# CULTURE
# OF ANIMAL CELLS

# CULTURE OF ANIMAL CELLS

## A MANUAL OF BASIC TECHNIQUE AND SPECIALIZED APPLICATIONS

### Sixth Edition

## R. Ian Freshney

Cancer Research UK Centre for Oncology and Applied Pharmacology
Division of Cancer Sciences and Molecular Pharmacology
University of Glasgow

WILEY-BLACKWELL

A John Wiley & Sons, Inc., Publication

*Library of Congress Cataloging-in-Publication Data:*

Freshney, R. Ian.
  Culture of animal cells : a manual of basic technique and specialized applications, / R. Ian Freshney. – 6th ed.
      p. cm.
  Includes index.
  ISBN 978-0-470-52812-9 (cloth)
  1. Tissue culture–Laboratory manuals. 2. Cell culture–Laboratory manuals. I. Title.
  QH585.2.F74 2010
  571.6′381–dc22
                            2010007042

Printed in the United States of America

10 9 8 7 6 5 4 3 2 1

This book is dedicated to all of the many friends and colleagues whose help and advice over the years has enabled me to extend the scope of this book beyond my own limited experience.

# Contents

# 21. Cytotoxicity, 365

# 22. Specialized Cells, 383

**Companion Website**
A companion resources site for this book is available at:
www.wiley.com/go/freshney/cellculture

# List of Figures

# List of Color Plates

# List of Protocols

# Preface and Acknowledgements

When the first edition of this books was published in 1983, although cell culture was an established technique it was still largely a research tool with a relatively small following. There was still an element of distrust that cell culture could deliver information relevant to processes in vivo. Largely because of the requirements of molecular genetics and virology the use of cell culture expanded into a major industrial process for the generation of biopharmaceuticals. Now the field is expanding further and entering other exiting areas of stem cell research and regenerative medicine. Perhaps one of the most exciting aspects of current progress in the field is that we can now grasp the "holy grail" of working with fully functional specialized cells in culture. A combination of selective culture conditions and manipulation of gene expression has meant that not only can we isolate and culture specialized cells, we can buy them "off the shelf," and we can evoke a plasticity in gene expression in both primitive stem cells and mature cells previously thought to be committed to their fate.

This book is the sixth edition of *Culture of Animal Cells: A Manual of Basic Technique and Specialized Applications.* Those who have used the previous edition will notice the extended title as some of the topics dealt with cannot be regarded as basic techniques. The book also has acquired a new chapter on stem cells, reflecting the current upsurge in interest in this area. Chapter 2, Training Programs, which is designed to enhance the use of this book as a teaching manual in addition to its role as a reference text, is now moved to the third to last chapter, on the assumption that instructors and trainees or students should have spent some time on the earlier chapters first, before attempting the exercises.

The number of color plate pages has been extended and, in combination with Figure 16.2, the book now provides photographs of around 40 different cell lines, including primary cultures, equipment, and processes. There are four new plates, two of stem cells and two of specialized cells (Courtesy of Cell Applications, Inc.). I am greatly indebted to Yvonne Reid and Greg Sykes of ATCC, Peter Thraves of ECACC, and many others for kindly providing illustrations. I hope that the color plates, in particular, will encourage readers to look at their cells more carefully and become sensitive to any changes that occur during routine maintenance.

For most of the book, I have retained the emphasis of previous editions and focused on basic techniques with some examples of more specialized cultures and methods. These techniques are presented as detailed step-by-step protocols that should give sufficient information to carry out a procedure without recourse to the prime literature. There is also introductory material to each protocol explaining the background and supplementary information providing alternative procedures and applications. Some basic biology is explained in Chapter 2, but it is assumed that the reader will have a basic knowledge of anatomy, histology, biochemistry, and cell and molecular biology. The book is targeted at those with little or no previous experience in tissue culture, including technicians in training, senior undergraduates, graduate students, postdoctoral workers, and clinicians with an interest in laboratory science. Those working in the biotechnology industry, including cell production, screening assays, and quality assurance, should also find this book of value.

The specialized techniques chapter 27, no longer contains protocols in molecular techniques as there are many other sources of these [e.g., Sambrook and Russell, 2006; Ausubel et al., 2009], and it is also an area in which I am not

**xxvii**

well versed. Similarly Chapter 26 on scale-up serves as an interface with biotechnology and provides some background on systems for increasing cell yield, but takes no account of full-scale biopharmaceutical production and downstream processes. The section on automation has been extended with more examples of the use of robotics in cell culture.

Protocols are given a distinct appearance from the rest of the text. Reagents that are specific to a particular protocol are detailed in the materials sections of the protocols and the recipes for the common reagents, such as Hanks's BSS or trypsin, are given in Appendix I at the end of the book. Details of the sources of equipment and materials are given in Appendix II. The suppliers' list (Appendix III) has been updated, but addresses, telephone and fax numbers, and email addresses are not provided, and only the website is given, on the assumption that all necessary contact information will be found there. Suppliers are not cited in the text unless for a specialized item.

Abbreviations used in the text are listed separately after this preface. Conventions employed throughout are D-PBSA for Dulbecco's PBS without $Ca^{2+}$ and $Mg^{2+}$ and UPW for ultrapure water, regardless of how it is prepared. Concentrations are given in molarity wherever possible, and actual weights have been omitted from the media tables on the assumption that very few people will attempt to make up their own media but will, more likely, want to compare constituents, for which molar equivalents are more useful.

As always, I owe a great debt of gratitude to the authors who have contributed protocols, and to others who have advised me in areas where my knowledge is imperfect, including Robert Auerbach, Bob Brown, Mike Butler, Kenneth Calman, Roland Grafström, Richard Ham, Rob Hay, Stan Kaye, Nicol Keith, John Masters, Wally McKeehan, Rona McKie, Stephen Merry, Jane Plumb, Peter Vaughan, Paul Workman, the late John Paul, and members of the staff at ECACC, including Isobel Atkins, Jim Collins, David Lewis, Chris Morris, and Peter Thraves. I am fortunate in having had the clinical collaboration of

David I. Graham, David G. T. Thomas, and the late John Maxwell Anderson. In the early stages of the preparation of this book I also benefited from discussions with Don Dougall, Peter del Vecchio, Sergey Federoff, Mike Gabridge, Dan Lundin, John Ryan, Jim Smith, and Charity Waymouth. I am eternally grateful to Paul Chapple who first persuaded me that I should write a basic techniques book on tissue culture. The original illustrations were produced by Jane Gillies and Marina LaDuke, although many of these have now been replaced due to the demands of electronic publishing. Some of the data presented were generated by those who have worked with me over the years including Sheila Brown, Ian Cunningham, Lynn Evans, Margaret Frame, Elaine Hart, Carol McCormick, Alison Mackie, John McLean, Alistair McNab, Diana Morgan, Alison Murray, Irene Osprey, Mohammad Zareen Khan, and Natasha Yevdokimova.

I have been fortunate to receive excellent advice and support from the editorial staff of John Wiley & Sons. I would also like to acknowledge, with sincere gratitude, all those who have taken the trouble to write to me or to John Wiley & Sons with advice and constructive criticism on previous editions. It is pleasant and satisfying to hear from those who have found this book beneficial, but even more important to hear from those who have found deficiencies, which I can then attempt to rectify. I can only hope that those of you who use this book retain the same excitement that I feel about the future prospects emerging in the field.

I would like to thank my daughter Gillian and son Norman for all the help they gave me in the preparation of the first edition, many years ago, and for their continued advice and support. Above all, I would like to thank my wife, Mary, for her hours of help in compilation, proofreading, and many other tasks; without her help and support, the original text would never have been written and I would never have attained the necessary level of technical accuracy that is the keynote of a good tissue culture manual.

*R. Ian Freshney*

# Abbreviations

| | | | |
|---|---|---|---|
| AEC | animal ethics committee | D-PBSB | Dulbecco's phosphate-buffered saline, solution B ($Ca^{2+}$ and $Mg^{2+}$) |
| AFP | α-fetoprotein | | |
| ANSI | American National Standards Institute | DSMZ | Deutsche Sammlung von Mikroorganismen und Zellkulturen (German Collection of Microorganisms and Cell Cultures) |
| ACDP | Advisory Committee on Dangerous Pathogens | | |
| ATCC | American Type Culture Collection | | |
| BMP | bone morphogenetic protein | DT | population doubling time |
| bp | base pairs (in DNA) | EBRA | European Biomedical Research Association |
| BFU-E | erythroid burst-forming units | EBSS | Earle's balanced salt solution |
| BPE | bovine pituitary extract | EBV | Epstein–Barr virus |
| BUdR | bromodeoxyuridine | EC | European Community |
| BSA | bovine serum albumin | EC | embryonal carcinoma |
| BUdR | bromodeoxyuridine | ECACC | European Collection of Animal Cell Cultures (now European Collection of Cell Cultures) |
| CAM | chorioallantoic membrane | | |
| CAM | cell adhesion molecule | ECGF | endothelial cell growth factor |
| CDC | Centers for Disease Control | EDTA | ethylenediaminetetraacetic acid |
| CCD | charge-coupled device | EGF | epidermal growth factor |
| CCTV | closed-circuit television | EGTA | ethylene glycol tetraacetic acid |
| cDNA | complementary DNA | EM | electron microscope |
| CE | cloning efficiency | ES | embryonic stem (cell) |
| CFC | colony-forming cells | FACS | fluorescence-activated cell sorter |
| CFC-GEMM | granulocyte, erythrocyte, macrophage, and megakaryocyte colony-forming cells | FBS | fetal bovine serum |
| | | FCS | fetal calf serum |
| CFC-mix | mixed colony-forming cells | FDA | Federal Drug Administration (USA) |
| CM | conditioned medium | FGF | fibroblast growth factor |
| CMC | carboxymethylcellulose | FITC | Fluorescein isothiocyanate |
| CMF | calcium- and magnesium-free saline | $G_1$ | gap one (of the cell cycle) |
| CMRL | Connaught Medical Research Laboratories | $G_2$ | gap two (of the cell cycle) |
| DEPC | diethyl pyrocarbonate | G-CSF | granulocyte colony stimulating factor |
| DMEM | Dulbecco's modification of Eagle's medium | GLP | good laboratory practice |
| DMSO | Dimethyl sulphoxide | GM-CFC | granulocyte and macrophage colony-forming cells |
| DNA | deoxyribonucleic acid | GM-CSF | granulocyte and macrophage colony stimulating factor |
| DoH | Department of Health (UK) | | |
| D-PBSA | Dulbecco's phosphate-buffered saline solution A (without $Ca^{2+}$ and $Mg^{2+}$) | GMP | good manufacturing practice |
| | | H&E | hemalum and eosin |

| | | | |
|---|---|---|---|
| HAT | hypoxanthine, aminopterin, and thymidine | NCAM | neural cell adhesion molecule |
| HBS | HEPES buffered saline | NCI | National Cancer Institute |
| HBSS | Hanks's balanced salt solution | NEAA | nonessential amino acids |
| HC | hydrocortisone | NICE | National Institute for Clinical Excellence (UK) |
| hCG | human chorionic gonadotropin | | |
| HEC | hospital ethics committee | NIH | National Institutes of Health (USA) |
| HEPES | 4-(2-hydroxyethyl)−1-piperazineethanesulfonic acid | NIOSH | National Institute for Occupational Safety and Health |
| hES | human embryonic stem (cell) | NMR | nuclear magnetic resonance |
| HFEA | Human Fertilization and Embryology Authority (UK) | NRC | National Research Council (USA) |
| | | NS | neurospheres |
| HGPRT | hypoxanthine guanosine phosphoribosyl transferase | NSF | National Science Foundation (USA) |
| | | O.D. | optical density |
| HITES | hydrocortisone, insulin, transferrin, estradiol, and selenium | OHRP | Office for Human Research Protections (USA) |
| | | OHS | Occupational Health and Safety |
| HIV | Human immunodeficiency virus | OHSA | Occupational Safety and Health Administration (USA) |
| HMBA | Hexamethylene-*bis*-acetamide | | |
| HPA | Health Protection Agency (UK) | OLAW | Office of Laboratory Animal Welfare (USA) |
| HPV | human papilloma virus | PA | plasminogen activator |
| HS | horse serum | PBS | phosphate-buffered saline |
| HSE | Health and Safety Executive (UK) | PBSA | phosphate-buffered saline, solution A ($Ca^{2+}$ and $Mg^{2+}$ free) |
| HSRRB | Health Science Research Resources Bank | | |
| HSV | herpes simplex virus | PBSB | phosphate-buffered saline, solution B ($Ca^{2+}$ and $Mg^{2+}$) |
| HT | hypoxanthine/thymidine | | |
| HuS | human serum | PCA | perchloric acid |
| ICAM | Intercellular adhesion molecule | PCR | polymerase chain reaction |
| ICM | inner cell mass (of embryo) | PD | population doubling |
| IL-1, 2 etc. | interleukin-1, 2, etc. | PDGF | platelet-derived growth factor |
| IMDM | Iscove's modification of DMEM | PE | plating efficiency (in clonogenic assays) |
| iPS | Induced pluripotent stem (cell) | PE | PBSA/EDTA (trypsin diluent) |
| ITS | insulin, transferrin, selenium | PEG | polyethylene glycol |
| JCRB | Japanese Collection of Research Bioresources | PGA | polyglycolic acid |
| KBM | keratinocyte basal medium | PHA | phytohemagglutinin |
| kbp | kilobase pairs (in DNA) | PLA | polylactic acid |
| KGM | keratinocyte growth medium | PMA | phorbol myristate acetate |
| LI | labeling index | PTFE | polytetrafluoroethylene |
| LIF | leukemia inhibitory factor | PVP | polyvinylpyrrolidone |
| LTBMC | long-term bone marrow culture | PWM | pokeweed mitogen |
| LTC-IC | long-term culture initiating cells | QA | quality assurance |
| M199 | medium 199 | QC | quality control |
| MACs | mammalian artificial chromosomes | RCCS™ | Rotatory Cell Culture System™ |
| MACS | magnet-activated cell sorting | RITC | Rhodamine isothiocyanate |
| MCDB | Molecular, Cellular, and Developmental Biology (Department, University of Colorado, Boulder, USA) | RFLP | restriction fragment length polymorphisms |
| | | RNA | ribonucleic acid |
| | | RPMI | Rosewell Park Memorial Institute |
| MEF | mouse embryo fibroblasts | RT-PCR | reverse transcriptase PCR |
| MEM | minimal essential medium (Eagle) | S | DNA synthetic phase of cell cycle |
| mES | mouse embryonic stem (cell) | SD | saturation density |
| MRC | Medical Research Council (UK) | SGM | second-generation multiplex |
| MRI | magnetic resonance imaging | SIT | selenium, insulin, transferrin |
| mRNA | messenger RNA | SLTV™ | Slow Turning Later Vessel™ |
| MSC | mesenchymal stem cell | S-MEM | MEM with low $Mg^{2+}$ and no $Ca^{2+}$ |
| MTT | 3-(4,5-dimethylthiazol-2yl)2,5-diphenyltetrazolium bromide | SOP | standard operating procedure |
| | | SSC | sodium citrate/sodium chloride |
| | | STR | short tandem repeat (in DNA profiling) |
| | | STR | stirred tank reactor (in scale-up) |
| NASA | National Aeronautics and Space Administration | SV40 | simian virus 40 |
| NAT | noradrenalin transporter | SV40LT | SV40 gene for large T-antigen |
| NBCS | newborn-calf serum | | |

| | | | |
|---|---|---|---|
| TCA | trichloroacetic acid | t-PA | tissue plasminogen activator |
| $T_D$ | population doubling time | TPA | tetradecanoylphorbol acetate |
| TE | trypsin/EDTA | u-PA | urokinase-like plasminogen activator |
| TEB | Tris/EDTA buffer | UPW | ultrapure water |
| TEER | transepithelial electrical resistance | US NRC | US Nuclear Regulatory Commission |
| TGF | transforming growth factor | UV | ultraviolet |
| TK | thymidine kinase | VEGF | vascular endothelial growth factor |
| TOC | total organic carbon | ZEF | zebrafish embryo fibroblasts |

# CHAPTER 1

# Introduction

As tissue culture enters its second century since its inception [Harrison, 1907], it is reaching what is probably one of, if not the, most exciting times in its history. For the first time it is possible for genetic manipulation of commonly and easily cultured cells, such as skin fibroblasts, to allow their conversion into pluripotent stem (iPS) cells, capable of differentiating into a range of different cell types [Lewitzky & Yamanaka, 2007; Nakagawa et al., 2007; Yu et al., 2007]. Coupled with the use of a chemical inducer of transcriptional changes in the genome (valproic acid), the four genes previously required is reduced to two [Huangfu et al., 2008] and the possibility of creating iPS cells by biochemical induction, rather than genetic intervention, becomes a real possibility. Added to that is the demonstration that it may also be possible to induce transdifferentiation from one lineage to another [Kondo & Raff, 2000; Le Douarin et al., 2004], and the field opens up to a whole new scenario: instead of the need for complex selective culture techniques, simple culture procedures may be used to initiate a cell line and biochemical regulation may be used to convert it into a new phenotype, directly via regression to a stem cell or to other progenitor cell. The possibilities that this opens up for the study of the regulation of differentiation, the determination of errors that occur in abnormal differentiation [Ebert et al., 2009] and malignancy, the provision of screening systems for diagnosis and drug development with cell lines from known pathologies, and the creation of autografts by tissue engineering promise a further expansion of tissue culture technology and usage comparable to the biotechnology boom of the turn of the century.

## 1.1 HISTORICAL BACKGROUND

Tissue culture was devised at the beginning of the twentieth century [Harrison, 1907; Carrel, 1912] (Table 1.1) as a method for studying the behavior of animal cells free of systemic variations that might arise in vivo both during normal homeostasis and under the stress of an experiment. As the name implies, the technique was elaborated first with undisaggregated fragments of tissue, and growth was restricted to the radial migration of cells from the tissue fragment, with occasional mitoses in the outgrowth. As culture of cells from and within such primary explants of tissue dominated the field for more than 50 years [Fischer, 1925; Parker, 1961], it is not surprising that the name "tissue culture" has remained in use as a generic term despite the fact that most of the explosive expansion in the field in the second half of the twentieth century (Fig. 1.1) was made possible by the use of dispersed cell cultures.

Disaggregation of explanted cells and subsequent plating out of the dispersed cells was first demonstrated by Rous [Rous & Jones, 1916], although passage was more often by surgical subdivision of the culture by Fischer, Carrel, and others, to generate what were then termed cell strains. L929 was the first cloned cell strain, isolated by capillary cloning from mouse L-cells [Sanford et al., 1948]. It was not until the 1950s that trypsin became more generally used for subculture, following procedures described by Dulbecco to obtain passaged monolayer cultures for viral plaque assays [Dulbecco, 1952], and the generation of a single cell suspension by trypsinization, which facilitated the further

**TABLE 1.1.** Key Events in Development of Cell and Tissue Cultures

| Date | Event | Reference |
|---|---|---|
| 1907 | Frog embryo nerve fiber outgrowth in vitro | Harrison, 1907 |
| 1912 | Explants of chick connective tissue; heart muscle contractile for 2–3 months | Carrel, 1912; Burrows, 1912 |
| 1916 | Trypsinization and subculture of explants | Rous & Jones, 1916 |
| 1923 | Subculture of fibroblastic cell lines | Carrel & Ebeling, 1923 |
| 1925–26 | Differentiation of embryonic tissues in organ culture | Strangeways & Fell, 1925, 1926 |
| 1929 | Organ culture of chick long bones | Fell & Robison, 1929 |
| 1948 | Introduction of use of antibiotics in tissue culture | Keilova, 1948; Cruikshank & Lowbury, 1952 |
| 1943 | Establishment of the L-cell mouse fibroblast; first continuous cell line | Earle et al., 1943 |
| 1948 | Cloning of the L-cell | Sanford et al., 1948 |
| 1949 | Growth of virus in cell culture | Enders et al., 1949 |
| 1952 | Use of trypsin for generation of replicate subcultures | Dulbecco, 1952 |
| | Virus plaque assay | Dulbecco, 1952 |
| | Salk polio vaccine grown in monkey kidney cells | Kew et al., 2005 |
| | Establishment the first human cell line, HeLa, from a cervical carcinoma | Gey et al., 1952 |
| 1954 | Fibroblast contact inhibition of cell motility | Abercrombie & Heaysman, 1953, 1954 |
| 1955 | Cloning of HeLa on a homologous feeder layer | Puck & Marcus, 1955 |
| | Development of defined media | Eagle, 1955, 1959 |
| | Requirement of defined media for serum growth factors | Sanford et al., 1955; Harris, 1959 |
| Late 1950s | Realization of importance of mycoplasma (PPLO) infection | Coriell et al., 1958; Rothblat & Morton, 1959; Nelson, 1960 |
| | Nuclear transplantation | Briggs & King, 1960; Gurdon, 1960 |
| 1961 | Definition of finite life span of normal human cells | Hayflick & Moorhead, 1961 |
| | Cell fusion—somatic cell hybridization | Sorieul & Ephrussi, 1961 |
| 1962 | Establishment and transformation of BHK21 | Macpherson & Stoker, 1962 |
| | Maintenance of differentiation (pituitary & adrenal tumors) | Buonassisi et al., 1962; Yasamura et al., 1966; Sato & Yasamura, 1966 |
| 1963 | 3T3 cells & spontaneous transformation | Todaro & Green, 1963 |
| 1964 | Pluripotency of embryonal stem cells | Kleinsmith & Pierce, 1964 |
| | Selection of transformed cells in agar | Macpherson & Montagnier, 1964 |
| 1964–69 | Rabies, mumps, and Rubella vaccines in WI-38 human lung fibroblasts | Wiktor et al., 1964; Sokol et al., 1968 |
| 1965 | Serum-free cloning of Chinese hamster cells | Ham, 1965 |
| | Heterokaryons—man-mouse hybrids | Harris & Watkins, 1965 |
| 1966 | Nerve growth factor | Levi-Montalcini, 1966 |
| | Differentiation in rat hepatomas | Thompson et al., 1966 |
| | Colony formation by hematopoietic cells | Bradley & Metcalf, 1966; Ichikawa et al., 1966 |
| 1967 | Epidermal growth factor | Hoober & Cohen 1967 |
| | HeLa cell cross-contamination | Gartler, 1967 |
| | Density limitation of cell proliferation | Stoker & Rubin, 1967 |
| | Lymphoblastoid cell lines | Moore et al., 1967; Gerper et al., 1969; Miller et al., 1971 |
| 1968 | Retention of differentiation in cultured normal myoblasts | Yaffe, 1968 |
| | Anchorage independent cell proliferation | Stoker et al, 1968 |
| 1969 | Colony formation in hematopoietic cells | Metcalf, 1969; Metcalf, 1990 |
| 1970s | Development of laminar flow cabinets | Kruse et al., 1991; Collins & Kennedy, 1999 |
| 1973 | DNA transfer, calcium phosphate | Graham & Van der Eb, 1973 |
| 1975 | Growth factors | Gospodarowicz, 1974; Gospodarowicz & Moran, 1974 |

**TABLE 1.1.** (*Continued*)

| Date | Event | Reference |
|------|-------|-----------|
| | Hybridomas—monoclonal antibodies | Kohler & Milstein, 1975 |
| 1976 | Totipotency of embryonal stem cells | Illmensee & Mintz, 1976 |
| | Growth factor supplemented serum-free media | Hayashi & Sato, 1976 |
| 1977 | Confirmation of HeLa cell cross-contamination of many cell lines | Nelson-Rees & Flandermeyer, 1977 |
| | 3T3 feeder layer and skin culture | Green, 1977 |
| 1978 | MCDB selective, serum-free media | Ham & McKeehan, 1978 |
| | Matrix interactions | Gospodarowicz et al., 1978b; Reid & Rojkind, 1979 |
| | Cell shape and growth control | Folkman & Moscona, 1978 |
| 1980s | Regulation of gene expression | Darnell, 1982 |
| | Oncogenes, malignancy, and transformation | Weinberg, 1989 |
| 1980 | Matrix from EHS sarcoma (later Matrigel™) | Hassell et al., 1980 |
| 1983 | Regulation of cell cycle; cyclin | Evans et al., 1983; Nurse 1990 |
| | Immortalization by SV40 | Huschtscha & Holliday, 1983 |
| 1980–87 | Development of many specialized cell lines | Peehl & Ham, 1980; Hammond et al., 1984; Knedler & Ham, 1987 |
| 1983 | Reconstituted skin cultures | Bell et al., 1983 |
| 1984 | Production of recombinant tissue-type plasminogen activator in mammalian cells | Collen et al 1984 |
| 1990s | Industrial scale culture of transfected cells for production of biopharmaceuticals | Butler, 1991 |
| 1991 | Culture of human adult mesenchymal stem cells | Caplan, 1991 |
| 1998 | Tissue engineered cartilage | Aigner et al., 1998 |
| 1998 | Culture of human embryonic stem cells | Thomson et al., 1998 |
| 2000 | Human Genome Project—genomics, proteomics, genetic deficiencies, and expression errors | Dennis et al., 2001 |
| 2002 | Exploitation of tissue engineering | Atala & Lanza, 2002; Vunjak-Novakovic & Freshney, 2006 |
| 2007 | Reprogramming of adult cells to become pluripotent stem (iPS) cells | Yu et al. 2007 |
| 2008 | Induction of iPS cells by reprogramming with valproic acid | Huangfu et al. 2008 |

Note: See also Pollack [1981].

development of single cell cloning. Gey established the first continuous human cell line, HeLa [Gey et al., 1952]; this was subsequently cloned by Puck [Puck & Marcus, 1955] when the concept of an X-irradiated feeder layer was introduced into cloning. Tissue culture became more widely used at this time because of the introduction of antibiotics, which facilitated long-term cell line propagation, although many people were already warning against continuous use and the associated risk of harboring cryptic, or antibiotic-resistant, contaminations [Parker, 1961]. The 1950s were also the years of the development of defined media [Morgan et al., 1950; Parker et al., 1954; Eagle, 1955, 1959; Waymouth, 1959], which led ultimately to the development of serum-free media [Ham, 1963, 1965] (see Section 9.6).

Throughout this book the term *tissue culture* is used as a generic term to include organ culture and cell culture. The term *organ culture* will always imply a three-dimensional culture of undisaggregated tissue retaining some or all of the histological features of the tissue in vivo. *Cell culture* refers to a culture derived from dispersed cells taken from original tissue, from a primary culture, or from a cell line or cell strain by enzymatic, mechanical, or chemical disaggregation. The term *histotypic culture* implies that cells have been reaggregated or grown to recreate a three-dimensional structure with tissue-like cell density, for example, by cultivation at high density in a filter well, by perfusion and overgrowth of a monolayer in a flask or dish, by reaggregation in suspension over agar or in real or simulated zero gravity, or by infiltration of a three-dimensional matrix such as collagen gel. *Organotypic* implies the same procedures but recombining cells of different lineages, such as epidermal keratinocytes in combined culture with dermal fibroblasts, in an attempt to generate a *tissue equivalent*.

Harrison [1907] chose the frog as his source of tissue, presumably because it was a cold-blooded animal, and consequently incubation was not required. Furthermore because tissue regeneration is more common in lower vertebrates, he perhaps felt that growth was more likely to

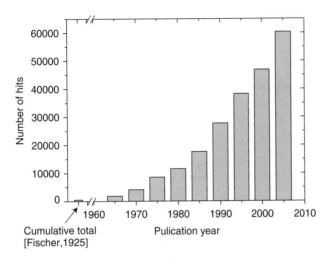

**Fig. 1.1. Growth of Tissue Culture.** Number of hits in PubMed for "cell culture" from 1965. The pre-1960 figure is derived from the bibliography of Fischer [1925].

occur than with mammalian tissue. Although his technique initiated a new wave of interest in the cultivation of tissue in vitro, few later workers were to follow his example in the selection of species. The stimulus from medical science carried future interest into warm-blooded animals, in which both normal development and pathological aberrations are closer to that found in humans. The accessibility of different tissues, many of which grew well in culture, made the embryonated hen's egg a favorite choice, but the development of experimental animal husbandry, particularly with genetically pure strains of rodents, brought mammals to the forefront as the favorite material. Although chick embryo tissue could provide a diversity of cell types in primary culture, rodent tissue had the advantage of producing continuous cell lines [Earle et al., 1943] and a considerable repertoire of transplantable tumors. The development of transgenic mouse technology [Beddington, 1992; Peat et al., 1992], together with the well-established genetic background of the mouse, has added further impetus to the selection of this animal as a favorite species.

The demonstration that human tumors could also give rise to continuous cell lines, such as HeLa [Gey et al., 1952], encouraged interest in human tissue, helped later by the classic studies of Leonard Hayflick on the finite life span of cells in culture [Hayflick & Moorhead, 1961] and the requirement of virologists and molecular geneticists to work with human material. The cultivation of human cells received a further stimulus when a number of different serum-free selective media were developed for specific cell types, such as epidermal keratinocytes, bronchial epithelium, and vascular endothelium (see Section 9.2.2). These formulations are now available commercially, although the cost remains high relative to the cost of regular media.

For many years the lower vertebrates and the invertebrates were largely ignored, although unique aspects of their development (tissue regeneration in amphibians,

metamorphosis in insects) make them attractive systems for the study of the molecular basis of development. More recently the needs of agriculture and pest control have encouraged toxicity and virological studies in insects, and developments in gene technology have suggested that insect cell lines with baculovirus and other vectors may be useful producer cell lines because of the possibility of inserting larger genomic sequences in the viral DNA and a reduced risk of propagating human pathogenic viruses. Furthermore the economic importance of fish farming and the role of freshwater and marine pollution have stimulated more studies of normal development and pathogenesis in fish. Procedures for handling nonmammalian cells have tended to follow those developed for mammalian cell culture, although a limited number of specialized media are now commercially available for fish and insect cells (see Section 27.5).

The types of investigation that lend themselves particularly to tissue culture are summarized in Fig. 1.2. These include basic studies on cellular metabolism, the regulation of gene expression and the cell phenotype at different stages of development, and the application of these studies to immunology, pharmacology, toxicology, and tissue regeneration and transplantation.

Initially the development of cell culture owed much to the needs of two major branches of medical research: the production of antiviral vaccines and the understanding of neoplasia. The standardization of conditions and cell lines for the production and assay of viruses undoubtedly provided much impetus to the development of modern tissue culture technology, particularly the production of large numbers of cells suitable for biochemical and molecular analysis. This and other technical improvements made possible by the commercial supply of reliable media and sera and by the greater control of contamination with antibiotics and clean-air equipment have made tissue culture accessible to a wide range of interests. Tissue culture is no longer an esoteric interest of a few but a major research tool in many disciplines and a huge commercial enterprise.

An additional force of increasing weight from public opinion has been the expression of concern by many animal-rights groups over the unnecessary use of experimental animals. Although most accept the idea that some requirement for animals will continue for preclinical trials of new pharmaceuticals, there is widespread concern that extensive use of animals for cosmetics development and similar activities may not be morally justifiable. Hence there is an ever-increasing lobby for more in vitro assays. The adoption in vitro assays, however, still requires proper validation and general acceptance. Although this seemed a distant prospect some years ago, the introduction of more sensitive and specifically targeted in vitro assays, together with a very real prospect of assaying for inflammation in vitro, has promoted an unprecedented expansion of in vitro testing (see Section 21.4).

The introduction of cell fusion techniques (see Section 27.6) and genetic manipulation [Maniatis et al.,

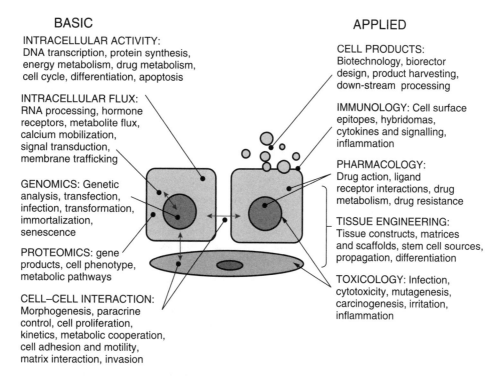

BASIC

**INTRACELLULAR ACTIVITY:**
DNA transcription, protein synthesis,
energy metabolism, drug metabolism,
cell cycle, differentiation, apoptosis

**INTRACELLULAR FLUX:**
RNA processing, hormone
receptors, metabolite flux,
calcium mobilization,
signal transduction,
membrane trafficking

**GENOMICS:** Genetic
analysis, transfection,
infection, transformation,
immortalization,
senescence

**PROTEOMICS:** gene
products, cell phenotype,
metabolic pathways

**CELL–CELL INTERACTION:**
Morphogenesis, paracrine
control, cell proliferation,
kinetics, metabolic cooperation,
cell adhesion and motility,
matrix interaction, invasion

APPLIED

**CELL PRODUCTS:**
Biotechnology, biorector
design, product harvesting,
down-stream processing

**IMMUNOLOGY:** Cell surface
epitopes, hybridomas,
cytokines and signalling,
inflammation

**PHARMACOLOGY:**
Drug action, ligand
receptor interactions, drug
metabolism, drug resistance

**TISSUE ENGINEERING:**
Tissue constructs, matrices
and scaffolds, stem cell sources,
propagation, differentiation

**TOXICOLOGY:** Infection,
cytotoxicity, mutagenesis,
carcinogenesis, irritation,
inflammation

*Fig. 1.2. Tissue Culture Applications.*

1978; Shih et al., 1979] established somatic cell genetics as a major component in the genetic analysis of higher animals, including humans. The technology has expanded rapidly and now includes sophisticated procedures for DNA sequencing, and gene transfer, insertion, deletion, and silencing. This technology has led to a major improvement in our understanding of how the regulation of gene expression and protein synthesis influence the expression of the normal and abnormal phenotype. The entire human genome has been sequenced in the Human Genome Project [Baltimore, 2001], and a new dimension added to expression analysis with multigene array technology [Iyer et al., 1999].

The insight into the mechanism of action of antibodies and the reciprocal information that this provided about the structure of the epitope, derived from monoclonal antibody techniques [Kohler & Milstein, 1975], was, like the technique of cell fusion itself, a prologue to a whole new field of studies in genetic manipulation. A vast new technology and a multibillion-dollar industry have grown out of the ability to insert exploitable genes into prokaryotic and eukaryotic cells. Cell products such as human growth hormone, insulin, interferon, and many antibodies are now produced routinely by genetically modified cells. The absence of post-transcriptional modifications, such as glycosylation, in bacteria suggests that mammalian cells may provide more suitable vehicles [Grampp et al., 1992], particularly in light of developments in immortalization technology (*see* Section 17.4).

The study of cell interactions and cell signaling in cell differentiation and development [Jessell and Melton, 1992;

Ohmichi et al., 1998; Balkovetz & Lipschutz, 1999] (*see also* Sections 2.2, 2.5, 16.7.1) have not only provided valuable fundamental information on mechanisms but have opened up whole new areas for tissue transplantation. Initial observations that cultures of epidermal cells form functionally differentiated sheets [Green et al., 1979] and endothelial cells may form capillaries [Folkman & Haudenschild, 1980] offered possibilities in homografting and reconstructive surgery using an individual's own cells [Limat et al., 1996; Tuszynski et al., 1996; Gustafson et al., 1998], particularly for severe burns [Gobet et al., 1997; Wright et al., 1998; Vunjak-Novakovic, 2006] (*see also* Section 25.4). With the ability to transfect normal genes into genetically deficient cells, it has become possible to graft such "corrected" cells back into the patient. Transfected cultures of rat bronchial epithelium carrying the β-gal reporter gene were shown to become incorporated into the rat's bronchial lining when they were introduced as an aerosol into the respiratory tract [Rosenfeld et al., 1992]. Similarly, cultured satellite cells were shown to be incorporated into wounded rat skeletal muscle, with nuclei from grafted cells appearing in mature, syncytial myotubes [Morgan et al., 1992]. Transfecting the normal insulin gene into β-islet cells cultured from diabetics, or even transfecting other cell types such as skeletal muscle progenitors [Morgan et al., 1992], would allow the cells to be incorporated into a low-turnover compartment and, potentially, give a long-lasting physiological benefit. Although the ethics of this type of approach seem less contentious, the technical limitations are still apparent.

Progress in neurological research has not had the benefit, however, of working with propagated cell lines from normal brain or nervous tissue, as the propagation of neurons in vitro has not been possible, until now, without resorting to the use of transformed cells (*see* Section 17.4). However, developments with human embryonal stem cell cultures [Thomson et al., 1998; Webber & Minger, 2004] suggest that this approach may provide replicating cultures that will differentiate into neurons and may provide useful and specific models for neuronal diseases [Ebert et al., 2008].

The prospect of transplantation of cultured cells has generated a whole new branch of culture, that of tissue engineering [Atala & Lanza, 2002; Vunjak-Novakovic & Freshney, 2006], encompassing the generation of tissue equivalents by organotypic culture (*see* Section 25.4), isolation and differentiation of human embryonal stem (ES) cells and adult totipotent stem cells such as mesenchymal stem cells (MSCs), gene transfer, materials science, construction and utilization of bioreactors, and transplantation technology. The technical barriers are steadily being overcome, bringing the ethical questions to the fore. The technical feasibility of implanting normal fetal neurons into patients with Parkinson disease has been demonstrated; society must now decide to what extent fetal material may be used for this purpose.

In vitro fertilization (IVF), developed from early experiments in embryo culture [Edwards, 1996], is now widely used [e.g., *see* Gardner & Lane, 2003] and has been accepted legally and ethically in many countries. The use of surplus embryos for research has also been accepted in some countries and will provide valuable material to further increase understanding of developmental processes and how to handle the cell lines generated. However, another area of development raising significant ethical debate is the generation of gametes in vitro from the culture of primordial germ cells isolated from testis and ovary [Dennis, 2003] or from ES cells. Oocytes have been cultured from embryonic mouse ovary and implanted, generating normal mice [Eppig, 1996; Obata et al., 2002], and spermatids have been cultured from newborn bull testes and cocultured with Sertoli cells [Lee et al., 2001]. Similar work with mouse testes generated spermatids that were used to fertilize mouse eggs, which developed into mature, fertile adults [Marh et al., 2003].

Tissue culture has also been used for diagnosis and toxicology. Amniocentesis (*see* Section 23.3.1) can reveal genetic disorders in the early embryo, although the polymerase chain reaction (PCR) and direct sampling are gradually replacing this, and the toxic effects of pharmaceutical compounds and potential environmental pollutants can be assayed in vitro (*see* Sections 22.3.1, 22.3.2, 22.4). In vitro toxicology has acquired greater importance in recent years due to changes in legislation regarding the usage of experimental animals, particularly in Europe.

## 1.2 ADVANTAGES OF TISSUE CULTURE

### 1.2.1 Control of the Environment

The two major advantages of tissue culture (Table 1.2) are the ability to control the physiochemical environment (pH, temperature, osmotic pressure, and $O_2$ and $CO_2$ tension), which has to be controlled very precisely, and the physiological conditions, which have to be kept relatively constant. However, the physiological environment cannot always be defined where cell lines still require supplementation of the medium with serum or other poorly defined constituents. These supplements are prone to batch variation and contain undefined elements such as hormones and other stimulants and inhibitors. The identification of some of the essential components of serum (*see* Table 8.5), together with a better understanding of factors regulating cell proliferation (*see* Table 9.4), has made the replacement of serum with defined constituents feasible (*see* Section 9.4). The role of the extracellular matrix (ECM) is important but similar to the use of serum—that is, the matrix is often necessary, but not always precisely defined. Prospects for defined ECM improve, however, as cloned matrix constituents become available [Kortesmaa et al., 2000; Belin & Rousselle, 2006; Braam et al., 2008; Dame & Verani, 2008; Domogatskaya et al., 2008] (see also Appendix II).

### 1.2.2 Characterization and Homogeneity of Samples

Tissue samples are invariably heterogeneous. Replicates, even from one tissue, vary in their constituent cell types. After one or two passages, cultured cell lines assume a homogeneous (or at least uniform) constitution, as the cells are randomly mixed at each transfer and the selective pressure of the culture conditions tends to produce a homogeneous culture of the most vigorous cell type. Hence, at each subculture, replicate samples are identical to each other, and the characteristics of the line may be perpetuated over several generations, or even indefinitely if the cell line is stored in liquid nitrogen. Because experimental replicates are virtually identical, the need for statistical analysis of variance is simplified. Furthermore the availability of stringent tests for cell line identity (*see* Section 15.4) and contamination (*see* Sections 12.1.1, 18.3, 18.6) means that preserved stocks may be validated for future research and commercial use.

### 1.2.3 Economy, Scale, and Mechanization

Cultures may be exposed directly to a reagent at a lower, and defined, concentration and with direct access to the cell. Consequently less reagent is required than for injection in vivo, where >90% may be lost by excretion and distribution to tissues other than those under study. Screening tests with many variables and replicates are cheaper, and the legal, moral, and ethical questions of animal experimentation are avoided. New developments in multiwell plates and robotics also have introduced significant economies in time and scale.

**TABLE 1.2.** Advantages of Tissue Culture

| Category | Advantages |
|---|---|
| Physicochemical environment | Control of pH, temperature, osmolality, dissolved gases |
| Physiological conditions | Control of hormone & nutrient concentrations |
| Microenvironment | Regulation of matrix, cell-cell interaction, gaseous diffusion |
| Cell line homogeneity | Availability of selective media; cell cloning |
| Characterization | Easily performed cytology, DNA profiling, immunostaining |
| Preservation | Stocks stored in liquid nitrogen |
| Validation & accreditation | Origin, history, purity authenticated and recorded |
| Replicates and variability | Easy quantitation and minimal statistical analysis |
| Reagent saving | Reduced volumes, direct access to cells, lower cost |
| Control of $C \times T$ | Ability to define dose, concentration, time |
| Mechanization | Available with microtitration and robotics |
| Scale | Number of replicates can be increased substantially |
| Time saving | Assay time reduced, at least, by an order of magnitude |
| Reduction of animal use | Cytotoxicity & screening of pharmaceutics, cosmetics, etc. |

## 1.2.4  In vitro Modeling of In vivo Conditions

Perfusion techniques allow the delivery of specific experimental compounds to be regulated in concentration, duration of exposure (*see* Table 1.2), and metabolic state. The development of histotypic and organotypic models, with a more accurate replication of the in vivo cell phenotypes, also increases the accuracy of in vivo modeling.

## 1.3  LIMITATIONS

### 1.3.1  Expertise

Culture techniques must be carried out under strict aseptic conditions because animal cells grow much less rapidly than many of the common contaminants, such as bacteria, molds, and yeasts (Table 1.3). Furthermore, unlike microorganisms, cells from multicellular animals do not normally exist in isolation and consequently are not able to sustain an independent existence without the provision of a complex environment simulating blood plasma or interstitial fluid. These conditions imply a level of skill and understanding on the part of the operator in order to appreciate the requirements of the system and to diagnose problems as they arise (*see* Chapters 2, 33). Also care must be taken to avoid the recurrent problem of cross-contamination and to authenticate stocks (*see* Sections 12.1.1, 15.2, 18.6). Hence tissue culture should not be undertaken casually to run one or two experiments, but requires proper training (*see* Chapter 28), strict control of procedures, and a controlled environment.

### 1.3.2  Quantity

A major limitation of cell culture is the expenditure of effort and materials that goes into the production of relatively little tissue. A realistic maximum per batch for most small laboratories (with two or three people doing tissue culture)

**TABLE 1.3.** Limitations of Tissue Culture

| Category | Examples |
|---|---|
| Necessary expertise | Sterile handling |
| | Avoidance of chemical contamination |
| | Detection of microbial contamination |
| | Awareness and detection of mis-identification |
| Environmental control | Isolation and cleanliness of workplace |
| | Incubation, pH control |
| | Containment and disposal of biohazards |
| Quantity and cost | Capital equipment for scale-up |
| | Medium, serum |
| | Disposable plastics |
| Genetic instability | Heterogeneity, variability |
| Phenotypic instability | Dedifferentiation |
| | Adaptation |
| | Selective overgrowth |
| Identification of cell type | Markers not always expressed |
| | Histology difficult to recreate and atypical |
| | Geometry and microenvironment changes cytology |

might be 1 to 10 g, wet weight, of cells. With a little more effort and the facilities of a larger laboratory, 10 to 100 g is possible; above 100 g implies industrial pilot–plant scale, a level that is beyond the reach of most laboratories but is not impossible if special facilities are provided, when kilogram quantities can be generated.

The cost of producing cells in culture is about 10 times that of using animal tissue. Consequently, if large amounts of tissue (>10 g) are required, the reasons for providing them by culture must be very compelling. For smaller amounts of tissue (∼10 g), the costs are more readily absorbed into routine expenditure, but it is always worth considering whether assays or preparative procedures can be scaled down. Microscale and nanoscale assays can often be quicker because of reduced manipulation times, volumes, and centrifuge times, for example, and so these assays are frequently more readily automated (*see* Sections 20.8, 21.3.5). Scaling down and automating assays enable more tests to be done, which in turn may require the cell preparation to be automated (*see* Section 26.4).

### 1.3.3   Dedifferentiation and Selection

When the first major advances in cell line propagation were achieved in the 1950s, many workers observed the loss of the phenotypic characteristics typical of the tissue from which the cells had been isolated. This effect was blamed on dedifferentiation, a process assumed to be the reversal of differentiation but later shown to be largely due to the overgrowth of undifferentiated cells of the same or a different lineage. The development of serum-free selective media (*see* Section 9.2.2) has now made the isolation of specific lineages possible, and it can be seen that under the right conditions, many of the differentiated properties of these cells may be restored (*see* Section 16.7).

### 1.3.4   Origin of Cells

If differentiated properties are lost, for whatever reason, it is difficult to relate the cultured cells to functional cells in the tissue from which they were derived. Stable markers are required for characterization of the cells (*see* Section 15.1); in addition the culture conditions may need to be modified so that these markers are expressed (*see* Sections 2.4, 16.7). Regrettably, many cell lines have been misidentified due to cross-contamination or errors in stock control in culture or in the freezer (*see* Sections 12.1.1, 15.2, 18.6). This makes it essential to have the technology, or access to it, to ensure authentication of each cell line that is used (*see* Section 15.2).

### 1.3.5   Instability

Instability is a major problem with many continuous cell lines, resulting from their unstable aneuploid chromosomal constitution. Even with short-term cultures of untransformed cells, heterogeneity in growth rate and the capacity to differentiate within the population can produce variability from one passage to the next (*see* Section 17.3).

### 1.4   MAJOR DIFFERENCES IN VITRO

Most of the differences in cell behavior between cultured cells and their counterparts in vivo stem from the dissociation of cells from a three-dimensional geometry and their propagation on a two-dimensional substrate. Specific cell interactions characteristic of the histology of the tissue are lost. As the growth fraction of the cell population increases, the cells spread out, become mobile, and, in many cases, start to proliferate. When a cell line forms, it may represent only one or two cell types, and many heterotypic cell–cell interactions are lost.

The culture environment also lacks the several systemic components involved in homeostatic regulation in vivo, principally those of the nervous and endocrine systems. Without this control, cellular metabolism may be more constant in vitro than in vivo, but may not be truly representative of the tissue from which the cells were derived. Recognition of this fact has led to the inclusion of a number of different hormones in culture media (*see* Sections 9.4.4, 10.4.3), and it seems likely that this trend will continue. The low oxygen tension due to the lack of oxygen transporter (hemoglobin) results in energy metabolism in vitro occurring largely by glycolysis; although the citric acid cycle is still functional, it plays a lesser role.

It is not difficult to find many more differences between the environmental conditions of a cell in vitro and in vivo (*see* Section 21.2), and this disparity has often led to tissue culture being regarded in a rather skeptical light. Still, although the existence of such differences cannot be denied, many specific pathways and specialized functions are expressed in culture, and as long as the limits of the model are appreciated, tissue culture can be a very valuable tool.

### 1.5   TYPES OF TISSUE CULTURE

There are three main methods of initiating a culture [Schaeffer, 1990] (*see* Fig. 1.3; Table 1.4; Appendix IV): (1) *Organ culture* implies that the architecture characteristic of the tissue in vivo is retained, at least in part, in the culture (*see* Section 25.2). Toward this end the tissue is cultured at the liquid–gas interface (on a raft, grid, or gel), which favors the retention of a spherical or three-dimensional shape. (2) In *primary explant culture* a fragment of tissue is placed at a glass (or plastic)–liquid interface, where, after attachment, migration is promoted in the plane of the solid substrate (*see* Section 11.3.1). (3) *Cell culture* implies that the tissue, or outgrowth from the primary explant, is dispersed (mechanically or enzymatically) into a cell suspension, which may then be cultured as an adherent monolayer on a solid substrate or as a suspension in the culture medium (*see* Sections 11.3, 12.4.5).

Because of the retention of histological interactions found in the tissue from which the culture was derived, organ cultures tend to retain the differentiated properties of that tissue. They do not grow rapidly (cell proliferation is limited to the periphery of the explant and is restricted mainly to embryonic tissue) and hence cannot be propagated; each experiment requires fresh explantations, which implies greater effort and poorer reproducibility of the sample than is

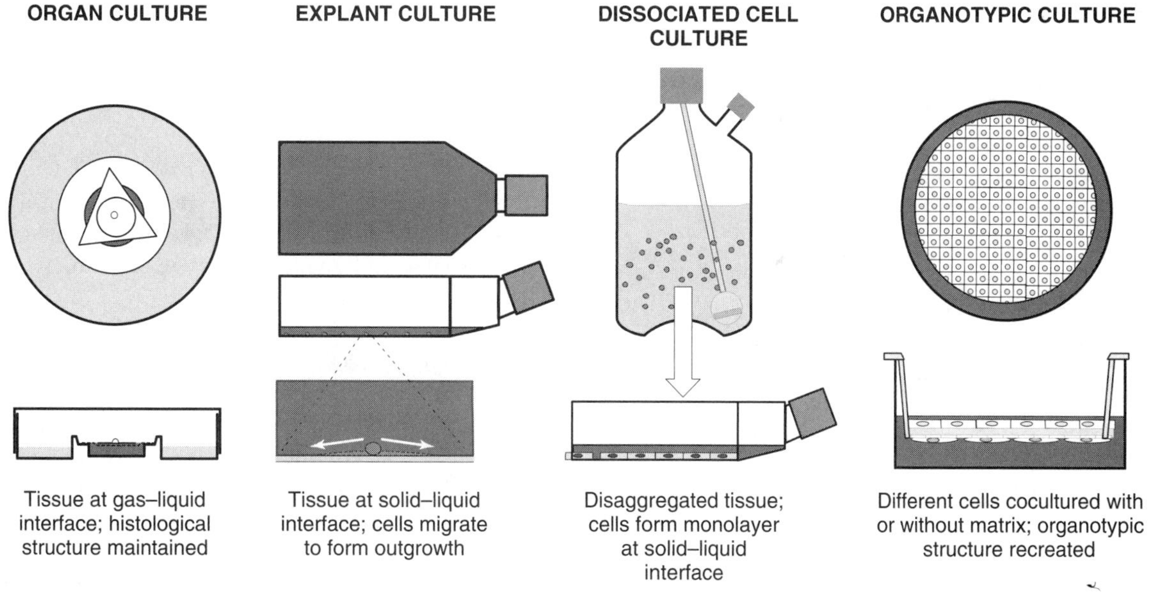

ORGAN CULTURE            EXPLANT CULTURE            DISSOCIATED CELL            ORGANOTYPIC CULTURE
                                                         CULTURE

Tissue at gas–liquid        Tissue at solid–liquid      Disaggregated tissue;       Different cells cocultured with
interface; histological     interface; cells migrate    cells form monolayer        or without matrix; organotypic
structure maintained        to form outgrowth           at solid–liquid             structure recreated
                                                         interface

*Fig. 1.3. Types of Tissue Culture.*

**TABLE 1.4.** Properties of Different Types of Culture

| Category | Organ culture | Explant | Cell culture | Organotypic culture |
|---|---|---|---|---|
| Source | Embryonic organs, adult tissue fragments | Tissue fragments | Disaggregated tissue, primary culture, propagated cell line | Primary culture or cell lines |
| Effort | High | Moderate | Low | Moderate |
| Characterization | Easy, by histology | Cytology and markers | Biochemical, molecular, immunological, and cytological assays | Histology, confocal microscopy, or MRI |
| Histology | Informative | Difficult | Not applicable | Informative |
| Biochemical differentiation | Possible | Heterogeneous | Lost, but may be reinduced | Often re-expressed |
| Propagation | Not possible | Possible from outgrowth | Standard procedure | Only after dissociation |
| Replicate sampling, reproducibility, homogeneity | High intersample variation | High intersample variation | Low intersample variation | Low intersample variation |
| Quantitation | Difficult | Difficult | Easy; many techniques available | May require image analysis |

achieved with cell culture. Quantitation is therefore more difficult, and the amount of material that may be cultured is limited by the dimensions of the explant ($\sim$1 mm$^3$) and the effort required for dissection and setting up the culture.

Cell cultures may be derived from primary explants or dispersed cell suspensions. Because cell proliferation is often found in such cultures, the propagation of cell lines becomes feasible. A monolayer or cell suspension with a significant growth fraction (*see* Section 20.11.1) may be dispersed by enzymatic treatment or simple dilution and reseeded, or subcultured, into fresh vessels (Table 1.5; *see also* Sections 12.1, 12.4). This constitutes a *subculture* or *passage*, and the daughter cultures so formed are the beginnings of a *cell line*.

The formation of a cell line from a primary culture implies (1) an increase in the total number of cells over several generations (population doublings) and (2) the ultimate predominance of cells or cell lineages with a high proliferative

**TABLE 1.5.** Subculture

| Advantages | Disadvantages |
| --- | --- |
| Propagation | Trauma of enzymatic or mechanical disaggregation |
| More cells | Selection of cells adapted to culture |
| Possibility of cloning | Overgrowth of unspecialized or stromal cells |
| Increased homogeneity | Genetic instability |
| Characterization of replicate samples | Loss of differentiated properties (may be inducible) |
| Frozen storage | Increased risk of misidentification or cross-contamination |

capacity, resulting in (3) a degree of uniformity in the cell population (*see* Table 1.5). The line may be characterized, and the characteristics will apply for most of its finite life span. The derivation of *continuous* (or "established," as they were once known) cell lines usually implies a genotypic change, or *transformation* (*see* Sections 3.8, 17.2), and the cell formation is usually accompanied by an increased rate of cell proliferation and a higher plating efficiency (*see* Section 17.5).

When cells are selected from a culture, by cloning or by some other method, the subline is known as a *cell strain*. A detailed characterization is then implied. Cell lines or cell strains may be propagated as an adherent monolayer or in suspension. *Monolayer* culture signifies that the cells are grown attached to the substrate. *Anchorage dependence* means that attachment to (and usually some degree of spreading onto) the substrate is a prerequisite for cell proliferation. Monolayer culture is the mode of culture common to most normal cells, with the exception of hematopoietic cells. *Suspension* cultures are derived from cells that can survive and proliferate without attachment *(anchorage independent);* this ability is restricted to hematopoietic cells, transformed cell lines, and transformed cells from malignant tumors. It can be shown, however, that a small proportion of cells that are capable of proliferation in suspension exists in many normal tissues (*see* Section 17.5.1). The identity of these cells remains unclear, but a relationship to the stem cell or uncommitted precursor cell compartment has been postulated. Cultured cell lines are more representative of precursor cell compartments in vivo than of fully differentiated cells, as most differentiated cells normally do not divide (*see* Sections 2.4, 16.3).

Because they may be propagated as a uniform cell suspension or monolayer, cell cultures have many advantages, in quantitation, characterization, and replicate sampling, but lack the retention of cell–cell interaction and cell–matrix interaction afforded by organ cultures. For this reason many workers have attempted to reconstitute three-dimensional cellular structures (*see* Sections 25.3, 25.4). Such developments have required the introduction, or at least redefinition, of certain terms. *Histotypic culture,* or *histoculture* (I use *histotypic culture*), has come to mean the high-density, or "tissue-like," culture of one cell type, whereas *organotypic* culture implies the presence of more than one cell type interacting, as the cells might, in the organ of origin. Organotypic culture has provided new prospects for the study of cell interaction among discrete, defined populations of homogeneous and potentially genetically and phenotypically defined cells and an opportunity to create differentiated populations of cells suitable for grafting.

In many ways some of the most exciting developments in tissue culture arise from recognizing the necessity of specific cell interaction in homogeneous or heterogeneous cell populations in culture. This recognition marks the transition from an era of fundamental molecular biology, in which many of the regulatory processes have been worked out at the cellular level, to an era of cell or tissue biology, in which that understanding is applied to integrated populations of cells, to a more precise elaboration of the signals transmitted among cells, and to the creation of fully functional tissues in vitro.

# CHAPTER 2

# Biology of Cultured Cells

## 2.1 THE CULTURE ENVIRONMENT

The validity of the cultured cell as a model of physiological function in vivo has frequently been criticized. Often the cell does not express the correct in vivo phenotype because the cell's microenvironment has changed. Cell–cell and cell–matrix interactions are reduced because the cells lack the heterogeneity and three-dimensional architecture found in vivo, and many hormonal and nutritional stimuli are absent. This creates an environment that favors the spreading, migration, and proliferation of unspecialized progenitor cells, rather than the expression of differentiated functions. The influence of the environment on the culture is expressed via five routes: (1) the nature of the substrate on or in which the cells grow—solid, as on plastic or other rigid matrix, semisolid, as in a gel such as collagen or agar, or liquid, as in a suspension culture; (2) the degree of contact with other cells; (3) the physicochemical and physiological constitution of the medium; (4) the constitution of the gas phase; and (5) the incubation temperature. The provision of the appropriate environment, including substrate adhesion, nutrient and hormone or growth factor concentration, and cell interaction, is fundamental to the expression of specialized functions [Alberts et al., 2008] (*see* Sections 16.1, 16.7).

## 2.2 CELL ADHESION

Most cells from solid tissues grow as adherent monolayers, and unless they have transformed and become anchorage independent (*see* Section 17.5.1), after tissue disaggregation or subculture they will need to attach and spread out on the substrate before they will start to proliferate (*see* Sections 12.4, 20.9.2). Originally it was found that cells would attach to, and spread on, glass that had a slight net negative charge. Subsequently it was shown that cells would attach to some plastics, such as polystyrene, if the plastic was appropriately treated with strong acid, a plasma discharge, or high-energy ionizing radiation. We now know that cell adhesion is mediated by specific cell surface receptors for molecules in the extracellular matrix (*see* Sections 7.1, 7.2, 16.7.3), so it seems likely that spreading may be preceded by the cells' secretion of extracellular matrix proteins and proteoglycans. The matrix adheres to the charged substrate, and the cells then bind to the matrix via specific receptors. Hence glass or plastic that has been conditioned by previous cell growth can often provide a better surface for attachment, and substrates pretreated with matrix constituents, such as fibronectin or collagen, or derivatives such as gelatin, will help more fastidious cells attach and proliferate.

With fibroblast-like cells the main requirement is for substrate attachment and cell spreading; usually the cells migrate individually at low densities. Epithelial cells may also require cell–cell adhesion for optimum survival and growth, and consequently they tend to grow in patches.

### 2.2.1 Cell Adhesion Molecules

Four major classes of transmenbrane proteins have been shown to be involved in cell–cell and cell–substrate adhesion (Fig. 2.1). (1) The classical cadherins are $Ca^{2+}$-dependent and are involved primarily in interactions between homologous cells, either via adherens junctions (cadherins E,

*Culture of Animal Cells: A Manual of Basic Technique and Specialized Applications, Sixth Edition*, by R. Ian Freshney
Copyright © 2010 John Wiley & Sons, Inc.

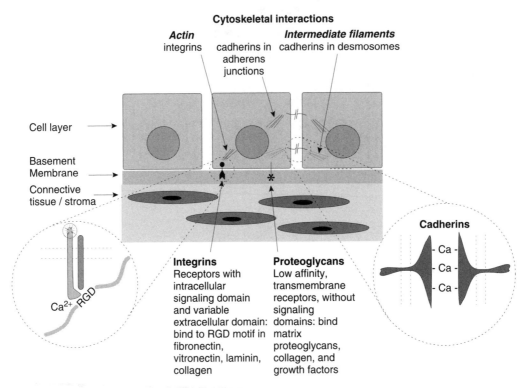

**Fig. 2.1. Cell Adhesion.** Diagrammatic representation of a layer of epithelial cells above connective tissue containing fibrocytes and separated from it by a basal lamina. CAMs and cadherins are depicted between like cells, and integrins and proteoglycans between the epithelial layer and the matrix of the basal lamina.

N, P, and VE) or desmosomes (desmoglein, desmocollin) [Alberts et al., 2008]. These proteins are homophilic; that is, homologous molecules in opposing cells interact with each other [Rosenman & Gallatin, 1991]. Cadherins E, N, P, and VE connect to the actin cytoskeleton and has a signaling, as well as structural, role acting via α- and β-catenins, vinculin, and α-actinin [Bakolitsa et al., 2004; Cavallaro & Christofori, 2004; Mège et al., 2006; Maddugoda et al., 2007]. Desmosomal cadherins connect via plakophilin, plakoglobin, and desmoplakin to the intermediate filament cytoskeleton and like the classical cadherins form cell–cell anchoring junctions that are capable of promoting sorting among cells to allow assembly into tissues. (2) The Ca$^{2+}$-independent cell–cell adhesion molecules (CAMs) such as NCAM, which is homophilic in neural synapses, and ICAM, which is heterophilic, interact with integrins in immunological synapses. (3) Integrins, which mediate cell–matrix adhesion, interact with molecules such as fibronectin, entactin, laminin, and collagen that bind to them via a specific motif usually containing the arginine-glycine-aspartic acid (RGD) sequence [Yamada & Geiger, 1997]. Each integrin comprises one α- and one β-subunit, whose extracellular domains are highly polymorphic, thus generating considerable diversity among the integrins. Integrins interact with the actin cytoskeleton via talin [Frame & Norman, 2008; Zhang et al., 2008], α-actinin or filamin, and vinculin [Janssen et al., 2006], in focal adhesions [Alberts et al., 2008] and signal to

the nucleus during adhesion and spreading, so allowing cells to enter the cycle. (4) Transmembrane proteoglycans also interact with matrix constituents such as other proteoglycans or collagen but not via the RGD motif. Some transmembrane and soluble proteoglycans may act as low-affinity growth factor receptors [Yevdokimova & Freshney, 1997; Forsten-Williams et al., 2008] and so stabilize, activate, and/or translocate the growth factor to the high-affinity receptor, participating in its dimerization [Schlessinger et al., 1995].

A fifth class of adhesion proteins are the claudins and occludens found in tight junctions. Like the cadherins they are homophilic transmembrane proteins that bind tightly to each other across the gap between adjacent cells. They interact with intracellular scaffold proteins, which in turn interact with desmosomes or adherens junctions, and this interaction helps establish cell polarity.

## 2.2.2 Intercellular Junctions
Although some cell adhesion molecules are diffusely arranged in the plasma membrane; others are organized into intercellular junctions. The role of junctions varies for mechanical junctions such as the desmosomes and adherens junctions, which hold epithelial cells together, tight junctions, which seal the space between cells such as between secretory cells in an acinus or duct or between endothelial cells in a blood vessel, and gap junctions, which allow ions, nutrients, and small signaling molecules such as

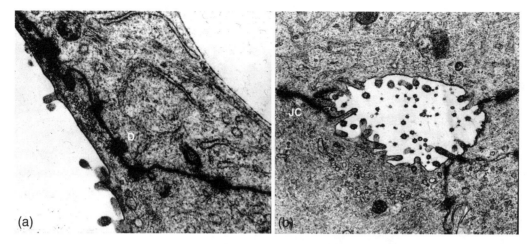

**Fig. 2.2. Intercellular Junctions.** Electron micrograph of culture of CA-KD cells, an early-passage culture from an adenocarcinoma secondary in brain (primary site unknown). Cells grown on Petriperm dish (Vivascience). (*a*) Desmosomes (D) between two cells in contact; mag. 28,000×. (*b*) Canaliculus showing tight junctions (T) and junctional complex (JC); mag. 18,500×. (Courtesy of Carolyn MacDonald.)

cyclic adenosine monophosphate (cAMP) to pass between cells in contact [Alberts et al., 2008]. Desmosomes may be distributed throughout the area of plasma membranes in contact (Fig. 2.2*a*), whereas adherens junctions are often associated with tight junctions the apical end of lateral cell contacts (Fig. 2.2*b*).

As epithelial cells differentiate in confluent cultures, they can form an increasing number of desmosomes and, if some morphological organization occurs, can form complete junctional complexes of adherens and tight junctions. The main role of tight junctions is to seal the intercellular space so that any molecules traveling from the apical to basal surface, and vice versa, must pass through the cell in a regulated fashion. Disaggregation of the tissue, or an attached monolayer culture, with protease will digest some of the extracellular matrix and may even degrade some of the extracellular domains of transmembrane proteins, allowing cells to become dissociated from each other. Epithelial cells are generally more resistant to disaggregation, as they tend to have tighter junctional complexes (desmosomes, adherens junctions, and tight junctions) holding them together, whereas mesenchymal cells, which are more dependent on integrin–matrix interactions for intercellular bonding, are more easily dissociated. Endothelial cells may also express tight junctions in culture, especially if left at confluence for prolonged periods on a preformed matrix, and can be difficult to dissociate. Homophilic binding of cadherins and integrin receptor binding to matrix constituents are both dependent on divalent cations $Ca^{2+}$ and $Mg^{2+}$. Hence chelating agents, such as EDTA, are often used to enhance disaggregation.

### 2.2.3 Extracellular Matrix

Intercellular spaces in tissues are filled with extracellular matrix (ECM), whose constitution is determined by the cell type (e.g., fibrocytes secrete type I collagen and fibronectin into the matrix), whereas epithelial cells produce laminin. Where adjacent cell types are different, such as at the boundary of the dermis (fibrocytes) and epidermis (keratinocytes), both cell types will contribute to the composition of the ECM, often producing a *basal lamina*. The complexity of the ECM is a significant component in the phenotypic expression of the cells attached to it, so a dynamic equilibrium exists in which the cells attached to the ECM control its composition, and in turn, the composition of the ECM regulates the cell phenotype [Kleinman et al., 2003; Zoubiane et al., 2003; Fata et al., 2004]. Hence a proliferating, migratory fibroblast will require a different ECM from a differentiating epithelial cell or neuron. Mostly cultured cell lines are allowed to generate their own ECM, but primary culture and propagation of some specialized cells, and the induction of their differentiation, may require exogenous provision of ECM [Lutolf & Hubbel, 2005; Blow, 2009].

ECM is comprised variously of collagen, laminin, fibronectin, hyaluronan, and proteoglycans such as betaglycan, decorin, perlecan, and syndecan-1, some of which bind growth factors or cytokines [Alberts et al., 2008]. It can be prepared by mixing purified constituents, such as collagen and fibronectin, by using cells to generate ECM and washing the producer cells off before reseeding with the cells under study (*see* Protocol 7.1), or by using a preformed matrix generated by the Engelberth-Holm-Swarm (EHS) mouse sarcoma, available commercially as Matrigel™ (*see* Section 7.2.1). Matrigel is often used to encourage differentiation and morphogenesis in culture, and it frequently generates a lattice-like network with epithelial (Fig. 2.3; Plate 12*c*) or endothelial cells.

At least two components of interaction with the substrate may be recognized: (1) adhesion, to allow the attachment and

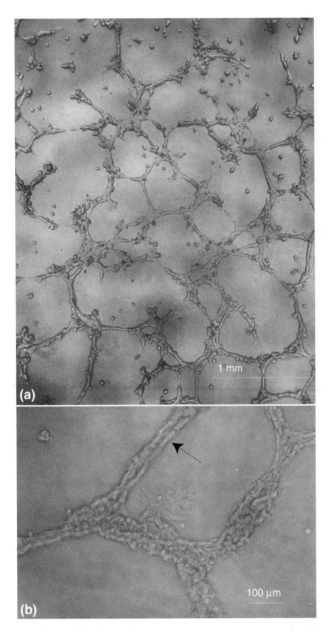

**Fig. 2.3. A549 Cells Growing on Matrigel.** Cultures of A549 adenocarcinoma cells growing on Matrigel. (*a*) Low-power shot showing lattice formation 24 h after seeding at $1 \times 10^5$ cells/mL. (*b*) Higher power, 3 days after seeding at $1 \times 10^5$ cells/nil. Arrow indicates possible tubular formation. (Courtesy of Jane Sinclair; *see also* Plate 12*c*.)

spreading that are necessary for cell proliferation [Folkman & Moscona, 1978], and (2) specific interactions, reminiscent of the interaction of an epithelial cell with basement membrane, with other ECM constituents, or with adjacent tissue cells, and required for the expression of some specialized functions (*see* Sections 2.4.1, 16.7.3). Rojkind et al. [1980], Vlodavsky et al. [1980], and others, explored the growth of cells on other natural substrates related to basement membrane. Natural matrices and defined-matrix macromolecules such as

Matrigel, Natrigel™, collagen, laminin, and vitronectin (BD Biosciences, Invitrogen) are available for controlled studies on matrix interaction.

The use of ECM constituents can be highly beneficial in enhancing cell survival, proliferation, or differentiation, but unless recombinant molecules are used [e.g., Braam et al., 2008; Dame & Varani, 2008; Domogatskaya et al., 2008], there is a significant risk of the introduction of adventitious agents from the originating animal (*see* Section 9.1). Recombinant collagen, fibronectin, and laminin fragments are available commercially (*see* Appendix II).

### 2.2.4 Cytoskeleton

Cell adhesion molecules are attached to elements of the cytoskeleton. The attachment of integrins to actin microfilaments via linker proteins is associated with reciprocal signaling between the cell surface and the nucleus [Fata et al., 2004]. Cadherins can also link to the actin cytoskeleton in adherens junctions, mediating changes in cell shape and morphogenesis [Maddugoda et al., 2007]. Desmosomes, which also employ cadherins, link to the intermediate filaments—in this case, cytokeratins—via an intracellular plaque, which has a structural as well as signaling role. Intermediate filaments are specific to cell lineages and can be used to characterize them (*see* Section 15.4.2; Plate 11*a–c*). The microtubules are the third component of the cytoskeleton; their role appears to be related mainly to cell motility and intracellular trafficking of microorganelles, such as the mitochondria and the chromatids at cell division.

### 2.2.5 Cell Motility

Time-lapse recording (*see* Section 27.3) demonstrates that cultured cells are capable of movement on a substrate. The most motile are fibroblasts at a low cell density (when cells are not in contact), and the least motile are dense epithelial monolayers. Fibroblasts migrate as individual cells with a recognizable polarity of movement. A lamellipodium, generated by polymerization of actin [Pollard & Borisy, 2003], extends in the direction of travel and adheres to the substrate, and the plasma membrane at the opposite side of the cell retracts, causing the cell to undergo directional movement. If the cell encounters another cell, the polarity reverses, and migration proceeds in the opposite direction. Migration proceeds in erratic tracks, as revealed by colloidal gold tracking [Scott et al., 2000], until the cell density reaches confluence, whereupon directional migration ceases. The cessation of movement at confluence, which is accompanied by a reduction in plasma membrane ruffling, is known as *contact inhibition* (*see* Section 17.5.2) and leads eventually to withdrawal of the cell from the division cycle. Myoblasts and endothelial cells migrate in a similar fashion and, like fibroblasts, may differentiate when they reach confluence, depending on the microenvironment.

Epithelial cells, unless transformed, tend not to display random migration as polarized single cells. When seeded at a low density, they will migrate until they make contact with another cell and the migration stops. Eventually cells accumulate in patches, and the whole patch may show signs of coordinated movement [Casanova, 2002].

## 2.3  CELL PROLIFERATION

### 2.3.1  Cell Cycle

The cell cycle is made up of four phases (Fig. 2.4). In the M phase (M = mitosis) the chromatin condenses into chromosomes, and the two individual chromatids, which make up the chromosome, segregate to each daughter cell. In the $G_1$ (Gap 1) phase, the cell either progresses toward DNA synthesis and another division cycle or exits the cell cycle reversibly ($G_0$) or irreversibly to commit to differentiation. It is during $G_1$ that the cell is particularly susceptible to control of cell cycle progression at a number of restriction points, which determine whether the cell will re-enter the cycle, withdraw from it, or withdraw and differentiate. $G_1$ is followed by the S phase (DNA synthesis), in which the DNA replicates. S in turn is followed by the $G_2$ (Gap 2) phase in which the cell prepares for re-entry into mitosis. Checkpoints at the beginning of DNA synthesis and in $G_2$ determine the integrity of the DNA and will halt the cell cycle to allow DNA repair or entry into apoptosis if repair is impossible. Apoptosis, or programmed cell death [Al-Rubeai & Singh, 1998], is a regulated physiological process whereby a cell can be removed from a population. Marked by DNA fragmentation, nuclear blebbing, and cell shrinkage (*see* Plate 17*c*, *d*), apoptosis can also be detected by a number of marker enzymes with kits such as Apotag (Oncor) or the COMET assay [Maskell & Green, 1995].

### 2.3.2  Control of Cell Proliferation

Entry into the cell cycle is regulated by signals from the environment. Low cell density leaves cells with free edges and renders them capable of spreading, which permits their entry into the cycle in the presence of mitogenic growth factors, such as epidermal growth factor (EGF), fibroblast growth factors (FGFs), or platelet-derived growth factor (PDGF) (*see* Sections 8.5.2, 9.4.5; Table 9.4), interacting with cell surface receptors. High cell density inhibits the proliferation of normal cells (though not transformed cells) (*see* Section 17.5.2). Inhibition of proliferation is initiated by cell contact and is accentuated by crowding, and the resultant change in the shape of the cell and reduced spreading with fewer focal adhesions.

Intracellular control is mediated by positive-acting factors such as the cyclins [McDonald & El-Deiry, 2000; Reed, 2003; Santamaria & Ortega, 2006] (*see* Fig. 2.4), which are upregulated by signal transduction cascades activated by phosphorylation of the intracellular domain of the receptor when it is bound to growth factor. Negative-acting factors such as p53 [Schwartz & Rotter, 1998], p16 and p21 [Caldon

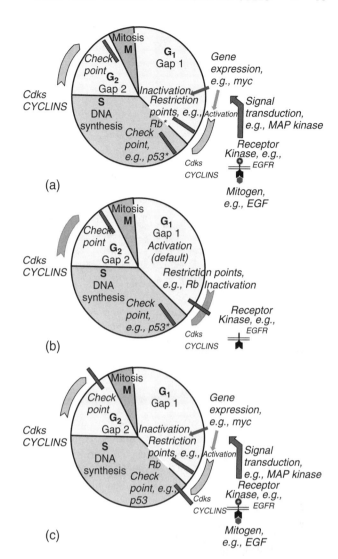

(a)

(b)

(c)

**Fig. 2.4. Cell Cycle.** The cell cycle is divided into four phases: $G_1$, S, $G_2$, and M. Progression round the cycle is driven by cyclins activated by cell division cycle kinases (Cdks), which in turn have been activated by regulatory genes, such as *myc*. Expression of positive-acting regulatory genes, such as *myc*, is induced by cytoplasmic signals initiated by receptor kinase following interaction with a mitogen, and transmitted via a signal transduction pathway, such as MAP kinase (*a*). The cell cycle is arrested at restriction points in $G_1$ by the action of Rb, and other cell cycle inhibitors in the absence of mitogens (*b*). When these are inactivated, usually by phosphorylation (Rb*), cells proceed round the cycle (*a*). The cell cycle can also be arrested at check points by cell cycle inhibitors such as and p53 if DNA damage is detected (*c*). Phosphorylation of p53 (p53*) allows the cycle to proceed (*a*).

et al., 2006], or the Rb gene product [Assoian & Yung, 2008] block cell cycle progression at restriction points or checkpoints [Planas-Silva & Weinberg, 1997] (*see* Fig. 2.4). The link between the extracellular control elements (both positive-acting, e.g., PDGF, and negative-acting, e.g., TGF-β) and intracellular effectors is made by cell membrane receptors and signal transduction pathways, often involving

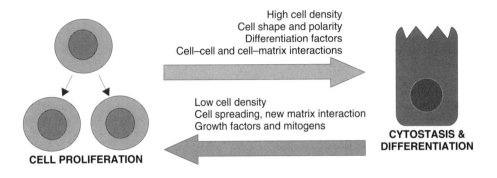

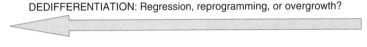

*Fig. 2.5. Differentiation and Proliferation.* Cells in culture can be thought to be in a state of equilibrium between cell proliferation and differentiation. Normal culture conditions (low cell density, mitogens in the medium) will favor cell proliferation, while high cell density and addition of differentiation factors will induce differentiation. The position of the equilibrium will depend on culture conditions. Dedifferentiation of the culture may be due to the effect of growth factors or cytokines inducing a more proliferative phenotype, reprogramming of gene expression, or overgrowth of a precursor cell type.

protein phosphorylation and second messengers such as cAMP, $Ca^{2+}$, and diacylglycerol [Alberts et al., 2008]. Much of the evidence for the existence of these steps in the control of cell proliferation has emerged from studies of oncogene and suppressor gene expression in tumor cells, with the ultimate objective of the therapeutic regulation of uncontrolled cell proliferation in cancer. The immediate benefit, however, has been a better understanding of the factors required to regulate cell proliferation in culture [McDonald & El-Deiry, 2000]. These studies have had other benefits as well, including the identification of genes that enhance cell proliferation, some of which can be used to immortalize finite cell lines (*see* Section 17.4).

## 2.4 DIFFERENTIATION

During early development the inner cell mass of the embryo differentiates into the three germ layers: (1) the *endoderm*, which gives rise to the epithelium of the gut and associated organs such as lung, liver, and pancreas, (2) the *ectoderm*, which give rise to the outer surface epithelia (epidermis, buccal epithelium, and outer cervical epithelium), and (3) the *mesoderm*, which give rise to the embryonic *mesenchyme* which in turn develops into connective tissue, supporting tissues such as bone, cartilage, muscle, vascular tissue (endothelium, smooth muscle, and pericytes), and the hematopoietic system. During neurulation the ectoderm also gives rise to the *neuroectoderm*, which in turn forms the neural system (central and peripheral neuron and glia), some neuroendocrine cells, and melanocytes. During organogenesis in the embryo, when the primitive organs start to form, these tissues derived from the primitive germ layers become associated in a process of

mutual induction of differentiation such that, for example, in the lung, the endodermally derived epithelial cells from a bud in the primitive gut, are induced to become tracheal, bronchial, and alveolar cells with secretory, lining, and respiratory functions, under the inductive influence of the associated mesenchyme, which in turn is induced by the endodermally derived cells to become fibrous and elastic connective tissue and smooth muscle. While most epithelial cells derive from the ectoderm or endoderm, some epithelial cells, such as kidney tubules and the mesothelium lining the body cavity, are mesodermal in origin.

Hence individual organs are comprised of tissues, often derived from different germ layers; for example, skin is made up of an outer epidermis (epithelial cells from the ectoderm) and an underlying dermis (from mesodermally derived mesenchyme). Tissues in turn are made up of individual cell type lineages; for example, the dermis contains connective tissue fibrocytes, vascular endothelial cells and smooth muscle cells, and the mesenchymal cells of the dermal papillae, among other cells. Each cell type can be traced back, via a series of proliferating cell stages, to an originating stem cell (*see* Section 2.7.1), forming a treelike structure. Each "branch" of that "tree" can be regarded as a *lineage*, as in a basal cell of the epidermis following a differentiation path to a mature cornified keratinocyte. Some lineages, such as the myeloid lineage of hematopoietic differentiation, may branch into sublineages (neutrophilic, eosinophilic, and basophilic), so lineage marker expression is also influenced by *differentiation*, namely the position of the cell in the lineage differentiation pathway (*see* Section 15.12).

As stated earlier (*see* Section 1.3.3), the expression of differentiated properties in cell culture is often limited by the promotion of cell proliferation (*see* also Section 16.3), which is necessary for the propagation of the cell line and the expansion of stocks. The conditions required for the induction of differentiation—a high cell density, enhanced cell–cell and cell–matrix interaction, and the presence of various differentiation factors (*see* Sections 16.1.1, 16.7.1)—may often be antagonistic to cell proliferation, and vice versa (Fig. 2.5). So, if differentiation is required, it may be necessary to define two distinct sets of conditions—one to optimize cell proliferation and one to optimize cell differentiation.

### 2.4.1  Maintenance of Differentiation

It has been recognized for many years that specific functions are retained longer when the three-dimensional structure of the tissue is retained, as in organ culture (*see* Section 25.2). Unfortunately, organ cultures cannot be propagated, must be prepared de novo for each experiment, and are more difficult to quantify than cell cultures. Re-creating three-dimensional structures by perfusing monolayer cultures (*see* Sections 25.3, 26.2.5) and culturing cells on or in special matrices, such as collagen gel, cellulose, or gelatin sponge, or other matrices (*see* Sections 2.2.3, 7.4.1, 7.4.2, 16.7.3) may be a better option. A number of commercial products, the best known of which is Matrigel™ (BD Biosciences), reproduce the characteristics of extracellular matrix but are undefined, although a growth factor-depleted version is also available (GFR Matrigel). The development and application of tissue engineering has placed great emphasis on this approach (*see* Sections 25.3, 25.4) and has led to the development of new tissue constructs incorporating new materials, such as polylactic acid (PLA), silk, and ceramics and, for some tissues, the use of dynamic stress [Vunjak-Novakovic & Freshney, 2006].

Although there is a degree of plasticity apparent in the phenotype of cells in culture (*see* Section 2.4.2), at the present state of the technology it is still important to select the correct lineage of cells when attempting to culture specialized cells (*see* Sections 13.1, 14.6, 22.1). If the correct precursors are grown, then induction of differentiation is more likely to be successfully induced (*see* Section 16.7). The ability to express the differentiated phenotype will also require propagation in the appropriate selective medium, usually in the absence of serum (*see* Section 9.2.2), before application of appropriate soluble inducers, such as hydrocortisone, retinoids, cytokines, or planar polar compounds (*see* Section 16.7.2).

### 2.4.2  Dedifferentiation

Historically the inability of cell lines to express the characteristic in vivo phenotype was blamed on *dedifferentiation*. According to this concept, differentiated cells lose their specialized properties in vitro, but it is often unclear whether (1) the wrong lineage of cells is selected in vitro, (2) undifferentiated cells of the same lineage (Fig. 2.5) overgrow terminally differentiated cells of reduced proliferative capacity, (3) the absence of the appropriate inducers (hormones, cell or matrix interaction) causes an adaptive, and potentially reversible, loss of expression of

differentiated properties (*see* Section 16.1.1), or (4) the differentiated cell reverts to a more primitive phenotype or even a stem cell. This last has been shown in the liver where, in response to partial hepatectomy, fully differentiated hepatocytes may dedifferentiate, proliferate, and redifferentiate when the liver mass is restored [Alison et al., 2004]. Also glial precursors, previously thought to be committed to become oligodendrocytes (OPCs), can be induced to revert to a common neural stem cell by treatment with BMP-2, BMP-4, and FGF-2, then reinduced to differentiate into either glia or neurons [Kondo & Raff, 2000]. Recent work has shown that adult fibroblasts can be induced to revert to a pluripotent stem cell by genetic [Nakagawa et al., 2007; Yu et al., 2007] or epigenetic [Huangfu et al., 2008] intervention (Fig. 2.7). This result suggests that given the right transcriptional inducers, many cells previously regarded as terminally differentiated may be induced to revert to progenitor or stem cell status. So dedifferentiation, so long out of fashion, reemerges as a possible, though perhaps less likely, cause of the undifferentiated status of cells in cultured cell lines. However, it is now clear that given the correct culture conditions, differentiated functions can be re-expressed (*see* Section 16.7). Surprisingly, in view of the concept that differentiation is dysfunctional in malignant cells, many transformed cell lines have provided the best model for the induction of differentiation (Table 2.1).

## 2.5  CELL SIGNALING

Cell proliferation, migration, differentiation, and apoptosis in vivo are regulated by cell–cell interaction, cell–matrix interaction, and nutritional and hormonal signals, as discussed above (*see* Section 2.4.1). Some signaling is contact mediated via cell adhesion molecules (*see* Section 2.2), but signaling can also result from soluble, diffusible factors. Signals that reach the cell from another tissue via the systemic vasculature are called endocrine, and those that diffuse from adjacent cells without entering the bloodstream are called paracrine. It is useful to recognize that some soluble signals arise in, and interact with, the same type of cell. I will call this homotypic paracrine, or *homocrine*, signaling (Fig. 2.8). Signals that arise in a cell type different from the responding cells are *heterotypic paracrine* and will be referred to simply as *paracrine* in any subsequent discussion. A cell can also generate its own signaling factors that bind to its own receptors or activate signal transduction pathways directly, and this is called *autocrine* signaling.

Although all of these forms of signaling occur in vivo, under normal conditions with basal media in vitro, only autocrine and homocrine signaling will occur. The failure of many cultures to plate with a high efficiency at low cell densities may be due, in part, to the dilution of one or more autocrine and homocrine factors, and this is part of the rationale in using conditioned medium (*see*

**TABLE 2.1.** Cell Lines with Differentiated Properties

| Cell type | Origin | Cell line | N | Species | Marker | Reference |
|---|---|---|---|---|---|---|
| Endocrine | Adrenal cortex | Y-1 | T | Mouse | Adrenal steroids | Yasamura et al., 1966 |
| Endocrine | Pituitary tumor | GH3 | T | Rat | Growth hormone | Buonassisi et al., 1962 |
| Endocrine | Hypothalamus | C7 | N | Mouse | Neurophysin; vasopressin | De Vitry et al., 1974 |
| Endothelium | Dermis | HDMEC | | Human | Factor VIII, CD36 | Gupta et al., 1997 |
| Endothelium | Pulmonary artery | CPAE | C | Cow | Factor VIII, ACE* | Del Vecchio & Smith, 1981 |
| Endothelium | Hepatoma | SK/HEP-1 | T | Human | Factor VIII | Heffelfinger et al., 1992 |
| Epithelium | Prostate | PPEC | N | Human | PSA | Robertson & Robertson, 1995 |
| Epithelium | Kidney | MDCK | C | Dog | Domes, transport | Gaush et al., 1966; Rindler et al., 1979 |
| Epithelium | Kidney | LLC–PKI | C | Pig | $Na^+$-dependent glucose uptake | Hull et al., 1976; Saier, 1984 |
| Epithelium | Breast | MCF-7 | T | Human | Domes, α-lactalbumin | Soule et al., 1973 |
| Glia | Glioma | MOG–G-CCM | T | Human | Glutamyl synthetase | Balmforth et al., 1986 |
| Glia | Glioma | C6 | T | Rat | Glial fibrillary acidic protein, GPDH | Benda et al., 1968 |
| Hepatocytes | Hepatoma | H4–11–E–C3 | T | Rat | Tyrosine amino-transferase | Pitot et al., 1964 |
| Hepatocytes | Liver | | T | Mouse | Aminotransferase | Yeoh et al., 1990 |
| Keratinocytes | Epidermis | HaCaT | C | Human | Cornification | Boukamp et al., 1988 |
| Leukemia | Spleen | Friend | T | Mouse | Hemoglobin | Scher et al., 1971 |
| Melanocytes | Melanoma | B16 | T | Mouse | Melanin | Nilos & Makarski, 1978 |
| Myeloid | Leukemia | K562 | T | Human | Hemoglobin | Andersson et al., 1979a, b |
| Myeloid | Myeloma | Various | T | Mouse | Immunoglobulin | Horibata & Harris, 1970 |
| Myeloid | Marrow | WEHI–3B D⁺ | T | Mouse | Morphology | Nicola, 1987 |
| Myeloid | Leukemia | HL60 | T | Human | Phagocytosis; Neotetrazolium Blue reduction | Olsson & Ologsson, 1981 |
| Myocytes | Skeletal muscle | C2 | C | Mouse | Myotubes | Morgan et al., 1992 |
| | | L6 | C | Rat | Myotubes | Richler & Yaffe, 1970 |
| Neuroendocrine | Pheochromo-cytoma | PC12 | T | Rat | Catecholamines; dopamine; norepinephrine | Greene & Tischler, 1976 |
| Neurons | Neuroblastoma | C1300 | T | Rat | Neurites | Lieberman & Sachs, 1978 |
| Type II pneumocyte or Clara cell | Lung carcinoma | A549 | T | Human | Surfactant | Giard et al., 1972 |
| | | NCI-H441 | T | Human | Surfactant | Brower et al., 1986 |
| Type II pneumocyte | Lung carcinoma | | I | Mouse | Surfactant | Wilkenheiser et al., 1991 |
| Various | Embryonal terato-carcinoma | F9 | T | Mouse | PA, laminin, type IV collagen | Bernstine et al., 1973 |

Note: Normal (N), continuous (C), immortalized (I), transformed (T); ACE, angiotensin II-converting enzyme.

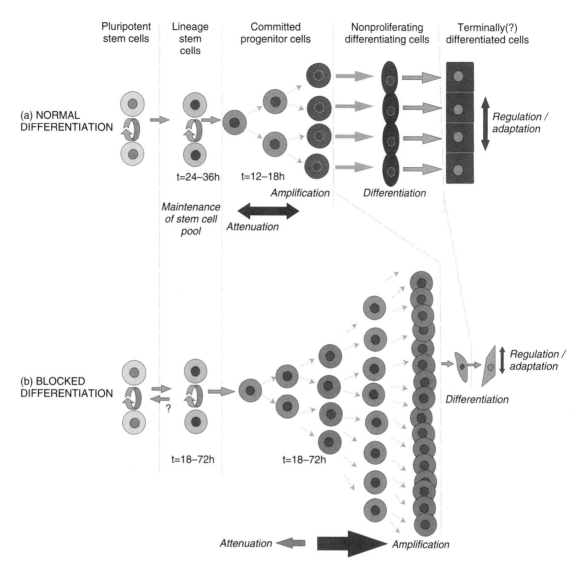

**Fig. 2.6. Differentiation from Stem Cells.** (a) In vivo, a small stem cell pool gives rise to a proliferating progenitor compartment that produces the differentiated cell pool. (b) In vitro, differentiation is limited by the need to proliferate, and the population becomes predominantly progenitor cells, although stem cells may also be present. Pluripotent stem cells (far left) have also been cultured from some tissues, but their relationship to the tissue stem cells is as yet unclear. Culture conditions select mainly for the proliferating progenitor cell compartment of the tissue or induce cells that are partially differentiated to revert to a progenitor status.

Section 13.2.2) or feeder layers (*see* Section 13.2.3) to enhance plating efficiency. As the maintenance and proliferation of specialized cells, and the induction of their differentiation, may depend on paracrine and endocrine factors, these must be identified and added to differentiation medium (*see* Section 16.7.2). However, their action may be quite complex as not only may two or more factors be required to act in synergy [e.g., McCormick & Freshney, 2000], but in trying to simulate cell–cell interaction by supplying exogenous paracrine factors, it is necessary to take into account that the phenotype of interacting cells, and hence the factors that they produce and the time frame in which they are produced, will change as a result of the interaction. Heterotypic

combinations of cells may be, initially at least, a simpler way of providing the correct factors in the correct matrix microenvironment, and analysis of this interaction may then be possible with blocking antibodies or antisense RNA.

## 2.6 ENERGY METABOLISM

Most culture media contain 4 to 20 mM glucose, which is used mainly as a carbon source for glycolysis, generating lactic acid as an end product. Under normal culture conditions (atmospheric oxygen and a submerged culture) oxygen is in relatively short supply. In the absence of an appropriate

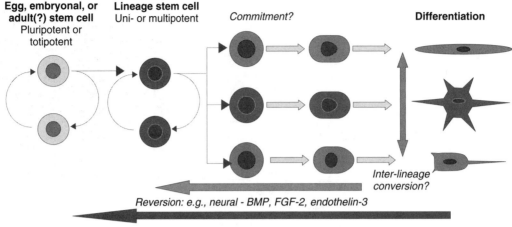

**Fig. 2.7. Commitment and Reversibility.** A lineage stem cell, such as a myeloid/erythroid stem cell, an epidermal stem cell, or a neural stem cell, gives rise to one or more lineages by a process of commitment to a particular pathway. However, this process is no longer regarded as irreversible, and reversion of committed precursors to a common lineage stem cell or to a pluripotent or even totipotent stem cell is possible.

carrier, such as hemoglobin, raising the $O_2$ tension will generate free radical species that are toxic to the cell, so $O_2$ is usually maintained at atmospheric levels. This results in anaerobic conditions and the use of glycolysis for energy metabolism [Danes & Paul, 1961], which is increased by insulin [Paul & Pearson, 1960]. However, the citric acid cycle remains active, and it has become apparent that amino acids, particularly glutamine, can be utilized as a carbon source by oxidation to glutamate by glutaminase and entry into the citric acid cycle by transamination to 2-oxoglutarate [Reitzer et al., 1979; Butler & Christie, 1994]. Deamination of the glutamine tends to produce ammonia, which is toxic and can limit cell growth, but the use of dipeptides, such as glutamyl-alanine or glutamyl-glycine, appears to minimize the production of ammonia and has the additional advantage of being more stable in the medium (e.g., Glutamax, Invitrogen).

## 2.7 ORIGIN OF CULTURED CELLS

Because most people working under standard conditions do so with finite or continuous proliferating cell lines, it is important to consider the cellular composition of the culture. The capacity to express differentiated markers under the influence of inducing conditions may mean either that the cells being cultured are mature and only require induction to continue synthesizing specialized proteins or that the culture is composed of precursor or stem cells that are capable of proliferation but remain undifferentiated until the correct inducing conditions are applied, whereupon some or all of the cells mature and become differentiated. It may be useful to think of a cell culture as being an equilibrium

between stem cells, undifferentiated precursor cells, and mature differentiated cells (*see* Fig. 2.6) and to suppose that the equilibrium may shift according to the environmental conditions. Routine serial passage at relatively low cell densities would promote cell proliferation and constrain differentiation, whereas high cell densities, low serum, and the appropriate hormones would promote differentiation and inhibit cell proliferation (*see* Fig. 2.5).

The source of the culture will also determine which cellular components may be present. Hence cell lines derived from the embryo may contain a higher proportion of stem cells and precursor cells and be capable of greater self-renewal than cultures from adults. In addition, cultures from tissues undergoing continuous renewal in vivo (epidermis, intestinal epithelium, and hematopoietic cells) may still contain stem cells that, under the appropriate conditions, will have a prolonged life span, whereas cultures from tissues that renew solely under stress (fibroblasts, muscle, glia) may contain only committed precursor cells with a limited life span. Thus the identity of the cultured cell is defined not only by its lineage in vivo but also by its position in that lineage (stem cell, precursor cell, or mature differentiated cell).

When cells are cultured from a neoplasm, they need not adhere to these rules. Thus a hepatoma from rat may proliferate in vitro and still express some differentiated features, but the closer the cells are to those of the normal phenotype, the more induction of differentiation may inhibit proliferation. The relationship between differentiation and cell proliferation may become relaxed but it is not lost—B16 melanoma cells still produce more pigment at a high cell density and at a low rate of cell proliferation than at a low cell density and a high rate of cell proliferation.

### 2.7.1 Initiation of the Culture

Primary culture techniques are described in detail later (*see* Section 11.1). Briefly, a culture is derived either by the outgrowth of migrating cells from a fragment of tissue or by enzymatic or mechanical dispersal of the tissue. Regardless of the method employed, primary culture is the first in a series of selective processes (Table 2.2) that may ultimately give rise to a relatively uniform cell line. In primary explantation (*see* Section 11.3.1), selection occurs by virtue of the cells' capacity to migrate from the explant, whereas with dispersed cells, only those cells that both survive the disaggregation technique and adhere to the substrate or survive in suspension will form the basis of a primary culture. If the primary culture is maintained for more than a few hours, a further selection step will occur. Cells that are capable of proliferation will increase, some cell types will survive but not increase, and yet others will be unable to survive under the particular conditions of the culture. Hence the relative proportion of each cell type will change and will continue to do so until, in the case of monolayer cultures, all the available culture substrate is occupied. It should be realized that primary cultures, although suitable for some studies such as cytogenetic analysis, may be unsuitable for other studies because of their instability and heterogeneity. Both cell population changes and adaptive modifications within the cells are occurring continuously throughout the culture period, making it difficult to select a time when the culture may be regarded as homogeneous or stable.

After *confluence* is reached (i.e., all the available growth area is utilized and the cells make close contact with one another), cells whose growth is sensitive to contact inhibition of cell motility and density limitation of cell proliferation (*see* Section 17.5.2) will stop dividing, while any transformed cells, which are insensitive to density limitation, will tend to overgrow. Keeping the cell density low (e.g., by frequent subculture) helps preserve the normal phenotype in cultures such as mouse fibroblasts where spontaneous transformants tend to overgrow at high cell densities [Todaro & Green, 1963].

Some aspects of specialized function are expressed more strongly in primary culture, particularly when the culture becomes confluent. At this stage the cells in the culture will show the closest morphological resemblance to the cells in the parent tissue and retain some diversity. Retention of the characteristics during subculture requires the development of selective conditions: (1) to retain the correct cell lineage, (2) to favor proliferation within this lineage, and (3) to allow for subsequent application of inducing conditions that will favor the expression of the differentiated phenotype.

### 2.7.2 Evolution of Cell Lines

After the first subculture, or passage (Fig. 2.9), the primary culture becomes known as a *cell line* and may be propagated and subcultured several times. With each successive subculture the component of the population with the ability to proliferate most rapidly will gradually predominate, and nonproliferating or slowly proliferating cells will be diluted out. This is most strikingly apparent after the first subculture, in which differences in proliferative capacity are compounded with varying abilities to withstand the trauma of trypsinization and transfer (*see* Section 12.1).

Although some selection and phenotypic drift will continue, by the third passage the culture becomes more stable and is typified by a rather hardy, rapidly proliferating cell. In the presence of serum and without specific selection conditions, mesenchymal cells derived from connective tissue fibroblasts or vascular elements frequently overgrow the culture. Although this has given rise to some very useful cell lines (e.g., WI-38 [Hayflick & Moorhead, 1961] and MRC-5 [Jacobs, 1970] human embryonic lung fibroblasts, BHK21 baby hamster kidney fibroblasts [Macpherson & Stoker, 1962], COS cells [Gluzman, 1981], CHO cells [Puck et al., 1958] (*see* Table 12.1), and perhaps the most famous of all, the L-cell, a mouse subcutaneous fibroblast treated with methylcholanthrene [Earle et al., 1943; Sanford et al., 1948]), this overgrowth represents one of the major challenges of tissue culture since its inception—namely how to prevent the overgrowth of the more fragile or slower growing specialized cells such as hepatic parenchyma or epidermal keratinocytes. Inadequacy of the culture conditions is largely to blame for this problem, and considerable progress has now been made in the use of selective media and substrates for the maintenance of many specialized cell lines (*see* Section 9.2.2; Chapter 22)

**TABLE 2.2.** Selection in Cell Line Development

| | Factors influencing selection | |
| Stage | Primary explant | Enzymatic disaggregation |
| --- | --- | --- |
| Isolation | Mechanical damage | Enzymatic damage |
| Primary culture | Adhesion of explant; outgrowth (migration), cell proliferation | Cell adhesion and spreading, cell proliferation |
| First subculture | Trypsin sensitivity; nutrient, hormone, and substrate limitations; proliferative ability | |
| Propagation as a cell line | Relative growth rates of different cells; selective overgrowth of one lineage Nutrient, hormone, and substrate limitations | |
| | Effect of cell density on predominance of normal or transformed phenotype | |
| Senescence; transformation | Normal cells die out; transformed cells overgrow | |

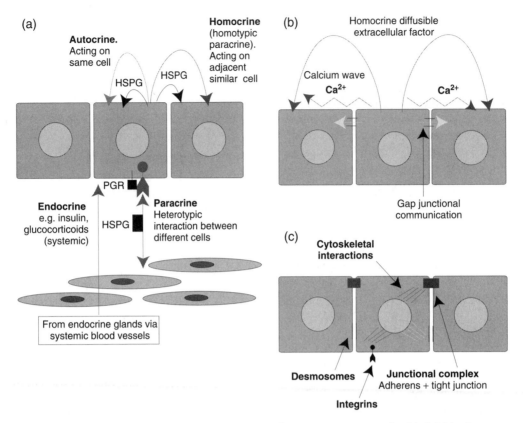

**Fig. 2.8. Cell Interaction and Signaling.** Routes of interaction among cells. (*a*) Soluble factors include endocrine hormones from the vasculature, paracrine factors from the stroma, homocrine factors from adjacent similar cells, and autocrine factors from the cell itself. Matrix, soluble, and cell-associated heparan sulfate proteoglycans (HSPG) and proteoglycan receptors (PGR) may help the activation, stabilization, and translocation of paracrine factors. (*b*) Uniformity of response in target tissue is improved by gap junctional communication, by calcium signaling, and possibly by homocrine factors from the stimulated cell. (*c*) Contact mediated effects also include adherens junctions and tight junctions (associated in junctional complexes) and desmosomes. These, along with integrins, signal via the cytoskeleton, enforcing position, shape, and polarity.

to the extent that many specialized cell types are available commercially (*see* Table 22.1; Appendix II).

## 2.7.3  Senescence

Normal cells can divide a limited number of times; hence cell lines derived from normal tissue will die out after a fixed number of population doublings. This is a genetically determined event involving several different genes and is known as *senescence*. It is thought to be determined, in part, by the inability of terminal sequences of the DNA in the telomeres to replicate at each cell division. The result is a progressive shortening of the telomeres until, finally, the cell is unable to divide further [Bodnar et al., 1998]. Exceptions to this rule are germ cells, stem cells, and transformed cells, which often express the enzyme telomerase, which is capable of replicating the terminal sequences of DNA in the telomere and extending the life span of the cells, infinitely in the case of germ cells and some tumor cells (*see also* Sections 17.4.1, 17.4.4).

## 2.7.4  Transformation and the Development of Continuous Cell Lines

Some cell lines may avoid senescence and give rise to continuous cell lines (*see* Fig. 2.9). The ability of a cell line to grow continuously probably reflects its capacity for genetic variation, allowing subsequent selection. Genetic variation often involves the deletion or mutation of the p53 gene, which would normally arrest cell cycle progression if DNA were to become mutated, and overexpression of the telomerase gene. Human fibroblasts remain predominantly euploid throughout their life span in culture and never give rise to continuous cell lines [Hayflick & Moorhead, 1961], whereas mouse fibroblasts (which are probably more correctly regarded as a more primitive mesodermal precursor cell) and cell cultures from a variety of human and animal tumors often become aneuploid in culture and frequently give rise to continuous cultures. Possibly the condition that predisposes most to the development of a continuous cell line is inherent genetic variation, so it is not surprising to find genetic instability perpetuated in continuous cell lines.

A common feature of many human continuous cell lines is the development of a subtetraploid chromosome number (Fig. 2.10). The alteration in a culture that gives rise to a continuous cell line is commonly called in vitro *transformation* (*see* Section 17.2) and may occur spontaneously or be chemically or virally induced (*see* Section 17.4). The word *transformation* is used rather loosely and can mean different things to different people. In this volume, *immortalization* means the acquisition of an infinite life span and *transformation* implies an additional alteration in growth characteristics (anchorage independence, loss of contact inhibition, and density limitation of growth) that will often, but not necessarily, correlate with tumorigenicity.

Continuous cell lines are usually *aneuploid* and often have a chromosome number between the diploid and tetraploid values (*see* Fig. 2.10). There is also considerable variation in chromosome number and constitution among cells in the population (*heteroploidy*) (*see also* Section 17.3.) It is not clear whether the cells that give rise to continuous lines are present at explantation in very small numbers or arise later as a result of the transformation of one or more cells. The second alternative would seem to be more probable on cell kinetic grounds, as continuous cell lines can appear quite late in a culture's life history, long after the time it would have taken for even one preexisting cell to overgrow. The

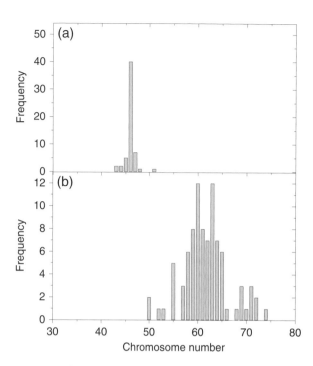

**Fig. 2.10. Chromosome Numbers of Finite and Continuous Cell Lines.** (*a*) A normal human glial cell line. (*b*) A continuous cell line from human metastatic melanoma.

possibility remains, however, that there is a subpopulation in such cultures with a predisposition to transform that is not shared by the rest of the cells.

The term transformation has been applied to the process of formation of a continuous cell line partly because the culture undergoes morphological and kinetic alterations and partly because the formation of a continuous cell line is often accompanied by an increase in tumorigenicity. A number of the properties of continuous cell lines, such as a reduced serum requirement, reduced density limitation of growth, growth in semisolid media, and aneuploidy (*see also* Table 17.1; Plate 14), are associated with *malignant transformations* (*see* Section 17.6). Similar morphological and behavioral changes can also be observed in cells that have undergone virally or chemically induced transformation.

Many (if not most) normal cells do not give rise to continuous cell lines. In the classic example, normal human fibroblasts remain euploid throughout their life span and at crisis (usually around 50 generations) will stop dividing, although they may remain viable for up to 18 months thereafter. Human glia [Pontén & Westermark, 1980] and chick fibroblasts [Hay & Strehler, 1967] behave similarly. Epidermal cells, on the other hand, have shown gradually increasing life spans with improvements in culture techniques [Rheinwald & Green, 1977; Green et al., 1979] and may yet be shown capable of giving rise to continuous growth. Such growth may be related to the self-renewal capacity of the tissue in vivo and successful propagation of the stem cells in vitro (*see* Sections 2.7, 23.1).

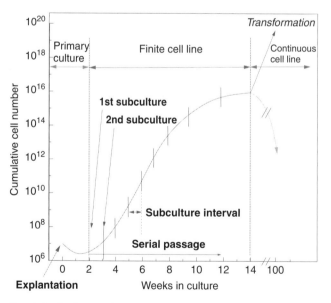

**Fig. 2.9. Evolution of a Cell Line.** The vertical (*y*) axis represents total cell growth (assuming no reduction at passage) for a hypothetical cell culture. Total cell number (cell yield) is represented on this axis on a log scale, and the time in culture is shown on the *x* axis on a linear scale. Although a continuous cell line is depicted as arising at 14 weeks, with different cells it could arise at any time. Likewise senescence may occur at any time, but for human diploid fibroblasts it is most likely to occur between 30 and 60 cell doublings, or 10 to 20 weeks, depending on the doubling time. Terms and definitions used are as in the glossary. (After Hayflick and Moorhead, 1961.)

# CHAPTER 3

# Laboratory Design, Layout, and Equipment

## 3.1 LAYOUT, FURNISHING, AND SERVICES

The need to maintain asepsis distinguishes the tissue culture laboratory from most others, so it is important it be dust free and have no through traffic. The introduction of laminar-flow hoods has greatly simplified the problem and allows the utilization of unspecialized laboratory accommodation, provided that the location is suitable (*see* Sections 3.2.2, 4.2.1, 5.2.1). Layout depends on the type and scale of the operations and the number of users (Figs. 3.1–3.4), but some general principles apply. The rooms should be designed for easy cleaning. Furniture should fit tightly to the floor or be suspended from the bench with a space left underneath for cleaning. Cover the floor with a coved vinyl, acrylic coating, or other dustproof finish, and allow a slight fall in the level toward a floor drain located outside the door of the room (i.e., well away from the sterile cabinets). This arrangement allows liberal use of water if the floor has to be washed, but more important, it protects equipment from damaging floods if stills, autoclaves, or sinks overflow.

If possible it is preferable for the tissue culture lab to be separated from the preparation, washup, and sterilization areas, while still remaining adjacent (*see* Figs. 3.3, 3.4). If you have a large tissue culture lab with a separate washup and sterilization facility, it will be convenient to have this on the same floor as, and adjacent to, the laboratory, with no steps to negotiate, so that carts or trolleys may be used. Across a corridor is probably ideal (*see* Fig. 3.4; *see also* Section 4.3). Try to imagine the flow of traffic—people, reagents, carts, and so on—and arrange for minimum conflict, easy and close access to stores, good access for replenishing stocks without interfering with sterile work, and easy withdrawal of soiled items.

If a conversion of existing facilities is contemplated, then there will be significant structural limitations; choose the location carefully to avoid space constraints and awkward projections into the room that will limit flexibility and air flow.

### 3.1.1 Requirements

Provision must be made for preparation and sterilization, aseptic handling, other activities within the culture area including centrifugation, cell counting, microscopy, incubation, and storage at room temperature, $4°C$, $-20°C$, and $-196°C$.

(1) *Number of users.* How many people will work in the facility, how long will they work each week, and what kinds of culture will they perform? These considerations determine how many laminar-flow hoods will be required (based on whether people can share hoods or will require a hood each for most of the day) and whether a large area will be needed to handle bioreactors, animal tissue dissections, or large numbers of cultures. As a rough guide, 12 laminar-flow hoods in a communal facility can accommodate 50 people with intermittent requirements; extended or continuous use will reduce the capacity proportionately.

(2) *Space.* What space is required for each facility? The largest area should be given to the culture operation, which

*Culture of Animal Cells: A Manual of Basic Technique and Specialized Applications, Sixth Edition*, by R. Ian Freshney
Copyright © 2010 John Wiley & Sons, Inc.

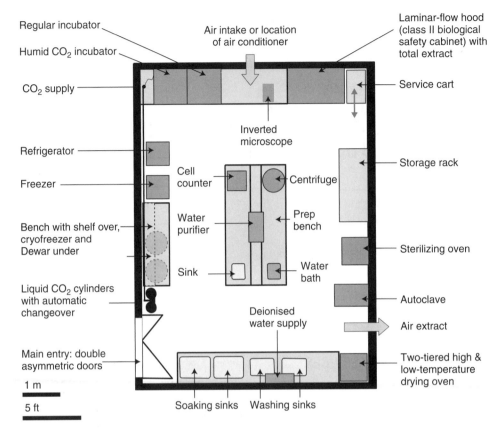

*Fig. 3.1. Small Tissue Culture Laboratory.* Suggested layout for simple, self-contained tissue culture laboratory for use by two or three persons. Dark-shaded areas represent movable equipment, lighter shaded areas fixed or movable furniture.

has to accommodate laminar-flow hoods, cell counters, centrifuges, incubators, microscopes, and using stocks of reagents, media, glassware, and plastics, and, if possible, a quarantine area (*see* Section 3.2). The second largest is for washup, preparation, and sterilization, the third is for storage, and the fourth is for incubation. A reasonable estimate is 4:2:1:1, in the order just presented.

(3) *Location of preparation area.* Facilities for washing up and for sterilization should be located (a) close to the aseptic area that they service and (b) on an outside wall to allow for the possibility of heat extraction from ovens and steam vents from autoclaves. Give your washup, sterilization, and preparation staff a reasonable visual outlook; they usually perform fairly repetitive duties, whereas the scientific and technical staff look into a laminar-flow hood and do not need a view. Opening windows can be a contamination hazard and sunlight can degrade culture medium.

(4) *Storage.* What is the scale of the work contemplated and how much storage space will this require for disposable plastics, and so on? What proportion of the work will be cell line work, with its requirement for storage in liquid nitrogen?

(5) *Access.* Make sure that doorways are wide enough and high enough and that ceilings have sufficient clearance to

allow the installation of equipment such as laminar-flow hoods (which may need additional space for ductwork), incubators, and autoclaves. Provide space for access for maintenance of equipment. Will people require access to the animal facility for animal tissue? If so, ensure that tissue culture is reasonably accessible to, but not contiguous with, the animal facility and that space is provided for a double change of lab coats.

(6) *Containment and sterility.* This is one of the more difficult problems because containment requires that the contents of the tissue culture room not escape to adjacent work areas, while asepsis requires that none of the contamination of surrounding areas enters the tissue culture. If the material being handled is potentially biohazardous, but the work needs to remain sterile, then some sort of buffer area will be required, such as the preparation area or the corridor (Fig. 3.5).

These questions will enable you to decide what size of facility you require and what type of accommodation—one or two small rooms (*see* Figs. 3.1, 3.2), or a suite of rooms incorporating washup, sterilization, one or more aseptic areas, an incubation room, a dark room for fluorescence microscopy and photomicrography, a refrigeration room, and storage (*see* Fig. 3.3). It is essential to have a dedicated

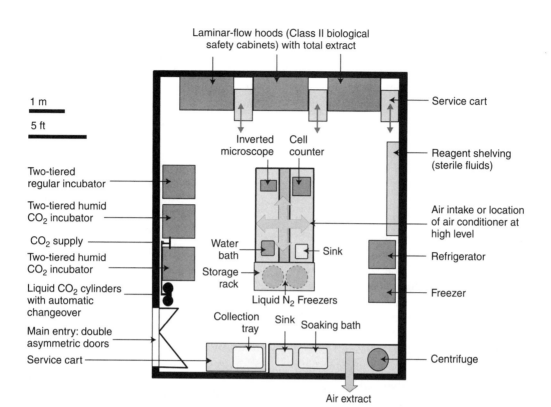

**Fig. 3.2. Medium-Sized Tissue Culture Laboratory.** Suitable for five or six persons, with washing-up and preparation facility located elsewhere. Dark-shaded areas represent movable equipment, light-shaded areas movable or fixed furniture.

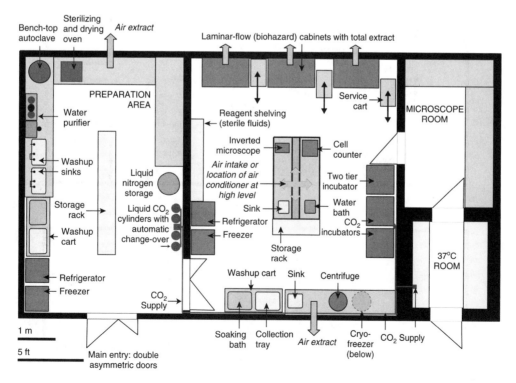

**Fig. 3.3. Tissue Culture Lab with Adjacent Prep Room.** Medium-sized tissue culture lab (see Fig 3.2), but with attached preparation area, microscope room, and 37°C room.

tissue culture laboratory with an adjacent preparation area, or a number of smaller ones with a common preparation area, rather than to have tissue culture performed alongside regular laboratory work with only a laminar-flow hood for protection. A separate facility gives better contamination protection, allows tissue culture stocks to be kept separate from regular laboratory reagents and glassware, and will, in any case, be required for containment if human or other primate cells are handled (*see* Section 6.8.1).

### 3.1.2 Services

Hot and cold water, power, combustible gas (domestic methane, propane, etc.), carbon dioxide, and compressed air will be required. Power is always underestimated, in terms of both the number of outlets and the amperage per outlet. Assess carefully the equipment that will be required, assume that both the number of appliances and their power consumption will treble within the life of the building in its present form, and try to provide sufficient power, preferably at or near the outlets (preferably located on a power track or buss) but at least at the main distribution board. Hot and cold water with sinks and drainage will be required in both the preparation (*see* Fig. 3.5) and the tissue culture areas (*see* Figs 3.1–3.4). Adequate floor drainage should be provided in the preparation/washup area, with a slight fall in floor level

from the tissue culture lab to the washup. Combustible gas may be required, but electricity is cleaner and generally easier to manage from a safety standpoint.

If possible, carbon dioxide should be piped into the facility. The installation will pay for itself eventually in the cost of cylinders of mixed gases for gassing cultures, and it provides a better supply, which can be protected, for gassing incubators (*see* Section 4.3.2). Gas-flow meters or electronic gas blenders (*see* Appendix II) can be installed at workstations to provide the correct gas mixture. Compressed air is generally no longer required at incubators, as $CO_2$ incubators regulate the gas mixture from pure $CO_2$, but will be required if a gas mixer is provided at each workstation. Compressed air is also used to expel cotton plugs from glass pipettes before washing and may be required for some types of glassware washing machine (e.g., Scientek 3000).

A vacuum line can be useful for evacuating culture flasks, but a collection vessel must be present with an additional trap flask, with a hydrophobic filter between the flasks, in order to prevent fluid, vapor, or some contaminant from entering the vacuum line and pump. Also the vacuum pump must be protected against the line being left open inadvertently; usually this can be accomplished via a pressure-activated foot switch that closes when no longer pressed. In many respects it is better to provide individual peristaltic pumps at each

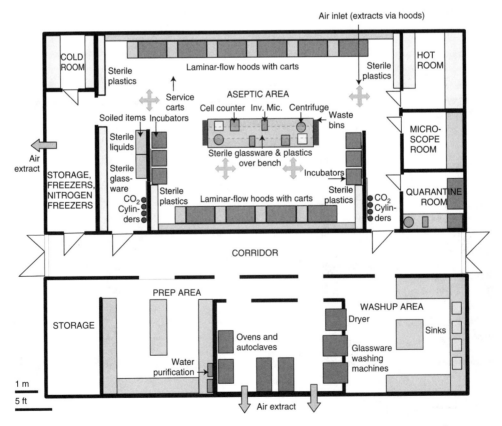

**Fig. 3.4. Large Tissue Culture Laboratory.** Suitable for 20 to 30 persons. Adjacent washing-up, sterilization, and preparation area. Dark-shaded areas represent equipment, light-shaded areas fixed and movable furniture.

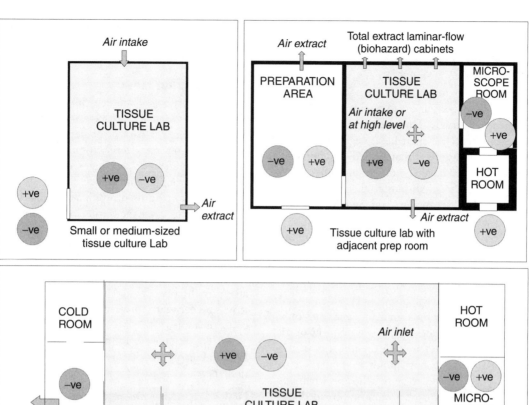

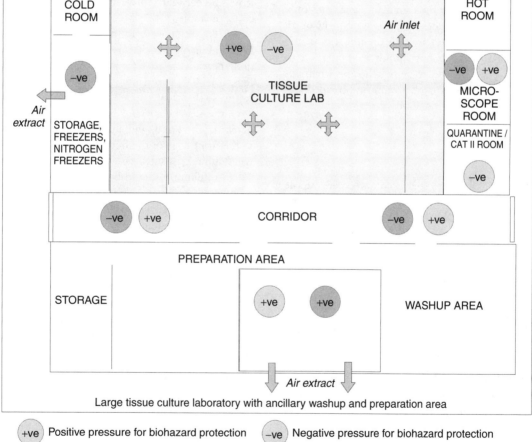

**Fig. 3.5. Air Pressure Balance.** Indications of relative pressure in different areas within an aseptic suite, dark-filled circles for maximum sterility, light-filled circles for biohazard protection.

workstation (*see* Figs. 4.1, 4.9), or one pump between two workstations.

### 3.1.3 Ventilation

Ideally a tissue culture laboratory should be at positive pressure relative to surrounding work areas, to avoid any influx of contaminated air from outside. However, if containment is required (*see* Section 6.8.1), it must be at negative pressure relative to the surrounding areas. To satisfy both requirements it may be preferable to have a positive-pressure buffer zone, receiving HEPA-filtered air outside the tissue culture laboratory, such as the Preparation Area and Microscope Room or the corridor (Fig. 3.5; see also Figs. 3.3, 3.4).

It is preferable to duct laminar-flow hoods to the exterior to improve air circulation and remove excess heat (300–500 W per hood) from the room. This also facilitates decontamination with formaldehyde, should it be required. Venting hoods to the outside will probably provide most of the air extraction required for the room, and it remains only to ensure that the incoming air, from a central plant or an air conditioner, does not interfere with the integrity of the airflow in the hood. Laminar-flow hoods are better left to run continuously, but if they are to be switched off when not in use, then an alternative air extract must be provided and balanced with the extract via the hoods.

## 3.2 LAYOUT

Six main functions need to be accommodated: sterile handling, incubation, preparation, washup, sterilization, and storage (Table 3.1). If a single room is used, create a "sterility gradient"; the clean area for sterile handling should be located at one end of the room, farthest from the door, and washup and sterilization facilities should be placed at the other end, with preparation, storage, and incubation in between. The preparation area should be adjacent to the washup and sterilization areas, and storage and incubators should be readily accessible to the sterile working area (*see* Figs. 3.1–3.4).

### 3.2.1 Sterile Handling Area

Sterile work should be located in a quiet part of the tissue culture laboratory and should be restricted to tissue culture (not shared with chemical work or with work on other organisms, e.g., bacteria, yeast, or protozoa), and there should be no through traffic or other disturbance that is likely to cause dust or drafts. Use a separate room or cubicle if laminar-flow hoods are not available. The work area, in its simplest form, should be a plastic laminate-topped bench, preferably plain white or neutral gray, to facilitate the observation of cultures and dissection, and to allow an accurate reading of pH when phenol red is used as an indicator. Nothing should be stored on the bench, and any shelving above should be used only in conjunction with sterile work (e.g., for holding pipettes and

instruments). The bench should be either freestanding (away from the wall) or sealed to the wall with a plastic sealing strip or mastic sealant.

### 3.2.2 Laminar Flow

The introduction of laminar-flow hoods with sterile air blown onto the work surface (*see* Section 4.2.1; Figs. 4.1, 5.3) affords greater control of sterility at a lower cost than providing a separate sterile room. Individual freestanding hoods are preferable, as they separate operators and can be moved around, but laminar-flow wall or ceiling units in batteries can be used. With individual hoods, only the operator's arms enter the sterile area, whereas with laminar-flow wall or ceiling units, there is no cabinet and the operator is part of the work area. Although this arrangement may give more freedom of movement, particularly with large pieces of apparatus (roller bottles, bioreactors), greater care must be taken by the operator not to disrupt the laminar flow, and it will be necessary to wear sterile caps and gowns to avoid contamination.

Select hoods that suite your accommodation— freestanding or bench top—and allow plenty of legroom underneath with space for pumps, aspirators, and so forth (*see* Figs. 3.1, 4.1). Freestanding cabinets should be on lockable castors so that they can be moved if necessary. Chairs should have the seat height and back angle adjustable to suit the height of the hood, and able to be drawn up close enough to the front edge of the hood to allow comfortable working well within it. A small cart, trolley, or folding flap (500 mm, 18 in. minimum) should be provided beside each hood for a notebook and materials that may be required but are not in immediate use. Trolleys can be pulled out for use and restocking. If desired, they can be exchanged for a clean freshly stocked trolley each time the operator changes.

Laminar-flow hoods should have a lateral separation of at least 500 mm (2 ft), to allow access for maintenance and to minimize interference in airflow between hoods. If hoods are opposed, there should be a minimum of 3000 mm (10 ft) between the fronts of each hood. Laminar-flow hoods should be installed as part of the construction contract as they will influence ventilation.

### 3.2.3 Service Bench

It may be convenient to position a bench for a cell counter, microscope, and other critical instruments, close to the sterile handling area and either dividing the area or separating it from the other end of the lab (*see* Figs. 3.1–3.4). The service bench should also provide for the storage of sterile glassware, plastics, pipettes, screw caps, and syringes, for example, in drawer units below and open shelves above. The bench may also be used for other accessory equipment, such as a small centrifuge whose contents should be readily accessible.

### 3.2.4 Quarantine and Containment

If sufficient space is available, designate a separate room as a quarantine and/or containment room (*see* Fig. 3.4). This

**TABLE 3.1.** Tissue Culture Facilities

| Minimum requirements | Desirable features | Useful additions |
|---|---|---|
| Sterile area, clean, quiet, and with no through traffic | Filtered air (air-conditioning) | Piped $CO_2$ and compressed air |
| Separate from animal house and microbiological labs | Service bench adjacent to culture area | Storeroom for bulk plastics |
| Preparation area | Separate prep room | Quarantine room |
| Washup area (not necessarily within tissue culture laboratory, but at least adjacent to it) | Hot room with temperature recorder | Containment room (could double as quarantine room) |
| Space for incubator(s) | Separate sterilizing room | Liquid $N_2$ storage tank ($\approx 500$ L) and separate storeroom for nitrogen freezers |
| Storage areas: | Separate cylinder store | Microscope room |
|   Liquids: ambient, 4°C, −20°C | | Darkroom |
|   Glassware (shelving) | | Vacuum line |
|   Plastics (shelving) | | |
|   Small items (drawers) | | |
|   Specialized equipment (slow turnover), cupboard(s) | | |
|   Chemicals: ambient, 4°C, −20°C (share with liquids, but keep chemicals in sealed container over desiccant) | | |
|   $CO_2$ cylinders | | |
|   Space for liquid $N_2$ freezer(s) | | |
| Sink | | |

is a separate aseptic room with its own laminar-flow hood (Class II microbiological safety cabinet), incubators, freezer, refrigerator, centrifuge, supplies, and disposal. This room must be separated by a door or air lock from the rest of the suite and be at negative pressure to the rest of the aseptic area. Newly imported cell lines or biopsies can be handled here until they are shown to be free of contamination, particularly mycoplasma (*see* Section 18.3.2; Protocols 19.2, 19.3), and proscribed pathogens such as HIV or hepatitis B. If local rules will allow, the same room can serve as a Level II containment room at different designated times. If used at a higher level of containment, it will also require a biohazard cabinet or pathogen hood with a separate extract and pathogen trap (*see* Section 6.8.2).

### 3.2.5  Incubation

The requirement for cleanliness is not as stringent as that for sterile handling, but clean air, a low disturbance level, and minimal traffic will give your incubation area a better chance of avoiding dust, spores, and the drafts that carry them. What type of incubation will be required is determined by size, temperature, gas phase, and proximity to the work space. Regular, nongassed incubators or a hot room may suffice, or a $CO_2$ and a humid atmosphere may be required. Generally, large numbers of flasks or large-volume flasks that are sealed are best incubated in a hot room, whereas open plates and dishes will require a humid $CO_2$ incubator.

*Incubators.* Incubation may be carried out in separate incubators or in a thermostatically controlled hot room (Fig. 3.6). Incubators are inexpensive and economical in terms of space if only one or two are required. These can be supplied for assembly (or disassembly) on site (Cellon)

allowing them to be withdrawn from use if not required. But as soon as you require more than three or four incubators, their cost becomes more than that of a simple hot room, and their use is less convenient. Incubators also lose more heat when they are opened and are slower to recover than a hot room. As a rough guide, you will need 0.2 m³ (200 L, 6 ft³) of incubation space with 0.5 m² (6 ft²) shelf space per person. Extra provision may need to be made for one or more humid incubators with controlled $CO_2$ (*see* Section 4.3.2).

*Hot room.* If you have the space within the laboratory area or have an adjacent room or walk-in cupboard readily available and accessible, it may be possible to convert the area into a hot room (*see* Fig. 3.6). The area need not be specifically constructed as a hot room, but it should be insulated to prevent cold spots being generated on the walls. If insulation is required, line the area with plastic laminate-veneered board, separated from the wall by about 5 cm (2 in.) of fiberglass, mineral wool, or fire-retardant plastic foam. Mark the location of the straps or studs carrying the lining panel in order to identify anchorage points for wall-mounted shelving if that is to be used. Use demountable shelving and space shelf supports at 500 to 600 mm (21 in.) to support the shelving without sagging. Freestanding shelving units are preferable, as they can be removed for cleaning the rack and the room. Allow 200 to 300 mm (9 in.) between shelves, and use wider shelves (450 mm, 18 in.) at the bottom and narrower (250–300 mm, 12 in.) ones above eye level. Perforated shelving mounted on adjustable brackets will allow for air circulation. The shelving must be flat and perfectly horizontal, with no bumps or irregularities.

Do not underestimate the space that you will require over the lifetime of the hot room. It costs very little more to equip

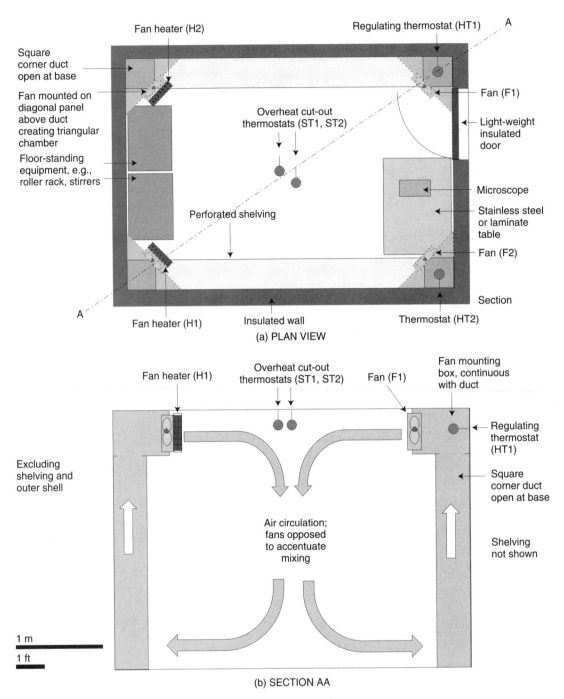

**Fig. 3.6. Hot Room.** Dual heating circuits and safety thermometers. (*a*) Plan view. (*b*) Diagonal section. Arrows represent air circulation. Layout and design were developed in collaboration with M. McLean of Boswell, Mitchell & Johnson (architects) and J. Lindsay of Kenneth Munro & Associates (consulting engineers).

a large hot room than a small one. Calculate costs on the basis of the amount of shelf space you will require; if you have just started, multiply by 5 or 10; if you have been working for some time, multiply by 2 or 4.

Wooden furnishings should be avoided as much as possible, as they warp in the heat and can harbor infestations. A small bench, preferably stainless steel or solid plastic laminate, should be provided in some part of the hot room. The bench should accommodate an inverted microscope, the flasks that you wish to examine, and a notebook. If you contemplate doing cell synchrony experiments or having to make any sterile manipulations at 37°C, you should also allow space for a small laminar-flow unit with a 300 × 300 or 450 × 450 mm (12–18 in.) filter size, mounted either on a wall or on a stand over part of the bench. Alternatively, a small laminar-flow hood around 1000 mm (3 ft) wide could be located in the room. The fan motor should be specified as for use in the tropics and should not run continuously. If it does run continuously, it will generate heat in the room and the motor may burn out.

Once a hot room is provided, others may wish to use the space for non−tissue-culture incubations, so the area of bench space provided should also take account of possible usage for incubation of tubes, shaker racks, and other such items. However, ban the use of other microorganisms, such as bacteria or yeast.

Incandescent lighting is preferable to fluorescent, which can cause degradation of the medium. Furthermore some fluorescent tubes have difficulty lighting up in a hot room.

The temperature of the hot room should be controlled within $\pm 0.5°C$ at any point and at any time, and depends on the sensitivity and accuracy of the control gear, the location of the thermostat sensor, the circulation of air in the room, the nature of the insulation, and the evolution of heat by other apparatus (stirrers, etc.) in the room.

*Heaters.* Heat is best supplied via a fan heater, domestic or industrial, depending on the size of the room. Approximately 2 to 3 kW per 20 m$^3$ (700 ft$^3$) will be required (or two heaters could be used, each generating 1.0−1.5 kW), depending on the insulation. The fan on the heater should run continuously, and the power to the heating element should come from a proportional controller.

*Air circulation.* A second fan, positioned on the opposite side of the room and with the airflow opposing that of the fan heater, will ensure maximum circulation. If the room is more than 2 × 2 m (6 × 6 ft), some form of ducting may be necessary. Blocking off the corners (*see* Fig. 3.6*a*) is often easiest and most economical in terms of space in a square room. In a long, rectangular room, a false wall may be built at either end, but be sure to insulate it from the room and make it strong enough to carry shelving.

*Thermostats.* Thermostats should be of the "proportional controller" type, acting via a relay to supply heat at a rate proportional to the difference between the room temperature and the set point. When the door opens and the room temperature falls, recovery will be rapid; on the other hand, the temperature will not overshoot its mark, as the closer it approaches the set point, the less heat is supplied.

Ideally there should be two separate heaters (H1 and H2), each with its own thermostat (HT1 and HT2). One thermostat (HT1) should be located diagonally opposite and behind the opposing fan (F1) and should be set at 37°C. The other thermostat (HT2) should be located diagonally opposite H2 and behind its opposing fan (F2) and should be set at 36°C (*see* Fig. 3.6*a*). Two safety override cutout thermostats should also be installed, one in series with HT1 and set at 38°C and the other in series with the main supply

to both heaters. If the first heater (H1) stays on above the set point, ST1 will cut out, and the second heater (H2) will take over, regulating the temperature on HT2. If the second heater also overheats, ST2 will cut out all power to the heaters (Table 3.2). Warning lights should be installed to indicate when ST1 and ST2 have been activated. The thermostat sensors should be located in an area of rapid airflow, close to the effluent from the second, circulating, fan for greatest sensitivity. A rapid-response, high-thermal conductivity sensor (thermistor or thermocouple) is preferred over a pressure-bulb type.

*Overheating.* The problem of unwanted heat gain is often forgotten because so much care is taken to provide heat and minimize loss. It can arise because of (1) a rise in ambient temperature in the laboratory in hot weather or (2) heat produced from within the hot room by apparatus such as stirrer motors, roller racks, and laminar-flow units. Try to avoid heat-producing equipment in the hot room. Induction-drive magnetic stirrers produce less heat than mechanically driven magnets, and drive motors for roller racks can sometimes be located outside the hot room. In tropical regions, or where overheating is a frequent problem, it may be necessary to incorporate cooling coils in the duct work of the heaters.

*Access.* If a proportional controller, good circulation, and adequate heating are provided, an air lock will not be required. The door should still be well insulated (with foam plastic or fiberglass), light, and easily closed, preferably self-closing. It is also useful to have a hatch leading into the tissue culture area, with a shelf on both sides, so that cultures may be transferred easily into the room. The hatch door should have an insulated core as well. Locating the hatch above the bench will avoid any risk of creating a "cold spot" on the shelving.

*Thermometer.* A temperature recorder should be installed and should be visible to the people working in the tissue culture room. The chart should be changed weekly. If possible, one high-level and one low-level warning light should be placed beside the chart or at a different, but equally obvious, location.

### 3.2.6 Preparation Area

*Media preparation.* The need for extensive preparation of media in small laboratories can be avoided if there is a reliable source of commercial culture media. Although a large enterprise (approximately 50 people doing tissue culture) may still find it more economical to prepare its own media, smaller laboratories may prefer to purchase readymade media. These laboratories would then need only to prepare reagents, such as salt solutions and ethylenediaminetetraacetic acid (EDTA), bottle these reagents and water, and package screw caps and other small items for sterilization. In that case, although the preparation area should still be clean and quiet, sterile handling is not necessary, as all the items will be sterilized.

**TABLE 3.2.** Hot-Room Thermostats

| Thermostat[a] | 37°C | <37°C | ≪37°C | >37°C<38°C | >37°C>38°C | >37°C>39°C |
|---|---|---|---|---|---|---|
| HT1 | O | I | I | O | O | O |
| ST1 | I | I | I | I | O | O |
| HT2 | O | O | I | O | O | O |
| ST2 | I | I | I | I | I | O |
| Warning light ST1 | | | | | ◆ | ◆ |
| Warning light ST2 | | | | | | ◆ |

*Note:* I = on; O = off; ◆ = pilot light illuminated.
[a]HT1, regulating thermostat for heater H1; HT2, regulating thermostat for heater H2; ST1, safety override cut out thermostat for H1; ST2, safety override cut-out thermostat for common supply to H1 and H2.

If reliable commercial media are difficult to obtain, the preparation area should be large enough to accommodate a coarse and a fine balance, a pH meter, and, if possible, an osmometer. Bench space will be required for dissolving and stirring solutions and for bottling and packaging various materials, and additional ambient and refrigerated shelf space will also be needed. If possible, an extra horizontal laminar-flow hood should be provided in the sterile area for filtering and bottling sterile liquids. Note that antibiotics and toxic reagents should not be handled in horizontal laminar flow due to the potential for inhalation of powder or aerosols. Incubator space must be allocated for quality control of sterility (i.e., incubation of samples of media in broth and after plating out).

Heat-stable solutions and equipment can be autoclaved or dry-heat sterilized at the nonsterile end of the preparation area. Both streams then converge on the storage areas (*see* Fig. 3.3).

*Washup.* Washup and sterilization facilities are best situated outside the tissue culture lab, as the humidity and heat that they produce may be difficult to dissipate without increasing the airflow above desirable limits. Autoclaves, ovens, and distillation apparatus should be located in a separate room if possible (*see* Figs. 3.3, 3.4), with an efficient extraction fan. The washup area should have plenty of space for soaking glassware and space for an automatic washing machine, should you require one. There should also be plenty of bench space for handling baskets of glassware, sorting pipettes, and packaging and sealing packs for sterilization. In addition you will need space for a pipette washer and dryer. If the sterilization facilities must be located in the tissue culture lab, place them nearest the air extract and farthest from the sterile handling area.

If you are designing a lab from scratch, then you can get sinks built in of the size that you want. Stainless steel or polypropylene are best, the former if you plan to use radioisotopes and the latter for hypochlorite disinfectants.

Sinks should be deep enough (450 mm, 18 in.) to allow manual washing and rinsing of your largest items without having to stoop too far to reach into them. They should measure about 900 mm (3 ft) from floor to rim (Fig. 3.7). It is better to be too high than too low—a short person can always stand on a raised step to reach a high sink, but a tall person will always have to bend down if the sink is too low. A raised edge around the top of the sink will contain spillage and prevent the operator from getting wet when bending over the sink. The raised edge should go around behind the taps at the back.

Each washing sink will require four taps: a single cold-water tap, a combined hot-and-cold mixer, a cold tap for a hose connection for a rinsing device, and a nonmetallic or stainless steel tap for deionized water from a reservoir above the sink (*see* Fig. 3.7). A centralized supply for deionized water should be avoided, as the piping can build up dirt and algae and is difficult to clean.

Trolleys or carts are often useful for collecting dirty glassware and redistributing fresh sterile stocks, but remember to allocate parking space for them.

### 3.2.7 Storage

Storage must be provided for the following items to ensure that sterile and nonsterile are kept separate and clearly labeled:

(1) Sterile liquids, at room temperature (salt solutions, water, etc.), at 4°C (media), and at −20°C or −70°C (serum, trypsin, glutamine, etc.)
(2) Sterile and nonsterile glassware, including media bottles and pipettes
(3) Sterile disposable plastics (e.g., culture flasks and Petri dishes, centrifuge tubes and vials, and syringes)
(4) Screw caps, stoppers, and other such sterile and nonsterile items
(5) Apparatus such as filters, sterile and nonsterile
(6) Gloves, plastic bags, and other disposable items
(7) Liquid nitrogen to replenish freezers; the liquid nitrogen should be stored in two ways:
  (a) in Dewars (25–50 L) under the bench, or
  (b) in a large storage vessel (100–150 L) on a trolley or in storage tanks (500–1000 L) permanently sited in a room of their own with adequate ventilation or, preferably, outdoors in secure, weatherproof housing (Fig. 3.8).

*Note.* Liquid-nitrogen storage vessels can build up

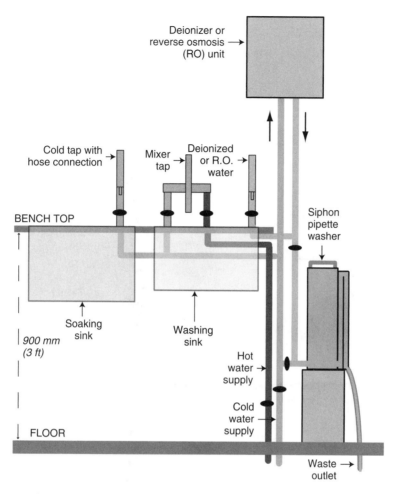

***Fig. 3.7. Washingup Sink and Pipette Washer.*** Suggested layout for soaking and washup sinks, with hot, cold, and deionized water supplies.

contamination, so they should be kept in clean areas. Using a perfused wall freezer will help prevent this (*see* Section 19.3.6).

Δ *Safety Note.* Adequate ventilation must be provided for the room in which the nitrogen is stored and dispensed, preferably with an alarm to signify when the oxygen tension falls below safe levels. The reason for this safety measure is that the filling, dispensing, and manipulating of freezer stocks is accompanied by the evaporation of nitrogen, which can replace the air in the room (1 L liquid $N_2 \approx 700$ L gaseous $N_2$).

(8) Cylinder storage for carbon dioxide, in separate cylinders for transferring to the laboratory as required

Δ *Safety Note.* The cylinders should be tethered to the wall or bench in a rack (*see* Fig. 6.3).

(9) A piped supply of $CO_2$ to be taken to workstations; or else the $CO_2$ supply can be piped from a pressurized tank of $CO_2$ that is replenished regularly (and must therefore be accessible to delivery vehicles). Which of the two means of storage you actually use will be based on your scale of operation and unit cost. As a rough guide, 2 to 3 people will only require a few cylinders, 10 to 15

will probably benefit from a piped supply from a bank of cylinders, and for more than 15 it will pay to have a storage tank.

Storage areas 1 to 6 should be within easy reach of the sterile working area. Refrigerators and freezers should be located toward the nonsterile end of the lab, as the doors and compressor fans create dust and drafts and may harbor fungal spores. Also refrigerators and freezers require maintenance and periodic defrosting, which creates a level and kind of activity best separated from your sterile working area.

The key ideal regarding storage areas is ready access for both withdrawal and replenishment of stocks, keeping older stocks to the front. Double-sided units are useful because they may be restocked from one side and used from the other.

Δ *Safety Note.* It is essential to have a lip on the edge of both sides of a shelf if the shelf is at a high level and glassware and reagents are stored on it. This prevents items being accidentally dislodged during use and when stocks are replenished.

Remember to allocate sufficient space for storage, as doing so will allow you to make bulk purchases, thereby

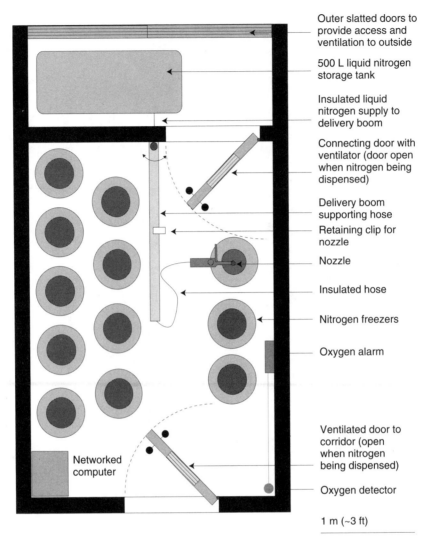

Outer slatted doors to provide access and ventilation to outside

500 L liquid nitrogen storage tank

Insulated liquid nitrogen supply to delivery boom

Connecting door with ventilator (door open when nitrogen being dispensed)

Delivery boom supporting hose

Retaining clip for nozzle

Nozzle

Insulated hose

Nitrogen freezers

Oxygen alarm

Ventilated door to corridor (open when nitrogen being dispensed)

Oxygen detector

Networked computer

1 m (~3 ft)

**Fig. 3.8. Liquid-Nitrogen Store and Cryostore.** The liquid-nitrogen store is best located on an outer wall with ventilation to the outside and easy access for deliveries. If the freezer store is adjacent, freezers may be filled directly from an overhead supply line and flexible hose. Doors are left open for ventilation during filling, and a wall-mounted oxygen alarm with a low-mounted detector sounds if the oxygen level falls below a safe level.

saving money, and, at the same time, reduce the risk of running out of valuable stocks at times when they cannot be replaced. As a rough guide, you will need 200 L (~8 ft³) of 4°C storage and 100 L (~4 ft³) of −20°C storage per person. The volume per person increases with fewer people. Thus one person may need a 250-L (10-ft³) refrigerator and a 150-L (6-ft³) freezer. Of course, these figures refer to storage space only, and allowance must be made for access and working space in walk-in cold rooms and deep freezer rooms.

In general, separate −20°C freezers are better than a walk-in −20°C room. They are easier to clean out and maintain, and they provide better backup if one unit fails.

You may also wish to consider whether a cold room has any advantage over refrigerators. No doubt, a cold room will give more storage per cubic meter, but the utilization of that space is important—how easy is it to clean and defrost, and how well can space be allocated to individual users? Several independent refrigerators will occupy more space than the equivalent volume of cold room, but may be easier to manage and maintain in the event of failure. However, a walk-in cold room may be required if cold preparation or isolation procedures are used (although cold benches may suffice). It is well worth considering budgeting for additional freezer and refrigerator space to allow for routine maintenance and unpredicted breakdowns.

# CHAPTER 4

# Equipment and Materials

## 4.1 REQUIREMENTS OF A TISSUE CULTURE LABORATORY

Unless unlimited funds are available, it will be necessary to prioritize the specific needs of a tissue culture laboratory: (1) essential—you cannot perform tissue culture reliably without this equipment; (2) beneficial—culture would be done better, more efficiently, quicker, or with less labor; and (3) useful—items that would improve working conditions, reduce fatigue, enable more sophisticated analyses to be made, or generally make your working environment more attractive (Table 4.1). In the following sections, items are presented under activity-based subject headings and their priorities given in Table 4.1.

The need for a particular piece of equipment is often very subjective—technical innovation, personal aspirations, merchandizing, and peer pressure. The real need is harder to define but is determined objectively by the type of work, the saving in time that the equipment would produce, the greater technical efficiency in terms of asepsis, quality of data, analytical capability, sample requirements, the saving in time or personnel, the number of people who would use the device, the available budget and potential cost benefit, and the special requirements of your own procedures.

The sources of individual pieces of equipment are listed in Appendix II and the Suppliers' details in Appendix III.

## 4.2 ASEPTIC AREA

### 4.2.1 Laminar-Flow Hood

It is possible to carry out aseptic procedures without laminar flow if you have appropriately isolated, clean accommodation with restricted access. However, it is clear that, for most laboratories, which are typically busy and overcrowded, the simplest way to provide aseptic conditions is to use a laminar-flow hood (see Section 5.4). Usually one hood is sufficient for two to three people. A horizontal-flow hood is cheaper and provides the best sterile protection for your cultures, but it is really suitable only for preparing medium (without antibiotics) and other nontoxic sterile reagents and for culturing nonprimate cells. It is particularly suitable for dissecting nonprimate material for primary culture. For potentially hazardous materials (any primate, including human, cell lines; virus-producing cultures; radioisotopes; and carcinogenic or toxic drugs), a Class II (Fig. 4.1) or Class III (Fig. 7.4) microbiological safety cabinet should be used. In practice, most laboratories now use a Class II microbiological safety cabinet as standard.

Δ *Safety Note.* It is important to familiarize yourself with local and national biohazard regulations before installing equipment because legal requirements and recommendations vary (see Section 6.8.1).

**TABLE 4.1.** Tissue Culture Equipment

| Item | Purpose | Requirement |
|---|---|---|
| **Tissue culture laboratory** | | |
| Laminar-flow hood (biological safety cabinet) | Maintain aseptic environment and containment | Essential (open bench or sterile room may suffice for small number of users in quiet area) |
| Trolleys or carts | Temporary storage at hoods | Nonessential but convenient |
| Incubator | Controlled temperature for incubation | Essential |
| Inverted microscope | Viewing and assessing cultures | Essential |
| Camera for inverted microscope(s) | Recording cell images; comparison with reference material | Useful for record purposes |
| Hemocytometer slides | Counting cells | Essential if electronic counter not available; still desirable for checking viability |
| Cell counter | Automatic cell counting | Preferable to hemocytometer for accuracy and lower error |
| Refrigerator | Local storage of media and reagents | Essential |
| Freezer | For $-20°C$ storage of unstable media, serum and reagents | Essential |
| Soaking bath or sink | Collection for washup | Essential |
| Pipette cylinder(s) | Collecting and disinfecting used pipettes | Essential |
| Peristaltic pump | Withdrawing fluid from flasks; filter sterilization | Preferable to vacuum pump for both but not essential if vacuum pump available |
| Bench centrifuge | Centrifuging cells to remove trypsin, preservative, or experimental additives | Essential |
| Humid, $CO_2$ incubator | Controlling humidity and $CO_2$ concentration | Essential for open plates or dishes (except for very small-scale activity where sealed chamber may suffice) |
| Liquid $CO_2$ cylinders, without siphon | Supply for $CO_2$ incubator | Essential unless piped $CO_2$ supply available |
| 5% $CO_2$ cylinder | Gassing flask cultures | Nonessential if $CO_2$ mixer available, permeable caps used in $CO_2$ incubator, or $CO_2$ not used |
| Piped $CO_2$ supply from cylinder store | Supply to hoods and incubators | Preferable to taking cylinders into aseptic suite, particularly if more than two incubators and two hoods |
| Automatic change-over device on $CO_2$ cylinders | Backup when first cylinder runs out | Essential if piped supply used |
| Pipettor(s) | Dispensing small volumes (5 μL 1 mL) accurately and reproducibly | Preferable to pipetting for small amounts |
| Pipette bulbs | Control uptake and dispensing by pipette | Some form of control essential; cheaper than controller but less accurate |
| Pipette controller(s) | Control uptake and dispensing by pipette | More accurate (though slower) than pipette bulbs |
| Liquid $N_2$ freezer | Preservation of seed and using stocks of cell lines | Essential for cell line work but not if limited to primary culture |
| Liquid $N_2$ storage Dewar | Local supply of liquid $N_2$ for freezer | Essential for cryostorage even if central supply is available |
| Slow-cooling device for cell freezing (see Section 19.3.4) | Control freezing rate for cells | Essential for cell freezing unless programmable controlled rate cooler available |
| Controlled-rate cooler | For cell freezing | Necessary for some cell types |
| Roller racks | For roller bottle culture | Option for monolayer scale-up |
| Magnetic stirrer | Maintaining uniform suspension of cells for suspension cultures or trypsinization for primary culture | Essential for large-scale suspension cultures and warm trypsinization (see Protocol 11.5) |
| Hot room | Incubation of large numbers of flasks, roller bottles, or stirrer cultures | Not required for small numbers of flasks or if all culture is in open plates or dishes |
| Portable temperature recorder | For checking hot room or incubators | Useful |

**TABLE 4.1.** (*Continued*)

| Item | Purpose | Requirement |
|---|---|---|
| *Wash-up* | | |
| Deep washing sink | Washing glassware | Not essential if only disposables used (although a small rinsing sink will still be required) |
| Pipette washer | Washing glass pipettes | Not required if using plastic pipettes |
| Pipette drier | Drying glass pipettes | Not required if disposable plastic pipettes are used; more convenient than oven for drying glass pipettes |
| Pipette plugger | Plugging glass pipettes | Not required if disposable plugged plastic pipettes are used |
| Glassware washing machine | Washing glassware and pipettes | Useful if a lot of glassware used |
| Drying oven(s), high and low temperature | Drying glassware and plastics | Essential if any washup being done |
| Trolleys or carts | Transfer of soiled glassware from culture area to washup | Beneficial, depending on scale of operation; small carts at hoods may be sufficient for small-scale operations but larger carts may become necessary if more than 2 hoods are being serviced |
| Plastics shredder/sterilizer | Disposal/recycling of plasticware | Useful, but use only after sterilization of used plastic |
| **Preparation** | | |
| Still or water purifier | Supply of ultrapure water | Essential if any media or reagents to be prepared or concentrates diluted |
| Balance | Weighing chemicals for reagent and medium preparation | Essential unless all media and reagents purchased ready for use |
| pH Meter | Measuring pH in prepared media and reagents | Essential if media or reagents being made up |
| Conductivity meter | Quality check on reagents | Useful; essential if making up own media |
| Osmometer | Quality check on reagents | Useful; essential if making up own media |
| Automatic dispenser | Dispensing liquids | Useful but not essential |
| Large refrigerator or walk-in cold room | Bulk long-term storage of media and reagents | Essential |
| **Sterilization** | | |
| Sterilizing oven | Sterilization of glassware, metals, and heat-resistant plastics; can double as drying oven | Essential unless only disposables used or if autoclave available |
| Steam sterilizer (autoclave, pressure cooker) | Sterilizing stable liquids and solutions | Essential unless all media and reagents purchased ready for use |
| Temperature recorders on sterilizing oven and autoclave and in hot room | Monitoring incubator and sterilizer performance | Essential if hot room or sterilizers used |
| Filter sterilization | Sterilizing heat-labile solutions | Essential if media made up from powder; otherwise, a selection of disposable units would suffice (*see* Table 4.2) |
| Polyethylene bag sealer | For packaging sterile items for long-term storage | Useful |
| *Ancillary equipment* | | |
| Phase-contrast, upright microscope | High-power observation of cells | Not required for routine culture |
| Fluorescence microscope (combined with above) | Visualizing staining with fluorochromes such as fluorescent antibodies (for fluorescent staining), DAPI or Hoechst 33285 (for mycoplasma detection) | Essential for mycoplasma testing by Hoechst method and for immunofluorescence staining |
| PCR thermal cycler | For PCR amplification of DNA or RT-PCR of RNA | Essential for mycoplasma testing by PCR |

(*continued overleaf*)

**TABLE 4.1.** (*Continued*)

| Item | Purpose | Requirement |
|---|---|---|
| Micromanipulators | Microinjection | Essential for nuclear transplantation or dye injection |
| CCD camera and time-lapse video equipment | Real time and time-lapse movies | Specialist requirement |
| Low-temperature ($\leqslant -70^\circ$C) freezer | Storing unstable reagents; freezing cryovials in insulated container | Dependent on reagents used and cell freezing practice; dry ice chest will suffice for latter |
| Computer | For freezer records and cell line database | Beneficial; facilitates record keeping and retrieval of images of cells |
| Colony counter | Counting colonies in plating efficiency and survival assays | Useful; very beneficial if large number of assays being done |
| Cell sizer (e.g., Casy, Coulter; Guava) | Measuring cell volume | Specialist requirement |
| Fluorescence-activated cell sorter | Cell sorting/separation; analysis of cell populations by variety of criteria | Specialist requirement |
| Confocal microscope | Analysis of fluorescence in thick specimens | Specialist requirement |
| Microtitration plate reader | Analysis of chromogenic endpoints in microtitration assays | Essential for assays such as MTT |
| Microtitration plate scintillation counter | Analysis of radioactivity in microtitration plates | Specialist requirement |
| High-capacity centrifuge (6 × 1 L) | Harvesting large-scale suspension cultures | Essential, if working on large scale (>2 L) |
| Centrifugal elutriator centrifuge and rotor | Large-scale cell separation | Specialist requirement |

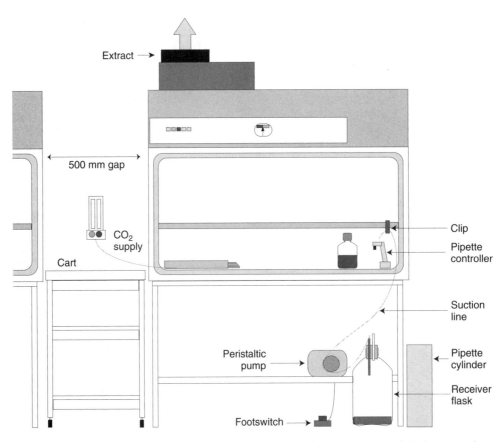

*Fig. 4.1. Laminar-Flow Hood.* A peristaltic pump, connected to a receiver vessel, is shown on the right side below the hood, with a foot switch to activate the pump. The suction line from the pump leads to the work area, and a delivery tube from a gas mixer provides a supply of $CO_2$ mixed in air.

The following check list should be considered when purchasing and installing a new hood:

(1) *Size.* A working surface of 1200 mm (4 ft) wide × 600 mm (2 ft) deep is usually adequate, unless specialized large equipment will be used in the hood.
(2) *Installation.* Make sure that the hood will go through the doorway to the laboratory. Check that there is sufficient headroom for venting to the room or for ducting to the exterior.
(3) *Servicing.* When in place, can the hood be serviced easily? (Ask the service engineer, not the salesperson!)
(4) *Functional efficiency.* Will the airflow from other cabinets, the room ventilation, or independent air-conditioning units interfere with the integrity of the work space of the hood? That is, will contaminated air spill in or aerosols leak out because of turbulence? Meeting this condition will require 3000 mm (10 ft) of face-to-face separation and a minimum of 500 mm (2 ft) of lateral separation. Expert testing by an engineer with experience in microbiological safety cabinets should be arranged immediately after installation, and with all other hoods and ventilation services operating.
(5) *Noise level.* Noisy hoods are more fatiguing.
(6) *Interior.* There should be access for cleaning both inside the working area and below the work surface in the event of spillage.
   (a) A divided work surface is easier to remove but can trap spillage in the crack between the sections.
   (b) A perforated work surface is more likely to allow spillage to go through; a solid work surface vented at the front and back is preferable.
   (c) The edges of the work surface, when lifted, should be smooth; if the edges are sharp, they can cause injury when cleaning out the hood.
(7) *Screen.* The front screen should be able to be raised, lowered, or removed completely, to facilitate cleaning and handling bulky culture apparatus. Remember, however, that a biohazard cabinet will not give you, the operator, or the culture the required protection if you remove the front screen.
(8) *Comfort.* Some cabinets have awkward ducting below the work surface, which leaves no room for your knees, lights or other accessories above that strike your head, or screens that obscure your vision. The person who will use the hood most should sit at it before purchasing in order to simulate normal use. Consider the following questions:
   (a) Can you get your knees under the hood while sitting comfortably and close enough to work, with your hands at least halfway into the hood?
   (b) Is there a footrest in the correct place?
   (c) Are you able to see what you are doing without placing strain on your neck?
   (d) Is the lighting convenient and adequate?

(e) Are you able to remove the work surface easily for cleaning?

## 4.2.2 Service Carts

A movable cart can be used to locate items for use at laminar flow hoods (Table 4.2). These carts conveniently fill the space between adjacent hoods and are easily removed for maintenance of the hoods. They can also be used to carry materials to and from the hoods and have basic items restocked by service staff. Larger carts are useful for clearing soiled glassware and used items from the aseptic area to the washup. They can be parked at a convenient location (*see* Figs. 3.3, 3.4, 4.1).

## 4.2.3 Sterile Liquid Handling—Pipetting and Dispensing

*Pipettes.* Pipettes should be of the blowout variety, wide tipped for fast delivery, and graduated to the tip, with the maximum point of the scale at the top rather than the tip. Pipettes are collected into disinfectant in pipette cylinders or hods, one per workstation.

Reusable glass or disposable graduates plastic pipettes can be used (*see below*) but Pasteur pipettes are best regarded as disposable and should be discarded, not into pipette cylinders, but into secure glassware waste. Alternatively, disposable plastic Pasteur pipettes can be used (Pastettes).

*Pipette cans.* It is worth considering using pipette cans even when using plastic pipettes, particularly if the pipettes are not individually wrapped. Cans keep the pipettes tidy in the hood and can be closed when not in use. Glass pipettes are usually sterilized in aluminum or nickel-plated steel cans. Square-sectioned cans, 75 × 75 × 300–400 mm, are preferable to round, as they stack more easily and will not roll about the work surface. Versions are available with silicone rubber lined top and bottom ends to avoid chipping glass pipettes during handling.

*Plastic versus glass pipettes.* Many laboratories use disposable plastic pipettes, which have the advantage of being prepacked and presterilized and do not have the safety problems associated with handling chipped or broken glass pipettes. Nor do they have to be washed, which is relatively difficult to do, or plugged, which is tedious. On the downside, they are very expensive and slower to use if singly packed. If instead they are bulk packed, there is a high wastage rate unless packs are shared, which is not recommended (*see* Section 5.3.5). Plastic pipettes also add a significant burden to disposal, particularly if they have to be disinfected first.

A large laboratory may find it more economical to use glass pipettes, even with the cost of hiring washup staff and energy costs in washing, drying, and sterilization. However, the convenience, safety of handling, and greater reliability (fewer washups and sterilization failures) tend to favor plastic disposables.

**TABLE 4.2.** Consumables

| Item | Size | Purpose |
|---|---|---|
| Pipettes | 1, 2, 5, 10, 25 mL | Dispensing and withdrawing medium, reagents, and cells |
| Culture flasks | 25, 75, 175 cm² | Primary culture and propagation |
| Petri dishes | 3.5, 5 or 6, 9 or 10 cm | Primary culture and propagation; cloning |
| Multiwell plates | 4-, 6-, 12-, 24-, 96-well | Replicate sampling |
| Sterile containers: | | Storage of sterile liquids |
| Sample pots | 50 mL | Storage of tissue |
| Universal | 30 mL | General purpose sample, media, tissue, and cell containers |
| Bijou | 5 mL | Storage of small samples |
| Centrifuge tubes | 15, 50, 250 mL | Centrifugation of cells |
| Bottles, glass or plastic | 100, 500 mL | Preparation and storage of media. |
| Cryovials | 1, 2 mL | Cryostorage (*see* Protocol 19.1) |
| Syringes | 1, 2, 5, 10, 25 mL | Withdrawing and dispensing viscous liquids and small volumes from vials |
| Syringe needles | 21–23 G | Withdrawing from septum vials |
| | 19 G | Dispensing liquids and cells |
| Filters: | | Sterilization |
| Syringe tip | 13, 25, 47 mm | Sterilizing small volumes |
| Bottle-top or flask | 47 mm | Sterilizing large volumes |
| Surgical gloves | Small, medium, and large | Operator protection from biohazards, solvents, and toxins |
| Lint-free swabs | 50 × 50 mm | Swabbing down worksurface |
| Paper towels | Various | Mopping up spillage |
| Disinfectants: | | |
| Isopropyl alcohol, or ethanol, 70% | | Swabbing down worksurface |
| Na-hypochlorite (Chloros or Chlorox) | | Disinfecting waste |

***Pipette cylinders.*** Pipette cylinders (sometimes known as pipette hods) should be made from polypropylene and should be freestanding and distributed around the lab, one per workstation, with sufficient numbers in reserve to allow full cylinders to stand for 2 h in disinfectant (*see* Section 6.8.5) before washing (glass) or disposal (plastic).

***Pipette controllers.*** Simple pipetting is one of the most frequent tasks required in the routine handling of cultures. Although a rubber bulb or other proprietary pipetting devices are cheap and simple to use, speed, accuracy, and reproducibility are greatly enhanced by a motorized pipette controller (Fig. 4.2), which may be obtained with a separate or built-in pump and can be mains operated or rechargeable. The major determinants in choosing a pipette controller are the weight and feel of the instrument during continuous use; it is best to try one out before purchasing it. Pipette controllers usually have a filter at the pipette insert to minimize the transfer of contaminants. Some filters are disposable, and some are reusable after resterilization (*see* Fig. 6.2 for the proper method of inserting a pipette into a pipetting device).

***Pipettors.*** These devices originated from Eppendorf micropipettes used for dispensing 10 to 200 μL. As the working range now extends up to 5 mL or more, the term "micropipette" is not always appropriate, and the instrument is more commonly called a pipettor (Fig. 4.3*a*). Only the

tip needs to be sterile, but the length of the tip then limits the size of vessels used. If a sterile fluid is withdrawn from a container with a pipettor, the nonsterile stem must not touch the sides of the container. Reagents volumes of 10 to 20 mL may be sampled in 5-μL to 1-mL aliquots from a sample tube such as a universal container or in 5- to 200-μL volumes from a bijou bottle or similar small vial but withdrawing liquids from larger containers will risk contamination unless extended length sterile tips are used. Pipettors are available with multiple tips (Fig. 4.3*b*) for use with microtitration plates.

It is assumed that the inside of a pipettor does not displace enough air to compromise sterility, but this may not always be the case. For example, if you are performing serial subculture of a stock cell line (as opposed to a short-term experiment with cells that will not be propagated beyond the experiment), the security of the cell line is paramount, and you must use a regular plugged pipette with a sterile length that is sufficient to reach into the vessel that you are sampling. If you are using a small enough container to preclude contact from the nonsterile stem, then it is permissible to use a pipettor, provided that the tip has a filter that prevents cross-contamination and minimizes microbial contamination. Otherwise, you run the risk of microbial contamination from the nonsterile stem or, more subtle and potentially more serious, cross-contamination from aerosol or fluid drawn up into the stem.

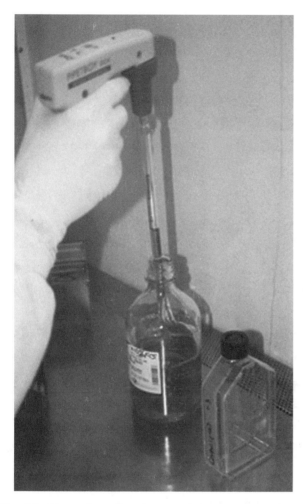

Fig. 4.2. *Pipette Controller.* Motorized pipetting device for use with conventional graduated pipettes.

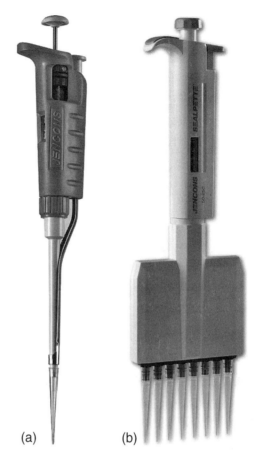

(a)  (b)

Fig. 4.3. *Pipettors.* (*a*) Variable-volume pipetting device. Also available in fixed volume. The pipettor is not itself sterilized but is used with sterilized plastic tips. (*b*) Multipoint pipettor with manifold to take 8 plastic tips; also available for 4 and 12. (Courtesy of VWR-Jencons.)

Routine subculture, which should be rapid and secure from microbial and cross-contamination but need not be very accurate, is best performed with conventional graduated pipettes. Experimental work, which must be accurate but should not involve further propagation of the cells used, may benefit from using pipettors.

Tips can be bought loose and can be packaged and sterilized in the laboratory, or they can be bought already sterile and mounted in racks ready for use. Loose tips are cheaper but more labor intensive. Prepacked tips are much more convenient but considerably more expensive. Some racks can be refilled and resterilized, which presents a reasonable compromise.

**Syringes.** Although it is not recommended that syringes and needles be used extensively in normal handling (for reasons of safety, sterility, and problems with shear stress in the needle when cells are handled), syringes are used for filtration in conjunction with syringe filter adapters, and with needles, the syringes may be required for extraction of reagents (drugs, antibiotics, or radioisotopes) from sealed vials.

**Large-volume dispensing.** A different approach to fluid delivery must be adopted with culture vessels exceeding 100 mL in the volume of medium. If only a few flasks are involved, a 100-mL pipette or a graduated bottle (Fig. 4.4) or media bag may be quite adequate, but if larger volumes (>500 mL) or a large number of high-volume replicates are required, then a peristaltic pump is preferable. Single fluid transfers of very large volumes (10–10,000 L) are usually achieved by preparing the medium in a sealed pressure vessel, sterilizing it by autoclaving and then displacing it by positive pressure into the culture vessel. It is possible to dispense large volumes by pouring, but this should be restricted to a single action with a premeasured volume (*see* Section 5.3.6).

**Repetitive dispensing.** Small-volume repetitive dispensing can be achieved by incremental movement of the piston in a syringe (Fig. 4.5*a*), or a repeated syringe action with a two-way valve connected to a reservoir, the Cornwall Syringe (Fig. 4.5*b*). Sticking of the valves can be minimized by avoiding the drying cycle after autoclaving and flushing the syringe out with serum-free medium or a salt solution

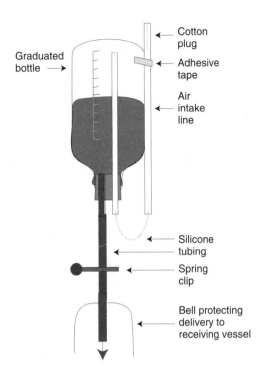

**Fig. 4.4. Graduated Bottle Dispenser.** Two-hole stopper inserted in the neck of a graduated bottle with a delivery line connected to a dispensing bell, a spring clip on the line, and an inlet line for balancing air. The stopper may be sterilized without the bottle and inserted into any standard bottle containing medium as required. (From an original design by Dr. John Paul.)

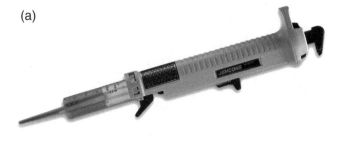

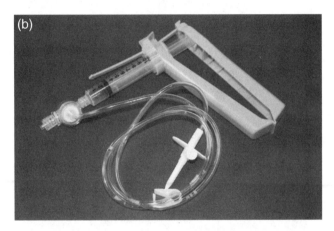

**Fig. 4.5. Syringe Dispensers.** (*a*) Stepping dispenser operated by incremental movement of syringe piston, activated by thumb button. (Repette, courtesy of VWR-Jencons.) (*b*) Repeating syringe dispenser with two-way valve connected to inlet tubing via an inline filter. (Cornwall Syringe, courtesy of Research Laboratory Supply.)

before and after use. A peristaltic pump can also be used for repetitive serial deliveries, and the advantage is that the pump can be activated via a foot switch, leaving the hands free (Fig. 4.6*a*). Care must be taken in setting up such devices to avoid contaminating the tubing at the reservoir and delivery ends. In general, they are worthwhile only if a very large number of flasks is being handled. Only the delivery tube is autoclaved, and accuracy and reproducibility can be maintained at high levels over a range from 10 to 100 mL. A number of delivery tubes may be sterilized and held in stock, allowing a quick changeover in the event of accidental contamination or change in cell type or reagent.

***Automation.*** Many attempts have been made to automate cell culture, but few devices or systems have the flexibility required for general laboratory scale use. However, robotic systems are being used increasingly for large-scale cell production (*see* Section 26.4). The introduction of microtitration plates (*see* Fig. 7.3) has brought with it many automated dispensers, plate readers, and other accessories (Fig. 4.7). Transfer devices, such as the Corning Transtar (Fig. 4.8), make it easier to seed from one plate to another. The range of equipment includes plate mixers and centrifuge carriers but is so extensive that it cannot be covered here. High-throughput screens, based on microtitration systems, are being adopted extensively in the pharmaceutical

industry (*see* Section 26.4.2) and represent one or the major exploitations of automation.

***Choice of system.*** Whether a simple manual system or a complex automated one is chosen, the choice is governed mainly by five criteria:

(1) Ease of use and ergonomic efficiency.
(2) Cost relative to time saved and increased efficiency.
(3) Accuracy and reproducibility in serial or parallel delivery.
(4) Ease of sterilization and effect on accuracy and reproducibility.
(5) Mechanical, electrical, chemical, biological, and radiological safety.

Δ **Safety Note.** Most pipetting devices tend to expel fluid at a higher rate than during normal manual operation and consequently have a greater propensity to generate aerosols. This must be kept in mind when using substances that are potentially hazardous.

***Aspiration pump.*** Suction from a peristaltic pump may be used to remove spent medium or other reagents from a

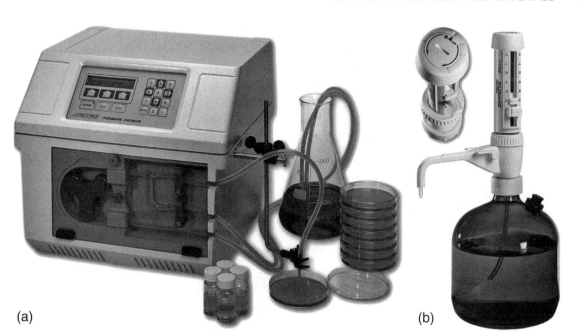

(a)  (b)

**Fig. 4.6. *Automatic Dispensers.*** (*a*) The Perimatic Premier, suitable for repetitive dispensing and dilution in the 1- to 1000-mL range. If the device is used for sterile operations, only the delivery tube needs to be autoclaved. (*b*) Zippette bottle-top dispenser, suitable for the 1- to 30-mL range; autoclavable. (Courtesy of VWR-Jencons.)

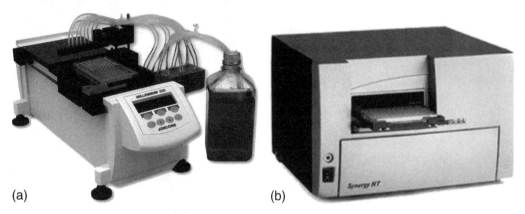

(a)  (b)

**Fig. 4.7. *Plate Filler and Plate Reader.*** (*a*) Automatic filling device for loading microtitration plates. The photo shows a nonsterile application, but the device can be used in sterile applications. (Courtesy of VWR-Jencons.) (*b*) Densitometer for measuring absorbance of each well; some models also measure fluorescence. (Courtesy of Biotek.)

culture flask (Fig. 4.9*a*), and the effluent can be collected directly into disinfectant (*see* Section 6.8.5) in a vented receiver (Fig. 4.9*b*, *c*), with minimal risk of discharging aerosol into the atmosphere if the outlet vent carries a cotton plug or micropore filter. The inlet line should extend further below the stopper than the outlet, by at least 5 cm (2 in.), so that waste does not splash back into the vent. Switch on the pump before inserting a pipette in the tubing (*see* Section 5.4) to keep effluent from running back into the culture flask. The pump tubing should be checked regularly for wear, and the pump should be operated by a self-canceling foot switch.

A vacuum pump downstream from the receiver may be used instead of a peristaltic pump, but a hydrophobic micropore filter will be required between the receiver and the pump to avoid the risk of waste entering the pump. Do not draw air through a vacuum pump from a receiver containing hypochlorite, as the free chlorine will corrode the pump and could be toxic. Also avoid vacuum lines; if they become contaminated with fluids, they can be very difficult to clean out.

### 4.2.4 Inverted Microscope

A simple inverted microscope is essential (Fig. 4.10*a*). It cannot be overemphasized that it is vital to look at cultures regularly to detect morphological changes (*see* Sections 12.4.1, 15.5; Fig. 12.1) and the possibility of microbiological

**Fig. 4.8. Transfer Device.** Transtar (Corning) for seeding, transferring medium, replica plating, and other similar manipulations with microtitration plates, enabling simultaneous handling of all 96 wells. (Reproduced by permission of Corning Life Sciences.)

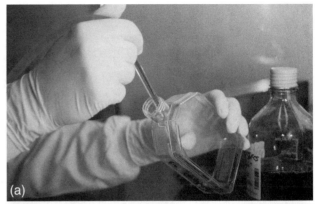

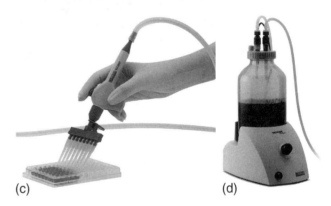

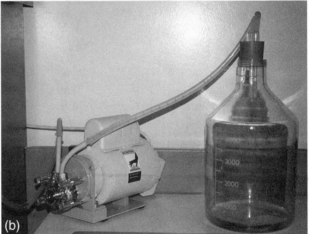

**Fig. 4.9. Aspiration of Medium.** (a) Pipette connected via tube to a peristaltic pump being used to remove medium from a flask. (b) Peristaltic pump on the suction line from the hood leading to a waste receiver. (c) Withdrawal of fluid from multiwell plate (same adapter can be used with regular pipettes and flasks), and (d) vacuum pump receiver (courtesy of Integra.)

contamination (*see* Section 18.3.1; Fig. 18.1). Make certain that the stage is large enough to accommodate large roller bottles, if required, between it and the condenser (*see* Section 26.2.2). It is worth getting a microscope with a phototube for digital recording or viewing linked to a monitor (*see* Fig. 4.10b), but it need not be a large and expensive research microscope. Long working-distance phase-contrast optics (condenser and objectives) are required to compensate for the thickness of plastic flasks. The increasing use of fluorescent tags (e.g., green fluorescent protein, GFP) for viewing live cells means that fluorescence optics may be considered as well.

A ring marker (Nikon) is a useful accessory to the inverted microscope. This device is inserted in the nosepiece in place of an objective and can be used to mark the underside of a dish to locate a colony or patch of cells. The colony can then be picked (*see* Section 13.4; Figs. 13.8, 13.9) or the development of a particularly interesting area in a culture followed.

### 4.2.5  CCD Camera and Monitor

Digital cameras and monitors have become a valuable aid to the discussion of cultures and the training of new staff or students (*see* Fig. 4.10b). Choose a high-resolution, but not high-sensitivity, camera, as the standard camera sensitivity is usually sufficient, and high sensitivity may lead to over-illumination. Black and white usually gives better resolution and is quite adequate for phase-contrast observation of living cultures. Color is preferable for fixed and stained specimens. A still digital camera is sufficient for record shots, but a charge-coupled-device (CCD) camera will allow real time

viewing and may be used for time-lapse recording (*see* Section 27.3).

### 4.2.6  Dissecting Microscope

Dissection of small pieces of tissue (e.g., embryonic organs or tissue from smaller invertebrates) will require a dissecting

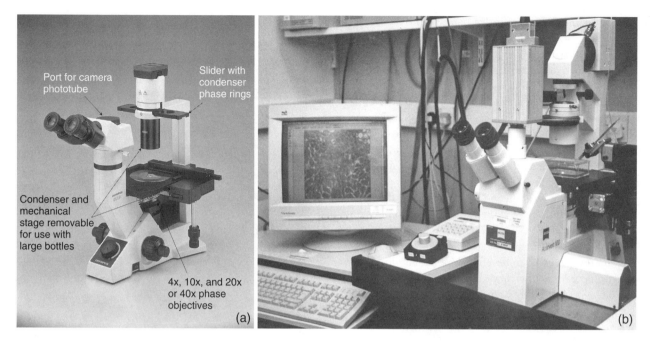

**Fig. 4.10. Inverted Microscope.** (*a*) Olympus CKX41 inverted microscope fitted with phase-contrast optics and trinocular head with port for attaching a digital camera. (Photo courtesy of Olympus, UK, Ltd.) (*b*) CCD camera attached to Zeiss Axiovert inverted microscope. Can be used for direct printing or for time-lapse studies when linked to a video recorder (*see* Section 27.3). Microinjection port on right. (Courtesy of Beatson Institute.)

microscope. The dissecting microscope is also useful for counting monolayer colonies and essential for counting and picking small colonies in agar.

### 4.2.7 Centrifuge

Periodically cell suspensions require centrifugation to increase the concentration of cells or to wash off a reagent. A small bench-top centrifuge, preferably with proportionally controlled braking, is sufficient for most purposes. Refrigeration is not necessary, although it can be used, set at room temperature, to prevent cell samples overheating. Cells sediment satisfactorily at 80 to 100 *g*; higher *g* may cause damage and promote agglutination of the pellet. A large-capacity refrigerated centrifuge, say, $4 \times 1$ L or $6 \times 1$ L, will be required if large-scale suspension cultures (*see* Section 26.1) are contemplated.

### 4.2.8 Cell Counting

Cultured cells may be counted by a variety of different direct and indirect methods (*see* Section 20.1).

***Hemocytometer slide.*** The simplest direct method uses an engraved graticule slide with a thick coverslip (*see* Section 20.1.1). It is the cheapest option and has the added benefit of allowing cell viability to be determined by dye exclusion (*see* Sections 20.1.1, 21.3.1; Fig. 20.1; Plate 17*a*; Protocols 20.1, 21.1). If used routinely, it is better to issue one slide per person each with multiple coverslips.

***Electronic cell counter.*** A cell counter (*see* Fig. 20.2) is a great advantage when more than two or three cell lines are carried and is essential for precise quantitative growth kinetics. Several companies now market models ranging in sophistication from simple particle counting up to automated cell counting, size analysis, and fluorescence emission (*see also* Section 20.1.2). Bench-top flow cytometers provide an alternative to electronic particle counting and will provide other parameters as well if required, such as viability by diacetyl fluorescein (DAF) uptake, apoptotic index, and DNA content.

***Cell sizing.*** Most midrange or top-of-the-range cell counters (*see* Fig. 20.2) will provide cell size analysis and the possibility of downloading data to a PC, directly or via a network.

### 4.3 INCUBATION AND CULTURE

#### 4.3.1 Incubator

If a hot room (*see* Section 3.2.5) is not available, it may be necessary to buy an equivalent dry incubator. Even with a hot room, it is sometimes convenient to have another incubator close to the hood for trypsinization. The incubator should be large enough, around 50 to 200 L (1.5–6 ft$^3$) per person, and should have forced-air circulation, temperature control to within $\pm 0.2°$C, and a safety thermostat that cuts off if the incubator overheats or, better, that regulates the incubator if the first thermostat fails. The incubator should be resistant to corrosion (e.g., stainless steel, although anodized aluminum is acceptable for a dry incubator) and easily cleaned. A double chamber, or two incubators stacked one above the other, independently regulated, is preferable to one large incubator because temperature control is generally better in a smaller incubator and if one half fails or needs to be cleaned, the other can still be used. In addition one can be used for frequent access and the other for limited access. Incubators are available which can be assembled or disassembled on site (Cellon), providing greater flexibility in location and meeting variable demands.

Many incubators have a heated water jacket to distribute heat evenly around the cabinet, thus avoiding the formation of cold spots. These incubators also hold their temperature longer in the event of a heater failure or cut in power. However, new high-efficiency insulation and diffuse surface heater elements have all but eliminated the need for a water jacket and make moving the incubator much simpler. (A water jacket generally needs to be emptied if the incubator is to be moved.)

Incubator shelving is usually perforated to facilitate the circulation of air. However, the perforations can lead to irregularities in cell distribution in monolayer cultures, with variations in cell density following the pattern of spacing on the shelves. The variations may be due to convection currents generated over points of contact relative to holes in the shelf, or they may be related to areas that cool down more quickly when the door is opened. Although no problem may arise in routine maintenance, flasks and dishes should be placed on a ceramic tile or metal tray in experiments where uniform density is important.

### 4.3.2   Humid CO$_2$ Incubator

Although cultures can be incubated in sealed flasks in a regular dry incubator or a hot room, some vessels, sucf as Petri dishes or multiwell plates, require a controlled atmosphere with high humidity and elevated CO$_2$ tension. The cheapest way of controlling the gas phase is to place the cultures in a plastic box, anaerobic jar, or culture chamber (Fig. 4.11), gas the container with the correct CO$_2$ mixture, and then seal it. If the container is not completely filled with dishes, include an open dish of water to increase the humidity inside the chamber.

CO$_2$ incubators (Fig. 4.12) are more expensive, but their ease of use and superior control of CO$_2$ tension and temperature justify the expenditure. A controlled atmosphere is achieved by using a humidifying tray (Fig 4.13) and controlling the CO$_2$ tension with a CO$_2$-monitoring device, which draws air from the incubator into a sample chamber, determines the concentration of CO$_2$, and injects pure CO$_2$ into the incubator to make up any deficiency. Air is circulated around the incubator by natural convection or by using a fan to keep both the CO$_2$ level and the temperature uniform. It is claimed that fan-circulated incubators recover faster

**Fig. 4.11. Culture Chambers.** Inexpensive alternatives to CO$_2$ incubator. Upper shelf, custom-made clear plastic box (Courtesy of Reeve Irvine Institute); lower shelf, anaerobic jar (BD Biosciences.)

after opening, although natural convection incubators can still have a quick recovery and greatly reduce the risks of contamination. Heated wall incubators also encourage less fungal contamination on the walls, as the walls tend to remain dry, even at high relative humidity. Some CO$_2$ controllers need to be calibrated every few months, but the use of gold wire or infrared detectors minimizes drift and many models reset the zero of the CO$_2$ detector automatically.

The size of incubator required will depend on usage, both the numbers of people using it and the types of cultures. Five people using only microtitration plates could have 1000 plates (~100,000 individual cultures) or 10 experiments each in a modest-sized incubator, while one person doing cell cloning could fill the incubator with a few experiments. Flask cultures, especially large flasks, are not an economical use of CO$_2$ incubators. They are better incubated in a regular incubator or hot room. If CO$_2$ is required, flasks can be gassed from a cylinder or CO$_2$ supply.

Frequent cleaning of incubators—particularly humidified ones—is essential (see Section 18.1.4), so the interior should dismantle readily without leaving inaccessible crevices or corners (see Fig. 4.12). Flasks or dishes, or boxes containing them, which are taken from the incubator to the laminar-flow hood, should be swabbed with alcohol before being opened (see Section 5.3.1).

### 4.3.3   Temperature Recorder

A recording thermometer with ranges from below $-70°$C to about $+200°$C will enable you to monitor frozen storage, the freezing of cells, incubators, and sterilizing ovens with one instrument fitted with a resistance thermometer or thermocouple with a long Teflon-coated lead.

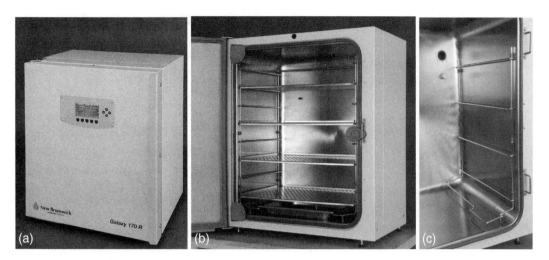

**Fig. 4.12. CO₂ Incubator.** Galaxy 170R fanless CO₂ incubator. (*a*) Exterior with LCD display panel. (*b*) Interior showing shelving and water tray in place; and (*c*) showing smooth easily cleaned interior with removable racking that does not penetrate the stainless steel lining. (*See also* Fig. 4.13.) (Courtesy of New Brunswick Scientific–RS Biotech.)

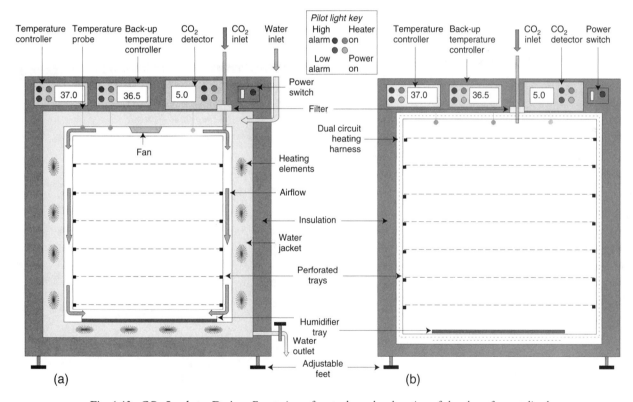

**Fig. 4.13. CO₂ Incubator Design.** Front view of control panel and section of chamber of two stylized humid CO₂ incubators. (*a*) Water-jacketed with circulating fan. (*b*) Dry-walled with no circulating fan (not representative of any particular makes).

Ovens, incubators, and hot rooms should be monitored regularly for uniformity and stability of temperature control. Recording thermometers should be permanently fixed into the hot room, sterilizing oven, and autoclave, and dated records should be kept to check regularly for abnormal behavior, particularly in the event of a problem arising.

### 4.3.4  Roller Racks

Roller racks are used to scale up monolayer culture (*see* Section 26.2.2). The choice of apparatus is determined by the scale (i.e., the size and number of bottles to be rolled). The scale may be calculated from the number of cells required, the maximum attainable cell density, and the surface area of the bottles (*see* Table 7.2). A large number of small bottles gives the highest surface area but tends to be more labor intensive in handling, so a usual compromise is bottles around 125 mm (5 in.) in diameter and various lengths from 150 to 500 mm (6–20 in.). The length of the bottle will determine the maximum yield but is limited by the size of the rack; the height of the rack will determine the number of tiers (i.e., rows) of bottles. Although it is cheaper to buy a larger rack than several small ones, the latter alternative (1) allows you to build up your racks gradually (having confirmed that the system works), (2) can be easier to locate in a hot room, and (3) will still allow you to operate if one rack requires maintenance (*see also* Section 26.2.2).

### 4.3.5  Magnetic Stirrer

A rapid stirring action for dissolving chemicals is available with any stirrer and may benefit from a built-in hotplate, but for enzymatic tissue disaggregation (*see* Section 11.3.3) or suspension culture (*see* Sections 12.4.6, 26.1), (1) the motor should not heat the culture (use the rotating-field type of drive or a belt drive from an external motor), (2) the speed must be controlled down to 50 rpm, (3) the torque at low rpm should still be capable of stirring up to 10 L of fluid, (4) the device should be capable of maintaining several cultures simultaneously, (5) each stirrer position should be individually controlled, and (6) a readout of rpm should appear for each position. It is preferable to have a dedicated magnetic stirrer for culture work (*see* Figs. 26.1, 26.2).

### 4.3.6  Culture Vessels

The choice of culture vessels is determined by (1) the yield (number of cells) required (*see* Section 7.3.1; Table 7.2), (2) whether the cell is grown in monolayer or suspension, and (3) the sampling regime (i.e., are the samples to be collected simultaneously or at intervals over a period of time?) (*see* Section 20.8). "Shopping around" will often result in a cheaper price, but do not be tempted to change products too frequently, and always test a new supplier's product before committing yourself to it (*see* Section 10.6.3).

Care should be taken to label sterile and nonsterile, tissue culture and non–tissue-culture grades of plastics clearly and to store them separately. Glass bottles with flat sides can be used instead of plastic, provided that a suitable washup and sterilization service is available. However, the lower cost tends to be overridden by the optical superiority, sterility, quality assurance, and general convenience of plastic flasks. Nevertheless, disposable plastics can account for approximately 60% of the tissue culture budget—even more than serum.

Petri dishes are much less expensive than flasks, though more prone to contamination and spillage. Depending on the pattern of work and the sterility of the environment, they are worth considering, at least for use in experiments if not for routine propagation of cell lines. Petri dishes are particularly useful for colony-formation assays in which colonies have to be stained and counted or isolated at the end of an experiment.

## 4.4  PREPARATION AND STERILIZATION

### 4.4.1  Washup

*Soaking baths or sinks.*  Soaking baths or sinks should be deep enough so that all your glassware (except pipettes and the largest bottles) can be totally immersed in detergent during soaking, but not so deep that the weight of the glass is sufficient to break smaller items at the bottom. A sink that is 400 mm (15 in.) wide ×600 mm (24 in.) long ×300 mm (12 in.) deep is about right (*see* Fig. 3.6).

*Glassware washing machine.*  Probably the best way of producing clean glassware is to have a reliable person doing the washing up, but when the amount gets to be too great, it may be worth considering the purchase of an automatic washing machine (Fig. 4.14). Several of these are currently available that are quite satisfactory. Look for the following principles of operation:

(1) A choice of racks with individual spigots over which you can place bottles, flasks, and other glassware. Open vessels such as Petri dishes and beakers will wash satisfactorily

**Fig. 4.14. Glassware Washing Machine.**  Glassware is placed on individual jets, which ensures thorough washing and rinsing (Betterbuilt). (Courtesy of Beatson Institute.)

in a whirling-arm spray, but narrow-necked vessels need individual jets. The jets should have a cushion or mat at the base to protect the neck of the bottle from chipping.

(2) The pump that forces the water through the jets should have a high delivery pressure, requiring around 2 to 5 hp, depending on the size of the machine.

(3) Water for washing should be heated to a minimum of 80°C.

(4) There should be a facility for a deionized water rinse at the end of the cycle. The water should be heated between 50°C and 60°C; otherwise, the glass may crack after the hot wash and rinse. The rinse should be delivered as a continuous flush, discarded, and not recycled. If recycling is unavoidable, a minimum of three separate deionized rinses will be required.

(5) Preferably, rinse water from the end of the previous wash cycle should be discarded and not retained for the pre-rinse of the next wash. Discarding the rinse water reduces the risk of chemical carryover when the machine is used for chemical and radioisotope washup as well as tissue culture.

(6) The machine should be lined with stainless steel and plumbed with stainless steel or nylon piping.

(7) If possible, a glassware drier should be chosen that will accept the same racks as the washer (*see* Fig. 4.14), so that they may be transferred directly, without unloading, via a suitably designed trolley.

*Pipette washer.* Reusable glass pipettes are easily washed in a standard siphon-type washer (*see* Section 10.3.2; Fig. 10.5). The washer should be placed just above floor level, rather than on the bench, to avoid awkward lifting of the pipettes and should be connected to the deionized water supply, as well as the regular cold water supply, so that the final few rinses can be done in deionized water. If possible, a simple changeover valve should be incorporated into the deionized water feed line (*see* Fig. 3.6).

*Pipette drier.* If a stainless steel basket is used in the washer, pipettes may subsequently be transferred directly to an electric drier. Alternatively, pipettes can be dried on a rack or in a regular drying oven.

*Drying oven.* This should be of large capacity, fan driven, and able to reach 100°C. In practice, the sterilizing oven (*see* Section 4.4.3) will double as a drying oven.

### 4.4.2 Preparation of Media and Reagents

*Water purifier.* Purified water is required for rinsing glassware, dissolving powdered media, and diluting concentrates. The first of these purposes is usually satisfied by deionized or reverse-osmosis water, but the second and third require ultrapure water (UPW), which demands a three- or four-stage process (Fig. 4.15; *see also* Fig. 11.10). The important principle is that each stage be qualitatively different; reverse osmosis may be followed by charcoal filtration, deionization, and micropore filtration (e.g., via a sterilizing filter; *see* Fig. 11.11), or distillation (with a silica-sheathed element) may be substituted for the first stage. Reverse osmosis is cheaper if you pay the fuel bills; if you do not, distillation is better and more likely to give a sterile product. If reverse osmosis is used, the type of cartridge should be chosen to suit the pH of the water supply, as some membranes can become porous in extreme pH conditions (check with supplier).

The deionizer should have a conductivity meter monitoring the effluent, to indicate when the cartridge must be changed. Other cartridges should be dated and replaced according to the manufacturer's instructions. A total organic carbon (TOC) meter can be used to monitor colloids (Fig. 4.15c).

Purified water should not be stored but should be recycled through the apparatus continually to minimize infection with algae or other microorganisms. All tubing or reservoirs in the system should be checked regularly (every 3 months or so) for algae, cleaned out with hypochlorite and detergent (e.g., Clorox or Chloros), and thoroughly rinsed in purified water before reuse.

Water is the simplest, but probably the most critical, constituent of all media and reagents, particularly serum-free media (*see* Chapter 9). A good quality-control measure is to check the plating efficiency of a sensitive cell line at regular intervals in medium made up with the water (*see* Section 10.6.3; Protocol 20.10).

*Balances.* Although most laboratories obtain media that are already prepared, it may be cheaper to prepare some reagents in house. Doing so will require a balance (an electronic one with automatic tare) capable of weighing items from around 10 mg up to 100 g or even 1 kg, depending on the scale of the operation. If you are a service provider, it is often preferable to prepare large quantities, sometimes 10× concentrated, so the amounts to be weighed can be quite high. It may prove better to buy two balances, coarse and fine, as the outlay may be similar and the convenience and accuracy are increased.

*Hot plate magnetic stirrer.* In addition to the ambient temperature stirrers used for suspension cultures and trypsinization, it may be desirable to have a magnetic stirrer with a hot plate to accelerate the dissolution of some reagents. Placing a solution on a stirrer in the hot room may suffice, but leaving solutions stirring at 37°C for extended periods can lead to microbial growth. So stable solutions are best stirred at a higher temperature for a shorter time.

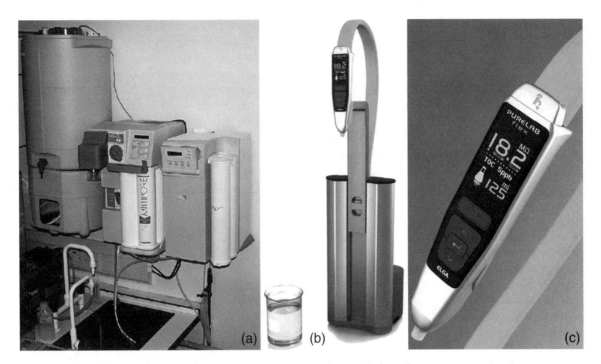

**Fig. 4.15. Water Purifier.** (*a*) High output. Tap water first passes through a reverse-osmosis unit on the right and then goes to the storage tank on the left. It then passes through carbon filtration and deionization (center unit) before being collected via a micropore filter (Millipore Milli-Q). Current models have integrated conductivity and TOC monitoring (Courtesy of Beatson Institute.) (*b*) Smaller self-contained bench-top unit (Elga Pureflex). (*c*) Handset with readout of resistivity and TOC. (*b, c,* courtesy of Elga.)

***pH Meter.*** A simple pH meter is required for the preparation of media and special reagents. Although a phenol red indicator is sufficient for monitoring pH in most solutions, a pH meter will be required when phenol red cannot be used (e.g., in preparing cultures for fluorescence assays or in estrogen binding assays where phenol red can interfere), in preparing stock solutions and for regular quality-control checks during preparation of media and reagents.

***Conductivity meter.*** When solutions are prepared in the laboratory, it is essential to perform quality-control measures to guard against errors (*see* Section 10.6.1). A simple check of ionic concentration can be made with a conductivity meter against a known standard, such as normal saline (0.15 M).

***Osmometer.*** One of the most important physical properties of a culture medium, and one that is often difficult to predict, is the osmolality. Although the conductivity is controlled by the concentration of ionized molecules, nonionized particles can also contribute to the osmolality. An osmometer (*see* Fig. 8.2) is therefore a useful accessory to check solutions as they are made up, to adjust new formulations, to compensate for the addition of reagents to the medium, and to act as a second line of quality control. Osmometers usually work by depressing the freezing point of a medium or elevating its vapor pressure. Choose one with a low sample volume (<1 mL) because you may want to measure a valuable or scarce reagent on occasion, and the accuracy (+10 mosmol/kg) may be less important than the value or scarcity of the reagent.

***Bottling—Automatic dispensers.*** Preparing media and reagents in bulk will usually require aliquotting into containers for sterilization and usage. Bottle-top dispensers (*see* Section 4.2.3; Fig. 4.6*b*) are suitable for volumes up to about 50 mL; above that level, gravity dispensing from a reservoir, which may be a graduated bottle (*see* Fig. 4.4) or plastic bag, is acceptable if accuracy is not critical. If the volume dispensed is more critical, then a peristaltic pump (*see* Figs. 4.6*a*, 10.12) is preferable. The duration of dispensing and diameter of tube determine the volume and accuracy of the dispenser. A long dispense cycle with a narrow delivery tube will be more accurate, but a wide-bore tube will be faster.

### 4.4.3 Sterilization

***Sterilizing oven.*** Although most sterilizing can be done in an autoclave, it is preferable to sterilize glassware by dry heat, avoiding the possibility of both chemical contamination from steam condensate and corrosion of pipette cans. Such sterilization, however, will require a high-temperature (160–180°C) fan-powered oven to ensure even heating throughout the load (*see also* Section 10.3.1). As

***Fig. 4.16. Bench-Top Autoclave.*** Simple, top-loading autoclave from Prestige Medical; left with lid closed, right with lid removed for filling. (Courtesy of Beatson Institute.)

with autoclaves, do not get an oven that is too big for the amount or size of glassware that you use. It is better to use two small ovens than one big one; heating is easier, more uniform, quicker, and more economical when only a little glassware is being used. You are also better protected during breakdowns.

***Steam sterilizer (autoclave).*** The simplest and cheapest sterilizer is a domestic pressure cooker that generates 100 kPa (1 atm, 15 lb/in.$^2$) above ambient pressure. Alternatively, a bench-top autoclave (Fig. 4.16) gives automatic programming and safety locking. A larger, freestanding model with a programmable timer, a choice of pre- and poststerilization evacuation, and temperature recording (Fig. 4.17) has a greater capacity, provides more flexibility, and offers the opportunity to comply with good laboratory practice (GLP).

A "wet" cycle (water, salt solutions, etc.) is performed without evacuation of the chamber before or after sterilization. Dry items (instruments, swabs, screw caps, etc.) require that the chamber be evacuated before sterilization, or the air replaced by downward displacement to allow efficient access of hot steam. The chamber should also be evacuated after sterilization to remove steam and promote subsequent drying; otherwise, the articles will emerge wet, leaving a trace of contamination from the condensate on drying. To minimize this risk when a "postvac" cycle is not available, always use deionized or reverse-osmosis water to supply the autoclave. If you require a high sterilization capacity (>300 L, 9 ft$^3$), buy two medium-size autoclaves rather than one large one so that during routine maintenance and accidental breakdowns you still have one functioning machine. Furthermore a medium-size machine will heat up and cool down more quickly and can be used more economically for small loads. Leave sufficient space around the sterilizer for maintenance and ventilation, provide adequate

air extraction to remove heat and steam, and ensure that a suitable drain is available for condensate.

Most small autoclaves come with their own steam generator (calorifier), but larger machines may have a self-contained steam generator, a separate steam generator, or the facility to use a steam line. If high-pressure steam is available on line, that will be the cheapest and simplest method of heating and pressurizing the autoclave; if not, it is best to purchase a sterilizer complete with its own self-contained steam generator. Such a sterilizer will be cheaper to install and easier to move. With the largest machines you may not have a choice, as they are frequently offered only with a separate generator. In that case you will need to allow space for the generator at the planning stage.

***Sterilization filters.*** Reusable apparatus is available for sterile filtration, the size depending on the scale of your operation. Filtration may be by positive pressure and will require a pump upstream from a pressure vessel, mostly for large volumes of 10 L or more, while smaller volumes can be handled with a smaller reservoir and a downstream peristaltic pump (*see* Section 10.5.2; Fig. 10.12). However, most laboratories now use disposable filters ranging in size from 25-mm syringe adapters through 47-mm in line, bottle-top adapters, or filter flasks. It is also wise to keep a small selection of larger sizes on hand.

## 4.5  STORAGE

### 4.5.1  Consumables

Stocks of culture vessels (*see* Section 7.3), tubes (Fig. 4.18), and other consumables (*see* Table 4.2) can be stored near hoods on shelving above adjacent benches or on carts. Shelf space is also required for stable liquids, such as PBS and water.

Temperature
probe

**Fig. 4.17. Freestanding Autoclave.** Medium-sized (300 L; 10 ft³) laboratory autoclave with square chamber for maximum load. The recorder on the top console is connected to a probe in the bottle in the center of the load. (Courtesy of Beatson Institute.)

### 4.5.2 Refrigerators and Freezers

Usually a domestic refrigerator or freezer is quite efficient and cheaper than special laboratory equipment. Domestic refrigerators are available without a freezer compartment ("larder refrigerators"), giving more space and eliminating the need for defrosting. However, if you require 400 L (12 ft³) or more storage, a large hospital, blood bank, or catering refrigerator may be better.

If space is available and the number of people using tissue culture is more than three or four, it is worth considering the installation of a cold room, which is more economical in terms of space than several separate refrigerators and is also easier to access. The walls should be smooth and easily cleaned, and the racking should be on castors to facilitate moving for cleaning. Cold rooms should be cleaned out regularly to eliminate old stock, and the walls and shelves should be washed with an antiseptic detergent to minimize fungal contamination.

Similar advice applies to freezers—several inexpensive domestic freezers will be cheaper, and just as effective as a specialized laboratory freezer. Most tissue culture reagents will keep satisfactorily at −20°C, so an ultradeep freeze is not essential. A deep-freeze room is not recommended—it is very difficult and unpleasant to clear out, and it creates severe problems in regard to relocating the contents if extensive maintenance is required.

Although autodefrost freezers may be bad for some reagents (enzymes, antibiotics, etc.), they are quite useful for most tissue culture stocks whose bulk precludes major temperature fluctuations and whose nature is less sensitive to

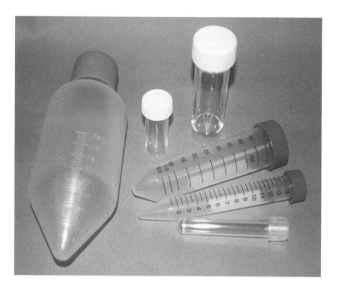

**Fig. 4.18. Tubes.** Centrifuge and samples tubes, available sterile but non−tissue-culture grade, although tissue culture grade tubes are available (BD Biosciences; Corning). Clockwise from the left: 250-mL centrifuge tube (Corning), 5-mL Bijou bottle (Sterilin), 30-mL universal container (Sterilin), 50-mL centrifuge tube (BD Biosciences), 15-mL centrifuge tube (BD Biosciences), and 5-mL sample tube (BD Biosciences).

severe cryogenic damage. Conceivably serum can deteriorate during oscillations in the temperature of an autodefrost freezer, but in practice it does not. Many of the essential constituents of serum are small proteins, polypeptides, and simpler organic and inorganic compounds that may be insensitive to cryogenic damage, particularly if solutions are stored in volumes >100 mL.

### 4.5.3 Cryostorage Containers

Details of cryostorage containers and advice on selection are given in Chapter 19 (*see* Section 19.3.6). In brief, the choice depends on the size and the type of storage system required. For a small laboratory, a 35-L freezer with a narrow neck and storage in canes and canisters (*see* Figs. 19.6*a*, 19.7*a*) or in drawers in a rack system (*see* Figs. 19.6*d*, 19.7*c*) should hold about 500 to 1000 ampoules (*see* Appendix II: Freezers, liquid nitrogen). Larger freezers will hold >10,000 ampoules and include models with walls perfused with liquid nitrogen, cutting down on nitrogen consumption, and providing safe storage without any liquid nitrogen in the storage chamber itself. It is important to establish, however, if selecting the perfused wall type of freezer, that adequate precautions have been taken to ensure that no particulate material, water, or water vapor can enter the perfusion system, as blockages can be difficult, or even impossible, to clear.

An appropriate storage vessel should also be purchased to enable a backup supply of liquid nitrogen to be held. The size of the vessel depends on (1) the size of the freezer, (2) the frequency and reliability of delivery of liquid nitrogen, and (3) the rate of evaporation of the liquid nitrogen. A 35-L

narrow-necked freezer using 5 to 10 L/wk will only require a 25-L Dewar as long as a regular supply is available. Larger freezers are best supplied on line from a dedicated storage tank (e.g., a 160-L storage vessel linked to a 320-L freezer with automatic filling and alarm, or a 500-L tank for a larger freezer or for several smaller freezers).

### 4.5.4 Controlled-Rate Freezer

Although cells may be frozen simply by placing them in an insulated box at −70°C, some cells may require different cooling rates or complex programmed cooling curves (*see* Section 19.3.4). A programmable freezer (e.g., Cryomed, Planer) enables the cooling rate to be varied by controlling the rate of injecting liquid nitrogen into the freezing chamber, under the control of a preset program (*see* Fig. 19.5). Cheaper alternatives for controlling the cooling rate during cell freezing are the variable-neck plug (Taylor Wharton), a specialized cooling box (Thermo−Nunc), a simple polystyrene foam packing container, or foam insulation for water pipes (*see* Figs. 19.2−19.4).

## 4.6 SUPPLEMENTARY LABORATORY EQUIPMENT

### 4.6.1 Computers and Networks

Whether or not a computer or terminal is located in the tissue culture laboratory itself, entering records for cell line maintenance (*see* Section 12.4.9), primary culture (*see* Section 11.3.11), and experiments facilitates later retrieval and analysis. Cell line data are best maintained on a computer database that can also serve as an inventory control for the nitrogen freezer (*see* Section 9.5.1). In larger laboratories stock control of plastics, reagents, and media can also be simplified. There is considerable advantage in networking, as individual computers can be backed up centrally on a routine basis, and information keyed in at one point can be retrieved elsewhere. For example, photographs recorded digitally in the tissue culture laboratory can be saved to a central server and retrieved in an office or writing area.

### 4.6.2 Upright Microscope

An upright microscope may be required, in addition to an inverted microscope, for chromosome analysis, mycoplasma detection, and autoradiography. Select a high-grade research microscope with regular bright-field optics up to 100× objective magnification, phase contrast up to at least 40× and preferably 100× objective magnification, and fluorescence optics with epi-illumination and 40× and 100× objectives for mycoplasma testing by fluorescence (*see* Protocol 18.2) and fluorescent antibody observation. A 50× water-immersion objective (e.g., from Leica) is particularly useful for observing routine mycoplasma wet preparations with Hoechst stain. A digital or CCD camera should also be fitted for photographic records of permanent preparations. A fully automatic fluorescence microscope is available from Olympus.

### 4.6.3 Low-Temperature Freezer

Most tissue culture reagents can be stored at 4°C or −20°C, but occasionally some drugs, reagents, or products from cultures may require a temperature of −70°C to −90°C, at which point most, if not all, of the water is frozen and most chemical and radiolytic reactions are severely limited. A −70°C to −90°C freezer is also useful for freezing cells within an insulated container (*see* Protocol 19.1). The chest type of freezer is more efficient at maintaining a low temperature with minimum power consumption, but vertical cabinets are much less extravagant in floor space and easier to access. If you do choose a vertical cabinet type, make sure that it has individual compartments—for example, six to eight in a 400-L (15-ft³) freezer, with separate close-fitting doors—and expect to pay at least 20% more than for a chest type.

Low-temperature freezers generate a lot of heat, which must be dissipated for them to work efficiently (or at all). Such freezers should be located in a well-ventilated room or one with air-conditioning such that the ambient temperature does not rise above 23°C. If this is not possible, invest in a freezer designed for tropical use; otherwise, you will be faced with constant maintenance problems and a shorter working life for the freezer, with all the attendant problems of relocating valuable stocks. One or two failures costing $1000 or more in repairs and the loss of valuable material soon cancel any savings that would be realized in buying a cheap freezer.

### 4.6.4 Confocal Microscope

Cytological investigations of fluorescently labeled cells often benefit from improved resolution when viewed by confocal microscopy (*see* Section 15.6). This technique allows the microscope to view an "optical section" through the specimen presenting the image in one focal plane and avoiding the interference caused by adjacent cells or organelles not in the same focal plane. The data are stored digitally and can be processed in a number of ways, including the creation of a vertical section through the sample (a so-called Z-section), particularly useful when viewing three-dimensional cultures such as filter wells or spheroids.

### 4.6.5 PCR Thermal Cycler

A number of ancillary techniques in cell line validation, such as mycoplasma detection (*see* Protocol 18.3) and DNA profiling (*see* Protocol 15.9), rely on amplification and detection of specific DNA sequences. If you plan to use these techniques, they utilize the polymerase chain reaction (PCR) and require a thermal cycler.

## 4.7 SPECIALIZED EQUIPMENT

### 4.7.1 Microinjection Facilities

Micromanipulators can be used to inject directly into a cell, for example, for nuclear transplantation or dye injection (*see* Fig. 4.10b).

### 4.7.2 Colony Counter

Monolayer colonies are easily counted by eye or on a dissecting microscope with a felt-tip pen to mark off the colonies, but if many plates are to be counted, then an automated counter will help. The simplest uses an electrode-tipped marker pen, which counts when you touch down on a colony. They often have a magnifying lens to help visualize the colonies. From there, a large increase in sophistication and cost takes you to a programmable electronic colony counter, which counts colonies using image analysis software. These counters are very rapid, can discriminate among colonies of different diameters, and can even cope with contiguous colonies (*see* Section 20.10.2).

### 4.7.3 Centrifugal Elutriator

The centrifugal elutriator is a specially adapted centrifuge that is suitable for separating cells of different sizes (*see* Section 14.2.2). The device is costly, but highly effective, particularly for high cell yields.

### 4.7.4 Flow Cytometer

This instrument can analyze cell populations according to a wide range of parameters, including light scatter, absorbance, and fluorescence (*see* Sections 14.4, 20.7). Multiparametric analysis can be displayed in a two- or three-dimensional format. In the analytic mode these machines are generally referred to as flow cytometers (*see* Section 20.7, e.g., BD Biosciences Cytostar), but the signals they generate can also be used in a fluorescence-activated cell sorter to isolate individual cell populations with a high degree of resolution (e.g., BD Biosciences FACStar). The cost is high ($100,000−200,000), and the best results are obtained with a skilled operator. There are also less expensive bench-top machines (Guava, Accuri) that can determine cell number and a variety of fluorescent parameters (*see* Section 20.1.4).

# CHAPTER 5

# Aseptic Technique

## 5.1 OBJECTIVES OF ASEPTIC TECHNIQUE

### 5.1.1 Risk of Contamination

Contamination by microorganisms remains a major problem in tissue culture. Bacteria, mycoplasma, yeast, and fungal spores may be introduced via the operator, the atmosphere, work surfaces, solutions, and many other sources (*see* Section 18.1; Table 19.1). Aseptic technique aims to exclude contamination by establishing a strict code of practice and ensuring that everyone using the facility adheres to it.

Contamination can be minor and confined to one or two cultures, can spread among several cultures and compromise a whole experiment, or can be widespread and wipe out your (or even the whole laboratory's) entire stock. Catastrophes can be minimized if (1) cultures are checked carefully by eye and on a microscope, preferably by phase contrast, every time that they are handled; (2) cultures are maintained without antibiotics, preferably at all times but at least for part of the time, to reveal cryptic contaminations (*see* Section 12.4.8); (3) reagents are checked for sterility (by yourself or the supplier) before use; (4) bottles of media or other reagents are not shared with other people or used for different cell lines; and (5) the standard of sterile technique is kept high at all times.

Mycoplasmal infection, invisible under regular microscopy, presents one of the major threats. Undetected, it can spread to other cultures around the laboratory. It is therefore essential to back up visual checks with a mycoplasma test, particularly if cell growth appears abnormal (*see* Section 18.3.2).

### 5.1.2 Maintaining Sterility

Correct aseptic technique should provide a barrier between microorganisms in the environment outside the culture and the pure, uncontaminated culture within its flask or dish. Hence all materials that will come into direct contact with the culture must be sterile, and manipulations must be designed such that there is no direct link between the culture and its nonsterile surroundings. It is recognized that this sterility barrier cannot be absolute without working under conditions that would severely hamper most routine manipulations. As testing the need for individual precautions would be time-consuming, procedures are adopted largely on the basis of common sense and experience. Aseptic technique is a combination of procedures designed to reduce the probability of infection, and the correlation between the omission of a step and subsequent contamination is not always absolute. The operator may abandon several precautions before the probability rises sufficiently that a contamination is likely to occur (Fig. 5.1). By then, the cause is often multifactorial, and consequently no simple single solution is obvious. If, once established, all precautions are maintained consistently, breakdowns will be rarer and more easily detected.

Although laboratory conditions have improved in some respects (with air-conditioning and filtration, laminar-flow facilities, etc.), the modern laboratory is often more crowded, and facilities may have to be shared. However, with rigid adherence to reasonable precautions, sterility is not difficult to maintain. It does, however, require that procedures be standardized among users of the facility and that the correct

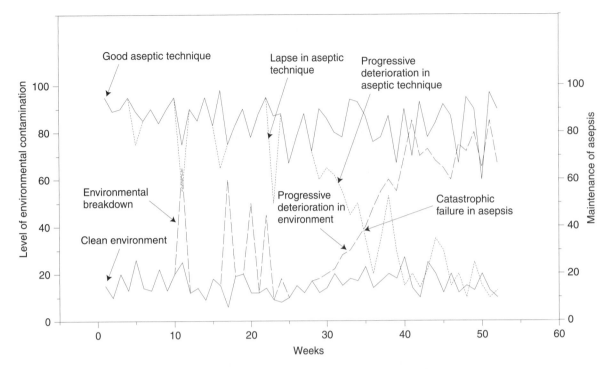

**Fig. 5.1. Probability of Contamination.** The solid line in the top graph represents variability in technique against a scale of 100 (right-hand axis), which represents perfect aseptic technique. The solid line in the bottom graph represents fluctuations in environmental contamination, with zero being perfect asepsis (left-hand axis). Both lines show fluctuations, the top one representing lapses in technique (forgetting to swab the work surface, handling a pipette too far down the body of the pipette, touching nonsterile surfaces with the tip of a pipette, etc.), the bottom one representing crises in environmental contamination (a high spore count, a contaminated incubator, contaminated reagents, etc.). As long as these lapses are minimal in degree and duration, the two graphs do not overlap. When particularly bad lapses in technique (dotted line) coincide with severe environmental crises (dashed line, e.g., at 10–11 weeks), where the dashed and dotted lines overlap briefly, the probability of infection increases. If the breakdown in technique is progressive (dotted line, 28–36 weeks), and the deterioration in the environment is also progressive (dashed line, 25–42 weeks), when the two lines cross, the probability of infection will be high, resulting in frequent, multispecific, and multifactorial contamination.

quality control measures and equipment checks be in place (*see* Section 6.10).

## 5.2  ELEMENTS OF ASEPTIC ENVIRONMENT

Conditions for achieving a clean area for cell culture have changed over the years (Fig. 5.2) mainly due to the introduction of antibiotics, laminar-flow cabinets, and filtered air-conditioning. The continuous use of antibiotics is neither necessary nor advisable as clean room air combined with laminar flow has made the creation of an aseptic environment much simpler to attain and more reliable.

### 5.2.1  Laminar Flow

The major advantage of working in a laminar-flow hood is that the work space is protected from dust and contamination by a constant, stable flow of filtered air passing over the work surface (Fig. 5.3). There are two main types of flow:

(1) horizontal, where the airflow blows from the side facing you, parallel to the work surface, and is not recirculated (*see* Fig. 5.3*a*), and (2) vertical, where the air blows down from the top of the hood onto the work surface and is drawn through the work surface and either recirculated or vented (*see* Fig. 5.3*b*). In most hoods 20% is vented and made up by drawing in air at the front of the work surface. This configuration is designed to minimize overspill from the work area of the cabinet. Horizontal-flow hoods give the most stable airflow and best sterile protection but will blow aerosols into your face and are unsuitable for handling material which is potentially biohazardous or toxic. Vertical-flow hoods give more protection to the operator, particularly when classified as a microbiological safety cabinet, Class II.

A Class II vertical-flow microbiological safety cabinet (MSC) should be used (*see* Section 6.8.2; Fig. 6.5*a*) if potentially hazardous material (human- or primate-derived cultures, virally infected cultures, etc.) is being handled. The

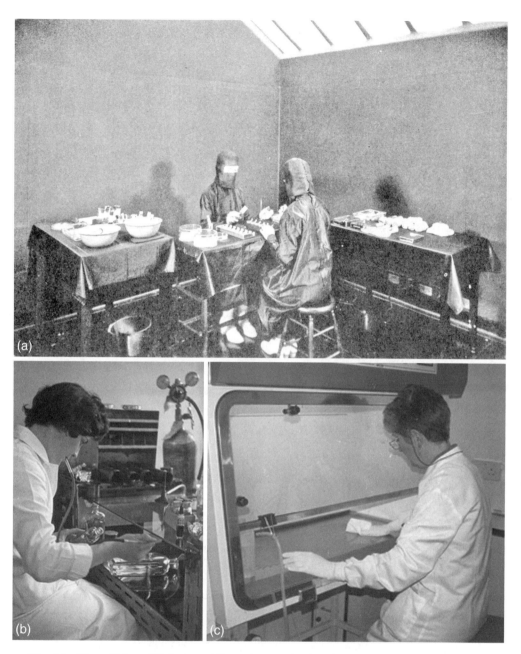

***Fig. 5.2. Tissue Culture Work Area.*** These photographs show how the layout of a tissue culture area has changed over the years. (*a*) One of Alexis Carrel's culture rooms in Rockefeller Institute in the 1930s [from Parker, 1938]. (*b*) Glass-topped table as used in John Paul's tissue culture room in the University of Glasgow Biochemistry Department in 1961. (*c*) Laminar-flow hood in use in Beatson Institute, Glasgow, in 1996.

best protection from chemical and radiochemical hazards is given by a chemical safety cabinet that has a carbon filter trap in the recirculating airflow or the effluent vented to outside the building (*see* Section 6.5.4). If known human pathogens are handled, a Class III pathogen cabinet with a pathogen trap on the vent is obligatory (*see* Section 6.8.2; Fig. 6.5*c*).

Laminar-flow hoods depend for their efficiency on a minimum pressure drop across the filter. When resistance builds up in the filter, the pressure drop increases, and the

flow rate of air in the cabinet falls. Below 0.4 m/s (80 ft/min), the stability of the laminar airflow is compromised, and sterility can no longer be maintained. The pressure drop can be monitored with a manometer fitted to the cabinet, but direct measurement of the airflow with an anemometer is preferable.

Routine maintenance checks of the primary filters are required (about every 3 to 6 months). With horizontal-flow hoods, primary filters may be removed (after switching off the

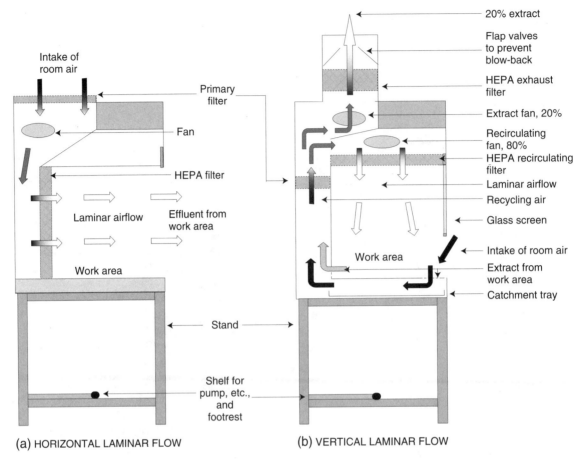

**Fig. 5.3. Airflow in Laminar-Flow Hoods.** Arrows denote direction of airflow. (*a*) Horizontal flow. (*b*) Vertical flow.

fan) and discarded or washed in soap and water. The primary filters in vertical-flow and biohazard hoods are internal and may need to be replaced by an engineer. They should be incinerated or autoclaved and discarded.

Every six months the main high-efficiency particulate air (HEPA) filter above the work surface should be monitored for airflow and holes, which are detectable by a locally increased airflow and an increased particle count. Monitoring is best done by professional engineers on a contract basis. Class II MSCs will have HEPA filters on the exhaust, which will also need to be changed periodically. Again, this should be done by a professional engineer, with proper precautions taken for bagging and disposing of the filters by incineration. If used for biohazardous work, cabinets should be sealed and fumigated before the filters are changed.

Regular weekly checks should be made below the work surface and any spillage mopped up, the tray washed, and the area sterilized with 5% phenolic disinfectant in 70% alcohol. Spillages should, of course, be mopped up when they occur. A regular check is imperative because occasionally they go unnoticed. Swabs, tissue wipes, or gloves, if dropped below the work surface during cleaning, can end up on the primary filter and restrict airflow, so take care during cleaning and

check the primary filter periodically with a mirror and torch if hidden by the ductwork.

Laminar-flow hoods are best left running continuously as this keeps the work area clean. Should any spillage occur, either on the filter or below the work surface, it dries fairly rapidly in sterile air, reducing the chance that microorganisms will grow.

Ultraviolet lights can be used to sterilize the air and exposed work surfaces in laminar-flow hoods between uses. Ultraviolet lights present a radiation hazard, particularly to the eyes, and will also lead to crazing of some clear plastic panels (e.g., Perspex) after six months to a year, particularly if used in conjunction with alcohol. Furthermore the effectiveness of the lights is doubtful because they do not reach crevices; alcohol or other liquid sterilizing agents are more effective as they will run into crevices by capillarity.

Δ *Safety Note.* If ultraviolet lights are used, protective goggles must be worn and all exposed skin covered.

### 5.2.2 Quiet Area

In the absence of a laminar-flow cabinet, a separate sterile room should be used for sterile work. If this is not possible, pick a quiet corner of the laboratory with little or no traffic and no other activity (*see* Section 3.2.1). With laminar flow, an area should be selected that is free of draughts from doors, windows, and so forth. The area should also have no through traffic and no equipment that generates air currents (e.g., centrifuges, refrigerators, and freezers); air conditioners and plenums should be positioned so that effluent air does not compromise the functioning of the hood (*see* Section 3.2). Activity in the area should be restricted to culture-related tissue, and animals and microbiological cultures should be excluded. The area should be kept clean and free of dust and should not contain equipment other than that connected with tissue culture. Nonsterile activities, such as sample processing, staining, or extractions, should be carried out elsewhere.

### 5.2.3  Work Surface

It is essential to keep the work surface clean and tidy. The following rules should be observed:

(1) Start with a completely clear surface.

(2) Swab the surface liberally with 70% alcohol (*see* Appendix I).

(3) Bring onto the surface only those items you require for a particular procedure.

(4) Between procedures, remove everything that is no longer required, and swab the surface down.

(5) Arrange your work area so that you have (a) easy access to all items without having to reach over one to get at another and (b) a wide, clear space in the center of the bench (not just the front edge!) to work on (Fig. 5.4*a*). If you have too much equipment too close to you, you will inevitably brush the tip of a sterile pipette against a nonsterile surface. Furthermore the laminar airflow will fail in a hood that is crowded with equipment (Fig. 5.4*b*).

(6) Do not allow your hands or any other nonsterile items (even the outside of a flask is nonsterile) to pass over an open flask or dish. Even when using horizontal flow, you should still work in a clear space with no obstructions between the central work area and the HEPA filter (Fig. 5.5).

(7) Work within your range of vision; for example, insert a pipette in a bulb or pipette controller with the tip of the pipette pointing away from you so that it is in your line of sight continuously and not hidden by your arm (*see* Fig. 5.9).

(8) Mop up any spillage immediately and swab the area with 70% alcohol.

(9) Remove everything when you have finished, and swab the work surface down again.

(10) Ensure that the space below the work surface is cleaned out regularly, at least once per week.

### 5.2.4  Personal Hygiene

There has been much discussion about whether hand washing encourages or reduces the bacterial count on the skin.

Regardless of this debate, washing will moisten the hands and remove dry skin that would otherwise be likely to blow onto your culture. Washing will also reduce loosely adherent microorganisms, which are the greatest risk to your culture. Surgical gloves may be worn and swabbed frequently, but it may be preferable to work without them (if no hazard is involved) and retain the extra sensitivity that this allows.

Caps, gowns, and face masks are required under good manufacturing practice (GMP) [Food and Drug Administration, 2007; Rules and Guidance for Pharmaceutical Manufacturers and Distributors, 2007] conditions but are not necessary under normal conditions, particularly when working with laminar flow. If you have long hair, tie it back. When working aseptically on an open bench, do not talk. Talking is permissible when you are working in a vertical laminar-flow hood, with a barrier between you and the culture, but should still be kept to a minimum. If you have a cold, wear a face mask, or, better still, do not do any tissue culture during the height of the infection.

### 5.2.5  Reagents and Media

Reagents and media obtained from commercial suppliers will already have undergone strict quality control to ensure that they are sterile, but the outside surface of the bottle they come in is not. Some manufacturers supply bottles wrapped in polyethylene, which keeps them clean and allows them to be placed in a water bath to be warmed or thawed. The wrapping should be removed outside the hood. Unwrapped bottles should be swabbed in 70% alcohol when they come from the refrigerator or from a water bath.

### 5.2.6  Cultures

Cultures imported from another laboratory carry high risk of being contaminated at the source or in transit. Imported cell lines should always be quarantined (*see* Sections 3.2.4, 18.1.8); that is, they should be handled separately from the rest of your stocks and kept free of antibiotics until they are shown to be uncontaminated. They may then be incorporated into your main stock. Antibiotics should not be used routinely as they may suppress, but not eliminate, some contaminations (*see* Section 8.4.7) and encourage poor technique.

## 5.3  STERILE HANDLING

### 5.3.1  Swabbing

Swab down the work surface with 70% alcohol before and during work, particularly after any spillage, and swab it down again when you have finished. Swab bottles as well, especially those coming from cold storage or a water bath, before using them, and also swab any flasks or boxes from the incubator. Swabbing sometimes removes labels, so use an alcohol-resistant marker. Isopropyl alcohol ("rubbing alcohol," IPA) can be used instead of ethanol or methanol and is available as a proprietary spray or as prepacked swabs.

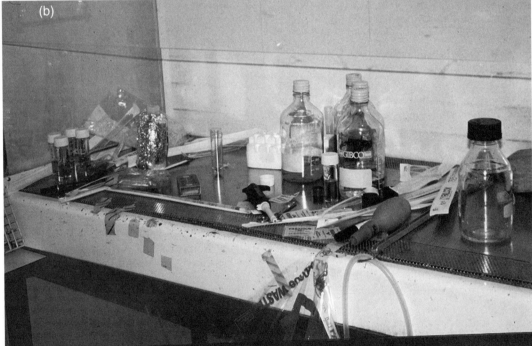

**Fig. 5.4. Layout of Work Area.** (*a*) Laminar-flow hood laid out correctly with pipettes on the left at the back, propped up for easy access and to allow airflow to the rear grill, medium to the left of the work area, culture flasks central and well back from the front edge, and pipette controller on the right. Positions may be reversed for left-handed workers. (*b*) Laminar-flow hood being used incorrectly. The hood is too full, and many items encroach on the air intake at the front, destroying the laminar airflow and compromising both containment and sterility.

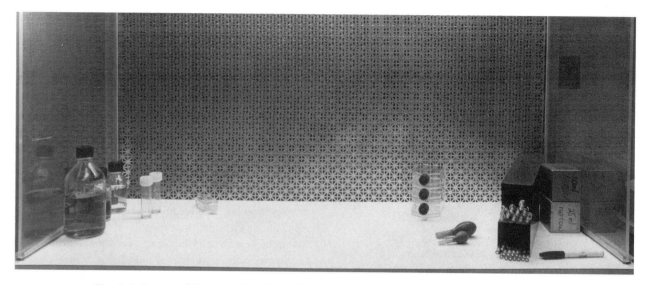

*Fig. 5.5. Layout of Horizontal Laminar-Flow Hood.* Correct layout for working in a horizontal laminar-flow hood. Positions may be reversed for left-handed workers.

## 5.3.2  Capping

Deep screw caps are preferred to stoppers, although care must be taken when washing caps to ensure that all detergent is rinsed from behind rubber liners. For this reason wadless polypropylene caps or disposables should be used. The screw cap should be covered with aluminum foil to protect the neck of the bottle from sedimentary dust, although the introduction of deep polypropylene caps (e.g., Duran) has made foil shrouding less necessary.

## 5.3.3  Flaming

When working on an open bench (Fig. 5.6), flame glass pipettes and the necks of bottles and screw caps before and after opening and closing a bottle, work close to the flame

where there is an updraft due to convection, and do not leave bottles open. Screw caps should be placed with the open side down on a clean surface and flamed before being replaced on the bottle. Alternatively, screw caps may be held in the hand during pipetting, avoiding the need to flame them or lay them down (Fig. 5.7).

Flaming is not advisable when you are working in a laminar-flow hood, as it disrupts the laminar flow, which in turn compromises both the sterility of the hood and its containment of any biohazardous material. An open flame can also be a fire hazard and can damage the HEPA filter or melt some of the plastic interior fittings. It is possible, however, to use flaming to burn off the alcohol used to sterilize instruments as long as the burner is outside the hood.

*Fig. 5.6. Layout of Work Area on Open Bench.* Items are arranged in a crescent around the clear work space in the center. The Bunsen burner is located centrally, to be close by for flaming and to create an updraft over the work area.

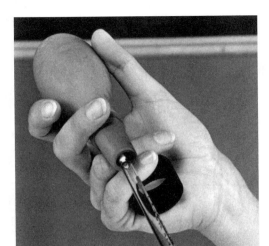

***Fig. 5.7. Holding Cap and Bulb.*** Cap may be unscrewed and held in the crook of the little finger of the hand holding the bulb or pipette controller.

Δ ***Safety Note.*** Care must be taken not to return the flamed instruments to the alcohol until they are fully extinguished.

### 5.3.4  Handling Bottles and Flasks

When working on an open bench, you should not keep bottles vertical when open; instead, keep them at an angle as shallow as possible without risking spillage. A bottle rack can be used to keep the bottles or flasks tilted. Culture flasks should be laid down horizontally when open and, like bottles, held at an angle during manipulations. When you are working in laminar flow, bottles can be left open and vertical, but do not let your hands or any other items come between an open vessel or sterile pipette and the HEPA filter. Flasks with angled necks facilitate pipetting when the flask is lying flat (Invitrogen–Gibco).

### 5.3.5  Pipetting

Standard glass or disposable plastic pipettes are still the easiest way to manipulate liquids. Syringes are sometimes used but should be discouraged as regular needles are too short to reach into most bottles. Syringing can also produce high shearing forces when you are dispensing cells, and the practice also increases the risk of self-inoculation. Wide-bore cannulae are preferable to needles but still not as rapid to use, except with multiple-stepping or repeating dispensers (*see* Fig. 4.5).

Pipettes of a convenient size range should be selected—1, 2, 5, 10, and 25 mL cover most requirements, although 100-mL disposable pipettes are available (BD Biosciences) and are useful for preparing and aliquotting media. Using fast-flow pipettes reduces accuracy slightly but gives considerable benefit in speed. If you are using glass pipettes and require only a few of each, make up mixed cans for sterilization and save space. Disposable plastic pipettes should be double wrapped and removed from their outer wrapping before

being placed in a hood. Unused pipettes should be stored in a dust-free container such as a pipette can.

Mouth pipetting should be avoided, as it has been shown to be a contributory factor in mycoplasma contamination and may introduce an element of hazard to the operator, such as with virus-infected cell lines and human biopsy or autopsy specimens or other potential biohazards (*see* Section 6.8.3). Inexpensive bulbs (Fig. 5.7) and electric pipette controllers are available (*see* Fig. 4.2); try a selection of these devices to find one that suits you and try holding a cap in the same hand. The instrument you choose should accept all sizes of pipette that you use without forcing them in and without the pipette falling out. Regulation of flow should be easy and rapid but at the same time capable of fine adjustment. You should be able to draw liquid up and down repeatedly (e.g., to disperse cells), and there should be no fear of carryover. The device should fit comfortably in your hand and should be easy to operate with one hand without fatigue.

Pipettors (*see* Fig. 4.3a) are particularly useful for small volumes (1 mL and less), although most makes now go up to 5 mL. Because it is difficult to reach down into a larger vessel without touching the inside of the neck of the vessel with the nonsterile stem of the pipettor, pipettors should only be used in conjunction with smaller flasks or by using an intermediary container, such as a universal container. Alternatively, longer tips may be used with larger volumes. Pipettors are particularly useful in dealing with microtitration assays and other multiwell dishes but should not be used for serial propagation unless filter tips are used.

Pipetting in tissue culture is often a compromise between speed and precision; speed is required to minimize deterioration during manipulations such as subculture, and precision is required for reproducibility during maintenance. However, an error of ±5% is usually acceptable, except under experimental conditions where greater precision may be required. Generally, using the smallest pipette compatible with the maneuver will give the greater precision required of most quantitative experimental work, while a larger pipette will allow quicker serial dispensing but with less accuracy.

All pipettes should have a cotton plug at the top to keep the pipette sterile during use. Plastic pipettes come with the plug already in place, but it needs to be inserted into glass pipettes after washing and drying and before sterilization. The plug prevents contamination from the bulb or pipette controller entering the pipette and reduces the risk of cross-contamination from pipette contents inadvertently entering the pipette controller. If the plug becomes wet, discard the pipette into disinfectant for return to the washup. Plugging pipettes for sterile use is a very tedious job, as is the removal of plugs before washing. Automatic pipette pluggers are available (*see* Fig. 10.6); they speed up the process, reduce the tedium, and blow out old plugs with compressed air.

Care must be taken to avoid contamination when setting up automatic pipetting devices and repeating dispensers (*see* Section 4.2.3). However, the increased speed in handling can

**TABLE 5.1.** Good Aseptic Technique

| Subject | Do | Don't |
|---|---|---|
| Laminar-flow hoods | Swab down before and after use. | Clutter up the hood. |
| | Keep minimum amount of apparatus and materials in hood. | Leave the hood in a mess. |
| | Work in direct line of sight. | |
| Contamination | Work without antibiotics. | Open contaminated flasks in tissue culture. |
| | Check cultures regularly, by eye and microscope. | Carry infected cells, even with antibiotics. |
| | Box Petri dishes and multiwell plates. | Leave contaminations unclaimed; dispose of them safely (see Section 18.4). |
| Mycoplasma | Test cells routinely. | Carry infected cells. |
| | | Try to decontaminate cultures. |
| Importing cell lines | Get from reliable source. | Get from a source far removed from originator. |
| | Quarantine incoming cell lines. | |
| | Check for mycoplasma. | |
| | Validate origin (even cells from originator). | |
| | Keep records. | |
| Exporting cell lines | Check for mycoplasma. | Send contaminated cell lines. |
| | Validate origin. | Pass on nonvalidated stock. |
| | Send data sheet. | |
| | Triple wrap. | |
| Glassware | Keep stocks separate. | Use for regular laboratory procedures. |
| Flasks | Pipette with flask sloped. | Have too many open at once. |
| | Gas from filtered $CO_2$ line. | Gas in $CO_2$ incubator unless with gas-permeable cap. |
| | Vent briefly at 37°C if sealed and stacked. | Stack too high (see Fig. 5.10). |
| Media and reagents | Swab bottles before placing them in hood. | Share among cell lines. |
| | Open only in hood. | Share with others. |
| | | Pour. |
| Pipettes | Use plugged pipettes. | Use the same pipette for different cell lines. |
| | Change if contaminated or plug wetted. | Share with other people. |
| | Use plastic for agar. | Overfill disposal cylinders. |

cut down on fatigue and on the time that vessels are open to contamination.

### 5.3.6 Pouring

Do not pour from one sterile container into another, unless the bottle you are pouring from is to be used once only to deliver all its contents (premeasured) in one single delivery. The major risk in pouring lies in the generation of a bridge of liquid between the outside of the bottle and the inside, permitting contamination to enter the bottle during storage or incubation.

## 5.4 STANDARD PROCEDURE

The essence of good aseptic technique embodies many of the principles of standard good laboratory practice (see Table 5.1). Keep a clean, clear space to work, and have on it only what you require at one time. Prepare as much as possible in advance so that cultures are out of the incubator for the shortest possible time and the various manipulations can be carried out quickly, easily, and smoothly. Keep everything in direct line of sight, and develop an awareness of accidental

contacts between sterile and nonsterile surfaces. Leave the area clean and tidy when you finish.

The two protocols that follow emphasize aseptic technique (see also Chapter 28, Ex. 1). Preparation of media and other manipulations are discussed in more detail under the appropriate headings (see Section 10.4).

---

### PROTOCOL 5.1. ASEPTIC TECHNIQUE IN VERTICAL LAMINAR FLOW

**Outline**
Clean and swab down work area, and bring bottles, pipettes, and other instruments (see Fig. 5.4a). Carry out preparative procedures first (preparation of media and other reagents), followed by culture work. Finally, tidy up and wipe over surface with 70% alcohol.

**Materials**

*Sterile (placed in hood):*

❑ Culture medium

❑ Pipettes, graduated, and plugged, in an assortment of sizes, 1, 5, 10, 25 mL
❑ Culture flasks

*Nonsterile:*

❑ Pipette controller or bulb (*see* Figs. 4.2, 5.7) in hood
❑ 70% alcohol in spray bottle (*see* Appendix I) in hood
❑ Lint-free swabs or wipes beside hood
❑ Absorbent paper tissues beside hood
❑ Pipette cylinder containing water and disinfectant (*see* Section 6.8.5) on floor beside hood
❑ Waste bin (for paper waste, swabs, and packaging) on floor beside hood on opposite side from pipette cylinder
❑ Suction line to aspirator below hood (*see* Fig. 4.9) or waste beaker (Fig. 5.8) in hood (both with disinfectant)
❑ Scissors
❑ Marker pen with alcohol-insoluble ink
❑ Notebook, pen, protocols

### Procedure

1. Swab down the work surface and all other inside surfaces of laminar-flow hood, including inside of the front screen, with alcohol and a lint-free swab or tissue.
2. Bring medium and reagents from cold store, water bath, or otherwise thawed from freezer; swab bottles with alcohol, and place those that you will need first in the hood.
3. Collect pipettes, and place at one side of the back of work surface in an accessible position (*see* Fig. 5.3*a*).
4. Open pipette cans and place lids on top or alongside, with the open side down, or stack individually wrapped pipettes, sorted by size, on a rack or in cans.
5. Collect any other glassware, plastics, and instruments that you will need, and place them close by (e.g., on a cart or an adjacent bench).
6. Slacken, but do not remove, caps of all bottles about to be used.
7. Remove the cap of the flask into which you are about to pipette, and the bottles that you wish to pipette from, and place the caps open side uppermost on the work surface, at the back of the hood and behind the bottle, so that your hand will not pass over them. Alternatively, if you are handling only one cap at a time, grasp the cap in the crook formed between your little finger and the heel of your hand (*see* Fig. 5.7), and replace it when you have finished pipetting.

8. Select a pipette:
   (a) For glass pipettes or bulk-wrapped plastic
      (i) Take a pipette from the can, lifting it parallel to the other pipettes in the can and touching them as little as possible, particularly at the tops (if the pipette that you are removing touches the end of any of the pipettes still in the can, discard it).
      (ii) Insert the top end of the pipette into a pipette controller or bulb, pointing the pipette away from you, and holding it well above the graduations, so that the part of the pipette entering the bottle or flask will not be contaminated (Fig. 5.9).
   (b) If pipettes individually wrapped plastic:
      (i) Open the pack at the top.
      (ii) Peel the ends back, turning them outside in.
      (iii) Insert the end of the pipette into the bulb or pipette controller.
      (iv) Withdraw the pipette from the wrapping without it touching any part of the outside of the wrapping, or the pipette touching any nonsterile surface (*see* Fig. 5.9).
      (v) Discard the wrapping into the waste bin.

Δ *Safety Note.* As you insert the pipette into the bulb or pipette controller take care not to exert too much pressure as pipettes can break if forced (*see* Section 6.5.3; Fig. 6.2).

9. The pipette in the bulb or pipette controller will now be at right angles to your arm. Take care that the tip of the pipette does not touch the outside of a bottle or the inner surface of the hood (*see* circled areas in Fig. 5.9); always be aware of where the pipette is. Following this procedure is not easy when you are learning aseptic technique, but it is an essential requirement for success and will come with experience.
10. Tilt the medium bottle toward the pipette so that your hand does not come over the open neck, withdraw 5 mL of medium, and transfer it a flask, also tilted.
11. Discard the pipette into the pipette cylinder containing disinfectant. Plastic pipettes can be discarded into double-thickness autoclavable biohazard bags.
12. Recap the flask.
13. Replace the cap on the medium bottle and flasks. Bottles may be left open while you complete a

particular maneuver, but should always be closed if you leave the hood for any reason.

*Note.* In vertical laminar flow, do not work immediately above an open vessel. In horizontal laminar flow, do not work behind an open vessel.

**14.** On completion of the operation, tighten all caps, and place flasks in incubator.

*Note.* If shortage of space requires you to stack flasks in the incubator you may need to release the pressure in the flasks after they have been in the incubator for about 30 min as the flasks may distort due to expansion of the gas phase, particularly with larger flasks (Fig. 5.10).

**15.** Remove all solutions and materials no longer required from the work surface, and swab down.

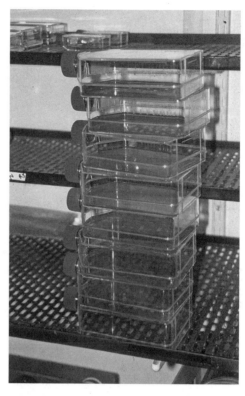

*Fig. 5.10. Tilting Flasks.* The air space inside a flask expands in the incubator or warm room. In large flasks, this causes the flask to bulge and will tilt the flasks, increasing the tilt with the height of the stack.

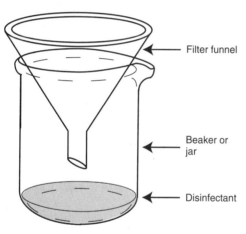

*Fig. 5.8. Waste Beaker.* Filter funnel prevents contents of beaker from splashing back.

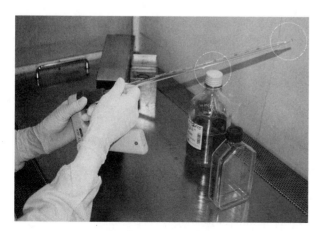

*Fig. 5.9. Inserting a Pipette in a Pipette Controller.* Pipette being inserted correctly with grip high on the pipette (above the graduations) and the pipette pointing away from the user. Circled areas mark potential risks, namely inadvertently touching the bottle or the back of the cabinet.

## PROTOCOL 5.2. WORKING ON THE OPEN BENCH

### Outline
Clean and swab down work area, and bring bottles, pipettes, and reagents (*see* Fig. 5.6). Carry out preparative procedures first. Flame articles as necessary and keep the work surface clean and clear. Finally, tidy up and wipe over surface with 70% alcohol.

### Materials

*Sterile or aseptically prepared:*

❑ Culture medium
❑ Pipettes, graduated, and plugged, in an assortment of sizes, 1, 5, 10, 25 mL
❑ Culture flasks

*Nonsterile:*

❑ Pipette controller or bulb (*see* Figs. 4.2, 5.7)
❑ 70% alcohol in spray bottle
❑ Lint-free swabs or wipes
❑ Absorbent paper tissues
❑ Pipette cylinder containing water and disinfectant (*see* Section 6.8.5) on floor to one side

❏ Waste bin (for paper waste, swabs, and packaging) on floor on opposite side from pipette cylinder
❏ Suction line to aspirator below hood (*see* Fig. 4.9) or waste beaker (Fig. 5.8) in hood (both with disinfectant)
❏ Bunsen burner (or equivalent) and lighter
❏ Scissors
❏ Marker pen with alcohol-insoluble ink
❏ Notebook, pen, protocols

### Procedure

1. Swab down bench surface with 70% alcohol.
2. Bring media and reagents, from cold store, water bath, or otherwise thawed from freezer, swab bottles with alcohol, and place those that you will need first on the bench in the work area, leaving the others at the side.
3. Collect pipettes and place at the side of the work surface in an accessible position (*see* Fig. 5.6).
   (a) If glass or bulk-wrapped plastic, open pipette cans and place lids on top or alongside, open side down.
   (b) If plastic, remove outer packaging and stack individually wrapped pipettes, sorted by size, on a rack or in cans.
4. Collect any other glassware, plastics, and instruments that you will need, and place them close by.
5. Flame necks of bottles, briefly rotating neck in flame, and slacken caps.
6. Select pipette:
   (a) If glass:
      (i) Take pipettes from can, lifting each parallel to the other pipettes in the can and touching as little as possible, particularly at the tops (if the pipette that you are removing touches the end of any of the pipettes still in the can, discard it),
      (ii) Insert the top end into a pipette aid, pointing pipette away from you and holding it well above the graduations, so that the part of the pipette entering the bottle or flask will not be contaminated (*see* Fig. 5.9).
   (b) If plastic:
      (i) Open the pack at the top
      (ii) Peel the ends back, turning them outside in
      (iii) Insert the end of the pipette into the bulb or pipette controller
      (iv) Withdraw the pipette from the wrapping without it touching any part of the outside of the wrapping, or the pipette

touching any nonsterile surface (*see* Fig. 5.9)
      (v) Discard the wrapping into the waste bin. Flame pipette (glass only) by pushing it lengthwise through the flame, rotate 180°, and pull the pipette back through the flame. This should only take 2 to 3 s, or the pipette will get too hot. You are not attempting to sterilize the pipette; you are merely trying to fix any dust that settled on it. It you have touched anything or contaminated the pipette in any other way, discard it into disinfectant for return to the washup facility; do not attempt to resterilize the pipette by flaming. Do not flame plastic pipettes.
7. Insert pipette in a bulb or pipette controller, pointing pipette way from you and holding it well above the graduations so that the part of the pipette entering the bottle or flask will not be contaminated (*see* Fig. 5.9).

Δ *Safety Note.* Take care not to exert too much pressure, as pipettes can break when being forced into a bulb (*see* Fig. 6.2).

8. The pipette in the bulb or pipette controller will now be at right angles to your arm. Take care that the tip of the pipette does not touch the outside of a bottle or pipette can. Always be aware of where the pipette is. Following this procedure is not easy when you are learning aseptic technique, but it is an essential requirement for success and will come with experience.
9. Holding the pipette still pointing away from you, remove the cap of your first bottle into the crook formed between your little finger and the heel of your hand (*see* Fig. 5.7). If you are pipetting into several bottles or flasks, they can be laid down horizontally on their sides. Work with the bottles tilted so that your hand does not come over the open neck. If you have difficulty holding the cap in your hand while you pipette, place the cap on the bench, open side down. If bottles are to be left open, they should be sloped as close to horizontal as possible in laying them on the bench or on a bottle rest.
10. Flame the neck of the bottle.

*Note.* If flaming a Duran bottle, do not use with pouring ring.

11. Tilt the bottle toward the pipette so that your hand does not come over the open neck.
12. Withdraw the requisite amount of fluid and hold.
13. Flame the neck or the bottle and recap.

**14.** Remove the cap of the receiving bottle, flame the neck, insert fluid, reflame the neck, and replace the cap.
**15.** When you have finished, tighten caps.
**16.** Remove from the work surface all solutions and materials no longer required, and swab down the work surface.

***Petri Dishes and Multiwell Plates.*** Petri dishes and multiwell plates are particularly prone to contamination because of the following factors:

(1) The larger surface area exposed when the dish is open.
(2) The risk of touching the rim of the dish when handling an open dish.
(3) The risk of carrying contamination from the work surface to the plate via the lid if the lid is laid down.
(4) Medium filling the gap between the lid and the dish due to capillarity if the dish is tilted or shaken in transit to the incubator.
(5) The higher risk of contamination in the humid atmosphere of a $CO_2$ incubator.

The following practices will minimize the risk of contamination:

(1) Do not leave dishes open for an extended period or work over an open dish or lid.
(2) When moving dishes or transporting them to or from the incubator, take care not to tilt, swirl, or shake them, to avoid the medium entering the capillary space between the lid and the base. Further precautions include:
(a) Use "vented" dishes (*see* Fig. 7.8).

***Fig. 5.11. Boxed Dishes.*** A transparent box, such as a sandwich box or cake box, helps protect unsealed dishes and plates, and flasks with slackened caps, from contamination in a humid incubator. This type of container should also be used for materials that may be biohazardous, to help contain spillage in the event of an accident. (Material known to be biohazardous would need a sealed container and a separate incubator.)

(b) If medium still lodges in this space, discard the lid, blot any medium carefully from the outside of the rim with a sterile tissue dampened with 70% alcohol, and replace the lid with a fresh one. (Make sure that the labeling is on the base!)
(3) Enclose dishes and plates in a transparent plastic box for incubation, and swab the box with alcohol when it is retrieved from the incubator (Fig. 5.11; *see* Section 5.7.7).

The following procedure is recommended for handling Petri dishes or multiwell plates.

---

**PROTOCOL 5.3. HANDLING DISHES OR PLATES**

***Materials***
As for Protocol 5.1 or 5.2, as appropriate.

***Procedure***

**1.** Place dish or plate on one side of work area.
**2.** Position medium bottle and slacken the cap.
**3.** Bring dish to center of work area.
**4.** Remove bottle cap and fill pipette from bottle.
**5.** Remove lid from dish.
**6.** Add medium to dish, directing the stream gently low down on the side of the base of the dish.
**7.** Replace lid.
**8.** Return dish to side of hood, taking care not to let the medium enter the capillary space between the lid and the base.
**9.** Discard pipette.
**10.** Tighten cap on medium bottle.
**11.** Remove from the work surface all solutions and materials no longer required, and swab down the work surface.

---

## 5.5 APPARATUS AND EQUIPMENT

All apparatus used in the tissue culture area should be cleaned regularly to avoid the accumulation of dust and to prevent microbial growth in accidental spillages. Replacement items, such as gas cylinders, must be cleaned before being introduced to the tissue culture area and no major movement of equipment should take place while people are working aseptically.

### 5.5.1 Incubators

Humidified incubators are a major source of contamination (*see* Section 18.1.4). They should be cleaned out at regular intervals (weekly or monthly, depending on the level of atmospheric contamination and frequency of access) by removing the contents, including all the racks or trays, and washing down the interior and the racks or shelves with a nontoxic detergent such as Decon or Roccall. Traces of detergent should then be removed with 70% alcohol, which should be allowed to evaporate completely before replacing the shelves and cultures.

A fungicide, such as 2% Roccall or 1% copper sulfate, may be placed in the humidifier tray at the bottom of the incubator to retard fungal growth, but the success of such fungicides is limited to the surface that they are in contact with, and there is no real substitute for regular cleaning. Some incubators have high-temperature sterilization cycles, but these are seldom able to generate sufficient heat for long enough to be effective and the length of time that the incubator is out of use can be inconvenient. Some incubators have micropore filtration and laminar airflow to inhibit the circulation of microorganisms, but it may be better to accept a minor increase in recovery time, eliminate the fan, and rely on convection for circulation (*see* Section 4.3.2).

### 5.5.2 Boxed Cultures

When problems with contamination in humidified incubators recur frequently, it is advantageous to enclose dishes, plates, and flasks, with slackened caps, in plastic sandwich boxes (*see* Fig. 5.11). The box should be swabbed before use, inside and outside, and allowed to dry in sterile air. When the box is subsequently removed from the incubator, it should be swabbed with 70% alcohol before being opened or introduced into your work area. The dishes are then carefully removed and the interior of the box swabbed before reuse. Flasks with gas-permeable caps (*see* Section 5.5.3; Fig. 7.8*b*) do not

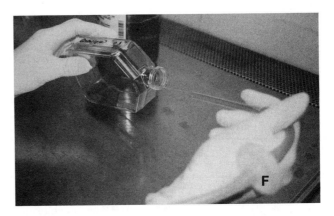

**Fig. 5.12. Gassing a Flask.** A pipette is inserted into the supply line from the $CO_2$ source, and 5% $CO_2$ is used to flush the air out of the flask without bubbling through the medium. The letter "F" indicates a micropore filter inserted in the $CO_2$ line.

require to be boxed but should be swabbed before placing in the hood.

### 5.5.3 Gassing with $CO_2$

It is common practice to place flasks, with the caps slackened, in a humid $CO_2$ incubator to allow for gaseous equilibration, but doing so does increase the risk of contamination. Flasks with gas-permeable caps allow rapid equilibration with the 5% $CO_2$ atmosphere without the risk of contamination. Alternatively, purge the flasks from a sterile, premixed gas supply (Fig. 5.12) and then seal them. This avoids the need for a gassed incubator for flasks and gives the most uniform and rapid equilibration.

# CHAPTER 6

# Safety, Bioethics, and Validation

## 6.1 LABORATORY SAFETY

In addition to the everyday safety hazards common to any workplace, the cell culture laboratory has a number of particular risks associated with culture work. Despite the scientific background and training of most people who work in this environment, accidents still happen, as familiarity often leads to a casual approach in dealing with regular, biological, and radiological hazards. Furthermore individuals who service the area often do *not* have a scientific background, and the responsibility lies with those who do have to maintain a safe environment for all who work there. It is important to identify potential hazards but, at the same time, to be rational and proportionate in identifying the risks. If a risk is not seen as realistic, then precautions will tend to be disregarded, and the whole safety code will be placed in disrepute.

It is essential that new personnel joining the tissue culture facility, from other departments or elsewhere, receive formal instruction in the safety issues relating to the tissue culture laboratory. This instruction should be accompanied by a printed document defining safety procedures and the role of the individual as well as that of the administration. This manual should also cover ethical issues (*see* Section 6.9) and the important aspects of quality assurance (*see* Section 6.10) and validation (*see* Section 6.11) essential for the reliable use of cell lines and the security of the tissue culture laboratory. A new member of the unit should be regarded as an apprentice, regardless of their seniority, until they demonstrate that they can work independently within the technical and regulatory procedures set down for the laboratory.

## 6.2 RISK ASSESSMENT

Risk assessment is an important principle that has become incorporated into most modern safety legislation. Determining the nature and extent of a particular hazard is only part of the process; the general environmental conditions are equally important in determining risk (Table 6.1). Such considerations as the amount of a particular material, the degree and frequency of exposure to a hazard, the scale of the operation, the procedures for handling materials and equipment, who uses them and their relevant training and experience, the type of protective clothing worn, ancillary hazards like exposure to heat, frost, and electric current, all contribute to the risk of a given procedure, although the nature of the hazard itself may remain constant.

A major problem that arises constantly in establishing safe practices in a biomedical laboratory is the disproportionate concern given to the more esoteric and poorly understood hazards, such as those arising from genetic manipulation, relative to the known and proved hazards of toxic and corrosive chemicals, solvents, fire, ionizing radiation, electrical shock, and broken glass. It is important that biohazards be categorized correctly [Caputo, 1996; CDC/OHS, 1999; HSE, 2008a], neither overemphasized

**TABLE 6.1.** Elements of Risk Assessment

| Category | Items affecting risk |
| --- | --- |
| **Operator** | |
| Experience | Level |
| | Relevance |
| | Background |
| Training | Previous |
| | New requirements |
| Protective clothing | Adequate |
| | Properly worn (buttoned lab coat) |
| | Laundered regularly |
| | Repaired or discarded when damaged |
| **Equipment** | |
| Age | Condition |
| | Adherence to new legislation |
| Suitability for task | Access, sample capacity, containment |
| Mechanical stability | Loading |
| | Anchorage |
| | Balance |
| Electrical safety | Connections |
| | Leakage to ground (earth) |
| | Proximity of water |
| Containment | Aerosols: |
| |    Generation |
| |    Leakage from hood ducting |
| |    Overspill from work area |
| | Toxic fumes |
| | Exhaust ductwork: |
| |    Integrity |
| |    Site of effluent and downwind risk |
| Heat | Generation |
| | Dissipation |
| | Effect on operator performance |
| Maintenance | Frequency |
| | Decontamination required? |
| Disposal | Route |
| | Decontamination required? |
| **Physical risks** | |
| Intense cold | Frostbite |
| | Numbing |
| Electric shock | Loss of consciousness |
| | Cardiac arrest |
| Fire | General precautions |
| | Equipment wiring, installation, maintenance |
| | Incursion of water near electrical wiring |
| | Fire drills, procedures, access to escape routes |
| | Solvent usage and storage (e.g., do not store ether in refrigerators) |
| | Flammable mixtures |
| | Identification of stored biohazards and radiochemicals |
| **Chemicals (including gases and volatile liquids)** | |
| Scale | Amount used |

**TABLE 6.1.** (*Continued*)

| Category | Items affecting risk |
| --- | --- |
| Toxicity | Poisonous |
| | Carcinogenic |
| | Teratogenic |
| | Mutagenic |
| | Corrosive |
| | Irritant |
| | Allergenic |
| | Asphyxiant |
| Reaction with water | Heat generation |
| | Effervescence |
| Reaction with solvents | Heat generation |
| | Effervescence |
| | Generation of explosive mixture |
| Volatility | Intoxication |
| | Asphyxiation |
| | Dissemination |
| Generation of powders and aerosols | |
| | Inhalation |
| Import, export, transportation | Breakage, leakage |
| Location and storage conditions | Access by untrained staff |
| | Illegal entry |
| | Weather, incursion of water |
| | Stability, compression, breakage, leakage |
| **Biohazards** | |
| Pathogenicity | Grade |
| | Infectivity |
| | Host specificity |
| | Stability |
| Scale | Number of cells |
| | Amount of DNA |
| Genetic manipulation | Host specificity |
| | Vector infectivity |
| | Disablement |
| Containment | Room |
| | Cabinet |
| | Procedures |
| **Radioisotopes** | |
| Emission | Type |
| | Energy |
| | Penetration, shielding |
| | Interaction, ionization |
| | Half-life |
| Volatility | Inhalation |
| | Dissemination |
| Localization on ingestion | DNA precursors, such as [$^3$H]thymidine |
| Disposal | Solid, liquid, gaseous |
| | Route |
| | Legal limits |

**TABLE 6.1.** (*Continued*)

| Category | Items affecting risk |
|---|---|
| **Special circumstances** | |
| Pregnancy | Immunodeficiency |
| | Risk to fetus, teratogenicity |
| Illness | Immunodeficiency |
| Immunosuppressant drugs | Immunodeficiency |
| Cuts and abrasions | Increased risk of absorption |
| Allergy | Powders (e.g., detergents) |
| | Aerosols |
| | Contact (e.g., rubber gloves) |
| **Elements of procedures** | |
| Scale | Amount of materials used |
| | Size of equipment & facilities, effect on containment |
| | Number of personnel involved |
| Complexity | Number of steps or stages |
| | Number of options |
| | Interacting systems and procedures |
| Duration | Process time |
| | Incubation time |
| | Storage time |
| Number of persons involved | Increased risk? |
| | Diminished risk? |
| Location | Containment |
| | Security and access |

nor underestimated, but the precautions taken should not displace the recognition of everyday safety problems.

## 6.3 STANDARD OPERATING PROCEDURES

Hazardous substances, equipment, and conditions should not be thought of in isolation but should be taken as part of a procedure, all the components of which should be assessed. If the procedure is deemed to carry any significant risk beyond the commonplace, then a standard operating procedure (SOP) should be defined, and all who work with the material and equipment should conform to that procedure. The different stages of the procedure—procurement, storage, operations, and disposal—should be identified, and the possibility must be taken into account that the presence of more than one hazard within a procedure will compound the risk or, at best, complicate the necessary precautions (e.g., how does one dispose of broken glass that has been in contact with a human cell line labeled with a radioisotope?).

## 6.4 SAFETY REGULATIONS

The following recommendations should not be interpreted as a code of practice but rather as advice that might

help in compiling safety regulations. The information is designed to provide the reader with some guidelines and suggestions to help construct a local code of practice, in conjunction with regional and national legislation and in full consultation with the local safety committee (Table 6.2). These recommendations have no legal standing and should not be quoted as if they do.

General safety regulations should be available from the safety office of the organization in which you work. In addition they are available from the Occupational Safety and Health Administration in the United States [OSHA, 2009] and in the United Kingdom from the Management of Health and Safety at Work Regulations [Management of Health and Safety at Work Regulations, 1999] (Table 6.3). These regulations cover all matters of general safety. The relevant regulations and recommendations for biological safety for the United States are contained in Biosafety in Microbiological and Biomedical Laboratories [CDC/OHS, 1999], a joint document prepared by the Centers for Disease Control and Prevention in Atlanta, Georgia, and the National Institutes of Health in Bethesda, Maryland. For the United

**TABLE 6.2.** Levels of Action

| Category | Action |
|---|---|
| Regulatory authority | Contact national, regional inspectors |
| Local safety committee | Appoint representatives |
| | Arrange meetings and discussion |
| Guidelines | Access local and national |
| | Generate local guidelines if not already done |
| Standard operating procedures (SOPs) | Define and make available |
| Protective clothing | Provide, launder, ensure that it is worn correctly |
| Containerization | Specify physical description (e.g., storage and packaging) |
| Containment levels | Specify chemical, radiological, biological levels |
| Training | Arrange seminars, supervision |
| Monitoring | Automatic smoke detectors, oxygen meter |
| Inspection | Arrange equipment, procedures, laboratory inspections by trained, designated staff |
| Record keeping | Safety officers and operatives to keep adequate records |
| Import and export | Regulate and record |
| Classified waste disposal | Define routes for sharps, radioactive waste (liquid and solid), biohazards, corrosives, solvents, toxins |
| Access | Limit to trained staff and visitors only |
| | Exclude children, except in public areas |

**TABLE 6.3.** Safety Regulations and Guidelines

| Topic | United States | United Kingdom (generally harmonized with EC Directives) |
|---|---|---|
| General | Occupational Safety and Health Administration [OSHA, 2009] | Management of Health and Safety at Work Regulations [1999] |
| Equipment | Occupational Safety and Health Administration [OSHA, 2009] | Provision and Use of Work Equipment [HSE, 1998] |
| Chemical | Occupational Safety and Health Administration [OSHA, 2009] National Institute for Occupational Safety and Health [CDC-NIOSH, 2009] | Control of Substances Hazardous to Health [Health and Safety Commission [HSE-COSHH]] |
| Biological | CDC NIH: Biosafety in Microbiological and Biomedical Laboratories [CDC/OHS, 1999] | Health & Safety Executive, Infections at Work and Genetically Modified Organisms [HSE, 2008a]; Advisory Committee on Dangerous Pathogens [ACDP, 2003, 2005] |
| Radiological | US Nuclear Regulatory Commission. Medical, Industrial, and Academic Uses of Nuclear Materials [U.S .NRC, 2008] | HSE Radiation Protection publications [HSE, 2008b] |
| Disposal | National Research Council [2009] Appropriate State Legislature | Department of Health [DoH, 2006] |

Kingdom, information and guidance on biological hazards at work is available from the Health and Safety Executive [HSE, 2009]. Safety guidelines in Europe are handled at the national level, and it is the responsibility of each country to ensure compliance with the appropriate EC Directives. The advice given in this chapter is general and should not be construed as satisfying any national or international legal requirement.

## 6.5   GENERAL SAFETY

Some aspects of general safety require particular emphasis in a tissue culture laboratory (Table 6.4).

### 6.5.1   Operator

It is the responsibility of the institution to provide the correct training in appropriate laboratory procedures and to ensure that new and existing members of staff are and remain familiar with safety regulations. It is the supervisor's responsibility to ensure that procedures are carried out correctly and that the correct protective clothing is worn at the appropriate times.

### 6.5.2   Equipment

A general supervisor should be appointed to be in charge of all equipment maintenance, electrical safety, and mechanical reliability. A curator should be put in charge of each individual piece of equipment to oversee the maintenance and operation of the equipment and to train others in its use. Particular risks include the generation of toxic fumes or aerosols from centrifuges and homogenizers, which must be contained either by the design of the equipment or by placing them in a fume cupboard.

The electrical safety of equipment is dealt with in the United States under Occupational Health and Safety [OHSA] and in the United Kingdom by the Health and Safety Executive [HSE, 1998].

### 6.5.3   Glassware and Sharp Items

The most common form of injury in performing tissue culture results from accidental handling of broken glass and syringe needles. Particularly dangerous are broken pipettes in a washup cylinder, which result from too many pipettes being forced into too small a container (Fig. 6.1). Glass Pasteur pipettes should not be inserted into a washup cylinder with other pipettes (see below, this section). Needles that have been improperly disposed of together with ordinary waste or forced through the wall of a rigid container when the container is overfilled are also very dangerous.

Accidental inoculation via a discarded needle or broken glass, or because of an accident during routine handling, remains one of the more acute risks associated with handling potentially biohazardous material. It may even carry a risk of transplantation when one handles human tumors [Southam, 1958; Scanlon et al., 1965; Gugel & Sanders, 1986], particularly in an immunocompromised host.

Pasteur pipettes should be discarded into a sharps bin (see Appendix II and www.cdc.gov/niosh/sharps1.html) and not into the regular washup as they break very easily and the shards are extremely hazardous. If reused, they should be handled separately and with great care. Disposable plastic Pasteur pipettes (e.g., Pastettes) are available but tend to have a thicker tip. Avoid using syringes and needles, unless they are needed for loading ampoules (use a blunt cannula)

**TABLE 6.4.** Safety Hazards in a Tissue Culture Laboratory

| Category | Item | Risk | Precautions |
|---|---|---|---|
| General | Broken glass | Injury, infection | Dispose of carefully in designated bin |
| | Pipettes | Injury, infection | Check glass for damage and discard; use plastic |
| | Sharp instruments | Injury, infection | Handle carefully; discard in sharps bin. |
| | Glass Pasteur pipettes | Injury, infection | Handle carefully; do not use with potentially biohazardous material; use plastic |
| | Syringe needles | Injury, infection | Minimize or eliminate use; discard into sharps bin |
| | Cables | Fire, electrocution, snagging, tripping | Check connections sound; clip together and secure in safe place |
| | Tubing | Leakage, snagging, tripping | Check and replace regularly; clip in place; keep away from passage floors |
| | Cylinders | Instability, leakage | Secure to bench or wall; check regularly with leak detector |
| | Liquid nitrogen | Frostbite, asphyxiation, explosion | Wear mask, lab coat, and gloves; do not store ampoules in liquid phase; if stored submerged, enclose when thawing |
| Burns | Autoclaves, ovens, & hot plates | Contact with equipment | Post warning notices |
| | | Handling items just sterilized | Provide gloves |
| Fire | Bunsen burners; flaming, particularly in association with alcohol | Fire, melting damage, burn risk | Keep out of hoods and do not place under cupboards or shelves; do not return flaming instruments to alcohol |
| | Manual autoclaves | Can burn dry and contents ignite | Install a timer and a thermostatic cut-out |
| | | | Ensure that a safety valve is present and active |
| Radiological | Radioisotopes in sterile cabinet | Emission, spillage, aerosols, volatility | Work on absorbent tray in Class II or chemical hazard hood; minimize aerosols |
| | Irradiation of cultures | Radiation dose | Use monitor, wear personal badge monitor and check regularly |
| Biological | Importation of cell lines and biopsies | Infection | Do not import from high-risk areas; screen cultures for likely pathogens |
| | Genetic manipulation | Infection, DNA transfer | Follow genetic manipulation guidelines |
| | Propagation of viruses | Infection | Observe CDC or ACDP guidelines; work in correct level of containment; minimize aerosols |
| | Position and maintenance of laminar-flow hoods | Breakdown in containment | Check airflow patterns, pressure drop across filter, and overspill from cabinet regularly |

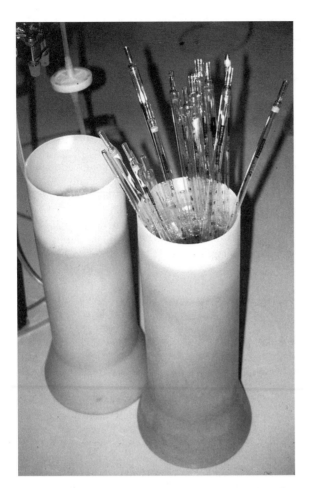

***Fig. 6.1. Overfilled Pipette Cylinder.*** Pipettes protruding from a pipette cylinder as a result of attempted insertion of pipettes after the cylinder is full; those protruding from the cylinder will not soak properly or be disinfected and are prone to breakage (if glass) when other pipettes are added.

or withdrawing fluid from a capped vial. When disposable needles are discarded, use a rigid plastic or metal container. Do not attempt to bend, manipulate, or resheath the needle. Provide separate hard-walled receptacles for the disposal of sharp items and broken glass, and do not use these receptacles for general waste.

Take care when you are fitting a bulb or pipetting device onto a glass pipette. Choose the correct size of bulb to guard against the risk of the pipette breaking at the neck and lacerating your hand. Check that the neck is sound, hold the pipette as near the end as possible, and apply gentle pressure with the pipette pointing away from your knuckles (Fig. 6.2). Although this is primarily a risk arising from the use of glass pipettes, even plastic pipettes can be damaged and break on insertion, so always check the top of each pipette before use.

### 6.5.4  Chemical Toxicity

Relatively few major toxic substances are used in tissue culture, but when they are, the conventional precautions should be taken, paying particular attention to the distribution

of powders and aerosols by laminar-flow hoods (*see* Section 6.8.2). Detergents—particularly those used in automatic washing machines—are usually caustic, and even when they are not, they can cause irritation to the skin, eyes, and lungs. Use liquid-based detergents in a dispensing device whenever possible, wear gloves, and avoid procedures that cause powdered detergent to spread as dust. Liquid detergent concentrates are more easily handled, but often are more expensive to achieve the same level of effectiveness. Chemical disinfectants such as hypochlorite should be used cautiously, either in tablet form or as a liquid dispensed from a dispenser. Hypochlorite disinfectants will bleach clothing, cause skin irritations, and even corrode welded stainless steel.

Specific chemicals used in tissue culture that require special attention are (1) dimethyl sulfoxide (DMSO), which is a powerful solvent and skin penetrant and can transport many substances through the skin [Horita and Weber, 1964] and even through some protective gloves (e.g., rubber latex or silicone, though little through nitrile), and (2) mutagens, carcinogens, and cytotoxic drugs, which should be handled in a safety cabinet. A Class II laminar-flow hood may be adequate for infrequent handling of small quantities of these substances, but it may be necessary to use a hood designed specifically for cytotoxic chemicals (*see* Fig. 6.5b). Mutagens, carcinogens, and other toxic chemicals are sometimes dissolved in DMSO, increasing the risk of uptake via the skin. Nitrile gloves provide a better barrier but should be tested for the particular agents in use.

The handling of chemicals is regulated by Occupational Health and Safety [OSHA, 2009] and by the Health and Safety Executive in the United Kingdom [HSE-COSHH]. Information and guidelines are also available from the National Institute for Occupational Safety and Health [CDC-NIOSH, 2009].

### 6.5.5  Gases

Most gases used in tissue culture ($CO_2$, $O_2$, $N_2$) are not harmful in small amounts but are nevertheless dangerous if handled improperly. They should be contained in pressurized cylinders that are properly secured (Fig. 6.3). If a major leak occurs, there is a risk of asphyxiation from $CO_2$ and $N_2$ and of fire from $O_2$. Evacuation and maximum ventilation are necessary in each case; if there is extensive leakage of $O_2$, call the fire department. An oxygen monitor should be installed near floor level in rooms where gaseous or liquid $N_2$ and $CO_2$ are stored in bulk, or where there is a piped supply to the room.

When glass ampoules are used, they are sealed in a gas oxygen flame. Great care must be taken both to guard the flame and to prevent inadvertent mixing of the gas and oxygen. A one-way valve should be incorporated into the gas line so that oxygen cannot blow back.

### 6.5.6  Liquid Nitrogen

Three major risks are associated with liquid nitrogen: frostbite, asphyxiation, and explosion (*see* Protocol 19.1.2). Because the temperature of liquid nitrogen is $-196°C$, direct contact with

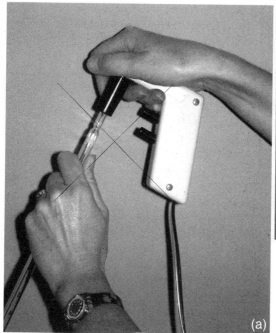

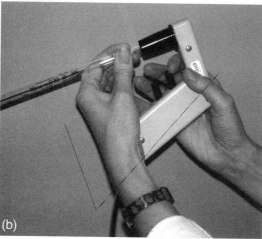

**Fig. 6.2. Safely Inserting a Pipette into a Pipetting Device.** *(a)* Wrong position: left hand too far down pipette, risking contamination of the pipette and exerting too much leverage, which might break the pipette; right hand too far over and exposed to end of pipette or splinters should the pipette break at the neck during insertion. Tip of pipette is also obscured by left hand and arm, risking contamination by contact with nonsterile surface. *(b)* Correct position: left hand farther up pipette with lighter grip; right hand clear of top of pipette, with tip of pipette in clear view.

**Fig. 6.3. Cylinder Clamp.** Clamps onto edge of bench or rigid shelf and secures gas cylinder with fabric strap. Fits different sizes of cylinder and can be moved from one position to another if necessary; clamp is available from most laboratory suppliers.

the liquid (via splashes, etc.), or with anything—particularly something metallic—that has been submerged in it, presents a serious hazard. Gloves that are thick enough to act as insulation, but flexible enough to allow the manipulation of ampoules, should be worn (*see* Cryoprotective Gloves, Appendix II). When liquid nitrogen boils off during routine use of the freezer, regular ventilation is sufficient to remove excess nitrogen, but when liquid nitrogen is being dispensed, or when a lot of samples are being inserted into the freezer, extra ventilation will be necessary. Remember, 1 L of liquid nitrogen generates nearly 700 L of gas. An oxygen monitor and alarm should be installed (*see* Fig. 3.8) and linked to the ventilation system, so that a nitrogen spillage reducing the oxygen concentration and triggering the alarm also increases the ventilation rate.

When an ampoule or vial is submerged in liquid nitrogen, a high-pressure difference results between the outside and the inside of the ampoule. If the ampoule is not perfectly sealed, liquid nitrogen may be drawn in, causing the ampoule to explode violently when warmed for thawing. Ampoules submerged in liquid nitrogen must be perfectly sealed; thawing of ampoules or vials that have been stored submerged in liquid nitrogen should always be performed in a container with a lid, such as a plastic bucket (*see* Protocol 19.2; Fig. 19.9), and a face shield or goggles must be worn. This problem can be avoided by storing the ampoules in the gas phase or in a perfused jacket freezer (*see* Section 19.3.6), which also reduces asphyxiation risks.

### 6.5.7 Burns

There are three main sources of risk from burns: (1) autoclaves, ovens, and hot plates; (2) handling of items that have just been removed from them; and (3) naked flames such as a Bunsen burner (*see* Table 6.4). Warning notices should be placed near all hot equipment, including burners, and items that have just been sterilized should be allowed to cool before removal from the sterilizer. Insulated gloves should be provided where hot items are being handled.

## 6.6 FIRE

Particular fire risks associated with tissue culture stem from the use of Bunsen burners for flaming, together with alcohol for swabbing or sterilization. Keep the two separate; always ensure that alcohol for sterilizing instruments is kept in minimum volumes in a narrow-necked bottle or flask that is not easily upset (Fig. 6.4). Alcohol for swabbing should be kept in a plastic wash bottle or spray and should not be used in the presence of an open flame. When instruments are

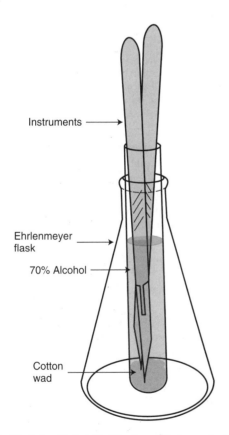

**Fig. 6.4. Flask for Alcohol Sterilization of Instruments.** The wide base prevents tipping, and the center tube reduces the amount of alcohol required so that spillage, if it occurs, is minimized. (From an original idea by M. G. Freshney.)

sterilized in alcohol and the alcohol is subsequently burned off, care must be taken not to return the instruments to the alcohol while they are still alight. If you are using this technique, keep a damp cloth nearby to smother the flames if the alcohol ignites.

## 6.7 IONIZING RADIATION

Three main types of radiation hazard are associated with tissue culture: ingestion, irradiation from labeled reagents, and irradiation from a high-energy source. Guidelines on radiological protection for the United States can be obtained from the Nuclear Regulatory Commission [NRC, 2008] and for the United Kingdom from HSE Radiation Protection Publications [HSE, 2008b].

### 6.7.1 Ingestion

Soluble radiolabeled compounds can be ingested by being splashed on the hands or via aerosols generated by pipetting or the use of a syringe. Tritiated nucleotides, if accidentally ingested, will become incorporated into DNA, and will cause radiolysis within the DNA due to the short path length of the low-energy $\beta$-emission from $^3H$. Radioactive isotopes of iodine will concentrate in the thyroid and may also cause local damage.

Work in a Class II hood to contain aerosols, and wear gloves. The items that you are working with should be held in a shallow tray lined with paper tissue or Benchcote to contain any accidental spillage. Use the smallest pieces of equipment (a pipettor with disposable plastic tips, small sample tubes, etc.) compatible with the procedure to generate minimum bulk when they are discarded into a radioactive waste container. Clean up carefully when you are finished, and monitor the area regularly for any spillage.

### 6.7.2 Disposal of Radioactive Waste

Procedures and routes for disposal of radioactive substances will be defined in local rules governing the laboratory; advice in setting up these rules can be obtained from the authorities given above. Briefly, the amount of radioactivity disposed of over a certain period will have an upper limit, disposal will be limited to certain designated sinks, and the amounts discarded will need to be logged in a record book at the site of disposal. Vessels used for disposal will then need to be decontaminated in an appropriate detergent, and the washings disposed of as radioactive waste. Disposal may need to take account of any biological risk, so items that are to be reused will first have to be biologically decontaminated in hypochlorite and then radioactively decontaminated in Decon or a similar detergent. Both solutions must then be regarded as radioactive waste.

### 6.7.3 Irradiation from Labeled Reagents

The second type of risk is from irradiation from higher energy $\beta$- and $\gamma$-emitters such as $^{32}P$, $^{125}I$, $^{131}I$, and $^{51}Cr$. Protection can be obtained by working behind a 2-mm-thick lead shield and storing the concentrated isotope in a lead pot. Perspex screens (5 mm) can be used with $^{32}$P at low concentrations for short periods. Work on a tray in a in a Class II hood (*see* Section 6.7.1).

### 6.7.4  Irradiation from High-Energy Sources

The third type of irradiation risk is from X-ray machines, high-energy sources such as $^{60}$Co, or ultraviolet (UV) sources used for sterilizing apparatus or stopping cell proliferation in feeder layers (*see* Section 13.2.3; Protocols 22.1, 22.4). Because the energy, particularly from X rays or $^{60}$Co, is high, these sources are usually located in a specially designed accommodation and are subject to strict control. UV sources can cause burns to the skin and damage to the eyes; they should be carefully screened to prevent direct irradiation of the operator, who should wear barrier filter goggles.

Consult your local radiological officer and code of practice before embarking on radioisotopic experiments. Local rules vary, but most places have strict controls on the amount of radioisotopes that can be used, stored, and discarded. Those wishing to use radioisotopes may be required to have a general medical examination, including storage of a blood sample, before starting work.

## 6.8  BIOHAZARDS

As for radioisotope use, those wishing to use potentially biohazardous material may require a general medical examination, including storage of a blood sample, before starting work. The need for protection against biological hazards [Caputo, 1996] is defined both by the source of the material and by the nature of the operation being carried out. It is also governed by the conditions under which culture is performed. Using standard microbiological technique on the open bench has the advantage that the techniques in current use have been established as a result of many years of accumulated experience. Problems arise when new techniques are introduced or when the number of people sharing the same area increases. With the introduction of horizontal laminar-flow hoods, the sterility of the culture was protected more effectively, but the exposure of the operator to aerosols was more likely. This led to the development of vertical laminar-flow hoods with an air curtain at the front (*see* Sections 5.2.1, 6.8.2) to minimize overspill from within the cabinet. These are now defined as Class II microbiological safety cabinets (*see* Section 6.8.2).

### 6.8.1  Levels of Biological Containment

Four biological safety levels have been defined by the National Institutes of Health (NIH) and the Centers for Disease Control and Prevention [CDC/OSH, 1999] (Table 6.5). These concern practices and the facilities and safety procedures that they require. UK guidelines also define four

levels of biological containment, although there are minor differences from the US classification (Table 6.6). These tables provide summaries only, and anyone undertaking potentially biohazardous work should contact the local safety committee and consult the appropriate national authority [e.g., CDC/OSH, 1999; HSE, 2008a].

### 6.8.2  Microbiological Safety Cabinets (MSCs)

Within the appropriate level of containment we can define three levels of handling determined by the type of safety cabinet used:

(1) *Minimal protection.* Open bench, depending on good microbiological technique. Again, this will normally be conducted in a specially defined area, which may simply be defined as the "tissue culture laboratory" but which will have Level 1 conditions applied to it.

(2) *Intermediate level of protection for potential hazards.* A vertical laminar-flow hood with front protection in the form of an air curtain and a filtered exhaust (biohazard hood or MSC, Class II; Fig. 6.5*a*) [NSF, 1993; British Standard BS5726-2005]. If recognized pathogens are being handled, hoods such as these should be housed in separate rooms, at containment levels 2, 3, or 4, depending on the nature of the pathogen. If there is no reason to suppose that the material is infected, other than by adventitious agents, then hoods can be housed in the main tissue culture facility, which may be categorized as containment Level 2, requiring restricted access, control of waste disposal, protective clothing, and no food or drink in the area (*see* Tables 6.5, 6.6.). All biohazard hoods must be subject to a strict maintenance program [Osborne et al., 1999], with the filters tested at regular intervals, proper arrangements made for fumigation of the cabinets before changing filters, and disposal of old filters made safe by extracting them into double bags for incineration.

(3) *Maximum protection from known pathogens.* A sealed pathogen cabinet with filtered air entering and leaving via a pathogen trap filter (biohazard hood or MSC, Class III; Fig. 6.5*c*). The cabinet will generally be housed in a separate room with restricted access and with showering facilities and protection for solid and liquid waste (*see* BSL 4 in Table 6.5 *and* Level 4 in Table 6.6), depending on the nature of the hazard.

Table 6.7 lists common procedures with suggested levels of containment. All those using the facilities, however, should seek the advice of the local safety committees and the appropriate biological safety guidelines (*see* Section 6.4) for legal requirements.

### 6.8.3  Human Biopsy Material

Issues of biological safety are clearest when known classified pathogens are being used, because the regulations covering

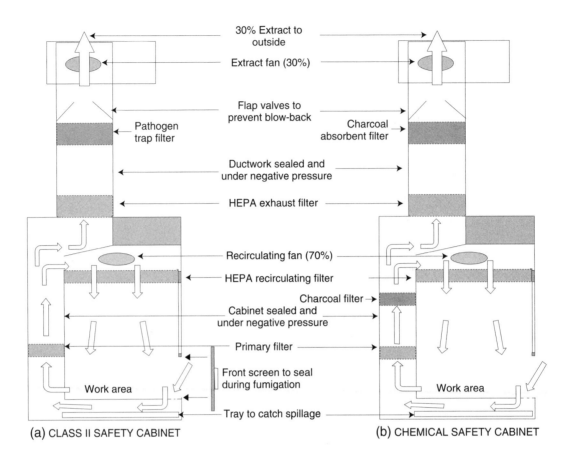

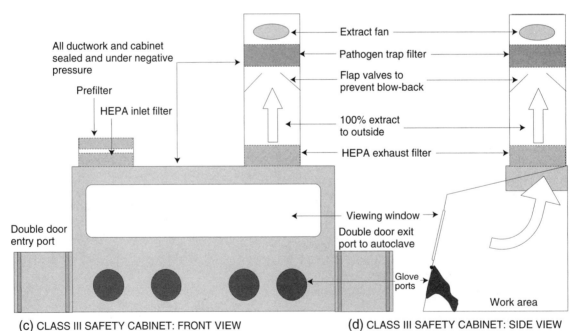

**Fig. 6.5. Microbiological Safety Cabinets.** *(a)* Class II vertical laminar flow, recirculating 80% of the air and exhausting 20% of the air via a filter and ducted out of the room through an optional pathogen trap. Air is taken in at the front of the cabinet to make up the recirculating volume and prevent overspill from the work area. *(b)* Class II chemical safety cabinet with charcoal filters on extract and recirculating air. *(c)* Class III nonrecirculating, sealed cabinet with glove pockets; works at negative pressure and with air lock for entry of equipment and direct access to autoclave, either connected or adjacent. *(d)* Side view of Class III cabinet.

**TABLE 6.5.** US Biosafety Levels

| | BSL 1 (SMP[a]) | BSL 2 (SP[b]) | BSL 3 (SP) | BSL 4 (SP) |
|---|---|---|---|---|
| Access | Access limited when work in progress. Controls against insect and rodent infestation. | Restricted. Predisposition assessed; immunization and baseline and periodic serum samples may be required. Hazard warnings posted when appropriate. | As BSL 2 + separate lab with two sets of self-closing doors. Doors closed when work in progress. Immunization or tests and baseline and periodic serum samples required. | As BSL 3 + Only designated personnel. Secure lockable doors. Changing space & shower. Door interlocks. Equipment and materials entry by double-ended autoclave or fumigation. No materials allowed except those required for the work being conducted. |
| Cleaning | Easy to clean; spaces between cabinets, equipment, etc.; impervious bench surfaces. | As BSL 1 + routine decontamination of work surfaces and equipment. | As BSL 2 + working on plastic-backed absorbent paper recommended. All room surfaces sealed and washable. | As BSL 3 + sealed joints, disinfectant traps on drains, HEPA filters on vents. Minimal surface area for dust. |
| Personal hygiene | Lab coats worn. No eating, drinking etc. No mouth pipetting. Wear gloves and protective eyewear (especially with contact lenses). Sink for hand washing. Remove gloves & wash hands on leaving. | As BSL 1 + provision for decontamination and laundering in house. Eyewash facility. | As BSL 2 + goggles and mask or face shield outside BSC. Respiratory protection when aerosol cannot be controlled. Solid-front gowns, removed before leaving lab. Automatic or elbow taps on sink. | As BSL 3 + change of clothing; clothing autoclaved. Shower before leaving. |
| Airflow and ventilation | Not specified. Windows that open should have screens. | As BSL 1. | Windows closed and sealed. Negative pressure, total extract, exhaust away from occupied areas or air intakes. | Dedicated and alarmed nonrecirculating ventilation system. Air exhaust through HEPA filters. Supply and extract interlocked. |
| Equipment | Not specified. | Routinely decontaminated and particularly before maintenance in house or away. | As BSL 2 + physical containment (e.g., sealed centrifuge cups and rotors). Any exhaust HEPA filtered. Vacuum lines protected by disinfectant traps and HEPA filters. Back-flow prevention devices. | As BSL 3. |

(*continued overleaf*)

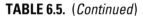

**TABLE 6.5.** (*Continued*)

| | BSL 1 (SMP[a]) | BSL 2 (SP[b]) | BSL 3 (SP) | BSL 4 (SP) |
|---|---|---|---|---|
| Sharps | Not specified. | Restricted to unavoidable use. Used or broken items into containers, decontaminated before disposal. | As BSL 2. | As BSL 2. |
| MSCs | Not required but aerosol generation minimized. | Class II. | Class II or III exhausting directly via HEPA filter. Class II may recirculate. | Class III or Class II with positive-pressure personnel suit and life support system. |
| Disinfection | Disinfectant available. Work surfaces decontaminated at least once per day. | As BSL 1 + procedures specified; autoclave nearby. | As BSL 2 + spills dealt with by trained staff. | As 3 + double-ended exit autoclave required, preferably from Class III cabinet. |
| Storage and transfer | Not specified. | Leak-proof container. | As BSL 2. | Viable materials leave in double-wrapped, nonbreakable containers via dunk tank or fumigation. |
| Disposal | Into disinfectant or by sealed container to nearby autoclave. | As BSL 1+ defined decontamination method. | As BSL 2 + decontamination within lab. | As BSL 3 + all effluent, excluding shower and toilet, and other materials disinfected before leaving via double-ended autoclave. Double-ended dunk tank for nonautoclavable waste. |
| Biosafety manual and training | Not specified. | Training required. | As BSL 2. | As BSL 3 + high proficiency in SMP. |
| Accidents and spills | Not specified. | Written report. Medical evaluation available. | As BSL 2 + spills dealt with by trained staff. | As BSL 3 + monitoring absence, care of illness, and quarantine. |
| Validation of facility | Not specified. | Not specified. | Not specified. | Safety of effluent. |

[a]Standard microbiological practices.
[b]Special practices, including handling agents of moderate potential hazard.

such pathogens are well established both by CDC/NIH in the United States [CDC/OSH, 1999] and in the United Kingdom by the Advisory Committee on Dangerous Pathogens [ACDP, 2003]. However, in two main areas there is a risk that is not immediately apparent in the nature of the material. One is in the development, by recombinant techniques such as transfection, retroviral infection, and interspecific cell hybridization, of new potentially pathogenic transgenes. Handling such cultures in facilities such as laminar-flow hoods introduces putative risks for which there are no epidemiological data available for assessment. Transforming viruses, amphitropic viruses, transformed human cell lines,

human–mouse hybrids, and cell lines derived from xenografts in immunodeficient mice, for example, should be treated cautiously until there are enough data to show that they carry no risk.

The other area of risk is the inclusion of adventitious agents in human or other primate biopsy or autopsy samples or cell lines [Grizzle & Polt, 1988; Centers for Disease Control, 1988; Wells et al., 1989; Tedder et al., 1995] or in animal products such as serum, particularly if those materials are obtained from parts of the world with a high level of endemic infectious diseases. When infection has been confirmed, the type of organism will determine

**TABLE 6.6.** UK Biological Containment Levels

|  | Level 1 | Level 2 | Level 3 | Level 4 |
|---|---|---|---|---|
| Access | Door closed when work in progress. | Restricted. | As 2 + separate lab with observation window. Door locked when lab unoccupied. | As 3 + controls against insect and rodent infestation. Changing space & shower. Door interlocks. Ventilated air lock for equipment. Telephone or intercom. |
| Space | Not specified. | 24 m³/person. | As 2. | As 2. |
| Cleaning | Easy to clean; impervious bench surfaces. | As 1 + routine decontamination of work surfaces. | As 2. | As 2. |
| Personal hygiene | Lab coats (side or back fastening) worn, stored, cleaned, replaced correctly. No eating, drinking, etc. No mouth pipetting. | As 1 + wash basin near exit to decontaminate hands. | As 2 + wear gloves; remove or replace before handling common items such as phone. | As 3 + change of clothing; clothing autoclaved. Shower before leaving. |
| Airflow | Negative pressure preferable. | Negative pressure required. | As 2. | Negative pressure ≥70 pascals (7 mm H₂0); air exhaust through two HEPA filters in series. Supply and extract interlocked. |
| Equipment | Not specified. | Not specified. | Should contain own equipment. | Must contain own equipment. |
| MSCs | Not required but aerosol generation minimized. | MSC or isolator required. | Class I or III (BS5725); Class II (BS5726). Exhaust to via HEPA filter to outside. | Class III. |
| Disinfection | Disinfectant available. | Procedures specified; autoclave nearby. | As 2 + lab sealable for decontamination. Autoclave preferably within lab. | As 3 + double-ended exit autoclave required, preferably from Class III cabinet. |
| Storage | Not specified. | Safe storage of biological agents. | As 2. |  |
| Disposal | Into disinfectant. | As 1+ waste labeled. Safe collection and disposal. | As 2. | As 3 + all effluent, including shower, disinfected. Double-ended dunk tank for nonautoclavable waste. |
| Accidents | Report. | As 1. | As 2. | As 3 + 2nd person present to assist in case of emergency. Respirators available outside. |
| Validation of facility | Not specified. | Not specified. | Not specified. | Required. |

**TABLE 6.7.** Biological Procedures and Suggested Levels of Containment

| Procedure | Containment level | Work space |
|---|---|---|
| Preparation of media | GLP | Open bench with standard microbiological practice, or horizontal or vertical laminar flow |
| Primary cultures and cell lines other than human and other primates | 1 | Open bench with standard microbiological practice, or horizontal or vertical laminar flow |
| Primary cultures and cell lines, other than human and other primates, that have been infected or transfected | 1 | Class II laminar-flow hood |
| Primary culture and serial passage of human and other primate cells | 2 | Class II laminar-flow hood |
| Interspecific hybrids or other recombinants, transfected cells, human cells, and animal tumor cells | 2 | Class II laminar-flow hood |
| Human cells infected with retroviral constructs | 3 | Class II laminar-flow hood |
| Virus-producing human cell lines and cell lines infected with amphitropic virus | 3 | Class II laminar-flow hood |
| Tissue samples and cultures carrying known human pathogens | 4 | Class III pathogen cabinet with glove pockets, filtered air, and pathogen trap on vented air |

Note: These are suggested procedures only and have no legal basis. Consult national legal requirements and local regulations before formulating proper guidelines.

the degree of containment, but even when infection has not been confirmed, the possibility remains that the sample may yet carry hepatitis B, human immunodeficiency virus (HIV), tuberculosis, or other pathogens as yet undiagnosed. Confidentiality frequently prevents HIV testing without the patient's consent, and for most adventitious infections the appropriate information will not be available. If possible, biopsy material should be tested for potential adventitious infections before handling. The authority to do so should be agreed on the consent form that the person donating the tissue will have been asked to sign (*see* example, Table 6.8), but the need to get samples into culture quickly will often mean that you must proceed without this information. Such samples should be handled with caution:

(1) Transport specimens in a double-wrapped container (e.g., a universal container or screw-capped vial within a second screw-top vessel, such as a polypropylene sample jar). This in turn should be enclosed in an opaque plastic or waterproof paper envelope, filled with absorbent tissue packing to contain any leakage, and transported to the lab by a designated carrier.

(2) Enter all specimens into a logbook on receipt, and place the specimens in a secure refrigerator marked with a biohazard label.

(3) Carry out dissection and subsequent culture work in a designated Class II biohazard hood, preferably located in a separate room from that in which routine cell culture is performed. This will minimize the risk of spreading contaminations, such as mycoplasma, to other cultures and will also reduce the number of people associated with the specimen, should it eventually be found to be infected.

(4) Avoid the use of sharp instruments (e.g., syringes, scalpels, glass Pasteur pipettes) in handling specimens. Clearly, this rule may need to be compromised when a dissection is required, but that should proceed with extra caution.

(5) Put all cultures in a plastic box with tape or labels identifying the cultures as biohazardous and with the name of the person responsible and the date on them (*see* Fig. 5.11).

(6) Discard all glassware, pipettes, and instruments, into disinfectant or into biohazard bags for autoclaving.

If appropriate clinical diagnostic tests show that the material is uninfected, and when it has been shown to be free of mycoplasma, the material may then be cultured with other stocks. However, if more than $1 \times 10^9$ cells are to be generated or if pure DNA is to be prepared, the advice of the local safety committee should be sought.

If a specimen is found to be infected, it should be discarded into double biohazard bags together with all reagents used with it, and the bags should then be autoclaved or incinerated. Instruments and other hardware should be placed in a container of disinfectant, soaked for at least 2 h, and then autoclaved. If it is necessary to carry on working with the material, the level of containment must increase, according to the category of the pathogen [CDC/OSH, 1999; ACDP, 2003].

### 6.8.4 Genetic Manipulation

Any procedure that involves altering the genetic constitution of cells or a cell line by transfer of nucleic acid will need to be authorized by the local biological safety committee. The current regulations may be obtained from the CDC or NIH

**TABLE 6.8.** Donor Consent Form

*CONSENT TO REMOVE TISSUE FOR DIAGNOSIS AND RESEARCH*

This form requests your permission to take a sample of your blood or one or more small pieces of tissue to be used for medical research. This sample, or cell lines or other products derived from it, may be used by a number of different research organizations, or it may be stored for an extended period awaiting use. It is also possible that it may eventually be used by a commercial company to develop future drugs. We would like you to be aware of this and of the fact that, by signing this form, you give up any claim that you own the tissue or its components, regardless of the use that may be made of it. You should also be aware of, and agree to, the possible testing of the tissue for infectious agents, such as the AIDS virus or hepatitis.

*I am willing to have tissue removed for use in medical research and development. I have read and understand, to the best of my ability, the background material that I have been given. (If the donor is too unwell to sign, a close relative should sign on his or her behalf.)*

Name of donor ..........................................    Name of relative ..........................................

Signature ................................................

Date ...................................................

---

This material will be coded, and absolute confidence will be maintained. Your name will not be given to anyone other than the person taking the sample.

*Do you wish to receive any information from this material that relates to your health?*    **Yes/No**

Signature ................................    Date .................

*Would you like, or prefer, that this information be given to your doctor?*    **Yes/No**

*If yes, name of doctor* ...................................................

Address of doctor ...................................................................
...................................................................
...................................................................

---

for the United States [CDC/OSH, 1999] and from the HSE for the United Kingdom [HSE, 2008a].

### 6.8.5 Disposal of Biohazardous Waste

Waste disposal is usually under local institutional control where each company or institution will have negotiated a code of practice with state or local regional government following national or federal advisory guidelines, the National Research Council in the United States [NRC, 2009] and the Department of Health in the United Kingdom [DoH, 2006]. Potentially biohazardous materials must be sterilized before disposal by placing them in unsealed autoclavable sacks and autoclaving, or by immersion in a sterilizing agent such as hypochlorite. Various proprietary preparations are available, such as Clorox or Chloros liquid concentrates and Precept or Haz-Tab tablets (*see* Appendix II: Disinfectants). Recommended concentrations vary according to local rules, but a rough guide can be obtained from the manufacturer's

instructions. Hypochlorite is often used at 300 ppm of available chlorine, but some authorities demand 2500 ppm (a 1:20 dilution of Chloros). Hypochlorite is effective and easily washed off those items that are to be reused, but is highly corrosive, particularly in alkaline solutions. It will bleach clothing and even corrode stainless steel (particularly at welded seams), so gloves and a lab coat or apron should be worn when handling hypochlorite, and soaking baths and cylinders should be made of polypropylene.

### 6.8.6 Fumigation

Some procedures may require that the MSC be sterilized after use, for example if high-grade pathogens are being used, or if the cabinet requires servicing. Fumigation is usually carried out with formaldehyde, requiring the cabinet to be switched off and sealed before fumigation is initiated with an electrically heated generator. The hood is switched on briefly to circulate the gas and then left for 1 h. After this time the hood is allowed to run overnight to exhaust the vapor, opening the front after about 10 min. Fumigation of cabinets can also be carried out with hydrogen peroxide (Bioquell), which is more easily dispersed after fumigation is complete.

## 6.9 BIOETHICS

In addition to potential biohazards, working with human and other animal tissue presents a number of ethical problems involving procurement, subsequent handling, and the ultimate use of the material [Hansson, 2009].

### 6.9.1 Animal Tissue

Most countries involved in biomedical research will now have in place regulations governing the use of experimental or other donor animals for the provision of tissue. These will apply to higher animals assumed to have sufficient brain capacity and organization to feel pain and distress, and generally will not apply to lower vertebrates such as fish, or to invertebrates. Usually a higher animal is assumed to be sentient any time after halfway through embryonic development and restrictions will apply to the method by which the animal is killed, or operated upon if it is to remain living, such that the animal suffers minimal pain or discomfort. Restrictions will also apply to the way the animal is housed and maintained, either in an animal house during husbandry or under experimental conditions, or in a veterinary hospital under clinical conditions. In each case control is usually exercised locally by an animal ethics committee and nationally by a governmentally or professionally appointed body. Legislation varies considerably from country to country, but the appointment of a local animal ethics committee (AEC) is usually the first step to making contact with the appropriate licensing authority, in the United States the Office of Laboratory Animal Welfare [OLAW, 2002] and the Home Office in the United Kingdom [Home Office, 2009], or via funding organizations such as the NIH Office of Biotechnology Activities [NIH OBA, 2009] or the Medical Research Council [MRC, 2008]. Information is also available through the European Biomedical Research Association [EBRA, 2009].

### 6.9.2 Human Tissue

Tissue will normally be collected under clinical conditions by an experienced medical practitioner, and the issues are more to do with the justification for taking the tissue and the uses to which it will be put. Local control will be exercised through the local hospital ethics committee (HEC), who will decide whether the work is reasonable and justified by the possible outcome. The HEC must be contacted before any experimental work with human tissue is initiated, and this is best done at the planning stage, as most granting authorities will require evidence of ethical consent before awarding funding. The US President has set up a Commission for the Study of Bioethical Issues [PCSBI, 2009] and guidance is also available from the Declaration of Helsinki [WMA, 2008]

and from the so-called Common Rule [OHRP, 2010]. The issue is also discussed at some length in Rebecca Skloot's fascinating book [Skloot, 2010]. Where embryological tissue is concerned the Office for Human Research Protections must be consulted in the United States [OHRP, 2002] and the Human Fertilization and Embryology Authority [HFEA, 2009] in the United Kingdom. Legislation varies across the European Union, but information is available on the internet via a newsletter [EuroStemCell, 2009].

There is also the question of ownership of the tissue, its contents such as DNA, any cell lines that are derived from it, and any products or marketable procedures that might ultimately be developed and sold for profit. The following issues need to be addressed:

(1) The patient's and/or relative's informed consent is required before taking tissue for research purposes, over and above any clinical requirement.

(2) A suitable form (*see* Table 6.8) should be drafted in a style readily understood by the patient or donor, requesting permission and drawing attention to the use that might be made of the tissue.

(3) Permission may be required from a relative if the donor is too unwell to be considered capable of a reasoned judgment.

(4) A short summary of your project should be prepared, in lay terms, explaining what you are doing, why, and what the possible outcome will be, particularly if it is seen to be of medical benefit.

(5) Confidentiality of the origin of the tissue must be ensured.

(6) Ownership of cell lines and their derivatives must be established.

(7) Authority may be needed for subsequent genetic modification of the cell lines.

(8) Patent rights from any commercial collaboration will need to be established.

(9) The donor will need to determine whether any genetic information derived from the tissue should be fed back to the patient directly or via an attending clinician.

(10) The donor will also be required to consent to screening of the tissue for adventitious pathogens and to say whether he or she wishes to be made aware of the outcome of the tests.

By far the easiest approach is to ask the donor and/or relatives to sign a disclaimer statement before the tissue is removed; otherwise, the legal aspects of ownership of the cell lines that might be derived and any future biopharmaceutical exploitation of the cell lines, their genes, and their products becomes exceedingly complex. Feedback of genetic information and evidence of a possible pathological infection such as HIV are more difficult problems; in the case of a patient in a hospital, the feedback is on a par with a diagnostic test and is most likely to be directed to the doctor, but in the case of a donor who is not hospitalized, you must ask the donor whether he or she wishes to know your findings and any implications that they might have. This may be done best via the donor's general practitioner. These factors are best dealt with by getting the donor to sign a consent form. Such a form may already have been prescribed by the HEC; if not, it will be necessary to prepare one (e.g., *see* Table 6.8) in collaboration with the HEC and other involved parties, such as clinical collaborators, patient support groups, and funding authorities. Further information is available from NIH in the United States [NIH, 2007] and in the United Kingdom from Nuffield [Nuffield Council on Bioethics, 2009] and from the Medical Research Council [MRC, 2009].

Perhaps the most controversial aspect of the consent process is the need for the donor to know something about what the tissue will be used for. This requires a brief description in lay terms that will neither burden nor confuse the donor. Tissue transplantation has such clear objectives that little explanation of the science is required, but some procedures, such as the examination of signal transduction anomalies in transformed cells, will require some generalization of the concept. Often a brief overview in simple terms given orally can be accompanied by a more detailed description, though still in lay terms, emphasizing the potential advantages but also identifying the ethical issues, such as the subsequent genetic modification of the cells or the transplantation of the cells into another individual after tissue engineering.

## 6.10  QUALITY ASSURANCE

Industrial and commercial laboratories are governed by strict regulations involving procedures and quality control. Most academic and research laboratories tend to have a more informal approach but would benefit from adherence to good laboratory practice [Good Laboratory Practice Regulations, 1999; FDA, 2003; OECD, 2009; U.S. EPA, 2009] including certain elements of quality assurance.

### 6.10.1  Procedures

It is difficult to introduce standard operating procedures (SOPs) into a research laboratory where individual groups may require variations of the procedure and the ability to modify it as the research progresses. However, it is recommended that a shared facility, like a tissue culture laboratory, should have defined SOPs for specific shared practices in the laboratory. A manual should be available, either printed or on the local Intranet, that shows the standard procedures for primary culture, subculture, cloning, cryopreservation, and so forth, from which individual users can derive their own protocols. Deviation from the SOP should be discouraged unless there is a sound valid scientific reason that does not compromise others using the facilities. Casual transfer of protocols by word of mouth, and the accidental deviations that result, should be avoided.

### 6.10.2  Quality Control (QC)

Regardless of the control over procedures it is essential that the preparation of reagents and media, the operation of the facilities, and the maintenance of the equipment are subject to routine testing. The general rule is that the person who prepares a reagent should be different from the person who carries out the QC, and the person who uses a piece of equipment should be different from the person who checks it. Specific aspects of QC are dealt with in later chapters (*see* Sections 10.6, 12.1.1, 15.2, 18.3).

## 6.11  VALIDATION

The proper use of cell lines, whether in research or commercial exploitation, requires that they be validated. In an industrial environment this will be a legal obligation if the ultimate product is to be accepted by the Federal Drug Administration (FDA) in the United States or the National Institute for Clinical Excellence (NICE) in the United Kingdom. Many of these criteria are defined by the International Conference on Harmonisation of Technical requirements for Registration of Pharmaceuticals for Human Use (ICH), e.g. Guidelines Q5A on viral safety and Q5D on derivation and characterisation [ICH, 2005]. However, in an academic research laboratory the requirement may be less well defined and the obligation left to individual conscience. Nevertheless, use of cell lines that are not properly validated reduces the reliability of the research and the likelihood that anyone will be able to repeat it (*see also* Section 12.1.1).

There are three major elements to validation:

(1) *Authentication.*  Is the cell line what it is claimed to be?
(2) *Provenance.*  How was the cell line derived and what has happened to the cell line since its original isolation?
(3) *Contamination.*  Is the cell line free from all known forms of microbial contamination?

### 6.11.1  Authentication

Several techniques are available to give a specific profile of the cell line (*see* Sections 15.2, 15.4). DNA profiling is probably the best but requires that DNA be available from the donor, or at least from an earlier generation of the cell line known to be authentic at the time it was preserved. DNA profiles can also be compared with reference collections (usually via cell banks; *see* Table 19.5) that, although they may not establish identity, will establish that the line is not cross-contaminated with another known cell line (*see also* Section 12.1.1). Failing this, a compilation of several characteristics will confirm the origin (species, tissue, etc.; *see* Section 15.4) beyond reasonable doubt. It should be remembered that one or more of the criteria that are used can be specific to the laboratory in which the cells are being used, sufficient for the purpose, though not necessarily readily transferable to another laboratory. The important issue is that some steps must be taken to authenticate the cell line before a major expenditure of time, effort, and funds is committed.

### 6.11.2 Provenance

Part of the validation process requires that there be a record of how a cell line was isolated and what has happened to it since isolation: maintenance regimens, contamination checks, decontamination procedures if used, properties expressed, genetic modification, spontaneous alterations, and so forth. Some knowledge of the provenance of the cell line, derived from the published literature or by word of mouth from a colleague, will have been the reason for selecting it in the first place but should be independently documented and added to as work progresses with the cell line. This means that proper records should be kept at all times (*see* Sections 11.3.1, 12.4.9, 19.3.8), detailing routine maintenance, significant experimental observations, and cryostorage. This does not have to be laborious, as the use of a spreadsheet or database will allow a new record to be made of a repeated procedure without having to rekey all the data except the date and that which is new or has changed. A cell line with a detailed and complete provenance gains value like a piece of antique furniture or a painting.

### 6.11.3 Contamination

However detailed your records or meticulous your experimental technique, the resulting work is devoid of value, or at least heavily compromised, if the cell line is shown to be contaminated with one or more microorganisms. Where the contamination is overt, it is less of a problem, as cultures can be discarded, but often it is cryptic because (1) the cells have been maintained in antibiotics (*see* Section 12.4.8), (2) routine testing for organisms such as mycoplasma has not been carried out (*see* Section 18.3.2), or (3) there is no routine test available for the organism as is the case with some viruses or prions (*see* Section 18.3.7). Contamination can be avoided (1) by observing proper aseptic technique (*see* Section 5.1); (2) by obtaining cell lines from a properly validated source such as a cell bank (*see* Table 19.5); (3) by culturing cells in the absence of antibiotics, even if only for part of the time (*see* Section 12.4.8 *and* Fig. 12.7); (4) by screening regularly for mycoplasma, namely by staining with Hoechst 33258 to detect any DNA-containing organism big enough to be resolved under a fluorescence microscope (*see* Protocol 18.2); or (5) by screening for the most common viruses using PCR or a commercial contract.

Cell lines that have been properly validated should be stored in liquid nitrogen and issued to end users as required (*see* Section 19.5.1). End users may store their own stock for the duration of a project; when these user stocks are no longer fully validated, they should not be passed on, and new users should revert to the validated stock.

# CHAPTER 7

# Culture Vessels and Substrates

## 7.1 THE SUBSTRATE

### 7.1.1 Attachment and Growth

The majority of vertebrate cells cultured in vitro grow as monolayers on an artificial substrate. Hence the substrate must be correctly charged to allow cell adhesion, or at least to allow the adhesion of cell-derived attachment factors, which will in turn allow cell adhesion and spreading. Although spontaneous growth in suspension is restricted to hematopoietic cell lines, rodent ascites tumors, and a few other selected cell lines, such as human small-cell lung cancer [Carney et al., 1981], many transformed cell lines can be made to grow in suspension and become independent of the surface charge on the substrate. However, most normal cells need to spread out on a substrate to proliferate [Folkman & Moscona, 1978; Ireland et al., 1989; Danen & Yamada, 2001; Frame & Norman, 2008; Zhang et al., 2008], and inadequate spreading due to poor adhesion or overcrowding will inhibit proliferation. Cells shown to require attachment for growth are said to be *anchorage dependent*; cells that have undergone transformation frequently become *anchorage independent* (*see* Section 17.5.1) and can grow in suspension (*see* Section 12.4.5) when stirred or held in suspension with semisolid media such as agar.

### 7.1.2 Common Substrate Materials

***Disposable plastic.*** Single-use sterile polystyrene flasks, Petri dishes, or multiwell plates provide a simple, reproducible substrate for culture. They are usually of good optical quality, and the growth surface is flat, providing uniformly distributed and reproducible monolayer cultures. As manufactured, polystyrene is hydrophobic and does not provide a suitable surface for cell attachment, so tissue culture plastics are treated by corona discharge, γ-irradiation, or chemically, to produce a charged, wettable surface. Because the resulting product can vary in quality from one manufacturer to another, samples from a number of sources should be tested by determining the growth rate and plating efficiency of cells in current use (*see* Protocols 20.7–20.10) in the appropriate medium containing limiting concentrations of serum or serum free. (High serum concentrations may mask imperfections in the plastic; *see* Section 22.2.4.)

To test a new substrate, grow the cells on it as a regular monolayer, with and without pretreating the surface (*see* Section 7.2.1), and then clone cells (*see* Protocol 20.10). PTFE can be used in a charged (hydrophilic) or uncharged (hydrophobic) form [Janssen et al., 2003; Lehle et al., 2003]; the charged form can be used for regular monolayer cells and organotypic culture (Biopore, Millipore; Transwell, Corning) and the uncharged for macrophages [von Briesen et al., 1990] and some transformed cell lines.

***Glass.*** This was the original substrate because of its optical properties and surface charge, but it has been replaced in most laboratories by plastic (usually polystyrene), which has greater consistency and superior optical properties. Glass is now rarely used, although it is cheap, is easily washed without losing its growth-supporting properties, can be sterilized readily by dry or moist heat, and is optically clear. Treatment with strong alkali (e.g., NaOH or caustic detergents) renders glass unsatisfactory for culture until it is neutralized by an acid wash (*see* Section 10.3.1). High optical quality glass is alkaline and often has a high lead content, which may reduce cell growth, hence slides and coverslips may need to be acid washed and/or coated for best results (*see* Section 7.2.1).

*Culture of Animal Cells: A Manual of Basic Technique and Specialized Applications, Sixth Edition*, by R. Ian Freshney

### 7.1.3 Alternative Substrates

*Other plastics.* Although polystyrene is by far the most common and cheapest plastic substrate, cells can also be grown on polyvinylchloride (PVC), polycarbonate, polytetrafluorethylene (PTFE; Teflon), Melinex, Thermanox (TPX), poly(methyl methacrylate) (PMMA; Plexiglas, Perspex, Lucite) [Gottwald et al., 2008], and a number of other plastics.

*Fibers.* Rayon, Nylon, poly-L-lactic acid (PLA), polyglycolic acid (PGA), and silk are often used for two- and three-dimensional constructs in tissue engineering (*see* Section 25.1.3; Fig. 25.2), particularly PLA, PGA, and silk, as they are biodegradable.

*Derivatization.* Substrates that are not naturally adhesive can be derivatized with the RGD tripeptide, usually as the pentapeptide GRGDS to allow interaction with integrins on the cell surface (*see* Section 2.2.1). EDC (1-ethyl-3-(3-dimethylaminopropyl) carbodiimide hydrochloride) and N-hydroxysulfosuccinimide (sulpho-NHS) have been used to derivatize processed silk to construct scaffolds for bone tissue engineering [Hofmann et al., 2006] and this treatment is potentially applicable to a number of different substrates [e.g. Lao et al., 2008] (*see also* Section 7.2.1 under Collagen).

*Metals.* Cells may be grown on stainless steel disks [Birnie & Simons, 1967] or other metallic surfaces [Litwin, 1973]. Observation of the cells on an opaque substrate requires surface interference microscopy, unless very thin metallic films are used. Westermark [1978] developed a method for the growth of fibroblasts and glia on palladium. Using electron microscopy shadowing equipment, he produced islands of palladium on agarose, which does not allow cell attachment in fluid media. The size and shape of the islands were determined by masks made by photo-etching, and the palladium was applied by shadowing under vacuum, as used in electron microscopy. Because the layer was very thin, it remained transparent.

## 7.2 TREATED SURFACES

### 7.2.1 Substrate Coating

*Conditioning.* Cell attachment and growth can be improved by pretreating the substrate [Barnes et al., 1984a]. A well-established piece of tissue culture lore has it that used glassware supports growth better than new [Paul, 1975]. If that is true, it may be due to etching of the surface or minute traces of residue left after culture. The growth of cells in a flask also improves the surface for a second seeding, and this type of conditioning may be due to collagen, fibronectin, or other matrix products [Crouch et al., 1987] released by the cells. The substrate can also be conditioned by treating it with spent medium from another culture [Stampfer et al., 1980], with serum, or with purified fibronectin or collagen (*see* Protocol 22.9).

*Polymers, nanoparticles, and photoetching.* McKeehan and Ham [1976a] found that it was necessary to coat the surface of plastic dishes with 1 mg/mL of poly-D-lysine before cloning in the absence of serum (*see* Sections 13.2.1). The same technique is also used to promote neurite outgrowth (*see* 22.4.1), so some effects of conditioning may be related to the surface charge. The D isomer is used in preference to the L as it is less readily digested by extracellular proteases, but both the D- and L-isomers have been used. Higher molecular weights become more viscous to handle but have more binding sites; both 100 kD (MP Biomedicals) and 500 kD (BD Biosciences) are available. Diamond nanoparticles have also been used to modify the substrate for the proliferation and differentiation of neural stem cells [Chen et al., 2010] and the configuration of the growth surface can also be altered by photoetching [Ploss et al, 2010].

*Collagen and gelatin.* Treatment of the substrate with denatured collagen improves the attachment of many types of cells, such as epithelial cells, and the undenatured gel may be necessary for the expression of differentiated functions (*see* Sections 2.2.3, 16.7.3, 22.2.1). Gelatin coating has been found to be beneficial for the culture of muscle [Richler and Yaffe, 1970] and endothelial cells [Folkman et al., 1979] (*see* Section 22.3.6), and it is necessary for some mouse teratomas. Coating with denatured collagen may be achieved by using rat tail collagen or commercially supplied alternatives (e.g., Vitrogen) and simply pouring the collagen solution over the surface of the dish, draining off the excess, and allowing the residue to dry. Because this procedure sometimes leads to detachment of the collagen layer during culture, a protocol was devised by Macklis et al. [1985] to ensure that the collagen would remain firmly anchored to the substrate, by cross-linking to the plastic with carbodiimide. Collagen can also be used in conjunction with fibronectin (*see* Protocol 22.9).

Collagen may also be applied as an undenatured gel (*see* Section 16.7.3), a type of substrate that has been shown to support neurite outgrowth from chick spinal ganglia [Ebendal, 1976] and morphological differentiation of breast cells [Nicosia & Ottinetti, 1990; Berdichevsky et al., 1992] and hepatocytes [Sattler et al., 1978; Fiorino et al., 1998], and to promote the expression of tissue-specific functions of a number of other cells in vitro (e.g., keratinocytes [Maas-Szabowski et al., 2002]; *see* Section 23.1.1). Diluting the concentrated collagen 1:10 with culture medium and neutralizing to pH 7.4 causes the collagen to gel, so the dilution and dispensing must be rapid. It is best to add the growth medium to the gel for a further 4 to 24 h to ensure that the gel equilibrates with the medium before adding cells. At this stage, fibronectin (25–50 µg/mL) or laminin (1–5 µg/mL), or both, may be added to the medium.

*Matrices.* Commercially available matrices (*see* Table 7.1), such as Matrigel™ (Becton Dickinson) from the Engelbreth–Holm–Swarm (EHS) sarcoma, contain laminin, fibronectin, and proteoglycans, with laminin predominating (*see also* Section 16.7.3). Other matrix products include Pronectin F (Protein Polymer Technologies), laminin, fibronectin, vitronectin entactin (UBI), heparan sulfate, EHS Natrix (BD Biosciences), ECL (US Biological), and Cell-tak (BD Biosciences). Some of these products are purified, if not completely chemically defined; others are a mixture of matrix products that have been poorly characterized and may also contain bound growth factors. If cell adhesion for survival is the main objective, and defined substrates are inadequate, the use of these matrices is acceptable, but if mechanistic studies are being carried out, they can only be an intermediate stage on the road to a completely defined substrate.

***Extracellular matrix.*** Although inert coating of the surface may suffice, it may yet prove necessary to use a monolayer of an appropriate cell type to provide the correct matrix for the maintenance of some specialized cells. Gospodarowicz et al. [1980] were able to grow endothelium on extracellular matrix (ECM) derived from confluent monolayers of 3T3 cells that had been removed with Triton X-100. This residual ECM has also been used to promote differentiation in ovarian granulosa cells [Gospodarowicz et al., 1980] and in studying tumor cell behavior [Vlodavsky et al., 1980].

---

### PROTOCOL 7.1. PREPARATION OF ECM

#### Outline
Remove a postconfluent monolayer of matrix-forming cells with detergent, wash flask or dish, and seed required cells onto residual matrix.

#### Materials
❑ Mouse fibroblasts (e.g., 3T3), MRC-5 human fibroblasts, or CPAE bovine pulmonary arterial endothelial cells (or any other cell line shown to be suitable for producing extracellular matrix)
❑ Sterile, ultrapure water (UPW) (*see* Section 10.4.1)
❑ Triton X-100, 1% in sterile UPW

#### Procedure
1. Set up matrix-producing cultures, and grow to confluence.
2. After 3 to 5 days at confluence, remove the medium and add an equal volume of sterile 1% Triton X-100 in UPW to the cell monolayer.
3. Incubate for 30 min at 37°C.
4. Remove Triton X-100 and wash residue three times with the same volume of sterile UPW.
5. Flasks or dishes may be used directly or may be stored at 4°C for up to 3 weeks.

---

### 7.2.2 Feeder Layers

Although matrix coating may help attachment, growth, and differentiation, some cultures of more fastidious cells, particularly at low cell densities [Puck & Marcus, 1955], require support from living cells (e.g., mouse embryo fibroblasts; *see* Protocol 13.3). This action is due partly to supplementation of the medium by either metabolite leakage or the secretion of growth factors from the fibroblasts (actually primitive mesenchymal cells and not fibroblasts), but may also be due to conditioning of the substrate by cell products. Feeder layers grown as a confluent monolayer may make the surface suitable, or even selective, for attachment for other cells (*see* Sections 13.2.3, 24.5.2; Protocols 22.1, 22.4). The survival and extension of neurites by central and peripheral neurons can be enhanced by culturing the neurons on a monolayer of glial cells, although in this case the effect may be due to a diffusible factor rather than direct cell contact [Seifert & Müller, 1984].

After a monolayer culture reaches confluence, subsequent proliferation causes cells to detach from the artificial substrate and migrate over the surface of the monolayers. The morphology of the cells may change (Fig. 7.1): The cells may become less well spread, more densely staining, and more highly differentiated. Apparently, and not too surprisingly, the interaction of a cell with a cellular underlay is different from the interaction of the cell with a synthetic substrate. The former can cause a change in morphology and reduce the cell's potential to proliferate.

### 7.2.3 Nonadhesive Substrates

Sometimes attachment of the cells is undesirable. The selection of virally transformed colonies, which are anchorage independent, can be achieved by plating cells in agar [Macpherson & Montagnier, 1964], as the untransformed cells do not form colonies readily in this matrix. There are two principles involved in such a system: (1) prevention of attachment at the base of the dish, where spreading and anchorage-dependent growth would occur, and (2) immobilization of the cells such that daughter cells remain associated with the colony, even if they are nonadhesive. Agar, agarose, or Methocel (methylcellulose of viscosity 4000 cP) is commonly used (*see* Section 13.3). The first two are gels, and the third is a high-viscosity sol. Because Methocel is a sol, cells will sediment slowly through it. It is therefore used with an underlay of agar (*see* Protocol 13.5). Dishes that are not of tissue culture grade can be used without an agar underlay, but some attachment and spreading may occur.

### 7.3 CHOICE OF CULTURE VESSEL

Some typical culture vessels are listed in Table 7.2. The anticipated yield of HeLa cells is quoted for each vessel; the yield from a finite cell line (e.g., diploid fibroblasts) would

**TABLE 7.1.** Matrix Materials

| Material | Composition | Source | Supplier |
|---|---|---|---|
| Matrigel | Laminin, fibronectin, collagen IV, proteoglycans, growth factors (growth factor depleted available) | EHS sarcoma | BD Biosciences |
| EHS Natrix | Laminin, fibronectin, collagen IV, proteoglycans, growth factors | Cell line from EHS sarcoma | BD Biosciences |
| Cell-Tak | Polyphenolic proteins | Mytilus edulis | BD Biosciences |
| Collagens (various) | Collagen I, II, III, IV | Human, bovine, rat tail | *See* Appendix II |
| ProNectin F | Protein polymer with multiple copies of the RGD containing epitope | Recombinant | Protein Polymer Technologies, Sanyo Chemical, MP Biomedicals, Sigma-Aldrich |
| Laminin | Attachment protein from basement membrane | Natural | *See* Appendix II |
| Laminin | Attachment protein from basement membrane | Recombinant | Kortesmaa et al. [2000]; [Belin & Rousselle 2006] |
| Fibronectin | Attachment protein from extracellular matrix | Natural | *See* Appendix II |
| Fibronectin | Attachment protein from extracellular matrix | Recombinant | Accurate Chemical & Scientific Co., AlpcoR & D |
| Heparan sulfate | Matrix proteoglycan | Natural | BD Biosciences, Sigma-Aldrich |
| ECL | Entactin–collagen IV–laminin | Natural | US Biological |
| Vitronectin | Attachment protein from extracellular matrix | Natural | BD Biosciences, Biosource International |
| ECM | Extracellular matrix proteins | Natural | *See* Matrix in Appendix II |

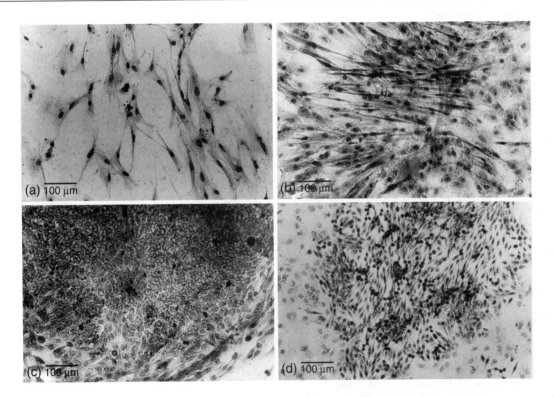

***Fig. 7.1. Morphology on Feeder Layers.*** Morphological alteration in cells growing on feeder layers: Fibroblasts from human breast carcinoma (*a*) growing on plastic and (*b*) growing on a confluent feeder layer of fetal human intestinal cells (FHI). Epithelial cells from human breast carcinoma growing (*c*) on plastic and (*d*) on same confluent feeder layer as in (*b*).

**TABLE 7.2.** Culture Vessel Characteristics

| Culture vessel | Replicates | mL | cm$^{2a}$ | Approximate cell yield (HeLa) |
|---|---|---|---|---|
| **Multiwell plates** | | | | |
| Microtitration | 96 | 0.1 | 0.3 | $1 \times 10^5$ |
| Microtitration | 144 | 0.1 | 0.3 | $1 \times 10^5$ |
| 4-well plate | 4 | 2 | 2 | $5 \times 10^5$ |
| 6-well plate | 6 | 2 | 10 | $2 \times 10^6$ |
| 12-well plate | 12 | 1 | 3 | $7.5 \times 10^5$ |
| 24-well plate | 24 | 1 | 2 | $5 \times 10^5$ |
| **Petri dishes** | | | | |
| 3.5-cm diameter | 1 | 2 | 8 | $2 \times 10^6$ |
| 5-cm diameter | 1 | 4 | 17.5 | $4 \times 10^6$ |
| 6-cm diameter | 1 | 5 | 21 | $5 \times 10^6$ |
| 9-cm diameter | 1 | 10 | 49 | $1 \times 10^7$ |
| **Flasks** | | | | |
| 10 cm$^2$ (T10) | 1 | 2 | 10 | $2 \times 10^6$ |
| 25 cm$^2$ (T25) | 1 | 5 | 25 | $5 \times 10^6$ |
| 75 cm$^2$ (T75) | 1 | 20 | 75 | $2 \times 10^7$ |
| 175 cm$^2$ (T175) | 1 | 50 | 175 | $5 \times 10^7$ |
| 225 cm$^2$ (T225) | 1 | 75 | 225 | $6 \times 10^7$ |
| Roller bottle | 1 | 200 | 850 | $2.5 \times 10^8$ |
| **Multisurface propagators** | | | | |
| Nunc Triple flask | 1 | 200 | 500 | $1 \times 10^8$ |
| Corning HyperFlask | 1 | 560 | 1720 | $5 \times 10^8$ |
| Nunclon Cell Factory | 1 | 200–8,000 | 632–25,284 | $2$–$75 \times 10^8$ |
| Corning CellStack | 1 | 150–8,000 | 636–25,440 | $2$–$75 \times 10^8$ |
| **Stirrer bottles** | | | | |
| 500 mL (unsparged) | 1 | 50 | | $5 \times 10^7$ |
| 5000 mL (sparged) | 1 | 4000 | | $4 \times 10^9$ |

$^a$These figures are approximate; actual areas will vary by source.

be about one-fifth to one-tenth of the HeLa figure. Several factors govern the choice of culture vessel, including (1) the cell mass required, (2) whether the cells grow in suspension or as a monolayer, (3) whether the culture should be vented to the atmosphere or sealed, (4) the frequency of sampling, (5) the type of analysis required, and (6) the cost.

### 7.3.1 Cell Yield

For monolayer cultures, the cell yield is proportional to the available surface area of the flask (Fig. 7.2). Small volumes and multiple replicates are best performed in multiwell plates (Fig. 7.3), which can have a large number of small wells (e.g., microtitration plates with 96 or 144 wells, 0.1 to 0.2 mL of medium, and 0.25-cm$^2$ growth area or 24-well "cluster dishes" with 1 to 2 mL medium in each well, 1.75-cm$^2$ growth area) up to 4-well plates with each well 5 cm in diameter and using 5 mL of culture medium (*see* Table 7.2). The middle of the size range embraces both Petri dishes (Fig. 7.4) and flasks ranging from 10 to 225 cm$^2$ (Fig. 7.5). Flasks are usually designated by their surface area (e.g., 25 or 175 cm$^2$, often abbreviated to T25 or T175, respectively), whereas Petri dishes are referred to by diameter (e.g., 3.5 or 9 cm).

Glass bottles are more variable than plastic because they are usually drawn from standard pharmaceutical supplies.

Glass bottles should have (1) one reasonably flat surface, (2) a deep screw cap with a good seal and nontoxic liner, and (3) shallow-sloping shoulders to facilitate harvesting of monolayer cells after trypsinization and to improve the efficiency of washing.

If you require large cell yields (e.g., $\sim 1 \times 10^9$ HeLa cervical carcinoma cells or $2 \times 10^8$ MCR-5 diploid human fibroblasts), then increasing the size and number of conventional bottles becomes cumbersome, and special vessels are required. Flasks with corrugated surfaces (Corning, Becton Dickinson) or multilayered flasks (Corning, Nunc) offer an intermediate step in increasing the surface area (Fig. 7.6). Cell yields beyond that require large multisurface propagators or roller bottles on special racks (*see* Section 26.2.2). Increasing the yield of cells growing in suspension requires only that the volume of the medium be increased, as long as cells in deep culture are kept agitated and sparged with 5% $CO_2$ in air (*see* Section 26.1).

### 7.3.2 Suspension Culture

Cells that grow in suspension can be grown in any type of flask, plate, or Petri dish that, although sterile, need not be treated for cell attachment. Stirrer bottles are used when agitation is required to keep the cells in suspension. These

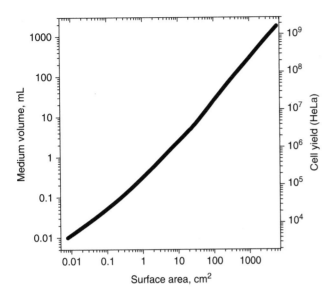

**Fig. 7.2. Cell Yield and Surface Area.** Relationship of volume of medium and cell yield to the surface area of a culture vessel. The graph is plotted on the basis of the volume of the medium for each size of vessel and is nonlinear, as smaller vessels tend to be used with proportionally more medium than is used with larger vessels. The cell yield is based on the volume of the medium and is approximate.

bottles are available in a wide range of sizes, usually in glass (Bellco, Techne). Agitation is usually by a top-driven suspended paddle or a pendulum containing a magnet, whose rotation is driven by a magnetic stirrer (Fig. 7.7; *see also* Figs. 12.7, 25.1). The rotational speed must be kept low, around 60 rpm, to avoid damage from shear stress. Generally,

the pendulum design is preferable for minimizing shear, although a paddle becomes preferable as the scale is increased. Suspension cultures can be set up as replicates or can be sampled repetitively from a side arm of the flask. They can also be used to maintain a steady-state culture by adding and removing medium continuously (*see* Section 26.1.1).

### 7.3.3 Venting

Multiwell and Petri dishes, chosen for replicate sampling or cloning, have loose-fitting lids to give easy access to the dish. Consequently they are not sealed and require a humid atmosphere with the $CO_2$ concentration controlled (*see* Section 8.2.2). As a thin film of liquid may form around the inside of the lid, partially sealing some dishes, vented lids with molded plastic supports inside should be used (Fig. 7.8*a*, arrow). If a perfect seal is required, some multiwell dishes can be sealed with self-adhesive film (*see* Appendix II: Plate Sealers).

Flasks may be vented by slackening the caps one full turn, when in a $CO_2$ incubator, to allow $CO_2$ to enter or to allow excess $CO_2$ to escape in excessive acid-producing cell lines. However, caps with permeable filters that permit equilibration with the gas phase are preferable as they allow $CO_2$ diffusion without risk of contamination (Fig. 7.8*b*). Solid, or "plug," caps should still be used in a non−$CO_2$ incubator or hot room.

### 7.3.4 Sampling and Analysis

Multiwell plates are ideal for replicate cultures if all samples are to be removed simultaneously and processed in the same way. If, instead, samples need to be withdrawn at different

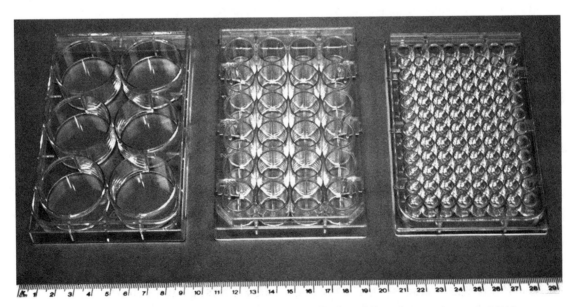

**Fig. 7.3. Multiwell Plates.** Six-well, 24-well, and 96-well (microtitration) plates. Plates are available with a wide range in the number of wells, from 4 to 144 (*see* Table 7.2 for sizes and capacities).

**Fig. 7.4. Petri Dishes.** Dishes of 3.5-, 5-, and 9-cm diameter. Square Petri dishes are also available, with dimensions 9 × 9 cm. A grid pattern can be provided to help in scanning the dish—for example, in counting colonies—but can interfere with automatic colony counting.

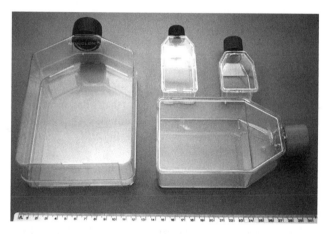

**Fig. 7.5. Plastic Flasks.** Sizes illustrated are 10 and 25 cm$^2$ (Falcon, BD Biosciences), 75 cm$^2$ (Corning), and 185 cm$^2$ (Nalge Nunc) (see Table 7.2 for representative sizes and capacities).

times and processed immediately, it may be preferable to use separate vessels (Fig. 7.9; *see also* Section 20.8). Individual wells in microtitration plates can be sampled by cutting and removing only that part of the adhesive plate sealer overlying the wells to be sampled. Alternatively, microtitration plates are available with removable wells for individual processing (Nunc). If you wish to use adherent cells, you should ensure that these wells are treated for tissue culture.

Low-power microscopic observation is performed easily on flasks, Petri dishes, and multiwell plates with the use of an inverted microscope. When using phase contrast, however, difficulties may be encountered with microtitration plates because of the size of the meniscus relative to the diameter of the well; even 24-well plates can only be observed satisfactorily by phase contrast in the center of the well. If microscopy will play a major part in your analysis, it may be advantageous to use a chamber slide (*see* Section 15.5.3;

**Fig. 7.6. Multisurface Flask.** The Nunc Triple-Flask with three 80-cm$^2$ growth surfaces that are seeded simultaneously. Although the growth surface is 240 cm$^2$, the shelf space is equivalent to a regular 80-cm$^2$ flask (*See also* Corning HyperFlask, Fig. 26.9). As the head space for gas phase is smaller, this flask is best used with a filter cap in a $CO_2$ incubator (arrow). (Photograph courtesy of Nalge Nunc International.)

Fig. 15.3). Large roller bottles give problems with some microscopes; it is usually necessary to remove the condenser, in which case phase contrast will not be available.

If processing of the sample involves extraction in acetone, toluene, ethyl acetate, or certain other organic solvents, then a problem will arise with the solubility of polystyrene. As this problem is often associated with organic solvents used in histological procedures, Lux (Bayer, MP Biomedicals) supplies solvent-resistant Thermanox (TPX) plastic coverslips, suitable for histology, that fit into regular multiwell dishes (which need not be of tissue culture grade). However, these coverslips are of poor optical quality and UV-impermeable so they should be mounted on slides with cells uppermost and a conventional glass coverslip on top.

Glass vessels are required for procedures such as hot perchloric acid extractions of DNA. Plain-sided test tubes or Erlenmeyer flasks (with no lip), used in conjunction with sealing tape or Oxoid caps, are quick to use and are best kept in a humid $CO_2$-controlled atmosphere. Regular glass scintillation vials, or "minivials," are also good culture vessels because they are flat bottomed. Once used with scintillation fluid, however, they should not be reused for culture.

### 7.3.5  Uneven Growth

Sometimes cells can be inadvertently distributed nonuniformly across the growth surface. Vibration, caused by opening and closing of the incubator, a faulty fan motor, or vibration from equipment can perturb the medium, which can result in resonance or standing waves in the flask, that, in turn, result in a wave pattern in the monolayer (Fig. 7.10) creates variations in cell density. Eliminating vibration and minimizing entry into the incubator will help reduce uneven growth. Placing a heavy weight in the tray or box with the plates and separating it from the shelf with plastic foam may also help alleviate the problem [Nielsen, 1989], but great care must be taken to wash and sterilize such foam pads, as they will tend to harbor contamination.

(a)                                    (b)

***Fig. 7.7. Stirrer Flasks.*** Four small stirrer flasks (Techne), 500-mL capacity, with 250 mL medium, on four-place stirrer rack (*see* also Figs. 12.5, 26.1, 26.2). (Courtesy of Sterilin.)

(a)

(b)

***Fig. 7.8. Venting Petri Dishes and Flasks.*** (*a*) Vented Petri dish. Small ridges, 120° apart, raise the lid from the base and prevent a thin film of liquid (e.g., condensate) from sealing the lid and reducing the rate of gas exchange. (*b*) Gas-permeable cap on 10-cm² flask (Falcon, BD Biosciences).

### 7.3.6 Cost

Cost always has to be balanced against convenience; for example, Petri dishes are cheaper than flasks with an equivalent surface area but require humid, $CO_2$-controlled conditions and are more prone to infection. They are, however, easier to examine and process.

Cheap soda glass bottles, although not always of good optical quality, are often better for culture than higher grade Pyrex, or optically clear glass, which usually contains lead. A major disadvantage of glass is that its preparation is labor intensive because it must be carefully washed and resterilized before it can be reused. Most laboratories now use plastic because of its convenience, optical clarity, and quality.

## 7.4 SPECIALIZED SYSTEMS

### 7.4.1 Permeable Supports

Semipermeable membranes are used as gas-permeable substrates and will also allow the passage of water and small molecules (<500–1000 da), a property that is exploited in some large-scale bioreactors (*see* Sections 25.3.2, 26.2.5). Growing cells on a water-permeable substrate increases diffusion of oxygen, $CO_2$, and nutrients. Attachment of cells to a natural substrate such as collagen may control phenotypic expression due to the interaction of integrin receptors on the cell surface with specific sites in the extracellular matrix (*see* Sections 2.2.1, 2.2.3, 16.7.3). The growth of cells on floating collagen [Michalopoulos and Pitot, 1975; Lillie et al., 1980] has been used to improve the survival of epithelial cells and promote terminal differentiation (*see* Sections 16.7.3, 25.3.8).

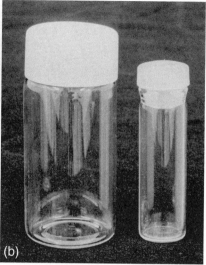

***Fig. 7.9. Screw–Cap Vials and Flasks.*** (*a*) Glass flasks are suitable for replicate cultures or storage of samples, particularly when plastic may not survive downstream processing. Screw caps are preferable to stoppers, as they are less likely to leak, and they protect the neck of the flask from contamination. (*b*) Scintillation vials are particularly useful for isotope incorporation studies but should not be reused for culture after containing scintillation fluid.

***Fig. 7.10. Nonrandom Growth.*** Examples of ridges seen in cultured monolayers in dishes and flasks, probably due to resonance in the incubator from fan motors or to opening and closing of the incubator doors. (Courtesy of Nunc.)

***Filter wells.*** The permeability of the surface to which the cell is anchored may induce polarity in the cell by simulating the basement membrane. Such polarity may be vital for full functional expression in secretory epithelia, and many other types of cells [Gumbiner & Simons, 1986; Chambard et al., 1987; Artursson & Magnusson, 1990; Mullin et al., 1997]. Permeable supports are available in the form of disposable filter well inserts of many different sizes, materials, and membrane porosities (*see* Appendix II: Filter Well Inserts). Also available are inserts precoated with collagen, laminin, or other matrix materials (e.g., Matrigel, BD Biosciences). Filter well inserts have been used extensively in studies of cell–cell interaction, cell–matrix interaction, differentiation and polarity, transepithelial permeability, and tissue modeling (*see* Section 25.3.6).

***Hollow fibers.*** Knazek et al. [1972] developed the technique of growing cells on the outer surface of bundles of plastic microcapillaries (Fig. 7.11; *see* Section 25.3.2). The plastic allows the diffusion of nutrients and dissolved gases from a medium perfused through the capillaries. Cells will grow to several cells deep on the outside of the capillaries, and an analogy with whole tissue is implied. Hollow fibers are also used in large-scale bioreactors (*see* Section 26.2.5).

### 7.4.2 Three-dimensional Matrices

It is apparent that many functional and morphological characteristics are lost during serial subculture, due to the loss of tissue architecture and cell–cell interaction (*see* Section 2.4.2). These deficiencies encouraged the exploration of three-dimensional matrices, such as collagen gel [Douglas et al., 1980], cellulose sponge (either alone or coated with collagen) [Leighton et al., 1968], or Gelfoam (*see* Section 25.3.1). Fibrin clots were one of the first media to be used for primary culture and are still used either as crude plasma clots (*see* Section 11.3.1) or as purified fibrinogen mixed with thrombin. Both systems generate a three-dimensional gel in which cells may migrate and grow, either on the solid–gel interface or within the gel [Leighton, 1991].

Three-dimensional matrices are used extensively in tissue engineering [Vunjak-Novakovic and Freshney, 2006] and can be inorganic, such as calcium phosphate, or organic, such as Gelfoam (*see* Section 25.3.1). As well as permitting cell attachment, proliferation, and differentiation, such matrices, or *scaffolds*, are required to degrade in vivo and be replaced by endogenous matrix.

Microcarriers made of polystyrene, Sephadex, polyacrylamide, and collagen or gelatin are available in bead form for the propagation of anchorage-dependent cells in suspension

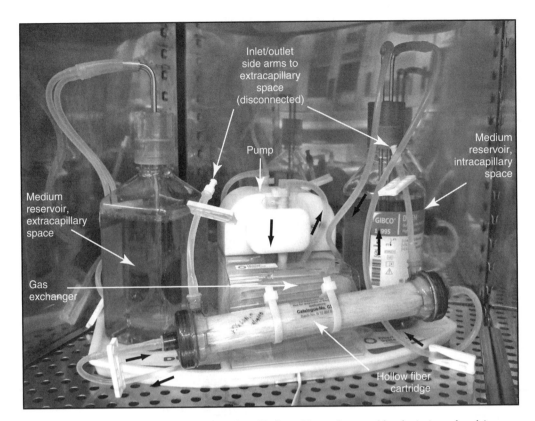

**Fig. 7.11. Hollow Fiber Culture.** A bundle of hollow fibers of permeable plastic is enclosed in a transparent plastic outer chamber and is accessible via either of the two side arms for seeding cells and collecting high molecular weight product. During culture the chamber is perfused from a reservoir, via a gas exchanger and peristaltic pump, down the center of the hollow fibers through connections attached to either end of the chamber. (Courtesy of FiberCell, Inc.; *see also* Fig. 26.6; Plate 24.)

(*see* Section 26.2.3; Table 26.1; Appendix II). Although technically three-dimensional, growth on many of these is functionally a two-dimensional monolayer, modified by the radius of curvature of the bead. However, some microcarriers are porous (*see* Sections 26.2.3, 26.2.4), so cells grow within the interstices of the matrix. Three-dimensional growth in

alginate beads occurs by encapsulation, rather than penetration of the matrix (*see* Section 25.3.5), allows a high level of cell–cell interaction and facilitates differentiation in chondrocytes (*see* Protocol 22.16) and neurons (*see* Section 25.3.5), and has been used to enhance antibody production by hybridomas [Zimmermann et al., 2003].

# CHAPTER 8

# Defined Media and Supplements

## 8.1 DEVELOPMENT OF MEDIA

Initial attempts to culture cells were performed in natural media based on tissue extracts and body fluids, such as chick embryo extract, plasma, serum, and lymph. With the propagation of cell lines (L929 cells, HeLa, etc.), the demand for larger amounts of a medium of more consistent quality led to the introduction of chemically defined media based on analyses of body fluids and nutritional biochemistry. Eagle's Basal Medium [Eagle, 1955] and subsequently Eagle's Minimal Essential Medium (MEM) [Eagle, 1959] became widely adopted, variously supplemented with calf, human, or horse serum, protein hydrolysates, and embryo extract. As more continuous cell lines became available, it was apparent that these media were perfectly adequate for the majority of them, and most of the succeeding developments were aimed at replacing serum (see Sections 9.2, 9.6), optimizing media for different cell types (e.g., RPMI 1640 for lymphoblastoid cell lines), or modifying for specific conditions (e.g., Leibovitz L15 to eliminate the need for adding $CO_2$ and $NaHCO_3$) [Leibovitz, 1963].

Isolation and propagation of cells of a specific lineage may require a selective serum-free medium (see Section 9.2.2), whereas cells grown for the formation of products, as hosts for viral propagation or for non–cell-specific molecular studies, rely mainly on Eagle's MEM [Eagle, 1959], Dulbecco's modification of Eagle's medium, DMEM [Dulbecco & Freeman, 1959], or, increasingly, RPMI 1640 [Moore et al., 1967], supplemented with serum. However, many industrial-scale production techniques now use serum-free media to facilitate downstream processing and reduce the risk of adventitious infectious agents (see Section 9.1). A popular compromise for many laboratories is a mixture of a complex medium, such as Ham's F12 [Ham, 1965], with one with higher amino acid and vitamin concentrations, such as DMEM. This alternative will horrify the purist, but it does generate a useful, all-purpose medium for primary culture as well as cell line propagation.

This chapter concentrates on the general principles of medium composition. It presents the widely used serum-supplemented media as examples, and Chapter 9 will focus on the design and use of serum-free media.

## 8.2 PHYSICOCHEMICAL PROPERTIES

### 8.2.1 pH

Most cell lines grow well at pH 7.4. Although the optimum pH for cell growth varies relatively little among different cell strains, some normal fibroblast lines perform best at pH 7.4 to pH 7.7, and transformed cells may do better at pH 7.0 to pH 7.4 [Eagle, 1973]. It was reported that epidermal cells could be maintained at pH 5.5 [Eisinger et al., 1979], but this level has not been universally adopted. In special cases it may prove advantageous to do a brief growth experiment (see Protocols 20.7–20.10), plating efficiency assay (see Section 20.10), or special function analysis (e.g., see Section 16.7) to determine the optimum pH.

Phenol red is commonly used as an indicator. It is red at pH 7.4 and becomes orange at pH 7.0, yellow at pH 6.5, lemon yellow below pH 6.5, more pink at pH 7.6, and purple at pH 7.8 (see Plate 22b). Because the assessment of color is highly subjective, it is useful to make up a set of standards using a sterile balanced salt solution (BSS) with phenol red at the correct concentration, in the same type of bottle, with the same headspace for air that you normally use for preparing a medium. The following example will use 25-cm² culture flasks as the final receptacles.

---

### PROTOCOL 8.1. PREPARATION OF pH STANDARDS

**Outline**
Prepare flasks of a series of sterile samples of medium or BSS adjusted to a range of pH from pH 6.5 to pH 7.8.

**Materials**

*Sterile:*

- Hanks's balanced salt solution (HBSS), or Eagle's MEM, 10× concentrate, without bicarbonate or glucose, with 20-mM HEPES
- Ultrapure water (UPW)
- 1 N NaOH
- Universal containers, 30 mL, 7
- Culture flasks, 25 cm² (T25), 7
- Pipettor tips, 100 μL

*Nonsterile:*

- pH meter
- Pipettor, 10–100 μL

**Procedure**

1. Make up the HBSS or medium at pH 6.5 or lower, and dispense 10 mL into each of seven labeled universal containers.
2. Allow to equilibrate with air for 30 min.
3. Using a pH meter, adjust the pH in the separate containers to 6.5, 6.8, 7.0, 7.2, 7.4, 7.6, and 7.8 with 1 N NaOH.
4. Filter sterilize 5 mL of each HBSS into separate labeled 25-cm² flasks (*see* Protocol 10.11), using the first 5 mL to flush out the filter before collecting the second 5 mL in the flask.
5. Cap the flasks securely.

---

### 8.2.2 CO₂ and Bicarbonate

Carbon dioxide in the gas phase dissolves in the medium, establishes equilibrium with $HCO_3^-$ ions, and lowers the pH. Because dissolved $CO_2$, $HCO_3^-$, and pH are all interrelated, it is difficult to determine the major direct effect of $CO_2$. The atmospheric $CO_2$ tension will regulate the concentration of dissolved $CO_2$ directly, as a function of temperature. This regulation in turn produces $H_2CO_3$, which dissociates according to the reaction

$$H_2O + CO_2 \Leftrightarrow H_2CO_3 \Leftrightarrow H^+ + HCO_3^- \qquad (1)$$

$HCO_3^-$ has a fairly low dissociation constant with most of the available cations, so it tends to reassociate, leaving the medium acid. The net result of increasing atmospheric $CO_2$ is to depress the pH, and the effect of elevated $CO_2$ tension can then be neutralized by increasing the bicarbonate concentration:

$$NaHCO_3 \Leftrightarrow Na^+ + HCO_3^- \qquad (2)$$

The increased $HCO_3^-$ concentration pushes equation (1) to the left until equilibrium is reached at pH 7.4. If another alkali (e.g., NaOH) is used instead, the net result is the same:

$$NaOH + H_2CO_3 \Leftrightarrow NaHCO_3 + H_2O$$
$$\Leftrightarrow Na^+ + HCO_3^- + H_2O \qquad (3)$$

The equivalent $NaHCO_3$ concentrations commonly used with different $CO_2$ tensions are listed in Tables 8.1, 8.2, and 8.3. Intermediate values of $CO_2$ and $HCO_3^-$ may be used, provided that the concentration of both is varied proportionately. Because many media are made up in acid solution and may incorporate a buffer, it is difficult to predict how much bicarbonate to use when other alkali may also end up as bicarbonate, as in equation (3). When preparing a new medium for the first time, add the specified amount of bicarbonate and then sufficient 1 N NaOH such that the medium equilibrates to the desired pH after incubation in a Petri dish at 37°C, in the correct $CO_2$ concentration, overnight. When dealing with a medium that is already at working strength, vary the amount of $HCO_3^-$ to suit the gas phase (*see* Table 8.1), and leave the medium overnight to equilibrate at 37°C. Each medium has a recommended bicarbonate concentration and $CO_2$ tension for achieving the

---

**TABLE 8.1.** Relationships among Bicarbonate, Carbon Dioxide, and HEPES

| Compound | Eagle's MEM Hanks's salts | Low $HCO_3^-$ + buffer | Eagle's MEM Earle's salts | DMEM |
|---|---|---|---|---|
| NaHCO₃ | 4 mM | 10 mM | 26 mM | 44 mM |
| CO₂ | Atmospheric and evolved from culture | 2% | 5% | 10% |
| HEPES[a] (if required) | 10 mM | 20 mM | 50 mM | — |

[a]If HEPES is used, the equivalent molarity of NaCl must be omitted and osmolality must be checked.

**TABLE 8.2.** Balanced Salt Solutions

| Component | M.W. | Earle's BSS g/L | Earle's BSS mM | Dulbecco's PBS Without Ca²⁺ and Mg²⁺ (D-PBSA) g/L | mM | With Ca²⁺ and Mg²⁺ g/L | mM | Hanks's BSS g/L | mM | Spinner salts (as in S-MEM) g/L | mM |
|---|---|---|---|---|---|---|---|---|---|---|---|
| **Inorganic salts** | | | | | | | | | | | |
| CaCl₂ (anhydrous) | 111 | 0.02 | 0.18 | | | 0.2 | 1.80 | 0.14 | 1.3 | | |
| KCl | 74.55 | 0.4 | 5.37 | 0.2 | 2.68 | 0.2 | 2.68 | 0.4 | 5.4 | 0.40 | 5.37 |
| KH₂PO₄ | 136.1 | | | 0.2 | 1.47 | 0.2 | 1.47 | 0.06 | 0.4 | | |
| MgCl₂ · 6H₂O | 203.3 | | | | | | | 0.1 | 0.5 | | |
| MgSO₄ · 7H₂O | 246.5 | 0.2 | 0.81 | | | 0.98 | 3.98 | 0.1 | 0.4 | 0.20 | 0.81 |
| NaCl | 58.44 | 6.68 | 114.3 | 8 | 136.9 | 8 | 136.9 | 8 | 136.9 | 6.80 | 116.4 |
| NaHCO₃ | 84.01 | 2.2 | 26.19 | | | | | 0.35 | 4.2 | 2.20 | 26.19 |
| Na₂HPO₄ · 7H₂O | 268.1 | | | 2.2 | 8.06 | 2.16 | 8.06 | 0.09 | 0.3 | | |
| NaH₂PO₄ · H₂O | 138 | 0.14 | 1.01 | | | | | | | 1.40 | 10.14 |
| **Total salt** | | | 147.9 | | 149.1 | | 154. | | 149.4 | | 158.9 |
| **Other components** | | | | | | | | | | | |
| D-glucose | 180.2 | 1 | 5.55 | | | | | 1 | 5.5 | 1.00 | 5.55 |
| Phenol red | 354.4 | 0.01 | 0.03 | | | | | 0.01 | 0.0 | 0.01 | 0.03 |
| **Gas phase** | | 5% CO₂ | | | | Air | | Air | | 5% CO₂ | |

correct pH and osmolality, but minor variations will occur in different methods of preparation.

With the introduction of Good's buffers (e.g., HEPES, Tricine) [Good et al., 1966] into tissue culture, there was some speculation that, as $CO_2$ was no longer necessary to stabilize the pH, it could be omitted. This proved to be untrue [Itagaki & Kimura, 1974], at least for a large number of cell types, particularly at low cell concentrations. Although 20-mM HEPES can control pH within the physiological range, the absence of atmospheric $CO_2$ allows equation (1) to move to the left, eventually eliminating dissolved $CO_2$, and ultimately $HCO_3^-$, from the medium. This chain of events appears to limit cell growth, although whether the cells require the dissolved $CO_2$ or the $HCO_3^-$ (or both) is not clear. Recommended $HCO_3^-$, $CO_2$, and HEPES concentrations are given in Table 8.1.

The inclusion of pyruvate in the medium enables cells to increase their endogenous production of $CO_2$, making them independent of exogenous $CO_2$, as well as $HCO_3^-$. Leibovitz L15 medium [Leibovitz, 1963] contains a higher concentration of sodium pyruvate (550 mg/L) but lacks $NaHCO_3$ and does not require $CO_2$ in the gas phase. Buffering is achieved via the relatively high amino acid concentrations. Because it does not require $CO_2$, L15 is sometimes recommended for the transportation of tissue samples. Sodium β-glycerophosphate can also be used to buffer autoclavable media lacking $CO_2$ and $HCO_3^-$ [Waymouth, 1979], and Gibco (Invitrogen) markets a $CO_2$-independent medium. If the elimination of $CO_2$ is important for cost saving, convenience, or other reasons, it might be

worth considering one of these formulations, but only after appropriate testing.

In sum, cultures in open vessels need to be incubated in an atmosphere of $CO_2$, whose concentration is in equilibrium with the sodium bicarbonate in the medium (*see* Tables 8.1, 8.2, 8.3). Cells at moderately high concentrations ($\geq 1 \times 10^5$ cells/mL) and grown in sealed flasks need not have $CO_2$ added to the gas phase, provided that the bicarbonate concentration is kept low ($\sim$4 mM), particularly if the cells are high acid producers. At low cell concentrations, however (e.g., during cloning), and with some primary cultures, it is necessary to add $CO_2$ to the gas phase of sealed flasks. When venting is required, to allow either the equilibration of $CO_2$ or its escape in high acid producers, it is necessary to leave the cap slack or to use a $CO_2$-permeable cap (*see* Fig. 7.8).

### 8.2.3 Buffering

Culture media must be buffered under two sets of conditions: (1) open dishes, where the evolution of $CO_2$ causes the pH to rise (*see* Section 8.2.2), and (2) overproduction of $CO_2$ and lactic acid in transformed cell lines at high cell concentrations, when the pH will fall. A buffer may be incorporated into the medium to stabilize the pH, but in (1) exogenous $CO_2$ may still be required by some cell lines, particularly at low cell concentrations, to prevent the total loss of dissolved $CO_2$ and bicarbonate from the medium. Despite its poor buffering capacity at physiological pH, bicarbonate buffer is still used more frequently than any other buffer because of its low toxicity, low cost, and nutritional benefit to the culture. HEPES, at 10 to 20 mM, is a much stronger

**TABLE 8.3.** Frequently Used Media

| Component | MEM [Eagle 1959] | DMEM [Dulbecco & Freeman, 1959] | F12 [Ham, 1965] | DMEM/F12 [Barnes & Sato, 1980] | α MEM [Stanners et al., 1971] | CMRL 1066 [Parker et al., 1957] | RPMI 1640 [Moore et al., 1967] | M199 [Morgan et al., 1950] | L15 [Leibovitz, 1963] | McCoy's 5A [McCoy et al., 1959] | MB 752/1 [Waymouth, 1959] |
|---|---|---|---|---|---|---|---|---|---|---|---|
| **Amino acids** | | | | | | | | | | | |
| L-Alanine | | | 1.0E-04 | 5.0E-05 | 2.8E-04 | 2.8E-04 | | 2.8E-04 | 2.5E-03 | 1.5E-04 | |
| L-Arginine | 6.0E-04 | 4.0E-04 | 1.0E-03 | 7.0E-04 | 6.0E-04 | 3.3E-04 | 1.1E-03 | 3.3E-04 | 2.9E-03 | 2.0E-04 | 3.6E-04 |
| L-Asparagine | | | 1.0E-04 | 5.0E-05 | 3.3E-04 | 3.8E-04 | 1.7E-03 | 3.0E-04 | 7.6E-05 | | |
| L-Aspartic acid | | | 1.0E-04 | 5.0E-05 | 2.3E-04 | 2.3E-04 | 1.5E-04 | 2.3E-04 | | 1.5E-04 | 4.5E-04 |
| l-Cysteine | | | 2.0E-04 | 1.0E-04 | 5.7E-04 | 1.5E-03 | | 5.6E-07 | 9.9E-04 | 2.0E-04 | 5.0E-04 |
| L-Cystine | 1.0E-04 | 2.0E-04 | | 1.0E-04 | 1.0E-04 | 8.3E-05 | 2.1E-04 | 9.9E-05 | | | 6.3E-05 |
| L-Glutamic acid | | | 1.0E-04 | 5.0E-05 | 5.1E-04 | 5.1E-04 | 1.4E-04 | 4.5E-04 | | 1.5E-04 | 1.0E-03 |
| L-Glutamine | 2.0E-03 | 4.0E-03 | 1.0E-03 | 2.5E-03 | 2.0E-03 | 6.8E-04 | 2.1E-03 | 6.8E-04 | 2.1E-03 | 1.5E-03 | 2.4E-03 |
| Glycine | | 4.0E-04 | 1.0E-04 | 2.5E-04 | 6.7E-04 | 6.7E-04 | 1.3E-04 | 6.7E-04 | 2.7E-03 | 1.0E-04 | 6.7E-04 |
| L-Histidine | 2.0E-04 | 2.0E-04 | 1.0E-04 | 1.5E-04 | 2.0E-04 | 9.5E-05 | 9.7E-05 | 1.0E-04 | 1.6E-03 | 1.0E-04 | 8.3E-04 |
| L-Hydroxyproline | | | | | | 7.6E-05 | 1.5E-04 | 7.6E-05 | | 1.5E-04 | |
| L-Isoleucine | 4.0E-04 | 8.0E-04 | 3.0E-05 | 4.2E-04 | 4.0E-04 | 1.5E-04 | 3.8E-04 | 1.5E-04 | 9.5E-04 | 3.0E-04 | 1.9E-04 |
| L-Leucine | 4.0E-04 | 8.0E-04 | 1.0E-04 | 4.5E-04 | 4.0E-04 | 4.6E-04 | 3.8E-04 | 4.6E-04 | 9.5E-04 | 3.0E-04 | 3.8E-04 |
| L-Lysine HCl | 4.0E-04 | 8.0E-04 | 2.0E-04 | 5.0E-04 | 4.0E-04 | 3.8E-04 | 2.2E-04 | 3.8E-04 | 5.1E-04 | 2.0E-04 | 1.3E-03 |
| L-Methionine | 1.0E-04 | 2.0E-04 | 3.0E-05 | 1.2E-04 | 1.0E-04 | 1.0E-04 | 1.0E-04 | 1.0E-04 | 5.0E-04 | 1.0E-04 | 3.4E-04 |
| L-Phenylalanine | 2.0E-04 | 4.0E-04 | 3.0E-05 | 2.2E-04 | 1.9E-04 | 1.5E-04 | 9.1E-05 | 1.5E-04 | 7.6E-04 | 1.0E-04 | 3.0E-04 |
| L-Proline | | | 3.0E-04 | 1.5E-04 | 3.5E-04 | 3.5E-04 | 1.7E-04 | 3.5E-04 | | 1.5E-04 | 4.3E-04 |
| L-Serine | | 4.0E-04 | 1.0E-04 | 2.5E-04 | 2.4E-04 | 2.4E-04 | 2.9E-04 | 2.4E-04 | 1.9E-03 | 2.5E-04 | |
| L-Threonine | 4.0E-04 | 8.0E-04 | 1.0E-04 | 4.5E-04 | 4.0E-04 | 2.5E-04 | 1.7E-04 | 2.5E-04 | 2.5E-03 | 1.5E-04 | 6.3E-04 |
| L-Tryptophan | 4.9E-05 | 7.8E-05 | 1.0E-05 | 4.4E-05 | 4.9E-05 | 4.9E-05 | 2.5E-05 | 4.9E-05 | 9.8E-05 | 1.5E-05 | 2.0E-04 |
| L-Tyrosine | 2.0E-04 | 4.0E-04 | 3.0E-05 | 2.1E-04 | 2.3E-04 | 2.2E-04 | 1.1E-04 | 2.2E-04 | 1.7E-03 | 1.2E-04 | 2.2E-04 |
| L-Valine | 4.0E-04 | 8.0E-04 | 1.0E-04 | 4.5E-04 | 3.9E-04 | 2.1E-04 | 1.7E-04 | 2.1E-04 | 8.5E-04 | 1.5E-04 | 5.6E-04 |
| **Vitamins** | | | | | | | | | | | |
| p-Aminobenzoic acid | | | | | | 3.6E-07 | 7.3E-06 | 3.6E-07 | | 7.3E-06 | |
| L-Ascorbic acid | | | | | 2.5E-04 | 2.8E-04 | | 2.8E-07 | | 3.2E-06 | 9.9E-05 |
| Biotin | | | 3.0E-08 | 1.5E-08 | 4.1E-07 | 4.1E-08 | 8.2E-07 | 4.1E-08 | | 8.2E-07 | 8.2E-08 |
| Calciferol | | | | | | | | 2.5E-07 | | | |
| Choline chloride | 7.1E-06 | 2.9E-05 | 1.0E-04 | 6.4E-05 | 7.1E-06 | 3.6E-06 | 2.1E-05 | 3.6E-06 | 7.1E-06 | 3.6E-05 | 1.8E-03 |
| Folic acid | 2.3E-06 | 9.1E-06 | 2.9E-06 | 6.0E-06 | 2.3E-06 | 2.3E-08 | 2.3E-06 | 2.3E-08 | 2.3E-06 | 2.3E-05 | 9.1E-07 |
| myo-Inositol | 1.1E-05 | 4.0E-05 | 1.0E-04 | 7.0E-05 | 1.1E-05 | 2.8E-07 | 1.9E-04 | 2.8E-07 | 1.1E-05 | 2.0E-04 | 5.6E-06 |
| Menadione | | | | | | | | 6.9E-08 | | | |
| Nicotinamide | 8.2E-06 | 3.3E-05 | 3.3E-07 | 1.7E-05 | 8.2E-06 | 2.0E-07 | 8.2E-06 | 2.0E-07 | 8.2E-06 | 4.1E-06 | 8.2E-06 |
| Nicotinic acid | | | | | | 2.0E-07 | | 2.0E-07 | | 4.1E-06 | |
| D-Ca pantothenate | 4.2E-06 | 1.7E-05 | 2.0E-06 | 9.4E-06 | 4.2E-06 | 4.2E-08 | 1.1E-06 | 4.2E-08 | 4.2E-06 | 8.4E-07 | 4.2E-06 |
| Pyridoxal HCl | 4.9E-06 | 2.0E-05 | 3.0E-07 | 1.0E-05 | 4.9E-06 | 1.2E-07 | | 1.2E-07 | | 2.5E-06 | |

| | 1 | 2 | 3 | 4 | 5 | 6 | 7 | 8 | 9 | 10 |
|---|---|---|---|---|---|---|---|---|---|---|
| Pyridoxine HCl | 2.7E-07 | | 3.0E-07 | 1.5E-07 | 1.2E-07 | 4.9E-06 | 1.2E-07 | 1.9E-07 | 2.4E-06 | 4.9E-06 |
| Riboflavin | | 1.1E-06 | 1.0E-07 | 5.8E-07 | 2.7E-08 | 5.3E-07 | 2.7E-08 | | 5.3E-07 | 2.7E-06 |
| Thiamin | 3.0E-06 | 1.2E-05 | 1.0E-06 | 6.4E-06 | 3.0E-08 | 3.0E-06 | 3.0E-08 | 2.4E-06 | 5.9E-07 | 3.0E-05 |
| Thiamin mono $PO_4$ | | | | | | | | | 4.8E-06 | |
| $\alpha$-Tocopherol | | | | | | | 2.3E-08 | | | |
| Retinol acetate | | | | | | | 3.5E-07 | | | |
| Vitamin $B_{12}$ | | | 1.0E-06 | 5.0E-07 | | 3.7E-09 | | | | 1.5E-07 |
| Antioxidants | | | | | | | | | | |
| Glutathione | | | | | 3.0E-05 | 3.0E-06 | 1.5E-07 | | 1.5E-06 | 4.5E-05 |
| **Inorganic salts** | | | | | | | | | | |
| $CaCl_2$ | 1.8E-03 | 1.8E-03 | 3.0E-04 | 1.1E-03 | 1.8E-03 | 1.3E-03 | 1.3E-03 | 1.3E-03 | 9.0E-04 | 8.2E-04 |
| KCl | 5.3E-03 | 5.3E-03 | 3.0E-03 | 4.2E-03 | 5.3E-04 | 5.3E-03 | 5.3E-03 | 5.3E-03 | 5.3E-03 | 2.0E-03 |
| $KH_2PO_4$ | | | | | | | 4.4E-04 | 4.4E-04 | | 5.9E-04 |
| $MgCl_2$ | | | | | 1.2E-01 | | | | | 1.2E-03 |
| $MgSO_4$ | 8.1E-04 | 8.1E-04 | 6.0E-04 | 7.0E-04 | 8.1E-04 | 4.0E-04 | 8.1E-04 | 1.6E-03 | 8.1E-04 | 8.1E-04 |
| NaCl | 1.16E-01 | 1.09E-01 | 1.28E-01 | 1.19E-01 | 1.16E-01 | 1.03E-01 | 1.16E-01 | 1.37E-01 | 1.10E-01 | 1.03E-01 |
| $NaHCO_3$ | 2.6E-02 | 4.4E-02 | 1.4E-02 | 2.9E-02 | 2.6E-02 | 2.3E-02 | 2.6E-02 | 2.6E-02 | 1.3E-02 | 2.7E-02 |
| $NaH_2PO_4$ | | 9.1E-04 | 4.5E-04 | 4.5E-04 | 1.0E-03 | 5.6E-03 | | | | |
| $Na_2HPO_4$ | 1.0E-03 | | 5.0E-04 | 5.0E-04 | | | 4.0E-04 | 1.6E-03 | 4.2E-03 | 2.1E-03 |
| **Trace elements** | | | | | | | | | | |
| $CuSO_4 \cdot 5H_2O$ | 1.6E-08 | 2.5E-07 | 1.6E-08 | 7.8E-09 | | | | | | |
| $Fe(NO_3)_3 \cdot 9H_2O$ | | | 1.2E-07 | 1.2E-07 | | | | | | |
| $FeSO_4 \cdot 7H_2O$ | 3.0E-06 | | 3.0E-06 | 1.5E-06 | | | | | | |
| $ZnSO_4 \cdot 7H_2O$ | 3.0E-06 | | 3.0E-06 | 1.5E-06 | | | | | | |
| **Bases, nucleosides, etc.** | | | | | | | | | | |
| Adenine $SO_4$ | | | | | | | 5.4E-05 | | | |
| Adenosine | | | | | 3.7E-05 | | | | | |
| AMP | | | | | | | 5.8E-07 | | | |
| ATP | | | | | | | 1.8E-05 | | | |
| Cytidine | | | | | 4.1E-05 | | | | | |
| Deoxyadenosine | | | 4.0E-05 | | 4.0E-05 | | | | | |
| Deoxycytidine | | | 3.8E-05 | | 4.2E-05 | | | | | |
| Deoxyguanosine | | | 3.7E-05 | | 3.7E-05 | | | | | |
| 2-Deoxyribose | | | | | | | 3.7E-06 | | | |
| DPN | | | 9.5E-06 | | | | | | | |
| FAD | | | 1.2E-06 | | | | | | | |
| Glucuronate, Na | | | 1.9E-05 | | | | | | | |
| Guanine | | | | | | | 1.6E-06 | | | |
| Guanosine | 3.5E-05 | | | | | | | | | |

(continued overleaf)

**TABLE 8.3.** (*Continued*)

| Component | MEM [Eagle 1959] | DMEM [Dulbecco & Freeman, 1959] | F12 [Ham, 1965] | DMEM/F12 [Barnes & Sato, 1980] | α MEM [Stanners et al., 1971] | CMRL 1066 [Parker et al., 1957] | RPMI 1640 [Moore et al., 1967] | M199 [Morgan et al., 1950] | L15 [Leibovitz, 1963] | McCoy's 5A [McCoy et al., 1959] | MB 752/1 [Waymouth, 1959] |
|---|---|---|---|---|---|---|---|---|---|---|---|
| Hypoxanthine | | | 3.0E-05 | 1.5E-05 | | | | 2.2E-06 | | | |
| 5-Me-deoxycytidine | | | | | | 4.1E-07 | | | | | |
| D-Ribose | | | 3.0E-06 | 1.5E-06 | | | | 3.3E-06 | | | |
| Thymidine | | | | | 4.1E-05 | 4.1E-05 | | | | | |
| Thymine | | | | | | 1.3E-06 | | 2.4E-06 | | | |
| TPN | | | | | | | | 2.7E-06 | | | |
| Uracil | | | | | | | | | | | |
| Uridine | | | | | 4.1E-05 | 1.8E-06 | | | | | |
| UTP | | | | | | | | | | | |
| Xanthine | | | | | | | | 2.0E-06 | | | |
| **Energy metabolism** | | | | | | | | | | | |
| Cocarboxylase | | | | | | 2.2E-06 | | | | | |
| Coenzyme A | | | | | | 3.3E-06 | | | | | |
| D-galactose | | | | | | | | | 5.0E-02 | | |
| D-glucose | 5.6E-03 | 2.5E-02 | 1.0E-02 | 1.8E-02 | 5.6E-03 | 5.6E-03 | 1.1E-02 | 5.6E-03 | | 1.7E-02 | 2.8E-02 |
| Sodium acetate | | | | | | 6.1E-04 | | 4.5E-04 | 5.0E-03 | | |
| Sodium pyruvate | | 1.0E-03 | 1.0E-03 | 1.0E-03 | 1.0E-03 | | | | | | |
| **Lipids and precursors** | | | | | | | | | | | |
| Cholesterol | | | | | | 5.2E-07 | | 5.2E-07 | | | |
| Ethanol (solvent) | | | | | | 3.5E-04 | | | | | |
| Linoleic acid | | 3.0E-07 | | 1.5E-07 | | | | | | | |
| Lipoic acid | | | 1.0E-06 | 5.1E-07 | | | | | | | 8.9E-05 |
| Tween 80 | | | | | 9.7E-07 | 1.8E-05 | | 1.8E-05 | | | |
| **Other components** | | | | | | | | | | | |
| Peptone, mg/mL | | | | | | | | | | 0.6 | |
| Phenol red | 2.7E-05 | 4.0E-05 | 3.2E-05 | 3.6E-05 | 2.9E-05 | 5.3E-05 | 1.3E-05 | 4.5E-05 | 2.7E-05 | 2.9E-05 | 2.7E-05 |
| Putrescine | | | 1.0E-06 | 5.0E-07 | | | | | | | |
| **Gas phase** | | | | | | | | | | | |
| $CO_2$ | 5% | 10% | 2% | 7% | 5% | 5% | 5% | 5% | Air | 5% | 5% |

Note: All concentrations are molar, and computer-style notation is used (e.g., 3.0E-2 = $3.0 \times 10^{-2}$ = 30 mM). Molecular weights are given for root compounds; although some recipes use salts or hydrated forms, molarities will, of course, remain the same. *Synonyms and abbreviations:* AMP, adenosine monophosphate; ATP, adenosine triphosphate; biotin = vitamin H; calciferol = vitamin $D_2$; FAD, flavine adenine dinucleotide; lipoic acid = thioctic acid; menadione = vitamin $K_3$; *myo*-inositol = L-inositol; nicotinamide = niacinamide; nicotinic acid = niacin; pyridoxine HCl = vitamin $B_6$; thiamin = vitamin $B_1$; α-tocopherol = vitamin E; retinol = vitamin $A_1$; TPN, triphosphopyridine nucleotide; UTP, uridine triphosphate; vitamin $B_{12}$ = cobalamin.

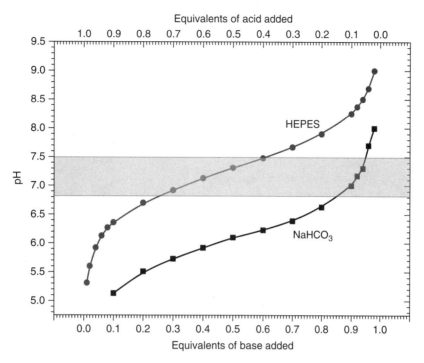

**Fig. 8.1. Buffering by HEPES and Bicarbonate.** The graphs show the effect of titrating with acid (top axis) or base (bottom axis) on the change in pH in the presence of either HEPES or sodium bicarbonate. It can be seen from the upper graph that pH is buffered more effectively in the physiological range (shaded horizontal bar) by HEPES than by bicarbonate (lower graph). [Transformed from Shipman, 1973.]

buffer in the pH 7.2 to 7.6 range (Fig. 8.1). It has been found that when HEPES is used with exogenous $CO_2$, the HEPES concentration must be more than double that of the bicarbonate for adequate buffering (*see* Table 8.1). A variation of Ham's F12 with 20-mM HEPES, 8-mM bicarbonate, and 2% $CO_2$ has been used successfully in the author's laboratory for the culture of a number of different cell lines. It allows the handling of microtitration and other multiwell plates outside the incubator without an excessive rise in pH and minimizes the requirement for HEPES, which is both toxic and expensive.

### 8.2.4 Oxygen

The other major constituent of the gas phase is oxygen. Whereas most cells require oxygen for respiration in vivo, cultured cells often rely mainly on glycolysis, a high proportion of which, as in transformed cells, may be anaerobic. Oxygen is still required, however. The use of $O_2$ carriers, such as perfluorocarbon [Radisic et al., 2006] or hemoglobin [Chen and Palmer, 2008], has had some success but is not widespread, and most cultures rely chiefly on dissolved $O_2$, which can be toxic due to the elevation in the level of free radicals. Providing the correct $O_2$ tension is therefore always a compromise between fulfilling the respiratory requirement and avoiding toxicity. Strategies involving both elevated and reduced $O_2$ levels have been employed, as has the incorporation of free radical scavengers, such as glutathione, 2-mercaptoethanol (β-mercaptoethanol),

or dithiothreitol, into the medium. Most such strategies have been derived empirically. Some incubators allow control of oxygen as well as $CO_2$ (e.g., New Brunswick Innova).

Cultures vary in their oxygen requirement, the major distinction lying between organ and cell cultures. Although atmospheric or lower oxygen tensions [Cooper et al., 1958; Balin et al., 1976] are preferable for most cell cultures, some organ cultures, particularly from late-stage embryos, newborns, or adults, require up to 95% $O_2$ in the gas phase [Trowell, 1959; De Ridder & Mareel, 1978]. This requirement for a high level of $O_2$ may be a problem of diffusion related to the geometry and gaseous penetration of organ cultures (*see* Section 25.2.1) but may also reflect the difference between differentiated and rapidly proliferating cells. Oxygen diffusion may also become limiting in porous microcarriers [Preissmann et al., 1997] (*see also* Section 26.2.4).

Most dispersed cell cultures prefer lower oxygen tensions, and some systems (e.g., human tumor cells in clonogenic assay [Courtenay et al., 1978], human embryonic lung fibroblasts [Balin et al., 1976], and mesenchymal stem cells [Ma et al., 2009]) prefer hypoxic conditions. McKeehan et al. [1976] suggested that the requirement for selenium in medium is related to oxygen toxicity, as selenium is a cofactor in glutathione synthesis. Oxygen tolerance—and selenium as well—may be provided by serum, so the control of $O_2$ tension is likely to be more critical in serum-free media.

Because the depth of the culture medium can influence the rate of oxygen diffusion to the cells, it is advisable to keep the depth of the medium within the range 2 to 5 mm ($0.2-0.5$ mL/cm$^2$) in static culture. Some types of cells (e.g., bronchial epithelium and keratinocytes) grown in filter well inserts appear to differentiate better when positioned at the air–liquid interface (*see* Section 25.3.6).

### 8.2.5 Osmolality

Most cultured cells have a fairly wide tolerance for osmotic pressure [Waymouth, 1970]. As the osmolality of human plasma is about 290 mOsm/kg, it is reasonable to assume that this level is the optimum for human cells in vitro, although it may be different for other species (e.g., around 310 mOsm/kg for mice [Waymouth, 1970]). In practice, osmolalities between 260 mOsm/kg and 320 mOsm/kg are quite acceptable for most cells but, once selected, should be kept consistent at ±10 mOsm/kg. Slightly hypotonic medium may be better for Petri dish or open-plate culture to compensate for evaporation during incubation. Changes in osmolality are generally achieved by altering the sodium chloride concentration, for example, to compensate for different bicarbonate concentrations (*see* Tables 8.2, 8.3) or to allow for the addition of HEPES.

Osmolality is usually measured by depression of the freezing point (Fig. 8.2), or elevation of the vapor pressure, of the medium. The measurement of osmolality is a useful quality-control step if you are making up the medium yourself, as it helps to guard against errors in weighing, dilution, and the like. It is particularly important to monitor osmolality if alterations are made in the constitution of the medium. The addition of HEPES and drugs dissolved in strong acids and bases and their subsequent neutralization can all markedly affect osmolality.

### 8.2.6 Temperature

The optimal temperature for cell culture is dependent on (1) the body temperature of the animal from which the cells were obtained, (2) any anatomic variation in temperature (e.g., the temperature of the skin and testis may be lower than that of the rest of the body), and (3) the incorporation of a safety factor to allow for minor errors in regulating the incubator. Thus the temperature recommended for most human and warm-blooded animal cell lines is 37°C, close to body heat, but set a little lower for safety, as overheating is a more serious problem than underheating.

Because of the higher body temperature in birds, avian cells should be maintained at 38.5°C for maximum growth. They will grow satisfactorily, however, if more slowly, at 37°C.

Cultured mammalian cells will tolerate considerable drops in temperature, can survive several days at 4°C, and can be frozen and cooled to −196°C (*see* Protocol 19.1). They cannot, however, tolerate more than about 2°C above normal (39.5°C) for more than a few hours and will die quite rapidly at 40°C and over.

**Fig. 8.2. Osmometer.** Löser osmometer. This model accepts samples of 100 μL. (Courtesy of Camlab, Ltd.)

Attention must be paid to the consistency of the temperature (within ±0.5°C) to ensure reproducible results. Doors of incubators or hot rooms must not be left open longer than necessary and large items or volumes of liquid, placed in the hot room to heat, should not be put near any cultures. The spatial distribution of temperature within the incubator or hot room must also be uniform (*see* Sections 3.2.5, 4.3); there should be no cold spots, and air should circulate freely. This means that a large number of flasks should not be stacked together when first placed in the incubator or hot room; space must be allowed between them for air to circulate. Also, as the gas phase within the flasks expands at 37°C, flasks, particularly large flasks and when stacked, must be vented by briefly slackening the cap 30 min to 1 h after placing the flasks in the incubator or warm room. Clearly, if vented caps are used in a $CO_2$ incubator, this is not necessary.

Poikilotherms (cold-blooded animals that do not regulate their blood heat within narrow limits) tolerate a wide temperature range, between 15°C and 26°C. Simulating in vivo conditions (e.g., for cold-water fish) may require an incubator with cooling as well as heating, to keep the incubator temperature below ambient. If necessary, poikilothermic animal cells can be maintained at room temperature, but the variability of the ambient temperature in laboratories makes this undesirable, and a cooled incubator is preferable.

A number of temperature-sensitive (ts) mutant cell lines have been developed that allow the expression of specific genes below a set temperature, but not above it [Wyllie et al., 1992; Varga et al., 2008]. These mutants facilitate studies on cell regulation, but also emphasize the narrow range within which one can operate, as the two discriminating temperatures are usually only about $2^{\circ}$C to $3^{\circ}$C apart. The use of ts mutants usually requires an incubator with cooling as well as heating, to compensate for a warm ambient temperature.

Apart from its direct effect on cell growth, the temperature will also influence pH due to the increased solubility of $CO_2$ at lower temperatures and, possibly, because of changes in ionization and the $pK_a$ of the buffer. The pH should be adjusted to 0.2 units lower at room temperature than at $37^{\circ}$C as it will rise during incubation. In preparing a medium for the first time, it is best to make up the complete medium and incubate a sample overnight at $37^{\circ}$C under the correct gas tension, in order to check the pH (*see* Section 8.2.2; Plate 22*b*).

### 8.2.7 Viscosity

The viscosity of a culture medium is influenced mainly by the serum content and in most cases will have little effect on cell growth. Viscosity becomes important, however, whenever a cell suspension is agitated (e.g., when a suspension culture is stirred) or when cells are dissociated after trypsinization. Any cell damage that occurs under these conditions may be reduced by increasing the viscosity of the medium with carboxymethylcellulose (CMC) or polyvinylpyrrolidone (PVP) [Cherry & Papoutsakis, 1990; *see also* Appendix I]. This becomes particularly important in low-serum concentrations, in the absence of serum, and in stirred bioreactor cultures (*see* Section 26.1), in which Pluronic F68 is often used, although its effect is probably multifactorial.

### 8.2.8 Surface Tension and Foaming

The effects of foaming have not been clearly defined, but the rate of protein denaturation may increase, as may the risk of contamination if the foam reaches the neck of the culture vessel. Foaming will also limit gaseous diffusion if a film from a foam or spillage gets into the capillary space between the lid and the base of a Petri dish or between a slack cap and the neck of a flask.

Foaming can arise in suspension cultures in stirrer vessels or bioreactors when 5% $CO_2$ in air is bubbled through medium containing serum. The addition of a silicone antifoam (Dow Chemical) or Pluronic F68 (Sigma), 0.01% to 0.1%, helps prevent foaming in this situation by reducing surface tension and may also protect cells against shear stress from bubbles. $CO_2$ should not be bubbled through the medium when gassing a flask as this may generate foam and can also spread aerosol from the flask.

## 8.3 BALANCED SALT SOLUTIONS

A balanced salt solution (BSS) is composed of inorganic salts and may include sodium bicarbonate and, in some cases, glucose. The compositions of some common balanced salt solutions are given in Table 8.2. HEPES buffer (5–20 mM) may be added to these solutions if necessary and the equivalent amount of NaCl omitted to maintain the correct osmolality. BSS forms the basis of many complete media, and commercial suppliers will provide Eagle's MEM with Hanks's salts [Hanks & Wallace, 1949] or Eagle's MEM with Earle's salts [Earle et al., 1943], indicating which BSS formulation was used; inclusion of Hanks's salts requires sealed flasks with a gas phase of air, whereas Earle's salts imply a higher bicarbonate concentration compatible with growth in 5% $CO_2$.

BSS is also used as a diluent for concentrates of amino acids and vitamins to make complete media, as an isotonic wash or dissection medium, and for short incubations up to about 4 h (usually with glucose present). BSS recipes are often modified—for instance, by omitting glucose or phenol red from Hanks's BSS or by leaving out $Ca^{2+}$ or $Mg^{2+}$ ions from Dulbecco's PBS [Dulbecco & Vogt, 1954]. PBS without $Ca^{2+}$ and $Mg^{2+}$ is known as PBS Solution A, and the convention D-PBSA will be used throughout this book to indicate Dulbecco's PBS without divalent cations. One should always check for modifications when purchasing BSS and should quote any modifications to the published formula in reports and publications.

The choice of BSS is dependent on both the $CO_2$ tension (Tables 8.1 and 8.2; *see* Section 8.2.2) and the intended use of the solution for tissue disaggregation or monolayer dispersal; in these cases $Ca^{2+}$ and $Mg^{2+}$ are usually omitted, as in Moscona's [1952] calcium- and magnesium-free saline (CMF) or D-PBSA (*see* Table 8.2). The choice of BSS also is dependent on whether the solution will be used for suspension culture of adherent cells. S-MEM, based on Eagle's Spinner salt solution, is a variant of Eagle's [1959] minimal essential medium (MEM), which is deficient in $Ca^{2+}$ in order to reduce cell aggregation and attachment (*see* Table 8.2).

HBSS, EBSS, and D-PBS rely on the relatively weak buffering of phosphate, which is not at its most effective at physiological pH. Paul [1975] constructed a Tris-buffered BSS that is more effective, but for which the cells sometimes require a period of adaptation and/or selection. HEPES (10–20 mM) is currently the most effective buffer in the pH 7.2 to pH 7.8 range (*see* Fig. 8.1), and Tricine in the pH 7.4 to pH 8.0 range, although both tend to be expensive if used in large quantities.

## 8.4 COMPLETE MEDIA

The term *complete medium* implies a medium that has had all its constituents and supplements added and is sufficient for the use specified. It is usually made up of a defined medium component, whose unstable constituents, such as glutamine and various supplements, such as serum, growth factors, or hormones, may be added just before use. Defined media range in complexity from the relatively simple Eagle's MEM [Eagle, 1959], which contains essential amino acids, vitamins, and salts, to complex media such as medium 199 (M199) [Morgan et al., 1950], CMRL 1066 [Parker et al., 1957], MB 752/1 [Waymouth, 1959], RPMI 1640 [Moore et al., 1967], and F12 [Ham, 1965] (Table 8.2) and a wide range of serum-free formulations (see Tables 9.1, 9.2). The complex media contain a larger number of different amino acids, including nonessential amino acids and additional vitamins, and are often supplemented with extra metabolites (e.g., nucleosides, tricarboxylic acid cycle intermediates, and lipids) and minerals. Nutrient concentrations are, on the whole, low in F12 (which was optimized by cloning) and high in Dulbecco's modification of Eagle's MEM (DMEM) [Dulbecco & Freeman, 1959; Morton, 1970], optimized at higher cell densities for viral propagation. Barnes and Sato [1980] used a 1:1 mixture of DMEM and F12 as the basis for their serum-free formulations to combine the richness of F12 and the higher nutrient concentration of DMEM. Although not always entirely rational, this combination has provided an empirical formula that is suitable as a basic medium for supplementation with special additives for many different cell types.

### 8.4.1 Amino Acids

The essential amino acids (i.e., those that are not synthesized in the body) are required by cultured cells, plus cysteine and/or cysteine, arginine, glutamine, and tyrosine, although individual requirements for amino acids will vary from one cell type to another. Other nonessential amino acids are often added as well, to compensate either for a particular cell type's incapacity to make them or because they are made but lost by leakage into the medium. The concentration of amino acids usually limits the maximum cell concentration attainable, and the balance may influence cell survival and growth rate. Glutamine is required by most cells, although some cell lines will utilize glutamate; evidence suggests that glutamine is also used by cultured cells as a source of energy and carbon [Butler & Christie, 1994; see also Section 3.6]. Glutamine is unstable in culture medium with a half-life of between 3 and 5 days, and it generates ammonia, which can be toxic [Hassell et al., 1991], as a by-product. Glutamax (Invitrogen) is an alanyl-glutamine dipeptide that is stable and bioavailable due to the action of dipeptidase.

### 8.4.2 Vitamins

Eagle's MEM contains only the water-soluble vitamins (the B group, plus choline, folic acid, inositol, and nicotinamide, but excluding biotin; see Table 8.3); other requirements presumably are derived from the serum. Biotin is present in most of the more complex media, including the serum-free recipes, and p-aminobenzoic acid (PABA) is present in

M199, CMRL 1066 (which was derived from M199), and RPMI 1640. All the fat-soluble vitamins (A, D, E, and K) are present only in M199, whereas vitamin A is present in LHC-9 and vitamin E in MCDB 110 (see Table 9.1). Some vitamins (e.g., choline and nicotinamide) have increased concentrations in serum-free media. Vitamin limitation—for example, by precipitation of folate from concentrated stock solutions—is usually expressed in terms of reduced cell survival and growth rates rather than maximum cell density. Like those of the amino acids, vitamin requirements have been derived empirically and often relate to the cell line originally used in their development; for example, Fischer's medium has a high folate concentration because of the folate dependence of L5178Y, which was used in the development of the medium [Fischer & Sartorelli, 1964].

### 8.4.3 Salts

The salts are chiefly those of $Na^+$, $K^+$, $Mg^{2+}$, $Ca^{2+}$, $Cl^-$, $SO_4^{2-}$, $PO_4^{3-}$, and $HCO_3^-$ and are the major components contributing to the osmolality of the medium. Most media derived their salt concentrations originally from Earle's (high bicarbonate; gas phase, 5% $CO_2$) or Hanks's (low bicarbonate; gas phase, air) BSS. Divalent cations, particularly $Ca^{2+}$, are required by some cell adhesion molecules, such as the cadherins [Yamada & Geiger, 1997]. $Ca^{2+}$ also acts as an intermediary in signal transduction [Alberts et al., 2008], and the concentration of $Ca^{2+}$ in the medium can influence whether cells will proliferate or differentiate (see Sections 16.7.2, 22.2.1). $Na^+$, $K^+$, and $Cl^-$ regulate membrane potential, whereas $SO_4^{2-}$, $PO_4^{3-}$, and $HCO_3^-$ have roles as anions required by the matrix and nutritional precursors for macromolecules, as well as regulators of intracellular charge.

Calcium is reduced in suspension cultures in order to minimize cell aggregation and attachment (see Table 8.2). The sodium bicarbonate concentration is determined by the concentration of $CO_2$ in the gas phase (see Section 8.2.2) and has a significant nutritional role in addition to its buffering capability.

### 8.4.4 Glucose

Glucose is included in most media as a source of energy. It is metabolized principally by glycolysis to form pyruvate, which may be converted to lactate or acetoacetate and may enter the citric acid cycle and be oxidized to form $CO_2$ and water. The accumulation of lactic acid in the medium, particularly evident in embryonic and transformed cells, implies that the citric acid cycle may not function entirely as it does in vivo.

### 8.4.5 Organic Supplements

A variety of other compounds, including proteins, peptides, nucleosides, citric acid cycle intermediates, pyruvate, and lipids, appear in complex media. Again, these constituents have been found to be necessary when the serum concentration is reduced, and they may help in cloning and in maintaining certain specialized cells, even in the presence of serum.

## 8.4.6  Hormones and Growth Factors

Hormones and growth factors are not specified in the formulas of most regular media, although they are frequently added to serum-free media (*see* Sections 8.5.2, 8.5.3, 9.4.4, 9.4.5).

## 8.4.7  Antibiotics

Antibiotics were originally introduced into culture media to reduce the frequency of contamination. However, the use of laminar-flow hoods, coupled with strict aseptic technique, makes antibiotics unnecessary. Indeed, antibiotics have a number of significant disadvantages:

(1)  They encourage the development of antibiotic-resistant organisms.
(2)  They hide the presence of low-level, cryptic contaminants that can become fully operative if the antibiotics are removed, the culture conditions change, or resistant strains develop.
(3)  They may hide mycoplasma infections.
(4)  They have antimetabolic effects that can cross-react with mammalian cells.
(5)  They encourage poor aseptic technique.

Hence it is strongly recommended that routine culture be performed in the absence of antibiotics and that their use be restricted to primary culture or high cost large-scale labor-intensive experiments. If conditions demand the use of antibiotics, then they should be removed as soon as possible, or, if they are used over the long term, parallel cultures should be maintained free of antibiotics (*see* Section 12.4.8) as a quality-control measure.

A number of antibiotics used in tissue culture are moderately effective in controlling bacterial infections (Table 8.4). However, a significant number of bacterial strains are resistant to antibiotics, either naturally or by selection, so the control that they provide is never absolute. Fungal and yeast contaminations are particularly hard to control with antibiotics; they may be held in check, but are seldom eliminated (*see* Section 18.5.1).

## 8.5  SERUM

Serum contains growth factors, which promote cell proliferation, and adhesion factors and antitrypsin activity, which promote cell attachment. Serum is also a source of minerals, lipids, and hormones, many of which may be bound to protein (Table 8.5). The sera used most in tissue culture are bovine calf, fetal bovine, adult horse, and human serum. Calf (CS) and fetal bovine (FBS) serum are the most widely used, the latter particularly for more demanding cell lines and for cloning. Human serum is sometimes used in conjunction with some human cell lines, but it needs to be screened for viruses, such as HIV and hepatitis B. Horse serum is preferred to calf serum by some workers, as it can be obtained from a closed donor herd and is often more consistent from batch to batch. Horse serum may also be less likely to metabolize polyamines, due to lower levels of polyamine oxidase; polyamines are mitogenic for some cells [Hyvonen et al., 1988; Kaminska et al., 1990].

## 8.5.1  Protein

Although proteins are a major component of serum, the functions of many proteins in vitro remain obscure; it may be that relatively few proteins are required other than as carriers for minerals, fatty acids, and hormones. Those proteins for which requirements have been found are albumin

**TABLE 8.4.** Antibiotics Used in Tissue Culture

| Antibiotic | Concentration, µg/mL (unless otherwise stated) | | Activity against |
| --- | --- | --- | --- |
| | Working | Cytotoxic | |
| Amphotericin B (Fungizone) | 2.5 | 30 | Fungi, yeasts |
| Ampicillin | 2.5 | | Bacteria, gram positive and gram negative |
| Ciprofloxacin | 100 | | Mycoplasma |
| Erythromycin | 50 | 300 | Bacteria, gram positive and gram negative; mycoplasma |
| Gentamicin | 50 | >300 | Bacteria, gram positive and gram negative; mycoplasma |
| Kanamycin | 100 | 10 mg/mL | Bacteria, gram positive and gram negative; mycoplasma |
| MRA (MP Biomedicals) | 0.5 | | Mycoplasma |
| Neomycin | 50 | 3000 | Bacteria, gram positive and gram negative |
| Nystatin | 50 | | Fungi, yeasts |
| Penicillin-G | 100 U/mL | 10,000 U/mL | Bacteria, gram positive |
| Polymixin B | 50 | 1 mg/mL | Bacteria, gram negative |
| Streptomycin SO$_4$ | 100 | 20 mg/mL | Bacteria, gram positive and gram negative |
| Tetracyclin | 10 | 35 | Bacteria, gram positive and gram negative |
| Tylosin | 10 | 300 | Mycoplasma |

Source: After Paul [1975] and Sigma-Aldrich, www.sigmaaldrich.com/life-science/core-bioreagents/learning-center/antibiotic-selection.html (*See also* Table 18.3.).

**TABLE 8.5.** Constituents of Serum

| Constituent | Range of concentration[a] |
|---|---|
| **Proteins and polypeptides** | 40–80 mg/mL |
| Albumin | 20–50 mg/mL |
| Fetuin[b] | 10–20 mg/mL |
| Fibronectin | 1–10 µg/mL |
| Globulins | 1–15 mg/mL |
| Protease inhibitors: $\alpha_1$-antitrypsin, $\alpha_2$-macroglobulin | 0.5–2.5 mg/mL |
| Transferrin | 2–4 mg/mL |
| **Growth factors** | |
| EGF, PDGF, IGF-I and -II, FGF, IL-1, IL-6 | 1–100 ng/mL |
| **Amino acids** | 0.01–1.0 µM |
| **Lipids** | 2–10 mg/mL |
| Cholesterol | 10 µM |
| Fatty acids | 0.1–1.0 µM |
| Linoleic acid | 0.01–0.1 µM |
| Phospholipids | 0.7–3.0 mg/mL |
| **Carbohydrates** | 1.0–2.0 mg/mL |
| Glucose | 0.6–1.2 mg/mL |
| Hexosamine[c] | 6–1.2 mg/mL |
| Lactic acid[d] | 0.5–2.0 mg/mL |
| Pyruvic acid | 2–10 µg/mL |
| **Polyamines** | |
| Putrescine, spermidine | 0.1–1.0 µM |
| **Urea** | 170–300 µg/mL |
| **Inorganics** | 0.14–0.16 M |
| Calcium | 4–7 mM |
| Chlorides | 100 µM |
| Iron | 10–50 µM |
| Potassium | 5–15 mM |
| Phosphate | 2–5 mM |
| Selenium | 0.01 µM |
| Sodium | 135–155 mM |
| Zinc | 0.1–1.0 µM |
| **Hormones** | 0.1–200 nM |
| Hydrocortisone | 10–200 nM |
| Insulin | 1–100 ng/mL |
| Triiodothyronine | 20 nM |
| Thyroxine | 100 nM |
| **Vitamins** | 10 ng–10 µg/mL |
| Vitamin A | 10–100 ng/mL |
| Folate | 5–20 ng/mL |

[a]The range of concentrations is very approximate and is intended to convey only the order of magnitude. Data are from Evans and Sanford [1978] and Cartwright and Shah [1994].
[b]In fetal serum only.
[c]Highest in human serum.
[d]Highest in fetal serum.

[Iscove & Melchers, 1978; Barnes & Sato, 1980], which may be important as a carrier of lipids, minerals, and globulins [Tozer & Pirt, 1964]; fibronectin (cold-insoluble globulin), which promotes cell attachment [Yamada & Geiger, 1997; Hynes, 1992], although probably not as effectively as cell-derived fibronectin; and $\alpha_2$-macroglobulin, which inhibits trypsin [de Vonne & Mouray, 1978]. Fetuin in fetal serum enhances cell attachment [Fisher et al., 1958], and transferrin [Guilbert & Iscove, 1976] binds iron, making it less toxic and bioavailable. Other proteins, as yet uncharacterized, may be essential for cell attachment and growth.

Protein also increases the viscosity of the medium, reducing shear stress during pipetting and stirring, and may add to the medium's buffering capacity.

## 8.5.2 Growth Factors

Natural clot serum stimulates cell proliferation more than serum from which the cells have been removed physically (e.g., by centrifugation). This increased stimulation appears to be due to the release of growth factors, particularly platelet-derived growth factor (PDGF), from the platelets during clotting. PDGF [Antoniades et al., 1979; Heldin et al., 1979] is one of a family of polypeptides with mitogenic activity and is probably the major growth factor in serum. PDGF stimulates growth in fibroblasts and glia, but other platelet-derived factors, such as TGF-β, may inhibit growth or promote differentiation in epithelial cells [Lechner et al., 1981].

Other growth factors to consider (*see* Table 9.4) are fibroblast growth factors (FGFs) [Gospodarowicz, 1974], epidermal growth factor (EGF) [Cohen, 1962; Carpenter & Cohen, 1977; Gospodarowicz et al., 1978a], endothelial cell growth factors such as vascular endothelial growth factor (VEGF) and angiogenin [Hu et al., 1997; Folkman & d'Amore, 1996; Joukov et al., 1997; Folkman et al., 1979; Maciag et al., 1979], and insulin-like growth factors IGF-I and IGF-II [le Roith & Raizada, 1989], which have been isolated from whole tissue or released into the medium by cells in culture. These growth factors have varying degrees of specificity [Hollenberg & Cuatrecasas, 1973] and are probably present in serum in small amounts [Gospodarowicz & Moran, 1974]. Many of these are available commercially (*see* Appendix II) as recombinant proteins, some of which also are available in long-form analogues (Sigma) with increased mitogenic activity and stability.

## 8.5.3 Hormones

Insulin promotes the uptake of glucose and amino acids [Kelley et al., 1978; Stryer, 1995] and may owe its mitogenic effect to this property or to activity via the IGF-1 receptor. IGF-I and IGF-II bind to the insulin receptor but also have their own specific receptors, to which insulin may bind with lower affinity. IGF-II also stimulates glucose uptake [Sinha et al., 1990]. Growth hormone may be present in serum—particularly fetal serum—and, in conjunction with the somatomedins (IGFs), may have a mitogenic effect.

Hydrocortisone is also present in serum—particularly fetal bovine serum—in varying amounts, and it can promote cell attachment [Ballard & Tomkins, 1969; Fredin et al., 1979] and cell proliferation [Guner et al., 1977; McLean et al., 1986] (*see also* Sections 23.1.1, 23.1.3, 23.1.4), but under certain conditions (e.g., at high cell density) may be cytostatic [Freshney et al., 1980a, b] and can induce cell differentiation [Moscona & Piddington, 1966; Ballard, 1979; McLean et al., 1986; Speirs et al., 1991; McCormick et al., 1995, 2000].

## 8.5.4 Nutrients and Metabolites

Serum may also contain amino acids, glucose, oxo (keto) acids, nucleosides, and a number of other nutrients and intermediary metabolites. These may be important in simple media but less so in complex media, particularly those with higher amino acid concentrations and other defined supplements.

## 8.5.5 Lipids

Linoleic acid, oleic acid, ethanolamine, and phospho-ethanolamine are present in serum in small amounts, usually bound to proteins such as albumin.

## 8.5.6 Minerals

Serum replacement experiments [Ham & McKeehan, 1978] have also suggested that trace elements and iron, copper, and zinc may be bound to serum protein, probably albumin. McKeehan et al. [1976] demonstrated a requirement for selenium, which probably helps to detoxify free radicals as a cofactor for GSH synthetase and is to be found in most serum-free formulations (*see* Section 9.4).

## 8.5.7 Inhibitors

Serum may contain substances that inhibit cell proliferation [Harrington & Godman, 1980; Liu et al., 1992; Varga Weisz & Barnes, 1993]. Some of these may be artifacts of preparation (e.g., bacterial toxins from contamination before filtration, or antibodies that cross-react with surface epitopes on the cultured cells), but others may be physiological negative growth regulators, such as TGF-β [Massague et al., 1992]. Heat inactivation removes complement from the serum and reduces the cytotoxic action of immunoglobulins without damaging polypeptide growth factors, but it may also remove some more labile constituents and is not always as satisfactory as untreated serum.

## 8.6 SELECTION OF MEDIUM AND SERUM

All 12 media described in Table 8.2 were developed to support particular cell lines or conditions. Many were developed with L929 mouse fibroblasts or HeLa cervical carcinoma cells, and Ham's F12 was designed for Chinese hamster ovary (CHO) cells; all now have more general applications and have become classic formulations. Among them, data from suppliers would indicate that RPMI 1640, DMEM, and MEM are the most popular, making up about 75% of sales. Other formulations seldom account for more than 5% of the total; most constitute 2% to 3%, although blended DMEM/F12 comes closer, with over 4%.

Eagle's minimal essential medium (MEM) was developed from Eagle's basal medium (BME) by increasing the range and concentration of the constituents. For many years, Eagle's MEM had the most general use of all media. Dulbecco's modification of BME (DMEM) was developed for mouse fibroblasts for transformation and virus propagation studies. It has twice the amino acid concentrations of MEM, has four times the vitamin concentrations, and uses twice the $HCO_3^-$ and $CO_2$ concentrations to achieve better buffering. αMEM [Stanners et al., 1971] has additional amino acids and vitamins, as well as nucleosides and lipoic acid; it has been used for a wide range of cell types, including hematopoietic cells. Ham's F12 was developed to clone CHO cells in low-serum medium; it is also used widely, particularly for clonogenic assays (see Protocol 21.3) and primary culture, often combined with DMEM (see Protocols 11.7–11.9).

CMRL 1066, M199, and Waymouth's media were all developed to grow L929 cells serum free but have been used alone or in combination with other media, such as DMEM or F12, for a variety of more demanding conditions. RPMI 1640 and Fischer's media were developed for lymphoid cells—Fischer's specifically for L5178Y lymphoma, which has a high folate requirement. RPMI 1640 in particular has quite widespread use, often for attached cells, despite being designed for suspension culture and lacking calcium. L15 medium was developed specifically to provide buffering in the absence of $HCO_3^-$ and $CO_2$. It is often used as a transport and primary culture medium for this reason, but its value was diminished by the introduction of HEPES and the demonstration that $HCO_3^-$ and $CO_2$ are often essential for optimal cell growth, regardless of the requirement for buffering.

Information regarding the selection of the appropriate medium for a given type of cell is usually available in the literature in articles on the origin of the cell line or the culture of similar cells. Information may also be obtained from the source of the cells. Cell banks, such as ATCC and ECACC, provide information on media used for currently available cell lines, and data sheets can be accessed from their websites (see Table 8.6; see also Table 19.5; Appendix III). Failing this, the choice is made either empirically or by comparative testing of several media, as for selection of serum (see Section 8.6.2).

Many continuous cell lines (e.g., HeLa, L929, and BHK21), primary cultures of human, rodent, and avian fibroblasts, and cell lines derived from them can be maintained on a relatively simple medium such as Eagle's MEM, supplemented with calf serum. More complex media may be required when a specialized function is being expressed (see Section 16.7) or when cells are subcultured at a low seeding density ($<1 \times 10^3$/ mL), as in cloning (see Section 13.2).

Frequently the more demanding culture conditions that require complex media also require fetal bovine serum rather than calf or horse serum, unless the formulation specifically allows for the omission of serum.

Some suggestions for the choice of medium and several examples of cell types and the media used for them are given in Table 8.6 [see also Mather, 1998]. If information is not available, a simple cell growth experiment with commercially available media and multiwell plates (see Protocols 20.8) can be carried out in about two weeks. Assaying for clonal growth (see Protocol 20.10) and measuring the expression of specialized functions may narrow the choice further. You may be surprised to find that your best conditions do not agree with those mentioned in the literature; reproducing the conditions found in another laboratory may be difficult because of variations in preparation, the impurities present in reagents and water, and differences between batches of serum. It is to be hoped that as serum requirements are reduced and the purity of reagents increases, the standardization of media will improve.

Finally, you may have to compromise in your choice of medium or serum because of cost. Autoclavable media are available from commercial suppliers (see Appendix II). They are simple to prepare from powder and are suitable for many continuous cell strains. They may need to be supplemented with glutamine for most cells and usually require serum. The cost of serum should be calculated on the basis of the volume of the medium when cell yield is not important, but if the objective is to produce large quantities of cells, one should calculate serum costs on a per cell basis. Thus, if a culture grows to $1 \times 10^6$/mL in serum A and $2 \times 10^6$ mL in serum B, serum B becomes the less expensive by a factor of two, given that product formation or some other specialized function is the same.

If fetal bovine serum seems essential, try mixing it with calf serum. This may allow you to reduce the concentration of the more expensive fetal serum. If you can, leave out serum altogether, or reduce the concentration, and use a serum-free formulation (see Section 9.4).

### 8.6.1 Batch Reservation

Considerable variation may be anticipated between batches of serum. Such variation results from differing methods of preparation and sterilization, different ages and storage conditions, and variations in animal stocks from which the serum was derived, including different strains and disparities in pasture, climate, and other environmental conditions. It is important to select a batch, use it for as long as possible, and replace it, eventually, with one as similar to it as possible.

Serum standardization is difficult, as batches vary considerably, and one batch will last only about six months to a year, stored at $-20°C$. Select the type of serum that is most appropriate for your purposes, and request batches to test from a number of suppliers. Most serum suppliers will reserve a batch until a customer can select the most suitable

**TABLE 8.6.** Selecting a Suitable Conventional Medium

| Cells or cell line | Medium | Serum |
|---|---|---|
| 3T3 cells | MEM, DMEM | CS |
| Chick embryo fibroblasts | Eagle's MEM | CS |
| Chinese hamster ovary (CHO) | Eagle's MEM, Ham's F12 | CS |
| Chondrocytes | Ham's F12 | FB |
| Continuous cell lines | Eagle's MEM, DMEM | CS |
| Endothelium | DMEM, M199, MEM | CS |
| Fibroblasts | Eagle's MEM | CS |
| Glial cells | MEM, DMEM/F12 | FB |
| Glioma | MEM, DMEM/F12 | FB |
| HeLa cells | Eagle's MEM | CS |
| Hematopoietic cells | RPMI 1640, αMEM | FB |
| Human diploid fibroblasts | Eagle's MEM | CS |
| Human leukemia | RPMI 1640 | FB |
| Human tumors | DMEM/F12, L15, RPMI 1640 | FB |
| Keratinocytes | αMEM | FB |
| L cells (L929, LS) | Eagle's MEM | CS |
| Lymphoblastoid cell lines (human) | RPMI 1640 | FB |
| Mammary epithelium | DMEM/F12 , RPMI 1640 | FB |
| MDCK dog kidney epithelium | DMEM, DMEM/F12 | FB |
| Melanocytes | M199 | FB |
| Melanoma | DMEM, MEM, DMEM/F12 | FB |
| Mouse embryo fibroblasts | Eagle's MEM | CS |
| Mouse leukemia | Fisher's, RPMI 1640 | FB, HoS |
| Mouse erythroleukemia | DMEM/F12, RPMI 1640 | FB, HoS |
| Mouse myeloma | DMEM, RPMI 1640 | FB |
| Mouse neuroblastoma | DMEM, DMEM/F12 | FB |
| Neurons | DMEM | FB |
| NRK rat kidney fibroblasts | MEM, DMEM | CS |
| Rat minimal-deviation hepatoma (HTC, MDH) | Swim's S77, DMEM/F12 | FB |
| Skeletal muscle | DMEM, F12 | FB, HoS |
| Syrian hamster fibroblasts (e.g., BHK 21) | MEM, GMEM, DMEM | CS |

Note: *See also* Table 9.5.
*Abbreviations:* CS, calf serum; FB, fetal bovine serum; HoS, horse serum. SF12 is Ham's F12 plus Eagle's essential amino acids and nonessential amino acids as in DMEM, available as 100× stock (*see* Appendix II, etc.). Further recommendations on the choice of medium can be found in Barnes et al. [1984(a–d)] and Mather [1998].

one (provided that this does not take longer than three weeks or so). When a suitable batch has been selected, the supplier is requested to hold the appropriate volume for up to one year, to be dispatched on demand. The other suppliers should also be informed so that they may return the rejected batches to their stocks.

## 8.6.2  Testing Serum
The quality of a given serum is assured by the supplier, but the firm's quality control is usually performed with one of a number of continuous cell lines. If your requirements are more demanding, then you will need to do your own testing. There are four main parameters for testing serum.

*Plating efficiency.* During cloning the cells are at a low density and hence are at their most sensitive, making this a very stringent test. Plate the cells out at 10 to 100 cells/mL, and look for colonies after 10 days to two

weeks. Stain and count the colonies (*see* Protocol 20.10), and look for differences in plating efficiency (survival) and colony size (cell proliferation). Each serum should be tested at a range of concentrations from 2% to 20%. This approach will reveal whether one serum is equally effective at a lower concentration, thereby saving money and prolonging the life of the batch, and will show up any toxicity at a high serum concentration.

*Growth curve.* A growth curve should be plotted for cell growth in each serum (*see* Protocols 20.7–20.9) so that the lag period, doubling time, and saturation density (cell density at "plateau") can be determined. A long lag implies that the culture has to adapt to the serum. Short doubling times are preferable if you want a lot of cells quickly, and a high saturation density will provide more cells for a given amount of serum and will be more economical. Performance of growth curves is labor intensive but can be simplified by using multiwell plates and a chromogenic endpoint such as the MTT assay (*see* Section 21.3.5). Alternatively, continuous in situ monitoring by image analysis (Incucyte, Chip-Man; *see* Section 24.4.3) allows rapid analysis of growth curves without the need to sample and count.

***Preservation of cell culture characteristics.*** Clearly, the cells must do what you require of them in the new serum, whether they are acting as host to a given virus, secreting a specific cell product, differentiating, or expressing a characteristic sensitivity to a given drug. Hence batch comparison may also need laboratory specific functional assays.

***Sterility.*** Serum from a reputable supplier will have been tested and shown to be free of microorganisms. However, in the unlikely event that a sample of serum is contaminated but has escaped quality control, the fact that it is contaminated should show up in mycoplasma screening (*see* Sections 18.3.2–18.3.6). So the selected batch should be included in routine screening before it is finally accepted.

### 8.6.3 Heat Inactivation

Serum is heat inactivated by incubation for 30 min at 56°C. It may then be dispensed into aliquots and stored at −20°C. Originally heating was used to inactivate complement for immunoassays, but it may achieve other effects not yet documented. Often heat-inactivated serum is used because of the adoption of a previous protocol, without any concrete evidence that it is beneficial. Claims that heat inactivation removes mycoplasma are probably unfounded, although heat treatment may reduce the titer for some mycoplasma.

## 8.7 OTHER SUPPLEMENTS

In addition to serum, tissue extracts and digests have traditionally been used as supplements to tissue culture media (*see also* Section 9.4.2).

### 8.7.1 Amino Acid Hydrolysates

Many such supplements are derived from microbiological media. Bactopeptone, tryptose, and lactalbumin hydrolysate (BD Biosciences) are proteolytic digests of beef heart or lactalbumin and contain mainly amino acids and small peptides. Bactopeptone and tryptose may also contain nucleosides and other heat-stable tissue constituents, such as fatty acids and carbohydrates. Sterility is easily achieved as they are autoclavable (*see* Section 10.5.1).

### 8.7.2 Embryo Extract

Embryo extract is a crude homogenate of 10-day-old chick embryo that is clarified by centrifugation (*see* Appendix I). The crude extract was fractionated by Coon and Cahn [1966] to give fractions of either high or low molecular weight. The low-molecular-weight fraction promoted cell proliferation, whereas the high-molecular-weight fraction promoted pigment and cartilage cell differentiation. Although Coon and Cahn did not fully characterize these fractions, more recent evidence suggests that the low-molecular-weight fraction may contain peptide growth factors and the high-molecular-weight fraction proteoglycans and other matrix constituents.

Embryo extract was originally used as a component of plasma clots (*see* Sections 2.7.1, 11.3.1) to promote cell migration from the explant and has been retained in some organ culture techniques (*see* Section 25.2). It should always be frozen and thawed at least twice to ensure that there is no carryover of live cells from the embryo.

### 8.7.3 Conditioned Medium

Puck and Marcus [1955] found that the survival of low-density cultures could be improved by growing the cells in the presence of feeder layers (*see* Section 13.2.3). In that instance the effect was due to soluble factors as the clones and feeder layer were kept separate. However, in current usage where clones and feeder are usually not separated, the effect is probably due to a combination of conditioning of the substrate by extracellular matrix constituents and conditioning of the medium by metabolites and growth factors [Takahashi & Okada, 1970]. Hauschka & Konigsberg [1966] showed that the conditioning of culture medium that was necessary for the growth and differentiation of myoblasts was due to collagen released by the feeder cells. Using feeder layers and conditioning the medium with embryonic fibroblasts or other cell lines remains a valuable method of culturing difficult cells [e.g., *see* Stampfer et al., 1980]. However, conditioning medium adds undefined components and should be eliminated if possible after the active constituents are determined.

# CHAPTER 9

# Serum-Free Media

Although many cell lines are still propagated in medium supplemented with serum, in many instances cultures may now be propagated in serum-free media (Tables 9.1, 9.2). Historically the need to standardize media among laboratories, provide specialized media for specific cell type, and eliminate variable natural products led to the development of more complex media, such as M199 of Morgan et al. [1950], CMRL 1066 of Parker et al. [1957], NCTC109 [Evans et al., 1956], Waymouth's MB 572/1 [1959], NCTC135 [Evans & Bryant, 1965], and Birch and Pirt [1971] for L929 mouse fibroblast cells, and Ham's F10 [Ham, 1963] and F12 [Ham, 1965] clonal growth media for Chinese hamster ovary (CHO) cells. Serum-free media were also developed for HeLa human cervical carcinoma cells [Blaker et al., 1971; Higuchi, 1977].

Although a degree of cell selection was probably involved in the adaptation of continuous cell lines to serum-free conditions, the MCDB series of media [Ham & McKeehan, 1978] (see also Table 9.1), Sato's DMEM/F12-based media [Barnes & Sato, 1980], and others based on RPMI 1640 [Carney et al., 1981; Brower et al., 1986] (see also Table 9.2), demonstrated that serum could be reduced or omitted without apparent cell selection if appropriate nutritional and hormonal modifications were made to the media [Barnes et al., 1984a–d; Cartwright & Shah, 1994; Mather, 1998]. These also provided selective conditions for primary culture of particular cell types. Specific formulations, such as MCDB 110 [Bettger et al., 1981], were derived to culture human fibroblasts [Ham, 1984], many normal and neoplastic murine and human cells [Barnes & Sato, 1980], lymphoblasts [Iscove & Melchers, 1978], and several different primary cultures [Mather & Sato, 1979a,b; Sundqvist et al., 1991; Gupta et al., 1997; Keen & Rapson, 1995; Vonen et al., 1992] in

the absence of serum, with, in several cases, some protein added [Tsao et al., 1982; Benders et al., 1991]. This list now covers a wide range of cell types, and many of the media are available commercially (see Appendix II). In addition the need to remove animal proteins from the in vitro production of biopharmaceuticals and concern for some of the safety issues invloved [Merten, 1999] has generated a number of formulations for continuous cell lines such as CHO and hybridomas [Froud, 1999; Ikonomou et al., 2003; Shah, 1999].

## 9.1 DISADVANTAGES OF SERUM

Using serum in a medium has a number of disadvantages:

(1) *Physiological variability.* The major constituents of serum, such as albumin and transferrin, are known, but serum also contains a wide range of minor components that can have a considerable effect on cell growth (see Table 8.5) and response to test substances (see Fig. 21.4). These components include nutrients (amino acids, nucleosides, sugars, etc.), proteins, peptide growth factors, hormones, minerals, and lipids, whose concentrations and actions have not been fully determined.

(2) *Shelf life and consistency.* Serum varies from batch to batch, and at best a batch will last one year, perhaps deteriorating during that time. It must then be replaced with another batch that may be selected as similar, but will never be identical, to the first batch.

(3) *Quality control.* Changing serum batches requires extensive testing to ensure that the replacement is as close as possible to the previous batch. This can involve

**TABLE 9.1.** Examples of Serum-Free Media Formulations

| Medium | Mol. wt. | MCDB 110 | MCDB 131 | MCDB 170 | MCDB 202 | MCDB 302 | MCDB 402 | WAJC 404 | MCDB 153 | Iscove's | LHC-9 |
|---|---|---|---|---|---|---|---|---|---|---|---|
| Cell type | | Human lung fibroblasts | Human vascular endothelium | Mammary epithelium | Chick embryo fibroblasts | CHO cells | 3T3 cells | Prostatic epithelium[1] | Keratinocytes | Lymphoid cells | Bronchial epithelium |
| Reference | | [Bettger et al., 1981] | [Knedler & Ham, 1987; Gupta et al., 1997] | [Hammond et al., 1984] | [McKeehan & Ham, 1976b] | [Hamilton & Ham, 1977] | [Shipley & Ham, 1983] | [McKeehan et al., 1984; Chaproniere & McKeehan, 1986] | [Peehl & Ham, 1980] | [Iscove & Melchers, 1978] | [Lechner & LaVeck, 1985] |
| Component | Mol. wt. | | | | | | | | | | |
| **Amino acids** | | | | | | | | | | | |
| L-Alanine | 89 | 1.0E-04 | 3.0E-05 | 1.0E-04 | 1.0E-04 | | 1.0E-04 | 1.0E-04 | 2.8E-04 | 1.0E-04 | |
| L-Arginine | 211 | 1.0E-03 | 3.0E-04 | 3.0E-04 | 3.0E-04 | | 3.0E-04 | 1.0E-03 | 1.0E-03 | 4.0E-04 | 2.0E-03 |
| L-Asparagine | 132 | 1.0E-04 | 1.0E-04 | 1.0E-03 | 1.0E-04 | 1.0E-03 | 1.0E-04 | 1.0E-04 | 1.0E-04 | 1.9E-04 | 1.0E-04 |
| L-Aspartic acid | 133 | 1.0E-04 | 1.0E-04 | 1.0E-04 | 1.0E-04 | 1.1E-04 | 1.0E-05 | 3.0E-05 | 3.0E-05 | 2.3E-04 | 3.0E-05 |
| L-Cysteine | 176 | 5.0E-05 | 2.0E-04 | 7.0E-05 | 2.0E-04 | 1.0E-04 | | | 2.4E-04 | | 2.4E-04 |
| L-Cystine | 240 | | | 2.0E-04 | 2.0E-04 | 1.0E-04 | 4.0E-04 | 2.4E-04 | | 2.9E-04 | |
| L-Glutamic acid | 147 | 1.0E-04 | 3.0E-05 | 1.0E-04 | 1.0E-04 | 1.0E-04 | 1.0E-05 | 1.0E-04 | 1.0E-04 | 5.1E-04 | 1.0E-04 |
| L-Glutamine | 146 | 2.5E-03 | 1.0E-02 | 2.0E-03 | 1.0E-03 | 3.0E-03 | 5.0E-03 | 6.0E-03 | 6.0E-03 | 4.0E-03 | 6.0E-03 |
| Glycine | 75 | 3.0E-04 | 3.0E-05 | 1.0E-04 | 1.0E-04 | 1.0E-04 | 1.0E-04 | 1.0E-04 | 1.0E-04 | 4.0E-04 | 1.0E-04 |
| L-Histidine | 210 | 1.0E-04 | 2.0E-04 | 1.0E-04 | 1.0E-04 | 1.0E-04 | 2.0E-03 | 8.0E-05 | 8.0E-05 | 2.0E-04 | 1.6E-04 |
| L-Isoleucine | 131 | 3.0E-05 | 5.0E-04 | 1.0E-04 | 1.0E-04 | 3.0E-05 | 1.0E-03 | 1.5E-05 | 1.5E-05 | 8.0E-04 | 3.0E-05 |
| L-Leucine | 131 | 1.0E-04 | 1.0E-03 | 3.0E-04 | 3.0E-04 | 1.0E-04 | 2.0E-03 | 5.0E-04 | 5.0E-04 | 8.0E-04 | 1.0E-03 |
| L-Lysine HCl | 183 | 2.0E-04 | 1.0E-04 | 2.0E-04 | 2.0E-04 | 2.0E-04 | 8.0E-04 | 1.0E-04 | 1.0E-04 | 8.0E-04 | 2.0E-04 |
| L-Methionine | 149 | 3.0E-05 | 1.0E-04 | 3.0E-05 | 3.0E-05 | 3.0E-05 | 2.0E-04 | 3.0E-05 | 3.0E-05 | 2.0E-04 | 6.0E-05 |
| L-Phenylalanine | 165 | 3.0E-05 | 3.0E-05 | 3.0E-05 | 3.0E-05 | 3.0E-05 | 3.0E-04 | 3.0E-05 | 3.0E-05 | 4.0E-04 | 6.0E-05 |
| L-Proline | 115 | 3.0E-04 | 2.0E-04 | 5.0E-05 | 5.0E-05 | 3.0E-04 | | 3.0E-04 | 3.0E-04 | 3.5E-04 | 3.0E-04 |
| L-Serine | 105 | 1.0E-04 | 1.0E-04 | 3.0E-04 | 3.0E-04 | 1.0E-04 | 1.0E-04 | 6.0E-04 | 6.0E-04 | 4.0E-04 | 1.2E-03 |
| L-Threonine | 119 | 1.0E-04 | 3.0E-04 | 3.0E-04 | 3.0E-04 | 1.0E-04 | 5.0E-04 | 1.0E-04 | 1.0E-04 | 8.0E-04 | 2.0E-04 |
| L-Tryptophan | 204 | 1.0E-05 | 1.0E-04 | 3.0E-05 | 3.0E-05 | 1.0E-05 | 1.0E-05 | 1.5E-05 | 1.5E-05 | 7.8E-05 | 3.0E-05 |
| L-Tyrosine | 181 | 3.5E-05 | 2.0E-05 | 5.0E-05 | 5.0E-05 | 4.4E-05 | 2.0E-04 | 1.5E-05 | 1.5E-05 | 4.6E-04 | 3.0E-05 |
| L-Valine | 117 | 1.0E-04 | 1.0E-03 | 3.0E-04 | 3.0E-04 | 1.0E-04 | 2.0E-03 | 3.0E-04 | 3.0E-04 | 8.0E-04 | 6.0E-04 |
| **Vitamins** | | | | | | | | | | | |
| Biotin | 244 | 3.0E-08 | 3.0E-08 | 3.0E-08 | 3.0E-08 | 3.0E-08 | 3.0E-08 | 6.0E-08 | 6.0E-08 | 5.3E-08 | 6.0E-08 |
| Choline chloride | 140 | 1.0E-04 | | 1.0E-04 | 1.0E-04 | 1.0E-04 | 1.0E-04 | 1.0E-04 | 1.0E-04 | 2.0E-05 | 2.0E-04 |

| | MW | | | | | | | | | |
|---|---|---|---|---|---|---|---|---|---|---|
| Folic acid | 441 | 1.8E-06 | 9.1E-06 | 1.8E-06 | 1.8E-06 | 3.0E-06 | 1.0E-08 | — | 1.0E-08 | 1.0E-09 |
| Folinic acid | 512 | 1.0E-04 | 4.0E-05 | 1.0E-04 | 1.0E-04 | 1.0E-04 | 1.0E-04 | — | 1.0E-06 | 1.0E-04 |
| myo-Inositol | 180 | 3.0E-07 | 3.3E-05 | 3.0E-07 | 3.0E-07 | 3.0E-07 | 5.0E-05 | — | 4.0E-05 | 5.0E-05 |
| Nicotinamide | 122 | 1.0E-06 | 1.7E-05 | 1.0E-06 | 1.0E-06 | 1.0E-06 | 5.0E-05 | — | 5.0E-05 | 5.0E-05 |
| Pantothenate | 238 | — | 2.0E-05 | — | — | — | 5.0E-05 | — | 5.0E-05 | 5.0E-05 |
| Pyridoxal HCl | 204 | — | — | — | — | — | — | — | — | — |
| Pyridoxine HCl | 206 | 3.0E-07 | — | 3.0E-07 | 3.0E-07 | 3.0E-07 | 1.0E-04 | — | 3.0E-07 | 3.0E-07 |
| Riboflavin | 376 | 1.0E-07 | 1.1E-06 | 1.0E-07 | 1.0E-07 | 1.0E-07 | 1.0E-06 | — | 3.0E-07 | 3.0E-07 |
| Thiamin HCl | 337 | 1.0E-06 | 1.2E-05 | 1.0E-06 | 1.0E-06 | 1.0E-06 | 1.0E-04 | — | 1.0E-06 | 1.0E-06 |
| α-Tocopherol | 430 | — | — | — | — | — | — | — | — | 1.4E-07 |
| Retinoic acid | 300 | 3.3E-07 | — | — | — | — | — | — | — | — |
| Retinol acetate | 329 | — | — | — | — | — | — | — | — | 4.2E-07 |
| Vitamin B$_{12}$ | 1355 | 3.0E-07 | 9.6E-09 | 3.0E-07 | 3.0E-07 | 1.0E-08 | 1.0E-07 | 1.0E-07 | 1.0E-08 | 1.0E-07 |
| **Antioxidants** | | | | | | | | | | |
| Dithiothreitol | 154 | — | — | — | — | — | — | — | — | 6.5E-06 |
| Glutathione | 307 | — | — | — | — | — | — | — | — | 6.5E-07 |
| **Inorganic salts** | | | | | | | | | | |
| CaCl$_2$ | 147 | 1.1E-04 | 1.5E-03 | 3.0E-05 | 1.3E-04 | 1.6E-03 | 6.0E-04 | 2.0E-03 | 1.6E-03 | 1.0E-03 |
| KCl | 75 | 1.5E-03 | 4.4E-03 | 1.5E-03 | 4.0E-03 | 1.5E-03 | 3.0E-03 | 3.0E-03 | 4.0E-03 | 5.0E-03 |
| KNO$_3$ | 160 | — | 7.5E-07 | — | — | — | 1.6E-08 | — | — | — |
| MgCl$_2$ | 203 | — | — | 6.0E-04 | — | — | 6.0E-04 | — | — | — |
| MgSO$_4$ | 247 | 2.2E-02 | 8.1E-04 | — | 6.0E-04 | 6.0E-04 | 6.1E-10 | 1.5E-03 | 1.0E-02 | 1.0E-03 |
| NaCl | 58 | — | 7.7E-02 | 1.2E-01 | 1.2E-01 | 1.2E-01 | 1.3E-01 | 1.2E-01 | 1.1E-01 | 1.1E-01 |
| NaHCO$_3$ | 84 | — | 3.6E-02 | 1.4E-02 | 1.4E-02 | 1.4E-02 | 1.4E-02 | 1.4E-02 | 1.4E-02 | —² |
| Na$_2$HPO$_4$ | 120 | 2.0E-03 | 1.0E-03 | 2.0E-03 | 2.0E-03 | 5.0E-04 | 1.2E-03 | 5.0E-04 | 5.9E-04 | 3.0E-03 |
| **Trace elements** | | | | | | | | | | |
| CuSO$_4$ · 5H$_2$O | 160 | 1.0E-08 | — | 1.1E-08 | 1.0E-09 | 5.0E-09 | 1.0E-09 | 1.0E-09 | 7.5E-09 | 1.0E-09 |
| FeSO$_4$ | 278 | 5.4E-04 | — | 5.0E-06 | 5.0E-06 | 5.0E-06 | 3.0E-06 | 5.0E-06 | 1.0E-06 | 5.0E-06 |
| MnSO$_4$·H$_2$O | 169 | 1.0E-09 | — | 1.0E-09 | 1.0E-09 | 1.0E-09 | — | 5.0E-10 | 1.2E-09 | 1.0E-09 |
| (NH$_4$)$_6$Mo$_7$O$_{24}$ | 1236 | 1.0E-09 | — | 1.0E-09 | 1.0E-09 | 1.0E-09 | 1.0E-08 | 1.0E-09 | 3.0E-09 | 1.0E-09 |
| NiCl$_2$ | 238 | 5.0E-10 | — | 5.0E-10 | 5.0E-10 | 5.0E-10 | — | 5.0#-12 | 4.2E-10 | 5.0E-10 |
| H$_2$SeO$_3$ | 129 | 3.0E-08 | — | 3.0E-08 | 3.0E-08 | 3.0E-08 | 1.3E-08 | 3.0E-08 | 3.0E-08 | 3.0E-08 |
| Na$_2$SiO$_3$ | 122 | 5.0E-07 | 1.0E-07 | 5.0E-07 | 5.0E-07 | 5.0E-07 | — | 5.0E-07 | — | 5.0E-07 |
| SnCl$_2$ | 190 | 5.0E-10 | — | 5.0E-10 | 5.0E-10 | 5.0E-10 | — | 5.0E-12 | 2.3E-05 | 5.0E-10 |
| NH$_4$VO$_3$ | 117 | 5.0E-09 | — | 5.0E-09 | 5.0E-09 | 5.0E-09 | 1.0E-08 | 5.0E-09 | 5.1E-09 | 5.0E-09 |
| ZnSO$_4$ · 7H$_2$O | 288 | 4.8E-07 | — | 5.0E-07 | 5.0E-07 | 5.0E-07 | 3.0E-06 | 1.0E-07 | 1.0E-09 | 5.0E-07 |
| **Lipids and precursors** | | | | | | | | | | |
| Cholesterol | 387 | — | — | — | — | — | — | — | — | 7.6E-06 |
| Ethanolamine | 61 | — | 1.0E-04 | — | — | 1.0E-04 | — | 1.0E-04 | 1.0E-04 | — |
| Linoleic acid | 280 | — | — | — | — | — | 3.0E-07 | 2.0E-07 | 2.0E-07 | — |

(continued overleaf)

**TABLE 9.1.** (*Continued*)

| Medium | Mol. wt. | MCDB 110 | MCDB 131 | MCDB 170 | MCDB 202 | MCDB 302 | MCDB 402 | WAJC 404 | MCDB 153 | Iscove's | LHC-9 |
|---|---|---|---|---|---|---|---|---|---|---|---|
| Cell type | | Human lung fibroblasts | Human vascular endothelium | Mammary epithelium | Chick embryo fibroblasts | CHO cells | 3T3 cells | Prostatic epithelium[1] | Keratinocytes | Lymphoid cells | Bronchial epithelium |
| Reference | | [Bettger et al., 1981] | [Knedler & Ham, 1987; Gupta et al., 1997] | [Hammond et al., 1984] | [McKeehan & Ham, 1976b] | [Hamilton & Ham, 1977] | [Shipley & Ham, 1983] | [McKeehan et al., 1984; Chaproniere & McKeehan, 1986] | [Peehl & Ham, 1980] | [Iscove & Melchers, 1978] | [Lechner & LaVeck, 1985] |
| Component | Mol. wt. | | | | | | | | | | |
| Lipoic acid | 206 | 1.0E-08 | 1.0E-08 | 1.0E-08 | 1.0E-08 | 9.8E-07 | 1.0E-08 | 1.0E-06 | 1.0E-06 | | 1.0E-06 |
| Phosphoethano-lamine | 141 | | | 1.0E-04 | | | | | | | 5.0E-07 |
| Soya lecithin, µg/ml | | 6 | | | | | | | | | |
| Soybean lipid, µg/ml | | | | | | | | | | 50 | |
| Sphingomyelin, µg/ml | | 1 | | | | | | | | | |
| **Hormones and growth factors** | | | | | | | | | | | |
| EGF, ng/ml | | 30 | | 10 | | | | 25 | 25 | | 5 |
| Epinephrine | 183 | | | | | | | | | | 2.7E-06 |
| Hydrocortisone[3] | 362 | 5.0E-07 | | 1.4E-07 | | | | 5.0E-07 | 1.4E-07 | | 2.0E-07 |
| Insulin, µg/ml | | 1 | | 5 | | | | 10 | 5 | | 5 |
| Prolactin, µg/ml | | | | 5 | | | | | | | |
| PGE1 | 355 | 2.5E-08 | | | | | | 2.5E-08 | 2.5E-08 | | |
| Triiodo-thyronine | 673 | | | | | | | | | | 1.0E-08 |
| **Nucleosides, etc.** | | | | | | | | | | | |
| Adenine SO4 | 184 | 1.0E-05 | | 1.0E-06 | | | 1.0E-06 | 1.8E-04 | 1.8E-04 | | 1.8E-04 |
| Hypoxanthine | 136 | 1.0E-06 | | 1.0E-06 | | 3.0E-05 | 1.0E-06 | | | | |
| Thymidine | 242 | 3.0E-07 | | 3.0E-07 | | | | 3.0E-06 | 3.0E-06 | | 3.0E-06 |

| | | | | | | | | | | |
|---|---|---|---|---|---|---|---|---|---|---|
| **Energy metabolism** | | | | | | | | | | |
| D-glucose | 180 | 4.0E-03 | 5.6E-03 | 8.0E-03 | 8.0E-03 | 1.0E-02 | 5.5E-03 | 6.0E-03 | 6.0E-03 | 2.5E-02 | 6.0E-03 |
| Phosphoenol-pyruvate | 190 | 1.0E-05 | | | | | | | | | |
| Sodium acetate $3H_2O$ | 136 | | | | | 1.0E | | 3.7E-03 | 3.7E-03 | 3.7E-03 | 3.7E-03 |
| Sodium pyruvate | 110 | 1.0E-03 | 1.0E-03 | 1.0E-03 | 5.0E-04 | 1.1E+02 | | 5.0E-04 | 5.0E-04 | 1.0E-03 | 5.0E-04 |
| **Other components** | | | | | | | | | | | |
| Cholera toxin | ~ 90,000 | | | | | | | 2.0E-10 | | | |
| HEPES, Na salt | 260 | 3.0E-02 | | 3.0E-02 | 3.0E-02 | | | 2.8E-02 | 2.8E-02 | 2.5E-02 | 2.3E-02 |
| Phenol red | 376 | 3.3E-06 | 3.3E-05 | 3.3E-06 | 3.3E-06 | 3.3E-06 | 3.3E-05 | 3.3E-05 | 3.3E-05 | 4.0E-05 | 3.3E-06 |
| Putrescine 2HCl | 161 | 1.0E-09 | 1.2E-09 | 1.0E-09 | 1.0E-09 | 1.0E-06 | 1.0E-09 | 1.0E-06 | 1.0E-06 | | 1.0E-06 |
| **Protein supplements** | | | | | | | | | | | |
| BPE, µgP/ml[4] | | 70 | | | | | | 25 | | | 35 |
| BSA, mg/ml | | | | | | | | | | 0.5–10 | |
| Dialyzed FBS, µgP/ml[3] | | | | 5 | | | | | 1 | | |
| Transferrin, $Fe^{3+}$ saturated, µg/ml | | | | | | | | | | 30–300 | 10 |
| **Gas phase** | | | | | | | | | | | |
| $CO_2$ | 44 | 2% | 5% | 2% | 2% | 5% | 5% | 5% | 5% | 10% | 5% |

[1] See also complete PFMR-4A [Peehl, 2002].

[2] Although no bicarbonate is specified in the formulation, 20 mM NaOH was used to neutralize the medium in a gas phase of 2% $CO_2$, resulting in 10 mM bicarbonate at pH 7.4.

[3] Soluble analogs of hydrocortisone, such as dexamethasone, can be used.

[4] Ovine prolactin can be substituted for BPE. Most concentrations are molar, and computer-style notation is used, e.g., 3.0E-2 = $3.0 \times 10^{-2}$ = 30 mM. Molecular weights are given for root compounds; although some recipes use salts or hydrated forms, molarities will, of course, remain the same. The units are given in the component column where molarity is not used.

Abbreviations: BPE, bovine pituitary extract; BSA, bovine serum albumin; EGF, epidermal growth factor; FBS, fetal bovine serum. See text for references.

**TABLE 9.2.** Examples of Serum-Free Media; Supplemented Basal Media

| Medium | HITES | ACL-3 | | N3 | G3 | K-1 | K-2 | | | |
|---|---|---|---|---|---|---|---|---|---|---|
| Reference | [Carney et al., 1981] | [Brower et al., 1986] | [Masui et al., 1986b] | [Bottenstein, 1984] | [Michler-Stuke & Bottenstein, 1982] | [Taub, 1984] | [Taub, 1984] | [Chopra & Xue-Hu, 1993] | [Robertson & Robertson, 1995] | [Naeyaert et al., 1991] |
| Cell type | Human small-cell lung carcinoma | Human non-small-cell lung carcinoma | Human lung adeno-carcinoma | Human neuroblastoma, LA-N-1 | Rat glial cells | MDCK (dog kidney) | LLC-PK, (pig kidney) | Human parotid | Human prostate | Human melanocytes |
| **Basal medium** | RPMI 1640 | RPMI 1640 | DMEM/F12 | DMEM/F12 | DMEM | DMEM/F12 | DMEM/F12 | MCDB 153 | αMEM/F12 | M199 |
| **Supplements** | | | | | | | | | | |
| Arg VP, µU/ml | | | | | | | 10 | | | |
| BPE, µgP/ml | | 5.0 | | | | | | 25 | 25 | |
| BSA, mg/ml | | | | | 0.3 | | | | | |
| Cholera toxin | | | | | | | | | | 1.0E-09 |
| Cholesterol | | | | | | | 1.0E-08 | | | |
| EGF, ng/ml | | 10 | | | | | | 10 | 10 | 1 |
| Epinephrine | 1.0E-08 | | | | | | | | | |
| Estradiol | | | 5.0E-07 | | | | | | | |
| Ethanolamine | | | | | | | | | 1.0E-03 | |
| FGF-2 (basic FGF), ng/ml | | | | | | | | | | 10 |
| Glucagon, µg/ml | | | 0.2 | | | | | | | |
| Glutamine (additional) | | 2.0E-03 | | | | | | | 2.0E-03 | |
| Hydrocortisone | 1.0E-08 | 5.0E-08 | | | | 5.0E-08 | 2.0E-07 | 1.4E-06 | 1.0E-06 | 1.4E-06 |
| Insulin, µg/ml | 5.0 | 20.0 | 5.0 | 5.0 | 0.5 | 5.0 | 25.0 | 5.0 | 5.0 | 10.0 |
| Na pyruvate (additional) | | 5.0E-04 | | | | | | | | |
| Na$_2$SeO$_3$ | 3.0E-08 | 2.5E-08 | 2.5E-08 | 3.0E-08 | | | | | | |
| Phosphoethanolamine | | | | | | | | | 1.0E-04 | |
| Progesterone | | | | 2.0E-08 | 2.0E-07 | | | | | |
| Prolactin | | | | | | | | | | |
| Prostaglandin E$_1$ | | | | | | 7.0E-08 | | | | |
| Putrescine | | | | 1.0E-04 | 1.0E-07 | | | | 5 | |
| Transferrin, Fe$^{3+}$ saturated, µg/ml | 100 | 10 | | 50 | 100 | 5 | 10 | | 5 | 10 |
| Triiodothyronine | | 1.0E-10 | 5.0E-10 | | 4.9E-07 | 5.0E-12 | 1.0E-09 | | | 1.0E-09 |
| Thyroxine | | | | | 4.5E-07 | | | | | |

Most concentrations are molar, and computer-style notation is used, e.g., 3.0E-2 = $3.0 \times 10^{-2}$ = 30 mM. Molecular weights are given for root compounds; although some recipes use salts or hydrated forms, molarities will, of course, remain the same. The units are given in the component column where molarity is not used. Soluble analogs of hydrocortisone, such as dexamethasone, can be used. *Abbreviations:* Arg VP, arginine vasopressin; BPE, bovine pituitary extract; BSA, bovine serum albumin; EGF, epidermal growth factor; FBS, fetal bovine serum; FGF, fibroblast growth factor.

several tests (for growth, plating efficiency, and special functions; *see* Section 9.7.2) on a number of different cell lines.

(4) *Specificity.* If more than one cell type is used, each type may require a different batch of serum, so several batches must be held on reserve simultaneously. Coculturing different cell types will present an even greater problem.

(5) *Availability.* Periodically the supply of serum is restricted because of drought in the cattle-rearing areas, the spread of disease among the cattle, or economic or political reasons. This can create problems at any time, restricting the amount of serum available and the number of batches to choose from, but can be particularly acute at times of high demand. Today demand is increasing, and it will probably exceed supply unless the majority of commercial users are able to adopt serum-free media. Although an average research laboratory may reserve 100 to 200 L of serum for a year, a commercial biotechnology laboratory can use that amount or more in a week.

(6) *Downstream processing.* The presence of serum creates a major obstacle to product purification and may even limit the pharmaceutical acceptance of the product.

(7) *Contamination.* Serum is frequently contaminated with viruses, many of which may be harmless to cell culture but represent an additional unknown factor outside the operator's control [Merten, 1999]. Fortunately, improvements in serum sterilization techniques have virtually eliminated the risk of mycoplasma infection from sera from most reputable suppliers, but viral infection remains a problem, despite claims that some filters may remove viruses (Pall Gelman). Because of the risk of spreading bovine spongiform encephalitis among cattle, cell cultures and serum shipped to the United States or Australia require information on the country of origin and the batch number of the serum. Serum derived from cattle in New Zealand probably has the lowest endogenous viral contamination, as many of the viruses found in European and North American cattle are not found in New Zealand. In addition, with the continued use of antibiotics in cattle rearing, there is also the risk of contamination from, for example, tetracycline [Steeb, L., *pers comm*], which can be toxic or alter gene transcription in some gene constructs.

(8) *Cost.* Cost is often cited as a disadvantage of serum supplementation. Certainly serum constitutes the major part of the cost of a bottle of medium (more than 10 times the cost of the chemical constituents), but if it is replaced by defined constituents, the cost of these may be as high as that of the serum. However, as the demand for such items as transferrin, selenium, and insulin rises, the cost is likely to come down with increasing market size, and serum-free media will become relatively cheaper. The availability of recombinant growth factors and proteins, coupled with market demand, may help to reduce their intrinsic cost.

(9) *Growth inhibitors.* As well as its growth-promoting activity, serum contains growth-inhibiting activity, and although stimulation usually predominates, the net effect of the serum is an unpredictable combination of both inhibition and stimulation of cell proliferation. Although substances such as PDGF may be mitogenic to fibroblasts, other constituents of serum can be cytostatic. Hydrocortisone, present at around $1 \times 10^{-8}$ M in fetal serum, is cytostatic to many cell types, such as glia [Guner et al., 1977] and lung epithelium [McLean et al., 1986], at high cell densities (although it may be mitogenic at low cell densities), and TGF-β, released from platelets, is cytostatic to many epithelial cells.

(10) *Standardization.* Standardization of experimental and production protocols is difficult, both at different times and among different laboratories, because of batch-to-batch variations in serum.

## 9.2 ADVANTAGES OF SERUM-FREE MEDIA

All of the problems above can be eliminated by removal of serum and other animal products. Serum-free media have in addition two major positive benefits.

### 9.2.1 Definition of Standard Medium
Given pure constituents are used, a given medium formulation can be standardized regardless of where it is used and by whom. Not only does this allow easier validation of industrial processes, it also means research labs can replicate conditions to repeat and confirm experimental data.

### 9.2.2 Selective Media
One of the major advantages of the control over growth-promoting activity afforded by serum-free media is the ability to make a medium selective for a particular cell type (*see* Tables 9.1, 9.2). Fibroblastic overgrowth can be inhibited in breast and skin cultures by using MCDB 170 [Hammond et al., 1984] and 153 [Peehl and Ham, 1980], melanocytes can be cultivated in the absence of fibroblasts and keratinocytes [Naeyaert et al., 1991], and separate lineages and even stages of development may be selected in hematopoietic cells by choosing the correct growth factor or group of growth factors (*see* Sections 23.3.7, 23.3.8). Many of these selective media are now available commercially along with cultures of selected cell types (*see* Tables 9.5, 22.1).

### 9.2.3 Regulation of Proliferation and Differentiation
Add to the ability to select for a specific cell type the possibility of switching from a growth-enhancing medium for propagation to a differentiation-inducing medium by altering the concentration and types of growth factors and other inducers.

## 9.3 DISADVANTAGES OF SERUM-FREE MEDIA

Serum-free media are not without disadvantages:

(1) *Multiplicity of media.* Each cell type appears to require a different recipe, and cultures from malignant tumors may vary in requirements from tumor to tumor, even within one class of tumors. Although this degree of specificity may be an advantage to those isolating specific cell types, it presents a problem for laboratories initiating or maintaining cell lines of several different origins.

(2) *Selectivity.* Unfortunately, the transition to serum-free conditions, however desirable, is not always as straightforward as it seems. Some media may select a sublineage that is not typical of the whole population, and even in continuous cell lines, some degree of selection may still be required.

(3) *Reagent purity.* The removal of serum also requires that the degree of purity of reagents and water and the degree of cleanliness of all apparatus be extremely high, as the removal of serum also removes the protective, detoxifying action that some serum proteins may have. Although removing this action is no doubt desirable, it may not always be achievable, depending on resources.

(4) *Cell proliferation.* Growth is often slower in serum-free media, and fewer generations are achieved with finite cell lines.

## 9.4 REPLACEMENT OF SERUM

The essential factors in serum have been described (*see* Section 8.5) and include (1) adhesion factors such as fibronectin; (2) peptides, such as insulin, PDGF, and TGF-β, that regulate growth and differentiation; (3) essential nutrients, such as minerals, vitamins, fatty acids, and intermediary metabolites; and (4) hormones, such as insulin, hydrocortisone, estrogen, and triiodothyronine. All these constituents regulate membrane transport, differentiation, and the constitution of the cell surface. Although some of these constituents are included in the formulation of serum-free media, others are not and may require addition and optimization.

### 9.4.1 Commercially Available Serum-Free Media

Around 50 suppliers (*see* Table 9.5; Appendix II) now make serum-free media [*see also* SEFREC, 2005; Focus on Alternatives, 2009]. Some are defined formulations, such as MCDB 131 (Sigma) for endothelial cells and LHC-9 (Invitrogen) for bronchial epithelium, whereas others are proprietary formulations, such as BD CHO Medium (BD Biosciences) and Hybridoma AGT (Mediatech). Many are designed primarily for culture of hybridomas, when the formation of a product that is free of serum proteins is clearly important, but others are applicable to other cell types. Although evidence exists in the literature for the use of proprietary media with specific cell types (*see* Table 9.5), commercial recipes are often a trade secret, and you can only rely on the supplier's advice or, better, screen a number of media over several subcultures with your own cells. The latter can be an extensive exercise but is justified if you are planning long-term work with the cells. Searching SEFREC [2005] and Focus on Alternatives [2009] might help in creating a short list to test.

It is important to determine the quality control performed by commercial suppliers of serum-free media. The ideal situation is for the medium to have been tested against the cells that you wish to grow (e.g., keratinocyte growth medium should have been tested on keratinocytes). Some suppliers, such as Clonetics (Lonza), Cascade, Cell Applications, ECACC, and PromoCell, will supply the appropriate cells with the medium, ensuring the correct quality control, but others may have performed quality control with common continuous cell lines, in which case you will have to do your own quality control before purchasing the medium.

### 9.4.2 Serum Substitutes

A number of products have been developed commercially to replace all or part of the serum in conventional media (Table 9.3; *see also* Appendix II: Serum Substitutes). Although they may offer a degree of consistency not obtainable with regular sera, variations in batches can still occur, and the constitution of the products is not fully defined. Serum substitutes may be useful as an ad hoc measure or for purposes of economy, but they are not a replacement for serum-free media. Nutridoma (BCL), ITS (BD Biosciences), ITS Premix, TCM, TCH (MP Biomedicals), SIT (Sigma), and Excell-900 (JRH Biosciences) are defined supplements aimed at replacing serum, partially or completely [Cartwright & Shah, 1994]. Sigma markets the MegaCell series of media, which are MEM, DMEM, RPMI 1640, and MEM or DMEM/F12 mixtures supplemented with growth factors and additional amino acids, which enable a substantial reduction in the concentration of serum.

**TABLE 9.3.** Serum Substitutes

| Brand name | Supplier |
|---|---|
| Biotain-MPS | Lonza |
| CPSR | Sigma |
| Excell-900 | Sigma (JRH Biosciences) |
| Ex-cyte | Bayer |
| ITS Premix | MP Biomedicals |
| Nutridoma | Roche Applied Sciences |
| Onco-CYTE | Millipore (Chemicon) |
| SerXtend | DuPont (NEN) |
| SIT | Sigma |
| TCM, TCH | Celox, MP Biomedicals |
| Ultroser | Invitrogen |
| Ventrex | Sigma (JRH Biosciences) |

The choice of serum-free media and serum replacements is now so large and diverse that it is not possible to make individual recommendations for specific tasks. The best approach is to check the literature, contact the suppliers, obtain samples of those products that seem most relevant from previous reports, and screen the products in your own assays (e.g., for growth, survival, or special functions).

Amino acid hydrolysates (*see* Section 8.7.1) are used to help reduce or eliminate serum in large-scale cultures for biotechnology applications. They have the advantage that they can be sterilized by autoclaving, but have an undefined constitution.

### 9.4.3 Serum-Free Subculture

Propagating attached cells in serum-free conditions may require modification of the subculture protocol as serum contains significant components not necessarily nutrient or growth factor related.

*Adhesion factors.* When serum is removed, it may be necessary to treat the plastic growth surface with fibronectin (25–50 μg/mL) or laminin (1–5 μg/mL), added directly or via the medium [Barnes et al., 1984a] (*see* Protocol 22.9). Pretreating the plastic with poly-D-lysine (1 mg/mL) was shown to enhance the survival of human diploid fibroblasts [McKeehan & Ham, 1976a; Barnes et al., 1984a] (*see also* Section 7.2).

*Protease inhibitors.* After trypsin-mediated subculture, the addition of serum inhibits any residual proteolytic activity. Consequently protease inhibitors such as soya bean trypsin inhibitor or 0.1 mg/mL aprotinin (Sigma) must be added to serum-free media after subculture. Furthermore, because crude trypsin is a complex mixture of proteases, some of which may require different inhibitors, it is preferable to use pure trypsin (e.g., Sigma Gr. III) followed by a trypsin inhibitor. Alternatively, one may wash cells by centrifugation to remove trypsin, although it may still be advisable to include a trypsin inhibitor in the wash.

*Trypsin and other proteases.* Special care may be required when trypsinizing cells from serum-free media, as the cells are more fragile and may need purified crystalline trypsin, and to be chilled to 4°C, to reduce damage [McKeehan, 1977]. Alternative sources of proteases are available to keep animal products from coming in contact with cells. Purified porcine trypsin may be replaced with recombinant trypsin (TrypLE™, Invitrogen; TrypZean, Sigma). Nonmammalian proteases are also available (e.g., Accutase™ or Accumax™, Sigma). Pronase, Dispase, Liberase (Roche), and collagenase are bacterial proteases not neutralized by trypsin inhibitors and will require removal by centrifugation. It is possible that Pronase can be inactivated by dilution without subsequent neutralization in serum-free conditions [McKeehan, personal communication]. Pronase is very effective but can be toxic

to some cells; Dispase and collagenase will not give a single cell suspension if there are epithelial cells present.

### 9.4.4 Hormones

Hormones that have been used to replace serum include growth hormone (somatotropin) at 50 ng/mL, insulin at 1 to 10 U/mL, which enhances plating efficiency in a number of different cell types, and hydrocortisone, which improves the cloning efficiency of glia and fibroblasts (*see* Tables 9.1, 9.2; Section 13.2.1) and has been found necessary for the maintenance of epidermal keratinocytes and some other epithelial cells (*see* Section 22.1.1). Recombinant insulin is available (Genway Biotech). Barnes & Sato [1980] described 10-pM triiodothyronine ($T_3$) as a necessary supplement for MDCK (dog kidney) cells, and it has also been used for lung epithelium [Lechner & LaVeck, 1985; Masui et al., 1986b]. Various combinations of estrogen, androgen, or progesterone with hydrocortisone and prolactin at around 10 nM can be shown to be necessary for the maintenance of mammary epithelium [Klevjer-Anderson & Buehring, 1980; Hammond et al., 1984; Strange et al., 1991; Lee et al., 1996].

Other hormones with activities not usually associated with the cells they were tested on were found to be effective in replacing serum, such as follicle-stimulating hormone (FSH) with B16 murine melanoma [Barnes & Sato, 1980] and bombesin-like peptide (gastrin-releasing peptide, GRP) with bronchial epithelium [DeMichele et al., 1994] and small-cell lung carcinoma [Hohla et al., 2007].

### 9.4.5 Growth Factors

The family of polypeptides that has been found to be mitogenic in vitro is now quite extensive (Table 9.3) and includes the heparin-binding growth factors (including the FGF family), EGF, PDGF [Barnes et al., 1984a, c], IGF-I and -II, and the interleukins [Thomson, 1991] that are active in the 1 to 10 ng/mL range. Growth factors and cytokines tend to have a wide-ranging specificity [Barnes et al., 1984d] beyond those tissues in which activity was first demonstrated. Keratinocyte growth factor (KGF) [Aaronson et al., 1991], besides showing activity with epidermal keratinocytes, will induce proliferation and differentiation in prostatic epithelium [Planz et al., 1998; Thomson et al., 1997]. Hepatocyte growth factor (HGF) [Kenworthy et al., 1992] is mitogenic for hepatocytes but is also morphogenic for kidney tubules [Furue & Saito, 1997; Montesano et al., 1997; Balkovetz & Lipschutz, 1999]. Growth factors and cytokines acquire their specificity by virtue of the fact that their production is localized and that they have a limited range. Most act as paracrine factors (they are active on adjacent different cells) and not by systemic distribution in the blood.

Growth factors may act synergistically or additively with each other or with other hormones and paracrine factors, such as prostaglandin $F_{2\alpha}$ and hydrocortisone [Westermark & Wasteson, 1975; Gospodarowicz, 1974]. For example, the action of interleukin 6 (IL-6) and oncostatin M on A549 cells is dependent on dexamethasone, a synthetic hydrocortisone analogue[McCormick et al., 1995, 2000]. The action is due to the production of a heparan sulfate proteoglycan (HSPG) [Yevdokimova & Freshney, 1997] by the A549 cells. The requirement for heparin or HSPG was first observed with FGF [Klagsbrun & Baird, 1991], but it may be a more general phenomenon; for example, β-glycan has been shown to be involved in the cellular response to TGF-β [Lopez-Casillas et al., 1993]. Some growth factors are dependent on the activity of a second growth factor before they act [Phillips & Christofalo, 1988]; for example, bombesin alone is not mitogenic in normal cells but requires the simultaneous or prior action of insulin or one of the IGFs [Aaronson et al., 1991].

### 9.4.6 Nutrients in Serum

Iron, copper, and a number of minerals have been included in serum-free recipes, although evidence that some of the rarer minerals are required is still lacking. Selenium ($Na_2SeO_3$), at around 20 nM, is found in most formulas, and there appears to be some requirement for lipids or lipid precursors such as choline, linoleic acid, ethanolamine, or phosphoethanolamine.

### 9.4.7 Proteins and Polyamines

The inclusion in medium of proteins such as bovine serum albumin (BSA), 0.5 to 10 mg/mL, or tissue extracts often increases cell growth and survival but adds undefined constituents to the medium and retains the problem of adventitious infectious agents. Tissue extracts include bovine pituitary extract used in conjunction with keratinocyte serum-free media, but it may be possible to replace this with defined recombinant growth factors. BSA, fatty acid free, is used at 1 to 10 mg/mL. Transferrin, at around 5 to 300 ng/mL, is required as a carrier for iron and may also have a mitogenic role. Recombinant transferrin is now available (Genway Biotech), or can be substituted with lactoferrin, also recombinant, (US Biological). Putrescine has been used at 100 nM.

### 9.4.8 Viscosity

One of the actions of serum is to increase the viscosity of the medium. This is particularly important in stirred suspension culture as it helps to minimize shear stress. Carboxymethylcellulose (CMC) at 1.2 to 30 mg/mL [Telling & Ellsworth, 1965], polyvinylpyrrolidone (PVP) at 1 mg/mL [Vireque et al., 2009], and the nonionic surfactant Pluronic F-68 (polyoxyethylene and polyoxypropylene) at 1 mg/mL [Cherry & Papoutsakis, 1990] have all been used to minimize mechanical damage.

## 9.5 SELECTION OF SERUM-FREE MEDIUM

### 9.5.1 Cell or Product Specificity

If the reason for using a serum-free medium is to promote the selective growth of a particular type of cell, then that reason will determine the choice of medium, such as MCDB 153 for epidermal keratinocytes, LHC-9 for bronchial epithelium, HITES for small-cell lung cancer, and MCDB 130 for endothelium (*see* Tables 9.1, 9.2, 9.4). If the reason is simply to avoid using serum with continuous cell lines, such as CHO cells or hybridomas, in order to reduce the likelihood of viral or serum proteins in the cell product, then the choice will be wider, and there will be several commercial sources to choose from (Table 9.4). When a cell line is obtained from the originator or a reputable cell bank, the supplier will recommend the appropriate medium, and the only reason to change will be if the medium is unavailable or is incompatible with other stocks. If possible, it is best to stay with the originator's recommendation, as this may be the only way to ensure that the line exhibits its specific properties. Table 9.4 summarizes the availability and selection of the serum-free media listed in Tables 9.1 and 9.2 and makes a few additional suggestions [*see also* Mather, 1998; Barnes et al., 1984a–d; Ham & McKeehan, 1979].

### 9.5.2 Adaptation to Serum-Free Media

Many continuous cell lines, such as HeLa, CHO-K1, or mouse myelomas, may be adapted to growth in the absence of serum. This often involves a prolonged period of selection of what may be a minority component of the cell population, so it is important to ensure that the properties of the cell line are not lost during this period of selection. If a myeloma is to be used to generate a hybridoma, or a CHO cell used for transfection, the selection of a serum-free line should be done before fusion or transfection, to minimize the risk of loss of properties during selection.

Adaptation to serum-free medium is usually carried out over several serial subcultures, with the serum concentration being reduced gradually at, or before, each subculture, or at a frequency determined by the rate of adaptation. Once stable cell proliferation is established at one serum concentration, subculture cells into a lower concentration until stable growth is reestablished and then dilute the serum again. In suspension cultures, this is done by monitoring the viable cell count and diluting the cell suspension accordingly, keeping the minimal cell concentration higher than for normal subculture. For monolayer cultures, reduce the serum concentration a few days before subculture, and then subculture into the new low serum concentration. During the adaptation process it may be necessary to supplement the medium with factors known to replace serum (*see* Sections 9.4, 9.6).

## 9.6 DEVELOPMENT OF SERUM-FREE MEDIUM

There are two general approaches to the development of a serum-free medium for a particular cell line or primary culture. The first is to take a known recipe for a related cell type, with or without 10% to 20% dialyzed serum, and alter the constituents individually or in groups, against a range of serum concentrations including zero. Clonal growth experiments will then show any sparing effects of a potential

**TABLE 9.4.** Growth Factors and Mitogens

| Name and synonyms | Abbreviation | Mol. mass. (KDa) | Source* | Function |
|---|---|---|---|---|
| Acidic fibroblast gf; aFGF; heparin binding gf 1, HBGF-1; endothelial cell gf (ECGF); myoblast gf (MGF) | FGF-1 | 13 h | Bovine brain; pituitary | Mitogen for endothelial cells |
| Activin; TGF-β family | | g | Gonads | Morphogen; stimulates FSH secretion |
| Amphiregulin | AR | 19, 21, 43 | MCF-7 conditioned medium | Autocrine EGF-like gf for keratinocytes |
| Angiogenin | | 16 | Fibroblasts, lymphocytes, colonic epithelial; cells | Angiogenic; endothelial mitogen |
| Astroglial growth factor-1; member of acidic FGFs | AGF-1 | 14 | Brain | Mitogen for astroglia |
| Astroglial growth factor-2; member of basic FGFs | AGF-2 | 14 | Brain | Mitogen for astroglia |
| Basic fibroblast gf; bFGF; HBGF-2; prostatropin | FGF-2 | 13 h | Bovine brain; pituitary | Mitogen for many mesodermal and neuroectodermal cells; adipocyte and ovarian granulosa cell differentiation |
| Brain-derived neurotrophic factor | BDNF | 28 | Brain | Neuronal viability |
| Cachectin | TNF-α | 17 | Monocytes | Catabolic; cachexia; shock |
| Cholera toxin | CT | 80–90 | Cholera bacillus | Mitogen for some normal epithelia |
| Ciliary neurotrophic factor; member of IL-6 group | CNTF | | Eye | |
| Connective tissue growth factor; IGFBP8 | CTGF | 38p | Peritoneal mesothelium; mesangial cells | Fibroblast mitogen; angiogenic; matrix production |
| Endothelial cell growth factor; acidic FGF family | ECGF | h | Recombinant | Endothelial mitogenesis |
| Endothelial growth supplement; mixture of endothelial mitogens | ECGS | | Bovine pituitary | Endothelial mitogenesis |
| Endotoxin | | | Bacteria | Stimulates TNF production |
| Epidermal growth factor, Urogastrone | EGF | 6 | Submaxillary salivary gland (mouse) human urine; guinea pig prostate | Active transport; DNA, RNA, protein, synthesis; mitogen for epithelial and fibroblastic cells; synergizes with IGF-1 and TGF-β |
| Erythropoietin | EPO | 34–39 g | Juxtaglomerular cells of kidney | Erythroid progenitor proliferation and differentiation |
| Eye-derived growth factor-1; member of basic FGFs | EDGF-1 | 14 | Eye | |
| Eye-derived growth factor-2; member of acidic FGFs | EDGF-2 | 14 | Eye | |

(*continued overleaf*)

**TABLE 9.4.** (*Continued*)

| Name and synonyms | Abbreviation | Mol. mass. (KDa) | Source* | Function |
|---|---|---|---|---|
| Fibroblast gf-3; product of *int*-2 oncogene | FGF-3 | 14 h | Mammary tumors | Mitogen; morphogen; angiogenic |
| Fibroblast gf-4; product of *hst*/KS3 oncogene | FGF-4 | 14 h | Embryo; tumors | Mitogen; morphogen; angiogenic |
| Fibroblast gf-5 | FGF-5 | 14 h | Fibroblasts; epithelial cells; tumors | Mitogen; morphogen; angiogenic |
| Fibroblast gf-6; product of *hst*-2 oncogene | FGF-6 | 14 h | Testis; heart; muscle | Mitogenic for fibroblasts; morphogen |
| Fibroblast gf-9 | FGF-9 | 14 h | Recombinant | |
| Fibroblast gf-10 | FGF-10 | 14 h | | Trophoblast invasion; stimulate uPA and PAI-1; alveolar epithelial mitogen |
| Granulocyte colony-stimulating factor; pluripoietin; CFS-β | G-CSF | 18–22 | | Granulocyte progenitor proliferation and differentiation |
| Granulocyte/macrophage colony-stimulating factor CSA; human CSFα | GM-CSF | 14–35 g | | Granulocyte/macrophage progenitor proliferation |
| Heparin-binding EGF-like factor | HB-EGF | | Recombinant | |
| Hepatocyte gf, HBGF-8; Scatter factor | HGF | h | Fibroblasts | Epithelial morphogenesis; hepatocyte proliferation |
| Heregulin; *erb*B2 ligand | HRG | 70 | Breast cancer cells; recombinant | Mammary and other epithelial cell mitogen |
| Immune interferon; macrophage-activating factor (MAF) | IFN-γ | 20–25 | Activated lymphocytes | Antiviral; activates macrophages |
| Inhibin; TGF-β family | | 31 g | Ovary | Morphogen; inhibits FSH secretion |
| Insulin | Ins | 6 | β Islet cells of pancreas | Glucose uptake and oxidation; amino acid uptake; glyconeogenesis |
| Insulin-like gf 1; somatomedin-C; NSILA-1 | IGF-1 | 7.6 | Liver | Mediates effect of growth hormone on cartilage sulfation; insulin-like activity |
| Insulin-like gf 2; MSA in rat | IGF-2 | 7 | BRL-3A cell-conditioned medium | Mediates effect of growth hormone on cartilage sulfation; insulin-like activity |
| Interferon-α1; leukocyte interferon | IFNα1 | 18–20 | Macrophages | Antiviral; differentiation inducer; anticancer |
| Interferon-α2; leukocyte interferon | IFN-α2 | 18–20 | Macrophages | Antiviral; differentiation inducer; anticancer |
| Interferon-β; fibroblast interferon | IFN-β1 | 22–27 g | Fibroblasts | Antiviral; differentiation inducer; anticancer |
| Interferon β2; fibroblast interferon; IL-6, BSF-2 (*see also* IL-6) | IFN-β2 | 22–27 g | Activated T-cells; fibroblasts; tumor cells | Keratinocyte differentiation; PC12 differentiation (*see also* IL-6) |
| Interferon γ; immune interferon | IFNγ | | Activated lymphocytes | Antiviral, macrophage activator; antiproliferative on transformed cells |

**TABLE 9.4.** (*Continued*)

| Name and synonyms | Abbreviation | Mol. mass. (KDa) | Source* | Function |
|---|---|---|---|---|
| Interleukin-1; lymphocyte-activating factor (LAF); B-cell-activating factor (BAF); hematopoietin-1 | IL-1 | 12–18 | Activated macrophages | Induces IL-2 release |
| Interleukin-2; T-cell gf (TCGF) | IL-2 | 15 | CD4+ve lymphocytes (NK); murine LBRM-5A4 and human Jurkat FHCRC cell lines | Supports growth of activated T-cells; stimulates LAK cells |
| Interleukin-3; multipotential colony-stimulating factor; mast cell growth factor | IL-3 | 14–28 g | Activated T-cells; WEHI-3b myelomonocytic cell lines | Granulocyte/macrophage production and differentiation |
| Interleukin-4; B-cell gf; BCGF-1; BSF-1 | IL-4 | 15–20 | Activated CD4+ve lymphocytes | Competence factor for resulting B-cells; mast cell maturation (with IL-3) |
| Interleukin-5; T-cell-replacing factor (TRF); eosinophil-differentiating factor (EDF) BCGF-2 | IL-5 | 12–18 g | T-lymphocytes | Eosinophil differentiation; progression factor for competent B-cells |
| Interleukin-6; Interferon β−2; B-cell-stimulating factor (BSF-2); hepatocyte-stimulating factor; hybridoma-plasmacytoma gf | IL-6 | 22–27 g | Activated T-cells macrophage/monocytes; fibroblasts; tumor cells | Acute phase response; B-cell differentiation; keratinocyte differentiation; PC12 differentiation |
| Interleukin-7; hematopoietic growth factor; lymphopoietin 1 | IL-7 | 15–17 g | Bone marrow stroma | Pre- and pro-B-cell growth factor |
| Interleukin-8; monocyte-derived neutrophil chemotactic factor (MDNCF); T-cell chemotactic factor; neutrophil-activating protein (NAP-1) | IL-8 | 8–10 h | LPS monocytes; PHA lymphocytes; endothelial cells; IL-1- and TNF-stimulated fibroblasts and keratinocytes | Chemotactic factor for neutrophils, basophils, and T-cells |
| Interleukin-9; human P-40; mouse T-helper gf; mast-cell-enhancing activity (MEA) | IL-9 | 30–40 | CD4 + ve T-cells; stimulated by anti-CD4 antibody PHA or PMA | Growth factor for T-helper, megakaryocytes, mast cells (with IL-3) |
| Interleukin-10; cytokine synthesis inhibitory factor (CSIF) | IL-10 | 20 | | Immune suppressor |
| Interleukin-11; adipogenesis inhibitory factor (AGIF) | IL-11 | 21 | | Stimulates plasmacytoma proliferation and T-cell-dependent development of Ig-producing B-cells |

(*continued overleaf*)

**TABLE 9.4.** (*Continued*)

| Name and synonyms | Abbreviation | Mol. mass. (KDa) | Source* | Function |
|---|---|---|---|---|
| Interleukin-12; cytotoxic lymphocyte maturation factor | IL-12 | 40, 35 subunits | | Activated T-cell and NK cell growth factor; induces IFN-γ |
| Keratinocyte gf, FGF-7 | KGF | 14 h | Fibroblasts | Keratinocyte proliferation and differentiation; prostate epithelial proliferation and differentiation |
| Leukemia inhibitory factor; HILDA; member of IL-6 group | LIF | 24 | SCO cells | Inhibits differentiation in embryonal stem cells |
| Lipopolysaccharide | LPS | 10 | Gram-positive bacteria | Lymphocyte activation |
| Lymphotoxin | TNF-β | 20–25 | Lymphocytes | Cytotoxic for tumor cells |
| Macrophage inflammatory protein—1α | MIP—1α | 10 | Macrophages | Hematopoietic stem cell inhibitor |
| Monocyte/macrophage colony-stimulating factor CSF-1 | M-CSF | 47–74 | B- and T-cells, monocytes, mast cells, fibroblasts | Macrophage progenitor proliferation and differentiation |
| Müllerian inhibition factor | MIF | | Testis | Inhibition of Müllerian duct; inhibition of ovarian carcinoma |
| Nerve gf, β | β NGF | 27 | Male mouse submaxillary salivary gland | Trophic factor; chemotactic factor; differentiation factor; neurite outgrowth in peripheral nerve |
| Neurotrophin-3 | NT-3 | | | Stimulation of neurite outgrowth |
| Oncostatin M; member of IL-6 group | OSM | 28 | Activated T-cells and PMA-treated monocytes | Differentiation inducer (with glucocorticoid); fibroblast mitogen |
| Phytohemagglutinin | PHA | 30 | Red kidney bean (*Phaseolus vulgaris*) | Lymphocyte activation |
| Platelet-derived endothelial cell growth factor; similar to gliostatin | PD-ECGF | ~ 70 | Blood platelets, fibroblasts, smooth muscle | Angiogenesis; endothelial cell mitogen; neuronal viability; glial cytostasis |
| Platelet-derived growth factor | PDGF | 30 | Blood platelets | Mitogen for mesodermal and neuroectodermal cells; wound repair; synergizes with EGF and IGF-1 |
| Phorbol myristate acetate; TPA; phorbol ester | PMA | 0.617 | Croton oil | Tumor promoter; mitogen for some epithelial cells and melanocytes; differentiation factor for HL-60 and squamous epithelium |
| Pokeweed mitogen | PWM | | Roots of pokeweed (*Phytolacca americana*) | Monocyte activation |
| Stem cell factor; mast cell growth factor; steel factor; *c-kit* ligand | SCF | 31 g | Endothelial cells, fibroblasts, bone marrow, Sertoli cells | Promotes first maturation division of pluripotent hematopoietic stem cell |

**TABLE 9.4.** (*Continued*)

| Name and synonyms | Abbreviation | Mol. mass. (KDa) | Source* | Function |
|---|---|---|---|---|
| Transferrin | Tfn | 78 | Liver | Iron transport; mitogen |
| Transforming growth factor α | TGF-α | 6 | | Induces anchorage-independent growth and loss of contact inhibition |
| Transforming growth factor β (six species) | TGF-β1–6 | 23–25 dimer | Blood platelets | Epithelial cell proliferation inhibitor; squamous differentiation inducer |
| Vascular endothelial gf | VEGF | | Kidney | Angiogenesis; vascular endothelial cell proliferation |

*Sources described are some of the original tissues from which the natural product was isolated. In many cases the natural product has been replaced by cloned recombinant material that is available commercially (*see* Appendix II). *Abbreviation* (other than in column 2): gf, growth factor; g, glycosylated; h, heparin binding; HBGF, heparin-binding growth factor; p, perlecan binding. Some of the information in this table was taken from Barnes et al. [1984a], Lange et al. [1991], Jenkins [1992], and Smith et al. [1997].

serum replacement (i.e., less serum will be required to achieve the same clonal growth in the presence of the replacement). The optimum concentration of the compound is determined at the minimum serum concentration, which will still allow satisfactory clonal growth. This was the approach adopted by Ham and coworkers [Ham, 1984] and generally will provide optimal conditions. If a group of compounds is found to be effective in reducing serum supplementation, the active constituents may be identified by the systematic omission of single components and then the concentrations of the essential components optimized [Ham, 1984]. However, this is a very time-consuming and laborious process, involving growth curves and clonal growth assays at each stage, and it is not unreasonable to expect to spend at least three years developing a new medium for a new type of cell.

The time-consuming nature of the first approach has led to the second approach: supplementing existing media such as RPMI 1640 [Carney et al., 1981] or combining media such as Ham's F12 with DMEM [Barnes & Sato, 1980] and restricting the manipulation of the constituents to a shorter list of substances, again by determining the optimum concentration at a limiting serum concentration. Among the substances usually tested are selenium, transferrin, albumin, insulin, hydrocortisone, estrogen, triiodothyronine, ethanolamine, phospho-ethanolamine, growth factors (EGF, FGF, PDGF, endothelial growth supplement, etc.), prostaglandins (PGE$_1$, PGF$_{2\alpha}$), and any other substances that may have special relevance. (*See* Section 9.4; Table 9.4.) Selenium, transferrin, and insulin will usually be found to be essential for most cells, whereas the requirements for the other constituents will be more variable.

## 9.7 PREPARATION OF SERUM-FREE MEDIUM

A number of recipes for serum-free media are now available—many of them commercially (*see* Appendix

II)—for particular cell types [Cartwright & Shah, 1994; Mather, 1998; *see also* Tables 9.1, 9.2; Chapter 23]. The procedure for making up serum-free recipes is similar to that for preparing regular media (*see* Section 10.4.4; *see also* [Waymouth, 1984]). Ultrapure reagents and water should be used and care taken with solutions of $Ca^{2+}$ and $Fe^{2+}$ or $Fe^{3+}$ to avoid precipitation. Metal salts tend to precipitate in alkaline pH in the presence of phosphate, particularly when the medium or salt solution is autoclaved, so cations in stock solutions should be kept at a low pH (below 6.5) and maintained phosphate free. They should be sterilized by autoclaving or filtration (*see* Section 10.5). It is often recommended that divalent cations be added last, immediately before using the medium. Otherwise, the constituents are generally made up as a series of stock solutions, minerals and vitamins at 1000×, tyrosine, tryptophan, and phenylalanine in 0.1 N HCl at 50×, essential amino acids at 100× in water, salts at 10× in water, and any other special cofactors, lipids, and so forth, at 1000× in the appropriate solvents. These are combined in the correct proportions and diluted to the final concentration, and then the pH and osmolality are checked (*see* Section 10.6).

Growth factors, hormones, and cell adhesion factors are best added separately just before the medium is used, as they may need to be adjusted to suit particular experimental conditions.

## 9.8 ANIMAL PROTEIN-FREE MEDIA

There is increasing pressure from regulatory authorities to remove all animal products from contact with cultured cells used in the production of biopharmaceuticals. Trypsin can be replaced with recombinant trypsin or a nonvertebrate protease (*see* Sections 9.4.3, 12.4) and growth factors with recombinant growth factors. BSA can often be replaced by

**TABLE 9.5.** Selecting a Serum-Free Medium

| Cells or cell line | Serum-free medium (see Table 9.1) | Refs. (see also Tables 9.1 and 9.2) | Commercial suppliers (of specified media or alternatives) |
|---|---|---|---|
| A549 | PeproGrow-1 | | PeproTech |
| Adipocytes | | [Rodriguez et al., 2004] | PromoCell; Stratech; Zen-Bio |
| BHK 21 | HyQ PF CHO; HyG PF CHO MPS; PC-1 | [Pardee et al., 1984] | Lonza; MP Biomedicals |
| Bronchial epithelium | LHC-9 | (See Protocol 22.9) | Biosource; Clonetics (Lonza); PromoCell |
| Chick embryo fibroblasts | MCDB 201, 202 | [McKeehan & Ham, 1976b] | Sigma |
| CHO | MCDB 302; PC-1 | [Hamilton & Ham, 1977] | Lonza; Sigma; Invitrogen; PeproTech; PromoCell |
| Chondrocytes | Supplemented DMEM/F12 | [Adolphe, 1984] (see Protocol 22.16) | PromoCell |
| Continuous cell lines | Eagle's MEM, M199, MB752/1, CMRL 1066, MCDB media, DMEM:F12 + supplements | [Waymouth, 1984] | Lonza; Invitrogen; JRH Biosciences; ICN; Sigma |
| Corneal epithelial cells | MCDB 153 | (See Protocol 22.2) | Lonza, KGM; Cascade |
| COS-1,7 | | [Doering et al., 2002] | Lonza |
| Endothelial cells | MCDB 130, 131 | [Knedler & Ham, 1987; Gupta et al., 1997; Hoheisel et al., 1998] | Cell Applications; Lonza; Cascade; PAA; PromoCell; Sigma |
| Fibroblasts | MCDB 110, 202, 402 | [Bettger et al., 1981; Shipley & Ham, 1983] | Lonza; Cascade; PromoCell; Sigma; Stratech; Zen-Bio |
| Glial cells | Michler-Stucke | [Michler-Stuke & Bottenstein, 1982] | Lonza; Invitrogen |
| Glioma | SF12 (Ham's F12 with extra essential and nonessential amino acids) | [Frame et al., 1980; Freshney, 1980] | Lonza; Invitrogen |
| HEK293 | HEKTOR; HEK; CDM4HEK293 | | Cell Culture Services; Hyclone; Millipore—CellGro; PeproTech |
| HeLa cells | | [Blaker et al., 1971; Bertheussen, 1993] | Lonza; PeproTech |
| Hematopoietic cells | αMEM; Iscove's | [Stanners et al., 1971; Iscove & Melchers, 1978] | Sigma; Roche Diagnostics |
| Hepatocytes | BD Hepato; Hepatozyme; Hepatocyte Growth Medium | | BD Biosciences; Invitrogen; Promocell |
| HL-60 | | [Li et al., 1997] | Lonza |
| HT-29 | | [Oh et al., 2001] | Lonza |
| Hybridomas | Ultradoma; Maxicell; Ex-Cell (and many others) | | Atlanta Biologicals; BD Biosciences; Invitrogen; Hyclone; Lonza; Sigma |
| Human diploid fibroblasts | MCDB 110, 202; PC-1 | [Bettger et al., 1981; Ham, 1984] | Cascade; Lonza; PromoCell |
| Human leukemia and normal leukocytes | Iscove's | [Breitman et al., 1984] | Lonza; Hyclone; MP Biomedicals; Invitrogen; Sigma |
| Human tumors | Brower; HITES; Masui; Bottenstein N3 | [Brower et al., 1986; Carney et al., 1981; Masui et al., 1986b; Bottenstein, 1984; Chopra et al., 1996] | |

**TABLE 9.5.** (*Continued*)

| Cells or cell line | Serum-free medium (*see* Table 9.1) | Refs. (*see also* Tables 9.1 and 9.2) | Commercial suppliers (of specified media or alternatives) |
|---|---|---|---|
| Hepatocytes, liver epithelium | Williams E, L15 | [Williams & Gunn, 1974; Mitaka et al., 1993] | Lonza; Sigma |
| Hybridomas | Iscove's | [Iscove & Melchers, 1978; Murakami, 1984] | Sigma; Invitrogen; Lonza; MP Biomedicals; Roche; Irvine Scientific; Metachem; PromoCell |
| Insect cells | | [Ikonomou et al., 2003] | Sigma; Millipore—CellGro |
| Keratinocytes | MCDB 153 | [Peehl & Ham, 1980; Tsao et al., 1982; Boyce & Ham, 1983] (*see* Protocol 22.1) | Invitrogen; Cascade; Cellntech; Lonza; Millipore; PromoCell; Sigma |
| L cells (L929, LS) | NCTC109; NCTC135 | [Birch & Pirt, 1970, 1971; Higuchi, 1977] | Sigma |
| Lymphoblastoid cell lines (human) | Iscove's | [Iscove & Melchers, 1978] | JRH Biosciences; CellGenix; GE Heathcare (Amersham) |
| Mammary epithelium | MCDB 170 | [Hammond et al., 1984] (*see* Protocol 22.3) | Cascade; Lonza; AthenaES |
| MDCK dog kidney epithelium | K-1; PC-1 | [Taub, 1984] | Lonza |
| LLC-PK, pig kidney | K-2 | [Taub, 1984] | |
| Melanocytes | Gilchrest | (*See* Protocol 22.22) | Lonza; Cascade; PromoCell |
| Melanoma | Gilchrest | [Halaban, 2004]; (*See* Protocol 22.21) | Lonza; Cascade; PromoCell |
| Mouse embryo fibroblasts; 3T3 cells | MCDB 402 | [Shipley & Ham, 1983; Ham, 1984] | Biosource |
| Mouse leukemia | | [Murakami, 1984] | |
| Mouse erythroleukemia | SF12 (Ham's F12 with extra essential and nonessential amino acids); Iscove's | [Frame et al., 1980; Freshney, 1980; Iscove & Melchers, 1978] | MP Biomedicals; Invitrogen; Sigma |
| Mouse myeloma | | [Murakami, 1984] | Sigma; Invitrogen; Peprotech |
| Mouse neuroblastoma | MCDB 411; DMEM:F12/N1 | [Agy et al., 1981; Bottenstein, 1984] | |
| Neurons | DMEM:F12/N3; B27/Neurobasal; N2, | [Bottenstein, 1984; Brewer, 1995] | Invitrogen |
| Osteoblasts | | [Shiga et al., 2003] | OGM, Lonza; PromoCell |
| Prostate | WJAC 404; | [Peehl, 2002] (*See* Table 9.1) | Lonza |
| Renal | REGM | | Epithelial, Lonza |
| | MsGM | | Mesangial, Lonza |
| Skeletal myoblasts | | [Goto et al., 1999] | PromoCell |
| Smooth muscle cells | | | Cascade; Lonza; PromoCell |
| Stem Cells | ESGRO; CellGro; StemLine; StemSpan (and many others) | | Millipore; Clonagen; CellGenix; Sigma; Stem Cell Technologies |
| Urothelium | MCDB 153; KSFMc; KGM-2, | [Southgate et al., 2002] | Lonza |
| Vero | PFEK-1; PF-Vero; MP-Vero; Ex-cell Vero | | Autogen Bioclear; Clonagen; Hyclone; MP Biomedicals; Sigma |
| **Matrix-coating products** | | | BD Biosciences; MP Biomedicals; Invitrogen; Biosource; Sigma |

Further recommendations on the choice of medium can be found in McKeehan [1977], Barnes et al. [1984a–d], and Mather [1998]; (*see also* Tables 9.1 and 9.2 *and* Appendix II) supplementation with those factors, such as lipids, hormones, minerals, and growth factors (*see* Section 9.4.4, 9.4.5), normally bound to BSA in serum, as it has yet to be established that BSA has a role in itself. Both transferrin and insulin are available as recombinant proteins (e.g., from Genway Biotech). Adaptation to protein-free medium may require further selection.

## 9.9  CONCLUSIONS

However desirable serum-free conditions may be, there is no doubt that the relative simplicity of retaining serum, the specialized techniques required for the use of some serum-free media, the considerable investment in time, effort, and resources that go into preparing new recipes or even adapting existing ones, and the multiplicity of media required if more than one cell type is being handled all act as considerable deterrents to most laboratories to enter the serum-free arena. There is also no doubt, however, that the need for consistent and defined conditions for the investigation of regulatory processes governing growth and differentiation, the pressure from biotechnology to make the purification of products easier, and the need to eliminate all sources of potential infection will eventually force the adoption of serum-free media on a more general scale. But first, recipes must be found that are less "temperamental" than some current recipes and that can be used with equal facility and effectiveness in different laboratories.

# CHAPTER 10

# Preparation and Sterilization

## 10.1 PREPARATION OF REAGENTS AND MATERIALS

All stocks of chemicals and glassware used in tissue culture should be reserved for that purpose alone. Traces of heavy metals or other toxic substances can be difficult to remove and are detectable only by a gradual deterioration in the culture. It follows that separate stocks imply separate washing of glassware. The requirements of tissue culture washing are higher than for general glassware. Although the level of soil may be lower, a special detergent may be necessary (*see* Section 10.3.4), and chemical contamination from regular laboratory glassware must be avoided. The almost universal adoption of single-use disposable plasticware has removed the problem of maintaining glass flasks suitable for culture vessels. Nevertheless, whenever glass is used, either for storage or for culture, the problem of chemical contaminants leaching out into media or reagents remains, and absolute cleanliness is therefore essential.

## 10.2 STERILIZATION OF APPARATUS AND LIQUIDS

All apparatus and liquids that come in contact with cultures or other reagents must be sterilized. The choice of method depends largely on the stability of the item at high temperatures (Table 10.1). In general, items that have a high resistance to heat, such as metals, glass, and thermostable plastics (e.g., PTFE), are best sterilized by dry heat. This is one of the simplest and most effective methods of sterilization, provided that all parts of the load reach the correct temperature for the required period. Sterilization by steam, or autoclaving, is also highly effective and can be applied to heat-stable liquids, such as water, salt solutions, and some specially formulated media. Penetration of steam through the packaging is essential in order that the correct temperature is achieved for effective sterilization (Fig. 10.1). Irradiation can be used to sterilize heat-labile materials, such as plastics, with $\gamma$-irradiation, usually with $^{60}$Co or $^{135}$Cs, being the most effective. As the dose required is quite high (25 kGy), this is usually done at a central facility, which can also often provide electron beam sterilization. Ethylene oxide can also be used for heat-labile plastics but tends to be adsorbed onto the plastic and may take up to several days to be released. Other chemical sterilants include sodium hypochlorite and formaldehyde, but these are more often used as decontamination agents (*see* Section 6.8.5) for liquid and plastics disposal. Formaldehyde is also used for fumigation (*see* Section 6.8.6).

Heat-labile liquids are sterilized by micropore filtration (*see* Section 10.5.2), which may be by positive pressure, through filters ranging in size from syringe tip to in-line disks of up to 293-mm diameter or pleated cartridge filters of up to 1000 cm$^2$ (*see* Table 10.5), or by negative pressure using a vacuum flask with filter attached (*see* Protocol 10.12). Where there is a suitable vacuum line or pump, this is a simple and effective method of filtration, as the filtrate goes straight into the vessel in which it will be stored. Negative-pressure filtration will tend to increase the pH as it draws off dissolved $CO_2$ from the medium so bottles may need to be gassed with 5% $CO_2$ before storage.

The preferred methods of sterilization of specific items (Tables 10.2, 10.3) are described in the following sections.

*Culture of Animal Cells: A Manual of Basic Technique and Specialized Applications, Sixth Edition*, by R. Ian Freshney
Copyright © 2010 John Wiley & Sons, Inc.

**TABLE 10.1.** Methods of Sterilization

| Method | Conditions | Materials | Limitations |
|---|---|---|---|
| **Dry heat** | 160°C, 1–2 h | Heat stable: metals, glass, PTFE | Some charring may occur, e.g., of indicating tape and cotton plugs |
| **Steam** | 121°C, 15–20 min | Heat-stable liquids: water, salt solutions, autoclavable media; moderately heat-stable plastics: silicones, polycarbonate, nylon, polypropylene | Steam penetration requires steam-permeable packaging; large fluid loads need time to heat up |
| **Irradiation** | | | |
| γ-Irradiation | 25 kGy | Plastics, organic scaffolds, heat-sensitive reagents and pharmaceuticals | Chemical alteration of plastics can occur; macromolecular degradation |
| Electron beam | 25 kGy | Plastics, organic scaffolds, heat-sensitive reagents and pharmaceuticals | Needs high-energy source; not suitable for average laboratory installation |
| Microwave | 5 min full power | Aqueous solutions and gels such as agar | Only useful for small volumes; usually just for melting agar |
| Short-wave UV | 254 nM, 50–100 W, 30 min | Flat surfaces, circulating air | Will not reach shadow areas. Spores resistant |
| **Chemical** | | | |
| Ethylene oxide | 1 h | Heat-labile plastics | Items must be ventilated for 24–48 h; leaves toxic residue |
| Hypochlorite | 300–2500 ppm 30 min | Contaminated solutions; plastics | Needs extensive washing. May leave residue |
| 70% Alcohol | Soak for 1 h | Dissecting instruments (combined with flaming); some plastics | Does not kill spores; fire risk with flaming; precast Perspex or Lucite may shatter if immersed in alcohol |
| **Filtration** | 0.1- to 0.2-μm porosity | All aqueous solutions; particularly suitable for heat-labile reagents and media; specify low protein binding for growth factors, etc | Not suitable for some solvents, e.g., DMSO; slow with viscous solutions |

## 10.3  APPARATUS

### 10.3.1  Glassware

If the glass surface is to be used for cell propagation, it must not only be clean but also carry the correct charge. Caustic alkaline detergents render the surface of the glass unsuitable for cell attachment and require subsequent neutralization with 0.1 M HCl or $H_2SO_4$, but neutral detergents do not alter the glass surface and can be removed more easily. The most effective washing procedure is as follows:

(1) Do not let soiled glassware dry out. A sterilizing agent, such as sodium hypochlorite, should be included in the water used to collect soiled glassware:
  (a) To remove any potential biohazard.
  (b) To prevent microbial growth in the water.
(2) Select a detergent that is effective in the water of your area, that rinses off easily, and that is nontoxic (see Section 10.3.4).
(3) Before drying the glassware, make sure that it has been thoroughly rinsed in tap water followed by deionized, reverse osmosis, or distilled water.

(4) Dry glassware inverted so that it drains readily.
(5) Sterilize the glassware by dry heat to minimize the risk of depositing toxic residues from steam sterilization.

Sterilization procedures are designed not just to kill replicating microorganisms but also to eliminate the more resistant spores. Moist heat is more effective than dry heat; however, it does carry a risk of leaving a residue. Dry heat is preferable, but a minimum temperature of 160°C maintained for at least 1 h is required. Moist heat (for fluids and perishable items) should be maintained at 121°C for 15 to 20 min (see Tables 10.1–10.3). For moist heat to be effective, steam penetration must be ensured, which means that the sterilization chamber must be evacuated before steam injection or the air must be completely replaced with steam by downward displacement (see Fig. 10.1).

Inserting Thermalog indicators (see Appendix II), as well as a temperature probe from a recording thermometer, into a sample of the load, centrally located, monitors both temperature and humidity during sterilization. The temperature of the chamber effluent is often used, but this does not accurately reflect the temperature of the load.

**TABLE 10.2.** Sterilization of Equipment and Apparatus

| Item | Sterilization |
|---|---|
| Ampoules for freezer, glass | Dry heat[a] |
| Ampoules for freezer, plastic | Autoclave[b] (usually bought sterile) |
| Apparatus containing glass and silicone tubing | Autoclave |
| Disposable tips for micropipettes | Autoclave in autoclavable trays or nylon bags |
| Filters, reusable | Autoclave; do not use prevacuum or postvacuum; remove air by displacement |
| Glassware | Dry heat |
| Glass bottles with screw caps | Autoclave with cap slack |
| Glass coverslips | Dry heat |
| Glass slides | Dry heat |
| Glass syringes | Autoclave (separate piston if PTFE) |
| Instruments | Dry heat |
| Magnetic stirrer bars | Autoclave |
| Pasteur pipettes, glass | Dry heat |
| Pipettes, glass | Dry heat |
| Plexiglas, Perspex, Lucite | 70% EtOH (can cause plastic to crack) |
| Polycarbonate | Autoclave |
| Repeating pipettes or syringes | Autoclave (separate PTFE pistons from glass barrels) |
| Screw caps | Autoclave |
| Silicone grease (for isolating clones) | Autoclave in glass Petri dish |
| Silicone tubing | Autoclave |
| Stoppers, rubber and silicone | Autoclave |
| Test tubes | Dry heat |

[a]Dry heat, 160°C for 1 h minimum.
[b]Autoclave, 100 kPa (1 bar, 15 lb/in.²), 121°C for 20 min.

Recording thermometers have the advantage that they will create a permanent record that can be archived. Indicators (e.g., Thermalog) additionally provide a visual confirmation of temperature and humidity and can be used to monitor several parts of the load simultaneously (*see also* Fig. 10.3). Both are recommended.

## PROTOCOL 10.1. PREPARATION AND STERILIZATION OF GLASSWARE

### Materials
☐ Disinfectant: hypochlorite, 500 to 1000 ppm available chlorine at minimum, when diluted in detergent (e.g., Precept tablets, Clorox or Chloros;

*see* Appendix II); heavily soiled material may require up to 5000 ppm available chlorine (WHO, 1989)
☐ Detergent (e.g., 7X™ or Decon®)
☐ Soaking baths
☐ Bottle brushes
☐ Stainless steel baskets (to collect washed and rinsed glassware for drying)
☐ Aluminum foil
☐ Sterility indicators (Alpha Medical for dry heat or steam sterilization; Thermalog indicators for steam sterilization only)
☐ Sterile-indicating tape or tabs (Alpha Medical; Bennett). This is different from the sterile-indicating tape used in autoclaves, as the sterilizing temperature is higher in an oven. Most autoclave tapes tend to char and release traces of volatile material from the adhesive, which can leave a deposit on the oven or even the glassware.
☐ Sterilizing oven, fan assisted, capable of reaching 160°C, and preferably with recording thermometer and flexible probe

### Procedure

Collection and washing of glassware (Fig. 10.2)

1. Immediately after use, collect glassware into detergent containing disinfectant. It is important that glassware does not dry before soaking, or cleaning will be much more difficult.
2. Soak overnight in detergent.
3. Rinse:
   (a) Brush glassware the following morning, and rinse thoroughly in four complete changes of tap water followed by three changes of deionized water. A sink-rinsing spray is a useful accessory; otherwise, bottles must be emptied and filled completely each time. Clipping bottles in a basket will help to speed up this stage.
   (b) Machine rinses should be done without detergent. If done on a spigot header (*see* Section 4.4.11), this can be reduced to two rinses with tap water and one with deionized or reverse-osmosis water.
4. After rinsing thoroughly, invert bottles, and other glassware, in stainless steel wire baskets and dry upside down.
5. Cap bottles with aluminum foil when cool, and store.

Sterilization of glassware

1. Attach a small square of sterile-indicating tape or other indicator label to glassware, and date.

2. Place glassware in an oven with fan-circulated air and temperature set to 160°C.
3. Ensure that the center of the load reaches 160°C:
    (a) Place a sterility indicator in a bottle or typical item in the middle of the load.
    (b) If using a recording thermometer, place the sensor in a bottle or typical item in the middle of the load.
    (c) Do not pack the load too tightly; leave room for circulation of hot air.
4. Close the oven, check that the temperature returns to 160°C, seal the oven with a strip of tape with the time recorded on it (or use automatic locking and recorder), and leave for 1 h.
5. After 1 h, switch off the oven and allow it to cool with the door closed. It is convenient to put the oven on an automatic timer so that it can be left to switch off on its own overnight and be accessed in the morning. This precaution allows for cooling in a sterile environment and also minimizes the heat generated during the day, when it is hardest to deal with.
6. Use glassware within 24 to 48 h.

Keep organic matter out of the oven. Do not use paper tape or packaging material, unless you are sure that it will not release volatile products on heating. Such products will eventually build up on the inside of the oven, making it smell when hot, and some deposition may occur inside the glassware being sterilized (*see above doors* Fig. 10.7).

Alternatively, bottles may be loosely capped with screw caps and foil, tagged with autoclave tape, and autoclaved for 20 min at 121°C with a prevacuum and a postvacuum cycle (*see* Section 4.4.4). Then the caps may be tightened when the bottles have cooled down. Caps must be very slack (loosened one complete turn) during autoclaving, so as to allow steam to enter the bottle and to prevent the liner (if one is used) from being sucked out of the cap and sealing the bottle. If a bottle becomes sealed during sterilization in an autoclave, sterilization will not be complete (Fig. 10.3; Plate 22a). Unfortunately, misting often occurs when bottles are autoclaved, and a slight residue may be left when the mist evaporates. Also the bottles risk becoming contaminated as they cool, by drawing in nonsterile air before they are sealed. Dry-heat sterilization is better, autoclaving the caps separately (*see* Section 10.3.1), because it allows the bottles to cool down within the oven before they are removed.

## 10.3.2 Glass Pipettes

Both glass and plastic pipettes are used in tissue culture. Plastic pipettes have the advantage that they are single use and disposable, avoiding the need for washup and sterilization. Glass pipettes are significantly cheaper but have to be deplugged and replugged each time they are used. Glass

pipettes must also be washed carefully in the pipette washer (Fig. 10.4), so that they will not retain soil and become vulnerable to subsequent blockage.

Δ *Safety Note.* Glass pipettes are prone to damage, and chipped ends present a severe hazard to both users and the washup staff. Discard or repair them.

---

**PROTOCOL 10.2. PREPARATION AND STERILIZATION OF GLASS PIPETTES**

### Materials
- ☐ Pipette cylinders (to collect used pipettes)
- ☐ Disinfectant: hypochlorite, 500 ppm available chlorine at minimum, when diluted in detergent (e.g., Precept tablets, Clorox, or Chloros)
- ☐ Detergent (e.g., 7X or Decon)
- ☐ Stainless steel baskets (to collect washed and rinsed pipettes for drying)
- ☐ Sterility indicators (*see* Appendix II)
- ☐ Pipette cans: square aluminum or stainless steel with silicone cushions at either end; square cans do not roll on the bench (Thermo Fisher)
- ☐ Sterile-indicating tape or tabs (*see* Protocol 10.1)
- ☐ Sterilizing oven, fan assisted, capable of reaching 160°C, and preferably with recording thermometer and flexible probe

### Procedure

Collection and washing

1. Place water with detergent and a disinfectant in pipette cylinder.
2. Discard pipettes, tip first, into cylinder immediately after use (Fig. 10.5):
    (a) Do not accumulate pipettes in the hood or allow pipettes to dry out.
    (b) Do not put pipettes that have been used with agar or silicones (water repellent, antifoam, etc.) in the same cylinder as regular pipettes. Use disposable pipettes for silicones, and either rinse agar pipettes after use in hot tap water or use disposable pipettes.
3. Soak pipettes overnight or for a minimum of 2 h. If usage of pipettes is heavy, replace cylinders at intervals when full, and soak for 2 h before entering rinse cycle.
4. After soaking, remove plugs with compressed air.
5. Transfer pipettes to pipette washer (Figs. 10.4, 10.5; *see also* Fig. 4.6), tips uppermost.
6. Rinse in tap water by siphoning action of pipette washer for a minimum of 4 h or in an automatic washing machine with a pipette adapter, but without detergent.

7. Turn valve to setting for deionized (DW) or reverse-osmosis (ROW) water (*see* Fig. 4.15, 10.9), or wait until last of the tap water finally runs out, turn off the tap water, and empty and fill washer three times with DW or ROW. (Use automatic deionized rinse cycle in glassware washing machine.)
8. Transfer pipettes to pipette dryer or drying oven, and dry with tips uppermost.
9. Plug with cotton (Fig. 10.6).
10. Sort pipettes by size and store dust free.

Sterilization

1. Place pipettes in pipette cans. (It is useful to have both ends of the cans labeled with the size of the pipettes.)
2. Fill each can with one size of pipette.
3. Fill a few cans with an assortment of 1-, 2-, 10-, and 25-mL pipettes (e.g., four of each size).
4. Attach sterile-indicating tape, bridging the cap to the can, and stamp date on tape.
5. Sterilize by dry heat for 1 h at 160°C. Use the smallest amount of tape possible, or replace with temperature indicator tabs, which are small and are made of less volatile material. The temperature should be measured in the center of the load, to ensure that this, the most difficult part to heat, attains the minimum sterilizing conditions. Leave spaces between cans when loading the oven, to allow for circulation of hot air (Fig. 10.7).
6. Remove pipette cans from oven, allow to cool, and transfer to tissue culture laboratory. If you anticipate that pipettes will lie idle for more than 48 h, seal cans around the cap with adhesive tape.

## 10.3.3 Screw Caps

There are two main types of caps that are in common use for glass bottles: (1) aluminum or phenolic plastic caps with synthetic rubber or silicone liners and (2) wadless polypropylene caps that are reusable (Duran); the latter are deeply shrouded and have ring inserts for better sealing and to improve pouring (although pouring is not recommended in sterile work). The following precautions should be observed:

(1) Wadless caps are preferable; if using phenolic or aluminum caps with liners, the liners must be removed during washing.
(2) Polypropylene caps will seal only if screwed down tightly on a bottle with no chips or imperfections on the lip of the opening. Discard bottles with chipped necks.
(3) Do not leave aluminum caps or any other aluminum items in alkaline detergents for more than 30 min, as they will corrode.

(4) Do not put glassware together with caps in the same detergent bath, or the aluminum may contaminate the glass.
(5) Avoid detergents that are made for machine washing, as they are highly caustic.

---

### PROTOCOL 10.3. PREPARATION AND STERILIZATION OF SCREW CAPS

#### *Materials*
- Disinfectant: hypochlorite, 500 ppm available chlorine (e.g., Precept tablets)
- Detergent (e.g., 7X or Decon)
- Soaking baths
- Stainless steel baskets (to collect caps for washing and drying)
- Sterility indicators (e.g., Thermalog; *see also* Appendix II)
- Glass Petri dishes (for packaging)
- Autoclavable plastic film (*see* Appendix II) or paper sterilization bags
- Sterile-indicating autoclave tape
- Autoclave, with recording thermometer with flexible probe that can be inserted in load

#### *Procedure*

Collection and washing

*Metal or phenolic caps with liners:*
1. Soak 30 min (maximum) in detergent.
2. Rinse thoroughly for 2 h. (Make sure all caps are submerged.) Liners should be removed and replaced after rinsing, which may be carried out in either of two ways:
   (a) In a beaker (or pail) with running tap water led by a tube to the bottom. Stir the caps by hand every 15 min.
   (b) In a basket or, better, in a pipette-washing attachment. Rinse in an automatic washing machine, but do not use detergent in the machine.

Polypropylene caps
3. These may be washed and rinsed by hand as just described (extending the detergent soak if necessary). Because these caps may float, they must be weighted down during soaking and rinsing. For automatic washers, after soaking in detergent, use pipette-washing attachment and normal cycle without machine detergent.

Stoppers
4. Shrouded caps are preferred to stoppers, but if the latter are required, use silicone or heavy metal-free white rubber stoppers in preference to those

made of natural rubber. Wash and sterilize as for caps. (There will be no problem with flotation in washing and rinsing.)

**Sterilization**

5. Place caps in a glass Petri dish with the open side down.
6. Wrap Petri dish containing caps in cartridge paper or steam-permeable nylon film, and seal with autoclave tape (Fig. 10.8).
7. Prepare similar package with sterility indicator enclosed and insert in the middle of the load.
8. Autoclave for 20 min at 121°C and 100 kPa (15 psi) (Fig. 10.9).

## 10.3.4 Selection of Detergent

When most culture work was done on glass, the quality (charge, chemical residue) of the glass surface was critical. As most cell culture is now carried out on disposable plastic, the major requirements for cleaning glassware are that (1) the detergent be effective in removing residue from the glass and (2) no toxic residue be left behind to leach out into the medium or other reagents. Some tissue culture suppliers will provide a suitable detergent that has been tested with tissue culture (e.g., 7X, MP Biomedicals, or Decon for manual washing), but often machine detergents will come from a general laboratory supplier or the supplier of the machine. The washing efficiency of a detergent can be determined by washing heavily soiled glassware (e.g., a

**TABLE 10.3.** Sterilization of Liquids

| Solution | Sterilization | Storage |
|---|---|---|
| Agar | Autoclave[a] or boil | Room temperature |
| Amino acids | Filter[b] | 4°C |
| Antibiotics | Filter | −20°C |
| Bacto-peptone | Autoclave | Room temperature |
| Bovine serum albumin | Filter (use stacked filters) | 4°C |
| Carboxylmethyl cellulose | Steam, 30 min[c] | 4°C |
| Collagenase | Filter | −20°C |
| DMSO | Self-sterilizing; dispense into aliquots in sterile tubes | Room temperature; keep dark, avoid contact with rubber or plastics (except polypropylene) |
| Drugs | Filter (check for binding; use low-binding filter, e.g., Millex-GV, if necessary) | −20°C |
| EDTA | Autoclave | Room temperature |
| Glucose, 20% | Autoclave | Room temperature |
| Glucose, 1–2% | Filter (low concentrations; caramelizes if autoclaved) | Room temperature |
| Glutamine | Filter | −20°C |
| Glycerol | Autoclave | Room temperature |
| Growth factors | Filter (low protein binding) | −20°C |
| HEPES | Autoclave | Room temperature |
| HCl, 1 M | Filter | Room temperature |
| Lactalbumin hydrolysate | Autoclave | Room temperature |
| Methocel | Autoclave | 4°C |
| NaHCO$_3$ | Filter | Room temperature |
| NaOH, 1 M | Filter | Room temperature |
| Phenol red | Autoclave | Room temperature |
| Salt solutions (without glucose) | Autoclave | Room temperature |
| Serum | Filter; use stacked filters | −20°C |
| Sodium pyruvate, 100 mM | Filter | −20°C |
| Transferrin | Filter | −20°C |
| Tryptose | Autoclave | Room temperature |
| Trypsin | Filter | −20°C |
| Vitamins | Filter | −20°C |
| Water | Autoclave | Room temperature |

[a]Autoclave, 100 kPa (15 lb/in.$^2$), 121°C for 20 min.
[b]Filter, 0.2-μm pore size.
[c]Steam, 100°C for 30 min.

bottle of serum or a medium containing serum that has been autoclaved). One simply uses the normal washup procedure; a visual check will then show which detergents have been effective.

The presence of a toxic residue is best determined by cloning cells (*see* Protocol 20.10), such as on a glass Petri dish that has been washed in the detergent, rinsed as previously indicated (*see* Protocol 10.1), and then sterilized by dry heat. A plastic Petri dish should be used as a control. This technique can also be used if you are anxious about residue left on glassware after autoclaving.

### 10.3.5  Miscellaneous Equipment

*Cleaning.* All new apparatus and materials (silicone tubing, filter holders, instruments, etc.) should be soaked in detergent overnight, thoroughly rinsed, and dried. Anything that will corrode in the detergent—such as mild steel, aluminum, copper, or brass—should be washed directly by hand without soaking (or with soaking for 30 min only, using detergent if necessary), brushed, and then rinsed and dried.

Used items should be rinsed in tap water and immersed in detergent immediately after use. Allow them to soak overnight, and then rinse in deionized water and dry in the oven. Again, do not expose materials that might corrode to detergent for longer than 30 min. Aluminum centrifuge buckets and rotors must never be allowed to soak in detergent.

Particular care must be taken with items treated with silicone grease or silicone fluids. These items must be treated separately and the silicone removed, if necessary, with carbon tetrachloride. Silicones are very difficult to remove if they are allowed to spread to other apparatus, particularly glassware.

*Packaging.* Ideally all apparatus being sterilized should be wrapped in a covering that will allow steam to penetrate (*see* Figs. 10.1, 10.8) but will be impermeable to dust, microorganisms, and mites. Proprietary bags are available with sterile-indicating marks that show up after sterilization, and semipermeable transparent nylon film, in rolls of flat tubes of different diameters, can be made up into bags with sterile-indicating tape (*see* Appendix II, Autoclavable bags and film). Although expensive, such film can be reused several times before becoming brittle.

Tubes and orifices should be covered with tape and paper or nylon film before packaging, and needles or other sharp points should be shrouded with a glass test tube or other appropriate guard.

*Sterilization.* The type of sterilization used will depend on the material (*see* Table 10.1). Metallic items are best sterilized by dry heat. Silicone rubber (which should be used in preference to natural rubber), PTFE, polycarbonate, cellulose acetate, and cellulose nitrate filters, for example, should be autoclaved for 20 min at 121°C and 100 kPa (1 bar, 15 lb/in.$^2$) with preevacuation and postevacuation

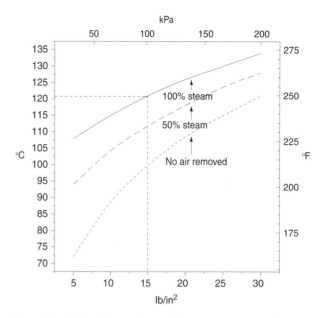

**Fig. 10.1. *Effect of Humidity on Temperature in Autoclave.*** The top curve shows that only when all the air is removed that 121°C can be achieved at 100 kPa (15 lb/in$^2$). Even when 50% of air is removed it still requires 150 kPa (23 lb/in$^2$) to reach 121°C. (Data from Breach, 1968.)

steps in the cycle, except when filters are sterilized in a filter assembly (*see* Protocol 10.4). In small bench-top autoclaves and pressure cookers, make sure that the autoclave boils vigorously for 10 to 15 min before pressurizing to displace all the air. (Take care that enough water is put in at the start to allow for evaporation.) After sterilization the steam is released and the items are removed to dry off in an oven or rack. Ensure proper records are taken for each preparation and sterilization run (Table 10.4).

Δ *Safety Note.* To avoid burns, take care in releasing steam and handling hot items. Wear elbow-length insulated gloves, and keep your face well clear of escaping steam when you open doors and lids. Use safety locks on autoclaves.

### 10.3.6  Reusable Sterilizing Filters

Although most laboratories now use disposable filter assemblies, some large-scale procedures may require stainless steel filter housings (Tables 10.5). Filter assemblies should be made up and sterilized in accordance with Protocol 10.4.

---

**PROTOCOL 10.4. STERILIZING FILTER ASSEMBLIES**

**Materials**

*Nonsterile:*

❑ Filter holder (Table 10.4).
❑ Micropore filters to fit holder

❑ Prefilter, glass fiber, if required
❑ Steam-permeable nylon film
❑ Sterile-indicating autoclave tape
❑ Autoclave

### Procedure

1. After thorough washing in detergent (*see* Section 10.3.6), rinse assembly in water, followed by deionized water, and dry.
2. Insert support grid in filter and place filter membrane on grid. If membrane is made of polycarbonate, apply wet to counteract static electricity.
3. Place prefilters, glass fiber, and other prefilters as required (*see* Section 10.5.3), on top of filter.
4. Reassemble filter holder, but do not tighten up completely. (Leave about one whole turn on bolts.)
5. Cover inlet and outlet of filter with aluminum foil.
6. Pack filter assembly in sterilizing paper or steam-permeable nylon film, and close assembly with sterile-indicating tape.
7. Autoclave at 121°C and 100 kPa (1 bar, 15 lb/in.$^2$) for 30 min with no preevacuation or postevacuation; use upward or downward displacement for 10 to 15 min to remove air before starting the sterilization cycle. (Use "liquids cycle" in automatic autoclaves.)
8. Remove and allow to cool.
9. Do not tighten filter holder completely until the filter is wetted at the beginning of filtration (*see* Protocol 10.14).

***Alternative methods of sterilization.*** Many plastics cannot be exposed to the temperature required for autoclaving or dry-heat sterilization. To sterilize such items, immerse them in 70% alcohol for 30 min and dry them off under UV light in a laminar-flow hood. Care must be taken with some plastics (e.g., Plexiglas, Perspex, Lucite), as they will depolymerize in alcohol or when they are exposed to UV light. Ethylene oxide may be used to sterilize plastics, but two to three weeks are required to clear it completely from the plastic surface after sterilization. The best method for sterilizing plastics is γ-irradiation, at a level of 25 kGy. Items should be packaged and sealed; polythene may be used and sealed by heat welding.

## 10.4  REAGENTS AND MEDIA

The ultimate objective in preparing reagents and media is to produce them in a pure form (1) to avoid the accidental inclusion of toxic substances, (2) to enable the reagent to be totally defined and the functions of its constituents to be fully understood, and (3) to reduce the risk of microbial contamination.

Most reagents and media can be sterilized either by autoclaving, if they are heat stable—such as water, salt solutions, and amino acid hydrolysates—or by membrane filtration, if they are heat labile. For autoclaving, solutions should be dispensed into borosilicate glass or polycarbonate and kept sealed to avoid evaporation and chemical pollution from the autoclave. If soda glass bottles are used, they are better left with the caps slack to minimize breakage. The evolution of vapor will help to prevent steam entering from the autoclave, but the level of the liquid will need to be restored with sterile UPW later. As with sterilizing apparatus, sterile indicators (e.g., Thermalog) and the probe from a recording thermometer should be placed in a mock sample in the center of the load.

Media and reagents supplied on line to large-scale culture vessels and industrial or semi-industrial bioreactors can be sterilized on line by short-duration ultrahigh-temperature treatment (Alfa-Laval). Adapting this process to media production might allow increased automation and ultimately reduce costs.

### 10.4.1  Water

Water used in tissue culture must be of a very high purity, particularly with serum-free media (*see also* Section 4.4.2). Because water supplies vary greatly, the degree of purification required may vary. Hard water will need a conventional ion-exchange water softener on the supply line before entering the purification system, but this will not be necessary with soft water.

There are four main approaches to water purification: reverse osmosis, distillation, deionization, and carbon filtration. For ultrapure water (UPW), the first stage is usually reverse osmosis, but it can be replaced by distillation (Fig. 10.9; *see also* Fig. 5.17). Distillation has the advantage that the water is heat sterilized, but it is more expensive because of power consumption and the need to clean out the boiler regularly. If glass distillation is used for the first stage, the still should be electric and automatically controlled, and the heating elements should be made of borosilicate glass or silica sheathed. Reverse osmosis depends on the integrity of the filtration membrane; hence the effluent must be monitored. The type of reverse-osmosis cartridge used is determined by the pH of the water supply (see manufacturer's specification for details). If the costs of both power for distillation and replacement membranes are deducted directly from your budget, reverse osmosis will probably work out cheaper, but if power is supplied free or is costed independently of usage, then distillation will be cheaper. The product of the first stage is semipurified water (SPW), useful for rinsing glassware and preparing some laboratory reagents (*see* Fig. 10.9).

The second stage is carbon filtration, which will remove both organic and inorganic colloids. The third stage is high-grade mixed-bed deionization to remove ionized inorganic material. The product of the second and third stages is purified water (PW) and can be used for some laboratory reagents,

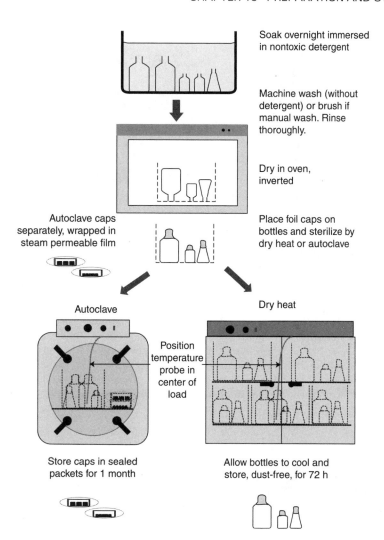

Soak overnight immersed in nontoxic detergent

Machine wash (without detergent) or brush if manual wash. Rinse thoroughly.

Dry in oven, inverted

Autoclave caps separately, wrapped in steam permeable film

Place foil caps on bottles and sterilize by dry heat or autoclave

Autoclave

Dry heat

Position temperature probe in center of load

Store caps in sealed packets for 1 month

Allow bottles to cool and store, dust-free, for 72 h

**Fig. 10.2. Washing and Sterilizing Glassware.** Sterilization conditions: autoclave at 121°C for 15 min; oven, minimum of 160°C for 1 h (*see* Protocol 10.1). Caps are sterilized separately from bottles to avoid condensation forming in bottles if autoclaved with caps in place (*see* Fig. 10.3).

D-PBSA, and trypsin. The final stage is micropore filtration to remove any microorganisms acquired from the system and to trap any resin that may have escaped from the deionizer. To minimize pollution during storage, the water should be collected directly from the final-stage micropore filter without being stored. If the water is recycled continuously from the micropore filter to the reservoir (*see* Fig. 10.9), with the supply from the first stage turned off (e.g., overnight), the stored water gradually "polishes" (i.e., increases in purity). This generates ultrapure water (UPW) suitable for preparing media and should be collected from the filtered outlet (right-hand side in Fig. 10.9) first thing in the morning.

The quality of the deionized water should be monitored by its conductivity (the inverse of resistivity) at regular intervals, and the cartridge should be changed when an increase in conductivity is observed. The ISO 3696 standard sets resistivity of type I water at a level $\geq 10$ M$\Omega$/cm (conductivity $<0.1$ $\mu$S/cm) at 25°C. The total organic

carbon (TOC) should be $\leq 10$ parts per billion (ppb). Conductivity meters are usually supplied with water purification systems; a TOC meter may need to be purchased separately (*see* Appendix II) but is included with some systems (*see* Fig. 4.15c). Distillation can precede deionization, providing sterile water to the deionization stage. An ultrafiltration stage can be inserted between deionization and micropore filtration to produce pyrogen-free water.

Sterilize water in suitable aliquots (e.g., 450 mL for media preparation from 10× concentrates) by autoclaving at 121°C and 100 kPa (15 lb/in.$^2$, 1 bar) for 20 min. Borosilicate glass (Pyrex) and polycarbonate bottles should be sealed during sterilization. If soda glass is used, it may break if sealed, so bottles are left with the caps slack and 10% extra volume per bottle is added to allow for evaporation. Caps are tightened when bottles are cool.

Protocol 10.5 is also used in Exercise 6 (*see* Section 28.2).

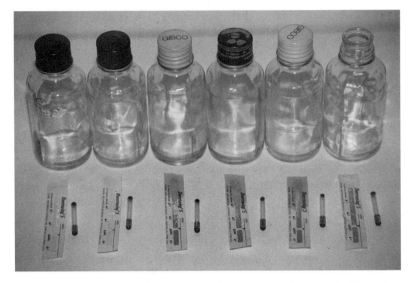

***Fig. 10.3. Sterilizing Capped Bottles.*** These bottles were autoclaved with Thermalog sterility indicators inside. Thermalog turns blue with high temperature and steam, and the blue area moves along the strip with time at the required sterilization conditions. The cap on the leftmost bottle was tight, and each succeeding cap was gradually slacker, until finally no cap was used on the bottle farthest to the right. The farthest left bottle is not sterile because no steam entered it. The second bottle is not sterile either, because the liner drew back onto the neck and sealed it. The next three bottles are all sterile, but the brown stain on the indicator shows that there was fluid in them at the end of the cycle. Only the bottle at the far right is sterile and dry. The glass indicators (Browne's tubes) all implied that their respective bottles were sterile. (*See also* Plate 22a.)

### PROTOCOL 10.5. PREPARATION AND STERILIZATION OF ULTRAPURE WATER (UPW)

***Equipment and Materials***

*Nonsterile:*

❑ Graduated glass borosilicate (e.g., Pyrex) or autoclavable plastic (polycarbonate) screw-cap bottles (ensure that there is sufficient head space for later additions)
❑ Screw caps to fit
❑ Sterility indicator strips
❑ Sterile-indicating tape
❑ Marker pen or preprinted labels
❑ Water purification equipment
❑ Autoclave
❑ Log (record of preparation and sterilization); book or computer database (Table 10.5)

***Procedure***

1. Create entry in log book or database (*see* Table 10.4); label bottles with date, contents, and batch number. (A label printer attached to the computer will generate labels automatically.)
2. Early in the morning after overnight recycling, run about 50 mL of water to waste from purifier, check conductivity (or resistivity) and total organic carbon (TOC) on respective meters, and enter in log.
3. If water is within specified limits (resistivity $\geq 10$ M$\Omega$/cm at 25°C, TOC $\leq 10$ ppb), collect ultrapure water directly into labeled bottles.
4. Fill to the specified mark (e.g., 430–450 mL) if to be used for diluting 10× concentrated medium (*see* Protocol 10.8). If to be autoclaved open, add 10% extra.
5. Place sterility indicator in one bottle (to be discarded when checked after autoclaving).
6. Seal bottles with screw caps. If using soda glass (i.e., if bottle is liable to breakage), then leave caps slack and add 10% extra water as in step 4.
7. Place bottles in autoclave with bottle containing sterility indicator in center of load.
8. Close autoclave and check settings: 121°C, 100 kPa (15 lb/in.², 1 bar), for 20 min with postvacuum deselected.
9. Start sterilization cycle.
10. On completion of cycle, check printout to confirm that correct conditions have been attained for the correct duration and enter in log.
11. Allow load to cool to below 50°C.

12. Open autoclave and retrieve bottles. If caps were slack during autoclaving, tighten when bottles reach room temperature.

*Note.* Bottles that have been open during autoclaving should not be sealed while they are still warm. This can cause the liner in some caps to be drawn into the bottle as it cools and the contents contract. Also there will be a rapid intake of air into the bottle when it is first opened, and this can cause contamination. Bottles that have been open during autoclaving should be allowed to cool to room temperature in a sterile atmosphere, such as in the autoclave or in horizontal laminar flow, before the caps are tightened.

13. Check sterility indicator to confirm that sterilization conditions have been achieved and enter in log (*see* Table 10.4).
14. Place bottles in short-term storage at room temperature.
15. Replenish culture room stocks as required, confirming by examination of sterile-indicating tape that the bottles have been through sterilization cycle.

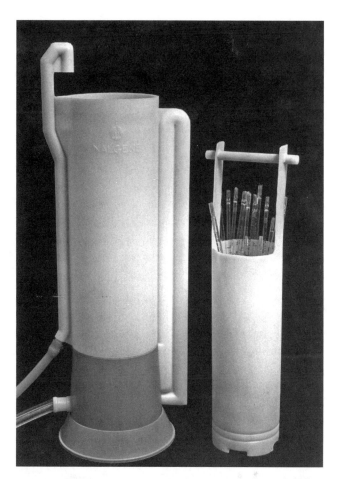

**Fig. 10.4. Siphon Pipette Washer.** Connected to main water supply. A slow fill establishes a siphoning action and drains the main chamber, which then refills. This process is repeated many times over 4 to 6 h. The pipettes shown here are as collected from use. Before the pipette basket is placed in the washer, the plugs are blown out and the pipettes inverted, as in Fig. 10.5.

## 10.4.2 Maintenance of Water Purifier

Institute a program of monitoring the system regularly by checking the record of conductivity and TOC of water at each outlet weekly and visually checking the distillation boiler, tubing, and connections regularly. When a pattern emerges indicating the expected useful life of the reverse osmosis, deionization, and charcoal cartridges, institute a program of replacement that preempts these deteriorations. Similarly, by observation, it can be determined how quickly the distillation boiler (if used) scales up. Clean as follows:

(1) Dismantle the glass boiler.
(2) Clean it in 1 M HCl and rinse thoroughly.
(3) Discard first boiling from still.
(4) Test first collection of water in plating efficiency assay (*see* Protocol 20.10).

The tubing and connections can also be contaminated with algae or other microorganisms. This can usually be detected visually and the tubing and connectors cleaned as follows:

(1) Disconnect and clean tubing.
(2) Soak in hot detergent.
(3) Rinse and soak in 1 M HCl.
(4) Rinse thoroughly in deionized water.
(5) Reinstall and discard first batch of water.

(6) Test first collection of water in plating efficiency assay (*see* Protocol 20.10).

The frequency with which these operations must be carried out will vary from one laboratory to another depending on the quality of the main water supply and can be established by keeping a record. Once a pattern is established, it should be written into standard procedure.

## 10.4.3 Balanced Salt Solutions

The formulation of BSS has been discussed previously (*see* Section 8.3). The formula for Hanks's BSS [after Paul, 1975] contains magnesium chloride in place of some of the sulfate originally recommended (*see* Table 8.2); it should be autoclaved below pH 6.5 to prevent calcium and magnesium phosphates from precipitating and should be neutralized just before use. Similarly Dulbecco's phosphate–buffered saline (D-PBS) is made up without calcium and magnesium (D-PBSA), which are made up separately (D-PBSB) and added just before use if required. D-PBS is often used without the addition of the $Ca^{2+}$ and $Mg^{2+}$ component, and in that form it should be referred to as D-PBSA or D-PBS without $Ca^{2+}$ and $Mg^{2+}$; the "D-PBSA" convention is used throughout this book. "D-PBS" is also used to distinguish Dulbecco's formulation from simple phosphate-buffered saline (PBS), which lacks potassium chloride and is just isotonic sodium chloride with phosphate buffer and should not be regarded as a balanced salt solution. Which formulation is being used should always be made clear in reports and publications.

Most balanced salt solutions contain glucose, and because glucose can caramelize on autoclaving, the salt solution is best omitted during preparation and pressure sterilization and added later. If glucose is prepared as a $100\times$ concentrate (200 g/L), caramelization during autoclaving is reduced, and it can be used at 5 to 25 mL/L BSS to give 1 to 5 g/L. Alternatively, complete balanced salt solutions can be prepared as for complete defined media (*see* Protocol 10.8) and sterilized by filtration (*see* Protocols 10.11–10.14). Protocol 10.6 is given for D-PBSA, commonly used as a rinsing solution and solvent for trypsin or ethylenediaminetetraacetic acid (EDTA), but can also be used for any BSS lacking glucose and bicarbonate. (For training, *see* Chapter 28, Exercise 7.)

---

**PROTOCOL 10.6. PREPARATION AND STERILIZATION OF D-PBSA**

**Outline**

Dissolve powder with constant mixing, make up to final volume, check pH and conductivity, dispense into aliquots, and autoclave.

**Materials**

*Nonsterile:*

- ❑ D-PBS powder (Solution A, lacking $Ca^{2+}$ and $Mg^{2+}$, e.g., Sigma D5652) or tablets (Oxoid Br 14a)
- ❑ Ultrapure water (UPW; *see* Section 10.4.1)
- ❑ Container: clear glass or clear plastic aspirator with tap outlet at base, Erlenmeyer flask or bottle with peristaltic metering pump and tubing
- ❑ Magnetic stirrer and PTFE-coated follower
- ❑ Bottles for storage, graduated; borosilicate glass
- ❑ Conductivity meter or osmometer
- ❑ pH meter
- ❑ Autoclave tape or sterile-indicating tabs
- ❑ Autoclave
- ❑ Logbook or access to computer database

**Procedure**

1. Add UPW to container.

---

2. Place container on magnetic stirrer and set to around 200 rpm.
3. Open packet of D-PBSA powder, or count out appropriate number of tablets, and add slowly to container while mixing.
4. Stir until completely dissolved.
5. Check pH and conductivity of a sample and enter in record.
   (a) pH should not vary more than 0.1 pH unit (pH may vary with different formulations).
   (b) Conductivity should not vary more than 5% from 150 μS/cm. Osmolality can be used as an alternative to conductivity (or in addition), and should show similar consistency between batches.

*Note.* It is important to check these parameters for consistency, as a quality control measure, to ensure that there has been no mistake in preparation. Any adjustments, such as to the osmolality, should be made after the quality control checks have been made.

   (c) Discard sample; do not add back to main stock.
6. Dispense contents of container into graduated bottles.
7. Cap and seal bottles.
8. Attach a small piece of autoclave tape or sterile-indicating tab and date.
9. Sterilize by autoclaving for 20 min at 121°C and 100 kPa (1 bar, 15 lb/in.$^2$) in sealed bottles.
10. Store at room temperature.
11. Complete records in log book or computer database.

---

## 10.4.4 Preparation and Sterilization of Media

During the preparation of complex solutions, care must be taken to ensure that all of the constituents dissolve and do not get filtered out during sterilization and that they remain in solution after autoclaving or storage. Concentrated media are often prepared at a low pH (between 3.5 and 5.0) to keep all the constituents in solution, but even then some precipitation may occur. If the constituents are properly resuspended, they will usually redissolve on dilution, but if the precipitate has been formed by degradation of some of the constituents of the medium, then the quality of the medium may be reduced. If a precipitate forms, the performance of the medium should be checked by cell growth and cloning and an appropriate assay of special functions (*see* Section 8.6.2), or the medium should be replaced.

Commercial media are supplied as (1) working-strength solutions ($1\times$) with or without glutamine; (2) $10\times$ concentrates, usually without $NaHCO_3$ and glutamine,

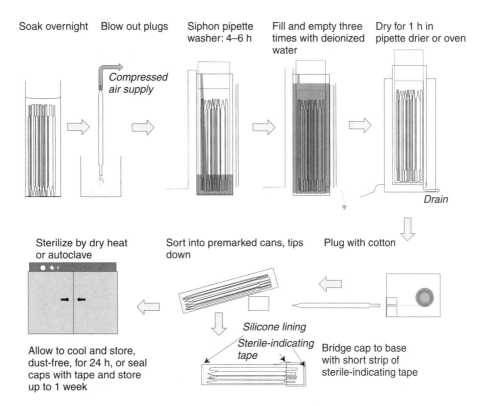

Soak overnight | Blow out plugs | Siphon pipette washer: 4–6 h | Fill and empty three times with deionized water | Dry for 1 h in pipette drier or oven

*Compressed air supply*

*Drain*

Sterilize by dry heat or autoclave | Sort into premarked cans, tips down | Plug with cotton

Allow to cool and store, dust-free, for 24 h, or seal caps with tape and store up to 1 week

*Silicone lining*

*Sterile-indicating tape*

Bridge cap to base with short strip of sterile-indicating tape

**Fig. 10.5. *Washing and Sterilizing Pipettes.*** Plugs are removed by compressed air (top left) and the pipettes are washed by automatically repeated emptying and filling of the pipette washer before transferring to a drier (top right). Plugs are reinserted (bottom right; *see also* Fig. 10.6) and the pipettes are placed in cans, sealed with sterile-indicating tape or marker and sterilized by hot air for a minimum of 1 h (*see also* Fig. 10.7).

which are available as separate concentrates; or (3) powdered media, with or without $NaHCO_3$ and glutamine. Powdered media are the cheapest and not a great deal more expensive than making up medium from your own chemical constituents if you include time for preparation, sterilization, and quality control and for the cost of raw materials of high purity and the cost of overheads such as power and wages. Powdered media are quality controlled by the manufacturer for their growth-promoting properties but not, of course, for sterility. They are mixed very efficiently by ball milling, so, in theory, a pack may be subdivided for use at different times. However, in practice, it is better to match the size of the pack to the volume that you intend to prepare, because once the pack is opened, the contents may deteriorate and some of the constituents may settle.

Tenfold concentrates cost two to three times as much per liter of working-strength medium as powdered media but save on sterilization costs. Buying media at working strength is the most expensive (4−5 times the cost of a 10× concentrate) but is the most convenient, as no further preparation is required other than the addition of serum, if that is required, and possibly glutamine. Protocol 10.7 describes the preparation of complete medium from 1× stock. (For training exercise, *see* Chapter 28, Exercise 8.)

**PROTOCOL 10.7. PREPARATION OF MEDIUM FROM 1× STOCK**

***Outline***
Check the formulation; if complete, it may be used directly, after adding serum if that is required (*see* Sections 8.6, 12.3.2). If the formulation is incomplete (e.g., lacking glutamine), add the appropriate stock concentrate.

***Note.*** A supplement (e.g., serum or antibiotics) is a component that is added to the medium and is not in the original formulation. It needs to be indicated in any record or publication. Other additions (e.g., glutamine or $NaHCO_3$) are part of the formulation and are not supplements. They need not be indicated in records or publications unless their concentrations are changed.

***Materials***
❑ Pipettes, and other items as listed for aseptic technique (*see* Protocols 5.1, 5.2)
❑ Medium stock, such as Eagle's MEM with Hanks's salts, 1×

☐ Glutamine, 200 mM (will need to be thawed)
☐ Serum (will need to be thawed), newborn or fetal bovine
☐ Antibiotics (if required; **not** recommended for routine use):
Penicillin in BSS or D-PBSA, 10,000 U/mL
Streptomycin in BSS or D-PBSA, 10 mg/mL

### Procedure

1. Check specification of medium, and determine what additions are required (e.g. glutamine).
2. Take medium to hood with any other supplement or addition that is required.
3. Remove any outer wrapping and swab bottle with 70% alcohol.
4. Uncap bottles.
5. Transfer the appropriate volume of each addition to the stock bottle to make the correct dilution; for example, for glutamine, 200 mM, dilute 100×, and antibiotics, dilute 200×.
6. Add serum to give final concentration required. This should be, for example, 11 mL to 100 mL to give 10%, but frequently 10 mL is added to 100 mL and called "10%." The first option is correct.
   (a) Use a different pipette for each addition.
   (b) Move each new stock to the opposite side of the hood after it has been added, so that you will know that it has been used.
   (c) Remove all additives or supplements from the hood when the medium is complete.
7. This protocol specifies MEM with Hanks's salts, which has a low bicarbonate concentration (4 mM). If MEM with Earle's salts, or another high-bicarbonate (23 mM) medium, is used then gas the air space with 5% $CO_2$. Do not bubble gas through any medium containing serum, as the medium will froth out through the neck, risking contamination.
8. Recap bottles.
9. Alter labeling to record additions, date and initial the bottle.
10. Return medium to 4°C or use directly.
11. If using a new medium for the first time, pipette an aliquot into a flask or Petri dish, and incubate for at least 1 h (preferably overnight) under your standard conditions, to ensure that the pH equilibrates at the correct value. If it does not, readjust the pH of the medium and repeat, or else alter the $CO_2$ concentration.
12. As soon as the medium is used for a particular cell line, label the bottle accordingly and do not use for any other cell line.

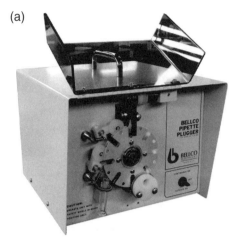

(a)

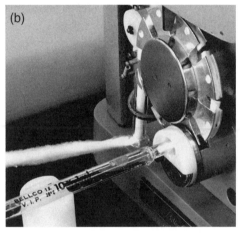

(b)

**Fig. 10.6. Semiautomatic Pipette Plugger (Bellco).** (*a*) Pipette plugger (Bellco) with front cover lifted (front cover should always be down in operation). (*b*) A strand of cotton is fed through the loading aperture, the appropriate length is inserted in the end of the pipette, and a section is cut off (photo taken on older model). Different thicknesses of cotton are required for different sizes of pipette.

**Note.** Changing the $CO_2$ concentration of the incubator will affect all other culture media in the same incubator. Changing the $CO_2$ concentration should be regarded as a one-time adjustment and should not be used for batch-to-batch variations, which are better controlled by adding sterile acid or alkali.

Preparing medium from 10× concentrate represents a good compromise between the economy of preparing medium from powder (*see* Protocol 10.9) and the ease of using a 1× preparation (*see* Protocol 10.7). (For training exercise, *see* Section 28.3, Exercise 10.)

### PROTOCOL 10.8. PREPARATION OF MEDIUM FROM 10× CONCENTRATE

#### Outline

Sterilize aliquots of deionized distilled water of such a size that one aliquot, when made up to full-strength medium, will last from one to three weeks. Add concentrated medium and other constituents, adjust the pH, and use the solution or return it to the refrigerator.

### Materials

□ Pipettes, and other items as listed for aseptic technique (*see* Protocols 5.1, 5.2)

### Sterile:

□ Premeasured aliquot of UPW
□ Medium, 10× concentrate, such as Eagle's MEM with Hanks's or Earle's salts
□ Glutamine (will need to be thawed), 200 mM
□ $NaHCO_3$, 7.5% (0.89 M)
□ Bovine serum, newborn or fetal (will need to be thawed)
□ Antibiotics if required (**not** recommended for routine use):
 Penicillin, 10,000 U/mL
 Streptomycin, 10 mg/mL
□ HEPES (if required), 1.0 M
□ NaOH, 1 M

### Procedure

1. Thaw serum and glutamine and bring to the hood.
2. Swab any bottles that have been in a water bath before placing in the hood.
3. Dilute 10× concentrated medium 1:10 in UPW.
4. Add glutamine, 1:100.
5. $NaHCO_3$:
 (a) Add 0.45 mL per 100 mL MEM with Hanks's salts for culture in a gas phase of air
 or
 (b) Add 2.6 mL per 100 mL MEM with Earle's salts for culture in a gas phase of 5% $CO_2$.
6. Add HEPES if required to increase the buffering capacity:
 (a) Final concentration 10 mM (1M stock 1:100) for MEM with Hanks's salts. This allows the flask to be vented to atmosphere for some cell lines at a high cell density if a lot of acid is produced.
 or
 (b) Final concentration of 20 mM (1M stock, 1:50) for MEM with Earle's salts but with only 1.0 mL $NaHCO_3$ per 100 mL medium and 2% $CO_2$ in the gas phase.

 *Note.* If HEPES is added check the osmolality; it may be necessary to add additional water and accept the minor dilution of nutrients that this will cause.

7. Add 1 M NaOH to give pH 7.2 at 20°C. When incubated, the medium will rise to pH 7.4 at 37°C,

but this figure may need to be checked by a trial titration the first time the recipe is used (*see* below, Protocol 10.7; Section 28.3, Exercise 10).

If any other constituents are required, and these additional constituents are made up in water, it will be necessary to remove an equivalent volume from the amount of water already present in the bottle before adding the main constituents. If the additional constituents are in isotonic salt,

**Fig. 10.7. Sterilizing Oven.** Pipette cans are stacked with spaces between to allow circulation of hot air. Brown staining on front of oven shows evidence of volatile material from sterile-indicating tape, a problem when using tape in a hot oven.

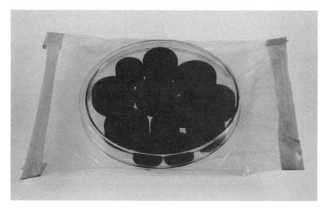

**Fig. 10.8. Packaging Screw Caps for Sterilization.** The caps are enclosed in a glass Petri dish, which is then sealed in an autoclavable nylon bag.

they can be added to the final volume of medium. Because it is isotonic, serum can be added to the final volume, although doing so will dilute the nutrients from the medium.

Always equilibrate and check the pH at 37°C, as the solubility of $CO_2$ decreases with increased temperature and the p$K_a$ of the HEPES will change.

***Adjusting pH.*** The amount of alkali needed to neutralize 10× concentrated medium (which is made up in acid to maintain the solubility of the constituents) may vary from batch to batch and from one medium to another, and in practice, titrating the medium to pH 7.4 at 37°C can sometimes be a little difficult. When making up a new medium for the first time, add the stipulated amount of $NaHCO_3$ and allow samples with varying amounts of alkali to equilibrate overnight at 37°C in the appropriate gas phase. Check the pH the following morning, select the correct amount of alkali, and prepare the rest of the medium accordingly.

***Sodium bicarbonate.*** The bicarbonate concentration is important in establishing a stable equilibrium with atmospheric $CO_2$. However, regardless of the amount of bicarbonate used, if the medium is at pH 7.4 and 37°C, the bicarbonate concentration at each concentration of $CO_2$ will be as previously described (*see* Section 8.2.2 and Table 8.1). Some media are designed for use with a high bicarbonate concentration and elevated $CO_2$ in the atmosphere (e.g., Eagle's MEM with Earle's salts), whereas others have a low bicarbonate concentration for use with a gas phase of air

(e.g., Eagle's MEM with Hanks's salts; *see* Tables 8.1, 8.2). If a medium is changed and its bicarbonate concentration altered, it is important to make sure that the osmolality is still within an acceptable range. The osmolality should always be checked (*see* Section 8.2.5) when any significant alterations are made to a medium that are not in the original formulation.

***Additions to medium.*** If your consumption of medium is fairly high (>200 L/year) and you are buying the medium readymade, then it may be better to get extra constituents included in the formulation, as this practice will work out to be cheaper. HEPES in particular is very expensive to buy separately. Glutamine is often supplied separately, as it is unstable; it is best to buy it separately and store it frozen. The half-life of glutamine in medium at 4°C is about 3 wk and at 37°C about 3 to 5 days. Some dipeptides of glutamine (e.g., Glutamax, Invitrogen) have increased stability, while retaining the bioavailability of the glutamine.

***Quality control.*** Once a batch of medium has been tested for its growth-promoting and other properties, if it is found to be satisfactory, it need not be tested each time it is made up to working strength. The sterility of a medium made by diluting a 10× concentrate should be checked, however, by incubating an aliquot of the complete medium at 37°C for 48 h before use (*see* Section 10.6.2).

### 10.4.5 Powdered Media

Instructions for the preparation of powdered media are supplied with each pack. Choose a size that you can make

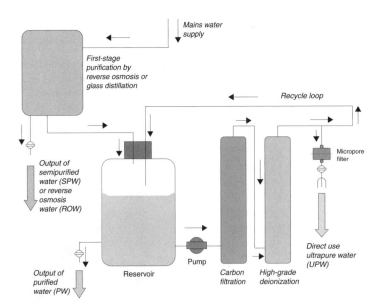

***Fig. 10.9. Water Purification.*** Tap water is fed to a storage container via reverse osmosis or glass distillation. This semipurified water is then recycled back to the storage container via carbon filtration, deionization, and micropore filtration. Reagent-quality water is available at all times from the storage reservoir; media–quality water is available from the micropore filter supply (at right of diagram). If the apparatus recycles continuously, then water of the highest purity will be collected first thing in the morning for the preparation of medium (*see also* Fig. 4.15).

**TABLE 10.4.** Suggested Layout for Log

| Batch | Initials | Date | Reagent | Meter readings | | | | Sterilization | | | |
|---|---|---|---|---|---|---|---|---|---|---|---|
| | | | | MΩ/cm[1] | pH | Osmol[2] | TOC[3] | Min[4] | °C | Tape[5] | Indic[6] |
| 12345 | RIF | 26/2/09 | HBSS | – | 6.5 | 290 | – | 15 | 121 | +ve | +ve |

| Storage | | | Quality control | | | |
|---|---|---|---|---|---|---|
| Location | °C | Date | Sterility test | Cell growth | PE | Other |
| Cold rm. | 4 | 27/2/09 | 37°C OK | — | — | — |

[1]Resistivity; conductivity, μS/cm, can also be used.
[2]mOsmol/kg.
[3]Total organic carbon in ppm.
[4]Time starts when temperature reaches 121°C.
[5]Sterile indicating tape; +ve confirms that appropriate marking has appeared.
[6]Indicator placed inside vessel at center of load; +ve confirms that color change or position of indicator strip confirms conditions met.

**TABLE 10.5.** Filter Size and Fluid Volume

| Filter size or designation | Disposable (D) or reusable (R) | Approximate volume that may be filtered | |
|---|---|---|---|
| | | Crystalloid | Colloid |
| 25 mm, Millex | D | 1–100 mL | 1–20 mL |
| 47 mm or Sterivex cartridge | R, D | 0.1–1 L | 100–250 mL |
| 90 mm | R | 1–10 L | 0.2–2 L |
| Millipak-20 | D | 2–10 L | 200 mL–2 L |
| Millipak-40 | D | 10–20 L | 2–5 L |
| Millipak-60 | D | 20–30 L | 5–7 L |
| Millipak-100 | D | 30–75 L | 7–10 L |
| Millipak-200 | D | 75–150 L | 10–30 L |
| Millidisk | D | 30–300 L | 5–50 L |
| 142 mm | R | 10–50 L | 1–5 L |
| 293 mm | R | 50–500 L | 5–20 L |

Note: Examples in the table are quoted from Millipore catalog. Similar products are available from Pall Gelman, Nalge Nunc, Sartorius, and a number of other suppliers. (*See* Appendix II.)

up all at once and use the medium within three months. Select a formulation lacking glutamine. If other unstable constituents are present, they also should be omitted and added later as a sterile concentrate just before use. Although BSS can be sterilized by autoclaving (*see* Protocol 10.6), this requires the omission of glucose and bicarbonate. If BSS is to be sterilized complete, follow the procedure for powdered medium (Protocol 10.9). This protocol is also presented as part of a training exercise (*see* Section 28.3, Exercise 6).

### PROTOCOL 10.9. PREPARATION OF MEDIUM FROM POWDER

#### Outline
Dissolve the entire contents of the pack in the correct volume of UPW, using a magnetic stirrer and adding the powder gradually with constant mixing. When

all the constituents have dissolved completely, the medium should be filtered immediately and not allowed to stand, in case any of the constituents precipitate or microbial contamination appears. The pH should be adjusted after the final constituents (e.g., glutamine, NaHCO$_3$, or serum) have been added to the medium.

#### Equipment and Materials

*Sterile:*

☐ Graduated bottles for medium
☐ Caps for bottles
☐ Universal containers for contamination control sampling

*Nonsterile:*

☐ Ultrapure water
☐ Powdered medium, such as MEM with Earle's salts, without glutamine

❑ Graduated Erlenmeyer flask or bottle, capacity plus head space
❑ Magnetic stirrer and PTFE-coated follower, 50 mm
❑ Conductivity meter

**Procedure**

1. Add appropriate volume of ultrapure water to container.
2. Add magnetic follower.
3. Place container on magnetic stirrer and set to around 200 rpm.
4. Open packet of powder and add contents slowly to container while mixing.
5. Stir until powder is completely dissolved.
6. Check pH and conductivity of a sample and enter in record:
    (a) pH should be within 0.1 unit of expected level for particular medium.
    (b) Conductivity should be within 2% of expected value for particular medium.
    (c) Discard sample; do not add back to main stock.
7. Sterilize by filtration (*see* Section 10.5.2; Protocols 10.11–10.14).
8. Cap and seal bottles.
9. Store at 4°C.
10. Add serum and glutamine, and correct pH to 7.4 as required, just before use.

---

For people using smaller amounts (<1.0 L/wk) or several different types of medium, smaller volumes may be prepared, complete with glutamine, and filtered directly into storage bottles with a bottle-top filter sterilizer or filter flasks (*see* Section 10.5.2 and Protocol 10.12). With this and other negative-pressure filtration systems, some dissolved $CO_2$ may be lost during filtration, and the pH may rise so it may be necessary to purge the head space of the bottles with 5% $CO_2$. Provided that the correct amount of $NaHCO_3$ is in the medium to suit the gas phase (*see* Section 8.2.2), the medium will reequilibrate in the incubator, but this should be confirmed the first time the medium is used.

For large-scale requirements (>10 L/wk), medium can be prepared in a pressure vessel, checked by sampling at intervals with a large pipette to determine whether solution is complete, and sterilized by positive pressure through an in-line disposable or reusable filter into a receiver vessel (*see* Section 10.5.2; Protocol 10.14).

### 10.4.6 Customized Medium

If there is a requirement to vary the formulation of the medium, or if it is to be made up in house from individual constituents, it is convenient to make up a number of concentrated stocks—amino acids at 50× or 100×, vitamins

at 1000×, and tyrosine and tryptophan at 50× in 0.1 M HCl, glucose at 200 g/L, and single-strength BSS. The requisite amount of each concentrate is then mixed to provide a combined stock concentrate, filtered through a sterilizing filter of 0.2-μm porosity, and diluted with high- or low-bicarbonate BSS (*see* Section 8.2.2; Table 8.2) for use. Care will be required to ensure that all constituents have dissolved and remain in solution before filtration.

---

**PROTOCOL 10.10. PREPARATION OF CUSTOMIZED MEDIUM**

**Outline**
Prepare stock concentrates (derived from Table 9.3 or 10.1) and store frozen. Thaw and blend as required. Sterilize by filtration and store until required.

**Materials**
❑ Amino acid concentrate, 100× in water, stored frozen
❑ Tyrosine and tryptophan, 50× in 0.1 M HCl
❑ Vitamins, 1000× in water
❑ Glucose, 100× (200 g/L in BSS)
❑ Additional solutions (e.g., trace elements and nucleosides; not lipids, hormones, or growth factors), which should be added just before using the medium
❑ Storage bottles, selected for optimum aliquot size for subsequent dilution (*see* step 5 of following procedure)

**Procedure**

1. Thaw solutions and ensure that all solutes have redissolved.
2. Blend constituents in correct proportions:
    (a) Amino acid concentrate 100 mL.
    (b) Tyrosine and tryptophan 200 mL.
    (c) Vitamins 10 mL.
    (d) Glucose 100 mL.
3. Mix and sterilize by filtration (*see* Protocols 10.12, 10.13).
4. Store frozen.
5. For use, dilute 41 mL of concentrate mixture with 959 mL sterile 1 × BSS.
6. Adjust to pH 7.4.
7. Store at 4°C.
8. Add serum or other supplements, such as hormones, growth factors, and lipids, just before using medium. If metals are used as trace elements, it is also better to add these just before using the medium, as they can precipitate in the presence of phosphate in concentrated stocks.

The advantage of this type of recipe is that it can be varied; extra nutrients (oxo-acids, nucleosides, minerals, etc.) can be added or the major stock solutions altered to suit requirements, but in practice, this procedure is so laborious and time-consuming that few laboratories make up their own media from basic constituents, unless they wish to alter individual constituents regularly. The reliability of commercial media depends entirely on the application of appropriate quality-control measures. Any laboratory carrying out its own preparation must make sure that appropriate quality-control measures are used (*see* Section 10.6.1).

There are now many reputable suppliers of standard formulations of media (*see* Appendix II), many of whom will supply specialized, serum-free formulations and custom media. It is important to ensure that the quality control these suppliers employ is relevant to the medium and the cells you wish to propagate. You might buy MCDB 153, which is tested on HeLa cell colony formation, but this test is of little relevance if you wish to grow primary keratinocytes.

## 10.5  STERILIZATION OF MEDIA

### 10.5.1  Autoclavable Media

Some commercial suppliers offer autoclavable versions of Eagle's MEM and other media. Autoclaving is much less labor intensive, is less expensive, and has a much lower failure rate than filtration. The procedure to follow is supplied in the manufacturer's instructions and is similar to that described earlier for D-PBSA (*see* Protocol 10.6). The medium is buffered to pH 4.25 with succinate, in order to stabilize the B vitamins during autoclaving, and is subsequently neutralized. Glutamine is replaced by glutamate or glutamyl dipeptides, or is added sterile after autoclaving. BSS is autoclavable without glucose, and various amino acid hydrolysates, such as tryptose phosphate broth and other microbiological media, are also sterilizable by autoclaving.

### 10.5.2  Sterile Filtration

Filtration through 0.1- to 0.2-$\mu$m microporous filters is the method of choice for sterilizing heat-labile solutions (Figs. 10.10–10.14). Numerous kinds of filters are available, made from many different materials, including polyethersulfone (PES), nylon, polycarbonate, cellulose acetate, cellulose nitrate, PTFE, and ceramics, and in sizes from syringe-fitting filters (Fig. 10.11*a*, *b*) through small and intermediate in-line filters (Fig. 10.11*c*, *f*; Fig. 10.12) or flask and bottle-top filters (Fig. 10.11*e*, *f*) to multidisk and cartridge filters (Fig. 10.13). Low-protein-binding filters are available from most suppliers (e.g., Durapore, Millipore); PES filters are generally found to be faster flowing. Polycarbonate filter membranes are absolute filters with an array of holes of a uniform porosity; the number of holes per unit area increases

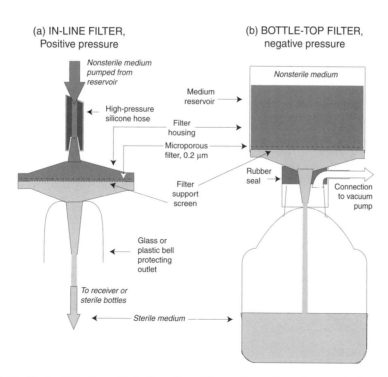

**Fig. 10.10. Sterile Filtration.** (*a*) In-line filter. Nonsterile medium from peristaltic pump (*see* Fig. 10.12) or pressure vessel (*see* Fig. 10.13). (*b*) Bottle-top filter or filter flask (designs are similar) for connection to vacuum pump. Medium added to upper chamber and collected in lower. Lower chamber can be used for storage.

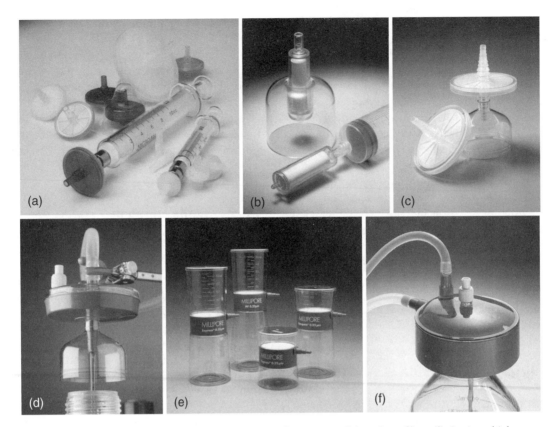

***Fig. 10.11. Disposable Sterilizing Filters.*** (*a*) Millex 25-mm disk syringe filter. (*b*) Sterivex high capacity; Luer fitting but can be attached via hose. (*c*) Millex 50-mm, in-line filter, with and without bell. (*d*) Steripak large in-line filter with bell (lowered in use to cover neck of bottle). (*e*) Strericup with storage vessels; also available as Stericap bottle-top filter. (*f*) Bottle-top filter drawing from separate reservoir (*see* Fig. 10.10 with built-in reservoir). (*a–d*) Positive pressure. (*e, f*) Negative pressure. (Photographs courtesy of Millipore Ltd., UK.)

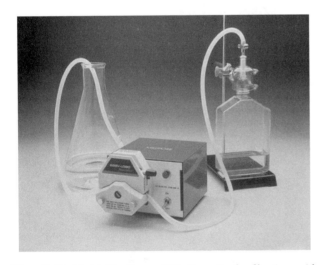

***Fig. 10.12. Peristaltic Pump Filtration.*** Sterile filtration with peristaltic pump between nonsterile reservoir and sterilizing filter (Courtesy of Millipore, Ltd., UK. This setup is no longer supplied by Millipore, who have introduced the Millipore Mobius FlexReady, designed for use on a larger scale, but a different peristaltic pump could be used.)

as the size diminishes, to maintain a uniform flow rate. Most other filters are of the mesh variety and filter by entrapment; they generally have a faster flow rate and reduced clogging, but will compress at high pressures.

Filtration may be carried out by negative pressure or positive pressure (*see* Figs. 10.10, 10.11), from a pressurized container (*see* Fig. 10.13) or with a peristaltic pump (*see* Fig. 10.12). The choice is based on the volume being processed and which type of pump is available, vacuum, peristaltic, or positive pressure (Fig. 10.14). High-pressure filtration from a pressure vessel is faster than a peristaltic pump and may benefit from using a collecting receiver as the flow cannot be regulated as easily, whereas a peristaltic pump can be switched on and off with more instantaneous effect during collection. High pressure also tends to compact the filter and is unsuitable for viscous solutions such as serum. Negative-pressure filtration is often simpler (*see* Figs. 10.10*b*, 10.11*e, f*), particularly for small-scale operations, and will collect directly into storage vessels. It may cause an elevation of the pH, however, because of the release of $CO_2$.

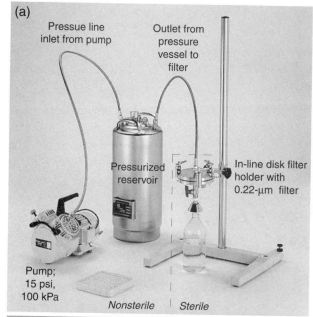

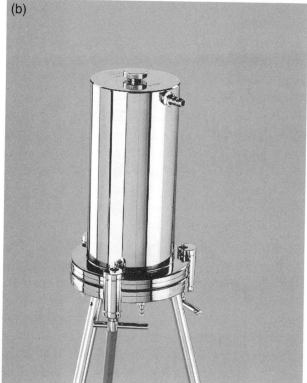

**Fig. 10.13. Large-Scale In-line Filter Assembly.** (*a*) In-line filter assembly supplied from a pressurized reservoir (center) and connected via a sterile reusable 142-mm filter holder to a receiver flask (right). Substituting a larger receiver with a tap outlet would allow collection of the entire contents of the pressure vessel for later dispensing into sterile containers (*see* Fig. 10.14*c*). Only the filter assembly and the receiver flask need to be sterilized. (*b*) Filter holder for cartridge filter for large-scale sterile filtration in place of disc filter in (*a*) or for use as a prefilter (*see* Fig. 10.14*c*). (Courtesy of Sartorius Stedim.)

*Disposable filters.* Disposable filter holder designs include simple disk filters, hollow-fiber units, and cartridges (*see* Fig. 10.11). Syringe-tip filters are generally used for low-volume filtration (for 2–100 mL) and vary in size from 13 to 50 mm in diameter (*see* Fig. 10.11*a*). Intermediate-size filters (for 50 ml–5 L) can be used in line with a peristaltic pump (*see* Fig. 10.12), or as bottle-top filters used with a vacuum line and a regular medium bottle (Fig. 10.11*f*). Intermediate-size filters can also be purchased as complete filter units for attaching to a vacuum line, with an upper chamber for the nonsterile solution and a lower chamber to receive the sterile liquid and to use for storage (Fig. 10.11*e*). Large-capacity cartridge filters (for 20–500 L) are usually operated in line under positive pressure from a pressure vessel (Fig. 10.13) that can be used to make up the medium as it is or lined with a media bag (*see* Appendix II). Although disposable filters are more expensive than reusable filters, they are less time-consuming to use and give fewer failures.

*Reusable filters.* Reusable filter holders may use membranes (Fig. 10.13a) or cartridges (Fig. 10.13b) and are sterilized by autoclaving (*see* Protocol 10.4). They are usually connected to a pressure reservoir (*see* Fig. 10.15) or a peristaltic pump and operate under positive pressure.

Protocols 10.11 through 10.14 feature disposable filters in small sizes and a reusable filter for larger volumes.

---

**PROTOCOL 10.11. STERILE FILTRATION WITH SYRINGE-TIP FILTER**

**Outline**

Fill a syringe with solution, attach filter, and expel solution though filter into sterile container.

**Materials**

*Sterile:*

❑ Plastic syringe, 10- to 50-mL capacity
❑ Syringe-tip filter
❑ Receiver vessel (e.g., a universal container)

*Nonsterile:*

❑ Solution for sterilization (5–100 mL)

**Procedure**

1. Swab down hood and assemble materials.
2. Fill syringe with solution to be sterilized.
3. Uncap receiver vessel.
4. Unpack filter and attach to tip of syringe, holding the sterile filter within the bottom half of the packaging while attaching to syringe.
5. Expel solution through filter into receiver vessel. Only moderate pressure is required.
6. Cap receiver vessel.
7. Discard syringe and filter.

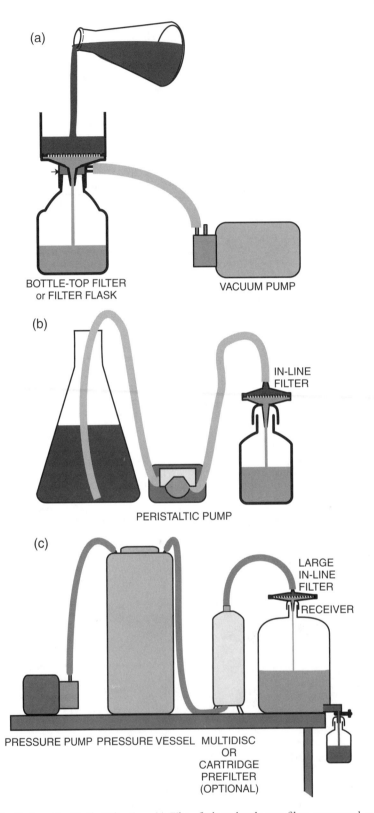

**Fig. 10.14. Options for Sterile Filtration.** (*a*) Filter flask or bottle-top filter connected to vacuum pump; popular laboratory scale setup for 1 to 10 L of medium. (*b*) In-line filter fed from large reservoir by peristaltic pump (*see* Fig. 10.12). Suitable for volumes up to 100 L, depending on size of filter. (*c*) Large-scale filtration with positive pressure pump, pressure vessel, prefilter (cartridge or multidisc, optional), and a final large-capacity sterilizing filter (*see* Fig. 10.14, but without the prefilter). Suitable for 100 to 10,000 L depending on sizes of components. Semi-industrial to industrial scale. Not to scale.

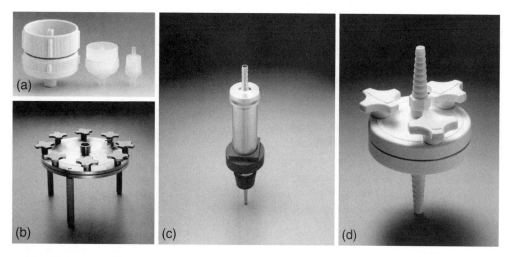

**Fig. 10.15. Reusable Filters.** (*a*) Swinnex polypropylene, in-line, Luer fitting. (*b*) Stainless steel housing, 293 mm high-capacity disk-type filter. (*c*) 47-mm filter with reservoir. (*d*) 47 mm in line with hose connections. (Courtesy of Millipore, Ltd., UK.)

The syringe may be refilled several times by returning the filter to the lower half of the sterile packaging, detaching it from the syringe, refilling the syringe, and reattaching the filter. If the back-pressure increases, take a new filter.

If a vacuum line or pump is available, volumes between 50 and 500 mL can be filtered conveniently by negative pressure into an integral filter flask (*see* Figs. 10.10*b*, 10.11*e*) or with a bottle-top filter and regular medium bottle (*see* Fig. 10.11*f*). Protocol 10.12 is presented for use in conjunction with Exercise 6 (*see* Section 28.2).

**PROTOCOL 10.12. STERILE FILTRATION WITH VACUUM FILTER FLASK**

**Outline**
Attach vacuum pump to outlet, pour medium into top chamber of filter unit, switch on pump and draw solution through to lower chamber; cap and store.

**Materials**

*Sterile:*

❏ Filter flask (*see* Figs. 10.10*b*, 10.11*e*; Table 10.5)
❏ Cap for lower chamber (if chamber to be used for storage)
❏ Sample tube or universal container for sterility test

*Nonsterile:*

❏ Medium for sterilization
❏ Vacuum pump or vacuum line
❏ Thick-walled connector tubing from pump or line, to fit filter flask connection tube

**Procedure**

1. Bring medium to·be sterilized and filter flask to hood.
2. Connect side outlet of filter flask to vacuum pump (located outside hood, e.g. on floor).
3. Remove cap from bottle and lid from top chamber of filter flask.
4. Pour nonsterile medium into top chamber.
5. Switch on pump.
6. Unpack cap for lower chamber, ready for use.
7. When liquid has all been drawn into lower chamber, switch off pump and detach filter housing and top chamber.
8. Transfer 10 mL from lower flask to universal container or equivalent tube.
9. Gas the head space in the lower chamber and the universal container with 5% $CO_2$.
10. Cap the lower chamber and universal.
11. Label lower chamber and universal with name of medium, date, and your initials.
12. Store lower chamber at 4°C until required.
13. Incubate the 10-mL sample in the universal at 37°C for 1 week and check for contamination. Do not release medium for use until shown to be free of contamination.
14. Complete records after filtration and after completion of sterility test.

Generally, this method of filtration has a low risk of contamination, so sampling for a contamination check may be omitted in due course.

Filter flasks are available with from 150- to 1000-mL receiver capacity. If a larger volume is to be filtered by negative pressure and dispensed into aliquots, it may be better to use one of several bottle-top filters, which may be set up to filter material directly into standard medium bottles (*see* Fig. 10.11*f*, also available with upper reservoir). Alternatively, solutions may be filtered by positive pressure, either from a pressurized reservoir (*see* Fig. 10.13) or via a peristaltic pump (*see* Fig. 10.12), passing through a sterile in-line membrane or cartridge filter equipped with a bell to protect the receiver vessel from contamination (*see* Figs. 10.10, 10.11*b*, *c*). Protocol 10.13 uses a small in-line filter and can be used in conjunction with Exercise 6 (*see* Section 28.2).

---

## PROTOCOL 10.13. STERILE FILTRATION WITH SMALL IN-LINE FILTER

### Outline
Pump medium from reservoir through a peristaltic pump and dispense into bottles through a sterilizing filter.

### Materials

*Sterile:*

❏ In-line filter with bell (*see* Fig. 10.11*c*, *d*), such as 47 mm
❏ Graduated medium bottles, foil capped, sterilized by dry heat (*see* Protocol 10.1)
❏ Caps, autoclaved (*see* Protocol 10.3)
❏ Sample tube or universal container for sterility test

*Nonsterile:*

❏ Medium for sterilization (*see* Protocol 10.9)
❏ Peristaltic pump (*see* Fig. 10.12) preferably with foot switch
❏ Silicone tubing to fit pump and inlet to filter
❏ Clamp stand to hold filter

### Procedure

1. Bring equipment to hood, swab as appropriate, and place in hood.
2. Feed tubing through peristaltic pump.
3. Insert upstream end into medium to be sterilized.
4. Unpack filter and connect to outlet from peristaltic pump.
5. Clamp filter in clamp stand at a suitable height such that bottles for receiving medium can be positioned below the filter with the neck shrouded and removed easily when ready to be filled. Use one of the sterile medium bottles to set up the filter if necessary, but do not use this bottle, ultimately, for sterile collection.

6. Switch on pump and collect about 20 mL into bottle used for setup.
7. Remove setup bottle, cap it and number it "1."
8. Remove foil from first medium bottle and place under filter bell.
9. Switch on pump.
10. Fill bottle to mark.
11. Switch off pump.
12. Remove bottle, cap it, number it, and replace with fresh sterile medium bottle.
13. Repeat steps 7 through 11, filling each bottles to mark, and collecting the remainder in the last bottle.
14. Gas the head space in each bottle with 5% $CO_2$.
15. Replace the aluminum foil over the cap and neck of each bottle to keep it free from dust during storage.
16. Label, date, and initial bottles.
17. Place bottles containing medium at 4°C for storage (*see* Section 10.6.4).
18. Place the first and last bottles, sealed, at 37°C and incubate for 1 week (*see* Section 10.6.2).
19. Complete records.
20. Store medium at 4°C until required; do not release for use until sterility test is complete. If any contamination is found in the test samples, refilter the whole batch or discard it.

---

Filtration with positive pressure can be scaled up by increasing the filter size and the size of the medium reservoir. The peristaltic pump is often replaced by adding positive pressure to the medium reservoir in a pressure vessel (*see* Fig. 10.13*a*). Protocol 10.14 is suitable for up to 50 L, but the volume can be increased up to industrial scale by selecting the appropriate pressure vessel and filter, often a pleated cartridge or other multisurface filter contained in a cylindrical housing (Fig. 10.13*b*).

---

## PROTOCOL 10.14. STERILE FILTRATION WITH LARGE IN-LINE FILTER

### Outline
Pump medium from pressure vessel through sterile filter into sterile receiver flask with outlet. Dispense from receiver into storage flasks.

### Materials
(*see* Figs. 10.13, 10.15, 10.16)

*Sterile:*

❏ Filter (e.g., 90-mm membrane and reusable filter holder; *see* Protocol 10.4; Table 10.4;

Fig. 10.13) or disposable disk (e.g., Millipore Millex, Fig. 10.11c) or cartridge (e.g., Pall Gelman Capsule, Fig. 10.13e).

❑ Receiver vessel for filtrate with outlet at the base with silicone tubing and glass bell for delivery and spring clip or valve on receiver outlet
❑ Silicone tubing from filter outlet to sterile receiver
❑ Medium bottles, foil capped, dry heat sterilized (see Protocol 10.1)
❑ Caps for medium bottles, autoclaved (see Protocol 10.3)

*Nonsterile:*

❑ Pressure vessel, 5 to 50 L.
❑ Pump, 100 kPa (15 lb/in.$^2$)
❑ Clamp stand and clamps to secure filter (unless filter holder has legs) and outlet bell if separate
❑ Pressure tubing from pump to pressure vessel and pressure vessel to filter

### Procedure

1. Place sterile filter holder and receiver flask in the hood and the pressure vessel on the floor or bench alongside the hood, but not in the hood.
2. Unwrap the filter holder and receiver and place in position.
3. Connect the outlet from the filter to the receiver vessel.
4. Secure the receiver at a suitable height such that medium bottles can be positioned below the outlet with the neck shrouded by the bell and removed easily when the bottle is filled. Use one of the sterile medium bottles to set up if necessary, but do not use this bottle, ultimately, for sterile collection.
5. Connect the inlet to the pressure vessel to the pump and the outlet to the filter inlet.
6. Decant the medium into the pressure vessel and close the lid.
7. For reusable filters, turn on the pump just long enough to wet the filter. Stop the pump and tighten up the filter holder. (This step is not required for disposable filters.)
8. Switch on the pump to deliver 100 kPa (15 lb/in.$^2$). When the receiver starts to fill, draw off aliquots into medium stock bottles of the desired volume.
9. Cap the bottles as each one is taken from the filter bell.
10. Replace the foil over the cap.
11. Label, date, and initial bottles.
12. Depending on the number and size of bottles filled, remove sample bottles from the beginning,

middle, and end of the run, and incubate at 37°C for 1 week to check for contamination.
13. Store medium at 4°C. Do not release for general use until quality control has been performed.
14. Complete records after filtration and after completion of sterility check.

Positive pressure is recommended for optimum performance of the filter and to avoid the removal of $CO_2$, which results from negative-pressure filtration. Positive pressure may also be applied by using a peristaltic pump (see Fig. 10.12) in line between the nonsterile reservoir and a disposable in-line filter, such as the Millipak (Millipore), which may be used instead of the reusable assembly. The only preparation and sterilization required is for the medium bottles as the disposable filter is bought sterile and no receiver is necessary (the flow can be interrupted easily by switching off the peristaltic pump).

### 10.5.3 Serum

Preparing serum is one of the more difficult procedures in tissue culture because of variations in the quality and consistency of the raw materials and because of the difficulties encountered in sterile filtration due to particulate material, colloids, and viscosity. Moreover serum is also one of the most costly constituents of tissue culture, accounting for 20% to 30% of the total budget if it is bought from a commercial supplier. Buying sterile serum is certainly the best approach from the point of view of consistency and quality control, but Protocol 10.15 is suggested if serum has to be prepared in the laboratory. The underlying principle is that a graded series of filters, which need not be sterile, are used to remove particulate material before the serum is passed through a sterilizing filter of 0.1 μm at a low pressure (Fig. 10.16).

---

**PROTOCOL 10.15. COLLECTION AND STERILIZATION OF SERUM**

### Outline
Collect blood, allow it to clot, and separate the serum. Sterilize serum through filters of gradually reducing porosity. Bottle and freeze filtered serum.

### Collection
Arrangements may be made to collect whole blood from a slaughterhouse. The blood should be collected directly from the bleeding carcass and not allowed to lie around after collection. Alternatively, blood may be withdrawn from live animals under proper veterinary supervision. The latter alternative, if performed consistently on the same group of animals, gives a more reproducible serum but a lower volume

for a greater expenditure of effort. If the procedure is done carefully, blood may be collected aseptically.

### Clotting

Allow the blood to clot by having it stand overnight in a covered container at 4°C. This so-called natural-clot serum is superior to serum that is physically separated from the blood cells by centrifugation and defibrination, as platelets release growth factors into the serum during clotting. Separate the serum from the clot, and centrifuge the serum at 2,000 $g$ for 1 h to remove sediment.

### Sterilization

Serum is usually sterilized by filtration through a sterilizing filter of 0.1-μm porosity, but because of its viscosity and high particulate content, the serum should be passed through a graded series of fiberglass or other prefilters before passing through the final sterilizing filter (*see* Fig. 10.16). Only the last filter, a 142- to 350-mm in-line disk filter or equivalent disposable filter (e.g., Millipak 200), needs to be sterile. The prefilter assemblies may be stainless steel with replaceable cartridges, disk filter units, or a single bonded unit (Pall Gelman). The last is easiest to use, but more difficult to clean and reuse.

### Materials

(For sterilizing volumes of 5 to 20 L)

*Sterile:*

☐ Sterilizing filter: 200 mm, 0.1-μm porosity (e.g., Millipak, Millipore). A porosity of 0.2 μm is sufficient for antibacterial and fungal sterilization, but 0.1 μm is required to remove mycoplasma.
☐ Sterile receiving vessel with outlet at base
☐ Sterile bottles with caps and foil
☐ Universal containers for sterility testing

*Nonsterile:*

☐ Peristaltic pump and tubing
☐ Clamp stand and clamps
☐ Nonsterile prefilters:
  Fiberglass disposable filter or 142-mm reusable filter (e.g., Pall Gelman)
  5-μm porosity disposable or reusable filter (e.g., Pall Gelman Versapore)
  1.2-μm porosity disposable or reusable filter (e.g., Millipore Opticap)
  0.45-μm porosity disposable or reusable filter (e.g., Millipore Millipak)

**Note.** The preceding filters are suggestions only; contact your supplier and request a series of filters and prefilters that will suit your serum requirements. If it is a once-only activity, choose disposable filters;

if collection will be repeated regularly, it will be more economical to employ reusable filter holders for the prefilter stages, while still using a disposable filter for sterilization, for added security.

### Procedure

1. Insert appropriate nonsterile filters into nonsterile prefilter holders (if reusable holders are being used).
2. Connect one or more prefilters in line and upstream from a sterile disposable or reusable filter holder (*see* Fig. 10.16) that contains a 0.1-μm porosity filter and is connected to a sterile receiver via the peristaltic pump.
3. Place the intake of the pump into the serum container.
4. Switch on the pump, and check any reusable filter holders for leakage as they are wetted; switch off the pump, and tighten filter holders as necessary.
5. Restart the pump and continue filtering, checking for leaks or blockages. Increasing the flow rate will increase the rate of filtration but may cause the filters to become packed or clogged.
6. Collect aliquots in sterile bottles, leaving at least 20% headspace to allow for expansion on freezing.
7. Collect samples at the beginning, middle, and end of the run to check sterility.
8. Cap, label, and number the bottles.
9. Replace the foil over the cap and neck for storage.
10. Store serum at −20°C until quality control is completed (*see* Section 10.6.1).
11. Complete records after filtration and after completion of sterility check.

**Small-scale serum processing.** If small amounts (<1 L) of serum are required, then the process is similar to Protocol 10.14 but can be scaled down. After clot retraction (*see* Protocol 10.15: Clotting), small volumes of serum may be centrifuged (5−10,000 $g$) and then filtered through a series of disposable filters (e.g., 50-mm Millipore Millex or Pall Gelman Acrodisc) and, finally, through a 50-mm, 0.1-μm porosity sterile disposable filter.

For very small volumes (10−20 mL), centrifuge at 10,000 $g$, and filter the serum directly through a graded series of syringe-tip disposable 25-mm filters (e.g., Acrodisc, Pall Gelman; Millex, Millipore), finishing with a 0.1-μm sterilizing filter (e.g., Millex).

**Storage.** Bottle the serum in sizes that will be used up within 2 to 3 weeks after thawing. Freeze the serum as rapidly as possible, and if it is thawed, do not refreeze it unless further prolonged storage is required.

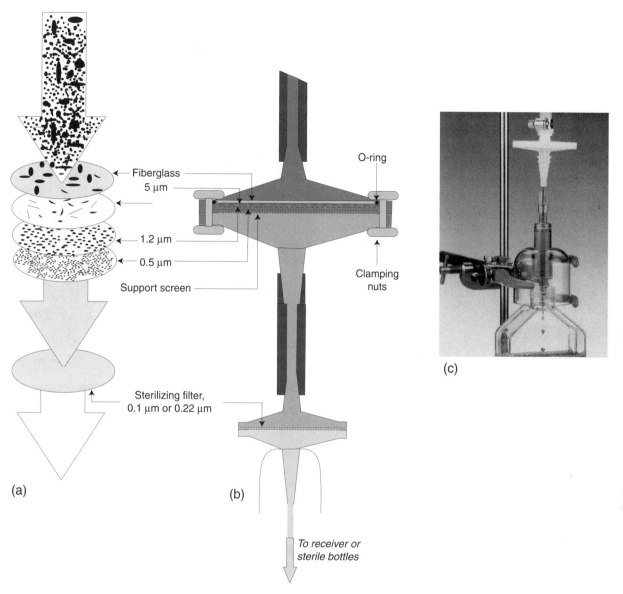

**Fig. 10.16. Prefiltration.** Prefiltration for filtering colloidal solutions (e.g., serum) or solutions with high particulate content. Several prefilters can be connected in series, and only the final filter needs to be sterile. (*a*) Diagrammatic representation of principle; medium passing through filters of gradually reducing porosity. (*b*) Filter series stacked in a nonsterile reusable filter holder, leading to a sterile disposable 0.1- or 0.22-μm in-line membrane filter. (*c*) One disposable prefilter (nonsterile or sterile) inserted upstream of a final 0.22-μm or 0.1-μm high-capacity Sterivex sterilizing filter.

Serum is best used within 6 to 12 months of preparation if it is stored at −20°C, but more prolonged storage may be possible at −70°C; usually, however, the bulk of serum stocks makes this impractical. Polycarbonate or high-density polypropylene bottles will eliminate the risk of breakage if storage at −70°C is desired. Regardless of the temperature of the freezer or the nature of the bottles, do not fill them completely; allow for the expansion of water during freezing.

***Human serum.*** Pooled outdated human blood or plasma from a blood bank can be used instead of or in addition to bovine or equine blood. It should be already

sterile and not require filtration. Titrate out the heparin or citrate anticoagulant with $Ca^{2+}$, allow the blood to clot overnight, and then separate and freeze the serum.

Δ ***Safety Note.*** Care must be taken with human donor serum to ensure that it is screened for hepatitis, HIV, tuberculosis, and other adventitious infections.

***Quality control.*** Use the same procedures as for medium (*see* Section 10.6).

A major problem that is emerging with the use of serum is the possibility of viral infection. When the possibility of bacterial infection was first appreciated, it was relatively easy to devise filtration procedures to filter out anything above 1.0 μm, and eventually a porosity of 0.45 μm became standard. Subsequently it was learned that mycoplasma would pass through filters as low as 0.2 μm, and commercial suppliers of serum lowered the exclusion limits of their filters to 0.1 μm. This reduction in size appears to have virtually eliminated mycoplasma from serum batches used in culture, but the problem of viral contamination remains. Filtering out virus would seem to be a much more significant task, but some companies (e.g., Pall Gelman) claim that it may be possible. Alternatively, serum can be irradiated to eliminate viruses but the dose required is quite high and there is a danger of radiation damage to some constituents. Collecting serum from areas of low indigenous infection and screening before processing are better alternatives.

*Dialysis.* For certain studies the presence of constituents of low molecular weight (amino acids, glucose, nucleosides, etc.) may be undesirable. These constituents may be removed by dialysis through conventional dialysis tubing.

---

## PROTOCOL 10.16. DIALYSIS OF SERUM

### Outline
Dialyze sterile serum against three changes of BSS and resterilize.

### Materials

*Sterile:*
- ❑ Serum to be dialyzed
- ❑ HBSS at 4°C
- ❑ Measuring cylinder
- ❑ Bottles and caps
- ❑ Sterilizing filter, 0.1 μm
- ❑ Prefilters: 5.0-μm fiberglass, 1.2-, 0.45-, and 0.22-μm membrane filters (e.g., Durapore, Millipore; *see* Protocol 10.14)

*Nonsterile:*
- ❑ Dialysis tubing
- ❑ Beaker with ultrapure water
- ❑ Bunsen burner
- ❑ Tripod with wire gauze

### Procedure

1. Boil five pieces of 30-mm × 500-mm dialysis tubing in three changes of distilled water.
2. Transfer tubing to sterile Hanks's balanced salt solution (HBSS), and allow to cool.
3. Wearing sterile gloves or using artery clamps, tie double knots at one end of each tube.
4. Half-fill each dialysis tube with serum (20 mL).

---

5. Express air and knot the open end of the tube, leaving a space of about half the tube length between the serum and the knot.
6. Place the tubing in 5 L of HBSS, and stir on a magnetic stirrer overnight at 4°C.
7. Change HBSS and repeat step 6 twice.
8. Collect serum into a measuring cylinder, and note the volume collected. (If the volume is less than the starting volume of the serum, add HBSS to return to the starting volume. If the volume is greater than the starting volume of the serum, make due allowance when adding to the medium later.)
9. Sterilize serum through a graded series of filters (*see* Protocol 10.15).
10. Bottle and freeze the serum.
11. Check sterility by incubating small sample for one week at 37°C.
12. Complete records after filtration and after sterility check.

---

If dialyzed serum is to be used frequently or for a prolonged period it may be preferable to employ positive pressure with a membrane filter with a low-molecular-weight cutoff (Millipore, Amicon). As these molecular filters concentrate as well as eliminate low-molecular-weight compounds, it is important to note the volume before filtration and restore it afterward with sterile water or BSS, if the serum is to remain equivalent to normal serum.

### 10.5.4 Preparation and Sterilization of Other Reagents

Individual recipes and procedures are given in Appendix I. On the whole, most reagents are sterilized by filtration if they are heat labile and by autoclaving if they are heat stable (*see* Table 10.3). Filters with low binding properties (e.g., Millex-GV) are available for sterilizing of proteins and peptides.

## 10.6 CONTROL, TESTING, AND STORAGE OF MEDIA

### 10.6.1 Quality Control

A medium that is prepared in the laboratory needs to be tested before use. If the medium is purchased readymade as a 1× working-strength solution, then it should be possible to rely on the quality control carried out by the supplier, other than any special requirements that you have of the medium. Likewise, if a 10× concentrate is used, the growth and sterility testing will have been done, and the only variable will be the water used for dilution. Provided that the conductivity and level of total organic carbon fall within specifications (*see* Section 10.4.1) and no major changes have been made in the supply of water, most laboratories will accept this compliance as adequate quality control.

However, if medium is prepared from powder, it will have been sterilized in the laboratory, and quality control will be required to confirm sterility. Nevertheless, if all the constituents have dissolved, you may be prepared to accept the quality control of the supplier regarding the medium's growth-promoting activity. Media made up from basic constituents will require complete quality control, involving both sterility testing and culture testing.

## 10.6.2 Sterility Testing

*Bubble point.* When positive-pressure filtration is complete and all the liquid has passed through the filter, raise the pump pressure until bubbles form in the effluent from the filter. This is the *bubble point* and should occur at more than twice the pressure used for filtration (see manufacturer's instructions). If the filter bubbles at the sterilizing pressure (100 kPa, 15 psi) or lower, then it is perforated and should be discarded. In that case any filtrate that has been collected should be regarded as nonsterile and refiltered. Single-use, disposable filters rarely fail the bubble point test, but reusable filters can fail, so they should be checked after every filtration run.

*Incubation.* Collect samples at the beginning, middle, and end of the run. If the bottles are small, this can be done by removing bottles. If the bottles are large, rather than wasting medium, collect samples into smaller containers at intervals during filtration. Remember, however, if you are bottling in 1000-mL sizes, taking 1-mL samples will reduce the sensitivity by a factor of $1 \times 10^3$; if you are not prepared to sacrifice whole bottles at this size, then you should at least sample 100 mL. It is best not to withdraw samples from individual bottles, as this will both increase the risk of contaminating the bottles and reduce the volume in the bottle used for sampling, thereby altering the dilutions of subsequent additions to that bottle. It is preferable to collect smaller samples at intervals during filtration.

Incubate samples according to either of the following procedures:

(1) Incubate samples of medium at 37°C for 1 week. If any of the samples become cloudy, discard them and resterilize the batch. If there are signs of contamination in the other stored bottles, the whole batch should be discarded.
(2) For a more thorough test, and when the solution being filtered does not have its own nutrients, take samples, as described in this section, and dilute one-third of each into nutrient broths (e.g., L-broth, beef heart hydrolysate, and thioglycollate). Divide each sample in two, and incubate one at 37°C and one at 20°C for 10 days, with uninoculated controls. If there is any doubt after this incubation, mix and plate out aliquots from the broths on nutrient agar and incubate at 37°C and 20°C.

*Downstream secondary filtration.* Place a demountable 0.45-μm sterile filter in the effluent line from the main sterilizing filter. Any contamination that passes because of failure in the first filter will be trapped in the second. At the end of the run, remove the second filter and place the filter on nutrient agar. If colonies grow, discard or refilter the medium. This method has the advantage that it monitors the entire filtrate, and not just a small fraction of it, although it does not avoid risks of contamination during bottling and capping.

*Autoclaved solutions.* Sterility testing of autoclaved stocks is much less essential, provided that proper monitoring (of the temperature and the time spent at the sterilizing temperature) of the center of the load of the autoclave is carried out (*see* Section 10.4).

## 10.6.3 Culture Testing

Media that have been produced commercially will have been tested for their capability of sustaining the growth of one or more cell lines. (If they have not, then you should change your supplier!) However, under certain circumstances, you may wish to test your own media for quality: (1) if it has been made up in the laboratory from basic constituents, (2) if any additions or alterations are made to the medium, (3) if the medium is for a special purpose that the commercial supplier is not able to test, and (4) if the medium is made up from powder and there is a risk of losing constituents during filtration.

Medium can become contaminated with toxic substances during filtration. For example, some filters are treated with traces of detergent to facilitate wetting, and the detergent may leach out into the medium as it is being filtered. Such filters should be washed by passing D-PBSA or BSS through them before use or by discarding the first aliquot of filtrate. Polycarbonate filters (e.g., Nuclepore) are wettable without detergents and are preferred by some workers, particularly when the serum concentration in the medium is low.

There are three main types of culture test: (1) plating efficiency, (2) growth curve at regular passage densities and up to saturation density, and (3) the expression of a special function (e.g., differentiation in the presence of an inducer, viral propagation, the formation of a specific product, or the expression of a specific antigen). All these tests should be performed on the new batch of medium with your regular medium as a control.

*Plating efficiency.* The plating efficiency test (*see* Protocol 20.10) is the most sensitive culture test, detecting minor deficiencies and low concentrations of toxins that are not apparent at higher cell densities. Ideally it should be performed with a limiting concentration of serum, which may otherwise mask deficiencies in the medium. To determine this concentration, do an initial plating efficiency test in different concentrations of serum and select a concentration such that the plating efficiency is about half that of the usual concentration but still gives countable colonies. A clonal growth assay will not always detect insufficiencies in the amount of particular constituents unless colony size is measured. For example, if the concentration of one or more amino acids is low, it may not affect the plating efficiency but could influence the mean colony size.

***Growth curve.*** A growth curve (*see* Protocols 20.7–20.9) gives three parameters of measurement: (1) the lag phase before cell proliferation is initiated after subculture, indicating whether the cells are having to adapt to different conditions; (2) the doubling time in the middle of the exponential growth phase, indicating the growth-promoting capacity of the medium; and (3) the maximum cell concentration attainable indicating whether there are limiting concentrations of certain nutrients. In cell lines whose growth is not sensitive to density (e.g., continuous cell lines; *see* Section 18.5.2), the terminal cell density indicates the total yield possible and usually reflects the total amino acid or glucose concentration. Remember that a medium that gives half the terminal cell density costs twice as much per cell produced. Growth curves can be performed by counting cells at intervals (*see* Protocols 20.7–20.9), by image analysis of the growing cultures (e.g. Incucyte or Cell-IQ; *see* Section 20.8.1), or by colorimetric tests (*see* Section 21.3.5).

***Special functions.*** If you are testing special functions, a standard test from the experimental system you are using (e.g., a virus titer in the medium after a set number of days) should be performed on the new medium alongside the old one.

A major requirement of these tests is that they should be initiated well in advance of the exhaustion of the current stock of medium so that proper comparisons may be made and there is time to have fresh medium prepared if the medium fails quality control.

***Records.*** All quality control tests should be recorded in a logbook or computer database, along with the other details relating to the preparation of this batch, and the name of the tester. The person supervising preparation and sterilization should review these records and determine failure rates and trends, to check for the need to alter procedures.

### 10.6.4 Storage

Opinions differ as to the shelf life of different media. As a rough guide, media made up without glutamine should last 6 to 9 months at 4°C. Once glutamine, serum, or antibiotics are added, the storage time is reduced to 2 to 3 weeks. Hence media that contain labile constituents should either be used within 3 weeks of preparation or be stored at −20°C (*see also* Section 28.2, Exercise 6). Substitution of glutamine with a glutamyl dipeptide such as Glutamax will enhance shelf life and stability in culture. Stored media can also show precipitation. If this does not redissolve on heating it is likely to be a chemical alteration of one of the constituents (e.g., metal ions combining with phosphate) and may not be reversible.

Some forms of fluorescent lighting will cause riboflavin, tryptophan, and tyrosine to deteriorate into toxic by-products, mainly peroxides [Wang & Nixon, 1978; Edwards et al., 1994]. While part of the toxic effect on cells may be due to photochemical deterioration, there is a direct mutagenic effect on cells [Bradley & Sharkey, 1977]. Thus incandescent lighting should be used in cold rooms where media are stored and in hot rooms where cells are cultured, and the light should be extinguished when the room is not occupied. Bottles of medium should not be exposed to fluorescent lighting for longer than a few hours; a dark freezer is recommended for long-term storage.

# CHAPTER 11

# Primary Culture

A primary culture is that stage of the culture after isolation of the cells but before the first subculture after which it becomes a cell line (*see* Section 2.7 and Appendix IV). Although there is formal acceptance for this definition of "primary culture" and "cell line" [Schaeffer, 1990], the advent of commercial supplies of early passage culture of specialized cells (*see* Appendix II) has seen the increasing use of the phrase "primary cells" or "primary cell lines." One has to be clear what is meant when purchasing these cells. In some cases they may be dissociated cells from a primary culture that will become a cell line when the recipient seeds them into culture, but in other cases they may be growing cultures already passaged from a primary culture and technically a cell line (usually a finite cell line) or frozen cells from a primary culture, or, more likely, from the first or second passage. This distinction may seem a little pedantic, but it is important to know the stage of the culture as it will affect its uniformity (*see* Section 2.7) and the number of generations left to it before senescence, namely the life span of the culture. If subcultured from the primary culture, the correct term is *early passage cell line* and not "primary cell line," which is a contradiction in terms.

## 11.1 INITIATION OF A PRIMARY CELL CULTURE

There are four stages to consider: (1) acquisition of the sample, (2) isolation of the tissue, (3) dissection and/or disaggregation, and (4) culture after seeding into the culture vessel. After isolation, a primary cell culture may be obtained either by allowing cells to migrate out from fragments of tissue adhering to a suitable substrate or by disaggregating the tissue mechanically or enzymatically to produce a suspension of cells, some of which will ultimately attach to the substrate. It appears to be essential for most normal untransformed cells, with the exception of hematopoietic cells and stem cells, to attach to a flat surface in order to survive and proliferate with maximum efficiency. Transformed cells, however, particularly cells from transplantable animal tumors, are often able to proliferate in suspension (*see* Section 17.5.1).

### 11.1.1 Enzymes Used in Disaggregation

The enzymes used most frequently for tissue disaggregation are crude preparations of trypsin, collagenase, elastase, pronase, Dispase, DNase, and hyaluronidase, alone or in various combinations, such as Elastase and DNase for type II alveolar cell isolation [Dobbs & Gonzalez, 2002], collagenase with Dispase [Booth & O'Shea, 2002], and collagenase with hyaluronidase [Berry & Friend, 1969; Seglen, 1975]. There are other, nonmammalian enzymes, such as Trypzean (Sigma), a recombinant, maize-derived, trypsin, TrypLE (Invitrogen), recombinant microbial, and Accumax and Accumax (Innovative Cell Technologies), and Liberase Blendzyme 3 or Liberase TM (Roche), also available for primary disaggregation. Crude preparations are often more successful than purified enzyme preparations. This is because the former contain other proteases as contaminants; the latter are nevertheless generally less toxic and more specific in their action. Trypsin and pronase give the most complete disaggregation but may damage the cells. Collagenase and Dispase, in contrast, give incomplete disaggregation, but are less harmful. Hyaluronidase can be used in conjunction with collagenase to digest the extracellular matrix, and DNase is used to disperse DNA released from lysed cells; DNA tends to impair proteolysis and promote reaggregation (*see* Table 12.5). Care should be taken when combining enzymes as some may inactivate others. For example DNase should be added after trypsin has been removed, as the trypsin may degrade the DNase.

As the specific activity of enzymes may vary make sure that each batch has the same activity or adjust the concentration to achieve the same activity (see Appendix I).

### 11.1.2 Common Features of Disaggregation

Although each tissue may require a different set of conditions, certain requirements are shared by most of them:

(1) Fat and necrotic tissues are best removed during dissection.

(2) The tissue should be chopped finely with sharp scalpels to cause minimum damage.

(3) Enzymes used for disaggregation should be removed subsequently by gentle centrifugation.

(4) The concentration of cells in the primary culture should be much higher than that normally used for subculture because the proportion of cells from the tissue that survives in primary culture may be quite low.

(5) A rich medium, such as Ham's F12, is preferable to a simple medium, such as Eagle's MEM, and if serum is required, fetal bovine often gives better survival than does calf or horse. Isolation of specific cell types will probably require selective serum-free media (*see* Section 9.2.2; Chapter 22).

(6) Embryonic tissue disaggregates more readily, yields more viable cells, and proliferates more rapidly in primary culture than does adult tissue.

## 11.2 ISOLATION OF THE TISSUE

Before attempting to work with human or animal tissue, make sure that your work fits within medical ethical rules or current legislation on experimentation with animals (*see* Section 6.9.1). For example, in the United Kingdom, the use of embryos or fetuses beyond 50% gestation or incubation is regulated under the Animal Experiments (Scientific Procedures) Act of 1986. Work with human biopsies or fetal material usually requires the consent of the local ethical committee and the patient and/or his or her relatives (*see* Section 6.9.2).

Δ *Safety Note.* Work with human tissue should be carried out at Containment Level 2 in a Class II biological safety cabinet (*see* Section 6.8.3).

An attempt should be made to sterilize the site of the resection with 70% alcohol if the site is likely to be contaminated (e.g., skin). Remove the tissue aseptically and transfer it to the tissue culture laboratory in dissection BSS (DBSS) or collection medium (*see* Appendix I) as soon

as possible. Do not dissect animals in the tissue culture laboratory, as the animals may carry microbial contamination. If a delay in transferring the tissue is unavoidable, it can be held at 4°C for up to 72 h, although a better yield will usually result from a quicker transfer.

Embryonic or fetal animals that are more than half term may require specified methods of humane killing before dissection. In the United Kingdom, guidelines are available from the Home Office [Home Office, 2005] and in the United States the Office of Laboratory Animal Welfare [OLAW, 2002]. Where guidelines are not available, seek advice from your local animal ethics committee. Usually separating the embryo from the fetal membranes or placenta followed by decapitation is regarded as humane, particularly if the gravid uteri or eggs are placed on ice beforehand.

### 11.2.1 Mouse Embryo

Mouse embryos are a convenient source of cells for undifferentiated mesenchymal cell cultures. These cultures are often referred to as "mouse embryo fibroblasts" and used as feeder layers (*see* Sections 13.2.3, 23.1). Full term is about 19 to 21 days, depending on the strain. The optimal age for preparing cultures from a whole disaggregated embryo is around 13 days, when the embryo is relatively large (Figs. 11.1, 11.2) but still contains a high proportion of undifferentiated mesenchyme, which is the main source of the culture. However, isolation and handling embryos beyond 50% full-term may require a license so 9- or 10-day embryos may be preferable. Although the amount of tissue recovered from these embryos will be substantially less, a higher proportion of the cells will grow. Most individual organs, with the exception of the brain and the heart, begin to form at about the 9th day of gestation but are difficult to isolate until about the 11th day. Dissection of individual organs is easier at 13 to 14 days, and most of the organs are completely formed by the 18th day. Stages in mouse embryo development are described in the Edinburgh Mouse Atlas Project [EMAP, 2003].

---

**PROTOCOL 11.1. ISOLATION OF MOUSE EMBRYOS**

**Outline**
Remove uterus aseptically from a timed pregnant mouse and dissect out embryos.

**Materials**

*Sterile:*

❑ DBSS: dissection BSS (BSS with a high concentration of antibiotics; *see* Appendix I) in 25- to 50-mL screw-capped tube or universal container

❑ BSS, 50 mL in a sterile beaker (used to cool instruments after flaming)

❑ Petri dishes, 9 cm

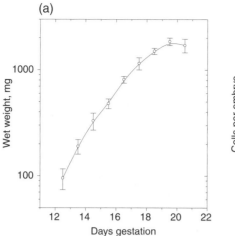
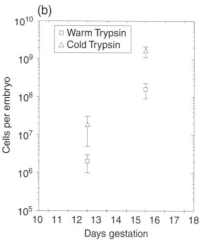

**Fig. 11.1.  *Total Wet Weight and Yield of Cells per Mouse Embryo*.** (*a*) Total wet weight of embryo without placenta or membranes, mean ± standard deviation (from Paul et al., 1969). (*b*) Viable cell yield per embryo after incubation in 0.25% trypsin at 37°C for 4 h with no intermediate harvesting (squares) or after soaking in 0.25% trypsin at 4°C for 5 h and incubation at 37°C for 30 min (triangles; Protocol 11.6).

❑ Pointed forceps
❑ Pointed scissors

*Nonsterile:*

❑ Small laminar-flow hood
❑ Timed pregnant mice (*see* step 1 of this protocol)
❑ Alcohol, 70%, in wash bottle
❑ Alcohol, 70%, to sterilize instruments (*see* Fig. 6.4)
❑ Bunsen burner

△ *Safety Note.* When sterilizing instruments by dipping them in alcohol and flaming them, take care not to return the instruments to alcohol while they are still alight!

**Procedure**

1. *Induction of estrus.* If males and females are housed separately, then put together for mating, estrus will be induced in the female 3 days later, when the maximum number of successful matings will occur. This process enables the planned production of embryos at the appropriate time. The timing of successful matings may be determined by examining the vaginas each morning for a hard, mucous plug.
2. *Dating the embryos.* The day of detection of a vaginal plug, or the "plug date," is noted as day zero, and the development of the embryos is timed from this date. Day 13 is optimal for primary culture.
3. Kill the mouse by cervical dislocation (UK Schedule I procedure [Home Office, 2005]), and swab the ventral surface liberally with 70% alcohol (Fig. 11.3*a*).

4. Tear the ventral skin transversely at the median line just over the diaphragm (Fig. 11.3*b*), and, grasping the skin on both sides of the tear, pull in opposite directions to expose the untouched ventral surface of the abdominal wall (Fig. 11.3*c*).
5. Cut longitudinally along the median line of the exposed abdomen with sterile scissors, revealing the viscera (Fig. 11.3*d*). At this stage, the two horns of the uterus, filled with embryos, are obvious in the posterior abdominal cavity (Fig. 11.3*e*).
6. Dissect out the uterus into a 25-mL or 50-mL screw-capped vial containing 10- or 20-mL DBSS (Fig. 11.3*f*).

*Note.* All of the preceding steps should be done outside the tissue culture laboratory; a small laminar-flow hood and rapid technique will help to maintain sterility. Do not take live animals into the tissue culture laboratory, as the animals may carry contamination. If an animal carcass must be handled in the tissue culture area, make sure that the carcass is immersed in alcohol briefly, or thoroughly swabbed, and disposed of quickly after use.

7. Take the intact uteri to the tissue culture laboratory, and transfer them to a fresh Petri dish of sterile DBSS (Fig. 11.3*g*).
8. Dissect out the embryos:
   (a) Tear the uterus with two pairs of sterile forceps, keeping the points of the forceps close together to avoid distorting the uterus and bringing too much pressure to bear on the embryos (Fig. 11.3*g, h*).

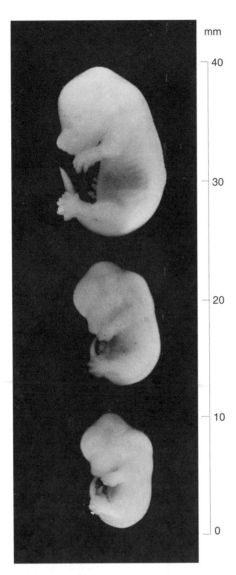

**Fig. 11.2. Mouse Embryos.** Embryos from the 12th, 13th, and 14th days of gestation. The 12-day embryo (bottom) came from a small litter (three) and is larger than would normally be found at this stage.

**(b)** Free the embryos from the membranes (Fig. 11.3*i*) and placenta and place them to one side of the dish to bleed.
9. Transfer the embryos to a fresh Petri dish. If a large number of embryos is required (i.e., more than four or five litters), it may be helpful to place the dish on ice (for subsequent dissection and culture; *see* Protocols 12.4–12.8).

## 11.2.2 Chick Embryo

Chick embryos are easier to dissect, as they are larger than mouse embryos at the equivalent stage of development. Like mouse embryos, chick embryos are used to provide predominantly mesenchymal cell primary cultures for cell proliferation analysis, to provide feeder layers, and as a substrate for viral propagation. Because of their larger size, it is easier to dissect out individual organs to generate specific cell types, such as hepatocytes, cardiac muscle, and lung epithelium. As with mouse embryos, the use of chick embryos may be subject to animal legislation and working with embryos that are more than half-term may require a license.

---

### PROTOCOL 11.2. ISOLATION OF CHICK EMBRYOS

#### Outline
Remove embryo aseptically from the egg and transfer to dish.

#### Materials

*Sterile:*

❑ DBSS: dissection BSS (BSS with a high concentration of antibiotics; *see* Appendix I) in 25- to 50-mL screw-capped tube or universal container
❑ BSS, 50 mL in a sterile beaker (used to cool instruments after flaming)
❑ Small beaker, 20 to 50 mL or egg cup
❑ Forceps, straight and curved
❑ Petri dishes, 9 cm

*Nonsterile:*

❑ Embryonated eggs, 10th day of incubation
❑ Alcohol, 70%
❑ Swabs
❑ Humid incubator (no additional $CO_2$ above atmospheric level)

#### Procedure

1. Incubate the eggs at 38.5°C in a humid atmosphere, and turn the eggs through 180° daily. Although hens' eggs hatch at around 20 to 21 days, the lengths of their developmental stages are different from those of mouse embryos. For a culture of dispersed cells from the whole embryo, the egg should be taken at about 8 days, and for isolated-organ rudiments, at about 10 to 13 days.
2. Swab the egg with 70% alcohol, and place it with its blunt end facing up in a small beaker (Fig. 11.4*a*).
3. Crack the top of the shell (Fig. 11.4*b*), and peel the shell off to the edge of the air sac with sterile forceps (Fig. 11.4*c*).
4. Resterilize the forceps (i.e., dip them in alcohol, burn off the alcohol, and cool the forceps in sterile BSS), and then use the forceps to peel off the white shell membrane to reveal the chorioallantoic

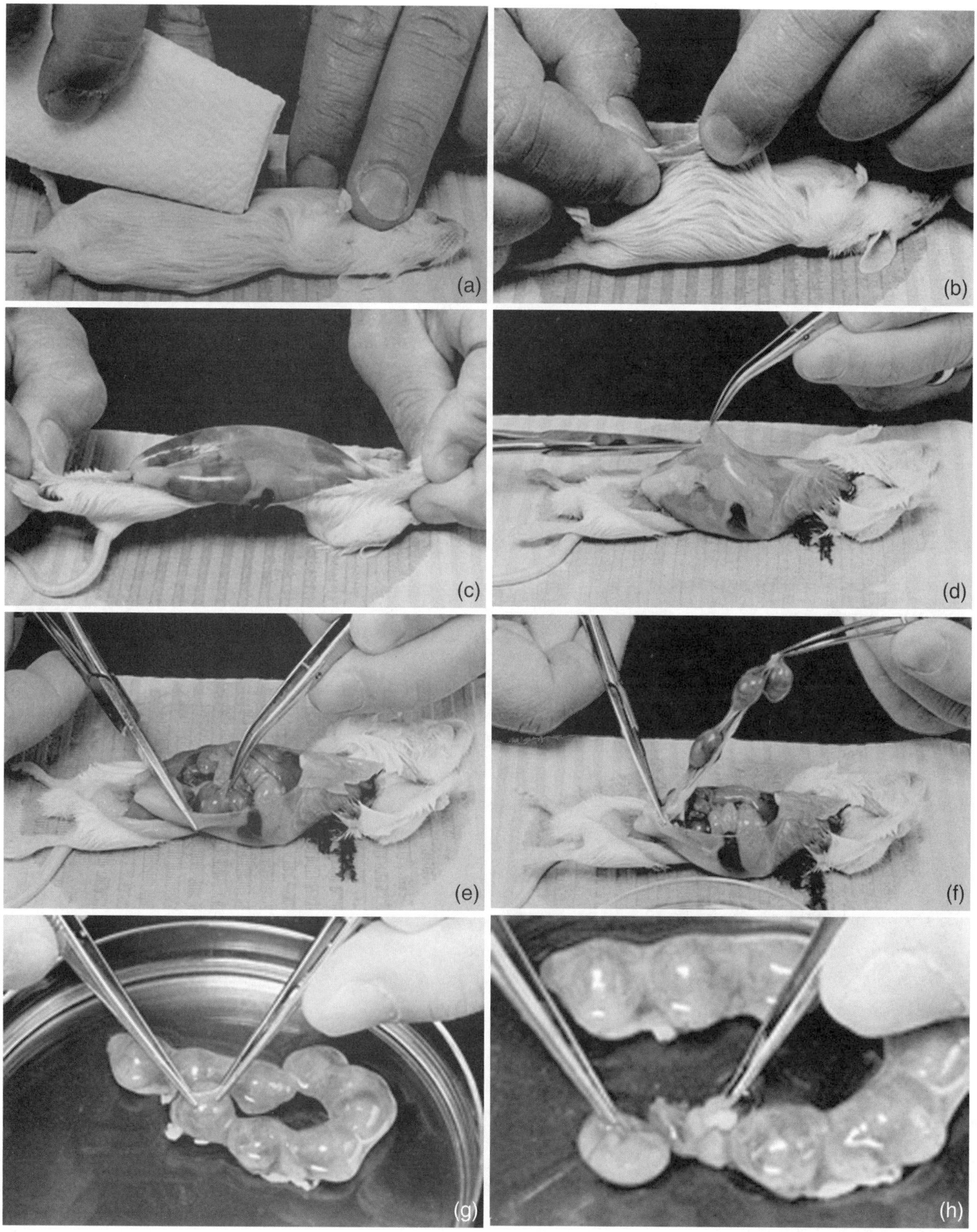

***Fig. 11.3. Mouse Dissection.*** Stages in dissection of a pregnant mouse for the collection of embryos (*see* Protocol 11.1). (*a*) Swabbing the abdomen. (*b*, *c*) Tearing the skin to expose the abdominal wall. (*d*) Opening the abdomen. (*e*) Revealing the uterus in situ. (*f*) Removing the uterus. (*g*, *h*) Dissecting the embryos from the uterus. (*i*) Removing the membranes. (*j*) Removing the head (optional). (*k*) Chopping the embryos. (*l*) Transferring pieces to trypsinization flask (for warm trypsinization, *see* Protocol 11.5). (*m*) Transferring the pieces to a small Erlenmeyer flask (for cold trypsinization, *see* Protocol 11.6). (*n*) Flask on ice.

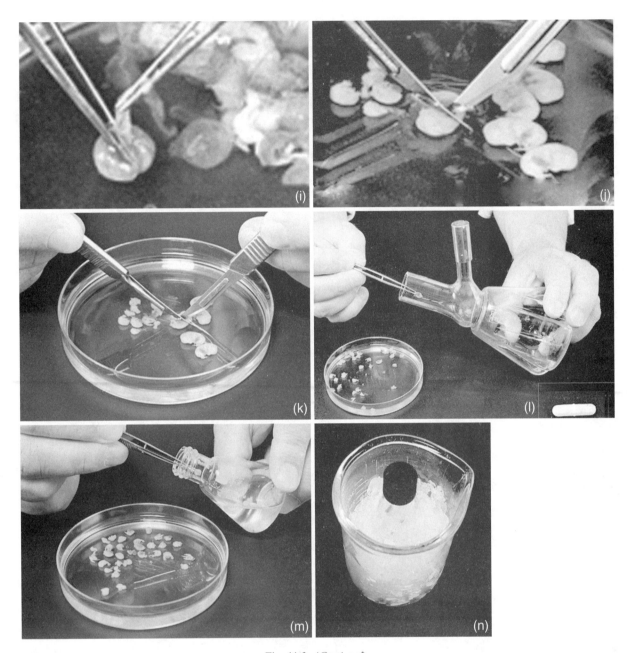

**Fig. 11.3.** (*Continued*)

membrane (CAM) below, with its blood vessels (Fig. 11.4*d*, *e*).

5. Pierce the CAM with sterile curved forceps (Fig. 11.4*f*), and lift out the embryo by grasping it gently under the head (Fig. 11.4*g*, *h*). Do not close the forceps completely, or else the neck will sever; place the middle digit under the forceps and use the finger pad to restrict the pressure of the forefinger (*see* Fig. 11.4*g*).

6. Transfer the embryo to a 9-cm Petri dish containing 20-mL DBSS (Fig. 11.4*i*). (For subsequent dissection and culture, *see* Protocol 11.7.)

## 11.2.3  Human Biopsy Material

Handling human biopsy material presents certain problems that are not encountered with animal tissue. It usually is necessary to obtain consent (1) from the hospital ethical committee, (2) from the attending physician or surgeon, and (3) from the donor or patient or the patient's relatives (*see* Section 6.9.2; see also WMA [2008]; PCSBI [2009]; OHRP [2010].). Furthermore biopsy sampling is usually performed for diagnostic purposes, and hence the needs of the pathologist must be met first. This factor is less of a problem if extensive surgical resection or nonpathological tissue (e.g., placenta or umbilical cord) is involved.

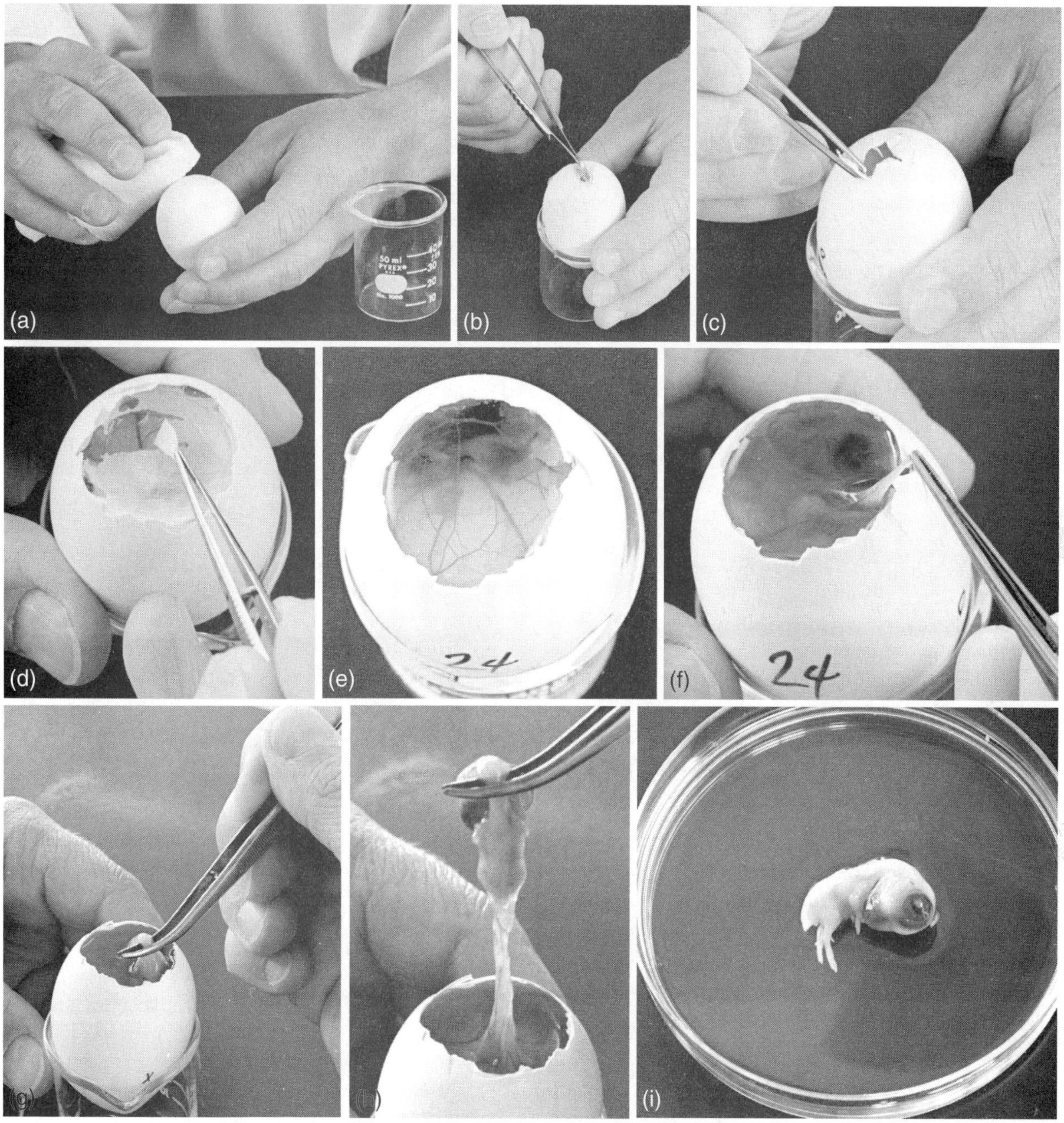

**Fig. 11.4. Removing a Chick Embryo from an Egg.** Stages in the extraction of the whole chick embryo from an egg. (*a*) Swabbing the egg with alcohol. (*b*) Cracking the shell. (*c*) Peeling off the shell. (*d*) Peeling off the shell membrane. (*e*) Chorioallantoic membrane (CAM) and vasculature revealed. (*f*) Removing CAM with forceps. (*g*) Grasping the embryo round the neck. (*h*) Withdrawing the embryo from the egg. (*i*) Isolated 10-day embryo in Petri dish.

The biopsy is often performed at a time that is not always convenient to the tissue culture laboratory, so some formal collection or storage system must be employed for times when you or someone on your staff cannot be there. If delivery to your lab is arranged, then there must be a system for receiving specimens, recording details of the source, tissue of origin, pathology, and so forth (*see* Section 11.3.11), and alerting the person who will perform the culture that the specimens have arrived; otherwise, valuable material may be lost or spoiled.

Δ *Safety Note.* Human biopsy material carries a risk of infection (*see* Section 6.8.3), so it should be handled under

Containment Level 2 in a Class II biohazard cabinet, and all media and apparatus must be disinfected after use by autoclaving or immersion in a suitable disinfectant (*see* Section 6.8.5). The tissue should be screened for adventitious infections such as hepatitis, HIV, and tuberculosis unless the patient has already been tested for these infections.

---

**PROTOCOL 11.3. HANDLING HUMAN BIOPSIES**

*Outline*
Consult with hospital staff, provide labeled container(s) of medium, and arrange for collection of samples from operating room or pathologist.

*Materials*

*Sterile:*

❑ Specimen tubes (15–30 mL) with leakproof caps about one-half full with culture medium containing antibiotics (*see* Appendix I: Collection Medium) and labeled with your name, laboratory address, and telephone number.

*Procedure*

1. Provide containers of collection medium, clearly labeled, to the anteroom of the operating theater or to the pathology laboratory.
2. Make arrangements to be alerted when the material is ready for collection.
3. Collect the containers after surgery, or have someone send them to you immediately after collection and inform you when they have been dispatched.
4. Transfer the sample to the tissue culture laboratory. The sample should be triple wrapped (e.g., in a sealed tube within a sealed plastic bag full of absorbent tissue, in case of leakage, within a padded envelope with your name, address, and telephone number on it; *see* Section 20.6). Usually, if kept at 4°C, biopsy samples survive for at least 24 h and even up to 3 or 4 days, although the longer the time from surgery to culture, the more the samples are likely to deteriorate.
5. Log receipt of sample as a numbered entry in a hand-written record book for subsequent transfer to a computerized database, or key into database directly on receipt. The log number should become incorporated into the subsequent culture designation. Linked to the pathology number or patient number, this should be the only connection to the patient or donor; the patients name or initials should not be used and remain confidential to the clinical staff.

---

6. *Decontamination.* Although most surgical specimens are sterile when removed, problems may arise with subsequent handling. Superficial specimens (skin biopsies, melanomas, etc.) and gastrointestinal tract specimens are particularly prone to contamination even when a disinfectant wash is given before skin biopsy and a parenteral antibiotic is given before gastrointestinal surgery. It may be advantageous to consult a medical microbiologist to determine which flora to expect in a given tissue and then choose your antibiotics for collection and dissection accordingly. If the surgical sample is large enough (i.e., 200 mg or more), then a brief dip (i.e., 30 s–1 min) in 70% alcohol will help to reduce superficial contamination without harming the center of the tissue sample.

---

## 11.3 TYPES OF PRIMARY CULTURE

Several techniques have been devised for the disaggregation of tissue isolated for primary culture. These techniques can be divided into (1) purely mechanical techniques, involving dissection with or without some form of maceration, and (2) techniques utilizing enzymatic disaggregation (Fig. 11.5). Primary explants are suitable for very small amounts of tissue; enzymatic disaggregation gives a better yield when more tissue is available, and mechanical disaggregation works well with soft tissues and some firmer tissues when the size of the viable yield is not important, or loosely adherent cells are removed from a more fibrous stroma.

### 11.3.1 Primary Explantation

The primary explant technique was the original method developed by Harrison [1907], Carrel [1912], and others for initiating a tissue culture. As originally performed, a fragment of tissue was embedded in blood plasma or lymph, mixed with heterologous serum and embryo extract, and placed on a coverslip that was inverted over a concavity slide. The clotted plasma held the tissue in place, and the explant could be examined with a conventional microscope. The heterologous serum induced clotting of the plasma, and the embryo extract and serum, together with the plasma, supplied nutrients and growth factors and stimulated cell migration from the explant. This technique is still used but has been largely replaced by the simplified method described in Protocol 11.4. (*See also* Section 28.4, Exercise 19, for training exercise.)

---

**PROTOCOL 11.4. PRIMARY EXPLANTS**

*Outline*
The tissue is chopped finely and rinsed, and the pieces are seeded onto the surface of a culture flask

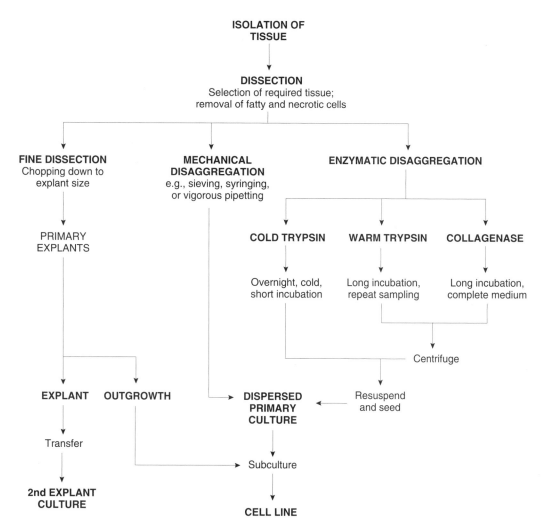

**ISOLATION OF TISSUE**

**DISSECTION**
Selection of required tissue;
removal of fatty and necrotic cells

**FINE DISSECTION**
Chopping down to
explant size

**MECHANICAL DISAGGREGATION**
e.g., sieving, syringing,
or vigorous pipetting

**ENZYMATIC DISAGGREGATION**

PRIMARY
EXPLANTS

**COLD TRYPSIN**

**WARM TRYPSIN**

**COLLAGENASE**

Overnight, cold,
short incubation

Long incubation,
repeat sampling

Long incubation,
complete medium

Centrifuge

**EXPLANT**   **OUTGROWTH**

**DISPERSED PRIMARY CULTURE**

Resuspend
and seed

Transfer

Subculture

**2nd EXPLANT CULTURE**

**CELL LINE**

*Fig. 11.5. Options for Primary Culture.* Multiple paths to obtaining a cell line; center and left, by mechanical disaggregation, right, by enzymatic disaggregation. An explant may be transferred to allow further outgrowth to form, while the outgrowth from the explant may be subcultured to form a cell line.

or Petri dish in a small volume of medium with a high concentration (i.e., 40–50%) of serum, such that surface tension holds the pieces in place until they adhere spontaneously to the surface (Fig. 11.6a). Once this is achieved, outgrowth of cells usually follows (Fig. 11.6b, c; Plates 1a, 2b).

### Materials

#### Sterile or aseptically prepared:

❑ Tissue sample
❑ Growth medium (e.g., 50:50 DMEM:F12 with 20% fetal bovine serum)
❑ DBSS, 100 mL
❑ Petri dishes, 9 cm, non–tissue–culture grade
❑ Forceps
❑ Scalpels

❑ Pipettes, 10 mL with wide tips
❑ Centrifuge tubes, 15 or 20 mL, or universal containers
❑ Culture flasks, 25 cm², or tissue-culture-grade Petri dishes, 5 to 6 cm. The size of flasks and volume of growth medium depend on the amount of tissue: roughly five 25-cm² flasks per 100 mg of tissue.

### Procedure

1. Transfer tissue to fresh, sterile DBSS, and rinse.
2. Transfer the tissue to a second dish; dissect off unwanted tissue, such as fat or necrotic material, and transfer to a third dish.

*Note.* Clean healthy tissue with little blood may not need these two transfers and can be dissected in the first dish,

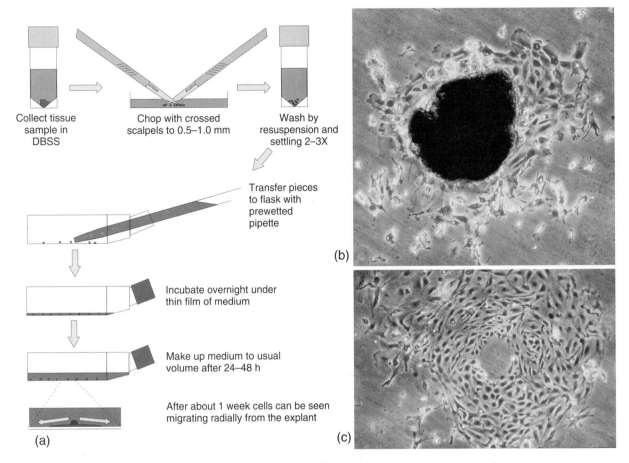

**Fig. 11.6. Primary Explant Culture.** (*a*) Schematic diagram of stages in dissection and seeding primary explants. (*b*) Primary explant culture from mouse squamous skin carcinoma; explant and early stage of outgrowth about 3 days after explantation (*see also* Plate 2*b*). (*c*) Outgrowth after removal of explant, about 7 days after explantation. 10× objective.

after transfer from the transport medium, which will have acted as the first wash.

3. Chop finely with crossed scalpels (*see* Fig. 11.6*a*, top) into about 1-mm cubes.
4. Transfer by pipette (10–20 mL, with wide tip) to a 15- or 50-mL sterile centrifuge tube or universal container. (Wet the inside of the pipette first with BSS or medium, or else the pieces will stick.)
5. Allow the pieces to settle.
6. Wash by resuspending the pieces in DBSS, allowing the pieces to settle, and removing the supernatant fluid. Repeat this step two more times. This step may be omitted if there is little blood or necrotic tissue.
7. Transfer the pieces (remember to wet the pipette) to a culture flask, with about 20 to 30 pieces per 25-cm² flask.
8. Remove most of the fluid, and add 1-mL growth medium per 25-cm² growth surface. Tilt the flask gently to spread the pieces evenly over the growth surface.
9. Cap the flask, and place it in an incubator or hot room at 37°C for 18 to 24 h.
10. Add 1 mL of medium the following day.
11. Make up the medium volume gradually to 5 mL per 25 cm² over the next 3 to 5 days.
12. Change the medium weekly until a substantial outgrowth of cells is observed (*see* Fig. 11.6*b*).
13. Once an outgrowth has formed, the remaining explant may be picked off with a scalpel (Fig. 11.6*c*) and transferred by prewetted pipette to a fresh culture vessel. (Then return to step 7.)
14. Replace the medium in the first flask until the outgrowth has spread to cover at least 50% of the growth surface, at which point the cells may be subcultured (*see* Protocol 12.3).

This technique is particularly useful for small amounts of tissue, such as skin biopsies, for which there is a risk of losing cells during mechanical or enzymatic disaggregation. Its disadvantages lie in the poor adhesiveness of some tissues and the selection of the more migratory cells in the outgrowth. In practice, however, most cells, particularly embryonic, migrate out successfully.

***Attaching explants.*** Both adherence and migration may be stimulated by placing a glass coverslip on top of the explant, with the explant near the edge of the coverslip, or the plastic dish may be scratched through the explant to attach the tissue to the flask [Elliget & Lechner, 1992] (*see* Protocol 22.9). Attachment may also be promoted by treating the plastic with polylysine or fibronectin (*see* Sections 7.2.12, 9.4.5, 13.2.1), extracellular matrix (*see* Protocol 7.1), or feeder layers (*see* Protocol 13.2.3). Historically plasma clots have been used to promote attachment. Place a drop of plasma on the plastic surface, and embed the explant in it. This should induce the plasma to clot in a few minutes, whereupon medium can be added. Alternatively, purified fibrinogen and thrombin can be used [Nicosia & Ottinetti, 1990].

## 11.3.2  Enzymatic Disaggregation

Cell–cell adhesion in tissues is mediated by a variety of homotypic interacting glycopeptides (cell adhesion molecules, or CAMs) (*see* Section 2.2.1), some of which are calcium dependent (cadherins) and hence are sensitive to chelating agents such as EDTA or EGTA. Integrins, which bind to the arginine-glycine-aspartic acid (RGD) motif in extracellular matrix, also have $Ca^{2+}$-binding domains and are affected by $Ca^{2+}$ depletion. Intercellular matrix and basement membranes contain other glycoproteins, such as fibronectin and laminin, which are protease sensitive, and proteoglycans, which are less so but can sometimes be degraded by glycanases such as hyaluronidase or heparinase. The easiest approach is to proceed from a simple disaggregation solution to a more complex solution (*see* Table 12.5) with trypsin alone or trypsin/EDTA as a starting point, adding or substituting other proteases to improve disaggregation, and deleting trypsin if necessary to increase viability. There is a risk including proteases with cocktails of enzymes as the resulting proteolysis make inactivate some of the enzymes. DNase, for example, will not work if added to trypsin, and needs to be added sequentially after the trypsin has been inactivated.

The choice of which trypsin grade to use has always been difficult, as there are two opposing trends: (1) the purer the trypsin, the less toxic it becomes, and the more predictable its action; (2) the cruder the trypsin, the more effective it may be, because of the presence of other proteases. In practice, a preliminary test experiment may be necessary to determine the optimum grade for viable cell yield, as the balance between sensitivity to toxic effects and disaggregation ability may be difficult to predict. Crude trypsin is by far the most common enzyme used in tissue disaggregation [Waymouth, 1974], as it is tolerated quite well by many cells and is effective for many tissues. Residual activity left after washing is neutralized by the serum of the culture medium, or by a trypsin inhibitor

(e.g., soya bean trypsin inhibitor) when serum-free medium is used. In general, increasing the purity of an enzyme will give better control and less toxicity with increased specificity but may result in less disaggregation potency.

Mechanical and enzymatic disaggregation of the tissue avoids problems of selection by migration and yields a higher number of cells that are more representative of the whole tissue in a shorter time. However, just as the primary explant technique selects on the basis of cell migration, dissociation techniques will select protease- and mechanical stress-resistant cells. Embryonic tissue disperses more readily and gives a higher yield of proliferating cells than newborn or adult tissue. The increasing difficulty in obtaining viable proliferating cells with increasing age is due to several factors, including the onset of differentiation, an increase in fibrous connective tissue and extracellular matrix, and a reduction of the undifferentiated proliferating cell pool. When procedures of greater severity are required to disaggregate the tissue (e.g., longer trypsinization or increased agitation), the more fragile components of the tissue may be destroyed. In fibrous tumors, for example, it is very difficult to obtain complete dissociation with trypsin while still retaining viable carcinoma cells.

## 11.3.3  Warm Trypsin

It is important to minimize the exposure of cells to active trypsin in order to preserve maximum viability. Hence, when whole tissue is being trypsinized at $37°C$, dissociated cells should be collected every half hour, and the trypsin should be removed by centrifugation and neutralized with serum in medium. (*See also* Section 28.4, Exercise 19 for training exercise.)

---

**PROTOCOL 11.5. TISSUE DISAGGREGATION IN WARM TRYPSIN**

### Outline
The tissue is chopped and stirred in trypsin for a few hours. The dissociated cells are collected every half hour, centrifuged, and pooled in medium containing serum (Fig. 11.7).

### Materials

*Sterile or aseptically prepared:*

- ❑ Tissue, 1 to 5 g
- ❑ DBSS, 50 mL (*see* Appendix I)
- ❑ Trypsin (crude), 2.5% in D-PBSA or normal saline

**Note.** As trypsin batches may vary in specific activity, it may be necessary to test a series of different concentrations with a particular tissue before selecting the correct one.

- ❑ D-PBSA, 200 mL
- ❑ Growth medium with serum (e.g., DMEM/F12 with 10% fetal bovine serum)

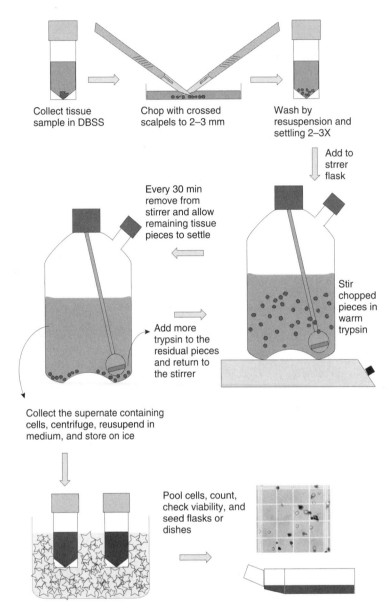

**Fig. 11.7. Warm Trypsin Disaggregation.** Coarsely chopped tissue is stirred in trypsin until fully disaggregated, with dissociated cells collected at intervals, centrifuged, resuspended in medium, and stored on ice to be pooled later.

❑ Culture flasks, 5 to 10 flasks per g tissue (varies depending on cellularity of tissue)
❑ Petri dishes, 9 cm, non–tissue–culture grade
❑ 50-mL centrifuge tubes, or universal containers, 2
❑ Trypsinization flask: 250-mL Erlenmeyer flask (preferably indented as in Fig. 11.3*l*) or stirrer flask (*see* Fig. 20.1)
❑ Magnetic follower, autoclaved in a test tube
❑ Curved forceps
❑ Pipettes (Pasteur, 2 mL, 10 mL)
❑ Pipettor, 100 µL, adjustable

*Nonsterile:*

❑ Naphthalene Black, 1% or Trypan Blue, 0.4%

❑ Magnetic stirrer
❑ Hemocytometer or cell counter

**Procedure**

1. Transfer the tissue to fresh, sterile DBSS in 9-cm Petri dish, and rinse.
2. Transfer the tissue to a second dish; dissect off unwanted tissue, such as fat or necrotic material, and transfer to a third dish.

**Note.** Clean healthy tissue with little blood may not need these two transfers and can be dissected in the first dish, after transfer from the transport medium, which will have acted as the first wash.

3. Chop with crossed scalpels (*see* Fig. 11.7) into roughly 3-mm cubes.

4. Transfer the tissue with curved forceps to a centrifuge tube or universal container.

5. Allow the pieces to settle.

6. Wash the tissue by resuspending the pieces in DBSS, allowing the pieces to settle, and removing the supernate. Repeat this step two more times. These repeated washes may be omitted if there is little blood or necrotic tissue.

7. Resuspend the pieces in the last wash and transfer to the empty trypsinization flask with a wide bore pipette or by pouring.

8. Remove most of the residual fluid in the trypsinization flask, and add 180 mL of D-PBSA.

9. Add 20 mL of 2.5% trypsin.

10. Add the magnetic follower to the flask.

11. Cap the flask, and place it on the magnetic stirrer in an incubator or hot room at 37°C.

12. Stir at about 100 rpm for 30 min at 37°C.

13. After 30 min, collect disaggregated cells as follows:

    **(a)** Allow the pieces to settle.

    **(b)** Pour off the supernate into a centrifuge tube and place it on ice. Carefully wipe off any medium running down the outside of side arm from which you have poured with a lint-free swab and 70% alcohol.

    **(c)** Add fresh trypsin to the pieces remaining in the flask, and continue to stir and incubate for a further 30 min.

    **(d)** Centrifuge the harvested cells from step 11(b) at approximately 500 *g* for 5 min.

    **(e)** Resuspend the resulting pellet in 10 mL of medium with serum, and store the suspension on ice.

14. Repeat step 11 until complete disaggregation occurs or until no further disaggregation is apparent (usually 3–4 h).

15. Collect and pool the chilled cell suspensions.

16. Remove any large remaining aggregates by filtering through sterile muslin or a proprietary sieve (e.g., *see* Fig. 11.8) or by allowing the aggregates to settle.

17. Count the cells by hemocytometer or electronic cell counter (*see* Section 12.5). As the cell population will be very heterogeneous, electronic cell counting may require confirmation with a hemocytometer count to check viability (*see* Section 21.3.1) and the degree of aggregation.

18. Dilute the cell suspension to $1 \times 10^5$ to $1 \times 10^6$/mL in growth medium, and seed as many flasks as are required, with approximately $2 \times 10^4$ to $2 \times 10^5$ cells/cm². When the survival rate

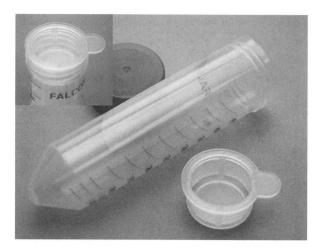

**Fig. 11.8. Cell Strainer.** Disposable polypropylene filter and tube for straining aggregates from primary suspensions (BD Biosciences). Can also be used for disaggregating soft tissues (*see also* Fig. 11.13).

is unknown or unpredictable, a cell count is of little value (e.g., in tumor biopsies, in which the proportion of necrotic cells may be quite high). In this case set up a range of concentrations from about 5 to 25 mg of tissue per mL of medium.

19. Change the medium at regular intervals (2–4 days as dictated by depression of pH). Check the supernate for viable cells before discarding it, as some cells can be slow to attach or may even prefer to proliferate in suspension.

The warm trypsin technique is useful for the disaggregation of large amounts of tissue in a relatively short time, particularly for chopped whole mouse embryos or chick embryos. It does not work as well with adult tissue, in which there is a lot of fibrous connective tissue, and mechanical agitation can be damaging to some of the more sensitive cell types, such as epithelium. If reaggregation is found after centrifugation and resuspension, incubate in DNase, 10 to 20 μg/mL, for 10 to 20 min, and recentrifuge.

### 11.3.4 Trypsinization with Cold Preexposure

One of the disadvantages of using trypsin to disaggregate tissue is the damage that may result from prolonged exposure to the tissue to trypsin at 37°C hence the need to harvest cells after 30-min incubations in the warm trypsin method rather than have them exposed for the full time (i.e., 3–4 h) as required to disaggregate the whole tissue. A simple method of minimizing damage to the cells during exposure is to soak the tissue in trypsin at 4°C for 6 to 18 h to allow penetration of the enzyme with little tryptic activity (Table 11.1). Following this procedure, the tissue will only require 20 to 30 min at 37°C for disaggregation [Cole & Paul, 1966]. (*See also* Section 28.4, Exercise 19 for training exercise.)

**TABLE 11.1.** Cell Yield from 12.5-Day Embryo by Warm and Cold Trypsinization

| Duration and temperature of trypsinization | | After trypsinization | | After 24 h in culture | | |
|---|---|---|---|---|---|---|
| 4°C | 37°C | Cells recovered per embryo × 10⁻⁷ | % Viability by dye exclusion (Trypan Blue) | Total no. of viable cells × 10⁻⁷ | Recovered, % of total seeded | Viability, % of viable cells seeded |
| 0 h | 4 h | 1.69 | 86 | 1.45 | 47.2 | 54.9 |
| 5.5 h | 0.5 h | 3.32 | 60 | 1.99 | 74.5 | 124 |
| 24 h | 0.5 h | 3.40 | 75 | 2.55 | 60.3 | 80.2 |

## PROTOCOL 11.6. TISSUE DISAGGREGATION IN COLD TRYPSIN

### Outline

Chop tissue and place in trypsin at 4°C for 6 to 18 h. Incubate after removing the trypsin, and disperse the cells in warm medium (Fig. 11.9).

### Materials

*Sterile or aseptically prepared:*

- Tissue, 1 to 5 g
- Growth medium (e.g., DMEM/F12 with 10% FBS)
- DBSS
- 0.25% crude trypsin in serum-free RPMI 1640 or MEM/Stirrer Salts (S-MEM)

*Note.* As trypsin batches may vary in specific activity, it may be necessary to test a series of different concentrations with a particular tissue before selecting the correct one.

- Petri dishes, 9 cm, non-tissue-culture grade
- Forceps, straight and curved
- Scalpels
- Glass Erlenmeyer flask, 25 or 50 mL, screw capped, centrifuge tube, or universal container
- Culture flasks, 25 or 75 cm²
- Pipettes (Pasteur, 2 mL, 10 mL)

*Nonsterile:*

- Viability stain: Naphthalene Black, 1% or Trypan Blue, 0.4%
- Ice bath

### Procedure

1. Transfer the tissue to fresh, sterile DBSS in a 9-cm Petri dish, and rinse.
2. Transfer the tissue to a second dish and dissect off unwanted tissue, such as fat or necrotic material.
3. Transfer to a third dish and chop with crossed scalpels (*see* Fig. 11.9) into about 3-mm cubes. Embryonic organs, if they do not exceed this size, are better left whole.

*Note.* Clean healthy tissue with little blood may not need these two transfers and can be dissected in the first dish, after transfer from the transport medium, which will have acted as the first wash.

4. Transfer the tissue with curved forceps to a glass Erlenmeyer, centrifuge tube, or universal container.
5. Allow the pieces to settle.
6. Wash the tissue by resuspending the pieces in DBSS, allowing the pieces to settle, and removing the supernate. Repeat this step two more times. This step may be omitted if there is little blood or necrotic tissue.
7. Carefully remove the residual fluid.
8. Add 10 mL/g of tissue of 0.25% trypsin in RPMI 1640 or S-MEM at 4°C.
9. Place the mixture at 4°C for 6 to 18 h.
10. Remove and discard the trypsin carefully, leaving the tissue with only the residual trypsin.
11. Place the tube at 37°C for 20 to 30 min.
12. Add warm medium, approximately 1 mL for every 100 mg of original tissue, and gently pipette the mixture up and down until the tissue is completely dispersed.
13. If some tissue does not disperse, then the cell suspension may be filtered through sterile muslin or stainless steel mesh (100–200 μm), or a disposable plastic mesh strainer (Fig. 11.8), or the larger pieces may simply be allowed to settle. When there is a lot of tissue, increasing the volume of suspending medium to 20 mL for each gram of tissue will facilitate settling and subsequent collection of cells in the supernate. Two to three minutes should be sufficient to get rid of most of the larger pieces. If there are strands of tissue which do not sediment easily, it is possible that these are aggregated with DNA; DNase, 10 to 20 μg/mL may be added for 10 to 20 min to disperse these.
14. Determine the cell concentration in the suspension by hemocytometer or electronic

15. Dilute the cell suspension to $1 \times 10^5$ to $1 \times 10^6$/ mL in growth medium, and seed as many flasks as are required, with approximately $2 \times 10^4$ to $2 \times 10^5$ cells/cm$^2$. When the survival rate is unknown or unpredictable, a cell count is of little value (e.g., in tumor biopsies, for which the proportion of necrotic cells may be quite high). In this case, set up a range of concentrations from about 5 to 25 mg of tissue per mL.

16. Change the medium at regular intervals (2–4 days as dictated by depression of pH). Check the supernate for viable cells before discarding it, as some cells can be slow to attach or may even prefer to proliferate in suspension.

cell counter (*see* Section 20.1), and check viability (*see* Protocol 21.1). The cell population will be very heterogeneous; electronic cell counting will initially require confirmation with a hemocytometer, which also allows for determination of viability.

The cold trypsin method usually gives a higher yield of viable cells, with improved survival after 24-h culture (*see* Figs. 11.1, 11.10; Table 11.1), and preserves more different cell types than the warm method (*see* Plates 2*d,e* and 3). Cultures from mouse embryos contain more epithelial cells when prepared by the cold method, and erythroid cultures from 13-day fetal mouse liver respond to erythropoietin after this treatment, but not after the warm trypsin method or mechanical disaggregation [Cole & Paul, 1966; Conkie, personal communication]. The cold trypsin method is also convenient, as no stirring or centrifugation is required and the incubation at 4°C may be done unattended overnight.

### 11.3.5 Chick Embryo Organ Rudiments
The cold trypsin method is particularly suitable for small amounts of tissue, such as embryonic organs. Protocol 11.7 gives good reproducible cultures from 10- to 13-day chick embryos with evidence of several different cell types characteristic of the tissue of origin. This protocol forms a good exercise for teaching purposes (*see* Section 28.4, Exercise 19).

---

**PROTOCOL 11.7. CHICK EMBRYO ORGAN RUDIMENTS**

**Outline**
Dissect out individual organs or tissues, and place them, preferably whole, in cold trypsin overnight. Remove the trypsin, incubate the organs or tissue

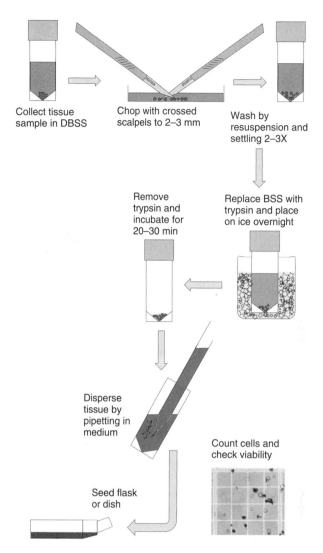

**Fig. 11.9. Cold Trypsin Disaggregation.** *See also* Plate 2*a, d, e,* and Plate 3.

briefly, and disperse them in culture medium. Dilute and seed the cultures.

**Materials**

*Sterile:*

❑ DBSS
❑ Crude trypsin (Difco 1:250 or equivalent) 0.25% in RPMI 1640 or S-MEM on ice; lower concentrations may be used with purer grades of trypsin (e.g., 0.05–0.1% Sigma crystalline or Worthington Grade IV)

*Note.* As trypsin batches may vary in specific activity, it may be necessary to test a series of different concentrations with a particular tissue before selecting the correct one.

❑ Culture medium (e.g., DMEM/F12 with 10% FBS), minimum of 12 mL per tissue

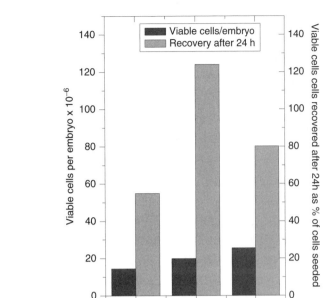

**Fig. 11.10. Warm and Cold Trypsinization.** Yield of viable cells per 12.5-day embryo increases by cold trypsinization up to 24 h at 4°C, but recovery after 24 h culture is greatest with shorter cold trypsinization (>100%, implying cell proliferation), perhaps because some of the cells released by longer cold trypsinization are not proliferative. (*See also* Table 11.1.)

---

❏ Petri dishes, 9 cm, non-tissue-culture grade
❏ Culture flasks, 25 cm² (2 per tissue)
❏ Scalpels (No. 11 blade for most steps)
❏ Iridectomy knives for fine dissection
❏ Curved and straight fine forceps
❏ Pipettes (Pasteur, 2 mL, 10 mL)
❏ Test tubes, preferably glass, 10 to 15 mL, with screw caps

*Nonsterile:*

❏ Embryonated hen's eggs, 10- to 13-day incubation
❏ Ice bath
❏ Binocular dissecting microscope

**Procedure**

1. Remove the embryo from the egg as described previously (*see* Protocol 11.2), and place it in sterile DBSS.
2. Remove the head (Fig. 11.11*a, b*).
3. Remove an eye and open it carefully, releasing the lens and aqueous and vitreous humors (Fig. 11.11*c, d*).
4. Grasp the retina in two pairs of fine forceps and gently peel the pigmented retina off the neural retina and connective tissue (Fig. 11.11*e*). (This step requires a dissection microscope for 10-day embryos. A brief exposure to 0.25% trypsin

in 1 mM EDTA will allow the two tissues to separate more easily.) Put the tissue to one side.
5. Pierce the top of the head with curved forceps, and scoop out the brain (Fig. 11.11*f*). Place the brain with the retina at the side of the dish.
6. Halve the trunk transversely where the pink color of the liver shows through the ventral skin (Fig. 11.11*g*). If the incision is made on the line of the diaphragm, then it will pass between the heart and the liver; but sometimes the liver will go to the anterior instead of the posterior half.
7. Gently probe into the cut surface of the anterior half, and draw out the heart and lungs (Fig. 11.11*h*; tease the organs out, and do not cut until you have identified them). Separate the heart and lungs and place at the side of the dish.
8. Probe the posterior half, and draw out the liver, with the folds of the gut enclosed in between the lobes (Fig. 11.11*i*). Separate the liver from the gut and place each at the side of the dish.
9. Fold back the body wall to expose the inside of the dorsal surface of the body cavity in the posterior half. The elongated lobulated kidneys should be visible parallel to and on either side of the midline.
10. Gently slide the tip of the scalpel under each kidney and tease the kidneys away from the dorsal body wall (Fig. 11.11*j*). (This step requires a dissection microscope for 10-day embryos.) Carefully cut the kidneys free, and place them on one side.
11. Place the tips of the scalpels together on the midline at the posterior end, and, advancing the tips forward, one over the other, express the spinal cord as you would express toothpaste from a tube (Fig. 11.11*k*). (This step may be difficult with 10-day embryos.)
12. Turn the posterior trunk of the embryo over, and strip the skin off the back and upper part of the legs (Fig. 11.11*l*). Collect and place this skin on one side.
13. Dissect off muscle from each thigh, and collect this muscle together (Fig. 11.11*m*).
14. Transfer all of these tissues, and any others you may want, to separate test tubes, add 1 mL of ice-cold trypsin, and place these tubes on ice. Make sure that the tissue is immersed in the trypsin.
15. Leave the test tubes for 6 to 18 h at 4°C.
16. Carefully remove the trypsin from the test tubes without disturbing the tissue; tilting and rolling the tube slowly will help.
17. Incubate the tissue in the residual trypsin for 15 to 20 min at 37°C.

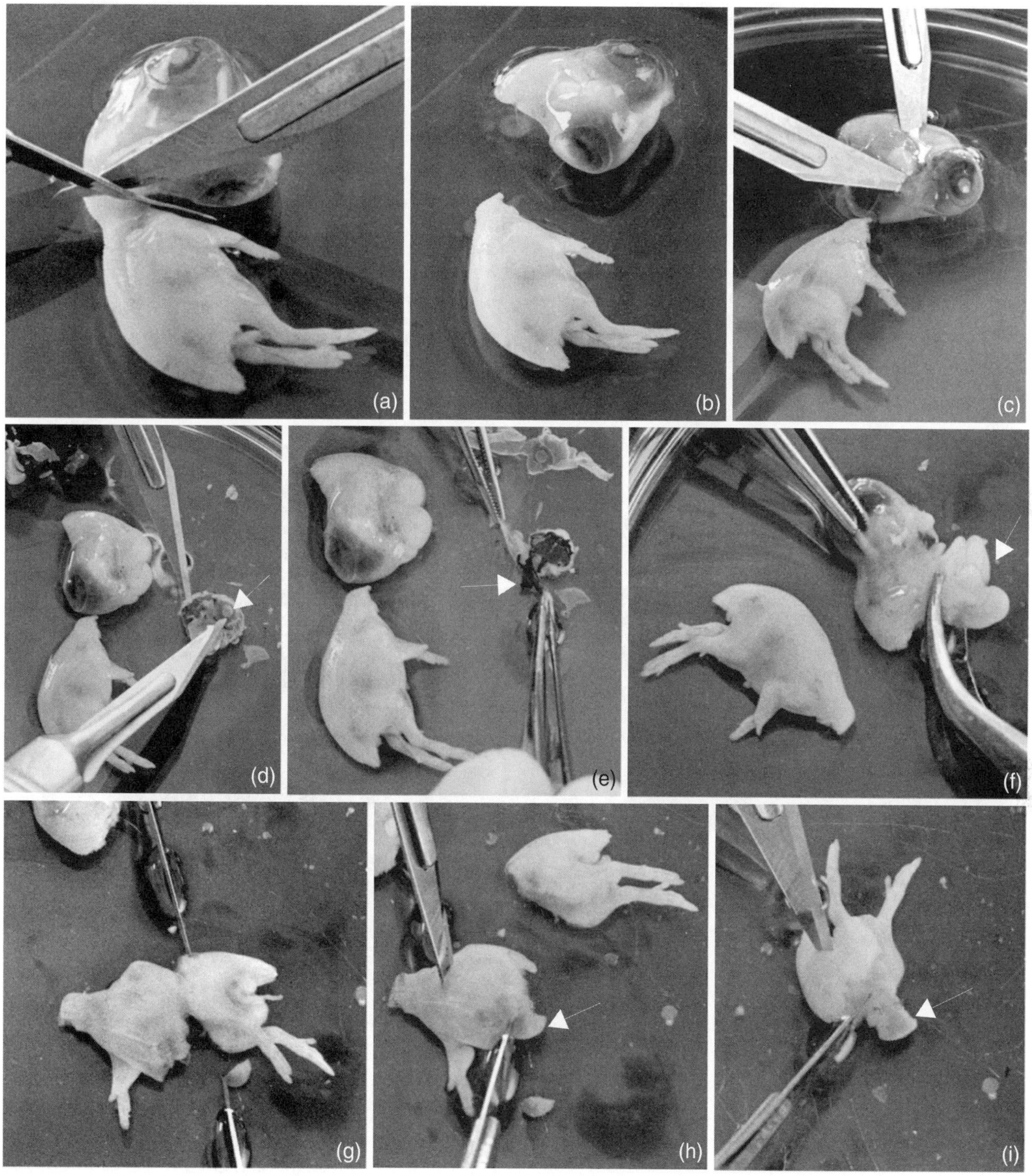

**Fig. 11.11. Dissection of a Chick Embryo.** (*a*, *b*) Removing the head. (*c*) Removing the eye. (*d*) Dissecting out the lens. (*e*) Peeling off the retina. (*f*) Scooping out the brain. (*g*) Halving the trunk. (*h*) Teasing out the heart and lungs from the anterior half. (*i*) Teasing out the liver and gut from the posterior half. (*j*) Inserting the tip of the scalpel between the left kidney and the dorsal body wall. (*k*) Squeezing out the spinal cord. (*l*) Peeling the skin off the back of the trunk and hind leg. (*m*) Slicing muscle from the thigh. (*n*) Organ rudiments arranged around the periphery of the dish. From the right, clockwise, we have the following organs: brain, heart, lungs, liver, gizzard, kidneys, spinal cord, skin, and muscle.

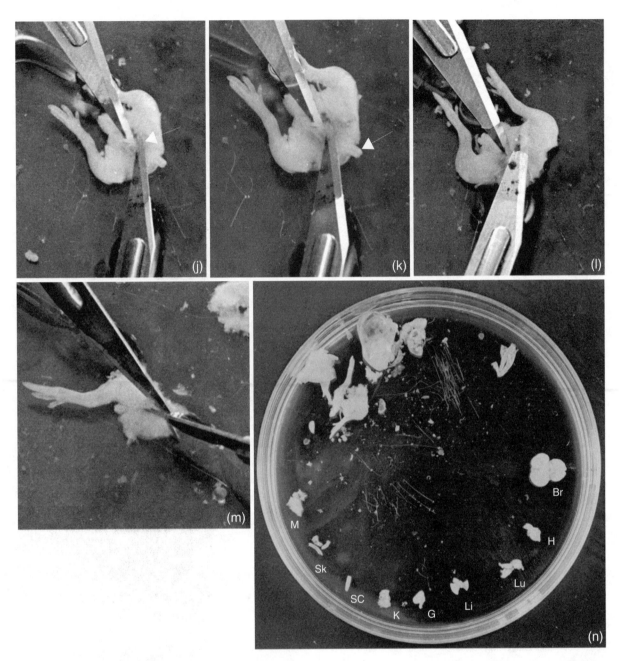

***Fig. 11.11.*** (*Continued*)

18. Add 4 mL of medium to each of two 25-cm² flasks for each tissue to be cultured.
19. Add 2 mL of medium to tubes containing tissues and residual trypsin, and pipette up and down gently to disperse the tissue.
20. Allow any large pieces of tissue to settle.
21. Pipette off the cells in the supernate into the first flask, mix, and transfer 1 mL of diluted suspension to the second flask. This procedure gives two flasks at different cell concentrations and avoids the need to count the cells. Experience

will determine the appropriate cell concentration to use in subsequent attempts.
22. Change the medium as required (e.g., for brain, it may need to be changed after 24 h, but pigmented retina will probably last 5–7 days), and check for characteristic morphology and function.

After 3 to 5 days, contracting cells may be seen in the heart cultures, colonies of pigmented cells in the pigmented retina culture, and the beginning of myotubes in skeletal muscle cultures. Cultures may be fixed and stained (*see* Protocol 15.2) for future examination.

### 11.3.6 Other Enzymatic Procedures

Disaggregation in trypsin can be damaging (e.g., to some epithelial cells) or ineffective for very fibrous tissue (e.g., such as fibrous connective tissue), so attempts have been made to utilize other enzymes. Because the extracellular matrix often contains collagen, particularly in connective tissue and muscle, collagenase has been the obvious choice [Freshney, 1972 (colon carcinoma); Speirs et al., 1996 (breast carcinoma); Chen, T. C., et al., 1989 (kidney); Booth & O'Shea, 2002 (gut); Heald et al., 1991 (pancreatic islet cells)] (*see* Sections 22.2.5–23.2.7; Table 12.5). Other bacterial proteases, such as Pronase [Schaffer et al., 1997; Glavin et al., 1996 and Dispase (Boehringer-Mannheim) [Compton et al., 1998; Inamatsu et al., 1998, have also been used with varying degrees of success. Liberase (Roche) is a cocktail of enzymes (neutral protease with collagenase or thermolysin) that has been used for isolation of hepatocytes from liver (*see* Protocol 22.6A) and islet cells from pancreas [Anazawa et al., 2009]. The participation of carbohydrate in intracellular adhesion has led to the use of hyaluronidase [Berry & Friend, 1969] and neuraminidase in conjunction with collagenase. Other proteases continue to appear on the market (*see* Section 11.1). With the selection now available, screening available samples is the only option if trypsin, collagenase, Dispase, Pronase, hyaluronidase, and DNase, alone and in direct or sequential combinations, do not prove to be successful.

### 11.3.7 Collagenase

This technique is very simple and effective for many tissues: embryonic, adult, normal, and malignant. It is of greatest benefit when the tissue is either too fibrous or too sensitive to allow the successful use of trypsin. Crude collagenase is often used and may depend, for some of its action, on contamination with other nonspecific proteases. More highly purified grades are available if nonspecific proteolytic activity is undesirable, but they may not be as effective as crude collagenase.

---

**PROTOCOL 11.8. TISSUE DISAGGREGATION IN COLLAGENASE**

*Outline*

Place finely chopped tissue in complete medium containing collagenase and incubate. When tissue is disaggregated, remove collagenase by centrifugation, seed cells at a high concentration, and culture (Fig. 11.12).

*Materials*

Sterile:

❑ Collagenase (2000 units/mL), Worthington CLS or Sigma 1A (screen batches)

❑ Culture medium (e.g., DMEM/F12 with 10% FBS)
❑ DBSS
❑ Pipettes, 1 mL, 10 mL
❑ Petri dishes, 9 cm, non-tissue-culture grade
❑ Culture flasks, 25 cm$^2$
❑ Centrifuge tubes or universal containers, 15 to 50 mL, depending on the amount of tissue being processed
❑ Scalpels

*Nonsterile:*

❑ Centrifuge

*Procedure*

1. Transfer the tissue to fresh, sterile DBSS, and rinse.
2. Transfer the tissue to a second dish and dissect off unwanted tissue, such as fat or necrotic material. This step may be omitted if there is no fat and necrotic tissue.
3. Transfer to another dish and chop finely with crossed scalpels (*see* Fig. 11.12) into about 1-mm cubes.
4. Transfer the tissue by pipette (10–20 mL, with wide tip) to a 15- or 50-mL sterile centrifuge tube or universal container. (Wet the inside of the pipette first with DBSS, or else the pieces will stick.)
5. Allow the pieces to settle.
6. Wash the tissue by resuspending the pieces in DBSS, allowing the pieces to settle, and removing the supernate. Repeat this step two more times.
7. Transfer 20 to 30 pieces to one 25-cm$^2$ flask and 100 to 200 pieces to a second flask.
8. Drain off the DBSS, and add 4.5 mL of growth medium with serum to each flask.
9. Add 0.5 mL of crude collagenase, 2000 units/mL, to give a final concentration of 200 units/mL collagenase.
10. Incubate at 37°C for 4 to 48 h without agitation. Tumor tissue may be left up to 5 days or more if disaggregation is slow (e.g., in scirrhous carcinomas of the breast or the colon), although it may be necessary to centrifuge the tissue and resuspend it in fresh medium and collagenase before that amount of time has passed if an excessive drop in pH is observed (i.e., <pH 6.5).
11. Check for effective disaggregation by gently moving the flask; the pieces of tissue will "smear" on the bottom of the flask and, with gentle pipetting, will break up into single cells and small clusters (Fig. 11.12b).
12. With some tissues (e.g., lung, kidney, and colon or breast carcinoma), small clusters of

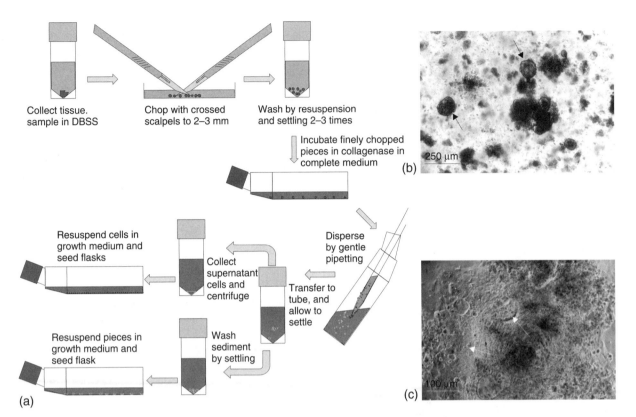

**Fig. 11.12. Tissue Disaggregation by Collagenase.** (*a*) Schematic diagram of dissection followed by disaggregation in collagenase. (*b*) Cell clusters from human colonic carcinoma after 48 h dissociation in crude collagenase (Worthington CLS grade), before removal of collagenase. Arrows point to probable epithelial clusters. (*c*) Same as (*b*), but after removal of collagenase, further disaggregation by pipetting, and culture for 48 h. The clearly defined rounded clusters (black arrows) in (*b*) form epithelium-like sheets (white arrows) in (*c*), some still three-dimensional, some spreading as a sheet, and the more irregularly shaped clusters produce fibroblasts. (*See also* Plate 2*c*).

epithelial cells can be seen to resist the collagenase and may be separated from the rest by allowing them to settle for about 2 min. If these clusters are further washed with DBSS by resuspension and settling and the sediment is resuspended in medium and seeded, then they will form islands of epithelial cells. Epithelial cells generally survive better if they are not completely dissociated.

13. When complete disaggregation has occurred, or when the supernatant cells are collected after removing clusters by settling, centrifuge the cell suspension from the disaggregate and any washings at 50 to 100 *g* for 3 min.

14. Discard the supernatant DBSS or medium, resuspend and combine the pellets in 5 mL of medium, and seed in a 25-cm² flask. If the pH fell during collagenase treatment (to pH 6.5 or less by 48 h), then dilute the suspension two- to threefold in medium after removing the collagenase.

15. Replace the medium after 48 h.

Some cells, particularly macrophages, may adhere to the first flask during the collagenase incubation. Transferring the cells to a fresh flask after collagenase treatment (and subsequent removal of the collagenase) removes many of the macrophages from the culture. The first flask may be cultured as well, if required. Light trypsinization will remove any adherent cells other than macrophages.

Disaggregation in collagenase has proved particularly suitable for the culture of human tumors [Pfragner & Freshney, 2004], mouse kidney, human adult and fetal brain, liver (*see* Protocol 22.6), lung, and many other tissues, particularly epithelium [Freshney & Freshney, 2002]. The process is gentle and requires no mechanical agitation or special equipment. With more than 1 g of tissue, however, it becomes tedious at the dissection stage and can be expensive, because of the amount of collagenase required. It will also release most of the connective tissue cells, accentuating the problem of fibroblastic outgrowth, so it may need to be followed by selective culture (*see* Section 9.2.2) or cell separation (*see* Chapter 14).

The discrete clusters of epithelial cells produced by disaggregation in collagenase (*see* step 11 of Protocol 11.8) and by the cold trypsin method (*see* Plates 2, 3) can be selected

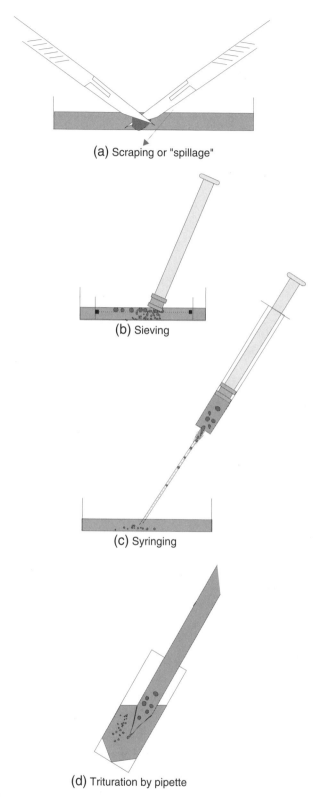

**Fig. 11.13.** *Mechanical Disaggregation.* (*a*) Scraping or "spillage." Cutting action, or abrasion of cut surface, releases cells. (*b*) Sieving. Forcing tissue through sieve with syringe piston. (Falcon Cell Strainer can be used; *see* Fig. 11.8.) (*c*) Syringing. Drawing tissue into syringe through wide-bore needle or canula and expressing. (*d*) Trituration by pipette. Pipetting tissue fragments up and down through wide-bore pipette.

under a dissection microscope and transferred to individual wells in a microtitration plate, alone or with irradiated or mitomycin C-treated feeder cells (*see* Sections 13.2.3, 22.2.1, 22.2.4).

## 11.3.8  Mechanical Disaggregation

The outgrowth of cells from primary explants is a relatively slow process and can be highly selective. Enzymatic digestion is rather more labor intensive, although, potentially, it gives a culture that is more representative of the cells in the tissue. As there is a risk of proteolytic damage to cells during enzymatic digestion, many people have chosen to use mechanical disaggregation: For example, *spillage:* collecting the cells that spill out when the tissue is carefully sliced and the slices scraped [Lasfargues, 1973]; *sieving:* pressing the dissected tissue through a series of sieves for which the mesh is gradually reduced in size; or *syringing:* forcing the tissue fragments through a syringe (with or without a wide-gauge needle) [Zaroff et al., 1961] or simply pipetting it repeatedly (*see* Fig. 11.13). This procedure gives a cell suspension more quickly than enzymatic digestion but may cause mechanical damage. Spillage (Fig. 11.13*a*) and sieving (Fig. 11.13*b*) are probably the gentlest mechanical methods, while pipetting (Fig. 11.13*d*) and, particularly, syringing (Fig. 11.13*c*), are most likely to generate shear stress. Protocol 11.9 is one method of mechanical disaggregation that has been found to be moderately successful with soft tissues such as brain.

---

**PROTOCOL 11.9. MECHANICAL DISAGGREGATION BY SIEVING**

**Outline**
The tissue in culture medium is forced through a series of sieves for which the mesh is gradually reduced in size until a reasonable suspension of single cells and small aggregates is obtained. The suspension is then diluted and cultured directly.

**Materials**

*Sterile:*

❏ Growth medium (e.g., DMEM/F12 with 10% FBS)
❏ Forceps
❏ Sieve (Fig. 11.13*b*), or graded series of sieves from 100 μm down to 20 μm, or Falcon Cell Strainer (BD Biosciences; *see* Fig. 11.8)
❏ Petri dishes, 9 cm
❏ Centrifuges tube, 50 mL
❏ Scalpels
❏ Disposable plastic syringes (2 or 5 mL)
❏ Culture flasks

**Procedure**

1. After washing and preliminary dissection of the tissue (*see* steps 1 and 2 of Protocol 11.5), chop the tissue into pieces about 3 to 5 mm across, and place a few pieces at a time into a stainless steel or polypropylene sieve of 1-mm mesh in a 9-cm Petri dish (*see* Fig. 11.13*b*) or 50 mL centrifuge tube (*see* Fig. 11.8).
2. Force the tissue through the mesh into medium by applying gentle pressure with the piston of a disposable plastic syringe. Pipette more medium through the sieve to wash the cells through it.
3. Pipette the partially disaggregated tissue from the Petri dish into a sieve of finer porosity, perhaps 100-μm mesh, and repeat step 2.
4. The suspension may be diluted and cultured at this stage, or it may be sieved further through 20-μm mesh if it is important to produce a single-cell suspension. In general, the more highly dispersed the cell suspension, the higher the sheer stress required and the lower the resulting viability.
5. Seed the culture flasks at $2 \times 10^5$, $1 \times 10^6$, and $2 \times 10^6$ cells/mL by diluting the cell suspension in medium.

Only soft tissues, such as spleen, embryonic liver, embryonic and adult brain, and some human and animal soft tumors, respond well to this technique. Even with brain, for which fairly complete disaggregation can be obtained easily, the viability of the resulting suspension is lower than that achieved with enzymatic digestion, although the time taken may be very much less. When the availability of tissue is not a limitation and the efficiency of the yield is not important, it may be possible to produce, in a shorter amount of time, as many viable cells with mechanical disaggregation as with enzymatic digestion, but at the expense of very much more tissue.

### 11.3.9 Separation of Viable and Nonviable Cells

When tissue is disaggregated and seeded into primary culture, only a proportion of the cells are capable of surviving and generating a primary culture (*see* Section 2.7). Some cells may not be capable of attachment but yet viable; others are nonviable, necrotic or apoptotic. If it is important to do so, the proportion of necrotic and apoptotic cells may be determined by viability staining and flow cytometry (*see* Sections 20.1.4, 21.1). Normally nonviable cells are removed at the first change of medium. With primary cultures maintained in suspension, nonviable cells are gradually diluted out when cell proliferation starts. If necessary, however, nonviable cells may be removed from the primary disaggregate by centrifuging the cells on a mixture of Ficoll and sodium metrizoate (e.g., Hypaque or Triosil) [Vries et al., 1973]. This technique is similar to the preparation of lymphocytes from peripheral blood (*see* Protocol 27.1). The viable cells collect at the interface between the medium and the Ficoll/metrizoate, and the dead cells form a pellet at the bottom of the tube.

---

**PROTOCOL 11.10. ENRICHMENT OF VIABLE CELLS**

**Outline**

Up to $2 \times 10^7$ cells in 9 mL of medium may be layered on top of 6 mL of Ficoll–Hypaque in a 25 to 50-mL screw-capped centrifuge tube. The mixture is then centrifuged, and viable cells are collected from the interface.

**Materials**

*Sterile:*

- ❑ Cell suspension with as few aggregates as possible
- ❑ Clear centrifuge tubes or universal containers
- ❑ D-PBSA
- ❑ Ficoll–Hypaque or equivalent (*see* Appendix II) (Ficoll/metrizoate, adjusted to 1.077 g/cc)
- ❑ Growth medium
- ❑ Syringe with blunt cannula or square-cut needle, Pasteur pipettes, or pipettor

*Nonsterile:*

- ❑ Hemocytometer or cell counter
- ❑ Centrifuge

**Procedure**

1. Allow major aggregates in the cell suspension to settle.
2. Layer 9 mL of the cell suspension onto 6 mL of the Ficoll–Hypaque mixture. This step should be done in a wide, transparent centrifuge tube with a cap, such as the 25-mL Sterilin or Nunclon universal container, or in the clear plastic Corning 50-mL tube, using double the aforementioned volumes.
3. Centrifuge the mixture for 15 min at 400 *g* (measured at the center of the interface).
4. Carefully remove the top layer without disturbing the interface.
5. Collect the interface carefully with a syringe, Pasteur pipette, or pipettor.

Δ **Safety Note.** If you are using human or other primate material, do not use a sharp needle or glass Pasteur pipette.

6. Dilute the mixture to 20 mL in medium (e.g., DMEM/F12/10FB).
7. Centrifuge the mixture at 70 *g* for 10 min.

**TABLE 11.2.** Data Record for Primary Culture

Date .......................... Time .................... Operator ...........................................

| | | Record |
|---|---|---|
| **Origin of tissue** | Species | |
| | Race or strain | |
| | Age | |
| | Sex | |
| | Path. # or animal tag # | |
| | Tissue | |
| | Site | |
| | Stored tissue/DNA location | |
| | | |
| **Pathology** | Normal/benign/malignant | |
| | Histology | |
| | | |
| **Disaggregation agent** | Trypsin, collagenase, etc. | |
| | Concentration | |
| | Duration | |
| | Diluent | |
| | | |
| **Cell count** | Concentration after resuspension ($C_I$) | |
| | Volume ($V_I$) | |
| | Yield ($Y = C_I \times V_I$) | |
| | Yield per g (wet weight of tissue) | |
| | | |
| **Seeding** | Number ($N$) and type of vessel (flask, dish, or plate wells) | |
| | Final concentration ($C_F$) | |
| | Volume per flask, dish, or well ($V_F$) | |
| | | |
| **Medium** | Type | |
| | Supplier and batch no. | |
| | Serum type and concentration | |
| | Supplier and batch no. | |
| | Other additives | |
| | $CO_2$ concentration | |
| | | |
| **Matrix coating** | Fibronectin, Matrigel, collagen, etc. | |
| | | |
| **Subculture** | Recovery at 1st subculture, cells/flask | |
| | % (cells recovered ÷ cells seeded) | |
| | Cell line designation | |
| | | |

8. Discard the supernate, and resuspend the pellet in 5 mL of growth medium.
9. Repeat steps 7 and 8 in order to wash cells free of density medium.
10. Count the cells with a hemocytometer or an electronic counter.
11. Seed the culture flask(s).

This enrichment procedure can be scaled up or down and works with lower ratios of density medium to cell suspension (e.g., 5 mL of cell suspension over 1 mL of density medium).

## 11.3.10 Primary Culture in Summary

The disaggregation of tissue and preparation of the primary culture make up the first, and perhaps most vital, stage in the culture of cells with specific functions. If the required cells are lost at this stage, then the loss is irrevocable. Many different cell types may be cultured by choosing the correct techniques (*see* Section 11.3; Chapter 22). In general, trypsin is more severe than collagenase but is sometimes more effective in creating a single-cell suspension. Collagenase does not dissociate epithelial cells readily, but this characteristic can be an advantage for separating the epithelial cells from stromal cells and maintaining viability in epithelial clusters. Mechanical disaggregation is much quicker than enzymatic procedures but damages more cells. The best approach is to try out the techniques described in Protocols 12.4 through 12.9 and select the method that works best in your system. If none of those methods is successful, try using additional or alternative enzymes, such as Pronase, Dispase, Liberase, Accutase, and DNase, at different concentrations, and consult the literature.

## 11.3.11 Primary Records

Regardless of the technique used to produce a culture, it is important to keep proper records of the culture's origin and derivation, including the species, sex, and tissue from which it was derived, any relevant pathology, and the procedures used for disaggregation and primary culture (Table 11.2). If you are working under good laboratory practice (*see* Section 6.10) conditions, then such records are not just desirable, but obligatory. Records can be kept in a notebook or another hard copy file of record sheets, but it is best at this stage to initiate a record in a computer database; this record then becomes the first step in maintaining the *provenance* of the cell line. The database record may never proceed beyond the primary culture stage, but, looking at the opposite extreme, it could be the first stage in creating an accurate record of what will become a valuable cell line. If the record becomes irrelevant, it can always be deleted, but it cannot, with any accuracy, be created later if the cell line assumes some importance. Records of primary cultures that are unsuccessful are nevertheless data that allow calculation of your success rate.

As it may be necessary to authenticate the origin of a cell line at a later stage in its life history, particularly if cross-contamination is suspected, it is important to save a sample of tissue, blood, or DNA from the same individual at the time of isolation of tissue for the primary culture. This sample can then be used as reference material for DNA profiling (*see* Sections 15.8.2, 15.8.3).

# CHAPTER 12

# Subculture and Cell Lines

## 12.1 SUBCULTURE AND PROPAGATION

The first *subculture* represents an important transition for a culture. The need to subculture implies that the primary culture has increased to occupy all of the available substrate. Hence cell proliferation has become an important feature. Although the primary culture could have a variable growth fraction after the first subculture (*see* Section 20.11.1), depending on the types of cells present in the culture, the growth fraction is usually high (80% or more). From a very heterogeneous primary culture, containing many of the cell types present in the original tissue, a more homogeneous cell line emerges. In addition to its biological significance, this process has considerable practical importance, as the culture can now be propagated, characterized, and stored, and the potential increase in cell number and the uniformity of the cells open up a much wider range of experimental possibilities (*see* Table 1.5). A large number of cell lines are now widely available through accredited cell banks (Table 12.1; *see also* Section 19.5, Appendix II).

### 12.1.1 Cross-contamination and Misidentification

However, there are a number of less desirable consequences of generating a cell line. While propagation and cryopreservation extends the lifetime of a culture and its availability, it also increases the risk of cross-contamination. Whenever more than one cell line is maintained in a laboratory, there is always a risk that cells from one cell line will be accidentally introduced into the other, and if its growth rate is faster, it will overgrow and eventually replace the original cell line

(*see also* Sections 15.2, 18.6). Such contamination could be due to poor pipetting techniques, sharing media and pipettes among cell lines, or the generation of aerosols when flasks or media bottles from more than one cell line are open simultaneously. Contamination accidents may also occur at subculture or during cryopreservation from mislabeling, seeding the wrong flask, or poor inventory control in the freezer leading to a cell line becoming misidentified. Presently there is an accumulation of around 350 documented cell lines that are not what they are claimed to be (Table 12.2; *see also* Appendix V, www.wiley.com/go/freshney/cellculture, and www.hpacultures.org.uk/collections/ecacc.jsp).

The history of the cell line HeLa makes an interesting story, particularly its origin and biology, and the bioethical, social, and legal implications of its widespread use [Skloot, 2010]. Cross-contamination was first suspected from research into enzyme polymorphisms [Gartler, 1967] and subsequently confirmed by chromosome analysis [Nelson-Rees & Flandermeyer, 1977] when it was found that the majority of continuous cell lines in current use in the 1960s and 1970s had been contaminated with HeLa cells. Regrettably, this problem has not diminished as one might have expected from the impact of these earlier revelations but has instead grown in scale, has been confirmed by wide-scale DNA profiling by cell banks and independent laboratories, and has been shown to involve many other victim and villain cell lines [Nelson-Rees et al., 1981; MacLeod et al., 1999; Stacey et al., 2000; Nardone, 2008; Hughes et al., 2007; Lacroix, 2008; Capes-Davis et al., 2010]. That new instances happen occasionally may be unfortunate but understandable; that the same contaminated cell lines that were described in the decade 1967 to 1977 are still in widespread

*Culture of Animal Cells: A Manual of Basic Technique and Specialized Applications, Sixth Edition*, by R. Ian Freshney
Copyright © 2010 John Wiley & Sons, Inc.

**TABLE 12.1.** Commonly Used Cell Lines

| Cell line | Morphology | Origin | Species | Age | Ploidy | Characteristics | Reference |
|---|---|---|---|---|---|---|---|
| **Finite, from normal tissue** | | | | | | | |
| IMR-90 | Fibroblast | Lung | Human | Embryonic | Diploid | Susceptible to human viral infection; contact inhibited | Nichols et al., 1977 |
| MRC-5 | Fibroblast | Lung | Human | Embryonic | Diploid | Susceptible to human viral infection; contact inhibited | Jacobs, 1970 |
| MRC-9 | Fibroblast | Lung | Human | Embryonic | Diploid | Susceptible to human viral infection; contact inhibited | Jacobs et al., 1979 |
| WI-38 | Fibroblast | Lung | Human | Embryonic | Diploid | Susceptible to human viral infection | Hayflick & Moorhead, 1961 |
| **Continuous, from normal tissue** | | | | | | | |
| 3T3-A31 | Fibroblast | | Mouse BALB/c | Embryonic | Aneuploid | Contact inhibited; readily transformed | Aaronson & Todaro, 1968 |
| 3T3-L1 | Fibroblast | | Mouse Swiss | Embryonic | Aneuploid | Adipose differentiation | Green & Kehinde, 1974 |
| BEAS-2B | Epithelial | Lung | Human | Adult | | | Reddel et al., 1988 |
| BHK21-C13 | Fibroblast | Kidney | Syrian hamster | Newborn | Aneuploid | Transformable by polyoma | Macpherson & Stoker, 1962 |
| BRL 3A | Epithelial | Liver | Rat | Newborn | | Produce IGF-2 | Coon, 1968 |
| C2 | Fibroblastoid | Skeletal muscle | Mouse | Embryonic | | Myotubes | Morgan et al., 1992 |
| C7 | Epithelioid | Hypothalamus | Mouse | | | Neurophysin; vasopressin | De Vitry et al., 1974 |
| CHO-K1 | Fibroblast | Ovary | Chinese hamster | Adult | Diploid | Simple karyotype | Puck et al., 1958 |
| COS-1, COS-7 | Epithelioid | Kidney | Pig | Adult | | Good hosts for DNA transfection | Gluzman, 1981 |
| CPAE | Endothelial | Pulmonary-artery endothelium | Cow | Adult | Diploid | Factor VIII, Angiotensin II converting enzyme | Del Vecchio & Smith, 1981 |
| HaCaT | Epithelial | Keratinocytes | Human | Adult | Diploid | Cornification | Boukamp et al., 1988 |
| HEK293 | Epithelial | Kidney | Human | Embryonic | Aneuploid | Readily transfected | Graham et al., 1977 |
| L6 | Fibroblastoid | Skeletal muscle | Rat | Embryonic | | Myotubes | Richler & Yaffe, 1970 |
| LLC-PKl | Epithelial | Kidney | Pig | Adult | Diploid | $Na^+$-dependent glucose uptake | Hull et al., 1976; Saier, 1984 |
| MDCK | Epithelial | Kidney | Dog | Adult | Diploid | Domes, transport | Gaush et al., 1966; Rindler et al., 1979 |
| NRK49F | Fibroblast | Kidney | Rat | Adult | Aneuploid | Induction of suspension growth by TGF-α, β | De Larco & Todaro, 1978 |
| STO | Fibroblast | | Mouse | Embryonic | Aneuploid | Used as feeder layer for embryonal stem cells | Bernstein, 1975 |
| Vero | Fibroblast | Kidney | Monkey | Adult | Aneuploid | Viral substrate and assay | Hopps et al., 1963 |
| **Continuous, from neoplastic tissue** | | | | | | | |
| A2780 | Epithelial | Ovary | Human | Adult | Aneuploid | Chemosensitive with resistant variants | Tsuruo et al., 1986 |
| A549 | Epithelial | Lung | Human | Adult | Aneuploid | Synthesizes surfactant | Giard et al., 1972 |
| A9 | Fibroblast | Subcutaneous | Mouse | Adult | Aneuploid | Derived from L929; lacks HGPRT | Littlefield, 1964b |

| Cell line | Morphology | Tissue | Species | Age | Ploidy | Characteristics | Reference |
|---|---|---|---|---|---|---|---|
| B16 | Fibroblastoid | Melanoma | Mouse | Adult | Aneuploid | Melanin | Nilos & Makarski, 1978 |
| C1300 | Neuronal | Neuroblastoma | Rat | Adult | Aneuploid | Neurites | Liebermann & Sachs, 1978 |
| C6 | Fibroblastoid | Glioma | Rat | Newborn | Aneuploid | Glial fibrillary acidic protein, GPDH | Benda et al., 1968 |
| Caco-2 | Epithelial | Colon | Human | Adult | Aneuploid | Transports ions and amino acids | Fogh, 1977 |
| EB-3 | Lymphocytic | Peripheral blood | Human | Juvenile | Diploid | EB virus +ve | Epstein & Barr, 1964 |
| Friend | Suspension | Spleen | Mouse | Adult | Aneuploid | Hemoglobin | Scher et al., 1971 |
| GH1, GH2, GH3 | Epithelioid | Pituitary tumor | Rat | Adult | Aneuploid | Growth hormone | Buonassisi et al., 1962; Yasamura et al., 1966 |
| H4-11-E-C3 | Epithelial | Hepatoma | Rat | Adult | Aneuploid | Tyrosine aminotransferase | Pitot et al., 1964 |
| HeLa | Epithelial | Cervix | Human | Adult | Aneuploid | G6PD Type A | Gey et al., 1952 |
| HeLa-S3 | Epithelial | Cervix | Human | Adult | Aneuploid | High plating efficiency; will grow well in suspension | Puck & Marcus, 1955 |
| HEP-G2 | Epithelioid | Hepatoma | Human | Adult | Aneuploid | Retains some microsomal metabolizing enzymes | Knowles et al., 1980 |
| HL-60 | Suspension | Myeloid leukemia | Human | Adult | Aneuploid | Phagocytosis | Olsson & Ologsson, 1981 |
| HT-29 | Epithelial | Colon | Human | Adult | Aneuploid | Neotetrazolium Blue reduction; Differentiation inducible with NaBt | Fogh & Trempe, 1975 |
| K-562 | Suspension | Myeloid leukemia | Human | Adult | Aneuploid | Hemoglobin | Andersson et al., 1979 a, b |
| L1210 | Lymphocytic |  | Mouse | Adult | Aneuploid | Rapidly growing; suspension | Law et al., 1949 |
| L929 | Fibroblast |  | Mouse | Adult | Aneuploid | Clone of L-cell | Sanford et al., 1948 |
| LS | Fibroblast |  | Mouse | Adult | Aneuploid | Grow in suspension; derived from L929 | Paul & Struthers, personal communication |
| MCF-7 | Epithelial | Pleural effusion from breast tumor | Human | Adult | Aneuploid | Estrogen receptor +ve, domes, α-lactalbumin | Soule et al., 1973 |
| MCF-10 | Epithelial | Fibrocystic mammary tissue | Human | Adult | Near diploid | Dome formation | Soule et al., 1990 |
| MOG-G-CCM | Epithelioid | Glioma | Human | Adult | Aneuploid | Glutamyl synthetase | Balmforth et al., 1986 |
| P388D1 | Lymphocytic |  | Mouse | Adult | Aneuploid | Grow in suspension | Dawe & Potter, 1957; Koren et al., 1975 |
| S180 | Fibroblast |  | Mouse | Adult | Aneuploid | Cancer chemotherapy screening | Dunham & Stewart, 1953 |
| SK-HEP-1 | Endothelial | Hepatoma, endothelium | Human | Adult | Aneuploid | Factor VIII | Heffelfinger et al., 1992 |
| WEHI-3B D+ | Suspension | Marrow | Mouse | Adult | Aneuploid | IL-3 production | Nicola, 1987 |
| ZR-75-1 | Epithelial | Ascites fluid from breast tumor | Human | Adult | Aneuploid | ER-ve, EGFr+ve | Engel et al., 1978 |

**TABLE 12.2.** Commonly-used Cross-contaminated Cell Lines (*see* Appendix V and www.hpacultures.org.uk/collections/ecacc.jsp for extended list)

| Cell line | Claimed species | Claimed cell type | Contaminating cell line | Actual species | Actual cell type | Reference |
|---|---|---|---|---|---|---|
| ALVA-31, 41 | Human | Prostatic carcinoma | PC-3 | Human | Prostatic carcinoma | Varella-Garcia et al., 2001 |
| BrCa 5 | Human | Breast carcinoma | HeLa | Human | Cervical adenocarcinoma | Nelson-Rees et al., 1981 |
| CaOV | Human | Ovarian carcinoma | HeLa | Human | Cervical adenocarcinoma | Nelson-Rees & Flandermeyer, 1976 |
| CHANG liver | Human | Embryonic liver epithelium | HeLa | Human | Cervical adenocarcinoma | Nelson-Rees & Flandermeyer, 1976 |
| COLO-818 | Human | Melanoma | COLO-800 | Human | Melanoma | Macleod et al., 1999 |
| Det98 | Human | Sternal marrow | HeLa | Human | Cervical adenocarcinoma | Nelson-Rees & Flandermeyer, 1976 |
| ECV-304 | Human | Normal endothelium | T-24 | Human | Bladder carcinoma | Dirks et al., 1999 |
| EJ-1 | Human | Bladder carcinoma | T24 | Human | Bladder carcinoma | Azari et al., 2007 |
| Girardi Heart | Pig | Adult heart | HeLa | Human | Cervical adenocarcinoma | Nelson-Rees & Flandermeyer, 1976 |
| HBL-100 | Human | Breast transformed but nontumorigenic cells | unknown | Human | Unknown; not female | ATCC |
| HBT-3 | Human | Breast carcinoma | HeLa | Human | Cervical adenocarcinoma | Nelson-Rees & Flandermeyer, 1976 |
| HEK | Human | Embryonic kidney | HeLa | Human | Cervical adenocarcinoma | Nelson-Rees & Flandermeyer, 1976 |
| HEp-2 | Human | Larynx epidermoid carcinoma | HeLa | Human | Cervical adenocarcinoma | Chen, 1988 |
| INT 407 | Human | Embryonic intestinal epithelium | HeLa | Human | Cervical adenocarcinoma | Nelson-Rees & Flandermeyer, 1976 |
| KB | Human | Orolaryngeal carcinoma | HeLa | Human | Cervical adenocarcinoma | Gartler, 1967; Lavappa et al., 1976; Nelson-Rees et al., 1981; Ogura et al., 1993 |
| L132 | Human | Embryonic lung epithelium | HeLa | Human | Cervical adenocarcinoma | Nelson-Rees & Flandermeyer, 1976 |
| McCoy | Human | Synovial cell | Strain L | Mouse | Connective tissue | Nelson-Rees & Flandermeyer, 1976 |
| MDA-MB-435 | Human | Breast carcinoma | M14 | Human | Melanoma | Ellison et al., 2003; Christgen & Lehmann, 2007; Rae et al., 2007 |
| MCF-7ADR | Human | Breast carcinoma (adriamycin resistant MCF-7) | OVCAR-8 | Human | Ovarian carcinoma | Liscovitch & Ravid, 2006 |
| MOLT-15 | Human | T-Cell leukemia | CTV-1 | Human | Monocytic leukemia | MacLeod et al., 1999 |
| NPA87 | Human | Thyroid cancer | M14/MDA-MB-435S | Human | Melanoma | Schweppe et al., 2008 |
| PC-93 | Human | Prostatic carcinoma | HeLa | Human | Cervical adenocarcinoma | van Bokhoven et al., 2003 |
| PPC-1 | Human | Prostatic carcinoma | PC-3 | Human | Prostatic carcinoma | Varella-Garcia et al., 2001 |
| TE-671 | Human | Medulloblastoma | RD | Human | Rhabdomyosracoma | Stratton et al., 1989; Chen et al., 1989 |
| U-373 MG | Human | Glioblastoma | U-251 MG | Human | Glioblastoma | Ishii et al., 1999 |
| WISH | Human | Newborn amnion epithelium | HeLa | Human | Cervical adenocarcinoma | Nelson-Rees & Flandermeyer, 1976 |

use is an appalling condemnation of the way that cell culture is taught and the lack of perception even among more senior scientists [Buehring et al., 2004]. Using standardized STR profiling, an international database of DNA profiles has been created listing cell lines [ASN-0002, 2010] including some of those previously reported as cross-contaminated [Cape-Davis et al., 2010].

The following practices help avoid cross-contamination:

(1) Obtain cell lines from a reputable cell bank that has performed the appropriate validation of the cell line (*see* Sections 6.11, 15.2, 19.2.1; Appendix II: Cell Banks), or perform the necessary authentication yourself as soon as possible, preferably by DNA STR profiling [Masters et al., 2001] (*see* Protocol 15.9; *see also* Appendix II).

(2) Do not have culture flasks of more than one cell line, or media bottles used with them, open simultaneously.

(3) Handle rapidly growing lines, such as HeLa, on their own and after other cultures.

(4) Never use the same pipette for different cell lines.

(5) Never use the same bottle of medium, trypsin, or other substances, for different cell lines. Dedicate one set of medium and other reagents to each cell line.

(6) Do not put a pipette back into a bottle of medium, trypsin, or other substances, after it has been in a culture flask containing cells.

(7) Add medium and any other reagents to the flask first, and then add the cells last.

(8) Do not use unplugged pipettes, or pipettors without plugged tips, for routine maintenance.

(9) Check the characteristics of the culture regularly, and suspect any sudden change in morphology, growth rate, or other phenotypic properties. Cross-contamination or its absence may be confirmed by DNA STR profiling (*see* Protocol 15.9), karyotype analysis [Nelson-Rees and Flandermeyer, 1977] (*see* Section15.7; Protocols 15.7, 23.6), or isoenzyme analysis [O'Brien et al., 1980] (*see* Protocol 15.10).

(10) When initiating a new cell line, it is important to retain tissue or DNA from the donor and to do comparative DNA STR profiling (or DNA barcoding for nonhuman cell lines) [Cooper et al., 2007] between the cell line and its presumed source as soon as it becomes established. Other cell lines in use in the laboratory should be included as controls.

## 12.1.2  Mycoplasma Contamination

The second major negative consequence of propagating cell lines, and particularly continuous cell lines, is the harboring of cryptic contaminations, most often mycoplasma [McGarrity, 1976]. While infection of a primary culture or an early-passage cell line often leads to rapid degeneration and loss of the culture, infection of continuous cell lines seems to be better tolerated and often goes undetected. The main sources of infection are from the operator, other infected cell lines or tissues, or natural substances like serum or trypsin. Although improved sterilization procedures have reduced infections arising from natural products, the operator and imported cell lines and other materials remain a serious problem (*see* Section 18.3.2). It is absolutely critical to have a mycoplasma detection program in operation to ensure (1) that cell lines being propagated remain clear of infection and (2) that any new material (tissue, cell lines, biologicals) introduced into the laboratory do not bring infection with them (*see* Protocols 18.2, 18.3). As for avoidance of cross-contamination, validated stocks obtained from a reputable cell bank will help prevent accidental importation of infected cell lines. Yet, however careful the procedures used in the laboratory, it does not pay to forgo regular testing; the consequences of a major outbreak far exceed the costs of performing regular tests.

## 12.1.3  Terminology

Once a primary culture is subcultured (or *passaged*), it becomes known as a *cell line*. This term implies the presence of several cell lineages of either similar or distinct phenotypes. If one cell lineage is selected, by cloning (*see* Protocol 13.1), by physical cell separation (*see* Chapter 14), or by any other selection technique, to have certain specific properties that have been identified in the bulk of the cells in the culture, this cell line becomes known as a *cell strain* (*see* Appendix IV). If a cell line transforms in vitro, it gives rise to a *continuous cell line* (*see* Sections 2.7.4, 17.4), and if selected or cloned and characterized, it is known as a *continuous cell strain*. It is vital at this stage to confirm the identity of the cell lines and exclude the possibility of cross-contamination (*see* Section 12.1.1). Continuous cell lines have a number of advantages but some disadvantages (Table 12.3).

The first subculture gives rise to a *secondary* culture, the secondary to a *tertiary*, and so on, although, in practice, this nomenclature is seldom used beyond the tertiary culture. In Hayflick's work and others with human diploid fibroblasts [Hayflick & Moorhead, 1961], each subculture divided the culture in half (i.e., the *split ratio* was 1:2), so the passage number was the same as the generation number. However, the numbers need not be the same. The *passage number* is the number of times that the culture has been subcultured, whereas the *generation number* is the number of doublings that the cell population has undergone, given that the number of doublings in the primary culture is very approximate. When the split ratio is 1:2, as in Hayflick's experiments, the passage number is approximately equal to the generation number. However, if subculture is performed at split ratios greater than 1:2, the generation number, which is the significant indicator of culture age, will increase faster than the passage number based on the number of doublings that the cell population has undergone since the previous subculture (*see* Section 12.4.3 and Fig. 12.4). None of these approximations takes account of cell loss through necrosis, apoptosis, or differentiation or premature aging and withdrawal from cycle, which probably take place at every growth cycle between each subculture.

**TABLE 12.3.** Properties of Finite and Continuous Cell Lines

| Properties | Finite | Continuous (transformed) |
|---|---|---|
| Ploidy | Euploid, diploid | Aneuploid, heteroploid |
| Transformation | Normal | Immortal, growth control altered, and tumorigenic |
| Anchorage dependence | Yes | No |
| Contact inhibition | Yes | No |
| Density limitation of cell proliferation | Yes | Reduced or lost |
| Mode of growth | Monolayer | Monolayer or suspension |
| Maintenance | Cyclic | Steady state possible |
| Serum requirement | High | Low |
| Cloning efficiency | Low | High |
| Markers | Tissue specific | Chromosomal, enzymatic, antigenic |
| Special functions (e.g., virus susceptibility, differentiation) | May be retained | Often lost |
| Growth rate | Slow ($T_D$ of 24–96 h) | Rapid ($T_D$ of 12–24 h) |
| Yield | Low | High |
| Control parameters | Generation number; tissue-specific markers | Stain characteristics |

### 12.1.4 Naming a Cell Line

New cell lines should be given a code name or designation, for example, normal human brain—NHB; a cell strain or cell line number (if several cell lines were derived from the same source, then NHB1, NHB2, etc.); and, if cloned, a clone number (NHB2-1, NHB2-2, etc.). It is useful to keep a logbook or computer database file where the receipt of biopsies or specimens is recorded before initiation of a culture. The accession number in the logbook or database file, perhaps linked to an identifier letter code, can then be used to establish the cell line designation; for example, LT156 would be lung tumor biopsy number 156. This method is less likely to generate ambiguities, such as the same letter code being used for two different cell lines, and gives automatic reference to the record of accession of the line or tissue. Rules of confidentiality preclude the use of a donor's initials in naming a cell line. When referenced in publications or reports, it is helpful to prefix the cell line designation with a code indicating the laboratory in which it was derived (e.g., WI for Wistar Institute, NCI for National Cancer Institute, SK for Sloan-Kettering) [Federoff, 1975]. In publications or reports the cell line should be given its full designation the first time it is mentioned and in the Materials and Methods section of the paper, and the abbreviated version can then be used thereafter. Confusion can arise in naming cell lines, particularly when short names are used; for example, EJ-1 appears as a bladder cancer cell line (which is actually T24), a bacteriophage of Streptococcus pneumonia [Goh et al., 2007], and a large B-cell lymphoma cell line [Goy et al., 2003]. Cell banks deal with this problem by giving each cell line an accession number; when reporting on cell lines acquired from a cell bank, you should give this accession number in the Materials and Methods section. Punctuation can also give rise to problems when one is searching for a cell line in a database, so always adhere to a standard syntax, and do not use apostrophes or spaces.

Other problems have arisen with designations such as 3T3, which is meant to identify the regimen of maintenance ($3 \times 10^5$ cells per 5-cm Petri dish every 3 days) rather than the actual name of the cells. This resulted in several cell lines called 3T3, 3T3-Swiss, NIH/3T3, and BALB/3T3 being the most popular because they were all maintained by the same regime. The derivation of cloned substrains, such as 3T3-L1 and BALB/3T3-A31, has added uniqueness to the names of these valuable cell strains, but the lesson to be learned is that the name of a cell line should be unique and verified by its DNA profile.

### 12.1.5 Culture Age

Cell lines with limited culture life spans are known as *finite* cell lines and behave in a fairly reproducible fashion (*see* Section 2.7.2). They grow through a limited number of cell generations, usually between 20 and 80 cell population doublings, before senescence (*see* Sections 2.7.3, 17.4.1). The actual number of doublings depends on species and cell lineage differences, clonal variation, and culture conditions, but it is consistent for one cell line grown under the same conditions. It is therefore important that reference to a cell line should express the approximate generation number or number of population doublings since explantation, which will be approximate because the number of generations that have elapsed in the primary culture is difficult to assess.

Continuous cell lines (*see* Table 12.1) have escaped from senescence control, so the generation number becomes less important and is usually replaced with the number of passages since last thawed from storage becomes more important. In addition, because of the increased cell proliferation rate and saturation density (*see* Section 18.5), split ratios become much greater (1:20–1:100) and cell concentration at subculture becomes much more critical (*see* Section 12.4.4).

For finite cell lines, the number of population doublings should be estimated and indicated after a forward slash, as in NHB2/2, and increases by one for a split ratio of 1:2 (NHB2/2, NHB2/3, etc.) by two for a split ratio of 1:4 (NHB2/2, NHB2/4, etc.). When dealing with a continuous cell line a "p" number at the end is often used to indicate the number the number of passages since the last thaw from the freezer, for example, HeLa-S$_3$/p4.

## 12.2 CHOOSING A CELL LINE

Apart from specific functional requirements, there are a number of general parameters to consider in selecting a cell line:

(1) *Finite or continuous.* Is there a continuous cell line that expresses the right functions? A continuous cell line generally is easier to maintain, grows faster, clones more easily, produces a higher cell yield per flask, and is more readily adapted to serum-free medium (*see* Table 12.3).

(2) *Normal or transformed.* Is it important whether the line is malignantly transformed or not? If it is, then it might be possible to obtain an immortal line that is not tumorigenic, such as 3T3-Swiss cells or BHK21-C13.

(3) *Species.* Is species important? A nonhuman cell line will have fewer biohazard restrictions and have the advantage that the tissue from which it was derived may be more accessible.

(4) *Growth characteristics.* What do you require in terms of growth rate, yield, plating efficiency, and ease of harvesting? You will need to consider the following parameters:
  (a) Population-doubling time (*see* Section 20.9.7).
  (b) Saturation density—yield per flask (*see* Section 20.9.3).
  (c) Plating efficiency (*see* Section 20.10).
  (d) Growth fraction (*see* Section 20.11.1).
  (e) Ability to grow in suspension (*see* Table 12.6; Section 17.5.1).

(5) *Availability.* If you have to use a finite cell line, are there sufficient stocks available, or will you have to generate your own line(s)? If you choose a continuous cell line, are authenticated stocks available? (*See* 6 below; *see also* Appendix II.)

(6) *Validation.* How well characterized is the line (*see* Section 6.11), if it exists already, or, if not, can you do the necessary characterization (*see* Chapter 15)? Is the line authentic (*see* Sections 12.1.1, 15.2, 18.6)? It is vital to eliminate the possibility of cross-contamination before embarking on a program of work with a cell line, as so many cross-contaminations have been reported (*see* Table 12.2; Appendix V).

(7) *Phenotypic expression.* Can the line be made to express the right characteristics (*see* Section 16.7)?

(8) *Control cell line.* If you are using a mutant, transfected, transformed, or otherwise abnormal cell line, is there a normal equivalent available, should it be required?

(9) *Stability.* How stable is the cell line (*see* Section 17.3; Plate 7)? Has it been cloned? If not, can you clone it, and how long would this cloning process take to generate sufficient frozen and usable stocks?

## 12.3 ROUTINE MAINTENANCE

Once a culture is initiated, whether it is a primary culture or a subculture of a cell line, it will need a periodic medium change (*feeding* or *refreshing*) followed eventually by subculture if the cells are proliferating. In nonproliferating cultures, the medium will still need to be changed periodically, as the cells will still metabolize and some constituents of the medium will become exhausted or will degrade spontaneously. Intervals between medium changes and between subcultures vary from one cell line to another, depending on the rate of growth and metabolism. Rapidly growing transformed cell lines, such as HeLa, are usually subcultured once per week, and the medium should be changed four days later. More slowly growing, particularly nontransformed, cell lines may need to be subcultured only every two, three, or even four weeks, and the medium should be changed weekly between subcultures (*see also* Sections 12.3.2, 12.4.1).

### 12.3.1 Significance of Cell Morphology

Whatever procedure is undertaken, it is vital that the culture be examined carefully to check the status and confirm the absence of contamination (Table 12.4; *see also* Section 19.3.1; Fig. 19.1). The cells should also be checked for any signs of deterioration, such as granularity around the nucleus, cytoplasmic vacuolation, and rounding up of the cells with detachment from the substrate (Fig. 12.1). Such signs may imply that the culture requires a medium change, or may indicate a more serious problem, such as inadequate or toxic medium or serum, microbial contamination, or senescence of the cell line (*see* Section 29.1). Medium deficiencies can also initiate apoptosis (*see* Sections 2.3.1, 21.1; Plate 17*c, d*). During routine maintenance the medium change or subculture frequency should aim to prevent such deterioration, as it is often difficult to reverse.

Familiarity with the cell's morphology may also allow you to spot the first sign of cross-contamination or misidentification. It is useful to have a series of photographs of cell types in regular use (*see* Fig. 15.2; Plates 9, 10, 25, 26), taken at different cell densities (preferably defined cell densities, e.g., by counting the number of cells/cm$^2$) to refer to when handling cultures, particularly when a new member of staff is being introduced to culture work. A training exercise in cell culture observation is provided in Chapter 28 (*see* Section 28.2, Exercise 7).

**TABLE 12.4.** Examination of Cells during Routine Maintenance

| Status | Criterion | Action indicated |
|---|---|---|
| **Appearance** | Morphology | Confirm against archival photographs at same cell density. |
| | Density | Compare with expected density at this stage of culture. |
| | | If higher or lower, check and adjust seeding concentration. |
| | Mitoses | Indicates proliferation. No action required unless absent, then check growth conditions. |
| | Deterioration | Check for contamination (*see* Section 18.3). |
| | | Check growth conditions and correct if necessary. |
| | | Discard and replace from stock or from freezer. |
| | Alteration, Transformation | Discard and replace from frozen stock. |
| | Heterogeneity (evidence of mixed cell types) | Compare against archival photographs. |
| | | Check identity and authenticity (*see* Section 15.2). |
| | Contamination | Discard culture and medium and reagents used with it (*see* Section 18.4). |
| | | Identify and eliminate source if repeated or widespread (*see* Section 18.5.4). |
| **Maintenance** | Cell density/concentration | If high, subculture. If moderate (50–70% confluent), check pH and feed if necessary. |
| | pH of medium | If low, feed or subculture, depending on cell density/concentration. |
| | | If high, check flasks for leaks, incubator $CO_2$ concentration, evidence of contamination. |
| | | If all cultures high, check medium preparation. |

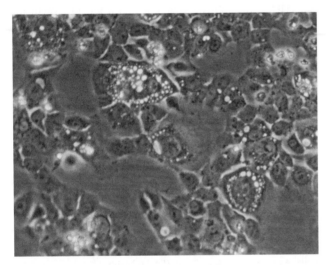

***Fig. 12.1. Unhealthy Cells.*** Vacuolation and granulation in bronchial epithelial cells (BEAS-2B) due, in this case, to medium inadequacy. The cytoplasm of the cells becomes granular, particularly around the nucleus, and vacuolation occurs. The cells may become more refractile at the edge if cell spreading is impaired.

## 12.3.2 Replacement of Medium

Four factors indicate the need for the replacement of culture medium:

(1) *A drop in pH.* The rate of fall and absolute level should be considered. Most cells stop growing as the pH falls from pH 7.0 to pH 6.5 and start to lose viability between pH 6.5 and pH 6.0, so if the medium goes from red through orange to yellow, the medium should be changed (*see* Plate 22*b*). Try to estimate the rate of fall. A culture at pH 7.0 that falls 0.1 pH units in one day will not come to harm if left a day or two longer before feeding, but a culture that falls 0.4 pH units in one day will need to be fed within 24 to 48 h and cannot be left over a weekend without feeding.

(2) *Cell concentration.* Cultures at a high cell concentration exhaust the medium faster than those at a low concentration. This factor is usually evident in the rate of change of pH, but not always.

(3) *Cell type.* Normal cells (e.g., diploid fibroblasts) usually stop dividing at a high cell density (*see* Section 17.5.2), because of cell crowding, shape change, growth factor depletion, and other reasons. The cells block in the $G_1$ phase of the cell cycle and deteriorate very little, even if left for two to three weeks or longer. Transformed cells, continuous cell lines, and some embryonic cells, however, deteriorate rapidly at high cell densities unless the medium is changed daily or they are subcultured.

(4) *Morphological deterioration.* This factor must be anticipated by regular examination and familiarity with the cell line (*see* Section 12.3.1). If deterioration is allowed to progress too far, it will be irreversible, as the cells will tend to enter apoptosis (*see* Sections 2.3.1, 21.1).

***Volume, depth, and surface area.*** The usual ratio of medium volume to surface area is 0.2 to 0.5 mL/cm$^2$ (*see also* Section 20.9.3). The upper limit is set by gaseous diffusion through the liquid layer, and the optimum ratio depends on the oxygen requirement of the cells. Cells with a high $O_2$ requirement do better in shallow medium (e.g., 2 mm), and those with a low requirement may do better in deep medium (e.g., 5 mm). If the depth of the medium is greater than 5 mm, then gaseous diffusion may become limiting. With monolayer cultures, this problem can be overcome by rocking or rolling the bottle (*see* Section 26.2.2) or by perfusing the culture with medium and arranging for gas exchange in an intermediate reservoir (*see* Section 25.3.2, 26.2.5). Stirring the medium improves oxygenation with cells growing in suspension up to around a depth of 5 cm, after which it will be necessary to sparge the medium (*see* Section 26.1).

***Holding medium.*** A *holding medium* may be used when stimulation of mitosis, which usually accompanies a medium change, even at high cell densities, is undesirable. Holding media are usually regular media with the serum concentration reduced to 0.5% or 2% or eliminated completely, or serum-free media without growth factors. This omission inhibits mitosis in most untransformed cells. Transformed cell lines are unsuitable for this procedure, as either they may continue to divide successfully or the culture may deteriorate, because transformed cells do not block in a regulated fashion in $G_1$ of the cell cycle (*see* Section 2.3.1).

Holding media are used to maintain cell lines with a finite life span without using up the limited number of cell generations available to them (*see* Section 2.7.3). Reduction of serum and cessation of cell proliferation also promote expression of the differentiated phenotype in some cells [Maltese & Volpe, 1979; Schousboe et al., 1979]. Media used for the collection of biopsy samples can also be referred to as holding media.

## 12.3.3  Standard Feeding Protocol

Protocol 12.1 is designed to accompany Exercise 9, using medium prepared from Exercise 8 (*see* Section 28.3). The cells and media are specified, but can easily be changed to suit individual requirements.

---

**PROTOCOL 12.1. FEEDING A MONOLAYER CULTURE IN FLASKS**

***Outline***
Examine the culture by eye and on an inverted microscope. If indicated (e.g. by a fall in pH), remove the spent medium and add fresh medium. Return the culture to the incubator.

***Materials***

*Sterile:*

- ❑ Cell cultures
- ❑ Growth medium
- ❑ Pipettes, graduated, and plugged in an assortment of sizes, 1, 5, 10, 25 mL, in a square pipette can (glass pipettes will have been sterilized in the can; individually wrapped plastic pipettes can also be placed in a can for protection and convenience
- ❑ Unplugged pipettes for aspirating medium if pump or vacuum line is available

*Nonsterile:*

- ❑ Pipette controller or bulb (*see* Figs. 4.2, 5.7)
- ❑ Tubing to receiver connected to vacuum line or to receiver via peristaltic pump (*see* Fig. 4.9)
- ❑ Alcohol, 70%, in spray bottle
- ❑ Lint-free swabs or wipes
- ❑ Absorbent paper tissues
- ❑ Cylinder or jar containing water and disinfectant for pipette disposal (*see* Fig. 4.1; Section 6.8.5)
- ❑ Marker pen with alcohol-insoluble ink
- ❑ Notebook, pen, protocols

***Procedure***

1. Prepare the hood by ensuring it is clear and swabbing it with 70% alcohol.
2. Bring the reagents and materials necessary for the procedure, swab bottles with 70% alcohol, and place items required immediately in the hood (*see* Protocol 5.1).
3. Examine the culture carefully for signs of contamination or deterioration (*see* Figs. 12.1, 18.1).
4. Check the previously described criteria—pH and cell density or concentration—and based on your knowledge of the behavior of the culture, decide whether or not to replace the medium. If feeding is required, proceed as follows.
5. Take the culture to the sterile work area.
6. Uncap the flask.
7. Take a sterile pipette and insert it into a bulb or pipetting aid, or select an unplugged pipette and connect it to a vacuum line or pump.
8. Withdraw the medium, and discard it into a waste beaker or aspirate the medium via a suction line.
9. Discard the pipette.
10. Uncap the medium bottle.
11. Take a fresh pipette and add the same volume of fresh medium as was removed, prewarmed to 37°C if it is important that there be no check in cell growth, and recap the bottle.
12. Discard the pipette.
13. Recap the flask and the medium bottle.

***Note.*** If the culture is maintained under 5% $CO_2$ use a permeable cap for a $CO_2$ incubator or gas the air space in the flask with 5% $CO_2$ for around 10 s before recapping for regular incubators or hot room.

14. Return the culture to the incubator ($CO_2$ or not as appropriate).
15. Complete the record of observations and feeding on a record sheet or lab book.
16. Clear away all pipettes, glassware, and other movable items, and swab down the work surface.

When a culture is at a low density and growing slowly, it may be preferable to half-feed it—that is, to remove only half of the medium at step 8 and replace it in step 11 with the same volume as was removed.

---

### PROTOCOL 12.2. FEEDING A MONOLAYER CULTURE IN PLATES OR DISHES

**Outline**
As for Protocol 12.1 but with additional precautions to avoid contamination due to the nature of the culture vessels.

**Materials**
As for Protocol 12.1

**Procedure**
Removing medium:

1. Stack dishes or plates carefully on one side of work area.
2. Switch on aspiration pump if to be used.
3. Select unplugged pipette and insert in aspiration line or plugged pipette is aspiration line not being used.
4. Lift first dish or plate to center of work area.
5. Remove lid and place behind dish, open side up.
6. Grasp the dish as low down on the base as you can, taking care not to touch the rim of the dish or to let your hand come over the open area of the dish or lid.

*Note.* With practice, you may be able to open the lid sufficiently and tilt the dish to remove the medium without removing the lid completely. This is quicker and safer.

7. Tilt dish and remove medium. Discard into waste beaker with funnel (*see* Fig. 5.8) if aspiration pump not available.
8. Replace lid.
9. Move dish to other side of work area from the untreated dishes in the initial stack.
10. Repeat procedure with remaining dishes or plates.
11. Discard pipette and switch off pump.

Adding medium or cells:

12. Position necessary bottles and slacken the cap of the one you are about to use.
13. Bring dish to center of work area.
14. Remove bottle cap and fill pipette from bottle.
15. Remove lid and place behind dish.
16. Add medium to dish, directing the stream gently low down on the side of the base of the dish.
17. Replace lid. (Again, with practice, you may be able to lift the lid and add the medium without laying the lid down, as you did when you removed the medium.)
18. Return dish to the side where the dishes were originally stacked, taking care not to let the medium enter the capillary space between the lid and the base.
19. Repeat with second dish, and so on.
20. Discard pipette.
21. Return dishes to CO$_2$ incubator.

---

If you are adding medium to a flask, dish, or plate with cells already in it, then the bottle of medium used must be designated for the cell line in use and not used for any other. Preferably, dispense sufficient medium into a separate bottle, and use only for this procedure.

## 12.4 SUBCULTURE

When a cell line is subcultured, the regrowth of the cells to a point ready for the next subculture usually follows a standard pattern (Fig. 12.2). A *lag period* after seeding is followed by a period of exponential growth, called the *log phase*. When the *cell density* (cells/cm$^2$ substrate) reaches a level such that all of the available substrate is occupied, or when the *cell concentration* (cells/mL medium) exceeds the capacity of the medium, growth ceases or is greatly reduced (*see* Fig. 15.2b, d, f, h, j, l; Plate 4d). Then either the medium must be changed more frequently or the culture must be divided. For an adherent cell line, dividing a culture, or *subculture* as it is called, usually involves removal of the medium and dissociation of the cells in the monolayer with trypsin, although some loosely adherent cells (e.g., HeLa-S$_3$) may be subcultured by shaking the bottle, collecting the cells in the medium, and diluting as appropriate in fresh medium in new bottles. Some exceptional cell monolayers cannot be dissociated in trypsin and require the action of alternative proteases, such as Pronase, Dispase, and collagenase (Table 12.5). Of these proteases, Pronase is the most effective but can be harmful to some cells. Dispase and collagenase are generally less toxic than trypsin but may not give complete dissociation of epithelial cells. Make sure that each batch of trypsin that you use has the same activity as the specific activity of trypsin may vary requiring a different concentration to achieve the same activity (*see* Appendix I). Other proteases, such as Accutase, Accumax (invertebrate proteases), Liberase, SplitKit (plant protease), and Trypzean or TrypLE (recombinant trypsins), are available, and their efficacy should be tested where either there is a problem with standard disaggregation protocols or there is a need to avoid mammalian (e.g., porcine trypsin) or bacterial (e.g., Pronase) proteases. The severity of the treatment required depends on the cell type, as does the sensitivity of the cells to proteolysis, and a protocol should be selected with the least severity that is compatible with the generation of a single-cell suspension of high viability.

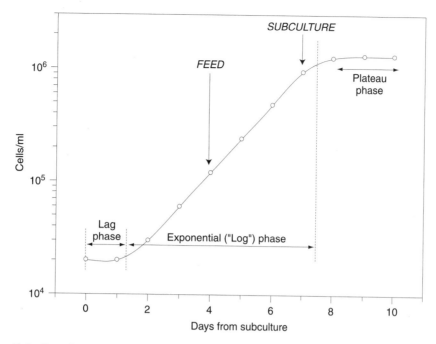

**Fig. 12.2. Growth Curve and Maintenance.** Semilog plot of cell concentration versus time from subculture, showing the lag phase, exponential phase, and plateau, and indicating times at which subculture and feeding should be performed (*see also* Section 20.9.2; Fig. 20.8).

With human embryonal stems cells, enzymatic dispersal can lead to the loss of stemness. Subculture must be done mechanically by subdividing a colony of cells and subculturing the pieces (*see* Protocol 23.4).

The attachment of cells to each other and to the culture substrate is mediated by cell surface glycoproteins and $Ca^{2+}$ (*see* Section 2.2). Other proteins, and proteoglycans, derived from the cells and from the serum, become associated with the cell surface and the surface of the substrate and facilitate cell adhesion. Subculture usually requires chelation of $Ca^{2+}$ and degradation of extracellular matrix and, potentially, the extracellular domains of some cell adhesion molecules.

## 12.4.1  Criteria for Subculture

The need to subculture a monolayer is determined by the following criteria:

(1) *Density of culture.* Normal cells should be subcultured as soon as they reach confluence. If left more than 24 h beyond this point, they will withdraw from the cycle and take longer to recover when reseeded. Transformed cells should also be subcultured on reaching confluence or shortly after; although they will continue to proliferate beyond confluence, they will start to deteriorate after about two doublings, and reseeding efficiency will decline. Some epithelial cell lines, such as Caco-2, need to be subcultured before they reach confluence as they become too difficult to trypsinize after confluence.

(2) *Exhaustion of medium.* Exhaustion of the medium (*see* Section 12.3.2) usually indicates that the medium requires replacement, but if a fall in pH occurs so rapidly that the medium must be changed more frequently, then subculture may be required. Usually a drop in pH is accompanied by an increase in cell density, which is the prime indicator of the need to subculture.

(3) *Time since last subculture.* Routine subculture is best performed according to a strict schedule, so that reproducible behavior is achieved and monitored. If cells have not reached a high enough density (i.e., they are not confluent) by the appropriate time, then increase the seeding density, or if they reach confluence too soon, then reduce the seeding density. Determination of the correct seeding density and subculture interval is best done by performing a growth curve (*see* Section 20.9.2). Once this routine is established, the recurrent growth should be consistent in duration and cell yield from a given seeding density. Deviations from this pattern then signify a departure from normal conditions or indicate deterioration of the cells. Ideally, a cell concentration should be found that allows for the cells to be subcultured after 7 days, with the medium being changed after 4 days.

(4) *Requirements for other procedures.* When cells are required for purposes other than routine propagation, they also have to be subcultured in order to increase the stock or to change the type of culture vessel or medium. Ideally this procedure is done at the regular subculture time, when it is known that the culture is performing routinely, what the reseeding conditions are, and what outcome can be expected. However, demands for cells do not always fit the established routine for maintenance, and

**TABLE 12.5.** Cell Dissociation Procedures

| Procedure | Pretreatment | Dissociation agent | Medium | Applicable to |
|---|---|---|---|---|
| Shake-off | None | Gentle mechanical shaking, rocking, or vigorous pipetting | Culture medium | Mitotic or other loosely adherent cells |
| Scraping | None | Cell scraper | Culture medium | Cell lines for which proteases are to be avoided (e.g., receptor or cell surface protein analysis); can damage some cells and rarely gives a single-cell suspension |
| Trypsin[a] alone | Remove medium completely | 0.01–0.5% Crude trypsin; usually 0.25% | D-PBSA, CMF, or saline citrate | Most continuous cell lines |
| Prewash + trypsin | D-PBSA | 0.25% Crude trypsin | D-PBSA | Some strongly adherent continuous cell lines and many early-passage cells |
| Prewash + trypsin | 1 mM EDTA in D-PBSA | 0.25% Crude trypsin | D-PBSA | Strongly adherent early-passage cell lines |
| Prewash + trypsin | 1 mM EDTA in D-PBSA | 0.25% Crude trypsin | D-PBSA + 1 mM EDTA | Many epithelial cells, but some can be sensitive to EDTA; EGTA can be used |
| Trypsin + collagenase | 1 mM EDTA in D-PBSA | 0.25% Crude trypsin; 200 U/mL crude collagenase | D-PBSA + 1 mM EDTA | Dense cultures and multilayers, particularly with fibroblasts |
| Dispase | None | 0.1–1.0 mg/mL Dispase | Culture medium | Removal of epithelium in sheets (does not dissociate epithelium) |
| Pronase | None | 0.1–1.0 mg/mL Pronase | Culture medium | Provision of good single-cell suspensions, but may be harmful to some cells |
| Liberase | None | A cocktail of Liberase with other proteases; 0.0025% | D-PBSA, CMF or medium | Has been used mainly for primary disaggregation of liver and pancreas (see Protocol 22.6A) |
| DNase | Trypsin or other proteases; DNase added after they have been removed. | 2–10 µg/mL crystalline DNase | Culture medium | Use of other dissociation agents which damage cells and release DNA |

[a]Digestive enzymes are available (Difco, Worthington, Roche Diagnostics, Sigma-Aldrich) in varying degrees of purity. Crude preparations—such as Difco trypsin, 1:250, or Worthington CLS-grade collagenase—contain other proteases that may be helpful in dissociating some cells but may be toxic to other cells. Start with a crude preparation, and progress to purer grades if necessary. Purer grades are often used at a lower concentration (µg/mL), as their specific activities (enzyme units/g) are higher. Purified trypsin at 4°C has been recommended for cells grown in low-serum concentrations or in the absence of serum [McKeehan, 1977] and is generally found to be more consistent. Batch testing and reservation, the same as for serum, may be necessary for some applications.

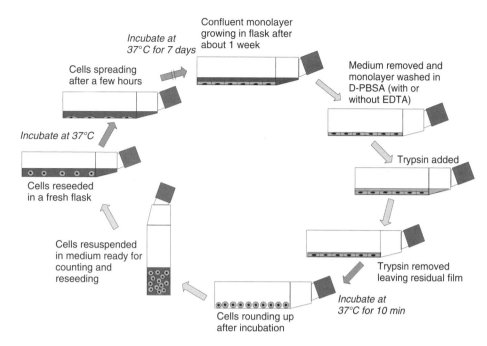

*Fig. 12.3. Subculture of Monolayer.* Stages in the subculture and growth cycle of monolayer cells after trypsinization (*see also* Plates 4, 5).

compromises have to be made. Nevertheless, cells should not be subcultured while still within the lag period, and should always be taken between the middle of the log phase and the time at which they will enter the plateau phase (unless there is a specific requirement for plateau-phase cells, in which case they will need frequent feeding or continuous perfusion before harvesting).

***Handling different cell lines.*** Different cell lines should be handled separately, with a separate set of media and reagents. If they are all handled at the same time, there is a significant risk of cross-contamination, particularly if a rapidly growing line, such as HeLa, is maintained alongside a slower growing line (*see* Sections 6.11, 12.1.1, 15.2, 18.6).

***Labeling.*** Always label flasks and other culture vessels before cells are added to help prevent misidentification. Labeling should be on the side of a flask so that viewing on the microscope is not impaired, and on the edge of the base of a Petri dish or multiwell plate (in case the lid gets separated from the base).

## 12.4.2  Typical Subculture Protocol for Cells Grown as a Monolayer

Protocol 12.2 describes trypsinization of a monolayer (Fig. 12.3; Plates 7–12) after an EDTA prewash to remove traces of medium, divalent cations, and serum (if used). This procedure can be carried out without the prewash, or with only D-PBSA as a prewash, and with 1 mM EDTA in the trypsin if required, depending on the type of cell (*see* Table 12.5).

Protocol 12.3 can be adapted for use in training (*see* Section 28.3, Exercise 13).

---

### PROTOCOL 12.3. SUBCULTURE OF MONOLAYER CELLS

#### Outline
Remove the medium and rinse the monolayer. Expose the cells briefly to trypsin; then remove the trypsin. Incubate the cells then disperse in medium, count, dilute, and reseed the subculture.

#### Materials

*Sterile:*

❑ Monolayer culture at late log phase (just reaching confluence)
❑ Growth medium
❑ Trypsin, 0.25% in D-PBSA (*see* Table 12.5) at 4°C
❑ D-PBSA with 1 mM EDTA
❑ Pipettes, graduated, and plugged in an assortment of sizes, 1, 5, 10, 25 mL, in a square pipette can as in Protocol 12.1
❑ Unplugged pipettes for aspirating medium if pump or vacuum line is available
❑ Universal containers or 50-mL centrifuge tubes
❑ Culture flasks

*Nonsterile:*

❑ Pipette controller or bulb (*see* Figs. 4.2, 5.7)

❑ Waste beaker or tubing to receiver connected to vacuum line or to receiver via peristaltic pump (*see* Figs. 4.9, 5.8)
❑ Alcohol, 70%, in spray bottle
❑ Lint-free swabs or wipes
❑ Absorbent paper tissues
❑ Pipette cylinder or jar containing water and disinfectant (*see* Sections 5.8.8, 7.8.5)
❑ Marker pen with alcohol-insoluble ink
❑ Notebook, pen, protocols
❑ Hemocytometer or electronic cell counter

### Procedure

1. Prepare the hood, and bring the reagents and materials to the hood (*see* Section 6.5).
2. Examine the cultures carefully for signs of deterioration or contamination (*see* Figs. 12.1, 19.1).
3. Check the criteria (*see* Section 12.7.1), and based on your knowledge of the behavior of the culture, decide whether or not to subculture. If subculture is required, proceed as follows.
4. Take the culture flasks to a sterile work area, and remove and discard the medium (*see* Protocol 12.1, steps 7–9). Handle each cell line separately, repeating this procedure from this step for each cell line handled.
5. Add D-PBSA/EDTA prewash (0.2 mL/cm$^2$) to the side of the flasks opposite the cells so as to avoid dislodging cells, rinse the prewash over the cells, and remove by pipette. This step is designed to remove traces of serum that would inhibit the action of the trypsin and deplete the divalent cations, which contribute to cell adhesion.
6. Add trypsin (0.1 mL/cm$^2$) to the side of the flasks opposite the cells. Turn the flasks over and lay them down. Ensure that the monolayer is completely covered. Leave the flasks stationary for 15 to 30 s. (A shorter exposure may be necessary if the trypsin is at room temperature.)
7. Tilt the flasks to remove the trypsin from the monolayer and quickly check that the monolayer is not detaching. Using trypsin at 4°C helps to prevent premature detachment, if this turns out to be a problem.
8. Withdraw all but a few drops of the trypsin.
9. Incubate, with the flasks lying flat, until the cells round up (Figs. 12.3, 12.4; Plates 8–11); now when the bottle is tilted, the monolayer should slide down the surface. (This usually occurs after 5–15 min at 37°C.) Do not leave the flasks longer than necessary, but do not force the cells to detach before they are ready to do so, or else clumping may result.

*Note.* In each case the main dissociating agent, be it trypsin or EDTA, is present only briefly, and the incubation is performed in the residue after most of the dissociating agent has been removed. If you encounter difficulty in getting cells to detach and, subsequently, in preparing a single-cell suspension, you may employ alternative procedures (*see* Table 12.5).

10. Add medium (0.1–0.2 mL/cm$^2$), and disperse the cells by repeated pipetting over the surface bearing the monolayer.
11. Finally, pipette the cell suspension up and down a few times, with the tip of the pipette resting on the bottom corner of the bottle, taking care not to create foam. The degree of pipetting required will vary from one cell line to another; some cell lines disperse easily, whereas others require vigorous pipetting in order to disperse them. Almost all cells incur mechanical damage from shearing forces if pipetted too vigorously. Primary suspensions and early-passage cultures are particularly prone to damage, partly because of their greater fragility and partly because of their larger size, but continuous cell lines are usually more resilient and require vigorous pipetting for complete disaggregation. Pipette the suspension up and down sufficiently to disperse the cells into a single-cell suspension. If this step is difficult, apply a more aggressive dissociating agent or DNase (*see* Table 12.5).

*Note.* A single-cell suspension is desirable at subculture to ensure an accurate cell count and uniform growth on reseeding. It is essential if quantitative estimates of cell proliferation or of plating efficiency are being made and if cells are to be isolated as clones.

12. Count the cells with a hemocytometer or an electronic counter (*see* Section 20.1), and record the cell counts.
13. Dilute the cell suspensions to the appropriate seeding concentration:
    (a) By adding the appropriate volume of cell suspension to a premeasured volume of medium in a culture flask
    or
    (b) By diluting the cells to the total volume required and distributing that volume among several flasks.

*Note.* Procedure (a) is useful for routine subculture when only a few flasks are used and precise cell counts and reproducibility are not critical, but procedure (b) is preferable when setting up several replicates, because the total number of manipulations is reduced and the concentrations of cells in each flask will be identical.

14. If the cells are grown in elevated $CO_2$ in a regular incubator or hot room, gas the flask by blowing the correct gas mixture (usually 5% $CO_2$) from a premixed cylinder, or a gas blender, through a filtered line into the flask above the medium (*see* Fig. 5.12). Do not bubble gas through the medium, as doing so will generate bubbles, which can denature some constituents of the medium and increase the risk of contamination. Alternatively, if culture will be in a $CO_2$ incubator, use a gas-permeable cap. If the normal gas phase is air, as with Eagle's medium with Hanks's salts, no or minimal gassing is required and the flask is sealed.

15. Cap the flasks, and return them to the incubator. Check the pH after about 1 h. If the pH rises in a medium with a gas phase of air, then return the flasks to the aseptic area and gas the culture briefly (1–2 s) with 5% $CO_2$. As each culture will behave predictably in the same medium, you eventually will know which cells to gas when they are reseeded, without having to incubate them first. If the pH rises in medium that already has a 5% $CO_2$ gas phase, either increase the $CO_2$ to 7% or 10% or add sterile 0.1 M HCl.

*Note.* The procedure in step 15 should not become a long-term solution to the problem of high pH after subculture. If the problem persists, then reduce the pH of the medium at the time it is made up, and check the pH of the medium in the incubator or in a gassed flask (*see* Sections 10.4.4, 10.4.5).

16. As the expansion of air inside plastic flasks causes larger flasks to swell and prevents them from lying flat, the pressure should be released by briefly slackening the cap 30 min after placing the flask in the incubator. Alternatively, this problem may be prevented by compressing the top and bottom of large flasks before sealing them (care must be taken not to exert too much pressure and crack the flasks). Incubation restores the correct shape as the gas phase expands. If working in 5% $CO_2$, using permeable caps will avoid this problem.

## 12.4.3 Growth Cycle and Split Ratios

Routine passage leads to the repetition of a standard growth cycle (Fig. 12.4*a; see also* Fig. 12.2). It is essential to become familiar with this cycle for each cell line that is handled, as

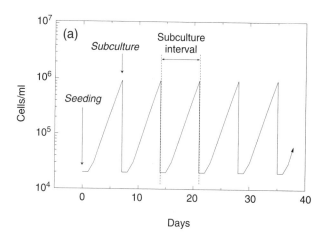

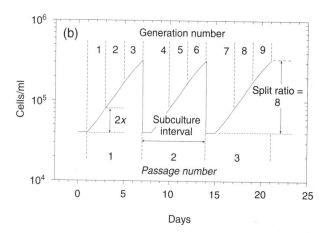

*Fig. 12.4. Serial Subculture.* (*a*) Repetition of the standard growth cycle during propagation of a cell line. If the cells are growing correctly, then they should reach the same concentration (peaks) after the same time in each cycle, given that the seeding concentration (troughs) and subculture interval remain constant. (*b*) Generation number and passage. Each subculture represents one passage, but the generation number (in this case 3 per passage) depends on the split ratio (8 for 3 generations per passage).

this cycle controls the seeding concentration, the duration of growth before subculture (the *subculture interval*), the duration of experiments, and the appropriate times for sampling to give greatest consistency. Cells at different phases of the growth cycle behave differently with respect to cell proliferation, enzyme activity, glycolysis and respiration, synthesis of specialized products, and many other properties (*see* Sections 20.9.2, 25.1.1).

For finite cell lines, it is convenient to reduce the cell concentration at subculture by 2-, 4-, 8-, or 16-fold (i.e., a split ratio of 2, 4, 8, and 16, respectively), making the calculation of the number of population doublings (PDs) easier (i.e., respectively, a split ratio of 2 corresponds to 1 PD, 4 to 2 PD, 8 to 3 PD, and 16 to 4 PD); for example, a culture divided 8-fold requires three doublings to return to the same cell density (Fig. 12.4*b*). A fragile or slowly growing line should be split 1:2, whereas a vigorous, rapidly growing normal cell line can be split 1:8 or 1:16 and some continuous cell lines may be split 1:50 or 1:100. Once a cell line becomes continuous (usually taken as beyond 150 or 200 generations), the cell concentration is the main parameter and the culture should be cut back to between $1 \times 10^4$ and $1 \times 10^5$ cells/mL. The split ratio, or dilution, is also chosen to establish a convenient subculture interval, perhaps 1 or 2 weeks, and to ensure that the cells are not diluted below the concentration that permits them to reenter the growth cycle within a reasonable lag period (24 h or less) and prevents them entering plateau before the next subculture.

When handling a cell line for the first time, or when using an early-passage culture with which you have little experience, it is good practice to subculture the cell line to a split ratio of 2 or 4 at the first attempt, noting the cell concentrations as you do so. As you gain experience and the cell line seems established in the laboratory, it may be possible to increase the split ratio—that is, to reduce the cell concentration after subculture—but always keep one flask at a low split ratio when attempting to increase the split ratio of the rest.

Even when a split ratio is used to determine the seeding concentration at subculture, the number of cells per flask should be recorded after trypsinization and at reseeding so that the growth rate can be estimated at each subculture and the consistency can be monitored (*see* Protocols 20.7–20.9). This is an important component of quality control for the culture (*see* Section 6.10). Otherwise, minor alterations will not be detected for several passages. If constraints of time or equipment prevent cell counting at each subculture you must, at least, compare the cell density before subculture with a reference photograph.

### 12.4.4 Cell Concentration at Subculture

The ideal method of determining the correct seeding density is to perform a growth curve at different seeding concentrations (*see* Protocols 20.7–20.9) and thereby determine the minimum concentration that will give a short lag period and early entry in rapid logarithmic growth (i.e., a short population-doubling time) but will reach the top of the exponential phase at a time that is convenient for the next subculture. As a rule, most continuous cell lines subculture satisfactorily at a seeding concentration of between $1 \times 10^4$ and $5 \times 10^4$ cells/mL, finite fibroblast cell lines subculture at about the same concentration, and more fragile cultures, such as endothelium and some early-passage epithelia, subculture at around $1 \times 10^5$ cells/mL. For a new culture, start at a high seeding concentration and gradually reduce until a convenient growth cycle is achieved without any deterioration in the culture.

### 12.4.5 Propagation in Suspension

Protocol 12.2 refers to the subculture of monolayers because this is the manner in which most primary cultures and cell lines grow. However, cells that grow continuously in suspension,

**TABLE 12.6. Monolayer and Suspension Cultures Compared**

| Monolayer | Suspension |
|---|---|
| **Culture requirements** | |
| Cyclic maintenance | Steady state |
| Trypsin passage | Dilution |
| Limited by surface area | Volume (gas exchange) |
| **Growth properties** | |
| Contact inhibition | Homogeneous suspension |
| Cell interaction | |
| Diffusion boundary | |
| **Useful for** | |
| Cytology | Bulk production |
| Mitotic shake-off | Batch harvesting |
| *In situ* extractions | |
| Continuous product harvesting | |
| **Applicable to** | |
| Most cell types, including primaries | Only transformed cells |

either because they are nonadhesive (e.g., many leukemias and murine ascites tumors, e.g., Fig. 12.6) or because they have been kept in suspension mechanically, or selected, may be subcultured like bacteria or yeast. Suspension cultures have a number of advantages (Table 12.6). For example, trypsin treatment is not required, so subculture is quicker and less traumatic for the cells, and scale-up is easier (*see also* Section 26.1). Replacement of the medium (feeding) is not usually carried out with suspension cultures, and instead the culture is (1) diluted and expanded, (2) diluted and the excess discarded, or (3) the bulk of the cell suspension is withdrawn and the residue is diluted back to an appropriate seeding concentration. In each case a growth cycle will result, similar to that for monolayer cells, but usually with a shorter lag period.

Cells that grow spontaneously in suspension can be maintained in regular culture flasks, which need not be tissue culture treated (although they must be sterile, of course). The rules regarding the depth of medium in static cultures are as for monolayers—2 to 5 mm to allow for gas exchange. When the depth of a suspension culture is increased—such as when it is expanded—the medium requires agitation, which is best achieved with a suspended rotating magnetic pendulum, with the culture flask placed on a magnetic stirrer (Fig. 12.5; *see also* Section 26.1). Roller bottles rotating on a rack can also be used to agitate suspension cultures (*see* Table 7.2; Figs. 26.13, 26.14).

### 12.4.6 Subculture of Cells Growing in Suspension

Protocol 12.3 describes routine subculture of a suspension culture into a fresh vessel. A continuous culture can be maintained in the same vessel, but the probability of contamination gradually increases with any buildup of minor

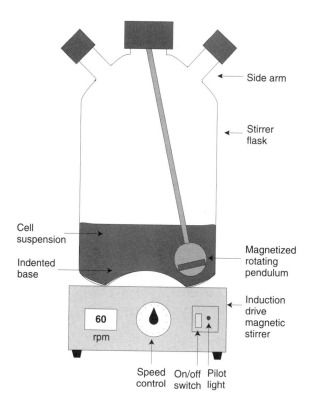

Side arm

Stirrer flask

Cell suspension

Indented base

Magnetized rotating pendulum

Induction drive magnetic stirrer

**60** rpm

Speed control · On/off switch · Pilot light

**Fig. 12.5. Stirrer Culture.** A small stirrer flask, based on the Techne design, with a capacity of 250 to 1000 mL. The cell suspension is stirred by a rotating pendulum, which rotates in an annular depression in the base of the flask.

spillage on the neck of the flask during dilution. The criteria for subculture are similar to those for monolayers:

(1) *Cell concentration*, which should not exceed $1 \times 10^6$ cells/mL for most suspension–growing cells.
(2) *pH*, which is linked to cell concentration, and declines as the cell concentration rises.
(3) *Time since last subculture*, which, as for monolayers, should fit a regular schedule.
(4) *Cell production requirements* for experimental or production purposes.

Protocol 12.3 can be adapted for use in training (*see* Section 28.3, Exercise 12).

---

## PROTOCOL 12.4. SUBCULTURE OF SUSPENSION CELLS

### Outline

Withdraw a sample of the cell suspension, count the cells, and seed an appropriate volume of the cell suspension into fresh medium in a new flask, restoring the cell concentration to the initial seeding level.

### Materials

*Sterile:*

❑ Starter culture, flask or stirrer culture
❑ Growth medium
❑ Pipettes, graduated, and plugged in an assortment of sizes, 1, 5, 10, 25 mL, in a square pipette can as in Protocol 12.1
❑ Unplugged pipettes for aspirating medium if pump or vacuum line available
❑ Universal containers or 50-mL centrifuge tubes
❑ Stirrer flask with magnetic pendulum stirrer (Techne, Bellco; *see* Figs. 7.7, 12.5)
❑ Culture flasks or small stirrer flask

*Nonsterile:*

❑ Pipette controller or bulb (*see* Figs. 4.2, 5.7)
❑ Waste beaker or tubing to receiver connected to vacuum line or to receiver via peristaltic pump (*see* Figs. 4.9, 5.8)
❑ Alcohol, 70%, in spray bottle
❑ Lint-free swabs or wipes
❑ Absorbent paper tissues
❑ Pipette cylinder containing water and disinfectant (*see* Fig. 4.1)
❑ Marker pen with alcohol-insoluble ink
❑ Notebook, pen, protocols
❑ Hemocytometer or electronic cell counter
❑ Magnetic stirrer platform

### Procedure

1. Prepare the hood, and bring the reagents and materials to the hood to begin the procedure (*see* Section 5.4).
2. Examine the culture by eye for signs of contamination or deterioration (fall in pH, aggregation, superficial scum, fungal mycelia or spores). This step is more difficult with suspension cultures than with monolayer cells, as the cells are in suspension making the medium cloudy already.
3. Examine on the microscope; cells that are in poor condition are indicated by shrinkage, an irregular outline, and/or granularity. Healthy cells should look clear and hyaline, with the nucleus visible on phase contrast, and are often found in small clumps in static culture.
4. Take the culture to the sterile work area and remove a sample to count the cells.
5. Based on the previously described criteria (*see* Section 12.7.1) and on your knowledge of the behavior of the culture, decide whether or not to subculture. If subculture is required, proceed as follows.
6. Mix the cell suspension, and disperse any clumps by pipetting the cell suspension up and down.

7. Add medium to the stirrer flask to a maximum depth of 5 cm. If a greater volume of medium is required, $CO_2$ must be supplied via a sparging tube (*see* Fig. 26.2).
8. Add a sufficient number of cells to give a final concentration of $1 \times 10^5$ cells/mL for slow-growing cells (36–48 h doubling time) or $2 \times 10^4$/ mL for rapidly growing cells (12–24 h doubling time).
9. Gas the air space of the stirrer flask with 5% $CO_2$.
10. Cap the stirrer flask and place on magnetic stirrers set at 60 to 100 rpm, in an incubator or hot room at 37°C. Take care that the stirrer motor does not overheat the culture. Insert a polystyrene foam mat under the bottle if necessary. Induction-driven stirrers generate less heat and have no moving parts.
11. If the cells grow spontaneously in suspension, you may wish to add medium to another regular culture flask or a small stirrer flask and seed cells at the same concentration. This will provide a seed culture for the next stirrer culture.

Suspension cultures have a number of advantages (*see* Table 12.6). The production and harvesting of large quantities of cells may be achieved without increasing the surface area of the substrate (*see* Section 26.1). Furthermore, if dilution of the culture is continuous and the cell concentration is kept constant, then a steady state can be achieved (*see* Section 26.1.1); this steady state is not readily achieved in monolayer cultures. Maintenance of monolayer cultures is essentially cyclic, with the result that growth rate and metabolism vary, depending on the phase of the growth cycle.

### 12.4.7 Standardization of Culture Conditions

Standardization of culture conditions is essential for maintaining phenotypic stability. Although some conditions may alter because of the demands of experimentation, development, and production, routine maintenance should adhere to standard, defined conditions, including replacement from frozen stocks at regular intervals (*see* Section 19.5.2)

*Medium.* The type of medium used will influence the selection of different cell types and regulate their phenotypic expression (*see* Sections 8.6, 9.2.2, 9.2.3, 16.2, 16.7). Consequently, once a medium has been selected, standardize on that medium, and preferably on one supplier, if the medium is being purchased readymade.

*Serum.* The best method of eliminating serum variation is to convert to a serum-free medium (*see* Section 9.2). If this is not feasible, serum substitutes (*see* Section 9.4.2) may offer greater consistency and are generally cheaper than serum or growth factor supplementation, although they do not offer the control over the physiological environment afforded by a defined serum-free medium. If serum is required, select a batch (*see* Section 8.6.1), and use that batch throughout each stage including cryopreservation.

*Plastics.* Most of the leading brands of culture flasks and dishes will give similar results, but there may be minor

**TABLE 12.7.** Data Record, Feeding

Date ..................... Time ................. Operator ...........................................................................................

| | Date: | | | | |
|---|---|---|---|---|---|
| **Cell line** | Designation | | | | |
| | Primary or subculture | | | | |
| | Generation or pass no. | | | | |
| | | | | | |
| **Status** | Phase of growth cycle | | | | |
| | Appearance of cells | | | | |
| | Density of cells | | | | |
| | pH of medium (approx.) | | | | |
| | Clarity of medium | | | | |
| **Medium** | Type | | | | |
| | Batch no. | | | | |
| | Serum type and concentration | | | | |
| | Batch no. | | | | |
| | Other additives | | | | |
| | $CO_2$ concentration | | | | |
| **Other parameters** | | | | | |

**TABLE 12.8.** Data Record, Subculture

Date ......................... Time .................. Operator ................................................

| | **Date:** | | | | |
|---|---|---|---|---|---|
| **Cell line** | Designation | | | | |
| | Generation or pass no. | | | | |
| **Status before subculture** | Phase of growth cycle | | | | |
| | Appearance of cells | | | | |
| | Density of cells | | | | |
| | pH of medium (approx.) | | | | |
| | Clarity of medium | | | | |
| **Dissociation agent** | Prewash | | | | |
| | Trypsin | | | | |
| | EDTA | | | | |
| | Other | | | | |
| | Mechanical | | | | |
| **Cell count** | Concentration after resuspension ($C_I$) | | | | |
| | Volume ($V_I$) | | | | |
| | Yield ($Y = C_I \times V_I$) | | | | |
| | Yield per flask | | | | |
| **Seeding** | Number ($N$) & type of vessel (flask, dish, or plate wells) | | | | |
| | Final concentration ($C_F$) | | | | |
| | Volume per flask, dish, or well ($V_F$) | | | | |
| | Split ratio ($Y \div C_F \times V_F \times N$), or number of flasks seeded $\div$ number of flasks trypsinized, where the flasks are of same size | | | | |
| **Medium/serum** | Type | | | | |
| | Batch no. | | | | |
| | Serum type and concentration | | | | |
| | Batch no. | | | | |
| | Other additives | | | | |
| | $CO_2$ concentration | | | | |
| **Matrix coating** | e.g., fibronectin, Matrigel, collagen | | | | |
| **Other parameters** | | | | | |
| | | | | | |
| | | | | | |
| | | | | | |

variations due to the treatment of the plastic for tissue culture (*see* Section 7.1.2). Hence it is preferable to use one type of flask or dish and supplier. Plastic flasks, dishes, and plates for suspension growth, though sterile, need not be tissue culture treated.

## 12.4.8  Use of Antibiotics

The continuous use of antibiotics encourages cryptic contaminations, particularly mycoplasma, and the development of antibiotic-resistant organisms (*see* Section 8.4.7). It may also interfere with cellular processes under investigation.

However, there may be circumstances for which contamination is particularly prevalent or a particularly valuable cell line is being carried, and in these cases antibiotics may be used. If they are used, then it is important to maintain some antibiotic-free stocks in order to reveal any cryptic contaminations; these stocks can be maintained in parallel, and stock may be alternated in and out of antibiotics (*see* Fig. 12.7) until antibiotic-free culture is possible. It is not advisable to adopt this procedure as a permanent regime, and if a chronic contamination is suspected, the cells should be discarded or the contamination eradicated (*see* Sections 18.4, 18.5), and then you may revert to antibiotic-free maintenance.

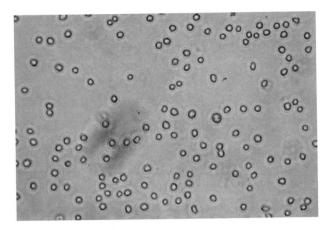

*Fig. 12.6. HL-60 Cells Growing in Suspension.*

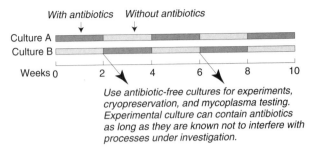

Use antibiotic-free cultures for experiments, cryopreservation, and mycoplasma testing. Experimental culture can contain antibiotics as long as they are known not to interfere with processes under investigation.

*Fig. 12.7. Parallel Cultures and Antibiotics.* A suggested scheme for maintaining parallel cell cultures with and without antibiotics, such that each culture always spends part of the time out of antibiotics.

### 12.4.9 Maintenance Records

Keep details of routine maintenance, including feeding and subculture (Tables 12.7, 12.8), and deviations or changes should be added to the database record for that cell line. Such records are required for GLP (*see* Section 6.10, 6.11), as for primary culture records (*see* Section 11.3.11), but are also good practice in any laboratory [Balls et al., 2006]. If standard procedures are defined, then entries need to say only "Fed" or "Subcultured," with the date and a note of the cell numbers. Any comments on visual assessment and any deviation from the standard procedure should be recorded as well.

This set of records forms part of the continuing provenance of the cell line, and all data, or at least any major event—for example, if the medium supplier is changed, the line is cloned, transfected, decontaminated, or changed to serum-free medium—should be entered in the database.

# CHAPTER 13

# Cloning and Selection

It can be seen from the preceding two chapters that a major recurrent problem in tissue culture is the preservation of a specific cell type and its specialized properties. Although environmental conditions undoubtedly play a significant role in maintaining the differentiated properties of specialized cells in a culture (*see* Sections 16.1, 16.7), the selective overgrowth of unspecialized cells of the wrong lineage remains a major problem. In addition many continuous cell lines are very heterogeneous (*see* Section 17.3), so it is desirable to isolate cloned lines expressing the correct properties before embarking on a program of work with them.

## 13.1  CELL CLONING

The traditional microbiological approach to the problem of culture heterogeneity is to isolate pure cell strains by cloning. Although this technique is relatively easy for continuous cell lines, its success in most primary cultures is limited by poor cloning efficiencies. Nevertheless, the cloning of primary cultures can be successful, as shown with Sertoli cells [Zwain et al., 1991], juxtaglomerular [Muirhead et al., 1990] and glomerular [Troyer & Kreisberg, 1990] cells from kidney, oval cells from liver [Suh et al., 2003], satellite cells from skeletal muscle [Zeng et al., 2002; McFarland et al., 2003; Hashimoto, 2004], and in the separation of different lineages from adult stem cell populations [Young et al., 2004].

A further problem of cultures derived from normal tissue is that they survive only for a limited number of generations (*see* Sections 2.7.3, 17.4.1), and by the time that a clone has produced a usable number of cells, it may already be near to senescence (Fig. 13.1). Although cloning of continuous cell lines is more successful than cloning finite cell lines, considerable heterogeneity may still arise within the clone as it is grown up for use (*see* Section 17.3; Fig. 17.1; Plate 7). Nevertheless, cloning may help reduce the heterogeneity of a culture.

Cloning is also used as a survival assay (*see* Sections 20.10, 23.3.2) for optimizing growth conditions (*see* Sections 9.6, 10.6.3) and for determining chemosensitivity and radiosensitivity (*see* Protocol 21.3). Cloning of attached cells may be carried out in Petri dishes, multiwell plates, or flasks, and it is relatively easy to discern individual colonies. Micromanipulation is the only conclusive method for determining genuine clonality (i.e., that a colony was derived from one cell), but when symmetrical colonies are derived from a single-cell suspension, particularly if colony formation is monitored at the early stages, then it is probable that the colonies are clones.

Cloning can also be carried out in suspension by seeding cells into a gel, such as agar or agarose, or a viscous solution, such as Methocel, with an agar or agarose underlay. The stability of the gel, or viscosity of the Methocel, ensures that daughter cells do not break away from the colony as it forms. Even in monolayer cloning, some cell lines, such as HeLa-S$_3$ and CHO, are poorly attached, and cells can detach from colonies as they form and generate daughter colonies, which will give an erroneous plating efficiency. This can be minimized by cloning in Methocel without an underlay, and allowing the cells to sediment on to the

*Culture of Animal Cells: A Manual of Basic Technique and Specialized Applications, Sixth Edition*, by R. Ian Freshney
Copyright © 2010 John Wiley & Sons, Inc.

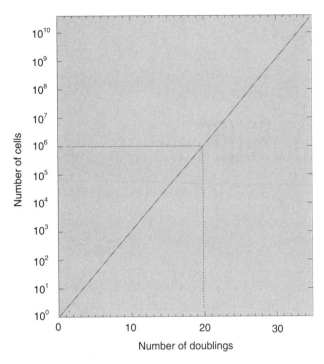

**Fig. 13.1. Clonal Cell Yield.** Relationship of the cell yield in a clone to the number of population doublings; for example, 20 doublings are required to produce $10^6$ cells.

plastic growth surface. Hematopoietic cells are usually cloned in suspension; depending on the cells and growth factors used, the colony generates undifferentiated cells with high repopulation efficiency, in vivo or in vitro, or may mature into colonies of differentiated hematopoietic cells with very little repopulation efficiency. Cloning then becomes an assay for reproductive potential and stem cell identity.

Continuous cell lines generally have a high plating efficiency in monolayer and in suspension because of their transformed status, whereas normal cells, which may have a moderately high cloning efficiency in monolayer, have a very low cloning efficiency in suspension because of their need to attach and spread out to enter the cell proliferation cycle. This distinction has allowed suspension cloning to be used as an assay for transformation (*see* Section 17.5.1).

Dilution cloning [Puck & Marcus, 1955] is the technique that is used most widely. It is based on the observation that cells diluted below a certain density form discrete colonies. Protocol 13.1 can be adapted for training (*see* Section 28.4, Exercise 20).

---

### PROTOCOL 13.1. DILUTION CLONING

#### Outline
Seed the cells at low density and incubate until colonies form (Fig. 13.2). Stain the cells (for plating efficiency and survival assays, *see* Protocols 15.3,

---

20.10, 21.3; Plate 6a, e) or isolate them and propagate into a cell strain if they are being used for selection (*see* Protocol 13.6).

#### Materials

*Sterile or aseptically prepared:*

- ❑ Cells (e.g., CHO-K1), 25 cm² flask, late log phase
- ❑ Medium: Ham's F12, 5% $CO_2$ equilibrated, 10% FBS
- ❑ Trypsin, 0.25%, 1:250, or equivalent (*see* Appendix I)
- ❑ Pipettes, 1, 5, 10, and 25 mL
- ❑ Petri dishes, 6 cm, 1 pack
- ❑ Tubes, or universal containers, for dilution

*Nonsterile:*

- ❑ Hemocytometer or electronic cell counter

#### Procedure

1. Trypsinize the cells (*see* Protocol 12.3) to produce a single-cell suspension. Under-trypsinizing will produce clumps and over-trypsinizing will reduce the viability of the cells, but it is fundamental to the concept of cloning that the cells be singly suspended. It may be necessary when cloning a new cell line for the first time to try different lengths of trypsinization and different recipes (*see* Table 12.5), to give the optimum plating efficiency from a good single-cell suspension.

2. While the cells are trypsinizing, number the dishes (on the side of the base), and measure out medium for the dilution steps. Up to four dilution steps may be necessary to reduce a regular monolayer accurately to a concentration suitable for cloning.

3. When the cells round up and start to detach, disperse the monolayer to a single-cell suspension in 5 mL of medium containing serum or trypsin inhibitor.

4. Count the cells, and dilute the cell suspension to $1 \times 10^5$ cells/mL, then to a concentration that will give roughly 50 colonies per Petri dish (Table 13.1). For CHO-K1 this will be 10 cells/mL, and the dilution steps will be as follows:

    (a) Dilute trypsinate to $1 \times 10^5$/mL (approximately 1:10 or 1:20, depending on the number of cells in the flask.

    (b) Dilute 200 μL of the $1 \times 10^5$/mL suspension to 20 mL (1:100) to give $1 \times 10^3$ cells/mL.

    (c) Dilute 200 μL of the $10^3$/mL suspension to 20 mL (1:100) to give 10 cells/mL. If you wish to add a variable to the cloning conditions, such as a range of serum concentrations, different sera, or a growth factor, prepare a range of tubes at this stage and add 200 μL of the $1 \times 10^3$ cells per mL suspension to each of them separately.

5. If cloning the cells for the first time, choose a range of 10, 20, 50, 100, 200, and 2000 cells/mL (*see* Protocol 20.10) to determine the plating efficiency.

6. Seed 3 Petri dishes each with 5 mL of medium containing cells from the final dilution stage. It is also advisable to seed dishes from the 2000-cells/mL suspension as controls to confirm that cells were present in case no clones form at the lower concentration.

7. Place the dishes in a transparent plastic sandwich or cake box.

8. Put the box in a humid $CO_2$ incubator or gassed sealed container (2–10% $CO_2$; *see* Section 9.3.2).

9. Leave the cultures untouched for 1 week for the colonies to form:

   (a) For plating efficiency assay (*see also* Protocol 20.10), stain and count the colonies (*see* Protocol 15.3; Plate 6).

   (b) For clonal selection, isolate individual colonies (*see* Protocol 13.6).

10. If no colonies are visible, replace medium and continue to culture for another week. Feed the dishes again and culture them for a third week if necessary. If no colonies appear by 3 weeks, then it is unlikely that they will appear at all.

*Feeding.*   As the density of cells during cloning is very low, the need to feed the dishes after one week is debatable. Feeding mainly counteracts the loss of nutrients (e.g., glutamine), which are unstable, replaces growth factors that have degraded or been depleted, and compensates for evaporation. However, it also increases the risk of contamination, so it is reasonable to leave dishes for two weeks without feeding. If it is necessary to leave the dishes for a third week, then the medium should be replaced, or at least half of it.

*Non-uniform distribution.*   The preferential formation of colonies at the center of the plate may result from incorrect seeding, either from seeding the cells into the center of a plate that already contains medium or from swirling the plate such that the cells tend to focus in the center. However, lack of uniformly may also be due to resonance in the incubator (*see* Section 7.3.5).

*Microtitration plates.*   If the prime purpose of cloning is to isolate colonies, then seeding into microtitration plates can be an advantage (Fig. 13.3). When the clones grow up, isolation is easy, although the plates have to be monitored at the early stages in order to mark which wells genuinely have single clones. The statistical probability of a well having a single clone can be increased by reducing the seeding density to a level such that only 1 in 5 or 10 wells would be expected to have a colony. For example, from Table 13.1, 100 cells/mL at 10% plating efficiency would give 10 colonies/mL, or 1 colony/0.1 mL, as added to a microtitration plate—that is 1 colony/well. If the seeding concentration is reduced to 10 cells/mL, then theoretically only 1 in 10 wells will contain a colony, and the probability of wells containing more than one colony is very low.

## 13.2  STIMULATION OF PLATING EFFICIENCY

When cells are plated at low densities, the rate of survival falls in all but a few cell lines. This does not usually present a severe problem with continuous cell lines, for which the plating efficiency seldom drops below 10%, but with primary cultures and finite cell lines, the plating efficiency may be quite low—0.5% to 5%, or even zero. Numerous attempts have been made to improve plating efficiencies, based on the assumption either that cells require a greater range of nutrients at low densities, because of loss by leakage, or that cell-derived diffusible signals or conditioning factors are present in high-density cultures and are absent or too dilute at low densities. The intracellular metabolic pool of a leaky cell in a dense population will soon reach equilibrium with the surrounding medium, while that of an isolated cell never will. This principle was the basis of the capillary technique of Sanford et al. [1948], by which the L929 clone of L-cells was first produced. The confines of the capillary tube

**TABLE 13.1.** Relationship of Seeding Density to Plating Efficiency

| Expected plating efficiency (%) | Optimal cell number to be seeded | | | |
| --- | --- | --- | --- | --- |
| | Per mL | Per cm$^2$ | Per dish, 6 cm | Per dish, 9 cm |
| 0.1 | $1 \times 10^4$ | $2 \times 10^3$ | 40,000 | 100,000 |
| 1.0 | $1 \times 10^3$ | 200 | 4,000 | 10,000 |
| 10 | 100 | 20 | 400 | 1,000 |
| 50 | 20 | 4 | 80 | 200 |
| 100 | 10 | 2 | 40 | 100 |

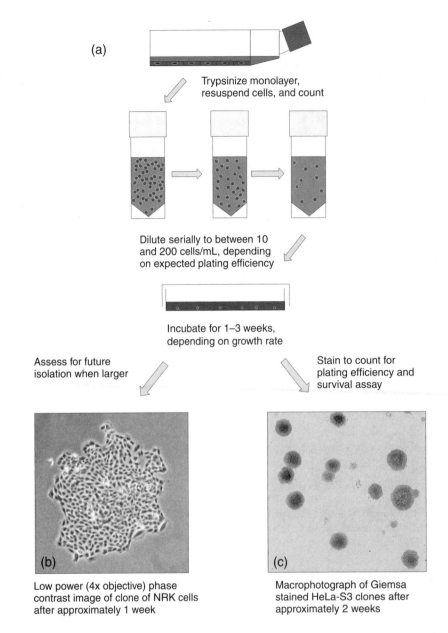

(a)

Trypsinize monolayer,
resuspend cells, and count

Dilute serially to between 10
and 200 cells/mL, depending
on expected plating efficiency

Incubate for 1–3 weeks,
depending on growth rate

Assess for future
isolation when larger

Stain to count for
plating efficiency and
survival assay

(b)

Low power (4x objective) phase
contrast image of clone of NRK cells
after approximately 1 week

(c)

Macrophotograph of Giemsa
stained HeLa-S3 clones after
approximately 2 weeks

**Fig. 13.2. Dilution Cloning.** Cells from a trypsinized monolayer culture (a) are counted and diluted
sufficiently to generate isolated colonies. When clones form (*b*; usually larger for optimal isolation),
they may be isolated (*see also* Figs. 13.8, 13.9). If isolation is not required and the cloning is being
performed for quantitative assay (*see* Protocols 20.10, 21.3), then the colonies are fixed, stained (*c*),
and counted. (*See also* Plate 6*a, e.*)

allowed the cell to create a locally enriched environment
that mimicked a higher cell concentration. In microdrop
techniques developed later, the cells were seeded as a
microdrop under liquid paraffin, again maintaining a relatively
high cell concentration, keeping one colony separate from
another, and facilitating subsequent isolation. As media
improved, however, plating efficiencies increased, and Puck
and Marcus [1955] were able to show that cloning cells by
simple dilution (as described in Protocol 13.1) in association
with a feeder layer of irradiated mouse embryo fibroblasts (*see*
Protocol 13.3) gave acceptable cloning efficiencies, although

subsequent isolation required trypsinization from within a
collar placed over each colony (*see* Protocol 13.6).

It has also been speculated that clonogenic cells in a
population of cultured cells represent the stem cells of that
population; hence the number may be quite low in normal
adult tissues. If that were so, then recloning from a clone
should give a higher plating efficiency, but this is generally not
the case. It may be that stem cells are in equilibrium with the
rest of the population and that equilibrium is reestablished
in a cloned strain or simply that clonogenic cells are not
synonymous with stem cells.

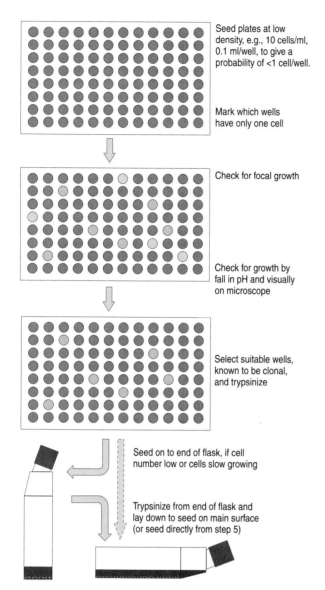

Seed plates at low density, e.g., 10 cells/ml, 0.1 ml/well, to give a probability of <1 cell/well.

Mark which wells have only one cell

Check for focal growth

Check for growth by fall in pH and visually on microscope

Select suitable wells, known to be clonal, and trypsinize

Seed on to end of flask, if cell number low or cells slow growing

Trypsinize from end of flask and lay down to seed on main surface (or seed directly from step 5)

**Fig. 13.3. Cloning in Microtitration Plates.** Cells are seeded at a low enough concentration to give a probability of <1 cell/well, so that some wells will have 1 cell/well, which can be checked by visual examination on the microscope a few hours after plating. Those wells are marked and followed, and when a fall in pH indicates growth, this is confirmed by microscopic observation and the colony is isolated by trypsinization.

## 13.2.1  Conditions That Improve Clonal Growth

(1) *Medium.* Choose a rich medium, such as Ham's F12, or a medium that has been optimized for the cell type in use, e.g., MCDB 110 Ham, 1984 for human fibroblasts, Ham's F12 or MCDB 302 for CHO [Ham, 1963; Hamilton and Ham, 1977] (*see* Sections 8.6, 9.5, Tables 9.1, 9.2; Chapter 22).

(2) *Serum.* When serum is required, fetal bovine is generally better than calf or horse. Select a batch for cloning experiments that gives a high plating efficiency during tests (*see* Protocol 20.10; Plate 6e).

(3) *Hormones.* Insulin, $1 \times 10^{-10}$ IU/mL, has been found to increase the plating efficiency of several cell types [Hamilton and Ham, 1977]. Dexamethasone, $2.5 \times 10^{-5}$ M (10 μg/mL) a soluble synthetic hydrocortisone analogue, improves the plating efficiency of chick myoblasts and human normal glia, glioma (Fig. 13.4a), fibroblasts, and melanoma and gives increased clonal growth (colony size) if removed five days after plating [Freshney et al., 1980a, b]. Lower concentrations (e.g., $1 \times 10^{-7}$ M) have been found to be preferable for epithelial cells, such as lung carcinoma cells (Fig. 13.4b) (*see* Sections 22.2.1, 22.2.3, 22.2.6).

(4) *Intermediary metabolites.* Oxo-acids (previously known as keto-acids)—such as pyruvate or α-oxoglutarate (α-ketoglutarate) [Griffiths and Pirt, 1967; McKeehan and McKeehan, 1979] and nucleosides in α-MEM [Stanners et al., 1971]—have been used to supplement media and are already included in the formulation of a rich medium, such as Ham's F12. Pyruvate is also added to Dulbecco's modification of Eagle's MEM [Dulbecco and Freeman, 1959; Morton, 1970].

(5) *Carbon dioxide.* $CO_2$ is essential for obtaining maximum cloning efficiency for most cells. Although 5% $CO_2$ is usually used, 2% is sufficient for many cells and may even be slightly better for human glia and fibroblasts. HEPES (20 mM) may be used with 2% $CO_2$, protecting the cells against pH fluctuations during feeding and in the event of failure of the $CO_2$ supply. Using 2% $CO_2$ also cuts down on the consumption of $CO_2$. At the other extreme, Dulbecco's modification of Eagle's basal medium (DMEM) is normally equilibrated with 10% $CO_2$ and is frequently used for cloning myeloma hybrids for monoclonal antibody production (although perhaps not always at the higher bicarbonate/$CO_2$ concentration). The concentration of bicarbonate must be adjusted if the $CO_2$ tension is altered, so that equilibrium is reached at pH 7.4 (*see* Table 8.1).

(6) *Treatment of substrate.* Polylysine improves the plating efficiency of human fibroblasts in low serum concentrations McKeehan and Ham, 1976a (*see* Section 7.2.1):

(a) Add 1 mg/mL of poly-D-lysine in UPW to the plates (~5 mL/25 cm²).

(b) Remove and wash the plates with 5 mL of D-PBSA per 25 cm². The plates may be used immediately or stored for several weeks before use.

Fibronectin also improves the plating of many cells [Barnes and Sato, 1980]. The plates may be pretreated with 5 μg/mL of fibronectin in growth medium.

(7) *Trypsin.* Purified (twice recrystallized) trypsin used at 0.05 μg/mL may be preferable to crude trypsin, but opinions vary. McKeehan [1977] noted a marked improvement in plating efficiency when trypsinization (using pure trypsin) was carried out at 4°C. The introduction of recombinant trypsin, TrypZean™ (Sigma) and TrypLE™ (Invitrogen) also gives the opportunity to

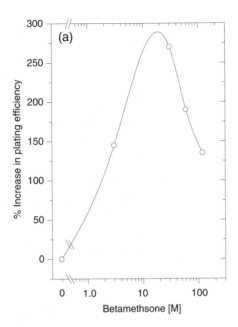
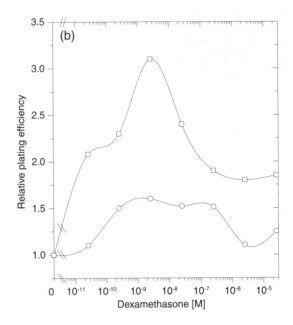

**Fig. 13.4.** *Effect of Glucocorticoids on Cloning.* Hydrocortisone analogs enhance the plating efficiency of several cell types. (*a*) Betamethasone and a human glioma cell line; (*b*) dexamethasone with SKMES-1 (squares) and A549 (circles) human non–small-cell lung carcinoma cells.

use a more highly purified trypsin of nonanimal origin; Invitrogen claim that Gibco TrypLE™ enhances plating efficiency in A549 (with serum) and MDCK cells (serum free). Be aware that the specific activity of trypsin may vary meaning that the concentration will need to be adjusted accordingly (*see* Appendix I).

### 13.2.2 Conditioned Medium

Medium that has been used for the growth of other cells acquires metabolites, growth factors, and matrix products from these cells and is known as *conditioned medium.* A number of different cells have been used for conditioning, such as mouse embryo fibroblasts (*see* Protocols 11.1, 11.6) for several cell types, and 5637 bladder carcinoma, Mo-T T-cell leukemia, and WEHI-3 mouse monocytic leukemia for hematopoietic precursors [Drexler, 2004]. Conditioned medium can improve the plating efficiency of some cells if it is diluted into the regular growth medium.

---

**PROTOCOL 13.2. PREPARATION OF CONDITIONED MEDIUM**

**Outline**
Harvest medium from homologous cells, or a different cell line, from the late log phase. Filter, and dilute with fresh medium as required.

---

**Materials**

*Sterile or aseptically prepared:*

☐ Cells for conditioning: same cell line, another cell line (e.g., 3T3 cells), or mouse embryo fibroblasts (*see* Protocols 11.1, 11.5, 11.6)
☐ Cloning medium: Ham's F12, with 10% FBS, or as appropriate for the cells to be cloned.
☐ Sterilizing filter: 0.45 m or 0.22 μm, filter flask

**Procedure**

1. Grow conditioning cells to 50% of confluence.
2. Change the medium, and incubate for a further 48 h.
3. Collect the medium.
4. Centrifuge the medium at 1,000*g* for 10 min.
5. Filter the medium through a 0.45-μm sterilizing filter. (The medium may need to be clarified first by prefiltration through 5-μm and 1.2-μm filters; *see* Protocol 11.14).
6. Store the medium frozen at −20°C.
7. Thaw the medium before use, and add it to cloning medium: 1 part conditioned medium to 2 parts cloning medium.

*Variations*

(1) *Conditioning cell type.* Mouse embryo fibroblasts (MEFs), Buffalo rat liver (BRL), or STO are often used with stem cells (*see* Protocols 23.2, 23.23), 3T3–L1 for keratinocytes, tumor stromal fibroblasts for tumor cells, 5637 cells for leukemic cells (*see* Protocol 24.5).

(2) *Phase of growth cycle.* Condition with plateau cells rather than log phase. Start conditioning just before confluence.

(3) *Duration of conditioning.* Extend conditioning period to 4 days.

(4) *Collection.* Multiple collections may be made from confluent cells and pooled.

Centrifugation, freezing and thawing, and filtration steps all help avoid the risk of carrying any cells over from the conditioning cells. If the same cells are used for conditioning as for cloning, then this problem is less important, but better cloning may be obtained by using a different cell line or primary mouse fibroblasts. If a cell strain is derived by this method, then its identity must be confirmed (e.g., *see* Protocols 16.10, 16.11, 16.12) to preclude cross-contamination from the conditioned medium.

## 13.2.3 Feeder Layers

The reason that some cells do not clone well may be related to their inability to survive at low cell densities. One way to maintain cells at clonogenic densities but, at the same time, mimic high cell densities is to clone the cells onto a growth-arrested feeder layer (Fig. 13.5). The feeder cells may provide nutrients, growth factors, and matrix constituents that enable the cloned cells to survive more readily.

---

**PROTOCOL 13.3. PREPARATION OF FEEDER LAYERS**

*Outline*
Seed homologous or heterologous cells—such as from mouse embryo, rendered nonproliferative by irradiation or drug treatment—at medium density before the cloning of test cells.

*Materials*

*Sterile or aseptically prepared:*

❑ Primary culture of 13-day mouse embryo fibroblasts (*see* Protocols 11.6)
❑ Culture medium for cells to be cloned
❑ Mitomycin C, 100 μg/mL stock, in HBSS or serum-free medium

*Nonsterile:*

❑ X-ray or $^{60}$Co source capable of delivering 30 Gy in 30 min or less (instead of mitomycin C)

---

*Procedure*

1. Trypsinize embryo fibroblasts, from the primary culture and reseed the cells at $1 \times 10^5$ cells/mL.
2. After 3 to 5 days (when cells are subconfluent and still dividing) block further proliferation by one of the following four methods:
   **(a)** By irradiation:
   (i) Expose cultures, in situ in flasks, to 60 Gy from an X-ray machine or source of γ-radiation such as $^{60}$Co.
   (ii) Expose trypsinized suspension, to 60 Gy from an X-ray machine or source of γ-radiation such as $^{60}$Co, and reseed at $1 \times 10^5$ cells/mL. Alternatively, the irradiated cell suspension may be stored at 4°C for up to 5 days.
   **(b)** By treatment with mitomycin C:
   (i) Add mitomycin C, 0.25 μg/mL final concentration, to subconfluent cells ($\sim 1 \times 10^5$ cells/mL), incubate at 37°C overnight ($\sim 18$ h) [Macpherson & Bryden, 1971] and replace the medium.
   (ii) Add at a final concentration of 20 μg/mL to a trypsinized suspension of $1 \times 10^7$ cells/mL (giving 2 μg/$10^6$ cells), incubate for 1 h at 37°C, wash by centrifugation ($4 \times 10$ mL) to remove the mitomycin C, and reseed at $1 \times 10^5$ cells/mL [Stanley, 2002] or store at 4°C for up to 5 days.
3. After a further 24 to 72 h in culture, trypsinize the cells and reseed, or reseed stored cells directly, in fresh medium at $5 \times 10^4$ cells/mL ($1 \times 10^4$ cells/cm$^2$).
4. Incubate the culture for a further 24 to 48 h, and then seed the cells for cloning.

---

Macpherson & Bryden [1971] maintained that the ratio of mitomycin C to cells should be equal to 2 μg/$1 \times 10^6$ cells; this is achieved in both options above. It is important that the cells enter the cell cycle after the mitomycin C treatment or the DNA damage may be repaired. Treatment of a high-density monolayer in the plateau phase of growth will almost certainly show resistance. In any case, it is essential that any cloned substrain isolated by this method be checked for identity to prevent the possibility of resistant feeder cell overgrowth.

When conditions are correct, the feeder cells will remain viable for up to three weeks, but will eventually die out and are not carried over if the colonies are isolated. Other

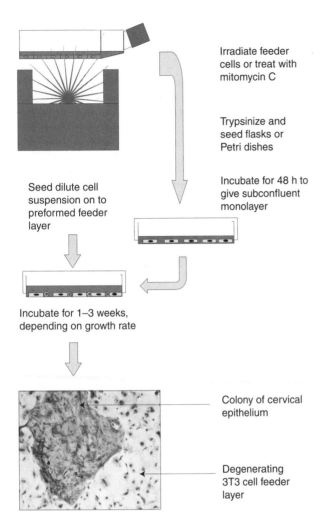

Irradiate feeder cells or treat with mitomycin C

Trypsinize and seed flasks or Petri dishes

Incubate for 48 h to give subconfluent monolayer

Seed dilute cell suspension on to preformed feeder layer

Incubate for 1–3 weeks, depending on growth rate

Colony of cervical epithelium

Degenerating 3T3 cell feeder layer

**Fig. 13.5. Feeder Layers.** Cells are irradiated and trypsinized (or may be trypsinized first and then irradiated in suspension, or treated with mitomycin C) and seeded at a low density to enhance cloning efficiency. (Photo courtesy of M. G. Freshney.)

cell lines or homologous cells may be used to improve the plating efficiency, but heterologous cells have the advantage that if clones are to be isolated later, chromosome analysis, isoenzyme analysis, or DNA barcoding will rule out accidental contamination from the feeder layer (see Section 15.4.1). Other cell lines that have been used for feeder cells include 3T3, MRC-5, and STO cells. Early-passage mouse embryo cells probably produce more matrix components than do established cell lines, but screening different cells is the only way to be sure which type of cell is best for a particular application. Mitomycin C treated feeder cells can be stored at 4°C for up to one week [Wigley, C., personal communication] or can be stored frozen (see Protocols 19.1, 19.2).

Cells may vary in their sensitivity to irradiation or mitomycin C, so a trial run should be carried out before cloning is attempted to ensure that none of the feeder cells survives (see also Section 28.4, Exercise 20, Experimental Variation 3). Even then, it is advisable to seed two or three feeder layer plates with feeder cells alone, to act as controls, when cloning on a feeder layer.

## 13.3 SUSPENSION CLONING

Some cells, particularly hematopoietic stem cells and virally transformed fibroblasts, clone readily in suspension. To hold the colony together and prevent mixing, the cells are suspended in agar or Methocel and plated on an agar underlay or into dishes that are not treated for tissue culture. Protocol 13.4 was submitted by Mary Freshney, Cancer Research UK Beatson Institute, Garscube Estate, Switchback Road, Bearsden, Glasgow, G61 1BD, Scotland, UK (see Fig. 13.6; Protocol 23.9).

---

### PROTOCOL 13.4. CLONING IN AGAR

#### Outline
Agar is liquid at high temperatures, but gels at 37°C. Cells are suspended in warm agar medium and, when incubated after the agar gels, form discrete colonies that may be isolated easily (see Fig. 13.10).

#### Materials

*Sterile:*

- Noble agar, Difco
- Medium at double strength (i.e., Ham's F12, RPMI 1640, Dulbecco's MEM, or CMRL 1066). Prepare the medium from 10× concentrate to half the recommended final volume, and add twice the normal concentration of serum if required.
- Fetal bovine serum (if required)
- Growth medium, 1×, for cell dilutions
- Sterile ultrapure water (UPW)
- Sterile conical flask
- Pipettes, including sterile plastic disposable pipettes for agar solutions
- Universal containers, bijoux bottles, or centrifuge tubes for dilution
- Petri dishes, 3.5-cm, non–tissue-culture grade

*Nonsterile:*

- Bunsen burner and tripod

□ Water bath at 55°C
□ Water bath at 37°C
□ Electronic cell counter or hemocytometer
□ Tray

*Note.* Before preparing the medium and the cells, work out the cell dilutions and label the Petri dishes. For an assay to measure the cloning efficiency of a cell line, prepare to set up three dishes for each cell dilution. Convenient cell numbers per 3.5-cm dish are 1,000, 333, 111, and 37—namely serial one-third dilutions of the cell suspension. If any growth factors, hormones, or other supplements are to be added to the dishes, they should be added to the 0.6% agar underlay.

### Procedure

1. Number or label the Petri dishes on the side of the base. It is convenient to place them on a tray.
2. Prepare 2× medium containing 40% FBS, and keep it at 37°C.
3. Weigh out 1.2 g of agar.
4. Measure 100 mL of sterile UPW into a sterile conical flask and another 100 mL into a sterile bottle. Add the 1.2 g of agar to the flask. Cover the flask, and boil the solution for 2 min. Alternatively, the agar may be sterilized in the autoclave in advance, but if subsequently stored, it will still need to be boiled or microwaved in order to melt it for use.
5. Transfer the boiled agar and the bottle of sterile UPW to a water bath at 55°C.
6. Prepare a 0.6% agar underlay by combining an equal volume of 2× medium and 1.2% agar (Fig. 13.6). Keep the underlay at 37°C. If any growth factors, hormones, or other supplements are being used, they should be added to the underlay medium at this point.

*Note.* If a titration of growth factors is being carried out or a selection of different factors is being used, add the required amount to the Petri dishes before the underlay is added.

7. Add 1 mL of 0.6% agar medium to the dishes, mix, and ensure that the medium covers the base of the dish. Leave the dishes at room temperature to set (*see* Fig. 13.6).
8. Prepare the cell suspension, and count the cells.
9. Prepare 0.3% agar medium, and keep it at 37°C. This medium may be prepared by diluting 2×

medium at 37°C with 1.2% agar at 55°C and UPW at 55°C in the respective proportions of 2:1:1 (*see* Fig. 13.6).
10. Prepare the following cell dilutions,
    (a) $1 \times 10^5$/mL.
    (b) Dilute $1 \times 10^5$/mL by 1/3 to give $3.3 \times 10^4$/mL.
    (c) Dilute $3.3 \times 10^4$/mL by 1/3 to give $1.1 \times 10^4$/mL.
    (d) Dilute $1.1 \times 10^4$/mL by 1/3 to give $3.7 \times 10^3$/mL.
11. Label four bijoux bottles or tubes one for each dilution and pipette 40 μL of each cell dilution, including the $1 \times 10^5$/mL concentration, into the respective container. Add 4 mL of 0.3% agar medium at 37°C to each container, mix, and pipette 1 mL from each container onto each of three Petri dishes (Fig. 13.6). This will give final concentrations as follows:
    (a) $1 \times 10^3$/mL/dish
    (b) 330/mL/dish
    (c) 110/mL/dish
    (d) 37/mL/dish

*Note.* Always be sure that the agar medium for the top layer has had adequate time to cool to 37°C before adding the cells to it.

12. Allow the solution in the Petri dishes to gel at room temperature.
13. Put the Petri dishes into a clean plastic box with a lid, and incubate them at 37°C in a humid incubator for 10 days.

Agarose, which has reduced sulfated polysaccharides, can be substituted for agar. Some types of agarose have a lower gelling temperature and can be manipulated more easily at 37°C. They are gelled at 4°C and then are returned to 37°C.

Because of the complexity of handling melted agar with cells, and the impurities that may be present in agar, some laboratories prefer to use Methocel, which is a viscous solution and not a gel [Buick et al., 1979]. Methocel has a higher viscosity when warm, and because it is a sol and not a gel, cells will sediment through it slowly. It is therefore essential to use an underlay with Methocel. Colonies form at the interface between the Methocel and the agar (or agarose) underlay, placing themselves in the same focal plane and making analysis and photography easier.

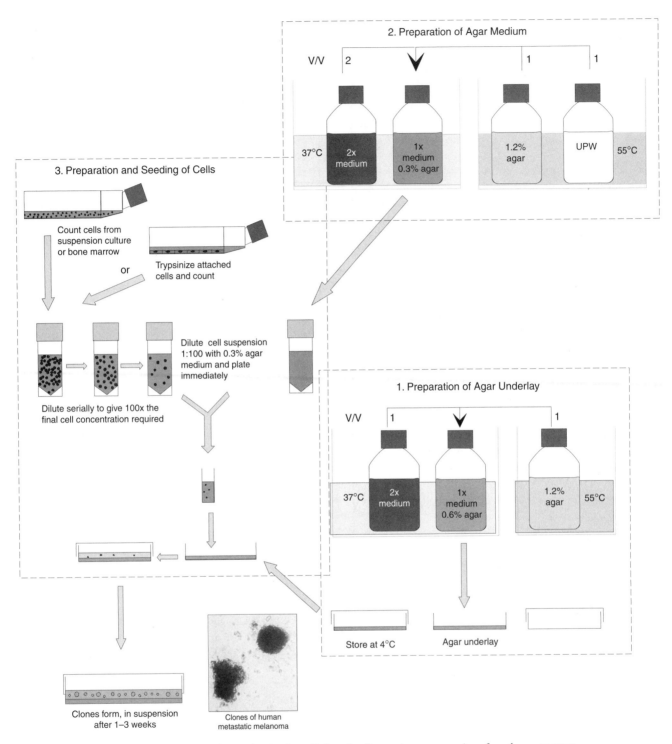

***Fig. 13.6. Cloning in Suspension in Agar.*** Cultured cells or primary suspensions from bone marrow or tumors, suspended in agar or low-melting-temperature agarose, which is then allowed to gel, form colonies in suspension. Use of an underlay prevents attachment to the base of the dish. (1) Preparation of agar underlay: agar, 1.2%, at 55°C is mixed with 2× medium at 37°C and dispensed immediately into dishes, where it is allowed to gel at room temperature or 4°C. (2) Preparation of agar medium: agar, 1.2%, and UPW are maintained at 55°C and mixed with 2× medium to give 0.3% agar for cloning. The use of low-melting-point agarose allows all solutions to be maintained at 37°C, but this agarose can be more difficult to gel. (3) Cells grown in suspension, derived from bone marrow, or trypsinized from an attached monolayer are counted and diluted serially, and the final product is diluted with agar or agarose, and seeded onto an agar underlay.

## PROTOCOL 13.5. CLONING IN METHOCEL

### Outline

Suspend the cells in medium containing Methocel, and seed the cells into dishes containing an agar or agarose underlay (Fig. 13.7).

### Materials

#### Sterile:

- ❑ Noble agar, Difco, or agarose
- ❑ Medium at double strength (i.e., Ham's F12, RPMI 1640, Dulbecco's MEM, or CMRL 1066). Prepare the medium from 10× concentrate to half the recommended final volume, and add twice the normal concentration of serum (if required)
- ❑ Fetal bovine serum (if required)
- ❑ Growth medium, 1×, for cell dilutions
- ❑ 1.6% Methocel, 4 Pa-s (4,000 centipoises), in UPW; place on ice (see Appendix I: Methocel)
- ❑ Sterile ultrapure water (UPW)
- ❑ Sterile conical flask
- ❑ Pipettes, including sterile plastic disposable pipettes for agar solutions
- ❑ Universal containers, bijoux bottles, or centrifuge tubes
- ❑ 3.5-cm Petri dishes, non–tissue-culture grade
- ❑ Syringes without needles to dispense Methocel (because of its viscosity Methocel tends to cling to the inside of pipettes, making dispensing difficult and inaccurate)

#### Nonsterile:

- ❑ Bunsen burner and tripod
- ❑ Water bath at 55°C
- ❑ Water bath at 37°C
- ❑ Electronic cell counter or hemocytometer
- ❑ Tray

### Procedure

1. Prepare agar underlays as in Protocol 13.4, steps 1 through 7.
2. Dilute the Methocel to 0.8% with an equal volume of 2× medium. Mix it well, and keep it on ice.
3. Trypsinize monolayer cells, or collect cells from suspension culture or bone marrow and count them.
4. Prepare the following cell dilutions:
   (a) $1 \times 10^5$/mL.
   (b) Dilute $1 \times 10^5$/mL by 1/3 to give $3.3 \times 10^4$/mL.
   (c) Dilute $3.3 \times 10^4$/mL by 1/3 to give $1.1 \times 10^4$/mL.
   (d) Dilute $1.1 \times 10^4$/mL by 1/3 to give $3.7 \times 10^3$/mL.
5. Methocel is viscous, so manipulations are easier to perform with a syringe without a needle.
6. Label four bijoux bottles or tubes one for each dilution, and pipette 40 μL of each cell dilution, including the $1 \times 10^5$/mL concentration, into the respective container. Add 4 mL of 0.8% Methocel medium to each container, and mix well with a vortex or, if the cells are known to be particularly fragile, by sucking the solution gently up and down with a syringe several times. Then use a syringe to add 1 mL from each container to each of three Petri dishes (See Fig. 13.7). This will give final concentrations as follows:
   (a) $1 \times 10^3$/mL/dish
   (b) 330/mL/dish
   (c) 110/mL/dish
   (d) 37/mL/dish
7. Incubate the dishes in a humid incubator until colonies form. Because the colonies form at the interface between the agar and the Methocel, fresh medium may be added, 1 mL per dish or well, after 1 week and then removed and replaced with more fresh medium after 2 weeks without disturbing the colonies.

Many of the recommendations that apply to medium supplementation for monolayer cloning also apply to suspension cloning. In addition sulphydryl compounds, such as mercaptoethanol (50 μM), glutathione (1 mM), and (α-thioglycerol (75 μM) [Iscove et al., 1980], are sometimes used. Macpherson [1973] found that the inclusion of DEAE dextran was beneficial for cloning, a finding that was later confirmed by Hamburger et al. [1978], who also found that macrophages enhanced the cloning of tumor cells, although others have found them to be detrimental. Courtenay et al. [1978] incorporated rat red blood cells into the medium and demonstrated that a low oxygen tension enhanced cloning. The problem may lie with the toxicity of free oxygen; with red blood cells present, it will be bound to hemoglobin. It may be possible to mimic this with perfluorocarbons [Lowe et al., 1998].

Most cell types clone in suspension with a lower efficiency than in monolayer, some cells by two or three orders of magnitude. The isolation of colonies is, however, much easier.

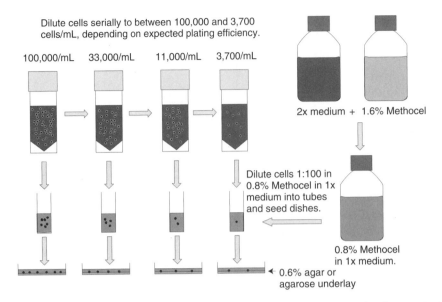

Dilute cells serially to between 100,000 and 3,700 cells/mL, depending on expected plating efficiency.

100,000/mL    33,000/mL    11,000/mL    3,700/mL

2x medium + 1.6% Methocel

Dilute cells 1:100 in 0.8% Methocel in 1x medium into tubes and seed dishes.

0.8% Methocel in 1x medium.

◄ 0.6% agar or agarose underlay

***Fig. 13.7. Cloning in Suspension in Methocel.*** A series of cell dilutions is prepared as for agar cloning, diluted 1:100 in Methocel medium, and plated into non-tissue-culture-grade dishes or dishes with an agar underlay.

## 13.4 ISOLATION OF CLONES

When cloning is used for the selection of specific cell strains, the colonies that form (*see* Plates 6) need to be isolated for further propagation. If monolayer cells are cloned directly into multiwell plates (*see* Protocol 13.1: Microtitration Plates), then colonies may be isolated by trypsinizing individual wells. It is, however, necessary to confirm the clonal origin of the colony during its formation by regular microscopic observation. If cloning is performed in Petri dishes, there is no physical barrier between colonies. This barrier must be created by removing the medium and placing a stainless steel or ceramic ring around the colony to be isolated (Fig. 13.8).

---

### PROTOCOL 13.6. ISOLATION OF CLONES WITH CLONING RINGS

#### Outline

The colony is trypsinized from within a porcelain, glass, PTFE, or stainless steel ring and transferred to one of the wells of a 24- or 12-well plate, or directly to a 25-cm² flask (Fig. 13.9).

#### Materials

*Sterile:*

❑ Cloning rings (*see* Appendix II); sterilize in a glass Petri dish by dry heat or autoclave
❑ Silicone grease; sterilize in a glass Petri dish by dry heat, 160°C for 1 h, or autoclave at 121°C for 15 min

---

❑ Yellow pipettor tips with filter, or plugged Pasteur pipettes with a bent end (Bellco #1273)
❑ Trypsin 0.25% in D-PBSA
❑ Growth medium
❑ Multiwell plate, 24-well, and/or 25-cm² flasks
❑ Sterile forceps

*Nonsterile:*

❑ Pipettor, 50 to 100 μL
❑ Felt-tip pen or, preferably, a Nikon ring marker or object marker that fits into the objective nosepiece of a microscope in place of one of the objectives

#### Procedure

1. Examine the clones, and mark those that you wish to isolate with a felt-tip marker on the underside of the dish, or use a ring marker.
2. Remove the medium from the dish, and rinse the clones gently with D-PBSA.
3. Using sterile forceps, take one cloning ring, dip it in silicone grease, and press it down on the dish alongside the silicone grease, to spread the grease around the base of the ring.
4. Place the ring around the desired colony.
5. Repeat steps 4 and 5 for two or three other colonies in the same dish.
6. Add sufficient 0.25% trypsin to fill the hole in the ring (0.1–0.4 mL, depending on the internal diameter of the ring).
7. Leave the trypsin for 20 s, and then remove it.
8. Close the dish, and incubate it for 15 min at 37°C.

9. Add 0.1 to 0.4 mL of medium to each ring.
10. Taking each clone in turn, pipette the medium up and down to disperse the cells, and transfer the medium to a well of a 24-well plate or to a 25-cm² flask standing on end. Use a separate pipette, or a separate pipettor tip, for each clone.
11. Wash out the ring with another 0.1 to 0.4 mL of medium, and transfer the medium to the same well or flask.

*Note.* The dish will dry out if left open for too long. Either limit the number of clones isolated or cover the dish between manipulations.

12. Make up the medium in the wells to 1.0 mL, close the plate, and incubate it. If you are using flasks; then add 1 mL of medium to each flask and incubate the flasks standing on end.
13. When the clone grows to fill the well, subculture to a 25-cm² flask, incubated conventionally with 5 mL of medium. If you are using the upended flask technique (*see* Fig. 13.9), then remove the medium when the end of a flask is confluent, trypsinize the cells, resuspend them in 5 mL of medium, and lay the flask down flat. Continue the incubation.

Flasks are available with a removable top film (Nunc; Midwest) that may be peeled off to allow harvesting of clones. Alternatively, where an irradiation source is available, clones may be isolated by shielding one with a lead disk and irradiating the rest of the monolayer with 30 Gy (*see* Protocol 13.7).

## PROTOCOL 13.7. ISOLATING CELL COLONIES BY IRRADIATION

### Outline
Invert the flask under an X-ray machine or 60Co source, screening the desired colony with lead.

Δ *Safety Note.* X-ray machines and 60Co sources must be used under strict supervision and with appropriate monitoring to safeguard your own exposure and that of others (*see* Section 6.7.4). Contact your local radio-protection officer before setting up this type of experiment.

### Materials
*Sterile:*

❑ Growth medium
❑ 0.25% trypsin
❑ D-PBSA

*Nonsterile:*

❑ X-ray or cobalt γ-source
❑ Pieces of lead cut from a 2-mm-thick sheet and of a size from about 2 to 5 mm in diameter

### Procedure
1. Select the desired colony, and mark it with a felt-tip pen or a Nikon ring marker.
2. Select a piece of lead of appropriate diameter.
3. Take the flask to the radiation source.
4. Invert the flask under the source.
5. Cover the colony with a 2-mm-thick piece of lead.
6. Irradiate the flask with 30 Gy.
7. Return the flask to the sterile area.
8. Remove the medium, trypsinize the cells, and allow the cells to reestablish in the same bottle, using the irradiated cells as a feeder layer.

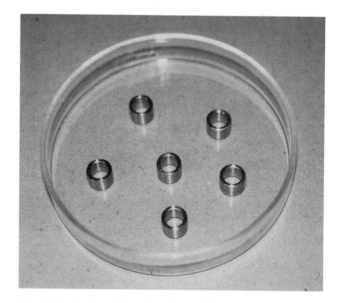

**Fig. 13.8. Cloning Rings.** Stainless steel rings, cut from tube, in 9-cm glass Petri dish. Porcelain (Fisher), thick-walled glass (Scientific Laboratory Supplies), or plastic (e.g., cut from nylon, silicone, or Teflon thick-walled tubing) can also be used. Whatever the material, the base must be smooth in order to seal with silicone grease onto the base of the Petri dish, and the internal diameter must be just wide enough to enclose one whole clone without the external diameter overlapping adjacent clones.

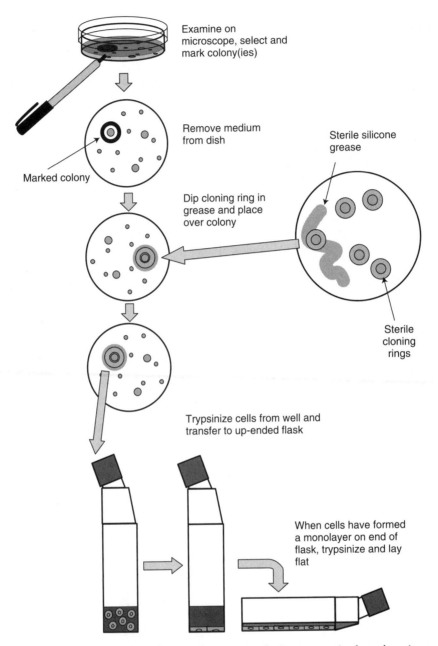

**Fig. 13.9. Isolation of Monolayer Clones.** The mature colonies are examined on the microscope, and suitable colonies are selected and marked. The medium is removed, and cloning rings, dipped in silicone grease, are placed around each colony, which is then trypsinized from within the ring.

If irradiation and trypsinization are carried out when the colony is about 100 cells in size, then the trypsinized cells will reclone, given a reasonably high cloning efficiency. Three serial clonings may be performed within six weeks by this method.

### 13.4.1 Other Isolation Techniques for Monolayer Clones

(1) Distribute small coverslips or broken fragments of coverslips on the bottom of a Petri dish. When plated out at the correct density, some colonies are found to be singly distributed on a piece of glass and may be transferred to a fresh dish or multiwell plate.

(2) Use the capillary technique of Sanford et al. [1948]. A dilute cell suspension is drawn into a sterile glass capillary tube (e.g., a 50-μL Drummond Microcap), allowing colonies to form inside the tube. The tube is then carefully broken on either side of a colony and transferred to a fresh plate. This technique was exploited by Echarti and Maurer [1989, 1991] for clonogenic assay of hematopoietic cells and tumor cells, for which the colony-forming efficiency can be quantified by scanning the capillary in a densitometer.

(3) Cells may be cloned in the OptiCell chamber, which is made up of two opposing growth surfaces of thin flexible plastic that may be cut with a scalpel or scissors. Provided that the outer surfaces are kept sterile, this can be used to cut out segments with colonies which are then trypsinized into a multiwell plate or flask.

## 13.4.2 Suspension Clones

The isolation of colonies growing in suspension is simple, but requires a dissection microscope.

---

### PROTOCOL 13.8. ISOLATION OF SUSPENSION CLONES

#### Outline
Draw the colony into a pipettor or Pasteur pipette, and transfer the colony to a flask or the well of a multiwell plate (Fig. 13.10).

#### Materials
*Sterile:*

❑ Growth medium in universal container
❑ Multiwell plates, 24 well
❑ Culture flask, 25 cm²
❑ Yellow pipettor tips with filter or plugged Pasteur pipettes

*Nonsterile:*

❑ Dissecting microscope, 20× to 50× magnification
❑ Pipettor, 100 µL
❑ Felt-tip pen or Nikon ring marker

#### Procedure

1. Examine the dishes on an inverted microscope, and mark the colonies with a felt-tip pen or a Nikon ring marker.
2. Pipette 1 mL of medium into each well of a 24-well plate.
3. Pick the colonies using a dissecting microscope:
4. Use a separate pipettor tip or Pasteur pipette for each colony.
5. Set the pipettor to 100 µL.
6. Draw approximately 50 µL of medium into the pipette tip, place the tip of the pipette against the colony to be isolated, and gently draw in the colony.
7. Transfer the contents of the pipette to a 24-well dish, and flush out the colony with medium. If the medium is made with Methocel, the colony will settle, and, if the cells are adherent they will attach, and grow out. Cells that normally grow in suspension will settle but, of course, will not

---

attach. If the medium is made from agar, you may need to pipette the colony up and down a few times in the well to disperse the agar. Clones may also be seeded directly into a 25-cm² plastic flask that is standing on end (*see* Protocol 13.6).

---

## 13.5 REPLICA PLATING

Bacterial colonies can be replated by pressing a moist pad gently down onto colonies growing on a nutrient agar plate and transferring the pad to a second, fresh, agar plate. Various attempts have been made to adapt this technique to cell culture, usually by placing a mesh screen or filter over monolayer clones and transferring it to a fresh dish after a few days [Hornsby et al., 1992]. For clones that have been developed in microtitration plates, there are a number of transfer devices available—such as, the Corning Transtar (*see* Fig. 4.8), which can be used with suspension cultures directly after agitating the culture or with monolayer cultures after trypsinization and resuspension.

## 13.6 SELECTIVE INHIBITORS

Manipulating the conditions of a culture by using a selective medium is a standard method for selecting microorganisms. Its application to animal cells in culture is limited, however, by the basic metabolic similarities of most cells isolated from one animal, in terms of their nutritional requirements. The problem is accentuated by the effect of serum, which tends to mask the selective properties of different media. Most selective media that have been shown to be generally successful have been serum-free formulations (*see* Section 9.2.2). A number of metabolic inhibitors, however, have had recurrent success. Gilbert and Migeon [1975, 1977] replaced the L-valine in the culture medium with D-valine and demonstrated that cells possessing D-amino acid oxidase would grow preferentially. Kidney tubular epithelia [Gross et al., 1992], bovine mammary epithelia [Sordillo et al., 1988], endothelial cells from rat brain [Abbott et al., 1992], and Schwann cells [Armati & Bonner, 1990] have been selected in this way. However this technique appears not to be effective against human fibroblasts [Masson et al., 1993].

Much of the effort in developing selective conditions has been aimed at suppressing fibroblastic overgrowth, which not only dominates other cells in the culture but may actively inhibit them [Halaban, 2004]. Kao & Prockop [1977] used *cis*-OH-proline, although it can prove toxic to other cells. Fry and Bridges [1979] found that phenobarbitone inhibited fibroblastic overgrowth in cultures of hepatocytes, and Braaten et al. [1974] were able to reduce the fibroblastic contamination of neonatal pancreas by treating the culture with sodium ethylmercurithiosalicylate. Fibroblasts also tend

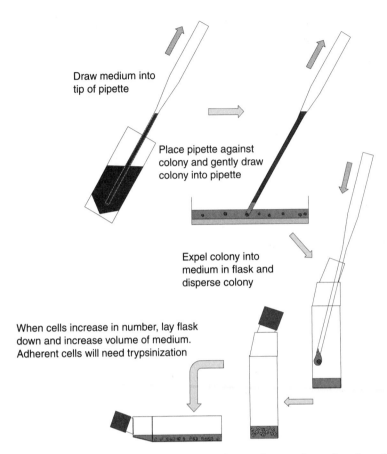

Draw medium into
tip of pipette

Place pipette against
colony and gently draw
colony into pipette

Expel colony into
medium in flask and
disperse colony

When cells increase in number, lay flask
down and increase volume of medium.
Adherent cells will need trypsinization

**Fig. 13.10. Isolation of Suspension Clones.** Mark the colony as for monolayer, then draw the colony into a Pasteur pipette (or pipettor tip). Transfer the colony to a culture flask, disperse it in medium, and incubate it. Make up medium when cells start to grow.

to be more sensitive to geneticin (G418) at 100 μg/mL [Halaban & Alfano, 1984; Levin et al., 1995].

One of the more successful approaches was the development of a monoclonal antibody to the stromal cells of a human breast carcinoma [Edwards et al., 1980]. Used with complement, this antibody proved to be cytotoxic to fibroblasts from several tumors and helped to purify a number of malignant cell lines. Cells may also be killed selectively with drug- or toxin-conjugated antibodies [e.g., Beattie et al., 1990]. However, selective antibodies are used more extensively in "panning" or magnetizable bead separation techniques (*see* Section 14.3).

Selective media are also commonly used to isolate hybrid clones from somatic hybridization experiments. HAT medium (*see* Appendix I), a combination of hypoxanthine, aminopterin, and thymidine, selects hybrids with both hypoxanthine guanine phosphoribosyltransferase and thymidine kinase from parental cells deficient in one or the other enzyme (*see* Section 27.6) [Littlefield, 1964a].

Transfected cells are also selected by resistance to a number of drugs, such as neomycin, its analogue geneticin (G418), hygromycin, and methotrexate, by including a resistance-conferring gene in the construct used for transfection (e.g.,

*neo* (aminoglycoside phosphotransferase), *hph* (hygromycin B phosphotransferase), or *dhfr* (dihydrofolate reductase; *see* Section 28.5). Culture in the correct concentration of the selective marker, determined by titration against the transfected and nontransfected controls, selects for stable transfectants. Selection with methotrexate has the additional advantage that increasing the methotrexate concentration leads to amplification of the *dhfr* gene and can coamplify other genes in the construct.

Negative selection is also possible by using the Herpes simplex virus (HSV) *TK* gene, which activates Ganciclovir (Syntex) into a cytotoxic product [Jin et al., 2003]. Transfected cells will be sensitized to the drug.

When a mixture of cells shows different responses to growth factors, it is possible to stimulate one cell type with the appropriate growth factor and then, taking advantage of the increased sensitivity of the more rapidly growing cells, to kill the cells selectively with irradiation or cytosine arabinoside (ara-c) (*see* Section 22.4.1). Alternatively, if an inhibitor is known or a growth factor is removed, which will take one population out of cycle, the remaining cycling cells can be killed with ara-c or irradiation.

## 13.7  ISOLATION OF GENETIC VARIANTS

Protocol 14.9 for the development of mutant cell lines that amplify the dihydrofolate reductase (DHFR) gene was contributed by June Biedler, Memorial Sloan–Kettering Cancer Center, New York, New York.

---

### PROTOCOL 13.9. METHOTREXATE RESISTANCE AND DHFR AMPLIFICATION

#### Principle

Cells exposed to gradually increasing concentrations of folic acid antagonists, such as methotrexate (MTX), over a prolonged period of time will develop resistance to the toxic effects of the drug [Biedler et al., 1972]. Resistance resulting from amplification of the DHFR gene generally develops the most rapidly, although other mechanisms—such as alteration in antifolate transport and/or mutations affecting enzyme structure or affinity—may confer part or all of the resistant phenotype.

#### Outline

Expose the cells to a graded series of concentrations of MTX for several weeks, periodically replacing the medium with fresh medium containing the same drug concentration. Select for subculturing those flasks in which a small percentage of cells survive and form colonies. Repeatedly subculture such cells in the same and in 2 to 10-fold higher MTX concentrations until the cells acquire the desired degree of resistance.

#### Materials

*Sterile:*

- ❑ Chinese hamster cells or rapidly growing human or mouse cell lines
- ❑ Methotrexate Sodium Parenteral (Wyeth Lederle)
- ❑ NaCl, 0.15 M
- ❑ Tissue culture flasks
- ❑ Pipettes
- ❑ Culture medium that does not contain thymidine and hypoxanthine (e.g., Eagle's MEM with 10% fetal bovine serum)

*Nonsterile:*

- ❑ Inverted microscope
- ❑ Liquid-nitrogen freezer

#### Procedure

1. Clone the parental cell line to obtain a rapidly growing, genotypically uniform population to be used for selection.

2. Dilute the MTX with sterile 0.15 M (0.85%) NaCl. The drug packaged for use in the clinic is in solution at 2.5 mg/mL.

3. Inoculate $2.5 \times 10^5$ cells into replicate 25-cm$^2$ flasks containing no drug or 0.01, 0.02, 0.05, and 0.1 μg/mL of MTX in complete tissue-culture medium. Adjust the pH of each solution to pH 7.4, and incubate the flasks at 37°C for 5 to 7 days.

4. Observe the cultures with an inverted microscope. Replace the medium with fresh medium containing the same amount of MTX in cultures showing clonal growth of a small proportion of cells amid a background of enlarged, substrate-adherent, and probably dying cells, and reincubate those cultures.

5. Allow the cells to grow for another 5 to 7 days changing the growth medium as necessary, but continuously exposing the cells to MTX. When the cell density has reached 2 to $10 \times 10^6$ cells/flask, subculture the cells at $2.5 \times 10^5$ cells/flask into new flasks containing the same and 2 to 10-fold higher drug concentrations.

6. After another 5 to 7 days, observe the new passage flasks as well as the cultures from the previous passage that had been exposed to higher drug concentrations; change the medium and select for viable cells as before.

7. Continue the selection with progressively higher drug concentrations at each subculture step until the desired level of resistance is obtained: 2 to 3 months for Chinese hamster cells with low to moderate levels of resistance, increase in DHFR activity, and/or transport alteration; 4 to 6 months or more for high levels of resistance and enzyme overproduction, for Chinese hamster, mouse, or fast-growing human cells.

8. Periodically freeze samples of the developing lines in liquid nitrogen (see Protocol 19.1).

---

***Analysis.*** Characterize resistant cells for levels of resistance to the drug in a clonal growth assay (*see* Protocol 21.3), for increase in activity or amount of DHFR by biochemical or gel electrophoresis techniques [Albrecht et al., 1972; Melera et al., 1980], and/or for increase in mRNA and copy number of the reductase gene by Northern, Southern, or dot blots [Scotto et al., 1986] with DHFR-specific probes to determine the mechanism(s) of resistance.

***Variations.*** Cell culture media other than Eagle's MEM can be used; the composition of the medium (e.g., folic acid content) can be expected to influence the rate and type of

MTX resistance development. Media containing thymidine, hypoxanthine, and glycine (*see* Tables 8.3, 9.1, 9.2) prevent the development of antifolate resistance and should be avoided. Cells can be treated with chemical mutagens before selection [Thompson & Baker, 1973]; this treatment may also alter the rate and type of mutant selection.

Selection can also be done with cells plated in the drug at low density in 10-cm tissue-culture dishes (with the isolation of individual colonies with cloning rings; *see* Protocol 13.6), using single cells in 96-well cluster dishes, or in soft agar, to enable the isolation of one or multiple clonal populations at each or any step during resistance development.

Cells can be made to be resistant to a number of other agents, such as antibiotics, other antimetabolites, and toxic metals, by similar techniques; differences in the mechanism of action or degree of toxicity of the agents, however, may require that treatment with the agent be intermittent rather than continuous and may increase the time necessary for selection.

Cell lines of different species or with slower growth rates, such as some human tumor cell lines, may require different (usually lower) initial drug concentrations, longer exposure times at each concentration, and smaller increases in the concentration between selection steps. Solubilization of MTX other than the Lederle product will require the addition of equimolar amounts of NaOH and sterilization through a 0.2-$\mu$m filter.

## 13.8  INTERACTION WITH SUBSTRATE

### 13.8.1  Selective Adhesion

Different cell types have different affinities for the culture substrate and attach at different rates. If a primary cell suspension is seeded into one flask and transferred to a second flask after 30 min, a third flask after 1 h, and so on for up to 24 h, then the most adhesive cells will be found in the first flask and the least adhesive in the last. Macrophages will tend to remain in the first flask, fibroblasts in the next few flasks, epithelial cells in the next few flasks, and, finally, hematopoietic cells in the last flask. A similar method has been used in the isolation of human chondroprogenitor cells from cartilage [Archer at al., 2007].

If collagenase in complete medium is used for primary disaggregation of the tissue (*see* Protocol 11.8), most of the cells that are released will not attach within 48 h unless the collagenase is removed. However, macrophages migrate out of the fragments of tissue and attach during this period and can be removed from other cells by transferring the disaggregate to a fresh flask after 48 to 72 h of treatment with collagenase. This technique works well during disaggregation of biopsy specimens from human tumors.

### 13.8.2  Selective Detachment

Treatment of a heterogeneous monolayer with trypsin or collagenase will remove some cells more rapidly than others. Periodic brief exposure to trypsin removed fibroblasts from cultures of fetal human intestine [Owens et al., 1974] and skin [Milo et al., 1980]. Lasfargues [1973] found that the exposure of cultures of breast tissue to collagenase for a few days at a

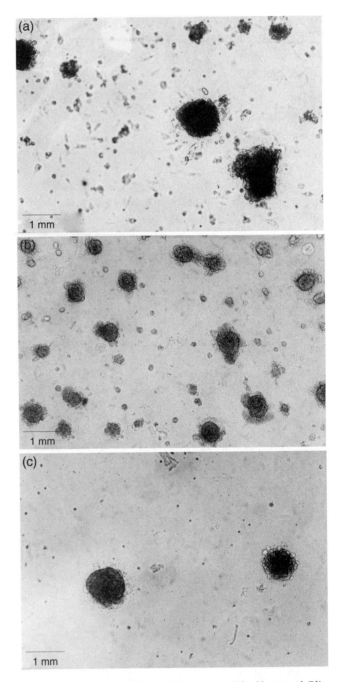

**Fig. 13.11.  Suspension Clones of Melanoma, Fibroblasts, and Glia.** Cells were plated out at $5 \times 10^5$ per 35-mm dish ($2.5 \times 10^5$ cells/mL) in 1.5% Methocel over a 1.25% agar underlay. Colonies were photographed after 3 weeks. (*a*) Melanoma; (*b*) human normal embryonic skin fibroblasts; (*c*) human normal adult glia.

time removed fibroblasts and left the epithelial cells. EDTA, however, may release epithelial cells more readily than it will release fibroblasts [Paul, 1975].

Dispase II (Boehringer Mannheim) selectively dislodges sheets of epithelium from human cervical cultures grown on feeder layers of 3T3 cells without dislodging the 3T3 cells (*see* Protocol 24.4). This technique may be effective in subculturing epithelial cells from other sources, selecting against stromal fibroblasts.

### 13.8.3 Nature of Substrate

Although several sources of ECM are now available (*see* Appendix II), the emphasis so far has been on promoting cell survival or differentiation, and little has been made of the potential for selectivity in "designer" matrices, although collagen has been reported to favor epithelial proliferation [Kibbey et al., 1992; Kinsella et al., 1992] and Matrigel also favors epithelial survival and differentiation [Bissell et al., 1987; Ghosh et al., 1991; Kibbey et al., 1992]. Because the constituents are now better understood, mixing various collagens with proteoglycans, laminin, and other matrix proteins could be used to create more selective substrates. The selective effect of substrates on growth may depend on differential rates both of attachment and of growth, or the net result of both. Collagen and fibronectin coating has been used to enhance epithelial cell attachment and growth [Lechner & LaVeck, 1985; Wise, 2002] (*see* Protocols 22.5, 22.9) and to support endothelial cell growth and function [Relou et al., 1998; Martin et al., 2004].

### 13.8.4 Selective Feeder Layers

As well as conditioning the substrate, feeder layers (*see* Section 13.2.3) can also be used for the selective growth of epidermal cells [Rheinwald & Green, 1975] (*see* Sections, 22.2.1, 22.2.4, 24.5.2) and for repressing stromal overgrowth in cultures of breast (*see* Fig. 24.2; Plate 6c, d) and colon carcinoma (*see* Protocol 24.4) [Freshney et al., 1982b]. The role of the feeder layer is probably quite complex; it provides not only extracellular matrix for adhesion of the epithelium but also positively acting growth factors and negative regulators that inactivate TGF-β [Maas-Szabowski & Fusenig, 1996]. Human glioma will grow on confluent feeder layers of normal glia, whereas cells derived from normal brain will not [MacDonald et al., 1985] (*see* Protocol 24.2).

### 13.8.5 Selection by Semisolid Media

The transformation of many fibroblast cultures reduces the anchorage dependence of cell proliferation [Macpherson & Montagnier, 1964] (*see* Section 17.5.1). By culturing the

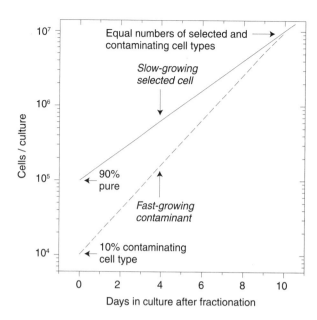

***Fig. 13.12. Overgrowth in Mixed Culture.*** Overgrowth of a slow-growing cell line by a rapidly growing contaminant. This figure portrays a hypothetical example, but it demonstrates that a 10% contamination with a cell population that doubles every 24 h will reach equal proportions with a cell population that doubles every 36 h after only 10 days of growth.

cells in agar (*see* Protocol 13.4) after viral transformation, it is possible to isolate colonies of transformed cells and exclude most of the normal cells. Normal cells will not form colonies in suspension with the high efficiency of virally transformed cells, although they will often do so with low plating efficiencies. The difference between transformed and untransformed cells is not as clear with early-passage tumor cell lines as plating efficiencies can be quite low; normal glia and fetal skin fibroblasts also form colonies in suspension with similar efficiencies (<1%) (Fig. 13.11).

Cell cloning and the use of selective conditions have a significant advantage over physical cell separation techniques (*see* Chapter 14), in that contaminating cells are either eliminated entirely by clonal selection or repressed by constant or repeated application of selective conditions. Even the best physical cell separation techniques still allow some overlap between cell populations, such that overgrowth recurs. A steady state cannot be achieved, and the constitution of the culture changes continuously. A 90%-pure culture of cell line A will be 50% overgrown by a 10% contamination with cell line B in 10 days, given that B grows 50% faster than A (Fig. 13.12). For continued culture, selective conditions are therefore required in addition to, or in place of, physical separation techniques.

# CHAPTER 14

# Cell Separation

Although cloning and using selective culture conditions are the preferred methods for purifying a culture (*see* Sections 9.2.2, 13.1, 13.6), there are occasions when cells do not plate with a high enough efficiency to make cloning possible or when appropriate selection conditions are not available. It may then be necessary to resort to a physical or immunological separation technique. Separation techniques have the advantage that they give a high yield more quickly than cloning, although not with the same purity.

The more successful separation techniques depend on differences in (1) cell density (specific gravity), (2) affinity of antibodies to cell surface epitopes, (3) cell size, and (4) light scatter or fluorescent emission as sorted by flow cytometry. The first two techniques involve a relatively low level of technology and are inexpensive, whereas the second two call for high technology with a significant outlay of capital. The most effective separations often employ two or more parameters to obtain a high level of purity.

## 14.1 CELL DENSITY AND ISOPYKNIC SEDIMENTATION

Separation of cells by density can be performed by centrifugation at low *g* with conventional equipment [Pretlow & Pretlow, 1989; Recktenwald, 1997; Calder et al., 2004; Al-Mufti et al., 2004]. The cells sediment in a density gradient to an equilibrium position equivalent to their own density (isopyknic sedimentation; Fig. 14.1). The density medium should be nontoxic and nonviscous at high densities (1.10 g/mL) and should exert little osmotic pressure in solution.

Serum albumin [Turner et al., 1967], dextran [Schulman, 1968], Ficoll (Thermo Fisher—Amersham) [Sykes et al., 1970], metrizamide (Nycomed) [Munthe-Kaas & Seglen, 1974], and Percoll (Thermo Fisher—Amersham) [Pertoft & Laurent, 1982] have all been used successfully (*see* also Protocol 27.1). Percoll (colloidal silica) and the radiopaque iodinated compounds metrizamide and metrizoate are among the more effective media currently used. A gradient may be generated (1) by layering different densities of Percoll with a pipette, syringe, or pump; (2) with a special gradient former (Fig. 14.2d); or (3) by high-speed spin (Fig. 14.3).

---

### PROTOCOL 14.1. CELL SEPARATION BY CENTRIFUGATION ON A DENSITY GRADIENT

**Outline**
Form a gradient, centrifuge cells through the gradient, collect fractions, dilute with medium, and culture (Fig. 14.1).

**Materials**

*Sterile:*

- ❑ Growth medium
- ❑ Growth medium + 20 % Percoll
- ❑ Centrifuge tubes, 25 mL
- ❑ D-PBSA
- ❑ Trypsin, 0.25 %
- ❑ Gradient former (*see* Fig. 14.2): (Thermo Fisher—Amersham, Buchler)

---

❏ Syringe or gradient harvester (Thermo Fisher—Amersham)

❏ Plates, 24 well or microtitration

*Nonsterile:*

❏ Refractometer or density meter

❏ Hemocytometer or cell counter

❏ Low-speed centrifuge

### Procedure

1. Prepare the density gradient
   **(a)** By layering:
      (i) Adjust the density of the Percoll medium to 1.10 g/cc and its osmotic strength to 290 mosmol/kg.
      (ii) Mix the Percoll and regular media in varying proportions to give the desired density range (e.g., 1.020–1.100 g/cc) in 10 or 20 steps.
      (iii) Layer one step over another, 2 mL or 1 mL per step, with a pipette, syringe or peristaltic pump, starting with the densest solution and building up a stepwise density gradient in a 25-mL centrifuge tube. It is also possible, and may be preferable, to layer from the bottom, starting with the least dense solution and injecting each layer of progressively higher density below the previous one, using a syringe or peristaltic pump.
      (iv) Gradients may be used immediately or left overnight.
   **(b)** With a gradient former: a continuous linear gradient may be produced by mixing, for example, 1.020 g/mL with 1.080 g/cc Percoll in a gradient former.
   **(c)** By centrifugation:
      (i) Place the medium containing Percoll at density 1.085 g/cc in a tube.
      (ii) Centrifuge at 20,000 $g$ for 1 h.
      (iii) Centrifugation generates a sigmoid gradient (*see* Fig. 14.3), whose the shape is determined by the starting concentration of Percoll, the duration and centrifugal force of the centrifugation, the shape of the tube, and the type of rotor.
2. Trypsinize cells and resuspend them in medium plus serum or a trypsin inhibitor. Check to make sure that the cells are singly suspended.
3. Using a syringe, pipettor, or fine-tipped pipette, layer up to $2 \times 10^7$ cells in 2 mL of medium on top of the gradient.
4. The tube may be allowed to stand on the bench for 4 h and will sediment under 1 $g$; or it may be centrifuged for 20 min at 100 to 1000 $g$. If the latter procedure is used, increase centrifuge speed gradually at start of run and do not apply brake at end of run.
5. Collect fractions with a syringe or a gradient harvester (Thermo Fisher; *see* Fig. 14.1). Fractions of 1 mL may be collected into a 24-well plate or of 0.1 mL into a microtitration plate. Samples should be taken at intervals for cell counting and for determining the density ($\rho$) of the gradient medium. Density may be measured on a refractometer or density meter.
6. Add an equal volume of medium to each well, and mix to ensure that the cells settle to the bottom of the well. Change the medium to remove the Percoll after 24 to 48 h incubation.

### Variations

*Position of cells.* Cells may be incorporated into the gradient during its formation by centrifugation. Although only one spin is required, spinning the cells at such a high $g$ force may damage them. In addition Percoll [Pertoft & Laurent, 1982], Isopaque [Splinter et al., 1978], and metrizamide [Freshney, 1976] may be taken up by cells, so it is preferable to layer cells on top of a preformed gradient.

*Other media.* Ficoll is one of the most popular media because, like Percoll, it can be autoclaved. Ficoll is a little more viscous than Percoll at high densities and may cause some cells to agglutinate. Metrizamide (Nycomed), a nonionic derivative of metrizoate, which is a radiopaque iodinated substance used in radiography (Isopaque, Hypaque, Renografin) and in lymphocyte purification (*see* Protocol 27.1), is less viscous than Ficoll at high densities [Rickwood & Birnie, 1975] but may be taken up by some cells (see above).

*Marker beads.* GE Healthcare (Pharmacia) markets colored marker beads of standard densities that may be used to determine the density of regions of the gradient.

Isopyknic sedimentation is quicker than velocity sedimentation at unit gravity (*see* Section 14.2.1) and gives a higher yield of cells for a given volume of gradient. It is ideal when clear differences in density ($\geq$ 0.02 g/cc) exist between cells. Cell density may be affected by uptake of the medium used to form the gradient, by the position of the cells in the growth cycle (plateau phase cells are denser), and by serum [Freshney, 1976]. Because high $g$ forces are not required, isopyknic sedimentation can be done on any centrifuge and can even be performed at 1 $g$.

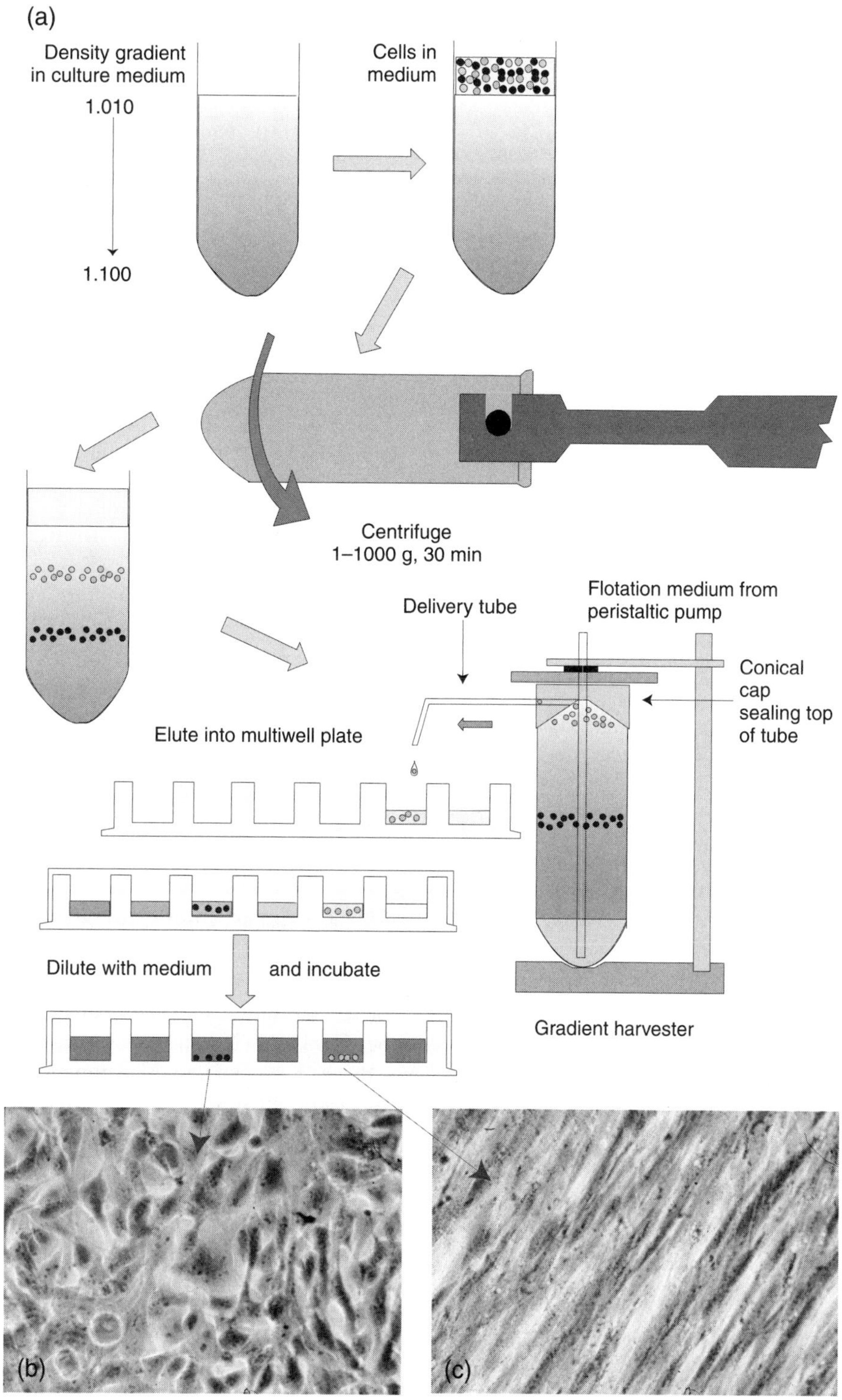

***Fig. 14.1.  Cell Separation by Density.*** (a) Cells are layered on to a preformed gradient (*see* Fig. 14.2) and the tube centrifuged. The tube is placed on a gradient harvester, and flotation medium (e.g., Fluorochemical FC43) is pumped down the inlet tube to the bottom of the gradient, displacing the gradient and cells upward and out through the delivery tube into a multiwell plate. The cells are diluted with medium (so that they will sink) and cultured. (b) Purified HeLa cells recovered from an artificial mixture of HeLa and MRC-5 finbroblasts. (c) Purified MRC-5 eluting at a lower density. (Gradient harvester after an original design by G. D. Birnie.)

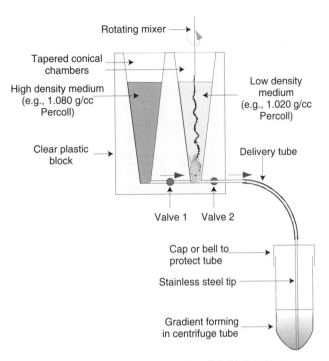

**Fig. 14.2. Gradient Former.** Two chambers are cut in a solid transparent plastic block and connected by a thin canal across the bases of the chambers and exiting to the exterior. A delivery tube with a stainless steel tip, long enough to reach the bottom of the centrifuge tube, is inserted in the outlet. With the valves closed, the left-hand chamber is filled with high-density medium and the right-hand chamber with low-density medium. Valve 1 is opened, the stainless steel tip is placed in the bottom of the tube, the stirrer is started, and then valve 2 is opened. As the solution in the left chamber runs into the right, it mixes and gradually increases the density, while the mixture is running out into the tube.

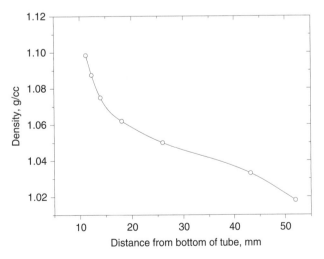

**Fig. 14.3. Centrifuge-Derived Gradient.** Gradient generated by spinning Percoll at 20,000 g for 1 h.

## 14.2 CELL SIZE AND SEDIMENTATION VELOCITY

Sedimentation of cells is also influenced by cell size (cross-sectional area), which becomes the major determinant of sedimentation velocity at 1 $g$ and a significant component at higher sedimentation rates at elevated $g$. The relationship between the particle size and sedimentation rate at 1 $g$, although complex for submicron-sized particles, is fairly simple for cells and can be expressed approximately as

$$v \approx \frac{r^2}{4}$$

[Miller & Phillips, 1969], where $v$ is the sedimentation rate in mm/h and $r$ is the radius of the cell in μm (*see* Table 20.3).

### 14.2.1 Unit Gravity Sedimentation
Layering cells over a serum gradient in medium will allow the cells to settle through the medium according to the equation above. However, unit gravity sedimentation is unable to handle large numbers of cells ($\sim 1 \times 10^6$ cells/cm$^2$ of surface area at the top of the gradient) and does not give particularly good separations unless the mean cell sizes are very different and the cell populations are homogeneous in size. It is usable where there are major differences in cell size, or when aggregates are being separated from single cells, such as after collagenase digestion (*see* Protocol 11.8).

### 14.2.2 Centrifugal Elutriation
Most cell separations based on cell size use either centrifugal elutriation (giving moderate resolution but a high yield) or a cell sorter (high resolution with a low yield; *see* Section 14.4). The centrifugal elutriator (Beckman Coulter) is a device for increasing the sedimentation rate and improving the yield and resolution of cell separation by performing the separation in a specially designed centrifuge and rotor (Fig. 14.4$a$) [Lutz et al., 1992]. Cells in the suspending medium are pumped into the separation chamber in the rotor while it is spinning. While the cells are in the chamber, centrifugal force tends to push the cells to the outer edge of the rotor (Fig. 14.4$b$). Meanwhile medium is pumped through the chamber such that the centripetal flow rate balances the sedimentation rate of the cells. If the cells were uniform, they would remain stationary in one position, but because they vary in size, density, and cell surface configuration, they tend to sediment at different rates. Because the sedimentation chamber is tapered, the flow rate increases toward the edge of the rotor, and a continuous range of flow rates is generated. Cells of differing sedimentation rates will therefore reach equilibrium at different positions in the chamber. The sedimentation

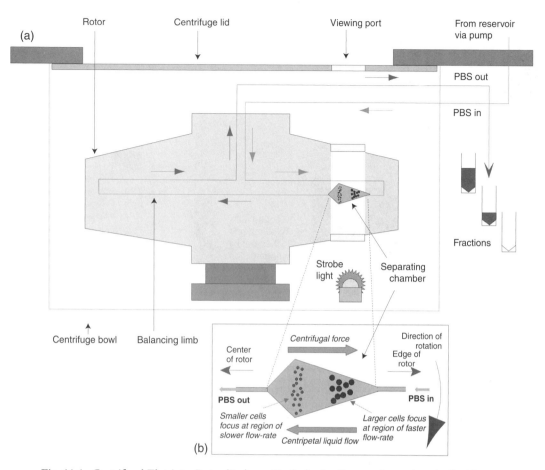

(a)

Rotor    Centrifuge lid    Viewing port    From reservoir via pump

PBS out

PBS in

Fractions

Centrifuge bowl    Balancing limb    Strobe light    Separating chamber

(b)

Centrifugal force

Center of rotor    Direction of rotation    Edge of rotor

**PBS out**    **PBS in**

*Smaller cells focus at region of slower flow-rate*    *Larger cells focus at region of faster flow-rate*

*Centripetal liquid flow*

***Fig. 14.4. Centrifugal Elutriator Rotor (Beckman Coulter).*** A cell suspension and carrier liquid enter at the center of the rotor and are pumped to the periphery and then into the outer end of the separating chamber. The return loop is via the opposite side of the rotor, to maintain balance.

(a)    Magnet    (b)    Cell pellet    (c)

***Fig. 14.5. Magnetic Sorting.*** Negative sort. Committed progenitor cells from bone marrow suspension bound to Dynal paramagnetic beads with antibodies to lineage markers. Lineage-negative (stem) cells are not bound and remain in the suspension ready for sorting by flow cytometry. (*a*) Inserting the tube into the magnetic holder. (*b*) Tube immediately after being placed in magnetic holder. (*c*) Tube 30 s after placement in magnetic holder. (*See also* Plate 23*a*).

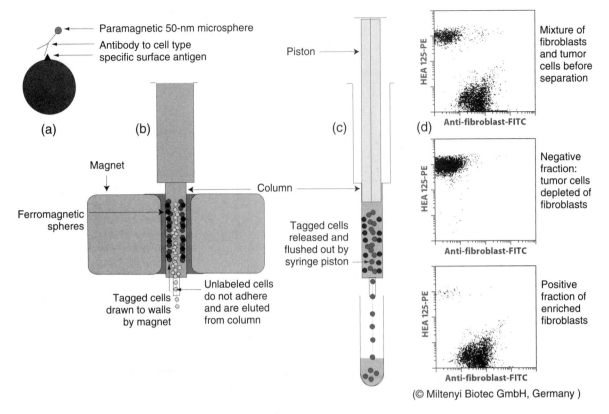

**Fig. 14.6. Magnetic Cell Sorting (MACS® Technology).** Positive sort. (*a*) Cells are preincubated with antibodies raised against a cell-type-specific surface antigen and conjugated to paramagnetic MACS MicroBeads. (*b*) When the cells are introduced into the column, cells bound to MicroBeads are retained in the magnetic field generated by magnet and column matrix; unlabeled cells go straight through. (*c*) The magnetically bound cells, released when the column is removed from the magnet, are flushed out with the piston. (*See also* Plate 23 *d, e*). (*d*) Dot plot from flow cytometry of tumor cell culture containing fibroblasts before MACS sorting (top), flow-through fraction after MACS sorting with Anti-Fibroblast MicroBeads, as in (*b*), enriched fibroblast fraction (bottom) after releasing from magnet. (*See also* Fig. 14.7*b, c.*) (*d,* © Miltenyi Biotec, used with permission.)

chamber is illuminated by a stroboscopic light and can be observed through a viewing port. When the cells are seen to reach equilibrium, the flow rate is increased, and the cells are pumped out into receiving vessels. The separation can be performed in a complete medium and the cells cultured directly afterward.

The procedure comprises four phases: (1) setting up and sterilizing the apparatus, (2) calibration, (3) loading the sample and establishing the equilibrium conditions, and (4) harvesting fractions. The details of the protocol are provided in the operating manual for the elutriator (Beckman Coulter). Equilibrium is reached in a few minutes, and the whole run may take 30 min. On each run, $1 \times 10^8$ cells may be processed, and the run may be repeated as often as necessary. The apparatus is, however, fairly expensive, and a considerable amount of experience is required before effective separations may be made. A number of cell types have been separated by this method [Lag et al., 1996; Recktenwald, 1997; Majore et al., 2009], as have cells of different phases of the cell cycle [Mikulits et al., 1997; Banfalvi, 2008].

## 14.3 ANTIBODY-BASED TECHNIQUES

There are a number of techniques that rely on the specific binding of an antibody to the cell surface. These include immune lysis by an antibody against unwanted cells such as fibroblasts in an epithelial population [Edwards et al., 1980], immune targeting of a cytotoxin [Beattie et al., 1990], fluorescence-activated cell sorting (*see* Section 14.4), immune panning (*see* Section 14.3.1), and sorting with antibody-conjugated magnetizable beads (magnetically activated cell sorting, MACS) [Saalbach et al., 1997] (*see* Section 14.3.2). These techniques all depend on the specificity of the selecting antibody and the presentation of the correct epitope on the cell surface of living cells, as confirmed by immune staining (*see* Section 15.11.1) or flow cytometry (*see* Sections 14.4, 15.11.2, 20.7.2).

### 14.3.1 Immune Panning

The attachment of cells to dishes coated with antibodies, a process called *immune panning*, has been used successfully

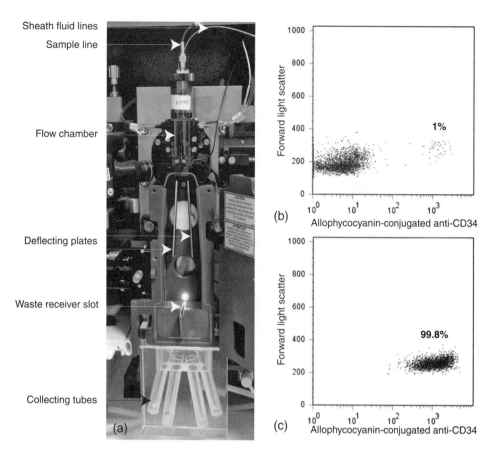

**Fig. 14.7. Fluorescence-Activated Cell Sorter (FACS).** (*a*) Close-up of flow chamber and separation compartment of cell sorter FACSAria (BD Biosciences; *see also* Fig. 14.8). Up to four fractions may be collected by deflection into tubes at bottom. (*b, c*) FACSAria used to analyze CD34$^+$ cell enrichment following CliniMACS (Miltenyi Biotech) selection from leukapheresis of peripheral blood analyzed on. (*b*) Percentage of CD34$^+$ cells, within a total mononuclear cell population, prior to enrichment. (*c*) Purified cells following CD34$^+$ selection. (Courtesy of Paul O'Gorman Leukaemia Research Centre, Glasgow.)

with a number of different cell types [Murphy et al., 1992; Fujita et al., 2004]. The vast majority of panning methods are derived from the work of Wysocki & Sata [1978] with lymphocyte subpopulations. A cell-type-specific antibody raised against a cell surface epitope is conjugated to the bottom of a Petri dish, and when the mixed cell population is added to the dish, the cells to which the antibody is directed attach rapidly to the bottom of the dish. The remainder can then be removed. Immune panning can be used positively, to select a specific subset of cells that can be released subsequently by mechanical detachment or light trypsinization, or negatively, to remove unwanted cells.

### 14.3.2 Magnetic Sorting

Magnetic sorting uses a specific antibody, raised against a cell surface epitope, conjugated to ferritin beads (Dynabeads, from Invitrogen—Dynal; Plate 23a–c) or microbeads (Miltenyi; Plate 23d, e). When the cell suspension is mixed with the beads and then placed in a magnetic field (Fig. 14.5), the cells that have attached to Dynabeads are drawn to the side of the separating chamber. The cells and beads are released when

the current is switched off, and the cells may be separated from the beads by trypsinization or vigorous pipetting. In the Miltenyi system, cells are immunologically bound to microparamagnetic beads that bind to ferromagnetic spheres in the separation column when placed in an electromagnetic field, and released when the column is removed from the magnetic field (Fig. 14.6). Several cell types have been separated by this method [Zborowski & Chalmers, 2007], including stem cell purification by negative sorting [Bertoncello et al., 1991], purification of undifferentiated human embryonic stem cells [Fong et al., 2009], isolation of kidney tubular epithelium [Carr et al., 1999], and isolation of CD34$^+$ cells from blood mononuclear cells (*see* Fig. 14.7b, c).

The use of immunomagnetic microbeads [Gaudernack et al., 1986] (*see* Fig. 14.6; Plate 23d,e) allows the cells to be cultured or processed through further sorting procedures, without the need to remove the beads (Miltenyi). The method is therefore readily applicable to a positive sort. Protocol 14.2 has been abstracted from the Miltenyi instruction sheet (*see also* Protocol 22.18D). Check Miltenyi website for latest versions of labeling and separation protocols (www.miltenyibiotec.com/).

---

### PROTOCOL 14.2. MAGNET-ACTIVATED CELL SORTING (MACS)

#### Outline

Buffy coat or another mixed-cell suspension is mixed with antibody-conjugated microbeads, diluted, and placed in a magnetic separation column. Cells bound to microbeads bind to ferromagnetic spheres in the column, while unbound cells flow through. Bound cells are released from the column when it is removed from the separator magnet and are purged from the column with a syringe piston.

#### Materials

*Sterile:*

☐ Buffer: D-PBSA, pH 7.2, with 0.5 % BSA and 2 mM EDTA

☐ Ficoll Hypaque (*see* Protocol 11.10)

☐ Nylon mesh or filter, 30 μm porosity (Miltenyi, #414:07)

☐ Magnetic cell separator: MiniMACS, MidiMACS, VarioMACS, or SuperMACS (Miltenyi)

☐ RS$^+$ or VS$^+$ column adaptors

☐ Positive selection column: MS$^+$/RS$^+$ for up to $1 \times 10^7$ cells; LS$^+$/VS$^+$ for up to $1 \times 10^8$ cells

☐ Magnetizable microbeads conjugated to antibody raised against cell surface antigen of cells to be collected

☐ Collection tube to match volume being collected: MS$^+$/RS$^+$1 mL; LS$^+$/VS$^+$5 mL

#### Procedure

*Labeling:*

1. Isolate peripheral blood mononuclear cells by the standard method (*see* Protocol 27.1), or prepare a cell suspension by trypsinization or an alternative procedure (*see* Protocols 11.5–11.9, 12.3, 12.4).
2. Remove dead cells with Ficoll-Hypaque (*see* Protocol 11.10).
3. Remove clumps by passing cells through nylon mesh or filter. (Wet the filter with a buffer before use.)
4. Wash the cells in the buffer by centrifugation.
5. Resuspend pellet from centrifugation in 80 μL of buffer per $1 \times 10^7$ total cells (80 μL minimum volume, even for $<1 \times 10^7$ cells).
6. Add 20 μL of MACS microbeads per $1 \times 10^7$ cells.
7. Mix and incubate suspension for 15 min at 6°C to 12°C. (For fewer cells, use the same volume.)

8. Dilute suspension by adding 10× to 20× the volume of buffer in step 5.
9. Centrifuge 300 *g* for 10 min.
10. Remove the supernate and resuspend cells plus microbeads in 500 μL of buffer per $1 \times 10^8$ total cells.

*Positive magnetic separation:*

11. Place the column in the magnetic field of the MACS separator.
12. Wash the column: MS$^+$/RS$^+$ 500 μL; LS$^+$/VS$^+$3 mL.
13. Apply the cell suspension to the column: MS$^+$/RS$^+$500 to 1000 μL; LS$^+$/VS$^+$1 to 10 mL.
14. Allow negative cells to pass through the column.
15. Rinse the column with buffer: MS$^+$/RS$^+$ 3 × 500 μL; LS$^+$/VS$^+$ 3 × 3 mL.
16. Remove the column from the separator and place the column on the collection tube.
17. Pipette buffer into the column (MS$^+$/RS$^+$1 mL, LS$^+$/VS$^+$5 mL), and flush out positive cells, using the plunger supplied with the column.
18. Count the cells, adjusting the concentration in the growth medium:
    (a) $1 \times 10^5$ to $1 \times 10^6$ cells/mL, and seed culture flasks for primary culture.
    (b) 10 to 1000 cells per mL, and seed Petri dishes for cloning.

---

## 14.4  FLUORESCENCE-ACTIVATED CELL SORTING

Fluorescence-activated cell sorting [Hoffman & Houck, 1997] operates by projecting a single stream of cells through a laser beam in such a way that the light scattered from the cells is detected by one or more photomultipliers and recorded (Figs. 14.8, 14.9). If the cells are pretreated with a fluorescent stain (e.g., propidium iodide or chromomycin A$_3$ for DNA) or a fluorescent antibody, the fluorescence emission excited by the laser is detected by a second photomultiplier tube. The information obtained is then processed and displayed as a dot plot (e.g., *see* Figs. 14.6*d*, 14.7*b,c* 20.7) or in a number of other formats [Shapiro, 2003; Applied Cytometry, 2008].

A *flow cytometer* is an analytical instrument that processes the output of the photomultipliers to analyze the constitution of a cell population (e.g., to determine the proportion of cells in different phases of the cell cycle, measured by a combination of DNA fluorescence and cell size measurements). A *fluorescence-activated cell sorter* (FACS; e.g., BD Biosciences FACSAria) is an instrument that uses the emission signals from each cell to sort the cell into one of four sample collection tubes and a waste reservoir. If specific coordinates are set to delineate sections of the display ("gating"), the cell

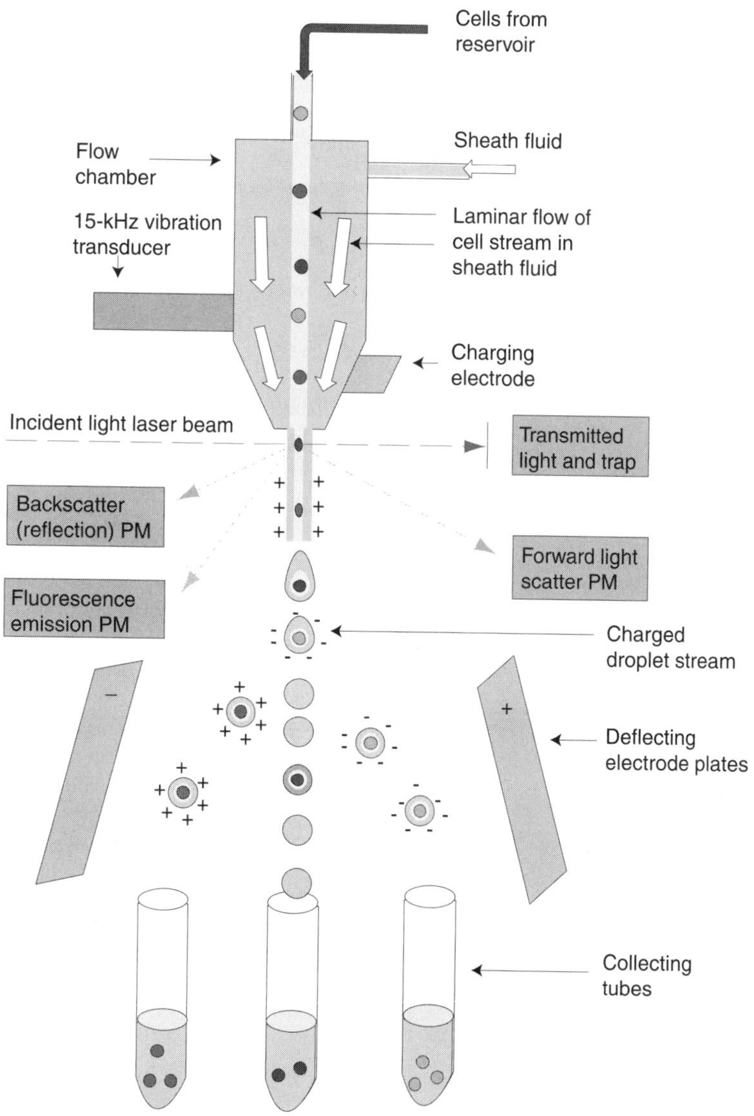

***Fig. 14.8. Flow Cytometry.*** Principle of operation of flow cytophotometer. Cell stream in D-PBSA enters at the top, and sheath liquid, also D-PBSA, is injected around the cell stream to generate a laminar flow within the flow chamber. As the cell stream exits the chamber, it cuts a laser beam, and the signal generated triggers the charging electrode, thereby charging the cell stream. The cell stream then breaks up into droplets, induced by the 15-kHz vibration transducer attached to the flow chamber. The droplets carry the charge briefly applied to the exiting cell stream and are deflected by the electrode plates below the flow chamber. The charge is applied to the plates with a sufficient delay to allow for the transit time from cutting the beam to entering the space between the plates (*see also* Fig. 14.7).

sorter will divert those cells with properties that would place them within these coordinates (e.g., high or low light scatter, high or low fluorescence) into the appropriate receiver tube, placed below the cell stream. The stream itself is broken up into droplets by a high-frequency vibration applied to the flow chamber, and the droplets containing single cells with specific attributes are charged as they leave the chamber. These droplets are deflected, left or right according to the charge applied, as they pass between two oppositely charged

plates. The charge is applied briefly and at a set time after the cell has cut the laser beam such that the droplet containing one specifically marked cell is deflected into the correct tube. The concentration in the cell stream must be low enough that the gap between cells is sufficient to prevent two cells from inhabiting one droplet. All cells having similar properties are collected into the same tube, and up to four sets of cells can be collected simultaneously by changing the polarity of the cell stream and deflecting the cells appropriately.

**TABLE 14.1.** Cell Separation Methods

| Method | Basis for separation | Equipment | Comments | Reference |
|---|---|---|---|---|
| Sedimentation velocity at 1g | Cell size | Custom-made separating funnel and baffles | Simple technique, but not very high resolution or yield | Miller & Phillips, 1969 |
| Isopyknic sedimentation | Cell density | Centrifuge | Simple and rapid | Pertoft & Laurent, 1982 (see Protocol 14.1) |
| MACS | Surface antibodies | Simple magnet and flow chamber | Specific, given a highly specific surface antibody | Zborowski & Chalmers, 2007 (see Protocol 14.2) |
| Immune panning | Surface antibodies | Antibody-coated dishes | Simple, low technology with precoated plates available, but also depends on specific surface antibody | Wysocki & Sata, 1978 |
| Centrifugal elutriation | Cell size, density, and surface configuration | Special centrifuge and elutriator rotor | Rapid, high cell yield, but quite complex process | Lutz et al., 1992 |
| FACS | Cell surface area, fluorescent markers, fluorogenic enzyme substrates, multiparameter | Flow cytometer | Complex technology and expensive; very effective; high resolution, but low yield | Hoffman & Houck, 1997 |
| Affinity chromatography | Cell surface antigens, cell surface carbohydrate | Sterilizable chromatography column | Elutriation of cells from columns is difficult, better in free suspension | Edelman, 1973 |
| Countercurrent distribution | Affinity of cell surface constituents for solvent phase | Shaker | Some cells may suffer loss of viability, but method is quite successful for others | Walter, 1977 |
| Electrophoresis in gradient or curtain | Surface charge | Curtain electrophoresis apparatus | | Kreisberg et al., 1977; Platsoucas et al., 1979 |

This method may be used to separate cells according to any differences that may be detected by light scatter (e.g., cell size) or fluorescence (e.g., DNA, RNA, or protein content; enzyme activity; specific antigens) and has been applied to a wide range of cell types. It has probably been utilized most extensively for hematopoietic cells [Yeung & Wai Eric So, 2009], for which disaggregation into the obligatory single-cell suspension is relatively simple, but has also been used for solid tissues (e.g., lung [Raiser & Kim 2009], skin [Nowak & Fuchs, 2009], and heart [Ieda et al., 2009]; see also Protocol 22.21). It is an extremely powerful tool but is limited by the cell yield (about $1 \times 10^7$ cells is a reasonable maximum number of cells that can be processed at one time). Although the more sophisticated machines with cell separation capability have a high capital cost and require a full-time skilled operator, less expensive bench-top machines are available for analytical use (e.g., Guava Technologies; Accuri).

## 14.5 OTHER TECHNIQUES

The many other techniques that have been used successfully to separate cells are too numerous to describe in detail. They are summarized below and listed in Table 14.1.

*Electrophoresis.* Performed either in a Ficoll gradient [Platsoucas et al., 1979] or by curtain electrophoresis; the second technique is probably more effective and has been used to separate kidney tubular epithelium [Kreisberg et al., 1977].

*Affinity chromatography.* Uses antibodies [Varon & Manthorpe, 1980; Au & Varon, 1979] or plant lectins [Pereira & Kabat, 1979] that are bound to nylon fiber [Edelman, 1973] or Sephadex (GE Healthcare—Amersham). This technique appears to be useful for fresh blood cells, but less so for cultured cells.

*Countercurrent distribution.* [Walter, 1975, 1977]. Utilized to purify murine ascites tumor cells with reasonable viability.

## 14.6 BEGINNER'S APPROACH TO CELL SEPARATION

It is best to start with a simple technique such as density gradient centrifugation or, if a specific cell surface phenotype can be predicted, positive or negative selection by MACS. If the resolution or yield is insufficient, then it may be necessary to employ FACS or centrifugal elutriation. Centrifugal elutriation is useful for the rapid sorting of large numbers of cells, but FACS will probably give the purest cell population, based on the combined application of two or more stringent criteria.

When a purification of several logs is required and selective culture is not an option, it will be necessary to employ at least a two-step fractionation, in a manner analogous to the purification of proteins. In many such procedures, density gradient separation is used as a first step, with panning or MACS as a second, and the final purification is performed by FACS, as has been used in the isolation of hematopoietic [Cooper & Broxmeyer, 1994] and embryonic stem cells [Fong et al., 2009].

# CHAPTER 15

# Characterization

## 15.1 THE NEED FOR CHARACTERIZATION

There are six main requirements for cell line characterization:

(1) Authentication, i.e., confirmation that the cell line is not cross-contaminated or misidentified (*see* Sections 6.11, 12.1.1, 15.2, 18.6; Table 12.2; Appendix V, www.wiley.com/go/freshney/cellculture.

(2) Confirmation of the species of origin.

(3) Correlation with the tissue of origin, which comprises the following characteristics:
 (a) Identification of the lineage to which the cell belongs.
 (b) Position of the cells within that lineage (i.e., the stem, precursor, or differentiated status).

(4) Determination of whether or not the cell line is transformed:
 (a) Is the cell line finite or continuous (*see* Sections 17.2, 17.4)?
 (b) Does it express properties associated with malignancy (*see* Section 17.5, 17.6)?

(5) Indication of whether the cell line is prone to genetic instability and phenotypic variation (*see* Sections 16.1.1, 17.3).

(6) Identification of specific cell lines within a group from the same origin, selected cell strains, or hybrid cell lines, all of which require demonstration of features unique to that cell line or cell strain.

## 15.2 AUTHENTICATION

Characterization of a cell line is vital, not only in determining its functionality but also in proving its authenticity; special attention must be paid to the possibility that the cell line has become cross-contaminated with an existing continuous cell line or misidentified because of mislabeling or confusion in handling (*see* Sections 6.11, 12.1.1, 18.6; Table 12.2; Appendix V). The demonstration that the majority of continuous cell lines in use in the United States in the late 1960s had become cross-contaminated with HeLa cells [Gartler, 1967; Nelson-Rees & Flandermeyer, 1977; Lavappa, 1978] first brought this serious problem to light, but the continued use of the lines 30 years later indicates that many people are still unaware, or are unwilling to accept, that many lines in common use (e.g., Hep-2, KB, Girardi Heart, WISH, Chang Liver) are not authentic [Capes-Davis et al., 2010] (*see also* Table 12.2; Appendix VI). The use of cell lines contaminated with HeLa or other continuous cell lines, without proper acknowledgment of the contamination, is still a major problem [Stacey et al., 2000; Drexler et al., 2003] and demonstrates that any vigorously growing cell line can overgrow another, more slowly growing line [Masters et al., 1988; van Helden et al., 1988; Christensen et al., 1993; Gignac et al., 1993; van Bokhoven et al., 2001a, b; Rush et al., 2002]. Although some of the work with these cell lines may remain perfectly valid, for example, if it is a molecular process of interest regardless of the origin of the cells, any attempt to correlate cell behavior with the tissue or its pathology is totally invalidated by cross-contamination. Unfortunately, this fact is frequently ignored by some journal editors, granting authorities, referees, and users of the cell lines. Characterization studies, particularly with continuous cell lines, have become primarily a process of authentication that is vital to the validation of the data derived from these cells.

*Culture of Animal Cells: A Manual of Basic Technique and Specialized Applications, Sixth Edition*, by R. Ian Freshney
Copyright © 2010 John Wiley & Sons, Inc.

Regardless of the intrinsic ability of the laboratory, DNA profiling has now become the major standard procedure for cell line identification, and a standard procedure with universal application has been defined [ASN-0002, 2010]. If you are unable or unwilling to carry out this authentication procedure yourself, then there are a number of commercial services (*see* Appendix II, Authentication) available; doing nothing is not an option. Multiple isoenzyme analysis by agarose gels (*see* Protocol 15.10) is also an option and inexpensive to set up in your own laboratory. Karyotype analysis is relatively simple and inexpensive to set up, though a little bit more difficult to interpret (*see* Sections 15.7, 23.3.1). Some of the methods in general use for cell line characterization are listed in Table 15.1 [*see also* Hay et al., 2000].

## 15.3 RECORD KEEPING AND PROVENANCE

When a new cell line is derived, either from a primary culture or from an existing cell line, it is difficult to predict its future value. Often it is only after a period of use and dissemination that the true importance of the cell line becomes apparent, and at that point details of its origin are required. However, by that time it is too late to collect information retrospectively. It is therefore vital that adequate records be kept from the time of isolation of the tissue, or of the receipt of a new cell line, detailing the origin, characteristics, and handling of the cell line. These records form the *provenance* of the cell line, and the more detailed the provenance, the more valuable the cell line (*see* Sections 11.3.11, 12.4.9). One of the most important elements of this provenance is the retention of tissue or DNA from the donor to provide reference material for subsequent confirmation of identity by DNA profiling.

The provenance of cell lines has become particularly important with their widespread dissemination through cell banks and personal contacts to research laboratories and commercial companies far removed from their origin. In particular, if a cell line becomes incorporated into a procedure that requires validation of its components, then the provenance of the cell line becomes important and its authentication crucial. Validation requires that the cell line be characterized on receipt, and periodically during use, and that these data be compatible with, and added to, the existing provenance, confirming the identity of the cell line and that it is fit for the use intended.

## 15.4 PARAMETERS OF CHARACTERIZATION

The nature of the technique used for characterization depends on the type of work being carried out; for example, if molecular technology is readily available, then DNA profiling (*see* Protocol 15.9), or analysis of gene expression (*see* Section 15.9) are likely to be of most use, whereas a cytology laboratory may prefer to use chromosome analysis (*see* Section 15.7) coupled with FISH and chromosome painting, and a laboratory with immunological expertise may prefer to use MHC analysis (e.g., HLA typing) coupled with lineage-specific markers. Combined with a functional assay related to your own interests, these procedures should provide sufficient data to authenticate a cell line as well as confirm that it is suited to your requirements.

### 15.4.1 Species Identification

Chromosomal analysis, otherwise known as *karyotyping* (*see* Protocols 15.7, 23.6), is one of the best traditional methods for distinguishing among species. Chromosome banding patterns can be used to distinguish individual chromosomes (*see* Section 15.7); chromosome painting, namely using combinations of specific molecular probes to hybridize to individual chromosomes (*see* Sections 15.7.1), adds further resolution and specificity to this technique. These probes identify individual chromosome pairs and are species specific. The availability of probes is limited to a few species at present, and most are either mouse or human, but chromosome painting is a good method for distinguishing between human and mouse chromosomes in potential cross-contaminations

**TABLE 15.1.** Characterization of Cell Lines and Cell Strains

| Criterion | Method | Reference | Protocol |
|---|---|---|---|
| DNA profile[a] | PCR of microsatellite repeats | Masters et al., 2001 | 15.11 |
| Karyotype[a] | Chromosome spread with banding | Rothfels & Siminovitch, 1958; Dracopoli et al., 2004 | 15.9 |
| Isoenzyme analysis[a] | Agar gel electrophoresis | Hay, 2000 | 15.12 |
| Genome analysis | Microarray | van Beers & Nederlof, 2006 | — |
| Gene expression analysis | Microarray | Sarang et al., 2003; Le Page et al., 2006; Staab et al., 2004 | — |
| Proteomics | Microarray | Barber et al., 2009 | — |
| Cell surface antigens | Immunohistochemistry | Hay, 2000, Burchell et al., 1983, 1987 | 15.13 |
| Cytoskeleton | Immunocytochemistry with antibodies to specific cytokeratins | Lane, 1982, Moll et al., 1982 | 15.13 |

[a]Most suited to authentication.

and interspecific hybrids. *Isoenzyme electrophoresis (see* Protocol 15.10) is also a good diagnostic test and is quicker than chromosomal analysis. A simple kit is available that makes this technique readily accessible. A combination of the two methods is often used and gives unambiguous results [Hay, 2000]. Analysis of polymorphisms in mitochondrial DNA, called *DNA barcoding*, is also becoming a promising approach to species recognition [Cooper et al., 2007; Pun et al., 2009; Aliabadian et al., 2009; Espiñeira et al., 2009].

### 15.4.2  Lineage or Tissue Markers

The progression of cells down a particular differentiation pathway toward a specific differentiated cell type can be regarded as a *lineage*, and as cells progress down this path they acquire *lineage markers* specific to the lineage and distinct from markers expressed by the stem cells. These markers often reflect the embryological origin of the cells from a particular germ layer (*see* Section 2.4). Lineage markers are helpful in establishing the relationship of a particular cell line to its tissue of origin (Table 15.1).

*Cell surface antigens.* These markers are particularly useful in sorting hematopoietic cells [Edvardsson & Olofsson, 2009] and have also been effective in discriminating epithelium from mesenchymally derived stroma with antibodies such as anti-EMA [Heyderman et al., 1979] and anti-HMFG 1 and 2 [Burchell & Taylor-Papadimitriou, 1989], distinguishing among epithelial lineages [Petersen et al., 2003; Labarge et al., 2007], and identifying neuroectodermally derived cells (e.g., with anti-A2B5) [Dickson et al., 1983].

*Intermediate filament proteins.* These are among the most widely used lineage or tissue markers [Lane, 1982; Ramaekers et al., 1982]. Glial fibrillary acidic protein (GFAP) for astrocytes [Bignami et al., 1980] (*see* Plate 11*b*) and desmin [Bochaton-Piallat et al., 1992; Brouty-Boyé et al., 1992] for muscle are the most specific, whereas cytokeratin marks epithelial cells [Lane, 1982; Moll et al., 1982] (*see* Plate 11*a*) and mesothelium [Wu et al., 1982]. Neurofilament protein marks neurons [Kondo & Raff, 2000] and some neuroendocrine cells [Bishop et al., 1988]. Vimentin (*see* Plate 11*a, c*), although usually restricted to mesodermally derived cells in vivo, can appear in other cell types in vitro.

*Differentiated products and functions.* Hemoglobin for erythroid cells, myosin or tropomyosin for muscle, melanin for melanocytes, and serum albumin for hepatocytes are examples of specific cell type markers, but like all differentiation markers, they depend on the complete expression of the differentiated phenotype (*see* Section 16.7).

Transport of inorganic ions, and the resultant transfer of water, is characteristic of absorptive and secretory epithelia [Abaza et al., 1974; Lever, 1986]; grown as monolayers, some epithelial cells will produce *domes*, which are hemicysts in the monolayer caused by accumulation of water on the underside of the monolayer [Rabito et al., 1980] (Fig. 15.1; *see* Plate 12*a, b*). Polarized transport can also be demonstrated in epithelial and endothelial cells using Boyden chambers or filter well inserts (*see* Section 25.3.6). Other tissue-specific functions that can be expressed in vitro include muscle contraction and depolarization of nerve cell membrane.

*Enzymes.* Three parameters are available in enzymic characterization: the constitutive level (i.e., in the absence of inducers or repressors), the induced or adaptive level (i.e., the response to inducers and repressors), and isoenzyme polymorphisms (*see* Sections 15.10.1, 15.10.2). Creatine kinase (CK) MM isoenzyme is found in muscle while the CKBB isoenzyme is characteristic of neuronal and neuroendocrine cells, as is neuron-specific enolase; lactic dehydrogenase is present in most tissues, but as different

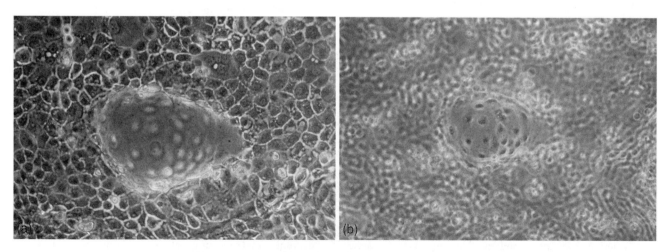

**Fig. 15.1. Domes.** (*a*) Dome, or hemicyst, formed in an epithelial monolayer by downward transport of ions and water, lower focus (on monolayer). (*b*) Upper focus (top of dome). (*See also* Plate 12*a, b*.)

**TABLE 15.2.** Enzymic Markers

| Enzyme | Cell type | Inducer | Repressor | Reference |
|---|---|---|---|---|
| Alkaline phosphatase | Type II pneumocyte (in lung alveolus) | Dexamethasone, oncostatin, IL-6 | TGF-β | Edelson et al., 1988; McCormick & Freshney, 2000 |
| Alkaline phosphatase | Enterocytes | Dexamethasone, NaBt | | Vachon et al., 1996 |
| Angiotensin-converting enzyme | Endothelium | Collagen, Matrigel | | Del Vecchio and Smith, 1981 |
| Creatine kinase BB | Neurons, neuroendocrine cells, SCLC | | | Gazdar et al., 1981 |
| Creatine kinase MM | Muscle cells | IGF-II | FGF-1,2,7 | Stewart et al., 1996 |
| DOPA-decarboxylase | Neurons, SCLC | | | Chung et al., 2006; Gazdar et al., 1980 |
| Glutamyl synthetase | Astroglia (brain) | Hydrocortisone | Glutamine | Hallermeyer & Hamprecht, 1984 |
| Neuron-specific enolase | Neurons, neuroendocrine cells | | | Hansson et al., 1984 |
| Nonspecific esterase | Macrophages | PMA, Vitamin $D_3$ | | Murao et al., 1983 |
| Proline hydroxylase | Fibroblasts | Vitamin C | | Pinnel et al. 1987 |
| Sucrase | Enterocytes | NaBt | | Pignata et al., 1994; Vachon et al., 1996 |
| Tyrosinase | Melanocytes | cAMP | | Park et al., 1993, 1999 |
| Tyrosine aminotransferase | Hepatocytes | Hydrocortisone | | Granner et al., 1968 |

isoenzymes, and a high level of tyrosine aminotransferase, inducible by dexamethasone, is generally regarded as specific to hepatocytes (Table 15.2) [Granner et al., 1968].

***Regulation.*** The level of expression of many differentiated products is under the regulatory control of environmental influences, such as nutrients, hormones, the matrix, and adjacent cells (*see* Section 16.7). Hence the measurement of specific lineage markers may require preincubation of the cells in, for example, a hormone such as hydrocortisone, specific growth factors, or growth of the cells on extracellular matrix of the correct type. Maximum expression of both tyrosine aminotransferase in liver cells and glutamine synthetase in glia requires induction with dexamethasone. Glutamine synthetase is also repressed by glutamine, so glutamate should be substituted in the medium 48 h before assay [DeMars, 1958]. While traditionally enzyme levels were determined by assays of activity, regulatory events are now more likely to be measured by specific gene expression or immunoassays of total enzyme protein.

***Lineage fidelity.*** Although many of the markers described above have been claimed as lineage markers, they are more properly regarded as tissue or cell type markers, as they are often more characteristic of the function of the cell than its embryonic origin. Cytokeratins occur in mesothelium and kidney epithelium, although both of these tissues derive from the mesoderm. Neuron-specific enolase and creatine kinase BB are expressed in neuroendocrine cells

of the lung, although these cells are now recognized to derive from the endoderm and not from neuroectoderm, as one might expect of neuroendocrine-type cells. Transformed epithelial cells can express vimentin in the presence or absence of cytokeratin expression, such as in epithelial-mesenchymal transition [Kokkinos et al., 2007; Gregory et al., 2008].

### 15.4.3 Unique Markers
Unique markers include specific chromosomal aberrations (e.g., deletions, translocations, polysomy), major histocompatibility (MHC) group antigens (e.g., HLA in humans), which are highly polymorphic, and DNA fingerprinting or SLTR DNA profiling (*see* Protocol 15.9). Enzymic deficiencies, such as thymidine kinase deficiency (TK⁻), and drug resistance, such as vinblastine resistance (usually coupled to the expression of the P-glycoprotein by one of the *mdr* genes that code for the efflux protein), are not truly unique, but they may be used to distinguish among cell lines from the same tissues but different donors.

### 15.4.4 Transformation
The transformation status forms a major element in cell line characterization and is dealt with separately (*see* Section 17.1).

## 15.5 CELL MORPHOLOGY

Observation of morphology is the simplest and most direct technique used to identify cells. It has, however, certain shortcomings that should be recognized. Most of these are related to the plasticity of cellular morphology in response to different culture conditions. For example, epithelial cells growing in the center of a confluent sheet are usually regular, polygonal, and with a clearly defined birefringent edge, whereas the same cells growing at the edge of a patch may be more irregular and distended and, if transformed, may break away from the patch and become fibroblast-like in shape (*see* Plate 1*a*, *c*).

Subconfluent fibroblasts from hamster kidney or human lung or skin assume multipolar or bipolar shapes (*see* Fig. 15.2*a*, *g*; Plate 8*a*) and are well spread on the culture surface, but at confluence they are bipolar and less well

spread (*see* Fig. 15.2*b*, *h*; Plates 8*b*, 10*d*, *e*). They also form characteristic parallel arrays and whorls that are visible to the naked eye (*see* Plate 10*c*). Mouse 3T3 cells (Fig. 15.2*s*, *t*) grow like multipolar fibroblasts at low cell density but become epithelioid at confluence (Fig. 15.2*v*, *w*; *see* Plate 10*b*). Alterations in the substrate [Gospodarowicz et al., 1978b; Freshney, 1980] and the constitution of the medium (*see* Section 16.7.2) can also affect cellular morphology. Hence comparative observations of cells should always be made at the same stage of growth and cell density in the same medium, and for growth on the same substrate.

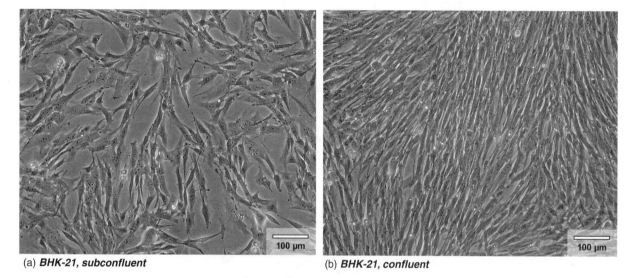

(a) *BHK-21, subconfluent*  (b) *BHK-21, confluent*

**Fig. 15.2. *Examples of Cell Morphology in Culture.*** (*a*) BHK-21 (baby hamster kidney fibroblasts) in log growth. The culture is not confluent, and the cells are well spread and randomly oriented (although some nonrandom orientation is beginning to appear). (*b*) BHK-21 cells at the end of log phase and entering plateau phase. (*c*) CHO-K1 cloned line of Chinese hamster ovary; some fibroblast-like, others more epithelioid. (*d*) CHO-K1 cells at high density; refractile and more elliptical or epithelioid with fewer spindle-shaped cells. (*e*) Vero cells in log phase; epithelial and forming sheets. (*f*) High-density population of Vero; postconfluent, dense sheet with smaller cell diameters. (*g*) Low-density (mid-log phase) MRC-5 human fetal lung fibroblasts; growth random, although some orientation beginning to appear. (*h*) Confluent population of MRC-5 cells with parallel orientation clearly displayed. (*i*) HEK 293 human embryonic kidney epithelial cell line, growing in sheets at mid-log phase. (*j*) High-density HEK 293 cells showing densely packed epithelial cells. (*k*) LNCaP clone FGC from a lymph node metastasis of prostate carcinoma; medium to high density. (*l*) High-density LNCaP cells forming aggregate. (*m*) HeLa cells from human cervical carcinoma. (*n*) HeLa-S3 clone of HeLa. (*o*) IMR-32 cells from human neuroblastoma. (*p*) Cos-7 cells from monkey kidney. (*q*) MDCK cells, Madin-Darby canine kidney. (*r*) BAE cells from bovine arterial endothelium. (*s*) Subconfluent Swiss 3T3 cells (Sw-3T3) from Swiss albino mouse embryo. (*t*) NIH-3T3 cells from NIH mouse embryo. (*u*) L-929, clone of mouse L-cells. (*v*) Sw-3T3 at confluence. Note low-contrast appearance implying very flat monolayer. (*w*) Postconfluent Sw-3T3. (*x*) Postconfluent Sw-3T3 showing a transformation focus of refractile cells overgrowing the monolayer, demonstrating why these cells should be subcultured well before they reach confluence. (*y*) Confluent STO mouse embryo fibroblasts, often used as feeder cells for embryonal stem cell culture. (*z*) Confluent EMT-6 mouse mammary tumor cell line; forms spheroids readily. (*aa*) Subconfluent A2780, human ovarian carcinoma. (*bb*) Caco-2, colorectal carcinoma, used in transepithelial transport studies. (*cc*) Hep-G2, human hepatoma, sometimes used to metabolize drugs or potential cytotoxins. (*dd*) HT-29 colorectal carcinoma; differentiation inducible with sodium butyrate. (*ee*) PC-12, rat adrenal pheochromocytoma; NGF induces neuronal phenotype. (*ff*) U-373 MG, human glioma; one of an extensive series of malignant and normal glial cell lines developed in Uppsala, Sweden. (Photos *a–q*, *s–v*, courtesy of ATCC; *r*, courtesy of Peter Del Vecchio; *y*, *z*, *aa–ff*, courtesy of ECACC; *see also* Plates 7–10, 25, 26.)

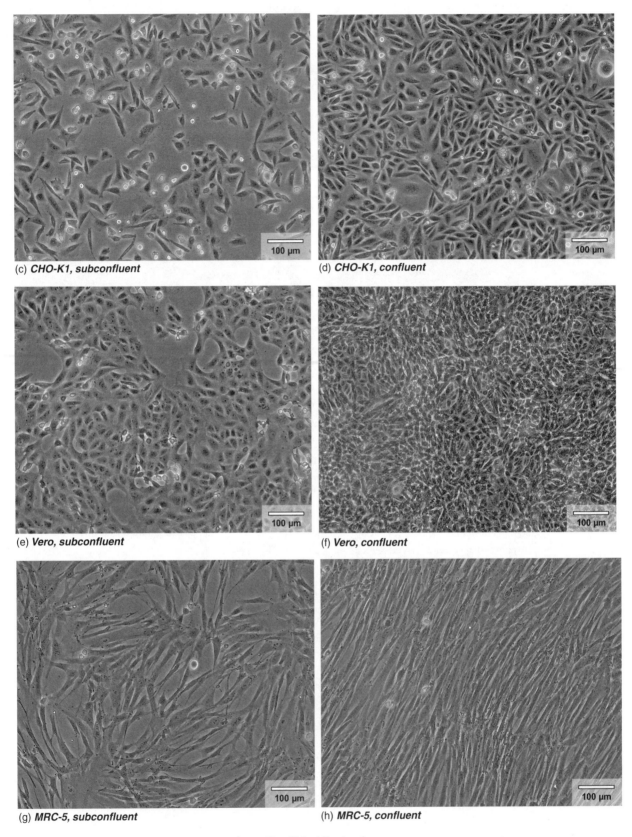

(c) *CHO-K1, subconfluent*

(d) *CHO-K1, confluent*

(e) *Vero, subconfluent*

(f) *Vero, confluent*

(g) *MRC-5, subconfluent*

(h) *MRC-5, confluent*

*Fig. 15.2.* (*Continued*)

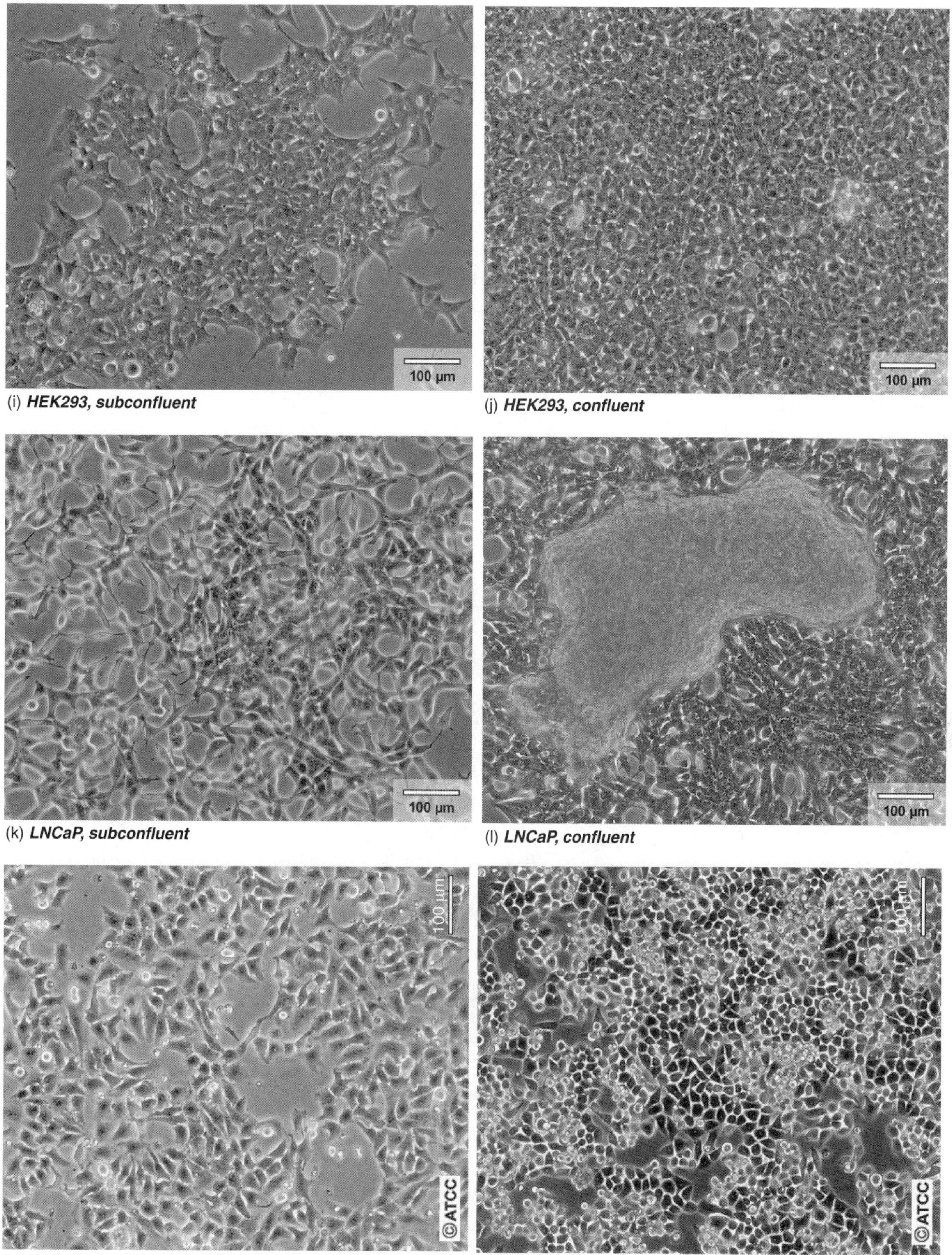

(i) *HEK293, subconfluent*

(j) *HEK293, confluent*

(k) *LNCaP, subconfluent*

(l) *LNCaP, confluent*

(m) *HeLa*

(n) *HeLa-S3*

**Fig. 15.2.** (*Continued*)

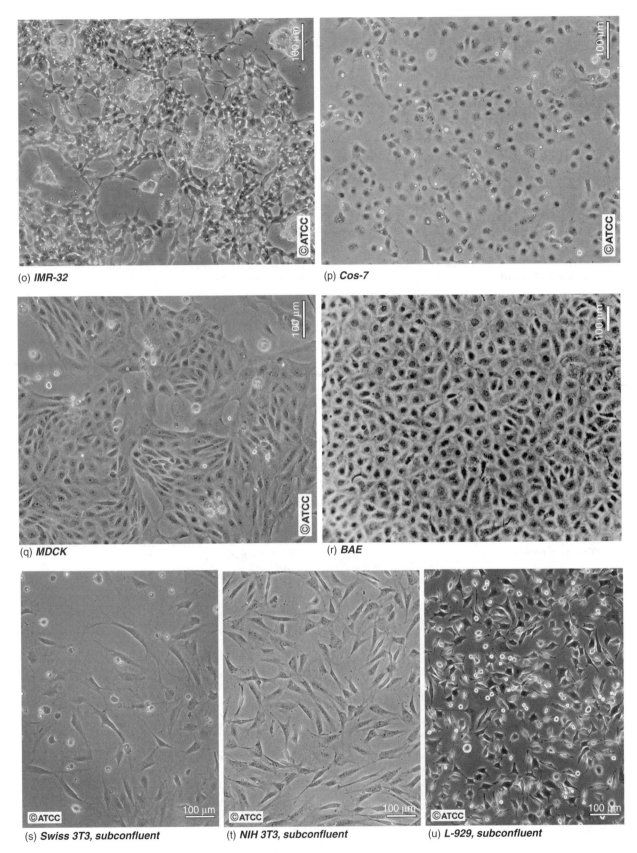

(o) *IMR-32*

(p) *Cos-7*

(q) *MDCK*

(r) *BAE*

(s) *Swiss 3T3, subconfluent*

(t) *NIH 3T3, subconfluent*

(u) *L-929, subconfluent*

*Fig. 15.2.* (*Continued*)

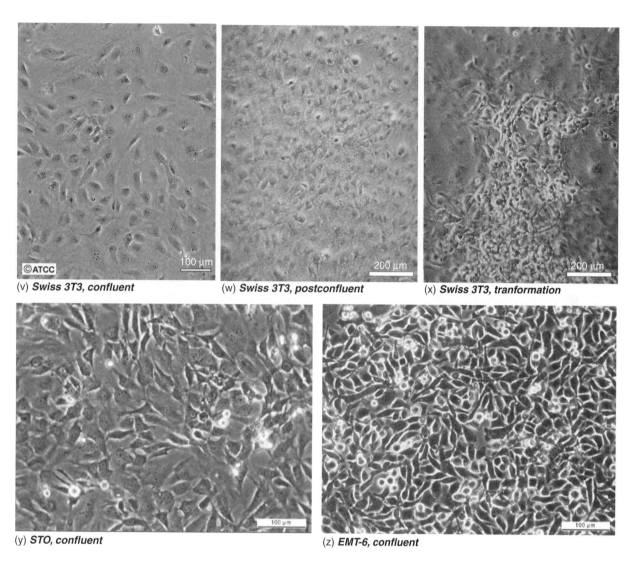

(v) *Swiss 3T3, confluent*     (w) *Swiss 3T3, postconfluent*     (x) *Swiss 3T3, tranformation*

(y) *STO, confluent*     (z) *EMT-6, confluent*

*Fig. 15.2.* (*Continued*)

The terms "fibroblastic" and "epithelial" are used rather loosely in tissue culture and often describe the appearance rather than the origin of the cells. Thus a bipolar or multipolar migratory cell, whose length is usually more than twice its width, would be called fibroblastic, whereas a monolayer cell that is polygonal, with more regular dimensions, and that grows in a discrete patch along with other cells, is usually regarded as epithelial (Fig. 15.2e, *f*, *i*, *j*, *m*, *n*, *q*). However, when the identity of the cells has not been confirmed, the terms "fibroblast-like" (or "fibroblastoid") and "epithelium-like" (or "epithelioid") should be used. Carcinoma-derived cells are often epithelium-like but more variable in morphology (Fig. 15.2m, *n*, *k*, *l*), and some mesenchymal cells like CHO-K1 and endothelium can also assume an epithelium-like morphology at confluence (Fig. 15.2d, *r*). Strictly speaking, precursor cells that are still capable of diving are called *blast* cells: for example, a *fibroblast*

is a proliferative precursor of a *fibrocyte*, a *myoblast* of a *myocyte*, and a *lymphoblast*, of a *lymphocyte*; these rules of terminology are not always observed.

Frequent brief observations of living cultures, preferably with phase-contrast optics, are more valuable than infrequent stained preparations that are studied at length. The former give a more general impression of the cell's morphology and its plasticity and also reveal differences in granularity and vacuolation that bear on the health of the culture. Unhealthy cells often become granular and then display vacuolation around the nucleus (*see* Fig. 12.1).

## 15.5.1 Microscopy

The inverted microscope is one of the most important tools in the tissue culture laboratory, but it is often used incorrectly. As the thickness of the closed culture vessel makes observation difficult from above, because of the long working distance, the culture vessel is placed on the stage, illuminated from above, and observed from below. As the thickness of the wall of the culture vessel still limits the working distance, the maximum objective magnification is usually limited to 40×. The use of phase-contrast optics, where an annular light path is masked by a corresponding dark ring in the objective and only diffracted light is visible, enables unstained cells to be viewed with higher contrast than is available by normal illumination. Because this means that the intensity of the light is increased, an infrared filter should be incorporated for prolonged observation of cells.

---

### PROTOCOL 15.1. USING AN INVERTED MICROSCOPE

#### Outline

Place the culture on the microscope stage, switch on the power, and select the correct optics. Focus on the specimen and center the condenser and phase ring, if necessary.

#### Materials

*Nonsterile:*

❑ Inverted microscope (*see* Section 4.2.4) with phase contrast on 10× and 20× or 40× objectives and condenser with appropriate phase rings
❑ Lens tissue

#### Procedure

1. Make sure that the microscope is clean. (Wipe the stage with 70% alcohol, and clean the objectives with lens tissue, if necessary.)
2. Switch on the power, bringing the lamp intensity up from its lowest to the correct intensity with the rheostat, instead of switching the lamp straight to bright illumination.
3. Check the alignment of the condenser and the light source by setting *Kohler Illumination*:
   (a) Deselect the phase contrast condenser if there is one.
   (b) Place a stained slide, flask, or dish on the stage.
   (c) Close down the field aperture.
   (d) Focus on the image of the iris diaphragm.
   (e) Center the image of the iris in the field of view.
4. Center the phase ring:
   (a) Reselect the correct phase contrast condenser.
   (b) If a phase telescope is provided, insert it in place of the standard eyepiece. Then focus on the phase ring, and check that the condenser ring (white on black) is concentric with the objective ring (black on white).
   (c) If no phase telescope is provided, then replace the stained specimen with an unstained culture

(living or fixed), refocus, and move the phase ring adjustment until optimum contrast is obtained.

5. Examine the cells (*see* Table 12.4). A 10× phase-contrast objective will give sufficient detail for routine examination. (A 4× phase-contrast objective will not give sufficient detail but may be used to determine pattern and select areas for investigation. Higher magnifications than 10× restrict the area scanned but may be used when a particularly interesting cell is seen.)
   (a) Note the growth phase (e.g., sparse, subconfluent, confluent, dense).
   (b) Note the state of the cells (e.g., clear and hyaline or granular, vacuolated).
6. Check the clarity of the medium (e.g., absence of debris, floating granules, signs of contamination), selecting a higher-magnification objective as required. (*See* Section 19.3.1; Fig. 19.1*a, c, e*; Plate 16*a, b, c*.)
7. Record your observations.
8. Turn the rheostat down and switch off the power to the microscope.
9. Return the culture to the incubator, or take it to a hood for sterile manipulations.

---

Culture vessels taken straight from the incubator will develop condensation on the inside of the top or lid. This condensation can be cleared in a flask by inverting the flask, running medium over the inside of the top of the flask (without allowing it to run into the neck), and standing the flask vertically for 10 to 20 s to allow the liquid to drain down. Clearing the condensation is more difficult for Petri dishes and may require replacement of the lid under sterile conditions. If the dish is correctly labeled (i.e., on one side of the base), this should not prejudice identification of the culture. Cultures in microtitration plates are particularly difficult to examine due to the diffraction caused by the meniscus.

It is useful to keep a set of photographs at different cell densities (*see* Fig. 15.2) for each cell line (*see* Protocol 15.6), prepared shortly after acquisition and at intervals thereafter, as a record in case a morphological change is subsequently suspected. Photographs of cell lines in regular use should be displayed above the inverted microscope. Photographic records can be supplemented with photographs of stained preparations and digital output from DNA profiling and stored with the cell line record in a database or stored separately and linked to the cell line database.

### 15.5.2 Staining

A polychromatic blood stain, such as Giemsa, provides a convenient method of preparing a stained culture. Giemsa stain is usually combined with May–Grünwald stain when staining blood, but not when staining cultured cells. Alone, it stains the nucleus pink or magenta, the nucleoli dark blue, and the cytoplasm pale gray-blue. It stains cells fixed in alcohol or formaldehyde but will not work correctly unless the preparation is completely anhydrous.

## PROTOCOL 15.2. STAINING WITH GIEMSA

### Outline
Fix the culture in methanol, stain it directly with undiluted Giemsa, and then dilute the stain 1:5. Wash the culture, and examine it wet.

### Materials

Nonsterile:

- ❑ D-PBSA
- ❑ D-PBSA:methanol, 1:1
- ❑ Methanol
- ❑ Undiluted Giemsa stain
- ❑ Deionized water

### Procedure
This procedure assumes that a cell monolayer is being used, but fixed cell suspensions (see Section 15.5.4) can also be used, starting at step 6.

1. Remove and discard the medium.
2. Rinse the monolayer with D-PBSA, and discard the rinse.
3. Add 5 mL of D-PBSA/methanol per 25 cm². Leave it for 2 min and then discard the D-PBSA/methanol.
4. Add 5 mL of fresh methanol, and leave it for 10 min.
5. Discard the methanol, and replace it with fresh anhydrous methanol. Rinse the monolayer, and then discard the methanol.
6. At this point the flask may be dried and stored or stained directly. It is important that staining be done directly from fresh anhydrous methanol, even with a dry flask. If the methanol is poured off and the flask is left for some time, water will be absorbed by the residual methanol and will inhibit subsequent staining. Even "dry" monolayers can absorb moisture from the air.
7. Add neat Giemsa stain, 2 mL per 25 cm², making sure that the entire monolayer is covered and remains covered.
8. After 2 min, dilute the stain with 8 mL of water, and agitate it gently for a further 2 min.
9. Displace the stain with water so that the scum that forms is floated off and not left behind to coat the cells. Wash the cells gently in running tap water until any pink cloudy background stain (precipitate) is removed, but stain is not leached out of cells (usually about 10–20 s).

10. Pour off the water, rinse the monolayer in deionized water, and examine the cells on the microscope while the monolayer is still wet. Store the cells dry, and rewet them to re-examine.

*Note.* Giemsa staining is a simple procedure that gives a good high-contrast polychromatic stain, but precipitated stain may give a spotted appearance to the cells. This precipitate forms as a scum at the surface of the staining solution, because of oxidation, and throughout the solution, particularly on the surface of the slide, when water is added. Washing off the stain by upward displacement, rather than pouring it off or removing slides from a dish, is designed to prevent the cells from coming in contact with the scum. Extensive washing at the end of the procedure is designed to remove any precipitate left on the preparation.

Fixed-cell preparations can also be stained with crystal violet. Crystal violet is a monochromatic stain and stains the nucleus dark blue and the cytoplasm light blue. It is not as good as Giemsa for morphological observations, but it is easier to use and can be reused. It is a convenient stain to use for staining clones for counting, as it does not have the precipitation problems of Giemsa and is easier to wash off, giving a clearer background and making automated colony counting easier.

## PROTOCOL 15.3. STAINING WITH CRYSTAL VIOLET

### Outline
Fix the culture in methanol, stain it directly with Crystal Violet, wash the culture, and examine.

### Materials

Nonsterile:

- ❑ D-PBSA
- ❑ D-PBSA/MeOH: 50% methanol in D-PBSA
- ❑ Methanol
- ❑ Crystal Violet, 0.1% w/v
- ❑ Filter funnel and filter paper of a size appropriate to take 5 mL of stain from each dish
- ❑ Bottle for recycled stain

### Procedure

1. Remove and discard the medium.
2. Rinse the monolayer with D-PBSA, and discard the rinse.
3. Add 5 mL of D-PBSA/MeOH per 25 cm². Leave it for 2 min and then discard the D-PBSA/MeOH.
4. Add 5 mL of fresh methanol, and leave it for 10 min.

5. Discard the methanol. Drain the dishes, and allow them to dry.
6. Add 5 mL of Crystal Violet per 25 cm², and leave for 10 min.
7. Place the filter funnel in the recycle bottle.
8. Discard the stain into the filter funnel.
9. Rinse the dish in tap water and then in deionized water, and allow the dish to dry.

### 15.5.3 Culture Vessels for Cytology: Monolayer Cultures

The following culture vessels have been found suitable for cytology:

(1) Regular 25-cm² flasks; may need to have the top removed for high power observation; flasks with detachable tops are available (Nunc).
(2) Petri dishes, 3.5 and 5 cm.
(3) Coverslips (glass or plastic; *see* Appendix II) in multiwell dishes or Petri dishes.
(4) Microscope slides in 9-cm Petri dishes or with attached single or multiwell chambers (*see* Fig. 15.3a, b; Appendix II: Chamber Slides).
(5) The OptiCell double membrane culture chamber (Thermo Fisher; Fig 15.3c).
(6) Petriperm dishes (Sartorius) a petri dish with a thin membrane base.
(7) Filter well inserts with cellulose acetate or polycarbonate filters.
(8) Thermanox, and PTFE-coated coverslips.

The last three vessels have all been used for EM cytology studies (some pretreatment of filters or PTFE may be required—e.g., gelatin, collagen, fibronectin, Matrigel, or serum coating; *see* Section 7.2.1).

### 15.5.4 Preparation of Suspension Culture for Cytology

Nonadherent cells must be deposited on a glass or plastic surface for cytological observations. The conventional technique for blood cells, the preparation of a blood smear, does not work well because cultured cells tend to rupture during smearing, although this can be reduced by coating the slide with serum and then spreading the cells in serum.

***Centrifugation.*** Cytological centrifuges have become the preferred option for creating monolayers from suspension cells. Centrifuges (usually called *cytocentrifuges*) are available with special slide carriers for spinning cells onto a slide (Fig. 15.4; *see* Appendix II: Cytocentrifuge). The carriers have sample compartments leading to an orifice that is placed

against a microscope slide and are located within a specially designed, enclosed, rotor. Coating the slides with serum helps to prevent the cells from bursting when they hit the slide.

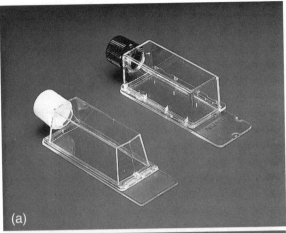
(a)

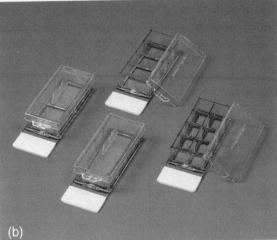
(b)

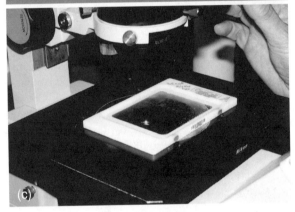
(c)

**Fig. 15.3. Culture Vessels for Cytology.** (*a*) Nunc slide flask; flask is detachable for processing. (*b*) Lab-Tek chamber slides: detachable plastic chambers on a regular microscope slide. One, 2, 4, 8, and 16 chambers per slide are available. (*c*) OptiCell culture chamber (Thermo Fisher) with upper and lower plastic membranes suitable for cell growth and observation. (*a*, *b*, courtesy of Nunc; *c*, courtesy of Dr. Donna Peehl.)

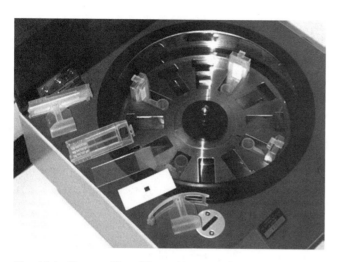

**Fig. 15.4. Cytocentrifuge.** View of interior of Cytotek centrifuge showing rotor and polypropylene slide carriers in place with one carrier disassembled on left-hand side (Sakura Finetek).

Some cytocentrifuges require that fixation be performed in situ.

Δ **Safety Note.** The rotor must be sealed during spinning if human or other primate cells are being used.

**Drop technique.** This procedure is the same as used for chromosome preparation (*see* Protocol 15.7), but without the colcemid and hypotonic treatments. Care must be taken to avoid clumping and the cell concentration must be adjusted to ensure the cells do not pile up. Cells at the edge of the spot may rupture.

**Filtration.** This technique is sometimes used in exfoliative cytology (see the instructions of the following manufacturers for further details: Pall Gelman, Millipore, Sartorius). This technique is suitable for sampling large volumes with a low cell concentration.

---

## PROTOCOL 15.4. PREPARATION OF SUSPENSION CELLS FOR CYTOLOGY BY CYTOCENTRIFUGE

### Outline
Dispense a small volume of concentrated cells into the sample compartment *in situ* and placed against a slide. Centrifuge the cells onto the slide.

### Materials

*Nonsterile:*

❑ Cell suspension, $5 \times 10^5$ cells/mL, in 50% to 100% serum
❑ Serum
❑ Methanol
❑ Cytocentrifuge (*see* Appendix II)
❑ Microscope slides
❑ Slide carriers for centrifuge
❑ Fan

### Procedure

1. Coat the slides in serum and drain.
2. Dry the slides with a fan.
3. Label the slides.
4. Place the slides on slide carriers, and insert the carriers in the rotor.
5. Place approximately 1 to $2 \times 10^5$ cells in 200 to 400 μL of medium in at least two sample blocks (*see* manufacturer's protocol).
6. Switch on the centrifuge, and spin the cells down onto the slides at 100 g for 5 min.
7. Dry off the slides quickly, and fix them in MeOH for 10 min.

---

## PROTOCOL 15.5. FILTRATION CYTOLOGY

### Outline
Gently draw a low-concentration cell suspension through a filter by vacuum. Wash the cells and fix it in situ.

### Materials

*Nonsterile:*

❑ D-PBSA, 20 mL
❑ Methanol, 50 mL
❑ Giemsa stain
❑ Mountant (DPX or Permount)
❑ Transparent filters (e.g., 25-mm TPX or polycarbonate, 0.5-μm porosity; Millipore, Pall-Gelman, Sartorius)
❑ Filter holder (e.g., Millipore, Pall-Gelman, Sartorius)
❑ Vacuum flask (Millipore)
❑ Cell suspension (~$10^6$ cells in 5–10 mL medium with 20% serum)
❑ Vacuum pump

### Procedure

1. Set up the filter assembly (Fig. 15.5) with a 25-mm-diameter, 0.5-μm-porosity transparent filter.
2. Draw the cell suspension onto the filter with a vacuum pump. Do not let all of the medium run through.
3. Gently add 10 mL of D-PBSA when the cell suspension is down to 2 mL.
4. Repeat step 3 when the D-PBSA is down to 2 mL.
5. Add 10 mL of methanol to the D-PBSA, and keep adding methanol until pure methanol is being drawn through the filter.

6. Switch off the vacuum before all of the methanol runs through.
7. Lift out the filter, and air dry it.
8. Stain the filter in Giemsa, rinse in water, and dry.
9. Mount the filter on a slide in DPX or Permount by pressing the coverslip down with a heavy weight to flatten the filter.

*Histology.* Regular monolayer or suspension cultures do not lend themselves readily to histological examination, but a number of techniques have been devised to make this feasible. Most of these are derived from 3D cultures (*see* Sections 25.2–25.4) but sectioning of monolayers and pelleted suspensions is possible by double embedding, where the cells are first embedded in agar or agarose, fixed, and re-embedded in wax or acrylic [Turnpenny & Hanley, 2007]. This technique can also be applied to transmission electron microscopy (e.g., *see* Fig. 2.2).

### 15.5.5 Photomicrography

Digital photography has greatly simplified photomicrography. A regular single lens reflex may be used with the appropriate K-mount and a photographic eyepiece. Choosing at least a 10-megapixel camera gives resolution that is more than adequate for publication. A charge-coupled device (CCD) camera will give the added advantage of constant monitoring on a VDU and connection to an intranet for file storage, printing, transmission, and accessibility.

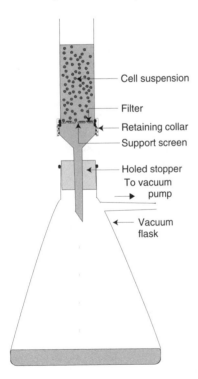

**Fig. 15.5. Filter Cytology.** Filter assembly for cytological preparation (e.g., Millipore).

---

### PROTOCOL 15.6. DIGITAL PHOTOGRAPHY ON A MICROSCOPE

#### *Outline*

Set up the microscope (*see* Protocol 15.1) and focus on the relevant area of the specimen. Check the connections of the camera to the computer, divert the light beam to the camera, focus, and name and save the image.

#### *Materials*

*Sterile or aseptically prepared:*

❑ Culture for examination (make sure that the culture is free of debris—e.g., change the medium on a primary culture before photography)

*Nonsterile:*

❑ Inverted microscope (*see* Sections 4.2.4, 4.2.5)
❑ Phase contrast on 10× and 20× or 40× objectives
❑ Phase condenser with phase rings for 10× and 20× or 40× objectives
❑ Trinocular head or other port with photo eyepiece and adapter for camera
❑ CCD camera or 35-mm single-lens reflex with K-mount
❑ Neutral density filters, 2× and 4×
❑ Lens tissue
❑ Record pad for images
❑ Micrometer slide

#### *Procedure*

1. Prepare culture for photography:
   (a) For flask cultures:
      (i) Take to a laminar flow hood, loosen the cap on the flask, and allow the culture to cool.
      (ii) Tighten the cap on the flask when the contents of the flask are at room temperature, in order to avoid any distortion of the flask as it cools.
      (iii) Rinse the medium over the inside of the top of the flask to remove any condensation that may have formed.
      (iv) Allow the film of medium to drain by standing the flask vertically for 10–20 s.
   (b) For dishes and plates:
      (i) Allow the culture to cool to room temperature in a laminar-flow hood.
      (ii) Replace the lid of the dish or plate (make sure that the sample is labeled on the base of the dish or plate).
2. Select the field and magnification (4× is best for clones or patterning of a monolayer, 10× for a

representative shot of cells, and 20× for cellular detail), avoiding imperfections or marks on the flask. (Always label the side of the flask or dish, and not the top or bottom.)

3. Check the focus:
    **(a)** Focus the binocular eyepieces to your own eyes, using the frame or graticule in the eyepiece.
    **(b)** Focus the microscope.
4. Switch to the monitor, and check the density and contrast of the image on screen, adjusting the light intensity with the rheostat on the microscope for a black-and-white record and with neutral density filters for color (although the image can later be reprocessed electronically if the color temperature is incorrect).
5. Refocus the microscope if necessary.
6. Save the image at a resolution appropriate to the image's ultimate use (1200 × 900 pixels minimum).
7. Turn down the light to avoid overheating the culture.

*Note.* An infrared filter may be incorporated to minimize overheating.

8. Repeat steps 3 through 6 with a micrometer slide at the same magnification to give the scale of magnification. If this image is processed in the same manner as the other shots, it can be used to generate a scale bar.
9. Return the culture to the incubator.
10. Complete a record manually, in the camera's datalog, or in a spreadsheet against the filename; otherwise, the images will be difficult to identify.
11. Saved images can be viewed, edited, and formatted with programs such as Adobe Photoshop and printed by high-quality inkjet, color laser printer, or thermal transfer printer. Publications usually require a minimum of 300 dots per inch (dpi) or 150 dots per cm.

## 15.6 CONFOCAL MICROSCOPY

Time-lapse records, or real-time records made on videotape, can be made via the confocal microscope with a conventional high-resolution video camera or CCD. These systems allow the recording of events within the cell depicted by the distribution and relocation of, and changes in the staining intensity of, fluorescent probes. Fluorescent imaging can be used to localize cell organelles, such as the nucleus or Golgi complex [Lippincott-Schwartz et al., 1991], to measure fluctuations in intracellular calcium [Cobbold & Rink, 1987], and to follow the penetration and movement within the cell of a drug [Neyfakh, 1987; Bucana et al., 1990]. Measurements can be made in three dimensions, as the excitation and detection system is capable of visualizing optical sections through the cell, and over very short periods of time. Confocal microscopy can also visualize cells within a three-dimensional culture system, as in filter well invasion assays [Brunton et al., 1997].

## 15.7 CHROMOSOME CONTENT

Chromosome content or *karyotype* is one of the most characteristic and best-defined criteria for identifying cell lines and relating them to the species and sex from which they were derived. Human, horse, and mouse karyotypes [UWMC Cytogenetics Laboratory, 2003] and the rat ideogram [Szpirer, 1996] are readily available. See also Mitelman [1995] for human and O'Brien et al. [2006] for other mammals. Chromosome analysis can also distinguish between normal and transformed cells (*see* Section 17.3.1) because the chromosome number is more stable in normal cells (except in mice, where the chromosome complement of normal cells can change quite rapidly after explantation into culture).

---

### PROTOCOL 15.7. CHROMOSOME PREPARATIONS

#### Outline
Fix cells arrested in metaphase and swollen in hypotonic medium. Drop the cells on a slide, stain, and examine (Fig. 15.6) [Rothfels & Siminovitch, 1958; Dracopoli et al., 2004].

#### Materials

*Sterile or aseptically prepared:*

☐ Culture of cells in log phase
☐ Colcemid, 10 µM in D-PBSA
☐ D-PBSA
☐ Trypsin, 0.25% crude

*Nonsterile:*

☐ Hypotonic solution: 40 mM KCl, 25 mM sodium citrate
☐ Acetic methanol fixative: 1 part glacial acetic acid plus 3 parts anhydrous methanol or ethanol, made up fresh and kept on ice
☐ Giemsa stain
☐ DPX or Permount mountant
☐ Ice
☐ Centrifuge tubes
☐ Pasteur pipettes
☐ Slides
☐ Coverslips, #00

❏ Slide dishes
❏ Low-speed centrifuge
❏ Vortex mixer

**Procedure**

1. Set up a 75-cm² flask culture at between $2 \times 10^4$ and $5 \times 10^4$ cells/mL $(4 \times 10^3$ and $1 \times 10^4$ cells/cm²) in 20 mL.

2. Approximately 3 to 5 days later, when the cells are in the log phase of growth, add 0.2 mL 10 μM colcemid to the medium already in the flask to give a final concentration of 0.1 μM.

3. After 4 to 6 h, remove the medium gently, add 5 mL of 0.25% trypsin, and incubate the culture for 10 min.

4. Centrifuge the cells in trypsin, and discard the supernatant trypsin.

5. Resuspend the cells in 5 mL of hypotonic solution, and leave them for 20 min at 37°C.

6. Add an equal volume of freshly prepared, ice-cold acetic methanol, mixing constantly, and then centrifuge the cells at 100 $g$ for 2 min.

7. Discard the supernatant mixture, "buzz" the pellet on a vortex mixer (e.g., hold the bottom of the tube against the edge of the rotating cup), and slowly add fresh acetic methanol with constant mixing.

8. Leave the cells for 10 min on ice.

9. Centrifuge the cells for 2 min at 100 $g$.

10. Discard the supernatant acetic methanol, and resuspend the pellet by "buzzing" in 0.2 mL of acetic methanol, to give a finely dispersed cell suspension.

11. Draw one drop of the suspension into the tip of a Pasteur pipette, and drop from around 30 cm (12 in) onto a cold slide. Tilt the slide and let the drop run down the slide as it spreads.

12. Dry off the slide rapidly over a beaker of boiling water, and examine it on the microscope with phase contrast. If the cells are evenly spread and not touching, then prepare more slides at the same cell concentration. If the cells are piled up and overlapping, then dilute the suspension two- to fourfold and make a further drop preparation. If the cells from the diluted suspension are satisfactory, then prepare more slides. If not, then dilute the suspension further and repeat this step.

13. Stain the cells with Giemsa:
    (a) Place the slides on a rack positioned over the sink.
    (b) Cover the cells completely with a few drops of neat Giemsa, and stain for 2 min.
    (c) Flood the slides with approximately 10 volumes of water.
    (d) Leave the slides for a further 2 min.
    (e) Displace the diluted stain with running water.

    *Note.* Do **not** pour the stain off the slides as it will leave a scum of oxidized stain behind; always displace the stain with water. Even when staining in a dish, the stain is never poured off but must be displaced from the bottom with water.

    (f) Finish by running the slides individually under tap water to remove any precipitated stain.
    (g) Check the staining under the microscope. If it is satisfactory, then dry the slide thoroughly and mount with a #00 coverslip in DPX or Permount.

**Variations**

*Metaphase block.* Vinblastine, 1 μM, may be used instead of colcemid. Duration of the metaphase block may be increased to give more metaphases for chromosome counting, or shortened to reduce chromosome condensation and improve banding (*see* Section 15.7.1).

*Collection of mitosis.* Some cells, such as CHO and HeLa, detach readily when in metaphase if the flask is shaken ("shake-off" technique), eliminating trypsinization. The procedure is as follows: add colcemid, remove the colcemid carefully, and replace it with hypotonic citrate/KCl; shake the flask to dislodge cells in metaphase either before or after incubation in hypotonic medium; and fix the cells as previously described.

*Hypotonic treatment.* Substitute 75-mM KCl alone or HBSS diluted to 50% with distilled water for the hypotonic citrate used in Protocol 15.9. The duration of the hypotonic treatment may be varied from 5 to 30 min to reduce lysis or increase spreading.

*Spreading.* There are perhaps more variations at this stage than any other, all designed to improve the degree and flatness of the spread. The procedure includes dropping cells onto a slide from a greater height (clamp the pipette, and mark the position of the slide, using a trial run with fixative alone); flame drying (dry the slide after dropping the cells, by heating the slide over a flame, or actually burn off the fixative by igniting the drop on the slide as it spreads; the latter may, however, make subsequent banding more difficult); making the slide ultracold (chill the slide on solid $CO_2$ before dropping the cells on to it); refrigerating the fixed-cell suspension overnight before dropping the cells onto a slide; dropping cells on a chilled slide (e.g., steep the slide in cold

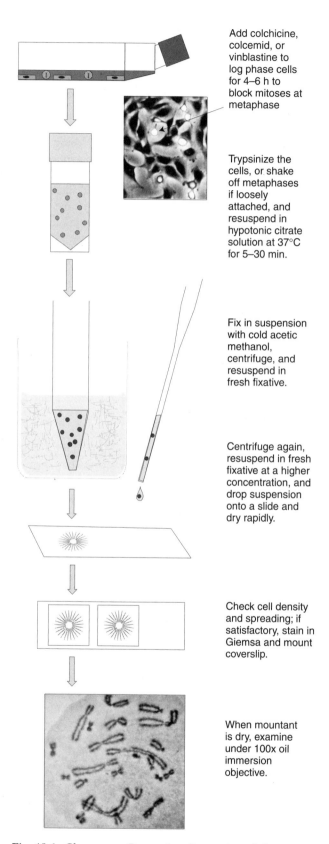

**Add colchicine, colcemid, or vinblastine to log phase cells for 4–6 h to block mitoses at metaphase**

**Trypsinize the cells, or shake off metaphases if loosely attached, and resuspend in hypotonic citrate solution at 37°C for 5–30 min.**

**Fix in suspension with cold acetic methanol, centrifuge, and resuspend in fresh fixative.**

**Centrifuge again, resuspend in fresh fixative at a higher concentration, and drop suspension onto a slide and dry rapidly.**

**Check cell density and spreading; if satisfactory, stain in Giemsa and mount coverslip.**

**When mountant is dry, examine under 100x oil immersion objective.**

**Fig. 15.6. Chromosome Preparation.** Preparation of chromosome spreads from monolayer cultures by the drop technique.

alcohol and dry it off) and then placing the slide over a beaker of boiling water; and tilting the slide or blowing the drop across the slide as the drop spreads (*see also* Protocols 23.6).

## 15.7.1 Chromosome Banding

This group of techniques [Dracopoli et al., 2004, Unit 4.3] was devised to enable individual chromosome pairs to be identified when there is little morphological difference between them [Wang & Fedoroff, 1972]. For Giemsa banding, the chromosomal proteins are partially digested by crude trypsin, producing a banded appearance on subsequent staining. Trypsinization is not required for quinacrine banding. The banding pattern is characteristic for each chromosome pair (Fig. 15.7). Other methods for banding include the following: Giemsa banding using trypsin and EDTA rather than trypsin alone (*see* Fig. 15.7*a*); Q-banding [Caspersson et al., 1968], which stains the cells in 5% (w/v) quinacrine dihydrochloride in 45% acetic acid, followed by rinsing the slide, and mounting it in deionized water at pH 4.5; and C-banding, which emphasizes the centromeric regions. The fixed preparations are pretreated for 15 min with 0.2 M of HCl and 2 min with 0.07 M NaOH and then are treated overnight with SSC (either 0.03 M sodium citrate, 0.3 M NaCl or 0.09 M sodium citrate, 0.9 M NaCl) before staining with Giemsa stain (*see* Protocol 23.6).

Techniques have been developed for discriminating between human and mouse chromosomes, principally to aid the karyotypic analysis of human-mouse hybrids. These methods include fluorescent staining with Hoechst 33258, which causes mouse centromeres to fluoresce more brightly than human centromeres [Hilwig & Gropp, 1972], and alkaline staining with Giemsa ("Giemsa-11") [Bobrow et al., 1972; Friend et al., 1976] (*see* Fig. 15.7*c*).

***Chromosome painting.*** Chromosome paints are available commercially from a number of sources (*see* Appendix II). The hybridization and detection protocols vary with each commercial source, but a general scheme is available [Ausubel, 2002, Unit 14.7; Dracopoli et al., 2004, Unit 4.6]. Karyotypic analysis is carried out classically by chromosome banding, using dyes that differentially stain the chromosomes (*see* Protocol 23.6). Thus each chromosome is identified by its banding pattern. However, traditional banding techniques cannot characterize many complex chromosomal aberrations. New karyotyping methods based on chromosome painting techniques—namely spectral karyotyping (SKY) and multicolor fluorescence in situ hybridization (M-FISH)—have been developed. These techniques allow the simultaneous visualization of all 23 human chromosomes in different colors. Furthermore visualization of the resulting fluorescence patterns by computer is more sensitive than the human eye. These techniques are proving to be highly successful in the identification of new chromosomal alterations that were previously unresolved by traditional approaches [Wienberg & Stanyon, 1997; Lim et al., 2004; McDevitt et al., 2007].

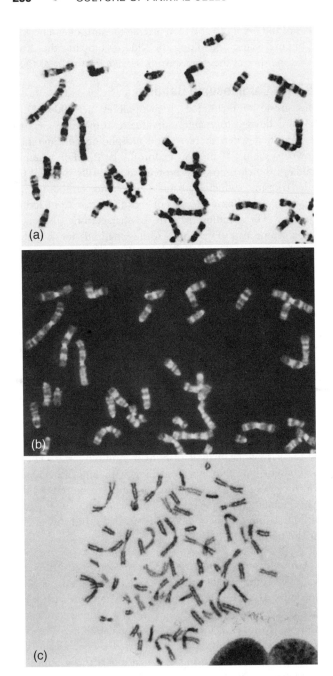

**Fig. 15.7. Chromosome Staining.** (*a*) Human chromosomes banded by the standard trypsin-Giemsa technique. (*b*) The same preparation as in (*a*), but stained with Hoechst 33258. (*c*) Human–mouse hybrid stained with Giemsa at pH 11. Human chromosomes are less intensely stained than mouse chromosomes. Several human-mouse chromosomal translocations can be seen. (Courtesy of R. L. Church.)

### 15.7.2 Chromosome Analysis

The following are methods by which the chromosome complement may be analyzed:

(1) *Chromosome count.* Count the chromosome number per spread for between 50 and 100 spreads. (The chromosomes need not be banded.) Closed-circuit television or a camera lucida attachment may help. You should attempt to count all of the mitoses that you see and classify them (a) by chromosome number or, if counting is impossible, (b) as "near diploid uncountable" or "polyploid uncountable." Plot the results as a histogram (*see* Fig. 2.10).

(2) *Karyotype.* Digitally photograph about 10 or 20 good spreads of banded chromosomes. Using Adobe Photoshop or an equivalent graphics program, cut the individual chromosomes and paste them into a new file where they can be rotated, trimmed, aligned, and sorted (Fig. 15.8). Image analysis can be used to sort chromosome images automatically to generate karyotypes (e.g., Leica CW4000).

Chromosome counting and karyotyping allow species identification of the cells and, when banding is used, distinguish individual cell line variations and marker chromosomes. However, karyotyping is time-consuming, and chromosome counting with a quick check on gross chromosome morphology may be sufficient to confirm or exclude a suspected cross-contamination.

## 15.8 DNA ANALYSIS

DNA content can be measured by propidium iodide fluorescence with a CCD camera or flow cytometry [Shapiro, 2003] (*see* Sections 14.4, 20.1.4), although the generation of the necessary single-cell suspension will, of course, destroy the topography of the specimen. DNA can be estimated in homogenates with Hoechst 33258 (*see* Protocol 20.3) and other DNA fluorochromes such as DAPI, propidium iodide, or Pico Green (Molecular Probes). Analysis of DNA content is particularly useful in the characterization of transformed cells that are often aneuploid and heteroploid (*see* Section 17.3).

### 15.8.1 DNA Hybridization

Hybridization of specific molecular probes to unique DNA sequences (Southern blotting) [Sambrook & Russell, 2006; Ausubel et al., 2009] can provide information about species-specific regions, amplified regions of the DNA, or altered base sequences that are characteristic to that cell line. Thus strain-specific gene amplifications, such as amplification of the dihydrofolate reductase (DHFR) gene, may be detected in cell lines selected for resistance to methotrexate [Biedler et al., 1972]; amplification of the MDR gene in vinblastine-resistant cells [Schoenlein et al., 1992]; overexpression of a specific oncogene, or oncogenes in transformed cell lines [Weinberg, 1989]; or deletion, or loss, of heterozygosity in suppressor genes [Witkowski, 1990; Marshall, 1991]. Although DNA aberrations can be detected in restriction digests of extracts of whole DNA, this is limited by the amount of DNA required.

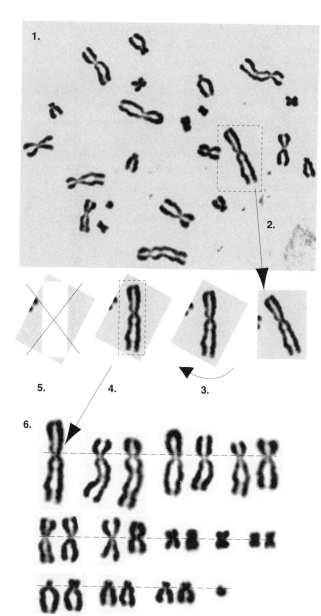

1. Prepare digital photographs of individual chromosome spreads.

2. Select and copy each chromosome separately and paste into a new file.

3. Rotate chromosome to vertical with short arms to top.

4. Cut chromosome and paste centromere on line, sorting by size and shape.

5. Discard residual background, including unwanted parts of adjacent chromosomes.

6. When all chromosomes are in place, adjust brightness and/or contrast to suppress background.

*Fig. 15.8. Karyotype Preparation.* Steps in the preparation of a karyotype from digital microphotographs of metaphase spreads. Chinese hamster cells recloned from the Y-5 strain of Yerganian and Leonard [1961] (acetic-orcein).

It is more common to use the polymerase chain reaction (PCR) with a primer specific to the sequence of interest, enabling detection in relatively small numbers of cells, such as would be available in a laboratory-scale experiment. Alternatively, specific probes can be used to detect specific DNA sequences by in situ hybridization [Ausubel, 2009], having the advantage of displaying topographical differences and heterogeneity within a cell population.

## 15.8.2  DNA Fingerprinting

DNA contains satellite DNA, nontranscribed regions that are highly repetitive, and of variable length, with minisatellite DNA having 1- to 30-kb repeats and microsatellite DNA having only 2 to 4 bases in repeating sequences. These regions are not highly conserved because they are not transcribed, and they give rise to regions of hypervariability. When the DNA is cut with specific endonucleases, sequences of interest may be probed with cDNAs that hybridize to these hypervariable regions [Jeffreys et al., 1985] or they may be amplified by PCR with specific primers. Electrophoresis reveals variations in fragment length in satellite DNA (restriction fragment length polymorphisms, RFLPs) that are specific to the individual from which the DNA was derived. When analyzed by polyacrylamide electrophoresis, each individual's DNA gives a unique hybridization pattern as revealed by autoradiography with radioactive or fluorescent probes. These patterns are known as DNA fingerprints and are cell line specific, except if more than one cell line has been derived from one individual or if highly inbred donor animals have been used [Stacey et al., 1992].

DNA fingerprints appear to be quite stable in culture, and cell lines from the same origin, but maintained separately in different laboratories for many years, still retain the same or very similar DNA fingerprints. DNA fingerprinting is a very powerful tool in determining the origin of a cell line, if the original cell line, or DNA from it or from the donor individual, has been retained. This emphasizes the need to retain a blood, tissue, or DNA sample when tissue is isolated for primary culture (*see* Section 11.3.11). Furthermore, if a cross-contamination or misidentification is suspected, this can be investigated by fingerprinting the cells and all potential contaminants (Fig. 15.9). DNA fingerprinting has confirmed

earlier isoenzyme and karyotypic [Gartler, 1967; Nelson-Rees & Flandermeyer, 1977; Lavappa, 1978; Nelson-Rees et al., 1981] data, indicating that many commonly used cell lines are cross-contaminated with HeLa (*see* Sections 12.1.1, 15.2, 18.6; Table 12.2; Appendix V).

### 15.8.3 DNA Profiling

For authentication of human cell lines, there is an ever increasing number of companies that provide identity testing services (*see* Appendix II, DNA Profiling). The author cannot vouch for the quality of service provided by any of these organizations, and, when approaching such companies, it is important to ask certain questions to assure yourself that they will provide the service you need. These questions include:

(1) Do they perform the testing themselves or do they outsource testing services?
(2) What is the specificity of the methods used for individual identification?
(3) Are the genetic markers used linked with those used by professional bodies or other expert centers?

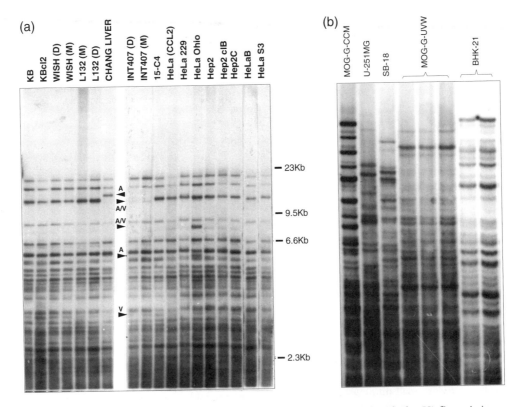

**Fig. 15.9. DNA Fingerprints.** Southern blots of cell line DNA digested with the *Hinf*I restriction enzyme and hybridized with the minisatellite probe 33.15 [Jeffreys et al., 1985]. (*a*) DNA fingerprints of HeLa cell-contaminated cell lines and subclones of the HeLa cell line. Banding patterns are identical in cases when master banks (M) and their derivative working or distribution cell banks (D) were analyzed. Fingerprint patterns were generally consistent between the cell lines, but some cases showed additional (A) or variable (V) bands (e.g., HeLa Ohio, INT 407, and Chang Liver). (Photograph courtesy of G. Stacey, NIBSC, UK.) (*b*) Four human glioma cell lines—MOG-G-CCM, U-251 MG, SB-18, and MOG-G-UVW (three separate freezings)—and duplicate lanes of BHK-21 controls. (Fingerprints courtesy of G. Stacey, NIBSC, UK.)

(4) Do they have experience in interpreting cell line data?

(5) Do they have in house expertise in the methods to assist in interpretation of results?

(6) Do they have accreditation by an appropriate professional or government body or do they have formal affiliation with an expert group or organization?

Cell banks such as ATCC, ECACC, DSMZ, and JCRB (*see* Table 19.5; Appendix II) will also be able to advise on such testing and may provide it as a service. An international standard protocol for DNA profiling has been published [ASN-0002, 2010].

Because microsatellite sequences are quite small, it has been possible to identify and quantify short tandem repeat (STR) sequences at specific loci. Second-generation multiplex (SGM) examines 7 different areas of the genome, and SGM Plus® examines 11, giving a discrimination potential of $1:10^9$. So far this only applies to human cell lines, but there is great potential for the generation of numerical data that would be interchangeable between different laboratories [Masters et al., 2001; Langdon, 2004]. DNA profiling is available as a service from a number of laboratories including LGC and ATCC. The following protocol has been submitted by Amber Meredith of LGC (www.lgc.co.uk).

## PROTOCOL 15.8. DNA STR PROFILING OF CELL LINES

### Background

Cross-contamination between cell lines is a longstanding and frequent cause of scientific misrepresentation. It has been estimated that up to 36% of cell lines are of a different origin than that claimed. STR profiling by the use of SGM Plus™ provides a quick and inexpensive way of identifying the STR profile of a cell line that is reproducible between laboratories [Masters et al., 2001].

### Outline

DNA is extracted from cell pellets with Qiagen® QIAamp® mini kits; the extracted DNA is amplified with SGM Plus™ and then run on an acrylamide gel with an ABI 377 DNA sequencer before being analyzed with GeneScan® and Genotyper® software.

### A. Extraction of Cell Pellets
### Materials

- QIAamp® DNA mini kit
- Microfuge tubes, 1.5 mL (e.g., Eppendorf Biopure Safelock tubes)
- Centrifuge (e.g., Eppendorf 5415 D or similar specification)
- Waterbath or oven capable of maintaining a temperature of 56°C
- Vortex mixer

- Absolute ethanol
- Phosphate buffer, 0.01 M (e.g., Sigma PBS tablets, 2 tablets dissolved in 400 mL of water)
- Before starting the procedure the following reagents must be prepared:
  (i) *Qiagen dissolved protease solvent.* Add the appropriate amount of protease solvent to the lyophilized Qiagen® protease. The solution is stable for up to two months when stored at 2°C to 8°C.
  (ii) *Qiagen AW1 and AW2 buffers.* Add the appropriate amount of absolute ethanol to the bottles containing the AW1 and AW2 concentrates. AW1 and AW2 buffers are stable at room temperature for a year.
  (iii) *Qiagen AL buffer (lysis buffer).* Before use, mix thoroughly by inversion.
  (iv) *Qiagen AE buffer (elution buffer)*
- Plastic disposable Pasteur pipettes (Liquipettes)

### Procedure
(Kit manufacturer's protocol with minor modifications) *If the cells are not suspended in culture medium, proceed from step 5. If the cell pellets have been received in culture medium:*

1. Centrifuge the samples at 20,000 g for 3 min.
2. Using a fine-point Liquipette, carefully remove the supernate and discard.
3. Resuspend the cells in 200 μL of PBS
4. Repeat steps 1 and 2 once more.
5. Resuspend the cells in 200 μL of PBS
6. Pipette 20 μL of Qiagen® protease into the bottom of an empty labeled 1.5-mL microfuge tube.
7. Add 200 μL of each sample into its corresponding tube containing the protease.
8. Add 200 μL of AL buffer into each tube.
9. Mix each tube by vortexing for approximately 15 s.
10. Incubate the samples at 56°C for 10 min.
11. After this time, settle the contents of each tube by centrifuging for approximately 3 s at 1000 g.
12. Carefully uncap each tube and add 200 μL of absolute ethanol.
13. Mix by vortexing each tube for 15 s, and centrifuge at 1000 g for 3 s to pellet the contents.
14. Carefully transfer the mixture into a correspondingly labeled QIAamp® spin column, which has been placed in a 2-mL collection tube.
15. Close the cap of the column and spin at 6000 g for 1 min.
16. Place the spin column into a clean collection tube and discard the tube containing the filtrate.

17. Carefully open the lid of the spin column and add 500 μL of AW1 buffer.
18. Repeat steps 15 and 16.
19. Carefully open the lid of the spin column and add 500 μL of AW2 buffer.
20. Close the cap and centrifuge at 20,000 g for 3 min.
21. As the AW2 buffer can have a detrimental effect on downstream applications, it is advisable to replace the collection tube and spin the column for a further minute at 20,000 g to ensure that all of the AW2 buffer has been discarded.
22. Place the column into a correspondingly labeled 1.5-mL microfuge tube.
23. Open the lid of the column and add 200 μL of AE buffer.
24. Incubate at 56°C for 1 min; then centrifuge at 6000 g for a further minute. Discard the column and cap the tube.
25. Quantify the DNA in each sample, such as with Picogreen (Molecular Probes).
26. If the samples are not to be amplified immediately, they should be stored frozen.

### B. Amplification with SGM Plus™

#### Reagents and Materials
❑ Components of SGM Plus™ PCR mix (AMPF/STR® PCR reaction mix, AMPF/STR® SGM Plus™ primer set, Amplitaq gold® DNA polymerase)
❑ Pure sterile water (e.g., Sigma tissue culture grade water)
❑ Thermocycler (e.g., ABI 9700)

#### Procedure
1. With reference to the ABI protocol, prepare sufficient mix for the samples to be amplified. Per sample, this comprises:
   (a) AMPF/STR® PCR reaction mix, 21 μL.
   (b) AMPF/STR® SGM Plus™ primer set, 11 μL.
   (c) Amplitaq gold® DNA polymerase, 1 μL.
2. Per sample, aliquot 30 μL of the mix either into individual PCR tubes or into a thin-walled microtitration plate, seal the plate with adhesive foil.
3. Remove the samples from the freezer and allow to defrost.
4. Mix the samples and spin down to settle the contents of each tube.
5. Add approximately 1 ng of DNA to the appropriate aliquot of SGM Plus™ PCR; mix and make up the volume added to 20 μL by adding the appropriate volume of pure sterile water.
6. Coamplify a sample of known genotype to act as a control (e.g., the kit control).

7. Place the samples onto a thermocycler programmed with the appropriate cycling details, as follows, and start the run:
   (a) Hold at 95°C for 11 min ±5 s.
   (b) Hold at 94 ± 1°C for 1 min ±2 s.
   (c) Hold at 59 ± 1°C for 1 min ±2 s.
   (d) Hold at 72 ± 1°C for 1 min ±2 s.
8. Repeat steps 7(b) to 7(d) 28 times.
9. Hold at 60 ± 1°C for 45 min ±5 s.
10. Hold at 4°C until required.

### C. Capillary Electrophoresis Using the ABI 31XX DNA Sequencer

#### Reagents and Materials
❑ ABI 31XX DNA Sequencer or equivalent (Fig. 15.10).
❑ Ice or electronic cold plate (e.g., Camlab Ice Cube)
❑ Thermocycler machine (e.g., ABI 9700)
❑ SGM Plus™ Gel loading mixture: Formamide, GeneScan 500 (GS500 ROX) internal size standard.

#### Procedure
1. With reference to the technical manual, set up the ABI 31XX DNA sequencer for a capillary run.
2. Remove an aliquot of HI-DI Formamide from the freezer and allow to thaw; vortex for 10 s and pulse spin.
3. Remove the GS 500 from the fridge, vortex for 10 s and pulse spin.
4. Remove the SGM+ ladder from the fridge, vortex for 10 s and pulse spin.
5. Prepare sufficient formamide/GS 500 reagent for the number of samples being processed. Each capillary run of 16 samples should contain at least one known positive sample and one ladder.
6. Into a 96-well Optical microtitration plate, add 9 μl of the Formamide/GS 500 mix into each of the wells to be used.
7. To each appropriate well, add 1 μl of amplified PCR product or allelic ladder.
8. Place the optical plate onto a thermocycler heated to 95°C, for 3 min before immediately transferring to a tray of ice or an electronic cooling block set to 0°C for a further 3 min.
9. After this time, insert the optical plate between the base and top of the plate assembly, ensuring that the rubber septa lines up with the openings on the plate.
10. Ensure that the instrument has sufficient polymer for the run and that it is topped up with running buffer.
11. Place the plate assembly onto the auto sampler and start the run.
12. Each capillary run takes approximately 40 min.

### D. Analysis of the Cell Line Results

**Materials**
- ❑ GeneMapper ID®
- ❑ PC

**Procedure**

1. Transfer the saved run from the 31XX onto the desktop of the PC.
2. From the File menu, select 'Add samples to project', and select the relevant run folder from the desk top.
3. Press 'Add to list'. Double click on the folder name to display the individual files, then press 'Add'.
4. The samples should appear in the samples window of the project and the status column should contain a green triangle for each sample.
5. Select the relevant sample type from the drop down menu and assign to each sample. Only one ladder per run can be designated the Allelic Ladder sample type.
6. Select the relevant analysis, table, and size standard settings for the run.
7. Once all the settings are updated, click on the green triangle on the toolbar and when prompted save the project with the required name. The software will then analyze each sample in turn.
8. Check the raw data and the size standard for each sample.
9. Review the ladder to ensure that the types are correctly assigned.
10. Select each profile in turn. Check through each profile and click off the type that is not a peak.
11. Print each profile using the 'portrait' mode.
12. Once the data have been printed, export the project to the desktop.
13. A comparison can then be made of the analyzed profile against the published profile for each cell line sample.

**Fig. 15.10. DNA Sequencer.** ABI 377 DNA sequencer using automated electrophoretic analysis on polyacrylamide gel to separate and quantify amplified STR sequences. (Photo courtesy of A. Meredith, LGC Promochem.)

DNA profiling has been used most extensively with human cell lines where the primers are most commonly available and the extension of this to other animal species is still somewhat limited (Fig. 15.11). Speciation can be achieved however using the so-called "barcode region" of the cytochrome oxidase I gene [Cooper et al., 2007] as well as by isoenzyme analysis (*see* Section 15.10.1).

## 15.9 RNA AND PROTEIN EXPRESSION

Genetic analysis can be performed on the total genome by microarray analysis. This represents a significant commitment in funding and technical support and it is unlikely that a laboratory would set this up just to identify cell lines. However, if the technology is accessible, it provides a valuable tool. Gene expression analysis also can be quantified by reverse transcriptase PCR (RT-PCR) of gene transcripts on a large scale by microarray gene expression analysis (Affymetrix) [Kiefer et al., 2004; Staab et al., 2004]. An example is given in Plate 24. The data in this plate compare three different cell types, for which this profiling exercise clearly indicates differences. Proteins can be detected by microarray analysis using antibodies [Barber et al., 2009; Simara et al., 2009]. While genome analysis may help in cell line identification, expression analysis will depend on the phenotypic status of the cells, which will vary with the growth conditions.

## 15.10 ENZYME ACTIVITY

Specialized functions in vivo are often expressed in the activity of specific enzymes, some of which may be expressed in vitro (*see* Table 15.2). Unfortunately, many enzyme activities are lost or greatly reduced in vitro (*see* Sections 3.4.2, 16.1.1) and are no longer available as markers of tissue specificity—for example, hepatocytes arginase activity within a few days of culture. However, some cell lines do express specific enzymes, such as tyrosine aminotransferase in the rat hepatoma HTC cell lines [Granner et al., 1968]. When looking for specific marker enzymes, the constitutive (uninduced) level and the induced level should be measured and compared with a number of control cell lines. Glutamyl synthetase activity, for instance, characteristic of astroglia in brain, is increased several-fold when the cells are cultured in the presence of glutamate instead of glutamine [DeMars, 1957; McLean et al., 1986]. Common inducers of enzyme activity are glucocorticoid hormones, such as dexamethasone;

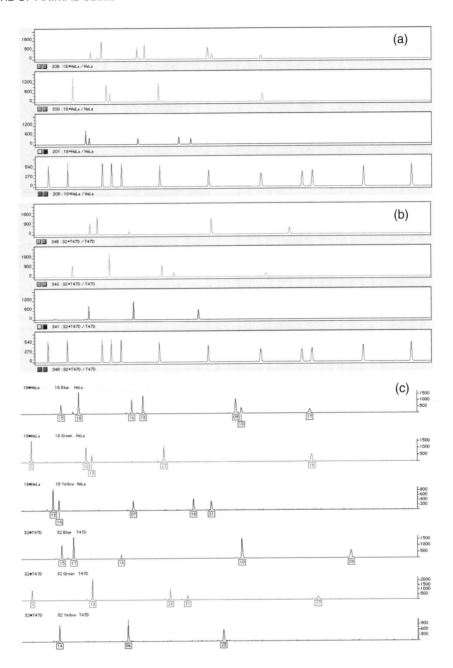

**Fig. 15.11. DNA Profiling.** (*a, b*) HeLa and T47D cell lines analyzed with Genescan® software to allocate sizes to the peaks detected by the ABI 377 DNA sequencer during gel electrophoresis. (*c*) HeLa and T47D cell lines analyzed with Genotyper® software to allocate allelic types to the peaks sized during analysis with Genescan® software. (Courtesy of A. Meredith, LGC Promochem.)

polypeptide hormones, such as insulin and glucagon; and alteration in substrate or product concentrations in the medium, as in the aforementioned example with glutamyl synthetase.

### 15.10.1 Isoenzymes

Enzyme activities can also be compared qualitatively between cell strains by using enzyme protein polymorphisms among species, and sometimes among races, individuals, and tissues within a species. These enzymes, called *isoenzymes* or *isozymes*,

may be separated chromatographically or electrophoretically, and the distribution patterns (zymograms) may be found to be characteristic of species or tissue (Fig. 15.12). Nims et al. [1998] determined that interspecies cell line cross-contamination can be detected with isoenzyme analysis if the contaminating cells represent at least 10% of the total cell population.

Electrophoresis media include agarose, cellulose acetate, starch, and polyacrylamide. In each case a crude enzyme extract is applied to one point in the gel, and a potential

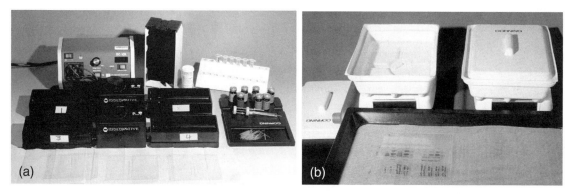

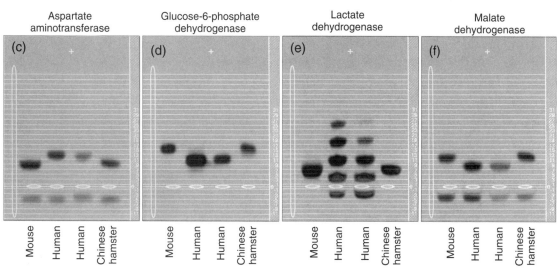

*Fig. 15.12. Isoenzyme Electrophoresis.* Analysis of four isoenzymes by the Authentikit (Innovative Chemistry) gel electrophoresis system. (*a*) A four-tank setup with power pack at rear, reagents on right, and three precast gels in the foreground. (*b*) Staining and washing trays (Corning) with developed gel in foreground. (*c–f*) Images from developed electropherograms. (Photos and electropherograms courtesy of ATCC.)

difference is applied across the gel. The different isoenzymes migrate at different rates and can be detected later by staining with chromogenic substrates. Stained gels can be read directly by eye and photographed, or scanned with a densitometer.

Protocol 15.10 for isoenzyme analysis using the Authentikit agarose gel system has been contributed by Jane L. Caputo and Yvonne Reid, American Type Culture Collection, 10801 University Blvd., Manassas, Virginia 20110, USA.

### 15.10.2 Isoenzyme Electrophoresis with Authentikit

The species of origin of a cell line can be determined with the Authentikit gel electrophoresis system, which can be used to determine the mobility of seven isoenzymes: nucleoside phosphorylase (NP; E.C. 2.4.2.1), glucose-6-phosphate dehydrogenase (G6PD; E.C. 1.1.1.49), malate dehydrogenase (MD, E.C. 1.1.1.37), lactate dehydrogenase (LD; E.C. 1.1.1.27), aspartate aminotransferase (AST, E.C. 2.6.1.1), mannose-6-phosphate isomerase (MPI, E.C. 5.3.1.8), and

peptidase B (Pep B, E.C. 3.4.11.4). This system allows easy screening of up to six cell lines per gel for seven genetic markers in less than three hours [Hay, 2000]. In most cases the species of origin can be determined by using only four of the seven isoenzymes listed: nucleoside phosphorylase, glucose-6-phosphate dehydrogenase, malate dehydrogenase, and lactate dehydrogenase. Similarly interspecies cell line cross-contamination can be detected in most cases by using these four isoenzymes [Nims et al., 1998], although peptidase B was also needed to analyze a contamination of a mouse cell line with a hamster cell line.

---

### PROTOCOL 15.9. ISOENZYME ANALYSIS

**Outline**

Harvest the cells, wash them in D-PBSA, resuspend them at $5 \times 10^7$ cells/mL, and prepare the cell extract. Store the extract at $-70°$C. Prepare the gel apparatus.

Apply 1 to 2 µL of the cell extract, standard, and control to the agarose gels. Fill the chambers with water, place the agarose gels in the chambers, and run electrophoresis for 25 min. Apply enzyme reagents, incubate the gels for 5 to 20 min, rinse the gels, dry them, and examine the finished gels showing enzyme zones.

### Materials

*Sterile or aseptically prepared:*

- ❑ Cells
- ❑ Extracts (Innovative Chemistry, Inc.)
- ❑ Standard, L929 extract
- ❑ Control, HeLa-S3 extract
- ❑ Cell extraction buffer (Innovative Chemistry, Inc.) or Triton X-100 extract solution: 1:15, v/v, Triton X-100 in 0.9% NaCl, pH 7.1, containing 0.66 mM EDTA (store at 4°C)

*Nonsterile:*

- ❑ Authentikit apparatus and reagents (all from Innovative Chemistry, Inc.)
- ❑ Agarose gel films, SAB 8.6, one for each enzyme tested
- ❑ Incubation tray liners, one for each enzyme tested
- ❑ Incubation trays
- ❑ Stain and wash trays
- ❑ Temperature-controlled chamber cover and base
- ❑ Power supply (160 V DC) with timer
- ❑ Safety interconnector
- ❑ Incubation chamber/dryer
- ❑ Magnetic stirrer with 1-in.-long stir bar
- ❑ Buffer, SAB 8.6
- ❑ Enzyme Stabilizer Innovative Chemistry (cat. no. R578, 10 mL; cat. no. R-590, 5 mL)
- ❑ 1 × Triton (ATCC® Cat# MD-8022-TB)
- ❑ Enzymatic substrates, 1 vial of each:
- ❑ Nucleoside phosphorylase (NP)
- ❑ Glucose-6-phosphate dehydrogenase (G6PD)
- ❑ Malate dehydrogenase (MD)
- ❑ Lactate dehydrogenase (LD)
- ❑ Aspartate aminotransferase (AST)
- ❑ Mannose-6-phosphate isomerase (MPI)
- ❑ Peptidase B (Pep B)
- ❑ Sample applicator Teflon tips®
- ❑ Gel documentation form
- ❑ Enzyme migration data form
- ❑ Cell I.D. final analysis form
- ❑ Microliter syringe
- ❑ Microcentrifuge (Eppendorf, Brinkmann)
- ❑ Eppendorf tubes, 1.5 mL
- ❑ Pipettor, 1.0 mL
- ❑ Pipettor tips
- ❑ Pipettes, 5.0 mL
- ❑ Graduated cylinder, 100 mL
- ❑ Marking pen
- ❑ Deionized water
- ❑ Distilled water
- ❑ D-PBSA
- ❑ Protective gloves
- ❑ Container for disposal of pipettor tips

### Procedure

#### A. Preparation of Extract

1. Grow cells in the conventional manner to minimum of $5 \times 10^6$ viable cells.
2. Harvest cells according to the procedure recommended for the particular cell line.
3. Resuspend the cell pellet in D-PBSA, and count the number of viable cells.
4. Centrifuge the suspension at 300 $g$ for 5 to 10 min to pellet the cells and decant the supernate.
5. Repeat steps 3 and 4 for a total of three washes.
6. Add 12.5 µL of D-PBSA (use for $5 \times 10^6$ viable cells).
7. Vortex the tube for 1 to 2 s until cell pellet is broken up.
8. Add 12.5 µL of 2X Triton-EDTA to cell suspension.
9. Gently pipette up and down to shear the cells. DO NOT VORTEX!
10. Briefly centrifuge (microfuge) cells at 4°C at maximum speed for 2 min.
11. Add equal volume of enzyme stabilizer solution to the extract/lysate.
12. Mix the lysate by gently pipetting. DO NOT VORTEX!
13. Divide extract to the desired volume and store at −70°C

#### B. Setup of Electrophoresis Apparatus

1. Turn on the incubation chamber.
2. Place one tray liner in each incubator tray.
3. Add 6.0 mL of deionized water to each incubator tray, and place the trays at 37°C for 20 min.
4. Place 95 mL of SAB 8.6 buffer in each chamber of the electrophoresis cell base (a total of 190 mL of buffer per base). This buffer should not be reused, as the pH changes significantly during the electrophoretic run.
5. Connect the filled cell base to the safety interconnector, which is plugged into a voltage-regulated power supply. The power supply should be plugged into a grounded electrical outlet.
6. Fill each temperature-controlled electrophoresis cell cover with 500 mL of cold water (4–10°C).

The cell cover must be filled when it is in a vertical position, or else the water will run out. Cool the water with ice, or store a sufficient supply of cold water in a standard refrigerator. Replace this water before each run.

7. Place 500 mL of deionized or distilled water in each stain, and wash the tray with it. This water must not be reused.

## C. Electrophoresis Procedure

1. Remove agarose gel from cold room just before electrophoresis and carefully label each agarose gel to be used. This can be done with labels available from Innovative Chemistry, Inc., or by writing on the back of the gel with a permanent marking pen.

2. Place the gel on the bench, with the plastic nipples down. Orient the gel so that the sample application wells are closest to you. Wear gloves while handling agarose gels.

3. Label each well with the identity of the sample to be tested (standard, control, unknown 1, unknown 2, etc.). Six unknowns can be tested on each gel.

4. Gently peel the agarose gel from its hard plastic cover. Be careful to handle the gel only by its edges. Discard the hard plastic cover.

5. Add the cell extracts to the sample wells. Use the dispenser with a Teflon tip attached to dispense exactly 1 µL of cell extract into each well. Use a fresh tip for each sample. Dispense the standard extract into track 1, the control into track 2, and the unknowns in tracks 3 to 8. Only the drop of cell extract should touch the well, not the tip, in order to avoid damaging the agarose. If 2 µL of cell extract are required, allow the first 1 µL to diffuse into the agarose before applying the second.

6. Insert each loaded agarose film into a cell cover. The agarose side must face outward. Match the anode (+) side of the agarose film with the anode (+) side of the cell cover. It may be necessary to bend the agarose gel film slightly to insert it into the cell cover.

7. Place each cell cover on a cell base. The black end with the magnet inside must be nearest to the power supply. Turn on the power supply (160 V), and set the timer for 25 min.

8. Remove each reagent substrate vial from the refrigerator about 5 min before the end of the run, and allow it to warm to room temperature.

9. Add 0.5 to 1 mL of deionized water to each reagent vial immediately before use. Swirl the vial gently to dissolve the reagent.

10. When the electrophoresis is finished, remove the cell cover from the cell base and place it on absorbent paper.

## D. Staining Procedure

1. To remove the agarose gel film from the cell cover, grasp the film by its edges, squeeze inward, and remove it from the cover.

2. Place the agarose gel film, gel side up, on absorbent paper on a flat surface, with the wells placed toward you.

3. Carefully blot the residual buffer from both ends of the agarose film with a lint-free tissue.

4. Place a 5-mL pipette along the lower edge of the agarose gel film.

5. Pour the reconstituted substrate evenly onto the agarose film along the leading edge of the pipette.

6. With a single, smooth motion, push the pipette across the surface of the agarose. Drag the pipette back toward you, and push it across the surface one more time. Roll it off the end of the agarose, removing the excess substrate in the process. Be very careful not to damage the agarose. No pressure is necessary to perform this step.

7. Place the agarose gel film, agarose side up, into a prewarmed incubator tray, and place the tray in a 37°C incubator for approximately 5 to 20 min.

8. After incubation, wash the agarose gel film twice in 500 mL of double-distilled or deionized water for 15 min each time, with agitation provided by a magnetic stirring bar. After the first 15 min, remove each gel from the water, discard the water, add 500 mL of fresh water to the dish, and immerse the gel in the water. Cover the gel to protect it from light. Be sure that the gel is immersed and not floating on top of the wash water.

9. Remove the gel from the water, and place it on a drying rack in the drying chamber of the incubator or oven. Dry it for 30 min or until the agarose is dry. Alternatively, agarose gel films will dry at room temperature overnight.

10. To clean up, pour out the buffer from the cell base, and rinse the cell base with distilled water. Pour the water out of the cell cover, rinse the inside of the cell cover, and allow it to dry.

11. Evaluate your results. The bands are permanent, and the films can be kept for future reference. If a background staining develops over time, the gels were not sufficiently washed.

**TABLE 15.3.** Antibodies Used in Cell Line Recognition

| Cell type | Antibody to | Localization | Specificity | Comments |
|---|---|---|---|---|
| Anterior pituitary | hGH | Golgi and secreted | High | |
| Astroglia | GFAP | Intermediate-filament cytoskeleton | High | (*See also* Table 22.3) |
| B-cells | CD22 | Cell surface | High | Co-receptor for B-cell receptor (BCR) |
| B-cells, immature | CD20 | Cell surface | High | Not in plasma cells |
| Breast epithelium | α-Lactalbumin, casein | Golgi and secreted | High | Needs differentiation induction |
| Colorectal and lung adenocarcinoma | CEA | Cell surface and Golgi | Intermediate | Cell adhesion molecule |
| Endothelium | Factor VIII | Weibl-Palade bodies | High | Granular appearance diagnostic |
| | V-CAM | Cell surface | High | Cell adhesion molecule |
| Epithelium | L-CAM | Cell surface | High | Cell adhesion molecule |
| | Cytokeratin | Intermediate-filament cytoskeleton | High | Some antibodies also stain mesothelium; specific antibodies for different epithelia |
| | EMA | Cell surface | High | Same antigen as HMFG |
| | HMFG I & II | Cell surface | High | Same antigen as EMA |
| Fetal hepatocytes | AFP | Golgi and secreted | High | Also expressed by hepatomas |
| Hematopoietic stem cells | CD34 | Cell surface | High | Also in precursors but not mature cells |
| Hepatocytes | Albumin | Golgi and secreted | High | Needs differentiation |
| Mesodermal cells | Vimentin | Intermediate-filament cytoskeleton | Low | Also expressed in epithelial and glial cells in culture |
| Mesothelium | Mesothelial cell antigen | Cell surface | High | |
| Monocytes and macrophages | CD14 | Cell surface | High | |
| Myeloid cells | CD13 | Cell surface | High | |
| Myocytes | Desmin | Intermediate-filament cytoskeleton | High | |
| Neural and neuroendocrine cells | NSE | Cytoplasmic | Intermediate | Can be expressed in SCLC |
| Neural cells | N-CAM | Cell surface | Intermediate | Also expressed in some SCLC |
| Neuroendocrine lung and stomach | GRP | Golgi and secreted | Intermediate | Also expressed in some SCLC |
| Neuronal cells | Neorofilament protein | Intermediate-filament cytoskeleton | High | Also stains some SCLC |
| Oligodendrocytes | Myelin basic protein | Cell surface | High | Also stains peripheral neurons |
| | Gal-C | Cell surface | High | (*See also* Table 22.3) |
| Placental epithelium | hCG | Golgi and secreted | Low | Also expressed in some lung tumors |
| Prostatic epithelium | PSA | Golgi and secreted | High | |
| Stem cells | ABC Transporter | Transmembrane | Moderate | Also expressed in some drug-resistant cells |
| | SSEA-1 | Cell surface | High | |
| | NANOG | Transcription factor | High | |
| | OCT-4 | Transcription factor | High | |
| T-cells | CD3 | Cell surface | High | |
| T-cells and endothelium | I-CAM | Cell surface | Intermediate | |

Note: This list is of examples only and is not meant to be comprehensive, as there are now so many antigenic markers (*see* Appendix II, Antibodies).
*Abbreviations:* GFAP, glial fibrillary acidic protein; EMA, epithelial membrane antigen; HMFG, human milk fat globule protein; AFP, α-fetoprotein; CEA, carcinoembryonic antigen; SCLC, small-cell lung cancer; hCG, human chorionic gonadotropin; I-CAM, intercellular cell adhesion molecule; NSE, neuron specific enolase; PSA, prostate-specific antigen; Gal-C, galactocerebroside; GRP, gastrin-releasing peptide.

*Analysis of electropherogram.* Attach the dried gel to the gel documentation form; line up the sample wells at the origin, and measure them. The bright yellow backgrounds of the forms are millimeter lined and give maximum contrast and enhance the purple bands on the dry gel (*see* Fig. 15.12). Measure the enzyme migration distance by measuring from the middle of the application zone to the middle of the enzyme zone. Record measured distances for the standard, the control, and the unknowns on the enzyme migration data form. Transfer the enzyme migration data to the cell I.D. final analysis form to confirm the species identification. Detailed instructions for identifying unknowns are included with the forms. Keep these forms in your notebook as a permanent record of your results. A detailed discussion of each enzyme can be found in the *Handbook for Cell Authentication and Identification*, available from Innovative Chemistry, Inc.

*Alternative extraction techniques.* Cell extracts can be prepared by ultracentrifugation, freezing and thawing rapidly three times, or treatment with octyl alcohol [Macy, 1978]. A cell extraction buffer is also available from Innovative Chemistry, Inc.

The control and standard may be prepared by using the standard procedure (*see* Protocol 15.10) for preparation of extract. The standard is prepared from the mouse L929 cell line (ATCC CCL-1), and the control is prepared from the human HeLa cell line (ATCC CCL-2). O'Brien et al. [1980] reported that samples could be stored at $-70°C$ for up to a year without substantial loss of enzyme activity.

## 15.11 ANTIGENIC MARKERS

Immunostaining and ELISA assays are among the most useful techniques available for cell line characterization (Table 15.3) facilitated by the abundance of antibodies and kits available from commercial suppliers (*see* Appendix II). Regardless of the source of the antibody, however, it is essential to be certain of its specificity by using appropriate control material. This is true for monoclonal antibodies and polyclonal antisera alike; a monoclonal antibody is highly specific for a particular epitope, but the specificity of the expression of the epitope to a particular cell type must still be demonstrated.

### 15.11.1 Immunostaining

Antibody localization is determined by fluorescence, wherein the antibody is conjugated to a fluorochrome, such as fluorescein or rhodamine, or by the deposition of a precipitated product from the activity of horseradish peroxidase or alkaline phosphatase conjugated to the antibody [Polak & Van Noorden, 2003]. Immunological staining may be direct—in that the specific antibody is itself conjugated to the fluorochrome or enzyme and used to stain the specimen directly. Usually, however, an indirect method is used wherein the primary (specific) antibody is used in its native form to bind to the antigen in the specimen, followed by treatment with a second antibody, raised against the immunoglobulin of the first antibody. The second antibody may be conjugated to a fluorochrome and visualized on a fluorescence microscope (*see* Plates 11*a*, 15*d*, *e*, 20*c*–*f*), or to peroxidase and visualized on a conventional microscope, after development with a chromogenic peroxidase substrate (*see* Plates 11*b*–*d*, 12*a*, *b*).

Various methods have been used to enhance the sensitivity of detection of these methods, particularly the peroxidase-linked methods. In the peroxidase–anti-peroxidase (PAP) technique, a further amplifying tier is added by reaction with peroxidase conjugated to anti-peroxidase antibody from the same species as the primary antibody [Polak & Van Noorden, 2003]. Even greater sensitivity has been obtained by using a biotin-conjugated second antibody with streptavidin conjugated to peroxidase or alkaline phosphatase (GE Healthcare) or gold-conjugated second antibody (Janssen) with subsequent silver intensification.

---

**PROTOCOL 15.10. INDIRECT IMMUNOFLUORESCENCE**

**Outline**
Fix the cells, and treat them sequentially with first and second antibodies. Examine the cells by UV light.

**Materials**

*Sterile or aseptically prepared*

❑ Culture grown on glass coverslip, chambered slide, or polystyrene Petri dish

*Nonsterile*

❑ Freshly prepared fixative: 5% acetic acid in ethanol (place at $-20°C$)
❑ Primary antibody diluted 1:100 to 1:1000 in culture medium with 10% FBS
❑ Second antibody raised against the species of the first (e.g., if the first antibody was raised in rabbit, then the second should be from a different species, e.g., goat anti-rabbit immunoglobulin); the second antibody should be conjugated to fluorescein or rhodamine
❑ Swine serum or another blocking agent
❑ D-PBSA
❑ Mountant: 50% glycerol in D-PBSA and containing a fluorescence-quenching inhibitor (Vecta)

**Procedure**

1. Wash the coverslip with cells in D-PBSA, and place it in a suitable dish—such as a 24-well plate for a 13-mm coverslip.
2. Place the dish at $-20°C$ for 10 min, add cold fixative to it, and leave it for 20 min.

3. Remove the fixative, wash the coverslip in D-PBSA, add 1 mL of normal swine serum, and leave the dish at room temperature for 20 min.
4. Rinse the coverslip in D-PBSA, drain it on paper tissue, and place the coverslip, inverted, on a 50-μL drop of diluted primary antibody.
5. Place the coverslip at 37°C for 30 min at room temperature for 1 to 3 h or overnight at 4°C. For the 4°C incubation the antibody may be diluted 1:1000.
6. Rinse the coverslip in D-PBSA, and transfer it to the second antibody, diluted 1:20, for 20 min at 37°C.
7. Rinse the coverslip in D-PBSA, and mount it on a slide in 50% glycerol in D-PBSA with fluorescence bleaching retardant (Vecta).
8. Examine the slide on a fluorescence microscope.

### Variations

*Fixation.* For the cell surface or, particularly, for fixation-sensitive surface antigens, treat the cells with antibodies first, and then postfix as in step 2. When a glass substrate is used, cold acetone may be substituted for acid ethanol.

*Indirect peroxidase.* Substitute peroxidase-conjugated antibody for the fluorescent antibody at stage 6, and then transfer it to peroxidase substrate. (Diaminobenzidine-stained preparations can be dehydrated and mounted in DPX, but ethyl carbazole must be mounted in glycerol as in step 7.)

*PAP.* Use peroxidase-conjugated second antibody at stage 6, and then transfer the coverslip to diluted PAP complex (1:100) (most immunobiological suppliers; e.g., Dako, Vecta) in D-PBSA for 20 min. Rinse the coverslip, and add peroxidase substrate as for indirect peroxidase. Incubate, wash, and mount the coverslip.

*Cell surface markers.* Specific cell surface antigens are usually stained in living cells (at 4°C in the presence of sodium azide to inhibit pinocytosis), whereas intracellular antigens are stained in fixed cells, sometimes requiring light trypsinization to permit access of the antibody to the antigen.

HLA and blood-group antigens can be demonstrated on many human cell lines and serve as useful characterization tools. HLA polymorphisms are particularly valuable, especially when the donor patient profile is known [Pollack et al., 1981; Hurley & Johnson, 2001].

### 15.11.2 Immunoanalysis

Assay of cell extracts can be performed by ELISA assay on a microtitration plate array with antibody-coated wells. Assays are provided in kit form (e.g., Assay Designs; R&D) and allow quantitation of marker and product protein expression.

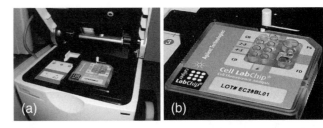

**Fig. 15.13. Agilent Immunoanalyzer.** Agilent 2100, which can handle the simultaneous analysis of multiple immunofluorescent stained samples. (*a*) Sample compartment; (*b*) sample holder. (Courtesy of PromoCell.)

Antigen expression analysis can also be automated and quantitated by flow cytometry (*see* Sections 14.4, 20.1.4), by analysis of multiple cell streams (Agilent 2100; Fig. 15.13), or by antibody microarray analysis (e.g., Affymetrix) [Barber et al., 2009].

The introduction of confocal microscopy provided both higher resolution and 3D analysis to immunostained material. As these systems are capable of restricting the image that is viewed and recorded to a very thin optical section, each focal plane is viewed with much greater clarity in lacking the interference of adjacent layers. Furthermore, as each optical section can be digitally stored, serial progression through a specimen provides data that can be analyzed automatically to general a so-called Z-section, a reconstructed vertical section through the specimen created at any point in the horizontal field. This enables cells to be viewed in suspension in collagen gels, Matrigel, or any other transparent gel (*see also* Section 17.6.3).

Fluorescence microscopy is now able to resolve with nanometric precision with probe-based super-resolution imaging using photoactivation localization microscopy (PALM) where excitation is brief and transient followed immediately by bleaching giving very high resolution as only a few fluorophores are activated and recorded at a time and lack the interference of surrounding fluorophores. A high resolution composite image is then synthesized from these brief transient emissions [Lippincott-Schwartz & Manley, 2009].

## 15.12 DIFFERENTIATION

Many of the characteristics described under antigenic markers or enzyme activities may also be regarded as markers of differentiation, and as such they can help to correlate cell lines with their tissue of origin as well as define their phenotypic status (*see also* Section 16.6; Chapter 22; Table 2.1). Although sometimes constitutively expressed (e.g., melanin in B16 melanoma or Factor VIII in endothelial cells), expression of differentiated lineage markers may need to be induced before detection is possible (*see* Section 16.7).

# CHAPTER 16

# Differentiation

## 16.1 EXPRESSION OF THE IN VIVO PHENOTYPE

The phenotype of cells cultured and propagated as a cell line is often different from that of the predominating cell type in the originating tissue (*see* Section 2.4.2). This is due to several factors that regulate the geometry, growth, and function in vivo, but that are absent from the in vitro microenvironment (*see* Section 16.1.1). *Differentiation* is the process leading to the expression of phenotypic properties characteristic of the functionally mature cell in vivo. This may be irreversible, such as the cessation of DNA synthesis in the erythroblast nucleus, neuron, or mature keratinocyte, or reversible, such as in the dedifferentiation of mature hepatocytes into precursors during liver regeneration. Some of the properties of the differentiated cell are adaptive, such as albumin synthesis in differentiated hepatocytes, which is often lost in culture but can be reinduced. Differentiation as used in this text describes the combination of *constitutive* (stably expressed without induction) and *adaptive* (subject to positive and negative regulation of expression) properties found in the mature cell. *Commitment* implies an irreversible transition from a stem cell to a particular defined lineage endowing the cell with the potential to express a limited repertoire of properties. The concept of commitment has nevertheless been called into question by the demonstration that some precursor cells can revert or convert to multipotent stem cells [Kondo & Raff, 2000, 2004; Le Douarin et al., 2004] (*see* Section 16.4) and even fully mature cells can be reprogrammed to form pluripotent stem cells (*see* Sections 16.5, 23.3.5).

*Terminal differentiation* implies that a cell has progressed down a particular lineage to a point at which the mature phenotype is fully expressed and beyond which the cell cannot progress. This stage may be reversible in some cells, such as fibrocytes, that can revert to a less differentiated phenotype, or even a stem cell, and resume proliferation, or irreversible in cells like erythrocytes, neurons, skeletal muscle, or keratinized squames.

### 16.1.1 Dedifferentiation

*Dedifferentiation* has been used to describe the loss of the differentiated properties of a tissue when it becomes malignant or when it is grown in culture. As dedifferentiation comprises complex processes with several contributory factors, including cell death, selective overgrowth, and adaptive responses, the term should be used with caution. When used correctly, dedifferentiation means the loss by a cell of the specific phenotypic properties associated with the mature cell. When dedifferentiation occurs, it may be either an adaptive process, implying that the differentiated phenotype may be regained given the right inducers (*see also* Section 2.4), or a selective process, implying that a precursor cell has been selected because of its greater proliferative potential. In either case the precursor cell may be induced to mature to the fully differentiated cell or even revert to a stem cell (*see* Section 16.5), given the right stimulus.

### 16.1.2 Lineage Selection

If the wrong lineage has been selected (e.g., stromal fibroblasts from liver instead of hepatocytes), no amount of induction can bring back the required phenotype. In the past this failure was often erroneously attributed to dedifferentiation but was more likely due to overgrowth of stromal fibroblasts induced by growth factors such as platelet-derived growth factor (PDGF) and inhibition of proliferation in epithelial cells by transforming growth factor β (TGF-β) as a result of culture in serum.

## 16.2 STAGES OF DIFFERENTIATION

There are two main pathways to differentiation in the adult organism. In constantly renewing tissues, like the epidermis, intestinal mucosa, and blood, a small population of stem cells, capable of self-renewal, gives rise, on demand, to precursor cells that will proliferate and progress toward terminal differentiation, losing their capacity to divide as they reach the terminal stages (*see* Fig. 2.6). This process gives rise to mature, differentiated cells that normally will not divide. Proliferation in the precursor cell compartment is regulated by feedback to generate the correct size of the differentiated cell pool.

In tissues that do not turn over rapidly, but replenish themselves in response to trauma, the resting tissue shows little proliferation; however, the mature cells may reenter division. In connective tissue, for example, cells such as fibrocytes may respond to a local reduction in cell density and/or the presence of one or more growth factors by losing their differentiated properties (e.g., collagen synthesis) and reentering the cell cycle. When the tissue has regained the appropriate cell density by division, cell proliferation stops and differentiation is reinduced. This type of renewal is rapid because a relatively large population of cells is recruited.

It is not clear whether the cells that reenter the cell cycle to regenerate the tissue are phenotypically identical to the bulk of the differentiated cell population, or whether they represent a subset of reversibly differentiated cells. In liver, which responds to damage by regeneration, it appears that the mature hepatocyte can reenter the cell cycle, whereas in skeletal muscle, where terminally differentiated cells cannot reenter the cycle, regenerating cells are derived from the satellite cells, which appear to form a quiescent stem cell population (*see* Section 22.3.3). Mature fibrocytes, blood vessel endothelial cells, and glial cells appear to be able to reenter the cell cycle to regenerate, but the possibility of a regenerative subset within the total population cannot be ruled out.

## 16.3 PROLIFERATION AND DIFFERENTIATION

As differentiation progresses, cell division is reduced and eventually ceases. In most cell systems, cell proliferation is incompatible with the expression of differentiated properties (*see* Fig. 2.5). Tumor cells can sometimes break this restriction, and in melanoma, for example, melanin continues to be synthesized while the cells are proliferating [Halaban, 2004]. Even in these cases, however, synthesis of the differentiated product increases when division stops.

There are severe implications for this relationship in culture, where expansion and propagation are often the main requirements, and it is therefore not surprising to find that the majority of cell lines do not express fully differentiated properties. This fact was noted many years ago by the exponents of organ culture (*see* Section 25.2), who set out to retain three-dimensional, high-cell-density tissue architecture and to prevent dissociation and selective overgrowth of undifferentiated cells. However, although of considerable value in elucidating cellular interactions regulating differentiation, organ culture has always suffered from the inability to propagate large numbers of identical cultures, particularly if large numbers of cells are required. In addition, the heterogeneity of the sample, assumed to be essential for the maintenance of the tissue phenotype, has in itself made the ultimate biochemical analysis of pure cell populations and of their responses extremely difficult.

Hence in recent years there have been many attempts to reinduce the differentiated phenotype in pure populations of cells by recreating the correct environment and, by doing so, to define individual influences exerted on the induction and maintenance of differentiation. This process usually implies the cessation of cell division and the creation of an interactive, high-density cell population, as in histotypic or organotypic culture. Cell−cell interaction has become a key feature in establishing engineered tissues (*see* Sections 16.7.1, 25.4.1, 25.4.2).

## 16.4 COMMITMENT AND LINEAGE

Progression from a stem cell to a particular pathway of differentiation traditionally implied an increase in commitment, with advancing stages of progression (*see* Fig. 2.6). A hematopoietic stem cell, after commitment to lymphocytic differentiation, would not change lineage at a later stage and adopt myeloid or erythrocytic characteristics. Similarly a primitive neuroectodermal stem cell, once committed to become a neuron, would not change to a glial cell. Commitment was regarded as the point between the stem cell and a particular precursor stage where a cell or its progeny can no longer transfer to a separate lineage. If such irreversible commitment exists it must occur much later than previously thought, as some precursor cells can revert to stem cells with multilineage potential, and even fully mature cells can be made to revert to stem cells status with appropriate genetic or epigenetic manipulation (*see* Sections 16.5, 23.3.5).

Many claims have been made in the past regarding cells transferring from one lineage to another. Perhaps the most substantiated of these claims is that of the regeneration of the amphibian lens by recruitment of cells from the iris [Clayton et al., 1980; Cioni et al., 1986]. As the iris can be fully differentiated and still regenerate lens, this claim has been proposed as *transdifferentiation*. Following the demonstration that mature cells can be reprogrammed into pluripotent stem cells (*see* Section 16.5) it now seems possible that, with the right transcription factors, direct conversion from one lineage to another is possible without reversion to a stem cell [Vierbuchen et al., 2010]. Transdifferentiation may occur in transformed cells but there are other possible explanations for phenotypic changes in tumors and cultures derived from them. For example, small-cell carcinoma of the lung has been found to change to squamous or large-cell carcinoma following relapse after the completion of chemotherapy. Whether this implies that one cell type, the Kulchitsky cell [de Leij et al., 1985], presumed to give rise to small-cell lung carcinoma, changed its commitment, or whether it implies that the tumor originally derived from a multipotent stem cell and on recurrence progressed down a different route, is still not clear [Gazdar et al., 1983; Goodwin et al., 1983; Terasaki et al., 1984]. Similarly advanced anaplastic squamous skin cancers can give rise to cells that appear mesenchymal [Oft et al., 2002] and this so-called epithelial-mesenchymal transition has been reported for other tumors [Gregory et al., 2008] and an important element of acquiring an invasive phenotype.

The K562 cell line was isolated from a myeloid leukemia but subsequently was shown to be capable of erythroid differentiation [Andersson et al., 1979b]. Rather than being a committed myeloid progenitor converting to erythroid, the tumor probably arose in the common stem cell known to give rise to both erythroid and myeloid lineages. For some reason, as yet unknown, continued culture favored erythroid differentiation rather than the myeloid features seen in the original tumor and early culture. In some cases, again in cultures derived from tumors, a mixed phenotype may be generated. For instance, the $C_6$ glioma of rat expresses both astrocytic and oligodendrocytic features, and these features may be demonstrated simultaneously in the same cells [Breen & De Vellis, 1974].

In general, these cases are unusual and are restricted to tumor cultures. Most cultures from normal tissues, although they may differentiate in different directions, do not alter to a different lineage. This raises the question of the actual status of cell lines derived from normal tissues. Most cultures are derived from (1) stem cells, or early progenitor cells, which may differentiate in one or more different directions; (2) late precursor cells, which may stay true to the lineage; or (3) differentiated cells, such as fibrocytes, which may dedifferentiate and proliferate, but still retain lineage fidelity (*see* Section 2.7). Some mouse embryo cultures, loosely called fibroblasts (e.g., the various cell lines designated 3T3), probably more correctly belong to category (1), as they can be induced to become adipocytes, muscle cells, osteocytes, and endothelium, as well as fibrocytes.

Cell lines are perhaps best regarded as a mixed population of stem cells, precursor cells, and differentiated cells, the balance being determined by soluble or contact mediated signals in the environment. In most propagated cell lines the majority of cells will have the precursor phenotype, but

may retain the plasticity to shift to either a stem cell or differentiated phenotype, given the correct signals.

There are now some well-described examples in which progenitor cells (e.g., the O2A common progenitor of the oligodendrocyte and type 2 astrocyte in the brain, which remains a proliferating progenitor cell in a mixture of PDGF and bFGF) will differentiate into an oligodendrocyte in the absence of growth factors or serum, or into a type 2 astrocyte in fetal bovine serum or a combination of ciliary neurotropic factor (CNTF) and FGF-2 [Raff et al., 1978; Raff, 1990]. However, the cells that have become oligodendrocyte precursors can still revert to a common progenitor phenotype when exposed to BMPs [Kondo & Raff, 2000, 2004]. Cardiac muscle cells remain undifferentiated and proliferative in serum and bFGF but differentiate in the absence of serum [Goldman & Wurzel, 1992], and primitive embryonal stem cell cultures differentiate spontaneously unless kept in the undifferentiated proliferative phase by bFGF, SCF (stem cell factor, Steel factor, kit ligand), and LIF (lymphocyte inhibitory factor) [Matsui et al., 1992].

Hence with the advent of more defined media and the identification of more differentiation-inducing factors, it is gradually becoming possible to define the correct inducer environment that will maintain cells in a stem-like, progenitor, or differentiated state.

## 16.5  STEM CELL PLASTICITY

Conventional stem cell theory predicts that the more primitive a stem cell, the greater is its potency. A *unipotent* stem cell is a cell that will give rise to only one lineage, for example, stem cells in the basal layer of the epidermis, which give rise to keratinocytes. A *multipotent* stem cell will generate more than two lineages, for example, the hemocytoblast in the bone marrow, which gives rise to granulocytes, lymphocytes, monocytes, megakaryocytes, mast cells, and erythrocytes. A *pluripotent* stem cell may give rise to several different lineages from different germ layers [Smith, 2006]. A *totipotent* stem cell will give rise to all known cell types [Mitalipov & Wolf, 2009] such as an embryonal stem cell (ES cell), which, by current dogma, is the only cell that can give rise to every cell lineage. Originating in the inner cell mass of the early embryo, it seems reasonable that ES cells should be totipotent. It has now been shown that mature cells from a differentiated compartment, such as fibroblasts from the dermis, may be reprogrammed to produce what have been called *induced pluripotent stem cells* (iPS cells) [Nakagawa et al., 2007; Yu et al., 2007; Aasen et al., 2008; Huangfu et al., 2008; Takahashi et al., 2007] that are capable of differentiating into several different lineages (*see* Fig. 2.7); it remains to be seen whether these are genuinely totipotent.

The etymology is confusing here as *multi*potent and *pluri*potent have similar meanings (*multi* = many, *pluri* = several) but in the current context multipotent appears to be reserved for a progenitor cell within a lineage (e.g., hematopoietic or neural) capable of making limited choices at a branch point in that lineage, such as a neural stem cell that may differentiate into a neuron, astrocyte, or oligodendrocyte (*see* Fig. 2.7), while pluripotent has been used to imply a more primitive stem cell capable of differentiating down several dissimilar lineages (endodermal, mesodermal, ectodermal, or neurectodermal), much like an embryonal stem cell.

The other traditional concept is that the greater the degree of commitment, the greater is the likelihood of the stem cell being located in a specific tissue; for example, hematopoietic stem cells are located in the bone marrow, enterocyte stem cells at the bottom of intestinal crypts, and keratinocyte stem cells in the epidermis. With this commitment and histological localization comes a reduction in potency, and one would expect that, for example, stem cells in the liver will only make liver cells (hepatocytes, bile duct cells). Furthermore tissues that do not regenerate, such as neurons in the central nervous system, would not be expected to possess stem cells. However, this tidy concept of lineage commitment, potency, and tissue localization is currently being questioned [Vescovi et al., 2002] by results that show that (1) tissues that are nonregenerative, such as neurons in the brain, do have stem cells [Pevny & Rao, 2003] and that (2) tissue localization does not necessarily mean lineage commitment and reduced potency, as liver stem cells can generate neurons [Deng et al., 2003], bone marrow stem cells can generate cardiac muscle [Mangi et al., 2003], muscle stem cells can generate hematopoietic cells [Cao et al., 2003], and neural stem cells can generate endothelium [Wurmser et al., 2004]. In some cases differentiation into an unpredicted lineage, such as bone marrow cells generating hepatocytes or cardiac muscle or neurons [Alison et al., 2004; Alvarez-Dolado et al., 2003], may be due to cell fusion [Greco & Recht, 2003], but others have claimed that cell fusion is not involved [Wurmser et al., 2004]. It the light of subsequent studies on iPS cells it seems more likely that these cells are being reprogrammed by being transferred from one microenvironmental niche to another.

The reprogramming of mature cells into iPS cells currently requires transfection with genes such as OCT4, NANOG, and SOX2 (*see* Section 23.3.5), but there is evidence that chemical regulators such as valproic acid (2-propylpentanoic acid), which changes chromatin rearrangement via inhibition of histone deacetylase, can reduce the number of transfected genes required to induce iPS cells [Huangfu et al., 2008]. Furthermore oligodendroglial precursors can be induced to differentiate and revert to a common neural stem cell capable of redifferentiating into mature astroglia, oligodendroglial, and neurons by treatment with growth factors such as BMP, FGF-2, again implicating chromatin remodeling [Kondo & Raff, 2004]. Similarly endothelin-3 can induce dedifferentiation in melanocytes and Schwann cells and generate a common precursor that can redifferentiate into glia and melanocytes [Dupin et al., 2000, 2003; Le Douarin et al., 2004]. This raises the interesting prospect that iPS cells may eventually be generated without the transfection of potentially harmful genes [Zhao & Daley, 2008]. Not only would these cells provide valuable homografts for transplantation and tissue repair, but they would also provide, from people with genetic abnormalities, a source of cells that could used to investigate the mechanism by which the phenotypic abnormalities arise.

## 16.6 MARKERS OF DIFFERENTIATION

Markers expressed early and retained throughout subsequent maturation stages are generally regarded as lineage markers—for example, intermediate filament proteins, such as the cytokeratins (epithelium) [Moll et al., 1982], or glial fibrillary acidic protein (astrocytes) [Eng & Bigbee, 1979; Bignami et al., 1980]. Markers of the mature phenotype representing terminal differentiation are more usually specific cell products or enzymes involved in the synthesis of these products—for example, hemoglobin in an erythrocyte, serum albumin in a hepatocyte, transglutaminase [Schmidt et al., 1985] or involucrin [Parkinson & Yeudall, 2002] in a differentiating keratinocyte, and glycerol phosphate dehydrogenase in an oligodendrocyte [Breen & De Vellis, 1974] (*see* Table 2.1). These properties are often expressed late in the lineage and are more likely to be reversible and under adaptive control by hormones, nutrients, matrix constituents, and cell–cell interaction (*see* Fig. 16.1).

As the genes encoding a large number of differentiated products have now been identified and sequenced [International Human Genome Sequencing Consortium, 2001], it is possible to look for the expression of differentiation marker proteins by RT-PCR [Ausubel et al., 2009] and microarray analysis [Kawasaki, 2004], which identifies the expression of specific mRNAs (*see* Plate 24). This method will not necessarily confirm synthesis of the final product, but it is possible to distinguish between very low levels (or no expression) and high levels of expression of specific gene transcripts. Expression of many genes can be screened simultaneously, a considerable advantage over other methods of determining product expression. Definition of the *proteome*, the total complement of proteins associated with a particular phenotype, can also be achieved by microarray analysis using antibody chips.

Differentiation should be regarded as the expression of one, or preferably more than one, marker associated with terminal differentiation. Although lineage markers are helpful in confirming cell identity, the expression of the functional properties of the mature cells is the best criterion for terminal differentiation and confirmation of cellular origin.

## 16.7 INDUCTION OF DIFFERENTIATION

There are five main parameters that control differentiation (Fig. 16.1): cell–cell interaction, cell–matrix interaction,

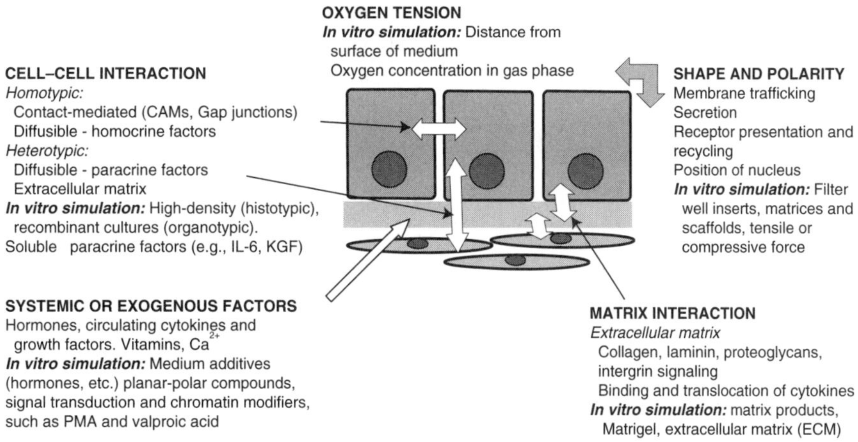

**OXYGEN TENSION**
*In vitro simulation:* Distance from
surface of medium
Oxygen concentration in gas phase

**CELL–CELL INTERACTION**
*Homotypic:*
    Contact-mediated (CAMs, Gap junctions)
    Diffusible - homocrine factors
*Heterotypic:*
    Diffusible - paracrine factors
    Extracellular matrix
*In vitro simulation:* High-density (histotypic),
    recombinant cultures (organotypic).
    Soluble  paracrine factors (e.g., IL-6, KGF)

**SHAPE AND POLARITY**
Membrane trafficking
Secretion
Receptor presentation and
recycling
Position of nucleus
*In vitro simulation:* Filter
    well inserts, matrices and
    scaffolds, tensile or
    compressive force

**SYSTEMIC OR EXOGENOUS FACTORS**
Hormones, circulating cytokines and
    growth factors. Vitamins, $Ca^{2+}$
*In vitro simulation:* Medium additives
(hormones, etc.) planar-polar compounds,
signal transduction and chromatin modifiers,
such as PMA and valproic acid

**MATRIX INTERACTION**
*Extracellular matrix*
    Collagen, laminin, proteoglycans,
    intergrin signaling
    Binding and translocation of cytokines
*In vitro simulation:* matrix products,
    Matrigel, extracellular matrix (ECM)

*Fig. 16.1. Regulation of Differentiation.* Parameters controlling the expression of differentiation, with some of the ways to reproduce these conditions in vitro.

cell shape and polarity, oxygen tension, and soluble systemic factors.

## 16.7.1  Cell Interaction

*Homotypic.*  Homologous cell interaction occurs at high cell density. It may involve gap junctional communication [Finbow & Pitts, 1981], in which metabolites, second messengers, such as cyclic AMP, diacylglycerol (DAG), $Ca^{2+}$, or electrical charge may be communicated between cells. This interaction probably harmonizes the expression of differentiation within a population of similar cells, rather than initiating its expression. (*See also* Section 25.1.)

Homotypic cell–cell adhesion molecules, such as CAMs or cadherins, which are calcium-dependent, provides another mechanism by which contacting cells may interact. These adhesion molecules promote interaction primarily between like cells via identical, reciprocally acting, extracellular domains (*see* Section 2.2.1), and they appear to have signal transduction potential via phosphorylation of the intracellular domains [Doherty et al., 1991; Gumbiner, 1995].

*Heterotypic.*  Heterologous cell interaction—such as between mesodermally and endodermally or ectodermally derived cells—is responsible for initiating and promoting differentiation. During and immediately after gastrulation in the embryo, and later during organogenesis, mutual interaction between cells originating in different germ layers promotes differentiation [Yamada et al., 1991; Hirai et al., 1992; Hemmati-Brivaniou et al., 1994; Muller et al., 1997]. For example, when endodermal cells form a diverticulum from the gut and proliferate within adjacent mesoderm, the mesoderm induces the formation of alveoli and bronchiolar ducts and is itself induced to become elastic tissue [Hardman et al., 1990; Caniggia et al., 1991].

The extent to which this process is continued in the adult is not clear, but evidence from epidermal maturation suggests that a reciprocal interaction, mediated by growth factors such as KGF and GM-CSF and cytokines such as IL-1α and IL-1β,

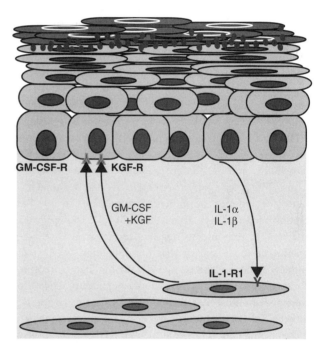

*Fig. 16.2. Reciprocal Paracrine Interaction.* Schematic illustration of the reciprocal paracrine pathways of keratinocyte growth regulation in organotypic cocultures with fibroblasts involving IL-1, KGF, and GM-CSF as well as their receptors [*see also* Maas-Szabowski et al., 2000; Szabowski et al. 2000]. (Figure adapted from Maas-Szabowski et al., 2002.)

with the underlying dermis is required for the formation of keratinized squames with fully cross-linked keratin (Fig. 16.2) [Maas-Szabowski et al., 2002]. In this context the recreation of a complex microenvironment becomes important for skin regeneration (*see* Section 22.2.1).

### Paracrine growth factors

*Positively acting.*  Factors derived from cell lineages different from their target are *heterotypic* paracrine factors (see Section 2.5), such as KGF in prostate [Planz et al., 1998], alveolar maturation factor in the lung [Post et al., 1984], and other potential candidates, including IL-6 and oncostatin M in lung maturation [McCormick et al., 1995, McCormick & Freshney, 2000], glia maturation factor in brain [Keles et al., 1992] (*see* Section 16.7.1; Plates 12*d*, 13*e*, *f*), and interferons [e.g., *see* Pfeffer & Eisenkraft, 1991].

Some growth factors act as morphogens [Gumbiner, 1992], such as epimorphin [Hirai et al., 1992; Radisky et al., 2003] and hepatocyte growth factor/scatter factor (HGF/SF) [Kinoshita & Miyajima, 2002]. HGF is one of the family of heparin-binding growth factors (HBGF) (*see* Table 9.4), shown to be released from fibroblasts, such as MRC-5 [Kenworthy et al., 1992]. Epimorphin and HGF induce tubule formation in the MDCK continuous cell line from dog kidney [Orellana et al., 1996; Montesano et al., 1997], salivary gland epithelium [Furue & Saito, 1997], and mammary epithelium [Soriano et al., 1995]. KGF, produced by dermal and prostatic fibroblasts induces epidermal [Aaronson et al., 1991; Gumbiner, 1992; Maas-Szabowski et al., 2002] and prostatic epithelial morphogenesis [Planz et al., 1998; Thomson et al., 1997]. Growth factors such as FGF-1, -2, and -3, KGF (FGF-7), TGF-β, and activin are also active in embryonic induction [Jessell & Melton, 1992].

HGF and KGF are both produced only by fibroblasts, bind to receptors found only on other cells (principally epithelium), and are classic examples of paracrine growth factors. The release of paracrine factors may be under the control of systemic hormones in some cases. It has been demonstrated that type II alveolar cells in the lung produce surfactant in response to dexamethasone in vivo. In vitro experiments have shown that this induction of surfactant synthesis is dependent on the steroid binding to receptors in the stroma, which then releases a peptide to activate the alveolar cells [Post et al., 1984]. Similarly the response of epithelial cells in the mouse prostate to androgens is mediated by the stroma [Thomson et al., 1997]. KGF has been shown to be at least one component of the interaction in the prostate [Yan et al., 1992].

The differentiation of the intestinal enterocyte, which is stimulated by hydrocortisone, also requires underlying stromal fibroblasts [Kédinger et al., 1987], and in this case modification of the extracellular matrix between the two cell types is implicated [Simon-Assmann et al., 1986]. Dexamethasone-dependent matrix modification is also implicated in the response of alveolar type II cells to paracrine factors (*see* Section 16.7.3) [Yevdokimova & Freshney, 1997]. Differentiation in the hematopoietic system is under the control of several positively acting lineage-specific growth factors, such as IL-1, IL-6, G-CSF, and GM-CSF (*see* Tables 9.4, 23.3), the last of which is also dependent on the matrix for activation [de Wynter et al., 1993]. Matrix heparan sulfates are also implicated in FGF activation [Klagsbrun & Baird, 1991; Fernig & Gallagher, 1994].

***Negatively acting.*** Factors such as MIP-1α maintain the stem cell phenotype [Graham et al., 1992], and similarly PDGF and FGF-2 promote the growth and self-renewal of the O-2A progenitor cell but inhibit its differentiation [Bögler et al., 1990]. Other negative regulators of differentiation include LIF in ES cell differentiation [Smith et al., 1988] and TGF-β in alveolar type II cell differentiation [Torday & Kourembanas, 1990; McCormick & Freshney, 2000; McCormick et al., 1995].

## 16.7.2 Systemic Factors

***Physiological inducers (Table 16.1).*** Systemic physiological regulators that induce differentiation include the following: (1) *Hormones* are secreted by a distant organ or tissue and reach the target tissue via the vasculature in vivo (i.e., *endocrine* factors). This category includes hydrocortisone, glucagon, and thyroxin (or triiodothyronine). (2) *Vitamins* such as vitamin $D_3$ [Jeng et al., 1994; Rattner et al., 1997] and retinoic acid [Saunders et al., 1993; Hafny et al., 1996; Ghigo et al., 1998] are derived from the diet and may be modified by metabolism. (3) Inorganic ions, particularly $Ca^{2+}$—high $Ca^{2+}$ promotes keratinocyte differentiation [Cho & Bikle, 1997] (*see* Section 22.2.1 for an example). This probably relates to the role of calcium in cell interaction (cadherins are calcium-dependent), intracellular signaling, and the membrane flux of calcium in so-called calcium waves that propagate signals from one responding cell to adjacent cells of the same lineage (*see* Fig. 2.8). Along with gap junctional communication, and possibly homocrine factors and heparan sulfate, this helps generate a coordinated response.

Other physiological inducers, the paracrine factors (*see* Section 16.7.1), are assumed to act directly, cell to cell, without the need for vascular transmission. However, some of these factors (FGF, PDGF, interleukins) are found in the blood, so they may also have a systemic role.

***Nonphysiological inducers.*** Rossi and Friend [1967] observed that mouse erythroleukemia cells treated with dimethyl sulfoxide (DMSO), to induce the production of Friend leukemia virus, turned red because of the production of hemoglobin (*see* Plate 13*a*, *b*), since confirmed by an increase in globin gene expression [Conkie et al., 1974] (*see* Plate 13*c*, *d*). Subsequently it was demonstrated that many other cells—such as neuroblastoma, myeloma, and mammary carcinoma—also responded to DMSO by differentiating (*see* Sections 22.2.7, 2.2.8). Many other compounds have now been added to this list of nonphysiological inducers: hexamethylene bisacetamide (HMBA); *N*-methyl acetamide; benzodiazepines, whose action may be related to that of DMSO; and a range of cytotoxic drugs, such as methotrexate, cytosine arabinoside, and mitomycin C (Table 16.2). Sodium butyrate has also been classed with these nonphysiological inducers of differentiation, but there is some evidence that butyrate occurs naturally, such as in the gut, and regulates normal enterocyte differentiation [Häner et al., 2010].

**TABLE 16.1.**  Soluble Inducers of Differentiation: Physiological

| Inducer | Cell type | Reference |
|---|---|---|
| **Steroid and related** | | |
| Hydrocortisone | Glia, glioma | McLean et al., 1986 |
| | Lung alveolar type II cells | Rooney et al., 1995; McCormick et al., 1995 |
| | Hepatocytes | Granner et al., 1968 |
| | Mammary epithelium | Marte et al., 1994 |
| | Myeloid leukemia | Sachs, 1978 |
| Retinoids | Tracheobronchial epithelium | Kaartinen et al., 1993 |
| | Endothelium | Lechardeur et al., 1995; Hafny et al., 1996 |
| | Enterocytes (Caco-2) | McCormack et al., 1996 |
| | Embryonal carcinoma | Mills et al., 1996 |
| | Melanoma | Lotan & Lotan, 1980; Meyskens & Fuller, 1980 |
| | Myeloid leukemia | Degos, 1997 |
| | Neuroblastoma | Ghigo et al., 1998 |
| **Peptide hormones** | | |
| Melanotropin | Melanocytes | Goding and Fisher, 1997 |
| Thyrotropin | Thyroid | Chambard et al., 1983 |
| Erythropoietin | Erythroblasts | Goldwasser, 1975 |
| Prolactin | Mammary epithelium | Takahashi et al., 1991; Rudland, 1992; Marte et al., 1994 |
| Insulin | Mammary epithelium | Marte et al., 1994; Rudland, 1992 |
| **Cytokines** | | |
| Nerve growth factor | Neurons | Levi-Montalcini, 1979 |
| Glia maturation factor, CNTF, PDGF, BMP2 | Glial cells | Keles et al., 1992; ; Raff, 1990; Kondo & Raff, 2000, 2004 |
| Epimorphin | Kidney epithelium | Hirai et al., 1992 |
| Fibrocyte-pneumocyte factor | Type II pneumocytes | Post et al., 1984 |
| Interferon-α, β | A549 cells | McCormick et al., 1995 |
| | HL60, myeloid leukemia | Kohlhepp et al., 1987 |
| Interferon-γ | Neuroblastoma | Wuarin et al., 1991 |
| CNTF | Type 2 astrocytes | Raff, 1990 |
| IL-6, OSM | A549 cells | McCormick et al., 1995; McCormick & Freshney, 2000 |
| BMP | 10T 1/2 | Shea et al., 2003 |
| KGF | Keratinocytes | Aaronson et al., 1991 |
| | Prostatic epithelium | Thomson et al., 1997; Yan et al., 1992 |
| HGF | Kidney (MDCK) | Bhargava et al., 1992; Li et al., 1992 |
| | Hepatocytes | Montesano et al., 1991 |
| TGF-β | Bronchial epithelium, melanocytes | Masui et al., 1986a; Fuller and Meyskens, 1981 |
| Endothelin | Melanocytes | Aoki et al., 2005 |
| **Vitamins** | | |
| Vitamin E | Neuroblastoma | Prasad et al., 1980 |
| Vitamin $D_3$ | Monocytes (U937) | Yen et al., 1993 |
| | Myeloma | Murao et al., 1983 |
| | Osteoblasts | Vilamitjana-Amedee et al., 1993 |
| | Enterocytes (IEC-6) | Jeng et al., 1994 |
| Vitamin K | Hepatoma | Bouzahzah et al., 1995 |
| | Kidney epithelium | Cancela et al., 1997 |
| Retinoids (see above in this table) | | |
| **Minerals** | | |
| $Ca^{2+}$ | Keratinocytes | Boyce & Ham, 1983 |

**TABLE 16.2.** Soluble Inducers of Differentiation: Nonphysiological

| Inducer | Cell type | Fate | Reference |
|---|---|---|---|
| **Planar-polar compounds** | | | |
| DMSO | Murine erythroleukemia | Immature erythrocytes | Rossi & Friend, 1967; Dinnen & Ebisuzaki, 1990 |
| | Myeloma | Granulocytes | Tarella et al., 1982 |
| | Neuroblastoma | Neurons | Kimhi et al., 1976 |
| | Mammary epithelium | Secretory epithelium | Rudland, 1992 |
| | Hepatocyte precursors, HepaRG hepatoma | Hepatocytes | Mitaka et al., 1993; Hino et al., 1999; Gripon et al., 2002 |
| Sodium butyrate | Erythroleukemia | Immature erythrocytes | Andersson et al., 1979b |
| | Colon cancer | Absorptive epithelium | Häner et al., 2010 |
| N-methyl acetamide | Glioma | Astrocyte | McLean et al., 1986 |
| N-methyl formamide, dimethyl formamide | Colon cancer | Absorptive epithelium | Dexter et al., 1979 |
| HMBA | Erythroleukemia | Immature erythrocytes | Osborne et al., 1982; Marks et al., 1994 |
| Butylated hydroxyanisole | Adipose-derived stem cells | Neurons | Safford & Rice, 2007 |
| Benzodiazepines | Erythroleukemia | Immature erythrocytes | Clarke and Ryan, 1980 |
| **Cytotoxic drugs** | | | |
| Genistein | Erythroleukemia | Immature erythrocytes | Watanabe et al., 1991 |
| Cytosine arabinoside | Myeloid leukemia | Granulocytes | Takeda et al., 1982 |
| Mitomycin C; anthracyclines | Melanoma | Melanocytes | Raz, 1982 |
| Methotrexate | Colorectal carcinoma | Absorptive & mucin-secreting epithelium | Lesuffleur et al., 1990 |
| **Chromatin modifiers** | | | |
| Valproic acid | Adipose-derived stem cells | Neurons | Safford & Rice, 2007 |
| | PC12 cells | Neurite extension | Kamata et al., 2007 |
| Azacytidine | NG108-15 neuronal cells | Cholinergic neurons | Aizawa et al., 2009 |
| **Signal transduction modifiers** | | | |
| Isobutylmethylxanthine | Adipose-derived stem cells | Adipocytes | Safford & Rice 2007 |
| Forskolin | Adipose-derived stem cells | Neurons | Safford & Rice, 2007 |
| PMA | Bronchial epithelium | Squamous epithelium | Willey et al., 1984; Masui et al., 1986a |
| | Mammary epithelium | | Wada et al., 1994 |
| | Colon (HT29, Caco-2) | Ductal morphogenesis | Velcich et al., 1995; Pignata et al., 1994 |
| | Monocytic leukemia (U937) | Monocytes | Hass et al., 1993 |
| | Erythroleukemia (K562) | Immature erythrocytes | Kujoth and Fahl, 1997 |
| | Neuroblastoma | Neurite outgrowth | Spinelli et al., 1982 |

*Abbreviations:* HMBA, Hexamethylene bisacetamide; PMA (TPA), phorbol myristate acetate.

The action of these compounds is unclear but may be mediated by changes in membrane fluidity (particularly for polar solvents, like DMSO, and anesthetics and tranquillizers); by their influence as lipid intercalators of signal transduction enzymes, such as protein kinase C (PKC) and phospholipase D (PLD), which tend to relocate from the soluble cytoplasm to the endoplasmic reticulum when activated; or by alterations in DNA methylation [Laurent et al., 2010] or histone acetylation [Keenen & de la Serna, 2008], also implicated in the action of 5-azacytidine [Aizawa et al., 2009] and valproic acid [Safford & Rice, 2007; Huangfu et al., 2008].

The induction of differentiation by polar solvents, such as DMSO, may be phenotypically normal, but the induction by cytotoxic drugs may also induce gene expression unrelated to differentiation [McLean et al., 1986].

Tumor promoters, such as phorbol myristate acetate (PMA), have been shown to induce squamous differentiation in bronchial mucosa, although not in bronchial carcinoma [Willey et al., 1984; Masui et al., 1986a, b; Saunders et al., 1993]. Although these tumor promoters are not normal regulators in vivo, they bind to specific receptors and activate signal transduction, such as by the activation of PKC [Dotto et al., 1985].

### 16.7.3 Cell–Matrix Interactions

Surrounding the surface of most cells is a complex mixture of glycoproteins and proteoglycans that is highly specific for each tissue, and even for parts of a tissue. Recreation of this complex microenvironment, involving cell–cell and cell–matrix interactions has been shown to be vital in the expression of the mature keratinocyte phenotype in the reconstruction of skin equivalents [Boehnke et al., 2007] and the maintenance of the stem cell compartment [Muffler et al., 2008] (see Section 22.2.1). Collagen has been found to be essential for the functional expression of many epithelial cells [Burwen & Pitelka, 1980; Flynn et al., 1982; Berdichevsky et al., 1992] and for endothelium to mature into capillaries [Folkman & Haudenschild, 1980]. The RGD motif (arginine-glycine-aspartic acid) in matrix molecules appears to be the peptide sequence which interacts with the receptor [Yamada & Geiger, 1997]. Small polypeptides containing this sequence effectively block matrix-induced differentiation, implying that the intact matrix molecule is required [Pignatelli & Bodmer, 1988].

Attempts to mimic matrix effects by use of synthetic macromolecules have been partially successful in using poly-D-lysine to promote neurite extension in neuronal cultures (see Section 22.4.1), but it seems that there is still a great deal to learn about the specificity of matrix interactions. It is unlikely that charge alone is sufficient to mimic the more complex signals demonstrated in many different types of matrix interaction, but charge alterations probably allow cell attachment and spreading, and under these conditions the cells may be capable of producing their own matrix.

It has been shown that endothelial cells [Kinsella et al., 1992; Garrido et al., 1995] and many epithelial cells differentiate more effectively on Matrigel [Kibbey et al., 1992; Darcy et al., 1995; Venkatasubramanian et al., 2000; Portnoy et al., 2004], a matrix material produced by the Engelberth–Holm–Swarm (EHS) sarcoma and made up predominantly of laminin, but also of collagen and proteoglycans. A number of cell lines, such as A549 adenocarcinoma of lung, show apparent morphogenesis when grown on Matrigel (see Plate 12c). This technique is useful but has the problem of introducing another biological variable to the system. Defined matrices are required; although fibronectin, laminin, collagen, and a number of other matrix constituents are available commercially, the specificity probably lies largely in the proteoglycan moiety, within which there is the potential for wide variability, particularly in the number, type, and distribution of the sulfated glycosaminoglycans, such as heparan sulfate [Fernig & Gallagher, 1994].

The extracellular matrix may also play a role in the modulation of growth factor activity. It has been suggested that the matrix proteoglycans, particularly heparan sulfate

proteoglycans (HSPGs), may bind certain growth factors, such as GM-CSF [Damon et al., 1989; Luikart et al., 1990], and make them more available to adjacent cells. Transmembrane HSPGs may also act as low-affinity receptors for growth factors and transport these growth factors to the high-affinity receptors [Klagsbrun & Baird, 1991; Fernig & Gallagher, 1994]. A549 cells require glucocorticoid to respond to differentiation inducers such as OSM, IL-6, and lung fibroblast-conditioned medium. The glucocorticoid, in this case dexamethasone (DX), induces the A549 cells to produce a low-charge density fraction of heparan sulfate, which, when partially purified, was shown to substitute for the DX and activate OSM, IL-6, and fibroblast-conditioned medium [Yevdokimova & Freshney, 1997].

### 16.7.4 Polarity and Cell Shape

Studies with hepatocytes [Sattler et al., 1978] showed that full maturation required the growth of the cells on collagen gel and the subsequent release of the gel from the bottom of the dish with a spatula or bent Pasteur pipette. This process allowed shrinkage of the gel and an alteration in the shape of the cell from flattened to cuboidal, or even columnar. Accompanying or following the shape change, and also possibly due to access to medium through the gel, the cells developed polarity, visible by electron microscopy; when the nucleus became asymmetrically distributed, nearer to the bottom of the cell, an active Golgi complex formed and secretion toward the apical surface was observed.

A similar establishment of polarity has been demonstrated in thyroid epithelium [Chambard et al., 1983] with a filter well assembly. In this case the lower (basal) surface generated receptors for thyroid-stimulating hormone (TSH) and secreted triiodothyronine, and the upper (apical) surface released thyroglobulin. Studies with hepatocytes [Guguen-Guillouzo & Guillouzo, 1986] (see Section 22.2.6) and bronchial epithelium [Saunders et al., 1993] suggest that floating collagen may not be essential, but the success of filter well culture (see Section 25.3.6) confirms that access to medium from below helps establish polarity.

### 16.7.5 Oxygen Tension

Expression of fully keratinized squamous differentiation in skin requires the positioning of the epidermal cells in an organotypic construct at the air/liquid interface [Maas-Szabowski et al., 2002]. Likewise location of alveolar type II cells at the air/liquid interface is necessary for optimal type II differentiation [Dobbs et al., 1997], and tracheal epithelium will only become mucus secreting at the air–liquid interface, becoming squamous if grown on the bottom of the dish [Kaartinen et al., 1993; Paquette et al., 2003]. It is assumed that positioning the cells at the air–liquid interface enhances gas exchange, particularly facilitating oxygen uptake without raising the partial pressure and risking free radical toxicity. However, it is also possible that the thin film above mimics the physicochemical conditions in vivo (surface tension, lack of nutrients) as well as oxygenation. The presence of D-PBSA on the apical surface may actually be preferable to complete medium [Chambard et al., 1983, 1987].

## 16.8 DIFFERENTIATION AND MALIGNANCY

It is frequently observed that with increasing progression of cancer, histology of a tumor indicates poorer differentiation, and from a prognostic standpoint, patients with poorly differentiated tumors generally have a lower survival rate than patients with differentiated tumors. It has also been stated that cancer is principally a failure of cells to differentiate normally. It is therefore surprising to find that many tumors grown in tissue culture can be induced to differentiate (see Table 16.2). Indeed much of the fundamental data on cellular differentiation has been derived from the Friend murine leukemia, mouse and human myeloma, hepatoma, and neuroblastoma. Nevertheless, there appears to be an inverse relationship between the expression of differentiated properties and the expression of malignancy-associated properties, even to the extent that the induction of differentiation has often been proposed as a mode of therapy [Spremulli & Dexter, 1984; Freshney, 1985].

## 16.9 PRACTICAL ASPECTS

It is clear that given the correct environmental conditions, and assuming that the appropriate cells are present, partial, or even complete, differentiation is achievable in cell culture. As a general approach to promoting differentiation, as opposed to cell proliferation and propagation, the following may be suggested:

(1) Select the correct cell type by use of appropriate isolation conditions and a selective medium (see Section 9.2.2; Chapter 22).

(2) Grow the cells to a high cell density ($>1 \times 10^5$ cells/cm$^2$) on the appropriate matrix. The matrix may be collagen of a type that is appropriate to the site of origin of the cells, with or without fibronectin or laminin, or it may be more complex, tissue derived or cell derived (see Protocol 7.1), such as Matrigel (see Section 16.7.3) or a synthetic matrix (e.g., poly-D-lysine for neurons).

(3) Change the cells to a differentiating medium rather than a propagation medium; for example, for epidermis increase $Ca^{2+}$ to around 3 mM, and for bronchial mucosa increase the serum concentration (see Protocol 22.9). For other cell types this step may require defining the growth factors appropriate to maintaining cell proliferation and those responsible for inducing differentiation.

(4) Add differentiation-inducing agents, such as glucocorticoids; retinoids; vitamin D$_3$; DMSO; HMBA; prostaglandins; and cytokines, such as bFGF, EGF, KGF, HGF, IL-6, OSM, TGF-$\beta$, interferons, NGF, and melanocyte-stimulating hormone (MSH), as appropriate for the type of cell (see Tables 16.1, 16.2).

(5) Add the interacting cell type during the growth phase (2, above), the induction phase (3 and 4 above), or both phases. Selection of the correct cell type is not always obvious, but lung fibroblasts for lung epithelial maturation [Post et al., 1984; Speirs et al., 1991], glial cells for neuronal maturation [Seifert & Müller, 1984], and bone marrow stromal cells for hematopoietic cells (see Protocol 23.9) are some of the better-characterized examples.

(6) Elevating the culture in a filter well [e.g., see Chambard et al., 1983] may be advantageous, particularly for certain epithelia, as it provides access for the basal surface to nutrients and ligands, the opportunity to establish polarity, and regulation of the nutrient and oxygen concentration at the apical surface by adjusting the composition and depth of overlying medium.

Not all of these factors may be required, and the sequence in which they are presented is meant to imply some degree of priority. Scheduling may also be important; for example, the matrix generally turns over slowly, so prolonged exposure to matrix-inducing conditions may be important, whereas some hormones may be effective in relatively short exposures. Furthermore the response to hormones may depend on the presence of the appropriate extracellular matrix, cell density, or heterologous cell interaction.

# CHAPTER 17

# Transformation and Immortalization

Transformation is seen as a particular event or series of events that both depends on and promotes genetic instability. Many of the cell line's properties are altered as a result, including growth rate, mode of growth (attached or in suspension), specialized product formation, longevity, and tumorigenicity (Table 17.1).

## 17.1 ROLE IN CELL LINE CHARACTERIZATION

It is important that transformation characteristics be included when a cell line is validated to determine whether it originates from neoplastic cells or has undergone transformation in culture. The transformation status must be known when culturing cells from tumors, in order to confirm that the cells are derived from the neoplastic component of the tumor, rather than from normal equivalent cells, infiltrating fibroblasts, blood vessel cells, or inflammatory cells.

More than one criterion is necessary to confirm neoplastic status, as most of the aforementioned characteristics are expressed in normal cells at particular stages of development. The exceptions are gross aneuploidy, heteroploidy, and tumorigenicity, which, taken together, are regarded as conclusive positive indicators of malignant transformation. However, some tumor cell lines can be near euploid and nontumorigenic so additional criteria are required.

## 17.2 WHAT IS TRANSFORMATION?

In microbiology, where the term was first used in this context, *transformation* implies a change in phenotype that is dependent on the uptake of new genetic material. Although

this process is now achievable artificially in mammalian cells, it is called *transfection* or *DNA transfer* in this case to distinguish it from transformation. Transformation of cultured cells implies a spontaneous or induced permanent phenotypic change resulting from a heritable change in DNA and gene expression. Although transformation can arise from infection with a transforming virus, such as polyoma, or from transfection with genes such as mutant *ras*, it can also arise spontaneously, after exposure to ionizing radiation, or after treatment with chemical carcinogens.

Transformation is associated with *genetic instability* and three major classes of phenotypic change, one or all of which may be expressed in one cell strain: (1) *immortalization*, the acquisition of an infinite life span, (2) *aberrant growth control*, the loss of contact inhibition of cell motility, density limitation of cell proliferation, and anchorage dependence, and (3) *malignancy*, as evidenced by the growth of invasive tumors in vivo. The term *transformation* is used here to imply all three of these processes. The acquisition of an infinite life span alone is referred to as *immortalization* because it can be achieved without grossly aberrant growth control and malignancy, which are usually linked.

## 17.3 GENETIC INSTABILITY AND HETEROGENEITY

### 17.3.1 Genetic Instability

The characteristics of a cell line do not always remain stable. In addition to the phenotypic alterations already described (*see* Sections 2.4.2, 16.1.1), cell lines are prone to genetic instability. Normal human finite cell lines are

*Culture of Animal Cells: A Manual of Basic Technique and Specialized Applications, Sixth Edition,* by R. Ian Freshney
Copyright © 2010 John Wiley & Sons, Inc.

**TABLE 17.1.** Properties of Transformed Cells

| Property | Assay | Protocol, figure, or reference |
|---|---|---|
| ***Growth*** | | |
| Immortal | Grow beyond 100 pd | Kopper & Hajdu 2004; Meeker & De Marzo 2004 |
| Anchorage independent | Clone in agar; may grow in stirred suspension | (*see* Protocols 13.4, 13.5, 12.4) |
| Loss of contact inhibition | Microscopic observation; time lapse | (*see* Fig. 17.3; Plate 14) |
| Growth on confluent monolayers of homologous cells | Focus formation | (*see* Fig. 17.3) |
| Reduced density limitation of growth | High saturation density; high growth fraction at saturation density | (*see* Protocols 17.3, 20.7, 20.8, 20.12) |
| Low serum requirement | Clone in limiting serum | (*see* Protocols 20.10, 21.3) |
| Growth factor independent | Clone in limiting serum | (*see* Protocols 20.10, 21.3) |
| Production of autocrine growth factors | Immunostaining; clone in limiting serum with conditioned medium; receptor-blocking antibody or peptide inhibitor | (*see* Protocols 15.11, 20.10, 21.3) |
| Transforming growth factor production | Suspension cloning of NRK | (*see* Protocols 13.2, 13.5) |
| High plating efficiency | Clone in limiting serum | (*see* Protocols 20.10, 21.3) |
| Shorter population-doubling time | Growth curve | (*see* Protocols 20.7–20.9) |
| ***Genetic*** | | |
| High spontaneous mutation rate | Sister chromatid exchange | (*see* Protocol 21.5) |
| Aneuploid | Chromosome content | (*see* Protocol 15.7) |
| Heteroploid | Chromosome content | (*see* Protocol 15.7) |
| Overexpressed or mutated oncogenes | Southern blot; FISH, immunostaining; microarray analysis | Ausubel et al., 2009 (*see* Protocol 15.1; Plate 24) |
| Deleted or mutated suppressor genes | Southern blot; FISH, immunostaining; microarray analysis | Ausubel et al., 2009 (*see* Protocol 15.11; Plate 24) |
| Gene and chromosomal translocations | FISH, chromosome paints | — |
| ***Structural*** | | |
| Modified actin cytoskeleton | Immunostaining | (*see* Protocol 16.13) |
| Loss of cell surface-associated fibronectin | Immunostaining | (*see* Protocol 16.13) |
| Modified extracellular matrix | Immunostaining; DEAE chromatography | (*see* Protocol 15.11) Yevdokimova & Freshney, 1997 |
| Altered expression of cell adhesion molecules (CAMs, cadherins, integrins) | Immunostaining | (*see* Protocol 15.11) |
| Disruption in cell polarity | Immunostaining; polarized transport in filter wells | (*see* Protocol 15.11) Halleux & Schneider, 1994 |
| ***Neoplastic*** | | |
| Tumorigenic | Xenograft in nude or scid mice | Giovanella et al., 1974; Russo et al., 1993 |
| Angiogenic | CAM assay; filter wells, VGEF production | (*see* Plate 15) Ment et al., 1997; Buchler et al., 2004 |
| Enhanced protease secretion (e .g., plasminogen activator) | Plasminogen activator assay | (*see* Fig. 17.9) Whur et al., 1980; Boxman et al., 1995 |
| Invasive | Organoid confrontation; filter well invasion assay | (*see* Figs. 17.6, 17.7) Mareel et al., 1979; Brunton et al., 1997 |

usually genetically stable, but cell lines from other species, particularly the mouse, are genetically unstable and transform quite readily. Continuous cell lines, particularly from tumors of all species, are very unstable, not surprisingly, as it was this instability that allowed the necessary mutations for the cell line to become continuous, and deletion or alteration in DNA surveillance genes, such as p53, are usually implicated. Continuous cell lines are usually *heteroploid*, meaning they show a wide range in chromosome number among individual cells in the population (*see* Fig. 2.10*b*), implying substantial genetic diversity. When continuous cell lines are cloned, there is evidence of considerable morphological diversity (*see* Plate 7). Other evidence of heterogeneity in seen in WIL lung adenocarcinoma, stained for CEA, showing that some cells are positive and others negative (*see* Plate 12*a*) and that subclones of H4-II-E-C3 minimal deviation hepatoma [Pitot et al., 1964] differ in their constitutive and glucocorticoid-induced levels of tyrosine aminotransferase (Fig. 17.1). Consequently continuous cell strains, even after cloning, contain a range of genotypes that are constantly changing.

There are two main causes of genetic heterogeneity: (1) the spontaneous mutation rate appears to be higher in vitro, associated, perhaps, with the high rate of cell proliferation and

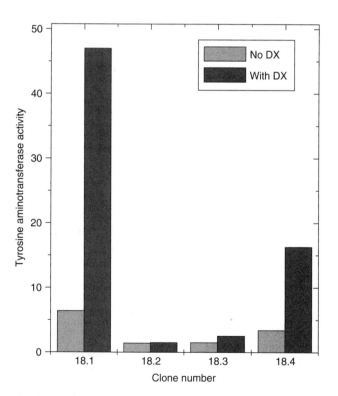

***Fig. 17.1. Clonal Variation.*** Variation in tyrosine aminotransferase activity among four subclones of clone 18 of a rat minimal-deviation hepatoma cell strain, H-4-II-E-C3. Cells were cloned; clone 18 was isolated, grown up, and recloned; and the second-generation clones were assayed for tyrosine aminotransferase activity, with and without pretreatment of the culture with dexamethasone. Light gray bars, basal level; dark gray bars. induced level. (Data J. Somerville.)

defective DNA surveillance genes, particularly p53, and (2) mutant cells are not eliminated unless their growth capacity is impaired.

### 17.3.2 Chromosomal Aberrations

Evidence of genetic rearrangement can be seen in chromosome counts (*see* Fig. 2.10) and karyotype analysis. Although the mouse karyotype is made up exclusively of small telocentric chromosomes, several metacentrics arise in many continuous murine cell lines due to Robertsonian fusion of the telomeres (Fig. 17.2*a*). Furthermore, although virtually every cell in the animal has the normal diploid set of chromosomes, this is more variable in culture. In extreme cases—such as continuous cell strains, such as HeLa-S$_3$—less than half of the cells will have exactly the same karyotype; i.e., they are heteroploid (*see also* Fig. 2.10).

Both variations in ploidy and increases in the frequency of individual chromosomal aberrations (Fig. 17.2*b*) can be found [Biedler, 1976; Croce, 1991], and the variations in chromosome number found in most tumor cultures (*see* Fig. 2.10; Protocol 15.7) [Murnane, 2006] are often reflected in abnormal DNA content by flow cytometry. The incidence of genetic instability and frequency of chromosomal rearrangement can be determined by the sister chromatid exchange assay [Bryant, 2004] (*see* Protocol 21.5; Plate 17*e*,*f*).

Most tumor culture show signs of *aneuploidy*, meaning deviations from the normal complement of chromosomes. Some specific aberrations are associated with particular types of malignancy [Ganmore et al., 2009], the first to be documented being the Philadelphia chromosome in chronic myeloid leukemia now known to be due to a reciprocal translocation between chromosomes 9 and 22. Subsequently translocations of the long arms of chromosomes 8 and 14 were found in Burkitt's lymphoma [Lebeau & Rowley, 1984]. Several other leukemias also express other translocations [Mark, 1971]. Meningiomas often have consistent aberrations, and small-cell lung cancer frequently has a 3p2 deletion [Wurster-Hill et al., 1984]. These aberrations constitute tumor-specific markers that can be extremely valuable in cell line characterization and confirmation of neoplasia [Mitelman et al., 2009].

### 17.4 IMMORTALIZATION

Most normal cells have a finite life span of 20 to 100 generations (*see* Section 2.7.4), but some cells, notably those from rodents and from most tumors, can produce continuous cell lines with an infinite life span. The rodent cells are karyotypically normal at isolation and appear to go through a crisis after about 12 generations; most of the cells die out in this crisis, but a few survive with an enhanced growth rate and give rise to a continuous cell line.

If continuous cell lines from mouse embryos (e.g., the various 3T3 cell lines) are maintained at a low cell density

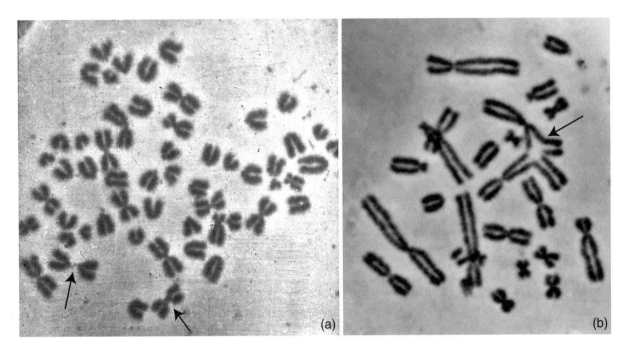

***Fig. 17.2. Chromosome Aberrations.*** Examples of aberrant recombinations. (*a*) P2 cells, a clone of L929 mouse fibroblasts, showing multiple telomeric fusions, with two marked by arrows. (*b*) Recombination event between two dissimilar chromosomes of the larger group in Y5 Chinese hamster cells. This cell would be unlikely to survive.

and are not allowed to remain at confluence for any length of time, they retain sensitivity to contact inhibition and density limitation of growth [Todaro & Green, 1963]. If, however, they are allowed to remain at confluence for extended periods, foci of cells appear with reduced contact inhibition, begin to pile up, and will ultimately overgrow (Fig. 17.3).

The fact that these cells are not apparent at low densities or when confluence is first reached suggests that they arise de novo, by a further transformation event. They appear to have a growth advantage, and subsequent subcultures will rapidly be overgrown by the randomly growing cell. This cell type is often found to be tumorigenic and loses its value as a feeder layer [Edington et al., 2004].

### 17.4.1 Control of Senescence

The finite life span of cells in culture is regulated by a group of 10 or more dominantly acting senescence genes, whose products negatively regulate cell cycle progression [Goldstein et al., 1989; Sasaki et al., 1996]. Somatic hybridization experiments between finite and immortal cell lines usually generate hybrids with a finite life span, suggesting that the senescence genes are dominant [Pereira-Smith & Smith, 1988; Nishizuka et al., 2001]. It is likely that one or more of these genes negatively regulate the expression of telomerase [Holt et al., 1996; Greider & Blackburn, 1996; Smith & de Lange, 1997; Bryan & Reddel, 1997], required for the terminal synthesis of telomeric DNA, which otherwise becomes progressively shorter during a finite life span, until the chromosomal DNA can no longer replicate. Telomerase

is expressed in germ cells and has moderate activity in stem cells, but is absent from somatic cells. Deletions and/or mutations within senescence genes, or overexpression or mutation of one or more oncogenes that override the action of the senescence genes, can allow cells to escape from the negative control of the cell cycle and re-express telomerase.

It has been assumed that immortalization is a multistep process involving the inactivation of a number of cell cycle regulatory genes, such as Rb and p53. The SV40 LT gene is often used to induce immortalization. The product of this gene, T antigen, is known to bind Rb and p53. By doing so, it not only allows an extended proliferative life span but also restricts the DNA surveillance activity of genes like p53, thereby allowing an increase in genomic instability and an increased chance of generating further mutations favorable to immortalization (e.g., the upregulation of telomerase or the downregulation of one of the telomerase inhibitors). Transfection of the telomerase gene with a regulatable promoter is sufficient to immortalize cells [Bodnar et al., 1998; Vaziri & Benchimol, 1998].

Immortalization per se does not imply the development of aberrant growth control and malignancy, as a number of immortal cell lines, such as 3T3 cells and BHK21-C13, retain contact inhibition of cell motility, density limitation of cell proliferation, and anchorage dependence, and are not tumorigenic. It must be assumed, however, that some aspects of growth control are abnormal and that there is a likely increase in genomic instability. Furthermore immortalized cell lines often lose the ability to differentiate,

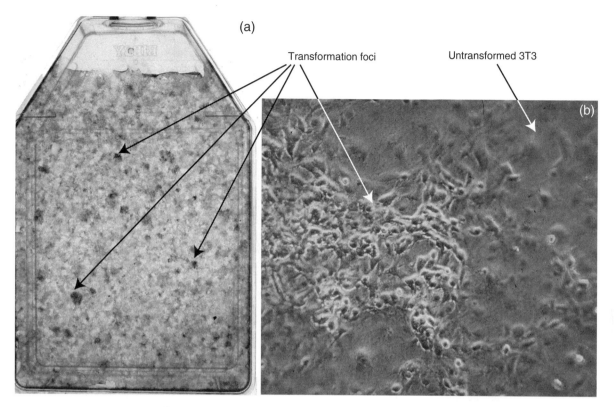

**Fig. 17.3. Transformation Foci.** A monolayer of normal, contact-inhibited NIH 3T3 mouse fibroblasts left at confluence for 2 weeks. (*a*) 75-cm² flask stained with Giemsa. (*b*) Phase-contrast image of focus of transformed cells overgrowing normal monolayer (10× objective). (*See also* Plate 14*c*.)

but there are reports of telomerase-induced immortalization of keratinocytes [Dickson et al., 2000] and skeletal muscle satellite cells [Wootton et al., 2003] without abrogation of p53 activity and retention of the ability to differentiate.

### 17.4.2 Immortalization with Viral Genes

A number of viral genes have been used to immortalize cells (Table 17.2). It has been recognized for some time that SV40 can be used to immortalize cells, and the gene responsible for this appears to be the large T (LT) gene [Mayne et al., 1996] (*see* Protocol 17.1). Other viral genes that have been used to immortalize cells are adenovirus E1a [Seigel, 1996], human papilloma virus (HPV) E6 and E7 [Peters et al., 1996; Le Poole et al., 1997], and Epstein–Barr virus (EBV; usually the whole virus is used) [Bolton & Spurr, 1996]. Most of these genes probably act by blocking the inhibition of cell cycle progression by inhibiting the activity of genes such as CIP-1/WAF-1/p21, Rb, p53, and p16, thus giving an increased life span, reducing DNA surveillance, and giving an enhanced opportunity for further mutations. Those genes that have been used most extensively are EBV for lymphoblastoid cells [Bolton & Spurr, 1996] and SV40LT for adherent cells such as fibroblasts [Mayne et al., 1996], keratinocytes [Steinberg, 1996], and endothelial cells [Punchard et al., 1996], and hTERT for mesenchymal stem cells (*see* Protocol 17.2) and a number of other cells.

Typically cells are transfected or retrovirally infected with the immortalizing gene before they enter senescence. This extends their proliferative life span for another 20 to 30 population doublings, whereupon the cells cease proliferation and enter *crisis*. After a variable period in crisis (up to several months), a subset of immortal cells overgrows. The fraction of cells that eventually immortalize can be $1 \times 10^{-5}$ to $1 \times 10^{-9}$.

### 17.4.3 Immortalization of Human Fibroblasts

The following introduction and Protocol 17.1 for the immortalization of fibroblasts has been condensed from Mayne et al. [1996].

By far the most successful and most frequently used method for deriving immortal human fibroblasts is through the expression of SV40 T antigen, which does not lead directly to immortalization but initiates a chain of events that results in an immortalized derivative appearing with a low probability, estimated at about 1 in $10^7$ [Shay & Wright, 1989; Huschtscha & Holliday, 1983]. SV40-transfected cells are selected directly under appropriate culture conditions, and the surviving cells are subcultured to give rise to a precrisis SV40-transformed cell population. These cells are cultivated continuously until they reach the end of their proliferative life span, when they inevitably enter crisis. They must then be nurtured with care, and sufficient cells must be cultured,

**TABLE 17.2.** Genes Used in Immortalization

| Gene | Insertion | Cell type | Reference |
|---|---|---|---|
| EBV: *ebna, lmp*1 | Infection | B-lymphocytes | Bolton & Spurr, 1996; Bourillot et al., 1998; Sugimoto et al., 2004 |
| SV40LT | Lipofection | Keratinocytes | Steinberg, 1996 |
| | Calcium phosphate transfection | Fibroblasts | Mayne et al., 1996 |
| | Calcium phosphate transfection | Astroglial cells | Burke et al., 1996 |
| | Adenovirus infection | Esophageal epithelium | Inokuchi et al., 1995 |
| | Microinjection | Rat brain endothelium | Lechardeur et al., 1995 |
| | Transfection | Prostate epithelium | Rundlett et al., 1992 |
| | Transfection | Mammary epithelium | Shay et al., 1993 |
| | Strontium phosphate transfection | Bronchial epithelium | De Silva et al., 1996 |
| | Strontium phosphate transfection | Mesothelial cells | Duncan et al., 1996 |
| HPV16 E6/E7 | Retroviral transfer | Cervical epithelium | Demers et al., 1994 |
| | Transfection | Keratinocytes | Bryan et al., 1995 |
| | Strontium phosphate transfection | Mesothelial cells | De Silva et al., 1994 |
| | Strontium phosphate transfection | Bronchial epithelium | De Silva et al., 1994 |
| | Retroviral infection | Ovarian surface epithelium | Tsao et al., 1995 |
| Ad5 E1a | | Epithelial cells | Douglas & Quinlan, 1994 |
| *htrt* | Transfection | Pigmented retinal epithelium | Bodnar et al., 1998 |
| | Transfection | Foreskin fibroblasts | Bodnar et al., 1998 |
| | Transduction | Bone marrow stem cells | Simonsen et al., 2002 |
| | Transduction | Keratinocytes | Dickson et al., 2000 |
| | Transduction | Myoblasts | Wootton et al., 2003 |

to give a reasonable chance for an immortalized derivative to appear.

The choice of T antigen expression vectors depends on the choice of the dominant selectable marker gene, which is the source of the promoter that drives T antigen expression and alternative forms of the T antigen itself. Although, in our experience [Mayne et al., 1996], selection for *gpt* is effective in human fibroblasts, G418 (*neo*) and hygromycin (*hygB*) are much more effective and easier to use. The majority of human SV40-immortalized fibroblast cell lines have been established with either SV40 virus or constructs, such as pSV3*neo*, that express T antigen from the endogenous promoter. We recommend the use of pSV3*neo* [Southern & Berg, 1982; Mayne et al., 1986] for the constitutive expression of T antigen.

Cells should be used between passages 7 and 15. Trypsinize the cells 24 to 48 h before transfection, and seed about 2 to $2.5 \times 10^5$ cells per 9-cm dish or about 5.5 to $6.8 \times 10^5$ cells per 175-cm$^2$ flask. Cells should be 70% to 80% confluent when transfected, in a final volume of medium of 10 mL/9-cm dish or 30 mL/175-cm$^2$ flask.

The calcium phosphate precipitation method relies on the formation of a DNA precipitate in the presence of calcium and phosphate ions. The DNA is first sterilized by precipitation in ethanol and resuspension in sterile buffer. It is then mixed carefully with calcium, and the resulting solution is added very slowly, with mixing, to a phosphate solution. When making the precipitate, it is important to note that optimal

gene transfer occurs when the final concentration of DNA in the precipitate is 20 µg/mL. The volume of DNA precipitate applied to the cultures should never exceed one-tenth of the total volume. It is necessary to leave the mixture to develop for 30 min before adding it to the cell cultures.

---

**PROTOCOL 17.1. FIBROBLAST IMMORTALIZATION**

**Materials**

*Sterile:*

☐ HEPES buffer: HEPES, 12.5 mM; pH 7.12
☐ 10 × CaHEPES: $CaCl_2$, 1.25 M; HEPES, 125 mM; pH 7.12
☐ 2 × HEPES-buffered phosphate (2 × HBP): $Na_2HPO_4$, 1.5 mM; NaCl, 280 mM; HEPES, 25 mM; pH 7.12
☐ NaOAc: NaOAc, 3 M; pH 5.5
☐ Tris-buffered EDTA (TBE): Tris·HCl, 2 mM; EDTA, 0.1 mM; pH 7.12
☐ Absolute ethanol
☐ Eagle's MEM/15% FCS
☐ G418 (Invitrogen): 20 mg/mL in HEPES buffer, pH 7.5, sterilized by filtration through a 0.2-µm membrane and stored in small aliquots at $-20°C$

❑ Hygromycin (Roche Applied Science): 2 mg/mL in UPW, filter sterilized with a 0.2-μm membrane and stored in small aliquots at −20°C

❑ SV40 T antigen DNA

### Procedure

1. Estimate the amount of DNA required for the transfection:

   (a) Use a maximum of 20 μg of T antigen vector (without carrier DNA) for each plate and 60 μg of the vector DNA per flask.

   (b) Include an additional 20 μg of DNA, as it is not always possible to recover the full expected amount after preparation of the precipitate.

2. Prepare a sterile solution of the vector DNA in a microcentrifuge tube:

   (a) Precipitate the DNA with one-tenth of a volume of 3.0 M NaOAc, pH 5.5, and 2.5 volumes of ethanol.

   (b) Mix well, ensuring that the entire inside of the tube has come into contact with the ethanol solution.

   (c) Leave the tube on ice briefly (∼5 min).

   (d) Centrifuge the tube for 15 min at 15,000 rpm in a microcentrifuge to collect the precipitate.

   (e) Gently remove the tube from the centrifuge, and open it in a laminar flow hood.

   (f) Remove the supernate by aspiration, taking care not to disturb the pellet. Ensure that the ethanol is well drained.

   (g) Allow the pellet to air dry in the cell culture hood until all traces of ethanol have evaporated.

   (h) Resuspend the DNA pellet in TBE to give a final concentration of 0.5 mg/mL. It may be necessary to vortex the tube in order to release the pellet from the side of the tube.

   (i) Incubate the tube at 37°C for 5 to 10 min, with occasional vortexing to ensure that the pellet is well resuspended.

**Note.** Do not use higher TBE concentrations for resuspending your DNA, as doing so can interfere with the formation of the DNA precipitate.

3. Prepare the DNA calcium phosphate precipitate:

   (a) Calculate the total volume of precipitate required. The final concentration of the DNA in the precipitate should be 20 μg/mL, and you will need 1 mL for each 9-cm plate and 3 mL for each 175-cm² flask. Remember to make an extra 1 mL of precipitate to ensure recovery of sufficient volume for all of your cultures, as some loss of volume will occur during preparation of the precipitate.

   (b) Dilute the DNA/TBE mix in 12.5 mM HEPES to give 20 μg/mL in the final mix.

   (c) Add 10 × CaCl₂, one-tenth of the volume of the final mix.

   (d) Add the solution dropwise, while mixing, to an equal volume of 2 × HBP. For example, for nine 9-cm dishes at 1 mL/dish, plus 1 mL to spare (i.e., 10 mL of mix), we have the following:

       (i) DNA/TBE, 0.5 mL

       (ii) 12.5 mM HEPES, 3.5 mL

       (iii) 10 × CaCl₂, 1.0 mL

       (iv) Add dropwise to 2 × HBP, 5.0 mL

       The final concentrations in the mix are as follows:

| | |
|---|---|
| DNA | 20 μg/mL |
| CaCl₂ | 0.125 M |
| Na₂PO₄ | 0.75 mM |
| NaCl | 140 mM |
| HEPES | 12.5 mM |

       The pH is 7.12

**Note.** Use plastic pipettes and tubes when preparing DNA calcium phosphate precipitates, as the precipitates stick very firmly to glass. For mixing, we recommend the use of two pipettes in two handheld pipette controllers, one for blowing bubbles of sterile air into the mixture and the other for carefully adding the DNA/calcium mix in a dropwise fashion to the phosphate solution; good mixing results in an even precipitate. Use 1- to 5-mL pipettes, depending on the volume of precipitate being made. As the DNA/calcium solution is added to the phosphate solution, a light, even precipitate will begin to form. This precipitate is quite obvious and gives a milky appearance when complete.

   (e) After all of the DNA/calcium solution has been added to the phosphate, replace the lid on the tube, invert the tube gently once or twice, and leave the tube to stand at room temperature for 30 min.

   (f) It is important to make a mock precipitate without DNA. This allows you to assess the effectiveness of your selection conditions. A mock precipitate can be made exactly as described for the regular precipitate, but the DNA for the mock precipitate is replaced with additional HEPES buffer.

4. Add 1 mL of the DNA or mock precipitate to each plate or 3 mL to each flask. Make sure that the volume of precipitate is no more than one-tenth

of the total volume of the culture medium already on the cells.

5. Leave the precipitate on the cells for a minimum of 6 h, but not more than overnight (~16 h). For fibroblasts from some individuals, exposure to calcium and phosphate for more than 6 h may be toxic.

6. Remove the calcium phosphate precipitate by aspiration. There is no need to wash the cells further, or, in our experience, to further treat the cells with either DMSO or glycerol.

7. Add medium to the cultures, and incubate them until 48 h from the start of the experiment; then add selective agents to the cultures. The agent that you add will depend on the vector used for transfection. Vectors carrying the *neo* gene confer resistance to G418 (Geneticin), and vectors carrying the *hyg b* gene confer resistance to hygromycin B. All fibroblasts, in our experience, require 100 to 200 μg/mL of G418 or 10 to 20 μg/mL of hygromycin B to kill the cells gradually over a period of a week.

8. Change the medium on the transfected plates:
   (a) Dispense the total volume of medium required into a suitably sized sterile bottle, and add G418 or hygromycin B from the concentrated stocks to give the correct final concentration.
   (b) Gently swirl the solution to mix it.
   (c) Aspirate the medium from the plates or flasks, and replace it with the selective medium.
   (d) Return the plates or flasks to the incubator.

9. Monitor the effects of the selective medium on a daily basis by examining the culture under the microscope. When a significant number of cells have lifted and died, replace the medium with fresh selective medium. Selection should be maintained at all times.

10. Continue to replace the medium until the background of cells has lifted and died. The mock-transfected plates that have not been transfected with DNA should have no viable cells remaining after 7 to 10 days. If there are cells remaining, then the selection has not worked adequately, and it may be necessary to raise the concentration of the selective agent.

11. Once the background of cells has died, it is no longer necessary to routinely change the medium on the cells. The cells should then be left undisturbed in the incubator for four to six weeks to allow the transfected cells to grow and form colonies.

12. Once colonies arise, pick out individual colonies by using cloning rings (*see* Protocol 13.6), or bulk the colonies together by trypsinizing the whole dish or flask.

13. Freeze aliquots of cells at the earliest opportunity and regularly thereafter. Once transfectants have been expanded into cultures and ample stocks frozen in liquid nitrogen, it is necessary to keep the culture going for an extended period of time until it reaches crisis. To minimize the risk from fungal contaminants, add amphotericin B (Fungizone, Invitrogen) at 2.5 μg/mL.

14. Subculture the cultures routinely until they reach the end of their in vitro life span. As the cells approach crisis, the growth rate often slows. As the cultures begin to degenerate and cell division ceases, it is no longer necessary to subculture the cells. However, if heavy cell debris begins to cling to the remaining viable cells, it may be advisable to trypsinize the cells to remove the debris. Either return all of the cells to the same vessel, or use a smaller vessel to compensate for the cell death that is occurring. In general, the cells grow and survive better if they are not too sparse. With patience, care, and the culture of sufficient cells from your freezer stocks, you should, in most cases, obtain a postcrisis line.

15. When healthy cells begin to emerge, allow the colonies to grow to a reasonable size before subculturing, and then begin to subculture the colonies again. Do not be tempted to put too few cells into a large flask.

16. Freeze an ampoule of cells at the earliest opportunity, and continue to build up a freezer stock before using the culture.

17. To check that your postcrisis line is truly immortal, we recommend selecting and expanding individual clones from the culture. This procedure has the additional benefit of providing a homogeneous culture derived from a single cell.

### Post-transfection care of cultures

(1) The level of selective agent is chosen to produce a gentle kill over a period of about a week. The majority of cells on the DNA-treated plates should die within seven days, and those cells that remain should be the successful transfectants.

(2) It is advisable to freeze ampoules of cells routinely, both to build up a stock of transfected cells and to save time if cultures are lost because of contamination.

(3) In most cases, crisis is a marked event, with the majority of cells showing signs of deterioration and a net loss of viable cells. On average, crisis lasts from 3 to 6 months, and the culture may deteriorate to the point where very few, if any, obviously healthy cells are present.

(4) Once a culture has entered crisis, you can then reliably predict the timing of crisis for parallel cultures stored in liquid nitrogen. This prediction allows one to retrieve ampoules from parallel cultures from the freezer and to build up a number of flasks sitting at the threshold of crisis. As these parallel cultures enter crisis and begin to lose viability, these flasks can be pooled. In some cases, there will be an adequate cover of cells in the flask but high levels of cellular debris. In these cases, replate the cells. The cells should be trypsinized and centrifuged, and the pellet should be returned to the original flask. The flask may be rinsed several times with trypsin to remove any adhering cell debris before returning the cells to it.

(5) The first sign of a culture emerging from crisis is usually the appearance of one or more foci of apparently healthy, robust cells with the typical appearance of SV40-transformed cells. On subculturing, these foci expand to give a healthy, regenerating culture. In many cases the early-emerging postcrisis cells grow poorly, but the growth properties improve with further subculturing, and subcloning helps select individual clones with better growth properties. It is important to freeze an ampoule of your new cell line as early as possible. Our working definition for an immortal line is that the culture has undergone a minimum of 100 population doublings post-transfection and has survived subsequent subcloning.

(6) As the appearance of an immortal derivative within these cultures is a relatively rare event, it is essential that any postcrisis cell lines that emerge are checked to ensure that they were derived from the original starting material and are not the result of cross-contamination from other immortal cell lines in the laboratory (*see* Sections 12.1.1, 15.2, 18.6; Protocol 15.9).

### 17.4.4 Telomerase-Induced Immortalization

Telomeres play an essential role in chromosome stability and determining cellular life span. Telomerase or terminal transferase is a ribonucleoprotein (RNP) complex that synthesizes telomere repeats in tissue progenitor cells and cancer cells. Active human telomerase is composed of three principal subunits, an RNA component (hTR), a protein catalytic subunit (hTERT), and dyskerin (DKC1). Very recently a holoenzyme subunit, TCAB1 (Telomerase Cajal body protein 1), has been identified, which is notably enriched in Cajal bodies, nuclear sides of RNP processing, and shown to be important for telomerase function as it controls telomerase trafficking and is required for telomere synthesis in human cancer cells [Venteicher et al., 2009]. The RNA subunit is ubiquitously expressed in both normal and malignant tissues, while hTERT is only expressed in cells and tissues such as tumors, germ line cells, and activated lymphocytes. The primary cause of senescence appears to be telomeric shortening, followed by telomeric fusion and the formation of dicentric chromosomes and subsequent apoptosis. Transfecting cells with the telomerase gene *htrt* extends the life span of the cell line, and a proportion of these

cells becomes immortal but not malignantly transformed [Bodnar et al., 1998, Simonsen et al., 2002]. As a high proportion of the *htrt*⁺ clones becomes immortal, this appears to be a promising technique for immortalization. Although the functionality of some of these lines has yet to be demonstrated, there are some encouraging reports of uncompromised differentiation, such as in keratinocytes [Dickson et al., 2000] and in myocytes [Wootton et al., 2003].

The preceding introduction and Protocol 17.2 were contributed by Nedime Serakinci, University of Southern Denmark, Institute of Regional Health Services, Telomere Aging Group. This protocol has been used successfully for various human stem cell lines and primary cultures such as mesenchymal, neuronal, muscle satellite, melanocyte, and mast cells.

---

## PROTOCOL 17.2. IMMORTALIZATION OF HUMAN STEM AND PRIMARY CELLS BY TELOMERASE

### *Materials*

General:

Sterile:

❑ Growth medium for human mesenchymal stem cells and mast cells: Dulbecco's modified Eagle's medium (DMEM) with high glucose, 4.5 g/L, and ʟ-glutamine, 2 mM, supplemented with 10% fetal bovine serum, 100 U/mL of penicillin, and 100 µg/mL streptomycin

❑ Alternative medium for mast cells: RPMI 1640 with high glucose, 4.5 g/L, and ʟ-glutamine, 2 mM, supplemented with 10% fetal bovine serum, 100 U/mL of penicillin and 100 µg/mL streptomycin

❑ Medium for human muscle satellite cells: Dulbecco's modified Eagle's medium with low glucose, and 25 mM HEPES, supplemented with 10% fetal calf serum, Glutamax I, 1:100 to give 2 mM

❑ Medium for melanocytes: Complete Melanocyte Medium M2 (Promocell)

❑ Polybrene, 8 mg/mL (Sigma)

❑ Universal containers

❑ Culture flasks, 25 cm², 75 cm²

### Production of retroviral vector

❑ Cell lines:
PG13 [Miller et al., 1991]
GP+E-86 [Markowitz et al., 1988]
AM12 [Markowitz et al., 1988]

❑ Retroviral vector with (pBABE-puromycin) and without selection marker (GC Sam)

❑ *htrt* DNA for transfection

❑ Tx buffer:

HEPES, 0.5 M, pH7.1...............................200 μL
NaCl, 5M.............................................100 μL
Na$_2$HPO$_4$/NaH$_2$PO$_4$, 1 M.......................3 μL
UPW.................................................1.7 mL
Total...............................................20 mL

□ Buffer A:
NaCl (5M)...........................................600 μL
EDTA (0,5M)..........................................40 μL
TrisHCl pH7.5 (0,5M)................................400 μL
UPW...................................................19 mL
Total.............................................20.04 mL

□ CaCl, 2.5 M

### Transduction of monolayer growing cells

[Christensen et al., 2000; Simonsen et al, 2002; Serakinci et al, 2007]

□ Cells: hMSC cells, neuronal stem cells, muscle satellite cells, or melanocytes
□ Flasks, 75 cm$^2$
□ Multiwell plates, 6-well

### Post-transduction

□ Materials for cryopreservation (*see* Protocol 19.1)
□ Materials for DNA profiling (*see* Protocol 15.9) or other authentication procedures (e.g., *see* Protocol 15.10)

### PCR reagents

□ DNA, 100 μg/mL
□ 10 × PCR Buffer (+ Mg)L (Qiagen)
□ ectoTERT primers for GC Sam TERT (sense: 5'-GGACCATCTCTAGACTGACG-3', antisense: 5'-GGAGCGCACGGCTCGGCAGC-3'; both primers 2 μM; expected PCR product size 100 bp)
□ ecto TERT primers for pBABE-puro TERT *(sense: 5'-CCGCCTCAATCCTCCCTTTAT-3', antisense: 5'-GGAGCGCACGGCTCGGCAGC–3'* (both primer 2 μM)(expected PCR product size 200 bp)
□ dNTP mix (10 mM) (Qiagen)
□ Q-solution (Qiagen; for amplification of GC-rich templates)
□ 10x PCR buffer (Qiagen)
□ DNA polymerase (Qiagen)
□ UPW

### *Protocol*

#### Production of retroviral vector

The retroviral vector with the hTERT gene, is packaged into the gibbon ape leukemia virus (GALV) packaging cell line PG13 [Miller et al., 1991] by a two-step procedure. First, use 20-μg/mL *htrt* DNA (Geron) to transfect packaging cell line GP+E-86 [Markovitz et al., 1988a, b], and then use the supernate to infect PG13 and/or AM-12 cells.

1. Seed $6.7 \times 10^5$ GP + E-86 cells in a small culture flask (25 cm$^2$).
2. Transfection of GP+E-86 cells:
   (a) Dispense 280 μL of the Tx-buffer per tube (universal containers).
   (b) Prepare DNA tubes

| Construct name | DNA conc. | 15 μg DNA | Buffer A | CaCl, 2.5 M | Total |
|---|---|---|---|---|---|
| | Z* μg/μL | X* μL | 280-30-X | 30 | 280 |

*X × Z = 15 μg.

   (c) Add the Tx buffer drop by drop to the tubes containing the DNA solution. Mix gently.
   (d) Incubate at room temperature for 30 min to allow precipitate formation.
   (e) Add fresh medium on to the GP+E-86 cells, 5 mL per 25-cm$^2$ flask.
   (f) Gently add the mix solution (Tx + DNA) onto the GP+E-86 cells and incubate 4 to 6 h at 37°C.
   (g) Wash the cells carefully 3 times with D-PBSA in the 25-cm$^2$ flask.
   (h) Add fresh medium to the cells and incubate at 37°C overnight. :
3. Change medium on the cells and add 2 mL fresh medium (instead of 5 mL) for virus production.
4. On the same day as step 3, seed $1 \times 10^4$ PG13 and/or AM-12 cells in each well of a 6-well plate.
5. On the following day, harvest the supernate from the transfected GP+E-86 cells and add Polybrene to a final concentration of 8 μg/mL (i.e., 1-μL/mL supernate).
6. Pass the supernate through a 0.45-μm filter and add 2 mL of filtrate to each well containing second packaging cells (e.g., PG13, AM-12 cells).
7. Centrifuge the plates at 32°C at 1000 *g*
8. Incubate at 37°C overnight.
9. Next day, change medium on cells.

#### Transduction of hMSCs, neuronal stem cells, muscle satellite, or melanocytes

1. Seed $2 \times 10^6$ transduced PG13-pBABEpuro hTERT or AM12-pBABEpuro hTERT or PG13-GC-Sam TERTor AM12-GC-Sam TERT cells in a 75-cm$^2$ culture flask.
2. Next day, add 6 mL of fresh medium to these packaging cells.
3. On the following day, plate the cells to be immortalized in 6-well plates at a concentration of approximately $2.5 \times 10^4$–$7.5 \times 10^4$ cells/mL.
4. Harvest the supernate from the packaging cells, add Polybrene to a final concentration of 8

μg/mL, and pass the supernate through a 0.45-μm filter.

5. Add 2 mL of filtered retroviral supernate to each well of the 6-well plates.

6. Centrifuge plates at 32°C at 1000 *g* and incubate at 37°C overnight.

7. Remove the retroviral supernatant and add fresh media to the cells.

### Transduction of suspension cells (e.g., mast cells)

1. Seed $2 \times 10^6$ transduced 2$^{nd}$ packaging cells in a 75-cm$^2$ culture flask.

2. Next day, add 6 mL of fresh medium to the second packaging cells.

3. Following day, plate the mass cells for transduction in 6-well plates at a concentration of approximately $7.5 \times 10^4 - 10 \times 10^4$ cells/mL.

4. Harvest the supernate from the packaging cells, add Polybrene to a final concentration of 8 μg/mL and pass the supernate through a 0.45-μm filter.

5. Centrifuge plates at 32°C at 1000 *g* for 3 min.

6. Carefully replace 1200 μL of media with 1200 μL of filtered retroviral supernate for each well of the 6-well plates.

7. Place the plates on a rotator/shaker with 100 rpm (to allow better and constant mixing of virus particles and cells at 37°C) and incubate overnight.

8. Next day centrifuge plates at 32°C at 1000 *g* for 3 mins.

9. Remove the retroviral supernatant and add fresh media to the cells according to the required cell.

### Post-transduction selection

When retroviral vector with selection marker (pBABE-puromycin) has been used start selection in cell-type-specific medium with appropriate puromycin concentration (for many primary and stem cells use 2 μg/ml puromycin as the final concentration).

### Post-transduction care of cultures

1. The cells that have not received hTERT gene should start to die at about passages 10 to 12; the remaining cells should be the cells that incorporated the hTERT gene and will be selected with subsequent passage.

2. It is advisable to freeze ampoules of cells routinely to build up a stock of transduced cells at different population doubling levels that match the PD growth curve, and to save time if cultures are lost due to contamination.

3. Use the following formula to derive the PD at each subculture and to establish a PD growth curve:

$$PD = \frac{\ln(N_{finish}/N_{start})}{\ln 2}$$

where PD is the number of population doublings, ln is the natural logarithm, $N_{start}$ is the number of cells initially seeded, and $N_{finish}$ is the total number of cells recovered at subculture. *Example (for $3.2 \times 10^6$ cells recovered from a seeding of $2 \times 10^5$):*

$$PD = \frac{\ln(3.2 \times 10^6/2 \times 10^5)}{\ln 2}$$

$$PD = \frac{\ln(16)}{\ln 2}$$

$$PD = \frac{2.7726}{0.6931} = 4$$

If the cells are split 1:4 at each subculture then $N_{finish}$ is multiplied by 4 for each time the cells have been subcultured. So, in the example above with 10 subcultures, $N_{finish}$ is $(3.2 \times 10^6) \times 4$, ten times, or $3.4 \times 10^{12}$.

*Example (for $3.2 \times 10^6$ cells recovered from an initial seeding of $2 \times 10^5$, after 10 subcultures at 1:4 split):*

$$PD = \frac{\ln(3.4 \times 10^{12}/2 \times 10^5)}{\ln 2}$$

$$PD = \frac{\ln(1.67 \times 10^7)}{\ln 2}$$

$$PD = \frac{16.6309}{0.6931} = 23$$

4. It is essential to check the authenticity of the immortalized cell line for a period after it is established to ensure that it was derived from the original starting material and is not the result of cross-contamination from other immortal cell lines in the laboratory (*see* Sections 12.1.1, 15.2, 18.6). This can be done, for example, by PCR against the transgene, hTERT (*see* below), or DNA profiling (*see* Protocol 15.9).

### PCR for hTERT

1. Thaw the PCR reagents and keep on ice.

2. Make a master mix and remember to include positive (hTERT positive other DNA or cDNA sample) and negative controls (no DNA template):

   **(a)** GC Sam hTERT PCR mastermix:

   DNA, 100 μg/mL.........................1 μL (100 ng)
   10× PCR Buffer (+ Mg)............................2 μL
   Sense primer (2 μM)...............................2 μL

Antisense primer (2 μM)..........................2 μL
dNTP mix (10 mM)...............................0.4 μL
DNA polymerase...................................0.1 μL
UPW.....................................................12.5 μL
Final volume..............20 mL (including DNA)

**(b)** pBabe-puro hTERT PCR mastermix:

DNA,100 μg/mL.................1–2 μL (100 ng)
10× PCR Buffer (+ Mg)........................2μL
Sense primer (2 μM)..............................2 μL
Antisense primer (2 μM).........................2 μL
Q-solution.............................................4 μL
dNTP mix (10 mM)...............................0.4 μL
HotStart Taq DNA polymerase...........0.1 μL
UWP.....................................................8.5 μL
Final volume..............20 mL (including DNA)

3. Place the tubes in PCR machine and run program: initial denaturation at 94°C for 3 min, denaturation at 94°C for 30 s, annealing at 59°C for 30 s, extension at 74°C for 1 min for 30 cycles; then 74°C for 10 min for GC Sam hTERT and run program for pBABe puro-hTERT initial denaturation 94°C for 7 min, denaturation at 95°C for 30 s, annealing at 58°C for 45 s, extension at 72°C for 30s for 33 cycles, then72°C for 10 min.
4. Run PCR products on 1.5% to 2% agarose gel.

## 17.4.5 Lymphocyte Immortalization

Early attempts to grow lymphocytes in culture [Moore et al., 1967] showed that immortalized lines could be derived from very dense cell pellets in a culture tube. The cells were from the B-lymphocyte lineage and were immortalized by endogenous Epstein Barr Virus (EBV). Immortalization of B-lymphocytes by infection with EBV is now a routine procedure with a very high success rate [Bolton & Spurr, 1996].

## 17.4.6 Transgenic Mouse

The transgenic mouse Immortomouse ($H-2K^b$-$tsA58$ SV40 large T) carries the temperature-sensitive SV40LT gene. A number of tissues from this mouse give rise to immortal cell lines, including colonic epithelium [Fenton & Hord, 2004], brain astroglia [Noble & Barnett, 1996], muscle [Ahmed et al., 2004], and retinal endothelium [Su et al., 2003].

## 17.5 ABERRANT GROWTH CONTROL

Cells cultured from tumors, as well as cultures that have transformed in vitro, show aberrations in growth control, such as growth to higher saturation densities [Dulbecco &

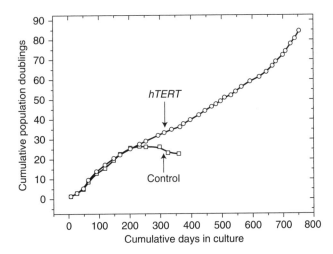

***Fig. 17.4. Cumulative Population Doublings (PD) of hTERT-Immortalized Cells.*** Cumulative PD were calculated for cultures of human mesenchymal stem cells (*see* Protocol 17.2) and plotted against time in culture for cells with ectopic expression of hTERT after retroviral transduction (circles) and control (nontransduced) cells (squares). (Data courtesy of N. Serakinci).

Elkington, 1973], clonogenicity in agar [Freedman & Shin, 1974], and growth on confluent monolayers of homologous cells [Aaronson et al., 1970]. These cell lines exhibit lower serum or growth factor dependence, usually form clones with a higher efficiency, and are assumed to have acquired some degree of autonomous growth control by overexpression of oncogenes or by deletion of suppressor genes. Growth control is often autocrine—meaning the cells secrete mitogens for which they possess receptors, such as TGF-α; or the cells express receptors, such as erbB, or stages in signal transduction that are permanently active and unregulated, such as hRas. Although immortalization does not necessarily imply a loss of growth control, many cells progress readily from immortalization to aberrant growth, perhaps because of genetic instability intrinsic to the immortalized genotype.

### 17.5.1 Anchorage Independence

Many of the properties associated with neoplastic transformation in vitro are the result of cell surface modifications [Hynes, 1973; Nicolson, 1976], such as changes in the binding of plant lectins [Laferte & Loh, 1992], in cell surface glycoproteins [Bruyneel et al., 1990; Carraway et al., 1992], and in cell adhesion molecules [Yang et al., 2004], many of which may be correlated with the development of invasion and metastasis in vivo. Fibronectin is lost from the surface of transformed fibroblasts [Hynes, 1973; Vaheri et al., 1976] due to alterations in integrin binding. This loss may contribute to a decrease in cell–cell and cell–substrate adhesion [Yamada, 1991; Reeves, 1992] and to a decreased requirement for attachment and spreading for the cells to proliferate.

Transformed cells may lack specific CAMs (e.g., L-CAM), which, when transfected back into the cell, regenerate the normal, noninvasive phenotype [Mege et al., 1989], and as such, they may be recognized as tumor suppressor genes. Other CAMs may be overexpressed, such as N-CAM in small-cell lung cancer [Patel et al., 1989], when the extracellular domain is subject to alternative splicing [Rygaard et al., 1992]. The expression of and degree of phosphorylation of integrins may also change [Persad & Dedhar, 2003; Lipscomb, & Mercurio, 2005], potentially altering cytoskeletal interactions, the regulation of gene transcription, the substrate adhesion of the cells, and the relationship between cell spreading and cell proliferation [Frame & Norman, 2008; Zhang et al., 2008].

In addition the loss of cell–cell recognition, a product of reduced cell–cell adhesion, leads to a disorganized growth pattern and the loss of contact inhibition of cell motility and density limitation of cell proliferation (*see* Section 17.5.1). Cells can grow detached from the substrate, either in stirred suspension culture or suspended in semisolid media, such as agar or Methocel. There is an obvious analogy between altered cell adhesion in culture and detachment from the tissue in which a tumor arises and the subsequent formation of metastases in foreign sites, but the rational basis for this analogy is not clear; new adhesions are clearly involved promoted by alterations in integrin binding [Dowling et al., 2008].

***Suspension cloning.*** Macpherson and Montagnier [1964] were able to demonstrate that polyoma-transformed BHK21 cells could be grown preferentially in soft agar, whereas untransformed cells cloned very poorly. Subsequently it has been shown that colony formation in suspension is frequently enhanced after viral transformation. The situation regarding spontaneous tumors is less clear, however, despite the fact that Shin and coworkers [Freedman and Shin 1974; Kahn & Shin, 1979] demonstrated a close correlation between tumorigenicity and suspension cloning in Methocel. Although Hamburger and Salmon [1977] showed that many human tumors contain a small percentage of cells (<1.0%) that are clonogenic in agar, a number of normal cells will also clone in suspension [Laug et al., 1980; Peehl & Stanbridge 1981; Freshney & Hart, 1982] (*see* Fig. 13.11). Because normal fibroblasts are among these cells, the value of this technique for assaying for the presence of tumor cells in short-term cultures from human tumors is in some doubt.

Techniques for cloning in suspension are described in Chapter 14 (*see* Protocols 13.4, 13.5). Variations with particular relevance to the assay of neoplastic cells lie in the choice of the suspending medium. It has been suggested [Neugut & Weinstein, 1979] that agar may allow only the most highly transformed cells to clone, whereas agarose (which lacks sulfated polysaccharides) is less selective. Montagnier [1968] was able to show that untransformed BHK21 cells, which would grow in agarose but not in agar, could be prevented from growing in agarose by the addition of dextran sulfate.

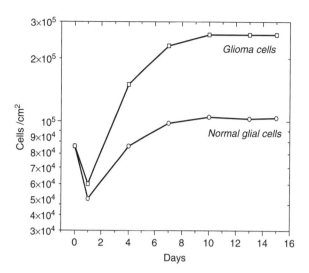

**Fig. 17.5. Density Limitation of Cell Proliferation.** The difference in plateaus (saturation densities) attained by cultures from normal brain (circles, solid line) and a glioma (squares, broken line). Cells were seeded onto 13-mm coverslips, and 48 h later the coverslips were transferred to 9-cm Petri dishes containing 20 mL of growth medium, to minimize exhaustion of the medium.

## 17.5.2 Contact Inhibition

The loss of contact inhibition may be detected morphologically by the formation of a disoriented monolayer of cells (*see* Plate 14a, b) or rounded cells in foci within the regular pattern of normal surrounding cells (*see* Figs. 17.3 and Plate 14c). Cultures of human glioma show a disorganized growth pattern and exhibit reduced density limitation of growth by growing to a higher saturation density than that of normal glial cell lines (Fig. 17.5) [Freshney et al., 1980a, b]. As variations in cell size influence the saturation density, the increase in the labeling index with [³H]thymidine at saturation density (*see* Protocol 20.11) is a better measure of the reduced density limitation of growth. Human glioma, labeled for 24 h at saturation density with [³H]thymidine, gave a labeling index of 8%, whereas normal glial cells gave 2% [Guner et al., 1977].

---

**PROTOCOL 17.3. DENSITY LIMITATION OF CELL PROLIFERATION**

***Outline***
Grow the culture to saturation density in nonlimiting medium conditions, and determine, autographically, the percentage of cells labeling with [³H]thymidine.

***Materials***

***Sterile or aseptically prepared:***

❑ Culture of cells ready for subculture

❑ Growth medium

❑ Maintenance medium (no serum or growth factors) containing 37 KBq/mL (1.0 μCi/mL) of [³H]thymidine, 74 GBq/mmol (2Ci/mmol)

❑ D-PBSA

❑ Trypsin, 0.25%

❑ Multiwell plates, 24-well, containing coverslips 13 mm in diameter

❑ Petri dishes, 9 cm (bacteriological grade, 1 per coverslip)

### Procedure

1. Trypsinize the cells and seed $1 \times 10^5$ cells/mL into a 24-well plate, 1 mL/well, each well containing a 13-mm-diameter coverslip.

2. Incubate the cells in a humidified $CO_2$ incubator for 1 to 3 days.

3. Transfer the coverslips to 9-cm bacteriological grade Petri dishes, each containing 20 mL of medium, and place the dishes in the $CO_2$ incubator.

4. Continue culturing, changing the medium every 2 days once the cells become confluent on the coverslips.

5. Trypsinize and count the cells from two coverslips every 3 to 4 days. As the cells become denser on the coverslip, it may be necessary to add 200 to 500 units/mL of crude collagenase to the trypsin in order to achieve complete dissociation of the cells for counting.

6. When cell growth ceases (i.e., two sequential counts show no significant increase), add 2.0 mL 37 KBq/mL (1.0 μCi/mL) of [³H]thymidine, 74 GBq/mmol (2 Ci/mmol), and incubate the cells for a further 24 h.

Δ **Safety Note.** Handle [³H]thymidine with care. Although it is a low-energy β-emitter, it localizes to DNA and can induce radiolytic damage. Wear gloves, do not handle [³H]thymidine in a horizontal laminar-flow hood but in a biohazard or cytotoxic drug handling hood (*see* Section 6.7.1), and discard waste liquids and solids by the appropriate route specified in the local rules governing the handling of radioisotopes.

7. Transfer the coverslips back to a 24-well plate, and trypsinize the cells for autoradiography (*see* Section 27.2). The cells may be fixed in suspension and dropped on a slide as for chromosome preparations (*see* Protocol 15.7 without the hypotonic treatment), centrifuged onto a slide with a cytocentrifuge (see Protocol 15.4), or trapped on filters by vacuum filtration (*see* Protocol 15.5).

*Note.* It is necessary to trypsinize high-density cultures for autoradiography because of their thickness and the weak penetration of β-emission from ³H. Labeled cells in the underlying layers will not be detected by the radiosensitive emulsion, due to the absorption of the β-particles by the overlying cells (the mean path length of β-particles in water is approximately 1 μm). If the cells remain as a monolayer at saturation density, this step may be omitted, and autoradiographs may be prepared by mounting the coverslips, cells uppermost, on a microscope slide.

*Analysis.* Count the number of labeled cells as a percentage of the total number of cells. Scan the autoradiographs under the microscope, and count the total number of cells and the proportion of cells labeled in representative parts of the slide (*see* Fig. 20.18).

*Variations.* Cells in DNA synthesis may also be labeled with bromodeoxyuridine (BUdR) and subsequently detected by antibody to BUdR-labeled DNA (Dako). Human cycling cells may be labeled with the Ki67 monoclonal antibody (Dako) against DNA polymerase, or with anti-PCNA against proliferating cell nuclear antigen (PCNA). Although there is generally good agreement between [³H]thymidine and BUdR labeling, Ki67 and PCNA will label more cells, as the antigen is present throughout the cycle and is not restricted to the S-phase.

Growth of cells at high density in nonlimiting medium can be achieved by growing the cells in a filter well, choosing a filter diameter substantially below that of the dish (e.g., the Corning 8-mm filter in a 24-well plate), and counting the number of cells in plateau, performing an autoradiograph or immunostaining with Ki67 or anti-PCNA.

### 17.5.3 Serum Dependence

Transformed cells have a lower serum dependence than their normal counterparts [Temin, 1966, Eagle et al., 1970], due, in part, to the secretion of growth factors by tumor cells [Todaro & DeLarco, 1978]. These factors have been collectively described as autocrine growth factors. Implicit in this definition is that (1) the cell produces the factor, (2) the cell has receptors for the factor, and (3) the cell responds to the factor by entering mitosis. Some of these factors may have an apparent transforming activity on normal cells (e.g., TGF-α) binding to the EGF receptor and inducing mitosis [Richmond et al., 1985], although, unlike true transformation (*see* Section 17.2), this type of transformation is reversible. These factors also cause nontransformed cells to adopt a transformed phenotype and grow in suspension [Todaro & DeLarco, 1978]. This effect can be assayed by treating NRK cells with conditioned medium from the test cell and cloning them in suspension (*see* Protocols 13.2, 13.4, 13.5).

Tumor cells can also produce many hemopoietic growth factors, such as interleukins 1, 2, and 3, along with colony-stimulating factor (CSF) [Fontana et al., 1984; Metcalf,

1990]. It has been proposed [Cuttitta et al., 1985] that some factors, such as gastrin-releasing peptide and vasoactive intestinal peptide (VIP), and human chorionic gonadotropin (hCG), hitherto believed to be ectopic hormones produced by lung carcinomas, may in fact be autocrine growth factors. Autocrine growth factors can be detected by immunostaining (*see* Protocol 15.11) or increased gene expression, but their value as transformation markers is limited, because many normal cells, such as glia, fibroblasts, and endothelial cells, produce autocrine factors when proliferating.

### 17.5.4 Oncogenes

Autonomous growth control is also achieved in transformed cells by oncogenes, expressed as modified receptors, such as the *erb*-B2 oncogene product, and the modified G protein, such as mutant *ras*, or by the overexpression of genes regulating stages in signal transduction (e.g., *src* kinase) or transcriptional control (e.g., *myc, fos,* and *jun*) [Bishop, 1991]. In many cases, the gene product is permanently active and is unable to be regulated [Alberts et al., 2008, p. 1243]. Overexpression of oncogenes can be detected by immunostaining (*see* Protocol 15.11), in situ hybridization, immunoblotting for the protein product, RT-PCR for mRNA, or microarray analysis. In some cases, the oncogene product (e.g., *erb*-B2, activated Ha-*ras*) can be distinguished from the normal product (e.g., EGF receptor, normal *ras*, respectively) qualitatively as well as quantitatively, by specific antibodies or gene sequencing.

## 17.6 TUMORIGENICITY

Transformation is a multistep process that often culminates in the production of neoplastic cells [Quintanilla et al., 1986; Alberts et al., 2008, p. 1209–1215]. However, cell lines derived from malignant tumors, presumably already transformed, can undergo further transformation with an increased growth rate, reduced anchorage dependence, more pronounced aneuploidy, and immortalization. This suggests that a series of steps, not necessarily coordinated or interdependent and not necessarily individually tumorigenic, is required for malignant transformation. Furthermore all cell lineages present within a tumor need not have the same transformed properties, and the same set of properties need not be expressed in every tumor. Progression may imply the expression of new properties or the deletion of old ones that may induce metastasis or even spontaneous remission.

There are therefore several steps in transformation, the sequence of which may be determined by environmental selective pressure. In vitro, where little restriction on growth is imposed, the events need not necessarily follow in the same sequence as in vivo.

### 17.6.1 Malignancy

Malignancy implies that the cells have developed the capacity to generate invasive tumors if implanted in vivo into an isologous host or if transplanted as a xenograft into an immune-deprived animal. Although the development of malignancy can be recognized as a discrete phenotypic event, it often accompanies the development of aberrant growth control, suggesting that some of the lesions responsible for aberrant growth control also cause malignancy. An obvious candidate for such lesions is a deficit in cell–cell interaction that deprives the cell of control of proliferation (density limitation of cell proliferation) and of motility control (contact inhibition).

Two approaches have been used to explore malignancy-associated properties: (1) cells have been cultured from malignant tumors and characterized; and (2) transformation in vitro with a virus or a chemical carcinogen, or transfection with oncogenes, has produced cells that were tumorigenic and that could be compared with the untransformed cells. The second approach provides transformed clones of the same lineage, which can be shown to be malignant, and these clones can be compared with untransformed clones, which are not malignant. Unfortunately, many of the characteristics of cells transformed in vitro have not been found in cells derived from spontaneous tumors. Ideally tumor cells and equivalent normal cells should be isolated and characterized. Unfortunately, there have been relatively few instances for which this arrangement has been possible, and even then, although the cells may belong to the same lineage, their position in that lineage is not always clear; thus comparison is not strictly justified. Although the bulk of cells in normal adult tissue will be differentiated and nonproliferative, those comprising a tumor will tend to be proliferating and undifferentiated, and this status will distinguish the tumor population from the normal, regardless of their malignancy. Furthermore it would appear that many, if not most, cells in a tumor do not have a prolonged life span in culture, and the population crucial to the advancement of the tumor, and consequently the main target for chemotherapy, may be quite small and equivalent to a stem cell population in normal tissue. As yet there is little conclusive evidence that these cells can be cultured, but an increasing amount of effort is being directed that way [Graziano et al., 2008; Vermeulen et al., 2008; Dick 2009].

### 17.6.2 Tumor Transplantation

The only generally accepted sign of malignancy is the demonstration of the formation of invasive or metastasizing tumors in vivo. Transplantable tumor cells ($\sim 1 \times 10^6$) injected into isogeneic hosts will produce invasive tumors in a high proportion of cases, whereas $1 \times 10^6$ normal cells of similar origin will not. Models have been developed, using immune-suppressed or immune-deficient host animals, to study the tumorigenicity of human tumor cells. The genetically athymic "nude" mouse [Giovanella et al., 1974] and thymectomized irradiated mice [Bradley et al., 1978a; Selby et al., 1980] have both been used extensively as hosts for xenografts. The take rate of the grafts varies, however, and many clearly defined tumor cell lines and tumor biopsies fail to produce tumors as xenografts; those that do take frequently fail to metastasize, although they may be invasive locally. Take rates can be improved by sublethal irradiation of the host nude mouse (30–60 Gy), by using asplenic athymic (*scid*) mice, or by implanting the cells in Matrigel™ [Pretlow et al., 1991]. Despite the frequency of false negatives, tumorigenesis remains a good indicator of malignancy.

### 17.6.3 Invasiveness

Tumorigenesis assays should always be accompanied by histology of the tumor to confirm its histopathological similarity to the original tumor and to demonstrate that it is invasive. However, if the cells are not tumorigenic, or if transplantation facilities are not available or are not considered desirable, then it is possible to utilize a number of in vitro assays. Some of these assays also provide models that are more readily quantified than in vivo assays.

*Chick chorioallantoic membrane.* The chorioallantoic membrane (CAM; *see* Fig. 11.4*d, e*) assay can be performed on chick embryos in ovo or on explanted CAM in vitro. Easty and Easty & Easty [1974] showed that invasion of the CAM could be demonstrated in organ culture, and others [Hart & Fidler, 1978] attempted, with some limited success, to construct a chamber capable of quantifying the penetration of tumor cells across the CAM. An advantage of the CAM assay in ovo is that it may also show angiogenesis (*see* Section 17.6.4; Plate 15*f, g*), and subsequent histology may reveal whether the tumor cells have penetrated the underlying basement membrane.

*Organoid confrontation.* Mareel et al. [1979] developed an in vitro model for invasion, using chick embryo heart fragments cocultured with reaggregated clusters of tumor cells (Fig. 17.6). Invasion appears to be correlated with the malignant origin of the cells, is progressive, and causes destruction of the host tissue. The application of this technique to human tumor cells shows a good correlation between malignancy and invasiveness in the assay [de Ridder & Calliauw, 1990]. This technique has been used extensively [e.g., Vanhoecke et al., 2005] but is difficult to quantify; it requires skilled histological interpretation.

*Filter well inserts.* A number of filter well techniques have been developed, based on the penetration of filters coated with Matrigel or some other extracellular matrix constituent (Fig. 17.7) [Repesh, 1989; Schlechte et al., 1990; Brunton et al., 1997; Lamb et al., 1997]. The degree of penetration into the gel, or through to the distal side of the filter, is rated as invasiveness and is determined histologically by the number of cells and the distance moved, or by prelabeling the cells with $^{125}$I and counting the radioactivity on the distal side of the filter or bottom of the dish. It is more readily quantified but lacks the presence of normal host cells

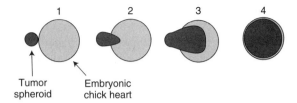

*Fig. 17.6. Chick Heart Assay.* Tumor spheroids (*see* Protocol 25.2) are cocultured with healed fragments of 8-day embryonic chick heart. (1) The spheroid adheres to the heart after a few hours; (2) after 24 to 48 h it starts to penetrate the chick heart. (3) It spreads within the heart fragment; (4) by 8 to 10 days it has completely replaced the heart tissue. (After Mareel et al., 1979).

in the barrier that are normally associated with invasion in vivo. The penetration into Matrigel is likely to be a measure of matrix degradation and reflects the production of proteases or glycosidases by the cells. It is not clear how cells that do not make their own degradative enzymes, but instead rely on the production of proteases induced in the stroma, will perform in these assays. Preliminary experiments in the author's laboratory showed that tumor cells seeded on top of a filter carrying an established monolayer of normal cells penetrated the filter and appeared in the lower compartment more so than normal cells seeded as controls but trypsinization before seeding the cells appeared to make both cell types invasive [Christie & Freshney, unpublished observations]. (*See also* Plate 19*e*.)

### 17.6.4 Angiogenesis

Tumor cells release factors, including VEGF [Joukov et al., 1997], FGF-2 [Thomas et al., 1997], and angiogenin [Hu et al., 1997], that are capable of inducing neovascularization [Folkman, 1992; Skobe & Fusenig, 1998]. Fragments of tumor, pellets of cultured cells, or cell extracts, implanted on the surface of the CAM of a hen's egg, promote an increase in vascularization that is apparent to the naked eye 6 to 8 days later (Plate 9). Because this assay is not readily quantified, the stimulation of cell migration [Bagley et al., 2003], production of vascular endothelial growth factor (VEGF) [Buchler et al., 2004], or morphogenesis [Chen et al., 1997; Ment et al., 1997] in filter well cultures of vascular endothelium have been used. Migration of endothelial cells from microcarriers (*see* Section 26.9) into a suitable matrix, such as fibrin [Nehls & Drenckhahn, 1995] or Matrigel [Crabtree & Subramanian, 2007] provides an assay for angiogenesis based on the migration and morphogenesis of endothelial cells.

The following introduction and Protocol 17.5 were contributed by Vasanta Subramanian, Department of Biology and Biochemistry, Building 4 South, University of Bath, Claverton Down, Bath BA2 7AY, England, UK.

Endothelial cells that line the vasculature are normally quiescent but can be induced to proliferate, migrate, and form new blood vessels when stimulated by angiogenic factors. This process is referred to as angiogenesis. Angiogenesis

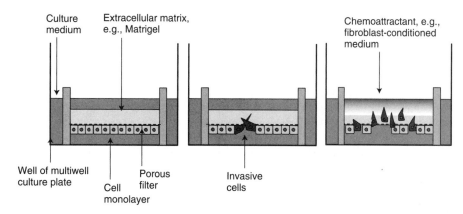

Culture medium | Extracellular matrix, e.g., Matrigel | Chemoattractant, e.g., fibroblast-conditioned medium

Well of multiwell culture plate | Cell monolayer | Porous filter | Invasive cells

*Fig. 17.7. Filter Well Invasion.* Cells plated on the underside of the filter migrate through the filter into growth factor-depleted Matrigel™ in the well of the filter insert, encouraged by the addition of a chemoattractant, such as fibroblast-conditioned medium, to the upper side of the Matrigel™. (After Brunton et al. [1997].)

occurs in disease conditions such as tumor growth and diabetic retinopathy as well as in some normal processes such as wound healing and embryonic development [Carmeliet, 2005]. Many of the assays for angiogenic factors and their inhibitors involve in vivo approaches such as the chick chorioallantoic membrane (CAM) assay and the rabbit corneal assay. These are technically difficult, tedious to perform, and involve the use of animals and embryos. Reproducibility can also be a problem in many of these in vivo assays.

The extracellular matrix (ECM) plays a major role in the regulation of angiogenesis [Adams and Watt, 1993]. It is possible to extract the ECM from Engelbreth–Holm–Swarm (EHS) mouse sarcoma cell lines [Kleinman et al., 1986] for use as a support for in vitro angiogenesis assays. Earlier in vitro assays for angiogenesis involved seeding endothelial cells on top of polymerized matrices such as Matrigel to form a two-dimensional culture system. Assaying growth factors or inhibitors of angiogenesis in such 2-D systems sometimes yields ambiguous results. However, cultured endothelial cells can form tubes that resemble endothelial vessels in 3-D matrices [Nehls and Drenckhahn, 1995] if provided the right culture conditions and thus provide an excellent alternative to in vivo assays.

Endothelial cells grown on Cytodex-3 microcarrier beads can be embedded in Matrigel (*see* Section 7.2.1) to provide an easy, rapid and reliable in vitro assay system for evaluating and measuring angiogenic activity [Crabtree and Subramanian, 2007]. This is a modification of the in vitro angiogenesis assay developed by Nehls and Drenckhahn [1995], which used fibrinogen to generate a fibrin matrix by the action of thrombin. This matrix was used for embedding the cell-coated Cytodex-3 beads. However, the polymerization of the fibrinogen into a gel is very variable and is sensitive to batch variations in the thrombin used for the polymerization. Matrigel consistently and reproducibly forms gels without the use of enzymes and thus is a preferred 3-D support. The modified angiogenesis assay using growth factor reduced Matrigel in conjunction with endothelial cell

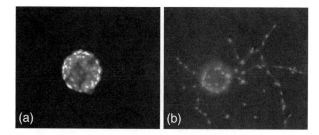

*Fig. 17.8. In vitro Angiogenesis Assay.* Endothelial cells grown on Cytodex beads induced to form tubes. (*a*) Endothelial cells forming a monolayer on the surface of the Cytodex beads (control) stained with DAPI. (*b*) Endothelial cells grown on beads and induced to form tubes by treatment with 10 ng/mL FGF. Stained with DAPI. (Courtesy of V. Subramanian. *See also* Plate 15*e*).

coated Cytodex-3 beads is reliable, easy to perform and the results can be easily quantified as the endothelial tubes radiate out from a focal point—the Cytodex3 bead (*see* Plate 15*e*) [Crabtree and Subramanian, 2007].

### PROTOCOL 17.4. IN VITRO ANGIOGENESIS ASSAY

#### *Outline*
Trypsinise monolayer of cells and seed on Cytodex-3 beads. Embed the cell-covered beads in Matrigel in 8-well chamber slides or 48-well plates. Incubate in culture medium. Treat with angiogenic factors, inhibitors drugs, and so on. Analyze after staining.

#### *Materials*

*Sterile or aseptically prepared:*

❑ Endothelial cell line(s) (*see* Section 22.2.9; Table 22.1; Appendix II)
❑ Cell culture medium with normal FBS as appropriate for the cell line (*see* Protocol

22.18) with 100 U/mL penicillin, 100 μg/mL streptomycin. Transferrin and Insulin (optional; Invitrogen)

❑ Reduced serum medium: same medium as above with 0.1% FBS. Use reduced serum medium as the basal medium for the angiogenic assay.

❑ NEAA (100× stock; Invitrogen)

❑ Fetal bovine serum (FBS; Invitrogen)

❑ Growth factor reduced (GFR), phenol red free Matrigel (BD Biosciences), freeze thawed 2 to 3 times prior to use (this is important as it reduces the background tube formation in the negative control)

❑ Human recombinant basic fibroblast growth factor (bFGF) (R&D Systems)

❑ Trypsin EDTA or any other disassociation agent appropriate for the cell line to be used

❑ Phosphate buffered saline (D-PBSA)

❑ Cytodex-3 microcarrier beads (GE Healthcare)

❑ Chamber slides, 8 well (Sterilin, Iwaki)

❑ Flasks: 25 cm² and 75 cm²

❑ Centrifuge tubes

❑ Plugged pipette tips (Thermo Fisher)

*Nonsterile:*

❑ Ice bucket

❑ Water bath

### Procedure

*A. Culture of cells on Cytodex-3 beads (48 h prior to embedding in Matrigel):*

1. Prepare and sterilize the Cytodex beads following the supplier's instructions.
2. Transfer 2 mL of Cytodex beads in PBSA to a 15-mL tube and allow to settle.
3. Add 5 mL of fresh D-PBSA to wash. Allow beads to settle and remove the D-PBSA.
4. Add 5 mL of complete medium with 10% FBS.
5. Allow the beads to settle and remove the medium.
6. Add 5 mL of fresh medium to the settled beads.
7. Equilibrate the Cytodex beads in 5 mL of the appropriate cell culture medium for 1 h at 37°C in 5% $CO_2$ (lids loosened on tubes) prior to the addition of cells.
8. Trypsinize an 80% to 90% confluent monolayer of endothelial cells from a 25-cm² flask and suspend in 6 mL of culture medium.
9. Count the cells.
10. Remove the equilibration medium from the beads, and add the cells to the Cytodex beads at about 30 cells per bead in 2 mL of complete medium in a 25-cm² tissue culture flask.
11. Incubate the flask in an upright position at 37°C in 5% $CO_2$ for 4 h with gentle agitation every 30 min.

12. At the end of the incubation, add a further 3 mL of medium and place the flask horizontally in the incubator and incubate at 37°C, 5% $CO_2$ for 48 h to allow the cells to grow to confluency on the Cytodex beads.

*B. Addition of cell-coated Cytodex-3 beads to Matrigel:*

1. Thaw Matrigel overnight at 4°C on ice (on the day prior to embedding cell coated beads in Matrigel).
2. Check that the beads are completely covered by the endothelial cells and that the cells look healthy (*see* Plate 18*c* for comparison).
3. Transfer the cell-coated Cytodex beads to a sterile 15-mL tube and allow to settle.
4. Aspirate the culture medium, and wash the beads once with 10 mL of D-PBSA and a further twice using reduced serum medium.
5. Incubate the beads for 1 h at 37°C in 5% $CO_2$ and reduced serum medium.
6. Label 8-well Iwaki chamber slides for the assay (48-well plates can be used instead), allowing at least 3 wells per sample. Include a negative control which is 0.1% FCS media alone and a positive control such as bFGF.
7. Count the beads and dilute to give 40 beads per 75 μL, and place on ice for 2 to 5 min; then add Matrigel in a ratio of 1:1 v/v. Mix gently to ensure uniform suspension of beads.
8. Add 150 μl of the Matrigel containing the cell-coated beads to each well of the 8-well chamber slides, and incubate at 37°C and 5% $CO_2$ to allow the Matrigel to set.
9. Once the Matrigel has set, add 500 μL of the reduced serum medium supplemented with the appropriate growth factors and inhibitors to be tested to each well.
10. Incubate at 37°C in 5% $CO_2$.

*C. Analysis:*

*Fixing and staining:*

1. Fix the Matrigel bead cell cultures in 4% buffered paraformaldehyde (pH 7.4) for 15 min at 4°C on day 5 or as required.
2. Stain with bis-benzimide (Hoechst 33258,20 μg/mL in PBSA for 1 h at room temperature) or with DAPI. This allows the number of cells in each process to be visualized by fluorescence microscopy (Fig. 17.8; Plate 15*e*)

*Quantification of cell migration and tube formation:*

3. Observe under the microscope daily for 5 days.
4. Count the number of Cytodex beads showing outgrowths (cellular processes), and express the results as percentage of microcarrier beads

showing cellular processes. A cellular process is classified as an outgrowth from the microcarrier bead that is longer than 150 μm and consists of 3 or more cells (Nehls and Drenckhahn, 1995). Cytodex-3 microcarrier beads have a diameter of 150 μm and serve as the internal standard for measuring the length of the outgrowth. The tube formation can also be confirmed by antibody staining for endothelial markers, such as factor VIII, PECAM-1, or endoglin (*see* Protocol 22.18).

5. Record the number of cellular processes present on each Cytodex bead using an inverted microscope.

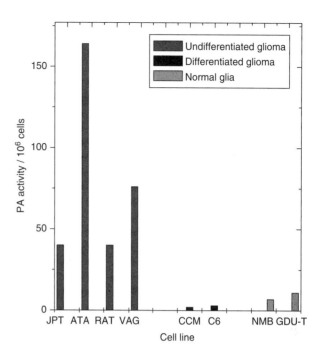

*Fig. 17.9. Plasminogen Activator.* PA produced by tumor cells in vitro (the units are arbitrary). The PA activities of the four gliomas—JPT, ATA, RAT, and VAG—were all higher than cells cultured from normal brain (NMB-C, GDU-T). It was also found that the only cells to produce the differentiated glial marker glial fibrillary acidic protein—CCM and C6—had the lowest PA of all. (Data courtesy of M. Frame).

*Applications.* The Matrigel bead assay for angiogenesis is a very reliable and reproducible assay as the Matrigel is more representative of the ECM than the individual components that have been used in the past. Matrigel sets easily and does not require enzymes for polymerization. The beads provide a focal point from which the tubes radiate out, making it easy to quantify the tube formation. This assay can be used to determine angiogenic activity, to assess the structure function relationships of angiogenic molecules, and to study the combination of factors necessary for inducing angiogenesis. This in vitro assay can be used to test inhibitors of angiogenesis. Gene expression studies and characterization of angiogenic inducers expressed during the process of angiogenesis can also be investigated using this assay system. It reduces the use of animals and embryos and hence raises fewer ethical problems. Since it is a three-dimensional system, it mimics the in vivo system quite closely.

## 17.6.5 Plasminogen Activator

Other products that tend to be increased in transformed cells are proteolytic enzymes [Mahdavi & Hynes, 1979], long since associated with theories of invasive growth [Liotta, 1987]. Because proteolytic activity may be associated with the cell surface of many normal cells and is absent on some tumor cells, an equivalent normal cell must be used as a control when using this criterion. Plasminogen activator (PA) is higher in some cultures from human glioma than in cultures from

normal brain [Hince & Roscoe, 1980; Frame et al., 1984] (Fig. 17.9), and other cultures have previously shown that PA is associated with many different tumors [de Vries et al., 1995; del Vecchio et al., 1993; Schwartz Albiez et al., 1991]. PA may be measured by clarification of a fibrin clot or by release of free soluble $^{125}$I from [$^{125}$I] fibrin [Unkless et al., 1974; Strickland & Beers, 1976]. In addition a simple chromogenic assay has been developed by Whur et al. [1980].

It has been proposed that, for some carcinomas, soluble urokinase-like PA (uPA) is elevated more than tissue-type PA (tPA) [Markus et al., 1980; Duffy et al., 1990], so it is informative to couple the chromogenic assay with zymogram analysis [Davies et al., 1993; Boxman et al., 1995] and then immunoblot to determine the proportion of each type of PA.

# CHAPTER 18

# Contamination

## 18.1 SOURCES OF CONTAMINATION

Maintaining asepsis is still one of the most difficult challenges to the newcomer to tissue culture. Awkwardness during early training can be overcome by experience, but in certain situations even the most experienced worker will suffer from contamination. There are several potential routes to contamination (Table 18.1), including failure in the sterilization procedures for solutions, glassware and pipettes, turbulence and particulates (dust and spores) in the air in the room, poorly maintained incubators and refrigerators, faulty laminar-flow hoods, the importation of contaminated cell lines or biopsies, and lapses in sterile technique. The last of these is probably the most significant.

### 18.1.1 Operator Technique

If reagents are sterile and equipment is in proper working order, contamination depends on the interaction of the operator's technique with environmental conditions. If the skill and level of care of the operator is high and the atmosphere is clean, free of dust, and still, contamination as a result of manipulation will be rare. If the environment deteriorates (e.g., as a result of construction work or a seasonal increase in humidity), or if the operator's technique declines (through the omission of one or more apparently unnecessary precautions), the probability of infection increases. If both happen simultaneously or sequentially, the results can be catastrophic.

Let us consider this conjunction of events in graphic form (*see* Fig. 5.1). The maintenance of good technique may be represented by the top graph, with occasional lapses shown as a downward peak, and the quality of the environment may be depicted by the bottom graph, with occasional sporadic increases in risk, such as a contaminated Petri dish opened accidentally in the area or dust generated from equipment maintenance, shown as an upward peak. Provided that the two curves (of good technique and a high-quality environment) are kept well apart, the coincidence of a lapse in technique and an environmental breakdown will be rare. If, however, there is a progressive decline in technique or in the environment, the frequency of contamination will increase, and if both conditions deteriorate, contamination will be regular and widespread.

### 18.1.2 Environment

It is fairly obvious that the environment in which tissue culture is carried out must be as clean as possible and free from disturbance and through traffic. Conducting tissue culture in the regular laboratory area should be avoided; a laminar-flow hood will not give sufficient protection from the busy environment of the average laboratory. A clean, traffic-free area should be designated, preferably as an isolated room or suite of rooms (*see* Sections 3.2.1, 5.2.2). Equipment brought into the sterile area from storage and air currents from doors, refrigerators, centrifuges, and the movement of operators, all increase the risk of contamination. Maintain a strict cleaning program for surfaces and equipment, and wipe down anything that is brought in with a sterilizing swab.

### 18.1.3 Use and Maintenance of Laminar-Flow Hood

The commonest example of poor technique is improper use of the laminar-flow hood. If the workspace becomes overcrowded with bottles and equipment (*see* Fig. 5.4b), the laminar airflow is disrupted, and the protective boundary layer between operator and room is lost. This in turn leads to the entry of nonsterile air into the hood and the release of potentially biohazardous materials into the room. In addition the risk of collision between sterile pipettes and nonsterile surfaces of bottles, for example, increases. One should bring into the hood only those items that are directly involved in the current operation.

Laminar-flow hoods also must be maintained regularly, and the integrity of the filters, ductwork, and cabinets should be checked at least twice a year by a competent engineer. The engineer should also check the containment of the workspace, to ensure that internal air does not spill out and outside air does not enter, both of which are dependent on the internal air velocity and outside turbulence.

### 18.1.4 Humid Incubators

A major source of contamination stems from the use of humid incubators (*see* Section 5.5.1). High humidity is not required unless open vessels are being used; sealed flasks are better kept in a dry incubator or a hot room (*see* Sections 3.2.5, 4.3.1). A low-$CO_2$ medium (e.g., based on Hanks's salts; *see* Section 8.2.2) can be used to avoid the need to gas flasks with $CO_2$. If there is a need to gas flasks with $CO_2$, this is better done from a cylinder or piped supply and the flasks sealed and placed in a normal incubator. Using permeable caps (*see* Section 5.5.3, 7.3.3) minimizes the risk of contamination but still exposes the flask to a higher risk atmosphere than in a dry incubator. Permeable caps may be sealed with a secondary rubber cap (BD Biosciences) for transfer to a nongassed incubator after equilibration with $CO_2$. If flasks are maintained in a $CO_2$ incubator with slack or permeable caps, it is possible to keep the incubator dry and use a different incubator for open plates. The $CO_2$ monitoring system will need to be recalibrated if the incubator is used dry, and the flasks will need to be checked for evaporation; preferably, tighten the caps after the pH equilibrates.

There are, however, many situations where a humid incubator must be used. Selecting an incubator with an interior that is readily accessible and easily cleaned (*see* Fig. 4.12) will reduce the risk of contamination. Cultures placed in the incubator can be enclosed in a plastic box (*see* Fig. 5.11), although equilibration of pH and temperature will take longer. Fan circulation in a $CO_2$ incubator shortens recovery time for both $CO_2$ and temperature, but at the cost of an increased risk of contamination (*see* Sections 4.3.2, 5.5.1); open plate cultures are better maintained in static air and frequency of access limited as much as possible. Having a dry, non-$CO_2$ incubator for sealed flasks and short procedures such as trypsinization will help limit access to an incubator used for open plates and cloning.

***Fungicides.*** Copper-lined incubators have reduced fungal growth but are usually about 20% to 30% more expensive than conventional ones. Placing copper foil in the humidifier tray also inhibits the growth of fungus, but only

in the tray, and will not protect the walls of the incubator. A number of fungal retardants are in common use, including copper sulfate, riboflavin, sodium dodecyl sulfate (SDS), and Roccall, a proprietary fungicidal cleaner used in a 2% solution. A comparison of colony formation in incubators with and without Roccall shows no toxic effect. Many of these retardants are detergents, so it is important not to have a $CO_2$ or an air line bubbling through the humidifier tray, or the liquid will foam. Remember, a fungicide will only protect the tray; there is no substitute for regular cleaning!

***Cleaning incubators.*** Cleaning should be carried out regularly with 1% Roccall (Pfizer) or an equivalent nontoxic antifungal cleaner. The frequency will depend on where the incubator is located; monthly may suffice for a clean area with filtered room air, but a shorter interval will be required for a rural site, where the spore count is higher, or during construction work or renovation. The frequency of access will also influence the buildup of fungal contamination. When the incubator is in use, any spillage must be mopped up immediately and contaminated cultures removed as soon as they are detected.

---

### PROTOCOL 18.1. CLEANING INCUBATORS

#### Outline
Remove cultures to an alternative incubator, switch off the empty incubator, wash it out with detergent and alcohol, switch on the heat, and allow the incubator to dry. Replenish the water in the tray and restore the $CO_2$.

#### Materials

*Nonsterile:*

- ❑ Water containing 2% Roccal-D or an equivalent fungal inhibitor, to refill the water tray
- ❑ Alternative $CO_2$ incubator or plastic box and sealing tape
- ❑ Detergent: 1% Roccal-D, or another fungicidal detergent
- ❑ Alcohol, 70%

#### Procedure

1. Remove all cultures to another $CO_2$ incubator, or enclose the cultures in a sealable container (e.g., a desiccator; *see* Fig. 4.11), gas them with $CO_2$, and place them in a regular incubator or hot room.
2. Switch off the incubator that is to be cleaned.
3. Remove all the shelves, the water tray, and any demountable panels from the incubator.
4. Wash the inside of the incubator with detergent solution; try to reach all corners and crevices.
5. Rinse with water.

6. Wipe the interior of the incubator with 70% alcohol.
7. Restore heat (not $CO_2$), and leave the door open until the chamber is dry.
8. Wash the shelves and panels in detergent, rinse them in water, and wipe them with 70% alcohol.
9. Return the panels and shelves to the incubator.
10. Run sterilization cycle if available and allow incubator to cool to normal working temperature.
11. Replace the water tray and fill it with sterile water containing 2% Roccal-D.
12. Close the door and restore $CO_2$.
13. When the temperature and $CO_2$ have stabilized, return the cultures to the incubator.

## 18.1.5 Cold Stores

Refrigerators and cold rooms also tend to build up fungal contamination on the walls in a humid climate, due to condensation that forms every time the door is opened, admitting moist air. The moist air increases the risk of deposition of spores on stored bottles; hence bottles should be swabbed with alcohol before being placed in the hood (*see* Section 5.5). The cold store should be cleared, and the walls and shelving should be washed down with disinfectant every few months.

## 18.1.6 Sterile Materials

There should be no risk of contamination from sterile plastics and reagents if the appropriate quality control is carried out, either in house (*see* Section 10.6) or by the supplier. However, failures in sterilization can occur if, for example, the load is too tightly packed, there is a cold spot in either the sterilizing oven or autoclave, packaging is punctured before or during storage, or a low-level contamination has escaped detection despite quality control (*see also* Section 29.4). Follow the correct procedures (*see* Section 10.6), make sure that all packing is correctly sealed, and monitor all sterilizer loads via probe in a replica container in the center of the load.

It is critical when new staff are introduced into the tissue culture laboratory that they are made familiar with sterilization procedures (*see* Section 28.2), even if they will not be called upon to carry out these procedures themselves. They should also be aware of the location of, and distinction between, sterile and nonsterile stocks (*see* Section 3.2.7). A simple error by a new recruit can cause severe problems that can last for several days before it is discovered.

## 18.1.7 Imported Cell Lines and Biopsies

Because cell lines and tissue samples brought into the tissue culture laboratory may be contaminated, they should be quarantined (*see* Section 18.1.8) until shown to be clear of contamination, at which point they, or their derivatives, can join other stocks in general use. Whenever possible, all cell lines should be acquired via a reputable cell bank, which will have screened for contamination. Cell lines from any other source, as well as biopsies from all animal and human donors, should be regarded as contaminated until shown to be otherwise.

## 18.1.8 Quarantine

Any culture that is suspected of being contaminated, and any imported material that has not been tested, should be kept in quarantine. Preferably, quarantine should take place in a separate room with its own hood and incubator, but if this is not feasible, one of the hoods that is in general use may be employed. In this case the hood should be used last thing in the day and should be sprayed and swabbed down with 70% alcohol containing 2% phenolic disinfectant after use. Then it should be withdrawn from service until the following day.

## 18.2 TYPES OF MICROBIAL CONTAMINATION

Bacteria, yeasts, fungi, molds, mycoplasmas, and occasionally protozoa can all appear as contaminants in tissue culture. Usually the species or type of infection is not important, unless it becomes a frequent occurrence. It is only necessary to note the general kind of contaminant (bacterial rods or cocci, yeast, etc.), how it was detected, the location where the culture was last handled, and the operator's name. If a particular type of infection recurs frequently, it may be beneficial to culture and identify it, and to try to trace its origin [Hay & Cour, 1997] taking care to keep this activity clear of the tissue culture facility. In general, rapidly growing organisms are less problematic as they are often overt and readily detected, whereupon the culture can be discarded. Difficulties arise when the contaminant is cryptic, either because it is too small to be seen on the microscope, such as mycoplasma, or slow growing such that the level is so low that it escapes detection. Use of antibiotics can be a common cause of cryptic contaminations remaining undetected (*see* Section 8.4.7).

## 18.3 MONITORING FOR CONTAMINATION

Potential sources of contamination are listed in Table 18.1, along with the precautions that should be taken to avoid them. Even in the best laboratories contaminations do arise, so the following procedure is recommended (*see also* Section 29.4):

(1) Check for contamination by eye and with a microscope at each handling of a culture. Check for mycoplasma every month (*see* Section 18.3.2).
(2) If it is suspected, but not obvious, that a culture is contaminated, and the fact cannot be confirmed in situ,

**TABLE 18.1.** Routes to Contamination

| Route or cause | Prevention |
| --- | --- |
| **Technique** | |
| *Manipulations, pipetting, dispensing, etc.* | |
| Nonsterile surfaces and equipment | Clear work area of items not in immediate use. |
| Spillage on necks and outside of bottles and on work surface | Swab regularly with 70% alcohol. Do not pour liquids. Dispense or transfer by pipette, autodispenser, or transfer device. If pouring is unavoidable: (1) do so in one smooth movement, (2) discard the bottle that you pour from, and (3) wipe up any spillage. |
| Touching or holding pipettes too low down, touching necks of bottles, inside screw caps | Hold pipettes above graduations. |
| | Do not work over open vessels. |
| Splash-back from waste beaker | Discard waste into a beaker with a funnel or, preferably, by drawing off the waste into a reservoir by means of a vacuum pump. |
| Sedimentary dust or particles of skin settling on the culture or bottle; hands or apparatus held over an open dish or bottle | Do not work over (vertical laminar flow and open bench) or behind and over (horizontal laminar flow) an open bottle or dish. |
| **Work surface** | |
| Dust and spillage | Swab the surface with 70% alcohol before during, and after work. |
| | Mop up spillage immediately. |
| **Operator hair, hands, breath, clothing** | |
| Dust from skin, hair, or clothing dropped or blown into the culture | Wash hands thoroughly or wear gloves. Wear a lint-free lab coat with tight cuffs and gloves overlapping them. |
| Aerosols from talking, coughing, sneezing, etc. | Keep talking to a minimum, and face away from work when you talk. |
| | Avoid working with a cold or throat infection, or wear a mask. |
| | Tie back long hair or wear a cap. |
| | Wear a lab coat different from the one you wear in the general lab area or animal house. |
| **Materials and reagents** | |
| *Solutions* | |
| Nonsterile reagents and media | Filter or autoclave solutions before using them. |
| Dirty storage conditions | Clean up storage areas and disinfect regularly. |
| Inadequate sterilization procedures | Monitor the performance of the autoclave with a recording thermometer or sterility indicator (*see* Protocol 10.5, Appendix II). |
| | Check the integrity of filters with a bubble-point or microbial assay after using them. |
| | Test all solutions after sterilization. |
| Poor commercial supplier | Test solutions; change suppliers. |
| *Glassware and screw caps* | |
| Dust and spores from storage | Shroud caps with foil. Wipe bottles with 70% alcohol before taking them into the hood. |
| | Replace stocks from the back of the shelf. Do not store anything unsealed for more than 24 h. |
| Ineffective sterilization (e.g., an overfilled oven or sealed bottles, preventing the ingress of steam) | Check the temperature of the load throughout the cycle. In the autoclave; keep caps slack on empty bottles. Stack oven and autoclave correctly (*see* Protocol 10.1). |
| *Instruments, pipettes* | |
| Ineffective sterilization | Sterilize items by dry heat before using them. |
| | Monitor the performance of the oven. |
| Contact with a nonsterile surface or some other material | Resterilize instruments. (Use 70% alcohol; burn and cool off the instruments.) |

**TABLE 18.1.** (*Continued*)

| Route or cause | Prevention |
|---|---|
| | Do not grasp any part of an instrument or pipette that will pass into a culture vessel. |
| ***Culture flasks and media bottles in use*** | |
| Dust and spores from incubator or refrigerator | Use screw caps instead of stoppers. Swab bottles before placing in hood. Box plates and dishes. |
| Dirty storage or incubation conditions. | Cover caps and necks of bottles with aluminum foil during storage or incubation. |
| | Wipe flasks and bottles with 70% alcohol before using them. |
| | Clean out stores and incubators regularly. |
| Media under the cap and spreading to the outside of the bottle | Discard all bottles that show spillage on the outside of the neck. |
| | Do not pour. |
| **Equipment and facilities** | |
| ***Room air*** | |
| Drafts, eddies, turbulence, dust, aerosols | Clean filtered air. |
| | Reduce traffic and extraneous activity. |
| | Wipe the floor and work surfaces regularly. |
| ***Laminar-flow hoods*** | |
| Perforated filter | Check filters regularly for holes and leaks. |
| Change of filter needed | Check the pressure drop across the filter. |
| Spillages, particularly in crevices or below a work surface | Clear around and below the work surface regularly. Let alcohol run into crevices. |
| ***Dry incubators*** | |
| Growth of molds and bacteria on spillages | Wipe up any spillage with 70% alcohol on a swab. |
| | Clean out incubators regularly. |
| ***$CO_2$, humidified incubators*** | |
| Growth of molds and bacteria on walls and shelves in a humid atmosphere | Clean out with detergent followed by 70% alcohol (*see* Protocol 18.1). |
| Spores, etc., carried on forced-air circulation | Enclose open dishes in plastic boxes with close-fitting lids (but do not seal the lids). |
| | Swab incubators with 70% alcohol before opening them. |
| | Put a fungicide or bacteriocide in humidifying water (but check first for toxicity). |
| ***Other equipment*** | |
| Dust on cylinders, pumps, etc | Wipe with 70% alcohol before bring in |
| ***Mites, insects, and other infestations in wooden furniture, or benches, in incubators, and on mice, etc., taken from the animal house*** | |
| Entry of mites, etc., into sterile packages | Seal all sterile packs. |
| | Avoid wooden furniture if possible; use plastic laminate, one-piece, or stainless steel bench tops. |
| | If wooden furniture is used, seal it with polyurethane varnish or wax polish and wash it regularly with disinfectant. |
| | Keep animals out of the tissue culture lab. |
| **Importation of biological materials** | |
| ***Tissue samples*** | |
| Infected at source or during dissection | Do not bring animals into the tissue culture lab. |
| | Incorporate antibiotics into the dissection fluid (*see* Section 11.3). |
| | Dip all potentially infected large-tissue samples in 70% alcohol for 30 s. |

(*continued overleaf*)

**TABLE 18.1.** (*Continued*)

| Route or cause | Prevention |
| --- | --- |
| ***Incoming cell lines*** | |
| Contaminated at the source or during transit | Handle these cell lines alone, preferably in quarantine, after all other sterile work is finished. Swab down the bench or hood after use with 2% phenolic disinfectant in 70% alcohol, and do not use it until the next morning. |
| | Check for contamination by growing a culture for two weeks without antibiotics. (Keep a duplicate culture in antibiotics at the first subculture.) |
| | Check for contamination visually, by phase-contrast microscopy and Hoechst stain for mycoplasma. Using indicator cells allows screening before first subculture. |

*Note:* No one-to-one relationship between prevention and cause is intended throughout this table; preventative measures are interactive and may relate to more than one cause.

clear the hood or bench of everything except your suspected culture and one can of Pasteur pipettes. Because of the potential risk to other cultures, this is best done after all your other culture work is finished. Remove a sample from the culture and place it on a microscope slide. (Kova Slides are convenient for this, as they do not require a coverslip.) Check the slide with a microscope, preferably by phase contrast. If it is confirmed that the culture is contaminated, discard the pipettes, swab the hood or bench with 70% alcohol containing a phenolic disinfectant, and do not use the hood or bench until the next day.

(3) Record the nature of the contamination.

(4) If the contamination is new and is not widespread, discard the culture, the medium bottle used to feed it, and any other reagent (e.g., trypsin) that has been used in conjunction with the culture. Discard all of these into disinfectant, preferably in a fume hood and outside the tissue culture area.

(5) If the contamination is new and widespread (i.e., in at least two different cultures), discard all media, stock solutions, trypsin, and so forth in current use.

(6) If the same kind of contamination has occurred before, check stock solutions for contamination (a) by incubation alone or in nutrient broth (*see* Section 10.6.2, Plate 16*d*) or (b) by plating out the solution on nutrient agar (Oxoid, Difco). If (a) and (b) prove negative, but contamination is still suspected, incubate 100 mL of solution, filter it through a 0.2-μm filter, and plate out filter on nutrient agar with an uninoculated control.

(7) If the contamination is widespread, multispecific, and repeated, check (a) the laboratory's sterilization procedures (e.g., the temperatures of ovens and autoclaves, particularly in the center of the load, the duration of the sterilization cycle), (b) the packaging and storage practices, (e.g., unsealed glassware should be resterilized every 24 h), and (c) the integrity of the aseptic room and laminar-flow hood filters.

(8) Do not attempt to decontaminate cultures unless they are irreplaceable.

### 18.3.1 Visible Microbial Contamination

Characteristic features of microbial contamination are as follows:

(1) A sudden change in pH, usually a decrease with most bacterial infections (Plate 16*a*), very little change with yeast (Plate 16*c*) until the contamination is heavy, and sometimes an increase in pH with fungal contamination.

(2) Cloudiness in the medium (*see* Plate 16*a*–*c*), sometimes with a slight film or scum on the surface or spots on the growth surface that dissipate when the flask is moved (Plate 16*b*).

(3) Under a 10× objective, spaces between cells will appear granular and may shimmer with bacterial contamination (Fig. 18.1*a*). Yeasts appear as separate round or ovoid particles that may bud off smaller particles (Fig. 18.1*b*). Fungi produce thin filamentous mycelia (Fig. 18.1*c*) and, sometimes, denser clumps of spores which may be blue or green. With toxic infection, some deterioration of the cells will be apparent.

(4) Under a 100× objective, it may be possible to resolve individual bacteria and distinguish between rods and cocci. At this magnification, the shimmering that is visible in some infections will be seen to be caused by mobility of bacteria. Some bacteria form clumps or associate with the cultured cells.

(5) With a slide preparation, the morphology of the bacteria can be resolved with a 100× objective, but this is not usually necessary. Microbial infection may be confused with precipitates of media constituents (particularly protein) or with cell debris, but can be distinguished by their regular morphology. Precipitates may be crystalline or globular and irregular and are not usually as uniform in size. Clumps of bacteria may be confused with

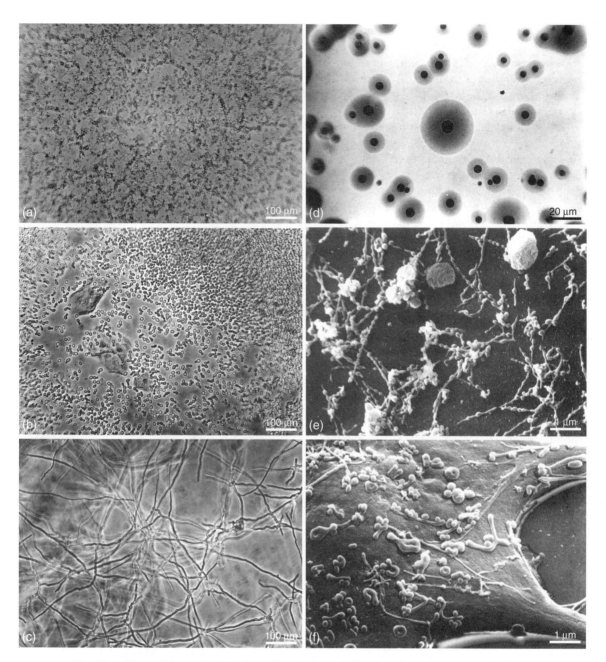

***Fig. 18.1. Types of Contamination.*** Examples of microorganisms found to contaminate cell cultures. (*a*) Bacteria. (*b*) Yeast. (*c*) Mold. (*d*) Mycoplasma colonies growing on special nutrient agar. (*e*, *f*). Scanning electron micrograph of mycoplasma growing on the surface of cultured cells. (*See* Plate 16*e*, *f* for mycoplasma stained with Hoechst 33258) *d*–*f*, courtesy of Dr. M. Gabridge; *a*–*c*, 10× objective.)

precipitated protein, but, particularly if shaken, many single or strings of bacteria will be seen. If you are in doubt, plate out a sample of medium on nutrient agar (*see* Section 10.6.2).

### 18.3.2  Mycoplasma

***Detecting mycoplasmal infections.*** Detection of mycoplasmal infections (Fig. 18.1*d*–*f*) is not obvious by routine microscopy, other than through signs of deterioration in the culture, and requires fluorescent staining, PCR, ELISA assay, immunostaining, autoradiography, or microbiological assay. Fluorescent staining of DNA by Hoechst 33258 [Chen, 1977] is one of the easiest and most reliable methods (*see* Protocol 18.2) and reveals mycoplasmal infections as a fine particulate or filamentous staining over the cytoplasm with a 50× or 100× objective (Plate 16*e*, *f*). The nuclei of the cultured cells are also brightly stained by this method and thereby act as a positive control for the staining procedure. Most other microbial contaminations will also show up with fluorescence staining, so low levels of contamination, or particularly small organisms such as micrococci, can be detected as well.

It is important to appreciate the fact that mycoplasmas do not always reveal their presence by means of macroscopic alterations of the cells or medium. Many mycoplasmal contaminants, particularly in continuous cell lines, grow slowly and do not destroy the host cells. However, they can alter the behavior and metabolism of the culture in many different ways [McGarrity, 1982; Doyle et al., 1990; Giron et al., Izutsu et al., 1996,1996; Paddenberg, et al., 1996; Jiang et al., 2008; Zhao et al., 2008]; for example, because mycoplasmas take up thymidine from the medium, infected cultures show abnormal labeling with [$^3$H]thymidine [Nardone et al., 1965] (see Fig. 27.2). Immunological studies can be totally frustrated by mycoplasmal contamination, as attempts to produce antibodies against the cell surface may raise antimycoplasma antibodies. There is therefore an absolute requirement for routine periodic assays to detect possible covert contamination of all cell cultures, particularly continuous cell lines.

*Monitoring cultures for mycoplasmas.* Superficial signs of chronic mycoplasmal infection include a diminished rate of cell proliferation, reduced saturation density [Stanbridge & Doersen, 1978], and agglutination during growth in suspension [Giron et al., 1996]. Acute infection causes total deterioration, with perhaps a few resistant colonies, although these and any resulting cell lines are not necessarily free of contamination and may carry a chronic infection.

## 18.3.3 Fluorescence Staining for Mycoplasma

The cultures are stained with Hoechst 33258, a fluorescent dye that binds specifically to DNA [Chen, 1977]. Because mycoplasmas contain DNA, they can be detected readily by their characteristic particulate or filamentous pattern of fluorescence on the cell surface and, if the contamination is heavy, in surrounding areas. Monolayer cell cultures can be fixed and stained directly, but it is preferable to use an indicator cell (i.e., another monolayer) known to be free of mycoplasma but also known to be a good host for mycoplasma and to spread well in culture, with adequate cytoplasm to reveal any adherent mycoplasma, such as Vero cells, 3T6, NRK, or A549. Cells growing in suspension will always need an indicator cell. Use of an indicator cell helps avoid problems with false positives arising from debris when cells are assayed just after thawing or from primary cultures, which you might not want to sacrifice anyway. If there is a lot of debris, the medium can be filtered through a sterile 5-μm filter or centrifuged at 100 g.

---

## PROTOCOL 18.2. FLUORESCENCE DETECTION OF MYCOPLASMA

### Outline

Culture cells in the absence of antibiotics for at least one week, transfer the supernatant medium to an early log phase indicator culture, incubate 3 to 5 days, fix and stain the cells, and look for fluorescence other than in the nucleus.

### Materials

*Sterile or aseptically prepared:*

❑ Supernatant medium, free of antibiotics, from 7-day monolayer or centrifuged suspension cell culture
❑ Indicator cells (e.g., 3T6, NRK, Vero, A549)
❑ Petri dishes, 6 cm

*Nonsterile:*

❑ Hoechst 33258 stain, 50 ng/mL in BSS without phenol red (BSS-PR) or D-PBSA
❑ BSS-PR: Hanks's BSS without Phenol Red
❑ D-PBSA: Dulbecco's PBS without $Ca^{2+}$ and $Mg^{2+}$
❑ Deionized water
❑ Fixative: freshly prepared acetic methanol (1:3, on ice)
❑ Buffered glycerol mountant with fade retardant (see Appendix I: Mycoplasma Reagents)

### Procedure

1. Seed indicator cells into Petri dishes without using antibiotics; seed enough to give 50% to 60% confluence in 4 to 5 days (e.g., $2 \times 10^4$ NRK or $1 \times 10^5$ A549, in 5 mL of medium).
2. Add 1.5 mL of medium from the test culture.
3. Incubate the culture until the indicator cells reach 50% to 60% confluence.

*Note.* Cultures must not reach confluence by the end of the assay or staining will be inhibited and the subsequent visualization of mycoplasma will be impaired.

4. Remove the medium and discard it.
5. Rinse the monolayer with BSS-PR or D-PBSA, and discard the rinse.
6. Add fresh BSS-PR or D-PBSA diluted 50:50 with fixative, rinse the monolayer, and discard the rinse.
7. Add pure fixative, rinse, and discard the rinse.
8. Add more fixative (~0.5 mL/cm$^2$), and fix for 10 min.
9. Remove and discard the fixative.
10. Dry the monolayer completely if it is to be stored. (Samples may be accumulated at this stage and stained later.)

**11.** If you are proceeding directly with staining, wash off the fixative with deionized water and discard the wash.

**12.** Add Hoechst 33258 in BSS-PR or D-PBSA, and stain 10 min at room temperature.

**13.** Remove and discard the stain.

**14.** Rinse the monolayer with water and discard the rinse.

**15.** Mount a coverslip in a drop of buffered glycerol mountant, and blot off any surplus from the edges of the coverslip.

**16.** Examine the monolayer by epifluorescence with a 330–380-nm excitation filter and an LP 440-nm barrier filter.

*Analysis.* Check for extranuclear fluorescence. Mycoplasmas give pinpoints or filaments of fluorescence over the cytoplasm and, sometimes, in intercellular spaces (*see* Plate 16e, *f*). The pinpoints are close to the limits of resolution with a 50× objective (0.1–1.0 μm) and are usually regular in size and shape. Not all of the cells will necessarily be infected, so as much as possible of the preparation should be scanned before the culture is declared to be uninfected.

Sometimes a light, uniform staining of the cytoplasm is observed, probably due to RNA. This fluorescence tends to fade on storage of the preparation, and examination the next day (after storing dry and in the dark) usually gives clearer results. This artifact never has the sharp punctuate or filamentous appearance of mycoplasma and can be distinguished fairly readily with further experience in observation.

If there is any doubt regarding the interpretation of the fluorescence test, it should be repeated after generating a further subculture of the test cells in the absence of antibiotics. If results are still equivocal, repeat with another assay, such as PCR (*see* Protocol 18.3) or MycoAlert (Lonza).

## 18.3.4 PCR for Mycoplasma

The following introduction, protocol, and some of the discussion in Section 18.3.5 have been contributed by Cord C. Uphoff and Hans G. Drexler of DSMZ, Braunschweig, Germany.

The polymerase chain reaction (PCR) provides a very sensitive and specific assay for the direct detection of mycoplasmas in cell cultures with low expenditure of labor, time, and cost, and with simplicity, objectivity of interpretation, reproducibility, and easy documentation of results. Several primer sequences are published for both single and nested PCR and with narrow or broad specificity for mycoplasma or eubacteria species. In most cases the 16S rDNA sequences are used as target sequences because this gene contains regions with more and less conserved sequences. This gene also offers the opportunity to perform a PCR with the 16S rDNA or an RT-PCR (reverse transcriptase-PCR) with the cDNA of the 16S rRNA.

Here we describe the use of a mixture of oligonucleotides for the specific detection of mycoplasmas. This approach reduces significantly the generation of false positive results due to possible contamination of the solutions used for sample preparation and the PCR run, and of other materials with airborne bacteria. One of the main problems concerning PCR reactions with samples from cell cultures is the inhibition of the *Taq* polymerase by unspecified substances. To eliminate those inhibitors, we strictly recommend that the sample DNA be extracted and purified by a conventional phenol-chloroform extraction, by a salt-precipitation protocol, or by the more convenient column- or matrix-binding extraction methods.

To confirm the error-free preparation of the sample and PCR run, appropriate control reactions have to be included in the PCR. These comprise internal control DNA for every sample reaction, and in parallel, positive and negative as well as water control reactions. The internal control consists of a DNA fragment with the same primer sequences for amplification but of a different size than the amplicon of mycoplasma-contaminated samples. This control DNA is added to the PCR mixture in a previously determined limiting dilution to demonstrate the sensitivity of the PCR reaction. The internal control DNA can be obtained from the authors. The performance of the PCR run highly depends on the employed Taq polymerase, buffer, and apparatus. Thus the protocol might be needed to be individually adjusted for optimal conditions. This can be performed with dilution series of the positive or the internal control DNA.

---

### PROTOCOL 18.3. DETECTION OF MYCOPLASMA BY PCR

#### *Outline*
Supernatant medium of adherent cells or of suspension cultures with settled cells is collected and the cellular particles are isolated by centrifugation. After the washing steps, the DNA is extracted and an aliquot is used for nucleic acid amplification with mycoplasma specific oligonucleotides. Appropriate positive and negative control reactions confirm the error-free run and verify the sensitivity of the PCR. The amplification products are visualized in an ethidium bromide stained agarose gel and subsequently documented.

#### *Materials*
❑ D-PBSA (phosphate-buffered saline): 137 mM NaCl, 2.7 mM KCl, 8.1 mM $Na_2HPO_4$, 1.5 mM $KH_2PO_4$, pH 7.2, and autoclaved 20 min at 121°C to sterilize the solution

❑ TAE (Tris acetic acid EDTA), 50×: 2 M Tris base, 5.71% glacial acetic (v/v), 100 mM EDTA. Adjust to about pH 8.5

❑ DNA extraction and purification system, such as phenol/chloroform extraction and ethanol precipitation, or DNA extraction kits applying salt-precipitation or DNA binding matrices

❑ *Taq* DNA polymerase with the appropriate 10× buffer (Platinum hot start *Taq* polymerase; Invitrogen, 10× buffer without MgCl$_2$)

❑ Loading buffer, 6×: 10 mM Tris-HCl (pH 7.6), 0.03% (w/v) bromophenol blue, 0.03% (w/v) xylene cyanol FF 60% glycerol (v/v), 60 mM EDTA

❑ Primers:

5′ primers (Myco–5′):

| cgc | ctg | agt | agt | acg | t**w**c | gc |
| tgc | ctg | **r**gt | agt | aca | ttc | gc |
| c**r**c | ctg | agt | agt | atg | ctc | gc |
| cgc | ctg | ggt | agt | aca | ttc | gc |

3′ primers (Myco–3′):

| gcg | gtg | tgt | aca | a**r**a | ccc | ga |
| gcg | gtg | tgt | aca | aac | ccc | ga |

(**r** = mixture of g and a; **w** = mixture of t and a)

Primer stock solutions: 100 μM in dH$_2$O, stored frozen at about −20°C. Working solutions: mix of forward primers at 5 μM each (Myco–5′) and mix of reverse primers at 5 μM each (Myco–3′) in dH$_2$O, aliquoted in small amounts (i.e., 25- to 50-μL aliquots), and stored frozen at −20°C.

❑ Internal control DNA: internal control may be prepared as published elsewhere [Uphoff & Drexler, 2002] or can be obtained from the authors. A limiting dilution should be determined experimentally by performing PCR with a dilution series of the internal control DNA.

❑ Positive control DNA: a 10-fold dilution of any mycoplasma-positive sample prepared as described below (or can be obtained from the authors)

❑ Deoxy-nucleotide triphosphate mixture (dNTP-mix): mixture contains 5 mM each of deoxyadenosine triphosphate (dATP), deoxycytidine triphosphate (dCTP), deoxyguanosine triphosphate (dGTP), and deoxythymidine triphosphate (dTTP) at 5 mM in H$_2$O, and stored as 50-μL aliquots at −20°C

❑ Magnesium chloride (50 mM)

❑ Agarose, 1.3% TAE gel.

❑ Thermal cycler

### Procedure

#### A. Sample Collection and Preparation of DNA:

1. Before collecting the samples, culture without any antibiotics for several days the cell line to be tested for mycoplasma contamination, or culture for at least two weeks after thawing. This should ensure that the titer of the mycoplasmas in the supernatant medium is within the detection limits of the PCR assay.

2. Collect 1 mL of the supernatant medium of adherent cells, or of cultures with settled suspension cells. Collecting the samples in this way will include some viable and/or dead cells, an advantage as some mycoplasma strains predominantly adhere to the eukaryotic cells or even invade them. The supernatant medium can be stored at 4°C for a few days or at −20°C for several weeks. After thawing, the samples should be processed immediately.

3. Centrifuge the supernatant medium at 13,000 *g* for 5 min, and resuspend the pellet in 1 mL of D-PBSA by vortexing.

4. Centrifuge the suspension again and wash one more time with D-PBSA as described in step 3.

5. After centrifugation, resuspend the pellet in 100 μL of D-PBSA by vortexing and then heat to 95°C for 15 min.

6. Immediately after lysing the cells, extract and purify the DNA by standard phenol/chloroform extraction and ethanol precipitation or other DNA isolation methods.

### B. PCR Reaction:

Established rules to avoid DNA carryover should be strictly followed: (1) separate the places where the DNA is extracted, where the PCR reaction is set up, and where the gel is run after the PCR; (2) store all reagents in small aliquots to provide a constant source of uncontaminated reagents; (3) avoid reamplifications; (4) reserve pipettes, tips, and tubes for their use in PCR only and irradiate pipettes frequently by UV light; (5) strictly follow the succession of the PCR set up described below; (6) wear gloves during the whole sample preparation and PCR setup; (7) include the appropriate control reactions, such as internal, positive, negative, and the water control reaction.

1. Set up two reaction pre-mixtures per sample to be tested with the following solutions:

   (a) *Sample only:* 1 μL dNTPs, 0.5 μL Myco–5′, 0.5 μL Myco–3′, 2.5 μL 10× PCR buffer, 1 μL MgCl$_2$, 0.2 μL hot start *Taq* DNA polymerase, 17.3 μL dH$_2$O.

   (b) *Sample and DNA internal standard:* 1 μL dNTPs, 0.5 μL Myco–5′, 0.5 μL Myco–3′, 2.5 μL 10× PCR buffer, 0.2 μl hot start *Taq* DNA polymerase, 16.3 μL dH$_2$O, 1 μL internal control DNA.

Prepare sufficient volumes of pre-mixtures for multiple samples, three additional samples, including three additional control reactions without internal control DNA (for the positive, negative, and the water controls) and two additional with the internal control DNA (for the positive and the negative controls), plus a surplus for pipetting.

2. Transfer 23 μL of each premixed stock to 0.2-mL PCR reaction tubes and add 2 μL UPW to the water control reactions.

3. Set aside all reagents used for the preparation of the premixed stocks.

4. Take out the samples of DNA to be tested and the positive control DNA. Do not handle the reagents and samples simultaneously.

5. Add 2 μL per DNA preparation to one reaction tube that contains no internal control DNA and to one tube containing the internal control DNA.

6. To perform the PCR, transfer the reaction mixtures to the thermal cycler and start one thermo cycle with the following parameters: step 1: 7 min at 95°C; step 2: 3 min at 72°C; step 3: 2 min at 65°C; step 4: 5 min at 72°C.

7. After this initial cycle for the activation of the Taq polymerase, perform 32 thermal cycles with the following parameters: step 1: 4 s at 95°C; step 2: 8 s at 65°C; step 3: 16 s at 72°C plus 1 s of extension time during each cycle.

8. Finish the reaction with a final amplification step at 72°C for 10 min and then cool the samples to room temperature.

9. Prepare a 1.3% agarose-TAE gel containing ethidium bromide, 0.3 μg/mL. Submerge the gel in 1× TAE and add 12 μL of the amplification product (10 μL reaction mixtures plus 2 μL of 6× loading buffer) to each well and run the gel at 10 V/cm.

10. Visualize the specific products on a suitable ultraviolet light screen and document the results.

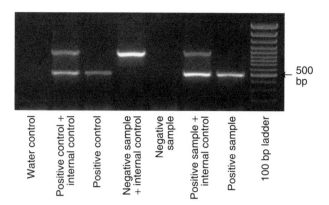

**Fig. 18.2. Mycoplasma Detection by PCR.** Ethidium bromide fluorescence of PCR products of infected, uninfected, and control cells, electrophoresed on 1.3% agarose. The 100-bp ladder consists of the following bands: 100, 200, 300, 400, 500 (strongly stained band), 600, 700, 800, 900, 1031, 1200, 1500, 2000, and 3000 bp. The wild-type mycoplasma bands are about 510 bp, and the internal control band is almost 1000 bp. (Courtesy of Cord Uphoff, DSMZ)

### Interpretation of results

(1) Ideally all samples containing the internal control DNA show a band at 986 bp. If the second run also shows no band for the sample and internal control, the whole procedure should be repeated.

(2) Mycoplasma-positive samples show a band at 502 to 520 bp, depending on the mycoplasma species (Fig. 18.2).

(3) Contaminations of reagents with mycoplasma-specific DNA or PCR product are revealed by a band in the water control and/or in the negative control sample. Weak mycoplasma-specific bands can occur after treatment of infected cell cultures with antimycoplasma reagents for the elimination of mycoplasmas or when other antibiotics

are applied routinely. In these cases the positive reaction might be due either to residual DNA in the culture medium derived from dead mycoplasma cells or to viable mycoplasma cells that are present at a very low titer.

The highly conserved regions of the 16S rRNA genes enable the selection of primers of wide specificity ( "universal primers") that will react with DNA of any mycoplasma and even with the DNA of other prokaryotes, but also of primers with a higher mycoplasmal specificity [Uphoff & Drexler, 1999]. It is important that no cross-reactions with the DNA of cell lines occur. The amplification is usually performed as a single-step PCR, but for higher sensitivity and specificity a nested PCR may be considered. Usually a very high sensitivity level is not required in routine diagnosis as a large number of organisms are present in the sample. The high sensitivity of PCR may cause problems in producing false-positive results due to contamination with target DNA. Another possible problem is false-negative data caused by the inhibition of the *Taq* polymerase by components in the samples. This problem can be overcome by performing a DNA preparation of the sample and by spiking an internal control DNA into the sample (see Protocol 18.3). Using a limiting dilution of the internal control DNA, the PCR reaction can easily be standardized. Ideally the internal control is amplified with the same primers as the wild type mycoplasma DNA fragment. A further development of the conventional PCR method is real-time PCR. This technique utilizes either SYBR green for the detection of amplified DNA or a third quenched fluorescently labeled oligonucleotide, which can hybridize specifically to the amplified product emanating from mycoplasmal DNA. If SYBR green is used, a melting point determination is required to distinguish mycoplasma-specific amplification products from unspecific amplicons. PCR protocols for the establishment of the tests in any laboratory [Uphoff & Drexler, 1999] as well as several PCR kits are commercially available (see Table 18.2).

### 18.3.5 Alternative Methods for Detecting Mycoplasma

*Biochemical.* Among other methods that have been reported for the detection of mycoplasmal infections are methods that detect mycoplasma-specific enzymes such as arginine deiminase or nucleoside phosphorylase *see* [Schneider & Stanbridge, 1975; Levine & Becker, 1977] and those that detect toxicity with 6-methylpurine deoxyriboside (Mycotect, Invitrogen). However, those assays were shown to produce false-negative results and should not be used as single mycoplasma detection assay [Uphoff et al., 1992]. The MycoAlert system (Lonza) detects ATP generated by carbamate kinase and/or acetate kinase that are active in arginine hydrolyzing and carbohydrate fermenting mycoplasmas, respectively. As these enzymes are also found in several other bacteria, the assay is not completely specific. The activity of the enzymes in the culture supernatant is measured by luminescence with luciferase. This assay is claimed to be very sensitive and quick (30 min) [Mariotti et al., 2008].

*Microbiological culture.* This is a very sensitive method, widely employed in quality control and validation procedures. However, it should only be used with isolation facilities and the appropriate background experience in microbiology, as these microorganisms are quite fastidious, and it will be necessary to culture live mycoplasma as a positive control. The cultured cells are seeded into mycoplasma broth [Doyle et al., 1990] (BD Biosciences), grown for 6 days, and plated out onto special nutrient agar [Hay, 2000] (BD Biosciences; Sigma-Aldrich). Colonies form in about 8 days and can be recognized by their size ($\sim$ 200 $\mu$m diameter) and their characteristic "fried egg" morphology-dense center with a lighter periphery (Fig. 18.1*d*). Mycoplasma detection kits and mycoplasma testing by microbiological culture are available (Table 18.2; *see also* Appendix II). The microbiological culture method is still one of the accepted industry standards, but it is much slower and more difficult to perform than the fluorescence technique and is best left to specialized laboratories with the appropriate quarantine facilities to handle the live mycoplasma needed for positive controls. Where a contamination is confirmed, specific monoclonal antibodies allow characterization of the mycoplasma (Roche; *see* Immunological Assays, below).

*Molecular hybridization.* Molecular probes specific to mycoplasmal DNA can be used to detect infections by conventional molecular hybridization techniques. A kit is available that uses biotin-labeled probes and streptavidin and enzymic amplification steps to produce a colorimetric endpoint (Mycoprobe, R&D). Mycoplasma Tissue Culture NI (MTC-NI, Gen-Probe) is a highly sensitive and specific detection assay that uses an acridinium ester labeled single-stranded DNA probe homologous to mycoplasmal 16S and 23S rRNA sequences. After the rRNA is released from the organism, the chemiluminescently labeled DNA probe combines with the target rRNA to form a stable DNA-RNA hybrid. The acridinium ester of the single-stranded

**TABLE 18.2.** Mycoplasma Detection Kits

| Kit | Principle | Supplier |
|---|---|---|
| MycoAlert | Luminescence detection of ATP from mycoplasma-specific enzymes | Lonza |
| Mycotect | Cytotoxicity generated by mycoplasma generated degradation of 6-methylpurine deoxyribose to 6-methylpurine | Invitrogen |
| ELISA | Elisa assay with polyclonal antibodies to *M. aginini, M. hyorhinis, A, laidlawii, and M. orale* | Roche |
| PCR | PCR of 16S rRNA | AppliChem Stratagene PAA |
| | PCR and real time PCR of 16s rRNA | Promokine Autogen Bioclear Minerva Biolabs Sigma |
| PCR-ELISA | PCR of 16S rRNA, subsequent detection of amplification products by ELISA | Roche |
| MycoProbe | Colorimetric detection of biotin-labeled probe for 16S rRNA in *M. hyorhinis, M. arginine, M. fermentans, M. orale, M. pirum, M. hominis, M. salivarium, A. laidlawii* | R & D |
| Mycoplasma Tissue Culture NI | DNA-RNA hybridization with acridinium ester labeled DNA and subsequent determination of the DNA-RNA hybrids by chemiluminescence | Gen-Probe |
| ImmuMark | Immunofluorescence assay with a monoclonal antibody | MP Biochemicals |

nonhybridized probe is then destroyed by addition of a hydrolysis buffer. Positive signals are then measured in a luminometer. The assay can be performed within one hour.

*Immunological assays.* Immunological methods use mycoplasma-specific polyclonal or monoclonal antibodies to detect mycoplasmas. The antibodies are commonly applied in immunofluorescence (Immumark Mycotest kit, MP Biochemicals) and enzyme-linked immunosorbent assays (ELISA). The species-specific immunoassays can also be used for the identification of the mycoplasma strains. The immunological assays have been shown to produce false-negative results in mycoplasma detection when only certain species are addressed. For example, the Mycoplasma Detection Kit from Roche does not detect *M. fermentans*, one of the most common mycoplasma species found in cell cultures [Uphoff et al., 1992]. A monoclonal antibody CCM-2 has been produced that recognizes a common antigen (elongation factor TU) shared by most mycoplasmas (Immumark Mycotest, MP Biochemicals). The staining is more specific than the fluorochrome staining and can thus be applied directly on the cells to be tested instead of using an indicator cell culture. However, the interpretation of the staining can sometimes be subjective and difficult to reproduce between observers.

### 18.3.6  Mycoplasma Detection Services
Most of the cells banks, such as ATCC, DSMZ, and ECACC, offer testing for mycoplasma using microbiological culture, DNA fluorescence staining, or PCR, or a combination of all three (*see also* Appendix II). Suspect cultures may be sent to them, and they will be tested for a fee. It is important when sending cultures for testing (1) to make sure it is labeled as potentially mycoplasma infected and (2) not to send a culture in advanced stages of deterioration as the cells will probably have died by the time the sample is received. Since the mycoplasma will have died also, you will get a false negative.

### 18.3.7  Viral Contamination
*Prevention of viral contamination.* Incoming cell lines, natural products, such as serum, in media, and enzymes such as trypsin, used for subculture, are all potential sources of viral contamination. A number of reagents are screened by manufacturers against a limited range of viruses, and claims have been made that the larger viruses can be filtered out during processing, but there is no certain way at present to eliminate viral contamination. The best way of avoiding it is to ensure that the products are collected from animals free from known virus infections. For this, you will need to rely on the quality control put in place by the supplier.

*Detection of viral contamination.* Screening with a panel of antibodies by immunostaining (*see* Protocol 15.11) or ELISA assays is one way of detecting viral infection. Alternatively, one could use PCR with the appropriate viral

primers, which has greater sensitivity. Some commercial companies (e.g., BioReliance) offer viral screening.

## 18.4  DISPOSAL OF CONTAMINATED CULTURES

It is important to ensure that all contaminated material is disposed of correctly. Culture vessels should be removed from the culture area, unopened if possible, and autoclaved. Open items, such as Petri dishes with the lids in place, and pipettes or other items that have come in contact with a contaminated culture should be immersed in hypochlorite disinfectant (Petri dishes can be opened while submerged). If only one of a series of similar cultures is contaminated, it is only necessary to discard the bottle of medium that was used with it, but if the contamination is widespread, then all medium as well as all other stock solutions and reagents, used with these cells, should be discarded into hypochlorite.

## 18.5  ERADICATION OF CONTAMINATION

### 18.5.1  Bacteria, Fungi, and Yeasts
The most reliable method of eliminating a microbial contamination is to discard the culture and the medium and reagents used with it, as treating a culture may be unsuccessful or lead to the development of an antibiotic-resistant microorganism. Decontamination should be attempted only in extreme situations, under quarantine, and with expert supervision. If unsuccessful, the culture and associated reagents should be autoclaved as soon as failure becomes obvious.

---

**PROTOCOL 18.4. ERADICATION OF MICROBIAL CONTAMINATION**

**Outline**
Wash the culture several times in a high concentration of antibiotics by rinsing the monolayer or by centrifugation and resuspension of nonadherent cells. Then grow the culture for three subcultures with, and three without, antibiotics. Test for contamination after each subculture.

**Materials**

*Sterile:*

❏ DBSS (*see* Appendix I: Dissection BSS)
❏ High-antibiotic medium (*see* Appendix I: Collection Medium; Table 9.4; *see* Section 18.5.2 for mycoplasma)
❏ Materials for subculture (*see* Protocol 13.3)

*Nonsterile:*

❏ Microscope, preferably with 40× and 100× phase-contrast objectives

❑ Materials for staining mycoplasma (*see* Protocol 18.2)
❑ Disinfectant for disposal (*see* Section 7.8.5)

**Procedure**

1. Collect the contaminated medium carefully. If possible, the organism should be tested for sensitivity to a range of individual antibiotics. If not, autoclave the medium or add disinfectant.
2. Wash the cells in DBSS (dilution can reduce the number of contaminants by two logs with each wash, unless they are adherent to the cells):
    (a) For monolayers, rinse the culture three times with DBSS, trypsinize, and wash the cells twice more in DBSS by centrifugation and resuspension.
    (b) For suspension cultures, wash the culture five times in DBSS by centrifugation and resuspension.
3. Reseed a fresh flask at the lowest reasonable seeding density.
4. Add high-antibiotic medium and change every 2 days.
5. Subculture in a high-antibiotic medium.
6. Repeat steps 1 through 4 for three subcultures.
7. Remove the antibiotics, and culture the cells without them for a further three subcultures.
8. Check the cultures by phase-contrast microscopy and Hoechst staining (*see* Protocol 18.2).
9. Culture the cells for a further two months without antibiotics, and check to make sure that all contamination has been eliminated (*see* Section 10.6.2).

The general rule remains that contaminated cultures are discarded and that decontamination is not attempted unless it is absolutely vital to retain the cell strain. In any event, complete decontamination is difficult to achieve, particularly with yeast, and attempts to do so may produce hardier, antibiotic-resistant strains.

### 18.5.2  Eradication of Mycoplasma

If mycoplasma contamination is detected in a culture, the first and overriding rule, as with other forms of contamination, is that the culture should be discarded for autoclaving or incineration. In exceptional cases (e.g., if the contaminated line is irreplaceable), one may attempt to decontaminate the culture. Decontamination should be done, however, only by an experienced operator, and the work must be carried out under conditions of quarantine to prevent contamination of other mycoplasma-free cell cultures.

Several agents are active against mycoplasma, including kanamycin, gentamicin, tylosin [Friend et al., 1966], polyanethol sulfonate [Mardh, 1975], and 5-bromouracil in combination with Hoechst 33258 and UV light [Marcus et al., 1980]. Coculturing with macrophages [Schimmelpfeng et al., 1968], animal passage [Van Diggelen et al., 1977], and cytotoxic antibodies [Pollock & Kenny, 1963] can also be effective in some cases. As some mycoplasma species can penetrate eukaryotic cells (e.g., *M. fermentans*), the eradication process has to include both intracellular and extracellular mycoplasmas. Thus the most successful agents have been antimycoplasmal antibiotics from the class of macrolides (e.g. tiamulin), fluoroquinolones (e.g. ciprofloxacin, enrofloxacin, sparfloxacin), and tetracyclines (e.g. minocyclin, doxycyclin) [Uphoff & Drexler, 2002]. Several products are available directly for the use in cell culture, such as MRA (Mycoplasma Removal Agent, MP Biomedicals), BM-Cyclin (Roche), and Plasmocin (Invivogen), and these antibiotics may be applied in single or in combination therapy. The more recently developed membrane-active peptides are also applied alone or in combination with other antibiotics. The activity of alamethicin, dermaseptin B2, gramicidin S, and surfactin lipopeptide antibiotics against mycoplasmas is reduced in the presence of serum and should be applied in cultures without serum or with reduced serum concentrations to up to 5%. Kits are available from Lonza (MycoZap) and Minerva Biolabs (Mynox Gold).

The antibiotic treatment of mycoplasma contamination in cell cultures is highly efficient and about three-quarters of treated cell cultures are cured with one of the mentioned antibiotics. Resistant mycoplasma strains can be eliminated subsequently with an antibiotic from another category. Nevertheless, resistance of the mycoplasma strain and loss of the culture due to cytotoxic effects of the antibiotic can occur. Thus it is of advantage to have several antibiotics from different classes at hand that can be used as alternatives. The prophylactic use of antibiotics can lead to the selection of resistant mycoplasma strains and is not recommended. Contaminated cultures should be treated as in Protocol 18.4, using the antibiotics mentioned in Table 18.3 at the recommended concentrations and treatment times. However, this operation should not be undertaken unless it is absolutely essential, and even then it must be performed in experienced hands and in isolation [Uphoff & Drexler 2004]. It is far safer to discard infected cultures.

**PROTOCOL 18.5. ERADICATION OF MYCOPLASMA CONTAMINATION**

**Outline**

Wash the culture several times in culture medium or D-PBSA by rinsing the monolayer or by centrifugation and resuspension of nonadherent cells. Then grow the culture for one to three weeks with, and at least two weeks without, antibiotics. Test for contamination after post-treatment cultivation.

**Materials**

*Sterile:*

❑ Cell culture specific medium
❑ D-PBSA
❑ Antibiotics (*see* Table 18.3 for concentrations)
❑ Materials for subculture (*see* Protocol 12.2)

*Nonsterile:*

❑ Microscope, preferably with 40× and 100× phase-contrast objectives
❑ Materials for mycoplasma detection (*see* Protocol 18.2)

**Procedure**

1. Collect the contaminated medium carefully, autoclave or add hypochlorite (*see* Section 18.4).
2. Wash the cells in culture medium or D-PBSA (dilution can reduce the number of contaminants by two logs with each wash, unless they are adherent to the cells):
   (a) For monolayers, rinse the culture three times with D-PBSA, trypsinize, and wash the cells twice more in D-PBSA by centrifugation and resuspension.
   (b) For suspension cultures, wash the culture in D-PBSA by centrifugation and resuspension.
3. Reseed one or more fresh flasks at high seeding densities.
4. Add antibiotic solution at the recommended concentration (see Table 18.3) and change the culture medium every two days.
5. Subculture in antibiotic medium for the recommended time (*see* Table 18.3). If antibiotics from different classes are applied alternately with BM-Cyclin, cells should be washed before a new antibiotic is added.
6. Remove the antibiotics, and culture the cells without them for at least a further two weeks.
7. Check the cultures by PCR or another mycoplasma detection assay (*see* Protocols 18.2, 18.3).
8. Culture the cells for a further two months without antibiotics, and check to make sure that all contamination has been eliminated (*see* Section 10.6.2).

### 18.5.3   Eradication of Viral Contamination

There are no reliable methods for eliminating viruses from a culture at present; disposal or tolerance are the only options.

### 18.5.4   Persistent Contamination

Many laboratories have suffered from periods of contamination that seems to be refractory to all the remedies suggested in Table 18.1. There is no easy resolution to this problem, other than to follow the previous recommendations in a logical and analytical fashion, paying particular attention to changes in technique, new staff, new suppliers, new equipment, and inadequate maintenance of laminar-flow hoods or other equipment (*see also* Section 29.4). Typically an increase in the contamination rate stems from deterioration in aseptic technique, an increased spore count in the atmosphere, poorly maintained incubators, a contaminated cold room or refrigerator, or a fault in a sterilizing oven or autoclave, the way that it is packed, or the monitoring of the sterilization cycle.

The constant use of antibiotics also favors the development of chronic contamination. Many organisms are inhibited, but not killed, by antibiotics. They will persist in the culture, undetected for most of the time, but resurfacing when conditions change. It is essential that cultures be maintained

**TABLE 18.3.** Antibiotic Treatments Suitable for the Eradication of Mycoplasmas from Cell Cultures

| Brand name | Generic names of constituents | Antibiotic category | Typical final concentration | Duration of treatment |
|---|---|---|---|---|
| Baytril | Enrofloxacin | Fluoroquinolone | 25 µg/ml | 1 week |
| Ciprobay | Ciprofloxacin | Fluoroquinolone | 10 µg/ml | 2 weeks |
| MRA | Undisclosed | Fluoroquinolone | 0.5 µg/ml | 1 week |
| Zagam | Sparfloxacin | Fluoroquinolone | 10 µg/ml | 1 week |
| Plasmocin | 1. Undisclosed | Fluoroquinolone | 25 µg/ml | 2 weeks |
| | 2. Undisclosed | Protein synthesis inhibition | | |
| BM-Cyclin | 1. Tiamulin | Macrolide | 10 µg/ml | 3 weeks |
| | 2. Minocycline | Tetracycline | 5 µg/ml | |
| MycoZap | 1. Undisclosed | Membrane-active peptide | --- | 3 passages |
| | 2. Undisclosed | --- | --- | |
| Mynox Gold | 1. Surfactin | Membrane-active peptide | --- | 1 passage |
| | 2. Undisclosed | --- | --- | 3 passages |

in antibiotic-free conditions for at least part of the time (*see* Fig. 12.7), and preferably all the time; otherwise, cryptic contaminations will persist, their origins will be difficult to determine, and eliminating them will be impossible.

A slight change in practices, the introduction of new personnel, or an increase in activity as more people use a facility can all contribute to an increase in the rate of contamination. Procedures must remain stringent, even if the

**TABLE 18.4. Cause and Prevention of Cross-contamination and Misidentification**

| Cause | Prevention |
|---|---|
| **Sourcing cell lines** | |
| Obtaining cells from nonvalidated source | Only use reliable resources like reputable cell banks. If not available from a reputable cell bank, carry out authentication on receipt and before use. |
| Passing unauthenticated cell lines on to others | Do not pass on unauthenticated cell lines; refer requests to the cell bank where they were obtained originally or ensure that the necessary authentication has been done. |
| **Handling** | |
| Handling rapidly growing cell lines along with other slower growing lines | Always handle rapidly growing cell lines last, when other cultures and their media have been put away. |
| Putting a pipette back into a medium or reagent bottle after being in contact with cells | Do not put a pipette back into a medium or reagent bottle after it has been in a culture flask with cells. |
| | Always put medium into culture vessel before cells |
| Transfer of aerosols between open cultures | Do not have cultures of two different cell lines open simultaneously. |
| Using unplugged pipettes | Always use plugged pipettes (except for aspiration after which they must be discarded). |
| Using pipettors with unplugged tips | Use plugged pipette tips when handling cells for further propagation. |
| **Quality assurance** | |
| Not checking cultures visually | Always check cultures on microscope before handling. Use reference photographs. |
| Not carrying out authentication checks | Carry out authentication check when introducing a new cell line (*see* Section 15.2). |
| **Sharing** | |
| Sharing media and/or pipettes with colleagues | Do not share media or pipettes. |
| Sharing a work space with a colleague | Do not share laminar-flow hoods or any other work space unless collaborating with the same cell line. |
| Using the same medium, reagents, and pipettes for different cell lines | Always keep a separate set of medium and other reagents (trypsin, D-PBSA, EDTA, etc.) for each cell line. |
| **Misidentification** | |
| Mislabeling at subculture | Always label flasks, plates, and dishes before seeding cells, and check at when cells are added. |
| Mislabeling at freezing | Label ampoules before adding cells and only handle one cell line at a time. |
| Mixing up ampoules on thawing | Use permanent labeling that will not deteriorate in freezer (*see* Section 19.3.5). |
| | Check label carefully before thawing. |
| Poor inventory control in freezer leading to thawing wrong ampoule | Back up written or computer records with color coding of ampoule caps or other device. |
| Seeding the wrong flask on thawing | Label flask and add medium before thawing ampoule. |
| | Do not thaw more than one cell line at a time. |
| Accidental transfer of ampoules from one cane to another in freezer | Enclose ampoule canes in sheath. |
| Accidental transfer of ampoules from one compartment to another in freezer | Do not remove an ampoule unless you are sure of its identity, and do not replace it without double-checking its correct location. |

reason is not always obvious to the operator, and alterations in routine should not be made casually. If strict practices are maintained, contamination may not be eliminated entirely, but it will be reduced considerably and detected early.

## 18.6   CROSS-CONTAMINATION

During the development of tissue culture, a number of cell strains have evolved with very short doubling times and high plating efficiencies. Although these properties make such cell lines valuable experimental material, they also make them potentially hazardous for cross-infecting other cell lines [Nelson-Rees & Flandermeyer, 1977; Lavappa, 1978; Nelson-Rees et al., 1981; Masters et al., 1986; Markovic & Markovic, 1998; Kneuchel & Masters, 1999; Dirks et al., 1999; MacLeod et al., 1999; van Bokhoven et al., 2003; Rush et al., 2002; Milanesi et al., 2003; Drexler et al., 2003; Nardone, 2008; Liscovitch & Ravid, 2007; Hughes et al., 2007; Lacroix, 2008; Schweppe et al., 2008] (*see also* Table 12.2; Appendix V and at www.wiley.com/go/freshney/cellculture). The extensive cross-contamination of many cell lines with HeLa and other rapidly growing cell lines is now clearly established [Stacey et al., 2000; MacLeod et al., 1999; Capes-Davis et al., 2010], but many operators are still unaware of the seriousness of the risk (*see also* Sections 12.1.1, 15.2). The responsibility lies with supervisors to impress upon new personnel the severity of the risks, with journal editors and referees to reject manuscripts, and with grant review bodies to reject grant proposals where there is no evidence of proper authentication of the cell lines used. Without the acceptance of these obligations, the situation can only get worse.

The probable causes of cross-contamination are listed in Table 18.4 along with probable causes of misidentificationas the two are not always distinguishable. It cannot be overemphasized that cross-contaminations exist and new instances can and do occur. It is essential that the appropriate precautions be taken (*see* Sections 12.1.1, 15.2) and that cell line characteristics be checked regularly. DNA profiling (*see* Protocol 15.9) is the preferred option and a standard set of conditions for the assay has been defined [ASN-0002, 2010].

## 18.7   CONCLUSIONS

- Check living cultures regularly for contamination by using normal and phase-contrast microscopy and for mycoplasmas by employing fluorescent staining of fixed preparations or by PCR.
- Do not maintain all cultures routinely in antibiotics. Grow at least one set of cultures of each cell line without antibiotics for a minimum of two weeks at a time, and preferably continuously, in order to allow cryptic contaminations to become overt.
- Do not attempt to decontaminate a culture unless it is irreplaceable, and then do so only under strict quarantine.
- Quarantine all new lines that come into your laboratory until you are sure that they are uncontaminated.
- Do not share media or other solutions among cell lines or among operators, and check cell line characteristics (*see* Section 15.3) regularly to guard against cross-contamination.
- New cell lines should be characterized, preferably by DNA profiling, as soon after isolation as possible. (*see* Protocol 15.9 and ASN-0002, [2010])

# CHAPTER 19

# Cryopreservation

As cell culture becomes established within a laboratory a number of cell lines may be developed or acquired. The use of each cell line adds to its provenance, and each one becomes a valuable resource. If unique, the cell line might be impossible to replace; at best, replacement would be expensive and time-consuming. It is therefore essential to protect this considerable investment by preserving the cell line.

## 19.1 RATIONALE FOR FREEZING

Cell lines in continuous culture are prone to variation due to selection in early-passage culture (*see* Section 2.7.2), senescence in finite cell lines (*see* Sections 2.7.3, 17.4.1), and genetic and phenotypic instability (*see* Section 16.1, 17.3) in continuous cell lines. In addition even the best-run laboratory is prone to equipment failure and contamination. Cross-contamination and misidentification (*see* Sections 12.1.1, 15.2, 18.6; Table 12.2; Appendix V and www.wiley.com/go/freshney/cellculture) also continues to occur with an alarming frequency [Capes-Davis et al., 2010]. There are many reasons therefore for freezing down a validated stock of cells; these reasons can be summarized as follows:

(1) Genotypic drift due to genetic instability.
(2) Senescence and the resultant extinction of the cell line.
(3) Transformation of growth characteristics and acquisition of malignancy-associated properties.
(4) Phenotypic instability due to selection and dedifferentiation.

(5) Contamination by microorganisms.
(6) Cross-contamination by other cell lines.
(7) Misidentification due to careless handling.
(8) Incubator failure.
(9) Saving time and materials by not maintaining lines other than those in current use.
(10) Need for distribution to other users.

## 19.2 CONSIDERATIONS BEFORE CRYOPRESERVATION

There are certain requirements that should be met before cell lines are considered for cryopreservation (Table 19.1). These concern the stage of development of the culture (primary, finite cell line, continuous cell strain, etc.), the definition of its maintenance conditions, the characterization of its properties including the confirmation of the identity of the cell line, and an estimate of the duration of its use, which will determine the size of the preserved stocks or bank.

### 19.2.1 Validation

Culture history and current characteristics should be recorded (*see* Section 15.3), particularly the specific properties that make the cell line unique and/or valuable and that may be forgotten during the life of the frozen stock. Cell lines should be shown to be free of contamination (*see* Section 18.3) and authentic (*see* Section 15.2) before cryopreservation. Although a few ampoules of a newly acquired cell line may be frozen before complete validation has been carried out (*see* Section 19.2.2), proper validation should be carried out before bulk stocks are frozen (*see* Section 6.11).

**TABLE 19.1.** Requirements before Freezing

| Status | Finite cell line | Freeze at early passage (<5 subcultures) |
|---|---|---|
| | Continuous cell line | Clone, select, and characterize; amplify |
| Standardization | Medium | Select optimal medium, and adhere to this medium |
| | Serum (if used) | Select a batch for use at all stages (*see* Section 8.6) |
| | Substrate | Standardize on one type and supplier, although not necessarily on one size or configuration |
| Validation | Provenance | Record details of origin, life history, and properties |
| | Authentication | Check cell line characteristics against origin and provenance (*see* Sections 15.1–15.3) |
| | Transformation | Determine transformed status (*see* Table 17.1) |
| | Contamination | Microbial (*see* Sections 18.3, 10.6.2) |
| | | Mycoplasma (*see* Protocols 18.2, 18.3) |
| | Cross-contamination and misidentification | Criteria to confirm identity (*see* Section 15.2) |

### 19.2.2 When to Freeze

Finite cell lines are grown to around the fifth population doubling in order to generate sufficient cells for freezing (*see* Section 19.5). Continuous cell lines should be cloned (*see* Sections 13.1, 13.4, 13.5), a characterized clone selected, and sufficient stocks grown for freezing.

## 19.3 PRINCIPLES OF CRYOPRESERVATION

### 19.3.1 Theoretical Background to Cell Freezing

Optimal freezing of cells for maximum viable recovery on thawing depends on minimizing intracellular ice crystal formation and reducing cryogenic damage from foci of high-concentration solutes formed when intracellular water freezes. This is achieved (1) by freezing slowly to allow water to leave the cell but not so slowly that ice crystal growth is encouraged, (2) by using a hydrophilic cryoprotectant to sequester water, (3) by storing the cells at the lowest possible temperature to minimize the effects of high salt concentrations on protein denaturation in micelles within the ice, and (4) by thawing rapidly to minimize ice crystal growth and generation of solute gradients formed as the residual intracellular ice melts. In certain cases, such as freezing embryos and embryonal stem (ES) cells, slow freezing is replaced by snap freezing in liquid nitrogen to minimize ice crystal formation by creating a glass (*vitrification*) rather than by slow removal of water (*see* Section 19.4).

### 19.3.2 Cell Concentration

Cells appear to survive freezing best when frozen at a high cell concentration. This is largely an empirical observation but may be related partly to the reduced viability on thawing requiring a higher seeding concentration and partly to improved survival at a high cell concentration if cells are leaky because of cryogenic damage. A high concentration at freezing also allows sufficient dilution of the cryoprotectant at reseeding after thawing so that centrifugation is unnecessary (at least for most cells). The number of cells frozen should be sufficient to allow for 1:10 or 1:20 dilution on thawing to dilute out the cryoprotectant but still keep the cell concentration higher than at normal passage; for example, for cells subcultured normally at $1 \times 10^5$/ mL, $1 \times 10^7$ should be frozen in 1 mL of medium, and, after thawing the cells, the 1 mL should be diluted to 20 mL of medium, giving $5 \times 10^5$ cells/mL (five times the normal seeding concentration). This dilutes the cryoprotectant from 10% to 0.5%, at which concentration it is less likely to be toxic. Cells normally seeded at $2 \times 10^4$/ mL can be diluted 1:100 to give $1 \times 10^5$/ mL and a cryoprotectant concentration of 0.1%.

### 19.3.3 Freezing Medium

The cell suspension is frozen in the presence of a cryoprotectant such as glycerol or dimethyl sulfoxide (DMSO) [Lovelock & Bishop, 1959]. Of these two, DMSO appears to be the more effective, possibly because it penetrates the cell better than glycerol. Concentrations of between 5% and 15% have been used, but 7.5% or 10% is more usual. It has been claimed that cells should be kept at 4°C after DMSO is added to the freezing medium and before freezing, but preliminary experiments by the author suggested that this did not improve survival after freezing and may even have reduced it [unpublished observations], perhaps by inhibiting intracellular penetration. However, this issue is still not fully resolved and is worth further experimentation (*see* Section 28.4, Exercise 18). There are situations where DMSO may be toxic or induce cells to differentiate after thawing (*see* Section 16.7.2), for example, with hematopoietic cell lines such as L5178Y or HL60 [M. Freshney, personal communication], and in these cases it is preferable either to use glycerol or to centrifuge the cells after thawing to remove the cryoprotectant. Otherwise, and for most monolayer cultures, residual cryoprotectant can be removed at the first medium change (as soon as the cells have attached).

DMSO should be colorless, and it needs to be stored in glass or polypropylene, as it is a powerful solvent and will leach impurities out of rubber and some plastics. Glycerol should be not more than one year old, as it may become toxic after prolonged storage. Other cryoprotectants have been suggested, such as polyvinylpyrrolidone (PVP) [Suzuki et al., 1995], polyethylene glycol (PEG) [Monroy et al., 1997], and hydroxyethyl starch (HES) [Pasch et al., 2000], but none has had the general acceptance of either DMSO or glycerol, although there may be some improvement with trehalose [Eroglu et al., 2000; Buchanan et al., 2004]. Many laboratories also increase the serum concentration in freezing medium to 40%, 50%, or even 100%.

## 19.3.4 Cooling Rate

Most cultured cells survive best if they are cooled at 1°C/min [Leibo & Mazur, 1971; Harris & Griffiths, 1977]. This is probably a compromise between fast freezing minimizing ice crystal growth and slow cooling encouraging the extracellular migration of water. The shape of the cooling curve is governed by (a) the ambient temperature, (b) any insulation surrounding the cells, including the ampoule, (c) the specific heat and volume of the ampoule contents, and (d) the latent heat absorption during freezing. When cells are frozen in an insulated container placed in an ultra-deep freeze, as in options (1) through (4) below, this results in a curve with a rapid cooling rate at the start. When the temperature differential is greatest, down to a minimum at the eutectic point, a slight rise as freezing commences, followed by a plateau as the latent heat of freezing is absorbed, and then a more rapid fall as freezing is completed, gradually slowing down as the temperature of the freezing chamber is reached

(Fig. 19.1). When cells are transferred to the liquid nitrogen freezer (or the end stage is reached in a programmable freezer) the temperature drops rapidly to between −180°C and −196°C.

A controlled cooling rate can be achieved in one of several different ways:

(1) Lay the ampoules on cotton wool in a polystyrene foam box with a wall thickness of about 15 mm. This box, plus the cotton wool, should provide sufficient insulation such that the ampoules will cool at 1°C/min when the box is placed at −70°C or −90°C in a regular deep freeze or insulated container with solid $CO_2$.

(2) Insert the canes in tubular foam pipe insulation, with a wall thickness of about 15 mm (Fig. 19.2), and place the insulation at −70°C or −90°C in a regular deep freeze or insulated container with solid $CO_2$.

(3) Place the ampoules in a freezer neck plug (Fig. 19.3), and insert the plug into the neck of the nitrogen freezer.

(4) Place the ampoules in a Nunc freezing container (Fig. 19.4), and place at −70°C or −80°C.

(5) Use a controlled-rate freezer programmed to freeze at 1°C/min (Fig. 19.5), with accelerated freezing through the eutectic point (see Fig. 19.1).

With any of the first four methods, such as the insulated tube method in cooling option (2) (see Fig. 19.2), the cooling rate will be an average of varying rates throughout the

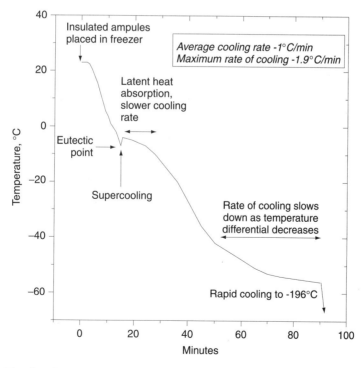

**Fig. 19.1. Freezing Curve.** Record of the fall in temperature in an ampoule containing medium, clipped with five other ampoules on an aluminum cane, enclosed in a cardboard tube, placed within a polyurea-foam tube (see Fig. 19.2), and placed in a freezer at −70°C.

**Fig. 19.2. Ampoules on Cane with Insulation for Slow Cooling.**
Plastic ampoules are clipped onto an aluminum cane (bottom), enclosed in a cardboard tube (middle), and placed inside an insulating foam tube (top). The insulating tube is plugged at either end with cotton or another suitable insulating material before placing it to cool slowly in a −70°C or −80°C freezer.

**Fig. 19.4. Nunc Cooler.** Plastic holder with fluid-filled base. The specific heat of the coolant in the base insulates the container and gives a cooling rate of about 1°C/min in the ampoules.

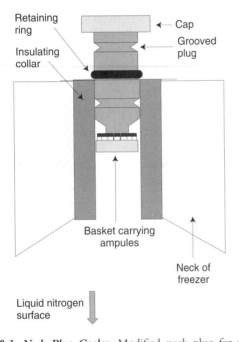

**Fig. 19.3. Neck Plug Cooler.** Modified neck plug for narrow-necked freezers, allowing controlled cooling at different rates (Taylor Wharton). Shown is the section of the freezer neck with the modified neck plug in place. The retaining ring is used to set the height of the ampoules within the neck of the freezer. The lower the height, the faster the cooling.

curve, and no attempt is made to control supercooling at the eutectic point or the duration of the plateau during latent heat absorption, both of which may impair survival. If recovery is low, it is possible to change the average cooling rate (i.e., by use of more or less insulation). Use of a programmable freezer (*see* Fig. 19.5) with a probe that senses the temperature of the ampoule and adds liquid nitrogen to the freezing chamber at the correct rate to achieve a preprogrammed cooling rate, and that can seed freezing as

the cell suspension reaches the eutectic, minimizes the stress of supercooling and can achieve a linear cooling rate throughout the range including a rapid transit through the eutectic point. Different cooling rates at different phases of the cooling curve, optimized experimentally to suit the cells being frozen, can be programmed into the cooling curve. Programmable coolers are, however, relatively expensive, compared to the simple devices described in cooling options (1) through (4), and have few advantages unless you have problems freezing particular cells and wish to vary the cooling rate [e.g., Foreman & Pegg, 1979] or alter the shape of the cooling curve.

With the insulated container methods, the cooling rate is proportional to the difference in temperature between the ampoules and the ambient air. If the ampoules are placed in a freezer at −70°C, they will cool rapidly to around −50°C, but the cooling rate falls off significantly after that (*see* Fig. 19.1). Hence the time that the ampoules spend in the −70°C freezer needs to be longer than the amount of time projected by a 1°C/min cooling rate, as the bottom of the curve is asymptotic. It is safer to leave the ampoules at −70°C overnight before transferring them to liquid nitrogen. Furthermore, when removed from the freezing device, they will heat up at a rate of about 10°C/min. It is critical that they do not warm up above −50°C, as they will start to deteriorate, so the transfer to the liquid nitrogen freezer must take significantly less than two minutes.

Alternatively, cells may be frozen rapidly so that the aqueous medium is vitrified (*see* Section 19.4).

### 19.3.5 Ampoules

Plastic ampoules are safer and more convenient than glass, but some repositories and cell banks prefer glass ampoules for seed stocks because the long-term storage properties of glass are well characterized and, when correctly performed, sealing is absolute. Plastic ampoules are usually polypropylene, and 1.2 mL is probably the most popular size (*see* Plate 22e). They may be labeled with a fine-tipped marker, or preprinted labels, both of which should be alcohol resistant and able to

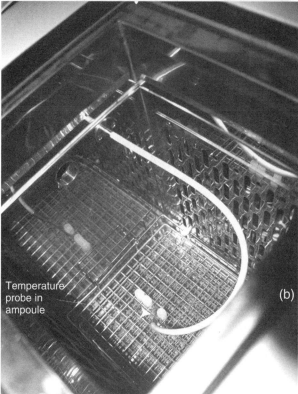

Temperature probe in ampoule

(a)

(b)

**Fig. 19.5. Programmable Freezer.** Ampoules are placed in an insulated chamber, and the cooling rate is regulated by injecting liquid nitrogen into the chamber at a rate determined by a sensor on the rack with the ampoules and a preset program in the console unit (Planer Biomed). (*a*) Control unit and freezing chamber (lid open). (*b*) Close-up of a freezing chamber with four ampoules, one with a probe in it.

withstand the low temperature of the freezer (*see* Appendix II: Cryomarkers; Cryolabels). The label should show the cell strain designation and, preferably, the date and user's initials, although the latter is not always feasible in the available space. If the cell line inventory is stored on a computer, labels can be printed automatically and barcoded if necessary.

Different colored caps also help identification (see Plate 22*e*). Remember, cell cultures stored in liquid nitrogen may well outlive you! They can easily outlive your stay in a particular laboratory. The record therefore should be readily interpreted by others and sufficiently comprehensive so that the cells may be of use to others (*see* Section 19.3.7) or discarded if not.

Plastic ampoules require the correct torsion for closing, as they will leak if too slack or too tight (due to distortion of the o-ring). It is worth practicing with a new batch to make sure (a) that they seal correctly and (b) that they withstand the low temperature; occasionally a faulty batch of ampoules may shatter on thawing. Plastic ampoules are of a larger diameter and taller than equivalent glass ampoules, so special canes must be used.

Δ *Safety Note.* Inexperienced users should avoid using glass ampoules as they have a serious risk of explosion when thawed. If glass ampoules are used, they must be perfectly and quickly sealed in a gas-oxygen flame. If sealing takes too long, the cells will heat up and die, and the air in the ampoule will expand and blow a hole in the top of the ampoule. If the ampoule is not perfectly sealed, it may inspire liquid nitrogen during storage in the liquid phase of the nitrogen freezer and will subsequently explode violently on thawing. If glass ampoules are used they should be stored in the vapor phase or in a perfused jacket freezer (*see* Section 19.3.6).

It is possible to check for leakage by placing ampoules in a dish of stain, such as 1% methylene blue in 70% alcohol, at 4°C for 10 min before freezing. If the ampoules are not properly sealed, the stain will be drawn into the ampoule, and the ampoule should be discarded.

### 19.3.6  Cryofreezers

Frozen cells are transferred rapidly to a cryofreezer when they are at or below −70°C. Storage in liquid nitrogen (Figs. 19.6, 19.7; *see also* Section 4.5.3) is currently the most satisfactory [Hay et al., 2000]. Cryofreezers differ in design depending on size of the access neck, storage system employed, and location of liquid nitrogen (Fig. 19.7).

*Neck size.* Canister storage systems tend to have narrow necks (Figs. 19.6*a*, *b*, 19.7*a*), which reduces the rate of evaporation of the liquid nitrogen but makes access a little awkward. Wide-necked freezers are chosen for ease of access and maximum capacity, usually with storage in sections within drawers, but tend to have a faster evaporation rate. However, it is possible to select a relatively narrow-necked freezer while still using a tray system for storage. Freezers are available with inventory control based on square-array storage trays; these trays are mounted on racks that are accessed by the same system as the cane and canister of conventional narrow-necked freezers (Figs. 19.6*d*, 19.7*b*), have equivalent holding times, and have the honeycomb storage array that many people prefer (*see* below).

*Storage system.* There are two mains types of storage used for 1.2-mL ampoules for cell culture work. The cane

**Fig. 19.6. Liquid–Nitrogen Freezers.** (*a*) Narrow-necked freezer with storage on canes in canisters. (*b*) Interior of the narrow-necked freezer, looking down on canes in canister, positioned in center as it would be for withdrawal. Normal storage position is under shoulder of freezer, just visible top right. (*c*) Wide-necked freezer with storage in triangular drawers. (*d*) Narrow-necked freezer with storage in square drawers (*see also* Fig. 19.7). (*e*) High-capacity freezer with offset access port open, revealing canes in canisters. (*f*) High-capacity freezers in a cell bank. Nearest freezer shows connections for monitoring and automatic filling. (*e, f,* photos courtesy of ATCC.)

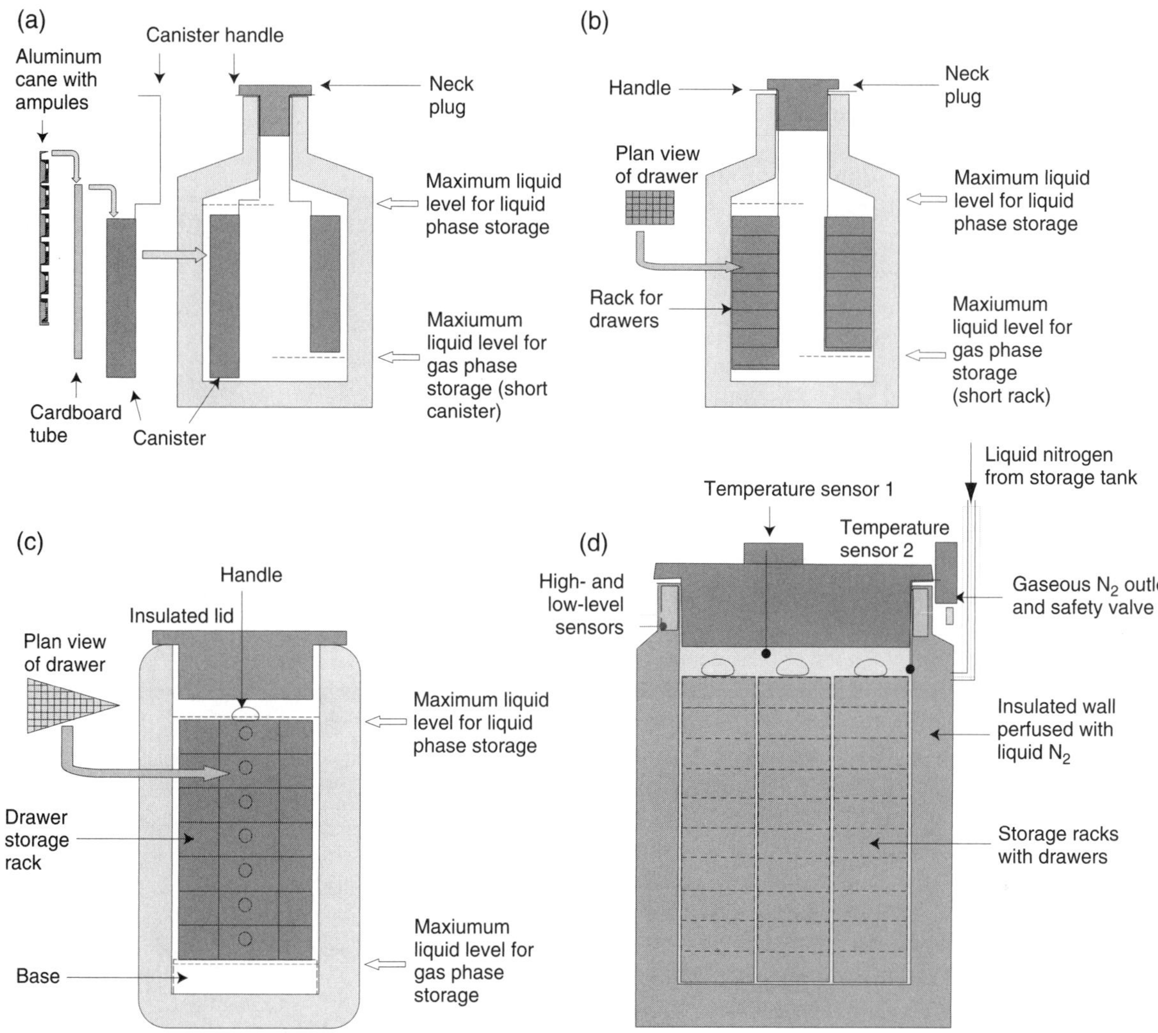

***Fig. 19.7. Nitrogen Freezer Design.*** Four main types of nitrogen freezers: (*a*) Narrow-necked with ampoules on canes in canisters (high capacity, low boil-off rate). (*b*) Narrow-necked with ampoules in square racks (moderate capacity, low boil-off rate). (*c*) Wide-necked with ampoules in triangular racks (high capacity, high boil-off rate). (*d*) Wide-necked with storage in drawers; piped liquid nitrogen perfused through freezer wall and level controlled automatically by high- and low-level sensors (top left).

system uses ampoules clipped on to an aluminum cane, inserted into a cardboard tube, and placed within cylindrical canisters in the freezer. It has the advantage that ampoules can be handled in multiples of six at a time, with all the ampoules on one cane being from the same cell line, making the transfer from the cooling device to the freezer easier and quicker. The canes can be colored and numbered, making location fairly easy, and one cane of ampoules can be partially withdrawn to retrieve an ampoule without exposing all the other ampoules to the warm atmosphere and without risking the error of replacing the ampoules in the wrong location. Storage in rectangular drawers is preferred by some users who feel that retrieval is easier, and individual ampoules can be identified by the drawer number and the coordinates within the drawer. It does mean, however, that the total contents of the drawer,

which can be from 20 to 100 ampoules, are exposed at one time when an ampoule is retrieved, the whole stack must be lifted out if you are accessing one of the lower drawers, and the risk of returning an ampoule to the wrong location is higher. Also, loading a large number of ampoules into the drawer must be done one ampoule at a time, risking delay and overheating, unless a whole drawer with a large batch of ampoules is frozen at one time in a programmable freezer.

***Location of liquid nitrogen.*** If the liquid nitrogen is located in the main body of the freezer, there is a choice of filling the freezer and submerging the ampoules, or only part-filling and storing the ampoules in the vapor phase. Storage in the liquid phase means that the container can be filled and the liquid nitrogen will therefore last longer, but risks uptake of nitrogen by leaky ampoules, which will then explode violently on thawing. There is also a greater likelihood of transfer of contamination between ampoules and the buildup of contamination from outside, carried in when material is introduced and concentrated by the constant evaporation of the liquid nitrogen in the tank. With the introduction of improved insulation and reduced evaporation, vapor-phase storage is preferable. It also eliminates the risk of splashing when the liquid nitrogen boils when something is inserted and reduces evaporation of nitrogen into the room air. There is, however, a gradient in the temperature from the surface of the liquid nitrogen up to the neck of approximately 80°C, from −190°C to around −110°C in gas phase storage [Rowley & Byrne, 1992], although the design and composition of the racking system may help eliminate this gradient.

Some freezers have the liquid nitrogen located within the wall of the freezer and not in the storage compartment. It is replenished by an automatic feed with high and low level controls (Fig. 19.7d), and evaporated nitrogen is released via a relief valve. This has the advantages of gas-phase storage, with the added advantage of a lower consumption of liquid nitrogen and elimination of the temperature gradient. However, the nitrogen level is not visible and cannot be measured by a dipstick, so complete reliance has to be made on electronic monitoring (see below). In addition any blockage of nitrogen flow within the freezer wall can be very difficult or even impossible to eliminate, so it is essential that the liquid nitrogen be filtered and that steps be taken to ensure that no water, or water vapor, enters the system, as ice can also block nitrogen flow.

Δ **Safety Note.** Biohazardous material *must* not be stored in the liquid phase and teaching and demonstrating should also not be done with liquid-phase storage. Above all, if liquid-phase storage is used, the user must be made aware of the explosion hazard of both glass and plastic and must wear a face shield or goggles.

***Monitoring and replenishing liquid nitrogen.*** The investment in the contents of a nitrogen freezer can be considerable and must be protected by a strict monitoring regime and electronic liquid level alarms. When the nitrogen is in the storage compartment, the level should be monitored at least once per week with a dipstick, and the level recorded on a chart. This should be done even if automatic filling is employed, as these systems can fail. Where the liquid nitrogen is totally enclosed, refilling is automatic based on level controls in the wall of the freezer but must still be backed up, preferably with two independent temperature recorders, both of which should sound an alarm, one if the temperature rises above −170°C and one above −150°C.

If liquid nitrogen storage is not available, the cells may be stored in a conventional freezer. The temperature in this freezer should be as low as possible; little deterioration has been found at −196°C [Green et al., 1967], but significant deterioration (5–10% per annum) may occur at −70°C.

### 19.3.7 Freezing Cultured Cells

The following protocol is based on the optimal conditions for most cultured cell lines but variations in the freezing rate or preservative may be required for some cells.

---

**PROTOCOL 19.1. FREEZING CELLS**

***Outline***
Grow the culture to late log phase, prepare a high concentration cell suspension in medium with a cryoprotectant, aliquot into ampoules, and freeze slowly (*see* Fig. 19.8).

***Materials***

*Sterile or aseptically prepared:*

☐ Culture to be frozen
☐ If monolayer: D-PBSA and 0.25% crude trypsin
☐ Growth medium (serum improves survival of the cells after freezing; up to 50%, or even pure, serum has been used. If serum is being used with serum-free cultures, it should be washed off after thawing)
☐ Cryoprotectant, free of impurities (*see* above): DMSO in a glass or polypropylene vial, or glycerol, fresh, and in a universal container
☐ Syringe, 1 to 5 mL, for dispensing glycerol if used (because it is viscous)
☐ Plastic ampoules, 1.2 mL, prelabeled with the cell line designation and the date of freezing

*Nonsterile:*

☐ Hemocytometer or electronic cell counter
☐ Canes or racks for storage (racks may already be in place in the freezer)
☐ Insulated container for freezing: polystyrene box lined with cotton wool or plastic foam insulation tube (*see* Fig. 19.2) or controlled rate freezer if available (*see* Fig. 19.5)
☐ Protective gloves, nitrile

***Procedure***

1. Make sure the culture satisfies the criteria for freezing (*see* Table 19.1), and check by eye and on microscope for:
   **(a)** Healthy appearance (*see* Section 12.4.1).
   **(b)** Morphological characteristics (*see* Section 15.5.1).
   **(c)** Phase of growth cycle (should be late log phase before entering plateau (*see* Section 12.4.3).
   **(d)** Freedom from contamination (*see* Section 18.3).
2. Grow the culture up to the late log phase, and if you are using a monolayer, trypsinize and count

the cells (*see* Protocol 13.2). If you are using a suspension, count and centrifuge the cells (*see* Protocol 12.3).

3. Resuspend at $2 \times 10^6$ to $2 \times 10^7$ cells/mL.

4. Dilute one of the cryoprotectants in growth medium to make freezing medium:

   (a) Add dimethyl sulfoxide (DMSO) to between 10% and 20%

   Δ *Safety Note.* DMSO can penetrate many synthetic and natural membranes, including *skin* and rubber gloves [Horita & Weber, 1964]. Consequently any potentially harmful substances in regular use (e.g., carcinogens) may well be carried into the circulation through the skin and even through rubber gloves. DMSO should always be handled with caution, particularly in the presence of any toxic substances.

   or

   (b) Add glycerol to between 20% and 30%.

5. Dilute the cell suspension 1:1 with freezing medium to give approximately $1 \times 10^6$ to $1 \times 10^7$ cells/mL and 5% to 10% DMSO (or 10–15% glycerol). It is not necessary to place ampoules on ice in an attempt to minimize deterioration of the cells. A delay of up to 30 min at room temperature is not harmful when using DMSO and is beneficial when using glycerol.

6. Dispense the cell suspensions into prelabeled ampoules, and cap the ampoules with sufficient torsion to seal the ampoule without distorting the gasket.

7. Place the ampoules on canes for canister storage (*see* Figs. 19.2, 19.6*a*, *b*, 19.7*a*), or leave them loose for drawer storage (*see* Figs. 19.6*c*, *d*, 19.7*b–d*).

8. Freeze the ampoules at 1°C/min by one of the methods described above (*see* Section 19.3.4). With the insulated container methods, this will take a minimum of 4 to 6 h after placing them at −70°C if starting from a 20°C ambient temperature (*see* Fig. 19.1), but preferably leave the ampoules in the container at −70°C overnight.

9. When the ampoules have reached −70°C or lower, check the freezer record before removing the ampoules from the −70°C freezer or controlled rate freezer, and identify a suitable location for the ampoules.

10. Transfer the ampoules to the liquid N₂ freezer, preferably not submerged in the liquid, placing the cane and tube into the predetermined canister or individual ampoules into the correct spaces in the predetermined drawer. This transfer must be done quickly (<2 min), as the ampoules will reheat at about 10°C/min, and the cells will deteriorate rapidly if the temperature rises above −50°C.

Δ *Safety Note.* Protective gloves and a face mask should be used when placing ampoules in or near liquid nitrogen.

11. When the ampoules are safely located in the freezer, complete the appropriate entries in the freezer index (*see* Tables 10.2, 19.3).

### 19.3.8 Freezer Records

Records should provide (1) an inventory showing what is in each part of the freezer, (2) an indication of free storage spaces, and (3) a cell strain index, describing the cell line, its designation, its origin, maintenance details, and freezing procedures, what its special characteristics are, and where it is located. This record may be kept on a conventional card index, but a computerized database will give superior data storage and retrieval, and can also print labels. This type of data can be provided by separate tables within the same database used for the provenance of the cell line (Tables 19.2, 19.3; *see also* Sections 11.3.11, 12.7, 12.8). Material stored on disks or tape must have backup copies on disk or tape or must have a hard-copy printout.

Using a computerized database requires that the curator of the freezers manages this database. If entries are to be made by users, user stocks can have both read and write access, whereas seed stock and distribution stock should be accessible only to the curator. Alternatively, the whole file can be read only and updated by the curator from paper entries on cards or a logbook.

### 19.3.9 Thawing Stored Ampoules

When required, cells are thawed and reseeded at a relatively high concentration to optimize recovery. The ampoule should be thawed as rapidly as possible to minimize intracellular ice crystal growth during the warming process. This can be done in warm water, in a bucket, or in a water bath, but if the ampoule has been submerged in liquid nitrogen during storage, the warming bath must be covered in case the ampoule has leaked and inspired liquid nitrogen, when it will explode violently on warming (Fig. 19.9).

The cell suspension should be diluted slowly after thawing as rapid dilution reduces viability. This gradual process is particularly important with DMSO, with which sudden dilution can cause severe osmotic damage and reduce cell survival by half. Most cells do not require centrifugation, as replacing the medium the following day will suffice for a monolayer or dilution for a suspension. However, some cells (often suspension-growing cells) are more sensitive to cryoprotectants, particularly DMSO, and must be centrifuged after thawing but still need to be diluted slowly in medium first.

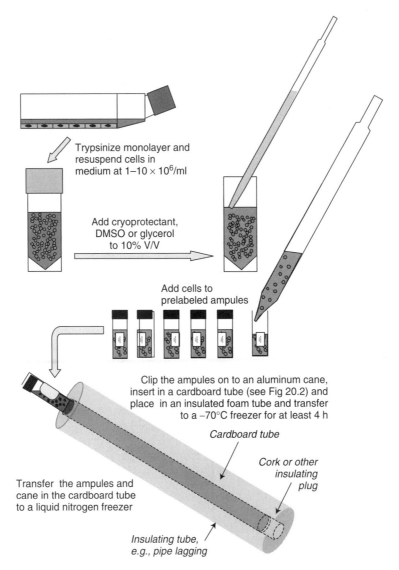

**Fig. 19.8. Freezing Cells.** Trypsinized cells, in medium with cryoprotectant, aliquoted into ampoules, which are then clipped on to an aluminum cane, inserted into a cardboard tube, and inserted into an insulated tube. The tube and contents are placed at −70°C or −80°C for 4 h or overnight, before transferring the cardboard tube containing the ampoules to a liquid nitrogen freezer with canister storage (*see* Fig. 19.7*a*).

---

### PROTOCOL 19.2. THAWING FROZEN CELLS

**Outline**
Thaw the cells rapidly, dilute them slowly, and reseed them at a high cell density (Fig. 19.9).

**Materials**
*Sterile:*

- ❑ Culture flask
- ❑ Centrifuge tube (if centrifugation is required)
- ❑ Growth medium

---

- ❑ Pipettes, 1 mL, 10 mL
- ❑ Pipettors, if used, must have filter tips
- ❑ Syringe and 19-g needle (if you are using glass ampoules)

*Nonsterile:*

- ❑ Naphthalene Black (Amido Black), 1%, or Trypan Blue, 0.4%.
- ❑ Alcohol, 70%
- ❑ Sterilized water at 37°C, 10 cm deep in a clean, alcohol-swabbed bucket with lid (if ampoule has been

stored submerged in liquid phase of the nitrogen) or to the top of a rack in a water bath (if ampoule has not been stored submerged in liquid nitrogen).
- ❑ Protective gloves and face mask
- ❑ Forceps
- ❑ Swabs

### Procedure

1. Check the index for the location of the ampoule to be thawed.
2. Collect all materials, prepare the medium, and label the culture flask.
3. Retrieve the ampoule from the freezer; check from the label that it is the correct one, and, if it has not been submerged in liquid nitrogen, place it in sterile water at 37°C in a beaker or water bath. If possible, avoid getting water up to the cap as this will increase the chance of contamination. A heating block is useful for this, though heat transfer may be slower.

Δ *Safety Note.* A closed lab coat and gloves must be worn when removing the ampoule from the freezer. If the ampoule has been stored **submerged** in liquid nitrogen, **a face shield, or protective goggles**, as well as a closed lab coat and gloves, must be worn. Ampoules, including plastic ampoules, stored in the liquid phase may inspire the liquid nitrogen and, on thawing, will explode violently. In this case **a plastic bucket with a lid must be used for thawing** to contain any explosion (*see* Fig. 19.9).

4. When the ampoule has thawed, double-check the label to confirm the identity of the cells; then swab the ampoule thoroughly with 70% alcohol, and open it in a laminar-flow hood.
5. Transfer the contents of the ampoule to a culture flask with a 1-mL pipette of pipettor with filter tip.
6. Add medium slowly to the cell suspension: 10 mL over about 2 min added dropwise at the start, and then a little faster, gradually diluting the cells and cryoprotectant.
   For cells that require centrifugation to remove the cryoprotectant:
   (a) Dilute the cells slowly, as in step 6, but in a centrifuge tube or universal container.
   (b) Centrifuge them for 2 min at 100 $g$.
   (c) Discard the supernatant medium with the cryoprotectant.
   (d) Resuspend the cells in fresh growth medium.
   (e) Seed flask for culture.

7. The dregs in the ampoule may be stained with Naphthalene Black or Trypan Blue to determine cell viability (*see* Protocol 21.3.1).
8. Check after 24 h:
   (a) For attached monolayer cells, confirm attachment and try to estimate percentage survival based on photographs of cells at the expected density (cells/cm$^2$) with full survival (*see* Sections 12.3.1, 16.4.5; Plate 4).
   (b) For suspension-growing cells, check appearance (clear cytoplasm, lack of granularity), and dilute to regular seeding concentration. This can be made more precise if the cells are counted and an estimate of viability is made (*see* Section 21.3.1), in which case the cells can be diluted to the regular seeding concentration of viable cells.

The dye exclusion viability and the approximate take (e.g., the proportion of cells attached after 24 h) should be recorded on the appropriate record card or file to assist in future thawing. One ampoule should be thawed from each new freezing, to check that the operation was successful.

### 19.3.10  Freezing Flasks

Whole flasks may be frozen by growing the cells to late log phase, adding 5% to 10% DMSO to the smallest volume of medium that will effectively cover the monolayer, and placing the flask in an expanded polystyrene container of 15-mm wall thickness [Ohno et al., 1991]. The insulated container is placed in a −70°C to −90°C freezer, and it will freeze at approximately 1°C/min. Survival is good for several months, as long as the flask in its container is not removed from the freezer. Twenty-four-well plates may also be frozen in the same manner [Ure et al., 1992] with about 150 μL freezing medium per well and can be used to store large numbers of clones during evaluation procedures.

### 19.4  VITRIFICATION

Although the slow cooling method is suitable for most cultured cell lines, some cells, such as preimplantation embryos and human embryonal stem cells, require rapid cooling and vitrification. It is possible that their three-dimensional structure, although very small, still presents diffusion limitations during the slow cooling process. Vitrification is the transformation of a liquid into a glass (a supercooled liquid) and is achieved by plunging the cells in a plastic capillary tube, or *straw*, into liquid nitrogen. The cryoprotectants DMSO and ethylene glycol are used with sucrose in stepwise increasing concentrations.

**TABLE 19.2.** Cell Line Record

| Cell line | Freeze date: | | | | | | | | | | | New card |
|---|---|---|---|---|---|---|---|---|---|---|---|---|
| | Location: | | | | | | | | | | | |

| | | |
|---|---|---|
| **Species:** Normal / Adult / Fetal / NB neoplastic | **Mycoplasma:** Method <br> Date of test          Result | **Freeze instructions:** <br>   Rate       Preservative       % |
| **Tissue:**   Site        Pass./ Gen. No | **Authentication:** | **Thaw instructions:** <br><br> Thaw rapidly to 37°C |
| **Author:**   Ref.        Mode of growth | Date of test          Method | Dilute      5 ml        10 mL <br> to: <br>              20 ml       50 mL |
| | **Normal Maintenance:** | Centrifuge to          Yes/No remove preservative? |
| **Special characteristics:** | Subculture  Seeding conc.       Agent frequency | **Special requirements** |
| | Medium      Type            Serum, change                       etc.... frequency                     % | |
| | Gas phase    Buffer          pH | |
| **Person completing card:**     Date | **Any other special conditions:** | **Biohazard precautions:** |

## 19.4.1  Cryopreservation of hES Cells

Human embryonic stem cells, like many other cells, can be stored long term in liquid nitrogen. While other stem cells have been frozen successfully by the conventional slow freezing (1°C/min) and rapid thawing (see Section 19.3.8), to date the most common process of preparing the hES cells for freezing is by vitrification. This method uses a stepwise increasing concentration of sucrose and DMSO freezing solutions to help preserve the cells and to reduce ice crystal formation within the cells (Fig. 19.10). The thawing process is very similar and uses a stepwise concentration to reduce sucrose and DMSO concentration, thereby thawing the cells slowly back into standard hES medium Protocols 19.3 and 19.4 are adapted from Cooke & Minger [2007].

**PROTOCOL 19.3. CRYOPRESERVATION OF hES CELLS BY VITRIFICATION**

***Reagents and Materials***

*Sterile or aseptically prepared:*

❑ Undifferentiated hES cells on MEF feeder layers
❑ hES-HEPES medium (see Section 23.1.4)
❑ 10% Vitrification solution:

    hES-HEPES Medium ............................2 mL
    Ethylene Glycol ..................................0.25 ml
    DMSO ..............................................0.25 ml

❑ 20% Vitrification solution:

    hES-HEPES medium ............................0.75 mL
    Sucrose, 1 M ....................................0.75 mL
    Ethylene glycol .................................0.5 mL
    DMSO ..............................................0.5 mL

❑ Multiwell plate, 4-well (see Fig. 19.10)
❑ Open pulled vitrification straws (see Fig. 19.10)
❑ Cryovial, 4.5 mL (see Plate 22e)
❑ Pulled glass pipettes (see Protocol 23.4)
❑ Pipette tips for use at 80 μL
❑ Syringe with wide bore needle (14 gauge)

*Nonsterile:*

❑ Pipettor, set at 80 μL
❑ Phase contrast dissecting microscope within a category II laminar-flow hood, with a heated stage set at 37°C
❑ Wide-necked vacuum flask filled with liquid nitrogen

❏ Long-handled forceps
❏ Appropriate safety equipment for handling liquid nitrogen

**Procedure**

1. Prepare a vitrification plate (*see* Fig. 19.10):
   Well 1: hES-HEPES medium, 1 mL.
   Well 2: empty.
   Well 3: 10% vitrification solution, 1 mL.
   Well 4: 20% vitrification solution, 1 mL.
   Upturned lid: 10 μL of 20% vitrification solution.
2. Pre-warm the plate in an incubator for 2 min prior to use.
3. Prelabel the 4.5 mL cryovial with the hES cell line name, passage number, date, and any other relevant information.
4. Carefully, using a wide-bore syringe needle gently heated over a gas burner, poke a hole in the top and bottom of the 4.5-mL cryovial to allow free flow of liquid nitrogen through the tube. This will help keep the cells consistently frozen during storage.
5. Cut the hES cell colonies into pieces roughly twice the size required for passaging (*see* Protocol 23.4). Small pieces of colony don't tend to survive the thawing process and therefore the correct size

for each hES cell line should be determined prior to bulk freezing.

6. Transfer 6 to 8 pieces of colony into well 1 for 60 s.
7. Using a pipettor transfer all of the colony pieces from well 1 into well 3, and time precisely for 60 s.
8. Transfer the colony pieces from well 3 to well 4 for precisely 30 s. Care must be taken with timing, as the DMSO solution is toxic to cells when not frozen.
9. Transfer the colony pieces into the 10-μL droplet.
10. Using a pipettor, place the vitrification straw on the end of the tip (*see* Fig. 19.10) and carefully, but quickly, draw up the solution containing the pieces of colony.
11. Carefully remove the pipette tip, and using a pair of long-handled forceps, submerge the straw at a slight angle into the liquid nitrogen to snap freeze the cells.
12. Once frozen (after around 5–10 s) place the straw in a 4.5-mL cryovial and place back in liquid nitrogen.
13. When full, transfer the 4.5-mL cryovial containing the straws into long-term liquid nitrogen storage for recovery later.

**TABLE 19.3.** Freezer Record

**Position:** Freezer no. ..................   Canister/section no ...................   Tube/drawer no .......................

**Cell strain/line:** ......................   Freeze date ..........................   Frozen by ............................

No. of ampoules frozen ...............   No. of cells/ampoule .................   in .................................. mL

Growth medium ......................   Serum ............ Conc. ............   Freeze medium .......................

Method of cooling ............................................................   Cooling rate ...........................

**Thawing record:**

| Thaw date | No. of ampoules | No. left | Seeding | | | Viability | | | Notes |
|---|---|---|---|---|---|---|---|---|---|
| | | | Conc. | Vol. | Medium | Dye exclusion | ~% attached by 24 h | Cloning efficiency | |
| | | | | | | | | | |
| | | | | | | | | | |
| | | | | | | | | | |
| | | | | | | | | | |
| | | | | | | | | | |
| | | | | | | | | | |

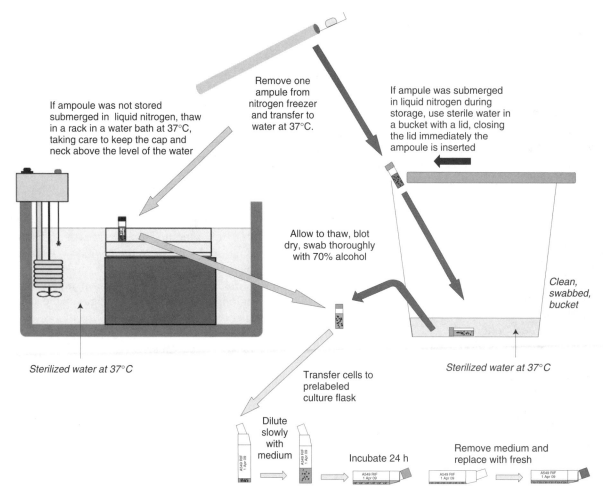

**Fig. 19.9. Thawing Cells.** Ampoules are removed from the freezer and thawed rapidly in warm water, under cover, to avoid the risk of explosion if the ampoules have been stored in the liquid phase (right-hand side of figure). Ampoules, which have not been submerged in liquid nitrogen, can be thawed in a water bath (left-hand side), with the water level kept below the cap and neck.

## 19.4.2 Thawing hES Cells

Thawing rates of hES cells can be very variable so it is advisable to have "practice runs" to determine the optimum size of colony to freeze to obtain maximum re-growth following thawing.

---

**PROTOCOL 19.4. THAWING hES CELLS CRYOPRESERVED BY VITRIFICATION**

### Reagents and Materials

*Sterile or aseptically prepared:*

❑ One straw of frozen hES cell colony pieces, preferably held in a transportable liquid nitrogen container
❑ Plate of MEFs (ideally inactivated 1–3 days prior to use)

---

❑ hES-HEPES medium (*see* Protocol 23.3)
❑ Sucrose solution, 0.2 M
    Sucrose, 1 M . . . . . . . . . . . . . . . . . . . . . . . . . . .1 mL
    hES-HEPES medium . . . . . . . . . . . . . . . . . . . .4 mL
❑ Sucrose solution, 0.1 M
    Sucrose, 1 M . . . . . . . . . . . . . . . . . . . . . . . . . .0.5 mL
    hES-HEPES . . . . . . . . . . . . . . . . . . . . . . . . . . .4.5 mL
❑ Multiwell plate, 4-well
❑ Pipette tips for 80 µL
❑ Nonsterile
❑ Pipettor, set at 80 µL
❑ Phase contrast dissecting microscope within a category II laminar flow hood, with a heated stage set at 37°C
❑ Insulated container filled with liquid nitrogen
❑ Long-handled forceps
❑ Appropriate safety equipment for handling liquid nitrogen

### Procedure

1. Prepare and pre-warm a thawing plate:
   Well 1, sucrose, 0.2 M, 1 mL
   Well 2, sucrose, 0.1 M, 1 mL
   Well 3 and well 4, hES-HEPES medium, 1 mL per well
2. Collect a cryovial from the long-term liquid nitrogen store and transfer into a portable liquid nitrogen container.
3. Using forceps, remove a single straw of hES cells and take to laminar flow hood.
4. Working quickly, place a finger over the top and submerge the narrowed end into well 1 containing the 0.2 M sucrose.
5. As soon as the frozen contents thaw, the hES cell clumps should be drawn into the sucrose solution (expel any remaining liquid with a pipettor).
6. Incubate the cells for precisely 60 s, before transferring into well 2 (0.1 M sucrose).
7. Incubate the cells in well 2 (0.1 M sucrose) for 60 s.
8. Transfer the cells into well 3 (hES-HEPES) for 5 min.
9. As the last step of thawing, transfer the cells into well 4 for 5 min.
10. Collect the hES cells with a pipettor and seed onto MEFs (*see* Protocol 23.4).

## 19.5 DESIGN AND CONTROL OF FREEZER STOCKS

As soon as a small surplus of cells becomes available, from subculturing a primary culture or newly acquired cell line, and it is shown to be free of contamination, a few ampoules should be frozen as what is called a *token freeze*. When the cell line has been propagated successfully, or a cloned cell strain selected with the desired characteristics, and its identity and freedom from contamination have been confirmed, then a *seed stock* should be stored frozen (Fig. 19.11; Table 19.4).

### 19.5.1 Freezer Inventory Control

The seed stock should be protected and not be made available for general issue. When an ampoule is thawed to check the viability of the seed stock, and if usage of the cell line is anticipated in the near future, cells can be grown up for a *distribution stock* and frozen. Ampoules from this stock are issued to individuals as required. Individual users requiring stocks over a prolonged period should then freeze down their own *user stocks*, which should be discarded when the work is finished. User stocks should never be passed on to another

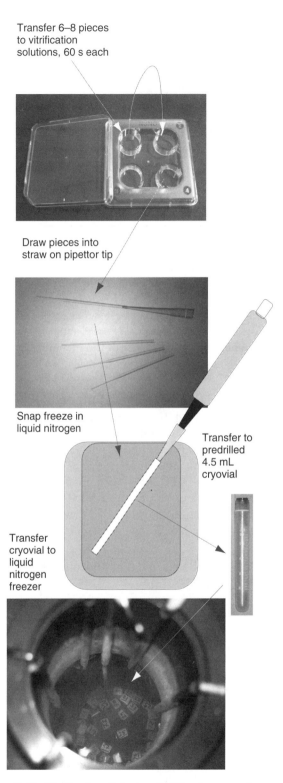

Transfer 6–8 pieces to vitrification solutions, 60 s each

Draw pieces into straw on pipettor tip

Snap freeze in liquid nitrogen

Transfer cryovial to liquid nitrogen freezer

Transfer to predrilled 4.5 mL cryovial

*Fig. 19.10. Vitrification.* Colonies of hES cells are cut into small pieces, transferred through vitrification solutions in a 4-well plate, drawn into a vitrification straw on a pipettor tip, snap-frozen in liquid nitrogen, and transferred in a predrilled 4.5-mL cryovial to a liquid nitrogen freezer (*see* Protocol 19.3). ( [After Cooke & Minger, 2007].)

**TABLE 19.4.** Acquisition and Storage of Cell Lines

| Stage | Source | Number of ampoules | Distribution | Validation |
|---|---|---|---|---|
| Token freeze | Originator | 1–3 | None | Provenance only |
| Seed stock | Original stock or token freeze | 12 | None (replenishment of distribution stock only) | Viability Authentication Transformation Contamination |
| Distribution stock | Test thaw from seed stock | 50–100 (or more as required) | Users, including other laboratories | Viability Contamination Authentication |
| User stock | Distribution stock | 20 (5-year project, culture stock replacement 4× per year) | None | Viability Contamination Authentication |

user as they will not have been fully validated; new users should request a culture or ampoule from the distribution stock. When the distribution stock becomes depleted, it may be replenished from the seed stock. When the seed stock falls below five ampoules, it should be replenished before any other ampoules are issued, and with the minimum increase in generation number from the first freezing.

### 19.5.2 Serial Replacement of Culture Stock

Stock cultures should be replaced from the freezer at regular intervals to minimize the effects of genetic drift and phenotypic variation. After a cell line has been in culture for two and a half months, thaw out another vial, check its characteristics, make sure that it is free from contamination (*see* Section 18.3), and expand it to replace existing stocks. Discard the existing stocks when they have been out of the freezer for three months, and move on to the new stock

(Fig. 19.10). Repeat this process every three months with cells that have a population-doubling time (PDT) of approximately 24 h; cell lines with shorter or longer PDTs may need shorter or longer replacement intervals, respectively.

### 19.6 CELL BANKS

Several cell banks exist (Table 19.5; *see also* Appendix II) for the secure storage and distribution of validated cell lines. Because many cell lines may come under patent restrictions, particularly hybridomas and other genetically modified cell lines, it has also been necessary to provide secure patent repositories with restricted access.

As a rule, it is preferable to obtain your initial seed stock from a reputable cell bank, where the necessary characterization and quality control will have been done.

**TABLE 19.5.** Cell Banks and Other Resources

| Cell/data bank | Description | Website |
|---|---|---|
| American Type Culture Collection (ATCC) | Cell and data bank | www.atcc.org |
| Coriell Cell Repositories (CCR) | Cell and data bank | http://ccr.coriell.org |
| Deutsche Sammlung von Mikroorganismen und Zellkulturen (DSMZ) | Cell and data bank | www.dsmz.de |
| European Collection of Cell Cultures (ECACC) | Cell and data bank | www.hpacultures.org.uk/collections/ecacc.jsp |
| Japanese Collection of Research Bioresources (JCRB) | Cell and data bank | http://cellbank.nibio.go.jp |
| Health Science Research Resources Bank (HSRRB) | Distribution service | www.jhsf.or.jp/English/index_e.html |
| NIH Stem Cell Unit (HIHSCU) | Stem cell bank and information service | http://stemcells.nih.gov/research/nihresearch/scunit |
| RIKEN | Gene and cell bank | www.riken.go.jp/engn |
| UK Stem Cell Bank (UKSCB) | Stem cell bank | www.ukstemcellbank.org.uk |
| Common Access to Biological Resources and Information (CABRI) | Consortium | www.cabri.org |

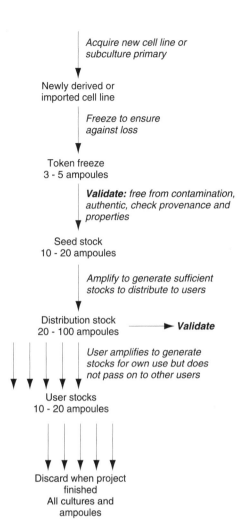

*Fig. 19.11. Cell Banking.* A newly acquired cell line, imported or subcultured from a primary, is frozen as soon as enough cells are available for a few ampoules. The stock is then grown up, characterized, shown to be free of contamination, and frozen as a validated seed stock. One ampoule of the seed stock is thawed, amplified, and revalidated and then frozen as distribution stock. Ampoules distributed to users can be used to generate user stock for the duration of a project but must then be discarded and not passed on to another user.

Furthermore it is highly recommended that you submit valuable cultures to a cell bank as this will protect you against loss and allow distribution of the cells to others unless you wish distribution restricted, which you can specify.

Most cell banks also make their catalog information available online, where they act as a major information resource. There are also several data banks that can be accessed online and that maintain information on cell lines held by subscribers in their own laboratories [Romano et al., 2009] (*see* Table 19.5). This type of data bank provides a vast increase in the amount of material that is potentially available. Nevertheless, it must be remembered that cell lines obtained from other laboratories will vary significantly in the amount of characterization that they have had and in the quality of their maintenance and must be subjected to proper authentication on receipt.

## 19.7 TRANSPORTING CELLS

Cultures may be transferred from one laboratory to another as frozen ampoules or as living cultures. In either case:

(1) Advise the recipient in advance of when the cells are to be shipped.
(2) Fax or email instructions on the following:
    (a) What to do on receipt.
    (b) Which medium and serum is required.
    (c) Any special supplements required.
    (d) Subculture regimen used.
(3) Tape the data sheet for the cells and a copy of the instructions to the outside of the package so that the recipient knows what to do before opening it.

### 19.7.1 Frozen Ampoules

Ship frozen ampoules in solid $CO_2$, in a thick-walled polystyrene foam container (Fig. 19.13*a*). The box should be about $30 \times 30 \times 30$ cm ($12 \times 12 \times 12$ in.) with a wall thickness of 5 cm (2 in.) and a central space of about $20 \times 20 \times 20$ cm ($8 \times 8 \times 8$ in.). You will need about 5 kg (12 lb) of solid $CO_2$ to fill the central space. Transfer the ampoule from the liquid nitrogen freezer to the solid $CO_2$ as quickly as possible, as it will warm up at about $10°C/min$ to $20°C/min$ and must not rise above $-50°C$. Place a sealable polypropylene centrifuge tube on solid $CO_2$ in an insulated box. Remove the ampoule from the solid $CO_2$, quickly wrap it in absorbent paper tissue (in case of leakage), and place it in the polypropylene tube and seal with a tight cap. Usually cells will remain frozen for up to three days if properly packed, but if they thaw slowly their viability will decline rapidly. The carrier must be informed when cells are shipped in solid $CO_2$.

On arrival, the ampoule should be thawed and seeded as normal (*see* Protocol 19.2).

### 19.7.2 Living Cultures

Alternatively, cells may be shipped as a growing culture. The cells should be at the mid- to late-log phase; confluent or postconfluent cultures will exhaust the medium more rapidly and may tend to detach in transit. The flask should be filled to the top with medium, taped securely around the neck with a stretch-type waterproof adhesive tape, and sealed in a small polythene bag. The bagged flask is packed within a rigid container filled with absorbent packing. This container is then sealed in a polythene bag and then in plastic foam or bubble wrap. Alternatively, there is an inflatable bag that is ideal for transporting cells (Air Packaging Technologies, Inc.). The flask is sealed in an inner envelope (Fig. 19.13*b*) and cushioned by the inflatable outer jacket. Place on the container a label that says "fragile" and, in large letters: DO NOT FREEZE!

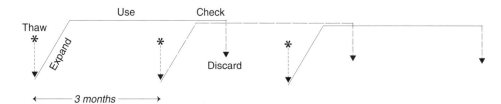

**Fig. 19.12. Serial Culture Replacement.** An ampoule is thawed and the cells grown up to the desired bulk for regular use and maintained for 3 months. Three months after the first thaw, a fresh ampoule is thawed, grown up, and, after checking characteristics against current stock, used to replace the current stock, which is then discarded. The cycle is repeated every 3 months, so that no culture remains in use for more than 3 months.

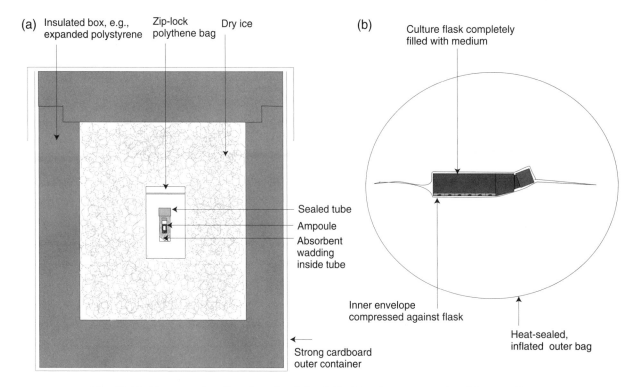

**Fig. 19.13. Transportation Containers for Cells.** Cells shipped as an ampoule of frozen cells or as a growing culture. (a) Rigid cardboard box with insulating expanded polystyrene lining containing dry ice with the ampoule, sealed in a tube with absorbent wadding and the tube sealed in a ziplock bag, placed in the center of the dry ice. (b) Double-skin plastic container with flask trapped in inner compartment by inflating the outer chamber (Air Box, Air Packaging Technologies).

On receipt, the flask is removed from the packing, swabbed thoroughly with 70% alcohol, and opened under sterile conditions. Most of the medium is removed from the flask, leaving only the normal amount for culture—for example, 5 mL for a 25-cm² flask—and the remainder kept for feeding the cells, if required. The culture can be weaned onto new medium when it is ready for the first feed, but keep the original shipping medium in case there are any problems of adaptation to the new batch of medium.

It is usually better to ship cells via a courier, who should be briefed as to the contents of the package and the urgency of delivery. International mailings will require negotiation with customs controls, and a competent nominated agent who is familiar with this type of importation can help move the package through customs. Cells shipped to the United States are required to be quarantined and tested by the Department of Agriculture, so it is better to arrange this shipment through one of the cell banks, ECACC or ATCC.

# CHAPTER 20

# Quantitation

Quantitation in cell culture is required for the characterization of the growth properties of different cell lines, for experimental analyses and to establish reproducible culture conditions for the consistency of primary culture and the maintenance of cell lines (*see* Sections 12.4.3, 12.4.4). Working under sterile conditions and the need to minimize the time that cultures spend out of the incubator often lead to a degree of compromise between speed and accuracy. Although primary culture and normal maintenance require a quantitative approach to ensure reproducibility, speed of handling is the key to good cell survival and growth, so in these situations reduced accuracy, in pipetting, for example, may be acceptable in pursuit of rapid handling. However, under experimental conditions, accuracy should dominate, even if the operations become a little slower. Although an error of roughly 10% may be acceptable in routine maintenance, this should be much lower under experimental conditions.

This chapter covers cell counting technology and a number of other assays used in quantifying cell proliferation, as well as other basic assays for determining cell bulk, such as DNA and protein estimations. Assays relating to cytotoxicity and cell survival are dealt with separately (*see* Section 21.3).

## 20.1   CELL COUNTING

Although estimates can be made of the stage of growth of a culture from its appearance under the microscope, the standardization of culture conditions and proper quantitative experiments are difficult to analyze and reproduce unless the cells are counted before and after, and preferably during, each experiment. There are a number of ways of determining cell number, both direct, such as hemocytometer, electronic counters, and flow cytometry, and indirect, such as staining with Crystal Violet or metabolically reduced MTT (Table 20.1).

### 20.1.1   Hemocytometer

The concentration of a cell suspension may be determined by placing the cells in an optically flat chamber under a microscope (Fig. 20.1). The cell number within a defined area of known depth (i.e., within a defined volume) is counted, and the cell concentration is derived from the count.

Protocol 20.1 can be adapted for use in training (*see* Section 28.3, Exercise 11).

---

**PROTOCOL 20.1. CELL COUNTING BY HEMOCYTOMETER**

***Outline***
Trypsinize a monolayer culture, or sample a suspension culture, prepare a hemocytometer slide, and add the cells to the counting chamber. Count the cells on a microscope and calculate the cell concentration.

---

## Materials

*Sterile:*

- ❑ D-PBSA
- ❑ Crude trypsin, 0.25%
- ❑ Growth medium
- ❑ Yellow pipettor tips

*Nonsterile:*

- ❑ Pipettor, 20 µL or adjustable 100 µL
- ❑ Hemocytometer (Improved Neubauer)
- ❑ Tally counter
- ❑ Microscope, preferably with phase contrast optics

## Procedure

1. Sample the cells:
   - (a) For a monolayer,
     - (i) Trypsinize the monolayer as for routine subculture (*see* Protocol 12.3) and resuspend in medium to give an estimated $1 \times 10^6$/mL. Where samples are being counted in a growth experiment, the trypsin need not be removed and the cells can be dispersed in the trypsinate and counted directly, or after diluting 50:50 with medium containing serum if the cells tend to reaggregate.
     - (ii) Mix the suspension thoroughly to disperse the cells, and transfer a small sample (~1 mL) to a vial or universal container.
   - (b) For a suspension culture,
     - (i) Mix the suspension thoroughly to disperse any clumps.
     - (ii) Transfer 1 mL of the suspension to a vial or universal container.

   A minimum of approximately $1 \times 10^6$ cells/mL is required for this method, so the suspension may need to be concentrated by centrifugation (at 100 *g* for 2 min) and resuspension it in a measured smaller volume.

2. Prepare the slide:
   - (a) Clean the surface of the slide with 70% alcohol, taking care not to scratch the semisilvered coating.
   - (b) Clean the coverslip, and wetting the edges very slightly, press it down over the grooves and semisilvered counting area (*see* Fig. 20.1*a–c*) with the edges of the coverslip extending beyond the outermost grooves. The appearance of interference patterns ("Newton's rings"—rainbow colors between the coverslip and the slide, like the rings formed by oil on water) indicates that the coverslip

is properly attached, thereby determining the depth of the counting chamber.

3. Mix the cell sample thoroughly, pipetting vigorously to disperse any clumps, and collect 20 µL into the tip of a pipettor.

4. Transfer the cell suspension immediately to the edge of the hemocytometer chamber, expel the suspension, and let it be drawn under the coverslip by capillarity. Do not overfill or underfill the chamber, or else its dimensions may change, due to alterations in the surface tension; the fluid should run only to the edges of the grooves.

5. Mix the cell suspension, reload the pipettor, and fill the second chamber if there is one.

6. Blot off any surplus fluid (without drawing from under the coverslip), and transfer the slide to the microscope stage.

7. Select a 10× objective, and focus on the grid lines in the chamber (*see* Fig. 20.1*f–h*). If phase contrast is not available and focusing is difficult because of poor contrast, close down the field iris, or make the lighting slightly oblique by offsetting the condenser.

8. Move the slide so that the field you see is the central area of the grid and is the largest area that you can see bounded by three parallel lines. This area is 1 mm². With a standard 10× objective this area will almost fill the field, or the corners will be slightly outside the field, depending on the field of view (*see* Fig. 20.1*f*).

9. Count the cells lying within this 1-mm² area, using the subdivisions (also bounded by three parallel lines) and single grid lines as an aid for counting. Count cells that lie on the top and left-hand lines of each square, but not those on the bottom or right-hand lines, to avoid counting the same cell twice. For routine subculture, attempt to count between 100 and 300 cells; the more cells that are counted, the more accurate the count becomes. For more precise quantitative experiments, 500 to 1000 cells should be counted.
   - (a) If there are very few cells (<100/mm²), count one or more additional squares (each 1 mm²) surrounding the central square mm.
   - (b) If there are too many cells (>1000/mm²), count only five small squares (each 0.04 mm² and bounded by three parallel lines) across the diagonal of the larger (1 mm²) square, giving a total of 0.2 mm².

10. If the slide has two chambers, move to the second chamber and do a second count. If not, rinse the slide and repeat the count with a fresh sample.

**TABLE 20.1.** Determination of Cell Number

| | Method | Advantages | Disadvantages |
|---|---|---|---|
| ***Direct*** | | | |
| Hemocytometer slide | Count number of cells in given area of marked slide under microscope and extrapolate back to stock cell suspension. | Cheap; can see cells, clumps, etc. Viability stain can be used | High statistical error Slow Low sensitivity |
| Electronic particle counter (Beckman Coulter Z1, Z2; Innovatis CASY) | Dilute cell sample in isotonic saline and count at prescribed setting. Based on resistance change when cell pass through narrow orifice. | Quick Large sample counted, low error Sizing possible | "Blind"; counts dead and live cells and clumps Coincident signals at high cell concentrations Orifice prone to blockage |
| Image analysis (Invitrogen Countess; Nexcelcom Cellometer; Labtech Digital Bio ADAM; Beckman Coulter Vi-CELL; Chemometec Nucleocounter; Innovatis Cedex) | Scan of microscope view of stained and unstained cells in special counting chamber. | Same principle as hemocytometer but faster and more accurate Cells visible on monitor | Needs single cell suspension |
| Flow cytometer (Guava; Accuri; Partec Cyflow CCA) | Count by light scatter, fixed or unfixed cell population, or by DNA fluorescent stain. | Many parameters can be measured simultaneously | Needs single cell suspension Technically more complex |
| ***Indirect*** | | | |
| MTT assay | Stain living cells with MTT and solubilize in DMSO. Read on ELISA plate reader. | Quick; readily automated. Only detects live cells | MTT formazan absorbance can vary |
| Sulphorhodamine | Stain fixed cells and read on fluorescence plate reader. | Sensitive, easy | Stains dead cells Need fluorescence detector |
| Crystal Violet stain | Stain fixed cells with CV and solubilise into EtOH. | Cheap, easy | Stains dead cells |
| Coomassie Blue staining | Stain with Coomassie Brilliant Blue and read on ELISA. | Quick, readily automated | Counts all cells (and even all protein) |

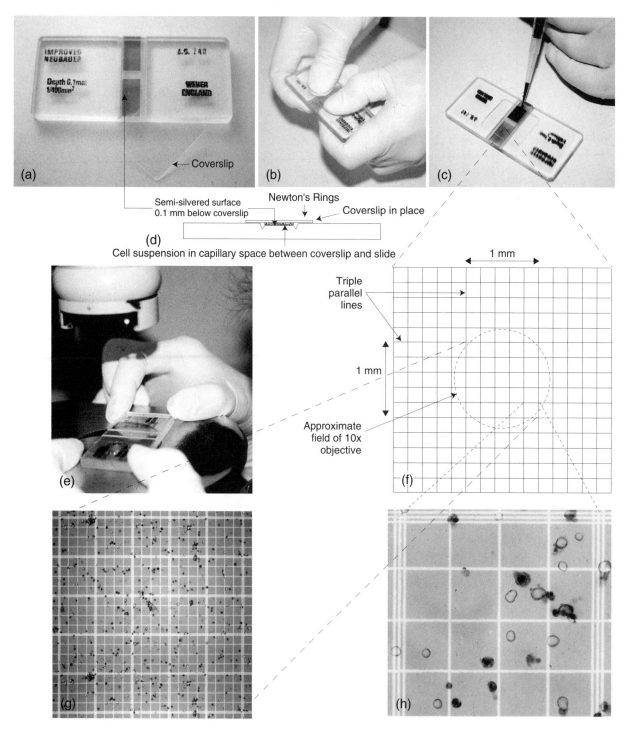

**Fig. 20.1. Using a Hemocytometer Slide.** (*a*) Hemocytometer slide (improved Neubauer) and coverslip before use. (*b*) Pressing coverslip down onto slide. (*c*) Adding a cell suspension to an assembled slide. (*d*) Longitudinal section of the slide, showing the position of the cell sample in a 0.1-mm-deep chamber. (*e*) Viewing slide on microscope. (*f*) Magnified view of the total area of the grid. The central area enclosed by the dotted circle is the area that would be covered by the average 10× objective. This area covers approximately the central 1 × 1 mm square of the grid. (*g*) Low-power (10× objective) microphotograph showing the 25 smaller 200 × 200-μm squares of a slide, which make up the central 1 × 1-mm square, loaded with cells pretreated with Naphthalene Black (Amido Black). (*h*) High-power (40× objective) microphotograph of one of the smaller 200-μm squares, bounded by three parallel lines and containing 16 of the smallest (50 × 50 μm) squares. Viable cells are unstained and clear, with a refractile ring around them; nonviable cells are dark and have no refractile ring.

*Analysis.* Calculate the average of the two counts, and derive the concentration of your sample using the formula

$$c = \frac{n}{v}$$

where $c$ is the cell concentration (cells/mL), $n$ is the number of cells counted, and $v$ is the volume counted (mL). For the Improved Neubauer slide, the depth of the chamber is 0.1 mm, and, assuming that only and all of the central 1 mm$^2$ is used, $v$ is 0.1 mm$^3$, or $1 \times 10^{-4}$ mL. The formula then becomes

$$c = \frac{n}{10^{-4}} \quad \text{or} \quad c = n \times 10^4$$

namely multiply your count by 10,000.

If the cell concentration was high and only the five diagonal squares within the central 1 mm$^2$ were counted (i.e., 1/5 of the total), this equation becomes

$$c = n \times 5 \times 10^4$$

If the cell concentration was low, and nine 1-mm$^2$ squares were counted, the expression then becomes

$$c = \frac{n \times 10^4}{9} \quad \text{or} \quad c = n \times 1.1 \times 10^3$$

This is the cell concentration in the suspension culture or created after trypsinization or centrifugation and resuspension. To determine the number of cells per culture flask, multiply by the volume of medium used to resuspend the cells (in this example 1 mL). To determine the concentration of cells in the culture before trypsinization, divide by the volume of medium in the original flask (5 mL in this case). To determine the density of cells in the culture (for monolayer cells only), divide by the surface area of the flask (25 cm$^2$ in this case). Cells per flask is the figure most useful in routine subculture, while cells/mL and cells/cm$^2$ are more appropriate for plotting a growth curve (*see* Protocol 20.7).

Hemocytometer counting is cheap and gives you the opportunity to see what you are counting. If the cells were previously mixed with an equal volume of a viability stain (*see* Protocol 21.1; Fig. 20.1*g*, *h*), a viability determination may be performed at the same time. (Remember to compensate for the additional dilution with viability stain to obtain an absolute count.) However, counting cells by hemocytometer is rather slow and prone to error, both in the method of sampling and in the total number of cells counted; it also requires a minimum of $1 \times 10^6$ cells/mL.

Most of the errors in this procedure occur by incorrect sampling and transfer of cells to the chamber. Make sure that the cell suspension is properly mixed before you take a sample, and do not allow the cells time to settle or adhere in the tip of the pipette before transferring them to the chamber. Ensure also that you have a single-cell suspension, as aggregates make counting inaccurate. Larger aggregates may enter the chamber more slowly or not at all. If aggregation cannot be eliminated during preparation of the cell suspension (*see* Table 12.5), lyse the cells in 0.1 M citric acid containing 0.1% crystal violet at 37°C for 1 h and then count the nuclei [Sanford et al., 1951].

## 20.1.2 Electronic Counting

There has been some diversification in the design of automatic cell counters since the original electrical resistance-based electronic counters were first introduced in the late 1950s. There are now three main types (Table 20.2): (a) the original resistance-based counters (e.g., Beckman Coulter Z1 and Z2 or the Innovatis CASY) based on the change in current generated when a cell passes through a narrow orifice, (b) image analysis of a microscope view of unstained or stained cells or nuclei in special counting chambers by visible light or fluorescence (e.g., Invitrogen Countess, Peqlab Cellometer oe Chemometic Nucleopcounter), and (c) bench-top flow cytometers that analyze a single cell stream for cell concentration and other parameters (e.g., Guava, Accuri). All perform best with a single cell suspension, obligatory in call counters of type (c) but with some compensation allowed in types (a) and (b).

*Resistance-based cell counter (see Fig. 20.2).* There are three main components of this electronic cell counter (*see* Fig. 20.3): (a) an orifice tube, with a 150-μm orifice, connected to a metering pump; (b) an amplifier, pulse-height analyzer, and scaler connected to two electrodes, one in the orifice tube and one in the sample beaker, and (c) an analog and a digital readout showing the cell count and a number of other parameters, such as cell volume and size distribution.

When the count is initiated, a measured volume of cell suspension is drawn through the orifice (*see* Fig. 20.3). As each cell passes through the orifice, it changes the resistance to the current flowing through the orifice by an amount proportional to the volume of the cell. This change in resistance generates a pulse (amp$^{-1}$) that is amplified and counted. Because the size of the pulse is proportional to the volume of the cell (or any other particle) passing through the orifice, a series of signals of varying pulse height are generated. The lower threshold (in cell diameter) is set to eliminate electronic noise and fine particulate debris, but to retain pulses derived from cells (*see* Calibration, below). The upper threshold is either set to infinity or is adjustable, depending on the model. These thresholds determine which range of particle sizes is counted. The display on the counter shows a histogram depicting the distribution of cell size. In some cases, the software is capable of correcting for clumping of cells and coincidence (i.e., when two or more cells go through the aperture at once).

Protocol 20.2 can be adapted for training (*see* Section 28.3, Exercise 11).

**TABLE 20.2.** Electronic Cell Counters

| Name | Supplier | Principle of action | Website |
|---|---|---|---|
| Z1, Z2 | Beckman Coulter | Electrical resistance change as cell passes through orifice | www.beckman.com |
| CASY | Innovatis; | Electrical resistance change as cell passes through orifice | www.innovatis.com; |
| | Sedna Scientific (UK) | | www.sednascientific.com |
| Cellometer | Nexcelom Biosciences; Peqlab | Fluorescence imaging of cells in optical chamber (PI and others), whole and lysed cells | www.nexcelom.com; www.peqlab.co.uk |
| NucleoCounter | Chemometec | Fluorescence and visible imaging (PI, Trypan Blue), whole and lysed cells (nuclei) | www.chemometec.com |
| Countess | Invitrogen | Image analysis with Trypan Blue | www.Invitrogen.com |
| ADAM | Digital-Bio; Labtech International | Fluorescence image analysis (PI) | www.digital-bio.com; www.labtech.co.uk |
| Cedex 2 | Innovatis | Image analysis of Trypan Blue exclusion | www.innovatis.com |
| Cellavista | Innovatis | Automated image analysis of Trypan Blue exclusion in MT plates | www.innovatis.com; www.sednascientific.com |
| C6 | Accuri | Flow cytometry of single cell stream; multiparametric | www.AccuriCytometers.com |
| CyFlow CCA | Partec | Flow cytometry of single cell stream; multiparametric | www.partec.com |
| Guava | Millipore | Flow cytometry of single cell stream; multiparametric | www.guavatechnologies.com |

---

## PROTOCOL 20.2. ELECTRONIC CELL COUNTING BY ELECTRICAL RESISTANCE

### Outline

Dilute a sample of cells in electrolyte (physiological saline or D-PBSA), place the diluted sample under the orifice tube, and count the cells by drawing 0.5 mL of the diluted sample through the counter.

### Materials

*Sterile or aseptically prepared:*

❑ Cell culture
❑ D-PBSA
❑ Crude trypsin 0.25%
❑ Growth medium

*Nonsterile:*

❑ Counting cups

### Procedure

1. Trypsinize the cell monolayer, or collect a sample from the suspension culture. The cells must be well mixed and singly suspended.
2. Dilute the sample of cell suspension to 1:50 in 20 mL of counting fluid in a 25-mL beaker or disposable sample cup. An automatic dispenser

(*see* Fig. 4.6) will speed up this dilution and improve reproducibility.

*Note.* Dispensing counting fluid rapidly can generate air bubbles that will be counted as they pass through the orifice. Consequently the counting fluid should stand for a few moments before counting. If the fluid is dispensed first and the cells added second, this problem is minimized.

3. Mix the suspension well, and place it under the tip of the orifice tube, ensuring that the orifice is covered and that the external electrode lies submerged in the counting fluid in the sample beaker.
4. Check the program settings:
   (a) Threshold setting(s) (minimum cell size, usually 7.0 μm).
   (b) Volume to be counted (usually 0.4 mL).
   (c) Background subtraction (if used).
   (d) Dilution settings (e.g., 50 if 0.4 mL is counted in 20 mL of D-PBSA).
5. Check the visual analog display:
   (a) To ensure that all cells fall within the threshold setting(s).
   (b) To check for viability or cell debris (indicated by a shoulder on the curve or histogram falling below the normal lower threshold setting).

**(c)** To check for aggregation (indicated by particles appearing above the normal size range).

6. Initiate the count sequence.

7. When the count cycle is complete, the size distribution will appear on the analog screen (Fig. 20.4). Switching to the digital screen will give the cell count per mL.

*Analysis.* If the background and dilution are set correctly, the cell concentration will be that of the starting suspension before dilution in counting fluid. Otherwise, if the counter is set to take 0.5 mL of a 1:50 dilution, the final count on the readout should be multiplied by 100 to give the concentration in cells/mL of the original cell suspension. However, some counters automatically compensate for the volume sampled, so check the instruction manual.

*Calibration.* Older counters required calibration by counting cells at increasing increments of the lower threshold (upper threshold set to infinity). The plot of these counts generated a plateau, whose center gave the correct threshold setting. In modern counters with an analog display, the lower

threshold may be set manually on the readout, usually at 7.0 μm, while the upper is set to 30 μm or infinity. Setting the lower threshold to 7.0 μm will include nonviable cells, whereas a higher setting (12 μm in Fig. 20.4) will exclude most of the nonviable cells. Nonviable cells have a smaller apparent diameter because the plasma membrane is leaky, the

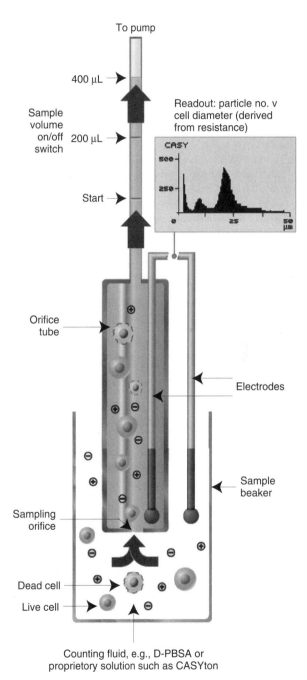

**Fig. 20.3. CASY Cell Counter Operation.** The electrolyte (e.g., D-PBSA or CASYton) in an orifice tube is connected to a pump and draws a measured volume of the cell sample from a beaker. A cell passing through the orifice alters the flow of current and generates a signal, the amplitude of which is proportional to the volume of the cell. (Courtesy of Schärfe Systems.)

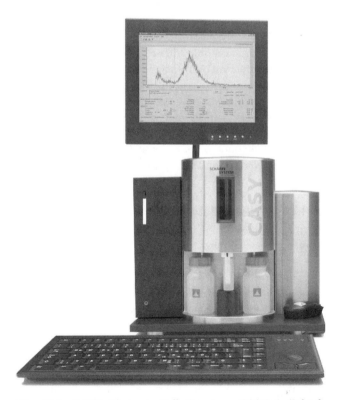

**Fig. 20.2. CASY Electronic Cell Counter.** CASY 1 (Schärfe Systems) cell counter, also suitable for cell sizing and discrimination between viable and nonviable cells, and single cells and aggregates. (Courtesy of Schärfe Systems.)

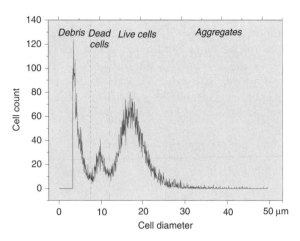

**Fig. 20.4. Analog Printout from CASY Electronic Cell Counter.** Cell number plotted against cell diameter gives a size distribution analysis that enables the lower threshold (vertical dashed line) to be set. In this display the upper threshold could be set to 30 μm or set to infinity. The peak below the vertical dashed line and the vertical dotted line represents nonviable cells that can be used with the viable cell count to determine percentage viability. (Data courtesy of Schärfe Systems.)

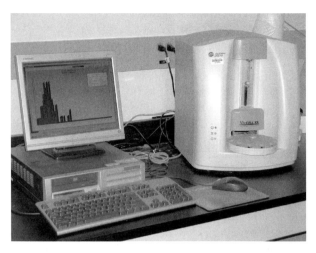

**Fig. 20.5. Beckman Coulter Vi-CELL.** An undiluted cell suspension is placed in a cup in the sample holder, sampled by the counter (right), mixed with Trypan Blue, and analyzed in a flow cell. The output appears on the display (left) as a graphic of cell size distribution or a microscope view of the cells with live cells circled in green and dead cells circled in red. (Courtesy of ATCC.)

cytoplasm has the same resistance as the electrolyte, and the resistance signal is generated by the nucleus. Counts at and above 30 μm indicate aggregation; aggregates can be partially excluded by setting an appropriate upper threshold (e.g., 25 or 30 μm in Fig. 20.4) or can be allowed for by using a statistical calculation available within the counter's operating program.

***Cell counting by image analysis.*** Image analysis is used to scan cells in an optical counting chamber and will discriminate between live and dead cells by recognizing Trypan Blue staining (e.g., Coulter Vi-CELL) or, in some models, fluorescence from propidium iodide (PI; e.g., Digital-Bio ADAM). This can be viewed on the display. Most of these counters can also generate an analog plot of cell size distribution, enabling upper and lower thresholds to be set (Fig. 20.5). A sample of cells, usually 20 μL to 1 mL, is injected into the chamber and placed in the counter. The range of cell concentrations that can be used range from $5 \times 10^4$ to $1 \times 10^7$ cells/mL for those using Trypan Blue, while those using PI can be used down to $5 \times 10^3$ cells/mL. After setting the correct thresholds, the count takes about 30 s. Counters using PI (e.g., Nucleocounter) can also estimate viability from the ratio of PI uptake in untreated cells divided by total PI uptake in lysed cells in the same sample.

Continuous monitoring of living cultures is also possible using image analysis of monolayers growing in flasks, Petri dishes, or multiwell plates (Incucyte, Chip-Man Cell-IQ). These systems work either by counting cells or determining residual available growth area and give continuous output of numerical data or microscope images which can be used to generate growth curves or time-lapse films (*see* Section 20.9.3).

Electronic cell counting is rapid and has a low inherent error because of the high number of cells counted. Although resistance counting is prone to misinterpretation, because cell aggregates, dead cells, and particles of debris of the correct size will all be counted, corrections are possible to exclude dead cells and aggregates and the CASY can make an approximate programmatic correction for aggregation. The cell suspension should still be examined carefully before dilution and counting on resistance counters; those working by image analysis give a visual display of the cell suspension being counted. Flow cytometers are, on the whole, a bit more complex to set up but allow the measurement of a larger range of parameters (*see* Sections 14.4, 20.1.4). Electronic cell counters are expensive but, if used correctly, are very convenient and give greater speed and accuracy to cell counting.

### 20.1.3 Stained Monolayers

There are occasions when cells cannot be harvested for counting or are too few to count in suspension (e.g., at low cell concentrations in microtitration plates). In these cases the cells may be fixed and stained in situ and counted by eye with a microscope. Because this procedure is tedious and subject to high operator error, isotopic labeling or the estimation of the total amount of DNA (*see* Protocol 20.3) or protein (*see* Protocol 20.4) is preferable, although these measurements may not correlate directly with the cell number—for example, if the ploidy of the cell varies. A rough estimate of the cell number per well can also be obtained by staining the cells with Crystal Violet and measuring the absorption on a densitometer. This method has also been used to calculate the number of cells per colony in clonal growth assays [McKeehan et al., 1977]. Staining cells with Coomassie

Blue, sulforhodamine B [Boyd, 1989; Skehan et al., 1990], or MTT [Plumb et al., 1989] also gives an estimation of the cell number, given that linearity has been demonstrated previously in a standard plot of absorption against cell number (*see* Protocol 21.4). MTT staining has the advantage that it stains only viable cells (*see* Table 20.1).

### 20.1.4   Flow Cytometry

Flow cytometry of a cell suspension [Shapiro, 2003; Applied Cytometry, 2008] (*see also* Section 14.4; Fig. 14.8),

while losing the relationship between cytochemistry and morphology, samples up to $1 \times 10^7$ cells, can measure multiple cellular constituents and activities [Kurtz & Wells, 1979; Klingel et al., 1994], and it enables correlation of these measurements with other cellular parameters, such as cell size, lineage, DNA content, or viability [Al-Rubeai et al., 1997]. While multiparametric analysis and cell sorting are best performed with one of the more elaborate machines, such as the FACStar, small bench-top flow cytometers can give useful information about the status of the cells in addition

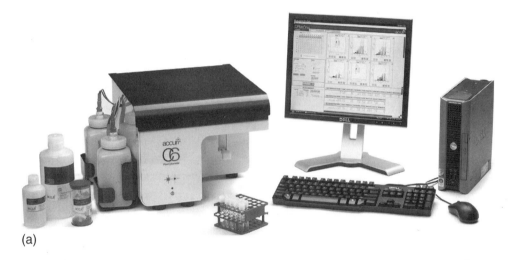

(a)

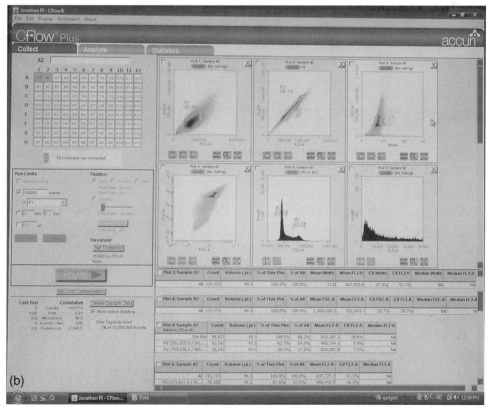

(b)

*Fig. 20.6. Accuri C6 Flow Cytometer.*   Upper panel, cytometer and associated PC. Lower panel, typical screen output. (Courtesy of Accuri Cytometers.)

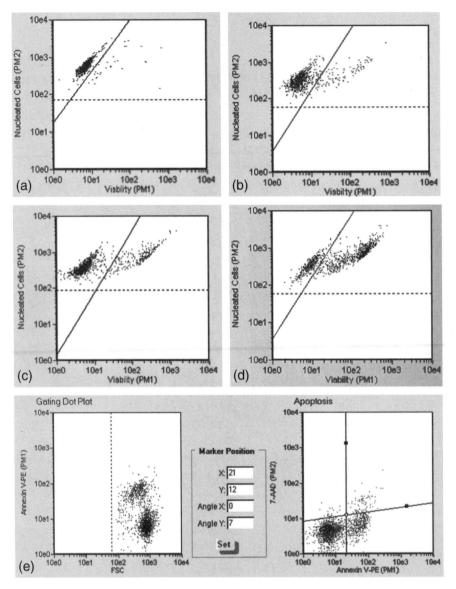

**Fig. 20.7. Output from Guava Flow Cytometer.** Dot plot from deteriorating cell culture. (a–d) Four panels show progressive accumulation of dead cells, stained with PI, in upper right quadrant. (e) Bottom two panels show apoptotic cells, staining with antibody to Annexin V (bottom right quadrant in right-hand image). (Courtesy of Edward Burnett, ECACC.)

to a cell count and are not much more expensive than an electronic cell counter (Fig. 20.6). Parameters such as viability (by propidium iodide uptake) or apoptosis (by annexin V immunostaining) can be readily quantified (Fig. 20.7).

## 20.2 CELL WEIGHT

Wet weight is seldom used unless very large cell numbers are involved because the amount of adherent extracellular liquid gives a large error. As a rough guide, however, there are about $2.5 \times 10^8$ HeLa cells (14–16 μm in diameter) per gram wet weight, about 8 to $10 \times 10^8$ cells/g for murine leukemias, such as L5178Y murine lymphoma, Friend murine

erythroleukemia, myelomas, and hybridomas (11–12 μm in diameter), and about $1.8 \times 10^8$ cells/g for human diploid fibroblasts (16–18 μm in diameter) (Table 20.3). Similarly dry weight is seldom used because salt derived from the medium contributes to the weight of unfixed cells, and fixed cells lose some of their low-molecular-weight intracellular constituents and lipids.

## 20.3 DNA CONTENT

In practice, besides the cell number, DNA and protein are the two most useful measurements for quantifying the amount of cellular material. DNA may be assayed by several fluorescence

**TABLE 20.3.** Comparison of Cell Size, Volume, and Mass

| Cell type | Diameter (μm) | Volume (μm$^3$) | Cells/g ×10$^{-6}$ Calculated | Measured |
|---|---|---|---|---|
| Murine leukemia (e.g., L5178Y or Friend) | 11–12 | 800 | 1250 | 1000 |
| HeLa | 14–16 | 1200 | 800 | 250 |
| Human diploid fibroblasts | 16–18 | 2500 | 400 | 180 |

methods, including reaction with DAPI [Brunk et al., 1979], PicoGreen (assay kit from Molecular Probes), or Hoechst 33258 [Labarca & Paigen, 1980]. The fluorescence emission of Hoechst 33258 at 458 nm is increased by interaction of the dye with DNA at pH 7.4 and in high salt to dissociate the chromatin protein. This method gives a sensitivity of 10 ng/mL but requires intact double-stranded DNA. DNA can also be measured by its absorbance at 260 nm, where 50 μg/mL has an optical density (O.D.) of 1.0. Because of interference from other cellular constituents, the direct absorbance method is useful only for purified DNA. Protocol 20.3 is a relatively simple and straightforward assay for DNA.

---

**PROTOCOL 20.3. DNA ESTIMATION BY HOECHST 33258**

*Outline*
Homogenize cells or tissue in buffer, and then sonicate the homogenate. Mix aliquots of the culture with H33258, and measure the fluorescence.

*Materials*
*Nonsterile:*

❑ Buffer: 0.05 M NaPO$_4$; 2.0 M NaCl, pH 7.4, containing $2 \times 10^{-3}$ M EDTA
❑ Hoechst 33258: in buffer, 1 μg/mL for DNA above 100 ng/mL and 0.1 μg/mL for 10 to 100 ng/mL

*Procedure*

1. Homogenize the cells in buffer, $1 \times 10^5$ cells/mL for 1 min, using a Potter homogenizer.
2. Sonicate the cells for 30 s.
3. Dilute the cells 1:10 in Hoechst 33258 and buffer.
4. Read fluorescence emission at 492 nm with excitation at 356 nm, using calf thymus DNA as a standard.

---

## 20.4   PROTEIN

The protein content of cells is widely used for estimating total cellular material and can be used in growth experiments or as a denominator in expressions of the specific activity of enzymes, the receptor content, or intracellular metabolite concentrations. The amount of protein in solubilized cells can be estimated directly by measuring the absorbance at 280 nm, with minimal interference from nucleic acids and other constituents. The absorbance at 280 nm can detect down to 100 μg of protein, or about $2 \times 10^5$ cells.

Colorimetric assays are more sensitive than measurements of UV absorption, and among these assays, the Bradford reaction with Coomassie Blue [Bradford, 1976] is one of the most widely used.

### 20.4.1   Solubilization of Sample

Because most assays rely on a final colorimetric step, they must be carried out on clear solutions. Cell monolayers and cell pellets may be dissolved in 0.5 to 1.0 M NaOH by heating them to 100°C for 30 min or leaving them overnight at room temperature. Alternatively, with 0.3 M NaOH and 1% sodium lauryl sulfate, the solution is complete after 30 min at room temperature.

### 20.4.2   Bradford Assay

The Bradford method is not dependent on specific amino acids and is quite sensitive, requiring 50 to 100,000 cells. Coomassie Blue undergoes a spectral change on binding to protein in acidic solution. Color is generated in one step after a short incubation and should be read within 30 min.

---

**PROTOCOL 20.4. PROTEIN ESTIMATION BY THE BRADFORD METHOD**

*Outline*
Dissolve protein, mix it with a color reagent, and read the O.D. after 10 min.

*Materials*

❑ Sodium lauryl sulfate (SLS, SDS), 3.5 mM in water or 0.3 M NaOH
❑ Coomassie Brilliant Blue G-250, 0.12 mM (0.01%), in 4.7% EtOH and 85% (w/v) phosphoric acid: dissolve 100 mg of Coomassie Blue in 50 mL of 95% EtOH; add 100 mL of 85% phosphoric acid; and dilute the solution to 1 L with water
❑ Protein standard solution (e.g., BSA, 10 μg/mL); on first setting up the assay and at intervals of 1 to 2

months, perform a standard curve with BSA or a similar standard protein (1–50 µg/mL)

**Procedure**

1. Solubilize the protein (1–20 µg) or cells (around $1 \times 10^6$) in 100 µL of 3.5 mM sodium dodecyl sulfate in water or 0.3 M NaOH.
2. Add 100 µL of reagent blank (SLS), 100 µL of test protein solution, and 100 µL of BSA standard to separate, triplicate tubes.
3. Add 1.0 mL of Coomassie Blue. Mix the solution, and let it stand for 10 min.
4. Read the tests and the BSA standard on a spectrophotometer or microplate reader at 595 nm against the reagent blank.

**Variations.** Reagents for the Bradford assay are available in a kit from Bio-Rad. Sulforhodamine B [Skehan et al., 1990] and bicinchonic acid (BCA) [Smith et al., 1985] can also be used to measure protein content and the sensitivity of the assays makes them very suitable for microtitration plate assays. A micro-BCA kit available from Pierce (*see* Appendix III) is very sensitive and suitable for small numbers of cells ($\sim 1 \times 10^3$).

## 20.5   RATES OF SYNTHESIS

### 20.5.1   DNA Synthesis

Measurements of DNA synthesis are often taken to be representative of the amount of cell proliferation (*see also* Section 21.3.5). [³H]thymidine ([³H]-TdR) or [³H]deoxycytidine is the usual precursor that is employed. Exposure to one of these precursors may be for short periods (0.5–1 h) for rate estimations or for longer periods (24 h or more) to measure accumulated DNA synthesis when the basal rate is low (e.g., in high-density cultures). [³H]-TdR should not be used for incubations longer than 24 h or at high specific activities as radiolysis of DNA will occur because of the short path length of β-emission ($\sim 1$ µm) from decaying tritium; the β-emission releases energy within the nucleus and causes DNA strand breaks. If prolonged incubations or high specific activities are required, [¹⁴C]-TdR or ³²P should be used.

---

**PROTOCOL 20.5. ESTIMATION OF DNA SYNTHESIS BY [³H]THYMIDINE INCORPORATION**

**Outline**

Label the cells with [³H]-TdR, extract the DNA, and determine the level of radioactivity by means of a scintillation counter.

---

**Materials**

*Sterile or aseptically prepared:*

❑ Cells at a suitable stage (grown in glass vials or test tubes if the hot 2 M PCA method is being used; otherwise, grown in conventional plastic)
❑ [³H]-TdR, 0.4 MBq/mL ($\sim 10$ µCi/mL), 100 µL for each mL of culture medium

*Note.* Some media—such as Ham's F10 and F12—contain thymidine, which will ultimately determine the specific activity of added [³H]-TdR. Allowance will have to be made for this factor when judging the amount of isotope to add. Although 40 KBq/mL ($\sim 1.0$ Ci/mL) of isotope may be sufficient for most media, 0.2 MBq/mL ($\sim 5$ µCi/mL) should be used with F10 or F12.

*Nonsterile:*

❑ Trichloroacetic acid (TCA), 0.6 M (on ice), 6 mL for each 1 mL of culture
❑ HBSS or D-PBSA, ice cold, 2 mL per mL of culture
❑ Perchloric acid, 2 M, or 0.3 M NaOH with 35 mM sodium lauryl sulfate (SLS, SDS), 0.5 mL per 1 mL of culture
❑ MeOH, 1 mL per 1 mL of culture
❑ Scintillation vials
❑ Scintillant (10× volume of perchloric acid or NaOH/SDS)

**Procedure**

1. Grow the culture to the desired density (usually the mid-log phase for maximum DNA synthesis or the plateau for density-limited DNA synthesis; *see* Section 17.5.2).
2. Add [³H]-TdR, 40 KBq/mL ($\sim 1.0$ µCi/mL), or 2 MBq/mmol ($\sim 50$ Ci/mol) in HBSS.
3. Incubate the cells for 1 to 24 h as required.
4. Remove the radioactive medium carefully, and discard it into the proper container for liquid radioactive waste.
5. Wash the cells carefully with 2 mL of HBSS or D-PBSA, and add 2 mL of ice-cold 0.6 M TCA for 10 min. Fix the cells in MeOH first if they are loosely adherent (*see* Protocol 20.6).
6. Repeat the trichloroacetic acid wash twice, for 5 min each time.
7. Add 0.5 mL of 2 M perchloric acid, place the solution on a hot plate at 60°C for 30 min, and then allow the solution to cool. Alternatively, add 0.5 mL of SLS in NaOH, and incubate the solution for 30 min at 37°C, or overnight at room temperature.
8. Collect the solubilized pellet, transfer it to the scintillant, and determine the radioactivity on a scintillation counter.

If you are using the perchloric acid method, the residue may be dissolved in alkali for protein determination (*see* Protocol 20.4). Replicate cultures should be set up to provide cell counts to allow calculation of the DNA synthesis related to cell number.

For suspension cultures, spin the cells at 100 *g* for 10 min in steps 4, 5, and 6. Mix the cells on a vortex mixer to disperse the pellet before each wash (in 0.6 M trichloroacetic acid and 2 M perchloric acid) in step 7. Spin the cells after step 7 at 1000 *g* for 10 min to separate the precipitate (for protein estimation, if required) and the supernatant (for scintillation counting).

△ *Safety Note.* [$^3$H]-TdR represents a particular hazard because it induces radiolytic damage in DNA. Take care to avoid accidental ingestion, injection, or inhalation of aerosols. When using [$^3$H]-TdR, work in an MSC, chemical hood, or on an open bench, but do not use horizontal laminar flow, or else aerosols will be blown directly at you.

*Analysis.* Incubation with an isotopic precursor can provide several different types of data, depending on the incubation conditions and subsequent processing. Incubation followed by a short wash in ice-cold BSS and extraction into 0.6 M TCA will give a measure of the uptake and, if carried out over a few minutes' duration, will give a fair measure of the unidirectional flux. In experiments measuring uptake, the incorporation of precursors into acid-insoluble molecules, such as protein and DNA, is assumed to be minimal because of the short incubation time, and only the acid-soluble pools are counted by extraction into cold 0.6 M TCA. In longer incubations (i.e., 2–24 h) it is assumed that the precursor pools become saturated. Equilibrium levels may be measured by cold trichloroacetic acid extraction, and the incorporation into polymers may be measured by extraction with hot 2 M of perchloric acid (DNA), cold dilute alkali (RNA), or hot 1.0 M of NaOH (protein).

## 20.5.2 Protein Synthesis

Colorimetric assays measure the total amount of protein present at any one time. Sequential observations over a period of time may be used to measure the net protein accumulation or loss (i.e., protein synthesized–protein degraded), while the rate of protein synthesis may be determined by incubating cells with a radioisotopically labeled amino acid, such as [$^3$H]leucine or [$^{35}$S]methionine, and measuring (e.g., by scintillation counting) the amount of radioactivity incorporated into acid-insoluble material per $1 \times 10^6$ cells or per milligram of protein over a set period of time.

△ *Safety Note.* Radioisotopes must be handled with care and according to local regulations governing permitted amounts, authorized work areas, handling procedures, and disposal (*see* Section 6.7.2).

---

### PROTOCOL 20.6. PROTEIN SYNTHESIS

#### Materials

*Sterile or aseptically prepared:*

❑ Cell cultures: for example, $1 \times 10^4$ to $1 \times 10^6$ cells/well, in a 24-well plate
❑ [$^3$H]leucine, 2 MBq/mL ($\sim$50 μCi/mL) in culture medium without serum (the specific activity is unimportant, as it will be determined by the leucine concentration in the medium)

*Nonsterile:*

❑ Sodium lauryl sulfate (SLS or SDS), 1% (35 mM) in 0.3 M NaOH
❑ Trichloroacetic acid (TCA)
❑ Scintillation vials
❑ Eppendorf tubes
❑ Scintillation fluid with a minimum of 10% water tolerance

#### Procedure

1. Incubate the culture to the required cell density.
2. Remove the culture from the incubator, and add a prewarmed solution of radioisotope in medium or BSS, diluting 1:10 (e.g., 100 μL per 1 mL/well).
3. Return the culture to the incubator as rapidly as possible.
4. Incubate the culture for 4 to 24 h.

*Note.* Different proteins turn over at different rates. This protocol is not aimed at any specific subset of proteins, but at the total protein in rapidly proliferating cells. When assaying protein synthesis in a cell line for the first time, check that the rate of synthesis is linear over the chosen incubation time. A lag may be encountered if the amino acid pool is slow to saturate.

5. Remove the culture from the incubator, and withdraw the medium carefully from the wells into a container designated for radioactive liquid waste. (For disposal *see* Section 6.7.2.)
6. Wash the cells gently with cold HBSS or D-PBSA.

*Note.* Some monolayer cultures—particularly some loosely adherent continuous cell lines, such as HeLa-S$_3$—may detach during washing. In this case, remove the isotope and add methanol to fix the monolayer. Leave the culture for 10 min, carefully remove the methanol, and dry the monolayer.

7. Replace the plates on ice. Add 0.6 M TCA, at 4°C, for 10 min, to remove any unincorporated precursor.

8. Repeat step 7 twice, but for only 5 min each time.
9. Wash the culture with MeOH, and then dry the plates.
10. Add 0.5 mL of 0.3 M NaOH, containing 1% SLS, and leave for 30 min at room temperature.
11. Mix the contents of each well, and transfer them to separate scintillation vials.
12. Add 5 mL of scintillant, and count on a scintillation counter.

*Note.* Biodegradable scintillation fluids—such as Ecoscint—are preferred over toluene- or xylene-based fluids, as the former are less toxic to handle and can be poured down the sink with excess water provided that the levels of radioactivity fall within the legal limits.

For suspension cultures, spin the cells at 1000 *g* for 10 min in step 5 to remove the medium, and at steps 6, 7, and 8. Also omit step 9.
Plates from step 9 can also be quantified directly on a phosphorimager.

## 20.6 PREPARATION OF SAMPLES FOR ENZYME ASSAY AND IMMUNOASSAY

As the amount of cellular material available from cultures is often too small for efficient homogenization, other methods of lysis are required to release soluble products and enzymes for assay. It is convenient either to set up cultures of the necessary cell number in sample tubes or multiwell plates (*see* Section 21.3.5; Protocol 20.8) or to trypsinize a bulk culture and place aliquots of cells into assay tubes. In either case the cells should be washed in HBSS or D-PBSA to remove the serum, and lysis buffer should be added to the cells. The lysis buffer should be chosen to suit the assay, but if the particular lysis buffer is unimportant, 0.15 M NaCl or D-PBSA may be used. If the product to be measured is membrane bound, add 1% detergent (Na deoxycholate, Nonidet P40) to the lysis buffer. If the cells are pelleted, resuspend them in the buffer by vortex mixing. Freeze and thaw the preparation three times by placing the sample tube in EtOH containing solid $CO_2$ ($\sim -90°C$) for 1 min and then in 37°C water for 2 min (longer for samples greater than 1 mL). Finally, spin the preparation at 10,000 *g* for 1 min (e.g., in an Eppendorf centrifuge), and collect the supernate for assay.

Alternatively, the whole extract may be assayed for enzyme activity, and the insoluble material may be removed by centrifugation later if necessary.

## 20.7 CYTOMETRY

### 20.7.1 In situ Labeling

Fluorescence labeling, either directly with a fluorescent dye (e.g., Hoechst 33258 or DAPI for DNA) or with a conjugated antibody for detection of an antigen or molecular probe, can measure the amounts of enzyme, DNA, RNA, protein, or other cellular constituents in situ with a CCD camera. This process allows qualitative and, when image analysis is used, quantitative analyses to be made, but is slow if large numbers of cells are to be scanned.

### 20.7.2 Flow Cytometry

Fluorescence labeled cell populations can also be quantified by flow cytometry where a very precise quantitative analysis may be made but in the absence of any structural relationships among the cells. Additional parameters that can be measured in a flow cytometer include forward and backward light scatter (influenced by cell size and surface configuration) and chromogenic enzyme substrates or products. Although flow cytometers used in cell separation (*see* Section 14.4) tend to be large and expensive machines, there are several low-cost bench-top analytical machines available, such as Guava and Accuri (*see* Section 20.1; Figs. 20.6, 20.7).

## 20.8 REPLICATE SAMPLING

Because in most cases cultured cells can be prepared in a uniform suspension, the provision of large numbers of replicates for statistical analysis is often unnecessary. Usually three replicates are sufficient, and for many simple observations (e.g., cell counts), duplicates may be sufficient.

Many types of culture vessel are available for replicate monolayer cultures (*see* Section 7.3.4), and the choice of which vessel to use is determined (a) by the number of cells required in each sample and (b) by the frequency or type of sampling. For example, if the incubation time is not a variable, replicate sampling is most readily performed in multiwell plates, such as microtitration plates or 24-well plates. If, however, samples are collected over a period of time (e.g., daily for 5 days), then the constant removal of a plate for daily processing may impair growth in the rest of the wells. In this case the replicates are best prepared in individual tubes or 4-well plates. Plain glass or tissue culture-treated plastic test tubes may be used, although Leighton tubes are superior, as they provide a flat growth surface. Alternatively, if the optical quality of the tubes is not critical, flat-bottomed glass specimen tubes and even glass scintillation vials may be good containers. If glass vials or tubes are used, they must be washed as tissue culture glassware (*see* Section 10.3.1); they cannot be used for tissue culture after use with scintillant.

Sealing large numbers of vials or tubes can become tedious, so many people seal tubes with vinyl tape rather than screw caps. Such tape can also be color coded to identify different treatments. Adhesive film may be used for sealing microtitration plates (*see* Appendix II: Plate Sealers). This reduces evaporation and contamination and gives a more even performance across the plate. It also means that individual wells or rows can be sampled without opening up the rest of the plate.

Handling suspension cultures is generally easier than dealing with monolayer cultures because the shape of the container and its surface charge are less important. Multiple sampling can also be performed on one culture when using suspension cultures. This sampling is done conveniently by sealing the bottle containing the culture with a silicone rubber membrane closure (Pierce) and then sampling with a syringe and needle. (Remember to replace the volume of culture removed with an equal volume of air.)

### 20.8.1  Data Acquisition

Analysis of data from cultured cells is not necessarily different from the way that data from any other system are handled. However, the production of large amounts of data in cell culture experiments is relatively easy, particularly when dealing with microtitration plates, with which several hundred data points can be generated without a great deal of effort, or even several thousand by using robotics. Handling tissue culture-derived data will depend on how they are generated, on the scale of the experiment, and the number of parameters. Although cell counting is the accepted method for generating data to construct a growth curve with one cell line under two or three sets of conditions, this does not lend itself to expansion to, say, determining multiple growth curves to measure the response of several cell lines to combinations of growth factors, except by using image analysis software in, for example, the Incucyte or Chip-Man Cell IQ. If a colorimetric end point is chosen, then absorbance (e.g., MTT assay; *see* Protocol 21.4) or fluorescence emission (e.g., sulforhodamine [Boyd, 1989]) can be used, the plates analyzed on a plate reader, and data reduction achieved by using the appropriate software.

A radiometric endpoint (e.g., [$^3$H]thymidine) used in conjunction with microtitration plates can be determined by simultaneous measurement of the whole plate in a microtitration plate scintillation counter (PerkinElmer). Similarly large numbers of sample tubes can be read automatically by γ or β counting, using robotic systems (Beckman Coulter).

### 20.8.2  Data Analysis

The ability to generate large amounts of data has been made possible by the creation of multiple replicate analysis systems, such as microtitration. As with ELISA analysis, the rate-limiting step is no longer the generation of the data but is instead its analysis. It is important therefore, when choosing a parameter of measurement to suit the culture system, that some thought be given to the amount of data to be generated and how those data will be handled. The easiest approach is to direct the data into a computer, either via a network or to a PC dedicated to that project. A number of companies now market computer programs that display and analyze data from microtitration plate assays (not just ELISA). These programs include titration curves, enzyme kinetics, and binding assays. With the necessary skills, and using a spreadsheet for importation of the data, you may be able to set up this type of program for yourself; alternatively, consultant advice is often available from the suppliers of plate readers (*see* Appendix II).

## 20.9  CELL PROLIFERATION

Measurements of cell proliferation rates are often used to determine the response of cells to a particular stimulus or toxin (*see* Section 21.3.3). Quantitation of culture growth is also important in routine maintenance, as it is a crucial element for monitoring the consistency of the culture and knowing the best time to subculture (*see* Section 12.4.4), the optimum dilution, and the estimated plating efficiency at different cell densities. Testing medium, serum, new culture vessels or substrates, and so forth, all require quantitative assessment. As with cell counting there are a number of different ways of determining cell proliferation using cell counting, plating efficiency, or labeling with radioisotopic precursors of DNA or antibodies to cell cycle specific proteins (Table 20.4). Counting and plating may be seen s direct, while labeling techniques are indirect as imply proliferation from the expression of metabolic markers.

### 20.9.1  Experimental Design

Knowledge of the growth state of a culture, and its kinetic parameters, is critical in the design of cell culture experiments. Cultures vary significantly in many of their properties between the lag phase, the period of exponential growth (log phase), and the stationary phase (plateau). It is therefore important to take account of the status of the culture both at the initiation of an experiment and at the time of sampling, in order to determine whether is it proliferating or not and, if it is, the duration of the population doubling time (PDT) and the cell cycle time. Cells that have entered the plateau phase have a greatly reduced growth fraction and a different morphology, may be more differentiated, and may become polarized. They tend to secrete more extracellular matrix and may be more difficult to disaggregate. Generally, cell

**TABLE 20.4.** Measurement of Cell Proliferation

| Term | Definition | Measurement |
|---|---|---|
| Growth curve | Semi-log plot of cell number against time from subculture | Cell counting (by trypsinization if necessary) by electronic counter or hemocytometer |
| Population doubling time | Time for a given cell (culture) population to double. Product of cell division, cell death and non-dividing cells | Cell counts during exponential phase of growth curve. <br> MTT assay of exponentially growing cells |
| Cell cycle time | Time from one point in the cell cycle until the cell reaches the same point again | % mitoses labeled with [$^3$H]-thymidine after 30-min incubation and sampled during 24–48 h cold thymidine chase <br> Flow cytometry with cells pulse-labeled with BUdR and stained with anti-BUdR at intervals during cold thymidine chase |
| Cell cycle analysis | Determination of duration of each phase of cell cycle | % labeled mitoses <br> Flow cytometry with cells pulse-labeled with BUdR |
| [$^3$H]Thymidine incorporation. | DPM acid insoluble [$^3$H]-TdR incorporated relative to DNA or protein | Scintillation counting of solubilized samples |
| Labeling index | % cells labeled with radioactive precursor or fluorescent antibody Ki67 or to BUdR or PCNA | [$^3$H]-thymidine incorporation into DNA followed by autoradiography <br> Stain with fluorescent antibody |
| Growth fraction | % cells capable of entering cycle | Prolonged label (24–48h) with [$^3$H]-thymidine, or antibody stain of cells which have incorporated BUdR over a prolonged period |
| Mitotic index | % cells in mitosis | Count mitoses as % of total population |
| Division index | % cells in cycle at any one time | Stain with antibodies to PCNA or Ki67 |
| Viability index | % cells excluding viability stain such as Trypan Blue, Naphthalene Black or erythrosin | Microscopic observation of unfixed cells stained with viability stain |
| Plating efficiency | Number of cells capable of forming discrete colonies following subculture at a low cell density | Trypsinise and plate out cells <br> Count colonies, above a defined threshold (usually 50 cells), after 10 days–3 weeks |
| Cloning efficiency | Number of cells capable of forming clones (colonies derived from one cell) following subculture at a low cell density | Trypsinise and plate out cells as single cells <br> Count colonies, above a defined threshold (usually 50 cells), after 10 days–3 weeks |
| Survival fraction | Fraction of colonies surviving relative to number of colonies in control | As for plating or cloning efficiency, express colony counts are a fraction of control |

cultures are most consistent and uniform in the log phase, and sampling at the end of the log phase gives the highest yield and greatest reproducibility.

It is also important to consider the effects of the duration of an experiment on the transition from one state to another. Adding a drug in the middle of the exponential phase and assaying later may give different results, depending on whether the culture is still in exponential growth when it is assayed or whether it has entered the plateau phase. Microtitration plate assays of cytotoxicity (*see* Section 21.3.5) are particularly susceptible to error if the culture reaches plateau during an assay; as the cells in those wells that are at the highest density reach plateau, cell proliferation decreases and there is an apparent shift in the 50% inhibitory point (ID$_{50}$) as more wells reach plateau (*see* Protocol 21.4). Hence scheduling treatment and sampling requires a detailed knowledge of the parameters of the growth cycle.

### 20.9.2 Growth Cycle

After subculture, cells progress through a characteristic growth pattern of lag phase, exponential, or log phase, and stationary, or plateau phase (Fig. 20.8; *see also* Section 12.4.3). The log and plateau phases give vital information about the cell line, the PDT during exponential growth, and the maximum cell density achieved in the plateau phase (i.e., the saturation density). The measurement of the PDT is used to quantify the response of the cells to different inhibitory or stimulatory culture conditions, such as variations in nutrient concentration, hormonal effects, or toxic drugs. It is also a good monitor of the culture during serial passage and enables the calculation of cell yields and the dilution factor required at subculture.

Single time points are unsatisfactory for monitoring growth if the shape of the growth curve is not known. A reduced cell count after, say, 5 days could be caused by

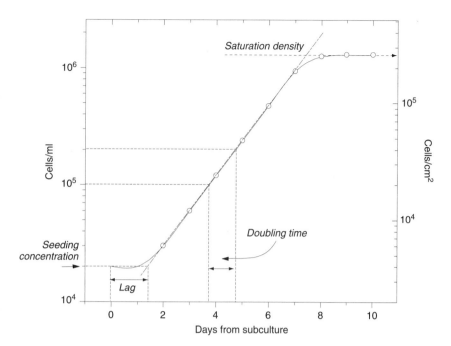

***Fig. 20.8. Growth Curve.*** Semilog plot of the increase in cell concentration (left axis) and cell density (right axis) after subculture. Replicate 25-cm² flask cultures are sampled daily and counted. It should be possible to draw a straight line through the part of the plot that represents the exponential phase and derive the population doubling time (PDT) from the middle region of this best-fit line. The time at the intercept of the line extrapolated from the exponential phase with the seeding concentration is the lag time, and the saturation density is found at the plateau (at least three linear points without an increase in cell concentration) at the top end of the curve.

a reduced growth rate of some or all of the cells; a longer lag period, implying adaptation or cell loss (it is difficult to distinguish between the two); or a reduction in saturation density (*see* Fig. 20.10). This is not to say that growth curves are of no value. They can be useful for testing media, sera, growth factors, and some drugs, and once the response being monitored is fully characterized and the type of response is predictable (e.g., a change in the PDT), then single time point observations can be made at a time point known to be in mid-log phase. Growth curves are particularly useful for the determination of the saturation density, although the amount of growth at saturation density should be assessed by the labeling index with [³H]thymidine (*see* Section 17.5.2; Protocols 20.11, 20.12).

The PDT derived from a growth curve should not be confused with the cell cycle or generation time. The PDT is an average figure that applies to the whole population, and it describes the net result of a wide range of division rates, including a rate of zero, within the culture. The *cell cycle time* or *generation time* is measured from one point in the cell cycle until the same point is reached again (*see* Section 20.12) and refers only to the dividing cells in the population, whereas the PDT is influenced by nongrowing and dying cells as well. PDTs vary from 12 to 15 h in rapidly growing mouse leukemias, like the L1210, to 24 to 36 h in many adherent continuous cell lines, and up to 60 or 72 h in slow-growing finite cell lines.

A new growth cycle begins each time the culture is subcultured and can be analyzed in more detail, as described in Protocol 20.7. Using flasks for a growth curve is more labor intensive; even limiting the number of replicates to two per day for 10 days, would require 20 flasks, with an additional 4 flasks for staining or to act as backup. A cell concentration of $2 \times 10^4$ cells/mL should be chosen for a rapidly growing line and $1 \times 10^5$ cells/mL for a slower growing finite cell line. Repeating the growth curve with higher or lower seeding concentrations should then allow the correct seeding concentration and subculture interval (*see* Section 12.4.4) to be established.

This protocol can be adapted for training (*see* Section 28.3, Exercise 15A).

## PROTOCOL 20.7. GROWTH CURVE WITH A MONOLAYER IN FLASKS

### Outline
Set up flasks, and count the cells at daily intervals until the culture reaches the plateau phase.

### Materials

*Sterile or aseptically prepared:*

❑ Monolayer cell culture, late log phase............................................1, 25 cm² flask

❏ Trypsin, 0.25%, crude, with 10 mM
EDTA..............................................................50 mL
❏ Growth medium.....................................200 mL
❏ D-PBSA (prewash and for cell counting) . . . 500 mL
❏ Flasks, 25 cm$^2$, or as appropriate.......................24

### Procedure

1. Trypsinize the cells as for a regular subculture (*see* Protocol 12.3).
2. Dilute the cell suspension to $2 \times 10^4$ cells/mL in 150 mL of medium.
3. Seed 24 25-cm$^2$ flasks.
4. Seal the flasks and place in an incubator at 37°C.

*Note.* A low-bicarbonate (4 mM) medium is specified for this protocol for the sake of simplicity, but if a high-bicarbonate medium is used (e.g., 23 mM), then either gas the flasks with 5% $CO_2$ or place in a $CO_2$ incubator with slack or gas–permeable caps.

5. After 24 h, remove the first two flasks from the incubator, and count the cells:
   (a) Remove the medium completely.
   (b) Add 2 mL of trypsin/EDTA to each flask.
   (c) Incubate the flasks for 15 min.
   (d) Disperse the cells in the trypsin/EDTA and transfer 0.4 mL of the suspension to 19.6 mL of D-PBSA (for Beckman Coulter Z1 or CASY); for other counters use 20 μL to 1 mL undiluted cell suspension as indicated in manufacturer's instructions.
   (e) Count the cells on an electronic cell counter.
6. Repeat sampling at 48 and 72 h, as in steps 5 and 6.
7. Change the medium at 72 h, or sooner, if indicated by a drop in the pH (*see* Protocol 12.1; Plate 22*b*).
8. Continue sampling daily for rapidly growing cells (i.e., cells with a PDT of 12–24 h), but reduce the frequency of sampling to every 2 days for slowly growing cells (i.e., cells with a PDT >24 h) until the plateau phase is reached.
9. Keep changing the medium every 1, 2, or 3 days, as indicated by the fall in pH.
10. Stain the cells in one flask at 2, 5, 7, and 10 days (*see* Section 15.5.2).

*Note.* A hemocytometer may be used to count the cells, but may be difficult to use for the lower cell concentrations at the start of the growth curve. If you use a hemocytometer, reduce the volume of trypsin to 0.5 mL, disperse the cells carefully, using a pipettor without frothing the trypsin, transfer the cells to the hemocytometer, and count enough squares to give a count in excess of 200 cells.

This type of assay is useful for comparing different media, supplements, and growth stimulants or inhibitors. However, for a quantitative assay for one or more variables, it is preferable to use multiwell plates. Protocol 20.8 can be used with 12-well plates in conjunction with Exercise 15B (*see* Section 28.3).

### PROTOCOL 20.8. GROWTH CURVE WITH A MONOLAYER IN MULTIWELL PLATES

#### Outline
Set up a series of multiwell plates with cultures at three different cell concentrations, and count the cells in one plate harvested at daily intervals until the culture reaches the plateau phase.

#### Materials

*Sterile or aseptically prepared:*

❏ Monolayer cell culture, late log phase.............................................1, 75 cm$^2$ flask
❏ Trypsin, 0.25%, crude, with 10 mM EDTA..............................................................50 mL
❏ Growth medium with 26 mM NaHCO$_3$.................................................300 mL
❏ D-PBSA (prewash and for cell counting)...............................................................500 mL
❏ Plates, 12 well...................................................10

*Nonsterile:*

❏ Plastic box or trays to hold the plates
❏ $CO_2$ incubator or $CO_2$ supply to purge the box with 5% $CO_2$

#### Procedure

1. Trypsinize the cells as for a regular subculture (*see* Protocol 12.3).
2. Dilute the cell suspension to $1 \times 10^5$ cells/mL, $3 \times 10^4$ cells/mL, and $1 \times 10^4$ cells/mL, in 25 mL of medium for each concentration.
3. Seed eight 12-well plates with 2 mL of the $1 \times 10^4$/mL cell suspension to each well of the top 4 wells, 2 mL of the $3 \times 10^4$/mL to each well of the second row, and 2 mL of the $1 \times 10^5$/mL to each well of the third row (Fig. 20.9). Add the cell suspension slowly from the center of the well so that it does not swirl around the well. Similarly, do not shake the plate to mix the cells, as the circular movement of the medium will concentrate the cells in the middle of the well.
4. Place the plates in a humid $CO_2$ incubator or a sealed box gassed with 5% $CO_2$.
5. After 24 h, remove the first plate from the incubator, and count the cells in three wells at each concentration:

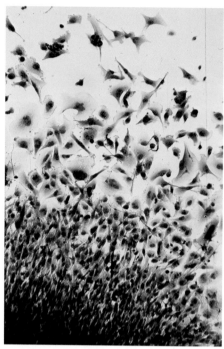

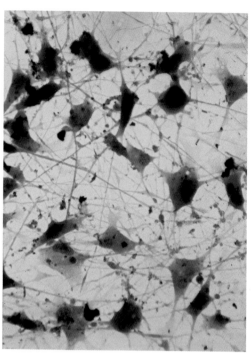

**(a) Human Lung Carcinoma.** *Outgrowth from primary explant from squamous cell carcinoma of human lung, MOG-L-DAN. Giemsa stained. 10x objective.*

**(b) Human glioma.** *Astroglial cells from an anaplastic astrocytoma. Giemsa. 40x objective.*

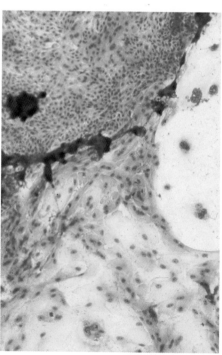

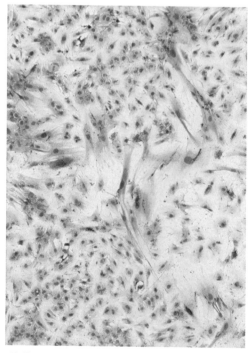

**(c) Human Cervical Intraepithelial Neoplasia (CIN).** *Primary culture from cervical biopsy subcultured on to 3T3 feeder layer of which a few degenerating cells remain, top right. Giemsa stained. 10x objective. (courtesy of M. G. Freshney.)*

**(d) Human Fibrosarcoma.** *Primary culture subcultured on to fetal intestinal feeder layer. The fibrosarcoma cells are spindle shaped and the feeder layer polygonal. Giemsa stained. 10x objective.*

*Plate 1. Primary Culture, Human.*

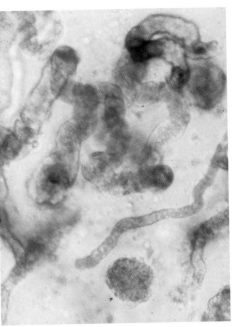

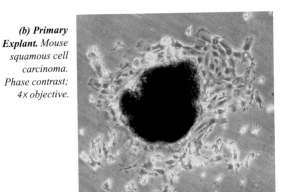

**(b) Primary Explant.** *Mouse squamous cell carcinoma. Phase contrast; 4x objective.*

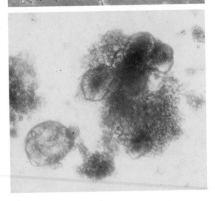

**(c) Collagenase Digest.** *Human colon carcinoma. Bright field; 4x objective.*

**(a) Mouse Kidney Tubules.** *Disaggregation of newborn mouse kidney by cold trypsin method. Connective tissue has dissociated, but fragments of tubules and glomeruli remain intact. Also seen with collagenase digestion. Normal bright field illumination; 4x objective.*

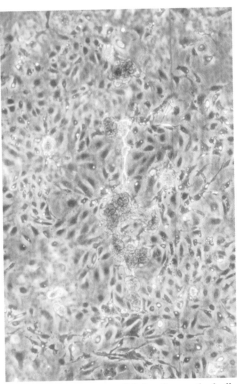

**(d) Outgrowth from Kidney Tubules.** *Outgrowth of cells from attached fragments following disaggregation by cold trysin method. Phase contrast; 10x objective.*

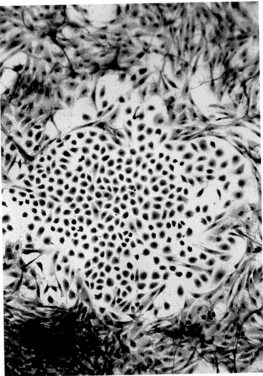

**(e) Newborn Rat Kidney.** *Primary culture from trypsinized (cold method) newborn rat kidney. Giemsa stained; 10x objective. (Courtesy of M. G. Freshney).*

*Plate 2. Primary Culture, Explant, Cold Trypsin, and Collagenase.*

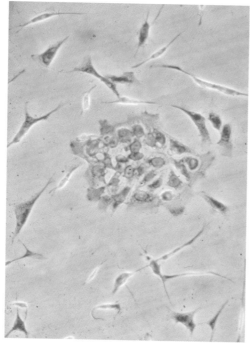

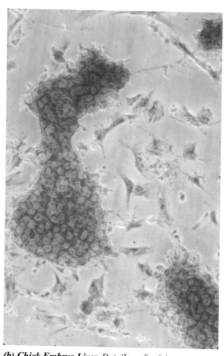

**(a) Chick Embryo Lung.** *Primary culture of 13-day chick embryo lung, 48 h after disaggregation by the cold trypsin method. Light Giemsa stain and phase contrast; 10x objective.*

**(b) Chick Embryo Liver.** *Details as for (a).*

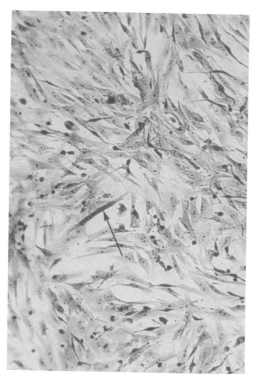

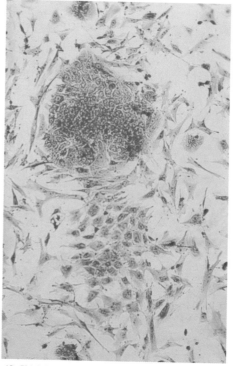

**(c) Chick Embryo Thigh Muscle.** *Culture details as for (a). Note multinucleate myotube (red arrow) resulting from fusion of myocytes (small dark-stained spindles). Giemsa stain, bright field; 10x objective.*

**(d) Chick Embryo Kidney.** *Culture details as for (a). Giemsa stain, bright field; 10x objective.*

**Plate 3. Primary Culture, Chick Embryo Organ Rudiments.**

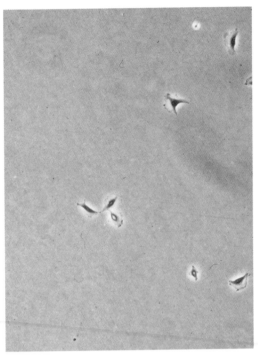

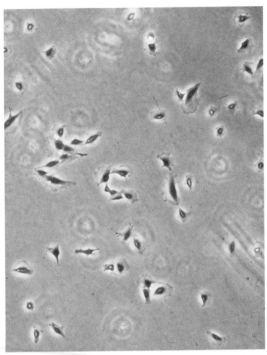

*(a) Newly subcultured monolayer.* NRK cells 24 h after subculture. Phase contrast; 10× objective.

*(b) Entering log phase.* NRK cells 48 h after subculture. Phase contrast; 10× objective.

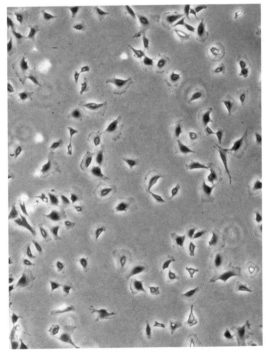

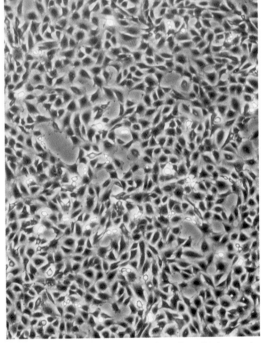

*(c) Mid-log phase cells.* NRK cells 3 days after subculture; ready for medium change. Phase contrast; 10× objective.

*(d) Late log phase cells.* NRK cells 7 days after subculture and ready for subculture again. Phase contrast; 10× objective.

*Plate 4. Phases of the Growth Cycle.*

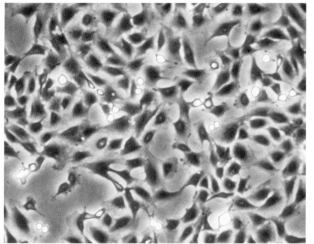

*(a) NRK Monolayer before Trypsinization. Phase contrast; 20× objective.*

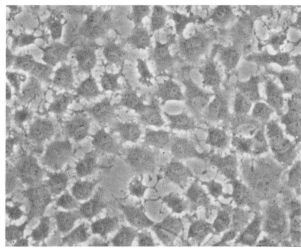

*(b) NRK Monolayer after D-PBSA/EDTA Prewash. Phase contrast; 20× objective.*

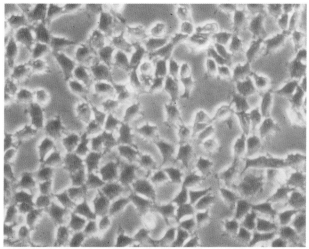

*(c) NRK Monolayer Immediately after Trypsin Removal. Phase contrast; 20× objective.*

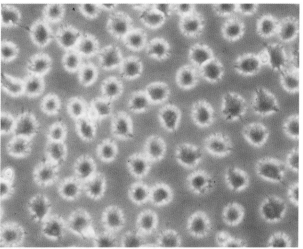

*(d) NRK Monolayer 1 min after Trypsin Removal. Phase contrast; 20× objective.*

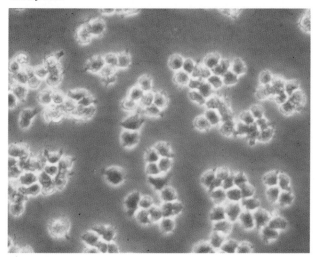

*(e) NRK Monolayer 5 min after Trypsin Removal. Phase contrast; 20× objective.*

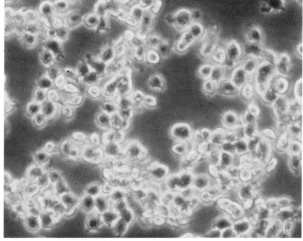

*(f) Fully Disaggregated Monolayer. 10 min after removal of trypsin and ready for dispersing and counting. Phase contrast; 20× objective.*

**Plate 5. Subculture by Trypsinization.**

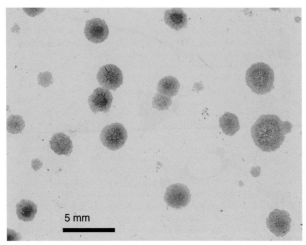

**(a) HeLa Clones.** *HeLa-S3 cloned by dilution cloning and stained with Giemsa after 3 weeks. Scale bar 5 mm.*

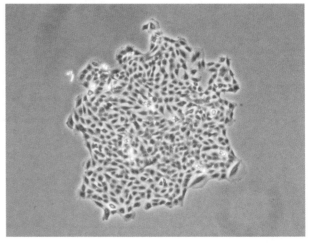

**(b) NRK Clone.** *Small clone of NRK cells, cloned by dilution cloning. Phase contrast, 4× objective.*

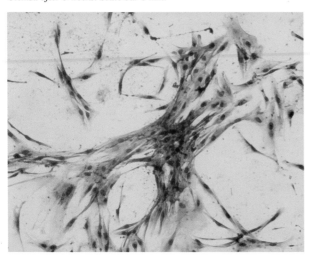

**(c) Breast Carcinoma, JUW, Cloned on Plastic.** *Microphotograph of secondary culture from human breast carcinoma, 4000 cells/cm², growing on plastic. Mainly fibroblasts. Giemsa stain. 10× objective.*

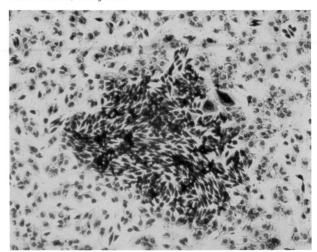

**(d) Breast Carcinoma on Feeder Layer.** *Microphotograph of epithelial colony from human breast carcinoma cells, 400 cells/cm², growing on confluent feeder layer of FHS74Int human fetal intestinal cells. Giemsa stain. 10× objective.*

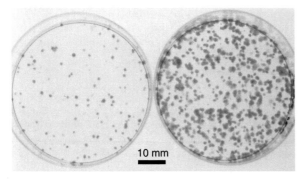

**(e) Effect of Serum on Plating Efficiency.** *Mv1Lu cells, transfected with myc oncogene, plated with 500 cells per 6-cm Petri dish, and fixed and stained 2 weeks later. 10% FBS (left), 20% FBS (right). Scale bar 10 mm. (Courtesy of M.Z. Khan.)*

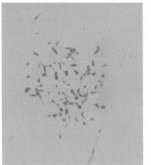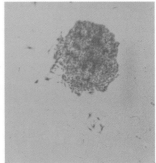

**(f) Effect of Glucocorticoid on Plating Efficiency.** *Early passage cell line from human glioma cloned in the absence (left) and presence (right) of 25 μM dexamethasone.*

**Plate 6. Cell Cloning.**

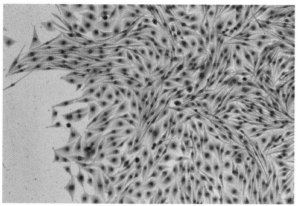

*(a) Clone of Continuous Glioma Cell Line, MOG-G-CCM.* Elliptical and spindle-shaped morphology. Giemsa stained. 10x objective.

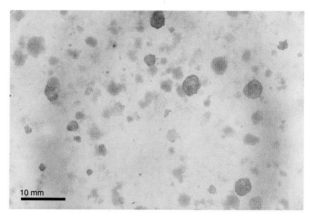

*(d) Cloned Culture of Early Passage Cell Line.* Human non-small cell lung carcinoma (NSCLC). Scale bar 10 mm.

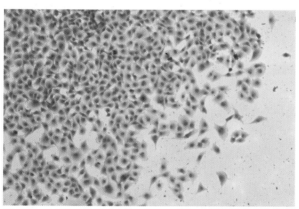

*(b) Different Clone of Same Continuous Glioma Cell Line, MOG-G-CCM.* Epithelioid morphology. Giemsa stained. 10x objective.

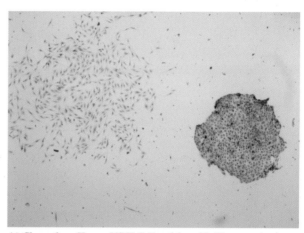

*(e) Clones from Human NSCLC.* Detail from (d). Giemsa stained. 4x objective.

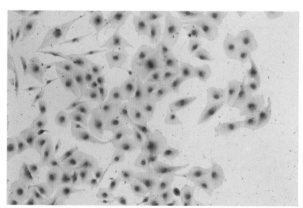

*(c) Third Clone of Continuous Glioma Cell Line, MOG-G-CCM.* Squamous epithelioid morphology. Same magnification as (a) and (b). Giemsa stained. 10x objective.

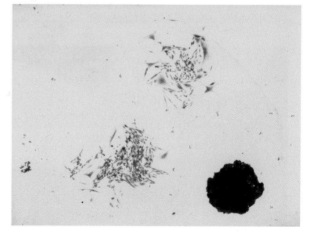

*(f) Cloned Culture of A2780.* Continuous cell line from human ovarian carcinoma. Giemsa stained. 4x objective.

*Plate 7. Cell Cloning, Morphological Diversity.*

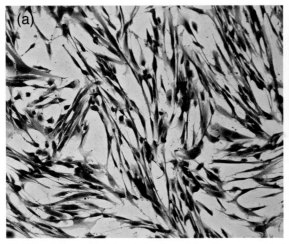

*(a) Normal Human Fetal Lung Fibroblasts.* Subconfluent normal fetal human lung fibroblasts. Giemsa stained; 10× objective.

*(b) Confluent Fibroblasts.* Dense culture of normal fetal human lung fibroblasts. Phase contrast; 10× objective.

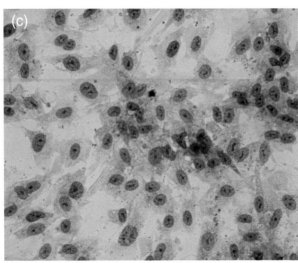

*(c) Human Umbilical Cord Endothelium.* Cell line from collagenase outwash of human umbilical vein. Giemsa stained. 20× objective.

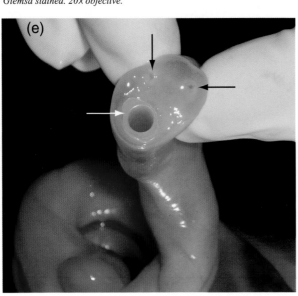

*(d) Human Melanocytes.* Secondary culture of TPA-treated epidermal melanocytes derived from African-American newborn foreskin. Note the dendritic morphology of the cells and their slender spindle shape. In contrast to the melanocytes derived from Caucasian newborn foreskin, these melanocytes display a high level of melanin granules (seen via phase contrast, ×320). (Courtesy of H.-Y. Park.)

*(e) View of Blood Vessels in Human Umbilical Cord.* Preparation for enzymatic isolation of endothelial cells from umbilical vein (see Protocol 22.18A). Black arrows, umbilical arteries. White arrow, umbilical vein with Luer adaptor inserted. (Courtesy of Charlotte Lawson; see also Fig. 22.3.)

*Plate 8. Finite Cell Lines and Cannulation of Human Umbilical Cord.*

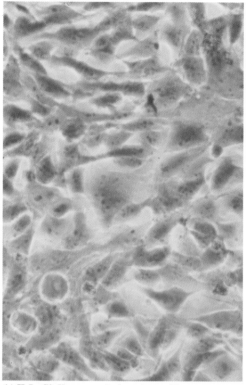

*(a) HeLa-S3. Human cervical carcinoma. Phase contrast. 40× objective.*

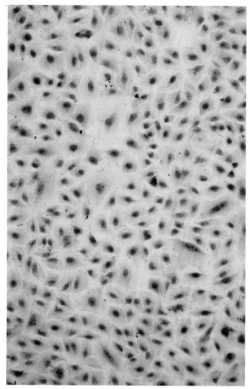

*(b) A549. Human lung adenocarcinoma. Giemsa stained. 10× objective.*

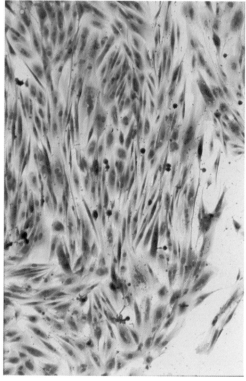

*(c) MOG-G-UVW. Human glioma. Giemsa stained. 10× objective.*

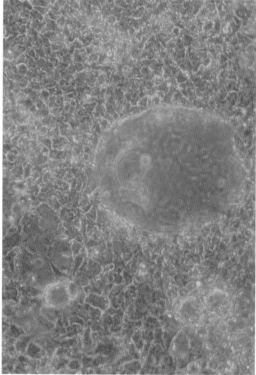

*(d) Caco-2. Human colorectal carcinoma showing dome formation. This cell line should be subcultured before it reaches this stage and while it is still subconfluent. Phase contrast; 10× objective.*

*Plate 9. Continuous Cell Lines from Human Tumors.*

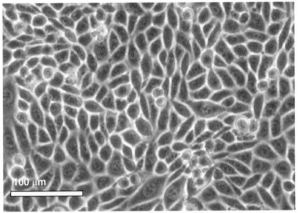

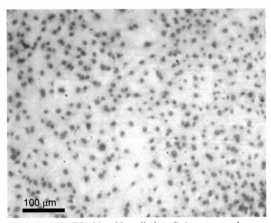

(a) CHO-K1. *Clone of CHO, continuous cell line from Chinese hamster ovary. Phase contrast.*

(b) Swiss 3T3. *Fibroblast-like cells from Swiss mouse embryo. Giemsa stained.*

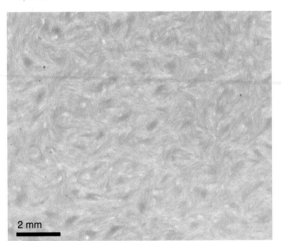

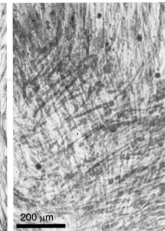

(c) BHK-21 Clone 13. *Baby hamster kidney fibroblasts after confluence showing typical swirling pattern formed by parallel arrays of cells. Giemsa stained.*

(d) BHK-21 Clone 13. *Baby hamster kidney fibroblasts approaching confluence and assuming parallel arrays. Giemsa stained.*

(e) High-Density BHK-21 C13. *Baby hamster kidney fibroblasts postconfluent and forming a second monolayer of more densely stained cells. Giemsa stained.*

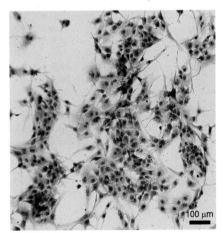

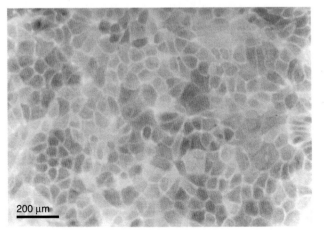

(f) MDCK. *Madin Darby canine kidney. Epithelial-like cell line **from dog kidney.***

(g) Mv1Lu. Mink Lung Epithelial Cell Line. *Stained by immunoperoxidase for cytokeratin (AE3 primary antibody; Photo courtesy of M.Z. Khan.)*

*Plate 10. Continuous Cell Lines from Normal, Nonhuman Animals.*

**(a) Normal Human Keratinocytes on 3T3 Feeder Layer.** *Keratinocyte pancytokeratin stained green, and vimentin in 3T3 cells, red. (Courtesy of Hans-Jurgen Stark.)*

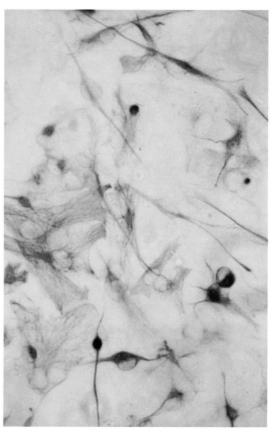

**(b) Human Glioma.** *MOG-G-CCM cells stained for GFAP by immunoperoxidase.*

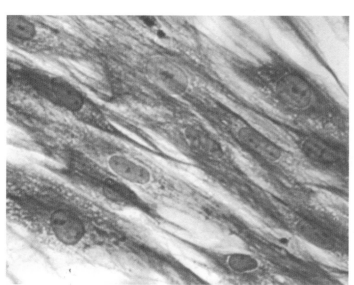

**(c) Breast Stromal Fibroblasts.** *Immunoperoxidase stained for vimentin (brown). (Courtesy of Valerie Speris.)*

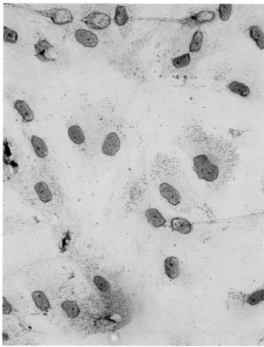

**(d) Human Umbilical Vein Endothelial cells.** *Factor VIII granular staining by immunoperoxidase.*

*Plate 11. Immunostaining.*

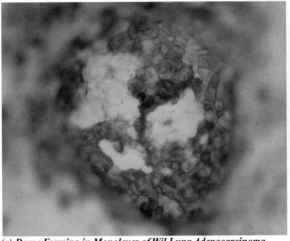

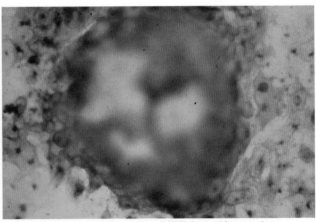

**(a) Dome Forming in Monolayer of Wil Lung Adenocarcinoma.**
*Mosaic of CEA-positive and CEA-negative cells. Focused on top of dome. Anti-CEA immonoperoxidase stained. 40x objective.*

**(b) Dome Forming in Monolayer of WIL Lung Adenocarcinoma.**
*Mosaic of CEA-positive and CEA-negative cells. Focused on monolayer. Anti-CEA immonoperoxidase stained. 40x objective.*

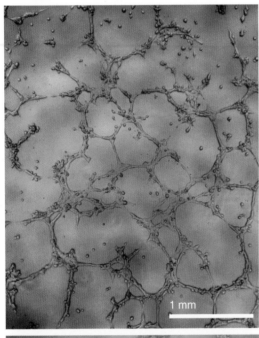

1 mm

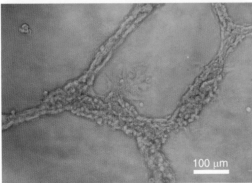

100 µm

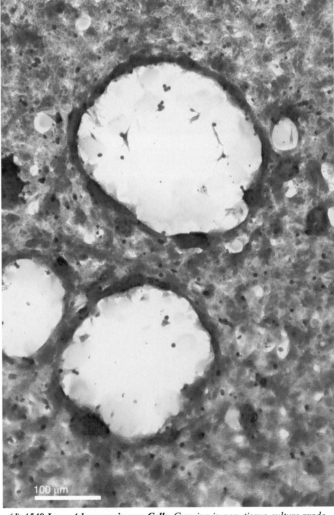

100 µm

**(c) A549 Lung Adenocarcioma Cells Growing on Matrigel.**
*Upper, 4x objective, lower 10x; bright field, unstained. (Courtesy of Jane Sinclair.)*

**(d) A549 Lung Adenocarcinoma Cells.** *Growing in non–tissue-culture grade Petri dish, with lung fibroblasts growing on a coverslip in center of dish. 10x objective; bright field.*

**Plate 12. Morphological Differentiation in Epithelial Cells.**

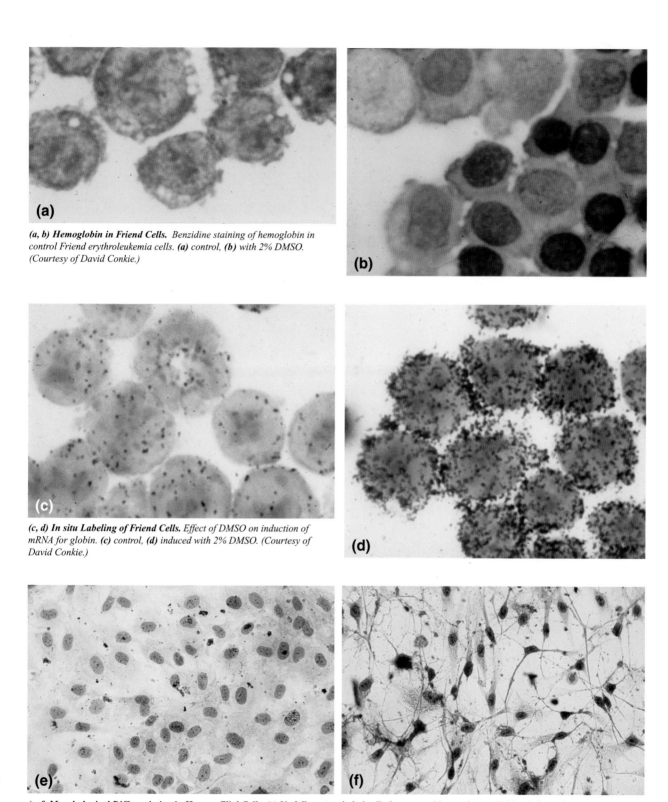

*(a, b) Hemoglobin in Friend Cells. Benzidine staining of hemoglobin in control Friend erythroleukemia cells. (a) control, (b) with 2% DMSO. (Courtesy of David Conkie.)*

*(c, d) In situ Labeling of Friend Cells. Effect of DMSO on induction of mRNA for globin. (c) control, (d) induced with 2% DMSO. (Courtesy of David Conkie.)*

*(e, f) Morphological Differentiation in Human Glial Cells. (e) Undifferentiated glial cells from normal human brain. (f) Morphological differentiation induced in glial cells from normal human brain by glia maturation factor. Giemsa stained.*

*Plate 13. Differentiation in Friend Cells and Human Glia.*

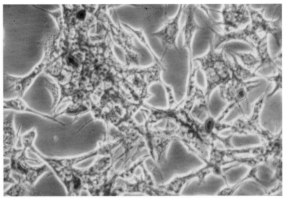

*(a) Contact Inhibition.* Late log-phase cultures of BHK21-C13 Cells tend to assume a parallel orientation and will not overgrow each other. Phase contrast. 10× objective.

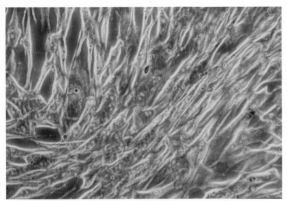

*(b) Loss of Contact Inhibition.* Polyoma-transformed clone BHK21-PyY cells that show no recognition of each other and grow randomly over each other. Phase contrast. 10× objective.

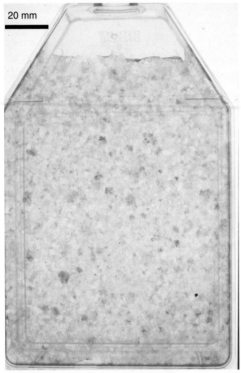

20 mm

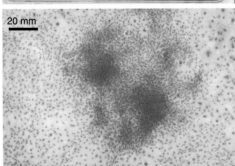

20 mm

*(c) Focus Formation in 3T3 Cells.* A monolayer of Sw-3T3 cells, left at confluence for 3 weeks, showed foci of transformed cells escaping contact inhibition. (top), Whole flask; (bottom) 4× objective. Giemsa stained.

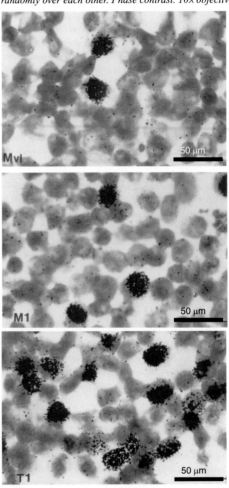

Mv1    50 µm

M1    50 µm

T1    50 µm

*(d) [³H]-thymidine Incorporation in Mv1Lu mink Lung Cells.* Mink lung cells, Mv1Lu, were labeled with [3H]-TdR for 1 h, fixed, and coated with autoradiographic emulsion (see Protocol 27.3), and stained with Giemsa. Mv1 (top) is the control cell line Mv1Lu, M1 (middle) was Mv1Lu transfected with the myc oncogene, and T1 (bottom) with mutant ras. T1 cells were tumorigenic and had a statistically significant ($p<0.001$) increase in labeling index compared to Mv1Lu, also seen with bromodeoxyuridine labeling [Khan et al., 1991]. (Courtesy of M. Z. Khan.)

*Plate 14. Properties of Transformed Cells.*

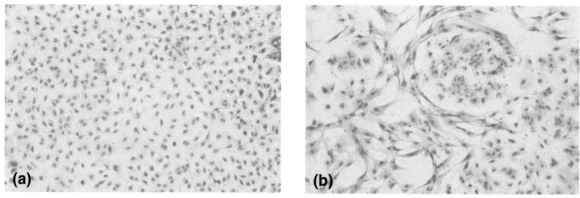

*(a, b) Infiltration of Normal Monolayer. (a) Contact inhibited confluent monolayer of normal human fetal instestinal cell line FHS74Int. (b) Human glioma cell Line WLY infiltrating FHS74Int. Normal glial cells do not infiltrate this confluent monolayer 10x objective; Giemsa stained.*

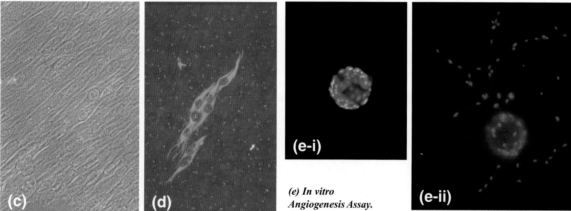

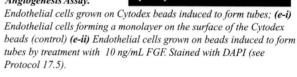

*(c, d) Non–Small Cell Carcinoma Cell Line Infiltrating Normal Fibroblasts. (c) Phase contrast. (d) Indirect immunofluorescence for cytokeratin revealing the tumor cells migrating in parallel with the fibroblasts. 40x objective.*

*(e) In vitro Angiogenesis Assay.*
*Endothelial cells grown on Cytodex beads induced to form tubes; (e-i) Endothelial cells forming a monolayer on the surface of the Cytodex beads (control) (e-ii) Endothelial cells grown on beads induced to form tubes by treatment with 10 ng/mL FGF. Stained with DAPI (see Protocol 17.5).*

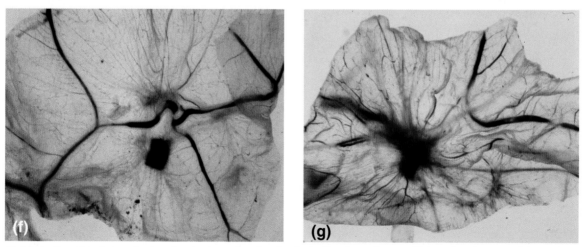

*(f, g) Induction of Angiogenesis. A crude extract was placed on the chorioallantoic membrane at 10 days of incubation, and the membrane was removed 2 weeks later. (f) Normal glial cell extract. (g) Crude extract of Walker 256 carcinoma cells. (Courtesy of Margaret Frame.)*

*Plate 15. More Properties of Transformed Cells.*

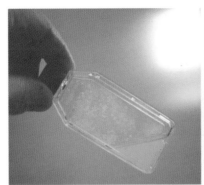

**(a) Contaminated Flask.** *Reduction in pH and cloudiness of medium.*

**(b) Floculated Contamination.** *Bacterial, but no drop in pH.*

**(c) Contaminated Medium Bottle.** *Cloudy with no drop in pH; yeast.*

**(d) Samples of Broth.** *L Broth with test samples of medium. Control (left) clear, Negative test (center), and positive control (right).*

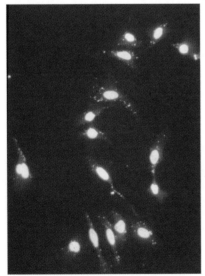

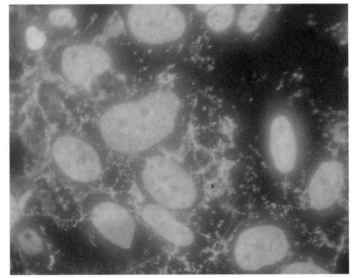

**(e) Mycoplasma, Low Power.** *Mycoplasma infected culture as revealed by Hoechst 33258 staining. 40× water-immersion objective.*

**(f) Mycoplasma, High Power.** *Hoechst-stained mycoplasma infected cells under 100× oil-immersion objective.*

*Plate 16. Examples of Contamination.*

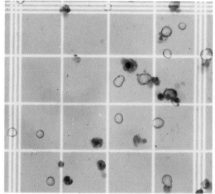 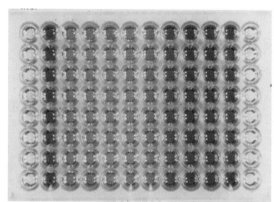

**(a) Dye Exclusion, Naphthalene Black.**
Hemocytometer slide 200-mm square with viable
(unstained) and nonviable (blue stained) cells. 40x
objective.

**(b) MTT Assay.** *Microtitration plate with cells stained with MTT
after 24 h in a range of concentrations of VP16. Extreme left- and
right-hand columns are blanks, 2nd from left and 2nd from right are
untreated controls. Column 3 to column 8, VP16 0 - 10 µM.*

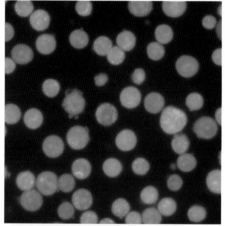 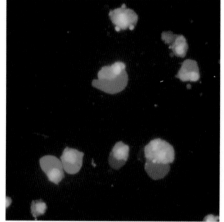

**(c, d) Apoptosis in HT29 cells.** *Treated with 0.25 µg/ml TRAIL (TNF- related apoptosis-inducing ligand) for 16 h and
stained with acridine orange.* **(c)** *Attached cells showing normal morphology.* **(d)** *Detached cells from* **(c)** *showing
chromatin condensation and fragmentation characteristic of apoptosis. 40x objective. (Courtesy of Angela Hague.)*

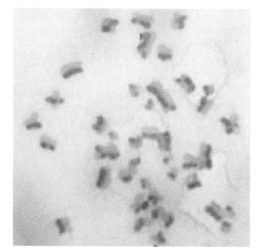 

**(e, f) Sister Chromatid Exchange (SCE).** *(e)* *Untreated cells: A2780/Cp70 with an additional human chromosome 2
transferred (A2780/cp70 +chr2).* *(f)* *Induced SCE. A2780/Cp70 cells treated with 10 µM Cisplatin for 1 h, showing
extensive SCE: dark Giemsa staining alternating between strands within individual chromosomes. (Courtesy of Robert
Brown & Maureen Illand.)*

**Plate 17. Viability and Cytotoxicity.**

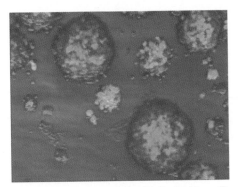

**(a) Transfected Mosaic Spheroids Derived from Human Glioma Cell Line MOG-G-UVW.** *The Spheroids, ranging in size from 100 to 500 μm diameter, are composed of mixtures of cells transfected with the GFP gene (green) and cells transfected with the NAT gene (red). (Courtesy of Marie Boyd and Rob Mairs; see Section 25.3.3.)*

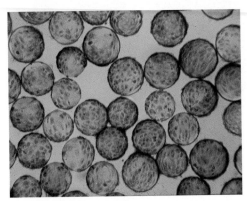

**(c) Microcarriers.** *Vero cells growing on microcarriers. (Courtesy of MP Biomedicals.)*

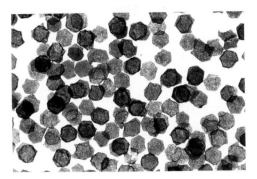

**(d) Hexagonal Microcarriers.** *Nunc Microhex Beads. (Courtesy of Nunc.)*

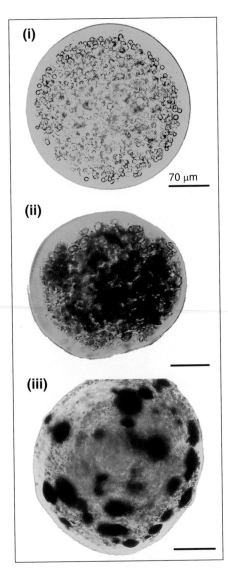

**(b) Alginate Encapsulation.** *Light microscopic images of cells encapsulated in alginate after 2 hours (i), 3 weeks (ii) and 4 months (iii) in vitro. Within the alginate beads both cell death and cell proliferation will occur, and for many cell lines multicellular spheroids will form inside the beads. (Courtesy of Tracy-Ann Read and Rolf Bjerkvig; see Section 25.3.5.)*

*Plate 18. Spheroids, Encapsulation, and Microcarriers.*

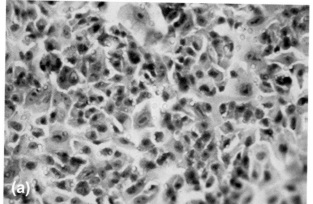

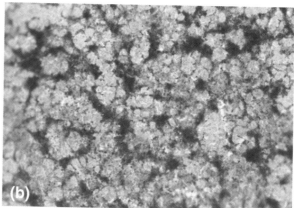

**(a) Culture on Filter Well Inserts.** *A549 human lung adenocarcinoma grown on polycarbonate filter. Holes in the filter (15 μm) are visible, particularly in top right-hand corner space. 40x objective. Giemsa stained.*

**(b) Human Fetal Lung Epithelial Cells.** *Cells growing on filter with human fetal lung fibroblasts on other side of filter. 4x objective. Giemsa stained.*

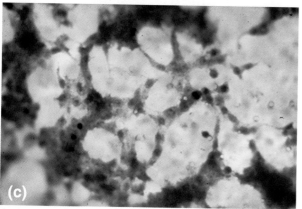

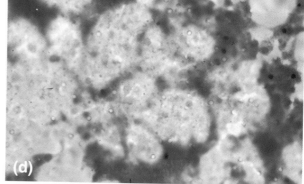

**(c, d) Lung Cell Coculture.** *Fetal lung epithelial cells and fibroblasts as in (b). (c) Focused on cells above filter. (d) Focused on cells below filter. 40x objective. Giemsa stained.*

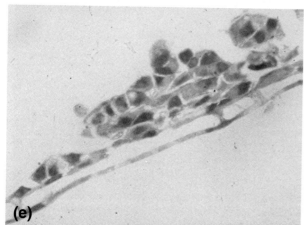

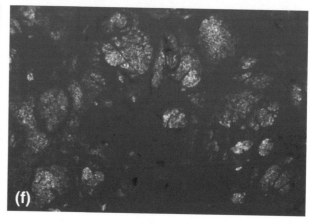

**(e) Section of Coculture.** *Section through filter with A549 cells and fibroblasts. Both cell types were seeded on top, but fibroblasts migrated through first, though some remained on the top. Three cells are seen travelling through the pores. 40x objective; H&E stain.*

**(f) Chick Embryo Lung.** *High-density primary culture of cold trypsin disaggregated 13d embryonic chick lung. 72 h after seeding at >1x106 cells/mL. 40x objective. Giemsa stained.*

**Plate 19. Organotypic Culture in Filter Wells.**

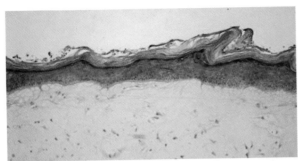

**(a) Keratinocyte and Dermal Fibroblast Cocultures.** *Organotypic cocultures after 2 weeks in vitro. 40x objective. H&E stain.*

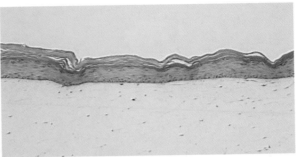

**(b) Organotypic Cultures Implanted In vivo.** *Organotypic cocultures in vivo 3 weeks after transplantation onto the nude mouse. 40x objective. H&E-stain.*

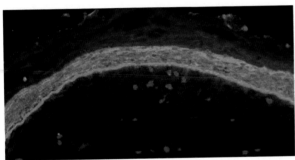

**(c) Integrin and Cytokeratin in Cocultures In vitro.** *Immunofluorescence for α6-integrin (green), keratin 10 (red) and DNA fluorescence (blue). 40x objective.*

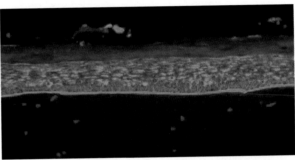

**(d) Integrin and Cytokeratin in Organotypic Cultures Implanted In vivo.** *Immunofluorescent staining for α6-integrin (green), keratin 10 (red) with DNA fluorescing blue. 40x objective.*

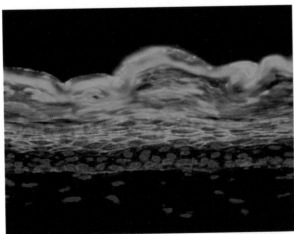

**(e) Involucrin and Collagen in Organotypic Cocultures at 2 weeks.** *Keratinocyte-fibroblast cocultures with collagen gel stained for involucrin (green) and the basement membrane component collagen VII (red). 40x objective.*

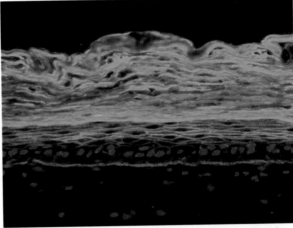

**(f) Involucrin and Collagen in Organotypic Cocultures at 3 weeks.** *Keratinocyte-fibroblast cocultures with collagen gel stained for involucrin (green) and the basement membrane component collagen VII (red). 40x objective.*

**Plate 20. Organotypic Culture of Skin.** (Courtesy of Hans-Jürgen Stark)

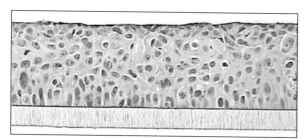

*(a) Exposed to Control Solution for 10 min.*

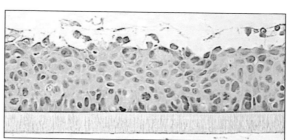

*(b) Exposed to 1% SDS for 10 min.*

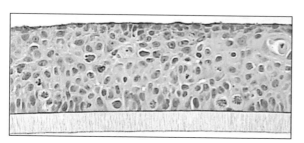

*(c) Control Solution for 20 min.*

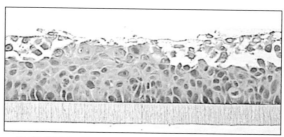

*(d) 1% SDS for 20 min.*

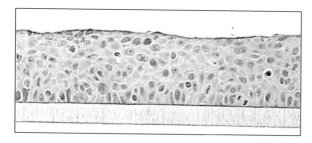

*(e) Control Solution for 60 min.*

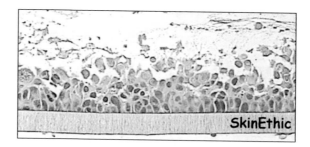

*(f) 1% SDS for 60 min.*

(*See also* Figs. 21.10, 25.7) (Examples, courtesy of SkinEthic.)

**Plate 21. In vitro Toxicity in Organotypic Model.**

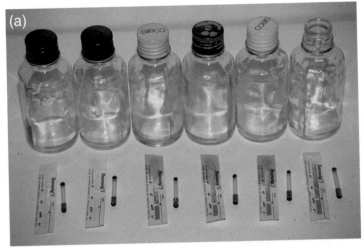

**(a) Sterilizing Bottles.** *These bottles were autoclaved with Thermalog sterility indicators inside. Thermalog turns blue with high temperature and steam, and the blue area moves along the strip with time at the required sterilization conditions. The cap on the leftmost bottle was tight, and each succeeding cap was gradually slacker, until, finally, no cap was used on the bottle furthest to the right. The leftmost bottle is not sterile because no steam entered it. The second bottle is not sterile either because the liner drew back onto the neck and sealed it. The next three bottles are all sterile, but the brown stain on the indicator shows that there was fluid in them at the end of the cycle. Only the bottle at the far right is sterile and dry.*

*The glass tube indicators (Browne's Tubes), which are temperature sensitive only, all indicated that the bottles were sterile.*

**(b) pH Standards.** *Phenol Red pH indicator in a standard set of solutions. Far left and far right are unacceptable and need immediate action (i.e., medium change, subculture or gassing.)*

pH6.5  pH7.0  pH7.4  pH7.6

**(c) Rotating Cell Culture System.** *Chamber rotates to create simulated zero gravity so that cells do not sediment and grow as aggregates within the chamber. (Courtesy of Synthecon.)*

**(d) BelloCell Culture System.** *Cells are grown on carriers in center of chamber and the bellows medium chamber alternately forces medium up over beads and back down into reservoir. (Courtesy of Bellco.)*

**(e) Color-coded Cryovials.** *Polypropylene cryostorage vials, 1.0, 1.8, 3.6, 4.6 mL. Colored inserts for caps helps to prevent misidentification. (Courtesy of Alpha Laboratories.)*

**Plate 22.  Medium Preparation, Culture Systems, and Cryovials.**

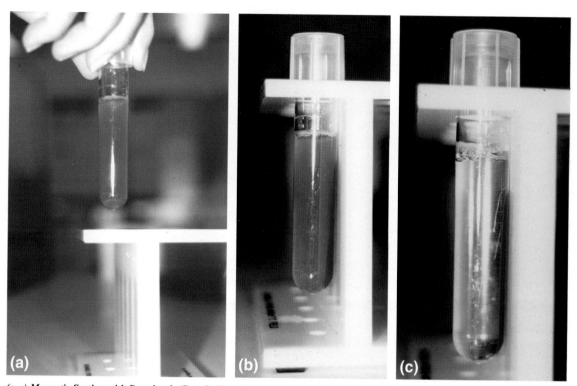

**(a–c) Magnetic Sorting with Dynabeads (Dynal).** *Negative sort; committed progenitor cells from bone marrow suspension bound to Dynal paramagnetic beads with antibodies to lineage markers. Lineage negative (stem) cells are not bound and remain in the suspension ready for sorting by flow cytometry.* **(a)** *Inserting the tube into the magnetic holder.* **(b)** *Tube immediately after being placed in magnetic holder.* **(c)** *Tube 30 s after placement in magnetic holder.*

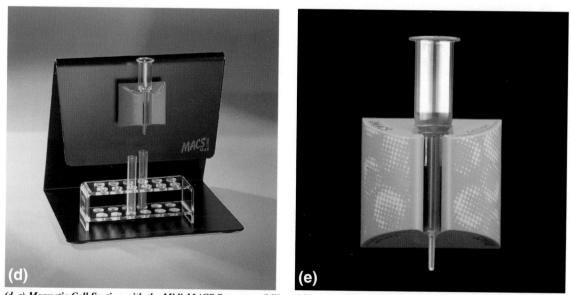

**(d, e) Magnetic Cell Sorting with the Midi-MACS Separator (Miltenyi Biotec).** **(d)** *Column, magnet, rack and stand.* **(e)** *Midi-MACS separator with LD column. (Courtesy of Miltenyi Biotec.)*

*Plate 23. Magnetically Activated Cell Sorting.*

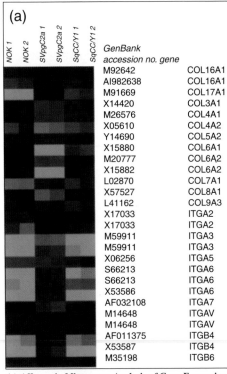

**(a)**

| NOK 1 | NOK 2 | SVpgC2a 1 | SVpgC2a 2 | SqCC/Y1 1 | SqCC/Y1 2 | GenBank accession no. | gene |
|---|---|---|---|---|---|---|---|
| | | | | | | M92642 | COL16A1 |
| | | | | | | AI982638 | COL16A1 |
| | | | | | | M91669 | COL17A1 |
| | | | | | | X14420 | COL3A1 |
| | | | | | | M26576 | COL4A1 |
| | | | | | | X05610 | COL4A2 |
| | | | | | | Y14690 | COL5A2 |
| | | | | | | X15880 | COL6A1 |
| | | | | | | M20777 | COL6A2 |
| | | | | | | X15882 | COL6A2 |
| | | | | | | L02870 | COL7A1 |
| | | | | | | X57527 | COL8A1 |
| | | | | | | L41162 | COL9A3 |
| | | | | | | X17033 | ITGA2 |
| | | | | | | X17033 | ITGA2 |
| | | | | | | M59911 | ITGA3 |
| | | | | | | M59911 | ITGA3 |
| | | | | | | X06256 | ITGA5 |
| | | | | | | S66213 | ITGA6 |
| | | | | | | S66213 | ITGA6 |
| | | | | | | X53586 | ITGA6 |
| | | | | | | AF032108 | ITGA7 |
| | | | | | | M14648 | ITGAV |
| | | | | | | M14648 | ITGAV |
| | | | | | | AF011375 | ITGB4 |
| | | | | | | X53587 | ITGB4 |
| | | | | | | M35198 | ITGB6 |

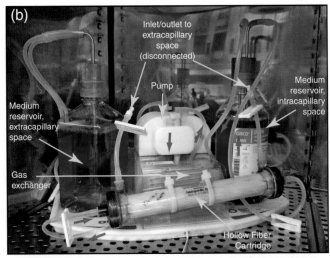

*(b) Hollow Fiber Culture. A bundle of hollow fibers of permeable plastic is enclosed in a transparent plastic outer chamber and is accessible via either of the two side arms for seeding cells and collecting high molecular weight product. During culture, the chamber is perfused (red arrows) from a reservoir, via a gas exchanger and peristaltic pump, down the center of the hollow fibers through connections attached to either end of the chamber. (Courtesy of FiberCell Inc; see also Fig. 26.6.)*

*(a) Affymetrix Microarray Analysis of Gene Expression. Expression of fibronectin, collagens, and integrin transcripts is detected by Affymetrix oligonucleotide microarray in normal (NOK), SV40 T-antigenimmortalized (SVpgC2a) and malignant (SqCC/Y1) human oral keratinocyte lines. GenBank gene abbreviations: COL16A1 = type XVI collagen 1 chain, etc.; ITGA2 = integrin 2 subunit, etc.; ITGB4 = integrin 4 subunit, etc. (From Sarang et al., 2003, by permission of the publisher and the authors.)*

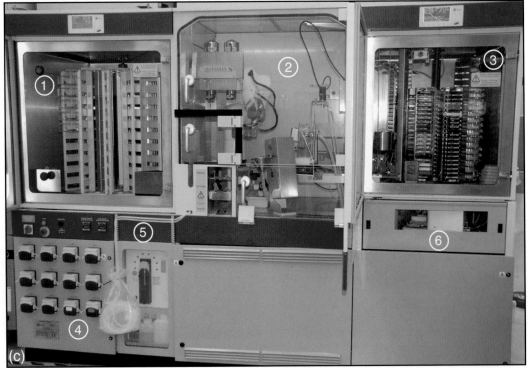

*(c) The CompacT SelecT Cell Culture Robotic System. Front aspect; (1) incubator chamber with flasks on top left of carousel, (2) main handling chamber with robotic arm in center, below left of number, capping/uncapping and dispensing unit to left of number and pipette container below number, (3) plate incubator chamber, (4) peristaltic pumps for media additions, (5) Cedex cell counter (Innovatis), (6) multiwell plate liquid handling unit. (Courtesy of The Automation Partnership.)*

***Plate 24. Automated Culture and Analysis.***

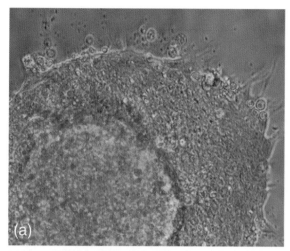

**(a) Mouse Embryonic Stem Cells.** *ES cell-like outgrowths from cultured epiblast. (Courtesy of Jason Wray & Jennifer Nichols; see Protocol 23.1.)*

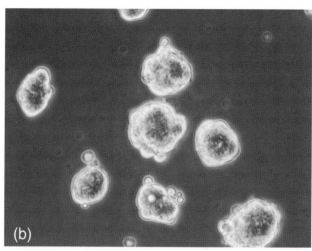

**(b) mES Colonies Growing in 2i Serum-Free Medium.** *(Courtesy of Jason Wray & Jennifer Nichols; see Protocol 23.1.)*

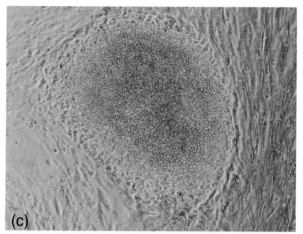

**(c) Human Embryonic Stem Cells.** *Colony of hES cells on feeder layer (See also Fig. 23.2.) (Reprinted from Cooke & Minger, 2007.)*

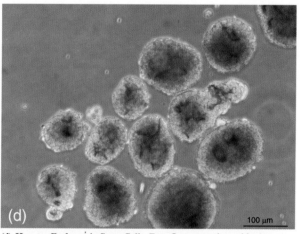

**(d) Human Embryonic Stem Cells.** *Free-floating embryoid bodies (EBs). (Reprinted from Cooke & Minger, 2007.)*

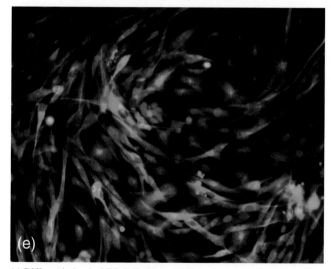

**(e) Differentiation in hES Cells.** *Neural differentiation after replating hES cell derived neurospheres on gelatin coated plastic. Immunofluorescent staining for nestin (green); nuclei stained with Hoechst 33342 (blue). (Jackson et al., 2007.)*

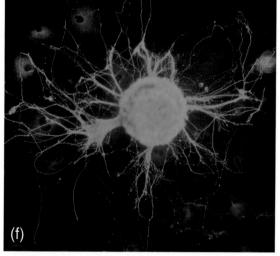

**(f) Differentiation in human embryonal carcinoma cells.** *Immunofluoescent staining for neural marker TUJ-1. (Przyborski, 2007.)*

*Plate 25. Embryo–Derived Stem Cells.*

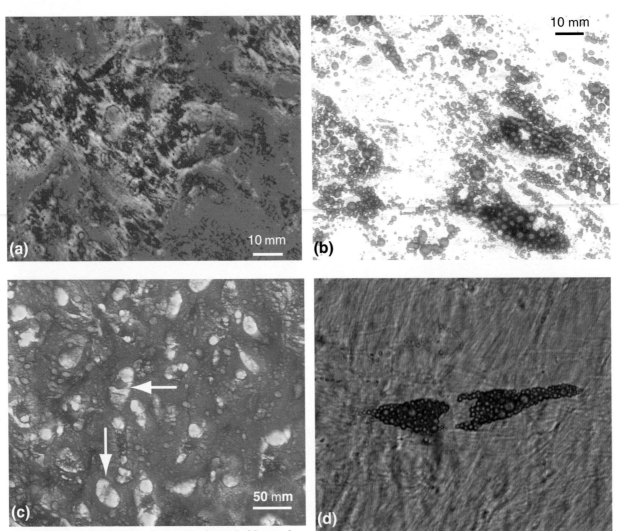

**(a–c) MSCs from Bone Marrow Stroma. (a)** *Alizarin Red S-stained monolayer showing osteogenic differentiation.* **(b)** *Oil Red O-stained monolayer showing adipogenic differentiation.* **(c)** *Toluidine Blue staining of 10 mm sections of pelleted cells. (Gregory & Prockop, 2007.) (See also Section 23.3.4.)*

**(d) Multipotent stem cells from dental pulp.** *Adipogenic differentiation shown by Oil Red O staining. (Sonoyama et al., 2007.)*

**Plate 26.  Juvenile and Adult Stem Cells.**

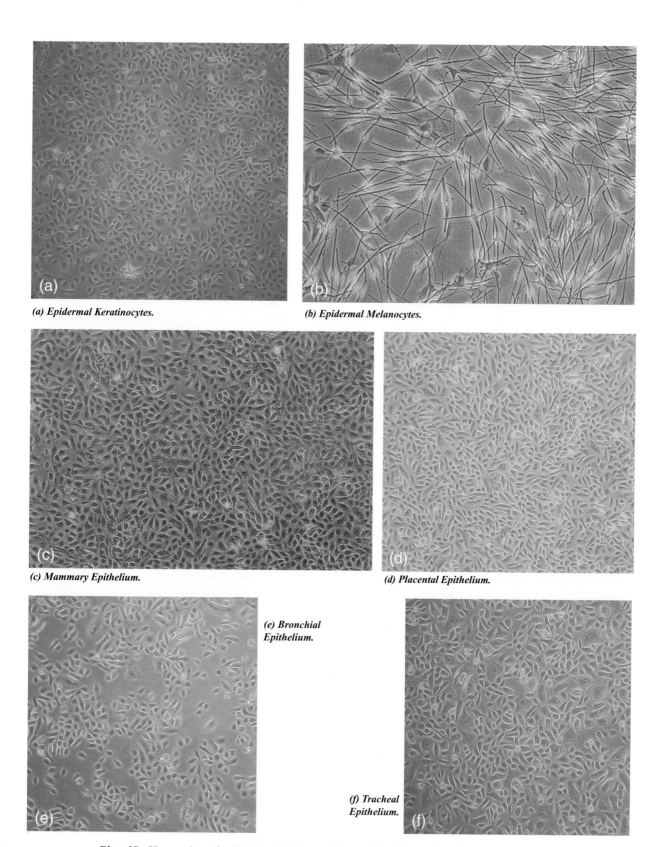

*(a) Epidermal Keratinocytes.*

*(b) Epidermal Melanocytes.*

*(c) Mammary Epithelium.*

*(d) Placental Epithelium.*

*(e) Bronchial Epithelium.*

*(f) Tracheal Epithelium.*

**Plate 27. Human Specialized Cells in Primary Culture (1).** *Cultures from first or second subcultures. Phase contrast, 10x objective (Courtesy of Cell Applications, Inc.)*

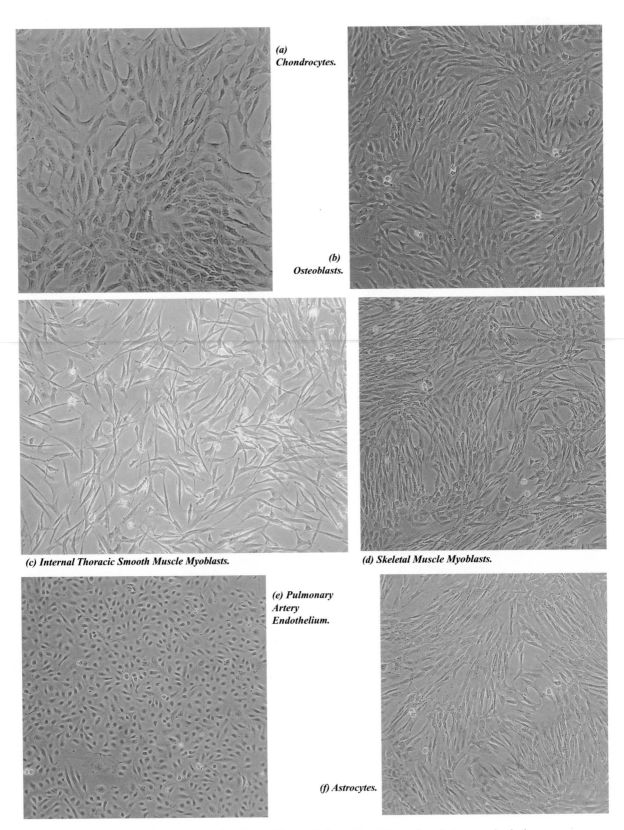

(a) Chondrocytes.

(b) Osteoblasts.

(c) Internal Thoracic Smooth Muscle Myoblasts.

(d) Skeletal Muscle Myoblasts.

(e) Pulmonary Artery Endothelium.

(f) Astrocytes.

**Plate 28. Human Specialized Cells in Primary Culture (2).** *Cultures from first or second subcultures.*
*Phase contrast, 10x objective (Courtesy of Cell Applications, Inc.)*

**(a)** Remove the medium completely from the three wells containing cells to be counted.

**(b)** Add 0.5 mL of trypsin/EDTA to each of the three wells.

**(c)** Incubate the plate for 15 min.

**(d)** Add 0.5 mL medium with serum, disperse the cells in the trypsin/EDTA/medium, and transfer 0.4 mL of the suspension to 19.6 mL of D-PBSA (for Beckman Coulter Z1 or Z2 or CASY); for other counters use 20 μL to 1 mL undiluted cell suspension as indicated in manufacturer's instructions.

**(e)** Count the cells on an electronic cell counter.

6. Stain the cells in the remaining wells at each cell density (*see* Section 15.5.2).

*Note.* A hemocytometer may be used to count the cells, but may be difficult to use for lower cell concentrations. If you use a hemocytometer, reduce the volume of trypsin to 0.1 mL, and disperse the cells carefully, using a pipettor without frothing the trypsin. Transfer the cells to the hemocytometer, and count enough squares to give a count in excess of 200 cells.

7. Repeat sampling at 48 and 72 h, as in steps 5 and 6.

8. Change the medium at 72 h, or sooner, if indicated by a drop in the pH (*see* Protocol 12.1; Plate 22*b*).

9. Continue sampling daily for rapidly growing cells (i.e., cells with a PDT of 12–24 h), but reduce the frequency of sampling to every 2 days for slowly growing cells (i.e., cells with a PDT >24 h) until the plateau phase is reached.

10. Keep changing the medium every 1, 2, or 3 days, as indicated by the fall in pH.

## 20.9.3 Analysis of Monolayer Growth Curves

(1) Calculate the number of cells per well, per mL of culture medium (*see* Fig. 20.8), and per cm$^2$ of available growth surface in the well as follows:

(a) *Primary count*: The count obtained from electronic counting or hemocytometer is the number of cells/mL of trypsinate.

(b) *Cells per flask or well*: Where 1 mL trypsin has been used for trypsinization, the primary count is the same as the number of cells per flask or well. If 2 mL trypsin was used, double the cell count to give cells per flask or well. If 0.5 mL trypsin was used, divide the primary count by 2 to give cells per flask or well.

(c) *Cells per mL of culture medium (cell concentration)*: Divide the number of cells per well or flask by the volume of medium used during culture.

(d) *Cells per cm$^2$ of growth surface (cell density)*: Divide the number of cells per well or flask by the surface area of the well or flask. As there are small variations among manufacturers, it is best to calculate the surface area for the culture vessel that you use. As a rough guide, however, each well is 2 cm$^2$ in a 24-well plate and 3.8 cm$^2$ in a 12-well plate. The flasks recommended for Protocol 20.7 have 25-cm$^2$ growth area.

(2) Plot the cell density (cells/cm$^2$) and the cell concentration (cells/mL), both on a log scale, against time on a linear scale (*see* Fig. 20.8). The scale on both vertical axes is the same but out of register by a factor that depends on the number of cm$^2$ per mL medium. In a 25-cm$^2$ flask, there are 5 mL medium covering 25 cm$^2$, 0.2 mL/cm$^2$ or 5 cm$^2$/mL medium, so the right-hand axis will be out of register with the left by a factor of 5, and $1 \times 10^5$ on the left axis will be equivalent to $2 \times 10^4$ on the right-hand axis.

(3) Determine the lag time, PDT, and plateau density (*see* Fig. 20.8; *see also* Section 20.11.1).

(4) Establish the appropriate starting density for routine passage from Protocol 20.7. Repeat the growth curve at different cell concentrations if necessary.

(5) Compare growth curves under different conditions (Fig. 20.10), and try to interpret the data (*see* legend for Fig. 20.10).

(6) Examine the stained cells at each density to:

(a) determine whether the distribution of cells in the flasks or wells is uniform and whether the cells are growing up the sides of the well, and

(b) observe differences in cell morphology as the density increases

Generating a growth curve can be quite labor intensive between setting up the replicate cultures, harvesting, counting, and analyzing the data. It is now possible to obtain image analysis software, operating with an inverted photomicroscope in a temperature regulated enclosure (Fig. 20.11), which will calculate the number of cells per field or residual unoccupied growth surface and generate a growth curve automatically. Replicate images are captured from multiple sites in a flask or multiwell plate, can be viewed directly or online from a distant location to assess the progress of a culture, can be used for time lapse observation, and generate the data necessary for constructing a growth curve (Fig. 20.12).

## 20.9.4 Medium Volume, Cell Concentration, and Cell Density

It is important when using multiwell plates and comparing data with culture flasks to remember that the volume of medium used in a multiwell plate is often proportionately higher than in a flask for a given surface area and the cell density will be higher for the same cell concentration. If $2 \times 10^4$ cells/mL are seeded in 5 mL into a 25-cm$^2$ flask, the cell density at seeding will be 4000 cells/cm$^2$

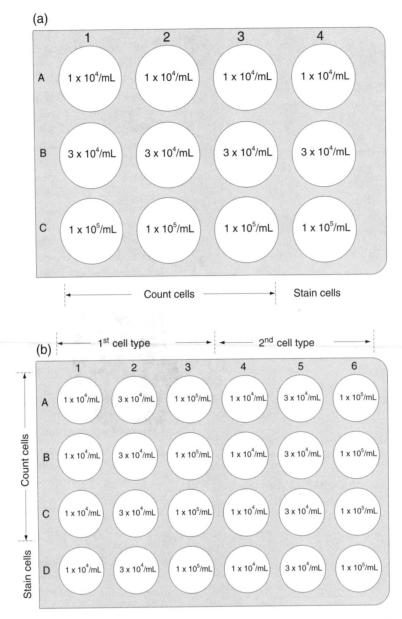

**Fig. 20.9. Layout of Multiwell Plates.** Layouts suggested for (*a*) 12-well plates with cells at three different concentrations and wells allocated for counting and staining and (*b*) 24-well plates with an additional variable, in this case the cell type.

(20,000 × 5 ÷ 25), whereas if the same cell concentration is seeded in 2 mL in a 12-well plate the cell density will be 10,500 cells/cm$^2$ (20,000 × 2 ÷ 3.8). This density is more than twice that of the flask, and the cells will reach plateau at least one day earlier. If an exact comparison is intended the ratio of medium to culture surface area must be the same, and a volume of 0.75 mL would be required to achieve the same cell density in a 12-well plate as in a 25-cm$^2$ flask for a given cell concentration. Unfortunately, such a low volume would cause uneven cellular distribution due to the shape of the meniscus, and cells would tend to concentrate at the edges of the wells. This problem increases as the wells get smaller

because the relative effect of the meniscus increases with a decrease in diameter of the well. A reasonable compromise is to ensure that the cell density (cells/cm$^2$) is the same, although the cell concentration will be less.

Multiwell plates are suitable for comparing different growth conditions, media, sera, or growth factors or cytotoxins, but if a growth curve is being used to establish conditions for routine maintenance, then the growth curve must be performed in the same vessels as being used for routine subculture (although the difference between a 25-cm$^2$ and a 75-cm$^2$ flask will be minimal given that the volume of medium per cm$^2$ remains the same).

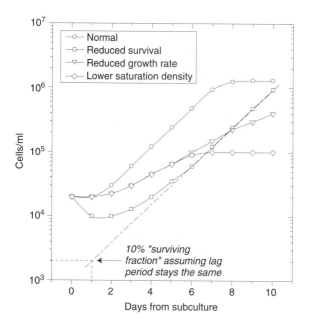

**Fig. 20.10. Interpretation of Growth Curves.** Changes in the shape of a growth curve can be interpreted in a number of different ways, but the labels in the key of this plot indicate what would normally be deduced from these curves.

It is also possible to perform a growth curve in microtitration plates, such as control plates to monitor cell numbers in a cytotoxicity assay (*see* Section 21.3.4). However, as the growth area per well is very small the cell numbers can be very low, particularly at the start of an experiment, four or eight wells may need to be pooled and counted for some forms of electronic counting. Alternatively, the cell count may be estimated from dye or reduced MTT absorbance (see Table 20.1 and Protocol 21.4, Steps 14 through 18).

### 20.9.5    Suspension Cultures

A growth curve can also be generated from cells growing in suspension, usually without replenishing the medium. The objectives are similar as for monolayer cells, namely to set conditions for routine maintenance or to assay differences in growth conditions. As trypsinization is not required, several samples can be harvested from the same vessel.

---

**PROTOCOL 20.9. GROWTH CURVE WITH CELLS IN SUSPENSION**

**Outline**
Set up a series of cultures at three different cell concentrations, and count the cells daily until they reach the plateau phase.

**Materials**

*Sterile or aseptically prepared:*

---

- ❑ Suspension cell culture
- ❑ Growth medium, 100 mL
- ❑ D-PBSA (for cell counting if required)
- ❑ Plates, 24 well

*Nonsterile:*

- ❑ Plastic box to hold the plates
- ❑ $CO_2$ incubator or 5% $CO_2$ supply to purge the box with 5% $CO_2$

**Procedure**

1. Add the cell suspension in growth medium to wells at a range of concentrations as for monolayer cultures (*see* Protocol 20.8).
2. Sample 0.4 mL (or as appropriate depending on the cell counter used) of the culture from triplicate wells at intervals, ensuring that the cells are well mixed and completely disaggregated.
3. Count the samples on electronic cell counter (*see* Section 20.1.2) or hemocytometer (*see* Section 20.1.1).
4. Calculate the cell concentration per sample, and plot on a log scale against time on a linear scale as for monolayer growth (*see* Section 20.9.3). Cell density does not apply to a suspension culture as they are not adherent to the growth surface.

---

***Variations.*** Seed two 75-cm² flasks with 20 mL of cell suspension in growth medium for each cell concentration, and sample each flask daily or as required. Mix the culture well before sampling it, and keep the flasks out of the incubator for the minimum length of time. Do not feed the cultures during the growth curve. Alternatively, set up a stirrer flask (Techne, Integra, Bellco) and sample daily. If a membrane closure is used on the side arm of the stirrer flask, then the flask can be sampled without removing it from the hot room by swabbing the membrane, upending the flask, and sampling via the side arm with a syringe and needle.

### 20.9.6    Phases of the Growth Cycle

The growth cycle (*see* Fig. 20.8) may be divided into three phases:

***Lag phase.*** This phase is the time after subculture and reseeding during which there is little evidence of an increase in the cell number. It is a period of adaptation during which the cell replaces elements of the cell surface and extracellular matrix lost during trypsinization, attaches to the substrate, and spreads out. During spreading the cytoskeleton reappears, an integral part of the spreading process. The activity of enzymes, such as DNA polymerase, increases, followed by the synthesis of new DNA and structural proteins. Some specialized cell products may disappear and not reappear until the cessation of cell proliferation at a high cell density.

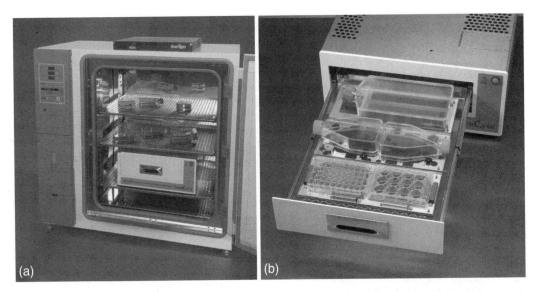

**Fig. 20.11. Incucyte.** Image analysis of cell cultures during growth. (*a*) Incubator with Incucyte in place; (*b*) Incucyte compartment open to show flasks. (Courtesy of Essen Instruments).

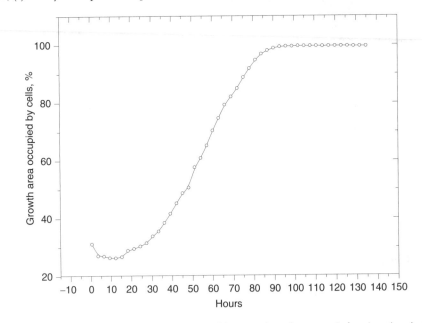

**Fig. 20.12. Incucyte Growth Curve.** Percentage of the growth surface occupied against time in hours (HCA-7 Col 29 human colon carcinoma cell line). (Courtesy of J. Cooper, ECACC.)

**Log phase.** This phase is the period of exponential increase in the cell number following the lag period and terminating one or two population doublings after confluence is reached. The length of the log phase depends on the seeding density, the growth rate of the cells, and the density that inhibits cell proliferation. In the log phase the growth fraction is high (usually 90–100%), and the culture is in its most reproducible form. It is the optimal time for sampling because the population is at its most uniform and the viability is high. However, the cells are randomly distributed in the cell cycle and, for some purposes, may need to be synchronized (*see* Section 27.4).

**Plateau phase.** Toward the end of the log phase, the culture becomes confluent—meaning all the available growth surface is occupied and all the cells are in contact with surrounding cells. After confluence the growth rate of the culture is reduced, and in some cases cell proliferation ceases almost completely after one or two further population doublings (*see* Section 17.5.2). At this stage the culture enters the plateau, or stationary, phase, and the growth fraction falls to between 0% and 10%. The cells may become less motile; some fibroblasts become oriented with respect to one another, forming a typical parallel array of cells. "Ruffling" of the plasma membrane is reduced, and the cell both occupies less surface area of substrate and presents less of its own surface to the medium. There may be a relative increase in the synthesis of specialized versus structural proteins, and the constitution and charge of the cell surface may be changed.

The cessation of motility, membrane ruffling, and growth following contact of cells at confluence was originally described by Abercrombie & Heaysman [1954] and was designated *contact inhibition*. It has since been realized that the reduction in the growth of normal cells after confluence is reached is not due solely to contact but may also involve reduced cell spreading [Stoker et al., 1968; Folkman & Moscona, 1978], buildup of inhibitors, and depletion of nutrients, particularly growth factors [Dulbecco & Elkington, 1973; Stoker, 1973; Westermark & Wasteson, 1975] in the medium [Holley et al., 1978]. This depletion can be quite local in a static monolayer, generating a diffusion boundary around the cells [Stoker, 1973] that can be overcome by irrigating the monolayer. The term *density limitation (of cell proliferation)* has been used to remove the implication that cell–cell contact is the major limiting factor [Stoker & Rubin, 1967], and the term *contact inhibition* is best reserved for those events resulting directly from cell contact (i.e., reduced cell motility and membrane ruffling, resulting in the formation of a strict monolayer and orientation of the cells with respect to each other).

Cultures of normal simple epithelial and endothelial cells stop growing after reaching confluence and remain as a monolayer. Most cultures, however, with regular replenishment of medium, will continue to proliferate (although at a reduced rate) well beyond confluence, resulting in multilayers of cells. Human embryonic lung and adult skin fibroblasts, which express contact inhibition of movement, will continue to proliferate, laying down layers of collagen between the cell layers until multilayers of six or more cells can be reached under optimal conditions [Kruse et al., 1970]. These fibroblasts still retain an ordered parallel array, however. Therefore the terms "plateau" and "stationary" are not strictly accurate and should be used with caution.

Cultures that have transformed spontaneously or have been transformed by virus or chemical carcinogens will usually reach a higher cell density in the plateau phase than their normal counterparts [Westermark, 1974] (Fig. 20.13). This higher cell density is accompanied by a higher growth fraction and the loss of density limitation of cell proliferation. The plateau phase for these cultures is an equilibrium between cell proliferation and cell loss. These cultures are often *anchorage independent* for growth—meaning they can easily be made to grow in suspension (*see* Section 17.5.1).

### 20.9.7 Derivatives from the Growth Curve

The construction of a growth curve from cell counts performed at intervals after subculture enables the measurement of a number of parameters that should be found to be characteristic of the cell line under a given set of culture conditions. The first of these parameters is the duration of

the lag period, or *lag time*, obtained by extrapolating a line drawn through the points for the exponential phase until it intersects the seeding concentration (*see* Fig. 20.8) and then reading off the elapsed time since seeding equivalent to that intercept. The second parameter is the *population doubling time* (PDT)—meaning the time taken for the culture to increase twofold in the middle of the exponential, or log, phase of growth. This parameter should not be confused with the generation time or cell cycle time (*see* Section 20.12), which are determined by measuring the transit of a population of cells from one point in the cell cycle until they return to the same point.

The last of the commonly derived measurements from the growth cycle are the *plateau level* and *saturation density*. The plateau level is the cell concentration (cells/mL of medium) in the plateau phase and is dependent on the cell type and the frequency with which the medium is replenished. The saturation density is the density of the cells (cells/cm$^2$ of growth surface) in the plateau phase. Saturation density and plateau level are difficult to measure accurately, as a steady state is not easily achieved in the plateau phase. Ideally the culture should be perfused, to avoid nutrient limitation or growth factor depletion, but a reasonable compromise is to grow the cells on a restricted area, say, a 15-mm diameter coverslip or filter well, in a 9-cm-diameter Petri dish with 20 mL of medium that is replaced daily (*see* Protocol 17.3). Under these conditions the limitation of growth by the medium is minimal, and the cell density exerts the major effect. A count of the cells under these conditions is a more accurate and reproducible measurement than a cell count in plateau under conventional culture conditions. Note that the term "plateau" does not imply the complete cessation of cell proliferation, but instead represents a steady state in which cell division is balanced by cell loss.

Although it is not appropriate to talk of "saturation density" in a suspension culture, nonadherent cells can still enter plateau because of exhaustion of the medium. Frequently, however, suspension cells at plateau phase will enter apoptosis quite quickly, show a marked fall in cell concentration, and may not generate a stable plateau. With normal cells, a steady state may be achievable by not replenishing the growth factors in the medium. In this case the cells are seeded and grown, and the plateau reached without changing the medium. Clearly, the conditions used to attain the plateau phase must be carefully defined.

### 20.10 PLATING EFFICIENCY

Colony formation at low cell density, or *plating efficiency*, is the preferred method for analyzing cell proliferation and survival (*see also* Protocol 21.3). This technique reveals differences in the growth rate within a population and distinguishes between alterations in the growth rate (colony size) and cell survival (colony number). It should be remembered,

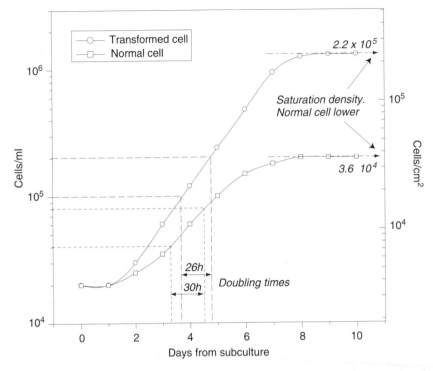

***Fig. 20.13. Saturation Density.*** Transformation produces an increase in the saturation density of transformed cells, relative to that found in the equivalent normal cells. This increase is often accompanied by a shorter PDT. (Data from normal human glia and glioma cell lines; Freshney et al., unpublished observations.)

however, that cells may grow differently as isolated colonies at low cell densities. In this situation fewer cells will survive, even under ideal conditions, and all cell interaction is lost until the colony starts to form. Heterogeneity in clonal growth rates reflects differences in the capacity for cell proliferation between lineages within a population, but these differences are not necessarily expressed in an interacting monolayer at higher densities when cell communication is possible.

When cells are plated out as a single-cell suspension at low cell densities (2–50 cells/cm²), they grow as discrete colonies (*see* Protocol 13.1; Plate 6). The number of these colonies can be used to express the plating efficiency:

$$\frac{\text{Number of colonies formed}}{\text{Number of cells seeded}} \times 100 = \text{Plating efficiency}$$

If it can be confirmed that each colony grew from a single cell, then this term becomes the *cloning efficiency*. Measurements of the plating efficiency are derived by counting the number of colonies over a certain size (usually around 50 cells) growing from a low inoculum of cells, and this term should not be used for the recovery of adherent cells after seeding at higher cell densities. Survival at higher densities is more properly referred to as the *seeding efficiency*:

$$\frac{\text{Number of cells attached}}{\text{Number of cells seeded}} \times 100 = \text{Seeding efficiency}$$

It should be measured at a time when the maximum number of cells has attached, but before mitosis starts. This time is a difficult point to define, as the window between maximum cell attachment and the initiation of mitosis may be quite narrow, and the events may even overlap. However, seeding time still provides a crude measurement of recovery in, for example, routine cell freezing or primary culture.

Protocol 20.10 for determination of plating efficiency can be adapted for training (*see* Section 28.4, Exercise 20), with, for example, varying concentrations of serum (*see* Plate 6e) or with and without a feeder layer (*see* Protocol 13.3).

---

### PROTOCOL 20.10. DETERMINATION OF PLATING EFFICIENCY

#### Outline
Seed the cells at low density, and incubate until colonies form (*see* Protocol 13.1); stain and count the colonies.

#### Materials

*Sterile:*

Culture of adherent cells
Growth medium............................................400 mL
Trypsin, 0.25%, crude....................................10 mL

Petri dishes, 6 cm............................................20
Tubes, or universal containers, for dilution......20

*Nonsterile:*

❏ Hemocytometer or electronic cell counter
❏ Fixative: anhydrous methanol.................100 mL
❏ D-PBSA...............................................200 mL
❏ Stain: Crystal Violet.............................100 mL
❏ Filter funnel and filter paper (to recycle the stain)

**Procedure**

1. Trypsinize the cells (*see* Protocol 12.3) to produce a single-cell suspension.
2. While the cells are trypsinizing,
   (a) Number the dishes on the side of the base.
   (b) Measure out medium for the dilution steps (Fig. 20.14). There should be more than enough medium for three replicates at each dilution.
3. When the cells round up and start to detach:
   (a) Disperse the monolayer in medium containing serum or a trypsin inhibitor.
   (b) Count the cells.
   (c) Dilute the cells to:
      (i) $2 \times 10^4$/mL for two 25-cm$^2$ flasks for routine maintenance.
      (ii) $2 \times 10^3$ cell/mL as top concentration for subsequent dilutions.
      (iii) Five further dilutions from (ii) to give 200, 100, 50, 20, and 10 cells/mL.
4. Seed the Petri dishes with 5 mL medium containing cells at each of the five concentrations in (iii). Seed two 6-cm Petri dishes at $2 \times 10^3$ cells/mL to act as controls in case the cloning is unsuccessful (to prove that there were cells present in the top dilution, at least).
5. Gas the flasks with 5% $CO_2$ and take to incubator.
6. Put the Petri dishes in a transparent plastic box and place in a humid $CO_2$ incubator, preferably one with limited access and reserved for cloning.
7. Incubate the dishes until colonies are visible to the naked eye (1–3 weeks).
8. Stain the colonies with Crystal Violet:
   (a) Remove the medium from the dishes.
   (b) Rinse the cells with D-PBSA, and discard the rinse.
   (c) Add 5 mL fresh D-PBSA and then add 5 mL methanol with gentle mixing (avoid colonies detaching).
   (d) Replace the 50:50 D-PBSA:methanol mixture with 5 mL fresh methanol, and fix the cells for 10 min.
   (e) Discard the methanol, and add Crystal Violet, neat, 2 to 3 mL per 6-cm dish, making

sure that the whole of the growth surface is covered.
   (f) Stain for 10 min.
   (g) Remove the stain, and return it to the stock bottle of stain via a filter.
   (h) Rinse the dishes with water and allow to dry.
9. Count the colonies in each dish, excluding those below 50 cells per colony. Magnifying viewers can make counting the colonies easier.

*Note.* It will be necessary to define a threshold above which colonies will be counted. If the majority of the colonies are between a hundred and a few thousands, then set the threshold at 50 cells per colony. In practice, this is a fairly natural threshold when counting by eye. However, if the colonies are very small (<100 cells), then set the threshold at 16 cells per colony. Below 16 cells, equivalent to 4 cell consecutive divisions, it would be hard to presume continued cell proliferation.

## 20.10.1   Analysis of Colony Formation

Calculate the plating efficiency (*see* calculation in the introduction to this section) at each seeding density. The plating efficiency should remain constant throughout the range of seeding densities (i.e., a plot of colony number against number of cells seeded should be linear). However, some cells may not plate well at very low densities, and the plating efficiency will fall. This can sometimes be minimized by using a feeder layer (*see* Section 13.2.3). If the plating efficiency falls at the higher concentrations, it implies that the cells are aggregating or colonies are coalescing. A seeding concentration that lies within the linear range of the plating efficiency curve (Fig. 20.15) should be selected for future assays.

The size distribution of the colonies may also be determined (e.g., to assay the growth-promoting ability of a test medium or serum; *see* Section 10.6.3) by counting the number of cells per colony by eye or estimating it by densitometry. To do so, after fixing and staining the colonies with Crystal Violet, measure absorption on a densitometer [McKeehan et al., 1977] or size the colonies on an automatic colony counter (*see* Section 20.10.2).

## 20.10.2   Automatic Colony Counting

If the colonies are uniform in shape and quite discrete, they may be counted on an *automatic colony counter* (*see* Appendix II: Colony Counters), which scans the plate with a CCD camera and analyzes the image to give an instantaneous readout of the number of colonies (Fig. 20.16). A size discriminator gives an analysis of the size of the colonies, based on the average colony diameter, but this is not always proportional to the cell number, as cells may pile up in the center of a colony.

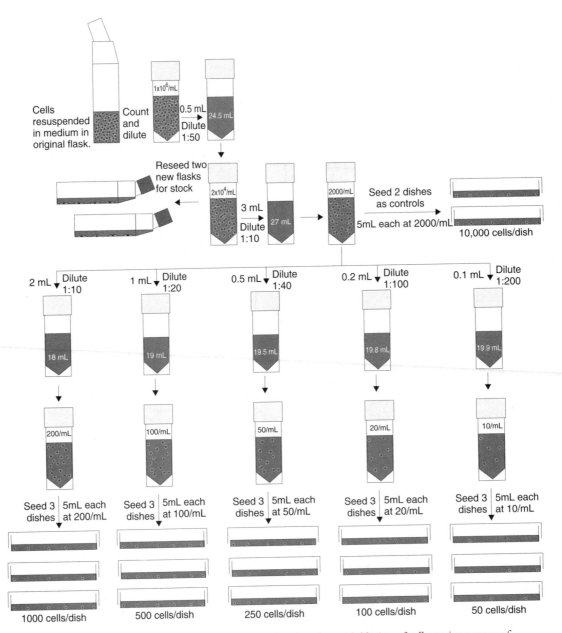

***Fig. 20.14. Diluting Cells for Cloning.*** Suggested regime for serial dilution of cells to give a range of seeding densities, suitable for cloning a cell line for the first time or establishing the linearity of plating efficiency versus seeding concentration (*see* Fig 20.15). Subsequently, when a suitable concentration has been selected, the cells may be diluted, with fewer steps, to the desired concentration.

Although expensive, these instruments can save a great deal of time and make colony counting more objective. However, they do not work well with colonies that overlap by more than 20% or have irregular outlines.

## 20.11 LABELING INDEX

Cells that are synthesizing DNA will incorporate [³H]-TdR (*see* Section 20.5.1). The percentage of labeled cells, determined by autoradiography (*see* Protocol 27.3), is known as the labeling index (LI) [e.g., *see* Westermark, 1974; Macieira-Coelho, 1973]. Measurement of the LI after a 30-min to 1-h labeling period with [³H]-TdR shows a large difference between exponentially growing cells (LI = 10–20%) and cells at the plateau phase (LI ∼ 1%). Normal Mv1Lu cells were shown to have a lower LI with [³H]-TdR than that of their neoplastic derivative transfected with mutant *ras* neoplastic cells [Khan et al., 1991] (*see* Fig. 20.17; Plate 14*d*). Because the LI is very low in the plateau phase, the duration of the [³H]-TdR labeling period may have to be increased to 24 h to show differences at saturation density.

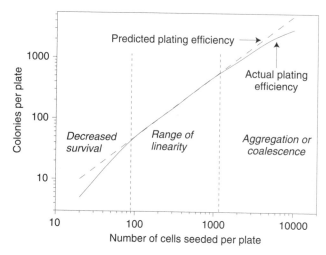

**Fig. 20.15. Linearity of Plating Efficiency.** If plating efficiency remains constant over a wide range of cell concentrations, the curve is linear (dashed line), whereas if there is poor survival at low densities or aggregation or coalescence at high densities, plating efficiency decreases (solid line).

---

## PROTOCOL 20.11. LABELING INDEX WITH [³H]THYMIDINE

### Outline

Grow cells to an appropriate density. Label the cells with [³H]-TdR for 30 min. Wash and fix the cells. Remove any unincorporated precursor from the cells and prepare autoradiographs.

### Materials

*Sterile or aseptically prepared:*

- ❏ Cell culture for assay
- ❏ Growth medium
- ❏ Trypsin, 0.25%, crude
- ❏ D-PBSA
- ❏ [³H]-TdR, 2.0 MBq/mL (∼50 µCi/mL), 75 GBq/mmol (∼2 Ci/mmol)

△ *Safety Note.* Handle [³H]-TdR with care, as it is radioactive and gentotoxic. Follow local guidelines for its use (*see* Section 6.7).

- ❏ Multiwell plate(s) containing 13-mm Thermanox coverslips (Nunc)

*Nonsterile:*

- ❏ Hemocytometer or electronic cell counter
- ❏ D-PBSA
- ❏ Acetic methanol (1 part glacial acetic acid to 3 parts methanol), ice cold, freshly prepared
- ❏ Microscope slides
- ❏ Mountant (e.g., DPX or Permount)

- ❏ Trichloroacetic acid (TCA), 0.6 M, ice cold
- ❏ Deionized water
- ❏ Methanol

### Procedure

1. Set up the cultures at $2 \times 10^4$ cells/mL to $5 \times 10^4$ cells/mL in 24-well plates containing coverslips.
2. Allow the cells to attach, start to proliferate (48–72 h), and grow to the desired cell density.
3. Add [³H]-TdR to the medium, 100 KBq/mL (∼5 µCi/mL), and incubate the cultures for 30 min.

*Note.* Some media—such as Ham's F10 and F12—contain thymidine (*see* Tables 8.3, 9.1, 10.2). In these cases the concentration of radioactive thymidine must be increased to give the same specific activity in the medium.

4. Remove the labeled medium, and discard it into a designated container for radioactive waste.
5. Wash the coverslips three times with D-PBSA. Lift the coverslips off the bottom of the wells (but not right out of them) at each wash to allow removal of the isotope from underneath.
6. Add 1:1 D-PBSA:acetic methanol, 1 mL per well, and then remove it immediately.
7. Add 1 mL of acetic methanol at 4°C to each well, and leave the cultures for 10 min.
8. Remove the coverslips, and dry them with a fan.
9. Mount the coverslip on a microscope slide with the cells uppermost.
10. Leave the mountant to dry overnight.
11. Place the slides in 0.6 M TCA at 4°C in a staining dish, and leave them for 10 min. Replace the TCA twice during this extraction, thereby removing unincorporated precursors.
12. Rinse the slides in deionized water, then in methanol, and dry the slides.
13. Prepare an autoradiograph (*see* Protocol 27.3).
14. When the autoradiograph has been exposed for the appropriate period (usually 1–2 weeks), develop, stain (*see* Protocol 27.3), and examine under microscope with 40× objective.
15. Count the percentage of labeled cells. To cover a representative area, follow the scanning pattern illustrated in Fig. 20.18.

### 20.11.1  Growth Fraction

If cells are labeled with [³H]-TdR for varying lengths of time up to 48 h, the plot of the LI against time increases rapidly over the first few hours and then flattens out to a very

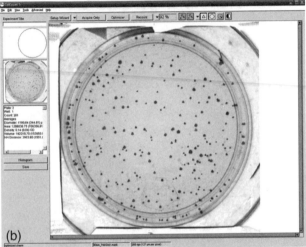

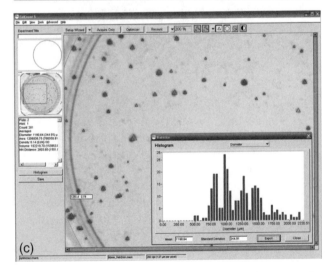

**Fig. 20.16. Automatic Colony Counter.** Automatic counter for colonies in plating efficiency and survival assays. (*a*) Counter with sample tray open and range of sample types; (*b*) screen image with colonies flagged; (*c*) histogram inset of size distribution histogram of selected colonies. (Courtesy of Oxford Optronix.)

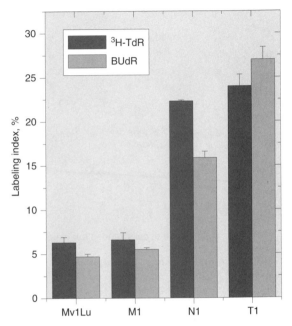

**Fig. 20.17. Labeling Index.** Mink lung cells, Mv1Lu, and oncogene-transfected derivatives were labeled with [³H]-TdR for 1 h, fixed, and coated with autoradiographic emulsion (*see* Protocol 27.3), stained with Giemsa (*see* Plate 14*d*), or labeled for 1 h with bromodeoxyuridine (BUdR), fixed, and stained by immunoperoxidase with an antibody directed against BUdR bound to DNA. Labeled nuclei were counted as percentage of the total in each case. Mv1Lu is the control cell line, M1 is Mv1Lu transfected with the *myc* oncogene, N1 is Mv1Lu cells transfected with normal human *ras*, and T1 with mutant human *ras*. T1 cells were tumorigenic and had a statistically significant ($p < 0.001$) increase in labeling index compared to Mv1Lu, also seen with bromodeoxyuridine labeling. (Tabular data from Khan et al., 1991.)

low gradient, almost a plateau (Fig. 20.19). The level of this plateau, read against the vertical axis, is the growth fraction of the culture—meaning the proportion of the cells in cycle at the time of labeling.

---

**PROTOCOL 20.12. DETERMINATION OF GROWTH FRACTION**

***Outline***
Label the culture continuously for 48 h, sampling at intervals for autoradiography.

***Procedure***
Protocol 20.10, except that at step 3, incubation should be carried out for 15 min, 30 min, and 1, 2, 4, 8, 24, and 48 h.

---

***Analysis.*** Count the number of labeled cells as a percentage of the total number of cells, using the scanning

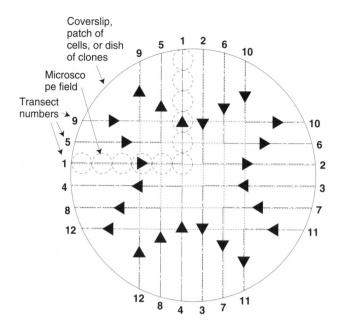

*Fig. 20.18. Scanning Slides or Dishes.* Scanning pattern for the analysis of cytological preparations on slides or dishes. Each dotted circle represents one microscope field, and the large circle represents the extent of the specimen (e.g., a coverslip, culture dish, well, or a spot of cells on a slide). Guide lines can be drawn with a nylon-tipped pen with a light, transparent ink.

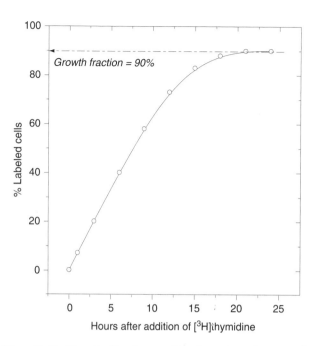

*Fig. 20.19. Growth Fraction.* To determine the growth fraction, cells are labeled continuously with [$^3$H]thymidine, and the percentage of labeled cells is determined at intervals by autoradiography (*see* Section 20.11.1).

pattern from Protocol 20.10. Plot the LI against time (*see* Fig. 20.18).

*Note.* Autoradiographs with $^3$H can be prepared only when the cells remain as a monolayer. If they form a multilayer, then they must be trypsinized after labeling, and slides must be prepared by the drop technique (*see* Protocol 15.7, without the hypotonic step) or by cytocentrifugation (*see* Protocol 15.4), as the energy of β-emission from $^3$H is too low to penetrate an overlying layer of cells.

The LI can also be determined by labeling cells with BUdR, which becomes incorporated into DNA. This effect can be detected subsequently by immunostaining with an anti-BUdR antibody (Dako). Results from this method are generally in agreement with the results of using [$^3$H]-TdR [Khan et al., 1991] (*see also* Fig. 20.17).

### 20.11.2 Mitotic Index

The mitotic index is the fraction or percentage of cells in mitosis and is determined by counting mitoses in stained cultures as a proportion of the whole population of cells. The scanning pattern should be as for counting labeled nuclei (*see* Fig. 20.18).

### 20.11.3 Division Index

A number of antibodies, such as Ki67 [Zhu & Joyce, 2004] or anti-PCNA [Katdare et al., 2004], are able to stain cells in the division cycle. These antibodies are raised against proteins expressed during the cell cycle but not

expressed in resting cells. Some of the antibodies are directed against DNA polymerase or cyclins. The cells are stained by immunofluorescence or immunoperoxidase (*see* Protocol 15.11), and the proportion of stained cells is determined cytologically (*see* Fig. 20.18) or by flow cytometry (*see* Section 15.11.2). This figure gives a higher index than that of either mitotic counting or DNA labeling, as cells stain throughout the cell cycle. It also gives a particularly useful indication of the growth fraction.

## 20.12 CELL CYCLE TIME

To determine the length of the cell cycle (generation time) and its constituent phases, cells are labeled continuously with BUdR, and the incorporation of the label is detected at intervals after labeling by immunofluorescence microscopy or flow cytometry with an antibody to DNA-bound BUdR. For flow cytometry, the cells are also stained with propidium iodide and analyzed for BUdR incorporation versus DNA content, to follow the progression of cells around the cycle [Dolbeare & Selden, 1994; Poot et al., 1994].

## 20.13 CELL MIGRATION

Cells in culture, particularly fibroblasts, are motile and can migrate significant distances across the substrate, dependent on cell density and the presence of stimulants such as growth factors. Motility is evidenced by ruffling of the cell membrane as visualized by time-lapse video. The amount of motility is difficult to quantify, but the quantification of cell migration can be achieved by detailed analysis of time-lapse video sequences (*see* Protocol 27.4) or by image analysis of tracks made by the cell's phagocytosis in dishes coated with colloidal gold [Kawa et al., 1997]. Migration can also be assayed by the movement of cells through a porous membrane (Cell Biolabs), as in chemotaxis assays in a Boyden chamber [Schor, 1994] or invasion assays in a filter well (*see* Section 17.6.3).

# CHAPTER 21

# Cytotoxicity

## 21.1 VIABILITY, TOXICITY, AND SURVIVAL

Once a cell is explanted from its normal in vivo environment, the question of viability, particularly in the course of experimental manipulations, becomes fundamental. Previous chapters have dealt with the status of the cultured material relative to the tissue of origin and how to quantify changes in growth and phenotypic expression. However, none of these data is acceptable unless the great majority of the cells are shown to be viable. Furthermore many experiments carried out in vitro are for the sole purpose of determining the potential cytotoxicity of compounds being studied, either because the compounds are being used as pharmaceuticals or cosmetics and must be shown to be nontoxic or because they are designed as anticancer agents and cytotoxicity may be crucial to their action.

New drugs, cosmetics, food additives, and so on, go through extensive cytotoxicity testing before they are released for use by the public. This testing usually involves a large number of animal experiments, although in Europe these experiments will be subject to new legislation [Cox & Chrisochoidis, 2003], introduced in 2009 for topical application and in 2013 for systemic application. There is much pressure, both humane and economic, to perform at least part of cytotoxicity testing in vitro. The introduction of specialized cell lines and interactive organotypic cultures, and the continued use of long-established cultures, may make this a reasonable proposition.

Toxicity is a complex event in vivo, where there may be direct cellular damage, as with a cytotoxic anticancer drug, physiological effects, such as membrane transport in the kidney or neurotoxicity in the brain, inflammatory effects, both at the site of application and at other sites, and other systemic effects. Currently it is difficult to monitor systemic and physiological effects in vitro, so most assays determine effects at the cellular level. Definitions of cytotoxicity vary [Kroemer et al., 2009], depending on the nature of the study and whether cells are killed or phenotypically altered. In addition cells may die by necrosis (*see* Fig. 12.1), apoptosis (*see* Plate 17*c*, *d*), self-digestion (*autophagy*); may cease proliferation (*cytostasis*); and/or may become terminally differentiated (e.g., cornification) [Galluzi et al., 2009]. Whereas demonstrating efficacy in an anticancer agent assay may require a *cytocidal* effect (cell killing), demonstrating the lack of toxicity of other pharmaceuticals may require a more subtle analysis of specific targets such as an alteration in gene transcription, cell signaling, or cell–cell interaction including those effects that may give rise to an inflammatory or allergic response.

Most assays oversimplify the events that they measure and are employed because they are cheap, easily quantified, and reproducible. However, it has become increasingly apparent that they are inadequate for modern drug development, which requires greater emphasis on specific molecular targets and precise metabolic regulation. Gross tests of cytotoxicity are still required, but there is a growing need to supplement them with more subtle tests of metabolic pathway regulation and signaling. Perhaps the most obvious of these tests is the induction of an inflammatory or allergic response, which need not imply cytotoxicity of the allergen and is still one of the hardest results to demonstrate in vitro.

The traditional approach to cytotoxicity has been to concentrate on cell growth or survival. Cell growth is generally taken to be the regenerative potential of cells, as measured by clonal growth (*see* Protocol 20.10), net change in population size (e.g., in a growth curve; *see* Protocols 20.7, 21.8), or a change in cell mass (total protein or DNA) or metabolic activity (e.g., DNA, RNA, or protein synthesis; MTT reduction). Other aspects will be considered later (*see* Section 21.6).

## 21.2 IN VITRO LIMITATIONS

It is important that any in vitro measurement can be interpreted in terms of the in vivo response of the same or similar cells, or at least that the differences that exist between in vitro and in vivo measurements are clearly understood.

### 21.2.1 Pharmacokinetics

The measurement of toxicity in vitro is generally a cellular event. For example, it would be very difficult to recreate the complex pharmacokinetics of drug exposure in vitro, and between in vitro and in vivo experiments there usually are significant differences in exposure time to and concentration of the drug, rate of change of the concentration, drug metabolism (activation and detoxification), tissue penetration, clearance, and excretion. Although it may be possible to simulate these parameters—for example, using multicellular tumor spheroids for drug penetration or timed perfusion to simulate concentration and time ($C \times T$) effects—most studies concentrate on a direct cellular response, thereby gaining simplicity and reproducibility.

### 21.2.2 Metabolism

Many nontoxic substances become toxic after being metabolized by the liver; in addition many substances that are toxic in vitro may be detoxified by liver enzymes. For in vitro testing to be accepted as an alternative to animal testing, it must be demonstrated that potential toxins reach the cells in vitro in the same form as they would in vivo. This proof may require additional processing by purified liver microsomal enzyme preparations [McGregor et al., 1988], coculture with activated hepatocytes [Guillouzo & Guguen-Guillouzo, 2008] (*see also* Appendix II: Hepatocytes), or hepatoma-derived cells such as Hep-G2 or HepaRG (*see* Section 22.2.8). Coculture may use transfilter 3-D culture or 2-D cellular microarrays [Khetani & Bhatia, 2008]. Genetic modification of the target cells with the introduction of genes for metabolizing enzymes under the control of a regulatable promoter [Macé et al., 1994] has also been used.

### 21.2.3 Tissue and Systemic Responses

The nature of the response must also be considered carefully. A toxic response in vitro may be measured by changes in cell survival (*see* Protocol 21.3) or metabolism (*see* Section 21.3.4), whereas the major problem in vivo may be a tissue response (e.g., an inflammatory reaction, fibrosis, kidney failure) or a systemic response (e.g., pyrexia, vascular dilatation). For in vitro testing to be more effective, models of these responses must be constructed, perhaps utilizing organotypic cultures reassembled from several different cell types and maintained in the appropriate hormonal milieu.

It should not be assumed that complex tissue and even systemic reactions cannot be simulated in vitro. Assays for inflammatory responses, teratogenic disorders, and neurological dysfunctions may be feasible in vitro, given the right tissue-engineered models and a proper understanding of cell–cell interaction and the interplay of endocrine hormones with local paracrine and autocrine factors.

## 21.3 NATURE OF THE ASSAY

The choice of assay will depend on the agent under study, the nature of the anticipated response, and the particular target cell. In vitro assays can be divided into five major classes:

(1) *Viability*. An immediate or short-term response, such as increased and uncontrolled membrane permeability or a perturbation of a particular metabolic pathway correlated with cell proliferation or survival.
(2) *Survival*. The long-term retention of self-renewal capacity (5–10 generations or more).
(3) *Metabolic*. Assays, usually microtitration based, of intermediate duration that can either measure a metabolic response (e.g., dehydrogenase activity; DNA, RNA, or protein synthesis) at the time of, or shortly after, exposure. Making the measurement two or three population doublings after exposure is more likely to reflect cell growth potential and may correlate with survival.
(4) *Genotoxicity and Transformation*. Survival in an altered state (e.g., one or more genetic mutations with resultant alterations in growth control or malignant transformation).
(5) *Irritancy*. A response analogous to inflammation, allergy, or irritation in vivo; as yet difficult to model in vitro, but may be possible to assay by monitoring cytokine release in organotypic cultures.

### 21.3.1 Viability

Viability assays are used primarily to measure the proportion of viable cells after a potentially traumatic procedure, such as primary disaggregation, cell separation, or cryopreservation, rather than to look at a long-term cytotoxic response.

Most viability tests rely on a breakdown in membrane integrity measured by the uptake of a dye to which the cell is normally impermeable, such as Trypan Blue, Erythrosin, Naphthalene Black (*see* Plate 17*a*), or propidium iodide (*see* Fig. 20.7), or the release of a dye normally taken up and retained by viable cells (e.g., diacetyl fluorescein or Neutral Red), or the release of lactate dehydrogenase by leaky cells

[Kendig & Tarloff 2006]. This effect presents immediately and does not always predict ultimate survival as dye exclusion tends to overestimate viability—for example, 90% of cells thawed from liquid nitrogen may exclude Trypan Blue, but only 60% prove to be capable of attachment 24 h later.

Note that routine assessment of viability at subculture can be uninformative regarding trypsinized adherent cells as most of the nonviable cells will be lost in the discarded medium and prewash before trypsinization. An accurate assessment of the viability status at subculture requires that all the cells be recovered from the medium and prewash and combined with the trypsinate. However, the viability of *reseeded* cells will be accurately determined without this recovery.

Protocol 21.1 can be adapted for training (*see* Sections 28.3, Exercises 11, 28.4, 18B, 19; *see also* Section 28.4).

---

## PROTOCOL 21.1. ESTIMATION OF VIABILITY BY DYE EXCLUSION

### Principle
Viable cells are impermeable to Naphthalene Black, Trypan Blue [Kaltenbach et al., 1958], propidium iodide [Darzynkiewicz & Gong, 1994], and a number of other dyes.

### Outline
Mix a cell suspension with stain, and examine it by low-power microscopy.

### Materials

*Sterile or aseptically prepared:*

❑ Cells for testing (e.g., flask for trypsinization, frozen vial to thaw, or primary disaggregate)
❑ Growth medium appropriate to cell type
❑ Trypsin, crude, 0.25%
❑ D-PBSA

*Nonsterile:*

❑ Hemocytometer
❑ Viability stain (e.g., 0.4% Trypan Blue or 1% Naphthalene Black in D-PBSA or HBSS)
❑ Pasteur pipettes
❑ Microscope
❑ Tally counter

### Procedure

1. Prepare a cell suspension at a high concentration (~$1 \times 10^6$ cells/mL) by trypsinization or by centrifugation and resuspension.
2. Take a clean hemocytometer slide and fix the coverslip in place (*see* Protocol 20.1; Fig. 20.1).
3. Mix one drop of cell suspension with one drop (Trypan Blue) or four drops (Naphthalene Black) of stain.

4. Load the counting chamber of the hemocytometer (*see* Protocol 20.1).
5. Leave the slide for 1 to 2 min before starting to count (do not leave any longer, or viable cells will deteriorate and take up the stain).
6. Place the slide on the microscope, and use a 10× objective to look at the counting grid (*see* Fig. 20.1; Plate 17a).
7. Count the total number of cells and the number of stained cells.
8. Wash the hemocytometer, and return it to its box.

---

***Analysis.*** Calculate the percentage of unstained cells. This figure is the percentage viability by this criterion. If the respective volumes of cell suspension and stain are measured accurately at step 3, then this method of viability determination can be incorporated into Protocol 20.1.

---

## PROTOCOL 21.2. ESTIMATION OF VIABILITY BY DYE UPTAKE

### Principle
Viable cells take up diacetyl fluorescein and hydrolyze it to fluorescein, to which the cell membrane of live cells is impermeable [Rotman & Papermaster, 1966]. Live cells fluoresce green; dead cells do not. Nonviable cells may be stained with propidium iodide and subsequently fluoresce red [Darzynkiewicz & Gong, 1994]. Viability is expressed as the percentage of cells fluorescing green. This method may be applied to CCD analysis or flow cytometry (*see* Section 20.7.2; Fig. 20.7).

### Outline
Stain a cell suspension in a mixture of propidium iodide and diacetyl fluorescein, and examine the cells by fluorescence microscopy or flow cytometry.

### Materials

*Sterile or aseptically prepared:*

❑ Single-cell suspension
❑ Fluorescein diacetate, 10 μg/mL, in HBSS without phenol red
❑ Propidium iodide, 500 μg/mL, in HBSS without phenol red

*Nonsterile:*

❑ Fluorescence microscope
❑ Filters:
  Fluorescein: excitation 450/590 nm, emission LP 515 nm

Propidium iodide: excitation 488 nm, emission 615 nm

**Procedure**

1. Prepare the cell suspension as for dye exclusion (*see* Protocol 21.1), but in medium without phenol red.
2. Add the fluorescent dye mixture at a proportion of 1 : 10 to give a final concentration of 1 μg/mL of diacetyl fluorescein and 50 μg/mL of propidium iodide.
3. Incubate the cells at 37°C for 10 min.
4. Place a drop of the cells on a microscope slide, add a coverslip, and examine the cells by fluorescence microscopy with excitation at 488 or 536 nm and detection at 562–588 or 617 nm respectively for propidium iodide and excitation at 494 nm and emission at 521 for fluorescein.

*Analysis.* Cells that fluoresce green are viable, whereas those that fluoresce red are nonviable. Viability may be expressed as the percentage of the total number of cells that fluoresce green. The stained cell suspension can also be analyzed by flow cytometry (see Section 20.7.2; Fig. 20.7).

*Neutral red uptake.* Living cells take up neutral red, 40 μg/mL in culture medium, and sequester it in the lysosomes. However, neutral red is not retained by nonviable cells. Uptake of neutral red is quantified by fixing the cells in formaldehyde and solubilizing the stain in acetic ethanol, and measuring absorbance on an ELISA plate reader at 570 nm [Borenfreund et al., 1990; Babich & Borenfreund, 1990]. Neutral red tends to precipitate, so the medium with stain is usually incubated overnight and centrifuged before use. This assay does not measure the total number of cells, but it does show a reduction in the absorbance related to loss of viable cells and is readily automated.

## 21.3.2 Survival

Although short-term tests are convenient and usually quick and easy to perform, they reveal only cells that are dead (i.e., permeable) at the time of the assay. Frequently, however, cells that have been subjected to toxic influences (e.g., irradiation, environmental toxins, and antineoplastic drugs) show an effect several hours, or even days, later. The nature of the tests required to measure viability in these cases is necessarily different because, by the time the measurement is made, the dead cells may have disappeared or some resistant cells may have recovered. Therefore long-term tests are used to demonstrate survival rather than short-term toxicity. Survival implies the retention of regenerative capacity and is usually measured by plating efficiency (*see* Protocol 20.10). Plating efficiency measures survival by demonstrating proliferative capacity for several cell generations, provided that the cells plate with a

high-enough efficiency that the colonies can be considered representative of the entire cell population. Although not ideal, a plating efficiency of over 10% is usually acceptable.

---

### PROTOCOL 21.3. CLONOGENIC ASSAY FOR ATTACHED CELLS

#### Outline
Treat the cells for 24 h with the experimental agent at a range of concentrations. Trypsinize the cells, seed them at a low cell density, and incubate them for 1 to 3 weeks. Stain the cells (*see* Plate 6a, e), and count the number of colonies (*see* Fig. 21.1).

#### Materials

*Sterile:*

- ❑ Growth medium
- ❑ D-PBSA
- ❑ Trypsin, crude, 0.25%
- ❑ Compound to be tested at 10× the maximum concentration to be used, dissolved in serum-free medium (check the pH and the osmolality of the test solution, and adjust if necessary)
- ❑ Flasks, 25 cm²
- ❑ Petri dishes, 6 or 9 cm, labeled on the side of the base

*Nonsterile:*

- ❑ D-PBSA
- ❑ Methanol
- ❑ Crystal Violet, 1%
- ❑ Hemocytometer or electronic cell counter

#### Procedure

1. Prepare a series of cultures in 25-cm² flasks, three for each of six agent concentrations, and three controls. Seed the cells at $5 \times 10^4$ cells/mL in 4.5 mL of growth medium, and incubate them for 48 h, by which time the cultures will have progressed into the log phase (*see* Section 20.9.2).
2. Prepare a serial dilution of the compound to give 2 mL at 10× of each of the final concentrations required:
   (a) If you are testing a compound for the first time, use 3- to 5-fold serial dilutions over a range of 3 to 5 logs.
   (b) If you can predict the approximate toxic concentration, then select a narrower, arithmetic range over one or two decades.
3. Add 0.5 mL of 10× concentrate to each of three flasks for each concentration, starting with control medium (no compound added but solvent present in amount used for toxin) and progressing from lowest to highest concentration.

4. Return the flasks to the incubator.
5. If the compound is slow acting or partially reversible, repeat step 3 twice; that is, expose the cultures to the agent for 3 days, replacing the medium and compound daily by changing the medium. With fast-acting compounds, 1-h exposure may be sufficient.
6. Remove the medium from each group of three flasks in turn (working from control then lowest concentration of compound to highest), trypsinize the cells, and count the cells in the controls.
7. Dilute and seed the cells into Petri dishes at the required density for clonal growth (*see* Protocols 13.1, 20.10), diluting all of the cultures by the same amount as the control. Work from the flasks exposed to the highest concentration of toxin down to the lowest and then the control.

8. Incubate the cultures until colonies form (usually 1–3 weeks).
9. Fix the cultures in absolute methanol, and stain them for 10 min in 1% Crystal Violet (*see* Protocol 15.3).
10. Wash the dishes in tap water, drain, and stand in an inverted position to dry.
11. Count the colonies with >50 cells (>5 doublings).

### Analysis of survival curve

(1) Calculate the plating efficiency at each drug concentration.
(2) Calculate the relative plating efficiency, which is the plating efficiency at each concentration as a fraction of the control—the *surviving fraction*.

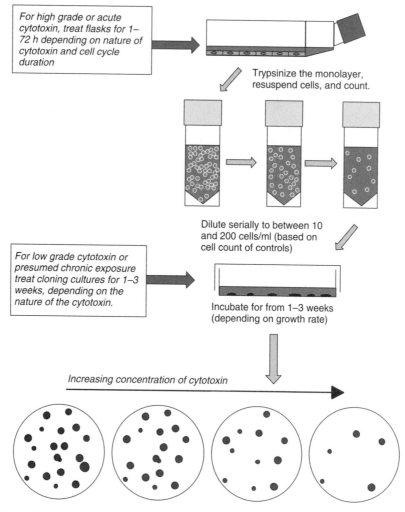

For high grade or acute cytotoxin, treat flasks for 1–72 h depending on nature of cytotoxin and cell cycle duration

Trypsinize the monolayer, resuspend cells, and count.

Dilute serially to between 10 and 200 cells/ml (based on cell count of controls)

For low grade cytotoxin or presumed chronic exposure treat cloning cultures for 1–3 weeks, depending on the nature of the cytotoxin.

Incubate for from 1–3 weeks (depending on growth rate)

*Increasing concentration of cytotoxin*

**Fig. 21.1. Clonogenic Assay for Adherent Cells.** Cells are trypsinized, counted, and diluted as for monolayer dilution cloning (*see* Protocol 13.1). The test substance can be added before trypsinization or after seeding for cloning (*see* Section 21.3.2). The colonies are fixed and stained when they reach a reasonable size for counting by eye but before they overlap.

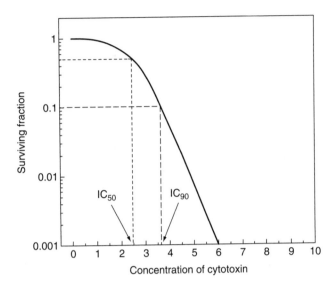

**Fig. 21.2. Survival Curve.** Semilog plot of the surviving fraction of cells (ratio of colonies forming from test cells to colonies forming from control cells) against the concentration of cytotoxin. Typically the curve has a "knee," and the $IC_{90}$ lies in the linear range of the curve. The $IC_{50}$, falling on the knee, is a less stable value.

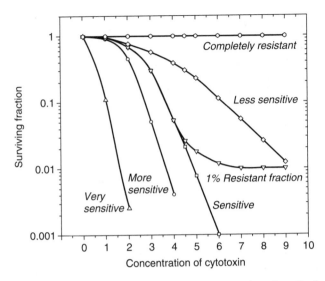

**Fig. 21.3. Interpretation of Survival Curves.** Semilog plot of cell survival against the concentration of cytotoxin. The slope increases with increasing sensitivity and decreases with reduced sensitivity until it becomes totally flat for complete resistance. Partial resistance as a resistant fraction is shown by the curve flattening out at the lower end.

(3) Plot the surviving fraction on a log scale against the concentration on a linear or log scale, depending on the concentration range used (*see* Fig. 21.2).

(4) Determine the $IC_{50}$ or $IC_{90}$, which is the concentration of compound promoting 50% or 90% inhibition of colony formation, respectively. As this is a semilog plot, the $IC_{90}$ is more appropriate, as it is more likely to fall on the linear part of the curve, whereas the $IC_{50}$ tends to fall on the knee of the curve, giving a less stable value.

(5) Analyze the curve for differences in sensitivity:

(a) *Slope of the curve and length of the knee.* A shallower slope and/or longer knee means reduced sensitivity; a steeper slope and/or shorter knee means increased sensitivity. Both the length of knee and the slope influence the $IC_{50}$ and the $IC_{90}$, although a more significant difference can be observed in the $IC_{90}$ (*see* Fig. 21.3).

(b) *Resistant fraction.* The fraction of resistant cells is indicated by a flattening of the lower end of the curve.

(c) *Total resistance* is indicated by the lack of any gradient on the curve.

(d) *Area under the curve:* Complex survival curves may be compared by calculating the area under the curve, but this is done for expediency and is not mathematically valid.

### Variable parameters in survival assay

**Concentration of agent.** A wide range of concentrations in log increments (e.g., 1 μM–1 mM, and control) should be used for the first attempt and a narrower range (log or linear),

based on the results from the first range, for subsequent attempts.

**Invariant agent concentrations.** Some conditions that are tested cannot easily be varied—such as the quality of medium, water, or an insoluble plastic. In these cases the serum concentrations can be varied. As serum may have a masking effect on low-level toxicity, an effect may only be seen in limiting serum (*see* Effect of Medium Constituents, below, *and* Fig. 21.4b).

**Duration of exposure to agent.** Some agents act rapidly, whereas others act more slowly. Exposure to ionizing radiation, for example, need last only a matter of minutes to achieve the required dose, whereas testing some cycle-dependent antimetabolic drugs may take several days to achieve a measurable effect. Duration of exposure ($T$) and drug concentration ($C$) are related, although $C \times T$ is not always a constant. Prolonging exposure, usually by replacing the drug daily, can increase sensitivity beyond that predicted by $C \times T$ because of cell cycle effects and cumulative damage.

**Time of exposure to agent.** When the agent is soluble and expected to be toxic, the procedure in Protocol 21.3 should be followed, but when the quality of the agent is unknown, stimulation is expected, or only a minor effect is expected (e.g., 20% inhibition rather than several-fold), the agent may be incorporated during clonal growth rather than at preincubation. Confirmation of anticipated toxicity—such as for a cytotoxic drug—requires a conservative assay with a minimal drug exposure, as compared to that in vivo, applied during culture before trypsinization for cloning.

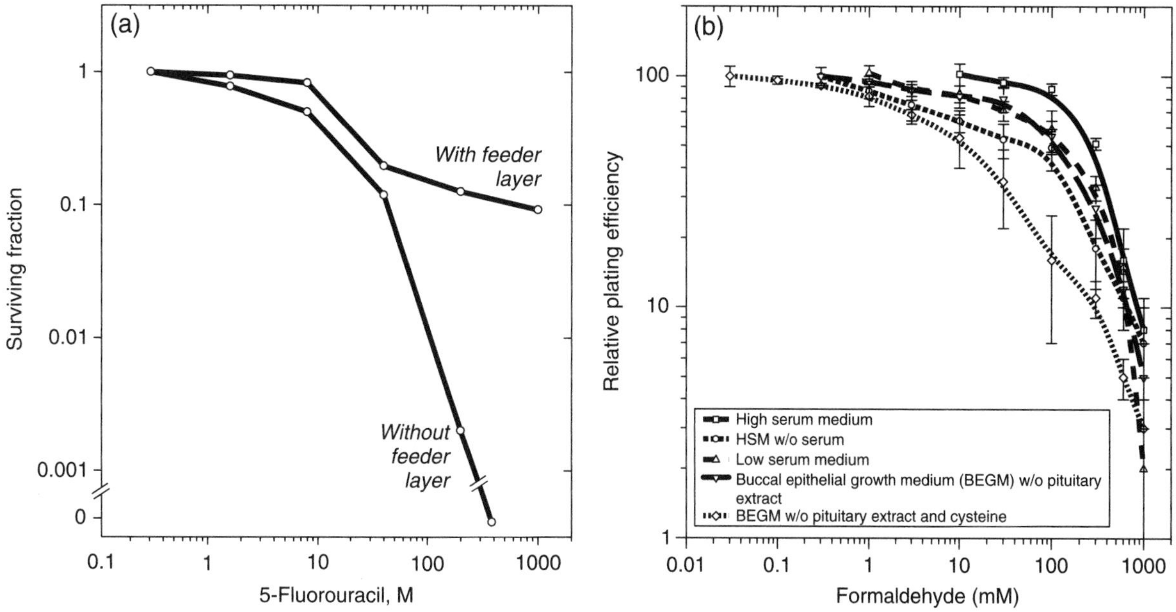

**Fig. 21.4. *Effect of Culture Conditions on Survival.*** (*a*) Cell density. Human glioma cells were plated out in the presence (dashed line) and absence (solid line) of a feeder layer after treatment with various concentrations of 5-fluorouracil. A 10% resistant fraction is apparent at $1 \times 10^{-4}$ M drug only in the presence of a feeder layer. In the absence of the feeder layer, the small number of colonies making up the resistant fraction were unable to survive alone. (*b*) Medium constituents. High serum medium (HSM; solid line and squares) compared with HSM without serum (dashed line and circles) shows about a 5-fold increase in the $IC_{50}$ with serum; removing cysteine decreases the $IC_{50}$ 10-fold (data from [Nilsson et al., 1998]).

Confirmation of the lack of toxicity—such as for tap water or a nontoxic pharmaceutical—requires a more stringent assay, with prolonged exposure added at seeding for clonal growth and maintained during clonal growth. The toxin may need to be replaced weekly; daily replacement would alone impair cloning efficiency.

***Cell density during exposure.*** The density of the cells during exposure to an agent can alter the response of the cells and the agent; for example, HeLa cells are less sensitive to the alkylating agent mustine at high cell densities [Freshney et al., 1975].

***Cell density during cloning.*** The number of colonies may fall at high concentrations of a toxic agent, but it is possible to compensate for this effect by seeding more cells so that approximately the same number of colonies form at each concentration. This procedure removes the risk of a low clonal density influencing survival and improves statistical reliability, but it is prone to the error that cells from higher drug concentrations are plated at a higher cell concentration, a factor that may also influence survival. It is preferable to plate cells on a preformed feeder layer, whose density $(5 \times 10^3$ cells/cm$^2$) greatly exceeds that of the cloning cells. This step ensures that the cell density is uniform regardless of clonal survival, which contributes little to the total cell density. Note that cloning on a feeder layer can sometimes

reveal a resistant fraction of cells that is not apparent without the feeder layer (Fig. 21.4*a*).

***Effect of medium constituents.*** The composition of the culture medium will also affect the way that cells respond to a toxin, partly because different media will have different effects on proliferation but also because individual constituents may affect the stability, binding, and metabolism of the toxin. Serum and cysteine, for example, reduce the toxicity of formaldehyde on buccal epithelium 5- and 10-fold respectively (Fig. 21.4*b*). Serum proteins may bind toxins, quite apart from their effect on cell growth and cysteine and other sulphydryls bind and detoxify reactive chemicals intracellularly and in the medium [Nilsson et al., 1998]. As cysteine auto-oxidizes in the medium, the age of the medium will add another variable [Grafström, *personal communication*].

***Colony size.*** Some agents are cytostatic (i.e., they inhibit cell proliferation) but not cytotoxic, and during continuous exposure they may reduce the size of colonies without reducing the number. In this case the size of the colonies should be determined by densitometry [McKeehan et al., 1977], automatic colony counting, or visually counting the number of cells per colony. For *colony counting*, the threshold number of cells per colony (e.g., 50 as in Protocol 21.9) is purely arbitrary, and it is assumed that most of the colonies are greatly in excess of this number. Colonies should be grown until they are quite large ($>1 \times 10^3$ cells), when the growth of larger colonies tends to slow down; smaller, but still viable, colonies tend to catch up with these larger colonies. For *colony sizing*, stain the cultures earlier, before the growth rate of larger colonies has slowed down, and score all of the colonies.

*Solvents.* Some agents to be tested have low solubilities in aqueous media, and it may be necessary to use an organic solvent to dissolve them. Ethanol, propylene glycol, and dimethyl sulfoxide have been used for this purpose, but may themselves be toxic to cells. Hence the minimum concentration of solvent should be used to obtain a solution. The agent may be made up at a high concentration in, for example, 100% ethanol, then diluted gradually with BSS and finally diluted into medium. The final concentration of solvent should be <0.5% and a *solvent control* must be included (i.e., a control with the same final concentration of solvent but without the agent being tested).

Take care when using organic solvents with plastics or rubber. It is better to use glass with undiluted solvents and to use plastic only when the solvent concentration is <10%.

Although calculating the plating efficiency is one of the best methods for testing cell survival rates, it should be remembered that plating efficiency only applies to the clonogenic component of the cell population, which may not be representative of the whole cell population. The question does not arise if controls plate with 100% efficiency; in practice, however, control plating efficiencies of 20% or less are more likely and the response is being measured in a subset of the total cell population.

### 21.3.3  Assays Based on Cell Proliferation

Cell counts after a few days in culture can also be used to determine the effect of various compounds on cell proliferation, but at least in the early stages of testing, a complete growth curve is required (*see* Protocols 20.7–21.9) because the interpretation of cell counts at a single point in time can be ambiguous (*see* Fig. 20.10, day 7). Growth curve analyses, using cell counting, are feasible only with relatively small numbers of samples, as they become cumbersome in a large screen, although automating growth curves in multiwell plates using image analysis (Incucyte; Chip-Man; *see* Section 20.9.3; Figs. 20.11, 20.12) can make this feasible.

In cases where there are many samples, a single point in time—such as the number of cells three to five days after exposure—can be used. The time should be selected as within the log phase, and preferably mid-log phase, of control cells. Any significant effect should be backed up with a complete growth curve over the whole growth cycle or by an alternative assay, such as a survival curve by clonogenic assay (*see* Protocol 21.3) or MTT assay (*see* Protocol 21.4).

### 21.3.4  Metabolic Cytotoxicity Assays

Plating efficiency tests are labor intensive and time-consuming to set up and analyze, particularly when a large number of

samples is involved (although this can be automated; *see* Section 26.4), and the duration of each experiment may be anywhere from two to four weeks. Furthermore some cell lines have poor plating efficiencies, particularly freshly isolated normal cells, so a number of alternatives have been devised for assaying cells at higher densities (e.g., in microtitration plates; *see* Section 21.3.5). None of these tests measures survival directly, however. Instead, the net increase in the number of cells (i.e., the growth yield; *see* Section 21.3.3), the increase in the total amount of protein or DNA, or continued metabolic activity, such as the reduction of a tetrazolium salt (MTT or XTT) to formazan or the synthesis of protein or DNA, is determined. Survival in these cases is defined as the retention of metabolic or proliferative ability by the cell population as a whole some time after removal of the toxic influence. However, such assays cannot discriminate between a reduction in metabolic or proliferative activity per cell and a reduced number of cells, and therefore any novel or exceptional observation should be confirmed by clonogenic survival assay.

### 21.3.5  Microtitration Assays

The introduction of multiwell plates revolutionized the approach to replicate sampling in tissue culture. These plates are economical to use, lend themselves to automated handling, and can be of good optical quality. The most popular are 96-well microtitration plates or *microplates* (*see* Plate 17*b*), each well having 28 to 32 mm$^2$ of growth area, 0.1 or 0.2 mL medium, and up to $1 \times 10^5$ cells. Microtitration offers a method by which large numbers of samples may be handled simultaneously, but with relatively few cells per sample. With this method, the whole population is exposed to the agent, and viability is determined subsequently, usually by measuring a metabolic parameter such as the ATP or NADH/NADPH concentration. Assay kits are available (*see* Appendix II: Cytotoxicity Assays).

The end point of a microtitration assay is usually an estimate of the number of viable cells, if the assay is done after the removal of the toxin. Although this result can be achieved directly by cell counts or by indirect methods, such as isotope incorporation, cell viability as measured by MTT reduction [Mosmann, 1983] is widely used as the endpoint [Cole, 1986; Alley et al., 1988]. MTT is a yellow water-soluble tetrazolium dye that is reduced by live, but not dead, cells to a purple formazan product that is insoluble in aqueous solutions. However, a number of factors can influence the reduction of MTT [Vistica et al., 1991]. The assay described in Protocol 21.4, provided by Jane Plumb of the Cancer Research UK Centre for Oncology and Applied Pharmacology, University of Glasgow, Scotland, UK, has been shown to give the same results as a standard clonogenic assay [Plumb et al., 1989] (*see also* Section 21.3.6). It illustrates the use of microtitration in the assay of anticancer drugs, but would be applicable, with minor modifications, to any cytotoxicity assay.

## PROTOCOL 21.4. MTT-BASED CYTOTOXICITY ASSAY

### Principle

Cells in the exponential phase of growth are exposed to a cytotoxic drug. The duration of exposure is usually determined as the time required for maximal damage to occur, but is also influenced by the stability of the drug. After removal of the drug, the cells are allowed to proliferate for two to three population-doubling times (PDTs) in order to distinguish between cells that remain viable and are capable of proliferation and those that remain viable but cannot proliferate. The number of surviving cells is then determined indirectly by MTT dye reduction. The amount of MTT-formazan produced can be determined spectrophotometrically once the MTT-formazan has been dissolved in a suitable solvent.

### Outline

Incubate monolayer cultures in microtitration plates in a range of drug concentrations (Fig. 21.5). Remove the drug, and feed the plates daily for two to three PDTs; then feed the plates again, and add MTT to each well. Incubate the plates in the dark for 4 h, and then remove the medium and MTT. Dissolve the water-insoluble MTT-formazan crystals in DMSO, add a buffer to adjust the final pH, and record the absorbance in a plate reader.

### Materials

*Sterile:*

❑ Growth medium
❑ Trypsin (0.25% + EDTA, 1 mM, in PBSA)
❑ MTT: 3-(4,5-dimethylthiazol-2-yl)-2,5-diphenylte-trazolium bromide (Sigma), 50 mg/mL, filter sterilized
❑ Sorensen's glycine buffer (0.1 M glycine, 0.1 M NaCl adjusted to pH 10.5 with 1 M NaOH)
❑ Microtitration plates (Iwaki)
❑ Pipettor tips, preferably in an autoclavable tip box
❑ Petri dishes (non–TC-treated), 5 cm and 9 cm or reservoir (Corning)
❑ Universal containers or tubes, 30 mL and 100 mL

*Nonsterile:*

❑ Plastic box (clear polystyrene, to hold plates)
❑ Multichannel pipettor
❑ Dimethyl sulfoxide (DMSO)
❑ DMSO dispenser (optional); such as Labsystems Microplate Dispenser (Cat No 5840 127, from Thermo Fisher; *see also* Fig. 4.7)

❑ ELISA plate reader (Molecular Devices, with SOFTmax PRO; *see also* Fig. 4.7; Appendix II: Plate Readers)
❑ Plate carrier for centrifuge (for cells growing in suspension; *see* Appendix II: Microtitration Plate Centrifugation)

### Procedure

Plating out cells

1. Trypsinize a subconfluent monolayer culture, and collect the cells in growth medium containing serum.
2. Centrifuge the suspension (5 min at 200 $g$) to pellet the cells. Resuspend the cells in growth medium, and count them.
3. Dilute the cells to 2.5 to $50 \times 10^3$ cells/mL, depending on the growth rate of the cell line, and allowing 20 mL of cell suspension per microtitration plate.
4. Transfer the cell suspension to a 9-cm Petri dish, and, with a multichannel pipette, add 200 µL of the suspension into each well of the central 10 columns of a flat-bottomed 96-well plate (80 wells per plate), starting with column 2 and ending with column 11, placing 0.5 to $10 \times 10^3$ cells into each well.
5. Add 200 µL of growth medium to the eight wells in columns 1 and 12. Column 1 will be used to blank the plate reader; column 12 helps maintain the humidity for column 11 and minimizes the "edge effect."
6. Put the plates in a plastic lunch box, and incubate in a humidified atmosphere at 37°C for 1 to 3 days, such that the cells are in the exponential phase of growth at the time that drug is added.
7. For nonadherent cells, prepare a suspension in fresh growth medium. Dilute the cells to 5 to $100 \times 10^3$ cells/mL, and plate out only 100 µL of the suspension into round-bottomed 96-well plates. Add drug immediately to these plates.

Drug addition

8. Prepare a serial fivefold dilution of the cytotoxic drug in growth medium to give eight concentrations. This set of concentrations should be chosen such that the highest concentration kills most of the cells and the lowest kills none of the cells. Once the toxicity of a drug is known, a smaller range of concentrations can be used. Normally three plates are used for each drug to give triplicate determinations within one experiment.

9. For adherent cells:

(a) Remove the medium from the wells in columns 2 to 11. This can be achieved with a hypodermic needle attached to a suction line.

(b) Feed the cells in the eight wells in columns 2 and 11 with 200 μL of fresh growth medium; these cells are the controls.

(c) Transfer the drug solutions to 5-cm Petri dishes, and add 200 μL to each group of four wells with a four-tip pipettor.

(d) Add the cytotoxic drug to the cells in columns 3 to 10. Only four wells are needed for each drug concentration, such that rows A through D can be used for one drug and rows E through H for a second drug.

10. For nonadherent cells, follow steps 9b–d but prepare the drug dilution at twice the desired final concentration; add 100 μL of diluted drug or control medium to the 100 μL of cells already in the wells.

11. Return the plates to the plastic box, and incubate them for a defined exposure period.

### Growth period

12. At the end of the drug exposure period, remove the medium from all of the wells containing cells, and feed the cells with 200 μL of fresh medium. Centrifuge plates containing nonadherent cells (5 min at 200 g) to pellet the cells. Then remove the medium, using a fine-gauge needle to prevent disturbance of the cell pellet.

13. Feed the plates daily for 2 to 3 PDTs.

### Estimation of surviving cell numbers

14. Feed the plate with 200 μL of fresh medium at the end of the growth period, and add 50 μL of MTT to all of the wells in columns 1 to 11.

15. Wrap the plates in aluminum foil, and incubate them for 4 h in a humidified atmosphere at 37°C. Note that 4 h is a minimum incubation time, and plates can be left for up to 8 h.

16. Remove the medium and MTT from the wells (centrifuge for nonadherent cells), and dissolve the remaining MTT-formazan crystals by adding 200 μL of DMSO to all of the wells in columns 1 to 11.

17. Add glycine buffer (25 μL per well) to all of the wells containing DMSO.

18. Record absorbance at 570 nm immediately, because the product is unstable. Use the wells in column 1, which contain medium and MTT but no cells, to blank the plate reader.

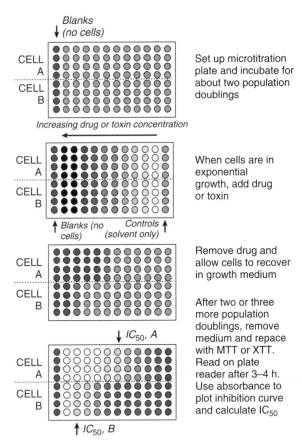

**Fig. 21.5. Microtitration Assay.** Stages in the assay of two different cell lines exposed to a range of concentrations of the same drug and then allowed to recover before the estimation of survival by the MTT reaction (*see* Protocol 21.4). The far left column has no cells and can be used as a blank to set the plate reader. This array is applicable when using plate sealers, when all wells are equivalent; however, with lids, there is a risk of an edge effect, probably due to evaporation, and it is better to leave the far left and far right columns blank (i.e., with medium only, as in Protocol 21.4), and some users leave the top and bottom rows blank as well. (*See also* Plate 17*b*).

### Analysis of MTT assay

(1) Plot a graph of the absorbance (*y*-axis) against the concentration of drug (*x*-axis).

(2) Calculate the $IC_{50}$ as the drug concentration that is required to reduce the absorbance to half that of the control. The mean absorbance reading from the wells in columns 2 and 11 is used as a control (columns 1 and 12 when plate sealers are used as in Fig. 21.5). The absorbance values in control columns should be the same. Occasionally they are not, however, and this is taken to indicate uneven plating of cells across the plate.

(3) The absolute value of the absorbance should be plotted so that control values may be compared, but the data can then be converted to a percentage-inhibition curve (Fig. 21.6) to normalize a series of curves.

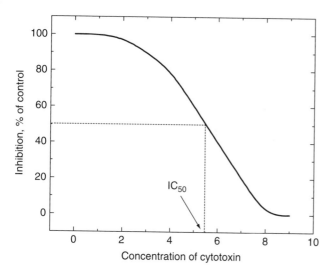

**Fig. 21.6. Percentage Inhibition Curve.** Test well values are calculated as a percentage of the controls and plotted against the concentration of cytotoxin. Typically a sigmoid curve is obtained, and ideally the $IC_{50}$ will lie in the center of the inflexion of the curve.

### Variations in MTT Assay

**Other applications.** A similar assay has also been used to determine cellular radiosensitivity [Carmichael et al., 1987b]. MTT can be used to determine the number of cells after a variety of treatments other than cytotoxic drug exposure, such as growth factor stimulation. However, in each case it is essential to ensure that the treatment itself does not affect the ability of the cell to reduce the dye and absorbance remains linear with cell number.

**Duration of exposure.** As with clonogenic assays (*see* Protocol 21.3), some agents may act more quickly, and the exposure period and recovery may be shortened. The cells must remain in exponential growth throughout (*see* Sections 12.4.3, 20.9.2), and the cell concentration at the end should still be within the linear range of the MTT spectrophotometric assay. When using a cell line for the first time, parallel plates should be set up for cell counts to generate a growth curve (*see* Protocol 21.4) and for MTT-formazan absorbance to ensure that absorbance is proportional to the number of cells. If the growth curve shows that the cells are moving into the stationary phase or the absorbance is nonlinear when plotted against cell concentration, shorten the assay and proceed directly to step 14 of Protocol 21.4.

Duration of exposure is related to the number of cell cycles that the cells have gone through during exposure and recovery. Cell cycle time will influence the choice between a short-form and long-form assay (Figs. 21.7, 21.8). With rapidly dividing cells, not only will the cell density increase more rapidly during exposure but, in addition, the response to cycle-dependent drugs will be quicker. When first trying an assay, it may be desirable to sample on each day of drug

exposure and recovery. If a stable $IC_{50}$ is reached earlier, then the assay may be shortened.

*End point.* Sulforhodamine, a fluorescent dye that stains protein, can also be used to estimate the amount of protein (i.e., cells) per well on a plate reader with fluorescence detection [Boyd, 1989]. It stains all cells and does not discriminate between live and dead cells. Labeling with [³H]thymidine (DNA synthesis), [³H] leucine [Freshney et al., 1975] or [³⁵S]methionine [Freshney & Morgan, 1978] (protein synthesis), or other isotopes can be substituted for MTT reduction. Quantitation is achieved by microtitration plate scintillation counting on a specially adapted scintillation counter (Perkin Elmer) or by preparing an autofluorogram and reading it on a densitometer [Freshney & Morgan, 1978].

In practice, it may not matter which criterion is used for determining viability or survival at the end of an assay; it is rather the design of the assay, such as duration of drug exposure and recovery, phase of the growth cycle (cell density, growth rate, etc.), that is more important. In a short assay with no or minimal recovery period, the endpoint must measure only viable cells (e.g., MTT), but in a longer assay the end point measures the difference between wells that have increased and those that have not, or have even decreased. In a monolayer assay, at least, nonviable cells will have been lost, and the increase or decrease relative to control wells is what is measured; whether by MTT, sulforhodamine, or isotope incorporation into DNA or protein becomes less important.

*Handling.* A variety of automated instruments are available to reduce the handling time required per sample, including autodispensers, diluters, cell harvesters, and programmable plate readers (*see* Fig. 4.7; Appendix II: Microtitration Equipment).

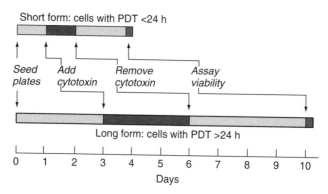

**Fig. 21.7. Assay Duration.** Pattern for short-form and long-form assays. The upper diagram represents an assay that is suitable for cell with a PDT < 24 h, and the bottom diagram represents an assay that is suitable for cells with a PDT > 24 h, although intermediate time scales are also possible.

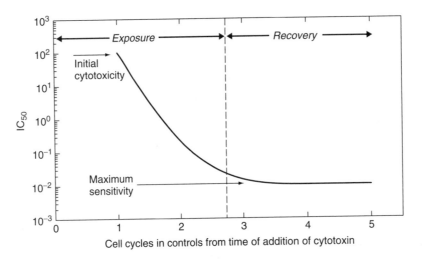

**Fig. 21.8. Time Course of the Fall in IC₅₀.** Idealized curve for an agent with a progressive increase in cytotoxicity with time, but eventually reaching a maximum effect after three cell cycles. Not all cytotoxic drugs will conform to this pattern [Freshney et al., 1975].

## 21.3.6 Comparison of Microtitration with Clonogenic Survival

The volume of medium required per sample for microtitration is less than one-fiftieth of that required for cloning, although the number of cells is approximately the same for both techniques. Microtitration assays are also shorter and more amenable to automated handling, data gathering, and analysis. Microtitration, however, is unable to distinguish between differential responses between cells within a population and the degree of response in each cell—for example, a 50% inhibition of a metabolic parameter could mean that 50% of the cells respond or that each cell is inhibited by 50%—but this becomes less important in an assay with a prolonged recovery period, where the relative increase by cell proliferation becomes the major criterion of survival.

A comparison of the $IC_{50}$ derived by microtitration and plating efficiency assays showed a good correlation between the two methods (Fig. 21.9) for the assay of antineoplastic drugs [Morgan et al., 1983; Plumb et al., 1989]. The correlation for $IC_{90}$ was not as tight.

A significant feature of microtitration assays, particularly with a photometric or radiometric end point, is the generation of large amounts of data, often in a format that is readily analyzed by computer (see Sections 20.8.1, 20.8.2). It is important, however, to scan the raw data as well as the data-reduced end point because computer analysis may make different assumptions or corrections to deal with aberrant data points, which are not apparent unless the raw data are available for scrutiny.

## 21.3.7 Drug Interaction

The investigation of cytotoxicity often involves the study of the interaction of different drugs; drug interaction is

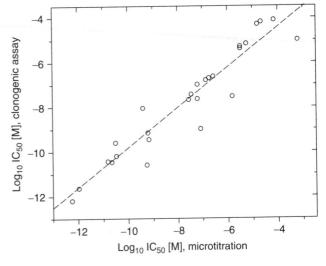

**Fig. 21.9. Correlation between Microtitration and Clonogenic Survival.** Measurement of the $IC_{50}$ values of a group of five cell lines from human glioma and six drugs (vincristine, bleomycin, VM-26 epidophyllotoxin, 5-fluorouracil, methyl CCNU, mithramycin). Most of the outlying points were derived from one cell line that later proved to be a mixture of cell types. The broken line is the regression, with the data points from the heterogeneous cell line omitted. Microtitration $IC_{50}$ was derived by [³⁵S]methionine incorporation [Freshney & Morgan, 1978; figure after Freshney et al., 1982a].

readily determined by microtitration systems, in which several different ratios of interacting drugs can be examined simultaneously. Analysis of drug interaction can be performed by using an isobologram to interpret the data [Chou, 2006]. A rectilinear plot implies an additive response, whereas a curvilinear plot implies synergy if the curve dips below the predicted line and antagonism if it goes above.

## 21.4 APPLICATIONS OF CYTOTOXICITY ASSAYS

### 21.4.1 Anticancer Drug Screening

Drug screening for the identification of new anticancer drugs can be a tedious and often inefficient method of discovering new active compounds. The trend is now more toward monitoring effects on specific molecular targets using high through-put screens. However, there have been attempts to improve screening by adopting rapid, easily automated assays, like those based on the determination of the number of viable cells by staining the cells with MTT [Mosmann, 1983; Carmichael et al., 1987a; Plumb et al., 1989]. To further cut down on manipulations, the MTT incubation step may be omitted and the end point determined by measuring the amount of total protein with sulforhodamine B [Boyd, 1989]. Although this method is quicker and easier than the MTT assay, it should be remembered that nonviable, and certainly nonreplicating, cells will still stain, so the assay should be confirmed when activity is detected, using a more reliable indicator such as clonogenicity or MTT reduction.

### 21.4.2 Predictive Drug Testing for Tumors

The possibility has often been considered that measurement of the chemosensitivity of cells derived from a patient's tumor might be used in designing a chemotherapeutic regime for the patient [Freshney, 1978]. This technique has never been exhaustively tested, although the results of small-scale trials were encouraging [Hamburger & Salmon, 1977; Bateman et al., 1979; Hill, 1983; Thomas et al., 1985; Von Hoff et al., 1986]. What is required is the development of reliable and reproducible culture techniques for neoplastic cells from the most common tumors (e.g., breast, lung, colon), such that cultures of pure tumor cells capable of cell proliferation over several cell cycles may be prepared routinely. As many of the cells within a tumor have a limited life span, the main targets for chemotherapy are the clonogenic populations with infinite repopulation capacity [Al-Hajj et al., 2004; Jones et al., 2004]. Advances in stem cell recognition (see Section 23.1) may make isolation of tumor stem cells feasible. It was hoped that the soft agar clonogenic assay [Hamburger & Salmon, 1977] might isolate transformed stem cells for assay but, although the technique seemed initially promising, isolated clones did not have long-term regenerative capacity. Routine isolation of tumor cell populations with long-term repopulation efficiency has yet to be achieved, but when it is, it may be possible to improve targeting and specificity of anticancer drugs, making predictive testing more meaningful. Assays might then be performed in a high proportion of cases, hopefully within two weeks of receipt of the biopsy and potentially effective drugs selected by genetic profiling.

The major problem, however, is one of logistics. The number of patients with tumors for which the correct target cells (1) will grow in vitro sufficiently to be tested, (2) can be expected to respond, and (3) will produce a response that can be followed up, is extremely small. Hence it has proved difficult to use any in vitro test as a predictor of response or even to verify the reliability of the assay. The correlation of insensitivity in vitro with nonresponders is high, but few clinicians would withhold chemotherapy because of an in vitro test, particularly when the agent in question would probably not be used alone. Ultimately it may be more profitable to isolate tumor stem cells and screen them by expression analysis for specific molecular targets and back this up with an in vitro cytotoxicity assay (see also Section 26.4.2).

### 21.4.3 Testing Pharmaceuticals

A number of pharmaceutical companies maintain a program of in vitro toxicity testing on the assumption that it might prove more economical and ethically acceptable than animal testing. Legislation enforcing the use of animal tests is difficult to introduce as the complexity of the wide range of effects seen in vivo is still very difficult to model in vitro. However, there is considerable political pressure to introduce such legislation, and this is driving large-scale comparative surveys to determine whether any of the many existing tests may be acceptable [Knight & Breheny, 2002; Vanparys, 2002; ECVAM, 2008; Hartung & Daston 2009].

## 21.5 GENOTOXICITY

Genotoxicity implies damage to DNA (mutagenesis) leading to permanent alteration in gene expression (transformation) which may lead to malignancy (carcinogenesis). DNA stress is often accompanied by increased expression of p53 and related genes, which can be used to assay genotoxicity (Gentronix). In vitro assays for transformation include demonstration of anchorage independence (see Protocols 13.4, 13.5), reduced density limitation of cell proliferation (see Protocol 17.3), and evidence of mutagenesis. (See also Table 17.1 and Genotoxicity Testing in Appendix II.) Mutagenesis can be assayed by sister chromatid exchange; this procedure is described in Protocol 21.5, which was contributed by Maureen Illand and Robert Brown when at the Cancer Research UK Centre for Oncology and Applied Pharmacology, University of Glasgow, Scotland.

### 21.5.1 Mutagenesis Assay by Sister Chromatid Exchange

Sister chromatid exchanges (SCEs) are reciprocal exchanges of DNA segments between sister chromatids at identical loci during the S-phase of the cell cycle. As SCEs are more sensitive indicators of mutagenic activity than chromosome breaks, they have become a major tool in mutagenesis research [Latt, 1981].

With the development of the thymidine analog bromodeoxyuridine (BUdR) and its subsequent use in DNA labeling experiments, the resolution of SCEs was greatly improved in comparison with previous methods, which involved the incorporation of radioactive nucleotides into replicating DNA [Taylor, 1958]. Later the fluorescence plus Giemsa (FPG) technique of [Perry and Wolf [1974] for the scoring of SCEs was enhanced, and for the first time, permanent staining of SCEs was demonstrated. Previously, during the scoring process, rapid bleaching of the fluorescent stain occurred [Latt, 1981].

The FPG method involves two distinct steps: (1) Cells are labeled with BUdR for two complete cycles and then treated with colcemid to block the cells in metaphase. After BUdR exposure, the DNA of one chromatid of each chromosome contains bromouracil in one strand, while the DNA of its sister chromatid contains bromouracil in both strands. (2) Chromosomes are then prepared from these cells and stained with the fluorescent dye Hoechst 33258, and then the BUdR is photodegraded with ultraviolet light; this is followed by Giemsa staining. These final steps highlight the differential incorporation of bromouracil into the sister chromatids. DNA that contains bromouracil quenches the fluorescence of Hoechst–DNA complexes. Therefore the chromatid containing bromouracil substituted in both strands fluoresces weakly and stains weakly with Giemsa, while the chromatid containing bromouracil in only one strand fluoresces more intensely, degrades the BUdR, and subsequently stains darkly with Giemsa (*see* Plate 17*e, f*). If any SCEs occur, this staining pattern produces what are called *harlequin chromosomes*.

---

### PROTOCOL 21.5. SISTER CHROMATID EXCHANGE

#### Outline

Trypsinize metaphase-arrested cells that have been labeled with BUdR for two cell cycles, incubate the cells in hypotonic buffer, and then fix the cells. Prepare slides of the cells, after treating the cells with Hoechst 33258, and photodegrade the chromosome spreads. Stain the chromosomes with Giemsa, and visualize on a light microscope under oil immersion.

#### Materials

*Sterile:*

❑ D-PBSA
❑ PE: 10 mM EDTA in D-PBSA
❑ Trypsin: 0.12% in PE
❑ BUdR (Sigma): 1 mM in sterile UPW
❑ Karyomax: Colcemid, 10 µg/mL (Invitrogen)
❑ Growth medium

*Nonsterile:*

❑ SSC, 2 × :1 : 10 dilution of 20 × SSC (*see* Appendix I)
❑ Hypotonic buffer: 0.075 M KCl
❑ Sorensen's buffer: phosphate buffer, 0.066 M, pH 6.8 (tablets from Merck)

❑ Methanol:acetic acid: 3:1, ice cold and freshly prepared
❑ Giemsa solution, 0.76%: Place 1 g of Giemsa powder (Merck) in 66 mL of glycerol and heat in a water bath at 56°C to 60°C for 11/2 to 2 h. Cool the solution, and add 66 mL of absolute alcohol.
❑ Giemsa, diluted to 3.5% in Sorensen's buffer, pH 6.8
❑ Hoechst 33258 (Sigma), 20 µg/mL, in UPW
❑ Latex photo-mountant or adhesive
❑ Xylene
❑ DPX mountant (Merck)
❑ Coverslips, 22 × 15 mm
❑ Coplin jar
❑ Slide rack (Thermo Fisher)
❑ Short wave UV lamp in irradiation box

Δ *Safety Note.* Hoechst 33258 is carcinogenic; weigh it out and dissolve it in a fume hood.

#### Pretreatment

1. Seed the cells at the appropriate density (e.g., $1 \times 10^6$ cells per 75-cm² flask), and incubate for 2 days at 37° C.
2. Add BUdR to the growth medium at a final concentration of 10 µM.
3. Incubate the cells in the dark at 37°C for a further 48 h (~2 cell cycles).
4. Add colcemid to the cells 1 to 6 h before harvesting, depending on the cycling time of the cells. For human cell lines, the final concentration of colcemid should be 0.01 µg/mL.

#### Harvesting cells

5. Wash the cells with D-PBSA, and trypsinize them with 1 mL of 0.12% trypsin in PE.
6. Resuspend the cells in 10 mL of growth medium. Transfer the cell suspension to 50-mL centrifuge tubes.
7. Centrifuge the suspension at 1200 *g* for 5 min.
8. Remove the supernate, leaving approximately 0.2 mL above the pellet. Flick the side of the tube to resuspend the pellet.
9. Slowly add 10 mL of hypotonic buffer (prewarmed to 37°C), and incubate the cell suspension for 10 to 15 min at room temperature.
10. Spin the cells in a benchtop centrifuge at 1200 *g* for 5 min.
11. Remove the supernate, leaving 0.2 mL above the pellet. Flick the side of the tube to resuspend the pellet.
12. Add 10 mL of ice-cold fixative, initially drop by drop, mixing well after each addition. Leave the tube on ice for 10 min.

13. Repeat steps 10 through 12 once more, letting the cells remain in fixative overnight at 4°C, to improve slide preparations.
14. Spin the fixed cells at 1200 *g* for 5 min, and resuspend the cells in 3 to 5 mL of methanol/acetic acid.
15. Store the cells at −20°C.

### Slide preparation

16. Slides should be clean and grease-free before use, so wipe them with absolute alcohol.
17. Using a short glass Pasteur pipette, take up approximately 500 μL of the fixed cells.
18. Hold the slide at a downward angle, and holding the Pasteur pipette at least 15 cm (6 in.) above the slide, drop 3 drops of the cell suspension onto the slide (*see* Protocol 15.7).
19. Air dry the slide in the dark.
20. Check the slide under phase contrast to ensure that the metaphase spreads are evenly distributed across the slide and that the chromosomes are well separated.

### Harlequin staining

21. Immerse the slides in a Coplin jar of Hoechst 33258 at a concentration of 20 μg/mL for 10 min. (Wear gloves, as Hoechst is toxic.)
22. Transfer the slides to a slide rack, and drop 500 μL of 2× SSC onto each slide.
23. Cover the slides with a 22 × 50-mm coverslip, and seal the edges with a temporary seal, such as latex photo-mountant, to prevent evaporation.
24. Place the slides in the slide rack, coverslips facing downwards, and place the slide rack on a shortwave UV box. Maintain a distance of approximately 4 cm between the slides and the UV source. The longer the slides are exposed to UV, the paler the pale chromatid will become; expose the slides for about 25 to 60 min.
25. Remove the coverslips from the slides, and wash the slides three times in UPW, 5 min per wash. Cover the slide holder with aluminum foil.
26. Air dry the slides in the dark.
27. Stain the slides in a Coplin jar containing 3.5% Giemsa solution in Sorensen's buffer, pH 6.8, for 3 to 5 min.
28. Carefully rinse the slides in tap water, and drain them with a paper tissue.
29. Air dry the slides on the bench for 1 h. Dip each slide into xylene, drop 4 drops of DPX mountant (Merck) onto the slide, and mount a 22 × 50-mm coverslip, expressing any air bubbles with tissue. (Carry out this final step in a fume hood, as xylene fumes are toxic. Also, wear gloves.)

30. Air dry the slides in a fume hood overnight.

### Analysis

31. Under the 40× objective of a light microscope, scan the slides for metaphase spreads.
32. Find an area on the slides where most of the metaphase spreads are located, and examine this area under oil immersion.
33. When no sister chromatid exchanges (SCEs) have occurred, each chromosome has one continuously staining pale chromatid and one continuously staining dark chromatid. One SCE has occurred when there is one area of dark staining and then light staining on one chromatid, and on the sister chromatid one area of light staining and then dark staining (*see* Plate 17*e*, *f*). Each point of the discontinuity in staining is scored as one SCE.
34. Count the number of SCEs per cell and also the number of chromosomes per cell.
35. Larger chromosomes usually have a greater number of SCEs than smaller ones, and the incidence of SCEs may vary from cell to cell. Therefore, scoring SCEs per chromosome is a more accurate measure of SCE rate. SCE score is calculated by the following formula:

$$\frac{\text{Mean number of SCEs cell}}{\text{Mean number of chromosomes/cell}}$$

Aim to score approximately 50 spreads per cell line being studied.

***Variations.*** Pulse labeling of cells with BUdR and subsequent staining, as described previously, can detect differences in early and late replicating regions of chromosomes during the cell cycle. When cells are labeled with BUdR at the latter part of the cell cycle, DNA that replicates early will stain darkly with Giemsa, due to very little BUdR incorporation, and for regions of the chromosome that are pulsed at the earlier stages of the cell cycle, only those regions that replicate their DNA early will stain faintly with Giemsa [Latt, 1973].

Additionally cells that have undergone only one cell cycle of continuous BUdR labeling show differential staining of chromatids only at certain bands (lateral asymmetry) because of the differences in thymine content of the DNA [Brito Babapulle, 1981]. After photodegradation of BUdR, the fluorescent dye acridine orange can also be used to stain SCEs. With this dye, green fluorescence is observed at regions that have double-stranded DNA and thus will have little BUdR incorporation, and red fluorescence is observed at regions that have single-stranded DNA, which will have incorporated the BUdR. Consequently the red fluorescence is equivalent to lighter staining with Giemsa, and the green fluorescence is equivalent to the darker staining with Giemsa [Karenberg & Freelander, 1974].

### 21.5.2 Carcinogenicity

The potential for in vitro testing for carcinogenesis is considerable [Berky and Sherrod, 1977; Grafström, 1990a,b; Zhu et al., 1991; Tweats et al., 2007], but this is one area in which in vivo testing is far from adequate; the models are poor, and the tests often take weeks, or even months, to perform. The development of a satisfactory in vitro test is hampered (1) by the lack of a universally acceptable criterion for malignant transformation in vitro and (2) by the inherent stability of human cells used as targets.

The most generally accepted tests so far assume that most carcinogens are mutagenic (*see* Section 17.3). This assumption is the basis of the Ames test [Ames, 1980], wherein bacteria are used as targets and activation can be carried out with liver microsomal enzyme preparations. This test has a high predictive value, but nevertheless, dissimilarities in uptake, susceptibility, and type of cellular response have led to the introduction of alternative tests using mammalian and human cells as targets, such as sister chromatid exchange (*see* Section 21.5.1).

Some of these tests are also mutagenesis assays, using suspensions of L5178Y lymphoma cells as targets [Cole et al., 1990] and the induction of mutations or reversion, or cytological evidence of sister chromatid exchange (*see* Protocol 21.5), as evidence of mutagenesis. Others [Styles, 1977] have used transformation as an end point, assaying clonogenicity in suspension (*see* Section 17.5.1) as a criterion for transformation. Critics of these systems say that both use cells that are already partially transformed as targets; even the BHK21-C13 cell used by some workers is a continuous cell line and may not be regarded as completely normal. Furthermore the bulk of the common cancers arise in epithelial tissues and not in connective tissue cells.

The demonstration of increased oncogene expression or amplification, or the presence of increased or altered oncogene products, may provide more reliable criteria, in some cases functionally related to the carcinogen. This is now possible with microarray analysis. Likewise the deletion or mutation of suppressor genes is open to molecular analysis, where deletions or mutations in the p53, Rb, p16, and L-CAM (E-cadherin) genes would cover a high proportion of malignant transformation events. It is now feasible to consider expression analysis and an alternative to mutagenesis in carcinogenicity testing [Nuwaysir et al., 1999; Desai et al., 2002; Vondracek et al., 2002].

### 21.6 INFLAMMATION

There is an increasing need for tissue culture testing to reveal the inflammatory responses that are likely to be induced by pharmaceuticals and cosmetics with topical application or by xenobiotics that may be inhaled or ingested and may be responsible for many forms of allergy. This is an area that is only at the early stages of development but bears great promise for the future. It is a sensitive topic in more ways than one. Animal rights groups are naturally incensed at the needless use of large numbers of animals to test new cosmetics that have little benefit except commercial advantage to the manufacturer, particularly when the testing of substances (e.g., shampoos) involves the Draize test, in which the compound is added to a rabbit's eye. More important, clinically, is the apparent increase in allergenic responses produced by pharmaceuticals and xenobiotics. These responses are little understood and poorly controlled, largely because of the absence of a simple reproducible in vitro test.

Since the advent of filter well technology, several models for skin (e.g., *see* Plate 20) and cornea have appeared [Braa and Triglia, 1991; Triglia et al., 1991; Fusenig, 1994b; Roguet et al., 1994; Kondo et al., 1997; Brinch & Elvig, 2001; Cantòn et al., 2010], utilizing the facility for coculture of different cell types that the filter well system provides. In these systems the interaction of an allergen or irritant with a primary target (e.g., epidermis) is presumed to initiate a paracrine response, which triggers the release of a cytokine from a second, stromal component (e.g., dermis) (Fig. 21.10). This cytokine can then be measured by ELISA technology to monitor the degree of the response. Although still in the early stages of development, kits for the measurement of irritant responses are available, for example, Epiderm (MatTek) [Koschier at al., 1997]; Episkin (Saduc) [Cohen et al., 1997]; SkinEthic [Brinch & Elvig, 2001]; review [Schäfer-Korting et al., 2008] (*see* Plate 21).

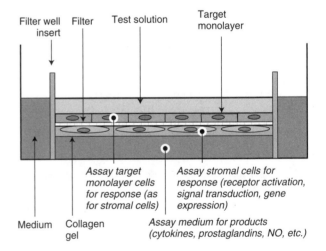

***Fig. 21.10. Organotypic Assay.*** Assay system for exposing one cell layer (e.g., epidermal keratinocytes) cocultured with another associated cell type (e.g., skin fibroblasts in collagen gel; *see* Plate 20) to an irritant and measuring the response by cytokine release (*see also* Plate 21).

Protocols for epidermal, buccal, and corneal culture suitable for modeling irritant responses are given in Chapter 22 (*see* Protocols 22.1, 22.2, 22.11).

It would seem that this type of system may be a major area of development, with the real prospect that allergen screening from patients' own skin in organotypic culture may become possible, analysis of similar cultures from GI tract may reveal allergens responsible for irritable bowel syndrome, and analysis of organotypic cultures from tracheal and bronchial cells may identify patient-related causes of asthma. In each case, and in many others, there is the possibility of specific mechanistic studies into the processes of abnormal cell interaction that typify many allergic and degenerative diseases.

# CHAPTER 22

# Specialized Cells

There is increasing interest in specific cell types to examine specialized developmental processes and their pathologies, to use in cell-based therapies, or for cell type specific screening. As a result of this interest there are now several commercial suppliers covering an extensive list of specialized cell types, both human and rodent (Table 22.1). These are often referred to in catalogs as "primary cells," implying that they are derived from primary cultures. However, this may not always be the case as some commercial cells may already have been subcultured and, technically, have become cell lines (though not "primary cell lines," which is incorrect terminology; *see* Section 2.7). These early passage cell lines are referred to here as *specialized cells* or *specialized cell lines* on the assumption that they have been selected to express the properties of specific cells constitutively or after appropriate induction (*see* Plates 27, 28).

For anyone wishing to culture specialized cells the question will always arise whether to attempt the isolation and selective culture yourself or to purchase the cells from a supplier. Different types of specialized cells (e.g., epidermal keratinocytes, melanocytes, endothelial cells, smooth muscle cells, dermal fibroblasts, melanocytes, mammary epithelium) are commercially available (*see* Appendix II). Skin cultures are also available for cytotoxicity and inflammation research from Episkin (Saduc), Epiderm (MatTek), and SkinEthic. These products are prepared in filter wells by combining keratinocytes with dermal fibroblasts and collagen supported by a nylon net called a "skin equivalent" (*see* Section 25.4.1). Other tissues—in particular, the cornea (SkinEthic)—have also been prepared in a similar way for toxicity studies (*see* Protocol 22.2; Plate 21).

The cost of these cells is significant, between $500 and $1000 per ampoule of frozen cells, but this cost has to be compared with the cost of setting up the necessary procedures yourself as well as the time and personnel commitment required. There may be situations where patient-specific or disease-specific tissue is required, or tissue with a particular genetic background or from a different species, that may make it preferable to undertake the isolation and culture yourself. In addition laboratories that are already doing specialized cell culture and using selective media may expand into another cell type without a major outlay. Accordingly a number of protocols are presented in this chapter to allow the reader to assess the complexity and outlay and make the choice of in-house preparation or purchase. In each case the name of one or more supplier is provided with the cell type (*see* Table 22.1; Appendix II: collectively under Specialized Cells and individually under cell type).

The generation of a specialized cell culture can be achieved in several ways: (1) by isolating differentiated cells or tissue for short-term nonregenerative culture (*see* Section 25.2), (2) by isolating precursor cells, expanding them in selective culture conditions (*see* below and Sections 9.2.2, 13.6), and then inducing them to differentiate (*see* Section 16.7), or (3) by isolating stem cells and creating cultures that either regenerate the stem cell population or progress toward one of several possible differentiated phenotypes (*see* Chapter 23). In each case the technology is different, and this chapter sets out to introduce some of the technology in approaches 1 and 2 for specific types of tissue.

**TABLE 22.1.** Commercial Suppliers of Specialized Cells

| Tissue | Cell type | Supplier[a] |
|---|---|---|
| Aorta | Endothelial | ATCC; Lonza |
| | Smooth muscle | ATCC |
| Blood | Stem cells | Promocell |
| Bladder | Microvascular endothelial cells | Lonza |
| Bone | Osteoblasts | Cell Applications; ECACC; Promocell; TCS Cellworks |
| Bone marrow | Stroma | Cell Applications; ECACC |
| | Hematopoietic stem cells | Befutur; ReachBio |
| | Mesenchymal stem cells (MSCs) | Befutur; Cell Applications; ECACC; Promocell; TCS Cellworks |
| Brain | Neural stem cells | Befutur; Neuromics |
| | Rat, mouse, & human neurons | Invitrogen; Millipore; Neuromics; TCS Cellworks |
| | Rat, mouse, & human astrocytes | Lonza; Millipore; Neuromics; TCS Cellworks |
| Breast | Epithelial | Cell Applications; ECACC; Invitrogen; Lonza |
| Cartilage | Chondrocyte | Cell Applications; ECACC; Cell Genix; Promocell; TCS Cellworks |
| Cornea | Epithelial | Invitrogen; Millipore; TCS Cellworks |
| | Keratocytes | TCS Cellworks |
| | Epithelial filter well constructs | SkinEthic |
| Esophagus | Epithelial | TCS Cellworks |
| Fat | Preadipocytes | BioPredic; Cell Applications; ECACC; Promocell; ReachBio; Zen Biologicals |
| | MSCs | Promocell; Zen Biologicals |
| Heart | Cardiac myocytes | Promocell; ReachBio |
| | Coronary artery endothelial cells; microvascular endothelial cells | Lonza |
| Human embryo | hES & mES cells | Befutur; Millipore |
| | Neural stem cells | Millipore |
| Kidney | Epithelium | ATCC; Lonza; TCS Cellworks |
| Liver | Hepatocytes | Invitrogen; Lonza Biologicals; Promocell; ReachBio; TCS Cellworks; Zen Biologicals |
| | Sinusoidal endothelial | TCS Cellworks |
| Lung | Epithelial | Cell Applications; ECACC; Lonza; TCS Cellworks |
| | Fibroblastic | Cell Applications; ECACC; TCS Cellworks |
| | Endothelial | Cell Applications; ECACC; Lonza |
| | Smooth muscle myoblast | Lonza |
| Lymphatics | Endothelial cells | Lonza |
| Mouse embryo | Fibroblasts | Millipore |
| | Stem (mES) cells | Thermo Fisher |
| Oral cavity | Keratinocytes | TCS Cellworks |
| | Gingival epithelial | Millipore |
| Peritoneum | Preadipocytes | Zen Biologicals |
| | Mesothelial | Zen Biologicals |
| Placenta | Epithelial | Cell Applications; ECACC |
| Prostate | Epithelial | ATCC; Lonza |
| | Fibroblasts | Lonza |
| | Smooth muscle myoblasts | Lonza |
| Skeletal muscle | Skeletal myoblasts | Cell Applications; ECACC; Lonza; Promocell; ReachBio; TCS Cellworks; Zen Biologicals |
| Skin | Keratinocytes | ATCC; BioPredic; Cell Applications; ECACC; Invitrogen; Lonza; Millipore; Promocell; TCS Cellworks; Zen Biologicals |
| | Fibroblasts | ATCC; BioPredic; Cell Applications; ECACC; Lonza; Promocell; Zen Biologicals |
| | Endothelial | Cell Applications; ECACC; Lonza |
| | Preadipocytes | Zen Biologicals |

**TABLE 22.1.** (*Continued*)

| Tissue | Cell type | Supplier[a] |
|---|---|---|
| | Melanocytes | ATCC; BioPredic; Cell Applications; ECACC; Invitrogen; Lonza; Promocell; TCS Cellworks |
| Synovium | Synoviocytes | Cell Applications; ECACC |
| Trachea | Epithelial | Cell Applications; ECACC; Lonza |
| Testis | Sertoli cells | Lonza |
| Umbilical cord | Endothelial | ATCC; Cell Applications; ECACC; Lonza; Millipore |
| | Smooth muscle myoblasts | Cell Applications; ECACC; Lonza |
| | MSCs | Promocell |
| Vascular | Endothelial | ATCC; BioPredic; Cell Applications; ECACC; Invitrogen; Lonza; Promocell |
| | Smooth muscle myoblasts | ATCC; Invitrogen; Lonza; Promocell; TCS Cellworks |
| | Fibroblasts | Invitrogen; TCS Cellworks |

[a]*See also* Appendix II for individual cell types; *see* Appendix III for web addresses.

## 22.1   CELL CULTURE OF SPECIALIZED CELLS

There are many reviews of culture techniques for specific cell types [Doyle et al., 1990–1999; Pollard & Walker, 1990; Freshney et al., 1994; Leigh & Watt, 1994; Freshney & Freshney, 1996; Ravid & Freshney, 1998; Haynes, 1999; Federoff & Richardson, 2001; Freshney & Freshney, 2002; Wise, 2002; Vunjak-Novakovic & Freshney, 2006] (*see also* Table 2.1).

The development of techniques for cell line immortalization [Freshney & Freshney, 1996] (*see also* Section 17.4) has meant that it has been possible to generate continuous cell lines from a number of finite lines from untransformed tissue [Chang et al., 1982; Freshney et al., 1994; Klein et al., 1990; Steele et al., 1992; Wyllie et al., 1992]. In many cases the differentiated properties are lost, but by using a switchable promoter (e.g., temperature sensitivity), it may prove possible to recover the differentiated phenotype. The development of the transgenic mouse carrying the large T gene of SV40 opened up a wide range of possibilities [Jat et al., 1991; Yanai et al., 1991; Morgan et al., 1994; Whitehead & Robinson, 2009] because cells cultured from these animals are already immortalized but still retain some differentiated functions. As these cells carry a temperature-sensitive promoter for the immortalizing gene, it is possible to recover the normal phenotype at the nonpermissive temperature, 37°C.

The protocols in this chapter are grouped by cell type and origin from a particular germ layer (*see* Section 2.4), starting with epithelial cells (Section 22.2; mostly ectodermal and endodermal but a few, e.g., kidney tubules, from mesoderm), then mesenchymal cells (derived from the mesoderm; Section 22.3), neuroectodermally derived cells (e.g., neurons, glia and melanocytes; Section 22.4), and hematopoietic cells (also mesodermal; Section 22.5). It is assumed in these protocols that the basic prerequisites of the tissue culture laboratory (*see* Chapters 3 & 4) will be available. Consequently items such as laminar-flow hoods, inverted microscopes, bench centrifuges, and water baths will not be among the materials listed.

## 22.2   EPITHELIAL CELLS

Epithelial cells are responsible for the recognized functions of many organs (e.g., polarized transport in the kidney and gut, digestive enzyme secretion in the pancreas, metabolism in the liver, gas exchange in the lung, and barrier protection in skin). Epithelial cells are also of interest as models of differentiation and stem cell kinetics (e.g., epidermal keratinocytes [Watt, 2001, 2002]) and are among the principal tissues in which common cancers arise. Consequently the culture of various epithelial cells has been a focus of attention for many years. The major problem in the culture of pure epithelium has been the overgrowth of the culture by stromal cells, such as connective tissue fibroblasts and vascular endothelium. Most of the variations in technique are aimed at preventing such overgrowth by nutritional manipulation of the medium or alterations in the culture substrate (Table 22.2) that promote the growth of the undifferentiated epithelium and, preferably, the stem cells. Subsequent modifications may then be employed to enhance epithelial differentiation, although usually at the expense of proliferation (*see* Section 16.3; Fig. 2.5).

Factors contained in serum—many of them derived from platelets—have a strong mitogenic effect on fibroblasts and tend to inhibit epithelial proliferation by inducing terminal differentiation. Consequently one of the most significant events in the isolation and propagation of specialized cell cultures has been the development of selective, serum-free media (*see* Section 9.2.2), supplemented with specific growth factors (*see* Table 9.4).

The isolation of epithelial cells from donor tissue is often best performed with collagenase (*see* Protocol 11.8), Dispase, or trypsin at 4°C, which disperse the stromal cells but leaves the epithelial cells in small clusters, allowing their separation by settling and favoring their subsequent survival.

### 22.2.1   Epidermis

The epidermis represents one of the most interesting models of epithelial differentiation, partly because of its

**TABLE 22.2.** Inhibition of Fibroblastic Overgrowth

| Method | Agent | Tissue | Reference |
|---|---|---|---|
| Selective detachment | Trypsin | Fetal intestine; cardiac muscle; epidermis | Owens et al., 1974; Milo et al., 1980 |
| | Collagenase | Breast carcinoma | Freshney, 1972; Lasfargues, 1973 |
| Confluent feeder layers | Mouse 3T3 feeder cells | Epidermis | Rheinwald & Green, 1975 |
| | Fetal human intestine feeder cells | Normal and malignant breast epithelium | Stampfer et al., 1980 |
| | | Colon carcinoma | Freshney et al., 1982b |
| Selective inhibitors | D-Valine | Kidney epithelium | Gilbert & Migeon, 1975, 1977 |
| | Cis-OH-proline | Cell lines | Kao & Prockop, 1977 |
| | Ethylmercurithiosalicylate | Neonatal pancreas | Braaten et al., 1974 |
| | Phenobarbitone | Liver | Fry & Bridges, 1979 |
| | Antimesodermal antibody | Squamous carcinomas | Edwards et al., 1980 |
| | | Colonic adenoma | Paraskeva et al., 1985 |
| | Geneticin | Melanocytes; melanoma | Levin et al., 1995 |
| Selective media (*see also* Section 9.2.1) | MCDB 153 | Epidermis | Boyce & Ham, 1983 |
| | MCDB 170 | Breast | Hammond et al., 1984 |
| | Low Ca$^{2+}$ | Epidermal melanocytes | Naeyaert et al., 1991 |

demonstration of developmental hierarchy by histology and partly because it was one of the first models to lend itself to 3T3 feeder layer-supported growth (*sees* Plate 11*a*). Numerous attempts have been made to use cultured epidermal keratinocytes as models for differentiation, as ideal tissue constructs for organotypic culture (*see* Plate 20) and studies on cell interaction Maas-Szabowski et al., 2002, and as grafts in burn or injury repair [Coolen et al., 2007]. Keratinocytes are available from a number of commercial suppliers (*see* Table 22.1; Plate 27*a*)

Protocol 22.1 for the culture of epidermal keratinocytes was contributed by N. E. Fusenig and H-J. Stark, Division of Carcinogenesis and Differentiation, German Cancer Research Center, Im Neuenheimer Feld 280, 6900 Heidelberg, Germany.

Because of the progress in basic cell culture technology and in our understanding of the culture requirements of the various epithelial tissues, keratinocytes of most stratified epithelia can now be grown and studied in cell culture. Mostly the squamous epithelia of the skin and their isolated epithelial cells, the keratinocytes, have been used to study their physiology and pathology in vitro. In addition cells of the oral mucosa [Tomakidi et al., 1997; Grafström, 2002], and of skin appendages such as the hair follicle [for a review, *see* Fusenig et al., 1994a, b], have been isolated and cultured under various conditions, and the reconstructed tissues have been formed in culture as well as in transplants [Limat et al., 1995; Maas-Szabowski et al., 2000, 2002].

Cells from different tissues and organs growing in primary or early subculture are commercially available through different providers (*see* Table 22.1; Appendix II;

Plate 27*a*), facilitating the use of normal cells for various applications such as for comparison of functional properties of cancer cells. Keratinocytes in early subculture provided commercially are usually derived from neonatal foreskin, often not discriminating between the ortho-keratinized cells of the outer leaflet and the mucous-type inner leaflet, thus limiting their comparability to keratinocytes derived from adult skin. Moreover, for specific research purposes, in particular dealing with cells of diseased skin such as psoriatic or malignant keratinocytes, isolation and culture of keratinocytes of specific adult human skin specimens is required.

As a first step separation of the epithelial compartment from the underlying connective tissue is usually done by enzymatic digestion with trypsin [e.g., Smola et al., 1993], Dispase type II (2.4 U/mL, Roche [Tomakidi et al., 1997]), or thermolysin [Germain et al., 1993]. The isolated epithelium is further dispersed by additional incubation in trypsin or, mechanically, by pipetting, after which the suspension is filtered through nylon gauze and propagated in a serum-free, low-calcium medium or on growth-arrested feeder cells by using different media formulations. Subpopulations of keratinocytes with stem cell characteristics can be isolated because of their selective attachment to basement membrane constituents [Bickenbach & Chism, 1998].

To avoid overgrowth by mesenchymal cells, the epithelial compartment has to be separated from the connective tissue and dispersed into single cells, which are then cultured on different substrata by using different media formulations. The most commonly used and reliable method is the feeder layer technique first established by Rheinwald & Green

[1975], who employed postmitotic (irradiated or mitomycin C-treated) 3T3 cells as mesenchymal feeders (*see* Plate 11*a*). Primary fibroblasts derived from dermis or submucosa and rendered postmitotic can be used instead [Limat et al., 1989; Tomakidi et al., 1997]. Feeder cells maintain a broad spectrum of physiological functions [Maas-Szabowski & Fusenig, 1996] even under arrested growth.

Alternatively, keratinocytes can be grown without feeder cells in serum-free media supplemented with pituitary extract or a variety of growth factors and other supplements, either at low (0.03–0.06 mM) $Ca^{2+}$ concentrations to inhibit stratification or at normal (1.2–1.4 mM) $Ca^{2+}$ concentrations to allow stratification and keratinization [Holbrook & Hennings, 1983]. Although maintenance at high cell density in subcultures, in the presence of feeder cells (e.g., irradiated 3T3 cells), or culture in a low–calcium, serum-free medium helps reduce fibroblast contamination, the mesenchymal cells are not completely eliminated by these procedures.

## PROTOCOL 22.1. EPIDERMAL KERATINOCYTES

### Outline

Separate the epidermis from the dermis enzymatically, disaggregate the keratinocytes, and seed them in a serum-free medium or on a growth-arrested feeder layer.

### Materials

#### Sterile:

- ❑ Prepared feeder layers, 3 days old (*see* Section 13.2.3; Protocol 22.4). Irradiate 3T3 at 30 Gy and human fibroblasts at 70 Gy, preferably in concentrated cell suspension.
- ❑ FAD: high-calcium medium for cultures with feeder layers; Ham's F12 and DMEM with additives, as formulated by Wu et al. [1982]: Ham's F12:DMEM, 1:3, supplemented with adenine ($1.8 \times 10^{-4}$ M) cholera toxin ($1 \times 10^{-10}$ M), EGF (1–10 ng/L), hydrocortisone (0.4 µg/mL, 1.1 µM), 5% to 10% FCS, penicillin (100 U/mL), and streptomycin (50 µg/mL)
- ❑ Keratinocyte growth medium (KGM): serum-free medium based on the original formulation of the medium MCDB 153 by Boyce and Ham [1983] (*see* Table 9.1; Appendix II), supplemented with bovine pituitary extract with low $Ca^{2+}$ (0.03–0.5 mM), which can be increased by adding $CaCl_2$
- ❑ Keratinocyte-defined medium (KDM): serum-free medium (*see* Appendix II) based on the original formulation of the medium MCDB 153 by Boyce and Ham [1983]; fully defined medium, tested as a fully competent medium in organotypic cultures [Stark et al., 1999]

- ❑ Supplemented KDM (SKDM):
  Keratinocyte-defined medium (KDM) without pituitary extract (Promocell; Lonza—Clonetics)
  Supplement with: insulin (5 µg/mL), hydrocortisone (0.5 µg/mL), rhEGF (0.1 ng/mL), transferrin (20 µg/mL), 0.1% highly purified bovine serum albumin (free of endotoxins and fatty acids, e.g., A-8806, Sigma), and L-ascorbic acid (50 µg/mL). $Ca^{2+}$ adjusted to 1.3 mM
  Primary keratinocyte cultures on uncoated plastic dishes in SKDM, maintained with low proliferative activity. Growth is improved when keratinocytes are plated on collagen (type 1)-coated dishes.
- ❑ Lifted collagen gels populated with fibroblasts. Organotypic cultures on such collagen gels in defined medium provide epidermal growth and differentiation indistinguishable from that obtained in FAD [Stark et al., 1999]
- ❑ D-PBSA
- ❑ EDTA (0.5%; ~15 mM)
- ❑ Glycerol (analytical grade)
- ❑ FAD medium with 20% FCS and 10% glycerol
- ❑ Enzymes:
  Thermolysin (T-7902, Sigma)
  Trypsin (1:250; 0.2 and 0.6%)
  Dispase (grade II, 2.4 U/mL, Roche Applied Science)
- ❑ Betadine solution (10% iodine)
- ❑ Cryopreservation ampoules
- ❑ Centrifuge tube, 50 mL
- ❑ Nylon gauze
- ❑ "Cell Strainer" (BD Biosciences)
- ❑ Petri dishes, bacteriological grade, 10 cm
- ❑ Scalpels, curved forceps

### Procedure

#### Specimens

1. Obtain foreskin (neonatal as well as juvenile), the most frequent laboratory source for human skin, or trunk skin obtained from surgery or postmortem (up to 48 h). Keratinocytes derived from foreskin seem to attach and proliferate better than cells obtained from adult skin.
2. Incubate skin biopsies for up to 30 min in Betadine solution (10% iodine), to prevent infection. (This does not visibly decrease keratinocyte cell viability.)
3. Rinse twice in D-PBSA for 10 min each.

#### Epidermal separation

Split-thickness skin is optimal

4. Slice full-thickness skin with a Castroviejo dermatome (Storz Instrument) set to 0.1 to 0.2 mm.

Alternatively, subcutaneous tissue and part of the dermis can be eliminated with curved scissors.

5. Dissect skin into $1 \times 2$-cm pieces with scalpels.
6. Rinse the tissue two to five times in D-PBSA.
7. Incubate tissue with protease by one of the following steps:
   (a) Trypsin. Float skin samples on 0.6% trypsin in D-PBSA (pH 7.4) for 20 to 30 min at 37°C. Alternatively (and this works particularly well also with full-thickness skin):
   (b) Cold trypsin. Float the samples on ice-cold 0.2% trypsin at 4°C for 15 to 24 h. The pH of the trypsin solution has to be monitored with phenol red to avoid a shift leading to altered enzyme activity and cell viability.
   (c) Dispase (grade II). Incubate with 2.4 U/mL for 30 min at 37°C.
   (d) Thermolysin (0.5 mg/mL) Germain et al., 1993. Incubate for 12 to 16 h at 4°C.
8. Monitor the separation of the epidermis carefully. When the first detachment of the epidermis is visible at the cut edges of skin samples, place the pieces (dermis side down) in 10-cm plastic Petri dishes and irrigate with 5 mL of complete culture medium including serum.
9. Peel off the epidermis with two fine curved forceps, and collect it in a 50-mL centrifuge tube containing 20 mL of complete culture medium.
   (a) When separation has been performed with trypsin, viable keratinocytes are detached from the epidermal parts by gently pipetting and sieving through nylon gauze of 100-μm mesh ("Cell Strainer," BD Biosciences).
   (b) To obtain single cells from epidermis separated by dispase or thermolysin, additional incubation with trypsin (0.2%) and 1.3 mM EDTA (0.05%), for 10 min at 37°C is required.
10. After splitting the skin in trypsin, gently scrape off the loosely attached epidermal cells on the remaining dermal part, and add the cells to the epidermal suspension. This can lead to a higher cell yield if the purity of the keratinocyte population is not a major issue, but the amount of mesenchymal cell contamination in the resulting keratinocyte cultures is significantly higher.
11. Wash the isolated epidermal cells twice in culture medium by centrifugation at 100 $g$ for 10 min, and count the total number of cells and the viable cells (those that do not absorb Trypan Blue; *see* Protocol 21.1).

### Primary culture

12. Seed cells at 37°C in FAD or KGM medium at desired densities:
   (a) In FAD with feeder cells (dishes seeded 3 days previously with postmitotic 3T3 cells [Rheinwald & Green, 1975] (*see* Protocol 13.3) or skin fibroblasts [Limat et al., 1989]), seed keratinocytes at 2 to $5 \times 10^4$ cells/cm$^2$.
   (b) For culture free of fibroblasts, seed primary keratinocytes at high density ($1-5 \times 10^5$ cells/cm$^2$) in FAD, and maintain them at high density in subcultures.
   (c) Alternatively, seed cells at low ($1 \times 10^3$ cells/cm$^2$) and high ($5 \times 10^4$ cells/cm$^2$) density in KGM in low Ca$^{2+}$ (0.03–0.1 mM), and subculture in the same medium.
   (d) Cells will attach and grow, but at a lower rate, in SKDM, so this medium is advised for experimental conditions when low proliferative activities are required.
13. When cells have attached (after 1–3 days), rinse cultures extensively with medium to eliminate nonattached dead and differentiated cells, and continue cultivation in either FAD or KGM. Stratification and slowing down of growth can be achieved by raising the Ca$^{2+}$ concentration in KGM.

### Subculture

14. Subculture as follows:
   (a) Cultures in FAD:
      (i) Incubate in 1.3 to 2.7 mM (0.05–0.1%) EDTA for 5 to 15 min to initiate cell detachment, which is visible by the enlargement of intercellular spaces.
      (ii) Incubate in 0.1% trypsin and 1.3 mM (0.05%) EDTA at 37°C for 5 to 10 min, followed by gentle pipetting, to completely detach the cells.
   (b) Cultures in KGM
      (i) EDTA pretreatment is not required, because of the low Ca$^{2+}$ concentration.
      (ii) Incubate in 0.1% trypsin with 1.3 mM EDTA, as with FAD cultures.

### Cryopreservation

Cryopreservation is often necessary to maintain large quantities of cells derived from the same tissue sample; the best results are reported when cells from preconfluent primary cultures are used.

(1) To freeze cells
   (a) Trypsinize cells as before, and centrifuge at 100 $g$ for 10 min.
   (b) Resuspend cells in complete culture medium with serum, and count.

(c) Dispense aliquots of $2 \times 10^6$ cells/mL in FAD medium with 20% FCS and 10% glycerol into cryopreservation tubes.

(d) Equilibrate at 4°C for 1 to 2 h.

(e) Freeze cells with a programmed freezing apparatus (Planer) at a cooling rate of 1°C per min. Alternatively, the cryopreservation tubes are inserted into cell freezing boxes filled with isopropanol (e.g., Cryo 1°C Freezing Container, cat. no. 5100-0001, Nunc), which allow an adequate cooling rate when put into a −80°C freezer.

(2) To recover cells

(a) Thaw cryotubes quickly in a 37°C water bath.

(b) Dilute cells 10-fold with medium.

(c) Centrifuge cells and resuspend them at an appropriate concentration in the desired culture medium, and seed culture vessel.

Human and mouse cells can be grown in all three media for several months. Although mouse cells can be subcultured only once or twice, human cells can be passaged four and seven times in FAD and KGM, respectively.

***Characterization of keratinocyte cultures.*** Cultured cells must be characterized for their epidermal (epithelial) phenotype to exclude contamination by mesenchymal cells. This is best achieved by using cytokeratin-specific antibodies for the epithelial cells (*see* Plates 11*a*, 20). Contaminating endothelial cells can be identified by antibodies against CD31 or factor VIII-related antigen. Identifying fibroblasts unequivocally is difficult because the use of antibodies against vimentin (the mesenchymal cytoskeletal element) is not specific; keratinocytes in vitro may initiate vimentin synthesis at frequencies that depend on culture conditions. As a practical assessment for mesenchymal cell contamination, cells should be plated at clonal densities ($1-5 \times 10^2$ cells/cm$^2$) on feeder cells, and clone morphology should be identified at low magnification after fixation and hematoxylin and eosin (H&E) staining of 10- to 14-day cultures. A more specific and highly sensitive method to identify contaminating fibroblasts is the analysis of expression of keratinocyte growth factor (KGF) by RT-PCR. Because this factor is produced in fibroblasts and not in keratinocytes, it represents a selective marker. Moreover KGF expression is enhanced by cocultured keratinocytes, so a minority of contaminating fibroblasts will be detected by this assay [Maas-Szabowski et al., 1999]. Individual fibroblasts in a keratinocyte monolayer may be detected by in situ hybridization with a cDNA of KGF. Human fibroblasts can be identified and quantitated by the fibroblast-specific antibody AS02 (Dianova) by immunoperoxidase technique [Saalbach et al., 1997].

***Variations.*** Keratinocytes can be obtained from the outer root sheaths (ORSs) of plucked scalp hair follicles by dissociating cells from the dissected follicle [Limat et al., 1989]. Up to $5 \times 10^3$ cells can be obtained from three hair follicles and plated on fibroblast feeder cells, resulting in about $1 \times 10^6$ cells within 15 days. Like interfollicular keratinocytes, ORS cells can be subcultured on feeder layer dishes and cryopreserved. These ORS-derived keratinocytes are similar in culture to interfollicular cells, form a regular neoepidermis when transplanted onto nude mice, and are used clinically to cover chronic wounds [Limat et al., 1996].

Keratinocytes can also be grown at clonal density to study clonal cell populations and their different proliferation potentials. For this purpose keratinocytes are cocultured with X-irradiated 3T3 cells [Rheinwald & Green, 1975], at reduced Ca$^{2+}$ concentrations with fibroblast-conditioned medium [Yuspa et al., 1981] or in defined serum-free medium [Boyce & Ham, 1983].

In order to provide conditions that are more in vivo like and to study the regulation of skin physiology by epithelial–mesenchymal interactions, organotypic coculture systems have been developed by seeding the cells on collagen gels (populated with fibroblasts; *see* Plate 20) or pieces of dermis and lifting the supports to the air–medium interface [Fusenig, 1994a]. Under these improved growth conditions keratinocytes express many aspects of growth and differentiation of the epidermis in vivo, including ultrastructural features and a complete basement membrane, features that are absent or less pronounced in submerged cultures on plastic [Smola et al., 1998]. Such organotypic cultures can now also be established and maintained for three weeks in defined SKDM medium, allowing the molecular interaction between epithelial and mesenchymal cells to be analyzed without disturbing influences from serum or other undefined tissue extracts [Stark et al., 1999].

Substantial improvement of epidermal tissue regeneration and differentiation is obtained by culturing keratinocytes on scaffold-enforced dermal equivalents composed of fibroblasts growing in a fibrin matrix within a scaffold of hyaluronic acid fibers (Hyalograft 3D) [Stark et al. 2004], as well as other scaffold materials. Moreover the authentic dermal matrix produced by the fibroblasts within the scaffold provides long-term culture periods of up to three months, enabling for the first time the observation of differentiation features in vitro reminiscent of epidermal homeostasis [Stark et al. 2006; Boehnke et al. 2007]. The effect of the scaffold is obviously not restricted to the hyaluronic acid based substrate of the scaffold but may as well be obtained with other materials [Boehnke et al., 2010]. Surprisingly, the dermal equivalents with authentic matrix produced a microenvironment comparable to stem cell niches in the adult skin enabling the re-expression of telomerase activity and stem cell features in the basal layer of the reformed epidermis in vitro [Muffler et al. 2008; Krunic et al. 2009].

Organotypic keratinocyte cultures are also available commercially (*see* Fig. 21.10; Plate 21; Section 21.6), either as cocultures with fibroblasts or as monocultures of keratinocytes growing on polycarbonate filter inserts (Falcon #3501), and are increasingly used for in vitro toxicity and biocompatibility testing of skin-related products [Fusenig, 1994b] (*see also* Section 21.6).

Optimal growth and differentiation of isolated keratinocytes are obtained under in vivo conditions when the cells are transplanted as intact cultures or in suspension onto nude mice [Fusenig 1994a]. This leads to an almost complete expression of differentiation characteristics of normal epidermis, and all biochemical and ultrastructural features of a completely keratinized epithelium, including the formation of a basement membrane [Breitkreutz et al., 1997].

## 22.2.2 Cornea

There have been a number of attempts to replace the rabbit's eye (Draize) test by using cultured corneal epithelium (*see* Section 21.6; Plate 21), so there has been considerable interest in culturing cornea [Dart, 2003]. The following introduction and Protocol 22.2 for the culture of normal human corneal cells in serum-free medium was provided by C. Cahn when at Gillette Medical Evaluation Laboratories, 401 Professional Drive, Gaithersburg, MD 20879.

The corneal epithelium contacts the external environment directly and is the first tissue compromised in ocular injury. Animal models are most frequently used to model the human ocular surface. The system to be described is being developed for in vitro toxicological investigations, which can complement in vivo studies.

The corneal epithelium in vivo is oxygenated by diffusion from the tear film; in addition it undergoes frequent mitosis. These two characteristics, coupled with the availability of donor corneal tissue, combine to make the corneal epithelium a good starting material for the generation of primary cultures.

Among the sources for the acquisition of corneal tissue is a sophisticated eye banking system that provides donor corneas for transplantation. Eye bank tissue is downgraded and made available to researchers after three days of storage, because of the limited in vitro viability of the endothelial cell layer. Although the endothelium is labile, the epithelium retains its generative capacity for extended periods of time in storage at 4°C, making expired tissue suitable as a source of viable epithelium.

---

### PROTOCOL 22.2. CORNEAL EPITHELIAL CELLS

**Outline**

Corneal tissue is placed on collagen and allowed to adhere, and the outgrowth is expanded and propagated in serum-free medium on fibronectin-collagen-coated surfaces.

---

**Materials**

*Sterile or aseptically prepared:*

❑ Keratinocyte serum-free medium (KGM, Lonza) containing 0.15 mM of calcium, human epidermal growth factor (0.1 ng/mL), insulin (5 µg/mL), hydrocortisone (0.5 µg/mL), and bovine pituitary extract (30 µg/mL)

❑ Eagle's Minimal Essential Medium

❑ D-PBSA: Dulbecco's phosphate-buffered saline without $Ca^{2+}$ and $Mg^{2+}$

❑ FBS

❑ Trypsin-EDTA: trypsin, 0.05%, EDTA, 0.5 mM

❑ Fibronectin/collagen (FNC): fibronectin, 10 µg/mL, collagen, 35 µg/mL, with bovine serum albumin (BSA), 100 µg/mL, added as a stabilizer

❑ Biocoat six-well plate precoated with rat tail collagen, type I (BD Biosciences)

**Procedure**

Primary cultures

1. Place donor corneas epithelial side up on a sterile surface, and cut them into 12 triangular-shaped wedges, using a single cut of the scalpel and avoiding any sawing motion. Careful handling of the cornea in this manner decreases damage to the collagen matrix of the stroma and minimizes the liberation of fibroblasts.

2. Turn each corneal segment epithelial side down, and place four segments in each well of a precoated six-well plate.

3. Press each segment down gently with forceps to ensure good contact between the tissue and the tissue culture surface. Allow the tissue to dry for 20 min.

4. Place one drop of KGM carefully upon each segment, and incubate the culture overnight at 37°C in 5% $CO_2$. Although the donor corneas received from the eye bank are stored in antibiotic-containing medium (either McCarey–Kauffman or Dexsol), all manipulations are performed under antibiotic-free conditions.

5. The next day, add 1 mL of KGM to each well. During the initial culture period, cells are observed to emigrate only from the limbal region of the cornea. No cells are observed to migrate away from the central cornea or the sclera. Fibroblast outgrowth is minimized by utilizing a serum-free medium that is low in calcium (0.15 mM) and that minimizes disruption of the collagen matrix.

6. Remove the tissue segment with forceps 5 days after the explantation of the donor

cornea slice, and add 3 mL of medium. After removal of the donor tissue, adherent cells continue to proliferate, and within 2 weeks from the time of establishment of the culture, confluent monolayers form, displaying the typical cobblestone morphology associated with epithelia. The yield is approximately 1 to $5 \times 10^6$ cells/cornea.

### Propagation

7. Following the initial outgrowth period, feed the cultures twice per week.
8. At 70% to 80% confluence, rinse the cells in D-PBSA, and release with trypsin/EDTA for 4 min at 37°C.
9. Stop the reaction with 10% FBS in D-PBSA.
10. Wash the cells (centrifugation followed by resuspension in KGM), count them, and plate at $1 \times 10^4$ cells/cm$^2$ onto tissue culture surfaces coated with FNC.
11. Incubate the culture at 37°C in 95% air and 5% $CO_2$.
12. Change the culture medium to fresh medium 1 day after trypsinization and reseeding.

Immediately after passage, cells appear more spindle-shaped, are refractile, and are highly migratory. Within seven days, control cultures become 70% to 80% confluent, continue to display a cobblestone morphology, and, if allowed to become postconfluent, retain the ability to stratify in discrete areas.

Although corneal epithelial cultures can be subcultured up to five times (approximately 9–10 population doublings), most of the proliferation occurs between the first and third passages. Approximate yields are 1 to $2 \times 10^6$ cells/cornea. Senescence always ensues by the fifth passage.

***Development of continuous cell lines.*** Lines of human corneal epithelium (HCE) with an extended life span have been developed using SV40LT immortalization (*see* Protocol 17.1) to provide a larger supply of cells for experimental purposes.

***Long-term storage of cells.*** Cells can be stored frozen in liquid nitrogen (*see* Protocol 19.1).

***Phenotypic development in vitro.*** Both primary cultures and HCE lines retain phenotypic characteristics of corneal epithelium in situ. They continue to synthesize collagenase and express EGF receptors and cornea-specific cytokeratins, although the level of expression under current culture conditions is less than that observed in situ. When corneal epithelium is cultured upon collagen membranes at

air–liquid interfaces, its morphology and barrier function are fairly well preserved. Stratified membranes develop that are able to inhibit the diffusion of Na-fluorescein. Air–liquid interface cultures survive for two weeks in vitro and have been used to investigate injury and repair mechanisms. A more complete three-dimensional tissue model may be constructed by supplying corneal fibroblasts in a collagen gel (*see* Sections 25.3.6, 25.4).

It is also possible to culture limbal, stromal, and endothelial stem cells from cornea [Du & Funderburgh, 2007]. Limbal stem cells will differentiate into stratified layers at the gas–liquid interface.

Human corneal epithelial cells propagated in vitro may provide a suitable model for exploring basic cell biological mechanisms as well as toxicological phenomena. Although the primary cultures are adequate for in vitro studies, HCE lines with an extended life span provide a reliable source of material that can be shared among laboratories.

### 22.2.3 Breast

Milk [Buehring, 1972; Ceriani et al., 1979] and reduction mammoplasty [Stampfer et al., 2002; Labarge et al., 2007] are suitable sources of normal ductal epithelium from the breast; the former gives purer cultures of epithelial cells, but they do not survive as long. Disaggregation in collagenase [Speirs et al., 1996] is preferred for primary disaggregation, growth on confluent feeder layers of fetal human intestine [Stampfer et al., 1980; Freshney et al., 1982b] represses stromal contamination of both normal and malignant tissue (*see* Plates 6c, d, 15b), and optimization of the medium [Stampfer et al., 1980; Smith et al., 1981; Hammond et al., 1984] enables serial passage and cloning of the epithelial cells. Cultivation in collagen gel allows three-dimensional structures to form that correlate well with the histology of the original donor tissue [Berdichevsky et al., 1992; Gomm et al., 1997] and, when cultured in Matrigel, can generate terminal ductal lobular units (TDLU) containing stem cells as well as differentiated acinar epithelium [LaBarge et al., 2007] (*see* Protocol 22.3). As with epidermis, cholera toxin [Taylor-Papadimitriou et al., 1980] and EGF [Osborne et al., 1980] stimulate the growth of epithelioid cells from normal breast tissue in vitro. The hormonal picture is more complex. Many epithelial cells survive better with insulin, (1–10 IU/mL) and hydrocortisone (∼10 nM), added to the culture. The differentiation of acinar breast epithelium in organ culture requires hydrocortisone, insulin, and prolactin [Darcy et al., 1995], and estrogen, progesterone, and growth hormone have also been shown to be required in cell culture [Klevjer-Anderson & Buehring, 1980]. There are a number of commercial sources for mammary epithelium (*see* Table 22.1; Appendix II; Plate 27c.)

Protocol 22.3 for the culture of cells from reduction mammoplasty was contributed by O. Petersen, M. LaBarge, and M. Bissell, and was originally published in Labarge et al. [2007].

(a)

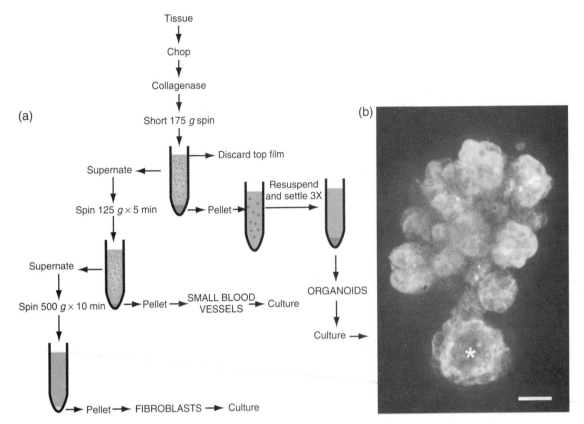

(b)

**Fig. 22.1. Isolation of Organoid Structures from Breast.** (*a*) Flowchart of isolation of different fractions by selective sedimentation and centrifugation following collagenase digestion. (*b*) A TDLU-like structure derived from a Muc-/ESA+/CD29hi D920 cell grown in collagen I/laminin-1 three-dimensional culture Labarge et al., 2007. The structure was stained for keratin-14 (red) and keratin-8 (green) expression, and imaged with a confocal microscope. The presented image is a reconstruction of several optical slices to give to appearance of three dimentions. There is a lumen indicated by a star. Scale bar = 50 μm. (*See also* Plate 26; Reproduced from LaBarge et al., 2007.)

## PROTOCOL 22.3. PREPARATION OF MAMMARY EPITHELIAL CELLS FROM REDUCTION MAMMOPLASTY SPECIMENS

### Outline

Specimens of reduction mammoplasty tissue are disaggregated in collagenase and the resulting crude digest separated by centrifugation and sedimentation into three main fractions (Fig. 22.1), each of which can be seeded on Vitrogen-coated 25-cm² flasks.

### Reagents and Materials

*Sterile or aseptically prepared:*

❑ DMEM/F12, 50:50
❑ CMD3 medium Petersen et al., 1990. Add to DMEM:F12, 1:1:
   Sodium selenite ................................. 2.6 ng/mL

| | |
|---|---|
| Epidermal growth factor (EGF) .......... | 100 ng/mL |
| Hydrocortisone ............................... | 0.5 μg/mL |
| (1.38 μM) | |
| Triiodothyronine ............................. | 10 nM |
| Fibronectin ..................................... | 100 ng/mL |
| Glutamine ...................................... | 2 mM |
| Transferrin ..................................... | 25 μg/mL |
| Dibutyryl cAMP ............................. | 10 nM |
| Phosphoethanolamine ...................... | 0.1 mM |
| Fetuin .......................................... | 20 μg/mL |
| Ascorbic acid ............................... | 0.06 mM |
| (10 μg/mL) | |
| Bovine serum albumin (fraction V) ..... | 0.01% |
| HEPES buffer .................................... | 10 mM |
| Gibco trace element mix ................... | 1:100 |
| Insulin .......................................... | 3.0 μg/mL |
| 17β-Estradiol.................................... | 0.1 nM |

Ethanolamine .................................0.1 mM
- ❏ Collagenase, 900 IU/mL in DMEM/F12
- ❏ Vitrogen
- ❏ Scalpels
- ❏ Culture flasks, 25 cm$^2$, Vitrogen-coated at 8 µg/cm$^2$

*Nonsterile:*

- ❏ Rotary shaker
- ❏ Swinging bucket centrifuge

**Procedure**

1. Transfer the tissue specimens to DMEM:F12 (1:1) immediately following surgery.
2. Using two scalpels (one in each hand) mince the tissue into roughly 2-mm cubes.
3. Digest the minced fragments in collagenase for 24 to 48 h on a rotary shaker (60 rpm) at 37°C.
4. Centrifuge the crude digest at 175 *g* for a few seconds:
   (a) Excess lipid and most cells with high lipid content float at the top of the supernate and should be discarded.
   (b) The supernate is comprised of smaller organoids of vascular origin and of single cells.
   (c) The pellet is comprised of larger organoids of both epithelial and vascular origin.
5. Resuspend the pellet in 10 mL of DMEM/F12. Within 30 s the larger epithelial organoids will sediment, leaving blood vessel fragments still suspended in the medium. These should be separated from the epithelial organoids as they will eventually settle to the bottom.
6. Isolate relatively pure fractions of epithelial organoids from the large blood vessels. Usually two additional rounds of resuspending the epithelial organoids in 10 mL of fresh media followed by sedimentation are sufficient.
7. Centrifuge the supernate from step (4) at 125 *g* for 5 min. The pellet will contain small blood vessels. Remove the supernate and centrifuge at 500 *g* for 10 min. The pellet will contain resident fibroblasts.
8. Seed the epithelial organoids, and other fractions if required, on to Vitrogen-coated 25-cm$^2$ flasks.
9. Maintain the epithelial organoids in CDM3 medium at 37°C, 5% $CO_2$, changing the medium three times per week.

These cultures can be used to isolate cloned populations of mammary stem cells and will form terminal ductal lobular units (TDLU) when grown in 3-D culture in Matrigel or collagen I with laminin-1 added [LaBarge et al., 2007].

Mammary epithelium is also available commercially (*see* Table 22.1; Appendix II; Plate 27*c*).

### 22.2.4 Cervix

Cervical keratinocytes can be grown in serial culture at clonal density by using a modification [Stanley & Parkinson, 1979] of the method described for epidermal keratinocytes [Rheinwald & Green, 1975]. Protocol 22.4 for the culture of epithelial cells from cervical biopsy samples was contributed by M. Stanley, Department of Pathology, University of Cambridge, UK.

---

**PROTOCOL 22.4. CERVICAL EPITHELIUM**

***Outline***

Single-cell suspensions from enzymatically disaggregated epithelium from punch or wedge cervical biopsies are inoculated into flasks or plates either with lethally inactivated Swiss 3T3 cells and grown in a medium supplemented with serum and growth factor (Protocol A) or grown in serum-free keratinocyte growth medium (KGM) without feeder support, by using modifications of the methods originally described for epidermal keratinocytes [Boyce & Ham, 1983]. When the keratinocyte colonies that arise contain 1000 cells or more, the cultures can be trypsinized and passaged again with the use of fibroblast feeder support.

***Materials***

*Sterile:*

- ❏ Transport medium: Dulbecco's modification of Eagle's medium (DMEM), supplemented with 10% fetal calf serum, 100 µg/mL of gentamycin sulfate, and 10 µg/mL of amphotericin
- ❏ Trypsin EDTA solution for disaggregation of epithelial cells: 0.25% trypsin (v/v), 0.25 mM (0.01%) EDTA as the disodium salt with pH 7.4 in D-PBSA
- ❏ Medium for Protocol A: Keratinocyte culture medium (KCM): DMEM supplemented with 10% fetal bovine serum, 0.5 µg/mL of hydrocortisone, and $1 \times 10^{-10}$ M cholera toxin:

  Fetal bovine serum: not all batches support growth adequately. Serum samples should be tested and a large batch of suitable quality bought.

  Epidermal growth factor (EGF): Sigma, 100 µg dissolved in 1 mL sterile UPW. Aliquot in 100-µL aliquots and store at −20°C. Prepare working stocks by diluting 100 µL of EGF at 100 µg/mL in 10 mL medium with serum. Store at 4°C. Dilute 1:100 in medium with serum for use at a final concentration of 10 ng/mL

Cholera toxin (CT; Sigma): add 1.18 mL of sterile UPW to 1 mg of CT to give a $10^{-5}$ M solution. Store at $+4°C$. Working stocks should be 100 μL of $10^{-5}$ M CT in 10 mL of medium with serum. Filter sterilize the culture and keep it at $+4°C$. Use at a final concentration of $10^{-10}$ M.

Hydrocortisone (Sigma): dissolve 1 mg of hydrocortisone in 1 mL of 50% ethanol/water (v/v). Dispense into 100-μL aliquots and keep at $-20°C$. Use at a final concentration of 0.5 μg/mL.

☐ Medium for Protocol B: keratinocyte growth medium: KGM (Lonza), or Keratinocyte-SFM (Invitrogen)

☐ Petri dishes: for tissue culture, 6-cm and 9-cm diameter

☐ Forceps, rat toothed

☐ Curved iris scissors

☐ Disposable scalpels, No. 22 blade

☐ Pipettes

☐ Centrifuge tubes

*Nonsterile:*

☐ Hemocytometer

☐ Radiation source (e.g., X rays or $^{60}Co$)

### Swiss 3T3 fibroblasts (Protocol A only)

1. Prepare a large master stock of cells and freeze in individual ampoules of $1 \times 10^6$ cells. Cells should not be used for more than 20 passages.

2. Grow 3T3s in DMEM/10% calf serum in 175-cm² tissue culture flasks. Inoculate cells at $1.5 \times 10^4$ cells/cm². Change the medium after 2 days. Subculture every 4 to 5 days.

3. To avoid low-level contamination, maintain one master flask of cells on antibiotic-free medium; these cells are then used at each passage to inoculate the flasks required for that week's feeder cells.

4. Inactivate Feeder layers by irradiation with 60 Gy (6000 rad), either from an X-ray or $^{60}Co$ source. Irradiated cells (XR-3T3) may be kept at 4°C for 3 to 4 days.

5. In the absence of a source of irradiation, inactivate feeder cells with mitomycin C.

6. Expose 3T3 cells growing in monolayer to mitomycin C at 20 μg/mL for 1 h at 37°C.

7. Trypsinize the treated cells, resuspend and wash the cell pellet twice with fresh medium with serum, resuspend the cells at a suitable concentration in complete medium, and use.

### Procedure A: With 3T3 Feeder Layers

#### Primary culture

1. Remove the cervical biopsy from the transport medium, and wash the cells two to three times with 5 mL of sterile D-PBSA containing gentamycin sulfate, 50 μg/mL, and amphotericin, 5 μg/mL.

2. Place the biopsy, epithelial surface down, on a sterile culture dish.

3. Using a disposable scalpel fitted with a No. 22 blade, cut and scrape away as much of the muscle and stroma as possible, leaving a thin, opaque epithelial strip.

4. Mince the epithelial strip finely with curved iris scissors.

5. Add 10 mL of trypsin/EDTA (prewarmed to 37°C) to the epithelial mince, and transfer the tissue to a sterile glass universal container containing a small plastic-coated magnetic stirrer bar.

6. Add a further 5 to 10 mL of trypsin/EDTA.

7. Place the universal container on a magnetic stirrer in an incubator or a hot room at 37°C, and stir slowly for 30 to 40 min.

8. Allow the suspension to stand at room temperature for 2 to 3 min.

9. Remove the supernate containing single cells, and filter it through a stainless steel or plastic mesh (e.g., Cell Strainer, BD Biosciences) into a 50-mL centrifuge tube (*see* Fig. 11.8).

10. Add 10 mL of complete medium to the filtrate.

11. Add a further 15 mL of warm trypsin/EDTA to the fragments in the universal container, and repeat steps 5 through 9 twice. Combine the trypsin supernates and spin in a bench centrifuge at 1000 rpm (80 *g*) for 5 min.

12. Remove the supernate, add 10 mL of complete medium to the pellet, resuspend the cells vigorously to give a single-cell suspension, and count the cells with a hemocytometer. Assess cell viability with Trypan Blue exclusion (*see* Protocol 21.1).

13. Dilute the cervical cell suspension with KCM, and plate cells out at $2 \times 10^4$ cells/cm² together with $1 \times 10^5$ cells/cm² of lethally inactivated 3T3 cells (i.e., for $1 \times 10^5$ cervical cells, $5 \times 10^5$ XR-3T3/60-mm dish).

14. Incubate the cultures at 37°C in 5% $CO_2$.

15. Seventy-two hours after the initial plating, replace the medium with a complete medium supplemented with EGF at 10 ng/mL. Check

the cultures microscopically to ensure that the feeder layer is adequate. Add further feeder cells if necessary.

16. Change the medium twice weekly; EGF should be present in the medium, except when the cells are initially plated. Keratinocyte colonies become visible on the microscope by days 8 to 12 and should be visible to the naked eye by days 14 to 16. Cultures should be passaged at this time.

### Subculture

17. Remove the medium from the cell layer, and remove the feeders by rinsing rapidly with 0.01% EDTA. Wash twice with D-PBSA.
18. To each culture dish, add enough prewarmed trypsin/EDTA to cover the cell sheet. Leave the cultures at 37°C until the keratinocytes have detached; check for detachment with a microscope. Do not leave the cells in trypsin for more than 20 min.
19. Remove the cell suspension from the plate and transfer it to a sterile centrifuge tube.
20. Rinse the growth surface with complete medium and add to the suspension. Mix and dispense the suspension with a 10-mL pipette.
21. Spin the cells at 80 to 100 $g$ for 5 min.
22. Remove the supernate, add 10 mL of complete medium, and resuspend the cells vigorously with a 10-mL pipette to achieve a single cell suspension.
23. Count the cells with a hemocytometer.
24. Cells may be replated on inactivated 3T3 cells and grown as just described or frozen for later recovery.

### Procedure B: Serum-Free Medium

#### Primary culture

1. Follow steps 1 through 11 of Protocol A to obtain a single-cell suspension in medium supplemented with serum.
2. Spin the suspension at 1000 rpm (80$g$) for 5 min. Remove the supernate and resuspend the cells in 10 mL of D-PBSA.
3. Spin the suspension again and wash the cells once more with D-PBSA. Resuspend the cells in 10 mL of keratinocyte growth medium (prepared as per the manufacturer's instructions), and seed into culture dishes or flasks at a density of $2 \times 10^5$ cells/cm$^2$ ($10^6$ cells/5-cm Petri dish, $4 \times 10^6$/9-cm Petri dish).
4. Incubate the cultures at 37°C in 5% $CO_2$.
5. Change the medium after 72 h and twice weekly thereafter. Cultures become confluent within 14 to 20 days and should be passaged at that time.

### Subculture

6. Follow steps 17 through 23 of Protocol A.
7. Spin the suspension at 80 to 100 $g$ for 5 min. Remove the supernate and resuspend the cells in 10 mL of D-PBSA.
8. Spin the suspension again and wash the cells once more with D-PBSA. Resuspend the cells in KGM and plate them onto culture dishes at $10^5$ cells/cm$^2$ ($5 \times 10^5$ cells/50-mm Petri dish, $2 \times 10^6$ cells/90-mm Petri dish).
9. Cells may also be frozen at this stage for recovery at a later date.

Cervical keratinocytes grown as just described can be used for a range of purposes, including investigations of papillomavirus carcinogenesis, differentiation studies, and examination of the response of these cells to mutagens and carcinogens.

## 22.2.5  Gastrointestinal Tract

The culture of normal epithelium from the gut lining has not been extensively reported, although there are numerous reports in the literature of continuous lines from human colon carcinoma (*see* Sections 24.6.1, 24.8.4). Owens et al. [1974] were able to culture cells from fetal human intestine as a finite cell line (FHS 74 Int), and extensive use has been made of the rat intestinal cell line IEC–6 [Rak et al., 1995; Bedrin et al., 1997]. Intestinal epithelium has also been isolated with a Wnt–dependent stem cell niche [Ootani et al., 2009].

Protocol 22.5 was condensed from Booth and O'Shea [2002]. The preparation of tissue specimens for cell culture is usually started within 1 to 2 h of removal from the patient. If this is impossible, fine cutting of the tissue into small pieces (1–2 mm) with scalpels and storage overnight at 4°C in washing medium (HBSS/PSG, *see* below) can also prove successful.

---

### PROTOCOL 22.5. ISOLATION AND CULTURE OF COLONIC CRYPTS

#### Outline
Finely chopped tissue is disaggregated in collagenase and dispase, and the crypts are separated by sedimentation through sorbitol.

#### Materials

*Sterile:*

❑ HBSS/PSG: Hanks's balanced salt solution with penicillin, 100 U/mL, streptomycin, 30 µg/mL, and gentamycin, 25 µg/mL
❑ Digestion mix: collagenase Type XI, 150 U/mL, Dispase, 40 µg/mL, FBS, 1%, in DMEM

❑ Growth medium: DMEM with 25 mM glucose, 6.8 mM pyruvate, 0.25 U/mL insulin, EGF, 50 ng/mL, FBS, 2.5%, and antibiotics as in HBSS above, buffered to pH 7.4 with sodium bicarbonate under 7.5% $CO_2$

❑ S-DMEM: growth medium with 2% sorbitol (0.11 M), sterilized by filtration

❑ All media should be at 37°C.

❑ Petri dishes, 9 cm

❑ Universal containers or 30- to 50-mL centrifuge tubes

❑ Multiwell plates, 24-well, collagen-coated:
   (i) Dilute stock Vitrogen 10-fold.
   (ii) Incubate plates with 0.5 mL/well for 3 h.
   (iii) Remove excess Vitrogen and dry plates in laminar-flow hood.
   (iv) Can be stored wrapped in cling film (Saran Wrap) for up to 1 week at 4°C.

❑ Flasks, 25 $cm^2$

❑ Scalpels, #4 holder, #22 blade

❑ Movette pipette (or similar wide-bore plastic pipette)

### Procedure

#### Isolation of crypts from human colon

1. Place tissue for culture in DMEM as soon as possible after excision from the patient. Preliminary studies indicate that as long as the tissue is quickly placed into DMEM, it will withstand a short period (up to 2 h) at 4°C, although a longer period is not advisable.

2. Wash the intact tissue twice in 1× HBSS/PSG to remove any contaminating factors.

3. Transfer the tissue to a clean Petri dish with fresh HBSS/PSG.

4. Remove as much muscle as possible and chop the tissue into pieces with two opposed scalpels (see Fig. 11.12).

5. Wash the chopped tissue pieces by transferring them to a 25-$cm^2$ flask containing 20 mL HBSS/PSG and pipette up and down with a Movette pipette.

6. Allow contents of the flask to settle and remove the HBSS/PSG.

7. Repeat the process until the HBSS/PSG remains almost clear. This usually takes about 5 repeats of the washing process.

8. Transfer the tissue to a Petri dish and remove any excess HBSS/PSG.

9. Chop the tissue with the scalpels until it has a fairly smooth consistency. Transfer the chopped tissue to a 25-cm flask with a wide-bore pipette (ensure that the inside of the pipette is already wet with HBSS/PSG to prevent sticking).

10. Allow pieces to settle and remove excess HBSS/PSG.

11. Add 25 to 30 mL of the digestion mix.

12. Incubate at 37°C for 1 h.

13. Replace the digestion mix.

14. Allow the tissue to settle.

15. Carefully remove the digestion mix from the flask and discard it (being sure not to remove any undigested tissue in the process).

16. Add fresh digestion mix and continue incubation at 37°C for a further 1 to 2 h.

17. Check the tissue every 15 to 30 min to ensure that crypts do not become overdigested. The isolation is ready when approximately 70% to 80% of the crypts have become liberated as individual free crypts.

*Note.* Do not wait until all crypts are liberated as this may result in a reduction in yield caused by overdigestion.

#### Sedimentation of human colonic crypts

Released crypts are purified by sorbitol sedimentation as follows. Large aggregates of fatty/mesenchymal tissue that rise to the top of the sedimentation tube should be discarded in the first steps because they may cause problems during the sedimentation process.

18. Divide the contents of the digestion flask between two 30-mL sterile universal containers and label "Set 1."

19. Top up to 25 mL with S-DMEM and remove any floating fatty or other mesenchymal tissue.

20. Allow the contents of the tubes to settle for a period of 1 min to allow any undigested matter to sediment.

21. Discard any fatty material floating on the surface.

22. Gently remove the supernatant suspension from the tubes and transfer to two clean tubes; label "Set 2."

23. Let the suspension settle for 30 s.

24. Transfer the contents to a further set of clean tubes; label "Set 3."

25. Top up with S-DMEM.

26. Centrifuge Set 3 tubes at 200 to 300 *g* for 4 min. (The slowest speed on a standard bench top centrifuge is usually about right.)

27. Discard the supernate.

28. Disperse the pellet by flicking/tapping the base of the tube.

29. Resuspend the crypts once more with S-DMEM and repeat the centrifugation step. This S-DMEM washing/centrifugation (Step 26) must be repeated about 5 times, or until the supernate is clear when the crypts are then ready to be plated out.

**30.** Before plating out the crypts examine the debris left behind in Set 1 and 2 tubes microscopically for the presence of crypts. If a high proportion of crypts remain, the material can be agitated and the initial sedimentation step repeated. If only undigested material remains, this can be discarded.

**31.** A critical plating density is required for good culture growth. Generally, the crypts appear to grow best when plated at a density of 800 to 1000 crypts/mL/well in a 24-well plate. To determine the number of crypts isolated:

**(a)** Resuspend the isolation in 10 mL of growth medium and agitate.

**(b)** Remove 100 μL and add to 900 μL growth medium.

**(c)** Take 100-μL sample from this and count the number of crypts present.

**(d)** Number of crypts/mL = Number of crypts counted ×100.

**(e)** Plate 800 to 1000 crypts per well in 1 mL growth medium.

**32.** Incubate the plates at 37°C in 7.5% $CO_2$ and allow the cells to attach for 2 days.

**33.** Remove the medium, containing all unattached material, from the plates and replace with fresh medium.

**34.** Supplement the medium at 5-day intervals with 0.5 mL fresh medium.

Crypts can be plated out on either collagen-coated plates or a 3T3 feeder layer. This method will allow routine culture for a period of about 1 week. After this period the epithelial cells deteriorate and the contaminating fibroblasts become prevalent. The cultures do not subculture well. Various cell recovery agents have been utilized, but so far we have had limited success subculturing intestinal epithelial cells. Not surprisingly, fibroblasts can be quite easily subcultured and are a potential source of feeder layer cells.

## 22.2.6  Liver

Although cultures from adult liver do not express all the properties of liver parenchyma, there is no doubt that selective isolation of the different cell types is possible and that the correct lineage of cells may be cultured. Because the main interest is devoted to hepatocytes, most efforts have been focused on conditions allowing improvement of their survival and function in vitro, including interaction with other cells in the liver [Corlu et al., 1997], so that functional hepatocytes isolated from liver tissue can be cultured under the correct conditions [Guguen-Guillouzo, 2002]. So far attempts at generating proliferating cell lines, mainly from human origin, have not been particularly successful, although primary hepatocytes are available from a number of sources

(*see* Table 22.1; Appendix II). However, the establishment of some differentiated cell lines has been reported such as Hep-G2 [Knowles et al., 1980] and recently great progress has been achieved with the new human hepatoma cell line, HepaRG, which is able to undergo a complete hepatocyte differentiation program and therefore represents a real alternative to human hepatocyte primary cultures (*see* Section 22.2.8).

### 22.2.7  Hepatocyte Primary Cultures

The following protocol for the culture of isolated adult hepatocytes was contributed by C. Guguen-Guillouzo, Université Rennes1, INSERM U522, Hôpital de Pontchaillou, Rue Henri le Guilloux, 35033 Rennes, France.

When perfused into the liver through the vessels and capillaries at an adequate flow rate, proteolytic enzymes such as collagenase, which are relatively noncytotoxic, will disrupt intercellular junctions and will digest the connective tissue framework within 15 min if the liver is previously cleared of blood and depleted of $Ca^{2+}$ by washing with calcium-free buffer [Berry and Friend, 1969]. Hepatocytes are selected from the cell suspension by two or three differential centrifugations at low speed.

---

**PROTOCOL 22.6A. ISOLATION OF RAT HEPATOCYTES**

**Outline**

Introduce a cannula in the portal vein or a portal branch, wash the liver with a calcium-free buffer (15 min), perfuse the liver with the enzymatic solution (15 min), collect and wash the cells, and count the viable hepatocytes [Guguen-Guillouzo and Guillouzo, 1986].

**Materials**

*Sterile:*

❑ L-15 Leibovitz medium

❑ Ham's F12 or Williams's E medium enriched with 0.2% bovine albumin (grade V, Sigma) and 10 μg/mL of bovine insulin (80–100 mL)

❑ Hepatocyte minimal medium (HMM): Eagle's MEM, 67.5%; medium 199, 32.5% (or Williams's E in place of both); bovine insulin, 5 μg/mL; BSA, 1 mg/mL; penicillin, 100 U/mL; streptomycin, 100 μg/mL; and 20% FBS

❑ HMM/SF: serum-free HMM with $1 \times 10^{-5}$ M hydrocortisone hemisuccinate

❑ Calcium-free HEPES buffer pH 7.65: 160.8 mM NaCl; 3.15 mM KCl: 0.7 mM $Na_2HPO_4 \cdot 12H_2O$, 33 mM HEPES, sterilization by 0.22-μm micropore filter (Millipore), and storage at 4°C (2 months)

❑ Collagenase solution: 0.025% collagenase (Sigma grade 1 or Roche #103578 or others); 0.075% $CaCl_2 \cdot 2H_2O$ in calcium-free HEPES buffer pH 7.65; preparation and sterilization by filtration just before use. As alternatives to collagenase, Liberase Blendzyme 3 or Liberase TM (Roche) can be used at a concentration 10 times lower, e.g. 0.0025%

❑ Nembutal (Abbot, 5%)

❑ Heparin (Roche)

❑ Tygon tubing (ID, 3.0 mm; OD, 5.0 mm)

❑ Disposable scalp vein infusion needles, 20G (Dubernard Hospital Laboratory, Bordeaux, France)

❑ Sewing thread for cannulation

❑ Graduated bottles and Petri dishes

❑ Surgical instruments (sharp, straight, and curved scissors and clips)

❑ Disposable syringes, 1-mL, 2

*Nonsterile:*

❑ Chronometer

❑ Peristaltic pump (10–200 rpm)

❑ Water bath

### Procedure

1. Warm the washing HEPES buffer and collagenase solution in a water bath (usually approximately 38–39°C to achieve 37°C in the liver). Oxygenation is not necessary.

2. Set the pump flow rate at 30 mL/min.

3. Anesthetize the rat (180–200 g) by intraperitoneal injection of Nembutal (150 µL/100 g), and inject heparin into the femoral vein (1000 IU).

4. Swab the underside of the rat with 70% alcohol.

5. Open the abdomen, place a loosely tied ligature around the portal vein approximately 5 mm from the liver, insert the cannula up to the liver, and ligate.

6. Rapidly incise the subhepatic vessels to avoid excess pressure, and start the perfusion with 500 mL of calcium-free HEPES buffer at a flow rate of 30 mL/min; verify that the liver whitens within a few seconds.

7. Perfuse 300 mL of the collagenase solution at a flow rate of 15 mL/min for 20 min. The liver becomes swollen.

8. Remove the liver and wash it with the calcium-free HEPES buffer.

9. After disrupting the Glisson capsule, disperse the cells in 100 mL of L-15 Leibovitz medium.

10. Filter the suspension through two layers of gauze or 60- to 80-µm nylon mesh, allow the viable cells to sediment for 20 min (usually at room temperature), and discard the supernate (60 mL) containing debris and dead cells.

11. Wash the cells three times by slow centrifugations (50 g for 40 s) to remove collagenase, damaged cells, and nonparenchymal cells.

12. Suspend isolated hepatocytes in HMM and proceed to count cell yield and determine viability.

13. Seed the cells at $7 \times 10^5$ viable cells/mL. Optionally a Percoll gradient can be performed (*see* Variations below *and* Protocol 14.1).

14. After 2 to 3 h, living cells will attach to plastic and begin to spread. Remove the dead cells with the supernatant medium.

15. After cell attachment, remove the FBS, and replace the medium with fresh HMM/SF. Renew every day thereafter.

If Liberase is chosen in place of collagenase, take into account that Liberase is 10 times more active than collagenase. When using biomatrix components for coating the dish (*see* Section 7.2.1), fibronectin, collagen or Matrigel are found to be most effective. FBS addition can be avoided completely. Growth factors (*see below* Variations: Hepatocyte Proliferation) can be added to the medium at cell seeding or early after cell spreading.

*Analysis.* Determine the cell yield and viability by the well-preserved birefringent shape or 0.2% Trypan Blue exclusion. The yield is usually 4 to $6 \times 10^8$ cells with a viability of more than 95%.

### Variations

*Enrichment of viable cells.* Fractionation on a Percoll gradient can greatly improve the viability of the hepatocyte suspension. It is performed on freshly isolated cells as follows: 10 mL of a 30% Percoll solution prepared in HMM are added to a sterile tube and 10 to 20 mL of the cell suspension are gently layered on top. A centrifugation of 5 min at 50 g allows the viable cells to sediment. The supernate is withdrawn and the cell pellet can be suspended in HMM for a new cell count and percentage viability calculation before seeding.

*Isolation of hepatocytes from other species.* The basic two-step perfusion procedure [Seglen, 1975] can be used for obtaining hepatocytes from various rodents, including mouse, rabbit, guinea pig, or woodchuck, by adapting the volume and the flow rate of the perfused solutions to the size of the liver. The technique has been adapted for the human liver [Guguen-Guillouzo et al., 1982] by perfusing a portion of the whole liver (usually with 1.5 L HEPES buffer and 1 L collagenase solution at 70 and 30 mL/min, respectively) or by taking biopsies (15–30 mL/min, depending on the size of the tissue sample). A complete isolation into a single cell suspension can be obtained by an additional collagenase incubation at 37°C under gentle stirring for 10 to 20 min (especially for human liver) [Guguen-Guillouzo and Guillouzo, 1986]. Fish hepatocytes can be obtained by cannulating the intestinal vein and incising the heart to avoid excess pressure. Perfusion is performed at room temperature at a flow rate of 12 mL/min.

*Maintenance and differentiation.* Isolated parenchymal cells can be maintained in suspension for 4 to 6 h and used for short-term experiments. When seeded at a concentration of $7 \times 10^5$ viable cells/mL of nutrient medium supplemented with 1 μM dexamethasone (or 50 μM hydrocortisone hemisuccinate) in plastic culture dishes, the cells survive for a few days. Presence of chromatin remodeling agents such as trichostatin A during perfusion and cell seeding [Henkens et al., 2007], or addition of 2% DMSO to the medium after hepatocyte attachment [Isom et al., 1985; Cable & Isom, 1997; Gilot et al., 2002], greatly favor their survival and differentiation. The use of extracellular matrices that prevent cell spreading such as Matrigel [Bissell MD et al., 1987] and the collagen sandwich configuration [Dunn et al., 1989] also improve hepatocyte functional stability (it is noteworthy that hepatocytes rapidly lose their specific differentiated functions when seeded on one collagen I layer). Higher stability (2 months) can be obtained by coculturing hepatocytes with rat liver epithelial cells presumed to derive from primitive biliary cells [Guguen-Guillouzo et al., 1983]. In these conditions cells secrete themselves a complex biomatrix composed of collagens I, III, and IV, fibronectin, and laminin, which deposits in a network around and above hepatocyte colonies [Corlu et al., 1991].

*Hepatocyte proliferation.* After cell attachment (4 h after seeding), the medium is renewed with the same medium without of fetal bovine serum and supplemented with 0.7 μM hydrocortisone hemisuccinate. EGF used at 50 ng/ml, TGF-α (both from Roche) at 20 ng/ml, in association or not with pyruvate at 20 mM, appear to be the primary mitogens for hepatocytes in vitro [Loyer et al., 1996; Garnier et al., 2009]. Extracellular matrix remodelling is required for multiple division cycles in rat hepatocytes and TNF-α is a good modulator of biomatrix remodeling [Serandour et al., 2005].

## 22.2.8 HepaRG Human Hepatocytes

Primary human hepatocytes are considered as the most pertinent in vitro model. Unfortunately, their unpredictability, scarce availability, and inter-donor basal variability greatly hamper their use. Human hepatic cell lines have been proposed as an alternative. Two main strategies are used to establish cell lines expressing liver specific functions. A lot of effort has been put into oncogenic immortalization of adult hepatocytes, but the results are quite disappointing. Probably the most powerful immortalized

cell line is the Fa2N-4 cell line originated from human hepatocytes transfected with the SV40 large T antigen gene [Mills et al., 2004]. In fact the most used human hepatocyte cell lines (e.g., Hep-G2, Hep3B, PLC/PRFs Huh7, HBG) are derived from liver tumors. The Hep-G2 line was very promising [Knowles et al., 1980], but cells have gradually lost most of the metabolic properties of normal liver. A subclone, Hep-G2/C3A, expressing higher drug metabolizing activities has been established [Kelly and Sussman, 2000]. HBG is a highly differentiated hepatoma cell line, but its slow growth hampers to some extend its application [Glaise et al., 1998]. The HUH7 line has preserved several characteristic functions that mainly make the cells very useful for viral hepatic studies [Lohmann et al. 1999].

Only the recently obtained human hepatoma HepaRG cell line has retained the expression of both liver-specific plasma proteins and liver-specific glycolytic enzymes, showing also high expression and inducibility of the major phase I and phase II detoxification enzymes [Aninat et al., 2006]. These cells were derived from a female suffering from liver carcinoma; they appear as a homogeneous cell population exhibiting limited karyotypic alterations and have the property of undergoing at confluence a complete program of hepatocyte differentiation [Gripon et al., 2002]. The sophistication of this cell model resides in its unique property of transdifferentiation, which makes the cells able to rapidly reverse in a few hours to undifferentiated hepatic bipotent progenitors when confluence is broken down. They start to divide actively until they reach confluence at which time they re-initiate differentiation programmes toward either hepatocyte or biliary cells, leading to the establishment, within two weeks in the presence of 2% DMSO, of a mixed population, approximately half of which are hepatocytes (Fig. 22.2*a*).

Following a defined culture protocol, the potential for differentiation of the cell line can be maintained and kept highly reproducible through several passages (up to 25 passages), so that the mixed population can be used for experiments. However, because of the difficulty of selectively counting the hepatocyte competent cells and quantifying their proper activities with high efficiency, a protocol has been designed to set conditions for obtaining pure hepatocyte cultures from confluent HepaRG cell populations. The aim of this protocol is to reproduce adult human hepatocyte primary culture potencies but without variability.

---

**PROTOCOL 22.6B. PURIFICATION OF HepaRG HUMAN HEPATOCYTES**

*Outline*
First, culture the HepaRG cells in conditions allowing the generation of a mixed population of cells where the colonies of polygonal granular hepatocytes are numerous (about half the population) and easily

detectable. Second, proceed to selective harvesting of these hepatocyte colonies and re-seed the cells at the appropriate density.

### Materials

*Sterile:*

❑ Williams's E medium (W's E) supplemented with 10% FBS, 100 units/ml penicillin, 100 µg/mL streptomycin, 5 µg/mL insulin, 2 mM glutamine, and 50 µM hydrocortisone hemisuccinate
❑ W's E/FBS: serum-free W's E medium with 50 µM hydrocortisone hemisuccinate
❑ W's E or W's E/FBS medium supplemented with 2% DMSO (GE Life Sciences—Pharmacia)
❑ Trypsin, 0.25%,EDTA, 1 mM, 1× solution (Invitrogen)
❑ D-PBSA buffer (pH 7.4)
❑ Conical plastic centrifuge tubes
❑ Graduated pipettes
❑ Flasks or Petri dishes

*Nonsterile:*

❑ Chronometer
❑ $CO_2$ incubator
❑ Refrigerated centrifuge
❑ Water bath
❑ Inverted microscope

### Procedure

#### HepaRG cell maintenance

1. Warm the D-PBSA, the trypsin, and the medium.
2. Perform passage on a 2-week-old subconfluent HepaRG culture.
3. Discard the medium, rinse the culture with D-PBSA, and incubate with 2 mL (for 25-cm² flasks) of trypsin at 37°C for 5 min.
4. Disperse the cells in the W's E medium and count.
5. Seed the cells in W's E medium at a density of $2.6 \times 10^4$ cells/cm².

#### HepaRG hepatocyte isolation and culture

6. Choose, preferentially, 4-week-old HepaRG culture (growing phase of 2 weeks in W's E medium + differentiation phase of 2 weeks in W's E supplemented with 2% DMSO).
7. Discard the medium and wash twice with D-PBSA.
8. Incubate the culture with 2 mL (for 25-cm² flasks) of trypsin solution diluted to 0.125%, at room temperature and for a maximum of 5 min.
9. Observe the selective detachment of the hepatocyte colonies using an inverted microscope

10. Proceed to gentle manual shaking for around 1 min to complete the hepatocyte detachment, with the biliary cells remaining attached and spread.
11. Block the trypsin by addition of W's E medium with FBS
12. Centrifuge to remove trypsin (50 *g* for 2 min)
13. Suspend the cell pellet in W's E medium containing 2% DMSO, prepare a homogeneous suspension by gentle repeated pipetting, and count the cells.
14. Seed isolated hepatocytes at a density of 7 to $9 \times 10^4$ cells/cm² depending on requirements.
15. Optionally an Optiprep gradient (*see* below) can be performed to improve the purity of the cell suspension before seeding.
16. After 24 h, living cells will attach to plastic and begin to spread.
17. Colonies of hepatocytes are formed after a further two days, (see Fig.22.2*b*)
18. Replace the medium with fresh W's E/FBS with or without DMSO according to requirements.

### Practical observations

*Maximum purity of hepatocyte suspension.* The use of the standard protocol described above, gives a purity of the hepatocyte suspension of around 90%. Improving this percentage is possible by adding to the protocol a step of Optiprep (Sigma) gradient separation, which gives >95% pure hepatocytes. This gradient separation is particularly recommended for experiments studying specific distribution of cell subpopulations using cytofluorimetric analyses. The Optiprep gradient can be performed as followed: 10 mL of a solution of 15% Optiprep prepared in W's E medium are added to a centrifuge tube and 20 mL of the cell suspension are gently layered over it. Centrifugation of 5 min at 80 *g* sediments the hepatocytes. The supernate is withdrawn and the cell pellet is suspended in W's E medium for a new cell count before seeding.

*Differentiation stability.* It should be noted that DMSO is crucial for getting pure hepatocyte cultures. At high concentration, DMSO potentiates differentiation stability of hepatocytes, whereas it is toxic for nondifferentiated cells and therefore it avoids reversion of hepatocytes to hepatic progenitors.

*Reproducibility of hepatocyte cultures.* Regarding the maintenance of the HepaRG cell line, the frequency of passages is every two weeks if the density of seeded cells is precisely calculated to be at $2.6 \times 10^4$ cells/cm² In those conditions the process of differentiation covers (1) two weeks for getting to confluence and two more weeks in presence of DMSO for completing the differentiation program; and (2)

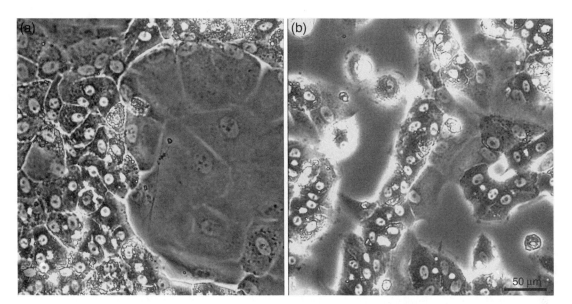

**Fig. 22.2.  HepaRG Cells.** Phase contrast of HepaRG living cells. (*a*) Four-week-old differentiated culture maintained for 2 weeks in presence of 2% DMSO. Note the mixed population with highly differentiated hepatocytes colonies bordered by flat biliary cells. (*b*) Two-day-old culture of selected HepaRG hepatocytes maintained in presence of 2% DMSO. Note the purity of the population and the hepatocyte organization in typical hepatic trabeculae.

reproducibility in the levels of functional properties of the cells is guaranteed from passage 5 to passage 22 only.

*Cryopreservation.*  Conservation of the cells in liquid nitrogen is performed in 10% DMSO/W's E medium.

## 22.2.9  Pancreas

Both acinar [Bosco et al., 1994] and islet cells [Vonen et al., 1992; Cao et al., 1997] have been grown from pancreatic tissue, and Liberase with collagenase or thermolysin (Roche; Serva) has been shown to be a suitable disaggregating agent for islet cell lines [Anazawa et al., 2009]. Cell lines have also been established from the "immortomouse" [Blouin et al., 1997]. The conversion of adenoma to carcinoma has generated particular interest [Perl et al., 1998].

Protocol 22.7 for the culture of pancreatic acinar cells was contributed by R. J. Hay and M. das Gracas Miranda, American Type Culture Collection, 10801 University Boulevard, Manassas, VA 20110–2209. Fractionated populations of pancreatic acinar epithelia are inoculated onto collagen-coated culture dishes, and two-dimensional aggregate colonies are allowed to develop for subsequent study [Hay, 1979; Ruoff & Hay, 1979].

### PROTOCOL 22.7. PANCREATIC EPITHELIUM

**Outline**

Dissociate guinea pig pancreatic tissue, filter the mixed cell suspension through cheesecloth or nylon sieves, and layer the suspension over a BSA solution. After three sequential centrifugation and fractionation steps, dispense the cell pellet, and seed the cells in collagen-coated culture vessels.

**Materials**

*Sterile:*

❑ F12K tissue culture medium with 20% bovine calf serum (F12K-CS20)
❑ HBSS
❑ HBSS without Ca²⁺ and Mg²⁺ (HBSS-DVC)
❑ Dextrose (Difco, no. 0155–174, BD Biosciences)
❑ Bovine serum albumin, fraction V (Sigma)
❑ Trypsin solution: trypsin 1:250 (Difco, BD Biosciences), 0.25% in citrate buffer; 3 g/L of trisodium citrate; 6 g/L of NaCl, 5 g/L dextrose; 0.02 g/L of phenol red; pH 7.6.
❑ Collagenase solution: dissolve collagenase (Invitrogen) in 1× HBSS-DVC (pH 7.2) to give 1800 U/mL. Adjust the solution to pH 7.2, and dialyze, using a 12-KDa exclusion membrane for 4 h at 4°C, against 1× HBSS-DVC containing 0.2% glucose. The HBSS is discarded, and this step is repeated for 16 to 18 h with fresh HBSS-DVC. Filter, sterilize, and store the solution in 10- to 20-mL aliquots at −70°C or below.
❑ Conical flask, 25 mL (Bellco), siliconized
❑ Pipettes, 5 or 10 mL, wide-bore (Bellco)
❑ Büchner funnel

❑ Dialysis tubing (Spectro-Por)
❑ Sidearm flask, 250 mL
❑ Polypropylene centrifuge tubes, 50 mL
❑ Cheesecloth
❑ Collagen-coated culture dishes (Biocoat, BD Biosciences)

*Nonsterile:*

❑ Agitating water bath (Backer model M5B#11–22-1)
❑ Centrifuge

### Procedure

1. Make up the dissociation fluid by mixing 1 part of collagenase with 2 parts of trypsin solution, and warm to 37°C.
2. Aseptically remove the entire pancreas (0.5–1.0 g), and place the organ in F12K-CS20.
3. Trim away mesenteric membranes and other extraneous matter, and mince the pancreas into 1- to 3-mm fragments.
4. Transfer the fragments to a siliconized, 25-mL conical flask in 5 mL of prewarmed dissociation fluid. Agitate the solution at about 120 rpm for 15 min at 37°C in a shaker bath. Repeat this dissociation step two or three times with fresh fluid, until most of the tissue has been dispersed.
5. After each dissociation, allow large fragments to settle, and transfer the supernate to a 50-mL polypropylene centrifuge tube with approximately 12 mL of cold F12K-CS20 to neutralize the dissociation fluid.
6. Spin the suspension at 600 rpm for 5 min, and resuspend the pellet in 5 to 10 mL of cold F12K-CS20. Keep the cell suspension in ice.
7. Pool the cell suspensions and pass through several layers of sterile cheesecloth in a Büchner funnel. Apply light suction by inserting the funnel stem into a vacuum flask during filtration. Take an aliquot for cell quantitation.
8. Generally, 1 to $2 \times 10^5$ cells/mg of tissue with 90% to 95% viability are obtained at this step.
9. Layer $5 \times 10^7$ to $5 \times 10^8$ cells from the resulting fluid onto the surface of 2 to 4 columns consisting of 35 mL of 4% cold BSA in HBSS-DVC (pH 7.2) in polypropylene 50-mL centrifuge tubes. Centrifuge the tubes at 600 rpm for 5 min. This step is critical to achieving a good separation of acinar cells from islet, ductal, and stromal cells of the pancreas. Discard the supernate, and repeat this fractionation step twice more, pooling and resuspending the pellets in cold F12K-CS20 after each step.
10. Collect the cell pellets in 5 to 10 mL of cold F12K-CS20. Take an aliquot for counting. Yields of 2 to $5 \times 10^4$ cells/mg of tissue are obtained, with 80% to 95% of the total being acinar cells. Inoculation densities can be $3 \times 10^5/$ $cm^2$. Cells adhere as aggregate colonies within 72 h.

*Variations.* After cells have adhered, F12K–C20 can be replaced by serum-free MEM with insulin (10 ng/mL), transferrin (5 ng/mL), selenium (9 ng/mL), and EGF (21 ng/mL). This medium supports pancreatic cell growth for at least 15 days without altering the cell morphology.

The preceding method has been applied to studies with guinea pig and human (transplant donor) tissues. The addition of irradiated human lung fibroblasts as a feeder layer produces a marked stimulation (up to 500%) in the incorporation of [3H]TdR and prolongs survival by at least a factor of two.

## 22.2.10 Kidney

The kidney is a structurally complex organ in which the system of nephrons and collecting ducts is made up of numerous functionally and phenotypically distinct segments. This segmental heterogeneity is compounded by a cellular diversity that has yet to be fully characterized. Some tubular segments possess several morphologically distinct cell types. In addition evidence points to rapid adaptive changes in cell ultrastructure that may correlate with changes in cell function [Stanton et al., 1981]. The structural and cellular heterogeneity presents a challenge to the cell culturist who is interested in isolating pure or highly enriched cell populations. The difficulty of the problem is further compounded in studies of the human kidney, with which form and access to the specimen may make some manipulations, such as vascular perfusion, difficult or impossible.

Several approaches have been used successfully to culture the cells of specific tubular segments. Density gradient methods (*see* Section 14.1) are now commonly used to isolate enriched populations of enzyme-digested tubule segments and are particularly effective in establishing proximal tubule cell cultures from experimental animals [Taub et al., 1989]. Specific nephron or collecting duct segments can also be isolated by microdissection and then explanted to the culture substrate. This method, developed with the use of experimental animals [Horster, 1979], has been applied to the culture of human kidneys [Wilson et al., 1985] and the cyst wall epithelium of polycystic kidneys [Wilson et al., 1986]. Immunodissection [Smith & Garcia-Perez, 1985] and immunomagnetic separation [Pizzonia et al., 1991] methods have also been developed to isolate specific nephron cell types on the basis of their expression of cell type-specific ectoantigens. These elegant methods hold considerable promise for the study of specific kidney cell types in health and disease, but as yet have been applied almost exclusively to studies on experimental animals. The limited (and unscheduled) availability and inconsistent form (e.g., excised pieces, damaged vasculature, lengthy postnephrectomy period) of human donor kidneys make progressive enzymatic dissociation a more practical means for isolating human kidney cells for culture.

A number of methods for the primary culture of human kidney tissue have been reported [Detrisac et al., 1984; States et al., 1986; McAteer et al., 1991] and commercial sources exist for early passage cultures (see Table 22.1; Appendix II). Protocol 22.8 and the preceding introduction were contributed by J. McAteer, S. Kempson, and A. Evan, Indiana University School of Medicine, Indianapolis, IN 46202-5120.

The primary culture of tissue fragments excised from the outer cortex of the human kidney provides a means of isolating cells that express many of the functional characteristics of the proximal tubule [Kempson et al., 1989]. Progressive enzymatic dissociation and crude filtration yields single cells and small aggregates of cells that, when seeded at high density, give rise to a heterogeneous epithelium-enriched population. Large numbers of cells can be harvested, making it practical to establish multiple replicate primary cultures or to propagate cells for frozen storage. Experience with the method shows that the functional characteristics of such primary and subcultured cells are reproducible for kidneys from different donors and that, with proper handling of specimens, good cultures can be derived from kidneys after even a lengthy postnephrectomy period.

---

## PROTOCOL 22.8. KIDNEY EPITHELIUM

### Outline
Tissue fragments are excised from the outer cortex, minced, washed, and incubated (with agitation) in a collagenase-trypsin solution. The tissue is periodically pipette triturated, and the dissociated cells are collected and pooled. The cell suspension is filtered through a size-limiting screen to remove undigested fragments. The cells are then washed free of enzyme, suspended in medium supplemented with serum, and seeded on culture-grade plastic.

### Materials

*Sterile:*

- Basal medium DMEM/F12: 50:50 mixture of DMEM and Ham's F12
- Fetal bovine serum (Sterile Systems; Hyclone)

- Complete culture medium: DMEM/F12 plus 10% fetal bovine serum
- BSS
- Collagenase (Worthington type IV), 0.1%, trypsin (1: 250, Sigma), 0.1% solution in 0.15 M NaCl
- DNase, 0.5 mg/mL in saline
- Trypsin-EDTA solution: trypsin (1:250), 0.1%, EDTA (culture grade, Sigma) 1 mM
- Scalpels, tissue forceps
- Nitex screen (160 μm) (Tetko)
- Culture dishes, flasks
- Tubes, sterile, 50 mL

*Nonsterile:*

- Orbital shaker (Bellco)

### Procedure

Tissue dissociation

1. Cut 5- to 10-mm-thick coronal slices of kidney, and wash the fragments in chilled basal medium.
2. Excise fragments from the outer cortex, and use crossed blades to mince the tissue into 1- to 2-mm pieces (see Section 11.3).
3. Transfer approximately 5 mL of tissue fragments to a 50-mL tube containing 20 mL of the warm collagenase-trypsin solution. Secure the tube to the platform of an orbital shaker within an incubator at 37°C.
4. Incubate the tissue with gentle agitation for 1 h.
5. Discard and replenish the enzyme solution.
6. Collect the supernate at 20-min intervals following the addition of 20 mL of basal medium containing 0.05 mg/mL DNase and gentle trituration through a 10-mL pipette (10 cycles).
7. Dilute the supernate that has been collected with an equal volume of complete culture medium, and hold the resulting suspension on ice.
8. Repeat this cell collection procedure 5 or more times, until the fragments are fully dispersed.
9. Pool the harvested supernates and dispense them into aliquots in 50-mL tubes.
10. Centrifuge at 100 g for 15 min, and resuspend each pellet in 45 mL of complete medium containing DNase.
11. Filter the suspension through Nitex cloth (160 μm).
12. The isolation protocol yields cell aggregates as well as single cells. Hence counting cells is unreliable. To propagate cells for subculture or cryopreservation, seed 15 mL into one 75-cm$^2$ flask.

**Subculture and cryopreservation**

13. Subculture NHK-C cells by routine methods (*see* Protocol 12.3). Dissociation with trypsin/EDTA produces a monodisperse suspension that is suitable for counting with a hemocytometer.
14. Seed culture-grade dishes at approximately $1 \times 10^5$ cells per $cm^2$.
15. Perform cryopreservation by routine methods using complete medium containing 10% dimethyl sulfoxide (*see* Protocol 19.1).

**Comments.** A prominent variable in this method is the quality of the kidney specimen itself. Procedures for procuring tissue are impossible to standardize. Donor tissue commonly includes segments of kidney collected at surgical nephrectomy (e.g., for renal cell carcinoma) and intact kidney tissue that is judged unsuitable for transplantation (e.g., because it has an anomalous vasculature). These specimens are collected under different conditions (e.g., with or without perfusion, with varying periods of warm ischemia, and for different durations of the postnephrectomy period). In addition differences in the age and health of the donor ensure that no two specimens are equivalent. Although not all kidneys yield satisfactory cultures, the foregoing protocol has been effective for a wide variety of specimens, including fresh surgical specimens and intact, perfused kidneys held on wet ice for nearly 100 h.

△ **Safety Note.** Work involving any tissue or fluid specimen of human origin requires precautions to avoid unprotected contact with blood-borne pathogens (*see* Section 6.8.3).

**Applications.** NHK-C cells exhibit predominantly the functional characteristics of the renal proximal tubule, including a parathyroid hormone (PTH)-inhibitable $Na^+$-dependent inorganic phosphate transport system [Kempson et al., 1989]. They also show phlorizin-sensitive $Na^+$-dependent hexose transport, exhibit a proximal tubule-like pattern of hormonal stimulation of cAMP (PTH responsive, vasopressin insensitive), and express several proximal tubule brush border enzymes (maltase, leucine aminopeptidase, gamma–glutamyl transpetidase). In addition NHK-C cells have been used to demonstrate the specificity of phosphonoformic acid, an inhibitor of $Na^+/P_i$ cotransport [Yusufi et al., 1986], and they serve as a model of oxidant injury in the renal tubule [Andreoli & McAteer, 1990].

### 22.2.11 Bronchial and Tracheal Epithelium

As for other epithelial cells there are now a number of commercial sources for early passage cultures of bronchial and tracheal epithelium (*see* Table 22.1, Appendix II; Plate 27*e, f*). Protocol 22.9 for the isolation and culture of normal human bronchial epithelial cells from autopsy tissue was contributed by M. A. LaVeck and J. F. Lechner, when at the National Institutes of Health, Bethesda, MD.

In the presence of serum, normal human bronchial epithelial (NHBE) cells cease to divide and, furthermore, terminally differentiate. The serum-free medium that has been optimized for NHBE cell growth does not support lung fibroblast cell replication, thus permitting the establishment of pure NHBE cell cultures [Lechner & LaVeck, 1985].

---

**PROTOCOL 22.9. BRONCHIAL AND TRACHEAL EPITHELIUM**

**Outline**
This procedure involves explanting fragments of large airway tissue in a serum-free medium (LHC-9) in order to initiate and subsequently propagate fibroblast-free outgrowths of NHBE cells; four subculturings and 30 population doublings are routine.

**Materials**

*Sterile:*

❏ Culture medium:
L-15 (Invitrogen)
LHC-9 [Lechner & LaVeck, 1985] (*see* Table 9.1)
HB [Lechner & LaVeck, 1985] (*see* Appendix I)
❏ Bronchial tissue from autopsy of noncancerous donors
❏ Plastic tissue culture dishes (6 and 10 cm)
❏ FN/V/BSA: Mixture of human fibronectin, 10 µg/mL (Collaborative Research, BD Biosciences), collagen (Vitrogen 100, 30 µg/mL; Collagen Corp.), and crystallized bovine serum albumin (BSA), 10 µg/mL (Bayer) in LHC basal medium (Biosource International) [Lechner & LaVeck, 1985]
❏ Trypsin, 0.02%, EGTA (Sigma), 0.5 mM, and polyvinylpyrrolidine (USB), 1% solution [Lechner & LaVeck, 1985]
❏ Scalpels 1621 (BD Biosciences)
❏ Surgical scissors
❏ Half-curved microdissecting forceps
❏ High-$O_2$ gas mixture (50% $O_2$, 45% $N_2$, 5% $CO_2$)

*Nonsterile:*

❏ Gloves (human tissue can be contaminated with biologically hazardous agents)
❏ Controlled-atmosphere chamber (Bellco no. 7741)
❏ Rocker platform (Bellco no. 7740)

*Procedure*

1. Coat a culture dish with 1 mL of the FN/V/BSA mixture per 6-cm dish, and incubate the dish in a humidified $CO_2$ incubator at 37°C for at least 2 h (not to exceed 48 h). Vacuum aspirate the mixture and fill the dish with 5 mL of culture medium.
2. Aseptically dissected lung tissue from noncancerous donors autopsied within the previous 12 h is placed into ice-cold L-15 medium for transport to the laboratory, where the bronchus is further dissected from the peripheral lung tissues.
3. Before culturing, scratch an area of one square centimeter at one edge of the surface of the 6-cm culture dishes with a scalpel blade.
4. Open the airways (submerged in the L-15 medium) with surgical scissors, and cut (slice, do not saw) the tissue with a scalpel into two pieces, 2 × 3 cm.
5. Using a scooping motion to prevent damage to the epithelium, pick up the moist fragments and place them epithelium side up onto the scratched area of the 6-cm dish. Remove the medium, and incubate the fragments at room temperature for 3 to 5 min to allow time for them to adhere to the scratched areas of the dishes.
6. Add 3 mL of HB medium to each dish and place them in a controlled-atmosphere chamber on a rocker platform.
7. Flush the chamber with a high-$O_2$ gas mixture.
8. Rock the chamber at 10 cycles per minute, causing the medium to flow intermittently over the epithelial surface.
9. Incubate rocking tissue fragments at 37°C, changing the medium and atmosphere after 1 day and at 2-day intervals for 6 to 8 days. This step improves subsequent explant cultures by reversing any ischemic damage to the epithelium that occurred from time of death of the donor until the tissue was placed in the ice-cold L-15 medium.
10. Before explanting, scratch seven areas of the surface of each 10-cm culture dish with a scalpel. Coat the surfaces of the scratched culture dishes with the FN/V/BSA mixture solution, and aspirate the surplus solution as before.
11. Cut the moist ischemia-reversed fragments into 7 × 7-mm pieces, and explant the pieces epithelium side up on the scratched areas.
12. Incubate the pieces at room temperature without medium for 3 to 5 min, as before.
13. Add 10 mL of LHC-9 medium to each dish, and incubate explants at 37°C in a humidified 5% air/$CO_2$ incubator.
14. Replace spent medium with fresh medium every 3 to 4 days.
15. After 8 to 11 days of incubation, when epithelial cell outgrowths radiate from the tissue explants more than 0.5 cm, transfer the explants to new culture dishes scratched and coated with FN/V/BSA to produce new outgrowths of epithelial cells. This step can be repeated up to seven times with high yields of NHBE cells.
16. Incubate the postexplant outgrowth cultures in LHC-9 medium for an additional 2 to 4 days before trypsinizing (with the trypsin/EGTA/PVP solution) for subculture or for experimental use.

## 22.2.12 Oral Epithelium

Oral keratinocytes have been cultured in medium similar to LHC-9 and MCDB153, again utilizing FN/C/BSA–coated culture dishes to promote attachment. There are commercial sources of oral keratinocytes (*see* Table 22.1; Appendix II). Protocol 22.10 has been abridged from Grafström [2002].

---

### PROTOCOL 22.10. ORAL KERATINOCYTES

*Materials*

*Sterile:*

- Transport medium: Leibowitz L-15 with 100 µg/mL gentamycin, 100 U/mL penicillin, 100 µg/mL, Fungizone, 1 µg/mL.
- Growth medium: modified MCDB 153 [Grafström, 2002]
- Antibiotic-supplemented growth medium: growth medium supplemented with gentamycin 100 µg/mL, penicillin-streptomycin 100 U/mL, and Fungizone, 1 µg/mL
- D-PBSA: Dulbecco's PBS lacking $Ca^{2+}$ and $Mg^{2+}$
- Trypsin, 0.17% in D-PBSA
- PET: trypsin, 0.025%, EGTA, 0.5 mM, polyvinylpyrrolidone (PVP, 40 KDa), 1%, in D-PBSA
- Scalpels, #11 blade
- Scissors, fine
- Forceps, fine, 2 pairs
- Petri dishes, tissue-culture grade, 5 or 10 cm
- Petri dishes, non–tissue-culture grade for dissectionsec, 3.5 cm, 10 cm
- Centrifuge tubes, 15 mL, conical
- Micropipette (e.g., Gilson), 100 µL
- FN/C/BSA-coated culture dishes, 5 cm (*see* Protocol 22.9)

*Procedure*

Primary culture

1. Obtain the tissue from surgery or early autopsy, place in cold transport medium, and transfer to the laboratory as soon as possible.
2. Transfer the tissue to a 10-cm dish and rinse with phosphate-buffered saline (D-PBSA).
3. Remove as much connective tissue as possible, including parts containing blood. If tissue specimens have a surface area of 1 cm$^2$, divide them into smaller pieces.
4. Place the specimens in a 3.5-cm dish and add enough 0.17% trypsin solution for coverage (1–2 mL), and incubate overnight at 4°C.
5. Hold each specimen with forceps; then peel and scrape away the epithelium with another pair of forceps into the trypsin solution.
6. Triturate the suspension carefully a few times, to further disaggregate the cells, and transfer the cell suspension to a centrifuge tube.
7. Rinse the dish with D-PBSA and subsequently add this rinsing solution to the cell suspension. Use the same volume for rinsing as used for the trypsin solution used for tissue digestion in step 4.
8. Remove an aliquot by micropipette for determination of cell yield.
9. Pellet the cells at 125 *g* at 4°C, preferably with a refrigerated centrifuge, and resuspend the cells in growth medium.
10. Dilute with additional growth medium as required, and then seed at $5 \times 10^3$ cells/cm$^2$ on FN/C/BSA-coated culture dishes. If cells are derived from autopsy material, use antibiotic-supplemented growth medium for the initial 3 to 4 days in culture.
11. Feed the cells with fresh medium after 24 h to remove possible cell debris and erythrocytes.
12. Feed the cells every second day.

Subculture

1. Rinse the cells once with D-PBSA.
2. Add PET solution, covering the cells completely with the solution (e.g., use 3 mL/10-cm dish).
3. Incubate at room temperature until the cells begin to round up and/or detach from the dish. Follow the cell detachment under the microscope (the procedure generally takes 5–10 min). The rate of detachment can be enhanced by addition of trypsin at higher concentrations or by increasing the temperature to 37°C.
4. When most cells have detached, carefully tap on the side of dish and pipette the trypsin solution

gently over the growth surface to mechanically enhance detachment.

5. When the cells have detached, add 5 to 10 mL of D-PBSA to the dish to inactivate the action of trypsin by dilution.
6. Transfer the cell suspension to centrifuge tubes and triturate gently to obtain an even distribution of the cells in the suspension.
7. Using a 1-mL pipette, take out a sample for a cell count and transfer to a hemocytometer chamber (*see* Protocol 20.1).
8. Determine the concentration of cells in the suspension.
9. Pellet the cells by centrifugation for 5 min at 125 *g* at 4°C.
10. Remove the supernate and tap the tube gently against the fingers until the pellet disperses.
11. Add growth medium as desired, disperse the cell suspension gently with a pipette, and add to each culture vessel.
12. Place all the vessels on a tray. Agitate the vessels gently by holding and moving the tray in your hands, alternating between different directions, to ensure even density of the cells in each vessel (simply swirling the dishes will focus the cells in the center of the dish). It is also important that the shelves are evenly fixed in the incubator where the vessels are placed, and that the incubator is free of vibration, to avoid an uneven distribution of cells during attachment.
13. Incubate the cells undisturbed for 4 to 24 h before changing the medium.

## 22.2.13  Prostate

A number of reports have used cells cultured from prostate tissue [Cronauer et al., 1997; Peehl, 2002] with a particular interest in regulating cell proliferation and differentiation by paracrine growth factors [Yan et al., 1992; Chopra et al., 1996; Thomson et al., 1997]. Prostatic epithelial, along with fibroblastic and smooth muscle cells, are available commercially (*see* Table 22.1; Appendix II). Protocol 22.11 for the primary culture of rat prostate epithelial cells was contributed by W. L. McKeehan, Texas A&M University, 2121 West Holcombe Boulevard, Houston, Texas 77030-3303A.

Isolated normal rat prostate epithelial cells are a valuable model for studying the cell biology of maintenance, growth, and functioning of normal prostate under investigator-controlled and –defined conditions [McKeehan et al., 1982, 1984; Chaproniere & McKeehan, 1986]. Normal cells serve as the control cell type out of which prostate adenocarcinoma cells arise. Conditions similar to those described here have also been useful for the primary culture of epithelial cells for transplantable rat tumors. Key features of this method are the specific support of epithelial cells by an improved nutrient medium and hormone-like growth factors [Yan et al., 1992], while concurrently inhibiting fibroblast outgrowth due to deficient or inhibitory properties of the medium.

---

## PROTOCOL 22.11. PROSTATIC EPITHELIUM

### Outline
Remove and prepare prostates for cell culture (30 min). Incubate prostatic tissue with collagenase (1 h), and collect single cells and small aggregates of cells (1 h). Inoculate and culture cells to confluent monolayer (7 days).

### Materials

*Sterile:*

- Growth medium WAJC404 [McKeehan et al., 1984; Chaproniere & McKeehan, 1986] (*see* Table 9.1)
- Media salt solution (MSS) [McKeehan et al., 1984]:
  NaCl .................................12 mM
  KCl ................................... 2 mM
  KH$_2$PO$_4$ ........................... 1 mM
  HEPES, pH 7.6........................... 30 mM
  Glucose ................................. 4 mM
  Phenol red ................................. 3.3 μM
- Fetal bovine or horse serum
- Penicillin
- Kanamycin
- Cholera toxin
- Dexamethasone
- Epidermal growth factor (EGF)
- Ovine or rat prolactin
- Insulin
- Partially purified [McKeehan et al., 1984; Chaproniere & McKeehan, 1986] or purified [Crabb et al., 1986] prostatropin (prostate epithelial cell growth factor)
- Collagenase, type I (Sigma)
- Petri dishes (glass or plastic), 60 mm
- Scissors
- Syringe and 14-G cannula
- Erlenmeyer flasks, 25 mL
- Wire screen, 1-mm mesh, made to fit a 50-mL conical centrifuge tube
- Conical centrifuge tubes, plastic, 50 mL
- Nylon screen filters of 250, 150, 100, and 40 μm to fit 50-mL conical tubes
- Multiwell plates, 24-well

*Nonsterile:*

- Shaking water bath at 37°C

- Centrifuge
- Hemocytometer or electronic cell counter
- Rats, 10 to 12 weeks old, male

### Procedure

1. Aseptically remove desired lobes of the prostate and place them in a sterile 60-mm Petri dish.
2. Trim fat from the lobes and weigh them.
3. Add 2 mL of collagenase at 675 U/mL in MSS containing 100 U/mL of penicillin and 100 μg/mL of kanamycin.
4. Mince the tissue with scissors to approximately 1-mm pieces, small enough to fit through a 14-G cannula.
5. Using a syringe and a 14-G cannula, transfer the minced tissue fragments to a sterile 25-mL Erlenmeyer flask. Add collagenase to 1 mL per 0.1 g of original wet tissue weight.
6. Incubate for 1 h at 37°C on a shaking water bath.
7. Aspirate the digested suspension three times through a 14-G cannula, and then pass the suspension through a coarse (1-mm) wire mesh screen fitted to 50-mL plastic conical tubes to remove debris and undigested material. Rinse the suspension with an equal volume of MSS containing 5% whole fetal calf serum or horse serum.
8. Collect cells by centrifugation at 100 g for 5 min at 4°C.
9. Resuspend the cell pellet in 5 mL of MSS plus 5% serum, and pass the suspension successively through nylon screen filters of mesh sizes 250, 150, 100, and 40 μm. Wash each screen with 5 mL of MSS plus 5% serum.
10. Collect cells by centrifugation, and resuspend them in 5 mL of nutrient medium WAJC404.
11. Count the cells, and adjust the concentration to 4 × 10$^6$ cells/mL.
12. Inoculate 50 μL containing 2 × 10$^5$ cells into each well of a 24-well plate (area of well ∼ 2 cm$^2$) containing 1 mL of medium WAJC404, 10 ng/mL of cholera toxin, 1 μM dexamethasone, 10 ng/mL of EGF, 1 μg/mL of prolactin, 5 μg/mL of insulin, 10 ng/mL of prostatropin, 100 U/mL of penicillin, and 100 μg/mL of streptomycin. Partially purified sources of prostatropin can be substituted, as described by McKeehan et al. [1984] and Chaproniere and McKeehan [1986].
13. Incubate the culture in a humidified atmosphere of 95% air and 5% CO$_2$ at 37°C. Change the medium at days 3 and 5. The cells should be near confluent by day 7.

*Variations.* The foregoing procedure can be applied to the culture of human prostate epithelial cells with modifications described by Chaproniere and McKeehan [1986].

*Paracrine control.* The purified growth factor prostatropin is identical to heparin-binding fibroblast growth factor type 1, previously called acidic FGF or HBGF-1 [Burgess & Maciag, 1989]. The new nomenclature, achieved by consensus, is now FGF-1. Molecular characterization of new members of the FGF ligand family, as well as the FGF receptor family in prostate cells, has revealed that prostate epithelial cells express a specific splice variant (FGF-R2IIIb) of one of the four FGF receptor genes. The specific receptor recognizes FGF-1 and stromal cell-derived FGF-7 (also called keratinocyte growth factor), but not FGF-2 (previously called bFGF or HBGF-2), described in Yan et al. [1992]. Prostate stromal cells express only the FGF-R2 gene, which recognizes FGF-1 and FGF-2, but not FGF-7. As stromal cells respond to both FGF-1 and FGF-2, prostatropin/FGF-1 can be substituted for FGF-7 in the procedure above to provide an additional selection for epithelial cells. FGF-2 will not substitute for FGF-1 or FGF-7. If purified FGF-1 is used in the protocol, its activity can be potentiated by the addition of 10 to 50 µg per mL of heparin.

## 22.3  MESENCHYMAL CELLS

Included under this heading are those cells that are derived from the embryonic mesoderm, but excluding the hematopoietic system (*see* Section 22.5). Mesenchymal cells include the structural (connective tissue, muscle, and bone) and vascular (blood vessel endothelium and smooth muscle) cells. Whereas epithelial cells are classed as *parenchyma*, the main functional cells within an organ, the connective tissue cells are classed as *stroma*, or supporting tissue.

### 22.3.1  Connective Tissue

Connective tissue cells are generally regarded as the weeds of the tissue culturist's garden. They survive most mechanical and enzymatic explantation techniques and may be cultured in many simple media, particularly if serum is present. They are available as commercially as *fibroblasts* (*see* Tables 22.1, 19.5). Although cells loosely called fibroblasts have been isolated from many different tissues and assumed to be connective tissue cells, the precise identity of cells in this class is not always clear. Fibroblast lines (e.g., 3T3 from the mouse) produce types I and III collagen and release it into the medium [Goldberg, 1977]. Although collagen production is not restricted to fibroblasts, synthesis of type I in relatively large amounts is characteristic of connective tissue. However, 3T3 cells can also be induced to differentiate into adipose cells [Kuriharcuch & Green, 1978; Smyth et al., 1993] and 10T½ cells, another mouse embryo cell line, can differentiate

into chondrocytes and bone [Shea et al., 2003] in response to bone morphogenetic proteins (BMPs). It appears that mouse embryo fibroblastic cell lines (MEFs), such as the 3T3 group of cell lines and 10T1/2 often used as feeder layers, are primitive mesodermal cells that may be induced to differentiate in more than one direction.

Human, hamster, and chick fibroblasts are morphologically distinct from mouse fibroblasts, as they assume a spindle-shaped morphology at confluence, producing characteristic parallel arrays of cells distinct from the pavement-like appearance of mouse fibroblasts. The spindle-shaped cell may represent a more highly committed precursor and may be more correctly termed a fibroblast. NIH3T3 cells may become spindle shaped if allowed to remain at high cell density.

It has also been suggested that fibroblastic cell lines may be derived from vascular pericytes, connective-tissue-like cells in the blood vessels, but in the absence of the appropriate markers, this is difficult to confirm. Clearly, cell lines loosely termed fibroblastic can be cultivated from embryonic and adult tissues, but these lines should not be regarded as identical or classed as fibroblasts without confirmation, by expression of the appropriate markers, that they can differentiate into fibrocytes. Collagen, type I, is one such marker. Thy I antigen has also been used [Raff et al., 1979; Saalbach et al., 1997], although it may also appear on some hematopoietic cells. Proline hydroxylase is also expressed in fibroblasts [Pinnel et al., 1987] and some fibroblasts produce KGF as a paracrine growth factor (*see* Characterization of keratinocyte cultures after Protocol 22.1), such as in prostate [Thomson et al., 1997] and intestine [Visco et al., 2009].

Cultures of fetal or adult fibroblasts can be prepared by conventional procedures such as primary explantation (*see* Protocol 11.4), warm (*see* Protocol 11.5) and cold (*see* Protocol 11.6) trypsinization, and collagenase digestion (*see* Protocols 12.8, 23.8). Fibroblast cultures from various sources are available commercially (*see* Table 12.1; Appendix II; *see also* Plate 8a, b).

### 22.3.2  Adipose Tissue

Although it may be difficult to prepare cultures from mature fat cells, differentiation may be induced in cultures of mesenchymal cells (e.g., mouse 3T3-L1) by maintaining the cells at a high density for several days [Kuriharcuch & Green, 1978; Vierick et al., 1996]. An adipogenic factor in serum appears to be responsible for the induction. Early-passage cultures of fat cells (adipocytes, preadipocytes) are available commercially (*see* Table 22.1; Appendix II). Primary cultures of fat cells can also be prepared from rat epididymus by collagenase digestion; Protocol 22.12 and the following introduction have been abridged from Quon [1998].

Adipose cells (adipocytes) are terminally differentiated, specialized cells whose primary physiological role has classically been described as an energy reservoir for the body. Adipocytes are a storage depot for triglycerides in times of energy excess and a source of energy in the form of free fatty acids released by lipolysis during times of energy need. Adipose cells as active regulators of carbohydrate and lipid metabolism has received increased attention [Flier, 1995; Reaven, 1995]. It is likely that specific abnormalities in adipose tissue can contribute directly to the pathogenesis of common diseases such as diabetes, hypertension, and obesity [Flier, 1995; Hotamisligil et al., 1995; Reaven, 1995; Walston et al., 1995; Zhang et al., 1994].

In Protocol 22.12 a modification of the isolation procedure originally described by Rodbell [1964] is optimized to obtain approximately 4 mL of packed adipose cells from the epididymal fat pads of four rats by flotation of disaggregated cells after centrifugation. This protocol can be scaled up or down as needed and has been used to provide insulin-responsive cells suitable for DNA transfer by electroporation.

---

## PROTOCOL 22.12. PRIMARY CULTURE OF ADIPOSE CELLS

### Materials

*Sterile:*

- DMEM-A: DMEM, pH 7.4, containing 25 mM of glucose, 2 mM of glutamine, 200 nM of $(R) - N^6$-(1-methyl-2-phenylethyl)adenosine (PIA, Sigma), 100 µg/mL of gentamycin, and 25 mM of HEPES. Prepare in 100-mL aliquots and warm to 37°C before use.
- DMEM-B: DMEM-A plus 7% BSA
- KRBH buffer: Krebs–Ringer medium, pH 7.4, containing 10 mM $NaHCO_3$, 30 mM HEPES (Sigma), 200 nM adenosine (Roche), and 1% (w/v) BSA (Intergen). To make 1 L of KRBH buffer, add 7 g of NaCl, 0.55 g of $KH_2PO_4$, 0.25 g of $MgSO_4 \cdot 7H_2O$, 0.84 g of $NaHCO_3$, 0.11 g of $CaCl_2$, 7.15 g of HEPES, 10 g of BSA, 1 mL of 200 µM adenosine, and UPW to a volume of 1 L. Dispense in aliquots of 100 mL. Warm to 37°C and adjust pH to 7.4 before use.
- KRBH-A buffer: KRBH with 5% BSA
- Collagenase, 20 mg/mL in 2 mL of KRBH buffer, sterilized by passage through a 0.22-µm filter
- Tissue-culture Petri dish, 6-cm diameter (BD Biosciences)
- Conical centrifuge tube, 50 mL (Corning)
- Polypropylene tubes, 17 × 100 mm
- Low-density polypropylene vial, 30 mL
- Nylon mesh filter, 250 µm, in holder or filter funnel
- Wide-bore pipette tips, 200 µL
- Scissors: two pairs of sharp dissecting scissors, 10 cm (4 in.); Perry forceps, 12.5 cm (5 in.)

*Nonsterile:*

- Male Sprague–Dawley rats (145–170 g, CD strain, Charles River Breeding Laboratories)
- Plastic box perfused with a gas mixture of 70% $CO_2$, 30% $O_2$
- Guillotine
- Pipettor for 200-µL tips

### Procedure

#### Dissection of epididymal fat pads

1. Anesthetize rats in a plastic box with a gas mixture of 70% $CO_2$ and 30% $O_2$.
2. Decapitate rats with a guillotine and exsanguinate.
3. Soak the bodies briefly in 70% ethanol.
4. Remove the epididymal fat pads while maintaining the highest level of sterility possible.
5. Cut through the skin on the lower abdomen with one pair of scissors to expose the peritoneum.
6. Using a second pair of scissors, open the peritoneum and pull up the testes with a pair of forceps.
7. Trim fat pads from epididymides, taking care to leave the blood vessels behind.
8. Transport tissue to the culture laboratory.

#### Collagenase digestion and washing of adipose cells

9. Add 4 g of fat pads (approximately equivalent to 8 fat pads) to a 30-mL low-density polypropylene vial containing 4 mL of KRBH buffer at 37°C.
10. Mince fat pads into pieces approximately 2 mm in diameter with scissors.
11. Add 1 mL of collagenase solution to the vial containing the minced fat pads, and incubate the pieces in a shaking water bath at 37°C for approximately 1 h, until the cell mixture takes on a creamy consistency.
12. After collagenase digestion, add 10 mL of KRBH buffer at 37°C to the vial.
13. Mix cells in the vial by swirling, and gently pass the cells through a 250-µm nylon mesh filter into a 50-mL conical tube.
14. Wash the cells by adding 30 mL of KRBH buffer at 37°C to the tube and centrifuging briefly at 200 g in a tabletop centrifuge. Remove infranate with a pipette. Note that adipose cells will be floating on top of the aqueous buffer.
15. Repeat the washing of cells by adding 40 mL of KRBH buffer, centrifuging, and removing infranate two additional times.
16. Wash cells twice with 40 mL of DMEM-A at 37°C.

17. After the final wash, resuspend the cells from the surface of the medium in DMEM-A at a cytocrit of approximately 40%.

**Primary culture of adipose cells**

18. Transfer 2 mL of the 40% cytocrit DMEM-A suspension, using 200-μL wide-bore pipette tips, into one 6-cm tissue culture dish.
19. Place the cells in a humid incubator at 37°C with 5% $CO_2$ for 1.5 h.
20. Add 5 mL of DMEM-B to each dish.

A glucose uptake assay [Quon, 1998] can be done on an aliquot of cells to check their viability.

### 22.3.3 Muscle

***Smooth muscle.*** Smooth muscle cells have been grown from vascular tissue [Subramanian et al., 1991] and cocultured with vascular endothelium [Jinard et al., 1997; Klinger & Niklason, 2006]. They are also available commercially (*see* Table 22.1; Appendix II; Plate 28*c*). Protocol 22.13 has been reproduced from Klinger and Niklason [2006].

---

**PROTOCOL 22.13. ISOLATION AND CULTURE OF SMOOTH MUSCLE CELLS**

### Materials

*Sterile:*

❑ Smooth muscle cell medium (SMCM): Dulbecco's Modified Eagle's medium (DMEM) supplemented with 20% fetal bovine serum, penicillin 100 U/mL, and streptomycin 100 μg/mL
❑ 0.25% trypsin/EDTA (*see* Appendix I)
❑ Petri dishes: 15 × 2.5 cm, 6 × 1.5 cm
❑ Scalpel and 10 surgical blade
❑ Dissection scissors
❑ Tissue forceps

### Procedure

1. Obtain thoracic aorta from young calves.
2. Immerse aorta in Hanks's saline.
3. Place on ice until ready to isolate cells.
4. Place aorta into a 15 × 2.5-cm Petri dish for dissection.
5. Incise the aorta longitudinally with dissection scissors.
6. Gently scrape the luminal surface of the aorta with the scalpel blade to remove endothelial cells.
7. Dissect the medial layer of the aorta free from the intimal and adventitial layers.

8. Cut the medial layer into segments of approximately 1 cm².
9. Place medial segments intimal side down in 6 × 1.5-cm Petri dishes.
10. Allow the segments to adhere to the dishes for approximately 10 min.
11. Add 1 mL of SMCM directly onto the medial segment in each dish.
12. Place medial segments in humidified incubator at 37°C and 10% $CO_2$ overnight.
13. The following day, add 5 mL fresh SMCM to each dish, being careful not to detach the medial segment from the dish.
14. Maintain the medial segments in an incubator for an additional 10 days, after which time the SMCs will have migrated off the segments and become established in two-dimensional culture.
15. Passage cells at subconfluence, using 0.25% trypsin/EDTA.

---

***Cardiac muscle and transplantation of skeletal myoblasts.*** Cardiac muscle can be isolated with the cold-trypsin method (Protocol 11.6) and will show contractions after about 3 days in culture [Deshpande et al., 1993]. Although human cardiac myocytes have also been grown [Goldman & Wurzel, 1992], transplantation of contractile and noncontractile cells from skeletal muscle into infarcted myocardium represents a new strategy for the treatment of acute myocardial infarctions; it has been performed in rabbit, rat, pig, dog, and sheep and shown to improve myocardial function in vivo [Léobon et al., 2003; Menasche, 2004; Taylor, 2001]. Because of these encouraging results, cell therapy, and especially cell transplantation, is evoked for several pathologies [Bhagavati & Xu, 2004].

Transplantation of autologous skeletal myoblasts into the postinfarction scar was recently applied to humans [Menasche et al., 2001]. These recent developments gave rise to a new interest in basic research on the molecular requirements for in vitro myogenic proliferation and differentiation [Neumann et al., 2003; Mal & Harter, 2003; Anderson & Wozniak, 2004; Li et al., 2004; Carvalho et al., 2004] together with a fast-developing interest in culture of stem cells giving myogenic cells [Bhagavati & Xu, 2004; Seruya et al., 2004; Bossolasco et al., 2004].

***Myogenic cells from skeletal muscle.*** It is possible to culture myogenic cells from adult skeletal muscle of several species under conditions in which the cells continue to express at least some of their differentiated traits. Called *satellite cells*, the myogenic cells partially mimic the first steps of skeletal muscle differentiation. They proliferate and migrate randomly on the substratum and then align and finally undergo a fusion process to form multinucleated myotubes [Richler & Yaffe, 1970; Campion, 1984; Hartley & Yablonka-Reuveni, 1990]. Although three to four passages can be performed by means of trypsinization, subculture is no longer possible once differentiation (i.e., fusion) has taken place. Early passage cultures of skeletal myoblasts are available commercially (*see* Table 22.1; Appendix II; Plate 28*d*).

Two different protocols are described below. Protocol 22.13 is a traditional method (digestion and mechanical trituration followed by seeding of isolated cells); Protocol 22.14 is culture based on single myofibers. Both can be used for animal cell culture, but the first can also be used for human cell culture. Nevertheless, the second protocol gives a closer simulation of the myogenic process in vivo. The first method is easy to perform and gives good results in terms of cell yield and differentiation capacities, especially if you use enough starting material; however, the percentage of myogenic cells obtained per unit of muscle tissue is very low, <1% of all cells present [Rosenblatt et al., 1995], calling into question whether the resultant culture is truly representative of the originating tissue. This method is described for primary cultures from human healthy muscle biopsies, which are highly enriched in myogenic cells (at least 85% positive for desmin immunostaining at day 10 after seeding). It can also be used for animal muscles (e.g., rat, mouse, rabbit, pig, or dog).

For single myofiber culture, muscles are removed individually and incubated in collagenase to digest the connective tissue. Gentle mechanical trituration then frees intact individual myofibers. Myofibers, together with associated satellite cells, can be derived from individual muscles. This permits comparisons between the responses of different muscle groups to experimental procedures or between muscles of different ages or genetic origins. Furthermore the myosin heavy chain type of individual myofibers can be determined by SDS gel electrophoresis, and so myofibers of different myosin heavy chain types can be compared [Rosenblatt et al., 1996].

Myogenic cultures can be initiated efficiently from single myofibers prepared from muscles of animals of any age, although a slight reduction is seen with increasing age. In animals suffering from chronic myopathies such as muscular dystrophy, there is some contamination by fibroblasts, probably due to an increase in connective tissue [Bockhold et al., 1998].

Protocol 22.14 and part of the preceding introduction were contributed by M. Malo, S. Bonavaud, Ph. Thibert, and G. Barlovatz-Meimon, Faculté des Sciences, Université Paris XII, Val de Marne, avenue General de Gaulle, F-94010 Creteil, France.

### PROTOCOL 22.14. CULTURE OF MYOBLASTS FROM ADULT SKELETAL MUSCLE

#### Outline

Following Pronase digestion, primary cultures can be grown easily in Ham's F12 medium supplemented with 20% fetal bovine serum (FBS).

Without modifying the culture conditions, these cells proliferate and differentiate by fusing to form multinucleated myotubes, confirming the myogenicity of the cultivated cells.

### Materials

*Sterile:*

- Transport medium: Ham's F12 (Seromed 08101) without $NaHCO_3$ and with 20 mM HEPES (Serva 25245), 75 mL
- Growth medium: Ham's F12 with 20% fetal bovine serum (GIBCO 011−06290), 200 mL
- Pronase solution: 0.15% pronase (Sigma P-6911) and 0.03% EDTA (Merck 8418) with 20 IU/mL of penicillin, and 20 μg/mL of streptomycin (0.4% of Eurobio PES 3000), in Ham's F12 or D-PBS, 100 mL
- D-PBSA, 100 mL
- Nylon mesh,100 μm
- Scalpels
- Long, fine scissors
- Petri dishes for dissectionsec, 9 cm
- Flasks, 25 $cm^2$, 4
- Centrifuge tubes, 20 mL, or universal containers, 5

*Nonsterile:*

- Hemocytometer

### Procedure

Dissociation

1. Trim off nonmuscle tissue from the biopsy with a scalpel, and rinse in D-PBSA.
2. Cut the biopsy into fragments parallel to the fibers and wash in D-PBSA before weighing the biopsy.
3. Place the fragments parallel to each other in the lid of a Petri dish, cut the fragments into thinner cylinders and then, finally, into 1-$mm^3$ pieces, without crushing the tissue. The final cutting can be done in a tube with long scissors, again avoiding crushing.
4. Rinse with D-PBSA and let the pieces settle; discard the supernate.
5. Digest the fragments for 1 h at 37°C in pronase solution. (Use 15 mL for a biopsy of 1–3 mg.) Shake the tube gently at 10- to 15-min intervals.
6. Triturate the culture with a pipette after incubation. The medium should become increasingly opaque as more and more cells are released.
7. Let the fragments settle to the bottom by gravity, forming pellet P1 and supernate S1.
8. Filter S1 through a 100-μm nylon mesh into a 20-mL centrifuge tube. Shake or pipette the supernate gently to resuspend the cells.

9. Centrifuge the tube 8 to 10 min at 350 $g$. Discard the supernate by aspiration.

10. Resuspend the pellet very, very gently by means of a rubber bulb pipette in precisely 10 mL of growth medium, and count the cells with a hemocytometer.

11. Dilute the suspension in growth medium to seed culture flasks with about $1.5 \times 10^4$ cells/mL. About 1 to $2 \times 10^5$ cells/g are obtained from healthy donor biopsies.

12. Add 15 mL of digestion medium to P1, and incubate the fragments for 30 min in a water bath at 37°C, with periodic shaking.

13. Pipette the suspension to disaggregate the cells and then filter the suspension through nylon mesh. Rinse the filter with 20 mL of growth medium.

14. Centrifuge the suspension for 8 to 10 min at 350 $g$, count the cells, and seed as before.

15. Transfer the flasks to a 37°C humidified incubator with 5% $CO_2$.

△ **Safety Note.** The rest of the biopsy and all tubes, pipettes, plates, and so forth, used in the procedure should be treated with hypochlorite before disposal (*see* Section 6.8.5).

## Maintenance of Cultures

16. Change the medium very gently 24 h after seeding and then every 3 to 4 days.

17. The development of these cultures is mainly toward differentiation. The timing of the three phases for human muscle cells is about 4 to 6 days for peak proliferation; then the cells align at about day 8, and around day 10 to 12 an increase in cell fusion and the formation of myotubes are observed. Nevertheless, one must keep in mind that some cells may differentiate earlier and that others will still proliferate when the majority of the culture is undergoing differentiation.

## Subculture

18. Add a small volume of EDTA gently to the cells and remove it immediately.

19. Add sufficient trypsin solution (0.25%) to form a thin layer over the cells.

20. When cells detach, add 5 to 10 mL of complete growth medium, pass the culture very gently in and out of a pipette, and then centrifuge the cells for 10 min at 350 $g$.

21. Count an aliquot and seed the cells at the chosen concentration.

*Proliferation and differentiation indexes.* The growth curves of human myogenic cell cultures obtained in Ham's F12 medium supplemented with 20% FBS show the three traditional phases (*see* Section 20.9.2): the lag phase, the exponential phase, and the plateau, which corresponds to the onset of fusion. The last, evaluated in terms of the number of nuclei incorporated into myotubes or in terms of a fusion index (the percentage of nuclei incorporated into myotubes relative to the total number of nuclei), commences usually around day 8 after plating and rises dramatically around day 10. According to the sample, this chronology can gain or lose one day.

Hence differentiation, expressed as the number of nuclei per myotube/cm$^2$, may be observed morphologically. But the differentiation process can also be monitored by the use of biochemical markers (e.g., sarcomeric proteins), enzymes involved in differentiation (e.g., creatine phosphokinase and its time-dependent muscle-specific isoform shift), or the appearance of α-actin [Buckingham, 1992].

A family of genes, the best known of which is MyoD, was shown to activate muscle-specific gene expression in myogenic progenitors. [Buckingham, 1992]. Myogenic cell differentiation either involves the activation of a variety of other genes with concurrent changes in cell surface adhesive properties [Dodson et al., 1990] or requires the cell surface plasminogen activator urokinase and its receptor [Quax et al., 1992].

*Variations and applications.* The explant–reexplant method has been developed by the group of Askanas [Askanas et al., 1990; Pegolo et al., 1990]; it has proved reliable for a long time, especially for studies on muscle defects.

When the aim is the study of the mechanism of differentiation, one has the choice between the explant and the enzymatic methods, followed by clonal or nonclonal culture or by three-dimensional cultivation. If the aim of the work is to obtain a large number of cells for biochemical or molecular studies, then the enzymatic method seems to be preferable.

Protocol 22.15 for single myofiber culture, the associated introduction, above, and the discussion following the protocol, were contributed by T. A. Partidge and L. Heslop, Muscle Cell Biology Group, MRC Clinical Sciences Centre, Hammersmith Hospital, Du Cane Road, London W12 ONN. T. A. Partidge's current address is Center for Genetic Medicine Research, Children's National Medical Center, 111 Michigan Avenue, Washington, DC 20010. This procedure is of value when it is necessary to extract all the satellite cells or it is important to avoid contamination with nonmyogenic cells. The single myofiber isolation technique can be performed on any muscle that can be removed with a large proportion of its fibers undamaged, preferably tendon to tendon. In the contributors' laboratory they routinely use murine extensor digitorum longus (EDL), soleus, and tibialis anterior (TA) muscles.

---

## PROTOCOL 22.15. SINGLE MYOFIBER CULTURE FROM SKELETAL MUSCLE

### Outline

Single myofibers are released by trituration of collagenase-treated whole skeletal muscle, and the outgrowth is collected after culture on Matrigel.

### Materials

#### Sterile:

- Matrigel (BD Biosciences) 1 mg/mL in DMEM
- Dulbecco's modified Eagle's medium (DMEM, GIBCO, Invitrogen)
- L-Glutamine, 200 mM (Sigma)
- Penicillin-streptomycin solution: penicillin $10^4$ U/mL, streptomycin 10 mg/mL in 0.9% NaCl (Sigma)
- Chick embryo extract (MP Biomedicals)
- Horse serum (GIBCO, Invitrogen)
- Fetal bovine serum (FBS; PAA Laboratories)
- Plating medium: DMEM with horse serum 10%, chick embryo extract 0.5%, L-glutamine 4 mM, penicillin 100 U/mL, and streptomycin 100 µg/mL
- Proliferation medium: DMEM with 20% FBS, 10% horse serum, chick embryo extract 1%, L-glutamine 4 mM, penicillin 100 U/mL, and streptomycin 100 µg/mL
- Collagenase type I (Sigma)
- Petri dishes, 9 cm
- Modified Pasteur pipettes (Pasteur pipettes are cut, using a diamond tipped pen to create apertures of varying sizes, and then fire-polished to prevent jagged edges damaging the myofibers.)
- Multiwell plates, 24 well, coated with 1 mg/mL Matrigel™ (Primaria®, BD Biosciences) in DMEM and incubated at 37°C for 30 to 60 min to allow the Matrigel™ to gel

#### Nonsterile:

- Diamond-tipped pen
- Shaking water bath at 35°C
- Transilluminating stereo dissecting microscope
- Fetal bovine serum (PAA Laboratories)

### Procedure

#### Isolation and primary culture

1. Dissect out the muscles immediately after killing the animal, taking care to handle the muscles by their tendons to minimize damage to the myofibers.

2. Place the muscles into 0.2% (w/v) type I collagenase in DMEM and incubate in a shaking water bath at 35°C for 60 to 90 min. Longer digestion times are required for larger muscles; as a general guide, an EDL muscle taken from a 6- to 8-week-old mouse requires 60 min.

3. Prepare Petri dishes to take myofibers by rinsing with horse serum to prevent myofibers from adhering to the surface of the dish, and add 20 mL DMEM.

4. After digestion, transfer the muscle to a Petri dish containing DMEM.

5. Under a transilluminating stereo dissecting microscope, liberate single muscle fibers by repeatedly triturating the muscle with a modified Pasteur pipette. Pipettes are rinsed in 10% horse serum in DMEM to prevent myofibers from adhering to the glass.

6. As myofibers are liberated, the diameter of the muscle decreases and Pasteur pipettes of progressively narrower aperture are used.

7. Once 20 to 30 myofibers have been liberated, the muscle bulk is transferred to a fresh Petri dish and more myofibers separated until sufficient numbers have been dissociated.

8. Larger muscles may require a second digestion step to free myofibers located in the core; after superficial myofibers have been dissociated, return the remaining muscle bulk to the collagenase for further digestion.

9. Plate myofibers on Matrigel-coated Primaria (BD Biosciences) tissue culture plates.

10. Place individual myofibers in the center of the well and allow them to attach to the Matrigel. To ensure purity of the preparation, it is important only to plate undamaged fibers, free of regions of hypercontraction that obscure contaminating adherent cells often associated with fragments of microvessel.

11. Add plating medium slowly to the well, taking care not to disturb the plated myofiber.

12. Place plates in an incubator at 37°C and 5% $CO_2$.

#### Proliferation

13. Satellite cells begin to migrate from the parent myofiber after 12 to 24 h in culture. By 3 to 4 days each myofiber will be surrounded by approximately 50 to 300 cells, the precise number of cells depends on a number of factors, including the muscle from which the myofiber originated [Rosenblatt et al., 1996] and the age of the donor mouse [Bockhold et al., 1998].

14. When a number of cells have migrated from the myofiber, remove the myofiber from the culture with a Pasteur pipette pulled to a fine tip in a flame.
15. Continue culture of the remaining cells in original well.
16. Subculture (*see* Protocol 12.3), if required, and plate out in proliferation medium.

*Differentiation.* Proliferating myogenic cells can be induced to differentiate into myotubes by reducing the serum content of the medium to 5% horse serum in DMEM.

The myogenicity of the cells migrating from myofibers can be tested by immunostaining for desmin, a muscle-specific intermediate filament protein. In our hands all cells migrating from 98% to 99% of the single myofibers were found to express desmin.

Single myofibers can be prepared from the muscles of the $H-2K^b$-tsA58 transgenic mouse to obtain myogenic cells capable of many rounds of division in vitro [Morgan et al., 1994]. The $H-2K^b$-tsA58 transgenic mouse harbors the tsA58 temperature-sensitive mutant of the simian virus 40 (SV40) large T antigen under the control of an inducible promoter ($H-2K^b$) [Jat et al., 1991]. The tsA58 protein is active at 33°C, and transcription can be stimulated by addition of mouse γ-interferon. Myogenic cells derived from these mice proliferate in the presence of mouse γ-interferon (20 units/mL) at 33°C. When switched to between 37°C and 39°C in the absence of γ-interferon, either in culture or on transplantation in vivo, they differentiate to form myofibers in a rapid and synchronous manner.

### 22.3.4 Cartilage

Chondrocytes have been cultured from several different sources and early passage cultures are available commercially (*see* Table 22.1; Appendix II; Plate 28a) Protocol 22.16 for culturing cells from articular cartilage was contributed by F. Lemare, UPRES-EA 1833 Paris-Descartes University, 8 rue Mechain, 75014 Paris, France, and Department of Clinical Pharmacy, Gustave Roussy Cancer Institute, 39 rue C. Desmoulins 94800 Villejuif, France, and S. Demignot of Laboratoire de Pharmacologie Cellulaire de l'Ecole Pratique des Hautes Etudes, Centre de Recherche des Cordeliers, 15 rue de L'Ecole de Médecine 75006 Paris, France.

Chondrocytes are highly specialized cells of mesenchymal origin that are responsible for synthesis, maintenance, and degradation of the cartilage matrix. A great deal of research in the field of rheumatology has been focused on understanding the mechanisms that induce metabolic changes in articular chondrocytes during osteoarthritis and rheumatoid arthritis. Articular chondrocytes in culture are a very useful tool, but cultured in a monolayer, they rapidly divide, become fibroblastic, and lose their biochemical characteristics [Benya et al., 1977; von der Mark, 1986]. As early as the first

passage, there is a gradual shift from the synthesis of type II collagen to types I and III collagens and from aggrecan to low-molecular-weight proteoglycans. In parallel with the phenotypic modulation, the metabolic response to interleukin−1β is quantitatively affected by subculture [Blanco et al., 1995; Lemare et al., 1998].

Among the various methods explored for maintaining the phenotype of chondrocytes [Adolphe and Benya, 1992], culture in alginate beads is now widely used, since this culture system leads to the formation of a matrix similar to that of native articular cartilage [Häuselmann et al., 1996; Petit et al., 1996]. The system not only maintains the expression of the differentiated phenotype but also restores it in dedifferentiated chondrocytes [Bonaventure et al., 1994; Lemare et al., 1998]. Another advantage over other three-dimensional methods is that cells can easily be recovered after the culture is completed, allowing protein and gene expression studies [Lemare et al., 1998].

### PROTOCOL 22.16. CHONDROCYTES IN ALGINATE BEADS

#### Outline
Culture in alginate beads is based on the gelation in calcium chloride of a suspension of chondrocytes in an alginate solution.

#### Materials
*Sterile:*

❑ Cartilage dissection medium (CDM): Ham's F12 medium (Invitrogen), supplemented with netilmycin (20 μg/mL), vancomycin (10 μg/mL), and ceftazidim (30 μg/mL) (all from Sigma), to be used for dissection of the joints and removal of cartilage, as well as for enzymatic digestion
❑ Growth medium: DMEM (glucose, 1 g/l)/Ham's F12 (1:1, v/v), gentamicin 4 μg/mL, supplemented with 10% FCS

*Note.* Different batches of serum have to be compared for their ability to support chondrocyte growth as well as the expression of the differentiated phenotype (e.g., assessed by immunolabeling of type II collagen). When one lot is selected, several months' supply should be put on reserve.

❑ Enzyme solutions for dissociation of cartilage:
  (i) Trypsin, 0.05%, porcine pancreatic (Sigma) in Ham's F12
  (ii) Collagenase, 0.3%, type 1 (Worthington, CLS1) in Ham's F12
  (iii) Collagenase 0.06% in Ham's F12 + 10% FCS (All solutions are sterilized using 0.22-μm filters.)

*Note.* New batches of collagenase should be tested for efficient cartilage dissociation by comparison with the previous batch.

❑ Trypsin-EDTA solution (for subculture): 0.1% w/v porcine pancreatic trypsin in PBSA with 5 mM (0.02% w/v) of EDTA

❑ Sodium alginate solution: 1.25% w/v sodium alginate (Fluka) in 20 mM HEPES, 0.15 M NaCl Preparation:

  (i) Using a heated stirrer, first dissolve the HEPES in 0.15 M NaCl and heat to ~ 60°C.

  (ii) Add the sodium alginate and continue stirring until it dissolves completely (>2 h).

  (iii) Carefully adjust the solution to pH 7.4 with 1 M NaOH; do not exceed pH 7.4, as the addition of acid (HCl or H2SO4) results in irreversible precipitation.

  (iv) Dispense the solution into aliquots and sterilize it by autoclaving at 121°C for 10 min. *Alternative preparation:* Sodium alginate dissolution in HEPES/NaCl can be achieved rapidly (~10 min) in a microwave oven, but care must be taken to avoid caramelization.

❑ Gelation solution: 102 mM $CaCl_2$, 5 mM HEPES, pH 7.4

❑ Solubilization solution: 50 mM EDTA, 10 mM HEPES, pH 7.4

❑ Magnet

### Procedure

#### Dissection

1. Prepare cultures from knee, shoulder, and hip joints. Fetal or young donors are preferable to adults, as they provide higher quantities of cells and take longer to senesce. One-month-old "Fauve de Bourgogne" or "New Zealand" rabbits are ideal. However, hip joints obtained from surgical resection from human adults are also suitable sources of chondrocytes.

2. Wash tissue pieces thoroughly with CDM before dissection. Dissection should begin without delay. Remove skin, muscle, and tendons from joints. Carefully take cartilage fragments from articulations that are free of connective tissue.

#### Dissociation

*Note.* To prevent the cartilage drying, carry out all dissociation steps in 10-cm non–TC-treated dishes containing CDM. The quantities of enzyme solutions indicated in the rest of the protocol are suitable for tibiofemoral and scapulohumeral articulations from one rabbit, but they must be adapted to the quantity of cartilage to be dissociated.

3. Using crossed scalpels (*see* Protocol 11.4), mince cartilage slices into 1-mm³ pieces.

4. Transfer cartilage fragments into a 30-mL flat-bottomed vial.

5. Add 10 mL of 0.05% trypsin solution, and incubate the fragments with moderate magnetic agitation for 25 min in a sealed vial at room temperature.

6. Remove the trypsin solution after the fragments sediment.

7. Add 10 mL of 0.3% collagenase solution, and incubate the culture with moderate magnetic agitation for 30 min in a sealed vial at room temperature.

8. Remove collagenase solution after the fragments sediment.

9. Add 10 mL of 0.06% collagenase solution.

10. Transfer the suspension into a 10-cm bacterial dish. Rinse the vial with another 10 mL of 0.06% collagenase solution, transfer the rinse into the dish, and incubate the cartilage fragments overnight at 37°C in an incubator with 5% $CO_2$.

11. Transfer the cell suspension into a 50-mL centrifuge tube and mix on a vortex mixer for a few seconds.

12. Remove residual material left after digestion by passing the digested material through a 70-μm nylon cell strainer (BD Biosciences; *see* Fig. 11.8).

13. Centrifuge the filtrate at 400 *g* for 10 min.

14. Resuspend the cell pellet in 20 mL of CDM, and count the cells with a hemocytometer.

15. Centrifuge the cells at 400 *g* for 10 min.

#### Entrapment

16. Resuspend the cell pellet in 1 mL of sodium alginate solution, and then dilute the suspension progressively in more sodium alginate solution, until a cellular density of $2 \times 10^6$ cells/mL is reached. This progressive dilution is necessary to obtain a homogeneous cell suspension in alginate.

17. Express the cell suspension in drops through a 21G needle into the gelation solution with moderate magnetic stirring, and allow the alginate to polymerize for 10 min to form beads.

18. Wash the beads 3 times in 5 vol NaCl, 0.15 M.

19. Distribute the beads, 5 mL ($1 \times 10^7$ cells), into 75-cm² flasks containing 20 mL of growth medium.

20. Incubate the culture at 37°C in a humidified atmosphere of 5% $CO_2$ and 95% air.

#### Recovery

21. Discard the medium.

22. Add 2 vol of solubilization solution to the beads.

23. Incubate the culture 15 min at 37°C.
24. Centrifuge the cells at 400 g for 10 min.
25. Resuspend the cell pellet in growth medium containing 0.06% collagenase.
26. Incubate the cells for 30 min at 37°C in a humidified atmosphere of 5% $CO_2$ and 95% air.
27. Centrifuge the cells at 400 g for 10 min.
28. Resuspend the cell pellet in serum-free DMEM/Ham's F12 medium, and count the cells with a hemocytometer.
29. Centrifuge the cells at 400 g for 10 min.
30. Repeat steps 28, 29, without counting the cells.

### Analysis of chondrocyte phenotype

**Collagen.** Type II collagen represents 85% to 90% of the total collagen synthesized by chondrocytes in cartilage or primary culture. Type II collagen is highly specific to chondrocytes and is considered as their main differentiation marker. The expression of type II collagen can be analyzed by indirect immunofluorescence using specific antibodies. These must be very carefully checked, however, to make sure that there is no cross-reactivity with type I collagen, which is expressed by dedifferentiated cells. The immunocytochemical study of collagen expression in chondrocytes cultured in alginate beads necessitates sectioning the beads as described in Lemare et al. [1998].

For a more precise analysis of the collagen phenotype, SDS-PAGE and two-dimensional CNBr peptide maps of purified collagen after [³H]-Proline incorporation have to be performed [Benya, 1981]. Concerning major collagens, SDS-PAGE analysis can reveal the presence of type III collagen and of the α2(I) chain of collagen I. However, it cannot separate the α1 chain of type I and type II (i.e., it cannot make a distinction between type I and type II trimer collagen), which is why two-dimensional CNBr peptide mapping is necessary to assert type II collagen expression.

The expression of these genetically distinct collagens can also be analyzed at the transcriptional level by real-time PCR [Murphy and Polak, 2004].

**Proteoglycans.** Alcian Blue staining at acidic pH is widely used as an indicator of the presence of sulfated aggrecan [Lemare et al., 1998]. As with immunocytochemical studies, this staining necessitates sectioning the beads. A more precise analysis of the different types of proteoglycans synthesized can be performed after $^{35}SO_4$ incorporation [Verbruggen et al., 1990]. As for collagen genes, the analysis of aggrecan and link protein gene expression can be performed by real-time PCR [Murphy and Polak, 2004].

**Variations and applications.** Several studies have shown that chondrocytes cultured in alginate beads represent an attractive alternative chondrocyte model [Grandolfo et al., 1993; Häuselmann et al., 1994, 1996; Petit et al., 1996]. The main advantage of this technique over other three-dimensional methods is that cells can easily be recovered after culture, allowing protein and gene expression studies [Lemare et al., 1998].

The number of available cells often becomes a limiting factor in studies of chondrocyte metabolism, as articular cartilage is difficult to obtain in large quantities and has a low cellularity. In this case the number of chondrocytes available can be amplified by first culturing the cells in a monolayer (two passages) and then restoring the differentiated properties of chondrocytes by cultivating the cells in alginate beads for two weeks [Bonaventure et al., 1994; Lemare et al., 1998]. In parallel with this phenotypic restoration, the metabolic response to Interleukin−1β of these cells was restored at a quantitative level similar to that of chondrocytes in a monolayer primary culture [Lemare et al., 1998]. This makes the culture in alginate beads a relevant model for the study of chondrocyte biology. Moreover alginate is a hydrogel that has potential for tissue engineering applications. For example, scaffolds filled with chondrocytes embedded in alginate gel have been proposed for cartilage tissue engineering and graft [Cancedda et al., 2003; Marijnissen et al., 2002; Masuda & Sah 2006; Selmi et al., 2008]. For these purposes hydrogels made of alginate and supplemented with another polymer such as hyaluronic acid, heparin, or agarose are currently under investigation. More recently bone marrow stromal cells have been shown to differentiate into chondrocytes and chondron-like structures when cultured in alginate beads and in the presence of a chondrogenic growth factor cocktail [Coleman et al., 2007]. In similar conditions chondrocytes from osteoarthritic cartilage can be stimulated to undergo 100-fold expansion and then redifferentiation [Hsieh-Bonassera et al., 2009].

## 22.3.5 Bone

Although bone is mechanically difficult to handle, thin slices treated with EDTA and digested in collagenase [Bard et al., 1972] give rise to cultures of osteoblasts that have some functional characteristics of the tissue. Antiserum against collagen has been used to prevent fibroblastic overgrowth without inhibiting the osteoblasts [Duksin et al., 1975]. Continuous cell lines have been obtained from osteosarcoma [Smith et al., 1976; Weichselbaum et al., 1976], but not from normal osteoblasts, although early passage lines are available commercially (see Table 22.1; Appendix II; Plate 28b). Mesenchymal stem cells (see Section 23.3.4) can be cultured and induced to form osteocytes, suitable for tissue engineering [Lennon & Caplan, 2005; Hofmann et al., 2005]

Protocol 22.17 for the culture of bone cells was contributed by E. Schwartz, Departments of Orthopedic Surgery and Physiology, Tufts University School of Medicine, 136 Harrison Avenue, Boston, MA 02111.

Bone culture suffers from the inherent problem that the hard nature of the tissue makes manipulation difficult.

However, conventional primary explant culture or digestion in collagenase and trypsin releases cells that may be passaged in the usual way.

---

## PROTOCOL 22.17. OSTEOBLASTS

### Outline

For explant culture, small fragments of tissue are allowed to adhere to the culture flask by incubation in a minimal amount of medium (*see* Protocol 12.4). The adherent explants are then flooded and the outgrowth is monitored. For disaggregated cell culture, trabecular bone is dissected down to 2- to 5-mm pieces and digested in collagenase and trypsin. Suspended cells are seeded into flasks in F12 medium.

### Materials

#### Sterile:

❑ Ham's F12 medium with 12% fetal bovine serum, 2.3 mM of $Mg^{2+}$, 100 U/mL of penicillin, and 100 µg/mL of streptomycin $SO_4$

❑ Trypsin solution to be used for cell passage: Dissolve 125 mg of trypsin (Sigma, type XI) in 50 mL of $Ca^{2+}$, $Mg^{2+}$-free Tyrode's solution. Adjust the resulting solution to pH 7.0 and filter sterilize it. Place aliquots of 5 and 10 mL in Pyrex tubes and store at −20°C.

❑ Digestion solution for the isolation of osteoblasts: Solution A: 8.0 g of NaCl, 0.2 g of KCl, 0.05 g of $NaH_2PO_4 \cdot H_2O$ in 100 mL of UPW

❑ Solution B (collagenase-trypsin solution): dissolve 137 mg of collagenase (Worthington, type I) and 50 mg of trypsin (Sigma, type III) in 10 mL of solution A. Adjust the combined solution to pH 7.2, and then bring the amount to 100 mL with UPW. Filter sterilize and distribute the solution into 10-mL aliquots that are stored at −20°C.

❑ Petri dishes, 9 cm

❑ Culture flasks 25 or 75 $cm^2$

❑ Scalpels

❑ Forceps

### Procedure A: Explant Cultures

1. Obtain bone specimens from the operating room.
2. Rinse the tissue several times at room temperature with sterile saline.
3. If the bone cannot be used immediately, cover the specimen with sterile Ham's F12 medium containing 12% fetal calf serum, penicillin, and streptomycin sulfate. The bone may be stored overnight at 4°C.

4. The next morning, rinse the tissue with Tyrode's solution containing penicillin (100 U/mL) and streptomycin sulfate (100 µg/mL).
5. Place the bone in a Petri dish and, with the use of scalpel and forceps, remove the trabeculae and place them in a second Petri dish.
6. Add 10 mL of Tyrode's solution over the excised trabeculae. Rinse the trabeculae several times with Tyrode's solution until blood and fat cells are removed.
7. To initiate explant cultures, prepare a 25-$cm^2$ flask by preincubating it with 2 mL of complete medium for 20 min to equilibrate the medium with the gas phase. Adjust the pH as necessary with $CO_2$, 4.5% $NaHCO_3$, or HCl.
8. Cut the trabeculae into fragments of 1 to 3 mm.
9. Remove the preincubation medium from the flask, and add 2.5 mL of fresh Ham's F12 medium to the flask.
10. Transfer between 25 and 40 fragments of trabeculae to the flask.
11. With the flask in an upright position, slide the explant pieces along the base of the flask with the aid of an inoculating loop, and distribute the explant pieces evenly.
12. Permit the flask to remain upright for 15 min at 37°C.
13. Slowly restore the flask into a normal horizontal position. The explant pieces will stick to the bottom of the flask.
14. Leave the flask in the horizontal position at 37°C for 5 to 7 days. After this period, check for outgrowth, and replace the medium in the flask with fresh medium. To prevent the explant pieces from becoming detached, lift the flask slowly into the vertical position before carrying it from the incubator to the hood or to the microscope.
15. To maintain cultures, change the medium twice per week.
16. When confluence is reached, remove the explant pieces, trypsinize the cell layer, isolate the cells by centrifugation, and seed cells into flasks or wells.

### Procedure B: Monolayer Cultures from Disaggregated Cells

1. Wash trabecular bone specimens repeatedly with Tyrode's solution to remove the fat and blood cells. The trabeculae are excised with scalpel and forceps under sterile conditions.
2. After collecting as much bone as possible, wash the remaining blood and fat cells away by rinsing the specimens three times with Tyrode's solution and cut into 2- to 5-mm fragments.

3. Wash the cut trabeculae with F12 medium containing fetal calf serum.

4. Place the pieces of bone in a small sterile bottle with a magnetic stirrer, and add 4 mL of digestion solution. (This amount should cover the bone specimens.)

5. Stir the solution containing bone fragments at room temperature for 45 min.

6. Remove the suspension of released cells and discard it, because these cells are most likely to contain fibroblasts.

7. Add a second aliquot of 4 mL of digestion solution to the bone fragments, and stir the mixture at room temperature for 30 min.

8. Collect the digestion solution from bone fragments, and centrifuge it for 2 min at 580 $g$ at room temperature.

9. After removing the supernate, suspend the cells in 4 mL of Ham's F12 medium with 20% fetal calf serum, and count the cells.

10. Centrifuge the suspension at 580 $g$ for 10 min, and resuspend the cells in 4 mL of complete medium. This suspension will become the inoculum.

11. Preincubate 75-cm$^2$ flasks for 20 min with 8 mL of complete F12 medium to equilibrate with the gas phase.

12. Remove the preincubation medium and add 2 mL of complete F12 medium.

13. Add 4 mL of medium containing the cell suspension. The inoculum should contain 6000 to 10,000 cells per cm$^2$ of surface area.

14. Finally, add another 6 mL of Ham's F12 medium, to give a total volume of 12 mL.

15. In the interim, add an additional 4 mL of digestion solution to the remaining pieces of bone, and repeat the digestion for 30 min. The released cells are harvested, and, if necessary, the digestion step is repeated several more times. With large amounts of bone, the digestion period can be increased to 1 to 3 h. Cell counts are performed after each digestion period, and the released cells are used to inoculate a different flask.

### Passage of cells in culture

1. Remove the pieces of explant.

2. Remove the medium and rinse the cell layer with D-PBSA, 0.2 mL/cm$^2$.

3. Add trypsin to the flask, 0.1 mL/cm$^2$, and incubate at 37°C until the cells have detached and separated from one another. Monitor cell detachment and separation on the microscope. In general, a 10-min incubation is sufficient.

4. Transfer the released cells to a centrifuge tube with an equal volume of Ham's F12 medium with 20% fetal calf serum (FCS).

5. Centrifuge the cells at 600 $g$ for 5 min.

6. Discard the supernate, and resuspend the cells in complete medium by gentle repeated pipetting.

7. Set one aliquot aside for the determination of cell concentration and another for DNA determination (see Protocol 20.3).

8. Inoculate the remaining cells into culture flasks or wells that have previously been equilibrated with medium. The cells should reattach within 24 h.

## 22.3.6 Endothelium

Endothelial cell culture can provide models for vascular disease, blood vessel repair, and angiogenesis in cancer, as well as providing cells for engineering blood vessels. Folkman and Haudenschild [1980] described the development of three-dimensional structures resembling capillary blood vessels derived from pure endothelial lines in vitro, which suggested that cultured endothelial cells still have the ability to reconstruct blood vessels. This has been borne out by later work [Klinger & Niklason, 2006] where smooth muscle cells have also been incorporated into the construct.

Growth factors, originally an angiogenesis factor(s) derived from Walker 256 cells in vitro, play an important part in maintaining cell proliferation and survival, so that secondary structures can be formed. Endothelial cultures are also good models for contact inhibition and density limitation of growth, as cell proliferation is strongly inhibited after confluence is reached [Haudenschild et al., 1976]. Endothelial cell cultures are also available commercially (see Table 22.1; Appendix II; Plate 28e).

The following introduction and Protocol 22.18 for the culture of large-vessel endothelial cells was contributed by C. Lawson, Royal Veterinary College, Royal College Street, London NW1 0UT, England, UK.

The inner surface of all blood vessels is composed of a single cell layer of endothelial cells. The endothelium provides a smooth anticoagulant surface, contributes to blood vessel contractility, and acts as a permeable barrier allowing the transport of nutrients and gases as well as influencing emigration of immune cells from the blood. Endothelial cells can be isolated from large blood vessels, including human umbilical veins [Jaffe et al., 1973] and bovine [Booyse et al., 1975] or porcine [Bravery et al., 1995] aorta. Microvascular endothelial cells can be isolated from bovine adrenal gland [Folkman et al., 1979], rat brain [Bowman et al., 1981], human skin [Davison et al., 1983], human adipose tissue [Kern et al., 1983], as well as human [McDouall et al., 1996] and murine [Marelli-Berg et al., 2000; Lidington et al., 2002] heart.

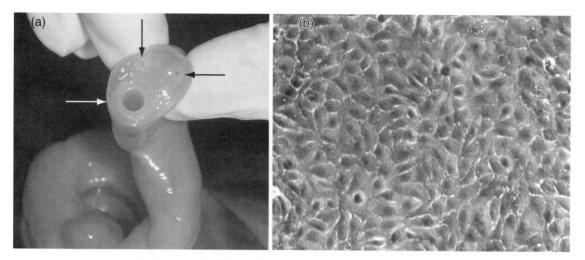

**Fig. 22.3. Vascular Endothelial Cells.** (*a*) View of blood vessels in human umbilical cord. Umbilical arteries (black arrows); umbilical vein (white arrow) with Luer adaptor inserted (*see also* Plate 8*e*). (*b*) View of human aortic endothelial cell monolayer under phase contrast microscope. 10× magnification.

## PROTOCOL 22.18. ISOLATION AND CULTURE OF VASCULAR ENDOTHELIAL CELLS

### Outline

Incubation of the blood vessel lumen with collagenase releases endothelial cells, which can be cultured for several population doublings on a gelatin-coated substrate in a suitable growth medium. Immunomagnetic separation (*see* Section 14.3.2; Figs. 14.5, 14.6) is used to isolate endothelial cells from disaggregated heart tissue.

### A. Isolation of Endothelial Cells from Human Umbilical Vein (HUVECS)

[Adapted from the method of Jaffe et al., 1973.]

### Common Materials for All Endothelial Procedures

*Sterile or aseptically prepared:*

❑ D-PBSA on ice
❑ HBSS: Hanks's balanced salt solution
❑ HBSS/PSG containing 300 U/mL penicillin, 300 μg/mL streptomycin (Sigma), and 50 μg/mL gentamycin (Invitrogen)
❑ Trypsin, 500 μg/mL, EDTA, 5 mM (0.2%) (Sigma cat. no. T4174)
❑ Culture flasks, 25 cm², coated with 0.5% gelatin
  (i) Dilute 2% solution of type B gelatin from bovine skin (Sigma, G1393) fourfold with D-PBSA
  (ii) Incubate in flask for 20 min at 37°C.
  (iii) Remove the gelatin just before addition of cells.

❑ Scalpel and #22 blade
❑ Needle
❑ Syringes, 20 mL
❑ Crocodile clips
❑ Artery clamps
❑ Forceps
❑ Sharp scissors

*Nonsterile:*

❑ Methanol, 70%
❑ Paper tissues
❑ Aluminum foil
❑ Cling film (or Saran Wrap)
❑ Cork board
❑ Extra strong thread

### Materials for HUVECs

❑ Human umbilical cords in HBSS/PSG
❑ HUVEC growth medium: M199 medium with Hanks's salts, 2 mM L-glutamine, and 25 mM HEPES supplemented with 20% fetal bovine serum (FBS), 150 units/mL penicillin/streptomycin and endothelial cell growth supplement (ECGS) from bovine brain (Roche, cat. no. 1033484, 75 mg; use at 20 μg/mL with heparin at 50 μg/mL or Sigma cat. no. E2759, 15 mg; use at 30 μg/mL with heparin at 10 μg/mL)
❑ Collagenase H: from *Clostridium histolyticum* (Roche cat. no. 1074059; use at 0.5 mg/mL in M199. Filter sterilize), 25 mL
❑ Newborn calf serum (NBS), cold, 5 mL
❑ Adaptors (Luer female to Luer male adaptor Portex cat. no. 700/180/700) 2

*Procedure*

Isolation of endothelial cells

1. Cover cork board with aluminum foil.
2. Place paper tissues on foil and soak with 70% methanol soaked tissues.
3. Squeeze down the length of the umbilical cord to expel any blood.
4. Cut 2 cm or more off the one end of the cord, including any clamp marks.
5. Insert Luer adaptor all the way into the vein (Fig. 22.3*a*; *see also* Plate 8).
6. Pin cord to cork board with needle near the vein but making sure not to pierce through to the lumen.
7. Using scalpel and forceps strip back the arteries about 2 cm.
8. Secure the connector in place with thread, using two very tight opposing knots
9. Clamping the other end of the cord shut, push up to 25 mL D-PBSA into the cord with a 20-mL syringe attached to the connector. With the vein distended under the pressure of the D-PBSA, check for any holes in the cord and clamp any holes with crocodile clips by nipping the exterior of the cord without closing the vein.
10. Empty cord of D-PBSA into waste pot.
11. Cut off about 2 cm from other end of cord and insert and secure connector as in step 8.
12. Using another 20-mL syringe, push approximately 25 mL collagenase into the cord until it becomes distended.
13. Wrap the cord in methanol-sprayed Clingfilm and leave at 37°C for 8 min.
14. Squeeze collagenase out of cord and withdraw back into syringe(s).
15. Remove syringes and empty collagenase into tube containing 5 mL newborn calf serum.
16. Replace syringe on one end of cord and use the other syringe to push 25 mL D-PBSA into the cord.
17. Squeeze or tap cord along its length for a few moments, then push D-PBSA back into syringes.
18. Add to contents of tube from step 15.
19. Centrifuge tube for 10 min, 210$g$, 4°C.
20. Resuspend pellet in 5 mL HUVEC growth medium.
21. Remove excess gelatin from flask and add seed cells from step 20.
22. The following day, remove the medium from the flask and discard.
23. Wash twice with D-PBSA and replace with fresh HUVEC growth medium.

Maintenance and culture

24. Remove half the medium and replace with fresh three times per week.
25. Passage, 1:2 split, trypsin/EDTA into gelatin coated flasks approximately once per week.
26. Do not use beyond 6th passage.

### B. Porcine Aortic Endothelial Cell (PAEC) Isolation

[Adapted from the method of Bravery et al., 1995.]

*Materials for PAECs*

❑ HBSS/PSGA: HBSS containing 300 units/mL penicillin/streptomycin, 50 μg/mL gentamycin, 2.5 μg/mL amphotericin B (Invitrogen)
❑ Fresh porcine aorta (collected aseptically if possible), transported in HBSS/PSGA
❑ PAEC growth medium: M199 medium with Hanks's salts, 2 mM L-glutamine, and 25 mM HEPES, supplemented with 150 U/mL penicillin/streptomycin, 50 μg/mL gentamycin, 2.5 μg/mL amphotericin B, 10% FBS, 10% NBBS
❑ Collagenase H (as in HUVEC Materials), 50 mL
❑ FBS in 50-mL tube, 5 mL

*Procedure*

1. Cover cork board with aluminum foil.
2. Lay aorta on methanol-sprayed tissue and pat dry.
3. On foil-covered board trim off excess tissue, leaving only branches—do not cut too short.
4. Tie off branches with thread and clamp thin end of aorta.
5. Pour in D-PBSA to fill; clamp top.
6. Clamp any leaks with sterile crocodile clips.
7. Discard D-PBSA.
8. Fill aorta with warm collagenase, reclamp, wrap in methanol-sprayed clingfilm, and incubate 8 min at 37°C.
9. Gently massage aorta to loosen cells.
10. Pour collagenase from aorta into 50-mL tube containing 5 mL FBS
11. Fill aorta with ice-cold D-PBSA, clamp, massage, add to tube of cells in step 10, and repeat this step.
12. Centrifuge cells at 500 $g$ for 6 min at 4°C.
13. Resuspend cell pellet in 5 mL medium, place in 25-cm$^2$ flask and incubate overnight at 37°C.
14. The following day remove medium from flask and discard.
15. Wash twice with D-PBSA and replace with fresh medium.

## Maintenance

16. Replace medium three times per week; if cells are less than 50% confluent remove and replace only half the medium; if cells are 50% to 80% confluent remove and replace all the medium.

17. When cells are 80% confluent passage with 1:3 split ratio using trypsin/EDTA.

18. Do not use beyond 7th passage.

### C. Murine Cardiac Endothelial Cell (MCEC) Isolation

[Adapted from the method of Marelli-Berg et al., 2000.]

#### Materials for MCECs

❑ Mouse hearts, 5, collected into HBSS/PSG (*see* Protocol 22.18A) containing 300 units/mL penicillin/streptomycin and 50 μg/mL gentamycin

❑ MCEC growth medium: DMEM with 25 mM glucose and 4 mM L-glutamine, supplemented with 10% FBS, 150 U/mL penicillin, 150 μg/mL streptomycin, and endothelial cell growth supplement (ECGS) from bovine brain

❑ Medium for nonendothelial cells: DMEM (1.0 g/L glucose) containing 15% FCS

❑ HBSS/PS: HBSS with 150 U/mL penicillin, 150 μg/mL streptomycin

❑ HBSS/FB: HBSS with 1% FBS

❑ CMF: HBSS without $Ca^{2+}$ and $Mg^{2+}$ with 150 U/mL penicillin, 150 μg/mL streptomycin

❑ Trypsin/EDTA (e.g., Sigma catalog)

❑ FBS

❑ Collagenase A: from *Clostridium histolyticum* (Roche cat. no. 103586); use at 1 mg/mL in HBSS. Make up fresh and filter sterilize, 10 mL

❑ Cell strainers, 70 μm mesh (BD Biosciences)

❑ Petri dishes, 9 cm, 2

❑ Irrelevant control antibody, such as OKT4 anti human CD4 (DAKO), 5 μg

❑ Rat anti-mouse endoglin (CD105) antibody; clone MY7/18 (BD Biosciences, Pharmingen), 5 μg

❑ Goat anti-rat Ig MicroBeads (Miltenyi cat. no. 130-048-501)

❑ MidiMacs separation unit (Miltenyi)

❑ LS MACS column (Miltenyi 130-042-401)

#### Procedure

1. Place hearts in Petri dish containing about 5 mL HBSS/PS.

2. Remove aorta, pulmonary artery, and other appendages from heart.

3. Cut heart in half and remove any large blood clots and connective tissue.

4. Place hearts in fresh Petri dish with CMF, wash and replace with fresh CMF.

5. Dice up material in the dish with scalpels.

6. Wash with CMF by resuspension and settling.

7. Scrape and pick up tissue on side of scalpel blade and transfer into universal containing 10 mL warmed freshly prepared collagenase A at 1 mg/mL.

8. Incubate for 30 min at 37°C with agitation.

9. Add CMF.

10. Disperse cells by shaking up tissue and allowing to settle. Aspirate off cell suspension and transfer to a fresh tube.

11. Repeat step 10 several times.

12. Spin cell suspension at 300 *g* for 8 min.

13. Carefully remove supernate from the loose cell pellet.

14. Resuspend cell pellet in 5 mL trypsin/EDTA, try to break up any clumps of cells by pipetting, and incubate for 10 min at 37°C.

15. Add 10 mL HBSS/FB containing 1% FS to disaggregated cells digest and filter through 70-μm cell strainer.

16. Wash tissue in cell strainer with 20 mL cold HBSS/FB.

17. Centrifuge cells in filtrate for 8 min at 300 *g*.

18. Aspirate supernate carefully, discard, and retain the cell pellet.

19. Add about 5 μg irrelevant antibody (e.g., OKT4/OKT8) to the cell pellet, flick the tube to disperse the pellet, and incubate on ice for 20 min to block Fc receptors.

20. Add 5 μg rat anti-mouse endoglin to cell pellet, flick the tube (clone MY7/18), and incubate at 4°C for 20 min.

21. Add 30 mL HBSS and centrifuge at 300 *g* for 8 min at 4°C.

22. Add 50 μL goat anti-rat beads in 200 μL HBSS.

23. Incubate for 15 min at 4°C.

24. Add 30 mL HBSS to wash and centrifuge at 300 *g* for 8 min at 4°C.

25. Place MidiMacs column on magnet and wash with 500 μL HBSS/1% FCS according to manufacturer's recommendations (*see also* Protocol 14.2).

26. Add 3 mL HBSS/1% FBCS to cell pellet and add to column.

27. Wash column 3× with 1 mL HBSS/FB followed by a final wash in 500 μL growth medium.

28. Remove column and place in 15-mL tube, add 4 mL MCEC growth medium, and put plunger in to collect bead bound cells. Move cell suspension to gelatin-coated 25-cm² flask

29. To keep nonendothelial cell population spin down washes from column and plate in 25-cm² flask in DMEM (1000 mg/L glucose) containing 15% FBS.

*Maintenance*

**30.** Replace medium twice per week; if cells are less than 50% confluent remove and replace only half the medium; if cells are 50% to 80% confluent remove and replace all of the medium.

**31.** When cells are 80% confluent, passage with a split ratio of 1:3 with trypsin/EDTA.

**32.** Do not use beyond 3rd passage.

### D. Human Cardiac Endothelial Cell (HCEC) Isolation

It is possible to isolate CEC from human left ventricle by following the method of McDouall et al. [1996].

#### Materials for HCEC

☐ Magnetic beads coated with antibodies to human MHC class II or human CD31 (Dynal)

☐ HCEC growth medium: M199 with 4 mM L-glutamine, supplemented with 10% FBS, 10% human male AB serum (Sigma), 150 U/mL penicillin, 150 μg/mL streptomycin, and ECGS as in Protocol A

☐ Collagenase II (Roche), 1 mg/mL

#### Procedure

**1.** Follow Protocol 22.18C.

**2.** Digest the tissue with collagenase II (Roche) at step 7.

**3.** Use magnetic beads coated with antibodies to human MHC class II or human CD31 (Dynal) to select endothelial cells as at step C20 and proceed as from step C23.

**4.** Maintain hCEC in HCEC growth medium.

*Identification of vascular endothelial cells.* Vascular endothelial cells appear polygonal and have a characteristic "cobblestone" morphology when cultured under static conditions (Fig. 22.3b). Fibroblast contamination should be easy to identify by their spindle-shaped morphology with a standard phase-contrast inverted microscope. Vascular endothelial cells can be identified by the production of factor VIII [Hoyer et al., 1973], angiotensin-converting enzyme [Del Vecchio & Smith, 1981], the uptake of acetylated low-density lipoprotein [Voyta et al., 1984], the presence of Weibel–Palade bodies [Weibel & Palade, 1964], and the expression of endothelium-specific cell surface antigens including PECAM-1 (CD31) [Parums et al., 1990; Albelda et al., 1990] and endoglin (CD105) [Gougos & Letarte, 1990]. Vascular endothelial cells will readily form tubes in Matrigel™ [McGuire & Orkin, 1987] (*see* Protocol 17.4).

*Variations.* Large-vessel endothelial cells can be also isolated from human or bovine aorta; from muscular arteries including pulmonary, coronary, and internal mammary arteries; and from veins, with protocols similar to those outlined above. Microvascular endothelial cells can be isolated from skin [Davison et al., 1983], brain [Bowman et al., 1981], and lung [Magee et al., 1994] and from specialized vascular beds including liver [Shaw et al., 1984], kidney [Green et al., 1992], and lymph node [Ager, 1987].

## 22.4 NEUROECTODERMAL CELLS

### 22.4.1 Neurons

Nerve cells will not survive well on untreated glass or plastic, but will demonstrate neurite outgrowth on collagen and poly-D-lysine. Neurite outgrowth is encouraged by the polypeptide nerve growth factor (NGF) [Levi-Montalcini, 1979] and other factors secreted by glial cells [Marchionni et al., 1993; Barnett, 2004; Miklic et al., 2004; Wu et al., 2004; Ernsberger, 2008]. Neural organization can be induced in vitro by growth on nanoparticles [Chen et al., 2010]. Cell proliferation has not been found in cultures of most neurons, even with cells from embryonic stages in which mitosis was apparent in vivo; however, studies with embryonal stem cells have shown that neuronal progenitor cells can be made to proliferate in vitro [Thomson et al., 1998] and recolonize in vivo [Snyder et al., 1992] (*see also* Plate 25e). Neural cells, including neurons, are available commercially (*see* Table 22.1; Appendix II).

Protocol 22.19 for the monolayer culture of cerebellar neurons was contributed by B. Engelsen and R. Bjerkvig, Department of Cell Biology and Anatomy, University of Bergen, Žrstadveien 19, N-5009 Bergen, Norway.

Cerebellar granule cells in culture provide a well-characterized neuronal cell population that is suited for morphological and biochemical studies [Drejer et al., 1983; Kingsbury et al., 1985]. The cells are obtained from the cerebella of 7- or 8-day-old rats, and nonneuronal cells are prevented from growing by the brief addition of cytosine arabinoside to the cultures.

---

### PROTOCOL 22.19. CEREBELLAR GRANULE CELLS

#### Outline

The cerebella from 4 to 8 neonatal rats are cut into small cubes and trypsinized for 15 min at 37°C in Hanks's BSS. The cell suspension is seeded in poly-L-lysine-coated culture wells or flasks.

#### Materials

*Sterile:*

☐ DMEM with
  Glucose.................................................30 mM
  L-glutamine............................................2 mM
  KCl.......................................................24.5 mM

Insulin (Sigma I-1882) ..................... 100 mU/L
*p*-Aminobenzoic acid (Sigma A-3659) ......... 7 μM
Gentamycin ...................................... 100 μg/mL
Fetal calf serum, heat inactivated ............... 10%
❑ Hanks's BSS with 3 g/L BSA (HBSS)
❑ Poly-L-lysine, mol. wt. >300, 000 (Sigma P-1524)
❑ Cytosine arabinoside (Cytostar, Upjohn; powder)
❑ Trypsin (type II): 0.025% in HBSS
❑ Silicone (Aquasil, Pierce 42799)
❑ Tissue culture Petri dishes, 3.5 cm
❑ Dissecting instruments: scalpels, scissors, and forceps
❑ Pasteur pipettes and 10-mL pipettes
❑ Test tubes, 12 and 50 mL

*Nonsterile:*

❑ Water bath
❑ Siliconization of Pasteur pipettes:
  (i) Dilute the Aquasil solution 0.1% to 1% in UPW.
  (ii) Dip the pipettes into the solution or flush out the insides of the pipettes.
  (iii) Air dry the pipettes for 24 h, or dry for several minutes at 100°C.
  (iv) Sterilize the pipettes by dry heat.
❑ Poly-L-Lysine treatment of culture dishes:
  (i) Dissolve the poly-L-lysine in UPW (10 mg/L).
  (ii) Sterilize the mixture by filtration.
  (iii) Add 1 mL of poly-L-lysine solution to each of the 35-mm Petri dishes.
  (iv) Remove the poly-L-lysine solution after 10 to 15 min, and add 1 to 15 mL of culture medium.
  (v) Place the culture dishes in the incubator (minimum 2 h) until the cells are to be seeded.

*Procedure*

1. Dissect out the cerebella aseptically and place them in HBSS.
2. Mince the tissue with scalpels into small cubes approximately 0.5 mm³.
3. Transfer the minced tissue to test tubes (12 mL) and wash the tissue three times in HBSS. Allow the tissue to settle to the bottom of the tubes between each washing.
4. Add 10 mL of 0.025% trypsin (in HBSS) to the tissue, and incubate the tube in a water bath for 15 min at 37°C.
5. Transfer the trypsinized tissue to a 50-mL centrifuge tube, and add 20 mL of growth medium to stop the action of the trypsin.
6. Disaggregate the tissue by trituration through a siliconized Pasteur pipette, until a single-cell suspension is obtained.

7. Let the cell suspension stay in the centrifuge tube for 3 to 5 min, allowing small clumps of tissue to settle to the bottom of the tube. Remove these clumps with a Pasteur pipette.
8. Centrifuge the single-cell suspension at 200 *g* for 5 min, and aspirate off the supernate.
9. Resuspend the pellet in growth medium, and seed the cells at a concentration of 2.5 to 3.0 × 10⁶ cells/dish.
10. After 2 to 4 days (best results usually are obtained after 2 days), incubate the cultures with 5 to 10 μM cytosine arabinoside for 24 h.
11. Change to regular culture medium.

*Characterization of neuronal cultures.* Neurons can be identified immunologically by using neuron-specific enolase antibodies or by using tetanus toxin as a neuronal marker. Astrocyte contamination can be quantified by using glial fibrillary acidic protein as a marker.

*Variations.* A single-cell suspension can be obtained by mechanical sieving through nylon meshes of decreasing diameter (Protocol 12.9) or by sequential trypsinization (i.e., a 3- to 5-min trypsin treatment). Instead of HBSS, Puck's solution, Krebs, or other buffers with glucose can be used.

## 22.4.2 Glial Cells

Glial cells have been cultured from avian, rodent, and human brains. Human adult normal astroglial lines from brain lines express glial fibrillary acidic protein (GFAP) [Burke et al., 1996], high-affinity γ-aminobutyric acid and glutamate uptake, and glutamine synthetase activity [Frame et al., 1980]. Oligodendrocytes do not readily survive subculture, but oligodendrocyte precursor cells from the optic nerve have been subcultured using PDGF and FGF-2 as mitogens [Bögler et al., 1990]. Schwann cells have been subcultured using cholera toxin as a mitogen [Raff et al., 1978; Brockes et al., 1979], but optimal growth has been found with the addition of glial growth factor and forskolin in medium containing 10% serum [Davis & Stroobant, 1990]. Cultured glial cells have had some success in transplant-mediated repair of spinal cord injury (*see* Section 25.4.2). Cultured astrocytes are available commercially (*see* Table 22.1; Appendix II; Plate 28f).

The following introduction and Protocol 22.20 have been abridged from Burke et al. [1996].

Astrocytes may be isolated from fetal or adult brain. For work on human tissue, we have modified the method published by McCarthy and de Vellis [1980] originally developed for the culture of rat pup cerebral tissue. Before obtaining human tissue samples, it is necessary to obtain ethical permission from your local governing body. Work on fetal tissue in the United Kingdom is further regulated under guidelines issued by the Human Fertilization and Embryology Authority [HFEA, 2009].

Brain tissue should be removed as soon as possible after death, under aseptic conditions, and transported to the laboratory in culture medium placed on ice. (Astrocytes will remain viable if left overnight at $4°C$.) Care should be taken to remove vascular tissue and membranes. The tissue is then cut into small pieces, if necessary, and gently triturated to produce a single cell suspension. The cell viability in this crude cell suspension varies from 30% to 50%. All subsequent manipulations with this cell suspension are performed at $4°C$. The crude cell suspension is washed several times in complete medium to remove any lipids present and is then passed through a 200-$\mu$m Nitex nylon membrane filter to remove any large particles that remain. A viable cell count is obtained in the presence of 0.4% Trypan Blue and the cells are plated at high density ($2 \times 10^5$ cells per cm$^2$) onto standard tissue culture plates in complete medium. The cultures should be left undisturbed for 3 days at $37°C$ in a humidified incubator with 5% $CO_2$ and the medium replaced regularly after 4 to 5 days.

Neuronal cells do not survive this isolation procedure, and after 7 to 9 days the plates are covered with a mixed glial population. Fibroblasts may contaminate these cultures, but the risk is reduced by careful removal of vascular tissue and membranes and by maintaining the cultures with inhibitors of fibroblast growth such as cis—4-hydroxy-L-proline or D-valine. On inspection, the mixed glial cultures are composed of a monolayer of astrocyte-like cells with oligodendrocytes resting on top. These oligodendrocytes can be efficiently removed by gently shaking the cultures on a rotary shaker overnight and removing the dislodged oligodendrocytes with the medium. The monolayer of astrocytes is subcultured by trypsinization and replated as purified astrocytes at a density of $1 \times 10^4$ cells/cm$^2$.

At least two populations of astrocytes can be distinguished on the basis of their morphological and biochemical properties. Type 1 astrocytes are polygonal in shape with a more diffuse GFAP staining pattern, while type 2 astrocytes are stellate in appearance and stain brightly for GFAP. The isolation method described below produces cultures containing both type 1 and type 2 astrocytes, but as the cells are passaged, fewer and fewer stellate cells are present. These astrocyte cultures will grow until confluence and can be passaged several times. After approximately 10 PD, the culture will no longer proliferate as it has come to the end of its natural life span. The cells are then characterized by a senescent appearance similar to that seen for fibroblasts, and though they no longer divide, they will remain viable for extended periods of time.

---

### PROTOCOL 22.20. PRIMARY CULTURE OF HUMAN ASTROCYTES

#### Outline
Brain is disaggregated by trituration and the cells seeded in medium containing cis-hydroxyproline to minimize fibroblastic overgrowth. Astrocytes are enriched by shaking off oligodendrocytes when a monolayer becomes established.

### Reagents and Materials

#### Sterile:

❑ Astrocyte culture medium (serum supplemented medium—SSM): DMEM/F12, 50:50, with 1.2 g NaHCO$_3$/L, and 15 mM HEPES, 100 U/mL penicillin, 100 $\mu$g/mL streptomycin, 1 mM sodium pyruvate 100 nM phorbol 12,13 dibutyrate, 1.5 mM cis-4-hydroxy-L-proline, 10% fetal calf serum.

**Note.** The basal medium, DMEM/F12, can be replaced with MEM/D-valine. This medium selectively inhibits the growth of fibroblasts without affecting astrocyte proliferation.

❑ Petri dishes, tissue culture grade, 9 cm
❑ Disposable Pasteur pipettes (e.g., Pastettes, Compettes), 2
❑ Centrifuge tubes, 2 × 50 mL
❑ Selection of cell culture flasks/dishes/multiwell plates
❑ Nitex (200 $\mu$m) monofilament nylon cloth

#### Nonsterile:

❑ Hemocytometer counting chamber

### Procedure

#### Isolation of Mixed Glial Cell Cultures

1. Remove brain tissue (usually by pathologist). May be stored overnight at $4°C$ in serum supplemented medium (SSM).

Δ **Safety Precautions.**
   (a) Ship brain samples on ice by private courier service and handle as if infected on arrival at the laboratory.
   (b) Work on the assumption that all human blood and tissue samples are a potential source of human pathogens.
   (c) Perform all manipulations in a Class II laminar-flow cabinet (MAT) in a room with restricted access.
   (d) Soak all glassware and materials in fresh disinfectant (chloros) immediately after use before further processing.

2. Transfer the brain tissue to a Petri dish and select pieces for processing. These pieces should be the most intact and undissociated fragments in the sample.

3. Remove any pieces of meninges and vascular tissue with forceps. Cut large pieces to smaller fragments, about 2 mm in diameter.
4. Clean the brain fragments by washing in medium.
5. Place the tissue pieces in a 30 to 50-mL centrifuge tube and add 10 mL SSM.
6. Invert several times to wash.
7. Centrifuge for 2 min at 400 g.
8. Repeat this wash cycle until the medium is clear.
9. After washing, dissociate the tissue by trituration. Add 10 mL of SSM to the tissue pieces and gently triturate using a wide-bore plastic bulb pipette until the fragments have been dissociated.
10. Remove lumps of tissue that have failed to dissociate by straining through a Nitex 200-μm filter.
11. Estimate the number of viable cells in this crude cell suspension by performing viable cell counts using 0.4% Trypan Blue.
12. Store cells on ice until seeding.

### Culture of Purified Astrocytes

13. Plate $2 \times 10^5$ viable cells/cm$^2$ in SSM in standard cell culture dishes.
14. Incubate at 37°C in a humidified environment with 5% $CO_2$. Do not disturb cultures for at least 2 days postseeding to allow the cells to settle and adhere to the plastic.
15. After 4 to 5 days postseeding, replace the medium and, thereafter, add fresh medium twice weekly.
16. Maintain the cultures in SSM until a confluent monolayer of mixed glial cells is generated (about 7–9 days postseeding). Phase-dark, process-bearing oligodendrocytes will be seen residing on top of the phase-light astrocyte monolayer.
17. Remove non-astrocytes from these mixed glial cultures:
    (a) Change the medium and allow it to equilibrate with the cells for 2 h in the incubator.
    (b) Remove the oligodendrocytes and place flask in a controlled environment in a shaking incubator.
    (c) Rotate the flasks gently at 200 rpm overnight. This treatment detaches the oligodendrocytes into the medium leaving the astrocyte monolayer attached to the flask.
    (d) Remove the oligodendrocytes with the culture medium and replace fresh medium into the flask.
18. Trypsinise the astrocyte monolayer and replate at $1 \times 10^4$ cells/cm$^2$ in SSM. The ensuing culture will be highly enriched for astrocytes (>95%), as

gauged by cytochemistry for the astrocyte marker, glial fibrillary acidic protein (GFAP) (*see* below).

***Characterization.*** The purity of the astrocyte preparation can be ascertained by immunohistochemical staining for the astrocyte specific marker, glial fibrillary acidic protein (GFAP). GFAP is a cytoskeletal protein found only in mature astrocytes. The isolation method described above routinely gives preparations that are 95% to 100% pure astrocytes as judged by staining for GFAP. Primary astrocyte cultures are relatively robust and can be routinely subcultured with trypsin and frozen in liquid nitrogen in a similar manner to fibroblasts.

Protocol 22.20 for the purification of rat olfactory ensheathing cells from the olfactory bulb by fluorescence-activated cell sorting was contributed by S. Barnett, Glasgow Biomedical Research Centre, Room B417, 120 University Place, Glasgow G12 8TA, Scotland, UK.

The ability to produce highly enriched populations of individual cell types is vital to the success of biological studies on the diverse elements that control cell growth, differentiation, and function. In this regard the fluorescence-activated cell sorter (FACS) is a powerful tool for the enrichment of relatively small populations of cells from a heterogeneous cell population. Initially the cell sorter was devised to characterize and sort cells of the hematopoietic system, for which a large number of cell-specific antibodies were available. With the increasing availability of antibodies that characterize both glia (Table 22.3) and neurons, it has now become possible to transfer this technique to the field of neurobiology Abney et al., 1983; Williams et al., 1985; Trotter & Schachner, 1988; Barnett et al., 1993; Franceschini & Barnett, 1996.

The olfactory bulb is a tissue of considerable interest because it is the only central nervous system (CNS) mammalian tissue that can support neuronal regeneration throughout life [Moulton, 1974; Graziadei & Monti Graziadei, 1979, 1980]. Once neuronal death occurs, either during the course of normal cell turnover or after injury, newly generated neurons are recruited from a pool of constantly proliferating putative stem cells in the olfactory epithelium [Schwob, 2002]. These neurons are able to produce axons that reinnervate the olfactory bulb and reestablish olfaction. It is thought that a population of glial cells present in the olfactory nerve layer of the olfactory bulb, termed olfactory ensheathing cells (OECs), supports or encourages the growth of olfactory axons within the CNS [Doucette, 1984, 1990]. Experiments have in fact demonstrated that OECs are capable of remyelinating experimentally created demyelinated lesions in rat spinal cord [Franklin et al., 1996; Imaizumi et al., 1998] and that they support the regeneration of cut axons [Li et al., 1997, 2003a, b; Ramon-Cueto et al., 1998]. Many reviews have now been written regarding the use of these cells in CNS

**TABLE 22.3.** Characterization of Glial Cells by Immunostaining

| Cell type | GFAP | Gal-C | O4 | A2B5 | NGF |
|---|---|---|---|---|---|
| O2-A | − | − | − | + | − |
| Type-1 astrocyte | + | − | − | − | − |
| Type-2 astrocyte | + | − | − | + | − |
| Oligodendrocyte | − | + | +[a] | +[a] | − |
| ONEC | +/−[c] | − | +[b] | − | + |

[a]Mature oligodendrocytes lose A2B5 and become GalC+ and O4+ only.
[b]ONECs lose O4 in culture.
[c]GFAP may be fibrous or amorphous. ONECs gain GFAP expression in culture.

repair [Franklin & Barnett, 1997, 2000; Raisman, 2001; Wewetzer et al., 2002; Barnett & Riddell, 2004; Barnett & Chang, 2004; Franssen et al., 2008]. It is now generally thought that it is more likely that the transplantation of OECs exerts a neuroprotective affect rather than promoting axonal regeneration across the injury site [Toft et al., 2007; Barnett & Riddell, 2007].

To study the biology of OECs in detail, it is necessary to separate them from the other major cell types that make up the olfactory bulb: neurons, CNS glial cells, oligodendrocytes and astrocytes, endothelial cells, and meningeal cells [Barnett & Chang, 2004].

The rat olfactory bulb fulfills the criteria for successful cell separation by FACS in that (1) it readily forms a single-cell suspension after enzymatic treatment and (2) the mouse monoclonal antibody O4-IgM; [Sommer & Schachner, 1981], in conjunction with anti-galactocerebroside antibody—anti-GalC [Ranscht et al., 1982], can be used to distinguish OECs from other cell types that are present [Barnett et al., 1993; Franceschini & Barnett, 1996]. The O4 antibody recognizes sulfatide, seminolipid, and preoligodendroblast antigen [Bansal et al., 1992] and labels many CNS and PNS glial cells, including Schwann cells, oligodendrocytes, and oligodendrocyte-type-2 astrocyte (O-2A) progenitor cells [Gard & Pfeiffer, 1990; Mirskey et al., 1990; Sommer & Schachner, 1981]. The olfactory bulb contains three cell populations that are labeled by means of the O4 antibody: OECs, oligodendrocytes, and O-2A progenitors [Barnett et al., 1993]. In contrast to OECs, oligodendrocytes and O-2A progenitors are also labeled by anti-GalC antibody, which can thereby be used as a negative selection marker. By using these two antibodies in conjunction, it has been found that 8% to 14% of the O4-positive cells are OECs and less than 1% of the total O4 positive cells are oligodendrocytes or O-2A progenitors. Moreover the characteristic highly branched or bipolar morphology of oligodendrocytes and O-2A progenitors, respectively, can easily be distinguished from the flat fibroblast-like morphology of the OECs.

The O4-positive FACS-purified OECs can be classified into two types of cells—Schwann cell-like and astrocyte-like, which are distinguished by a panel of neural markers [Doucette, 1990; Franceschini & Barnett, 1996; Wewetzer et al., 2002]. These two cell types may have different functional roles in vivo. The growth factors that regulate their growth and differentiation are currently being investigated. In the presence of conditioned medium from type-1 astrocytes (ACM), OECs exhibit good viability and growth for at least 14 days. The method for growing astrocytes in culture can be found in Noble & Murrey [1984] and Wolswijk & Noble [1989]. Further experiments demonstrated that an isoform of neuregulin (neu–differentiation factor β) was identified as the factor in ACM [Pollock et al., 1999]. More recently a combination of growth factors have been found to be potent OEC mitogens, which include a mix of ACM, FGF-2, forskolin, and neuregulin, in 5% fetal bovine serum [Alexander et al., 2002].

Protocol 22.20 describes a rapid method for purifying a minor population of glial cells from the olfactory bulb by using a cell surface antibody and FACS. These principles, however, can be applied to the purification of virtually any glial or neuronal population, provided that two main criteria are fulfilled: the researcher is able to produce a viable single-cell suspension, and suitable cell type-specific antibodies are available for identifying the cells of interest.

## PROTOCOL 22.21. OLFACTORY ENSHEATHING CELLS

### Materials

*Sterile or aseptically prepared:*

- Poly-L-lysine (Sigma), stock, 4 mg/mL, diluted 1:300 in sterile UPW for use at a final concentration of 13.3 μg/mL
- Collagenase type I (MP Biomedicals, cat no. 195109): stock, 2,000 U/mL in L-15
- Ca$^{2+}$- and Mg$^{2+}$-free Hanks's BSS (HBSS, Invitrogen)
- Leibowitz L-15 medium plus 25 μg/mL of gentamycin (MP Biomedicals)
- DMEM/FBS: DMEM (GIBCO) with 5% FBS

❑ DMEM/BS: DMEM (GIBCO) containing (all from Sigma) [Bottenstein & Sato, 1979]:
Glucose........................................ 25 mM (4.5 g/L)
Gentamycin (Invitrogen)............. 25 µg/mL
BSA Pathocyte (MP Biomedicals).......... 0.0286% (v/v)
Glutamine...................................... 2 mM
Bovine pancreas insulin................. 10 µg/mL
Human transferrin......................... 100 µg/mL
Progesterone................................. 0.2 µM
Putrescine.................................... 0.10 µM
L-Thyroxine................................ 0.45 µM
Selenium...................................... 0.224 µM
3,3′,5-Triiodo-L-thyronine............. 0.49 µM

❑ Monoclonal antibodies:
O4: MAB345 (Chemicon)
Anti-galactocerebroside Ranscht et al., 1982 (anti-GalC): MAB342 (Chemicon)

❑ Second antibodies: anti-mouse IgM-fluorescein and anti-mouse IgG3-phycoerythrein (Southern Biotechnology Associates or Cambridge Biosciences)

❑ Flasks, dishes and coverslips; coat all tissue culture plastics and glass coverslips with 13.3 mg/mL poly-L-lysine:
(i) Incubate for at least 1 h to a maximum of 24 h.
(ii) Wash the plastics and coverslips once with D-PBSA.
(iii) Air dry them before use.

❑ Petri dish
❑ Falcon 15-mL polypropylene conical tube with cap
❑ Forceps, curved and straight (size 7)
❑ Tubes, polystyrene, 12 × 75 mm, round bottom with caps (Falcon #2058; these tubes fit both the sample tubing and the sort-collecting port of the FACS).
❑ Scissors, curved, dissecting
❑ Scalpel

*Nonsterile:*

❑ Sprague–Dawley rats, 6 to 8 days postpartum. A greater yield of OECs is obtained from neonatal rats rather than older animals, although it is possible to make preparations from animals of any age.
❑ Dissecting board and pins
❑ Alcohol, 70%, in a spray bottle
❑ Fluorescence-activated cell sorter (FACS; BD Biosciences; *see* Section 14.4).

### Procedure

### Preparation of dissociated cell suspension of OECs

1. Kill the rats by decapitation. Use 15 to 20 animals for a FACS sort of 2 to 3 h to recover enough OECs for 10 to 15 coverslips (10,000 cells per coverslip).
2. Pin the head dorsal side up onto a dissecting board and spray with 70% ethanol.
3. Dip all dissecting instruments in 70% alcohol before use, and shake the instruments dry.
4. Remove the skin from the head by using sharp curved scissors, and make a circular cut to remove the top of the skull, revealing the brain and the two olfactory bulbs at the anterior tip of the cortex near the nose (Fig. 22.4).
5. Using curved forceps, gently release the olfactory bulbs from the brain, and place them in a Petri dish containing a drop of L-15 plus gentamycin.
6. Using a sterile scalpel blade, chop the olfactory bulbs into small pieces.
7. Place the pieces into a small vial containing 1 mL of stock collagenase and 1 mL of L-15.
8. Incubate the pieces of olfactory bulbs at 37°C for 30 to 45 min.
9. Centrifuge the resultant suspension at 100 *g* for 5 min and remove the supernate.
10. Resuspend the pelleted bulb tissue in 1 mL of $Ca^{2+}$- and $Mg^{2+}$-free HBSS.
11. Dissociate the bulb tissue gently through a 19G hypodermic needle followed by a 23G hypodermic needle. Glial cells are fragile; therefore, to produce a single-cell suspension, the olfactory bulb tissue must be dissociated gently, taking care not to produce air bubbles.
12. Add 5 mL of $Ca^{2+}$- and $Mg^{2+}$-free HBSS, and spin the suspension again at 100 *g* for 10 min. Remove the supernate and resuspend the cells in DMEM-FBS at a concentration of $5 \times 10^6$ cells/mL.

### FACS sorting of OECs

13. Remove as many of the O4$^+$ oligodendrocytes as possible from the sorted cell population. It is preferable to label them with a second antibody to galactocerebroside [anti-GalC; Ranscht et al., 1982; Barnett et al., 1993], thus allowing discrimination between oligodendrocytes and OECs. This way a dual-color sort can be carried out in which the unwanted cells, now labeled with both antibodies, can be ignored. On a FACSVantage the sort takes 1 h.
14. After tissue dissociation, resuspend $5 \times 10^6$ cells in 500 µL of a mixture of O4 and anti-GalC antibody hybridoma supernates, and incubate at 4°C for 30 to 45 min in a 15-mL centrifugation tube. There will be more than $5 \times 10^6$ cells, so prepare enough tubes for the number of cells. If the hybridoma supernate does not contain high concentrations of antibody, then an increase in

cell number will result in an apparent decrease in the percentage of O4$^+$ cells. A similar consideration applies to the anti-GalC hybridoma supernate. It is therefore always necessary to titrate antibodies by FACS analysis before use.

15. After 30 to 45 min, wash the cells twice in DMEM/FBS and resuspend in 500 μL of DMEM/FBS containing a 1:100 dilution of both goat anti-mouse IgM-fluorescein and goat anti-mouse IgG3-phycoerythrein. Incubate the cells for a further 30 min at 4°C.

16. Wash the cells twice in 15 mL of DMEM-FCS by centrifugation, and resuspend them at a concentration of $2 \times 10^6$ cells/mL for sorting in ice-cold DMEM/FBS media in cell-sorting tubes (BD Biosciences 2085). Keep the cell suspension on ice. The FACS profile for antibody-labeled cells is stable for at least 3 to 4 h when the cells are kept on ice.

17. This protocol describes the method used for sorting cells on a FACStar cell sorter. However, any type of cell sorter should produce similar results. Align the laser to the 488-nm excitation line. Fluorescence compensation must be set to negate fluorescence spillover into the individual fluorescent channels. A control sample of cells labeled with the second antibodies alone should be analyzed before the labeled sample, so that any nonspecific cell label can be accounted for and removed by decreasing the gain on the photomultiplier tube.

18. Set a cell sort window around the O4$^+$ GalC$^-$ cells. The average percentage of O4$^+$ olfactory bulb cells should be around $10 \pm 1\%$.

19. Sort the O4$^+$ GalC$^-$ cells into a polystyrene round-bottom tube containing 4 mL of DMEM/FBS.

20. After sorting, transfer the sorted cells to a 15-mL centrifugation tube, wash out the sort tube with medium, pool the washes, and fill the centrifuge tube with DMEM/FBS. Spin the centrifuge tube at 100 g for 15 min.

21. Remove the supernate and resuspend the cells in DMEM/BS. Optimum cell viability and growth for the OECs may be obtained in a 1:1 dilution of DMEM/BS:ACM [Barnett et al., 1993]. ACM is collected from monolayers of purified cortical astrocytes [Noble & Murrey, 1984] incubated for 48 h in DMEM/BS. Cell can also be grown in ACM (1:5), FGF2 (10 ng/mL) forskolin (10 μM) diluted in DMEM/BS then diluted 1:1 in DMEM containing 10% FBS [Alexander et al., 2000].

This protocol produces a highly enriched population of O4$^+$ olfactory bulb glial cells (>97%). These cells may be plated onto poly-L-lysine-coated coverslips, Petri dishes, or flasks and used for further investigations. Current evidence suggests that these cells do correspond to the OECs described in electron microscope studies [Doucette, 1984; Raisman, 1985], and research continues on their further characterization.

### 22.4.3 Endocrine Cells

The problems of culturing endocrine cells [O'Hare et al., 1978] are accentuated because the relative number of endocrine cells in a particular tissue relative to other parenchymal and stromal cells may be quite small. Sato and colleagues [Buonassisi et al., 1962; Sato & Yasumura, 1966] cultured functional adrenal and pituitary cells from rat tumors by mechanical disaggregation of the tumor [Zaroff et al., 1961] and regular monolayer culture. The functional integrity of the cells was retained by intermittent passage of the cells as tumors in rats [Buonassisi et al., 1962; Tashjian et al., 1968]. These lines are now fully adapted to culture and can be maintained without animal passage [Tashjian, 1979] and, in some cases, in fully defined media [Hayashi & Sato, 1976]. The Y-1 rat adrenal cell line [Sato & Yasumura, 1966] is still used and has been supplemented by a human line NCI-H295, also derived from a tumor [Gazdar et al., 1990; Rainey et al., 2004]. Normal adrenal cells have been difficult to propagate as a cell line, although primary cultures, dissociated in collagenase, have been used [Kanczkowski et al., 2009].

Fibroblasts have been reduced in cultures of pancreatic islet cells by treatment with ethylmercurithiosalicylate [Braaten et al., 1974], and islet cells have also been purified by density

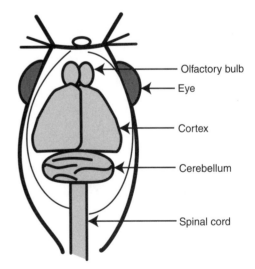

**Fig. 22.4. Olfactory Bulb Dissection.** Schematic diagram of the isolation of the olfactory bulbs from the brain of newborn rats. (Courtesy of Susan Barnett.)

gradient centrifugation [Prince et al., 1978] and by centrifugal elutriation [Bretzel et al., 1990]. The islet cells apparently produce insulin, but not as propagated cell lines. Short- and long-term cultures have also been generated from non-islet cells, capable of differentiating toward insulin-secreting cells in serum-free medium [Gao et al., 2003; Dorrell et al., 2008] and current emphasis is tending toward sourcing β-islet cells from stem cells [Evans-Molina et al., 2009].

Pituitary cells that produce pituitary hormones for several subcultures have been isolated from the mouse [Sato & Yasumura, 1966; de Vitry et al., 1974]. Although normal human pituitary cells do not survive well, and most pituitary adenoma cultures gradually lose the capacity to synthesize hormones, some cultures continue to produce growth hormone [e.g., [Bossis et al., 2004] and three-dimensional cultures of pituitary adenomas continue to secrete prolactin [Guiraud et al., 1991].

### 22.4.4 Melanocytes

The high substrate dependency of cultured keratinocytes has been utilized to obtain preferential melanocyte attachment and growth in a hormone-supplemented medium [Naeyaert et al., 1991]. The system circumvents the problem of keratinocyte contamination, while supporting good melanocyte proliferation with minimal supplementation of serum [Gilchrest et al., 1985; Wilkins et al., 1985] in the absence of chemotherapeutic agents or tumor promoters such as phorbol esters. These substances downregulate protein kinase C (PKC), and hence tyrosinase activity, curtailing the production of pigment [Park et al., 1993]. With conventional serum supplementation (5–20%), melanocyte growth is far better than reported in other systems, and fibroblast overgrowth can be prevented by reducing the $Ca^{2+}$ concentration to 0.03 mM [Naeyaert et al., 1991]. The system described here can be further modified by removing cholera toxin (which otherwise promotes melanocyte growth by greatly increasing intracellular levels of cAMP) and substituting dibutyryl cAMP. This medium then provides physiologic levels of both PKC and cAMP that permit studies of cAMP-modulating agents [Park et al., 1999] under conditions that are optimal for melanogenesis studies.

Human melanocyte cultures are also available commercially (*see* Table 22.1; Appendix II; Plate 27b).

The preceding introduction and Protocol 22.22 for the culture of human melanocytes were contributed by H-Y. Park, M. Yaar, and B. A. Gilchrest, Department of Dermatology, Boston University School of Medicine, 609 Albany Street, Boston, MA 02118–2394.

---

**PROTOCOL 22.22. CULTURE OF MELANOCYTES**

*Outline*

Epidermis, stripped from small fragments of skin after trypsinization, is dissociated in EDTA and cultured

---

in serum-free medium supplemented with basic FGF (FGF-2).

*Materials*

*Sterile:*

❑ D-PBSA
❑ Fetal bovine serum
❑ Melanocyte hormone-supplemented medium (MHSM):

|  | Vol. stock | Final conc. |
|---|---|---|
| Medium 199 (GIBCO 400–1200) | 93.1 mL | |
| EGF (BD Biosciences), 10 µg/mL stock | 0.1 mL | 10 ng/mL |
| Transferrin (Sigma), 10 mg/mL stock | 0.1 mL | 10 µg/mL |
| Insulin (Sigma), 10 mg/mL stock | 0.1 mL | 10 µg/mL |
| Triiodothyronine (Sigma), 1 µM stock | 0.1 mL | 1 nM |
| Hydrocortisone (Calbiochem), 0.28 mM stock | 0.5 mL | 1.4 µM |
| Cholera toxin (Calbiochem), 0.1 µM stock | 1.0 mL | 1 nM |

*Note.* Cholera toxin can be replaced by bovine pituitary extract (Clonetics), 35 µg/mL, stock 0.7 ng/mL final, and dibutyryl cAMP (Sigma), 1 mM stock, 100 µM final.

| | | |
|---|---|---|
| Basic FGF (Amgen), 200 ng/mL stock | 5.0 mL | 10 ng/mL |

❑ Trypsin/EDTA: Trypsin 0.25% in D-PBSA with 0.1% (3.5 mM) EDTA (GIBCO 610–5050)
❑ Soybean trypsin inhibitor (Sigma), 10 µg/mL
❑ Tissue culture dishes (Falcon, BD Biosciences), 10, 6, and 3.5 cm in diameter
❑ Centrifuge tubes, 15 and 50 mL

*Nonsterile:*

❑ Humidified incubator (37°C, 7% $CO_2$)
❑ Water bath
❑ Electronic cell counter or hemocytometer

*Procedure*

**Establishing the Primary Culture**

*Day 1*

1. Rinse the skin specimen in 70% alcohol and then twice in D-PBSA.
2. Transfer the tissue to a sterile 10-cm dish, epidermal side down.
3. With dissecting scissors, excise subcutaneous fat and deep dermis.

4. Cut the remaining tissue into 2 × 2-mm pieces with a scalpel, rolling the blade over the tissue. (Do not use a sawing action.)

5. Transfer the tissue fragments to cold 0.25% trypsin in a 15-mL centrifuge tube.

6. Incubate the tissue fragments at 4°C for 16 to 18 h (do not go beyond 24 h). For some donors, additional incubation at 37°C for 1 to 2 h may give higher melanocyte yields.

*Day 2*

7. Gently tap the centrifuge tube to dislodge fragments that have settled at the bottom, and then rapidly pour the contents of the tube into a 6-cm dish.

8. Using forceps, transfer the tissue fragments individually to a dry 10-cm dish, epidermal side down. Gently roll each tissue fragment against the dish. The epidermal sheet should adhere to the dish and allow a clean separation of the dermis with forceps. Discard the dermal fragments.

9. Transfer all epidermal sheets to a sterile 15-mL centrifuge tube containing 5 mL of 0.02% (0.7 mM) EDTA. Take care to place each epidermal sheet in the EDTA solution, not on the plastic wall of the centrifuge tube.

10. Gently vortex the tube to disaggregate the epidermal sheets into a single-cell suspension.

11. Centrifuge the cells for 5 min at 350 $g$, and then aspirate the supernate.

12. Resuspend the pellet in serum-free MHSM.

13. Count the cells with a hemocytometer, and inoculate 1 × 10$^6$ cells (approximately 2–4 × 10$^4$ melanocytes) per 3.5-cm dish in 2 mL of MHSM containing 5% FBS to facilitate attachment of the cells.

14. Refeed the culture with fresh MHSM containing growth factors and bovine pituitary extract twice weekly.

*Days 3 to 30*

15. At 24 h, the cultures will contain primary keratinocytes with scattered melanocytes (Fig. 22.5a). Keratinocyte proliferation should cease with several days, and colonies should begin to detach during the second week. By the end of the third week, only melanocytes should remain (Fig. 22.5b, c; Plate 8d). Most cultures attain near confluence and are ready to passage within 2 to 4 weeks.

Subculture

16. Gently rinse the culture dish twice with 0.02% (0.7 mM) EDTA.

17. Add 3 mL of 0.25% trypsin/0.1% (2.5 mM) EDTA, and incubate at 37°C. Examine the dish under phase microscopy every 5 min to detect cell detachment.

18. When most cells have detached, inactivate the trypsin with soybean trypsin inhibitor (10 μg/mL) or 1 mL of medium containing 10% calf serum. (Melanocytes maintained under serum-free conditions greatly benefit from a "serum kick" at the time of subculture.)

19. Pipette the contents of the dish to ensure complete melanocyte detachment.

20. Aspirate and centrifuge the cells for 5 min at 350 $g$.

21. Aspirate the supernate, resuspend the cells in a calcium-free melanocyte medium containing 5% FBS without calcium, and replate at 2 to 4 × 10$^4$ cells per 100-mm dish. FBS is required for good melanocyte attachment (>75%) to plastic dishes, but can be omitted if the dishes are coated with fibronectin or type I/III collagen [Gilchrest et al., 1985].

22. Refeed the culture twice a week with serum-free or serum-containing melanocyte medium. The serum-containing medium will give a greater melanocyte yield, but encourages the growth of fibroblasts. Reducing the calcium concentration to 0.03 mM has no effect on the proliferation of melanocytes but inhibits fibroblast overgrowth [Naeyaert et al., 1991].

***Confirmation of melanocytic identity.*** Melanocyte cultures may be contaminated initially with keratinocytes and at any time by dermal fibroblasts. Both forms of contamination are rare in cultures established and maintained by an experienced technician or investigator but are common problems for the novice. The cultured cells can be confirmed to be melanocytes with moderate certainty by frequent examination of the culture under phase microscopy, provided that the examiner is familiar with the respective cell morphologies (Fig. 22.5b, c). More definitive identification is provided by electron microscopic examination, DOPA staining, or immunofluorescent staining with Mel 5 antibody, directed against tyrosinase–related protein-1.

## 22.5 HEMATOPOIETIC CELLS

Hematopoietic cells have been grown in colony-forming assays (*see* Section 23.3.8), in long-term cultures from bone marrow, and as continuous cell lines Freshney et al., 1994; Klug & Jordan, 2002. Under the control of specific growth

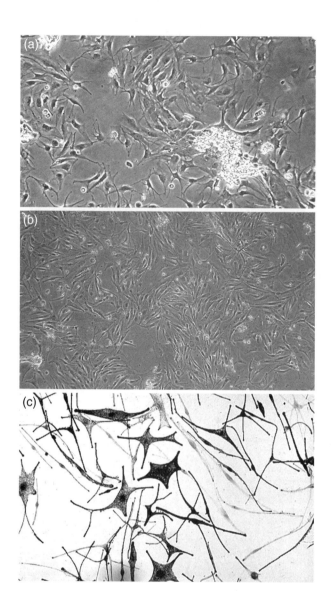

*Fig. 22.5. Melanocyte Cultures.* (a) Culture of Caucasian newborn foreskin-derived melanocytes 1 week after inoculation. Note the multiple keratinocyte colonies with central stratification and tightly apposed epithelial cells at the periphery. The melanocytes are the relatively small, dark dendritic cells, most of them in contact with the keratinocyte colonies by means of dendritic projections (seen via phase contrast, 100× magnification). (b) Ten-day-old primary cultures of Caucasian newborn epidermal melanocytes in medium lacking TPA. Many cells display branching dendrites, and other cells display bipolar to polygonal morphology (seen via phase contrast, 40× magnification). (c) Secondary culture of TPA-treated epidermal melanocytes derived from African-American newborn foreskin. Note the dendritic morphology of the cells and their slender spindle shape. In contrast to the melanocytes derived from Caucasian newborn foreskin, these melanocytes display a high level of melanin granules (seen via phase contrast, 320× magnification). (*See also* Plate 8*d*; photographs courtesy of Hee-Young Park.)

factors, hematopoietic stem cells, or CFU-A, can be recloned from agar cultures. MIP−1α appears to be one of the main factors that keeps the stem cells in the regenerative compartment. However, most suspension colonies, which contain cells of only one lineage, survive only as primary cultures that lose their repopulation efficiency, and hence cannot be subcultured. Granulocytic colonies are the most common, but under the appropriate conditions, other lineages can be produced (*see* Protocol 23.9).

A number of myeloid cell lines have been developed from murine leukemias [Horibata & Harris, 1970] and, like some of the human lymphoblastoid lines [Collins et al., 1977], have been shown to make globulin chains and, in some cases, complete α- and γ-globulins (*see* Section 27.7). Some of these lines can be grown in serum-free medium [Iscove & Melchers, 1978]. T-cell lines require T-cell growth factors, such as IL-2 [Gillis & Watson, 1981], and B-cell growth factors have also been described [Howard et al., 1981; Sredni et al., 1981; *see also* Freshney et al., 1994].

The first human lymphoblastoid cell lines were derived by culturing peripheral lymphocytes from blood at very high cell densities ($\sim$110$^6$/mL), usually in deep culture (>10 mm) Moore et al., 1967. Subsequently, immortalization was shown to be due to EBV, and reproducible techniques are now available for this procedure Bolton & Spurr, 1996.

Rossi and Friend [1967] demonstrated that a mouse RNA virus (the "Friend virus") could cause splenomegaly and erythroblastosis in infected mice. Cell cultures taken from minced spleens of these animals could, in some cases, give rise to continuous cell lines of erythroleukemia cells. All of these lines are transformed by what is now recognized as a complex of defective and helper viruses derived from Moloney sarcoma virus Ostertag & Pragnell, 1978, 1981. Some cell lines can produce a virus that is infective in vivo, but not in vitro, and the cells can also be passaged as solid tumors or ascites tumors in DBA2 or BALB-c mice.

Treating cultures of Friend cells with a number of agents, including DMSO, sodium butyrate, isobutyric acid, and hexamethylene bisacetamide, promotes erythroid differentiation [Friend et al., 1971; Leder & Leder, 1975]. Untreated cells resemble undifferentiated proerythroblasts, whereas treated cells show nuclear condensation, a reduction in cell size, and an accumulation of hemoglobin (*see* Section 16.7.2) to the extent that centrifuged cell pellets are red in color. Evidence for differentiation can be demonstrated by staining the cells for hemoglobin with benzedine (*see* Plate 13*a*, *b*) and in situ hybridization of globin-specific messenger RNA (*see* Plate 13*c*, *d*). The human leukemic cell line K562 can also be induced to differentiate with sodium butyrate and hemin, though not with DMSO [Andersson et al., 1979c].

Macrophages may be isolated from many tissues by collecting the cells that attach during enzymatic disaggregation. The yield is rather low, however, and a number of techniques have been developed to obtain larger numbers of macrophages. Mineral oil or thioglycollate broth

[Marques da Silva, 2009; Takahashi et al., 2009] may be injected into the peritoneum of a mouse, and 3 days later the peritoneal washings will contain a high proportion of macrophages. If necessary, macrophages may be purified by their ability to attach to the culture substrate in the presence of proteases (*see* Section 13.8.1). Macrophages can be subcultured only with difficulty, because of their insensitivity to trypsin. Methods have been developed using hydrophobic plastics (e.g., Petriperm dishes, Sartorius).

There are some reports of propagated lines of macrophages, mostly from murine neoplasia. Normal mature macrophages do not proliferate, although it may be possible to culture replicating precursor cells (*see* Protocols 23.9, 23.10).

## 22.6 GONADS

### 22.6.1 Ovary

Ovarian granulosa cells can be maintained and are apparently functional in primary culture [Orly et al., 1980], but specific functions are lost on subculture. A cell line started from Chinese hamster ovary, CHO-K1 [Kao & Puck, 1968] has been in culture for many years, but its lineage still has not been identified. Although epithelioid at some stages of growth, it undergoes a fibroblast-like modification when it is cultured in dibutyryl cyclic AMP [Ilsie & Puck, 1971]. Human surface epithelial cells have also been isolated and immortalized with hTERT (*see* Protocol 17.2) retaining functional pRb and p53 [Li et al., 2007]. Some surface epithelial cells have been shown to be capable of maturing into early oocytes [Virant-Klun et al., 2008].

### 22.6.2 Testis

A number of cell types (Sertoli cells, spermatogonial, spermatocytes) have been cultured from human testis to investigate the role of hormonal control of spermiogenesis and cell–cell interaction [Sá et al., 2008]. TM4 is a cell line from mouse Sertoli cells [Mather, 1979; Kato et al., 2009] and TM3 from mouse Leydig cells [Matsumoto et al., 2008]. Pluripotent stem cells have also been isolated from human testis [Conrad et al., 2008] (*see* Section 23.2) and used as model for investigation growth factor control of germ cell maintenance and spermiogenesis [Huleihel et al., 2007].

# CHAPTER 23

# Stem Cells, Germ Cells, and Amniocytes

## 23.1 STEM CELLS

Perhaps the most promising approach to culturing specialized cells that has emerged in recent years is via the use of stem cells, including embryonic stem cells, mesenchymal stem cells from bone marrow and other sources, or induced pluripotent stem cells (iPS cells) generated by modifying the expression of specific genes in normal adult cells. Culture of stem cells was developed first with hematopoietic cells, where the identification of lineage markers and the relative ease of disaggregation of hemopoietic tissues, such as bone marrow, made isolation and purification feasible [Yeung et al., 2009] (*see* Protocol 23.9). Isolation and culture of stem cells from solid tissues has been much more problematical, but as techniques develop, progress has been made in identifying and culturing cells, such as from mammary gland [Welm et al., 2003; LaBarge et al., 2007], lung [Tesei et al., 2009], intestine [Booth et al., 1999], liver [Alison et al., 2004], and skin [Nowak et al., 2009]. One problem has been the dearth of stem cell markers in solid tissues, but some progress has been made in this area [Zhou et al., 2001; Potten et al., 2003], opening up possibilities for cell sorting.

### 23.1.1 Embryonic Stem Cells

Embryonic stem cells from several species, but particularly mouse, have been used extensively to study differentiation [Martin & Evans, 1974; Martin, 1975, 1978; Rizzino, 2002; zur den et al., 2003] because they may develop into a variety of different cell types (muscle, bone, nerve, etc.). Those that have been through animal passage form teratomas, analogous to spontaneous human teratomas. Cells grown on feeder layers of, for example, STO mouse fibroblasts, will proliferate, but not differentiate, whereas, when the cells are grown on gelatin without a feeder layer or in nonadherent plastic dishes, nodules form that eventually differentiate. Leukemia inhibitory factor (LIF) appears to be one of the main regulatory factors that hold cells within the stem cell compartment, whereas retinoids, vitamin D3, and planar polar compounds induce lineage-specific differentiation in mouse and human [Draper et al., 2002] cells.

### 23.1.2 Derivation of Mouse Embryonic Stem Cells

The following introduction and Protocol 23.1 were contributed by J. P. Wray and J. Nichols, Wellcome Trust Centre for Stem Cell Research, University of Cambridge, Cambridge, UK.

Embryonic stem (ES) cells grow in culture as cell lines that can be maintained indefinitely. They are unique in their ability to differentiate into cells from each of the three germ layers—ectoderm, endoderm, and mesoderm—a property termed pluripotency [Smith, 2006]. This is clearly demonstrated by the ability of ES cells to contribute to all tissues of the adult mouse, including the germ line, upon reintroduction to the developing embryo [Bradley et al., 1984]. As genetic manipulations can be carried out and confirmed in culture, ES cells have become a powerful tool for the generation of genetically modified mice [Robertson et al., 1986].

The observation that embryos grafted into mice produce teratocarcinomas [Solter et al., 1970; Stevens, 1968, 1970], from which pluripotent stem cells termed embryonal carcinoma (EC) cells could be derived, gave rise to the

*Culture of Animal Cells: A Manual of Basic Technique and Specialized Applications, Sixth Edition*, by R. Ian Freshney
Copyright © 2010 John Wiley & Sons, Inc.

idea that it should be possible to derive pluripotent cell lines directly from the embryo. Coculture of EC cells with fibroblasts supported their growth [Martin & Evans, 1975], and under these conditions EC cells retained a high differentiation capacity. In 1981 Evans & Kaufman [1981] and Martin [1981] succeeded in establishing pluripotent cell lines by plating delayed-implantation blastocysts or inner cell masses (ICMs) isolated from the expanded late blastocyst by immunosurgery onto mitotically inactivated STO feeder layers. The resulting cell lines displayed characteristics of EC cells but retained a normal karyotype, whereas EC cells display aneuploidy. Martin termed the new cell lines *embryonic stem cells* to distinguish them from their tumor-derived counterparts.

These first ES cell lines were derived and cultured in the presence of serum on feeder layers. Similar methods are still commonly used to derive ES cells but are reliable for derivation only from inbred 129 strain mice and 129 hybrids. Techniques for ES cell derivation seek to expand the pluripotent cells of the preimplantation epiblast and restrict growth of cells from the extraembryonic tissues—the trophectoderm and the hypoblast. Derivation from the recalcitrant CBA strain has been reported, but microdissection is required to separate pluripotent epiblast cells from extraembryonic tissues [Brook & Gardner, 1997]. This technique is technically demanding, and alternative methods have now been devised to remove the extraembryonic tissues [Batlle-Morera et al., 2008]; or to prevent their formation [Ying et al., 2008].

---

## PROTOCOL 23.1. DERIVATION AND PRIMARY CULTURE OF MOUSE EMBRYONIC STEM CELLS

### Outline
Isolate delayed implantation blastocysts and plate on feeders. Disaggregate and replate epiblast outgrowths to establish embryonic stem cell lines.

### Materials

*Sterile or aseptically prepared:*

- Prepared feeder layers (*see* Protocol 13.3)
- ES cell culture medium: Glasgow minimum essential medium (GMEM) with
  Glutamine .............................................2 mM
  FBS (batch tested for EC cell culture) .. 10%
  Non-essential amino acids (NEAA), .... 1:100
  Sodium pyruvate, ................................ 1 mM
  2-Mercaptoethanol ............................ 0.1 mM
  Leukemia inhibitory factor (LIF; *see*
      Appendix II: ESGRO) ..................... 1000 U/mL
  Penicillin, ............................................... 50 U/mL
  Streptomycin ..................................... 50 µg/mL
  (Use of antibiotics is optional.)

- Feeder medium: as for ES cell culture medium but without LIF
- N2B27 (Stem Cell Sciences), serum-free culture medium [Nichols & Ying, 2006] (for formulation, *see* Appendix I)
- N2 supplement [Nichols & Ying 2006] (*see* Appendix I)
- Serum-free ES cell culture medium as described by Ying et al. [2003]: to N2B27 add LIF (1000 U/mL) and bone morphogenetic protein 4 (BMP4, recombinant human, 10 ng/mL, R and D systems, www.rndsystems.com).
- 2i ES cell culture medium as described in Ying et al. [2008]: To N2B27 add 1 µM PD0325901 and 3 µM CHIRON99021 (Division of Signal Transduction Therapy, DSTT, University of Dundee, www.lifesci.dundee.ac.uk).
- KSOM (Millipore—Chemicon).
- KSOM+2i: to KSOM add 1 µM PD0325901 and 3 µM CHIRON99021 (*DSTT as above*).
- PD184352 (*DSTT as above*).
- D-PBSA
- Gelatin (Sigma): 1% stock solution prepared in UHP water, autoclaved, and stored in aliquots at 4°C. To prepare the 0.1% working solution, warm 1% gelatin to 37°C until it liquefies and dilute 1:10 in sterile D-PBSA.
- Trypsin: 0.25% (Invitrogen), 1 mM EDTA, and 1% chick serum in D-PBSA. Stored in aliquots at −20°C.
- Tamoxifen (Sigma)
- Depo-Provera (Medroxyprogesterone 17-acetate, Sigma)
- Acid Tyrode's solution (Millipore—Chemicon)
- Rabbit anti-mouse antiserum (Sigma).
- Rat serum (as a source of complement, made in house, not heat inactivated, kept at −80°C) or guinea pig complement (Sigma)
- Poly-L-ornithine (Sigma)
- Laminin (Sigma)
- Tissue culture flasks and 4-well plates (Iwaki or Corning)
- Centrifuge tubes, 15 and 50 mL (Greiner)
- Disposable pipettes, 5, 10, and 25 mL
- Pasteur pipettes

*Nonsterile:*

- Humidified incubator set to 7% $CO_2$ and 37°C

### Procedure

1. Prepare the required number of plates with feeders:
   (a) Coat the plates by adding enough gelatin to cover the entire surface of each well and

incubate at room temperature for at least 20 min.

**(b)** Grow the required number of mouse embryonic fibroblasts (MEFs) until they have just reached confluence (*see* Protocol 13.3). Add mitomycin C (10 μg/mL, Sigma) to the culture medium and incubate for 2 to 3 h at 37°C.

**(c)** Wash cells twice in D-PBSA.

**(d)** Add trypsin (~1 ml for every 25 cm²) and incubate at 37°C for 5 min or until the cells detach.

**(e)** Add 4 trypsin volumes of feeder medium and pipette up and down to resuspend cells.

**(f)** Transfer the cell suspension to a centrifuge tube and centrifuge at 300 *g* for 3 min.

**(g)** Aspirate the supernate and resuspend the cell pellet in feeder medium.

**(h)** Count the cells using a hemocytometer.

**(i)** Aspirate gelatin from the coated plates and plate cells at about $5 \times 10^4$ cells/cm².

**(j)** Allow the feeder cells to attach overnight. Plates can be kept for up to 1 week at 37°C.

2. Recover delayed implantation blastocysts:

**(a)** Induce delayed implantation in mice by intraperitoneal injection of tamoxifen (10 μg per mouse) at 2.5 days *post coitum* (dpc).

**(b)** Inject Depo-Provera (1–3 mg/mouse) subcutaneously as a source of progesterone.

**(c)** Flush the embryos from the uterus 4 or 5 days after administration of tamoxifen and plate into pre-equilibrated 4-well plates containing feeders and ES cell culture medium. Up to 10 embryos can be placed in each well.

3. Disaggregate and expand outgrowths:

**(a)** Isolate outgrowths after 3 to 5 days by gently detaching each cell clump, using a mouth controlled finely drawn plugged Pasteur pipette whose tip diameter is just bigger than the outgrowth. Each outgrowth is handled separately at this point and in all subsequent manipulations.

**(b)** Wash each outgrowth twice by transferring it to drops of D-PBSA in a Petri dish before transferring it to a small (approximately 5 μL) drop of trypsin.

**(c)** Incubate the outgrowths in trypsin at 37°C for a few minutes until they begin to dissociate.

**(d)** Transfer each outgrowth to a new well of a 4-well plate containing feeders and ES cell medium, using a finely drawn Pasteur pipette. Carry over as little trypsin as possible.

**(e)** Triturate the outgrowths into clumps of 1 to 5 cells. Replace medium every 2 days; ES cell colonies should become identifiable after about 5 days (Fig. 23.1*a*).

4. Identify and expand ES cells:

**(a)** Identify ES cell colonies based on morphology. ES cells have a high nucleocytoplasmic ratio, prominent nucleoli (Fig. 23.1*b*) and form tight, refractile colonies (Fig. 23.1*a*).

**(b)** Using a pulled Pasteur pipette, manually dissect the ES cell colonies from surrounding differentiated cells and passage as described for blastocyst outgrowths. For the first passage place dissociated ES cells in a single well of a 96-well plate.

**(c)** Replace medium every 2 days until the well approaches confluency.

5. Expand ES cell cultures:

**(a)** Wash wells containing ES cell colonies once in D-PBSA.

**(b)** Add 30 μL trypsin and incubate at 37°C for 3 min or until colonies begin to detach from the plate.

**(c)** Add 170 μL ES cell medium and pipette up and down to dissociate clumps.

**(d)** Transfer the cell suspension to a centrifuge tube and centrifuge at 300 *g* for 3 min.

**(e)** Aspirate the supernatant medium, taking care not to disturb the cell pellet that may be too small to be clearly visible. Resuspend the pellet in 500 μL ES cell medium and transfer to a 4-well plate containing feeders.

**(f)** Passage ES cells to plates of increasing well size each time a well approaches confluence by repeating steps (a) through (e), adjusting volumes to reflect the increased surface area.

### Serum- and feeder-free derivation using LIF and BMP4 [Nichols & Ying, 2006]

1. Proceed as described above but plate blastocysts in gelatin-coated 4-well plates with N2B27 containing LIF (1000 U/ml).

2. Following disaggregation, culture in serum-free ES cell medium.

3. Use N2B27 without growth factors for trypsinization steps

### Improvements to serum-free derivation

The efficiency of ES cell derivation can be improved by removing the extraembryonic tissues.

1. Remove the trophectoderm by immunosurgery as described [Solter & Knowles, 1975]:

**(a)** Incubate blastocysts in N2B27 + 20% mouse antiserum for 1 h at 37°C.

**(b)** Rinse the blastocysts three times in pre-equilibrated N2B27.

**(c)** Transfer blastocysts to EITHER pre-equilibrated N2B27 containing 20% rat serum (add freshly thawed rat serum immediately before transfer of the embryos) or guinea pig complement. Incubate for 10 min at 37°C.

**(d)** Transfer blastocysts to pre-equilibrated N2B27 and incubate for 1 h at 37°C. At this point up to 10 blastocysts can be placed in the same well.

**(e)** Remove trophectoderm lysate by drawing the embryos through a finely drawn Pasteur pipette.

2. Remove the hypoblast as described [Batlle-Morera et al., 2008]:

**(a)** Incubate the ICMs isolated by immuno-surgery in serum-free ES cell medium in hanging drops for about 12 h at 37°C. To form hanging drops, place 10 μL drops of medium on the lids of Petri dishes, place an ICM in each, and invert the lid onto its base. Place 2 to 5 mL of medium or D-PBSA in the base to reduce evaporation. During this time the hypoblast grows to surround the epiblast forming a "rind and core" structure (Fig. 23.1c).

**(b)** Remove the hypoblast "rind" by repeatedly drawing the ICM through a drawn Pasteur pipette with a tip diameter wider than the epiblast but narrower than the entire ICM. Take care to remove all hypoblast cells.

**(c)** Place the isolated epiblasts in gelatinized 4-well plates in serum-free ES cell medium and incubate for 3 to 7 days until ES-like outgrowths can be seen (Fig. 23.1d).

**(d)** Disaggregate and expand as described above.

3. Decrease differentiation in epiblast outgrowths and expanding ES cell lines by adding the MEK inhibitor PD184352 (1 μM) to serum-free ES cell culture medium when the epiblasts have been isolated and in all subsequent steps. This is essential if deriving ES cells from non-129 strains of mice.

### Derivation in "2i" as described [Nichols et al., 2009]

Embryos cultured in 2i medium from the 8-cell stage do not form a hypoblast, allowing them to be placed in culture immediately after trophectoderm removal. The 2i medium allows derivation of ES cells from mouse strains that are recalcitrant to standard derivation protocols and also from rat [Buehr et al., 2008; Li et al., 2008].

To derive ES cell lines in 2i medium proceed as follows:

1. Flush embryos from oviducts at the 8-cell stage and place into pre-equilibrated organ culture dishes containing KSOM+2i. Fill the outer well of the dish with D-PBSA to humidify.

2. Incubate for 2 days at 37°C.

3. Transfer the embryos to a fresh organ culture dish containing pre-equilibrated 2i medium plus LIF. Fill the outer well of the dish with D-PBSA to humidify.

4. Incubate for 1 day at 37°C.

5. If the embryos have not hatched, remove the zona pellucida as follows:

**(a)** Using a pulled Pasteur pipette, place groups of embryos in a drop of acid Tyrode's. Carry over as little medium as possible.

**(b)** Observe under a microscope until the zona pellucida disappears.

**(c)** Using a pulled Pasteur pipette containing a little medium transfer the embryos to a fresh organ culture dish containing pre-equilibrated 2i medium plus LIF.

6. Remove the trophectoderm by immunosurgery as described above.

7. Place the isolated epiblasts into gelatin-coated 96-well plates containing 200 μl pre-equilibrated 2i medium plus LIF per well.

8. Incubate at 37°C for 5–7 days to allow the epiblast to expand.

9. Disaggregate and expand the cultured epiblasts. ES cell-like colonies should now form (Fig. 23.1e, f).

### Notes

(1) While expanding primary ES cell lines in serum-free conditions, particular care must be taken with enzymatic passaging. Minimize the time that the cells are exposed to trypsin, dilute trypsin in at least 10 volumes of N2B27, and after centrifugation take care to remove as much of the supernate as possible without disturbing the cell pellet, which may be loose.

(2) ES cells cultured in 2i medium form tightly packed colonies with a tendency to detach from gelatin-coated plates. Feeders or laminin-coated plates can be used to improve cell attachment.

### 23.1.3 Subculture and Propagation of Mouse Embryonic Stem Cells

The following introduction and Protocol 23.2 were contributed by J. P. Wray and A. Smith, Wellcome Trust Centre for Stem Cell Research, University of Cambridge, Cambridge, UK.

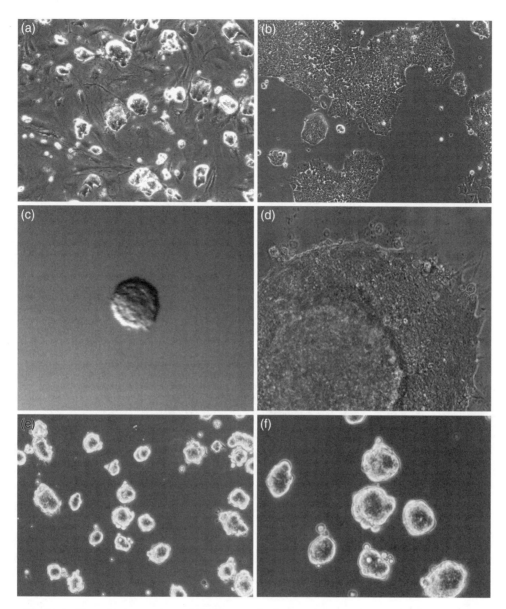

**Fig. 23.1.** *Mouse Embryonic Stem Cells.* (*a*) Growth of mES colonies on feeder layers with serum and LIF. (*b*) Subcultured mES cells in serum and LIF. (*c*) Isolated inner cell mass (ICM); "rind and core" structure of hypoblast growing round epiblast. (*d*) Outgrowths of mES cell-like from cultured epiblast (*see also* Plate 25*a*). (*e, f*) Growth of mES colonies in 2i medium, (*e*) 10× and (*f*) 20× magnification (*see also* Plate 25*b*).

Mouse embryonic stem (ES) cells can self-renew (divide to give rise to identical copies of themselves) in culture without apparent limit while retaining the ability to differentiate into cell types of all three germ layers. They are used as a model for the normally inaccessible processes of early development, as a source for the generation of specialized cells for in vitro and transplantation studies and as a tool for the genetic modification of mice [Smith, 2001].

Mouse ES cells were originally cultured on mitotically inactivated fibroblast feeder layers in the presence of serum [Evans & Kaufman, 1981; Martin, 1981]. These culture conditions are still widely used, but this complex environment has made it difficult to understand and to manipulate the environmental signals that regulate ES cell self-renewal and differentiation. ES cells are extremely sensitive to the composition of the culture medium and will undergo spontaneous differentiation in suboptimal culture conditions. In particular, serum batches must be tested for their ability to support ES cell culture. Refinements to the culture medium have allowed populations of ES cells with low levels of spontaneous differentiation to be cultured in the absence of feeder layers and serum. Addition of the cytokine leukemia inhibitory factor (LIF) removes the need for feeder layers [Smith et al., 1988; Williams et al., 1988] while serum–free medium can be used if it is supplemented with bone morphogenetic protein 4 (BMP4) [Ying et al., 2003]. Alternative culture conditions have also been developed that depend on the inhibition of mitogen activated protein kinase kinase-1 and -2 (MEK1/2) and glycogen synthase kinase 3-α and -β (GSK3α/β) using small molecules [Ying et al., 2008].

---

## PROTOCOL 23.2. PROPAGATION OF MOUSE EMBRYONIC STEM CELL LINES

### Outline

Detach and disaggregate ES cells by incubation with trypsin. Seed into a new flask and change medium every 1 to 2 days.

### Materials

*Sterile:*

- ❑ Prepared feeder layers (*see* Protocol 13.3)
- ❑ ES cell culture medium with selected serum batch tested with ES cells (*see* Protocol 23.1)
- ❑ Feeder medium: as for ES cell culture medium but without LIF
- ❑ N2B27, serum-free culture medium (*see* Protocol 23.1)
- ❑ N2 supplement (*see* Protocol 23.1)
- ❑ Serum-free ES cell culture medium (*see* Protocol 23.1)
- ❑ 2i ES cell culture medium (*see* Protocol 23.1)
- ❑ D-PBSA
- ❑ Gelatin (*see* Protocol 23.1)
- ❑ Trypsin (*see* Protocol 23.1)
- ❑ Tissue culture flasks (Iwaki or Corning)
- ❑ Centrifuge tubes, 15 and 50 mL (Greiner)
- ❑ Sterile, disposable pipettes, 5, 10, and 25 ml (Greiner)
- ❑ Pasteur pipettes, glass autoclaved.
- ❑ Humidified incubator set to 7% $CO_2$ and 37°C

### Procedure

This protocol assumes that cultures are maintained in 25-cm² flasks with 7% $CO_2$ and at 37°C. Scale volumes up/down for larger/smaller surface areas. Mouse ES cells should be passaged when 70% to 80% confluent. Under routine tissue culture regimes this will be every 2nd or 3rd day.

1. Prepare the required number of flasks with feeders (*see* Protocols 13.3, 23.1)
2. Warm D-PBSA, trypsin and ES cell medium to 37°C in a water bath.
3. Aspirate the culture medium and remove excess serum by washing adherent ES cells once in D-PBSA.

4. Add 1 ml trypsin and incubate cells at 37°C for 2 to 3 min or until cells can be detached from the flask with a sharp tap.
5. Add 9 mL ES cell culture medium and pipette up and down 4 to 5 times using a narrow-mouthed pipette to ensure a single cell suspension.
6. Transfer cell suspension to a centrifuge tube and centrifuge at 300 *g* for 3 min.
7. Aspirate the supernate taking care not to disturb the cell pellet.
8. Resuspend the pellet in ES cell culture medium and transfer the required number of cells to a pre-prepared tissue culture flask with feeders. For routine passaging, a split ratio of between 1:5 and 1:10 is normally used to give about 2 to 4 × 10⁴ cells/cm².
9. Add ES cell medium to a final volume of 10 mL and rock the flask (back and forth and side to side) to disperse the cells evenly.
10. Monitor the culture closely. Undifferentiated ES cells have a high nucleus:cytoplasm ratio and prominent nucleoli (Fig. 23.1*b*). Colonies growing on feeders typically assume a tightly packed, rounded-up morphology (Fig. 23.1*a*). Some differentiated cells will appear spontaneously but should not expand after passaging.
11. Replace the medium with fresh ES cell culture medium every 1 to 2 days.

### Feeder-free culture

Many 129 strain mouse ES cell lines are readily maintained in the absence of feeders. For feeder-free culture passage and feed as described above but coat tissue culture flasks only with gelatin (*see* Protocol 23.1, Step 1a).

*Note.* Feeder-free cultures may assume a more flattened, monolayer-like appearance than those cultured on feeder layers (Fig. 23.1*b*) and more differentiation may be evident. Feeder-free culture is particularly sensitive to the batch of serum used.

### Feeder- and serum-free culture

It is possible to maintain mouse ES cells in the absence of feeders and serum if the media are supplemented with both LIF and BMP4 [Ying et al., 2003]. For serum-free culture proceed as follows:

1. To transfer an established line into serum-free medium, first passage the cells and allow them to grow for 24 h in their usual culture conditions.
2. Wash once with D-PBSA and add 10 mL serum-free ES cell culture medium.

3. To passage remove medium and add 1 mL trypsin, incubate at room temperature for 1 to 2 min or until cells can be detached with a sharp tap.

4. Add 9 mL N2B27 and pipette up and down 4 to 5 times to ensure a single cell suspension.

5. Transfer cell suspension to a centrifuge tube and centrifuge at 300 $g$ for 3 min.

6. Aspirate the supernate; be sure to remove as much of the supernate as possible taking care not to disturb the cell pellet. The pellet may be only loosely attached in serum-free culture.

7. Resuspend the pellet in serum-free ES cell culture medium and transfer the required number of cells to a gelatin-coated flask. For routine passaging a split ratio of between 1:5 and 1:10 is normally used.

8. Add serum-free ES cell medium to a final volume of 10 mL and rock the flask (back and forth and side to side) to disperse the cells evenly.

9. Replace the culture medium daily.

### ES cell culture in 2i medium

Mouse ES cells can readily be cultured in serum-free medium supplemented with a combination of small molecule inhibitors [Ying et al., 2008]. For 2i culture proceed as follows:

1. To transfer an established line into 2i medium, first passage the cells and allow them to grow for 24 h in their usual culture conditions.

2. Wash once with D-PBSA and add 10 mL of 2i cell culture medium.

3. Replace culture medium daily and passage as described for serum-free culture. A degree of cell death and differentiation is normal during the first 1 to 2 passages in 2i.

*Notes*

(1) ES cells cultured in 2i medium are prone to detach from gelatin-coated plates. Substrate adhesion is enhanced by drying the gelatin coated plates for 15 to 60 min at room temperature after aspirating the gelatin. While adding medium to the flasks, take care not to pipette medium directly onto the adherent cells.

(2) ES cell growth in 2i medium is further enhanced by the addition of LIF to the medium.

(3) ES cells cultured in 2i medium form tightly packed, three-dimensional colonies (Fig. 23.1*e*, *f*). Cultures should be passaged well before reaching confluence to keep cells at the center of large colonies from becoming necrotic. During routine culture this will typically be every third day.

## 23.1.4 Primary Culture of Human Embryonic Stem Cells

Stem cell cultures have also been derived from human embryos and show differentiation down a number of different pathways, encouraging their use in tissue repair [Thomson et al., 1998; Rippon & Bishop, 2004]. Human teratocarcinomas cells such as NTera-2 [Paquet-Durand et al., 2004] and ES cells [Thomson et al., 1998] and human EG cells [Schamblott et al., 1998; Turnpenny & Hanley, 2007] can similarly be propagated in culture in the presence of LIF, FGF-2, and forskolin [Schamblott et al., 2002] with capacity to differentiate and significant potential for tissue engineering [Laslett et al., 2003; Rippon & Bishop, 2004]. Cultures derived from spontaneously generated embryoid bodies, isolated from EG cultures and disaggregated in collagenase and Dispase, give rise to EBD (embryoid body derived) cell lines that appear to be more easily maintained and have a longer life span in culture [Schamblott et al., 2002; Turnpenny & Hanley, 2007]. These cell lines are also capable of differentiation in a number of different directions and may have considerable potential for tissue engineering [Kerr et al., 2003], provided that genetic stability can be guaranteed [Lefort et al., 2008].

The following introduction and Protocol 23.3 for the derivation of human ES cells has been adapted from Cooke & Minger [2007]. Information on the use of human embryonic stem cells for research in the United States is available on the National Institutes of Health website (http://stemcells.nih.gov). In the United Kingdom, the Human Fertilization and Embryology Authority (HFEA: www.hfea.gov.uk) licenses all human embryonic stem cell derivation research projects.

During assisted reproduction treatment, good quality embryos are used in fertility treatment of the patients requiring treatment. Often the second-grade embryos are frozen for later patient use, and only then the poorest quality embryos that would not have been used for patient fertilization will be donated for research [Pickering et al., 2003]. Alternatively, high-quality embryos can be obtained from screening for known genetic disorders using preimplantation genetic diagnosis (PGD). The human embryo rapidly grows from the fertilized oocyte, reaching the blastocyst stage after approximately 5 to 6 days after fertilization. The inner cell mass (ICM) can be isolated by immunosurgery from the surrounding trophectoderm at the blastocyst stage using protease to lyse the trophectoderm [Bongso et al., 1994]. The ICM can also be isolated using a laser to avoid exposure of the embryo to animal products [De Vos & Steirteghem, 2001]. The ICM can then be cultured in vitro on MEFs until a putative stem cell line emerges.

## PROTOCOL 23.3. DERIVATION OF HUMAN EMBRYONIC STEM CELLS

### Outline

Explant the inner cell mass from a pronase-treated 5- to 6-day blastocyst onto a mouse embryo cell feeder layer after immunolysis of the trophectoderm.

### Reagents and Materials

*Sterile or aseptically prepared:*

❑ Day 5 to 6 human blastocyst obtained through an HFEA licence and with full patient consent as previously discussed. Full details of HFEA requirements can be found on the HFEA website (www.hfea.gov.uk).

❑ DMEM with Glutamax

❑ Buffalo rat liver (BRL) medium (BRL-M): DMEM, with 4.0 mM Glutamax, 4.5 g/L glucose, 1% NEAA, 0.1 mM 2-mercaptoethanol, 20% ES-grade FBS (Autogen Bioclear)

❑ BRL-conditioned medium (BRL-CM): BRL cells are grown to confluence in BRL-M and the medium replaced with fresh BRL-M collected 3 days later (*see also* Protocol 13.2)

❑ hES medium: BRL-CM:BRLM/20FB, 3:2, with 1000 U/mL LIF (ESGRO, Millipore)

❑ Pronase, 0.5%, in DMEM

❑ Anti-human antibody (Sigma), 30% to 50%

❑ Guinea pig complement diluted 1:1 with DMEM

❑ Single well of freshly inactivated MEFs seeded at $7.5 \times 10^4$ per well of a 4-well plate containing 500 μL of hES medium

❑ Wide-bore pipette

### Procedure

1. Allow the embryo to reach blastocyst stage, usually around day 5.

2. If embryo has not hatched from the *zona pellucida*, incubate at 37°C with 5 to 10 μL of 0.5% Pronase until the *zona* is dissolved.

3. Expose the *zona*-free blastocyst to 30% to 50% anti-human antibody diluted in DMEM with Glutamax for 10 min.

4. Following incubation, rinse the blastocyst briefly in hES medium to inactivate the anti-human antibody.

5. Incubate the blastocyst at 37°C for 5 to 15 min with 5 to 10 μL 20% guinea pig complement in order to lyse the trophectoderm. The embryo can be gently passed through a wide-bore pipette to help the process at this stage.

6. When the trophectoderm is fully lysed, gently remove the intact inner cell mass using a pipette and transfer immediately onto one well of the MEFs in 500 μL of hES medium.

7. The ICM attaches within 2 to 5 days; observe MEF plate daily for outgrowth. It can be left in situ for up to 15 days, having freshly inactivated MEFs added to the well as required (only when the colony is large enough to passage should it be transferred to a fresh MEF plate).

8. Monitor cells with stem cell-like morphology from the ICM appearing from the center of the colony (Fig. 23.1*a*).

9. When the colony reaches around 0.1 to 0.5 mm in size, dissecte it into 2 to 10 pieces, using a pulled glass pipette, and transfer the pieces onto 2 to 4 wells of fresh inactivated MEF plates. This process should be repeated every 5 to 7 days.

10. After the first few passages of the newly derived hES cell line, follow the protocols of hES cell propagation.

*Note.* As soon as there is more than a single colony, clumps of the nascent hES cell line should be frozen at every passage (*see* Protocol 19.4) until large numbers of cryo-straws from each cell line have been successfully frozen. Minimally 50% of the first 10 to 15 passages should be frozen until the cell line is well-established.

11. Feeding:
    (a) On day 1 following seeding, 250 μL fresh medium should be added without removing the old medium to allow the hES cells to attach to the MEFs. Cells should be double fed (i.e., 1000 μL of fresh hES medium per well) if they cannot be fed or supplemented the following day, for example, on a Saturday to support the cells until Monday.
    (b) After hES cells have attached, replace the medium with 500 μL of fresh hES medium per well.
    (c) On days in between feeding, 250 μL of fresh hES medium should be added to each well.

Colonies of 300 to 500 cells detached from the feeder layer and maintained in suspension form embryoid bodies (Fig. 23.2*b*; *see also* Plate 25*d*). On reattachment to the correct matrix, and in the presence of the correct growth factors, the colonies will differentiate into a number of different cells types, depending on the microenvironment.

## 23.1.5 Passaging hES Cells

Passaging hES cells involves the division of undifferentiated colonies into smaller pieces and transfer onto a new support

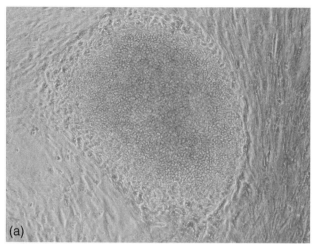

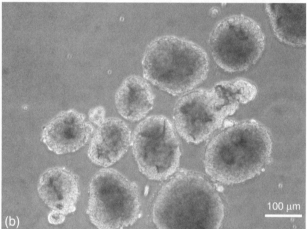

**Fig. 23.2. Human Embryonic Stem Cells.** (*a*) Colony of hES cells on feeder layer. Undifferentiated cells around the edge of the colony surround more differentiated cells with a brownish appearance in the center. The colony is surrounded by growth-inactivated mouse embryo feeder cells (see also Plate 25*c*). (*b*) Free-floating embryoid bodies (EBs), typically 300 to 500 cells per cluster, generated by detaching colonies from feeder layer and preventing further reattachment (see also Plate 25*d*). (From Cooke & Minger, 2007).

layer or matrix to allow further outgrowth of the cells. This culture method maintains the cells in an undifferentiated state by encouraging self-renewal. This can be performed using manual techniques such as cutting the colonies with a pulled glass pipette, or dislodging the colonies with glass beads or by using chemicals or enzymes such as Collagenase IV to dislodge the cells from the feeder layer. Manual passage reduces the potential for genetic alteration, which may be seen with repeated exposure to enzymes or chemicals [Suemori et al., 2006].

This protocol has also been condensed from Cooke & Minger [2007].

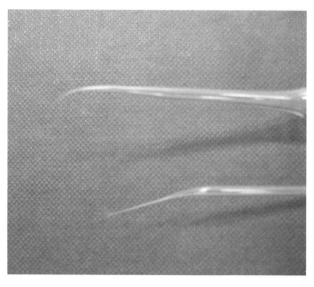

**Fig. 23.3. Pulled Glass Pipettes.** Various different pulled tips should be explored to find a preference, for example a straight-edged or rounded-edged pipette tip. (From Cooke & Minger, 2007).

## PROTOCOL 23.4. MANUAL PASSAGE OF hES CELLS

### Outline

Using pulled glass pipettes (Fig. 23.3), cut the hES cell colonies into smaller pieces, allowing for precise selection of undifferentiated cells within a colony, and reseed on to fresh MEFs.

### Reagents and Materials

*Sterile or aseptically prepared:*

❑ Undifferentiated hES cell colonies on MEF plates
❑ Fresh MEF plate (ideally within 1–3 days post inactivation)
❑ hES medium (*see* Protocol 23.3)
❑ Pipettor tips for 50 µL
❑ Glass Pasteur pipettes

*Nonsterile:*

❑ Pipettor 100 µL, adjustable, set at 50 µL
❑ Gas burner with naked flame or similar
❑ Dissecting phase contrast microscope preferably with a heated stage set at 37°C housed within a category II laminar-flow hood

### Procedure

1. All work must be done under sterile conditions using surgical masks, gloves, and lab coats.
2. Change the MEFM on the fresh MEF plates for 500 µL of hES medium.

3. Light the gas burner and use a blue flame to pull the glass pipettes. Gently rotate the glass pipette and pull when soft to produce a sealed narrow end no thicker than a human hair (Fig. 23.3).

4. Place all pulled pipettes immediately in the laminar-flow hood to keep sterile. Pipettes should only be pulled at the time of passaging to reduce the risk of contamination.

5. Remove hES cell plate from the incubator and place on the heated stage.

6. Select areas of the colonies that are undifferentiated by their appearance. Undifferentiated hES cells are small, round, and compact with a translucent color. Areas of brown cells that are spontaneously differentiating should be avoided (*see* Fig. 23.2*a*).

7. Using the tip of a pulled pipette, cut around the area of undifferentiated cells and cut into smaller equal chunks. Typically a colony of similar size to Fig. 23.2*a* should be cut into 6 to 10 pieces depending on the rate of re-growth of that particular hES line.

8. When the colony has been cut, lift the free-floating pieces with a pipettor and suitable tip.

9. Transfer the pieces of colony onto new MEF plates. As a guide, equally distribute around 4 to 6 smaller pieces within a fresh MEF well.

10. Carefully transfer the old and new hES plates back into the incubator. The pieces should be allowed to settle overnight without disturbing the plate, so that cells can attach and not clump in the center of the plate.

11. The following day, supplement the hES medium in both old and new plates with another 250 µL of hES medium to reduce the risk of aspirating unattached hES cells in a full feed.

*Note.* A fresh sterile pulled pipette should be used each time you enter a well to avoid contamination.

### 23.1.6  Pluripotent Stem Cells from Fish Embryos

Pluripotent stem cells can be isolated from many other species besides rodents and humans. The following protocols describe the generation primary cultures, the feeder cells required for maintaining them, and the cell lines that can be derived from them, retaining pluripotency. Protocol 26.6 for culturing cells from zebrafish has been abridged from Collodi [1998].

The zebrafish possesses many favorable characteristics that make it a popular nonmammalian model for studies of vertebrate development and toxicology [Powers, 1989; Driever et al., 1994]. Zebrafish reach sexual maturity in approximately three months, and females produce 100 to 200 eggs each week throughout the year. Embryogenesis is completed outside of the mother in 3 to 4 days, and the large, transparent embryos are amenable to experimental manipulations involving cell labeling or ablation techniques [Westerfield, 1993]. In vitro approaches utilizing embryo cell cultures have also been employed for the study of zebrafish development. Cell lines and long-term primary cultures, initiated from blastula, gastrula, and late-stage embryos, have been established [Collodi et al., 1992; Ghosh & Collodi, 1994; Sun et al., 1995a,b; pelenbosch et al., 1995].

***Cell lines derived from early-stage embryos.*** As differentiation occurs during zebrafish gastrulation, the cells in earlier stage embryos, such as the blastula, are pluripotent [Kane et al., 1992], and methods have been developed for the culture of cells from these early-stage embryos. The ZEM-2 cell line, initiated from mid-blastula-stage embryos, has been growing in culture for more than 300 generations in medium containing low concentrations of fetal bovine and trout sera, insulin, trout embryo extract, and medium conditioned by buffalo rat liver cells [Ghosh & Collodi, 1994].

A fibroblastic cell line, ZEF, has also been derived from early-stage embryos in medium supplemented with FBS and FGF. Once established, the line has been maintained in LDF medium (*see* Protocol 23.5) containing 10% FBS [Sun et al., 1995a], and ZEF cells have been utilized as a feeder layer for primary cultures of zebrafish embryo cells [Sun et al., 1995a, Bradford et al., 1994a].

***Primary cultures derived from early-stage embryos.*** In addition to the continuously growing embryo cell lines that are available, methods have been developed for the initiation of primary cultures derived from early zebrafish embryos [Sun et al., 1995b, Bradford et al., 1994a, b]. Primary cultures, derived from early gastrula-stage embryos, maintained a diploid chromosome number and exhibited a morphology characteristic of pluripotent ES cells when derived on a feeder layer of ZEF fibroblasts in medium containing FBS, trout serum, fish embryo extract, insulin, and leukemia inhibitory factor [Sun et al., 1995b; Bradford et al., 1994a].

***Cell lines derived from late-stage embryos.*** Late-stage zebrafish embryos (20–24 h postfertilization) have been used for the derivation of three fibroblastic cell lines: ZF29, ZF13 [Peppelenbosch et al., 1995], and ZF4 [Driever & Rangini, 1993]. The lines were derived in Leibowitz's L-15 (ZF29 and ZF13) or a mixture of Ham's F12 and Dulbecco's modified Eagle's media (ZF4) supplemented with FBS.

## PROTOCOL 23.5. CELL CULTURES FROM ZEBRAFISH EMBRYOS

### Outline

Cells isolated from the blastula may be cultured on a zebrafish embryo feeder layer and can give rise to continuous cell lines.

### Materials

*Sterile:*

☐ LDF basal medium:
Leibowitz's L-15 medium ........... 100 mL
DMEM ...................................... 70 mL
Ham's F-12 ............................... 30 mL
Sodium selenite ......................... 6 μM 200 μL
(Store refrigerated at 4°C.)

☐ LDF primary medium—LDF basal medium plus:
FBS ....................................... 1%
Trout serum ......................... 0.5%
Trout embryo extract ........... 40 μg of protein/mL
Insulin ................................ 10 μg/mL
Leukemia inhibitory factor, human,
recombinant ................... 10 ng/mL

☐ LDF maintenance medium—LDF basal medium plus:
FBS ....................................... 1%
Trout serum (Sea Grow, East Coast
Biologicals) ...................... 0.5%
Trout embryo extract ........... 40 μg of protein/mL
Insulin ................................ 10 μg/mL
BRL-conditioned medium .... 50%

☐ D medium—DMEM/F12/10FB: 50/50 DMEM/Ham's F12 with 10% fetal bovine serum (FBS)

☐ Trout embryo extract:

(i) Collect the embryos (Shasta Rainbow or other strains of trout, 28 days postfertilization, reared at 10°C; or zebrafish, 3 days postfertilization, reared at 28°C), and store them frozen at −80°C.

(ii) To prepare the extract, thaw the embryos (approximately 150 g) and homogenize them in 10 mL of LDF for 2 min on ice, using a Tissuemizer homogenizer (Teledyne Tekmar).

(iii) Pass the homogenate through several layers of cheesecloth to remove the chorions, and then centrifuge the homogenate at (20,000 g for 30 min at 4°C).

(iv) After centrifugation, collect the supernate, leaving behind the bright-orange lipid layer present on the surface.

(v) Transfer the supernate to a new tube, and centrifuge it as before (in step iii).

(vi) Collect the supernate, leaving behind any remaining lipid, and then ultracentrifuge it at 100,000 g for 60 min at 4°C.

(vii) After ultracentrifugation, collect the supernate, leaving behind the lipid layer. Dilute the supernate with LDF (1:10), and filter sterilize through a series of filters (1.2 μm, 0.45 μm, and 0.2 μm); *see* Protocol 10.15.

(viii) Store the extract frozen at −80°C in 0.5-mL aliquots.

(ix) To use the extract for cell culture, measure the concentration of protein (*see* Protocol 20.4) and then dilute the extract with LDF to the desired working concentration.

(x) Store the diluted extract refrigerated at 4°C for a maximum of two months.

☐ BRL cell-conditioned medium:

(i) Culture BRL cells (ATCC) at 37°C in 75-cm² flasks in DMEM/F12/10FB.

(ii) When the cultures become confluent, replace the FD medium with LDF supplemented with 2% FBS, and incubate the cells at 37°C.

(iii) After 5 days, remove the LDF, filter it, and store it frozen at −20°C.

(iv) Add fresh LDF to the BRL cultures, and repeat the process 5 days later. Conditioned LDF medium can be collected up to three times from the same flask before the cells must be split and allowed to grow again to confluence.

☐ Holtfreter's buffer:
NaCl ..................................... 70 g
KCl ....................................... 1.0 g
NaHCO$_3$ ............................. 4.0 g
CaCl$_2$ ................................ 2.0 g
UPW to ............................... 1000 mL
(Store at 4°C. Prepare a working solution by diluting 1:20 with UPW.)

☐ Pronase E, 0.5 mg/mL in Holtfreter's buffer

☐ Trypsin, 1%, EDTA, 1 mM, in D-PBSA

### A. Fibroblast Feeder Layers

### Outline

Prepare feeder layers of embryonic fibroblasts from gastrula-stage zebrafish [Sun et al., 1995a] by removing the chorion in pronase, culturing the cells in FGF-supplemented medium, and selecting the fibroblasts by differential trypsinization.

### Materials for Feeder Layers

*Sterile or aseptically prepared:*

☐ Embryos, 8 h postfertilization
☐ Pronase
☐ Trypsin
☐ FBS

❑ LDF primary medium with 10 ng/mL bovine FGF (*a* + *b* mixture)
❑ LDF basal medium with 10% FBS
❑ Flasks, 25 cm$^2$

### Procedure

1. Collect approximately 30 embryos (8 h postfertilization).
2. Remove the chorion by pronase treatment [Sun et al., 1995a].
3. Incubate the embryos in trypsin (1 min) while gently pipetting to dissociate the cells.
4. Add FBS (10% final concentration) to stop the action of the trypsin.
5. Collect the cells into a pellet by centrifugation (at 500 *g* for 10 min).
6. Resuspend the cell pellet in 5 mL of LDF primary medium containing FGF, and transfer the cells to a 25-cm$^2$ flask.
7. Allow the cells to attach and grow to confluence at 26°C.
8. When the cells are confluent, passage the culture in the same medium.
9. After 2 to 3 passages, a mixed population of epithelial and fibroblastic cells will be present in the culture, and the cells can be maintained in LDF basal with 10% FBS. Select the fibroblasts for further culture, by differential trypsinization:
   (a) Treat the culture with trypsin for 1 min to remove most of the fibroblasts, and leave the epithelial cells attached to the plastic.
   (b) Transfer the fibroblasts to another flask, and repeat this process when the culture becomes confluent.
   (c) Passage 2 or 3 times until the culture consists of a homogeneous population of fibroblasts.
10. Prepare feeder layers of growth-arrested fibroblasts:
   (a) Add the cells to the appropriate culture dish or flask, and allow the fibroblasts to grow into a confluent monolayer.
   (b) Add mitomycin C, 10 µg/mL, to the cultures, and incubate the cultures for 3 h at 26°C.
   (c) After being rinsed three times with LDF, the growth-arrested fibroblasts can be used as feeder layers for zebrafish embryo cell cultures.

### B. Primary Cultures:

#### Outline

Collect embryos, remove the chorion, disaggregate the embryos in trypsin, and grow primary cultures derived from zebrafish blastula- and early-gastrula-stage embryos on feeder layers of embryonic fibroblasts.

### Materials for Primary Culture

*Sterile:*

❑ LDF primary medium
❑ Holtfreter's buffer
❑ Dilute bleach, 0.1% in UPW
❑ Pronase E solution
❑ Feeder layers of embryonic fibroblasts
❑ Human recombinant leukemia inhibitory factor (LIF), 1 µg/mL

### Procedure

1. Harvest embryos at the midblastula or early gastrula stage, and rinse them several times with clean water.
2. After rinsing, transfer the embryos into 6-cm Petri dishes (50–100 embryos/dish), take them to a laminar-flow hood, and maintain them under aseptic conditions.
3. Soak the embryos for 2 min in dilute bleach, and rinse them several times in sterile Holtfreter's buffer.
4. Dechorionate the embryos by incubating them in 2 mL of Pronase E solution for about 10 min, and then gently swirl the embryos in the Petri dish to separate them from the partially digested chorion.
5. Tilt the dish to collect the embryos on one side, and gently remove 1.5 mL of the Pronase solution with a Pasteur pipette. To prevent the dechorionated embryos from adhering to the dish and rupturing, keep the dish tilted so that the embryos remain suspended in the remaining Pronase solution.
6. Gently rinse the embryos by adding 2 mL of Holtfreter's buffer and swirling gently.
7. Tilt the dish and remove most of the Holtfreter's buffer, leaving the embryos suspended in about 0.5 mL.
8. Repeat the rinse procedure 2 more times.
9. After the final rinse, leave the embryos suspended in 0.5 mL of Holtfreter's buffer, and add 2 mL of trypsin solution.
10. Incubate the embryos in the trypsin for 1 min and then dissociate the cells by gently pipetting 3 to 4 times.
11. Immediately transfer the cell suspension into a sterile polypropylene centrifuge tube, and add to the tube 200 µL of FBS to stop the trypsin.
12. Collect the cells by centrifugation at 500 *g* for 5 min and resuspend the pellet in LDF primary medium (without FBS or trout serum).

**13.** Seed the cells at $1 \times 10^4$ cells/cm$^2$ onto feeder layers of growth-arrested embryonic fibroblasts, contained in multiwell dishes or flasks.

**14.** Allow the cells to attach to the feeder layers (~15 min) before adding FBS and trout serum. Human recombinant leukemia inhibitory factor (10 ng/mL) is used in the medium in preference to BRL-conditioned medium [Sun et al., 1995a, b].

### C. Cell Lines:

#### Materials for Cell Lines

*Sterile or aseptically prepared:*

- ZEM-2 cells (or equivalent)
- LDF maintenance medium

#### Protocol

**1.** Grow cultures derived from early zebrafish embryos, such as ZEM-2, in LDF maintenance medium to about 70% confluence.

**2.** Incubate the cultures at 26°C in ambient air.

**3.** Change the medium approximately every 5 days.

**4.** Subculture by trypsinization (*see* Protocol 12.3).

Trout serum and trout embryo extract have also been shown to stimulate the growth of embryo cells from other fish species [Collodi & Barnes, 1990] and the zebrafish spleen cell line ZSSJ has been used a feeder layer for zebrafish ES cell culture [Xing et al., 2009].

## 23.2 GERM CELLS

Primordial germs cells can be isolated from the post gastrulation human embryo and grown on STO feeder layers [Turnpenny & Hanley, 2007]. They will differentiate into a number of cell types after removal of growth factors and the feeder layer. Germline cells capable of differentiating into somatic cells of all three germ layers have also been isolated from spermatogonial cells of human adult testis [Conrad et al., 2008]. Male gametes can be derived from human ES cells [Geijsen et al., 2004].

## 23.3 EXTRAEMBRYONIC CELLS

Stem cells have also been isolated from trophoblast [Chung et al., 2006; Douglas et al., 2009], placenta [Lokesh Battula et al., 2007], and amniotic fluid [Perin et al., 2007; Ditadi et al., 2009] from which amniocytes are explanted for prenatal chromosomal and molecular analysis. Trophoblast cells represent a commitment away from the ES cell phenotype determined by DNA methylation of the transcription factor Elf5, although this can be manipulated [Roper & Hemberger, 2009].

### 23.3.1 Culture of Amniocytes

The human fetal karyotype can be determined by culturing amniotic fluid cells obtained by amniocentesis. Amniocentesis can now be performed from 11 weeks of gestation onward, although most amniocenteses are still performed from 15 to 18 weeks of gestation.

Inborn errors of metabolism and other sex-linked or autosomal recessive and dominant conditions are diagnosed mainly on placental tissues obtained by chorionic villus sampling and using direct techniques (i.e., using uncultured material). In addition methods are now available for trisomy screening using quantitative fluorescence- (QF)- PCR. However, most prenatal diagnoses for Down's syndrome are still based on a complete chromosome analysis, and a full chromosome analysis from amniotic fluid requires the culture of cells.

The following protocol and the above introduction was provided by M. Griffiths, Regional Genetics Laboratory, Birmingham Women's Hospital, Edgbaston, Birmingham, B15, 2 TG, England, UK.

The principle of culturing amniotic fluid cells is based on separating the cells by centrifugation and setting up the cell suspension in a suitable culture vessel. A variety of approaches exist that use either closed or open systems to grow the cells in tubes, flasks, slide chambers, or on coverslips in Petri dishes. Good-quality, rapid results can be obtained with any of these techniques, but a closed-tube system has the advantage of robustness, simplicity, and minimal use of resources.

### PROTOCOL 23.6. CULTURE OF AMNIOCYTES

#### Outline

Incubated cell cultures in a plastic Leighton tube in a standard incubator at 37°C, and harvest the cells using a trypsin suspension method.

#### Materials

*Sterile or aseptically prepared:*

- ❑ Amniotic fluid sample, 10 to 15 mL
- ❑ Complete medium: 100 mL Ham's F10 medium with 20 mM HEPES buffer, 2 mM glutamine, 100 U/mL penicillin, and 100 µg/mL streptomycin with 10% fetal bovine serum or 2% Ultroser G (Invitrogen)
- ❑ D-PBSA
- ❑ Colcemid, 10 µg/mL: add 1 mL to 9 mL of sterile D-PBSA for a working solution, and then add 0.1 mL of this for each 1 mL of medium in the culture, to reach a final concentration of 0.1 µg/mL
- ❑ Thymidine (Sigma): make a stock solution of 15 mg/mL in D-PBSA, and filter sterilize it. Add

0.1 mL of the stock thymidine solution to 10 mL of medium. Change to use the thymidine medium before harvesting cells. The final concentration in the medium should be 0.15 mg/mL.

❑ Bromodeoxyuridine, BUdR (Sigma): a vial contains 250 mg; dissolve all of it in 25 mL of sterile D-PBSA, and filter sterilize the solution. Add 0.5 mL of the solution to 11.5 mL sterile of D-PBSA and 8 mL of diluted colcemid solution. Dispense the resultant solution into aliquots, and store them frozen. Add 0.1 mL of the BUdR-colcemid mix to each 1 mL of culture. The final concentrations should be 25 µg/mL of BUdR and 0.04 µg/mL of colcemid.

❑ Trypsin (Bacto, Difco): reconstitute the contents of the vial with 10 mL of UPW, following manufacturer's instructions to give a 5% w/v solution

❑ EDTA, 0.5 mM in isotonic saline (Invitrogen)

❑ Trypsin/EDTA (TE) solution for subculture and harvest: 2 mL of reconstituted Bacto trypsin in 100 mL of EDTA giving 0.1% w/v trypsin

❑ Universal containers for collecting samples of amniotic fluid

❑ Plastic, disposable, 10-mL pipettes, and a pipetting aid for setting up cultures

❑ Leighton tubes (flat sided, plastic) and caps (Nunc)

❑ Syringes, 20 mL; plastic and plastic mixing needles (Henley Medical) for feeding cultures

❑ Transfer pipettes/Pastettes, 1 mL and 3 mL, plastic, disposable

❑ Histopaque-1077 (Sigma); adapt the manufacturer's instructions

*Nonsterile:*

❑ Disinfectant (e.g., Virkon—Merck); follow the manufacturer's instructions

❑ Potassium chloride hypotonic solution, 0.075 M KCl in UPW; add 5.574 g of Analar potassium chloride to 1 L of UPW

❑ Fixative: Freshly mixed, 3 parts Analar methanol to 1 part Analar glacial acetic acid

❑ Transfer pipettes/Pastettes, 1 mL, plastic, disposable

❑ Glass microscope slides, precleaned (Berliner Glas KG, from Skan, through Richardsons of Leicester)

❑ Coplin jars

❑ Hydrogen peroxide solution, 5% v/v: Dilute 1 part 30% hydrogen peroxide with 5 parts water

❑ Saline solution, 0.9% w/v NaCl in UPW

❑ Trypsin/saline solution for banding: Add 1.4 mL of reconstituted Bacto trypsin (Difco, BD Biosciences) to 50 mL of saline giving 0.14% w/v trypsin

❑ Buffer, pH 6.8 (Gurr—Merck): Add one buffer tablet to 1 L of UPW

❑ Leishman stain (Gurr—Merck), usually diluted 1 part to 4 parts pH 6.8 buffer

❑ DPX slide mountant and coverslips

❑ Oven at 60°C

❑ Hot plate at 75°C

❑ Bright-field and phase-contrast upright microscope, to assess slide making, banding, and for chromosome analysis

### Procedure

Δ *Safety Note.*

(a) Wear gloves and a laboratory coat. All samples should be handled under sterile conditions in a class II microbiological safety cabinet (*see* Section 6.8.2)

(b) Discarded media, and hypotonic supernates (but not fixative) should be poured off into a disinfectant solution—such as Virkon or hypochlorite. Fixative should be discarded into sodium bicarbonate solution to neutralize the acid. After standing 2 h, both types of waste may be discarded into normal drainage, with plenty of running water.

(c) Bromodeoxyuridine (BUdR) is a known mutagen, and its use is optional. An appropriate local safety assessment (e.g., OSHA or COSHH; *see* Section 6.5.4) should be undertaken before proceeding with any work involving this chemical. Solutions and supernates containing significant amounts of the chemical may be discarded into sealable vessels containing vermiculite (or a similar absorbent material) and destroyed by incineration.

(d) Protocols involving the use of methanol and acetic acid for fixing and preparing slides should be carried out in a fume hood.

### Setting up and monitoring tube cultures:

1. Expect approximately 10 to 15 mL of amniotic fluid to be delivered to the laboratory in a plastic sterile universal container.

2. Use a sterile 20-mL syringe fitted with a sterile plastic mixing needle (or a 10-mL pipette) to divide the sample between three Leighton tubes. Label the tubes with patient and culture identification details. If the sample is small (a volume of 5 mL or less), set up only two cultures.

3. Centrifuge the tubes for 5 min at 150 *g*.

4. Pour the supernates back into their original container (for biochemical assay or

immunoassay—e.g., α-fetoprotein [AFP] estimation), or discard them into disinfectant solution (Virkon).

5. Resuspend the cell pellets in 1 mL of culture medium per Leighton tube.

6. Incubate the tubes at 37°C.

7. Leave the cultures undisturbed for 5 to 7 days, to allow the cells to settle and establish colonies.

8. Assess the cultures. Depending on the degree of cell growth, either add 0.5 mL of fresh medium or remove and discard the old medium, and add approximately 1 mL of fresh medium.

9. Reassess cell growth as necessary (every 2–4 days), changing the medium as appropriate until there is sufficient growth for a harvest, usually 7 to 12 days after the cultures were initiated. Change the medium on the day prior to harvesting, if possible.

*Notes.*

(a) Rapid cell growth can be encouraged by spreading colonies using trypsin (*see* Dispensing and Subculturing below). Overgrown cultures can be recovered by subculturing into several additional Leighton tubes or flasks if large quantities of cells are required.

(b) Bloodstained amniotic fluid samples may not grow as well as clear samples. One approach to working with such samples is to separate some of the amniocytes from the contaminating red blood cells, using density gradient centrifugation (e.g., Histopaque; see Protocol 12.10).

(c) Assessment of heavily bloodstained samples at 6 to 7 days is usually not possible without removing the bloodstained medium first. The medium can be collected into supplementary tubes, which may be discarded if the original tubes show growth, or can be incubated further if necessary.

(d) Supplementary tubes may be established at the first change of medium in any case when discarding the original suspension is undesirable.

## Dispersing and subculturing

10. Prewarm approximately 1.5 mL of sterile TE solution for each tube to be processed to 37°C.

11. Remove the medium from the culture tube, and rinse it once with 1 mL of sterile TE solution.

12. Disperse the culture:

(a) Add 0.2 mL of TE to the culture, and incubate the culture at 37°C for 1 to 2 min to detach the cells.

(b) Check the cells on an inverted microscope. When the cells are in suspension, add 1 mL of culture medium, and return the tube to the incubator. (The serum in the medium will inactivate the TE solution.)

13. Subculture:

(a) Add 0.5 mL of TE to the culture, and incubate it at 37°C for 1 to 2 min to detach the cells.

(b) Check the culture on an inverted microscope. When the cells are in suspension, add 1 to 2 mL of culture medium, and divide the suspension between an appropriate number of subculture tubes. Two to 8 subcultures may be seeded, depending on the initial cell density. It is usually worth varying the concentration of cells in each subculture tube, as doing so improves the chances of being able to harvest a tube at an optimal cell density.

(c) Top up each subculture with fresh medium to a final volume of 1 mL, and then incubate.

14. Check that the cells have resettled the next day, and change the medium.

## Routine tube harvesting

Harvests can be routinely carried out on primary cultures so long as other cultures are available as a backup. If only one culture remains, subculture it prior to harvesting the cells, or salvage it after the harvest.

15. When the cultures are ready to be harvested, add 0.1 mL of diluted colcemid to each culture.

16. Incubate the cultures for as long as is necessary to accumulate enough rounded-up mitotic cells. This is typically 2 to 3 h, but may be as little as half an hour for very active cultures, or more than 4 h for slow-growing cultures.

17. Remove the medium into Virkon solution, and drain the tube briefly onto a paper towel.

18. Add 2 mL of TE, and incubate the cultures at 37°C for 3 min to detach the cells.

19. When the cells are in suspension, add 7 mL of KCl hypotonic solution, and leave the cultures at 37°C for 5 min.

20. Centrifuge the tubes at 150 *g* for 5 min.

21. Carefully remove the supernate by pouring it into Virkon solution. Drain the tube briefly onto a paper towel. Flick the tube gently to resuspend the cells in the small amount of remaining liquid.

22. Slowly fix the cells using fresh fixative:

(a) Flick the tube, and add the first 1 to 2 mL of fixative drop by drop, continually agitating the cells to avoid cell clumping.

(b) Add a further 3 to 4 mL of fixative.

23. Centrifuge the tubes at 150 *g* for 5 min.

24. Remove the supernate fixative by pouring it into sodium bicarbonate solution.
25. Gently resuspend the cell pellet, and add 5 mL of fixative to it.
26. Change the fixative by repeating the centrifugation, pour-off, and refixation steps 9–11).
27. Centrifuge the cells again, pour off the supernate, resuspend the cell pellet in residual fixative, and make slides of the cells (*see* later stepsslide preparation).

*Notes.*

(a) Fixed cell suspensions can be stored at $-20°C$ at any of the fixed stages. Change the fix twice before making slides.

(b) Salvage harvests may be used as an alternative to subculture or when only a single culture remains. This method requires the addition of sterile colcemid and sterile TE in the initial stages of the harvest, followed by transfer of the cells to a separate centrifuge tube after the cells have detached but before the addition of hypotonic solution. Fresh medium can be added to the original culture tube; the medium should then be changed the next day, after the cells have resettled, to remove residual traces of trypsin and colcemid.

(c) If a harvest produces an unacceptably low yield of metaphases, then alternative strategies such as the following can be used:

(i) Subculture the cells into a flask that is supported at a slight angle to the horizontal while the cells settle. Tilting the flask in this manner ensures a leading edge of the cells that are always growing at an optimal rate across the whole width of the flask.

(ii) Expose the cells to a reduced concentration of colcemid overnight.

(d) Longer chromosomes for higher resolution analysis may be produced by using either thymidine synchronization or overnight exposure to colcemid in the presence of bromodeoxyuridine. However, both these approaches work best with cells in an exponential growth phase, requiring careful assessment of the growth.

### Thymidine synchronized harvesting

1. On the morning of the day before the cultures are to be harvested, change the medium in the cultures, using the thymidine-supplemented medium (to a final thymidine concentration of 0.15 mg/mL).
2. Early on the day of harvest, rinse out the thymidine, using prewarmed medium. Pour off the medium, and rinse the cells twice. Add 1 mL of medium (without thymidine), which releases the thymidine block. The time of release depends on the time you intend to harvest the culture.
3. Incubate the culture for 4 h; then add 0.1 mL of diluted colcemid.
4. Incubate the culture for a further 2 h, and then harvest as for routine tube harvests. (Start with step of Routine tube harvesting, earlier in this protocol.)

### Bromodeoxyuridine overnight colcemid harvests

1. Change the medium in the cultures to be harvested during the morning of the day before the harvest.
2. In the afternoon of the same day, add 0.1 mL of BUdR-colcemid solution for each 1 mL of culture medium (final concentrations in culture: BUdR, 25 μg/mL; colcemid, 0.04 μg/mL).
3. The next morning (after about 20 h of exposure), pour off the BUdR–colcemid medium into a sealable container filled with vermiculite. Add 2 mL of prewarmed TE to the culture tube, and incubate the tube for 3 min to detach the cells.
4. Check that the cells are in suspension, and then add 7 mL of 0.075 M KCl hypotonic solution. Incubate the culture for 5 min.
5. Centrifuge the culture at 150 g for 5 min.
6. Pour off the hypotonic supernate into the vermiculite container.
7. Gently resuspend the cell pellet, slowly fix the cells and continue the harvest, starting with step of Routine tube harvesting, earlier in this protocol.

### Slide preparation

1. Use cleaned slides. The slides may be purchased as precleaned slides; however, it may still be beneficial to add 1 drop of fresh fix to each slide and wipe the slide with a paper towel immediately prior to use. If it is possible to control the surrounding environment, make the slides at 20°C to 25°C and a relative humidity of 40% to 50%.
2. Place one drop of cell suspension onto each slide by dropping the suspension from a height of 1 to 3 cm. Allow the drop to dry naturally.
3. Assess the quality of spreading using a phase-contrast microscope. If the spreading is acceptable, make the rest of the slides, adjusting the cell density by adding extra drops of fix to the suspension if necessary.
4. In some circumstances, the spreading may need to be improved, and the following suggestions may be helpful:

(a) Breathe on the slide first, and then place one drop of cell suspension onto the slide from a height of 1 to 3 cm. Allow the slide to dry naturally.

(b) Place one drop of cell suspension onto the slide, either with or without breathing first. Leave the slide for a few seconds, and then place a drop of fresh fix on top before the first drop has dried. Allow the slide to dry.

(c) Place one drop of suspension onto the slide, either with or without breathing on the slide first. Leave the slide for several seconds, and then, as the drying surface becomes dimpled or Newton rings become visible, place a drop of fresh fix on top. Allow the slide to dry.

(d) If the quality of spreading is still unacceptable, place the fixed cell suspension in the freezer in fix overnight, and remake the slides the next day.

### G-Banding with trypsin

1. Pretreatment of the slide may be carried out prior to trypsin exposure but, depending on the age of the slide, may not be essential. Hydrogen peroxide pretreatment works well with fresh slides. Concentrated Hanks's salt solution works well with older slides. Coplin jars are suitable staining vessels. Slides that are not pretreated may also be used.

2. Any of the following pretreatment methods may be used:

   (a) Immerse the fresh slides (after drying for 1 h) in 5% hydrogen peroxide for 20 s to 2 min.

   (b) Incubate the fresh slides in an oven at 60°C for a few hours or overnight, to age the slides. Immerse the slides in 5× Hanks's BSS for 5 min.

   (c) Immerse 1 d or older slides in pH 6.8 buffer for 60 s.

3. Rinse the slides well in pH 6.8 buffer.

4. Immerse the slides in 0.14% trypsin in saline solution for 3 s to 2 min.

5. Rinse the slides well in pH 6.8 buffer.

6. Stain the slides with Leishman stain, freshly diluted 1:4 with pH 6.8 buffer, for 4 min.

7. Wash the slides with tap water, and drain them to dry or blot them dry with care.

8. Assess banding under a bright-field microscope at 400× magnification. If the slides are underbanded, they may be destained in pH 6.8 buffer or methanol and the procedure repeated. If the slides are overbanded, start the procedure

again with another slide and vary the exposure time to trypsin or change the pretreatment.

9. Leave the slides on a hot plate (75°C) for a few minutes to ensure that they are completely dry, and then mount them with a glass coverslip, using DPX.

## 23.3.2 Cells from Neonates and Juveniles

Stem cells have been isolate from umbilical cord [Sanchez-Ramos, 2002; Newman et al., 2003; Howe et al., 2009]. Hematopoietic stem cells and mesenchymal stem cells (MSCs) are recoverable from the umbilical vein capable of forming a number of cell lineages including hemopoietic and neuroglial [McGuckin et al., 2004], and MSCs are also found in Wharton's Jelly, the proteoglycan rich matrix surrounding the umbilical arteries and vein. Dental pulp from deciduous teeth is also a source of MSCs [Sonoyama et al., 2007], similar in characteristics to bone marrow MSCs (*see* Section 23.3.4). Cord and dental pulp represent noninvasive sources for multipotent stem cells without complex ethical implications.

## 23.3.3 Multipotent Stem Cells from the Adult

There are numerous reports of isolation of multipotent stem cells from a number of adult tissues, including bone marrow [Suva et al., 2004; Lokesh Battula et al., 2007; Ma et al., 2009], liver [Deng et al., 2003; Alison et al., 2004], brain [Vescovi et al., 2002; Greco & Recht, 2003; Rieske et al., 2009], and muscle [Cao et al., 2003]. The identification of multipotential stem cells in adult tissues opens up a wholly unexpected area in stem cell biology, previously locked into the concept that stem cell regeneration was tissue specific. It raises many exciting prospects for understanding commitment and differentiation in adult progenitor cells but also raises a number of major questions. If stem cells exist in the brain, why are neurons not replaced? If there is a potential for circulating stem cells to repopulate other tissues, why do satellite cells from skeletal muscle not repair cardiac myocyte injury unless introduced artificially? What is the biological significance, from an evolutionary standpoint, of regenerative capacity and pluripotency of stem cells in many tissues that is never used? The concept that stem cells can change their differentiative capacity, not just from one lineage to another (e.g., from astroglial to oligodendroglial, or from erythroid to myeloid) but from the derivative of one germ layer to the derivative of another (e.g., neuroectodermal to mesodermal [Wurmser et al., 2004; Rieske et al., 2009] or endodermal to neuroectodermal [Deng et al., 2003]) conflicts so strongly with the established paradigm of lineage fidelity that many have claimed the atypical development of stem cells at an ectopic site is due to fusion of the incoming stem cell with a resident progenitor cell [Alvarez-Dolado et al., 2003; Greco & Recht, 2003]. However, although cell fusion probably does account for differentiation of some ectopic stem cells, the

**TABLE 23.1.** Primary Antibodies for Characterizing Pluripotent Stem Cell Markers

| Name | Species | Company | Catalog number | Working dilution |
|---|---|---|---|---|
| NANOG | Goat polyclonal | R& D Systems | AF1997 | 1:20 |
| OCT4 | Goat polyclonal | Santa Cruz Biotechnology, Inc. | sc-8629 | 1:150 |
| SSEA1 | Mouse monoclonal | DSHB | MC-480 | 1:20 |
| SSEA3 | Rat monoclonal | DSHB | MC-631 | 1:20 |
| SSEA4 | Mouse monoclonal | DSHB | MC-813-70 | 1:20 |
| PGC surface marker | Mouse monoclonal | DSHB | EMA-1 | 1:20 |
| hTERT | Mouse monoclonal | Novocastra | NCL-L-hTERT | 1:50 |
| TRA-1-60 | Mouse monoclonal | Chemicon, Ltd. | MAB4360 | 1:50 |
| TRA-1-81 | Mouse monoclonal | Chemicon, Ltd. | MAB4381 | 1:50 |

Source: Reprinted from Turnpenny & Hanley, 2007.

evidence for stem cell plasticity seems convincing, not least by the proof that adult cells such as epidermal keratinocytes and dermal fibroblasts can be reprogrammed to behave like pluripotent stem cells (*see* Section 23.3.5). It may be possible that stem cells found in many tissues of the adult acquire their new potency from the microenvironment into which they are transplanted, rather than that were already pluripotent in their native site.

Markers that have been used to confirm pluripotent stem cell identity include NANOG, OCT4, SSEA1, 3, and 4, PGC, hTERT, TRA-1-60, TRA-1-81 (Table 23.1) and alkaline phosphatase.

### 23.3.4 MSCs from Human Bone Marrow

One of the sites that has been exploited most extensively is the bone marrow from which can be derived not only hematopoietic stem cells but also mesenchymal stem cells (MSCs) with the potential to differentiate into a number of cell types including adipose, muscle, cartilage, bone [Hofmann et al., 2006; Gregory & Prockop, 2007], and cardiomyocytes [Kawada et al., 2004]. Protocol 23.5 for the cultures of MSCs and the following introduction was contributed by C. Gregory and D. Prockop, Texas A & M Health Science Center, Institute for Regenerative Medicine, 5701 Airport Road, Module C, Temple, TX; it is adapted from Gregory & Prockop [2007], and is reproduced here with their agreement.

Iliac crest bone marrow aspirates are generally preferred for the isolation and expansion of hMSCs, although MSC-like cells have been recovered from trabecular bone [Sakaguchi et al., 2004, synovium Sakaguchi et al., 2005], adipose tissue [Zuk et al., 2002], and even exfoliated teeth [Miura et al., 2003]. A 2-mL bone marrow aspirate is adequate for the production of enough MSCs for most applications. The bone marrow can be stored in heparinized blood drawing tubes charged with 3 mL αMEM and stored on ice for up to 8 h prior to processing. Longer incubations at 4°C decrease the initial rate of propagation of MSCs and are therefore discouraged.

Human bone marrow is initially processed by enriching for the nucleated component of the bone marrow that contains the hematopoietic and mesenchymal stem cells by Ficoll-mediated discontinuous density gradient centrifugation. The bone marrow is then cultured on 15-cm tissue culture plates or, after recovery on tissue culture plates, the cells can be expanded in Nunc Cell Factories (*see* Protocol 26.2) with frequent washes and media changes. The non-adherent hematopoietic component of the culture is gradually washed away over a few days resulting in an exclusively adherent MSC culture.

### PROTOCOL 23.7. MSC PRODUCTION FROM HUMAN BONE MARROW

#### Outline

Fractionate bone marrow aspirate on Ficoll–Paque and seed the nucleate fraction into Petri dishes. Wash to remove hematopoietic cells and subculture when 60% confluent.

#### Reagents and Materials

*Sterile or aseptically prepared:*

❏ Bone marrow aspirate, 2 mL collected into 3 mL α-MEM
❏ Complete culture medium (CCM): a-MEM without ribonucleosides or deoxyribonucleosides, with 4 mM glutamine, 20% hybridoma qualified FBS, 100 U/mL penicillin, 100 μg/mL streptomycin
❏ D-PBSA
❏ Trypsin/EDTA: porcine trypsin, 0.25% in PBSA with 1 mM EDTA
❏ Hanks's balanced salt solution without calcium or magnesium (HBSS)
❏ Ficoll-Paque
❏ Polypropylene centrifuge tubes, 15 mL and 50 mL
❏ Plastic tissue culture Petri dishes 15 cm diameter

❑ Plastic micro-pipettor tips for dispensing 10 to 1000 µL

*Nonsterile:*

❑ Trypan Blue solution in 0.85% saline
❑ Microcentrifuge
❑ Refrigerated benchtop centrifuge with swinging bucket rotor
❑ Water bath set to 37°C
❑ Improved Neubauer hemocytometer
❑ Pipettors, Eppendorf P10, P20, P100, and P1000 or equivalent

### Procedure

1. Uncap the drawing tube of bone marrow and transfer to one 50 mL centrifuge tube. Make the volume up to 25 mL with room temperature HBSS.
2. To another 50 mL centrifuge tube, add 20 mL Ficoll-Paque and gently overlay the 25-mL cell solution on to the Ficoll. The interface between the HBSS and the Ficoll should not be disrupted.
3. Centrifuge at 1800 $g$ for 30 min at room temperature with the brake off.
4. After centrifugation, collect the white cell layer at the interface of the Ficoll and HBSS and transfer to a fresh 50-mL centrifuge tube.
5. Make the volume of the interface cell suspension up to at least 3 volumes with HBSS and centrifuge at 1000 $g$ for 10 min at room temperature with the brake on. Repeat wash.
6. Suspend the cell pellet in 30 mL CCM pre-warmed to 37°C.
7. Add 10 µL of the cell suspension to 10 µL of Trypan Blue and assess the viability with a hemocytometer; viability should be above 80%.
8. Transfer the 30-mL cell suspension to a 15-cm diameter tissue culture Petri dish and culture in a humidified incubator under 5% $CO_2$ for at least 15 h.
9. Remove the Petri dish from the incubator and remove the medium.
10. Add and remove 20 mL pre-warmed D-PBSA to wash the monolayer.
11. Repeat the wash procedure 3 times.
12. Replace with 30 mL of fresh pre-warmed CCM.
13. Repeat this wash and medium replenishment every second day for 6 days.
14. After 6 days, examine the monolayers with an inverted microscope. Adherent, fibroblast-like colonies of MSCs should be clearly visible in the Petri dish (Fig. 23.4*a*, *b*). In some cases there may be signs of hematopoietic contamination, but these cells will be depleted upon passaging the cells. When the culture is 50% to 60% confluent, proceed to subculture.

### Subculture of MSCs from human bone marrow

15. Inspect the MSC cultures. If the cultures consist of small, adherent, spindle-shaped fibroblastoid cells that are approximately 60% confluent (see Fig. 24.3*d*), proceed. If the monolayer is sparse (see Fig. 24.3*c*), continue to wash and replenish medium as described above.
16. Trypsinize the monolayer as follows:
    (a) Wash the monolayer with 20 mL of pre-warmed PBSA and add 5 mL of trypsin/EDTA.
    (b) Place the plate at 37°C for 2 min, and then inspect the monolayer at 10× magnification. The adherent cells should be in the process of detaching from the plastic substratum.
    (c) Replace plate at 37°C for 2 min; then inspect again. Repeat inspection until 90% of the MSCs have detached from the plastic.
    (d) Add 5 mL of CCM, transfer the 10 mL suspension to a 15 mL conical tube, and centrifuge for 10 min at 500 $g$.
    (e) After the centrifugation, remove the supernate from the cell pellet and resuspend the pellet in 1 to 2 mL warm PBSA per tube. If necessary, combine multiple resuspended pellets for a single cell count.
17. Add 10 µL of the cell solution to 10 µL of Trypan Blue and count with a hemocytometer. An adequate concentration for the cell suspension should be between 2 to $5 \times 10^5$ cells per mL with a viability > 80%.
18. Suspend the MSCs at a concentration of $7 \times 10^3$ (for a final density of 50 cells per cm$^2$) to $1 \times 10^4$ cells/mL (for a final density of 100 cells per cm$^2$) in pre-warmed CCM. The cells should be plated at low density to maintain the rapidly self-renewing, multipotential phenotype.
19. Prepare the appropriate number of plates by adding 25 mL of pre-warmed CCM to each of the 15-cm plates.
20. Seed the plates by adding 1 mL of the suspension prepared in step 18. Slide the plates from side to side, do not swirl, to distribute the cells evenly. Replace the plates in the incubator.
21. After 2 to 3 days of culture, inspect the plates and make an assessment of morphology. The MSCs should adopt a small, spindle-shaped morphology with frequent refractile doublets (e.g., *see* Fig. 24.3*b*). This is the sign of a healthy culture of MSCs.

22. Aspirate the medium from the plates, wash the MSCs with 20 mL pre-warmed PBSA, and replace with 25 mL of fresh pre-warmed CCM.
23. The number of subsequent expansion plates per passage is limited only by the number of cells available to seed the plates. The volumes quoted in the protocol above are suitable for a single 15-cm plate of MSCs. This can be expanded proportionally to accommodate multiple plates where necessary.

### 23.3.5 Induced Pluripotent Stem Cells

Further evidence of plasticity in the generation of stem cells comes from the rapidly accumulating body of data demonstrating that adult cells can be converted to pluripotent stem cells by altering gene expression, by transfection [Nakagawa et al., 2007], which can be reversible using an excisable lentivirus construct [Sommer et al., 2010], a piggyback transposon [Kaji et al., 2009; Woltjen et al., 2009], retroviral infection [Aasen et al., 2008; Woltjen et al., 2009], adenoviral reprogramming [Stadfeld et al., 2008], transfection of microRNAs [Judson & Blelloch, 2009], or

chemical manipulation [Huangfu et al., 2008; Lin et al., 2009]. Alterations in DNA methylation and histone acetylation [Kondo & Raff, 2004; Hrzenjak et al., 2006; Keenen et al., 2008; Boheler, 2009] are implied by most of the methods induced by the expression of transcription factors such as c-Myc, Klf4, Oct4, and Sox2.

The following introduction and Protocol 23.8 for reprogramming human dermal fibroblasts, using lentiviral infection to introduce the reprogramming factors, has been contributed by S. Sullivan, R. Jones, and G. P. Davey, School of Biochemistry and Immunology and Institute of Neuroscience, Trinity College, Dublin 2, Ireland.

Currently much interest surrounds the generation and use of human-induced pluripotent stem (iPS) cells for applications relevant to human therapy [Nishikawa et al., 2008]. In the long term these may include human cell replacement therapies, but in the short term such cells will be used primarily as sources of disease- or patient-tailored cells for drug screening (assaying activity, toxicity, efficacy) and for disease modeling (in this context defined as the process by which underpinnings of disease are investigated by in vitro models of disease using stem cell derivatives).

Successful reprogramming of human dermal fibroblasts depends on several key parameters, including avoidance of microbial contamination such as mycoplasma, high

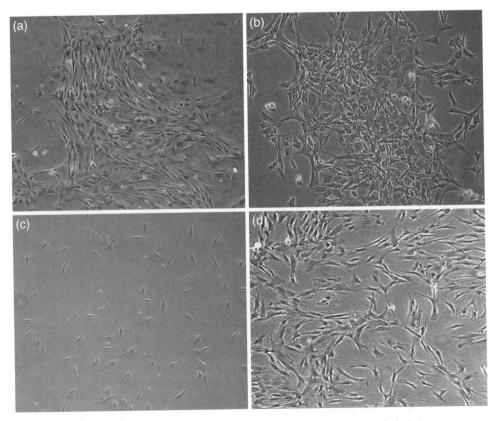

***Fig. 23.4. Bone Marrow-Derived MSCs.*** (*a, b*) Colonies formed after plating of whole bone marrow mononuclear cells. Morphology and optimal passaging density of MSCs. (*c*) An early passage culture of MSCs. (*d*) A monolayer at the appropriate density for passage. (From Gregory & Prockop, 2007).

proliferative rate of the fibroblasts in culture, generation of high titers of infective virus coding for the reprogramming factors (Oct-4, Sox2, Klf4, c-Myc), maintenance of a high and sustained level of these reprogramming factors during the cell reprogramming process, and extensive functional characterization of clones bearing a human embryonic stem cell morphology after following the reprogramming process.

---

### PROTOCOL 23.8. REPROGRAMMING HUMAN DERMAL FIBROBLASTS FOR THE GENERATION OF PLURIPOTENT STEM CELLS

#### A. Generation of Human Dermal Fibroblast Cell Lines:

#### Outline

Take a skin biopsy using a skin punch, digest the skin sample, remove the dermis and seed this into a gelatin-treated well of a 24-well plate. After 30 min add medium to the well and leave the dermal fibroblast explants to grow out for 11 to 15 days before passaging the cells.

#### Materials

*Sterile:*

- ❑ Collection medium: D-PBSA supplemented with penicillin (100 U/mL) and streptomycin (100 µg/mL)
- ❑ DFM: dermal fibroblast medium; DMEM supplemented with 10% fetal calf serum, penicillin (100 U/mL) and streptomycin (100 µg/mL).
- ❑ Trypsin, 0.25%, in D-PBSA
- ❑ Collagenase (Sigma, approx. 625 U/mL) in D-PBSA
- ❑ Gelatin, type A, porcine skin, 0.1% w/v (SIGMA cat. no. G1890)
- ❑ Centrifuge tube, 15 mL (BD Biosciences, Falcon)
- ❑ Skin punch, 4 mm
- ❑ Petri dishes for dissection, 9 cm
- ❑ Multiwell plates, 24-well, gelatin-coated (*see* Protocol 22.18A)
- ❑ Dissection kit: fine forceps and razor blades for manipulation of the skin sample and removal of the dermal layer after digestion

#### Procedure

Ethical approval for skin biopsy of patients (including informed consent; *see* Section 6.9.2) is required prior to collecting a human skin biopsy, which is usually done by a trained dermatologist or similar professional.

1. Swab the skin is swabbed with 70% ethanol or a similar antimicrobial prior to biopsy to reduce the risk of microbial contamination.
2. Collect the biopsy using a 4-mm skin punch to reduce pain and risk of infection.
3. Place the biopsy in a sterile 15-mL tube with 10 mL of collection medium and transfer to the tissue culture facility.
4. Remove the biopsy from the tube with sterile forceps and place in a 15-mL tube containing 2 mL collagenase.
5. Incubate in collagenase for 2 h at 37°C.
6. Transfer to a Petri dish and manipulate the skin to separate the dermal layer from the epidermis (*see also* Protocol 22.1) and transfer the dermis to a fresh dish.
7. Mince the dermis finely using two razor blades in a minimal amount of residual medium.
8. Transfer the resulting "pulp" to an empty well of a gelatinized gelatin-coated 24-well plate.
9. Leave to adhere to the well by incubating the plate in a $CO_2$ incubator for 30 min.
10. Gently remove the plate from the incubator and add 2 mL DFM to the well with the adherent pieces of dermis.
11. Culture the cells 37°C in 95% air, 5% $CO_2$, and 100% humidity. Fibroblast outgrowths should be ready to passage after 11 to 15 days in culture.
12. Subculture when approaching confluence, and use the cultures between passage 2 and 4.

***Note.*** If fibroblasts lines are to be purchased commercially, it is important that they are at a low passage number and have a good proliferative rate. Whether the fibroblast lines are derived in house or acquired commercially, they should be tested for mycoplasma prior to proceeding to the reprogramming step.

#### B. Generation of High Titers of Infective Virus Coding for iPS Factors:

*Background:*

A doxycycline inducible lentiviral system for generating human iPS cells has been developed by Maherali et al. [2008] and is now routinely used for making human iPS lines from somatic cells. The relevant constructs are freely available from Addgene (www.addgene.org; search term: hochedlinger). To generate iPS cells, skin fibroblasts are infected with lentivirus containing the constitutively active reverse tetracycline transactivator and the four reprogramming factors OCT4, SOX2, cMYC, and KLF4. For the process to work, it is necessary (1) to generate high titers of infective lentivirus and (2) to ensure a high expression level of the cells is maintained within the cells for several days. Controls for transfection of vectors into packaging lines, viral infectivity, and reprogramming factor induction should be carried out in parallel with actual transfections so that problems can be trouble-shot if they arise. It is imperative to use low-passage, nonstressed HEK293T cells that have been verified to be mycoplasma free. It should be noted that viral packaging lines like HEK293T are common sources of mycoplasma, so they should be quarantined on receipt and tested by regular testing for mycoplasma before release into general culture handling (*see* Sections 18.1.7, 18.1.8). Regular testing (e.g., with kits like MycoAlert (Lonza cat. no. LT07-118) is recommended (*see* Sections 18.3.2–18.3.6).

## Outline

HEK293T cells are used to generate lentivirus containing the constructs coding for iPS conversion and the supernatant medium is used to infect the skin fibroblasts to initiate reprogramming.

## Materials

*Sterile or aseptically prepared:*

❑ HEK293T cells (ATCC cat. no. CRL-11268)
❑ Plasmids: reprogramming constructs FUdeltaGW-rtTA; FU-tet-o-hKlf4; FU-tet-o-hOct4; FU-tet-o-hSox2; FU-tet-o-hc-Myc (Addgene)
❑ Lentiviral packaging constructs delta8.9 and vsv-g (Addgene)
❑ DMEM/10FB/PS: Dulbecco's modified Eagle's medium (DMEM) (Sigma cat. no. D6546), with 2 mM glutamine, supplemented with 100 U/mL penicillin, and 100 µg/mL streptomycin added from concentrated penicillin-streptomycin-glutamine (100x) (Invitrogen cat. no. 10378-016), and 10% FBS
❑ DMEM/10FB: Dulbecco's modified Eagle's medium (Sigma, cat. no. D6546), with 2 mM glutamine, supplemented with 10% FBS but no antibiotics
❑ hES cell growth medium: 80% Knockout DMEM (Invitrogen cat. no.10829), 20% KO Serum Replacement (Invitrogen cat. no.10828), 4 ng/mL bFGF, 2 mM Glutamax-I (Invitrogen cat. no. 35050), 0.055 mM 2-mercaptoethanol, and nonessential amino acids (Invitrogen cat. no.11140050).
❑ D-PBSA: phosphate-buffered saline without $Ca^{2+}$ and $Mg^{2+}$
❑ FBS (Hyclone cat. no. SH30070)
❑ Polybrene (Sigma cat. no. 10768-9)
❑ Endofree Plasmid Maxi Kit (Qiagen cat. no. 12362)
❑ Fugene 6 reagent (Roche Applied Science)
❑ TC plates Petri dishes, 10-cm (BD Biosciences, Falcon)

❑ Cell culture Petri dishes, 3.5 cm, gelatin-coated
❑ Conical centrifuge tubes, 50 mL (Falcon cat. no. 352070)
❑ Steriflip-GP filter (0.22-µm) unit (Millipore cat. no. SCGP00525)
❑ Ultracentrifuge tubes 17.0 mL (Beckman Coulter cat. no. 344061)

*Nonsterile equipment:*

❑ Beckman ultracentrifuge Beckman with SW28.1 rotor
❑ Fluorescence microscope

## Procedure

### Day 1: HEK293T cell seeding

1. Grow HEK293T cells in DMEM/10FB for several days in the absence of antibiotics and test for mycoplasma as previously outlined.
2. Trypsinize the cells (if clear) and count with a hemocytometer.
3. Seed at a density of $1 \times 10^6$ cells per 10-cm Petri dish. Tap dish to ensure uniform covering of cells on the dish.
4. Culture at 37°C in 5% $CO_2$.
5. Generate several mg of plasmid DNA for each of the plasmids coding for the reprogramming factors using Qiagen Endofree. To ensure the highest possible yield, make up the bacterial broth fresh before use and use a starter culture as per manufacturer's instructions.

### Day 2: Transfection of HEK293tT cells with reprogramming and viral packaging vectors

The cells should now be about 50% confluent and proliferating. It is important that a cell density higher than this not be used, as the medium will become acidic and result in a reduced yield of infective virus.

6. Add 10 mL of fresh DMEM/10FB/PS to each dish and transfect the cells with the reprogramming and viral packaging vectors, using FUGENE 6 reagent as per the manufacturer's instructions. Add the transfection mixture dropwise to each dish and swirl to insure maximum transfection of the HEK293T cells.
7. Transfect each plate with one reprogramming vector (FU-tet-o-hKlf4, FU-tet-o-hOct4, FU-tet-o-hSox2, FU-tet-o-hc-Myc) and the two viral packaging vectors delta8.9 and vsv-g with the mass ratio 4:3:2.

**Note.** The process appears to work best when the stochiometry of the four reprogramming factors is 1:1:1:1. If HEK293T cells are transfected with vectors with all four reprogramming factors, together the reprogramming frequency will be lower because some reprogramming factors will be produced at the expense of others. Consequently it is better to produce virus for each reprogramming factor separately, and then filter and mix the media from the four dishes (each transfected with a single reprogramming vector and two viral packaging vectors) at the very end.

### Day 3: Medium change

**8.** Observe the cells and confirm that they are not approaching confluency and that the medium is not turning too acidic.

**9.** Rinse cells in each dish with D-PBSA as per Fugene6 instructions.

**10.** Add 5 mL of fresh medium to each dish.

### Day 4: Harvest and concentrate virus particles from supernate

**11.** Collect and filter the medium from each dish through a 0.22-μm filter. Filtering is important as it prevents virus producing HEK293T cells being introduced into the fibroblast cultures now to be reprogrammed.

**12.** Concentrate the viral particles by ultracentrifugation using aerosol free ultracentrifuge tubes:

  **(a)** Centrifuge the supernate at 120,000 g for 1.5 h at 4°C using a swinging bucket rotor. In the Beckman SW28.1 rotor the capacity is 17 mL/tube with a total of 6 tubes, so the supernate obtained from 2 dishes (∼18 ml) can be concentrated in one tube.

  **(b)** Pour off the supernate. The pellet will be almost invisible.

  **(c)** Resuspend the viral pellet in 90 μL of D-PBSA.

**13.** Seed $1 \times 10^5$ skin fibroblasts in DFM on a gelatin-coated 3.5-cm dish.

### Day 5: Infection

**14.** Infect skin fibroblasts, seeded the previous day, with 2 mL DFM, supplemented with 6 μg/mL Polybrene, 10 μL rtTA, and 5 μL reprogramming factor (Oct4, Sox2, Klf4, c-Myc) viral preps.

### Day 6: Prepare feeder layer

**15.** Prepare 10-cm tissue culture dishes with good quality feeder layer (*see* Protocol 13.3).

### Day 7: Passage

**16.** Trypsinise the infected dermal fibroblasts, resuspend pellet in 10 mL DFM, and passage to feeder layers at a density of 90,000 infected dermal fibroblasts cells per 10-cm dish.

### Day 8: Add doxycycline addition (to induce reprogramming factor expression)

**17.** Remove old medium and add 2 mL of hES medium supplemented with doxycycline, 0.5 μg/mL.

**18.** Replace the medium with fresh medium supplemented with 0.5 μg/ml doxycycline every 3 days (switching to a lower concentration 0.25 μg/ml after day 9), and wait until colonies appear (about 3 weeks after infection). Once colonies appear in the doxycycline, stop supplementing the hES medium with doxycycline.

### Day 26 onward: Morphological assessment and picking of putative iPS clones

**19.** Check colonies on microscope. iPS colonies should resemble resemble human ES cells (see Fig. 23.5): flat colonies with pert, defined boarders, cells having prominent nucleoli and a high nucleus to cytoplasm ratio.

**20.** Stain for tissue nonspecific alkaline phosphatase [Turnpenny & Hanley, 2007] using Alkaline Phosphatase Staining (Stemgent, Kit 00–0009).

**21.** Immunostain (see Section 15.1.1; Protocol 15.11) putative iPS clones for hNanog to confirm reprogramming of other genes associated with pluripotency.

Morphological assessment will also need to be supplemented with functional assessment of putative iPS clones. Initial functional tests should be embryoid body formation in medium lacking bFGF, and differentiation as monolayers into the three germ layers.

## 23.3.6 Long-Term Bone Marrow Cultures from Mouse

The following introduction and Protocol 23.6 for the long-term culture from mouse bone marrow was contributed by E. Spooncer, Department of Biomolecular Sciences, UMIST, Sackville Street, Manchester M60 1QD, England, UK.

By culturing whole bone marrow, the relationship between the stroma and stem cells is maintained, and in the presence of the appropriate hematopoietic cell and stromal cell interactions, stem cells and specific progenitor cells can continue proliferating over several weeks [Dexter et al., 1984; Spooncer et al., 1992]. Progenitor cells from fresh marrow or long-term cultures may be assayed by clonogenic growth in soft agar [Heyworth & Spooncer, 1992] or in mice [Till & McCulloch, 1961].

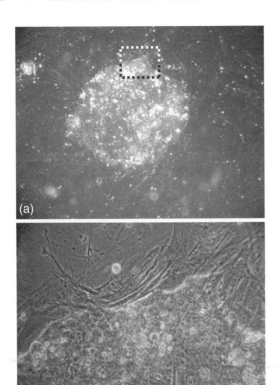

**Fig. 23.5. Colony of iPS Cells.** Typical iPS colony 25 days after infection. (*a*) Dark field, low magnification; (*b*) higher magnification of edge of colony outlined with dotted line.

---

## PROTOCOL 23.9. LONG-TERM HEMATOPOIETIC CELL CULTURES FROM MOUSE BONE MARROW

### Outline

Marrow is aspirated into growth medium and maintained as an adherent cell multilayer for at least 12, and up to 30, weeks. Stem cells and maturing and mature myeloid cells are released from the adherent layer into the growth medium. Granulocyte/macrophage progenitor cells can be assayed in soft gels (*see* Protocol 23.11).

### Materials

❑ All reagents must be pretested to check their ability to support the growth of the cultures.

*Sterile:*

❑ Fischer's medium (Invitrogen) supplemented with 50 U/mL of penicillin and 50 μg/mL of streptomycin and containing 16 mM (1.32 g/L) of $NaHCO_3$

❑ Growth medium: Fischer's as above, 100-mL aliquots supplemented with 1 μM hydrocortisone, sodium succinate, and 20% horse serum (hydrocortisone sodium succinate made up as 1 mM stock in Fischer's medium and stored at −20°C)

---

❑ Syringes, 1 mL, with 21G needles
❑ Gauze, swabs, scissors, forceps
❑ Tissue culture flasks, 25 cm$^2$

*Nonsterile:*

❑ Five mice: (C57Bl/6× DBA/2)F$_1$ bone marrow performs well in long-term culture, but marrow from some strains (e.g., CBA) does not [Greenberger, 1980]

### Procedure

1. Kill the donor mice by cervical dislocation.
2. Wet the fur with 70% alcohol and remove both femurs. Collect 10 femurs in a Petri dish containing Fischer's medium on ice. One femur contains 1.5 to 2.0 × 10$^7$ nucleated cells.
3. In a laminar-flow hood:
   (a) Clean off any remaining muscle tissue with gauze swabs.
   (b) Hold the femur with forceps and cut off the knee end. The 21G needle should fit snugly into the bone cavity.
   (c) Cut off the other end of the femur as close to the end as possible.
   (d) Insert the tip of the bone into a 100-mL bottle of growth medium, and aspirate and depress the syringe plunger several times until all the bone marrow is flushed out of the femur.
   (e) Repeat steps through (d) with the other nine bones.
4. Disperse the marrow to a suspension by pipetting the large marrow cores through a 10-mL pipette. There is no need to disaggregate small clumps of cells.
5. Dispense 10-mL aliquots of the cell suspension into 25-cm$^2$ tissue culture flasks, swirling the suspension often to ensure an even distribution of the cells in the 10 cultures.
6. Gas the flasks with 5% $CO_2$ in air and tighten the caps.
7. Incubate the cultures horizontally at 33°C.
8. Feed the cultures weekly:
   (a) Agitate the flasks gently to suspend the loosely adherent cells.
   (b) Remove 5 mL of growth medium, including the suspension cells; take care not to touch the layer of adherent cells with the pipette.
   (c) Add 5 mL of fresh growth medium to each flask; to avoid damage, do not dispense the medium directly onto the adherent layer.
   (d) Gas the cultures and replace them in the incubator.

*Analysis.* Cells harvested during feeding can be investigated by a range of methods, including morphology, CFC assays (*see* Protocol 23.11), and the in vivo CFU-S assay for stem cells [Till & McCulloch, 1961].

*Variations.* Mouse erythroid [Dexter et al., 1981], B-lymphoid [Whitlock et al., 1984], and human long-term cultures [Gartner & Kaplan, 1980; Coutinho et al., 1992] have been grown.

### 23.3.7 Long-Term Culture of Human Primitive Hemopoietic Cells

Primitive hemopoietic cells, with proliferative potential, can be maintained in culture for extended periods of time, typically several months. These culture conditions have been called long-term bone marrow culture (LTBMC) [Coulombel et al., 1983]. Briefly, LTBMC requires the formation of a supportive stromal layer that supplies the necessary microenvironment to allow the primitive haemopoietic cells to proliferate over time (*see also* Section 23.3.6). An application of LTBMC is an assay that measures the number of long-term culture initiating cells (LTC-IC) [Sutherland et al., 1991; Hogge et al., 1996]. This introduction and Protocol 23.10 have been contributed by A. Hamilton, Paul O'Gorman Leukaemia Research Centre, Section of Experimental Haematology, Faculty of Medicine, University of Glasgow, Gartnavel General Hospital, 1053 Great Western Road, Glasgow, G12 0YN. Scotland, UK.

---

**PROTOCOL 23.10. HUMAN LONG-TERM CULTURE-INITIATING CELL (LTC-IC) ASSAY**

**Outline**
The cells of interest are overlaid on pre-established, irradiated stromal feeder layers. After 5 weeks of culture, the contents of each plate are set up in a committed progenitor assay for a further 2 weeks. The number of colonies is then counted and the frequency of LTC-IC is determined.

**Materials**

*Sterile:*

❑ Sl/Sl fibroblasts, (ATCC, CRL-2453)
❑ M210-B4 fibroblasts, (ATCC, CRL-1972)
❑ DMEM (Sigma-Aldrich, D-6546)
❑ Iscove's modification of Dulbecco's medium (MDM), (Sigma-Aldrich, I-7633)
❑ RPMI 1640, (Sigma-Aldrich, R-0883)
❑ L-glutamine, 200 mM, (Sigma-Aldrich, G-7513)
❑ Penicillin, 100 U/mL/Streptomycin, 100 μg/mL, (Sigma-Aldrich, P-4458)

❑ Fetal bovine serum, (Invitrogen, 16050-098)
❑ RPMI10FB: RPMI 1640 with 10% fetal bovine serum
❑ IMDM2FB: Iscove's MDM with 2% fetal bovine serum
❑ Hygromycin B, (Sigma-Aldrich, H-0654)
❑ G418 disulphate salt solution, (Sigma-Aldrich, G8168)
❑ Hydrocortisone 21-hemisuccinate, (Stem Cell Technologies, 07904)
❑ MyeloCult long term culture medium (MC; Stem Cell Technologies, H5100)
❑ MethoCult methylcellulose medium, with cytokines, (Stem Cell Technologies, H4034)
❑ Sterile water for irrigation, (Baxter, UKF7114)
❑ Hanks' Balanced Salts Solution, $Ca^{2+}$, $Mg^{2+}$-free, (HBSS-CMF Sigma-Aldrich, H9394)
❑ Trypsin, 0.25%, EDTA, 0.54 mM, (Sigma-Aldrich, T-4049)
❑ Tissue culture flasks, 25-$cm^2$, 75-$cm^2$, with vented caps
❑ Syringes, 1 mL and 20 mL
❑ Syringe-tip filter units (0.22 μm)
❑ Needles, blunt-end, 16 gauge (Stem Cell Technologies)
❑ Centrifuge tubes (15 mL)
❑ Bijou containers
❑ Multiwell plates, 6-well, 3.5 cm, collagen type I coated (Sterilin/Iwaki)
❑ Suspension dishes: Petri dishes, sterile but not treated for cell attachment, 3.5 cm (Thermo—Nunc)
❑ Nunc Petri dishes, 10 cm

*Nonsterile:*

❑ Hemocytometer
❑ Centrifuge

**Procedure**

Maintenance of M2-10B4 fibroblast cell line

M2-10B4 cells are genetically modified to express the growth factors granulocyte colony stimulating factor (G-CSF) and interleukin-3 (IL-3), and are maintained in RPMI 1640 containing 10% fetal bovine serum (RPMI10FB) in tissue culture flasks (T 25 $cm^2$ or T 75 $cm^2$). Since M210-B4 producers grow as a monolayer, it is useful to express the initial concentration per surface area, rather than per mL. Cultures should be initiated at a concentration of $1 \times 10^4$/ $cm^2$. Therefore a 75 $cm^2$ tissue culture flask containing 10 mL of culture medium, should receive $0.75 \times 10^6$ cells.

Passage of M2-10B4 cells

1. Remove culture medium.

2. Add 2 mL of HBSS-CMF, rotate flask carefully and discard medium.
3. Add 2 mL of Trypsin-EDTA per 25 cm² flask.
4. Incubate 2 to 10 min at 37°C, or until adherent cells start to detach from the surface of the tissue culture flask.
5. Add 0.2 mL FBS to neutralize trypsin and mix by pipette to disperse cells.
6. Transfer cells to a 15-mL sterile tube, and fill with RPMI/10% FBS.
7. Centrifuge for 7 to 10 min at 300 g.
8. Wash cells once with RPMI10FB.
9. Transfer 1/50 to 1/100 of the volume to a new 25-cm² flask containing 8 mL of RPMI10FB.
10. Maintain the cultures at 37°C under 5% $CO_2$ in a humidified atmosphere.
11. To minimize the proliferation of untransduced wild type cells, the cultures should be fed on alternate weeks with medium supplemented with hygromycin B (final concentration: 62.5 μg/mL) and G418 (final concentration: 400 μg/mL). Filter sterilize the supplemented medium using a 0.22-μm syringe filter.
12. The cells should be passaged once they have reached confluency (i.e., every 7 to 10 days).

## Maintenance of Sl/Sl fibroblast cell line

Sl/Sl fibroblasts are cells originally established from Sl/Sl mouse embryos and were genetically engineered to express the growth factors, IL-3 and stem cell factor (SCF). Maintain the cell line in DMEM containing 10% fetal bovine serum (DMEM10FB) in 25-cm² or 75-cm² flasks. They are maintained in exactly the same way as M2-10B4 cells.

## Preparation of long-term culture medium

### Stock solution of hydrocortisone:

1. Thaw MyeloCult human myeloid long-term culture medium (MC) in a 37°C waterbath or overnight with refrigeration (4°C). Mix well. Store the medium for up to one month at 4°C.
2. Prepare hydrocortisone by dissolving hydrocortisone sodium hemisuccinate powder in MC to a final concentration of $1 \times 10^{-3}$ M.
3. Prepare a working stock by diluting $1 \times 10^{-3}$ M solution 1:10 to yield a final concentration of $1 \times 10^{-4}$ M.
4. Filter sterilize the hydrocortisone solution using a 0.22-μm syringe filter. Hydrocortisone should be freshly prepared each week.

### Use of hydrocortisone in culture medium:

5. Add 1 mL of $1 \times 10^{-4}$ M stock solution of hydrocortisone to every 100 mL of MC, to achieve a final, working concentration of $1 \times 10^{-6}$ M.

6. The medium (MC/HC) should be used for a maximum of one month from the day of preparation.

### Preparation of irradiated stromal feeder layers

In the LTC-IC assay, the cells of interest are overlaid on pre-established, irradiated stromal layers in microtitration plates. The M210-B4 and Sl/Sl producers are genetically modified murine, fibroblastic cell lines that can be used to provide the stromal support necessary for the propagation of LTC-IC.

### Establishment of irradiated stroma:

1. Trypsinise M2-10B4 and Sl/Sl cells from flasks as described, and wash twice in the appropriate culture medium.
2. Following the final wash, resuspend cells in 1 to 2 mL of MC/HC and perform a nucleated cell count.
3. Irradiate the cells at 80 Gy, using an appropriate source.

*Note.* It may be necessary to confirm that this dose of irradiation allows the stromal cells to support LTC-IC but is sufficient to inhibit cell proliferation.

4. Prepare a volume of each lot of irradiated cells at a concentration of $1.5 \times 10^5$/mL in MC/HC
5. Depending on the number of wells required, mix together equal volumes of both lots of irradiated cells, according to the example below:

*Example:*

(a) Calculate the number of wells required (e.g., six 6-well plates, i.e., 24 wells in total, using the two center wells for water only; *see* note in step below).
(b) Each well should receive 2 mL of cell suspension at $1.5 \times 10^5$/mL.
(c) The total volume of cell suspension would be $24 \times 2 = 48$ mL.

*Note.* Always prepare more than the required volume (in this case ~50 mL).

(d) For this example, the stock suspension of M2-10B4 cells would be irradiated at a concentration of $7.0 \times 10^6$/ mL.
(e) The following formula should be used to calculate the volume of stock suspension:

$$\frac{\begin{array}{c}\text{Required volume of cells (mL)} \\ \times \text{ Required concentration of cells} \\ (\times 10^6/\text{mL})\end{array}}{\text{Concentration of stock cells}(\times 10^6/\text{mL})}$$

For example:

$$\frac{25 \times 0.15}{7.0} = \frac{3.75}{7.0} = 0.54 \text{ mL}$$

Thus 0.54 mL of stock suspension of M2-10B4 cells should be added to 25 mL of MC/HC. Similarly, if the Sl/Sl cells were irradiated at a concentration of $5.5 \times 10^6$/mL, then the cells would be prepared as follows:

$$\frac{25 \times 0.15}{5.5} = \frac{3.75}{5.5} = 0.68 \text{ mL}$$

Thus 0.68 mL of stock suspension of Sl/Sl cells should be added to 25 mL of MC/HC.

6. Combine both of the volumes of stromal cells to give the mixed feeder cell suspension.
7. Plate 2 mL of the mixed feeder cells in MC/HC in collagen-coated tissue culture dishes. The collagen-coating on the tissue culture dish promotes the adherence of the stromal cell lines.

*Note.* Always add cells to the outer 4 wells of a 6-well plate, with 4 mL $dH_2O$ added to each of the middle two wells to aid humidification.

8. Incubate the cultures at 37°C in a humidified incubator (>95%) with 5% $CO_2$ in air; can be used for up to 10 days.
9. It is recommended that the cultures be incubated for a minimum of 24 h prior to the addition of test cells.

### Test cell suspensions

A variety of hemopoietic cells can be used to measure LTC-IC in this assay, for example, peripheral blood stem cells, whole human bone marrow cells, and umbilical cord blood. In addition purified populations of CD34+ cells can also be established in this system. These purified populations may be obtained by a number of technologies, including Isolex 300i (Baxter), CLINIMACS[plus−] (Miltenyi; *see* Fig. 13.7*b, c*), and the StemSep system (Stem Cell Technologies).

### Quantitation of LTC-IC using bulk cultures

#### Initiation of cultures:

1. Set up each lot of test cells in a minimum of 1 duplicate pair of irradiated mixed feeder stromal layers.
2. Carefully draw up the medium of each of the wells (*see* step 7 under Establishment of irradiated stroma, above), into a 5-mL disposable pipette,

and then add back a volume of 1 mL. To avoid disturbing the adherent layer, place the tip of the pipette against the side of the culture dish and remove/add the culture supernate very slowly.
3. Make up the deficit by adding 1 mL of fresh MC/HC. Again, add carefully and slowly against the side of the culture dish to avoid disturbing the adherent layer.
4. Prepare the bone marrow aspirate, peripheral blood or whole blood sample, using either Ficoll (*see* Protocol 27.1) or $NH_4Cl$ to isolate the mononuclear cells, or magnetic sorting (*see* Protocol 14.2; Fig. 14.7) to isolate the CD34+ cells.
5. Resuspend test cells (Table 23.2) in the required volume of MC/HC and add carefully to the dishes containing feeder cells. Routinely, a cell inoculum equivalent to 5000 CD34+ cells should be added. However, the use of three or four different initial test cell concentrations is recommended.

*Example:*

> Bone marrow cell sample:
> Cell count:          $87.4 \times 10^6$/ mL
> % CD34+ cells:     0.36
> Concentration of CD34+ cells:
>                $0.31 \times 10^6$ CD34+ cells/mL
> Volume containing 5000 CD34+ cells:
> $$\frac{5000}{0.31 \times 10^6} = 15.9 \text{ μL}$$

Thus, to each well of irradiated feeders, add 15.9 μL of cell suspension.

6. Once the test cells have been added, place the culture dishes within a plastic box with a lid. Return the cultures to the incubator and maintain at 37°C/5% $CO_2$ in a humidified atmosphere.

### Maintenance of cultures

7. At weekly intervals, remove one-half (approximately 1 mL) of the culture supernate, as described above.
8. Carefully replace this with an equal volume (1 mL) of fresh MC/HC, without disturbing the adherent layer, and gently swirl the dish to mix the contents.
9. When aspirating the culture supernate from multiple dishes, be sure to change the pipette after every aspiration to minimize any cross-contamination.
10. Examine the cultures periodically, using an inverted microscope to assess hematopoiesis and to detect any contamination.

11. At this weekly maintenance, always check the water levels in the middle wells of the plate, to ensure the humidity is kept high.

12. Continue this feeding regimen at weekly intervals for 5 weeks.

### Harvest of LTC-IC

After 5 weeks of incubation, LTC-IC cultures are harvested (both adherent and non-adherent cells) and the clonogenic progenitors are assayed in methylcellulose medium.

1. For each LTC-IC culture well, pipette the culture supernate (containing non-adherent cells) into a sterile 15-mL centrifuge tube (harvest tube). This should be a volume of approximately 2 mL.

2. Carefully rinse the adherent layer twice with 2 mL of HBSS-CMF (to remove any remaining MC/HC containing serum and loosely attached cells) and swirl gently.

3. Add 1 mL of trypsin-EDTA and place at 37°C in humidified incubator for a maximum of 10 min. At intervals, swirl the culture gently and examine for evidence of detachment of the adherent layer.

4. Detachment can be facilitated by repeatedly pipetting trypsin-EDTA solution over the surface of the well to ensure that all adherent cells are detached and to make a single cell suspension. Transfer this suspension to the harvest tube.

5. Immediately add 2 mL of IMDM2FB to the LTC-IC well and rinse by swirling gently. Transfer this wash volume to the harvest tube.

6. Add 2 mL of HBSS-CMF to the LTC-IC well and rinse by swirling gently. Transfer this wash volume to the harvest tube. The total volume in the harvest tube should now be about 9 mL.

7. Centrifuge for 10 min at 300 $g$.

8. Remove the supernate and gently resuspend the cells in the residual supernate.

9. Transfer the cell suspension to a bijou container and record the volume (normally < 200 µL).

10. Remove a sample and perform a nucleated cell count.

11. Measure the number of colony-forming cells by setting up the committed progenitor assay for each cell suspension.

### Committed progenitor assay

1. If sufficient cells have been harvested from the LTC-IC assay, then set up the committed progenitor assay at a cell concentration of $5 \times 10^4$ cells per dish (in duplicate).

2. Adjust cells to $5 \times 10^5$ cells per mL in IMDM2FB, and add 0.3 mL of cells to 3 mL of Methocult™ methylcellulose medium and vortex, until the contents are thoroughly dispersed.

3. Let stand for 2 to 3 min to allow bubbles to rise to the top.

4. Using a 1-mL syringe and blunt-end needle, take up a 1.1 mL aliquot from the culture tube and express it slowly into a 3.5-cm suspension culture dish. Repeat for a second 1.1 mL aliquot.

5. Rotate carefully to spread methylcellulose medium over the surface of the dish and ensure that the surface of the plates are completely and uniformly covered.

6. Place two dishes within a 10-cm Petri dish containing a third uncovered 35-mm dish with 6 mL of sterile water (to aid humidity and prevent the plates from drying out).

7. Incubate methylcellulose cultures for at least 14 days at 37°C in humidified incubator (>95%) with 5% $CO_2$ in air.

8. After the incubation period, score and record the total number of colonies per dish using an inverted microscope.

### Calculation of LTC-IC frequency

The number of LTC-IC present in the initial test cell suspension by dividing the total number of colony-forming cells (CFC) detected in the culture by the average number of clonogenic progenitors per LTC-IC for the standard conditions used. Alternatively values can be expressed as LTC-IC derived CFC per number of test cells.

*Example:*

A culture dish containing irradiated stromal feeder cells was initiated with $2 \times 10^6$ normal bone marrow cells, purified for the mononuclear cell fraction, and incubated for 5 weeks with weekly half medium exchanges.

Number of test cells set up in each dish: $2 \times 10^6$ bone marrow cells.

Total cells harvested after 5 weeks of culture: $1 \times 10^6$

Number of cells plated per committed progenitor assay: $5 \times 10^4$

Average number of CFC obtained per progenitor assay: 70

Average CFC output per LTC-IC (Hogge et al., 1996): 8

*Calculation:*

Total CFC: $5 \times 10^4$ cells yielded 70 CFC.

The entire LTC-IC culture yielded $1 \times 10^6$ cells; therefore it would have contained 1400 CFC

If it is presumed that 1 LTCIC produces (on average) 8 CFC, then the total LTC-IC: **175 LTC-IC derived CFC** (per $2 \times 10^6$ bone marrow cells).

**TABLE 23.2.** Recommended Concentration of Cells for Initiation of LTC-IC on Stromal Feeder Layers

| Cell type | Cells per well of 6-well plate |
|---|---|
| Bone marrow - NH$_4$Cl preparation | $2 \times 10^6$ |
| Bone marrow - Ficoll preparation | $1 \times 10^6$ |
| Bone marrow - CD34$^+$ selected | 3000–5000 |
| Peripheral blood - Ficoll preparation | $8 \times 10^6$ |
| Cord blood - Ficoll preparation | $5 \times 10^5$ |

## 23.3.8 Hematopoietic Colony-Forming Assays

Hematopoietic progenitor cells may be cloned in suspension in semisolid media in the presence of the appropriate growth factor(s) [Heyworth & Spooncer, 1992]. Pure or mixed colonies will be obtained, depending on the potency of the stem cells that are isolated. Assays for the detection of granulocyte and macrophage colony-forming cells (GM-CFC), erythroid burst-forming units (BFU-E), mixed colony-forming cells (CFC-mix), and granulocyte, erythrocyte, macrophage, and megakaryocyte colony-forming cells (CFC-GEMM) are described by Testa and Molineux [1993], and their place in routine hematopoietic cell culture technology is already well established [Metcalf, 1990]. The following introduction and Protocol 23.7 have been abridged from Freshney et al., [1994].

The efficiency of growth of colonies in these assays is increased by using Methocel instead of agar as the semisolid phase. This practice makes for a tighter colony that is easier to evaluate and count. Because Methocel is a high-viscosity liquid, and not a gel like agar, cells will sediment through it, albeit slowly, and plate out on the plastic base of the dish. This route places them all in one focal plane for subsequent observation. The colonies will form if grown in an atmosphere of 5% CO$_2$ in air, but this may be at the expense of adequate hematopoiesis, and ideally the gas phase should be 10% CO$_2$ and 5% O$_2$ in air [Bradley et al., 1978b]. If an incubator with this gas mixture is not available, a cylinder of mixed gases can be rented. Place the dishes in a plastic box with a lid with a hole in it. Seal the box with plastic tape, and gas via the hole before sealing it. A dish of water in the box will keep the atmosphere humid.

---

**PROTOCOL 23.11. HEMATOPOIETIC COLONY-FORMING ASSAYS**

**Outline**

Suspend bone marrow cells in agar or Methocel, and seed the cells into dishes with the appropriate growth factors.

---

**Materials**

*Sterile or aseptically prepared:*

❑ Bone marrow cells (*see* Protocol 23.9, steps–4 for preparation). Count the nucleated cells in a hemocytometer after staining them with methylene blue, or lyse the cells with Zapoglobin (Beckman Coulter) and count the nuclei on an electronic cell counter. Each femur will yield 1.0 to $1.5 \times 10^7$ cells.

❑ Methylcellulose, 4000 cP (Fluka)

*Preparation of methylcellulose:*

(i) Weigh out 7.2 g of Methocel, and add it to a 500-mL bottle containing a large magnetic stirrer bar.

(ii) Sterilize the Methocel by autoclaving.

(iii) Add 400 mL of sterile UPW heated to 90°C to wet the Methocel.

(iv) Stir the mixture at 4°C overnight to dissolve the Methocel. The solution is now $2 \times$ Methocel. It is more accurate to use a syringe (without a needle) than a pipette to dispense Methocel.

Methocel can be purchased already prepared as MethoCult (Stem Cell Technologies).

❑ Noble agar (Difco, BD Biosciences)

❑ Alpha MEM stock (Invitrogen)

*Preparation of Alpha medium stock solution:*

(i) Alpha medium, powder (GIBCO 10-L pack size)

(ii) MEM vitamin stock, 100×, 100 mL

(iii) Gentamycin sulfate, 200 mg. Stir the medium on a heated stirrer until it dissolves, and make to 3 L with UPW. (Do not allow the temperature to rise above 37°C.) Before final filtration through a 0.22-μm filter, it is advantageous to prefilter the medium through stacked filters of pore sizes 5, 1.2, 0.8, and 0.45 μm. Dispense the medium into 21-mL aliquots, and store it in premeasured volumes at −20°C.

*Preparation of Alpha medium, 2×:*

(i) Alpha medium stock solution, 21 mL

(ii) Fetal bovine serum (FBS), 25 mL

(iii) Glutamine (200 mM) 1 mL

(iv) NaHCO$_3$, 7.5%, 3 mL

(v) Mix the ingredients in a sterile bottle and equilibrate to 37°C.

❑ FBS

❑ Growth factors (Table 23.2), either recombinant (R&D) or from a conditioned medium

❑ Wehi-cell-conditioned medium (Wehi-CM): Wehi 3B is a mouse myelomonocytic cell line that when cultured, releases IL-3 (multi-CSF) into the medium [Bazill et al., 1983] (*see* Protocol 13.2 for preparation of conditioned medium).

❑ BSA, 10% in D-PBSA
❑ Petri dishes, 3 cm

*Nonsterile:*

❑ With agar, use two water baths, one at 37°C and the other at 55°C.
❑ Incubator: gas phase, 10% $CO_2$, 5% $O_2$, 85% $N_2$

**Procedure**

*A. BFU-E, CFC-mix, and CFU-GEMM:*

1. Mix an equal volume of alpha 2× medium to which 1% BSA has been added, with 2× Methocel to make the required amount of medium for the experiment. Keep the mixture cold; Methocel is more liquid when it is cold.
2. Set up cultures in triplicate, but make enough mix for 4 dishes, as Methocel clings to the side of tubes and some is always lost.
3. Add the required concentration of growth factors to the tube (Table 23.3).
4. Add $5 \times 10^4$ cells.
5. Increase the total volume to 4.4 mL with the addition of 1× medium, and mix the tubes on a vortex mixer.
6. Using a syringe, plate out 1 mL of medium into 3-cm non–tissue-culture-grade dishes.
7. Incubate the culture at 37°C in a humid atmosphere of 10% $CO_2$ and 5% $O_2$ in air for 8 to 15 days.

Identification of the colonies

BFU-E colonies can be either single colonies composed of very small cells or multicentric colonies (bursts), each with tightly packed very small pink or red cells. The incidence of these colonies is 40 to $80/10^5$ bone marrow cells.

CFC-mix colonies can be single, compact colonies, usually with a halo of cells of widely varying size. They may be multicentric, but the cell population is obviously heterogeneous. The incidence of colonies in this assay is 100 to $180/10^5$ bone marrow cells, of which only about 10% will be mixed colonies containing erythroid cells. If the erythroid cells are not red, it can be very difficult for the inexperienced eye to identify the colonies accurately. The colonies should be photographed and then picked out, after which cytocentrifuge preparations should be made that can be fixed and stained with 10% Giemsa, and help sought with identification of the cells [Heyworth & Spooncer, 1992].

*B. GM-CFC:*

1. Use 3 cm of non–tissue-culture-grade Petri dishes, lay out the dishes, and label them.
2. Make a 0.3% agar medium (*see* Protocol 13.4) and keep it at 37°C.
3. Prepare bone marrow cells (*see* Protocol 23.9). Count the nucleated cells after staining the cells with methylene blue, or lyse the cells with Zapoglobin (Beckman Coulter) and count the nuclei with an electronic cell counter. Each femur will yield 1.0 to $1.5 \times 10^7$ nucleated cells.
4. Add 0.1 ng/mL of rMurGM-CSF to each dish.
5. Add 1 mL of agar medium containing $7.5 \times 10^4$ cells and swirl gently to mix the cells and agar. Allow the agar to set at room temperature.
6. Alternatively, 0.8% Methocel medium can be used. This generally gives a tighter colony, which is easier to count.
7. Place the dishes in a clean plastic box.
8. Incubate the dishes for 6 days in a humidified incubator at 37°C in an atmosphere of 5% $CO_2$ in air.
9. Using an inverted microscope, count the colonies that contain more than 50 cells, and express the number as colonies/$10^5$ cells seeded.
10. The incidence of GM-CFC in normal bone marrow is 100 to $120/10^5$ cells.

**TABLE 23.3.** Addition of Cells and Growth Factors for Colony-Forming Assays

| Assay | IL-6 | Growth factors/mL | | | | |
| | | rMurGM-CSF[a] | rIL-3[b] | Epo | Cells | Incidence/$10^5$ |
|---|---|---|---|---|---|---|
| BFU-E | | 0.1 ng | 1 ng | 2 U | $5 \times 10^4$ | 40–80 |
| CFC-mix | | | 1 ng | 2 U | $5 \times 10^4$ | 100–180 |
| CFU-GEMM | 100 ng | | 1 ng | 2 U | $5 \times 10^4$ | 92–106 |

[a]10% AF1-19T CM can be substituted for GM-CSF Pragnell et al., 1988.
[b]10% Wehi-CM can be substituted for rIL-3 Bazill et al., 1983.

# CHAPTER 24

# Culture of Tumor Cells

## 24.1 PROBLEMS OF TUMOR CELL CULTURE

The culture of cells from tumors, particularly spontaneous human tumors, presents problems similar to those of the culture of specialized cells from normal tissue. The tumor cells must be separated from normal connective tissue cells, preferably by provision of a selective medium that will support tumor cells but not normal cells. Although the development of selective media for normal cells has advanced considerably (*see* Section 9.2.2; Chapter 22), progress in tumor culture has been limited by variation both among and within samples of tumor tissue, even from the same tumor type. It is often surprising to find that tumors that grow in vivo, largely as a result of their apparent autonomy from normal regulatory controls, fail to grow in vitro.

There are many possible reasons for the failure of some tumor cultures to survive. Their nutritional requirements may be different from those of the equivalent normal cells, or perhaps attempts to remove stroma may actually deprive the tumor cells of a matrix, nutrients, or signals necessary for survival. Alternatively, dilution of tumor cells to provide a sufficient amount of nutrients per cell may also dilute out autocrine growth factors produced by the cells. Strictly speaking, truly autocrine factors should be independent of dilution if they are secreted onto the surface of the cell and are active on the same cell, but it is possible that some so-called autocrine growth factors are in fact homocrine (*see* Section 2.5)—namely they act on adjacent similar cells and not only on the cell releasing them. Hence a closely interacting population is required. Interaction with certain types of stromal cells may provide paracrine interaction if they are able to make the requisite growth factors, either spontaneously or in response to the tumor cells, necessary for their survival.

It may be incorrect to assume that the growth factor dependence of a tumor cell is similar to that of the normal cells of the tissue from which it was derived. Tumor cells may produce endogenous autocrine growth factors, such as TGF-$\alpha$, and the provision of exogenous growth factors, such as EGF, may compete for the same receptor. Furthermore the response of a tumor cell to a growth factor, or hormone, will depend on what other growth factors are present, some of which may be tumor cell derived, and on the status of the cell. A normal cell, capable of expressing growth suppressor and senescence genes, may respond differently from a cell in which one or more of these genes are inactive or mutated, and in which antagonistic, growth-promoting oncogenes are overexpressed.

So there are many possible reasons for a tumor cell population responding differently to the nutritional and mitogenic environment optimized for normal cells of the same lineage. To confirm this difference, more information is required on the nutritional requirements of tumor cells, but given the heterogeneity of tumors (*see* Fig. 17.1; *see also* Plate 7), the task is a daunting one. As the potential therapeutic benefit to be derived from knowledge of the nutritional requirements of individual tumor cell lines is not likely to be great, greater emphasis has been placed on the response of tumor cells to growth factors and the differences in signal transduction.

It is probable that the bulk of the cells in a tumor have a limited life span, due to genetic aberration, terminal differentiation, apoptosis, or natural senescence, and only a few cells, analogous to a stem cell population in normal tissue, have the potential for continuous survival. Dilution into culture may reduce the number of these cells, as well as their interaction with other cells, such that survival is impossible. Cells from multicellular animals, unlike prokaryotes, do not survive readily in isolation. Even a tumor is still a multicellular organ and may require continuing cell interaction for survival. The lethality of the tumor to the host lies in its uncontrolled infiltration and colonial growth, but the origin of the bulk of the cell population may reside in a relatively small population of transformed stem cells [Jones et al., 2004; Al-Hajj et al., 2004]. This pool of stem cells may be so small that its dilution on explantation deprives it of some of its prerequisites for survival, particularly paracrine growth factors from stromal elements and homocrine interaction with other tumor subclones.

There has been continuing interest in the stem cell origin of cancer for many years, and it has often been proposed that the stem cell population is the main repopulating fraction after therapy and hence should be the main target for chemotherapy [Hamburger & Salmon, 1977; Vermeulen et al., 2008; Dick, 2009] particularly as one of the class of genes associated with drug resistance, the ABC transporter, is expressed in stem cells [Robey et al., 2009]. Now that stem cell markers have been better defined, evidence is accumulating for malignant stem cells in breast [Pontier & Muller, 2009], prostate [Hurt et al., 2007; Kasper, 2008], head and neck [Graziano, et al., 2008], and brain [Vescovi, et al., 2006].

In sum, the goal is either to create the correct, defined nutritional and hormonal environment or, failing that, to provide a sustaining environment, as yet undefined but nevertheless able to permit the survival of an appropriate or representative population. There has been a continuing trend to use serum and feeder layers in order to get tumor cells to grow, and only a few tumors have responded to serum-free culture. As many transformed cells are not inhibited by TGF-β, there has not been the same need to eliminate serum, other than to repress fibroblastic growth, which remains a major problem. The adaptation of medium designed for equivalent normal cells is still the most logical approach to obtaining cell lines from tumors, even if supplemented with minimal amounts of serum or conditioned medium from other cells [Dairkee et al., 1995].

## 24.2 SAMPLING

### 24.2.1 Selection of Representative Cells

In addition to preventing the overgrowth of connective tissue or vascular cells, both of which are stimulated to invade and proliferate by many tumors, tumor cell culture requires the separation of the transformed cells from the normal equivalent tissue cells, which may have similar characteristics. Furthermore, although any section of gastric epithelium may be regarded as representative of that particular zone of the gastric mucosa, tumor tissue, dependent as it is on genetic variation and natural selection for its development, is usually heterogeneous and composed of a series of often diverse subclones displaying considerable phenotypic diversity. Ensuring that cultures derived from this heterogeneous population are representative is difficult, and can never be guaranteed unless the whole tumor is used and survival is 100%. As these conditions are practically impossible to achieve, the average tumor culture is a compromise. Assuming that representative subpopulations have been retained and are able to interact, the corporate identity may be similar to the original tumor. Alternatively, if the stem cell compartment is the main target for therapy, this may be the key representative target population and may even be more homogeneous if the heterogeneity arises in those cells that are several cell generations ways from the stem cells.

The problem of selectivity is accentuated when sampling is carried out from secondary metastases, which often grow better, but may not be typical either of the primary tumor or of all other metastases. It is, however, interesting to speculate on the analogy between metastatic occurrences and ectopic development of stem cells (*see* Section 23.3.3).

In view of these practically overwhelming problems facing tumor culture, it is almost surprising that the field has produced any valid data whatsoever. In fact it has, and this may result from (1) the aforementioned autonomy of tumor populations, which might have allowed the proliferation of tumor cells under conditions in which normal cells would not multiply; (2) the increased size of the proliferative pool in tumors, which is larger than that of most normal tissues; (3) the ability of tumor cells to give rise to tumors as xenografts in immune-deprived mice and increased success in deriving culture from the xenografts; and (4) the propensity of malignantly transformed cells to give rise to continuous immortalized cell lines more frequently than normal cells. This last feature, more than any other, has allowed extensive research to be carried out on tumor cell populations, even on apparently normal differentiation processes, despite the uncertainty of their relationship to the tumor from which they were derived.

### 24.2.2 Preservation of Tissue by Freezing

It is often difficult to take advantage of a large biopsy and utilize all of the valuable material that it provides. It is possible in these cases to preserve the tissue by freezing.

## PROTOCOL 24.1. FREEZING BIOPSIES

### Outline
Chop the tumor, expose the pieces to DMSO, and freeze aliquots in liquid nitrogen.

### Materials

*Sterile:*

- Biopsy
- Plastic ampoules, 1.2 mL (Thermo—Nunc)
- DBSS (*see* Appendix I)
- Collection medium (*see* Appendix I)
- DMSO (self-sterilizing if placed in a sterile container)
- Instruments (scalpels, forceps, dishes, etc., as for primary culture)

### Procedure

1. After removing necrotic, fatty, and fibrous tissue, chop the tumor into about 3- to 4-mm pieces, and wash the pieces in DBSS, as for primary culture.
2. Place four or five pieces in each ampoule.
3. Add 1 mL of growth medium containing 10% DMSO to the pieces, and leave them for 30 min at room temperature.
4. Freeze the ampoules at 1°C/min (*see* Protocol 19.1), and transfer them to a liquid nitrogen freezer (to avoid explosion risk, do not submerge in liquid nitrogen).
5. To thaw an ampoule, place it in 37°C water (with appropriate precautions; *see* Protocol 19.2).
6. Swab the ampoule thoroughly in alcohol, open it, allow the pieces to settle, and remove half of the medium.
7. Replace the medium slowly with fresh, DMSO-free medium. Mix by gentle shaking, and allow to stand for 5 min.
8. Gradually replace all of the medium with DMSO-free medium, transfer the pieces to a Petri dish, and proceed as for regular primary culture, but allowing twice as much material per flask.

## 24.3 DISAGGREGATION

Some tumors, such as human ovarian carcinoma, some gliomas, and many transplantable rodent tumors, are readily disaggregated by purely mechanical means, such as pipetting and sieving (*see* Section 11.3.8), which may also help minimize stromal contamination, as stromal cells are often more tightly locked in fibrous connective tissue. Many of the common human carcinomas, however, are hard, or scirrhous, and the tumor cells are contained within large amounts of fibrous stroma, making mechanical disaggregation difficult, although scraping the cut surface of scirrhous tumors has been used successfully in the so-called spillage technique [Lasfargues, 1973; Oie et al., 1996], to release tumor cells from the fibrous stroma.

Enzymatic digestion has proved to be preferable to mechanical disaggregation in most cases. Although trypsin has often been used for this purpose, its effectiveness against fibrous connective tissue is limited, and it can reduce the seeding efficiency of the tumor cells [Lounis et al., 1994]; crude collagenase has been found to be more effective with several different types of tumor [Dairkee et al., 1997]. Enzymatic disaggregation also releases many stromal cells, requiring selective culture techniques for their elimination (*see* Section 24.5; Fig. 14.6d; Plate 23). Collagenase exposure may be carried out over several hours, or even days, in complete growth medium (*see* Protocol 11.8).

Extensive necrosis is also a problem of tumor tissue that is infrequently encountered with normal tissue. Usually the attachment of viable cells allows necrotic material to be removed on subsequent feeding, but if the amount of necrotic material is large and not easily removed at dissection, it may be advisable to use a Ficoll-metrizoate separation (*see* Protocol 11.10) to remove necrotic cells.

## 24.4 PRIMARY CULTURE

Some cells—such as macrophages—attach to the substrate during collagenase digestion but may be removed by transferring the disaggregated cell suspension to a fresh flask when the collagenase is removed. The adherent cells may be retained and cultured separately or irradiated or treated with mitomycin C and used as a feeder layer (*see* Section 13.2.3; Plates 6d, 11a, 15b; Protocols 22.1, 22.4). The reseeded cells will contain many stromal cells (principally fibroblasts and endothelium), some of which can be removed by a second transfer to a fresh vessel in 2 to 4 h, as tumor cells, particularly clusters of malignant epithelium, often take longer to attach. This method of removal by serial transfer is generally only partially successful, however, and it will usually require selective culture conditions for the complete removal of the stromal cells. It may still be advantageous to retain the stromal cell cultures to use as feeder layers and a source of DNA for profiling.

Physical separation techniques have also been used to remove stromal contaminants [Csoka et al., 1995; Oie et al.,

1996; *see also* Chapter 14], but in general, these methods are suitable only if the cells are to be used immediately, as stromal overgrowth usually follows in the absence of selective conditions. However, magnetic sorting (*see* Section 14.3.2; Plate 23) with antifibroblast microbeads (*see* Fig. 14.6*d*; Miltenyi Biotec) appears to be very effective.

Cloning as a method of purification has limitations, as tumor cells in primary culture often have poor plating efficiencies (<0.1%). Suspension cloning has been proposed in the past as a means of isolating not only tumor cells but tumor stem cells [Hamburger & Salmon, 1977]. However, by the time a clone has grown to sufficient numbers to be of potential analytical value, it could have changed considerably, and it could even have become heterogeneous, due to genetic instability. Cloned isolates from a tumor should be studied collectively, and even pooled, for a meaningful interpretation.

There has also been some difficulty in propagating cell lines from primary clones, particularly from clones isolated by the suspension method. It may be that although these cells are clonogenic, few of them really are stem cells, or, if they are, they mature spontaneously due to the suspension mode of growth and lose their regenerative capacity. Nevertheless, cell strains cloned directly from tumors would be valuable material for studying tumor clonal diversity and interaction, and they represent a key area of study for future investigation, particularly if selective conditions can be used to isolate tumor stem cells.

## 24.5   SELECTIVE CULTURE OF TUMOR CELLS

Three main approaches have been adopted to select tumor cells in primary culture: selective media (*see* Sections 9.2.2, 13.6), confluent feeder layers (*see* Section 24.5.2), and suspension cloning (*see* Sections 13.3, 13.8.5).

### 24.5.1   Selective Media

There are only a few media that have been developed as selective agents for tumor cells, because of their inherent problems of variability and heterogeneity. HITES [Carney et al., 1981; *see* Table 9.2] is one such medium and may owe its success to the production of peptide growth factors by small-cell lung cancer, for which the medium was developed. A proportion, but not all, of small-cell lung cancer biopsies will grow in pure HITES; others will survive with a low-serum supplement (e.g., 2.5%). HITES medium is modified RPMI 1640 with hydrocortisone, insulin, transferrin, estradiol, and selenium. Of these constituents, selenium, insulin, and transferrin are probably the most important and are found in many serum-free formulations (*see* Table 9.2). The NCI group also produced a selective medium for adenocarcinoma; reputedly suitable for lung, colon, and, potentially, many other carcinomas [Brower et al., 1986]. This medium is also based on RPMI 1640, supplemented with selenium, insulin, and transferrin, with the addition of hydrocortisone, EGF, triiodothyronine, BSA, and sodium pyruvate (*see* Table 9.2).

Other selective media have been used successfully with prostate [Uzgare et al., 2004], bladder [Messing et al., 1982], and mammary [Ethier et al., 1993] carcinoma.

Other types of selective media depend on the metabolic inhibition of fibroblastic growth and are not specifically optimized for any particular type of tumor (*see* Section 13.6). However, inhibitors have not been found to be generally effective, with the exception of the use of monoclonal antibodies against fibroblasts by Edwards et al. [1980] and Paraskeva et al. [1985]. These antibodies have proved useful in establishing cultures from laryngeal and colon cancer. Antibodies have also proved useful in either positively selecting epithelial cells or negatively sorting stromal cells from tumor cell suspensions by panning or magnetic sorting (*see* Sections 14.3, Fig. 14.6*d*; Plate 23).

### 24.5.2   Confluent Feeder Layers

The use of confluent feeder layers (see Figs. 13.5, 13.9; Plate 6*c, d*), has been applied successfully to many types of tumor. Confluent feeder layers of fetal human intestine, FHS74Int, have been used to grow epithelial cells from mammary carcinoma [Lan et al., 1981], with media conditioned by other cell lines, although later reports suggest that selective culture in MCDB 170 is a more reproducible approach [Hammond et al., 1984]. Feeder layers of mouse 3T3 or STO embryonic fibroblasts were used successfully with breast, colon, and basal cell carcinoma [Rheinwald & Beckett, 1981; Leake et al., 1987].

Feeder layer techniques rely on the prevention of fibroblastic overgrowth by a preformed monolayer of other contact-inhibited cells. They are not selective against normal epithelium, as normal epidermis and normal breast epithelium both form colonies on confluent feeder layers (Fig. 24.1). Results from glioma [MacDonald et al., 1985], however, suggest that selection against equivalent normal cells may be possible on a homologous feeder layer. Glioma grown on normal glial feeder layers should lose any normal glial contaminants. By the same argument, breast carcinoma seeded on confluent cultures of normal breast epithelium—such as from reduction mammoplasty (*see* Protocol 22.3)—could become free of any contaminating normal epithelium.

---

**PROTOCOL 24.2. GROWTH ON CONFLUENT FEEDER LAYERS**

*Outline*
Treat feeder cells in the mid-exponential phase with mitomycin C, and reseed the cells to give a confluent monolayer. Seed tumor cells, dissociated from the biopsy by collagenase digestion, or from a primary culture with trypsin, onto the confluent monolayer (*see* Fig. 24.1, 24.2). Colonies from epithelial tumors

may form in 3 weeks to 3 months. Fibrosarcoma and gliomas do not always form colonies, but may infiltrate the feeder layer and gradually overgrow [*see* Plates 1*d*, 15*b*].

### Materials

*Sterile:*

❑ Feeder cells (e.g., 3T3, STO, 10T1/2, or FHS74Int)
❑ Mitomycin C (Sigma), 1 mg/mL

*Note.* It is advisable to do a dose-response curve with mitomycin C when using feeder cells for the first time (*see* Protocol 21.3), to confirm the dose that allows the feeder layer to survive for 2 to 3 weeks but does not permit further replication in the feeder layer after about two doublings, at most; namely no resistant colonies form. This is usually 0.25 µg/mL for overnight exposure or 20 µg/mL for 1 h exposure.

❑ Growth medium
❑ Collagenase, 2000 U/mL, CLS grade (Worthington) or equivalent
❑ Trypsin, 0.25%, in PBSA
❑ Tumor biopsy or primary culture
❑ Forceps, fine curved
❑ Scalpels with #22 blades
❑ Petri dishes for dissection, as for primary culture

### Procedure

1. Grow up the feeder cells to 80% confluence in six 75-cm$^2$ flasks.
2. Add mitomycin C to give the appropriate final concentration, usually around 0.25 µg/mL.
3. Incubate the cells overnight (∼18 h) in mitomycin C.
4. Remove the medium with mitomycin C, and wash the monolayer with fresh medium.
5. Grow the cells for a further 24 to 48 h.
6. Trypsinize the cells, and reseed them in 25-cm$^2$ flasks at $5 \times 10^5$ cells/mL ($1 \times 10^5$ cells/cm$^2$). Incubate the cultures for 24 h.
7. If you are using biopsy material, the biopsy should be dissected and placed in collagenase (*see* Protocol 11.8) during step 2.
8. Remove the collagenase from the disaggregated tumor cells, either by repeated settling or centrifugation (*see* Protocol 11.8).
9. Resuspend the tumor cells and seed approximately 20 to 100 mg/flask, into two of the 25-cm$^2$ flasks, such that each flask holds 6 mL of suspension.
10. Remove 1 mL of the suspension from each flask, and add it to 4 mL of medium in each of two more flasks.

11. The third pair of flasks should be kept as controls to guard against feeder cells surviving the mitomycin C treatment.

If you are using a primary culture from a tumor biopsy, trypsinize or dissociate the cells in collagenase, 200 U/mL final (*see* Protocol 11.8), and seed onto the feeder layer at $1 \times 10^5$ cells/mL in two flasks and $1 \times 10^4$ cells/mL in two flasks. If the cells are from a glioma (Plate 15*b*) or fibrosarcoma (Plate 1*d*), colonies may not appear as the tumor cells migrate freely among the feeder cells, and the surviving tumor will be confirmed only by subculturing the cells without a feeder layer (by which time contaminating stromal cells should have been eliminated).

It is essential to confirm the species of origin of any cell line derived by this method, in order to guard against accidental contamination from resistant cells in the feeder layer. The species of origin can be confirmed by chromosomal analysis (*see* Protocol 15.7), isoenzyme electrophoresis (*see* Protocol 15.10), or polymorphism within the cytochrome oxidase gene [Cooper et al., 2007] if the feeder is of a different species from the primary culture. If feeder cells of the same species as the primary culture are used, it is necessary to do a DNA profile (*see* Protocol 15.9) of both the feeder layer cells and any culture that is generated and to compare the results with a portion of the biopsy or other tissue taken from the donor.

### 24.5.3 Suspension Cloning

The transformation of cells in vitro leads to an increase in their clonogenicity in agar (*see* Sections 13.3, 17.5.1); tumorigenicity has also been shown to correlate with cloning in Methocel [Freedman & Shin, 1974] (*see* Protocol 13.5). As cells may be cloned in suspension directly from disaggregated tumors [Hamburger & Salmon, 1977], or at least colonies may grow (they may not be clones) in preference to normal stromal cells, suspension cloning would seem to be a potentially selective technique. However, the colony-forming efficiency is often very low (often <0.1%), and it is not easy to propagate cells isolated from the colonies. Although this method has not generated cell lines, it has been used for drug screening with tumor biopsies (*see* Section 21.4.2). FACS or immunomagnetic positive sorting of clonogenic cells by expression of stem cell markers, such as Sca-1, Kit [Takahashi et al., 2004], and the ABC transporter [Zhou et al., 2001] (*see also* Table 23.1) and negative sorting for more differentiated lineage markers may allow enrichment of a putative stem cell pool, which may make a better target for drug screening and molecular drug targeting.

### 24.5.4 Xenografts

When cultures are derived from human tumors, the scarcity of material and the infrequency of rebiopsy make it difficult to

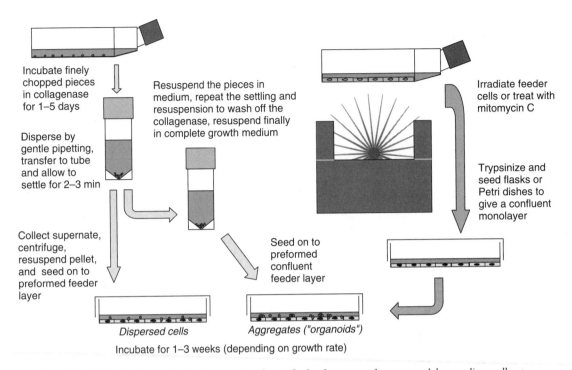

**Fig. 24.1. Confluent Feeder Layers.** Confluent feeder layers can be prepared by seeding cells at a high density or by allowing a normally contact inhibited cell to grow to confluence. Epithelial clusters from collagenase digestion form colonies when seeded onto confluent feeder layers, such as contact-inhibited fetal intestinal epithelium (FHS74Int; *see* Fig. 24.2; Plate 6c, d), or irradiated 3T3 or STO cells (*see* Section 13.8.4). Dispersed cells, although containing more stromal cells, can also form colonies on confluent feeder layers with significant restriction of stromal overgrowth. Selection is against stromal components, but not normal epithelium.

make several attempts to culture the same tumor. The growth of some tumors in immune-deprived animals [Rofstad, 1994] provides an alternative approach that makes much greater amounts of tumor available. It also favors selection and amplification of the tumor cell population as normally the human stromal component will not grow and is replaced by mouse stromal tissue. It has sometimes been found that cultures can be initiated more easily from xenografts than from the parent biopsy and tend not to be overgrown by stromal cells, but whether this is due to the availability of more tissue, enrichment of transformed cells, progression of the tumor, or modification of the tumor cells by the heterologous host (e.g., by murine retroviruses) is not clear.

Two main types of host are used: the genetically athymic nude mouse, which is T-cell deficient [Giovanella et al., 1974], and neonatally thymectomized animals that are subsequently irradiated and treated with cytosine arabinoside [Selby et al., 1980; Fergusson et al., 1980]. The first type of host is expensive to buy and difficult to rear, but maintains the tumor for longer. Thymectomized animals are more trouble to prepare, but cheaper and easier to provide in large numbers. They do, however, regain immune competence and ultimately reject the tumor after a few months. Take rates for tumors can be enhanced by using mice that are asplenic as well as athymic, genetically (e.g., *scid* mice) or by splenectomy,

or by sublethally irradiating nude mice. Implantation with fibroblasts or Matrigel has also been reported to improve tumor take [Topley et al., 1993].

If access to a nude mouse colony is available, or facilities exist for neonatal thymectomy and irradiation, xenografting should be considered as a first step in generating a culture. Although only a small proportion of tumors may take, the resulting tumor will probably be easier to culture, and repeated attempts at culture may be made with subsequent passage of the tumor in mice. However, particular care must be taken, as with isolation from mouse feeder layers, to ensure that the cell line ultimately surviving is human, and not mouse, by proper characterization with isoenzyme (*see* Section 15.10) and chromosome analysis (*see* Section 15.7) or mitochondrial DNA profiling (*see* Section 15.8.3).

## 24.6 DEVELOPMENT OF CELL LINES

### 24.6.1 Subculture of Primary Tumor Cultures

Primary cultures of carcinoma cells do not always take readily to trypsin passage, and many of the cells in the primary culture may not be capable of propagation because of a genetic or phenotypic aberration, terminal differentiation, or nutritional insufficiency. Nevertheless, some primary

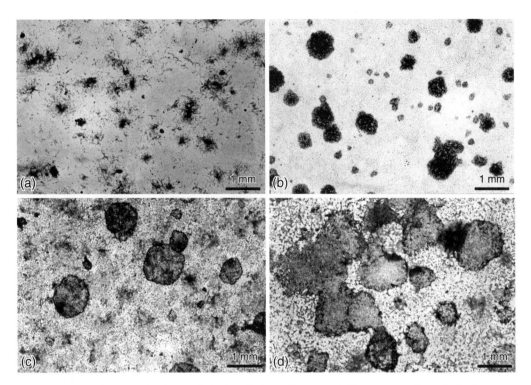

***Fig. 24.2. Selective Feeder Layers.*** Selective cloning of breast epithelium on a confluent feeder layer. (*a*) Colonies forming on plastic alone after seeding 4000 cells/cm$^2$ ($2 \times 10^4$ cells/mL) from a breast carcinoma culture. Small, dense colonies are epithelial cells, and larger, stellate colonies are fibroblasts. (*b*) Colonies of cells from the same culture, seeded at 400 cells/cm$^2$ (2000 cells/mL) on a confluent feeder layer of FHS74Int cells [Owens et al., 1974]. The epithelial colonies are much larger than those in (*a*), the plating efficiency is higher, and there are no fibroblastic colonies. (*c*) Colonies from a different breast carcinoma culture plated onto the same feeder layer. Note the different colony morphology with a lighter stained center and ring at the point of interaction with the feeder layer. (*d*) Colonies from normal breast culture seeded onto FHI cells (fetal human intestine; similar to FHS74Int). A few small, fibroblastic colonies are present in (*c*) and (*d*). [Smith et al., 1981; A. J. Hackett, personal communication.] (*See also* Plate 6*c, d*.)

cultures from tumors can be subcultured, opening up major possibilities. Evidence for tumor cells in the subculture implies that they have not been overgrown, may even have a faster growth rate than contaminating normal cells, and may be available for cloning or other selective culture methods (*see* Sections 9.2.2, 13.6, 24.5).

One of the major advantages of subculture is amplification. Expanded cultures can be cryopreserved and replicate cultures prepared for characterization and assay of specific parameters such as genomic alterations, changes in gene expression and metabolic pathway analyses, chemosensitivity, and invasiveness. Disadvantages of subculture include evolution away from the phenotype of the tumor due to the inherent genetic instability of the cells and selective adaptation of the cell line to the culture environment.

## 24.6.2 Continuous Cell Lines

One major criterion for the neoplastic origin of a culture is its capacity to form a continuous cell line (*see* Fig. 16*k–n,aa–ff*; Plate 9), which is usually aneuploid, heteroploid, insensitive to density limitation of growth, anchorage independent, and

often tumorigenic (*see* Section 17.6.2). The relationship of this cell line to the primary culture and the parent tumor is still difficult to assess, however, as such cells are not always typical of the tumor population. The cells of a continuous cell line may represent (1) further transformation stimulated by adaptation to culture, made possible by the unstable genotypic characteristics of tumor cells, or (2) a specific subset or stem cell population of the tumor. Currently the second possibility seems more likely [Petersen et al., 2003], as the emergence of a continuous cell line is often from colonies within the monolayer, suggesting that the continuous line arises from a minor immortalized subset of the tumor cell population, with cell culture merely providing the appropriate conditions for their expansion.

The capacity to form continuous cell lines is a useful criterion for a malignant origin, and some authors maintain that the characteristics of the cell lines (tumorigenicity, histology, chemosensitivity, etc.) still correlate with the tumor of origin [Tveit & Pihl, 1981]. In any event, these cell lines provide useful experimental material, although the time required for their evolution makes immediate clinical application difficult. Cancer cell lines are reviewed by Masters and Palsson in a series of books [Masters & Palsson, 1999–2000].

The uncertainty of the status of continuous cell lines remains, yet they have provided a valuable source of human cell lines for molecular and virological research. The question of whether they represent advanced stages of progression of a tumor whose development has been accelerated in culture, a cryptic stem cell population, or a purely in vitro artifact is still to be resolved. Continuous cell lines are certainly distinct from most early-passage tumor cultures, but they may still contain significant elements of the genotype of the parental cell from which they were derived. Their immortality is more likely to be due to the deletion or suppression of genes inducing senescence [Pereira-Smith & Smith, 1988; Goldstein et al., 1989; Holt et al., 1996; Sasaki et al., 1996] and to increased telomerase activity [Bryan & Reddel, 1997; Bodnar et al., 1998] than to overexpression of genes conferring malignancy per se.

## 24.7 CHARACTERIZATION OF TUMOR CELL CULTURES

### 24.7.1 Heterogeneity of Tumor Cultures

The isolation of cells from tumors may give rise to several different cell line types. Besides the neoplastic cells, connective-tissue fibroblasts, vascular endothelial and smooth muscle cells, infiltrating lymphocytes, granulocytes, and macrophages, as well as elements of the normal tissue in which the neoplasia arose, can all survive explantation. The hematopoietic components seldom form cell lines, although hematopoietic cell lines have been derived from small-cell carcinoma of the lung, causing serious confusion, because this carcinoma also tends to produce suspension cultures that can express myeloid markers [Ruff & Pert, 1984]. Macrophages and granulocytes are so strongly adherent and nonproliferative that they are generally lost at subculture. Smooth muscle does not propagate readily without the appropriate growth factors and selective medium, so the major potential contaminants of tumor cultures are fibroblasts, endothelial cells, and the normal equivalents of the neoplastic cells.

Of these contaminants, the major problem lies with the fibroblasts, which grow readily in culture and may also respond to tumor-derived mitogenic factors. Similarly endothelial cells, particularly in the absence of fibroblasts, may respond to tumor-derived angiogenesis factors and proliferate readily. The role of normal equivalent cells is harder to define, as their similarity to the neoplastic cells has made the appropriate experiments difficult to analyze. Characterization criteria should be chosen to exclude nontumor cells. For example, endothelial cells are factor VIII positive, contact inhibited, and sensitive to density limitation of growth; fibroblasts have a characteristic spindle-shaped morphology, are density limited for growth (though less so than endothelial

cells), have a finite life span of 50 generations or so, make type I collagen, and are rigidly diploid.

In general, the normal cell component, phenotypically equivalent to the tumor cells, is harder to identify and eliminate. The cells will be diploid, although some tumor cells may be close to diploid. They are usually anchorage dependent and will have a finite life span, although, again, there are cases of normal epithelial cell lines becoming continuous [Boukamp et al., 1988]. Moreover, if the cells are epithelial, they are more likely to be inhibited by serum TGF-β. The tumor cells are more likely to show genetic aberrations, such as oncogene amplification, translocations, and suppressor gene deletions, identifiable by FISH or PCR. The tumor cells are also likely to be angiogenic (see Section 17.6.4), show a higher expression of a urokinase-like plasminogen activator uPA (see Section 17.6.5), and be invasive (see Section 17.6.3).

General characteristics that may be used to identify tumor cells in culture are described in Chapter 17 (see Sections 17.3–17.6; Table 17.1). Although these characteristics are often expressed in continuous cell lines, their detection in early-passage cultures may be difficult because of the greater heterogeneity of this stage of culture. Detection of specific genetic abnormalities may be required, preferably by in situ analysis [e.g., Keith, 2003; Malvestiti, et al., 2009]. Similarly overexpressed oncogene products, such as mutant p53, or erb-B may be detected by immunostaining in cytological preparations or by flow cytometry.

It is essential, if a new cell line arises, that its DNA profile be shown to match DNA from the donor (from retained blood or tissue) and be distinct from any other continuous cell line growing concurrently in the laboratory.

From a behavioral aspect, the ability of neoplastic cells to grow on a preformed monolayer of the normal cells of same type is a good criterion for tumor cell identity and a potential model for separation. The normal cells also provide a feeder layer to sustain the tumor cells. Glioma, for example, will grow readily (better than on plastic in some cases) on a preformed monolayer of normal glial cells [MacDonald et al., 1985], but their normal counterparts will not; the same may be true for hepatoma cells grown on normal hepatocytes, skin carcinomas grown on normal epidermal keratinocytes, and mammary carcinoma grown on normal mammary epithelium.

Normal cells tend to have a low growth fraction at saturation density (see Sections 17.5.2, 20.11.1), whereas neoplastic cells continue to grow faster after reaching confluence. The maintenance of cultures at high density can sometimes provide conditions for overgrowth of the neoplastic cells (see Section 25.3.6; Protocol 17.3).

### 24.7.2 Histotypic Culture

Apart from organ culture (Section 25.2), which is not fundamentally different for tumor tissue than for normal tissue, histotypic methods with particular application to tumor culture are spheroid culture (see Protocol 25.3.3) and filter well culture (see Protocol 25.3.6).

***Spheroid formation.*** Normal stromal cells do not form spheroids or even become incorporated in tumor-derived spheroids. Hence cultures from tumors allowed to form spheroids on nonadhesive substrates, like agarose (*see* Protocol 25.2), will tend to overgrow their stromal component. Some cultures from breast and small-cell lung carcinoma can generate spheroids or irregular cellular organoids that float off and may be collected from the supernatant medium, leaving the stroma behind. However, the spheroids or organoids do not always appear soon after culture and can sometimes take weeks, or even months, to form, suggesting derivation from a minority cell population in the tumor. In other cases, such as mammary carcinoma, three-dimensional organoids may represent the normal epithelial component [Speirs, 2004]. Spheroid generation does not arise in all tumor cultures but has been described in neuroblastoma, melanoma, and glioma (*see* Section 25.3.3).

***Filter well inserts.*** Filter wells are designed to recreate the cell and matrix interactions of the tissue from which the cells were derived. Hence they provide an ideal model for the study of invasiveness (*see* Section 17.6.3), angiogenesis (*see* Section 17.6.4), and other abnormalities of cell interaction in cancer.

## 24.8  SPECIFIC TUMOR TYPES

The general protocols described in Chapter 11 (*see* Protocols 11.3–11.10), together with selective culture (*see* Section 24.7; Chapter 22), provide a good starting point for culturing most tumor types. In general, a reasonable approach to tumor culture is to combine collagenase digestion (*see* Protocol 11.8) with the tissue-specific approaches given in the protocols in Chapter 22, with or without the use of a feeder layer. Although serum-free conditions are often selective for cells from normal tissues, the nutritional and growth factor requirements of tumor cells may be more variable (*see* Section 24.5) and require the use of serum. Some specific examples of tumor culture are discussed briefly below.

### 24.8.1  Breast

Breast carcinoma can be cultured from collagenase digestion of biopsies [Leake et al., 1987; Dairkee et al., 1995, 1997; Speirs, 2004] and propagated on feeder layers or in MCDB 170. However, many of the conditions used to derive cultures from normal breast (growth factor supplementation, collagen coating) may not be optimal for mammary carcinoma [Ethier et al., 1993], so a variety of conditions may need to be tested in preliminary attempts and the neoplastic identity of the cells cultured confirmed.

Identification of breast tumor cells, as distinct from normal breast cells, will require detection of specific genetic lesions including erbB-2, c-*myc*, and fibroblast growth factor receptor (FGFR) 2 [Ethier et al., 1993; Ray et al., 2004] and elevated cyclin E levels [Willmarth et al., 2004]. Lineage can be confirmed by cytokeratin, EMA [Heyderman et al., 1979], and anti-HMFG 1 and 2 [Burchell and Taylor-Papadimitriou, 1989]; it has been proposed that keratin 19 is more likely to be expressed in tumor cells than in normal mammary cells in vitro [Taylor-Papadimitriou et al., 1989].

The following protocol has been contributed by V. Speirs, Breast Research Group, Leeds Institute of Molecular Medicine, St James's University Hospital, Leeds LS9 7TF, UK, adapted from Speirs [2004]. A schematic diagram of the breast tissue dispersal method is shown in Figure 24.3.

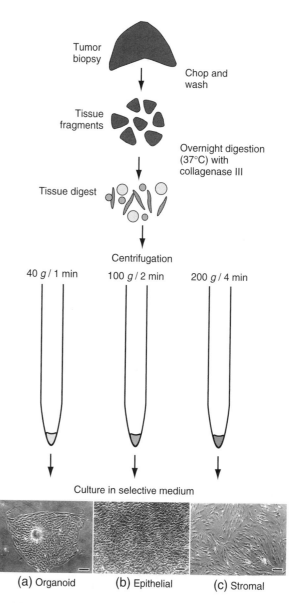

***Fig. 24.3. Fractionation of Breast Carcinoma Digest.*** A suspension of epithelial cell clusters, dispersed epithelial cells, and stromal cells can be segregated into fractions by differential centrifugation, where the epithelial clusters sediment at lower *g*, dispersed epithelial cells at intermediate *g*, and stromal cells at the highest *g* (*see* Protocol 24.3; *see also* Protocol 22.3; Fig. 22.1). (From Speirs, 2004.)

## PROTOCOL 24.3. CULTURE OF MAMMARY TUMOR CELLS

### Outline

Biopsy tissue is digested in collagenase, fractionated by low $g$ sedimentation, and cultured in serum-free selective medium.

### Reagents and Materials

- ☐ Antibiotic solution (ABC-PBSA): PBSA + 200 U/mL penicillin, 200 µg/mL streptomycin, 5 µg/mL Fungizone
- ☐ Collection medium (CM): DMEM + 200 U/mL penicillin, 200 µg/mL streptomycin, 5 µg/mL Fungizone
- ☐ Organoid medium (OM): DMEM supplemented with 100 U/mL penicillin, 100 µg/mL streptomycin, 2 mM glutamine, 10 mM HEPES, 0.075% BSA, 10 ng/mL cholera toxin, 0.5 µg/mL hydrocortisone, 5 µg/ml insulin, 5 ng/ml EGF (all Sigma-Aldrich)
- ☐ Complete culture medium (CCM): DMEM containing 2 mM glutamine, 100 U/mL penicillin, 100 µg/mL streptomycin, 10% heat inactivated fetal bovine serum (HIFBS)
- ☐ Primary culture medium (PCM): 3:1 mixture of OM and DMEM/10FB
- ☐ Collagenase: 0.1% type III in CCM
- ☐ Scalpels, #11 blades
- ☐ Universal containers or 50-mL centrifuge tubes

Unless otherwise stated, all culture reagents are from Invitrogen.

### Procedure

1. With approval from your local ethics committee, obtain breast tumors biopsies from pathology and transport to the laboratory in CM. If the tissue cannot be processed immediately (recommended), it may be stored refrigerated for up to 24 h without significant loss of viability.
2. In an MSC, wash tissue extensively in ABC-PBS.
3. Mince finely using crossed scalpels (#11 blades are recommended).
4. Wash twice more in ABC PBS.
5. Disaggregate for 18 to 20 h in 0.1% collagenase in a 37°C incubator. We have evaluated a number of different collagenase enzymes for this purpose and find that collagenase III gives a superior cell yield without compromising viability [Speirs et al., 1996].
6. Remove tissue from incubator and shake vigorously by hand. This will break up any remaining large clumps of partially digested tissue.

7. Centrifuge at 40 $g$ for 1 min using a swing-out rotor.
8. Carefully remove supernate and place in a fresh universal. The pellet contains single cells and small fragments of partially digested tissue, which we term the *organoid fraction*.
9. Retain and resuspend the pellet in 5 to 10 mL ABC PBS.
10. Re-centrifuge the supernate at 100 $g$ for 2 min.
11. Remove second supernate and transfer to a fresh universal. The pellet contains the *epithelial fraction*.
12. Resuspend in 5 to 10 mL ABC PBS and retain.
13. Re-centrifuge the second supernate at 200 $g$ for 4 min.
14. Remove and discard the third supernate. The pellet contains the *fibroblast fraction*, which should be retained and resuspended in 10 mL ABC-PBSA.
15. Pellet each fraction by centrifuging at 200 $g$ for 4 min.
16. Aspirate supernates and wash twice more with ABC-PBSA.
17. Plate the cells from the organoid and epithelial fractions separately in PCM.
18. After 24 h remove medium and replace with OM. Maintain in this medium for the duration of the culture. The components of OM selectively inhibit fibroblast overgrowth and encourage the growth of epithelial cells.
19. Seed stromal cells from step 16 in CCM and maintain in this throughout.

### 24.8.2 Lung

Both small-cell lung carcinoma (SCLC) and non–small-cell lung carcinoma (NSCLC) have been cultured successfully [Oie et al., 1996; for protocols, *see* Wu, 2004] with serum-free selective media—HITES for SCLC and ACL4 for NSCLC—mechanical spillage, and density gradient separation on Ficoll for isolating the cells. A substantial panel of these cell lines has been accumulated by the NCI, and some are available through the ATCC. The effects of matrix on oncogene and growth factor expression have also been studied [Pavelic et al., 1992] and have been used to facilitate culture of lung carcinoma cells from bone marrow micrometastases [Pantel et al., 1995] and selective media have been used to examine angiogenesis potential in short-term culture [Heinzman et al., 2008]. The assay of chemotherapeutic drugs in brain metastases from lung assayed in vitro shows considerable heterogeneity of response [Marsh et al., 2004].

A number of markers are available for the identification of SCLC cells in vitro, including bombesin-like immunoreactivity, DOPA-decarboxylase, N-*myc*, and creatine kinase

BB isoenzyme overexpression [Carney et al., 1985; Pedersen et al., 2003], but variant SCLC and non-SCLC lack these markers. Squamous cell lung cancer has been shown to overexpress EGFR, *erb*B-2, and TGF-α [Piyathilake et al., 2002], and other NSCLC cells lines have been shown to express abnormal properties in vitro, including expression of HER2/neu, TP53, and K-*ras*, which correlate with in the in vivo phenotype of the tumor from which they were derived [Wistuba et al., 1999].

### 24.8.3  Stomach

Solid tumors from gastric cancer can be disaggregated by mincing with scissors and pipetting to free tumor cell aggregates from fibrous stroma. Ascites can be purified by centrifugation on Ficoll/metrizoate [for protocols, *see* Park et al., 2004]. Enrichment for tumor cells is possible by harvesting tumor cell aggregates mechanically and subculturing, by scraping off fibroblasts, or by differential trypsinization. The morphology of the cell lines obtained range from well-flattened pavement-like monolayers, through more refractile attached cobblestone-like monolayers, to loosely attached aggregates (Fig. 24.4).

### 24.8.4  Colon

Serum-free conditions for the culture of some human colorectal cancer cell lines have been described [Murakami & Masui, 1980; Fantini et al., 1987], but these conditions are generally not suitable for newly isolated carcinoma cultures, which require serum. Colorectal carcinoma has been cultured from biopsies that have been taken from both primary tumors and metastases [Danielson et al., 1992; Paraskeva & Williams, 1992; Park & Gazdar, 1996] and have been used as a model for studying the control of epithelial-mesenchymal transition [Vincan et al., 2008]. As in lung carcinoma, some colorectal tumors occur with neuroendocrine properties, and some success has been reported on the use of HITES medium (*see* Table 9.2) with them [Lundqvist et al., 1991]. Density centrifugation on Percoll has been used to purify colonic carcinoma cells for primary culture in conventional medium (RPMI 1640 with 10% FB) [Csoka et al., 1995].

Protocols for culture of colon carcinoma are available in Whitehead [2004]. Protocol 24.4 has been adapted from Paraskeva and Williams [1992].

---

**PROTOCOL 24.4. CULTURE OF COLORECTAL TUMORS**

***Outline***
Adenomas are usually digested enzymatically and carcinomas just chopped with surgical blades. If a well-differentiated colorectal cancer does not release

---

tumor cells readily when the specimen is cut, it can be digested enzymatically.

***Reagents and Materials***

*Sterile:*

❑ Growth medium: DMEM containing 2 mM glutamine and supplemented with 20% fetal bovine serum (batch selection is essential), 1 µg/mL hydrocortisone sodium succinate, 0.2 U/mL insulin, 2 mM glutamine, 100 U/mL penicillin, 100 µg/mL streptomycin.

❑ Washing medium: growth medium with 5% FBS, 200 U/mL penicillin, 200 µg/mL streptomycin, 50 µg/mL gentamycin. Gentamycin is kept in the primary culture for at least the first week and then is removed.

❑ Digestion solution: DMEM and antibiotics, as described for the washing solution, with collagenase (1.5 mg/mL, Worthington type IV) hyaluronidase (0.25 mg/mL, Sigma type 1), and 2.5% to 5% FBS. Although we use Worthington collagenase, Sigma culture grades can also be tried.

❑ Dispase for subculture: prepare Dispase (a neutral protease, Roche Applied Science, grade 1) at 2 U/mL in DMEM containing 10% FBS, glutamine, penicillin, and streptomycin. Sterile filter the solution and store at −20°C. If a precipitate forms on thawing, the solution should be centrifuged and the active supernatant should be removed and used.

❑ Flasks, 25 cm², coated with collagen type IV, with Swiss 3T3 feeder cells at approximately 1 × 10⁴ cells/cm² (*see* Protocols 13.3, 22.4)

***Procedure***

*A. Enzyme Digestion:*

1. Wash tumor specimens four times in washing medium.

2. Mince in a small volume of the same medium, just enough to cover the tissue. (Do not allow the tissue to dry out!) Mince the tissues with crossed surgical blades or sharp scissors to fragments of approximately 1 mm³.

3. After cutting, wash the tissue again four times (the number of washings can be varied with experience, depending on whether contamination is a factor) by bench centrifugation (300 *g* for 3 min) and resuspension.

4. After washing the tumor fragments, put them into the digestion solution, and rotate at 37°C, usually overnight (approximately 12–16 h). The time the specimens are left in the solution is not critical because digestion is a mild process, but it is important not to let the digestion medium become

acid during the procedure. A low pH indicates that the specimens were left in the medium for too long or too much tissue was put in the volume of the digestion mixture. Approximately 1 cm$^3$ of tumor tissue is put into 20 to 40 mL of digestion solution.

### B. Nonenzymic Tissue Preparation:

Adenomas almost invariably need digestion with enzymes. However, with carcinomas, it is often found that during the cutting of the tumor with blades into 1-mm$^3$ pieces, small clumps of tumor cells are released from the tumor tissue into the washing medium. In this case the following procedure can be carried out:

1. Collect the washing medium containing released clumps of cells.
2. Separate into large and small clumps by allowing them to settle by gravity for a few minutes in a centrifuge tube.
3. Remove the supernatant phase.
4. Either put the remaining tissue pieces directly into culture or rotate the tissue gently for 30 to 60 min in washing medium to release more small clumps. The clumps of cells can then be collected, plated, and put into culture separately from the remaining larger pieces of tissue.
5. Wash all samples three times before putting them into culture.

### C. Standard Primary Culture Conditions:

1. Inoculate epithelial tubules and clumps of cells derived from tissue specimens into flasks in 4 mL of medium per collagen-coated 25-cm$^2$ flask of feeder cells.
2. Incubate at 37°C in a 5% $CO_2$ incubator.
3. Change the culture medium twice weekly.

*Note.* The tubules and cells start to attach to the substratum, and epithelial cells migrate out within 1 to 2 days. Most of the tubules and small clumps of epithelium attach within 7 days, but the larger organoids can take up to 6 weeks to attach, although they will remain viable all that time. The attachment of epithelium during primary culture and subculture is more reproducible and efficient when cells are inoculated onto collagen-coated flasks (Biocoat, BD Biosciences; *see also* Protocols 22.2, 22.9), and significantly better growth is obtained with 3T3 feeders than without. When the epithelial colonies expand to several hundred cells per colony, they become less dependent on 3T3 feeders, and no further addition of feeders is necessary.

### D. Subculture and Propagation:

Most colorectal adenoma primary cultures and adenoma-derived cell lines cannot be passaged by routine trypsin/EDTA procedures [Paraskeva et al., 1984, 1985]. As disaggregation of the cultured adenoma cells to single cells with 0.1% trypsin in 0.25 mM (0.1%) EDTA results in extremely poor growth, Dispase is used instead.

1. Add Dispase to the cell monolayer, just enough to cover the cells (~2.5 mL/25-cm$^2$ flask), and leave the solution to stand for 40 to 60 min for primary cultures and 20 to 40 min for cell lines.
2. Once the epithelial layers begin to detach (they do so as sheets rather than single cells), pipette to help detachment and disaggregation into smaller clumps.
3. Wash and reseed the cells under standard culture conditions. It may take several days for clumps to attach, so replace the medium carefully when feeding.

*Fibroblast contamination of colorectal tumor cultures.* One or a combination of the following techniques can be employed to deal with fibroblast contamination:

(1) Physically remove well-isolated fibroblast colonies by scraping with a sterile blunt instrument (e.g., a cell scraper). Care has to be taken to wash the culture up to six times to remove any fibroblasts that have detached, in order to prevent them from reattaching.

(2) Differential trypsinization can be attempted with the carcinomas [Kirkland & Bailey, 1986].

(3) Dispase preferentially (but not exclusively) removes the epithelium during passaging and leaves behind most of the fibroblastic cells attached to the culture vessel [Paraskeva et al., 1984]. During subculture, cells that have been removed with dispase can be preincubated in plastic Petri dishes for 2 to 6 h to allow the preferential attachment of any fibroblasts that may have been removed together with the epithelium. Clumps of epithelial cells still floating can be transferred to new flasks under standard culture conditions. This technique takes advantage of the fact that fibroblasts in general attach much more quickly to plastic than do clumps of epithelial cells, so that a partial purification step is possible.

(4) Use a conjugate between anti-Thy-1 monoclonal antibody and the toxin ricin [Paraskeva et al., 1985]. Thy-1 antigen is present on colorectal fibroblasts, but not colorectal epithelial cells; therefore the conjugate

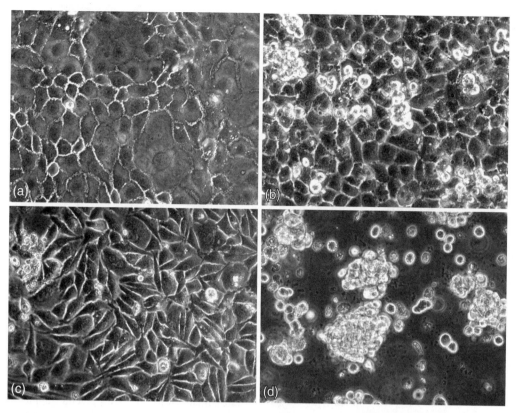

***Fig. 24.4.  Cell Lines from Gastric Carcinoma.*** (*a*) Well-flattened monolayer of SNU-216 gastric carcinoma cell line. (*b*) Pavement-like monolayer of SNU-484 gastric carcinoma cell line. (*c*) More refractile, cobblestone-like appearance of SNU-668 gastric carcinoma cell line. Cancer cells grow as adherent cultures, showing diffusely spreading growth of cultured tumor cells with fusiform or polygonal contours. (*d*) SNU-620 gastric carcinoma cell line. Cancer cells grow as both adherent and floating cell aggregates. (From Park et al., 2004.)

kills contaminating fibroblasts but shows no signs of toxicity toward the epithelium, whether derived from an adenoma or a carcinoma.

(5) Reduce the concentration of serum to about 2.5% to 5% if there are heavy concentrations of fibroblastic cells. It is worth remembering that normal fibroblasts have a finite growth span in vitro and that using any or all of the preceding techniques will eventually push the cells through so many divisions that any fibroblasts will senesce.

## 24.8.5  Pancreas

Cell lines from pancreatic primary tumors or metastases have been isolated and propagated in RPMI 1640 supplemented with fetal bovine serum (*see also* Protocol 22.7). The cell lines were adapted to protein-free medium for the examination of cell products [Yamaguchi et al., 1990]. Culture of pancreatic carcinoma has also been used to study the effect of genistein on angiogenesis [Buchler et al., 2004].

Pancreatic cell lines have been generated for studies in autocrine growth control by gastrin [Monstein et al., 2001]; no correlation was detected between the expression of gastrin and cholecystokinin (CCK) receptors, which normally bind gastrin. KCI-MOH1 is a cloned line of pancreatic adenocarcinoma isolated after passage in the SCID mouse [Mohammad et al., 1999]. Xenografted tumor was also used to establish the HPAC cell line for studies on glucocorticoid receptors [Gower et al., 1994]. Growth in xenografts—as well as proliferation, migration, invasiveness in Matrigel-coated filter well inserts, and soft agar cloning—is enhanced by coculture with stromal cells from pancreatic cancer [Hwang et al., 2008]. Protocols for pancreatic carcinoma culture by digestion in collagenase and plating onto collagen-coated dishes are available in Iguchi et al. [2004], who derived cell lines from primary lesions, ascites, and liver metastases. Two lines from liver metastases were passaged through nude mice by splenic injection (Fig. 24.5).

## 24.8.6  Ovary

A number of cell lines have been established from ovarian epithelial tumors (e.g., OAW series [Wilson et al., 1996], OVCAR-3 [Hamilton et al., 1983], A2780, [Tsuruo et al., 1986]), some in serum-free medium [Jozan et al., 1992] and others in serum-containing medium (e.g., OSE medium, 50:50 M199:MCDB105, supplemented with 15% FBS) used

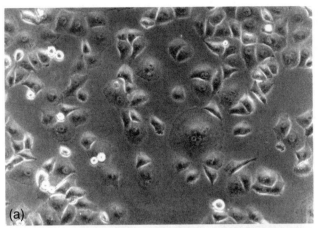

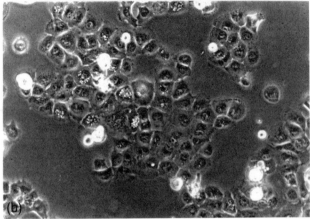

**Fig. 24.5. Cell Lines from Pancreatic Cancer.** Photomicrographs of cell lines KP-1N (a) and KP-3 (b) from liver metastases, xenografted in nude mice. Phase-contrast optics (125×). [From Iguchi et al., 2004.]

after collagenase digestion [Lounis et al., 1994]). Generation of cultures from ascites or pleural effusions appears to be easier than from solid tumor material [Verschraegen et al., 2003]. Density centrifugation on Percoll has also been used to purify ovarian carcinoma cells for primary culture, as used for colonic carcinoma [Csoka et al., 1995]. Protocols for culture of ovarian carcinoma are available in Wilson [2004].

### 24.8.7 Prostate

The matrix-assisted method, described above for lung carcinoma, has also proved to be successful in isolating cell lines from prostate tumors [Pantel et al., 1995]. Serum-free culture has also been used for initiating cultures from normal prostate and from benign and malignant tumors [Chopra et al., 1996; Peehl, 2002; Bright & Lewis, 2004] (see also Protocol 22.11). Long-term culture was possible after immortalization with a retroviral construct encoding the E6 and E7 transforming proteins of HPV16 [Bright & Lewis, 2004]. The molecular characterization of 21 prostatic carcinoma cell lines established 17 of these to be genuine

[van Bokhoven et al., 2003]. Using one of these, PC3, and human bone marrow endothelial cells, it was shown that dexamethasone inhibited angiogenesis and enhanced the antiangiogenic properties of docetaxel [Wilson et al., 2008].

Development of prostatic epithelium is known to be under paracrine control from the stroma by the FGF family of growth factors, including KGF [Thomson et al., 1997]. When cultures were prepared from normal, benign prostatic hyperplasia (BPH) and carcinoma, it was shown that FGF-17 was elevated twofold in cultures from BPH [Polnaszek et al., 2004]. Tumor angiogenesis has also been studied with cultures from normal and neoplastic prostate utilizing nitric oxide [Wang et al., 2003] and VEGF signaling [Shih et al., 2003].

### 24.8.8 Bladder

Culture of bladder carcinoma has shown that superficial tumors tend to have a limited life span, whereas cell lines from myoinvasive tumors often form continuous cell lines [Yeager et al., 1998]. Primary culture from transitional cell carcinomas of bladder have been used to study the role of FasL in immune protection of bladder cancer cells [Chopin et al., 2003] and a method has been described for culturing low-grade superficial urothelial carcinoma with a 63% success rate enhanced to 86% with glycine supplementation [Seifert et al., 2007]. Protocols for culture of bladder carcinoma can be found in Fu et al. [2004].

### 24.8.9 Skin

**Melanoma.** Pigment cells from skin can be cultured with the appropriate growth factors (see Protocol 22.22), and cultures can also be obtained from melanomas with a reasonable degree of success [Creasey et al., 1979; Mather & Sato, 1979a, b]. Primary melanomas are often contaminated with fibroblasts, but cloning on confluent feeder layers of normal cells (see Protocol 24.2) may be possible [Creasey et al., 1979; Freshney et al., 1982b]. Cell cultures derived by mechanical spillage can be freed of fibroblasts by treatment with 100 μg/mL Geneticin (G418) [Halaban, 2004]

MCDB 153, supplemented with FGF-2, insulin, transferrin, α-tocopherol, bovine pituitary extract, hydrocortisone, and 5% serum, with catalase and PMA added for the first two passages, has been used to grow melanocytes from normal skin, dysplastic nevi, and melanotic metastases [Levin et al., 1995]. An organotypic system for studying the interaction of melanoma with keratinocytes has been reported [Santiago-Walker et al., 2009]. Protocols for melanoma culture are available in Halaban [2004].

**Basal cell carcinoma.** Basal cell carcinoma (BCC) of the skin, the most common type of cancer in humans, is caused mainly by activation of the SONIC HEDGEHOG gene, and the effect of mutant suppressor gene PATCHED on this pathway has been studied in skin equivalent organotypic culture [Brellier et al., 2008]. The 3T3 cell feeder layer technique has proved successful for BCC of skin [Rheinwald & Beckett, 1981] (*see* Protocol 22.1). Oh et al. [2003] used primary cultures from BCC to study angiogenic potential and found a correlation with progression to aggressive disease. BCC cultures were also shown to be more sensitive to a cytotoxic effect of interferons than normal keratinocytes [Brysk et al., 1992].

***Squamous cell carcinoma (SCC).***  SCC and erythroplakias have also been cultured on 3T3 feeder layers [for protocols, *see* Edington et al., 2004] and have led to the establishment of valuable cell lines (the BICR series) representing different stages of malignancy and immortalization [Fitzsimmons et al., 2003; Gordon et al., 2003]. HaCaT cells [Boukamp et al., 1988] have been used as a model for SCC with and without Ras transfection [Fusenig & Boukamp, 1998]. Many SCC are oral, and it has been shown that their invasiveness may be regulated by annexin A5 [Wehder et al., 2009].

## 24.8.10  Cervix

Benign and malignant tumors may be established from cervical biopsies with the 3T3 feeder layer technique described for normal cervix [Stanley & Parkinson, 1979] (*see* Protocol 22.4; Plate 1*c*). Cell cultures have been used to study chromosomal instability and the integration of human papillomaviruses, which are implicated in the development of cervical cancer [Koopman et al., 1999], and in epithelial-mesenchymal transition [Lee et al., 2008].

Protocols for cervical tumor culture are available in Stern et al. [2004].

## 24.8.11  Glioma

Cultures of human glioma can be prepared by mechanical disaggregation, trypsinization, or collagenase digestion [Westermark et al., 1973; Pontén & Macintyre, 1968; Pontén, 1975; Freshney, 1980; for additional protocols, *see* Darling, 2004]. Protocol 11.8 can be used with glioma biopsies, omitting step 12, and gives a high success rate—about 80% for primary cultures and 60% for early passage cell lines [Freshney, 1980] (Fig. 24.6; *see also* Plates 7*a–c*, 9*c*). Early-passage cultures of human glioma cells will proliferate when plated onto confluent monolayers or normal glial cells, while normal glial cells will not [MacDonald et al., 1985], opening up a possible selective method to exclude normal glia from glioma cultures. However, most glioma cultures will outgrow the normal glia, especially if grown until they become continuous cell lines [Pontén & McIntyre, 1968], which is often achieved without any apparent crisis.

A number of gliomas have been cultured from rodents, among which the C6 deserves special mention [Benda et al., 1968]. This cell line expresses the astrocytic

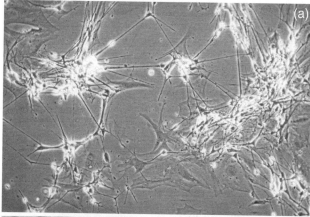

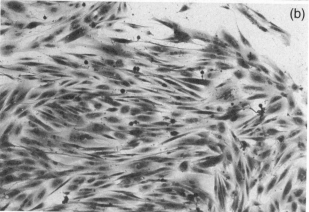

**Fig. 24.6.  *Cultures from Human Glioma.*** Two cell cultures from human anaplastic astrocytoma. (*a*) Primary culture from collagenase digest, showing typical astrocytic cells, which may be differentiated tumor cells or reactive glia. (*b*) Continuous cell line MOG-G-UVW, showing one of several morphologies found in cell lines from glioma. This example shows a pleomorphic fibroblast-like morphology; there are no typical multipolar astrocytic cells, often lost in continuous cell lines.

marker—glial fibrillary acidic protein—in up to 98% of cells [Freshney et al., 1980a] but still carries the enzymes glycerol phosphate dehydrogenase and 2′, 3′-cyclic nucleotide phosphorylase [Breen & De Vellis, 1974], both of which are oligodendrocytic markers. The line appears to be an interesting example of a precursor cell tumor that can express a dual phenotype.

Human glioma cultures have been used in predictive chemosensitivity testing [Thomas et al., 1985]; glucocorticoid sensitivity [Guner et al., 1977; Freshney, 1980b] and, more recently, retrovirus-mediated therapy [Rainov & Ren 2003]; and targeting with brain tumor-selective peptide ligands fused to cytoxins [Liu et al., 2003]. It has been claimed that a combination of monolayer culture and culture in neurospheres in serum-free medium helps preserve the in vivo genotype in resultant cell lines [Fael Al-Mayhani et al., 2009].

### 24.8.12 Neuroblastoma

Several lines of neuroblastoma (e.g., SK–N–BE(2) [Biedler and Spengler, 1976], [Tumilowicz et al., 1970]) have been isolated and are of particular interest, because of their potential for differentiation [Dimitroulakos et al., 1994] This has led to the examination of their potential role in neurotoxicity studies, although there appear to be significant differences in response to toxins compared to normal neurons [LePage et al., 2005]. Mouse neuroblastomas have been found to be suitable substrates for transmissible spongiform encephalopathies (TSEs) [Solassol et al., 2003].

### 24.8.13 Seminoma

Testicular seminomas have been cultured by using STO cells as a feeder layer and then have been supplemented with stem cell factor (SCF), LIF, and FGF-2, as used for embryonal stem cell cultures [Olie et al., 1995]. Although the cultures were heterogeneous, they did not give rise to primordial germ cell cultures.

### 24.8.14 Lymphoma and Leukemia

Routine generation of cell lines from lymphoma and leukemia has been difficult. Many of the continuous cell lines in existence were derived from Burkitt's lymphoma [Drexler & Minowada 2000] and shown to carry integrated EB viral genes. Protocols for lymphoma-leukemia culture can be found in Drexler [2004].

Protocol 24.5 was contributed by H. Drexler, DSMZ-German Collection of Microorganisms and Cell Cultures, Braunsweig, Germany, and adapted from Drexler [2004].

---

**PROTOCOL 24.5. ESTABLISHMENT OF CONTINUOUS CELL LINES FROM LEUKEMIA/LYMPHOMA**

#### Outline

Mononucler cells are isolated from disaggregated lymph node, peripheral blood, or bone marrow from leukemia/lymphoma (LL) patients by Ficoll–Hypaque and cultured at a high concentration in medium supplemented with 5637 cell conditioned medium.

#### Reagents and Materials

Sterile:

❑ Heparinized sample from bone marrow or peripheral blood, or mesh-disaggregated lymph node (*see* Protocol 11.9).
❑ 5637 CM: conditioned medium from 5637 cells (*see* Protocol 13.2) conditioned for 4 days from the point when cells are just subconfluent and pooled

---

from multiple collections; screened against M-07e cells [DSMZ catalog]
❑ Culture medium: RPMI 1640, IMDM, α-MEM, or McCoy's 5A supplemented with 20% FBS and 10% CM
❑ Culture flasks: 80 cm² (or 24-well or microtitration plates)

#### Procedure

1. Fractionate cell sample by centrifuging on a discontinuous Ficoll–Hypaque gradient and collecting the cells at the interface (*see* Protocol 27.1).
2. Adjust the concentration of the cell suspension to $2 \times 10^6$ to $5 \times 10^6$/ mL in culture medium.
3. Place 5 to 10 mL of the cell suspension in culture medium in an 80 cm² plastic culture flask. If 24-well-plates are used, add 1 to 2 mL cell suspension into each well. Add 100 to 200 µL of cell suspension into wells of 96-well flat-bottomed microtitration plates.
4. Place the cells in a humidified incubator at 37°C and 5% $CO_2$ in air. Alternatively, incubate the cells in a humidified 37°C incubator with 6% $CO_2$, 5% $O_2$, and 89% $N_2$.
5. Expand the cells by exchanging half of the spent culture medium volume with fresh culture medium plus 20% FBS and 10% 5637 CM (or with appropriate concentrations of growth factors) once a week, either by centrifugation or by settling.
6. After 4 h, some cells become adherent. These adherent cells appear to be the source of colony-stimulating factors for both normal and malignant cells. During the first two weeks, it is not necessary to remove these adherent cells from the culture unless there is a specific reason to do so, for example, because of the addition of a purified growth factor to the medium in order to obtain a unique type of cell line.
7. After 2 weeks, if the suspension cells grow very rapidly, the adherent cells can be removed simply by transferring the suspension cells into new culture vessels in order to reduce the potential for overgrowth of fibroblasts and normal lymphoblastoid cells.
8. During the first weeks, the neoplastic cells may appear to proliferate actively. If the medium becomes acidic quickly (yellow in the case of RPMI 1640 medium), change half of the volume of medium at 2 to 3 day intervals (seldom daily). If the number of the cells increases rapidly, readjust the cell concentration weekly to at least $1 \times 10^6$/ ml in fresh complete medium by dilution or subdivision

into new flasks or wells of the plate. The neoplastic cells from the majority of LL patients undergo as many as four doublings in two weeks, but after 2 to 3 weeks most malignant cells cease proliferating. Following a lag time of 2 to 4 weeks ("crisis period"), a small percentage of cells from the total population may still proliferate actively and may continue to grow, forming a cell line.

9. If the malignant cells continue to proliferate for more than two months, there is a high possibility of generating a new LL cell line. Then the task of characterizing the proliferating cells should be begun as soon as possible [Drexler, 2004]. Prior to the characterization of the cells, freeze ampoules of the proliferating cells containing a minimum of $3 \times 10^6$ cells/ampoule in liquid nitrogen (*see* Protocol 19.1) in order to avoid accidental loss of the cells.

10. Use limiting dilution of the cells in 96-well plates to generate monoclonal cell lines (*see* Sections 13.1, 13.4, 13.5). However, after prolonged culture in vitro, the cell line may effectively become monoclonal because of the outgrowth of selected cell clones. In most cases, it is not absolutely necessary to subclone the cell line by limiting dilution. In some types of LL cell lines, such as immature T- and precursor B-cell lines, it may be difficult, if not virtually impossible, to "clone" the cells.

# CHAPTER 25

# Three-Dimensional Culture

## 25.1 CELL INTERACTION AND PHENOTYPIC EXPRESSION

The historical divergence between maintenance of a fragment of explanted tissue and propagation of the cells that grew out from it led to the development of organ culture and cell culture (*see* Section 1.5), and it is cell culture that has become dominant. Now, although the potential uses of propagated cell lines are far from exhausted, many people are reverting to the notion that nutritional and hormonal supplementation are in themselves inadequate to recreate full structural and functional competence in a given cell population. The vital missing factor is cell interaction and the reciprocal signaling capacity that it entails both directly and in the creation of the correct matrix for a specific microenvironmental niche (*see* Section 22.2.1).

### 25.1.1 Effect of Cell Density

Cell−cell interaction is manifested at the simplest level when a cell culture reaches confluence and the constituent cells begin to interact more strongly with each other because of contact-mediated signaling, formation of junctional complexes, and increased potential for exchange of homocrine factors (*see* Sections 2.5, 16.7.1). The first noticeable effect is cessation of cell motility (contact inhibition) and withdrawal from cell cycle (density limitation of cell proliferation; *see* Section 17.5.2) in normal cells and reduced cell proliferation and increased apoptosis in transformed cells. Where cells have the capacity to differentiate, there is often an increase in the proportion of differentiated cells. C6 rat glioma cells show an increase in the percentage of GFAP-positive cells, (Fig. 25.1), secretory or absorptive epithelial cells form domes

(*see* Fig. 15.1; Plate 12*a, b*), and skeletal myocytes fuse to become multinucleated myotubes. Normal fibroblasts, although contact inhibited, will secrete more collagen after reaching confluence and will tend to multilayer, with the upper layer of cells smaller and darker staining, resembling fibrocytes (*see* Fig. 7.1*b*).

### 25.1.2 Reciprocal Interactions

Interacting populations of different cells have a reciprocal effect on their respective phenotypes (*see* Section 16.7.1; Fig. 16.2), and the resultant phenotypic changes lead to new interactions. Cell interaction is therefore not just a single event, but a continuing cascade of events. Similarly exogenous signals do not initiate a single event, as may be the case with homogeneous populations, but initiate a new cascade, as a result of the exogenously modified phenotype of one or both partners. For example, alveolar cells of the lung synthesize and release surfactant only in response to hormonal stimulation of adjacent fibroblasts [Post et al., 1984]; similarly the response of developing prostate epithelium to stromal signals is in turn activated by androgen binding to the stroma [Thomson et al., 1997] and inducing the release of KGF. Linser and Moscona [1980] separated the Müller cells of the neural retina from pigmented retina and neurons and demonstrated that glucocorticoid-induced differentiation did not occur unless the Müller cells (astroglia) were recombined with neurons from the retina. Neurons from other regions of the brain were ineffective.

Epithelium differentiates in response to matrix constituents that are often determined jointly by the epithelium on one side and connective tissue on the other, as may be the case with the interaction between epidermis and dermis in vitro

*Culture of Animal Cells: A Manual of Basic Technique and Specialized Applications, Sixth Edition*, by R. Ian Freshney
Copyright © 2010 John Wiley & Sons, Inc.

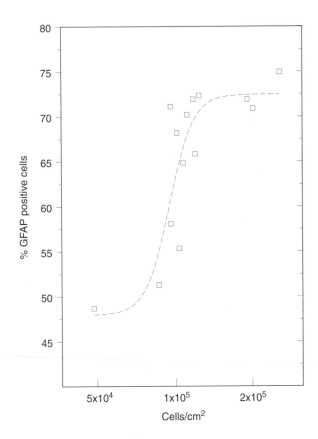

**Fig. 25.1. Effect of Cell Density on Expression of GFAP in C6 Cells.** Flask cultures were grown to different densities, fixed, and stained by immunoperoxidase for GFAP (*see* Plate 11*b*), and the percentage of stained cells was calculated from replicate counts in representative fields. (Courtesy of R. L. Shaw).

[Fusenig, 1994a; Limat et al., 1995]. Hence the whole integrated tissue may respond differently to simple ubiquitous signals, not because of the specificity of the signal or the receptor affinity but because of the quality of the microenvironment encoded in the juxtaposition of one cell type with a specific correspondent. As in human society the response of one individual to an exogenous stimulus is dictated as much by the spatial and temporal relationship of the individual with other individuals as by the endogenous makeup of the individual. Likewise a primitive neural crest cell may become a neuron, an endocrine cell, or a melanocyte depending on its ultimate location, its interaction with adjacent cells, and its response, mediated by neighboring cells, to hormonal stimuli.

In essence this preamble establishes that although some cell functions, such as cell proliferation, glycolysis, respiration, and gene transcription, can proceed in isolation, their regulation as related to a functioning multicellular organism ultimately depends on the interaction among cells of the appropriate lineage, the appropriate stage in that lineage, and on the interaction among cells of different lineages occupying the same microenvironment. This concept suggests that if you

want to study the biology of isolated cells, or use the cells as a substrate, conventional monolayer or suspension cultures may be adequate, but if you want to learn something of the integrated function, or dysfunction, of cells in vivo, a histotypic or organotypic model will be required.

### 25.1.3 Choice of Models

There are two major ways to approach this goal. One is to accept the cellular distribution within the tissue, explant it, and maintain it as an organ culture. The second is to purify and propagate individual cell lineages, study them alone under conditions of homologous cell interaction, recombine them, and study their mutual interactions. These approaches have given rise to three main types of three-dimensional culture: (1) *organ culture*, in which whole organs, or representative parts, are maintained as small fragments in culture and retain their intrinsic distribution, numerical and spatial, of participating cells; (2) *histotypic culture*, in which propagated cells are grown alone to high density in a three-dimensional matrix or are allowed to form three-dimensional aggregates in suspension; (3) *organotypic culture*, in which cells of different lineages are recombined in experimentally determined ratios and spatial relationships to recreate a component of the organ under study (Fig. 25.2).

Organ culture seeks to retain the original structural relationship of cells of the same or different types, and hence their interactive function, in order to study the effect of exogenous stimuli on further development [Lasnitzki, 1992; Hume et al., 1996; Kuslak and Marker 2007; Dame et al., 2008b]. This relationship may be preserved by explanting the tissue intact or recreated by separating the constituents and recombining them, as in the now-classic experiments of Grobstein and Auerbach and others in organogenesis [Auerbach & Grobstein, 1958; Cooper, 1965; Wessells, 1977] (*see also* Section 16.7.1). Organotypic culture represents the synthetic approach, whereby a three-dimensional, high-density culture is regenerated from isolated (and, preferably, purified and characterized) lineages of cells that are then recombined, after which their interaction is studied and their response to exogenous stimuli is characterized. The exogenous stimuli may be regulatory hormones, nutritional conditions, or xenobiotics. In each case the response is likely to be different from the responses of a pure cell type in isolation, grown at a low cell density.

### 25.2 ORGAN CULTURE

### 25.2.1 Gas and Nutrient Exchange

A major deficiency in tissue architecture in organ culture is the absence of a vascular system, limiting the size (by diffusion) and potentially the polarity of the cells within the organ culture. When cells are cultured as a solid mass of tissue, gaseous diffusion and the exchange of nutrients and metabolites is from the periphery, and the rate of this

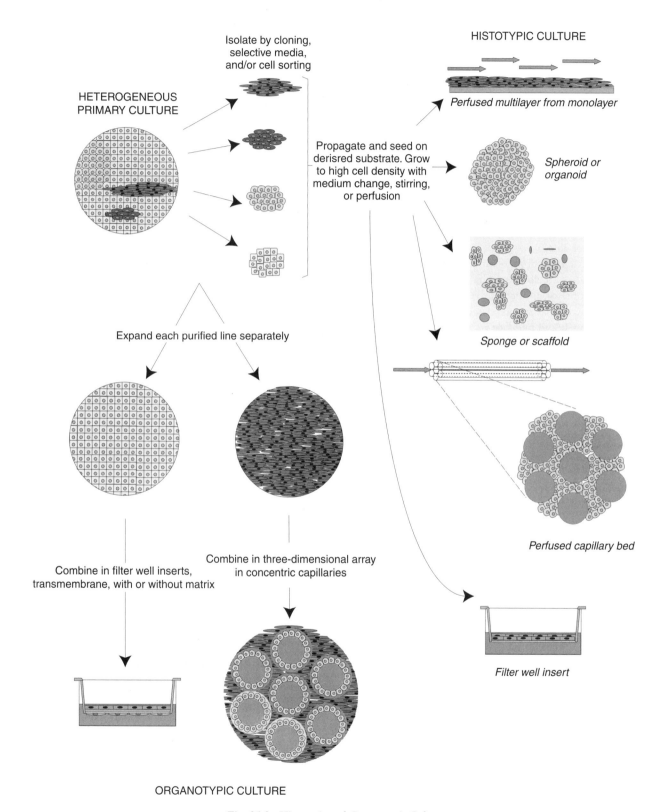

Isolate by cloning,
selective media,
and/or cell sorting

HETEROGENEOUS
PRIMARY CULTURE

HISTOTYPIC CULTURE

*Perfused multilayer from monolayer*

Propagate and seed on
derisred substrate. Grow
to high cell density with
medium change, stirring,
or perfusion

*Spheroid or
organoid*

*Sponge or scaffold*

*Perfused capillary bed*

Expand each purified line separately

Combine in filter well inserts,
transmembrane, with or without matrix

Combine in three-dimensional array
in concentric capillaries

*Filter well insert*

ORGANOTYPIC CULTURE

*Fig. 25.2.  Histotypic and Organotypic Culture.*

diffusion limits the size of the tissue. The dimensions of individual cells cultured in suspension or as a monolayer are such that diffusion is not limiting, but survival of cells in aggregates beyond about 250 $\mu$m in diameter ($\sim$5000 cell diameters) starts to become limited by diffusion, and at or above 1.0 mm in diameter ($\sim$2.5 $\times$ 10$^5$ cell diameters) central necrosis is often apparent. To alleviate this problem, organ cultures are usually placed at the interface between the liquid and gaseous phases, to facilitate gas exchange while retaining access to nutrients. Most systems achieve this by positioning the explant in a filter well insert (see Figs. 25.7, 25.8) or on a raft or gel exposed to the air (Fig. 25.3), but explants anchored to a solid substrate can also be aerated by rocking the culture, exposing it alternately to a liquid medium and a gas phase [Nicosia et al., 1983; Lechner & LaVeck, 1985] (see Protocol 22.9), or by using a roller bottle or rotating tube rack (see Protocol 26.3).

Anchorage to a solid substrate can lead to the development of an outgrowth of cells from the explant and resultant alterations in geometry, although this effect can be minimized by using a nonwettable surface. One of the advantages of culture at the gas–liquid interface is that the explant retains a spherical geometry if the liquid is maintained at the correct level. If the liquid is too deep, gas exchange is impaired; if it is too shallow, surface tension will tend to flatten the explant and promote outgrowth.

Increased permeation of oxygen can also be achieved by using increasing $O_2$ concentrations up to pure oxygen or by using hyperbaric oxygen. Certain tissues—such as thyroid [de Ridder & Mareel, 1978], prostate, trachea: [Dame et al., 2010] and skin [Lasnitzki, 1992], particularly from a newborn or an adult—may benefit from elevated $O_2$ tension, but often this benefit is at the risk of $O_2$-induced toxicity. As increasing the $O_2$ tension will not facilitate $CO_2$ release or nutrient-metabolite exchange, the benefits of increased oxygen may be overridden by other limiting factors.

### 25.2.2 Structural Integrity

The maintenance of structural integrity, above other considerations, was and is the main reason for adopting organ culture as an in vitro technique in preference to cell culture. Whereas cell culture utilizes cells dissociated by mechanical or enzymic techniques or spontaneous migration, organ culture deliberately maintains the cellular associations found in the tissue. Initially organ culture was selected to facilitate histological characterization, but ultimately it was discovered that certain elements of phenotypic expression were found only if cells were maintained in close association. The reasons for this are discussed above (see Section 25.1).

### 25.2.3 Growth and Differentiation

There is a relationship between growth and differentiation such that differentiated cells no longer proliferate (see Section 16.3). It is also possible that cessation of growth, regardless of cell density, may contribute to the induction of differentiation, if only by providing a permissive phenotypic state that is receptive to exogenous inducers of differentiation. Because of density limitation of cell proliferation and the physical restrictions imposed by organ culture geometry, most organ cultures do not grow, or if they do, proliferation is limited to the outer cell layers. Hence the status of the culture is permissive to differentiation and, given the appropriate cellular interactions and soluble inducers (see Section 16.7), should provide an ideal environment for differentiation to occur.

### 25.2.4 Limitations of Organ Culture

Experimental analysis of organ cultures depends largely on histological techniques, and they do not lend themselves readily to biochemical and molecular analyses. Biochemical monitoring requires reproducibility between samples, which is less easily achieved in organ culture than in propagated cell lines, because of sampling variation in preparing an organ culture, minor differences in handling and geometry, and variations in the ratios of cell types among cultures (see Table 1.4).

Organ cultures are also more difficult to prepare than replicate cultures from a propagated cell line and do not have the advantage of a characterized reference stock to which they may be related. Organ cultures cannot be propagated, and hence each experiment requires recourse to the original donor tissue, making the procedure labor intensive and prone to variation. Furthermore, as the population of reacting cells may be a minor component of the culture, attributing a molecular response to the correct cell type requires histological sectioning for autoradiographic, histochemical, immunocytochemical, or molecular in situ analysis.

Organ culture is essentially a technique for studying the behavior of integrated tissues rather than isolated cells. It is precisely in this area that a future understanding of the control of gene expression (and ultimately of cell behavior) in multicellular organisms may lie, but the limitations imposed by the organ culture system are such that recombinant systems between purified cell types may contribute more information at this particular stage. However, there is no doubt that organ culture has contributed a great deal to our understanding of developmental biology and tissue interactions and that it will continue to do so in the absence of adequate synthetic systems.

### 25.2.5 Types of Organ Culture

As techniques for organ culture have been dictated largely by the requirement to place the tissue at a location that allows optimal gas and nutrient exchange, most of these techniques put the tissue at the gas–liquid interface on semisolid gel substrates of agar [Wolff & Haffen, 1952] or clotted plasma [Fell & Robison, 1929], or on a raft of microporous filter, lens paper, or rayon supported on a stainless steel grid [Lasnitzki, 1992; Fig. 1.3] or adhering to a strip of Perspex or Plexiglas [Auerbach & Grobstein, 1958]. This type of geometry is now most easily attained with filter well inserts (see Protocol 25.3). Protocol 25.1 uses organ primordia from chick embryo but is applicable to many other types of tissue.

## PROTOCOL 25.1. ORGAN CULTURE

### Outline

Dissect out the organ or tissue, reduce it to 1 mm³, or to a thin membrane or rod, and place it on a support at the air–medium interface (e.g., filter well insert; *see* Figs. 25.7, 25.8). Incubate it in a humid $CO_2$ incubator, changing the medium as required.

### Materials

*Sterile or aseptically prepared:*

❑ Medium (e.g., M199), with or without serum
❑ Filter well inserts, non–tissue-culture treated (e.g., Corning Transwells polycarbonate #3402)
❑ Multiwell plates, 12 well (Corning)
❑ Pasteur pipettes
❑ Instruments for dissection

*Nonsterile:*

❑ Fertile hen's eggs at 8 days of incubation

### Procedure

1. Place the filter well inserts in the wells of a multiwell plate, and add sufficient medium to reach the level of the bottom of the filter (~1 mL).
2. Place the dishes in a humid $CO_2$ incubator to allow the pH of the medium to equilibrate at 37°C.
3. Prepare the tissue, or dissect out whole embryonic organs (e.g., femur or tibiotarsus of an 8-day chick embryo; *see* Protocols 11.2 *and* 11.7). The tissue must not be more than 1 mm thick, preferably less, in one dimension. (For example, 8-day embryonic tibiotarsus is perhaps 5 mm long, but only 0.5–0.8 mm in diameter. A fragment of skin might be 10 mm square but only 200 μm thick. Tissue that must be chopped down to size, such as liver or kidney, should be no more than 1 mm³.)
4. For short dissections (<1 h), HBSS is sufficient, but for longer dissections, use 50% serum in HBSS buffered with HEPES to pH 7.4.
5. Take the dishes from the incubator, and transfer the tissue carefully to filters. A Pasteur pipette is usually best for this task and can be used to aspirate any surplus fluid transferred with the explant, although care should be taken not to puncture the filter. Wet the inside of the pipette with medium before aspiration, to prevent fragments of tissue from sticking to the pipette.
6. Check the level of medium, making sure that the tissue is wetted, but not totally submerged, and return the dishes to the incubator.
7. Check after 2 to 4 h to ensure that a film of medium remains over the filter and explant, but that it is not deep enough to submerge the explant or cause it to float.
8. Incubate the dishes for 1 to 3 weeks, changing the medium every 2 or 3 days and sampling as required.

### Variations

(1) *Medium.* M199 or CMRL 1066 may be used with or without serum, and BGJ medium has been used for cartilage or bone [Biggers et al., 1961].
(2) *Type of support.* Organ cultures may be supported by a filter (e.g., polycarbonate) lying on top of a stainless steel grid in a center-well organ culture dish (Falcon #353037; *see* Fig. 25.3). However, filter well inserts have a number of advantages in terms of handling, range of sizes, materials, and matrix coatings (*see* Fig. 25.7; Table 25.1). Different types of tissue may be combined on opposite sides of a filter to study their interaction (*see* Section 16.7.1). Furthermore, with small filter well inserts (e.g., 6.5 mm, as in Corning #3402), the well formed on the top side of the filter assembly generates a meniscus of medium with a large surface area available for gas exchange. It is also possible to alter the configuration of the tissue by raising or lowering the level of medium in the dish or in the well; deeper medium gives a spherical explant, and shallower medium flattens the explant.
(3) $O_2$ *tension.* Embryonic cultures are usually best kept in air, but late-stage embryo, newborn, and adult tissues are better kept in elevated oxygen [Trowell, 1959; de Ridder & Mareel, 1978; Zeltinger & Holbrook, 1997].
(4) *Stirred cultures.* Stirred cultures of small tissue fragments in a gyratory shaker have been used for confrontational

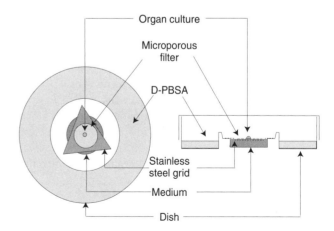

**Fig. 25.3. Organ Culture.** Small fragment of tissue on a filter laid on top of a stainless steel grid over the central well of an organ culture dish.

**TABLE 25.1.** Types of Filter Well Inserts

| Make | Name | Material | Qualities | Transparency | Porosity (μm) |
|---|---|---|---|---|---|
| Millipore | Millicel | Nitrocellulose | Mesh | Opaque | 0.45 |
| | | Polyolefin | Mesh | Transparent | 0.45 |
| Corning | Transwells | Polycarbonate | Absolute filter | Transparent | 5–8 |
| | | | | Translucent[a] | 0.45–1.0 |
| BD Biosciences | Inserts | Polyethylene terephthalate | Absolute filter | Transparent | 0.45–3 |
| Nunc | Anocel | Ceramic | Sieve | Transparent | 0.01 |
| Greiner Bio-One | Thinsert | Polyethylene terephthalate | Absolute filter | Transparent | 0.4–3.0 |
| | | | | Translucent | 0.4–8.0 |
| Earl-Clay | Ultraclone | Collagen | Mesh | Transparent | |

[a]The higher the pore frequency, the lower is the transparency. Low-porosity filters have a high pore frequency and are consequently less transparent.

cultures for assay of invasion [Mareel et al., 1979; Bjerkvig et al., 1986a,b] (*see* Protocol 25.4).

(5) *Rocking or rotated cultures.* The tissue is anchored to a substrate and subjected alternately to liquid culture medium and the gas phase by placing the culture vessel on a rocking platform (*see* Protocol 22.9) [Nicosia et al., 1983], or by anchoring the tissue to the wall of a rotating flask or tube (*see* Section 4.3.4; Protocol 26.3).

## 25.3 HISTOTYPIC CULTURE

Histotypic culture, in this context, is defined as high-density cell culture with the cell density approaching that of the tissue in vivo. Various attempts have been made to regenerate tissue-like architecture from dispersed monolayer cultures. Green and Thomas [1978] showed that human epidermal keratinocytes will form dermatoglyphs (i.e., friction ridges) if they are kept for several weeks without transfer, and Folkman and Haudenschild [1980] were able to demonstrate the formation of capillary tubules in cultures of vascular endothelial cells cultured in the presence of endothelial growth factor and medium conditioned by tumor cells. As cells reach a high density, medium nutrients will become limiting. To avoid this, the ratio of medium volume to cell number should remain approximately as it was in low-density culture. This can be achieved by seeding cells on a small coverslip in the center of a large non–tissue-culture grade dish or by use of filter well inserts, which give the opportunity for the formation of both high-density polarized cultures and heterotypic combinations of cell types to create organotypic cultures (*see* Protocol 25.3; Section 25.4; Plates 19–21). A high medium-to-cell ratio can also be maintained by perfusion (*see* Section 25.3.2).

### 25.3.1 Gel and Sponge Techniques

Leighton first demonstrated that both normal and malignant cells penetrate cellulose sponge [Leighton et al., 1968] facilitated by collagen coating; Gelfoam (a gelatin sponge matrix used in reconstructive surgery) can be used in place of cellulose [Sorour et al., 1975] and has been used in studies

of the effect of mechanical strain on lung development [Liu et al., 1995]. These systems require histological analysis and are limited in dimensions, like organ cultures, by gaseous and nutrient diffusion. The use of three-dimensional sponges and gels has increased significantly with the development of tissue engineering (*see* Section 25.4.2).

***Collagen gel.*** Collagen gel (native collagen, as distinct from denatured collagen coating) provides a matrix for the morphogenesis of primitive epithelial structures. Many different types of cell can be shown to penetrate such matrices and establish a tissue-like histology. Mammary epithelium forms rudimentary tubular and glandular structures when grown in collagen [Gomm et al., 1997], whereas breast carcinoma shows more disorganized growth [Berdichevsky et al., 1992]. The kidney epithelial cell line MDCK responds to paracrine stimulation from fibroblasts by producing tubular structures, but only in collagen gel [Kenworthy et al., 1992]. Collagen gel is also an important constituent of many organotypic models of skin (*see* Section 25.4.1; Plate 20), differentiated hepatocytes [Gomez-Lechon et al., 1998] and for demonstration of angiogenesis [Ment et al., 1997; Yang et al., 2004]. Neurite outgrowth from sympathetic ganglia neurons growing on collagen gels follows the orientation of the collagen fibers in the gel [Ebendal, 1976].

***Matrigel.*** Matrigel is a commercial product (BD Biosciences), derived from the extracellular matrix of the Engelbreth–Holm–Swarm (EHS) mouse sarcoma, that has been used for coating plastic (see Section 7.2.1) but can also be used in gel form. It is composed of laminin, collagen, fibronectin, and proteoglycans with a number of bound growth factors, although it can be obtained in a growth factor-depleted form. It has been used as a substrate for epithelial morphogenesis [Larsen et al., 2004], formation of capillaries from endothelial cells [Jain et al., 1997; Vouret-Craviari et al., 2004] (*see* Protocol 17.4), and in the study of malignant invasion [Brunton et al., 1997; De Wever et al., 2004]. It is, however, a complex and not completely defined matrix and can inhibit some morphogenetic events, such as hepatocyte growth factor (HGF)-induced tubulogenesis of MDCK cells [Williams & Clark, 2003].

## 25.3.2 Hollow Fibers

As medium supply and gas exchange become limiting at high cell densities, Knazek et al. [1972; Gullino and Knazek, 1979] developed a perfusion chamber from a bed of plastic capillary fibers, now available commercially (*see* Appendix II: Hollow Fiber Perfusion). The fibers are gas and nutrient permeable and support cell growth on their outer surfaces. Medium, saturated with 5% $CO_2$ in air, is pumped through the centers of the capillaries, and cells are added to the outer chamber surrounding the bundle of fibers (*see* Fig. 7.11; Fig. 26.8). The cells attach and grow on the outside of the capillary fibers, fed by diffusion from the perfusate, and can reach tissue-like cell densities. Different plastics and ultrafiltration properties give molecular weight cutoff points at 10, 50, or 100 kDa, regulating the diffusion of macromolecules.

It is claimed that cells in this type of high-density culture behave as they would in vivo. For example, in such cultures, choriocarcinoma cells release more human chorionic gonadotrophin [Knazek et al., 1974] than they would in conventional monolayer culture and colonic carcinoma cells produce elevated levels of CEA [Rutzky et al., 1979; Quarles et al., 1980]. However, there are considerable technical difficulties in setting up the chambers, and they are costly. Furthermore sampling cells from these chambers and determining the cell concentration are difficult operations. Overall, however, hollow fibers appear to present an ideal system for studying the synthesis and release of biopharmaceuticals and are now being exploited on a semi-industrial scale (*see* Section 26.2.5). They have also been used as a model for studying oxygen transport [Chen & Palmer, 2008].

## 25.3.3 Spheroids

When dissociated cells are cultured in a gyratory shaker, they may reassociate into clusters. Dispersed cells from embryonic tissues will sort during reaggregation in a highly specific fashion [Linser & Moscona, 1980]. Cells in these heterotypic aggregates appear to be capable of sorting themselves into groups and forming tissue-like structures.

Homotypic reaggregation also occurs fairly readily, and spheroids generated in gyratory shakers or by growth on agar (*see* Plate 18a) have been used as models for chemotherapy in vitro [Twentyman, 1980] and for the characterization of malignant invasion [Mareel et al., 1980]. As with organ cultures, the growth of spheroids is limited by diffusion, and a steady state may be reached in which cell proliferation in the outer layers is balanced by central necrosis (Fig. 25.4).

The following introduction and Protocol 25.2 for preparing multicellular tumor spheroids were contributed by M. Boyd and R. J. Mairs, Center for Oncology and Applied Pharmacology, Glasgow University, Cancer Research UK Beatson Laboratories, Garscube Estate, Bearsden, Glasgow G61 1BD, Scotland, UK.

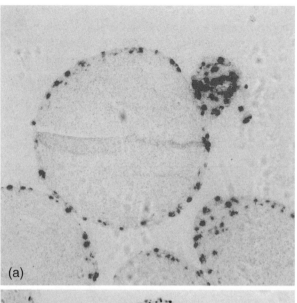

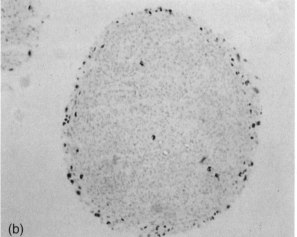

***Fig. 25.4. Dividing Cells in Spheroids.*** Sections through mature spheroids of approximately 600- to 800-μm diameter. (*a*) Autoradiograph labeled with [³H]thymidine, showing restriction of label to periphery. (*b*) Immunoperoxidase staining with anti-BUdR, showing similar restriction of label to periphery. (Courtesy of Ali Neshasterez; *see also* Plate 18a).

Multicellular tumor spheroids provide a proliferating model for avascular micrometastases. The three-dimensional structure of spheroids allows the experimental study of aspects of drug penetration and resistance to radiation or chemotherapy that are dependent on intercellular contact. Spheroids are also well suited to the study of "bystander effects" in experimental targeted or gene therapy [Boyd et al., 2002, 2004]. Human tumor spheroids are more easily developed from established cell lines or from xenografts than from primary tumors [Sutherland, 1988].

From a single cell suspension (trypsinized monolayer or disaggregated tumor), cells can be inoculated into magnetic stirrer vessels (Techne; *see* Fig. 7.7) and incubated to allow the formation of small aggregates over 3 to 5 days [Boyd et al.,

2001]. This procedure is optimal; however, the majority of cell lines do not form spheroids in this manner. Alternatively, aggregates may be formed from cell suspensions in stationary flasks, previously base-coated with agar. Aggregates may be left in the original flasks or transferred individually (by pipette) to multiwell plates, where continued growth over weeks will yield spheroids of maximum size, about 1000 μm [Yuhas et al., 1977; Sutherland, 1988].

---

## PROTOCOL 25.2. 3-D CULTURE IN SPHEROIDS

### Outline
Trypsinize monolayer cells, or disaggregate primary tissue, and seed the cells onto an agar-coated substrate. Transfer the aggregates to 24-well plates for analysis.

### Materials
*Sterile:*

- ❑ Noble agar (Difco, BD Biosciences)
- ❑ Growth medium
- ❑ Ultrapure water (UPW)
- ❑ Trypsin, 0.25%, in PBSA
- ❑ Flasks, 25 cm², or multiwell plates, 24 well
- ❑ Petri dishes, 9 cm

*Note.* When agar coating is used, all flasks, plates, and dishes should be sterile, but not necessarily tissue culture grade.

*Nonsterile:*

- ❑ Pi-pump (Schuco) or equivalent pipetting aid.

### Procedure
**Agar coating in 25-cm² flasks**

1. Add 1 g of Noble agar to 20 mL of UPW in a 100-mL borosilicate glass bottle with a loosely screwed-on cap.
2. Heat the agar in a water bath at 100°C for 10 min or until the agar has completely dissolved.
3. Add the contents of the bottle immediately to 60 mL of growth medium, previously heated to 37°C, and put 5-mL aliquots into each flask. Ensure that the agar is free from bubbles.
4. The agar will set at room temperature in ~5 min, giving a 1.25%-agar-coated flask.

**Agar coating in multiwell plates**

1. Add 0.5 g of agar to 10 mL of UPW, heat as in step 2 for 25-cm² flasks, and then add 40 mL of UPW.
2. Place 0.5 mL of the resulting solution in each well of a 24-well plate to give a base coat of 1% agar.

Accuracy and careful placement are important to ensure easy well-to-well focus of the microscope in subsequent viewing of spheroids.

**Spheroid initiation**

1. Trypsinize the confluent monolayer (for established lines, *see* Protocol 12.3) or disaggregate (for solid tumors, *see* Protocols 11.5, 11.6, 11.8) to give a single-cell suspension.
2. Neutralize the trypsin with medium containing serum (if necessary).
3. Count the number of cells, using an electronic cell counter or a hemocytometer.
4. Place $5 \times 10^5$ cells in 5 mL of growth medium in each agar-coated 25-cm² flask, and incubate the cultures. If the cells are capable of spheroid formation, small aggregated clumps (about 100–300 μm in diameter) will form spontaneously in 3 to 5 days.

For subsequent growth, spheroids should be transferred to new 25-cm² flasks or 24-well plates.

**Transfer to 25-cm² flasks**

1. Transfer the contents of the original flasks to conical centrifuge tubes or universal containers.
2. Allow the spheroids to settle, and remove single cells with the supernate.
3. Resuspend the spheroids in fresh medium, and transfer the suspensions to new agar-coated flasks, where growth will proceed by division of cells in the outer layer.

**Transfer to 24-well plates**

1. Transfer the contents of each 25-cm² flask into a 6-cm Petri dish.
2. Add 0.5 mL of medium to each agar-coated well of a 24-well plate.
3. Select individual spheroids of chosen dimensions under low-power magnification (40×) and, using a Pasteur pipette and a Pi-Pump or another pipetting aid with suitably fine control, transfer selected spheroids of similar diameter individually to the agar-coated wells of the 24-well plate.
4. Place the plate in a $CO_2$ incubator.
5. Replace the medium in the plate once or twice weekly (exchanging 0.5 mL each time), or add 0.5 mL of medium (without removing any medium) once or twice weekly, giving 2 mL/well after 2 to 4 weeks.

---

*Analysis.* Spheroid growth in wells or flasks may be quantified by regular (e.g., 2–3 times/week) measurement of the diameters of the spheroids, using a microscope eyepiece micrometer or graticule, or preferably by measurement of the cross-sectional area with an image analysis scanner. The most accurate growth curves are obtained when spheroids are grown in wells and are individually monitored.

Spheroids can also be utilized for clonogenic assay after treatment with test agents. Spheroids are collected in universal containers, washed with D-PBSA to remove residual medium, and then incubated with trypsin at 37°C for 5 min. Spheroids are then disaggregated to single cell suspension by passage through a syringe needle and counted, and the number of viable cells is assessed by Trypan Blue exclusion (*see* Protocol 21.1). Cells can then be utilized for conventional clonogenic assay [Boyd et al., 2001, 2004].

***Transfectant mosaic spheroids.*** Spheroids can be grown from populations of cells that have been transfected with different genes. The cells are first grown in monolayer and then are transfected with the transgene and subjected to selection for transgene-expressing cells. Mosaic spheroids are formed by the addition of both transfected and nontransfected monolayer cells in any desired proportions [Boyd et al., 2002, 2004]. The different cell populations are distributed throughout the resultant spheroids in approximately uniform mosaic patches, maintaining the same proportions of transfected to nontransfected cells as were added at the formation stage (Plate 18*a*).

***Applications.*** Spheroids have wide applications in the modeling of avascular tumor growth [Ward & King, 1997], the role of three-dimensional spatial configurations in gene expression in cell populations [Waleh et al., 1995; Dangles et al., 1997], and the assessment of cytotoxic treatment. Treatment end points include growth delay, determination of the proportion of spheroids sterilized ("cured") by treatment, and colony formation in monolayer after disaggregation of treated spheroids [Freyer & Sutherland, 1980, Boyd et al., 1999, 2004]. An important area is the use of spheroids to study the penetration of cytotoxic drugs, antibodies, or other molecules used in targeted therapy [Sutherland, 1988; Carlsson & Nederman, 1989]. This category represents a special application that is not possible in single cell suspensions or monolayer cultures. Spheroids have also proved useful in the study of cell killing by biologically targeted radionuclides [Mairs & Wheldon, 1996; Boyd et al., 2001; Fullerton et al., 2004]. Spheroid cultures have also been used in confrontation experiments to assess the invasiveness of spheroids derived from malignant cell populations that are grown in close proximity to normal cell cultures [de Ridder, 1997]; such techniques have also been used in nononcological studies of disease processes, such as studies of rheumatoid arthritis [Ermis et al., 1998]. Mosaic spheroids are a new variant form that has special applications in the assessment of bystander effects. For example, a current difficulty of gene therapy for cancer is the inefficiency of gene transfer procedures, leading to the requirement for bystander effects to eliminate cells in a tumor population that have not been transfected successfully.

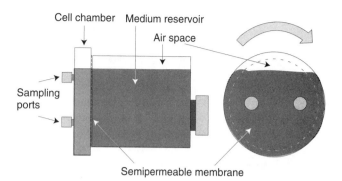

**Fig. 25.5. Rotating Chamber System.** Miniperm. The concentrated cell suspension in the left chamber is separated from the medium chamber by a semipermeable membrane. High-molecular-weight products remain with cells and can be harvested from the sampling ports, while replenishment of the medium is carried out via the right-hand port. Mixing is achieved by rotating the chamber on a roller rack. (Greiner Bio-one; Sigma.)

Mosaic spheroids mirror this situation in vitro and allow evaluation of different forms of the bystander effect, such as radiation cross fire when transfected cells are targeted with a radioactive agent [Boyd et al., 2002, 2004].

### 25.3.4 Rotating Chamber Systems

***Miniperm bioreactor.*** Mixing and aeration can also be achieved by rolling the culture vessel, either in a conventional roller bottle (*see* Protocol 26.3) or in two-compartment chambers (Fig. 25.5). If the cell suspension is limited to one small compartment, then the cell concentration can be quite high, as the cells are not diluted by the bulk of the medium. The product concentration (e.g., antibody) accumulates in the cellular compartment, while nutrients and waste products diffuse across the semipermeable membrane to and from the medium compartment. An example of this kind of design is the Miniperm, a two compartment cylinder with cells in the smaller compartment and medium in the larger, separated by a semipermeable membrane (Greiner Bio-one; Sigma; *see* Fig. 25.5). It is rotated to ensure mixing, and the medium can be sparged or replaced without disturbing the cells or product. The geometry of the chamber and the slow rotation tend to favor aggregate formation, and this may enhance product formation.

***Rotatory cell culture system (RCCS).*** Intrigued by the concept of growing cells in microgravity, in the 1980s NASA constructed a rotating chamber in which cells, growing in suspension, achieved simulated zero gravity with a slowly rotating chamber altering the sedimentation vector continuously (Fig. 25.6; *see also* Plate 22*c*). The cells remain stationary, are subject to zero shear force, and tend to form three-dimensional aggregates, spheroid-like structures, which may be more differentiated with enhanced product formation [NASA Bioreactors Workshop, 1997]. Gas exchange occurs from the cell-containing cylinder through a central silicone

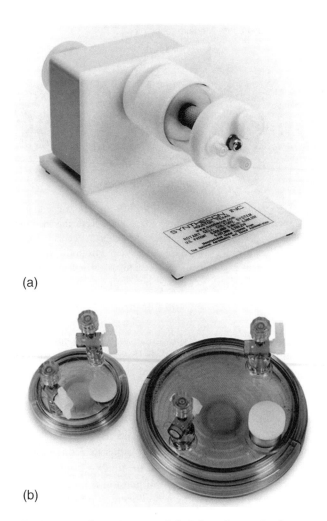

(a)

(b)

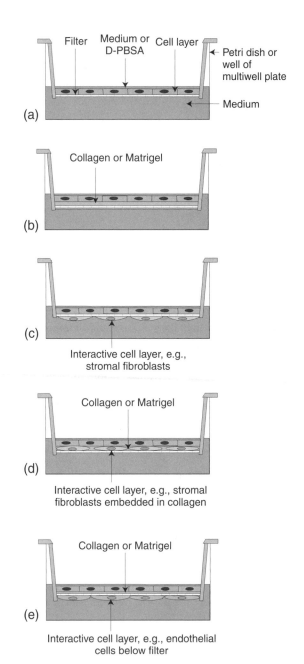

Filter    Medium or    Cell layer
            D-PBSA                    Petri dish or
                                      well of
                                      multiwell plate

(a)                                   Medium

Collagen or Matrigel

(b)

(c)

Interactive cell layer, e.g.,
stromal fibroblasts

Collagen or Matrigel

(d)

Interactive cell layer, e.g., stromal
fibroblasts embedded in collagen

Collagen or Matrigel

(e)

Interactive cell layer, e.g., endothelial
cells below filter

*Fig. 25.6. Synthecon Rotatory Cell Culture System.* In the rotary cell culture system (RCCS), cells are maintained in suspension by adjusting the rotation speed of a cylindrical culture chamber. A gas-permeable silicone membrane core gives the cells ample gas while disallowing shear causing bubbles. (*a*) Reusable cylindrical chamber on RCCS. (*b*) Disposable chambers. (*See* also Plate 22*c*.) The NASA-designed bioreactor is available from Synthecon, Inc., Houston, Texas, and distributors worldwide. (Courtesy of Synthecon, Inc.)

*Fig. 25.7. Filter Well Inserts.* Sectional diagram of a hypothetical filter well insert. (*a*) Monolayer grown on top of the filter. (*b*) Monolayer grown on matrix on top of a filter. (*c*) Interactive cell layer added to the underside of a filter. (*d*) Interactive cell layer added to the matrix. (*e*) Interactive cell layer added to the underside of the filter with matrix coating above. (*See also* Plates 19–21.)

membrane. When the rotation stops, the aggregates sediment and the medium can be replaced. This bioreactor is available from Synthecon, Inc., as a disposable RCCS unit or as a reusable STLV (slow turning lateral vessel).

In addition to its original purpose, to determine the effects of microgravity on cells with application to the space program [Freed & Vunjak-Novakovic, 2002], this culture vessel has also provided a suitable bioreactor for bulk culture of tissue engineering constructs [Dutt et al., 2003; Vunjak-Novakovic, 2006].

### 25.3.5 Immobilization of Living Cells in Alginate

The technique of encapsulating living cells within alginate beads has been widely used in experimental research into product formation—for example, hybridoma cells for

monoclonal antibody production, hormone-producing cells used in animal models for the treatment of diabetes mellitus [Zimmermann et al., 2003] and chondrocyte differentiation [Lemare et al., 1998] (*see also* Protocol 22.16; Plate 18*b*).

### 25.3.6 Filter Well Inserts

Filter well inserts are a commercialization of a filter-based culture system the origins of which go back to the 1950s and

**Fig. 25.8.** **Transwells.** Filter well inserts in a 12-well plate, with a filter well insert alongside. (Transwells: Corning.)

used in various forms since then. A filter substrate provides an environment for studying cell interaction, stratification, polarization, and tissue modeling. Polarity and functional integrity can be established as in thyroid [Chambard et al., 1983], intestinal [Halleux & Schneider, 1994], and kidney [Mullen et al., 1997] epithelium (*see* Section 16.7.4). Filter cultures allow generation of stratified epidermis [Limat et al., 1995; Kondo et al., 1997; Maas-Szabowski et al., 2000, 2002]. Others have used them to study invasion by granulocytes or malignant cells [McCall et al., 1981; Elvin et al., 1985; Repesh, 1989; Schlechte et al., 1990; Brunton et al., 1997].

One of the major advantages of filter well inserts is that they allow the culture of cells at very high, tissue-like densities, with ready access to medium and gas exchange, but in a multireplicate form. They also allow recombination of multiple different cell types with or without matix (*see* Fig. 25.7; Section 25.4). Filter well inserts are now available from several suppliers (*see* Appendix II) in a variety of translucent or transparent materials, including polycarbonate, PTFE, and polyethylene teraphthalate, and ranging in size from 6.5 mm to 9 cm, suitable for 24-well, 12-well, and 6-well plates, or larger dishes (Table 25.1; Figs. 25.7, 25.8; *see also* Plates 19–21). Filters can be obtained precoated with collagen, laminin, fibronectin, or Matrigel.

## PROTOCOL 25.3. FILTER WELL INSERTS

### Outline
Seed cells into filter well inserts, and culture the cells in excess medium in multiwell plates.

### Materials

*Sterile or aseptically prepared:*

❑ Approximately $0.5 \times 10^6$ cells per $cm^2$ of filter
❑ Growth medium, 1 to 20 mL per filter (depending on the vessel the housing filter)
❑ Filter well inserts
❑ Multiwell plates for filter inserts: 6, 12, or 24 well
❑ Forceps, curved

*Nonsterile:*

❑ Pipettor

### Procedure

1. Place the filter wells in the plate or dish.
2. Add medium, tilting the dish to allow the medium to occupy the space below the filter and to displace the air with minimum entrapment. Add medium until it is level with the filter (2.5 mL for a 6-well plate; 1.0 mL for a 24-well plate).
3. Level the dish, and add $2 \times 10^6$ cells in 2 mL of medium to the top of the filter for a 25-mm-diameter filter, or $5 \times 10^5$ in 200 µL of medium for a 6.5-mm-diameter filter, taking care not to perforate the filter.
4. Place the dish in a humid $CO_2$ incubator in a protective box (*see* Section 5.5.2). It is critical to avoid shaking the box, and the cultures should not be moved in the incubator, to avoid spillage and resultant contamination.
5. Monolayers should become established in 3 to 5 days, although 5 to 10 days or longer may be required for histotypic differentiation (e.g., polarized transport) to become established.
6. Cultures may be maintained indefinitely, replacing the medium or transferring the insert to a fresh well or dish every 3 to 5 days.

### Applications and analysis

(1) *Permeability.* Some epithelial cells (e.g., MDCK and Caco-2) and endothelial cells (e.g., from umbilical vein) form tight junctions several days after reaching confluence. This process is accelerated by precoating the membranes with collagen. Transepithelial permeability then becomes restricted to physiologically regulated transport through the cells, and pericellular transport falls to near zero. The process can be monitored by looking at dye (e.g., lucifer yellow), [$^{14}$C]methylcellulose, or [$^{14}$C]inulin transfer across the membrane, or by increasing transepithelial electrical resistance (TEER).

(2) *Polarized transport.* The addition of labeled glucose or amino acid to the upper compartment of the filter well insert will show unidirectional transport to the lower compartment. If the cells possess P-glycoprotein or some

other efflux transporter, cytotoxins (e.g., vinblastine) added to the lower compartment will be transported to the upper compartment, but not vice versa.

(3) *Penetration of cells through the filter.* Trypsinize and count each side of the filter in turn (trypsinized cells will not pass through even an 8-μm filter, as their spherical diameter in suspension exceeds this size), or fix the filter, embed, and section and examine by electron microscope or conventional histology. Visualization is possible in whole mounts by mounting the fixed, stained (Giemsa) filter on a slide in DPX under a coverslip under pressure, to flatten the filter. Differential counting can then be performed by alternately focusing on each plane.

(4) *Detachment of cells from the filter to the bottom of the dish.* Count the cells by trypsinization or scanning.

(5) *Partition above and below the filter.* Either count the cells as in (3), or prelabel the cells with rhodamine or fluorescein isothiocyanate (5 μg/mL for 30 min in a trypsinized suspension) and measure the fluorescence of solubilized cells (0.1% SDS in 0.3 N NaOH for 30 min) trypsinized from either side of the filter. Fluorescent labeling avoids problems with counting cells if they replicate in the lower compartment as fluorescence remains constant, regardless of replication.

(6) *Cellular invasion.* Precoat the filter with another barrier cell layer (normal fibroblasts, keratinocytes, endothelium, smooth muscle, etc.; *see* Chapter 22; Table 22.1), use microscopic examination to ensure that confluence is achieved, and then seed EDTA-dissociated test cells on top of a preformed layer ($1 \times 10^5 - 1 \times 10^6$ cells per filter). If the test cells are RITC or FITC labeled, then fluorescent measurements will reveal the appearance of the cells below the filter.

(7) *Matrix invasion.* Coat the filter with Matrigel, apply the cells above the Matrigel, and monitor the appearance of the cells below the filter [Repesh, 1989; Schlechte et al., 1990]. Alternatively, seed the cells onto the lower surface of the filter, coat the upper surface of the filter with Matrigel, and monitor the invasion of the cells into the Matrigel by confocal microscopy [Brunton et al., 1997].

### Variations

(1) *Depth of medium.* The depth of medium above the filter will regulate oxygen tension at the level of the cells. Keratinocytes or type II pneumocytes from lung alveoli will require little medium and a high oxygen tension, whereas enterocytes, such as Caco-2, may be better off submerged and with a lower oxygen tension.

(2) *Filter porosity.* 1-μm filters allow cell interaction and contact without transit across the filter, 8-μm filters allow live cells to cross the filter, and 0.2-μm filters probably do not allow cell contact. Low-porosity filters may be used to study cell interaction without permitting the cells to intermingle.

(3) *Transfilter combinations.* Invert the filter well insert, place upside down in a Petri dish, and load the underside

(now uppermost) with 0.5 mL, $2 \times 10^6$ cells/mL, of cell suspension. Place the lid on the top of the dish and touching the drop of cell suspension, before all of the medium drains through the filter. The depth of the dish will need to be about 0.5 to 1 mm higher than the height of the filter well insert so that a capillary space forms between the lid and the filter. Incubate the filter for 18 h. Capillarity will hold the medium and cells until the cells sediment onto the filter and attach [Brunton et al., 1997]. The next day, invert the filter and load the well with interacting cells or Matrigel, as described in (7) above (*see also* Protocol 25.3).

### 25.3.7 Cultures of Neuronal Aggregates

Aggregating cultures of fetal brain cells have been extensively used to study neural cell differentiation [Seeds, 1971; Trapp et al., 1981; Bjerkvig et al., 1986a; Jackson et al., 2007]. The aggregating cells follow the same developmental sequence as observed in vivo, leading to an organoid structure consisting of mature neurons, astrocytes, and oligodendrocytes. A prominent neuropil is also formed. In tumor biology, the aggregates can be used to study brain tumor cell invasion in vitro [Bjerkvig et al., 1986b].

The preceding introduction and Protocol 25.4 for aggregating cultures of brain cells have been contributed by Rolf Bjerkvig, Department of Cell Biology and Anatomy, University of Bergen, Žrstadveien 19, N-5009 Bergen, Norway.

---

### PROTOCOL 25.4. NEURONAL AGGREGATES

**Outline**

Remove brains from fetal rats at day 17 or 18 of gestation (consult local Animal Ethics Committee; *see* Section 6.9.1), and prepare the brains as a single cell suspension. Form brain aggregates by culturing in agar-coated wells in a multiwell plate. The cells in the aggregates form a mature organoid brain structure during a 20-day culture period.

**Materials**

*Sterile:*

❑ Dulbecco's modification of Eagle's medium, containing 10% heat-inactivated newborn calf serum; four times the prescribed concentration of nonessential amino acids; 2 mM L-glutamine; 100 U/mL penicillin, 100 μg/mL streptomycin

❑ Phosphate-buffered saline (PBS) with $Ca^{2+}$ and $Mg^{2+}$

❑ Trypsin type II (0.025% in D-PBSA)

❑ Agar (Difco)

- Multiwell tissue culture dishes (24-well plates; Nunc)
- Petri dishes, 10 cm
- Test tubes, 12 mL
- Scalpels, scissors, and surgical forceps
- Erlenmeyer flasks, 2, 100 mL
- To coat the wells with agar-medium, use the following procedure:
  - (i) Prepare a 3% stock solution (3 g of agar in 100 mL of D-PBSA) in an Erlenmeyer flask.
  - (ii) Heat the flask in boiling water until the agar is dissolved. Place an empty Erlenmeyer flask in boiling water, and add 10 mL of hot agar solution to it.
  - (iii) Slowly add warm complete growth medium to the flask until a medium-agar concentration of 0.75% is reached.
  - (iv) Add 0.5 mL of warm medium-agar solution to each well in the multiwell dish.
  - (v) Allow agar to cool and gel.
    The multiwell dishes can be stored in a refrigerator for 1 week.

*Nonsterile:*

- Water bath

**Procedure**

1. Dissect out, aseptically, the whole brains from a litter of fetal rats at day 17 or 18 of gestation, and place the tissue in a 10-cm Petri dish containing D-PBSA.
2. Mince the tissue into small cubes, about 0.5 cm$^3$, using crossed scalpels.
3. Transfer the tissue to a test tube, and wash it three times in D-PBSA. Allow the tissue to settle to the bottom of the tube between each washing.
4. Add 5 mL of trypsin solution to the tissue, and incubate in a water bath for 5 min at 37°C.
5. Disaggregate the tissue by trituration through a Pasteur pipette approximately 20 times.
6. Allow the tissue to settle for 3 min, and transfer the clump-free milky cell suspension to a test tube containing 5 mL of growth medium.
7. Add 5 mL of fresh trypsin to the undissociated tissue, and repeat the trypsinization and dissociation procedure twice more.
8. Spin the cell suspension at 200 $g$ for 5 min.
9. Aspirate and discard the supernate, resuspend the cells, and pool them in 10 mL of growth medium.
10. Count the cells, and add $3 \times 10^6$ cells in 1 mL to each agar-coated well.
11. Place the multiwell dish in a $CO_2$ incubator for 48 h.

12. Remove the aggregates to a sterile 10-cm Petri dish, and add 10 mL growth medium to the dish.
13. Transfer larger aggregates individually to new agar-coated wells by using a Pasteur pipette.
14. Change the medium every third day by carefully removing and adding new overlay medium.

During 20 days in culture, the aggregates will become spherical and develop into an organoid structure.

***Analysis.*** Fix and embed in paraffin or epon for histological or electron microscopic evaluation. Oligodendrocytes, astrocytes, and neurons are identifiable by transmission electron microscopy or by immunohistochemical localization of myelin basic protein, glial fibrillary acidic protein, and neuron-specific enolase, respectively.

***Variations.*** A single cell suspension can be obtained by mechanical sieving through steel or nylon meshes [Trapp et al., 1981] (*see* Section 11.3.8). Reaggregation cultures can also be obtained by using a gyratory shaker. Select a speed (about 70 rpm) such that the cells are brought into vortex, thereby greatly increasing the number of collisions between cells. This movement also prevents cell attachment to the culture flasks.

## 25.4  ORGANOTYPIC CULTURE

High density, three–dimensional culture involving the recombination of different cell lineages may be referred to as *organotypic culture*, a term that I am using to distinguish these reconstruction techniques from *organ culture* where the original cells are not dissociated. The key event that distinguishes these constructs from *histotypic* culture is the introduction of heterotypic cell interaction, including diffusible paracrine effects and signaling implicating the extracellular matrix. The relationship of the cells allows the generation of a structured microenvironment, cell polarity, and enhanced differentiation. Organotypic culture can be created by mixing cells randomly, and allowing them to interact and sort (*see* Protocol 25.4), as can happen spontaneously particularly with embryonic cells [Linser & Moscona, 1980], or the construct may be design to keep the interacting cells separate so that their interactions may be studied (*see* Section 25.3.6).

Usually organotypic culture will require a structural matrix similar to those described above (*see* Section 25.3) with filter well technology being used extensively (*see* Protocol 25.3).

### 25.4.1 Tissue Equivalents

The advent of filter well technology, boosted by its commercial availability, has produced a rapid expansion in the study of organotypic culture methods. Skin equivalents have been generated by coculturing dermis with epidermis (Mattek Epiderm, Episkin, SkinEthic; *see* Plate 21), with an intervening layer of collagen, or with dermal fibroblasts incorporated into the collagen [Limat et al., 1995; Maas-Szabowski et al., 2000, 2002; Stark et al., 2004, 2006] (*see* Plate 20), and models for paracrine control of growth and

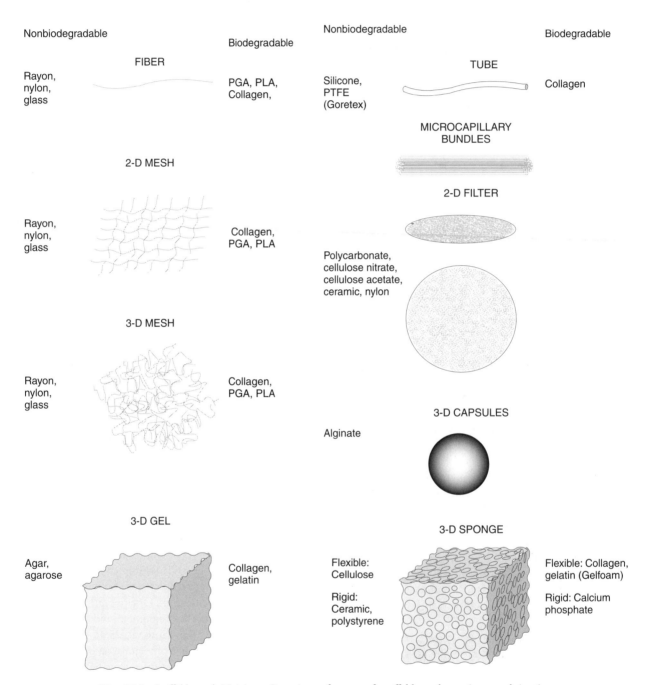

***Fig. 25.9. Scaffolds and Matrices.*** Overview of types of scaffolds and matrices used in tissue engineering constructs. Many different geometries have been employed including linear fibers or tubes (top), two-dimensional mesh screens or filters (center), and three-dimensional cubes or spheres (bottom). Although nondegradable materials have been used (left-hand labels in each column) the trend is toward biodegradable scaffolds (right-hand labels in each column) such as collagen, gelatin, polyglycolic acid (PGA) or polylactic acid (PLA), and calcium phosphate.

differentiation have been developed with cells from lung [Speirs et al., 1991], prostate [Thomson et al., 1997], and breast [Van Roozendahl et al., 1992].

The opportunity for heterotypic cell interaction has also opened up numerous opportunities for studying inflammation and irritation *in vitro* (*see* Plate 21; *see also* Section 21.6) and for creating other models for tissue interaction with increased in vivo relevance [Emura et al., 1997; Gomm et al., 1997]. Construction of tissue equivalent cultures has also made tissue replacement therapy possible. Skin equivalent cultures have been used in burn repair [Limat et al., 1989; Gobet et al., 1997; Wright et al., 1998], and now tissue engineering is being applied to the construction of tissue replacements for many different locations including cartilage, bone, ligament, cardiac and skeletal muscle, blood vessels, liver, and bladder [Atala & Lanza, 2002; Vunjak-Novakovic & Freshney, 2006].

### 25.4.2 Tissue Engineering

Just as organotypic culture needs cell interaction, constructs for tissue engineering often require similar interactions, as in the interaction between endothelium and smooth muscle in blood vessel reconstruction [Klinger & Niklason, 2006]. In addition to biological interactions, some constructs require physical forces; skeletal muscle needs tensile stress [Shansky et al., 1997; Powell et al., 2002; Shansky et al., 2006], bone [Mullender et al., 2004] and cartilage [Seidel et al., 2004] need compressive stress, and vascular endothelium in a blood vessel construct needs pulsatile flow [Niklason et al., 2001].

Engineering of tissue constructs depends on several components, depending on the tissue:

(1) Tissue cells of the correct lineage and at a proliferative, precursor stage

(2) Interactive cells, such as dermal fibroblasts in skin, smooth muscle cells in blood vessels, or glial cells in neural constructs

(3) A biodegradable scaffold to support the structure, such as polyglycolic acid (PGA), polylactic acid (PLA), or calcium phosphate (Fig. 25.9)

(4) Matrix, such as collagen, in place of the scaffold for soft tissues, or coating the scaffold to enhance cellular attachment

(5) Mechanical stress, namely tensile for muscle, compressive for cartilage and bone, and pulsatile for blood vessels

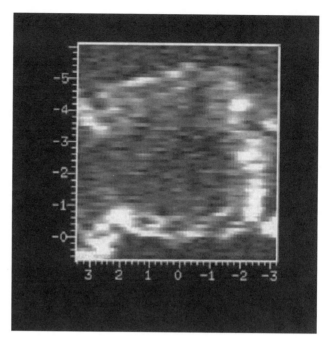

**Fig. 25.10. MRI of Cartilage Construct.** Brightness of image is proportional to cell density. Distances on grid are in mm. (Courtesy of Dr. Kevin Brindle; from Thelwall et al., 2001.)

Neural cells have also been used for tissue reconstruction, for example, spinal repair, but in these cases a cell suspension, rather than a construct, is injected into the injured site [Franklin et al., 1996; Franklin & Barnett 2000; Wewetzer et al., 2002; Totoiu et al., 2004; Groves et al., 1993].

### 25.5 IMAGING CELLS IN 3-D CONSTRUCTS

As microscopic observation becomes difficult when cells are incorporated into a scaffold in a three-dimensional organotypic construct, alternative methods must be used to visualize the status of the cells within the construct. This can be done by NMR if the bioreactor housing the constructs is placed within an NMR detector and the output displayed as an MRI [Neves et al., 2003] (Fig. 25.10), and the emission spectrum can be analyzed (*see* Fig. 26.22).

# CHAPTER 26

# Scale-up and Automation

There are two aspects to scaling up cell culture activities (1) increasing the mass of cells produced and (2) increasing the number of replicate cultures for screening in cell-based assays. The first implies increasing the volume of the culture, and the complexities of scale include nutrient supply, gas exchange, and the surface area for growth if the cells are attached, but the number of manipulations may not increase even if the increase in scale requires qualitative changes. Automation is required generally in the area of fluid transfer and process control over a prolonged culture period (days or even weeks). The second implies a numerical increase in the number of manipulations, based on the number of samples handled, placing the emphasis on automation via robotics to replicate the manipulations required, with less emphasis on process control, given the time scale is usually shorter (minutes or hours). Most of this chapter will deal with culture volume and geometry with a short section on automation and robotics at the end.

The meaning of mass scale-up varies according to the laboratory. For a basic research laboratory it may mean the difference between $1 \times 10^6$ and $1 \times 10^7$ cells per culture, readily achievable with conventional culture ware, particularly with the availability of multisurface flasks (BD Biosciences, Corning, Nunc; see Figs. 7.6, 26.9), and $1 \times 10^9$ to $1 \times 10^{10}$ cells that can be produced in simple stirrer cultures of around 1- to 10-L capacity (see Section 12.4.5). Larger scale cultures of $1 \times 10^{10}$ to $1 \times 10^{12}$ cells will require apparatus ranging from 10 to 100 L in a bench-top or floor standing laboratory-scale fermentor (e.g., Bellco BCS-36, New Brunswick Celligen, or Sartorius Wave Biorecator) and may be used for research applications or industrial pilot plant

which may increase to capacities of from 100 to 1000 L. Full-scale industrial production uses 5000- to 20,000-L bioreactors, but these are beyond the scope of this book.

The terms *stirrer culture* or *spinner culture* are synonymous and tend to be used for simple culture systems at the low end of the laboratory range of equipment. The terms *fermentor* and *bioreactor*, although not synonymous, have considerable overlap. The name *fermentor* derives from microbiological culture systems, designed originally for bacteria and yeast, and was used initially for laboratory equipment of around 50- to 100-L capacity, but the increase in scale in line with developments in the biotechnology industry, coupled with a greater diversity in design to cope with monolayer cultures as well as suspension, has led to the introduction of the name *bioreactor*. In tissue engineering, the term includes stirred culture of constructs in relatively small volumes.

The method employed to increase the scale of a culture depends on whether the cells proliferate in suspension or must be anchored to the substrate. Methods for suspension cells are generally simpler and will be dealt with first. Monolayer techniques, being dependent on increased provision of growth surface, tend to be more complex and will be dealt with later (see Section 26.2). Many of the monolayer techniques are also applicable to suspension cells.

## 26.1 SCALE-UP IN SUSPENSION

Scale-up of suspension cultures involves, primarily, an increase in the volume of the culture medium. Agitation of the medium is necessary when the depth exceeds 5 mm, to

**Fig. 26.1. Large Stirrer Flask.** Large stirrer flask (5 L) on a magnetic stirrer. Note the offset pendulum that makes an excursion in the annular depression in the base of the flask. The side arms are for sampling or perfusion with $CO_2$ in air, which would be required with this volume of medium (*see* Fig. 26.2) (Techne, courtesy of Sterilin.)

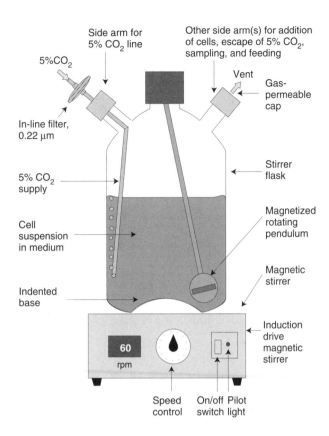

**Fig. 26.2. Large Stirrer Culture.** Diagram of a large stirrer flask suitable for volumes up to 8 L.

enhance gas exchange and prevent the cells from forming a dense sediment; above 5 to 10 cm (depending on the ratio of surface area to volume), sparging with $CO_2$ and air is required to maintain adequate gas exchange (Figs. 26.1, 26.2). Stirring is best done slowly with a magnet encased in a glass pendulum (Techne, Integra) or with a large-surface-area paddle (Bellco). The stirring speed should be between 30 and 100 rpm, sufficient to prevent cell sedimentation, but not so fast as to create shear forces that would damage the cells. Antifoam (Dow Chemical Co.) or Pluronic F68 (Sigma), 0.01% to 0.1%, should be included when the serum concentration is above 2%, particularly if the medium is sparged. In the absence of serum, it may be necessary to increase the viscosity of the medium with carboxymethyl cellulose (1–2%) (molecular weight $\sim10^5$). A simple laboratory-scale bioreactor for setting up a 4-L culture of cell growing in suspension is described in the following protocol.

---

### PROTOCOL 26.1. STIRRED 4-LITER BATCH SUSPENSION CULTURE

#### Outline
Grow a pilot culture of cells, and add it to a prewarmed stirrer flask or bioreactor containing medium already equilibrated with 5% $CO_2$ at 37°C. Stir the cell suspension slowly, with sparging, until the required cell concentration is reached, and then harvest the cells.

#### Materials

*Sterile or aseptically prepared:*

❑ Starter culture vessel (e.g., small stirrer flask; *see* Figs. 7.7, 12.5, or static flask if cell grow spontaneously in suspension)
❑ Growth medium, warmed to 37°C
❑ Antifoam (silicone: Dow Corning, Merck; Pluronic F68: Sigma)
❑ Prepared large stirrer flask (Bellco, Techne; *see* Figs. 26.1, 26.2) with:
  (i) A magnet enclosed in a glass pendulum, or a suitable alternative stirrer suspended from the top cap (e.g., Techne).
  (ii) One inlet port with a removable $CO_2$-permeable cap (*see* Fig. 7.8*b*), or a cap with a silicone diaphragm pierced by a wide-bore needle with a Luer connection fixed to a sterile in-line filter (e.g., Millex).

(iii) A second inlet port carrying a tube for the gas line (e.g., Bellco #1965), with a 2- to 3-mm internal diameter, that reaches almost to the bottom of the vessel but remains clear of the pendulum when it is stirring, and with the external entry terminating in a female Luer fitting guarded by aluminum foil.

(iv) Sterilize the stirrer culture vessel by autoclaving, at 100 kPa (15 lb/in.$^2$) for 20 min, fully assembled, but with the $CO_2$-permeable cap removed and sterilized separately.

(v) Alternatively, use a small commercially provided bioreactor (Bellco, New Brunswick) and follow the manufacturer's instructions.

❏ An in-line sterile Luer-fitting micropore filter, 25-mm diameter, 0.2-μm porosity, (e.g., Millex), to attach to the gas entry line at the time of use

*Nonsterile:*

❏ Magnetic stirrer or motor drive for paddle
❏ Supply of 5% $CO_2$, preferably from a metered supply at about 10 to 30 kPa (2–5 lb/in.$^2$)
❏ D-PBSA for counting cells
❏ Electronic cell counter or hemocytometer

### Procedure

1. Set up a starter culture (*see* Protocol 12.4; Fig. 12.5) with 200 mL medium, and seed it with cells at $5 \times 10^4$ to $1 \times 10^5$ cells/mL. Place the culture on the magnetic stirrer, rotating at 60 rpm, and incubate the culture until a concentration of $5 \times 10^5$ to $1 \times 10^6$ cells/mL is reached. (*Note.* Do not exceed $1 \times 10^6$ cells/mL as the cells may enter apoptosis.)
2. Set up the large stirrer flask or bioreactor as in Fig. 26.2.
3. Add 4 L of medium to the flask.
4. Add 0.4 mL of antifoam to the flask, using a disposable pipette or syringe.
5. Close the side arm with the $CO_2$-permeable cap.
6. Place the flask on the magnetic stirrer in a 37°C room or incubator.
7. Connect the 5%-$CO_2$ air line via the sterile, micropore in-line filter to the gas inlet Luer fitting.
8. Turn on the gas at a flow rate of approximately 10 to 15 mL/min.
9. Stir the culture at 60 rpm.
10. Incubate the large stirrer flask for about 2 h to allow the temperature and $CO_2$ tension to equilibrate.
11. Turn off the $CO_2$, and disconnect the flask from the $CO_2$ line.
12. Bring the large stirrer flask and the starter culture back to the laminar-flow hood.
13. Transfer the starter culture to the large stirrer flask by pouring in one single smooth action.

14. Return the large stirrer flask to the 37°C room or incubator.
15. Restart the magnetic stirrer at 60 rpm.
16. Reconnect the 5% $CO_2$ line, and adjust the gas flow to 10 to 15 mL/min. (Estimate the rate of flow by counting the bubbles if there is no flow meter on the line.)
17. Incubate the culture for 4 to 7 days.
18. Sample daily to check the cell proliferation rate:
    (a) Bring the stirrer flask back to the laminar-flow hood.
    (b) Remove the side-arm cap.
    (c) Flush out the contents of the sampling tube and then withdraw 5 to 10 mL of the cell suspension.
    (d) Return the stirrer flask to 37°C.
    (e) Count the cells, and check their viability by dye exclusion.
19. When the cell concentration reaches the desired level:
    (a) Turn off the gas and the magnetic stirrer.
    (b) Disconnect the stirrer flask from the gas supply.
    (c) Take the culture to the laboratory, and pour off the cells into centrifuge bottles.
    (d) Centrifuge the cells at 100 *g* for 10 min, and collect the cells (from the pellet) or the supernatant medium, as appropriate.

*Analysis.* Monitor the growth rate of the cells daily. For the best results the cells should not show a lag period of more than 24 h and should still be in exponential growth when harvested. Plot the cell counts daily, and harvest the cells at approximately $1 \times 10^6$ cells/mL. Suspension cultures tend to enter apoptosis if this concentration is exceeded.

### Variations

*Adding medium.* The following procedures are alternatives for adding medium to vessels larger than 5 L:

(1) Buy medium in media bags (*see* Appendix II) that can be hooked up alongside the stirrer flask, warmed to 37°C, and allowed to run in unattended.

(2) Sterilize the stirrer flask with a premeasured volume of UPW, and make up the medium from concentrates in situ. Mark the side of the flask to indicate the level of water, so that any water lost by evaporation during autoclaving can be made up.

(3) Use an autoclavable medium (*see* Appendix II), and autoclave it in situ (again restoring volume with sterile UPW to compensate for evaporation).

*Harvesting.* Pouring from a large stirrer flask can be difficult, unless there is a bottom outlet. However, bottom outlets tend to clog with dead cells and are best avoided. Harvesting is best done by pumping or displacement:

(1) Remove the in-line filter from the gas line and add silicone tubing and either siphon off the cells or use a peristaltic pump, tilting the vessel when the fluid is near to the bottom.
(2) Alternatively, disconnect the 5% $CO_2$ supply, replace the micropore filter with a flexible tube, attach the 5% $CO_2$ supply to the other port, and blow the cells out through the gas line. (*Note.* This requires that the gas line is open ended, i.e. without a sintered glass outlet, as in Fig. 26.2.)

### 26.1.1 Continuous Culture

If it is required that the cells be kept at a set concentration, the culture can be maintained as a *chemostat* or *biostat*. In this type of culture system the cells are grown to the mid-log phase (monitored by daily cell counts); a measured volume of cells is removed each day and replaced with an equal volume of medium. Alternatively, the cells may be run off continuously, at a constant rate, at mid-log phase, and medium added at the same rate. The latter will require a stirrer vessel with four ports, two for $CO_2$ inlet and outlet, one for medium inlet, and one for spent medium outlet, collecting into a reservoir (Fig. 26.3). The flow rate of medium may be calculated from the growth rate of the culture [Griffiths, 2000] but is better determined experimentally by serial cell counting at different flow rates of the medium. Cell counts can be performed on the bioreactor by disconnecting the $CO_2$ inlet and withdrawing a sample by a syringe with a Luer double female adapter. The flow rate may be regulated by a variable peristaltic pump on the inlet line.

Production of cells in bulk, in the 1- to 20-L range, is best done by the batch method described in Protocol 26.1.

A steady state is required for monitoring metabolic changes related to cell density but is more expensive in medium and is more likely to lead to contamination. However, if the operation is in the 50- to 1000-L range, then more investment and time are spent in generating the culture, and the batch method becomes more costly in time, materials, and down time. Continuous culture may be better in some cases, but, with the availability of disposable bioreactors (Sartorius Stedim; Xcellerex), where preparation and down time are significantly reduced, most emphasis is now placed on fed-batch stirred tank reactor (STR) processes.

Suspension cultures can also be scaled up by mixing in rotating bottles on a roller rack (*see* Fig. 26.13), as for monolayer cultures (*see* Protocol 26.3).

***Adherent cells.*** Anchorage-dependent cells cannot be grown in liquid suspension, except on microcarriers (*see* Protocol 26.4), but transformed cells (e.g., virally transformed or spontaneously transformed continuous cell lines, such as CHO or HeLa-S$_3$) can. Because these cells are still capable of attachment, the culture vessels will need to be coated with a water-repellent silicone (e.g., Repelcote), and the calcium concentration may need to be reduced. S-MEM medium (Invitrogen, Sigma) is a variation of Eagle's MEM with no calcium in the formulation and has been used for the culture of HeLa-S$_3$ and other cells in suspension. RPMI 160, which also has no calcium and a low magnesium concentration, was developed for growing lymphoblastoid cells in suspension and is also suitable for other suspension-growing cells.

### 26.1.2 Scale and Complexity

Standard bench-top stirrer cultures operate satisfactorily up to around 10 L for the bulk production of cells or soluble products in the medium. If, however, more attention must be

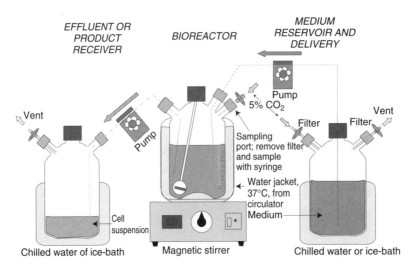

**Fig. 26.3. Biostat.** A modification of the suspension culture vessel in Fig. 26.1, with continuous matched input of fresh medium and output of cell suspension. The objective is to keep the culture conditions constant rather than to produce large numbers of cells (*see also* Figs. 26.17–26.19). Bulk culture, per se, is best performed in batches in the apparatus in Fig. 26.1 or Fig. 26.17.

paid to process control, then a controlled bioreactor should be used (Figs. 26.4, 26.5). These bioreactors have regulated input of medium and gas and provide the capability for data collection from oxygen, $CO_2$, and glucose electrodes in the culture vessel. They are regulated by a programmable control unit that records and outputs data and can be used to regulate gas and liquid input and output, stirring speed, temperature, and so on (*see* Section 26.3).

The capital cost and the preparation, setup, and turn round time for reusable bioreactors are significant, and this has led to the introduction of disposable bioreactors (Xcellerex; Sartorius; Millipore Mobius) where the sterile compartment is plastic, disposable, and fits within a nonsterile incubator jacket (*see* Fig. 26.5*b*) and is fully validated [Hacker et al., 2009].

### 26.1.3  Mixing and Aeration

Problems of increased scale in suspension cultures revolve around mixing and gas exchange. In contrast to simple stirrer cultures, most laboratory-scale bioreactors agitate the medium with a rotating turbine or paddle. Designs of these bioreactors vary, mostly to try to achieve maximum movement of liquid with minimum shear stress for the cells (*see* Fig. 26.4*a*). Successful designs usually employ a slowly rotating large-bladed paddle with a relatively high surface area [Griffiths, 2000] and a chemical additive designed to minimize the harmful effects of shear such as Pluronic F68 [Cherry & Papoutsakis, 1990; Nienow, 2006], carboxymethylcellulose (CMC), or polyvinylpyrrolidone (PVP). Sparging with $CO_2$ in air is required for deep cultures. Although a small bubble size promotes more rapid gas exchange it tends to be more damaging when they burst at the surface of the medium (minimized by Pluronic F68) and larger bubble sizes are generally preferred [Nienow, 2006].

*Wave bioreactors.*  Culture bags (see Appendix II), similar to those used for medium storage, can be used to scale up culture of suspension cells. They are gas permeable and can be agitated by rocking on trays on a flat rocking platform (Fig. 26.6; Sartorius). This also promotes gas exchange. The

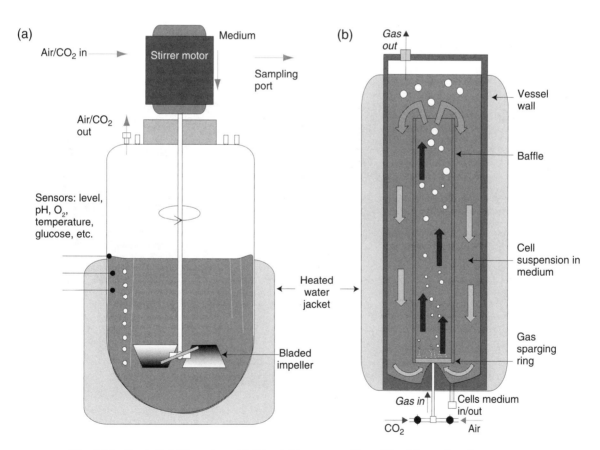

**Fig. 26.4. Controlled Bioreactors.** (*a*) Stirred bioreactor with paddle, inlet and outlet ports, gas sparging, and control sensors. (*b*) Air lift bioreactor consisting of two concentric cylinders, with the inner cylinder being shorter at both ends than the outer, thereby creating an outer and an inner chamber. The bottom of the inner chamber carries a sintered steel ring through which 5% $CO_2$ in air is bubbled. The bubbles rise, carrying the cell suspension with them. Air/$CO_2$ is vented from the top, and displacement ensures the return of the cell suspension down the outer chamber. (Modified from Griffiths, 2000.)

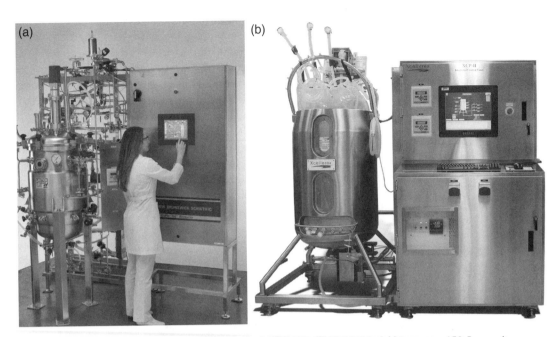

**Fig. 26.5. Large-Scale Bioreactors.** (*a*) Autoclavable reusable RightModel bioreactor, 150-L capacity. (Courtesy of New Brunswick Scientific.) (*b*) Single use XDR bioreactor where the culture bag is sterile and disposable; capacity 200 to 2000 L. (Courtesy of Xcellerex.)

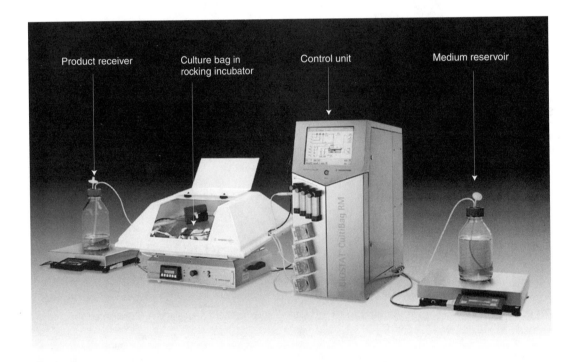

**Fig. 26.6. Wave Bioreactor.** A culture bag with a cell suspension in medium is rocked inside a rocking incubator enclosure, maintaining agitation of the cell suspension and enhancing gas exchange. Medium can be fed to the culture bag, regulated by the control unit, and cell product can be harvested to the receiver; both media supply and receiver can be media bags instead of bottles. (Courtesy of Sartorius Stedim.)

(a)

(b)

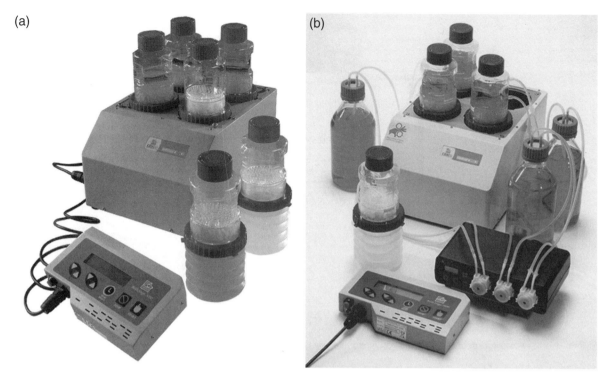

**Fig. 26.7. BelloCell Aerator Culture.** Cells are grown on carriers that are alternately submerged in medium and exposed for gas exchange by the pump action of the lower bellows compartment. (*a*) Four bioreactors on drive consol with control unit at front. Individual bioreactors at front show the BioNOC carriers cell compartment above central ring and concertina shape of bellows compartment. (*b*) Connected to reservoirs and peristaltic pump for continuous perfusion. (*See also* Plate 22*d*.) (Courtesy of Bellco, Inc.)

bag comes sterile, is disposable, and can be housed in a purpose built incubator/shaker. The unit cost and preparation time are low.

***Air lift fermentors.*** Large-scale fermentors sometimes use the air lift principle (Fig. 26.4*b*): 5% $CO_2$ in air is pumped into a porous steel ring at the base of the central cylinder, and bubbles stream up the center, carrying a flow of liquid with them, and are released at the top, while the medium is recycled to the bottom of the cylinder. This fairly simple type of fermentor has been used extensively in the biotechnology industry, up to capacities of 20,000 L but fed-batch STRs are now generally preferred [M. Butler, *personal communication;* Hacker et al., 2009].

***BelloCell aerator culture.*** This unusual device has a bellows medium compartment that alternately forces medium over cells anchored in porous matrices and withdraws it again, in a "breathing" motion of the bellows (Fig. 26.7; Bellco). Optimum mixing and aeration is claimed with minimum shear and very high densities may be obtained with cell output equivalent to 12 roller bottles [Ho et al., 2004].

***Perfused suspension culture.*** Hollow fiber and membrane perfusion systems both operate on the principle of

compartmentalization. The cells are retained in a low-volume compartment at a very high concentration, and medium is perfused through hollow fibers within the cell compartment (Fig. 26.8; *see also* Section 7.4.1). Regulation of gas exchange in the medium is external to the culture chamber. Hollow fibers are also available with double concentric spaces so that nutrient can be perfused down the center, product collected in the outer space, and cells grown in between the two concentric tubes (*see* Fig. 25.2).

## 26.2 SCALE-UP IN MONOLAYER

Although cells that grow in suspension are easier to manage, anchored cells, with their resultant potential for developing polarity, may yet represent a better model for posttranslational modification of proteins and the appropriate membrane flux, leading to their secretion from the cells in a bioactive form. Most normal, untransformed, cells grow in this configuration. For anchorage-dependent monolayer cultures, it is necessary to increase the surface area of the substrate in proportion to the number of cells and the volume of medium. This requirement has prompted a variety of different strategies, some simple and others complex.

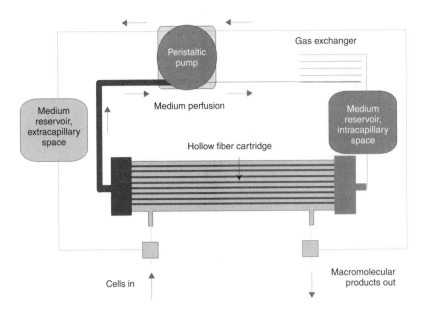

**Fig. 26.8. Hollow Fiber Perfusion.** Sectional diagram of hollow fiber perfusion system. Medium is circulated from a reservoir to the culture chamber by a peristaltic pump. Aeration and $CO_2$ exchange are via a gas-permeable tubing coil when the assembly is housed in a $CO_2$ incubator (*see also* Fig. 7.11.)

**Fig. 26.9. Corning Hyperflask.** Multisurface flask with 1720-$cm^2$ growth area and the same footprint as a 175-$cm^2$ flask. Arrow indicates air space between growth surfaces where gas exchange occurs with growth compartments. (Flask provided by The Automation Partnership.)

## 26.2.1 Multisurface Propagators

*Flasks.* Multisurface flasks are available as an intermediate stage of scale-up, such as Nunc Triple-Flask, (*see* Fig. 7.6) and Corning HyperFlask (Fig. 26.9).

*Stacked trays.* One of the simpler systems for scaling up monolayer cultures is to stack multiple tray units in tiers of up to 40 layers. The original design was the Nunc Cell Factory, and this has been joined by the similar Corning Cellstack (Fig. 26.10; also Table 7.2). These systems are made up of a welded stack of rectangular Petri dish-like units, with a total surface area of 600 to 24,000 $cm^2$, interconnected

at two adjacent corners by vertical tubes. Because of the positions of the apertures in the vertical tubes, medium can flow between compartments only when the unit is placed on end. When the unit is rotated and laid flat, the liquid in each compartment is isolated, although the apertures in the interconnecting tubes still allow connection of the gas phase. These bioreactors have the advantage that they are not different in the geometry or the nature of the substrate from a conventional flask or Petri dish.

The recommended method of using the Cell Factory is described in Protocol 26.2 (*see* Fig. 26.11). Handling the CellStack is similar.

---

### PROTOCOL 26.2. NUNC CELL FACTORY

**Outline**
Prepare a cell suspension in medium, and run the suspension into the chambers of the unit, lying on its long edge (*see* Fig. 26.11). Rotate through 90°, and then lay the unit flat and gas it with $CO_2$. Seal and incubate the unit.

**Materials**

Sterile:

- ❑ Monolayer cells
- ❑ Growth medium
- ❑ Trypsin, 0.25% crude
- ❑ D-PBSA
- ❑ Nunc Cell Factory, 10 chamber
- ❑ Silicone tubing and connectors

*Nonsterile:*

❑ Hemocytometer or electronic cell counter and counting fluid

**Procedure**

1. Trypsinize the cells (*see* Protocol 12.3), resuspend them, count the cells (*see* Protocols 20.1, 20.2), and dilute the suspension to $2 \times 10^4$ cells/mL in 2 L of medium.
2. Place the chamber on a long edge, and connect the medium delivery tube to a female:female Luer connector in the supply tube to the bottom port (*see* Fig. 26.11).
3. Run the cells and medium in through the supply tube. Medium in all chambers will reach the same level.
4. Rotate the chamber through 90° in the plane of the monolayer, so that the unit lies on the short edge, with the entry ports and supply tube at the top.
5. Disconnect the medium delivery tube from the medium reservoir at the Luer connection.
6. Rotate the chamber through 90° perpendicular to the plane of the monolayer, so that it lies flat on its base, with the culture surfaces horizontal.
7. Purge the unit with 5% $CO_2$ in air for 5 min, and then clamp off both the supply and the outlet. (The chamber may be gassed continuously if desired.)
8. Transport it to the incubator, tipping the medium away from the supply and vent ports.
9. To change the medium (or collect the medium), follow step 6 in reverse; remove the filter on the inlet tube, and then follow step 4 in reverse.
10. Swab Luer connection, open the clamp, and drain off the medium.
11. Replace the medium as in steps 2 through 8.
12. Collect the cells:
    (a) Remove the medium as in steps 9 and 10.
    (b) Add 500 mL of D-PBSA and then remove it.
    (c) Add 500 mL of trypsin at 4°C, and remove it after 30 s.
    (d) Incubate the cells with the residual trypsin for 15 min.
    (e) Add 500 mL medium, distribute evenly among the chambers, and rock to resuspend the cells.
    (f) Run the cells and medium off as in steps 9 and 10.
13. The residue may be used to seed the next culture, although this method makes it difficult to control the seeding density. It is better to discard the chamber and start fresh.

*Analysis*

*Observation.*  Monitoring cell growth in these chambers is difficult, so a single tray is supplied to act as a pilot culture. It is assumed that the single tray will behave as the multichamber unit.

*Product harvest.*  The supernatant medium can be collected repeatedly for virus or cell product purification. The collection of cells for analysis depends on the efficiency of trypsinization. This technique has the advantage of simplicity, though it can be expensive if the unit is discarded each time cells are collected. Harvesting of supernatant medium is primarily good for producing large numbers of cells ($1 \times 10^8 - 1 \times 10^{10}$).

***Multiarray disks, spirals, and tubes.***  Disks, spirals, and tubes have all been used to increase the surface area for monolayer growth, but few of these systems are now available commercially. Most matrix or multisurface propagators have gone toward perfusion (*see* Section 26.2.5), with the emphasis on product recovery. Corning markets a multisurface perfusion system, called CellCube (Fig. 26.12), that is a hollow polystyrene cube with multiple inner lamellae, perfused with oxygenated, heated medium. The inner lamellae are capable of supporting monolayer growth on both surfaces but growth may not be entirely random [Aunis et al., 2003].

## 26.2.2  Roller Culture

If cells are seeded into a round bottle or tube that is then rolled around its long axis on a roller rack (Fig. 26.13), the medium carrying the cells runs around the inside of the bottle. If the cells are nonadherent, they will be agitated by the rolling action but will remain in suspension in the medium. If the cells are adherent, they will gradually attach to the inner surface of the bottle and grow to form a monolayer (Fig. 26.14). This system has three major advantages over static monolayer culture: (1) the increase in utilizable surface area for a given size of bottle; (2) the constant, but gentle, agitation of the medium; and (3) the increased ratio of the medium's surface area to its volume, which allows gas exchange to take place at an increased rate through the thin film of medium over cells not actually submerged in the deep part of the medium.

---

**PROTOCOL 26.3. ROLLER BOTTLE CULTURE**

**Outline**

Seed a cell suspension in medium into a round bottle, and rotate the bottle slowly on a roller rack.

**Materials**

*Sterile or aseptically prepared:*

❑ Medium and medium dispenser
❑ D-PBSA
❑ Crude trypsin, 0.25%
❑ Monolayer culture
❑ Roller bottles

*Nonsterile:*

❑ Hemocytometer, or electronic cell counter and counting fluid
❑ Supply of 5% $CO_2$
❑ Roller rack (*see* Fig. 26.13)

**Procedure**

1. Trypsinize the cells, and seed them at the usual density for these cells.

*Note.* The gas phase is large in a roller bottle. With media based on Hanks's salts and a gas phase of air (*see* Table 8.1), it may therefore be necessary to blow a little 5% $CO_2$ into the bottle (e.g., for 2 s at 10 L/min). If the medium is $CO_2$/$HCO_3$-buffered, then the gas phase should be purged with 5% $CO_2$ (for 30 s–1 min at 20 L/min, depending on the size of the bottle; *see* Protocol 12.3).

2. Place the bottle on the roller rack, and rotate it at 20 rph until the cells attach (24–48 h).
3. Increase the rotational speed to 60 to 80 rph as the cell density increases.
4. To feed the cells or harvest the medium, take the bottles to a sterile work area, draw off the medium as usual, and replace it with fresh medium (*see* Protocol 12.1). A transfusion device (*see* Fig. 4.4) or media bag is useful for adding fresh medium, provided that the volume is not critical. If the volume of medium is critical, it may be dispensed by pipette or metered by a peristaltic pump (*see* Section 4.2.3).
5. Harvest the cells:
   (a) Remove the medium, rinse the cells with 50 to 100 mL of D-PBSA, and discard the D-PBSA.
   (b) Add 50 to 100 mL of trypsin at 4°C, and roll the bottle for 15 s by hand or on a rack at 20 rpm.
   (c) Draw off the trypsin, incubate the bottle for 5 to 15 min, and add medium to the bottle.
   (d) Rock and rotate the bottle and wash off the cells by pipetting.

*Analysis.* Monitoring cells in roller bottles can be difficult, but it is usually possible to see the cells on an inverted microscope. However, with some microscopes, the condenser needs to be removed, and with others, the bottle may not fit on the stage. So choose a microscope with sufficient stage accommodation.

The roller bottle system is probably the most economical for repeated harvesting of large numbers of cells or for collecting supernatant medium, although it is labor intensive and requires investment in a roller rack (*see* Fig. 26.13).

**Variations**

*Aggregation.* Some cells may tend to aggregate before they attach. This behavior is difficult to overcome, but may be reduced by lowering the initial rotational speed to 5, or even 2, rph, trying a different type or batch of serum, or precoating the surface of the roller bottle with fibronectin or polylysine (*see* Section 7.2.1).

*Size.* A range of bottles (around 500–1800 cm²), both disposable and reusable, is available (*see* Table 7.2; Appendix II: Roller Bottles). Some bottles are provided with a ribbed inner surface to increase the surface area available for cell growth (Corning).

*Volume.* The volume of the medium may be varied. A low volume will give better gas exchange and may be better for untransformed cells. Transformed cells, which are more anaerobic, grow faster, and produce more lactic acid, may be better in a larger volume of medium. The volumes given in Table 7.1 are mean values and may be halved or doubled as appropriate.

*Mechanics.* Roller racks are preferable for bottles (*see* Fig. 26.13*b*), as they are economical in space and allow easy observation of the bottles. Roller drums (Fig. 26.15) require more space for a given volume of culture bottle but can be useful for large numbers of smaller bottles or tubes. Small bottles and tubes can also be rotated on a roller rack by enclosing them in a cylindrical holder.

*Disposables.* Besides regular disposable roller bottles, it is possible to purchase presterilized sets of 20 bottles, equivalent to 200 standard roller bottles, which are single use and disposable. They come as a sterile validated pack, are connected to medium supply lines that fit in a peristaltic pump on the unit, and are inserted into a drum on a drive unit, which rolls into a $CO_2$ or standard incubator. Handling can be manual or robotically controlled (Rollercell 40, Cellon).

## 26.2.3 Microcarriers

Monolayer cells can be grown on microbeads 90 to 300 μm in diameter. These are made of plastic, glass, gelatin, or collagen [Griffiths, 2000] (Table 26.1). Usually there is a central core, which may be polystyrene, cellulose, or dextran, coated with a variety of different surfaces, including glass, tissue culture treated polystyrene, gelatin, or fibronectin. The surface charge is usually cationic but may vary. The microbeads are usually

***Fig. 26.10. Multisurface Propagators.*** (*a, b*) Nunc Cell Factory; stacked culture trays, each of 632 cm². (*a*) Two-chamber, 1264 cm², and 10-chamber, 6320 cm², cell factories. (*b*) Forty-chamber cell factory, 25,280 cm². (Courtesy of Nunc A/S.) (*c–e*) Corning CellStack. (*c*) Single-chamber, 636 cm², (*d*) 10-chamber, 6360 cm², and (*e*) forty-chamber, 254,440 cm², cell factories. (Courtesy of Corning Life Sciences.)

spherical, and some are macroporous so that cells can grow in the interstices of the bead as well as on the surface (e.g., Cytoline, Cytopore, and Cultispher; Fig. 26.16). Culturing monolayer cells on microbeads gives a maximum ratio of the surface area of the culture to volume of the medium, up to 90,000 cm²/L, depending on the size and density of the beads, and has the additional advantage that the cells may be treated as a suspension. Whereas the Nunc Cell Factory gives

an increase in scale with conventional geometry (*see* Section 26.2.1), microcarriers require a significant departure from the usual substrate design. However, this difference has relatively little effect at the microscopic level (*see* Plate 18*c*), as the cells are still growing on a smooth surface at the solid–liquid interface, although some cells are influenced by the radius of curvature and may prefer larger bead diameters or planar surfaces (*see* Plate 18*d*).

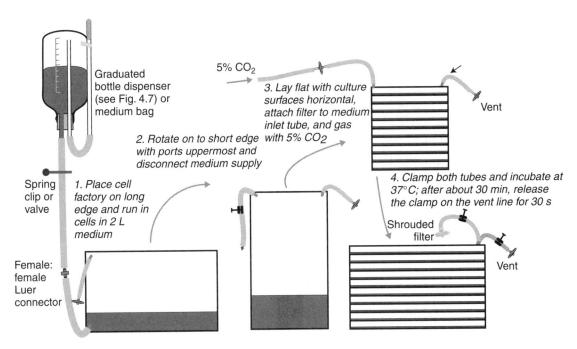

Graduated bottle dispenser (see Fig. 4.7) or medium bag

5% $CO_2$

*3. Lay flat with culture surfaces horizontal, attach filter to medium inlet tube, and gas with 5% $CO_2$*

Vent

*2. Rotate on to short edge with ports uppermost and disconnect medium supply*

Spring clip or valve

*1. Place cell factory on long edge and run in cells in 2 L medium*

*4. Clamp both tubes and incubate at 37°C; after about 30 min, release the clamp on the vent line for 30 s*

Shrouded filter

Female: female Luer connector

Vent

*Fig. 26.11. Filling Nunc Cell Factory.*

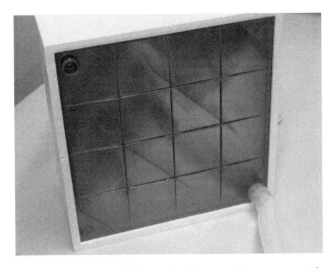

*Fig. 26.12. Corning CellCube.* Multisurface cell propagator with 6500-cm² available growth surface. (Courtesy of ATCC.)

The major difference created by microcarrier systems is in the mechanics of handling [Griffiths, 2000]. Efficient stirring without grinding the beads is essential and can be achieved with a suspended rotating pendulum (Techne; *see* Figs. 7.7, 26.1, 26.2) or paddle (Bellco), as for suspension cell culture, by rotating at 30 rpm. Technical literature is available from microcarrier suppliers describing the properties of the microcarriers, and to assist in setting up satisfactory cultures, a number of protocols for microcarrier systems have been published [e.g., Griffiths, 2000; Butler, 2004]. (*See also* Protocol 17.4.)

## PROTOCOL 26.4. MICROCARRIERS

### Outline
Seed the cells at a high cell and bead concentration, dilute, stir, and sample the cells as required.

### Materials

*Sterile:*

- Growth medium
- Microcarriers (*see* Table 26.1; Appendix II)
- Starter culture
- Stirrer flask (*see* Fig. 7.7, 26.1, 26.2; Appendix II)

*Nonsterile:*

- Magnetic stirrer (Techne, Bellco)

### Procedure

1. Suspend the beads at 2 to 3 g/L, in one-third of the final volume of medium required.
2. Trypsinize and count the cells.
3. Seed the cells at three to five times the normal seeding concentration into the bead suspension.
4. Stir the culture at 10 to 25 rpm for 8 h.
5. Add medium to reach the final volume, dictated by a bead concentration of 0.7 to 1 g/L.
6. Increase the stirring speed to approximately 60 rpm.
7. If the pH falls, feed the culture by switching off the stirrer for 5 min, allowing the beads to settle,

and then replacing one-half to two-thirds of the medium.

8. Harvest the cells:
   (a) Remove the medium.
   (b) Wash the cells by settling.
   (c) Trypsinize the beads with trypsin/EDTA.
   (d) Allow the beads to settle.
   (e) Spin down the cells.
   (f) Wash the cells by resuspension and centrifugation.

*Analysis.* Cell counting on beads can be difficult. Therefore the growth rate of the cells should be checked by determination of DNA (*see* Protocol 20.3) or protein (*see* Protocol 20.4) if nonproteinaceous beads are used, or by dehydrogenase activity if using the MTT assay (*see* Protocol 21.4) on a sample of the beads [Liu et al., 2004].

*Variations.* Most variations on this method arise from the choice of bead or design of the culture vessel and stirrer [Griffiths, 2000]. Bead density varies from 1.03 to 1.05 g/cc and influences the stirring speed, since higher speed is required for higher density beads. Composition, or coating, of the beads will also influence attachment, and more fastidious cells may prefer gelatin or collagen beads, which are actually soluble when digested by protease activity. Glass beads are the easiest to recycle.

Cells can also be cultured in hollow microspheres of sodium alginate (*see* Protocol 22.16).

### 26.2.4 Large Microcarriers

There are porous microcarriers that are larger (sometimes called *macrocarriers*), varying in size from around 1 × 2 mm (GE Healthcare Cytoline) to 6 mm diameter discs (New Brunswick FibraCel). Their macroscopic structure is made up of a number of different materials, such as treated polyester/polypropylene, polylactic acid (PLA), polyglycolic acid (PGA), collagen, or gelatin (Gelfoam) in a variety of different geometries (*see* Fig. 25.9). These can be loaded with cells and stirred in a bioreactor or perfused in a fixed-bed reactor (*see* Section 26.2.5; Figs. 26.7, 26.17, 26.18).

### 26.2.5 Perfused Monolayer Culture

Perfusion is frequently used to facilitate medium replacement and product recovery. The Cellcube (Corning) is a perfused, multisurface, single-use propagator with growth-surface areas from 21,250 to 85,000 cm$^2$ (*see* Fig. 26.12), with associated pumps, oxygenator, and system controller. Other perfusion systems use cells anchored to macrocarriers or beads in fixed-bed reactors (*see* below).

*Hollow fiber perfusion.* There are a number of hollow fiber perfusion systems in which adherent cells grow on the outer surface of the perfused microcapillary bundles. High-molecular-weight products concentrate in the outer space with the cells, while nutrients are supplied and metabolites removed via the inner space (*see* Section 25.3.2; Fig. 26.8). The potential of these systems for adherent cells lies in the re-creation of high, tissue-like cell densities, matrix interactions, and the establishment of cell polarity, all of which may be important for post-translational processing of proteins and for exocytosis. Although originally designed for attached cells, hollow fiber systems are also used extensively for suspension cells, such as hybridomas (*see* Section 27.7).

*Fixed-bed reactors.* Systems have also been developed with glass, matrix beads, or macrocarriers such as FibraCell (Figs. 26.7, 26.17, 26.18; Table 26.1), with the medium being perfused upward through the bed or percolating downward by gravity. The product is collected with the spent medium

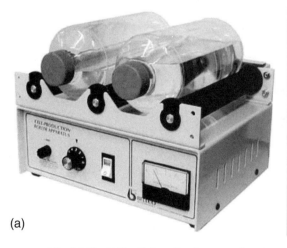

(a)          (b)

**Fig. 26.13. Roller Culture Bottles on Racks.** (*a*) Small bench-top rack. (Courtesy of Bellco.) (*b*) Large freestanding extendable rack. (Bellco; Courtesy of Beatson Institute.)

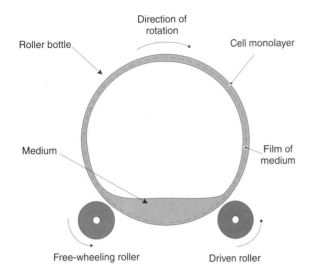

**Fig. 26.14. Roller Bottle Culture.** The cell monolayer (dotted line) is constantly bathed in liquid but is submerged only for about one-fourth of the cycle, enabling frequent replenishment of the medium and rapid gas exchange.

in a reservoir [Nehring et al., 2006; Pörtner & Platas Barradas, 2007].

***Microencapsulation.*** Sodium alginate behaves as a sol or gel, depending on the concentration of divalent cations. It will gel as a hollow sphere around cells in suspension in a high concentration of divalent cations (*see* Protocol 22.16; Plate 18*b*). Because the alginate acts as a barrier to high-molecular-weight molecules, macromolecules secreted by the cells are trapped within the vesicle, while nutrients, metabolites, and gas freely permeate the gel. The product and

**Fig. 26.15. Roller Drum Apparatus.** New Brunswick Scientific's TC-7 Roller Drum. Roller drums are used for roller culture of large numbers of small bottles or tubes. From an original idea by George Gey. (Courtesy of New Brunswick Scientific.)

cells are recovered by reducing the concentration of divalent cations. These gels have a low immunoreactivity and can be implanted in vivo.

## 26.3 PROCESS CONTROL

The progress of suspension cultures is monitored via pH, oxygen, $CO_2$, and glucose electrodes that read from the

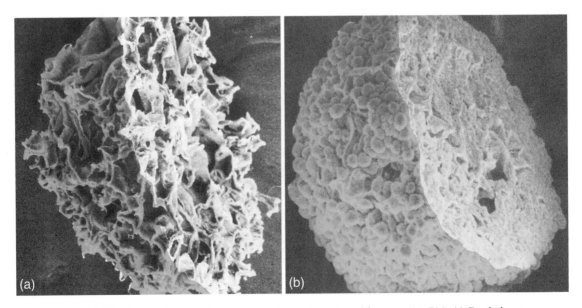

**Fig. 26.16. Cytopore Microcarriers.** Porous microcarriers viewed by scanning EM. (*a*) Bead alone; (*b*) bead coated with CHO cells. (Courtesy of GE Healthcare.)

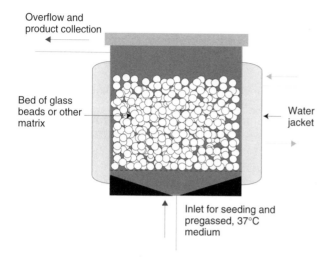

**Fig. 26.17. Fixed-Bed Reactor.** Cells grown on the surface of beads or macrocarriers are perfused with medium. The beads may be glass, plastic, or porous ceramic and are settled in a dense bed resting on a perforated base at the bottom of the culture vessel or, if of a lighter material, are restrained within a cage. Once the culture is established, the beads do not move, and medium percolates around them.

culture in situ, and by assaying the utilization of nutrients, such as glucose and amino acids, or the buildup of metabolites, such as lactate and ammonia, and products,

such as immunoglobulin from hybridomas (Fig. 26.19). The number of cells and other parameters, such as ATP, DNA, and protein, are determined in samples drawn from the culture and are used to calculate the total biomass. The temperature of the medium is regulated by preheating the input medium and by heating the surrounding water jacket regulated by feedback from the temperature probe. The flow rate of the medium is controlled, and can match the output to the sample line if the suspension is running as a biostat (*see* Section 26.1.1), and the stirring speed can be regulated, along with the viscosity of the medium, to reduce the shear stress.

There is a recurrent problem when monolayer cell cultures are scaled up, particularly in a fixed-bed or hollow fiber bioreactor: it is no longer possible to observe the cells directly, and both monitoring the progress of a culture by cell counting and determining the biomass become difficult. One approach to this problem [Brindle, 1998; Thelwall et al., 2001] uses nuclear magnetic resonance (NMR) to assay the contents of the culture chamber. By placing the cells in a perfused hollow fiber chamber within the magnetic field of an NMR spectroscope, a characteristic NMR spectrum is generated, enabling the identification and quantitation of specific metabolites (Fig. 26.20). NMR spectroscopes can also be used as imaging devices, producing a quasi-optical section through the chamber to reveal the distribution of the cells (*see* Fig. 25.10), and even for distinguishing between proliferating and nonproliferating zones of the culture.

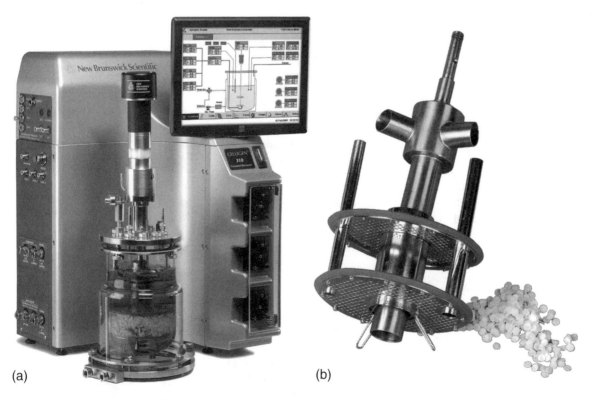

(a)  (b)

**Fig. 26.18. Celligen® 310 Bioreactor.** (*a*) Bioreactor chamber with FibraCel® macrocarriers in medium surrounded by water jacket connected to pump and control unit. (*b*) Impeller, and basket with FibraCel® disks lying alongside. (Courtesy of New Brunswick Scientific.)

**TABLE 26.1.** Microcarriers and Macrocarriers

| Name | Supplier | Composition | Specific gravity | Diameter, μm |
|---|---|---|---|---|
| Biosilon (Nunclon) | Sigma Krackeler | TC-treated polystyrene | 1.05 | 160–300 |
| Cultispher-G | Sigma | Gelatin, macroporous | N/A | 120–180 |
| Cultispher-G | Thermo Fisher | Gelatin, macroporous | N/A | 120–180 |
| Cultispher-GL | Thermo Fisher | Gelatin, macroporous | N/A | 150–330 |
| Cultispher-S | Sigma | Gelatin, macroporous | N/A | 120–180 |
| Cultispher-S | Thermo Fisher | Gelatin, macroporous | N/A | 120–180 |
| Cytodex 1 | GE Healthcare | DEAE-dextran | 1.03 | 130–220 |
| Cytodex 1 | Sigma | DEAE-dextran | 1.03 | 130–220 |
| Cytodex 2 | GE Healthcare | DEAE-dextran | 1.04 | 115–200 |
| Cytodex 2 | Sigma | DEAE-dextran | 1.04 | 133–215 |
| Cytodex 3 | GE Healthcare | Gelatin-coated dextran | 1.04 | 130–210 |
| Cytoline 1 | GE Healthcare | High-density polyethylene/silicate macroporous | 1.32 | 500–1000 × 1700–2500 |
| Cytoline 2 | GE Healthcare | High-density polyethylene/silicate macroporous | 1.03 | 500–1000 × 1700–2500 |
| Cytopore 1 | GE Healthcare | Macroporous DEAE-cellulose (low charge density) | 1.03 | 230 |
| Cytopore 2 | GE Healthcare | Macroporous DEAE-cellulose (high charge density) | 1.03 | 230 |
| FACT III, Collagen | SoloHill | Gelatin/polystyrene | 1.02, 1.04 | 90–212 |
| FibraCel | New Brunswick | Treated polyester/polypropylene | | 6000 diameter disc |
| Glass | SoloHill | Glass-coated polystyrene | 1.02, 1.04 | 90–150, 125–212 |
| Hillex | SoloHill | Modified polystyrene, cationic | 1.12 | 90–212 |
| Hillex II | SoloHill | Modified polystyrene, cationic | 1.11 | 160–180 |
| HyQ Spheres | Hyclone (Thermo Fisher)/SoloHill | Cross-linked polystyrene; uncoated, cationic, or with fibronectin or collagen coating | 1.02/1.12 | 90–212 |
| MicroHex 2D | Nunc (Thermo Fisher) | Polystyrene | 1.05 | 125 × 25 |
| Plastic | SoloHill | Cross-linked polystyrene | 1.02, 1.04 | 90–150, 125–212 |
| Plastic Plus | SoloHill | Cross-linked polystyrene, cationic | 1.02, 1.04 | 90–150, 125–212 |
| ProNectin | SoloHill | Cross-linked polystyrene, recombinant fibronectin | 1.02, 1.04 | 90–150, 125–212 |
| Rapidcell C | MP Biomedicals | Collagen-coated | 1.02, 1.03 | 90–210 |
| Rapidcell G | MP Biomedicals | Glass | 1.02, 1.03 | 90–210 |
| Rapidcell P | MP Biomedicals | Polystyrene | 1.02, 1.03 | 90–210 |
| SoloHill FACT III, Collagen | Sigma | Gelatin/polystyrene | 1.02, 1.04 | 90–212 |
| SoloHill ProNectin | Sigma | Cross-linked polystyrene, recombinant fibronectin | 1.02, 1.04 | 90–150, 125–212 |

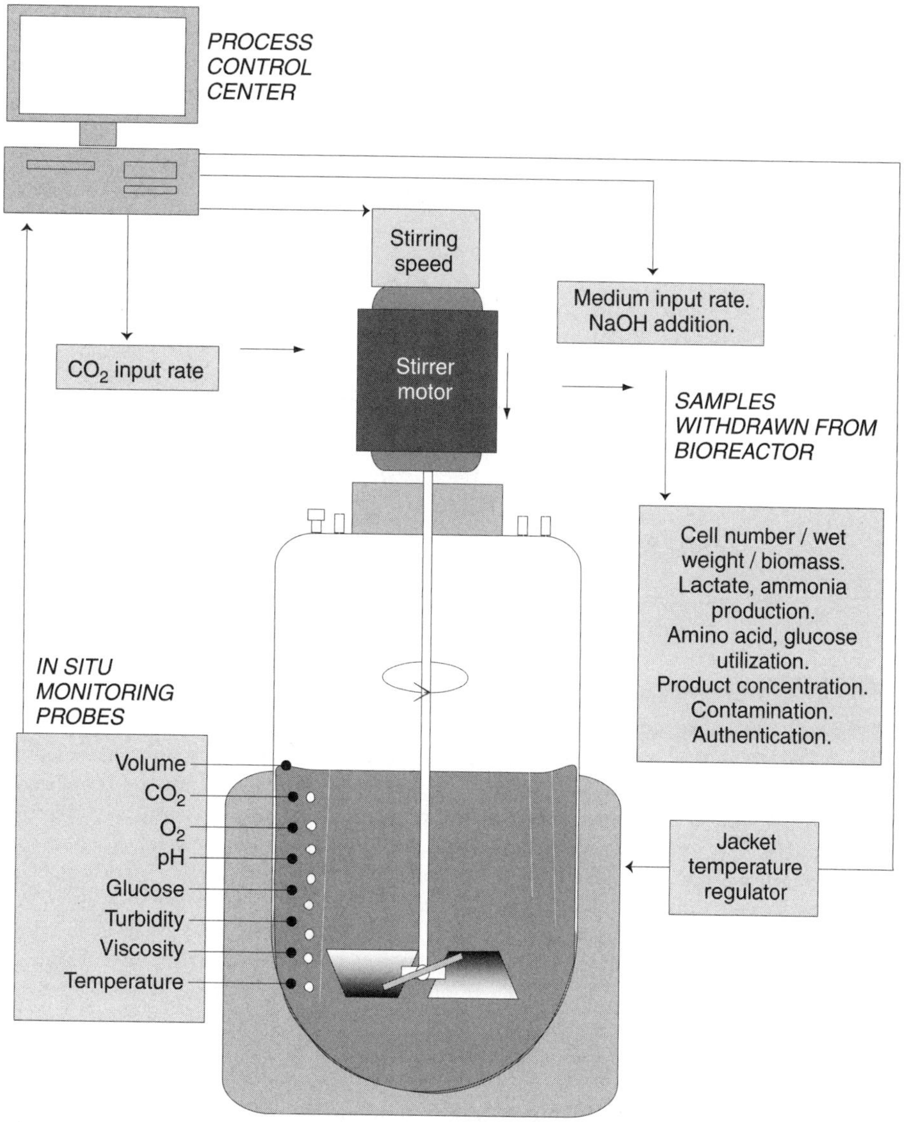

***Fig. 26.19. Bioreactor Process Control.*** Schematic representation of a paddle-stirred bioreactor with direct-reading probes on the left, feeding to a control unit (top left) that stores the data and also regulates conditions within the bioreactor. A sampling port on the right withdraws the cell suspension from the bioreactor for analysis.

## 26.4 AUTOMATION

Implicit in many of the procedures described for scale-up is the involvement of some degree of automation. Bioreactors can be fed and harvested automatically (*see* Figs. 26.3, 26.19), and many other devices are continuously monitored for physiochemical and physiological parameters with feedback used to adjust pH or nutrient concentrations (*see* Section 26.3). However, there has always been a need to automate routine culture and subculture procedures and the large-scale handling of multiple replicates for high-throughput screening. This has driven the introduction of robotics into cell culture, on the one hand, for large-scale

production in conventional culture flasks, and on the other, for handling and analyzing tens of thousands of microsamples per day.

### 26.4.1 Robotic Cell Culture

One of most widely used systems is the Cellmate (The Automation Partnership, TAP), which will handle regular culture flasks and roller bottles and subculture by enzymatic or mechanical disaggregation (Fig. 26.21; *see also* Plate 24). Versions can also clone cells and harvest the clones (CellCelector) and expand embryonic stem cell cultures for tissue engineering (CompacT SelecT). As all operations are carried out in a sealed environment, contamination is

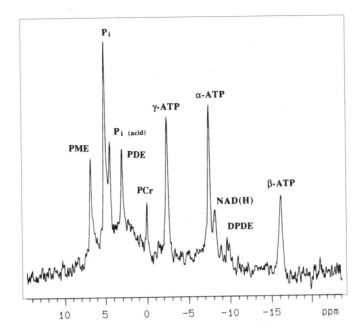

***Fig. 26.20. Analysis by NMR.*** Cell growth in hollow fibers, analyzed by NMR. [$^{31}$P] NMR spectrum of CHO cells growing in a hollow fiber reactor. The cells had been grown on macroporous beads in the extracapillary space of a specially constructed hollow fiber cartridge that could be accommodated within a 25-mm-diameter NMR probe. The cells were present at a density of approximately $7 \times 10^7$ cells/mL. *Abbreviations:* PME, phosphomonoesters, including phosphocholine and phosphoethanolamine; $P_i$, extracellular inorganic phosphorus; $P_i$(acid), an acidic (pH 6.7) extracellular $P_i$ pool within the bioreactor; PCr, phosphocreatine; γ-ATP, γ-phosphate of ATP; α-ATP, α-phosphate of ATP; PDE, phosphodiesters; DPDE, diphosphodiesters, including UDP-glucose; β-ATP, β-phosphate of ATP; NAD(H), nicotinamide-adenine dinucleotide (reduced). The chemical shift scale is referenced to phosphocreatine at 0.0 ppm. (*See also* Fig. 25.10) (Courtesy of Dr. Kevin Brindle.)

effectively prevented, and the system can work continuously as long as it is provided with fresh medium and seeding cultures. This frees up skilled staff to carry out more demanding operations.

The Compact Select system has been used in a number of applications [Gryseels, 2008] including drug discovery [Cawkill & Tucker, 2008; Lou et al., 2008] and stem cell production [Thomas et al., 2008]. It and a number of other robotics systems will handle multiwell plates and microplates, providing multiple replicates for cell-based high-throughput screening (*see* Fig. 26.21a, zones 3 and 6).

## 26.4.2 High-Throughput Screening

While in terms of culture volume and cell number most industrial cell culture output is used for production of biopharmaceuticals, much of the output in terms of number of individual cultures handled is directed toward the screening the response to experimental pharmaceuticals or cytotoxins of large numbers of increasingly small cell samples, usually in some sort of microplate or microscope slide array that can be accessed by microarray analysis [Fernandes et al., 2009; Kikuchi et al., 2009]. Cellular responses are recorded in situ and processed by downstream computer analysis. There are two major advantages of these systems: (1) very large numbers of samples with relatively few cells can be assayed simultaneously, and (2) data recording and reduction are instantaneous with minimal manual intervention.

There have been several successful attempts at automating growth curve generation in conventional cultures via in situ digital photo recording, as this is useful for general maintenance as well as analytical studies (*see* Section 20.9.3; Fig. 20.11) and can be used in conjunction with robotic

***Fig. 26.21. Robotic Cell Culture.*** The CompacT SelecT cell culture robotic system. (*a*) Front aspect:
(1) incubator chamber with flasks on top left of carousel; (2) main handling chamber with robotic arm
in center, below left of number, capping/uncapping and dispensing unit to left of number and pipette
container below number; (3) plate incubator chamber; (4) peristaltic pumps for media additions;
(5) Cedex cell counter (Innovatis); (6) multiwell plate liquid handling unit. (*See also* Plate 24.) (*b*)
Flask in gripper of robotic arm, showing flask being rocked, though it can also be shaken from side
to side. (*c*) uncapping, where cap remains on device until flask is returned for recapping. (*d*) Pipetting
cell suspension from front flask to rear flask. (Courtesy of The Automation Partnership.)

systems such as the CompacT SelecT. These systems are
also applicable to multiwell plates and microplates and can
perform scheduled observations on selected wells. There
are also a number of parallel mini bioreactor systems with

image monitoring, such as the PANsys 3000 (Applikon) and
Hexascreen (Hexascreen Culture Technologies), and even
down to nanoscale on a microfluidics platform [Barbulovic-
Nad et al., 2010]

# CHAPTER 27

# Specialized Techniques

This chapter covers a number of techniques that are ancillary to or dependent on cell culture, but that are not an essential element to initiation of a culture or its regular maintenance. They represent specialist interests or applications that may be of direct value to some readers, or simply provide background on tissue culture applications for others.

## 27.1 LYMPHOCYTE PREPARATION

### 27.1.1 Isolation by Density

Flotation on a combination of Ficoll and sodium metrizoate (Ficoll-Hypaque) is the most widely used technique [Boyum, 1968a, b] for separating lymphocytes from plasma and erythrocytes. It is suitable for applications such as PHA stimulation (*see* Protocol 27.2) for chromosome analysis, and can be used as a first step in the further purification of lymphocyte subclasses. It is also used in the isolation of stem cells from bone marrow (*see* Protocol 23.7) and in the enrichment of viable cells in primary cultures (*see* Protocol 11.10).

---

**PROTOCOL 27.1. PREPARATION OF LYMPHOCYTES**

**Outline**

Layer whole citrated or heparinized blood, depleted in red blood cells by dextran-accelerated sedimentation, on top of a dense layer of Ficoll-Hypaque. After centrifugation, most of the lymphocytes are found at the interface between the Ficoll-Hypaque and the plasma.

**Materials**

*Sterile:*

❑ Blood sample with heparin or citrate (concentration determined by blood sample collection container, which will already contain citrate or heparin)
❑ Clear centrifuge tubes or universal containers
❑ D-PBSA
❑ Ficoll-Hypaque, adjusted to 1.077 g/cc (*see* Appendixes I, II)
❑ Serum-free medium
❑ Syringe with blunt cannula, plastic Pasteur pipettes, or 1-mL pipettor with tips

*Nonsterile:*

❑ Hemocytometer or electronic cell counter
❑ Low-speed centrifuge

**Procedure**

1. Collect the blood sample in citrated or heparinized container and transport to laboratory.

△ *Safety Note.* Human blood may be infected with HIV, hepatitis, or other pathogens, and should be handled with great care (*see* Section 6.8.3).

---

2. Dilute 1:1 with D-PBSA, and layer 9 mL onto 6 mL Ficoll-Hypaque. This should be done in a wide, transparent centrifuge tube with a cap, such as a 25-mL universal container, or with double these volumes in a clear plastic 50-mL centrifuge tube.

3. Centrifuge the suspension for 15 min at 400 $g$ (measured at the center of the interface; see Appendix I).

4. Carefully remove the plasma/D-PBSA without disturbing the interface.

5. Collect the interface with a syringe (fitted with a blunt cannula), a plastic Pasteur pipette, or a 1-mL pipettor and dilute it to 20 mL in serum-free medium (e.g., RPMI 1640).

Δ **Safety Note.** Glass Pasteur pipettes and syringes with sharp needles should not be used with human blood. Instead, use a 1-mL pipettor or a syringe with a blunt cannula.

6. Centrifuge the diluted cell suspension from the interface at 70 $g$ for 10 min.

7. Discard the supernate, and resuspend the pellet in 2 mL of serum-free medium. If several washes are required—such as to remove serum factors—repeat resuspension in 20 mL of serum-free medium and centrifugation, two or three times more, before finally resuspending the pellet in 2 mL of serum-free medium.

8. Stain a sample of the cells with methylene blue, and count the nucleated cells on a hemocytometer. Alternatively, lyse a sample of the cells with Zapoglobin (Beckman Coulter), and count the nuclei on an electronic cell counter with a 70– 100-μm orifice tube or on a Nucleocounter (New Brunswick Scientific).

Lymphocytes will be concentrated at the interface, along with some platelets and monocytes. Some granulocytes may be found in the interface, although most will be found in the Ficoll–Hypaque, and in the pellet created in step 3. Monocytes and residual granulocytes can be removed from the interface fraction by taking advantage of their adherence to glass (beads or the surface of a flask) or to nylon mesh. Use a positive sort by MACS (see Section 14.3.2) or FACS (see Section 14.4) with specific lymphocyte subclass surface markers if purer preparations are required.

### 27.1.2 Blast Transformation

Lymphocytes in purified preparations or in whole blood may be stimulated with mitogens such as phytohemagglutinin (PHA), pokeweed mitogen (PWM), and antigen [Hume & Weidemann, 1980]. The resultant response may be used to quantify the immunocompetence of the cells. PHA stimulation is also used to produce mitoses for chromosomal analysis of peripheral blood [Dracopoli et al., 2004; McDevitt et al., 2007] (see Protocol 15.6).

---

**PROTOCOL 27.2. PHA STIMULATION OF LYMPHOCYTES**

**Outline**
Nucleated cells from Ficoll-Hypaque separation are exposed to a mitogen such as phytohemagglutinin.

**Materials**

*Sterile:*

❏ Medium: HEPES or $CO_2$-buffered DMEM, CMRL 1066, or RPMI 1640 supplemented with 10% autologous serum or fetal bovine serum
❏ Phytohemagglutinin (PHA), 50 μg/mL
❏ Colcemid, 0.01 μg/mL in BSS

*Nonsterile:*

❏ Test tubes or universal containers
❏ Microscope slides
❏ KCl, 0.075 M

**Procedure**

1. Using the washed interface fraction from step 7 of Protocol 27.1, incubate $2 \times 10^6$ cells/mL, 1.5 to 2.0 cm deep, in medium.

2. Add PHA, 5 μg/mL (final concentration), to stimulate mitosis from 24 to 72 h later.

3. Collect samples at 24, 36, 48, 60, and 72 h, and prepare drop preparations (see Protocol 15.6) or cytocentrifuge slides of the samples to determine the optimum incubation time (i.e., the peak mitotic index).

4. Add 0.001 μg/mL (final concentration) of colcemid for the 2 h during which the peak of mitosis is anticipated from observations made in step 3.

5. Centrifuge the cells after the colcemid treatment, resuspend the pellet in 0.075 M KCl for hypotonic swelling, and proceed as for chromosome preparation (see Protocols 15.6, 23.6).

---

## 27.2 AUTORADIOGRAPHY

This section is intended to cover microautoradiography of any small molecular precursor into a cold acid-insoluble macromolecule, such as DNA, RNA, or protein. Other variations may be derived from this text or found in the literature [Rogers, 1979; Baker, 2006]. Because autoradiography is used extensively to localize material in blots from electrophoresis as well as in microscope preparations, the two methods may be distinguished as macroautoradiography and microautoradiography, respectively (see Appendix IV).

**TABLE 27.1.** Isotopes Suitable for Autoradiography

| Isotope | Emission | Energy (mV) (mean) | T1/2 |
|---------|----------|--------------------|------|
| $^3$H | β$^-$ | 0.018 | 12.3 years |
| $^{55}$Fe | X rays | 0.0065 | 2.6 years |
| $^{125}$I | X rays | 0.035, 0.033 | 60 days |
| $^{14}$C | β$^-$ | 0.155 | 5570 years |
| $^{35}$S | β$^-$ | 0.167 | 87 days |
| $^{45}$Ca | β$^-$ | 0.254 | 164 days |

Isotopes suitable for microautoradiography are listed in Table 27.1. A low-energy emitter (e.g., $^3$H or $^{55}$Fe) in combination with a thin emulsion gives high intracellular resolution. Slightly higher energy emitters (e.g., $^{14}$C and $^{35}$S) give localization at the cellular level. Still higher energy isotopes (e.g., $^{131}$I, $^{59}$Fe, and $^{32}$P) give poor resolution at the microscopic level but are used for macroautoradiographs of chromatograms and blots from DNA, RNA, and protein electrophoresis, for which the absorption of low-energy emitters would limit the detection of incorporation. Low concentrations of higher energy isotopes ($^{14}$C and above), used in conjunction with thick nuclear emulsions, produce tracks that are useful in locating a few highly labeled particles (e.g., virus particles infecting a cell population or tissue).

Tritium is often used for autoradiography at the cellular and subcellular level, because the β-particles released have a mean range of about 1 μm in aqueous media, giving very good resolution. Tritium-labeled compounds are usually less expensive than the $^{14}$C- or $^{35}$S-labeled equivalents and have a longer half-life. Because of their low energy of emission, however, it is important that the radiosensitive emulsion be positioned in close proximity to the specimen, with nothing between the cell and the emulsion. Even in this situation only the incorporation in the top 1 μm of the specimen will be detected.

β-Particles entering the emulsion produce a latent image in the silver halide crystal lattice within the emulsion at the point where they stop and release their energy. The image may be visualized as metallic silver grains after treatment with an alkaline reducing agent (developer) and subsequent removal of the remaining unexposed silver halide by an acid photographic fixer ("hypo").

The latent image is more stable at low temperatures and in anhydrous conditions, so its sensitivity (signal relative to background) may be improved by exposure in a refrigerator over desiccant. This reduces the background silver grain formation by thermal activity. (Collaboration with N. Keith is gratefully acknowledged in the preparation of this protocol.)

---

# PROTOCOL 27.3. MICROAUTORADIOGRAPHY

## Outline
Incubate cultured cells with the appropriate isotopically labeled precursor (e.g., [$^3$H]thymidine to label DNA) and wash, fix, and dry the cells (Fig. 27.1). Perform any necessary extractions (e.g., to remove unincorporated precursors). Coat the specimen with emulsion in the dark and leave it to expose. After development silver grains can be seen in overlying areas where the radioisotope was incorporated (Fig. 27.2; *see* Plates 13*c, d,* 14*d*).

## Materials

### Setting up the culture

*Sterile:*

❑ Cells
❑ D-PBSA
❑ Trypsin
❑ Growth medium
❑ Coverslips or slides, and Petri dishes (may be non–tissue-culture grade if coverslips or slides are used) or plastic bottles

*Nonsterile:*

❑ Hemocytometer, or electronic cell counter and counting fluid

### Labeling with the Isotope

*Sterile:*

❑ Isotope
❑ HBSS

*Nonsterile:*

❑ Protective gloves
❑ Containers for disposal of radioactive pipettes
❑ Container for radioactive liquid waste

### Fixing and processing the cells

*Nonsterile:*

❑ Acetic methanol (1:3, ice cold, freshly prepared)
❑ DPX
❑ TCA, 0.6 N
❑ Deionized water
❑ Gelatin/chrome alum (subbing solution):
   (i) Add 0.2 g gelatin to 190 mL of UPW.
   (ii) Microwave for 2 min, cool, and filter.
   (iii) Dissolve 0.2 g chrome alum in 10 mL UPW.
   (iv) Add chrome alum to gelatin.

### Setting up autoradiographs

*Nonsterile:*

❑ Safelight: Kodak Wratten II or equivalent
❑ Emulsion (GE Healthcare/Amersham Hypercoat Emulsion LM-1, RPN 40)

**Note.** It is convenient to melt the emulsion and disperse it into aliquots in dipping vessels (GE Healthcare/

Amersham) suitable for the number of slides to be handled at one time. If the slides are sealed in a dark box, they may be stored at 4°C until required.

- ❑ Light-tight microscope slide boxes (Clay Adams, Raven)
- ❑ Silica gel (Fisher)
- ❑ Dark vinyl tape
- ❑ Black paper or polyethylene

**Note.** All glassware must be carefully washed and free of radioisotopic contamination. Plastic coverslips should be used in preference to glass, to minimize radioactive background. Be particularly careful to prevent spillages, and immediately mop up any that do occur. Wear gloves, and change them regularly—such as when you move from incubation (a high level of isotope) to handling washed, fixed slides (a low level of isotope).

### Processing autoradiographs

*Nonsterile:*

- ❑ Developer: Phenisol (Ilford), 20%
- ❑ Stop bath: 1% acetic acid
- ❑ Photographic fixer: 30% sodium thiosulfate
- ❑ Coverslips (#00)

### Staining

*Nonsterile:*

- ❑ Hematoxylin (filter before use) or Giemsa stain (*see* Protocol 15.2)
- ❑ Phosphate buffer, 0.01 M, pH 6.5
- ❑ Ethanol, 50%, 70%, and 100%
- ❑ Histoclear (Fisher)
- ❑ DPX (Merck)

### *Procedure*

#### Setting up the culture

1. Set up replicate monolayer cultures on coverslips, chamber slides (Nunc, Bellco), or Petri dishes.
2. Incubate the cultures at 37°C until the cells reach the appropriate stage for labeling.

#### Adding the isotope

3. Add the isotope, usually in the range of 0.1 to 10 µCi/mL (~4.0 KBq–0.4 MBq/mL), 100 Ci/mmol (~4 GBq/µmol), for 0.5 to 48 h as appropriate.

Δ *Safety Note.* Follow local rules for handling radioisotopes. Because such rules vary, no special recommendations are made here, other than the following: wear gloves; do not work in horizontal laminar flow; use a shallow tray with an absorbent liner to contain any accidental spillage; incubate the cultures in a box or tray labeled for use with

radioactivity; and regulate the disposal of radioactive waste according to local limits (*see* Section 6.7.2).

4. Remove all medium containing isotope, and wash the cells carefully in BSS, discarding the medium and washes.

Δ *Safety Note.* $^3$H-nucleosides are highly toxic because of their ultimate localization in DNA (*see* Section 6.7.1).

#### Fixing and processing the cells

5. Fix the cells in ice-cold acetic methanol for 10 min.
6. Prepare the slides:
    (a) Coverslips: Mount the coverslips on a slide with DPX or Permount, cells uppermost.
    (b) Cell suspensions (from growth in suspension or trypsinized): Centrifuge the cells onto a slide (*see* Protocol 15.4) or make drop preparations (*see* Protocol 15.7, steps 4–12).
    (c) Prepare several extra control slides for use in determining the correct duration of exposure. All preparations are referred to as "slides" from now on.
7. Extract acid-soluble precursors (when labeling DNA, RNA, or protein) with ice-cold 0.6 N TCA (3 × 10 min), and perform any other control extractions (e.g., with lipid solvent or enzymatic digestion).
8. Dip the slides in the gelatin/chrome alum subbing solution for 2 min.
9. Drain the slides vertically, and let them dry. (The slides may be stored dry at 4°C.)

#### Setting up the autoradiographs

10. Take the slides to the darkroom.
11. Under a dark-red safelight, melt the emulsion in a water bath at 46°C.
12. Mix the emulsion gently with a clean slide, taking care not to create bubbles.
13. Dip the slides:
    (a) Dip the slides in the emulsion for 5 s at 46°C, making sure that the cells are completely immersed. Note that the temperature is critical and will determine the thickness of the emulsion and, consequently, the resolution.
    (b) Withdraw the slides, and drain them for 5 s.
    (c) Dip the slides again in the emulsion for 5 s.
    (d) Withdraw the slides, and drain them vertically for 5 s.
    (e) Wipe the back of each slide with a paper tissue.
    (f) Allow the slides to dry flat on a tissue or piece of filter paper for 10 min.

**(g)** Place the slides on a rack in a light-tight box, and allow them to dry completely. Do not force the slides to dry; dry them slowly in humid conditions to avoid cracking the emulsion.

14. When the slides are dry (2–3 h), transfer them to light-tight microscope slide boxes with a desiccant, such as silica gel. Make sure that you do not touch the slides and that they do not touch each other or the desiccant.

15. Seal the boxes with dark vinyl tape (e.g., electrical insulation tape), wrap them in black paper or polythene, and place them in a refrigerator. Make sure that this refrigerator is not used for the storage of isotopes.

16. Leave the boxes at 4°C for 1 to 2 weeks. The exact time required will depend on the activity of the specimen and can be determined by processing one of the extra slides at intervals. Slides with prolonged exposure times have an increased background and are prone to latent image fade. It is better to increase the activity of the label than to increase the length of exposure.

### Processing the autoradiographs

17. Return the boxes to the darkroom and, under a dark-red safelight, unseal the boxes.

18. Allow the slides to come to atmospheric temperature and humidity (~2 min).

19. Prepare the solutions for development at 20°C.

20. Place the slides in the developer for 2.5 min with gentle intermittent agitation.

21. Wash the slides briefly in UPW and transfer to stop bath for a few seconds.

22. Wash the slides briefly in UPW and transfer to a photographic fixer for 5 min.

23. Rinse the slides in deionized water, and then place them in a hypo clearing agent (Kodak) for 1 min.

24. Wash the slides in cold running water, or with five changes of water over 5 min.

25. Dry the slides, and examine them on the microscope. Phase contrast may be used by mounting a thin glass (#00) coverslip in water. Remove the coverslip when you are finished examining the slides and before the water dries out, or else the coverslip will stick to the emulsion.

### Staining

26. To stain with hematoxylin:
    **(a)** Stain the slides in freshly filtered hematoxylin for 45 s.
    **(b)** Rinse the slides in running water for 2 min.
    **(c)** Dehydrate the slides in a succession of 50, 70, and 100% EtOH.
    **(d)** Clear the slides in Histoclear; mount coverslips in DPX. If a coverslip is used, it must be #00

with a minimum of mountant, to allow sufficient working distance for a 100× objective.

27. To stain with Giemsa:
    **(a)** Immerse the dry slides in neat Giemsa stain for 1 min.
    **(b)** Dilute the stain in situ 1:10 in 0.01 M phosphate buffer (pH 6.5) for 10 min.
    **(c)** Remove the staining solution by upward displacement with water. The slides should not be withdrawn or the stain poured off, or else the scum that forms on top of the stain will adhere to the specimen.
    **(d)** Rinse the slides thoroughly under running tap water until the color is removed from emulsion, but not from the cells.

*Qualitative analysis.* Determine the specific localization of grains—for example, over nuclei only or over one cell type rather than another.

*Quantitative analysis*

(1) *Grain counting*: Count the number of grains per cell, per nucleus, or localized elsewhere in the cells. This requires a low grain density—about 5 to 20 grains per nucleus, 10 to 50 grains per cell—no overlapping grains, and a low uniform background.

(2) *Labeling index*: Count the number of labeled cells as a proportion of the total number of cells (*see also* Protocol 20.11). The grain density should be higher for this assessment than for grain counting, to ease the recognition of labeled cells. If the grain density is high (e.g., ~100 grains per nucleus), set the lower threshold at, say, 10 grains per nucleus or per cell, but remember that low levels of labeling, significantly over background, may yet contain useful information, such as regarding DNA repair.

Microautoradiography is a useful tool for determining the distribution of isotope incorporation within a population, but it is less suited to total quantitation of isotope uptake or incorporation, for which scintillation counting is preferable.

*Variations*  Isotopes of two different energies—such as $^3$H and $^{14}$C—may be localized in one preparation by coating the slide first with a thin layer of emulsion, then coating that layer with gelatin alone, and finally coating the gelatin with a second layer of emulsion [Rogers, 1979]. The weaker β-emission from the $^3$H is stopped by the first emulsion and the gelatin overlay, while the higher energy β-emission from the $^{14}$C, having a longer mean path length of around 20 μm, will penetrate the upper emulsion.

Adams [1980] described a method for autoradiographic preparations from Petri dishes or flasks such that liquid emulsion is poured directly onto fixed preparations without the need for trypsinization.

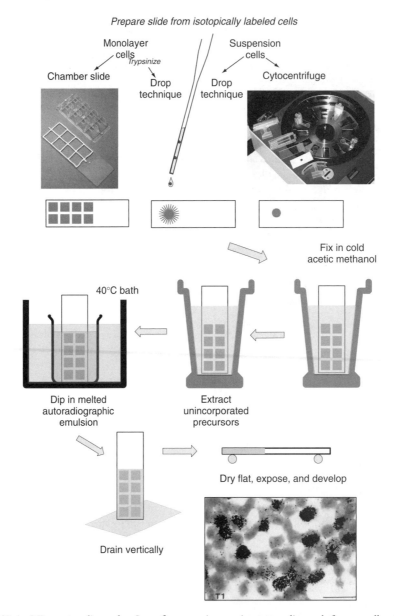

*Fig. 27.1. Microautoradiography.* Steps for preparing a microautoradiograph from a cell culture.

Soft β-emitters may also be detected in electron microscope preparations, using very thin films of emulsion or silver halide sublimed directly onto the section [Rogers, 1979].

Fluorescent and luminescent probes (GE Healthcare—Amersham) are now being used in place of radioisotopically labeled probes [Shaner et al., 2005]. The resolution with this method is often superior, quantitation is possible by confocal microscopy or a CCD camera, and disposal of reagents is environmentally friendly. However, the equipment for microscopic evaluation is expensive (~$50,000–$200,000).

## 27.3  TIME-LAPSE RECORDING

Time-lapse recording was developed primarily as a cinematography technique by which naturally slow processes can be observed at a greatly accelerated rate. At its inception the technique required a camera operated automatically by a signal from a timing device. Time-lapse cinemicroscopy for recording the behavior of cultured cells was started by Ronald Canti in 1920s [Canti, 1928]. In the 1950s Michael Abercrombie was the first to use the technique for behavior studies of tissue culture cells in a rigorously quantitative way [Abercrombie and Heaysman, 1953]. Video systems gradually replaced cine film, and "video microscopy" optimized their use in combination with light microscopes [Inoué and Spring, 1997]. The quality of video imaging progressively improved, and its digital form eventually superseded 16-mm film [Entwistle, 1998]. The microscopy

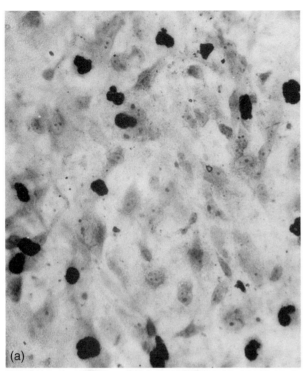

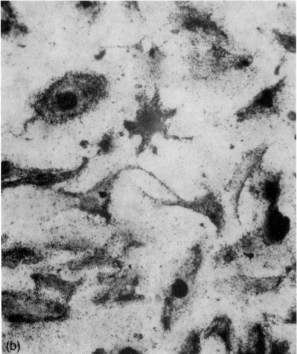

***Fig. 27.2. Microautoradiographs.*** Examples of [³H]thymidine incorporation into a cell monolayer. Normal glial cells were incubated with 0.1 μCi/mL (3.7 KBq/mL), 200 Ci/mmol (7.4 GBq/μmol), of [³H]thymidine for 24 h, washed, and processed as described in Protocol 27.3. (*a*) Typical densely labeled nuclei, suitable for determining the labeling index (*see* Protocol 20.11). (*b*) A similar culture, infected with mycoplasma showing [³H]thymidine incorporation in the cytoplasm. (*See also* Plates 13*c*, *d*, 14*d*).

technique itself, however, remains the critical link in the whole process. Development of green fluorescence protein (GFP) and other biofluorescent proteins [Shaner et al., 2005] that allow fluorescence labeling of specific subcellular structures in live cells revolutionized the time-lapse recording in tissue culture and dramatically increased its popularity. Detailed analysis of cell behavior in a limited number of recordings acquired by standard microscopy techniques is often performed manually or interactively on a frame-by-frame basis. Automated analysis of such recordings by digital image processing is generally difficult and is usually tailor developed for specific application [Zicha and Dunn, 1995]. Automatic analysis of fluorescence recordings tends to be easier than other commercially available microscopy techniques such as phase-contrast. Fully automatic analysis with unstained vertebrate tissue culture cells can be achieved with dual beam interference microscopy [Dunn and Zicha, 1993] or quadrature interferometry [Hogenboom et al., 1998]. These techniques are not currently commercially available but similar information can be derived by digital holographic microscopy available for example from Lyncee Tech (www.lynceetech.com). Automatic assessment of cell behavior is an important aspect of time-lapse analysis because large numbers of data are usually required in order to tackle the high intrinsic variability of the relevant phenomena.

This background information and the following protocol for time-lapse recording has been contributed by D. Zicha, Cancer Research UK London Research Institute, Lincoln's Inn Fields Laboratories, Light Microscopy, 44 Lincoln's Inn Fields, London, WC2A 3PX, UK.

### PROTOCOL 27.4. TIME-LAPSE VIDEO RECORDING

#### Outline
Prepare cells in a glass bottomed plastic Petri dish, which has better optical properties for microscopy techniques. The microscope is equipped with a video camera or higher quality scientific camera connected to a recording computer, and additional equipment provides the desired environmental conditions for the cell culture (temperature, $CO_2$, and humidity). During recording, individual exposures are triggered by the computer that also operates a shutter or shutters, eliminating unnecessary illumination of cells.

#### Materials

*Sterile:*

❏ Medium: e.g. DMEM without phenol red (in order to reduce background autofluorescence) with other standard supplements

❏ Petri dish: plastic or glass bottom (e.g., from MatTek) or a dish with a glass coverslip glued over a hole in its base

*Nonsterile:*

❑ Inverted microscope (e.g., Eclipse TE2000-E, Nikon)

❑ Environmental incubator providing 37°C, $CO_2$, and humidity (e.g., Cube/Box/Brick, available from Life Imaging Services; www.lis.ch)

❑ Computer-controlled shutter and filter wheel system (e.g., Lambda 10-3, from Sutter Instrument Co.; www.sutter.com)

❑ Camera (e.g., CoolSNAP cf (Photometrics, www.photomet.com/)

❑ Control computer with imaging software that records microscope images in time-lapse mode (e.g., Metamorph package from Molecular Devices; www.moleculardevices.com)

**Procedure**

1. Seed around 20,000 cells in a Petri dish (to achieve subconfluent culture for observation of individual cells) and allow the cells to settle at 37°C in a humidified atmosphere with 10% $CO_2$ overnight.

2. Switch on all the equipment, especially the heater, well in advance (40 min to 1 h is usually sufficient), to allow the temperature within the environmental incubator to stabilize.

3. Place the Petri dish with the cells into the stage incubator.

4. Use the appropriate microscopy technique—for example, phase contrast and fluorescence with Plan Fluor ELWD 20×/0.45 Ph1 DM objective lens. A higher power objective lens with a short working distance will require the glass bottom Petri dish.

5. To perform an alternating acquisition of phase-contrast and fluorescence images, with the Metamorph imaging system:

    (a) Calibrate the image size using a stage micrometer unless the information is already available.

    (b) Chose the Multi Dimensional Acquisition application in Apps.

    (c) Tick Timelapse and Multiple Wavelengths in Main tab.

    (d) Select Directory and specify Base Name in Saving tab.

    (e) Specify Number of time points and Time Interval in Timelapse tab; analysis of translocation in slowly moving, fibroblast-like, cells will give satisfactory results at 5-min lapse intervals. Observation of cell morphology that changes much faster will require lapse intervals of 1 min or even less. Also translocation of fast cells, such as neutrophil leukocytes, will need 1 min or shorter lapse intervals.

    (f) Specify Number of Wavelengths in Wavelengths tab as 2 and select appropriate Illumination configurations (shutters and positions of filter wheels) with Exposures, using Snap and Live buttons to check the images; exposure time of 0.1 s is usually sufficient for phase contrast, whereas weak fluorescence requires a longer exposure such as 0.4 s.

    (g) Preview button will help to arrange windows to be displayed during the acquisition.

    (h) Start acquisition using the Start button in the Preview or the Acquire button in the main Multi Dimensional Acquisition window.

    (i) Separate phase-contrast and fluorescence movie sequences can be viewed, after the acquisition, either independently or in combination using Review Multi Dimensional Data application.

6. Store the image sequences on a data server or other backup medium.

Δ **Safety Note.** Arc lamps used for fluorescence excitation emit UV radiation which can be dangerous. Follow the manufacturer's instructions for handling these light sources.

**Analysis** The imaging software Metamorph can be also used for the interactive or automatic measurements of positions, areas, and intensities in the images. It can be applied to fluorescence images for the evaluation of cell translocation using centroid positions, of cell spread area, of the intensity of the fluorescence signal, or of the shape factors in morphological studies.

**Variations** More extensive information on time-lapse recording of tissue culture cells can be found in [Dobbie & Zicha 2002]. The choice of medium depends on the requirements of the cells under observation. Medium with Hanks's salts has the advantage that its pH equilibrates automatically in a sealed culture chamber. Specialized chambers with perfusion option are also available (e.g., Confocal Imaging Chamber RC-30HV from Warner Instruments). Differential interference contrast (DIC) is a common alternative to phase contrast that improves the contrast of details at high resolution. Time-lapse analysis with multiwell plates has become popular with multifield recording, using motorized stages, which is supported by the commercially available acquisition software. It is no longer worth giving serious consideration to film-based methods, or even disc or tape time-lapse video recorders, because digital recording on a computer now provides greater convenience as well as other features—namely image processing, including contrast enhancements, background-noise subtraction, and support for analysis. Alternative time-lapse recording software packages for Mac and Windows OS include Volocity (Perkin Elmer) or Micro-Manager (www.micro-manager.org).

## 27.4  CELL SYNCHRONY

A number of techniques have been developed that follow the progression of cells through the cell cycle. The recorded cell population may be fractionated or blocked metabolically so that on return to regular culture conditions, the cells are all at the same phase and progress through the cycle in synchrony [Jackman & O'Connor, 2001].

### 27.4.1  Cell Separation

Techniques for cell separation have been described previously (*see* Chapter 14). Sedimentation at unit gravity can be used [Shall & McClelland, 1971; Shall, 1973], but centrifugal elutriation is preferable if a large number of cells ($>5 \times 10^7$) are required [Mikulits et al., 1997] (*see* Figs. 15.4, 15.5). Fluorescence-activated cell sorting (*see* Figs. 15.8, 15.9) can also be used, in conjunction with a nontoxic, reversible DNA stain, such as Hoechst 33342. The yield is lower for this method than for centrifugal elutriation ($\sim 1 \times 10^7$ cells or less), but the purity of the fractions is higher.

One of the simplest techniques for separating synchronized cells is mitotic shake-off: monolayer cells round up at metaphase and detach when the flask in which they are growing is shaken. This method works well with CHO cells [Tobey et al., 1967; Petersen et al., 1968] and some sublines of HeLa-S$_3$. Placing the cells at 4°C for 30 min to 1 h can also be used to synchronize cells in cycle [Rieder & Cole, 2002] and enhances the yield at mitotic shake-off [Miller et al., 1972].

### 27.4.2  Blockade

Two types of blocking have been used:

(1) *Metabolic inhibitors*: Inhibition of (a) DNA synthesis, thymidine, and (b) cell cycle progression, in G1 (lovastatin), S (aphidicolin, mimosine), and G2/M (nocodazole [Jackman & O'Connor, 2001]. The effects of these agents are variable because many are toxic, and blocking cells in cycle at phases other than G$_1$ can lead to deterioration of the cells and entry into apoptosis. Hence the culture will contain nonviable cells, cells that have been blocked in cycle but are viable and cells that have escaped the block [Yoshida & Beppu, 1990].

(2) *Nutritional deprivation (G$_1$ phase)*: Serum [Chang & Baserga, 1977] or isoleucine [Ley & Tobey, 1970] is removed from the medium for 24 h and then restored, whereupon transit through the cycle is resumed in synchrony [Yoshida & Beppu, 1990; Jackman & O'Connor, 2001].

A high degree of synchrony (e.g., >80%) is achieved only in the first cycle; by the second cycle, the degree of synchrony may be <60%, and by the third cycle, cell cycle distribution may be close to random. A chemical blockade is often toxic to the cells, and nutritional deprivation does not work well in many transformed cells. Physical fractionation techniques are probably most effective and do less harm to the cells (*see* Sections 14.2, 14.4).

## 27.5  CULTURE OF CELLS FROM POIKILOTHERMS

The approach to the culture of cells from cold-blooded animals (poikilotherms) has been similar to that employed for warm-blooded animals, largely because the bulk of present-day experience has been derived from culturing cells from birds and mammals. Thus the dissociation techniques for primary culture use proteolytic enzymes, such as trypsin, with EDTA as a chelating agent. Fetal bovine serum appears to substitute well for homologous serum or hemolymph (and is more readily available), but modified media formulations may improve cell growth. A number of these media are available through commercial suppliers (*see* Appendix II), and the procedure for using these media is much the same as for mammalian cells: Try those media and sera that are currently available, assaying for growth, plating efficiency, and specialized functions (*see* Sections 8.6, 9.5). As the development of media for many invertebrate cell lines is in its infancy, it may prove necessary to develop new formulations if an untried class of invertebrates or type of tissue is examined. Most of the accumulated experience so far relates to insects and mollusks.

Culture of vertebrate cells other than bird and mammal has also followed procedures for warm-blooded vertebrates, and so far there has not been a major divergence in technique. As this is a developmental area, certain basic parameters will still need to be considered to render culture conditions optimal, and if a new species is being investigated, optimal conditions for growth may need to be established—such as pH, osmolality (which will vary from species to species), nutrients, and mineral concentration. Temperature may be less vital, but it should be fixed within the appropriate environmental range and regulated within ±0.5°C; overheating is particularly damaging.

Two reviews cover some of the early developments in invertebrate tissue culture [Vago, 1971, 1972; Maramorosch, 1976], and later work on fish and invertebrates is reviewed in an issue of *In vitro Cellular and Developmental Biology, Animal* [Smagghe & Goodman, 2009]. Aquatic invertebrate cell culture is reviewed in Mothersill & Austin [2000].

### 27.5.1  Fish Cells

Fish cell culture has become increasingly popular, because of the growing commercial interest in fish farming and environmental marine and fresh water toxicology [Lee et al., 2009]. Lee et al. list 31 fish cell lines available from ATCC and several others that have been reported in the literature from a wide range of fish species, including goldfish, catfish, zebrafish, salmon, trout, carp, and fathead minnow. Fish cell lines have been used in studies of microsporidial infections [Monaghan et al., 2009], and a spleen cell line, ZSSJ, has been developed as a feeder layer for fish ES cells [Xing et al., 2009]. A protocol has been provided for the culture of ES cells from zebrafish embryos, including the preparation of fish embryo fibroblastic feeder layers (see Protocol 23.5).

## 27.5.2 Insect Cells

There has been considerable interest for some time in the culture of insect cells for studies of pest control and environmental toxicology [Smagghe et al., 2009]. However, the greatest increase in the usage of insect cell culture has resulted from the use of baculovirus for gene cloning [Midgley et al., 1998]. Baculovirus is often grown in Sf9 cells, a continuous cell line from the fall armyworm *Spodoptera frugiperda*. Protocol 27.5, for the culture of Sf9 cells, has been abridged from Midgley et al. [1998]. In this method, Sf9 cells are kept growing continuously in a magnetic spinner culture flask (see Protocol 12.4) or a flat-bottom flask with a magnetic stirrer bar mixing at about 80 rpm, ensuring that the stirrer is not a source of heat. Ideally the cells should be maintained at 27°C, but it is possible to grow them without an incubator in a room with constant temperature between 20°C and 28°C. $CO_2$ is not required for these media. Cells can be maintained in standard plastic tissue culture flasks, but as the cells attach to the surface of the flask, they must be detached for subculture by scraping or dislodging the cells with a jet of medium. However, this method will result in a lot of cell death because the cells attach quite tightly to plastic when grown in the presence of serum. The cells should have a population-doubling time of under 24 h.

---

### PROTOCOL 27.5. PROPAGATION OF INSECT CELLS

#### Outline
Disperse the cells mechanically from the monolayer, and propagate them in suspension at 27°C.

#### Materials

*Sterile or aseptically prepared:*

❑ Cells: Sf9 [Smith et al., 1983] (ATCC #CRL-1711) or lines derived from the cabbage looper *Trichoplusia ni* (Tn368, or BTI-TN-5B1-4; also known as "High Five," available from Invitrogen)
❑ Growth medium: EX-CELL 400 (JRH Biosciences) containing 2 mM L-glutamine, supplemented with 5% FBS, and 5 mL of penicillin/streptomycin solution (50 U/mL penicillin, 50 μmL streptomycin). Store the medium at 4°C in the dark, and always

---

warm it to room temperature before use. Sf9 cells are very sensitive to changes in growth medium, so for any change (e.g., to use serum-free EX-CELL 400), acclimatize the cells by gradually adding the new medium over a number of days.

❑ Dimethyl sulfoxide (DMSO), 10% in FBS
❑ Pluronic F68
❑ Culture flasks
❑ Spinner flask and magnetic stirrer
❑ Incubator at 27°C ($CO_2$ not required)

#### Procedure

*Routine maintenance*

1. Detach the cells from the flask culture by scraping or dispersing them with a jet of medium, or use cells grown in suspension in a spinner flask.
2. Count the cells by hemocytometer, and determine their viability by dye exclusion with Trypan Blue or Naphthalene Black.
3. Seed the spinner flask at 0.5 to $1 \times 10^6$ viable cells per mL (20–100 mL in a 500-mL spinner flask).
4. Incubate the cells at 20°C to 28°C. (27°C is optimal.)
5. Dilute the cells to 0.5 to $1 \times 10^6$ cells per mL every 48 to 72 h, or when there are about 4 to $5 \times 10^6$ cells/mL.
6. Transfer the cells to a clean flask every 3 to 4 weeks.
7. If the cells clump, try stirring them slightly faster, and add the surfactant Pluronic F-68 (0.5–1.0% v/v) to reduce shearing.

*Freezing cells for storage*

1. Count the cells, and centrifuge at 1000 rpm (~200 *g*) for 5 min.
2. Resuspend the pellet at $1 \times 10^7$ cells/mL in 10% DMSO in FBS.
3. Dispense the cells into aliquots, place the aliquots into ampoules, and chill the ampoules on ice for 1 h.
4. Pack the tubes into a Styrofoam container.
5. Freeze the tubes slowly (~1°C/min) overnight at −70°C.
6. Transfer the tubes to a liquid nitrogen freezer.
7. To recover the frozen cells, thaw the ampoules rapidly at 37°C. (If the cells are stored in the liquid phase, take care to thaw them in a covered vessel, to avoid risk of injury from explosion of the ampoule.)
8. Transfer the cells into a 25-cm² flask containing 5 mL of medium. Tip the flask to spread out the cells evenly.

9. Remove the medium after 2 to 3 h, when most of the cells should have attached, and add 5 mL of fresh medium.
10. Leave the cells 2 to 3 days to recover before detaching, and then transfer them to a stirrer flask or a larger plastic flask as described previously.

## 27.6  SOMATIC CELL FUSION

### 27.6.1  Cell Hybridization

Somatic cells fuse if cultured with inactivated Sendai virus or with polyethylene glycol (PEG) [Pontecorvo, 1975]. A proportion of the cells that fuse progress to nuclear fusion, and a proportion of these cells progress through mitosis, such that both sets of chromosomes replicate together and a hybrid is formed. In some interspecific hybrids—such as human-mouse—one set of chromosomes (the human) is gradually lost [Weiss & Green, 1967]. Thus genetic recombination is possible in vitro, and in some cases, segregation is possible as well.

As the proportion of viable hybrids is low, selective media are required to favor the survival of the hybrids at the expense of the parental cells. TK$^-$ and HGPRT$^-$ mutants (*see* Section 13.7) of the two parental cell types are used, and the selection is carried out in HAT medium (hypoxanthine, aminopterin, and thymidine) (Fig. 27.3) [Littlefield, 1964a]. Only cells formed by the fusion of two different parental cells (heterokaryons) survive, since the parental cells and fusion products of the same parental cell type (homokaryons) are deficient in either thymidine kinase or hypoxanthine guanine phosphoribosyl transferase. The parental cells and homokaryons cannot therefore utilize thymidine or hypoxanthine from the medium, and as aminopterin blocks endogenous synthesis of purines and pyrimidines, they are unable to synthesize DNA.

Protocol 27.6 for somatic cell fusion has been contributed by I. Hickey, Science Department, St. Mary's University College, Belfast, Northern Ireland, UK.

Although many cell lines undergo spontaneous fusion, the frequency of such events is very low. To produce hybrids in significant numbers, cells are treated with the chemical fusogen polyethylene glycol (PEG) [Pontecorvo, 1975]. Selection systems that kill parental cells but not hybrids are then used to isolate clones of hybrid cells.

---

### PROTOCOL 27.6. CELL HYBRIDIZATION

**Outline**

Bring the cells to be fused into close contact, either in suspension or in monolayers. Treat the cells with PEG briefly, to minimize cell killing. Usually the cells

---

are given a 24-h period to recover before selection for hybrids.

**Materials**

*Sterile:*

❏ PEG 1000 (Merck):
   (i) Autoclave the PEG to liquefy and sterilize it.
   (ii) Allow it to cool to 37°C, and then mix it with an equal volume of serum-free medium, prewarmed to 37°C.
   (iii) Adjust the pH to approximately 7.6 to 7.9, using 1.0 M NaOH.
   (iv) Store the solution at 4°C for up to 2 weeks.
❏ Complete growth medium
❏ Serum-free growth medium
❏ NaOH, 1.0 M
❏ Petri dishes, 5 cm
❏ Universal containers

**Procedure**

*A. Monolayer Fusion:*

1. Inoculate equal numbers of the two types of cells to be fused into 5-cm tissue culture dishes. Between $2.5 \times 10^5$ and $2.5 \times 10^6$ of each parental cell line per dish is usually sufficient.
2. Incubate the mixed culture overnight.
3. Warm the PEG solution to 37°C. It may be necessary at this point to readjust its pH, using NaOH.
4. Remove the medium thoroughly from the cultures and wash them once with serum-free medium.
5. Add 3.0 mL of the PEG solution and spread it over the monolayer of cells.
6. Remove the PEG solution after exactly 1.0 min, and rinse the monolayer three times with 10 mL of serum-free medium before returning the cells to complete medium.
7. Culture the cells overnight.
8. Add selection medium.

*B. Suspension Fusion:*

1. Centrifuge a mixture of $4 \times 10^6$ cells of each of the two parental cell lines at 150 *g* for 5 min at room temperature. Carry out centrifugation and subsequent fusion in 30-mL plastic universal containers or centrifuge tubes.
2. Resuspend the pellet in 15 mL of serum-free medium, and centrifuge again.
3. Aspirate off all of the medium.
4. Resuspend the cells in 1 mL of PEG solution by gently pipetting.
5. After 1.0 min, dilute the suspension with 9 mL of serum-free medium, and transfer half of the suspension to each of two universal containers

or centrifuge tubes containing a further 15 mL of serum-free medium.

**6.** Centrifuge the suspensions at 150 *g* for 5 min. Remove the supernate, and resuspend the cells in complete medium.

**7.** Incubate the cells overnight at 37°C.

**8.** Clone the cells in selection medium.

*Variations* A large number of variations of the PEG fusion technique have been reported. Although the procedure described here works well with a range of mouse, hamster, and human cells in interspecific and intraspecific fusions, it is unlikely to be optimal for all cell lines. Inclusion of 10% DMSO in the PEG solution has the advantage of reducing its viscosity and has been reported to improve fusion [Norwood et al., 1976]. Also the molecular weight of the PEG used need not be 1000 Da. Preparations with molecular weights from 400 to 6000 Da have been successfully used to produce hybrids.

*Selection of hybrid clones* The method of selection used in any particular instance depends on the species of origin of the two parental cell lines, the growth properties of the cell lines, and whether selectable genetic markers are present in either or both cell lines. Hybrids are most frequently selected with the HAT system: 0.1 mM hypoxanthine, 0.6 μM aminopterin, and 1.6 μM thymidine [Littlefield, 1964a]. This system can be used to isolate hybrids made between

pairs of mutant cell lines deficient in the enzymes thymidine kinase (TK⁻) and hypoxanthine guanosine phosphoribosyl transferase (HGPRT⁻), respectively. TK⁻ cells are selected by exposure to BUdR and HGPRT⁻ cells by exposure to thioguanine, following the procedures described by Biedler in Chapter 14 (*see* Protocol 13.9). When only one parent cell line carries such a mutation, HAT selection can still be applied if the other cell line does not grow, or grows poorly in culture (e.g., lymphocytes, senescing primary cultures).

Differential sensitivity to the cardiac glycoside ouabain is an important factor in the selection of hybrids between rodent cells and cells from a number of other species, including human. Rodent cells are resistant to concentrations of this antimetabolite up to 2.0 mM, whereas human cells are killed at 10 μM ouabain. The hybrids are much more resistant to ouabain than the human parental cells. If a rodent cell line that is HGPRT⁻ is fused to unmarked human cells, then the hybrids can be selected in medium containing HAT and low concentrations of ouabain.

Although many other selection systems have been reported, few have been widely used. Exogenous dominant markers such as G418 resistance can be used successfully. It must be stressed that whichever method is used to isolate clones of putative hybrid cells, confirmation of the hybrid nature of the cells must be obtained. This is usually done with cytogenetic (*see* Protocol 15.7) or molecular techniques. In certain cases, comparing the number of hybrids with the frequency of revertants may be the only way of making this confirmation.

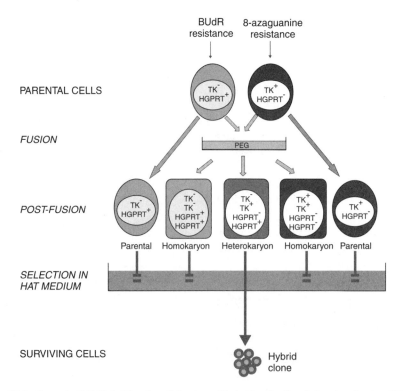

**Fig. 27.3. *Somatic Cell Hybridization.*** Selection of hybrid cells after fusion (*see* Section 27.7).

## 27.6.2  Nuclear Transfer

Genetic recombination experiments can also be carried out with isolated nuclei, but the major interest in this technique is related to cloning individual animals [Wolf et al., 1998] and examining the effect of nuclear transfer on gene expression [Gurdon & Melton, 2008] and stemness in heterokaryons of stem cells and somatic cells [Sumer et al., 2009]. Nuclei can be isolated by centrifuging cytochalasin B-treated cells and fusing the extracted nuclei to recipient whole cells or enucleated cytoplasts in the presence of PEG. However, in animal cloning experiments micromanipulation techniques are used to remove the nucleus from one cell and inject it into a fertilized, preimplantation egg from which the existing nucleus has been removed.

## 27.7  PRODUCTION OF MONOCLONAL ANTIBODIES

Monoclonal antibodies have become indispensable tools in research, diagnostics, and therapeutics. Since hybridoma technology was first introduced by Kohler and Milstein in 1975, monoclonal antibodies have replaced polyclonal antibodies in many different applications. The following introduction and Protocol 27.7 were contributed by J. Payne and T. Kuus-Reichel of Hybritech Incorporated, a subsidiary of Beckman Coulter Inc., 7330 Carroll Road, San Diego, CA 92121.

Hybridomas are produced by fusing a nonsecreting myeloma cell with an antibody-producing B-lymphocyte in the presence of polyethylene glycol (Fig. 27.4). The myeloma cell is deficient in hypoxanthine-guanine phosphoribosyl transferase (HGPRT) or thymidine kinase (TK), necessary for DNA synthesis, and cannot survive in selection medium containing hypoxanthine, aminopterin, and thymidine. Any unfused B-lymphocytes from the spleen cannot survive in culture for more than a few days. Any B-cell-myeloma hybrids should contain the genetic information from both parent cells and are thus able to survive in the HAT selection medium. They can be cultured indefinitely and will produce unlimited quantities of antibody. Supernates from surviving hybridomas are screened for antibody by ELISA. Those hybridomas selected are then subcloned to ensure that they are producing antibody that is specific for a single epitope. Antibody production can be scaled up in vivo as ascites in mice or in vitro as a suspension culture. Hybridomas also grow very well in various hollow fiber (*see* Section 25.3.2) and stirred bioreactor (*see* Section 26.1) systems.

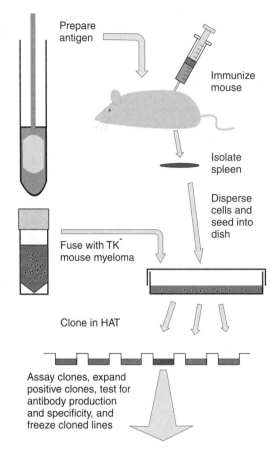

**Fig. 27.4. Production of Hybridomas.** Schematic diagram of the production of hybridoma clones capable of secreting monoclonal antibodies.

---

### PROTOCOL 27.7. PRODUCTION OF MONOCLONAL ANTIBODIES

#### Outline
Using polyethylene glycol (PEG), fuse spleen cells from an immunized mouse with myeloma cells

(P3.653). Select hybrid colonies (hybridomas) in HAT medium. Screen the supernates by ELISA 10 to 14 days after fusion (the ELISA screening protocol should be developed before the fusion), and expand, freeze, and subclone the desired hybridomas, to ensure monoclonality.

#### Materials

*Sterile or aseptically prepared:*

❑ Mice (Balb/c or A/J), 6 to 10 weeks old
❑ P3.653 myeloma: This myeloma cell line from ATCC does not secrete immunoglobulin and performs well in fusions.
❑ Antigen (125 μg per mouse is optimal): Small antigens can be conjugated to keyhole limpet hemocyanin (KLH, Sigma) to increase antigenicity.
❑ Adjuvants (alum or Freund's adjuvant; complete and incomplete, Sigma): A detailed description of these and other adjuvants can be found in Vogel and Powell [1995].

- TCD (T-Cell depletion) buffer: Hanks's balanced salt solution + 10 mM HEPES + 0.3% BSA
- NH$_4$Cl, 0.16 M
- Antimouse Thy 1.2 antibody (Accurate): For use, dilute 1:500 in TCD buffer, and filter sterilize.
- Rabbit complement, Low Tox M (Accurate): Reconstitute in 1 mL of cold UPW, dilute 1:12 in TCD buffer, and filter sterilize
- D-PBSA
- SFM (serum-free medium): MEM that has been stored at room temperature, with the cap of the container loosened to allow the release of CO$_2$; the pH must be alkaline.
- PEG (polyethylene glycol): Melt 10.5 mL of PEG 1450 (Sigma) in a 56°C water bath; add 19.5 mL of warm sterile MEM (pH 8.3 to pH 8.7); mix well; and allow the solution to equilibrate, with the cap of the container loosened, for 5 to 7 days before fusion.
- HAT stock (100×) (10 mM hypoxanthine, 40 μM aminopterin, 1.6 mM thymidine): 1.36 g of hypoxanthine, 729 μL of aminopterin (from 25 mg/mL stock), 0.387 g of thymidine, 0.022 g of glycine dissolved in 4 mL of 5 M NaOH + 26 mL UPW. Make up to 1 L with UPW. Filter sterilize.
- HAT medium: basal medium, such as MEM or RPMI, +10% FBS + 20% spleen-conditioned medium plus HAT stock (final dilution 1:100)
- SCM (spleen-conditioned medium): Tease 3 to 5 naive mouse spleens in PBSA. Count the cells, and resuspend them at 1 × 10$^6$ cells/mL in MEM + 10% horse serum. Transfer the suspension to a 500-mL spinner flask, and incubate it at 37°C in 5% CO$_2$ for 48 h. Remove the cells by centrifugation, and store the supernatant frozen.
- 8-Azaguanine stock, 10 mM (100×): 1.52 g of 8-azaguanine dissolved in 4 mL of 5 M NaOH + 21 mL of UPW. Make the solution up to 1 L with UPW. Filter sterilize.
- MEM, 10% fetal calf serum, 0.1 mM 8-azaguanine (for maintenance of P3.653 myeloma)
- HT stock (100×): 0.408 g of hypoxanthine, 0.1161 g of thymidine, 0.0067 g of glycine, dissolved in 2 mL of 5 M NaOH + 8 mL of UPW. Make the solution up to 300 mL with UPW. Filter sterilize.
- HT medium: HT stock diluted 100× in basal medium (as for HAT medium, above)
- Syringes, 1 mL with 25G needles
- Syringes, 1 mL, with 23G needles, ×2
- Dissecting instruments (scissors and forceps)
- Petri dishes, 60 × 15 mm
- Centrifuge tubes, 15 mL and 50 mL
- Multiwell plates, 24 well and 96 well
- Culture flasks

- 100-mL Nalgene bottle
- Pipette tips
- Reservoir for multipipettor (100 mL, Matrix Technologies; Corning Costar)

*Nonsterile:*

- Trypan Blue (Sigma)
- Multipipettor, 12 channel (Matrix Technologies)
- Inverted phase-contrast microscope
- Unopette microcollection system (BD Biosciences)
- Hemocytometer

### Procedure

*A. Immunization:*

1. Bleed the mice on day 0 before the initial injection, and check the serum for background antigen reactivity.
2. Immunize the mice (A/J or Balb/c) with antigen emulsified in Freund's adjuvant or mixed with a 1/10 volume of alum and vortexed. Give three injections of antigen intraperitoneally, according to the following schedule:

| Day | Amount of antigen | Adjuvant |
|-----|------|----------|
| 0 | 50 μg | Alum or complete Freund's adjuvant |
| 14 | 25 μg | Alum or incomplete Freund's adjuvant |
| 28 | 25 μg | Alum or D-PBSA |

3. Bleed the mice on day 35 and measure the serum titer of antibody by an ELISA assay.
4. Dilute the serum serially 1:4 after a 1:30 dilution, and up to 1:30,720.
5. Select mice with the highest ratio of serum titers to antigen for fusion.
6. Give the selected mice a final boost of 10 μg of antigen i.v. or 25 μg of antigen i.p. 3 days before fusion.

*B. Myeloma:*

1. It is convenient to perform fusions on a Thursday, with the mice receiving a final boost of antigen on a Monday.
2. Maintain the P3.653 myeloma cell line in MEM + 10% fetal calf serum + 8-azaguanine.
3. Dilute the P3.653 cells to 3.5 × 10$^5$ cells/mL each day for the three days before fusion.

*C. T-Cell Depletion:*

1. Bleed mice with appropriate serum titers, and sacrifice them by cervical dislocation.
2. Aseptically remove the spleens, and place them in a sterile Petri dish with 5 mL of sterile D-PBSA.

3. Gently tease the spleens with two 23G needles on 1-mL syringes. Teasing spleens roughly will result in a high concentration of fibroblasts.

4. Transfer the cells to a 15-mL conical tube, and allow clumps to settle.

5. Transfer spleen cells (without clumps) to a 50-mL conical tube, and, after a 1:100 dilution in a Unopette, count the cells with a hemocytometer.

6. Spin the cells at 200 $g$ for 8 min.

7. To lyse the red blood cells, resuspend the resultant pellet in 0.84% $NH_4Cl$ (10 mL/spleen), and incubate the suspension at 4°C for 15 min.

8. Underlayer the cell suspension with 14 mL of horse serum, and spin the solution at 450 $g$ for 8 min.

9. Resuspend the resultant pellet in 50 mL of TCD buffer, and spin the suspension at 200 $g$ for 8 min.

10. For T-cell depletion, resuspend the resultant cell pellet in diluted anti-Thy 1.2 at a final concentration of $1 \times 10^7$ cells/mL.

11. Incubate the suspension at 4°C for 45 min, and then spin it at 200 $g$ for 8 min.

12. Resuspend the resultant pellet in diluted rabbit complement.

13. Incubate the suspension at 37°C for 45 min, and then spin it at 200 $g$ for 8 min.

14. Count the cells by Trypan Blue exclusion on a hemocytometer. B-cell recovery should be 30% to 50%.

### D. Fusion:

1. Mix the myeloma and B-cells in a 50-mL centrifuge tube. One fusion can be done on a maximum of $1.2 \times 10^8$ spleen cells. Mix the spleen cells with P3.653 myelomas at a ratio of 4:1; thus, the maximum number of P3.653 cells per fusion is $3 \times 10^7$ cells.

2. Centrifuge the suspension at 200 $g$ for 8 min.

3. Break up the resultant pellet by tapping, and add 1 mL of PEG to the tube over 15 s.

4. Mix the suspension by gently swirling the tube for 75 s.

5. Add 1 mL of SFM over 15 s, and gently swirl the tube for 45 s.

6. Add 2 mL of SFM over 30 s, and swirl the tube for 90 s.

7. Add 4 mL of HAT medium over 30 s, and swirl the tube for 90 s.

8. Finally, add 8 mL of HAT medium over 30 s, and swirl tube for 90 s.

9. Add this volume (16 mL) to a sterile Nalgene bottle containing the calculated amount of HAT medium (125 mL if the maximum cell concentration has been used). 16 mL, containing $1.5 \times 10^8$ cells, from Step 1 in this section of the protocol plus 125 mL of HAT medium in the bottle makes 141 mL. With the wash in the next step (Step 10), the total volume is 150 mL and will result in a final concentration of $1 \times 10^6$ cells/mL.

10. Wash the 50-mL conical tube with 9 mL of HAT medium, and add this volume to the bottle.

11. Mix the contents of the bottle well, and transfer the cells to the sterile reservoir.

12. Using a 12-channel multipipettor, plate the cells at 200 µL/well into a sterile 96-well plate. The final concentration is then $2 \times 10^5$ cells/well.

### E. Selection of Hybridomas:

1. Feed the fusion plates 5 days after fusion, by aspirating most of the culture media from the wells and replacing it with 150 to 200 µL/well of fresh HAT medium.

2. Feed the plates twice per week.

3. Screen the clones for selection of positive hybridomas, usually two weeks after fusion, using ELISA.

4. After a further 48 h, retest those clones that tested positive in the previous step.

5. Expand the most productive hybridomas by culturing them in two wells of a 96-well plate in media containing 10% FBS and HT.

6. Retest the clones, expand the positive hybridomas to a 24-well plate, and wean them off HT medium, at which time 2 mL of culture supernatant should be harvested for screening. At this step, enough volume is harvested to perform several selection assays to ensure that the antibody is directed only at the antigen of interest.

7. Expand the hybridomas to be kept to 4 wells of a 24-well plate, and cryopreserve them (*see* Protocol 19.1).

8. Perform a second cryopreservation after expanding the hybridoma to a 75-cm² flask.

***Screening*** Take care in developing the screening strategy to obtain a monoclonal antibody with the characteristics that you want. Hybridoma culture supernates should be screened as early as feasible for desired reactivity patterns. After initial selection by ELISA for reactivity to the immunogen, the expanded culture supernatant should be tested in the application for which it was developed (Western blot, competitive immunoassay, flow cytometry, etc). A more detailed discussion of ELISA and other immunoassays can be found in Knott et al. [1997].

***Subcloning*** To ensure monoclonality, subclone hybridomas of interest. This can be done by serially diluting cells and plating the equivalent of 1 cell per 3 wells in a 96-well plate (*see* Section 13.1) or by sorting with an automated cell deposition unit (ACDU) on a FACStarplus (BD Biosciences) and plating at one cell per well. Subcloning can be done on top of a mouse spleen feeder layer plate. Feeder layers are prepared by teasing a naive mouse spleen and resuspending the cells at $1 \times 10^6$ cells/mL. The cells are then plated in a 96-well microtitration plate at a final concentration of $2 \times 10^5$ cells/well. After subcloning, colonies can usually be seen at day 5 and must be checked visually for monoclonality. Plates are fed with fresh medium beginning on day 7. Screening for positive hybridomas is usually done between days 10 and 14. Those clones selected are then expanded and frozen in the same way as the parental hybridoma.

***Antibody Production*** Concentrated antibody from clones of interest can be produced in vivo as ascites in IFA primed mice (Balb/c or nu/nu [Gillette, 1987]) or in vitro as a suspension culture. Several hollow fiber cell culture systems are also available (*see* Sections 25.3.2, 26.1.3, 26.2.5; Appendix II). When hybridomas are inoculated into a hollow fiber system, the cells are maintained in a compartment of the bioreactor, while fresh media and waste from the cells are recirculated. High concentrations of antibody are produced in the cell compartment, and culture supernate containing antibody can be harvested at multiple time points.

# CHAPTER 28

# Training Programs

## 28.1 OBJECTIVES

This book has been designed, primarily, as a source of information on procedures in tissue culture, with additional background material provided to place the practical protocols in context and explain the rationale behind some of the procedures used. There is a need, however, to assist those who are engaged in the training of others in tissue culture technique. Whereas an independent worker will access those parts of the book most relevant to his or her requirements, a student or trainee technician with limited practical experience may need to be given a recommended training program, based on previous experience and the supervisor's requirements. This chapter is intended to provide programs at basic, advanced, and specialized levels for an instructor to use or modify in the training of new personnel. Exercises presented in bold font in Table 28.1 are regarded as essential.

The programs are presented as a series of exercises in a standard format with cross-referencing to the appropriate standard protocols and background text and to ancillary protocols that are not part of the exercise but topically related. Cross-references are provided in the background section of each exercise and not repeated in the materials and procedures, as the student or trainee is expected to have read the relevant background material before embarking on the exercise and not be sourcing this information while carrying out the exercise. Protocol instructions are repeated in the exercises with suggestions for possible experimental modifications to make each exercise more interesting and, where possible, to generate data that the trainee can then analyze. Most are described with a minimal number of samples

to save manipulation time and complexity, so the trainee should be made aware of the need for a greater number of replicates in a standard experimental situation. The exercises are presented in a sequence, starting from the most basic and progressing toward the more complex, in terms of technical manipulation. They are summarized in Table 28.1, with those exercises that are regarded as indispensable presented in bold type. The basic and advanced exercises are assumed to be of general application and good general background, although available time and current laboratory practices may dictate a degree of selection.

Where more than one protocol is required, the protocol numbers are separated by a semicolon; where there is a choice, the numbers are separated by "or," and the instructor can decide which is more relevant or best suited to the work of the laboratory. It is recommended that all the basic and advanced exercises in Table 28.1 be attempted, and those in bold font be regarded as essential. The instructor may choose to be more selective in the specialized section.

Additional ancillary or related protocols are listed within each exercise. These do not form a part of the exercise but can be included if they are likely to be of particular interest to the laboratory or the student/trainee.

## 28.2 PREPARATIVE AND MANIPULATIVE SKILLS

The exercises that a trainee or student should attempt first are designed to help the trainee learn the necessary sterile manipulations and understand how sterile reagents and

**TABLE 28.1.** Training Programs

| Exercise Number | Procedures (those in bold font regarded as essential) | Training objectives | Based on Protocol |
|---|---|---|---|
| | *Basic preparative and manipulative skills* | | |
| 1. | **Pipetting and transfer of fluids** | Familiarization. Handling and accuracy skills. | 5.1, 5.2 |
| 2. | Washing and sterilizing glassware | Familiarization with support services. Appreciation of need for clean and nontoxic glass containers. | 10.1 |
| 3. | Preparation and sterilization of water | Appreciation of need for purity and sterility. Applications and limitations. Sterilization by autoclaving. | 10.5 |
| 4. | Preparation of D-PBS without phenol red | Constitution of salt solutions. Osmolality. Buffering and pH control. Sterilization of heat-stable solutions by autoclaving. | 10.6 |
| 5. | Preparation of pH standards; D-PBS with phenol red | Familiarization with use of phenol red as a pH indicator. | 8.1 |
| 6. | Preparation of stock medium and sterilization by filtration | Technique of filtration and appreciation of range of options. | **10.9**, 10.11–10.14 |
| | *Basic cell culture techniques* | | |
| 7. | Observation of cultured cells | Use of inverted microscope. Appreciation differences in cell morphology within and among cell lines. Use of camera and preparation of reference photographs. | **15.1** 15.6 |
| 8. | **Aseptic technique: preparing medium for use** | Aseptic handling. Skill in handling sterile reagents and flasks without contamination. Adding supplements to medium. | **5.1; 10.7** |
| 9. | Feeding a culture | Assessing a culture. Changing medium. | **12.1** |
| 10. | **Preparation of complete medium from 10× stock** | Aseptic handling. Constitution of medium. Control of pH. | **10.8** |
| 11. | **Counting cells by hemocytometer and an electronic counter** | Quantitative skill. Counting cells and assessment of viability. Evaluation of relative merits of two methods. | **20.1, 20.2** |
| 12. | Subculture of continuous cell line growing in suspension | Assessing a culture. Aseptic handling. Cell counting and viability. Selecting reseeding concentration. | **12.4** |
| 13. | Subculture of continuous cell line growing in monolayer | Assessing a culture. Aseptic handling. How to disaggregate cells. Technique of trypsinization. | **12.3** |
| 14. | Staining a monolayer cell culture with Giemsa | Cytology of cells. Phase-contrast microscopy. Fixation and staining. Photography. | 15.1, 15.2, 15.6 |
| 15. | Construction and analysis of growth curve | Replicate subcultures in multiwell plates. Cell counting. Selecting reseeding concentration. | 20.8, 20.1, 20.2 |

534

## Advanced culture techniques

| No. | Technique | Description | Reference |
|---|---|---|---|
| 16. | **Cell line characterization** | Confirmation of cell line identity. Increase awareness of overgrowth, misidentification, and cross-contamination. | **15.7 or 16.8 or 16.9** |
| 17. | **Detection of mycoplasma** | Awareness of importance of mycoplasma screening. Experience in fluorescence method or PCR for routine screening of cell lines for mycoplasma contamination. | **18.2 or 18.3** |
| 18. | **Cryopreservation** | How to freeze cells, prepare cell line and freezer inventory records, stock control. | **19.1. 19.2** |
| 19. | **Primary culture** | Origin and diversity of cultured cells. Variations in primary culture methodology. | **11.2; 11.6 or 11.7** |
| 20. | Cloning of monolayer cells | Technique of dilution cloning. Determination of plating efficiency. Clonal isolation. | 13.1, 20.10, 13.6 |

## Specialized culture techniques

| No. | Technique | Description | Reference |
|---|---|---|---|
| 21. | Cloning in suspension | Technique of dilution cloning in suspension. Isolation of suspension clones. | 13.4 or 13.5; 13.8 |
| 22. | Selective media | Demonstration of selective growth of specific cell types. | 22.1 or 22.2 |
| 23. | Cell separation | Isolation of cell type with desired phenotype by one of several separation methods. | 14.1 or 14.2 |
| 24. | Preparation of feeder layers | How to improve cloning efficiency. Selective effects of feeder layers. | 13.3 |
| 25. | Histotypic culture in filter well inserts | Familiarization with high-density culture. Potential for differentiation, nutrient transport, and invasion assay. | 25.4 |
| 26. | Cytotoxicity assay | Familiarization with high-throughput screening methods. Positive and negative effects. | 21.4 |
| 27. | Survival assay (Can be run as a component of Exercise 20 or as a separate exercise.) | Use of clonal growth to identify positive and negative effects on cell survival and proliferation. | 22.3 |

materials are prepared. Most of these early exercises are simple and straightforward to perform and are designed to enhance skills and provide a background to the later exercises. As there is a danger that these exercises may be seen as boring and distant from actual culturing of cells, the supervisor may choose to run them in parallel with the next section, Basic Cell Culture Techniques.

A tour of the tissue culture facilities is an essential introduction; this lets trainees meet other staff, determine their roles and responsibilities, and see the level of preparation that is required. The principles of storage should also be explained and attention drawn to the distinctions in location and packaging between sterile and nonsterile stocks, tissue culture grade and non-tissue-culture grade plastics, using stocks and backup storage, fluids stored at room temperature versus those stored at $4°C$ or $-20°C$. The trainee should know about replacement of stocks: what the shelf life is for various stocks, where replacements are obtained, who to inform if backup stocks are close to running out, and how to rotate stocks so that the oldest is used first. The degree to which the trainee will participate in preparatory procedures will depend on their ultimate role.

## Exercise 1  Sterile Pipetting and Transfer of Fluids

### *Purpose of Procedure*
To transfer liquid quickly, accurately, and aseptically from one container to another.

### *Applications*
Preparation of medium, feeding cultures, subculture, and all other manual liquid handling.

### *Training Objectives*
Skill in handling pipettes; appreciation of level of speed, accuracy, and reproducibility required. For example, in routine maintenance, speed and reproducibility are more important than accuracy, while in an experimental situation, accuracy may be more important.

*Supervision:* Continuous initially, then leave trainee to repeat exercise and record accuracy.

*Time:* 30 min to 1 h.

### *Background Information*
Sterile liquid handling (*see* Section 4.2.3); handling bottles and flasks (*see* Section 5.3.4); pipetting (*see* Section 5.3.5). **Standard protocols:** Aseptic technique (*see* Protocol 5.1 or Protocol 5.2).

*Demonstration materials or operations:* Instructor should demonstrate pipette handling: inserting in pipette controller and fluid transfer, and give some guidance on the compromise required between speed and accuracy. Instructor should also demonstrate fluid withdrawal by vacuum pump (if used in laboratory) and explain the mechanism and safety constraints.

## Instructions for Exercise 1
Adapted from Protocol 5.1. If laminar-flow hood not available, use Protocol 5.2.

### Outline
Clean and swab down work area, and bring bottles, flasks, pipettes, and other instruments. Transfer medium from bottle to flasks. Finally, tidy up and wipe over work surface with 70% alcohol.

### Materials
*Sterile (placed in hood):*
- ❏ Eagle's $1×$ MEM with Hanks's salts and $NaHCO_3$, without antibiotics, 100 mL
- ❏ Pipettes, graduated, and plugged, in an assortment of sizes, 1 mL, 5 mL, 10 mL, 25 mL; glass or bulk wrapped plastic in sterile square pipette cans, or individually wrapped plastic
- ❏ Culture flasks, 25 cm$^2$, preweighed ....................10
- ❏ Petri dishes, 5 cm, preweighed ........................10

*Nonsterile:*
- ❏ Pipette controller or bulb in hood
- ❏ Alcohol, 70%, in spray bottle in hood
- ❏ Lint-free swabs or wipes beside hood
- ❏ Absorbent paper tissues beside hood
- ❏ Pipette cylinder containing water and disinfectant on floor beside hood
- ❏ Waste bin (for paper waste, swabs, and packaging) on floor beside hood on opposite side from pipette cylinder
- ❏ Suction line to aspirator or waste beaker in hood (aspirator/waste beaker with disinfectant)
- ❏ Scissors
- ❏ Marker pen with alcohol-insoluble ink
- ❏ Notebook, pen, protocols

### Procedure
1. Swab down the work surface and all other inside surfaces of laminar-flow hood, including inside of front screen, with alcohol and a lint-free swab or tissue.
2. Bring media and reagents from cold store, water bath, or otherwise thawed from freezer, swab bottles with alcohol, and place those that you will need first in the hood.
3. Collect pipettes and place at one side of the back of work surface in an accessible.
4. Open pipette cans and place lids on top or alongside, with the open side down, or stack individually wrapped pipettes, sorted by size, on a rack or in cans.
5. Collect any other glassware, plastics, and instruments that you will need, and place them close by (e.g., on a cart or an adjacent bench).
6. Slacken, but do not remove, caps of all bottles about to be used.
7. Remove the cap of the flask into which you are about to pipette, and the bottles that you wish to

pipette from, and place the caps open side uppermost on the work surface, at the back of the hood and behind the bottle, so that your hand will not pass over them.

8. Select a pipette:

(a) If glass or bulk-wrapped plastic:

   (i) Take a pipette from the can, lifting it parallel to the other pipettes in the can and touching them as little as possible, particularly at the tops (if the pipette that you are removing touches the end of any of the pipettes still in the can, discard it).

   (ii) Insert the top end of the pipette into a pipette controller or bulb, pointing the pipette away from you and holding it well above the graduations, so that the part of the pipette entering the bottle or flask will not be contaminated.

(b) If individually wrapped plastic:

   (i) Open the pack at the top.

   (ii) Peel the ends back, turning them outside in.

   (iii) Insert the end of the pipette into the bulb or pipette controller.

   (iv) Withdraw the pipette from the wrapping without it touching any part of the outside of the wrapping, or the pipette touching any nonsterile surface.

   (v) Discard the wrapping into the waste bin.

Δ **Safety Note.** As you insert the pipette into the bulb or pipette controller take care not to exert too much pressure as pipettes can break if forced.

9. The pipette in the bulb or pipette controller will now be at right angles to your arm. Take care that the tip of the pipette does not touch the outside of a bottle or the inner surface of the hood; always be aware of where the pipette is. Following this procedure is not easy when you are learning aseptic technique, but it is an essential requirement for success and will come with experience.

10. Tilt the medium bottle toward the pipette so that your hand does not come over the open neck, and, using a 5-mL pipette, withdraw 5 mL of medium and transfer it to a T-25 flask, also tilted.

11. Repeat with a further 4 flasks. When you are pipetting into several bottles or flasks, they can be laid down horizontally on their sides. Ensure that the flasks remain well back in the hood and that your hand does not come over open necks. Record the time taken to add the medium to the 5 flasks.

12. Discard the pipette into the pipette cylinder containing disinfectant. Plastic pipettes can be discarded into double-thickness autoclavable biohazard bags.

13. Recap the flasks.

14. Repeat the procedure, using a 25-mL pipette to transfer 5 mL to each of 5 flasks from one filling of the pipette. Record the time taken to add the medium to the 5 flasks.

15. Replace the cap on the medium bottle and flasks. Bottles may be left open while you complete a particular maneuver, but should always be closed if you leave the hood for any reason.

**Note.** In vertical laminar flow, do not work immediately above an open vessel. In horizontal laminar flow, do not work behind an open vessel.

16. On completion of the operation, tighten all caps, remove all solutions and materials no longer required from the work surface, and swab down.

17. Weigh the flasks to check accuracy of medium dispensing.

18. Place at 37°C for 1 week, to check for possible contamination.

*Repeat exercise with 5- or 6-cm Petri dishes:*

1. Place dishes on one side of work area.

2. Position medium bottle and slacken the cap.

3. Bring dish to center of work area.

4. Remove bottle cap and fill 5-mL pipette from bottle.

5. Remove lid and place behind dish.

6. Add medium to dish, directing the stream gently low down on the side of the base of the dish.

7. Replace lid.

8. Discard pipette.

9. Return dish to side, taking care not to let the medium enter the capillary space between the lid and the base.

10. Repeat with 4 more dishes.

11. Repeat again with 5 dishes and 25-mL pipette.

12. Discard pipette.

13. Weigh dishes.

14. Place dishes in $CO_2$ incubator for 1 week to check for possible contamination.

### Data

(1) Calculate the mean weight of liquid in each flask.

(2) Note the range and calculate the error as a percentage of the volume dispensed.

### Analysis

(1) Compare the results obtained with each pipette and comment on the differences:

   (a) In accuracy

   (b) In time

(2) When would it be appropriate to use each pipette?

(3) What is an acceptable level of error in the accuracy of pipetting?

(4) Which is more important: absolute accuracy or consistency?

## Exercise 2   Washing and Sterilizing Glassware

### Purpose of Procedure
To clean and resterilize soiled glassware. This exercise is nonexperimental and should use the regular protocols (*see* below).

### Training Objectives
Appreciation of preparative practices and quality control measures carried on outside aseptic area.

*Supervision:* Nominated senior member of washup staff should take trainee through standard procedures.

*Time:* 20 to 30 min should be adequate for each session, but the time spent will depend on the degree of participation by the trainee in procedures as determined by his/her ultimate role and the discretion of the supervisor.

### Background Information
Preparation area (*see* Section 3.2.6); washup (*see* Section 4.4.1); glassware washing machine (*see* Fig. 4.14); sterilizer (*see* Section 4.4.3; Figs. 4.16, 4.17); washing and sterilizing apparatus (*see* Section 10.3). **Standard protocols:** Preparation of apparatus (*see* Protocols 10.1–10.4). **Ancillary protocols:** Sterilizing filter assemblies (*see* Protocol 10.4).

*Demonstration materials or operations:* Trainee should observe all steps in preparation and participate where possible; this may require repeated short visits to see all procedures. Trainee should see all equipment in operation, including stacking, quality control (QC), and safety procedures, although not operating the equipment, unless future duties will include washup and sterilization.
   Δ **Safety Note.** Chemical hazard from detergents; high temperature from autoclaves and ovens: follow standard laboratory procedures (*see* Sections 6.5.4, 6.5.7).

## Instructions for Exercise 2

### Outline
Collecting, rinsing, soaking, washing, and sterilizing glassware and pipettes.

### Equipment and Materials
As in regular use in preparation area (*see* Protocols 10.1–10.3).

### Procedure
As these have no experimental elements use the following standard protocols:
1. Preparation and Sterilization of Glassware (*see* Protocol 10.1).
2. Preparation and Sterilization of Pipettes (*see* Protocol 10.2).
3. Preparation and Sterilization of Screw Caps (*see* Protocol 10.3).

### Data
Trainee should become familiar with noting and recording QC data, such as numerical and graphical output from ovens and autoclaves.

## Exercise 3   Preparation and Sterilization of Water

### Purpose of Procedure
Provision of regular supply of pure, sterile water.

### Training Objectives
Appreciation of preparative practices carried on outside aseptic area. Knowledge of need for purity of water and process of preparation.

*Supervision:* Intermittent.

*Time:* 30 min.

### Background Information
Water (UPW; *see* Section 10.4.1; Figs. 4.15, 10.9). **Standard protocol:** Preparation and sterilization of ultra pure water (*see* Protocol 10.5). **Ancillary protocol:** Preparation of glassware (*see* Protocol 10.1).

*Demonstration materials or operations:* Preparation supervisor should discuss principles and operation of water purification equipment and demonstrate procedures for collection, bottling, sterilization, and QC. Trainee participation at discretion of supervisor and instructor.

## Instructions for Exercise 3

### Equipment and Materials
*Nonsterile:*
- ❏ Graduated glass borosilicate (e.g., Pyrex) or autoclavable plastic (polycarbonate) screw-cap bottles (ensure there is sufficient head space for later additions), 500 mL ......21
- ❏ Screw caps to fit ........................................21
- ❏ Sterility indicator strips ................................. 2
- ❏ Sterile-indicating tape
- ❏ Marker pen or preprinted labels
- ❏ Water purification equipment
- ❏ Autoclave
- ❏ Log (record of preparation and sterilization); book or computer database

### Procedure
1. Create entry in log book or database; label bottles with date, contents, and batch number. (A label printer attached to the computer will generate labels automatically.)
2. Early in the morning after overnight recycling, run about 50 mL of water to waste from purifier, check

conductivity (or resistivity) and total organic carbon (TOC) on respective meters, and enter in log book and in your notes.

3. If water is within specified limits (resistivity $\geq 10$ M$\Omega$/cm at 25°C, TOC $\leq 10$ ppb), collect ultrapure water directly into labeled bottle.

4. Prepare labeled bottles with deionized water and tap water, noting resistivity and TOC as above in log book and in your notes.

5. Fill to the specified mark (e.g., 430–450 mL if to be used for diluting 10× concentrated medium; (if to be autoclaved open, add 10% extra).

6. Place sterility indicator in one bottle (to be discarded when checked after autoclaving so can just be tap water).

7. Seal bottles with screw caps. (If using soda glass, i.e., if bottle is liable to breakage, then leave caps slack and add 10% extra as in step 4.)

8. Place bottles in autoclave with bottle containing sterility indicator in center of load.

9. Close autoclave and check settings: 121°C, 100 kPa (15 lb/in,$^2$, 1 bar), for 20 min with postvacuum deselected.

10. Start sterilization cycle.

11. On completion of cycle, check printout to confirm that correct conditions have been attained for the correct duration and enter in log and take a copy of the printout for your notes.

12. Allow load to cool to below 50°C.

13. Open autoclave and retrieve bottles. If caps were slack during autoclaving, tighten when bottles reach room temperature.

14. Repeat with deionized water (or reverse osmosis water) and tap water, noting resistivity and TOC at Step 3 but not rejecting if outside limits. Label and retain for Ex. 10.

**Note.** Sealing bottles that have been open during autoclaving and are still warm can cause the liner in some caps to be drawn into the bottle as it cools and the contents contract. Also there will be rapid intake of air into the bottle when it is first opened, and this can cause contamination. Bottles that have been open during autoclaving should be allowed to cool to room temperature in a sterile atmosphere, such as in the autoclave or in horizontal laminar flow, before the caps are tightened.

15. Check sterility indicator to confirm that sterilization conditions have been achieved and enter in log book and in your notes.

16. Place bottles in short-term storage at room temperature.

17. Keep and use to make medium from 10× stock (Exercise 10) to compare later by clonogenic assay (Exercise 20) or Growth Curve (Exercise 15).

### QC Data

*Acquisition:* Resistivity (or conductivity) meter on water purifier and total organic carbon (TOC) meter. Automatic printout from autoclave. Sterile tape on bottles. Sterility indicator in center bottle.

*Recording:* Enter appropriate readings and observations in logbook and in your notebook, along with copies of the printout.

### Analysis

Review logbook at intervals of 1 week, 1 month, and 3 months, and note trends or variability in water quality or sterilizer performance. Use to assay water quality when making up medium from powder (*see* Exercise 6) or 10× stock (*see* Exercise 10) by growth curve analysis (*see* Exercise 15) or clonogenic assay (*see* Exercise 20).

## Exercise 4   Preparation and Sterilization of Dulbecco's Phosphate-Buffered Saline (D-PBS) without Ca$^{2+}$ and Mg$^{2+}$ (D-PBSA)

### Purpose of Procedure
Preparation of isotonic salt solution.

### Applications
Diluent for concentrates such as 2.5% trypsin, prerinse for trypsinization, washing solution for cell harvesting or changing reagents. As the solution contains no calcium, magnesium, sodium bicarbonate, or glucose, it is not suitable for prolonged incubations or as a balanced salt solution in preparing medium.

### Training Objectives
Constitution of simple salt solution. Osmolality. Buffering and pH control. Sterilization of heat stable solutions by autoclaving.

*Supervision:* Continuous while preparing solution, then intermittent during QC steps. Continuous at start and completion of sterilization and interpretation of QC data.

*Time:* 2 h.

### Background Information
Balanced salt solutions (*see* Section 8.3; Table 8.2); buffering (*see* Section 8.2.3). **Standard protocol:** Preparation and sterilization of D-PBSA (*see* Protocol 10.6).

*Demonstration materials or operations:* Use of osmometer or conductivity meter. Supervised use of autoclave or bench-top sterilizer.

Δ **Safety issues.** Steam sterilizers present high risk of burns and possible risk of explosion (*see* Sections 6.5.2, 6.5.7). Simple bench-top autoclaves can burn dry and, consequently, present a fire risk, unless protected with an automatic, temperature-controlled cut-out.

## Instructions for Exercise 4

### Outline
Dissolve powder with constant mixing, make up to final volume, check pH and conductivity, dispense into aliquots, and autoclave.

### Materials
*Nonsterile:*
- ❏ D-PBS powder (Solution A, lacking $Ca^{2+}$ and $Mg^{2+}$, e.g., Sigma D5652), 1-L pack or tablets (Oxoid Br 14a) .................................................. 1
- ❏ Ultrapure water (UPW; *see* Section 10.4.1) ........... 1 L
- ❏ Container:
  Clear glass or clear plastic aspirator with tap outlet at base, 1 L............................................... 1
  or
  Erlenmeyer flask or bottle, peristaltic metering pump and tubing, 1 L ............................... 1
- ❏ Magnetic stirrer and PTFE-coated follower............. 1
- ❏ Bottles for storage, graduated; borosilicate glass, 100 mL .......................................... 10
- ❏ Conductivity meter or osmometer
- ❏ pH meter
- ❏ Autoclave tape or sterile-indicating tabs
- ❏ Autoclave

### Procedure
1. Add 1 L UPW to container.
2. Place container on magnetic stirrer and set to around 200 rpm.
3. Open packet of D-PBSA powder, or count out 10 tablets, and add slowly to container while mixing.
4. Stir until completely dissolved.
5. Check pH and conductivity of a sample and enter in record and in your notes.
6. pH should not vary more than 0.1 pH unit (pH may vary with different formulations).
7. Conductivity should not vary more than 5% from 150 µScm-1. Osmolality can be used as an alternative to conductivity (or in addition), and should show similar consistency between batches.

**Note.** It is important to check these parameters for consistency, as a quality control measure, to ensure that there has been no mistake in preparation. Any adjustments, such as to the osmolality, should be made after the quality control checks have been made.

8. Discard sample; do not add back to main stock.
9. Dispense contents of container into graduated bottles.
10. Cap, label (date, contents and your initials) and seal bottles.
11. Attach a small piece of autoclave tape or sterile-indicating tab and date.
12. Sterilize by autoclaving for 20 min at 121°C and 100 kPa (1 bar, 15 lb/in.²) in sealed bottles (*see* Fig. 10.3).

13. Store at room temperature.

### QC Data

*Acquisition:* Measure osmolality or conductivity and pH after dissolving constituents.

*Recording:* Enter details into log book and your notes with date and batch number.

## Exercise 5    Preparation of pH Standards

### Purpose of Procedure
To prepare a series of flasks, similar to those in current use in the laboratory, containing a simple medium or salt solution with phenol red, and adjusted to a pH range embracing the range normally found in culture.

### Applications
Assessment of pH during preparation of medium and before feeding or subculturing.

### Training Objectives
Familiarization with use of phenol red as a pH indicator. Sterilization with syringe filter and autoclave; comparison of effect on pH.

*Supervision:* Continuous at start, but minimal thereafter until operation complete.

*Time:* 2 h.

### Background Information
Physicochemical properties, pH (*see* Section 8.2 and Plate 22b). **Standard protocols:** Preparation of pH standards (*see* Protocol 8.1). Sterile filtration with syringe-tip filter (*see* Protocol 10.11).

*Demonstration materials or operations:* Use of pH meter. Principle, use, and range of syringe filters (*see* Fig. 10.11a, b).
△ **Safety issues.** None, as long as no needle is used on outlet.

## Instructions for Exercise 5

### Outline
Prepare flasks of a series of sterile samples of BSS adjusted to a range of pH from pH6.5 to pH7.8.

### Materials
*Sterile:*
- ❏ Hanks's BSS (HBSS) with phenol red, without bicarbonate or glucose, 1 L
- ❏ NaOH, 1 M ....................................... 100 mL
- ❏ Erlenmeyer flasks, 100 mL ............................. 7
- ❏ Culture flasks, 25 cm² ............................... 14
- ❏ Pipettor, 100–500 µL
- ❏ Pipettor tips, pink

❑ Syringe, 10 mL ........................................... 1
❑ Syringe-tip filter, 25 mm ............................... 1

*Nonsterile*:
❑ pH meter
❑ Magnetic stirrer and 1- to 2-cm follower

## Procedure

1. Dispense 100 mL HBSS into each of seven Erlenmeyer flasks labeled with the appropriate pH.
2. Allow to equilibrate with air for 30 min.
3. Place on magnetic stirrer and stir slowly.
4. Using a pH meter, adjust the pH in the separate Erlenmeyer flasks to 6.5, 6.8, 7.0, 7.2, 7.4, 7.6, and 7.8 with 1 M NaOH.
5. Filter sterilize 5 mL of each HBSS into separate labeled culture flasks (*see* Protocol 10.11), using the first 5 mL to flush out the filter, before collecting the second 5 mL in the flask.
6. Cap the culture flasks securely with solid caps.
7. Cap the Erlenmeyer flasks with the residue with aluminum foil and autoclave at 121°C for 20 min.
8. When cool, and in a laminar flow hood, dispense 5 mL from each Erlenmeyer flask into each of 7 labeled culture flasks, and cap the flasks with solid caps.

### Data

*Acquisition:*  Note colors in flasks and compare with pH standard chart for phenol red.

### Analysis

(1) Are the colors the same in the two series? If not, explain.
(2) Change to gas-permeable caps on culture flasks, leave overnight and compare the next day.

## Exercise 6   Preparation of Stock Medium from Powder and Sterilization by Filtration

### Purpose of Procedure
Preparation of complex solutions and sterilization of heat-labile reagents and media.

### Training Objectives
Technique of filtration and appreciation of range of options. Comparison of positive- and negative-pressure filtration.

*Supervision:*  Instruction on preparation of medium. Constant supervision during setup of filter, intermittent during filtration process, and continuous during sampling for quality control.

*Time:*  2 h.

### Background Information
Preparation of medium from powder (*see* Section 10.4.5); sterile filtration (*see* Section 10.5.2). **Standard protocols:** Preparation and sterilization of medium (*see* Protocols

10.9, 10.12); alternative procedures (*see* Protocols 10.11, 10.13, 10.14). **Ancillary protocols:** Preparation of customized medium (*see* Protocol 10.10); autoclavable media (*see* Section 10.5.1); reusable sterilizing filters (*see* Section 10.3.6); sterile filtration with syringe-tip filter (*see* Protocol 10.11); sterile filtration with large in-line filter (*see* Protocol 10.14); serum (*see* Section 10.5.3; Protocol 10.15).

*Demonstration materials or operations:*  Range of disposable filters and reusable filter assemblies, preferably the actual items, but if not, photographs may be used. Emphasize concept of filter size (surface area) and scale. Principles and advantages/disadvantages of positive-/negative-pressure filtration (*see* Section 10.5.2). Handling of filter, filtration, collection, and QC sampling should be demonstrated.

### Instructions for Exercise 6

#### Outline
Dissolve powder in UPW, filter-sterilize, bottle, and sample for sterility.

#### A. Preparation of medium from powder.
This is the first stage and should be followed by sterile filtration as soon as the solution complete. If medium has to be stored before filtration, it must be kept at 4°C for no longer than 24 h.

#### Equipment and Materials
*Sterile*:
❑ Graduated bottles for medium, 100 mL ................ 10
❑ Caps for bottles ....................................... 10
❑ Universal containers for contamination control sampling ............................................... 3

*Nonsterile*:
❑ Ultrapure water....................................... 1 L
❑ Powdered medium (e.g., MEM with Earle's salts, without glutamine) ..................................... 1-L pack
❑ Graduated Erlenmeyer flask or bottle, capacity plus head space, 1 L ........................................... 1
❑ Magnetic stirrer and PTFE-coated follower, 50 mm ................................................. 1
❑ Conductivity meter

#### Procedure
1. Add appropriate volume of ultrapure water to container.
2. Add magnetic follower.
3. Place container on magnetic stirrer and set to around 200 rpm.
4. Open packet of powder and add contents slowly to container while mixing.
5. Stir until powder is completely dissolved.
6. Check pH and conductivity of a sample and enter in record:
   (a) pH should be within 0.1 unit of expected level for particular medium.
   (b) Conductivity should be within 2% of expected value for particular medium.

7. Discard sample; do not add back to main stock.
8. Sterilize by filtration (*see* below and Section 10.5.2, Protocols 10.11–10.14).

### B. Sterile filtration with vacuum filter flask.

*Background:* CO$_2$ and bicarbonate (*see* Section 8.2.2); buffering (*see* Section 8.2.3); standard sterilization protocols (*see* Section 10.5).

#### Outline
Attach vacuum pump to outlet, pour medium into top chamber of filter unit, switch on pump, and draw solution through to lower chamber; cap and store.

#### Materials
*Sterile:*
- [ ] Filter flask 500 mL (*see* Fig. 10.10*b*; e.g., *see* Fig. 10.11*e*) ............................................... 1
- [ ] Cap for lower chamber (if chamber is used for storage) ................................................... 1
- [ ] Sample tube or universal container for sterility test ...................................................... 1

*Nonsterile:*
- [ ] Medium for sterilization ......................... 450 mL
- [ ] Vacuum pump or vacuum line
- [ ] Thick-walled connector tubing from pump or line, to fit filter flask inlet

#### Procedure
1. Take medium to be sterilized and filter flask to hood.
2. Connect side outlet of filter flask to vacuum pump (located outside hood, e.g., on floor).
3. Remove cap from bottle and lid from top chamber of filter flask.
4. Pour nonsterile medium into top chamber.
5. Switch on pump.
6. Unpack cap for lower chamber, ready for use.
7. When liquid has all been drawn into lower chamber, switch off pump and detach filter housing and top chamber.
8. Note pH in lower chamber, gas air space with 5% CO$_2$ if high, and cap.
9. Transfer 10 mL from lower flask to universal container or equivalent tube, label with date, contents, and your initials, and incubate at 37°C for 1 week to check for contamination.
10. Label lower chamber and with name of medium, date, and your initials.

### C. Sterile filtration with small in-line filter.

Alternative to Procedure B, above. Should be performed in addition to B for training purposes.

#### Outline
Pump medium from reservoir through a peristaltic pump and dispense into bottles through a sterilizing filter.

#### Materials
*Sterile:*
- [ ] In-line filter with bell (*see* Figs. 10.10*a*, 10.11*f*) .................................................. 47 mm
- [ ] Graduated 100-mL medium bottles, foil capped, sterilized by dry heat (*see* Protocol 10.1) ............... 7
- [ ] Caps, autoclaved (*see* Protocol 10.3) ................... 6
- [ ] Sample tube or universal container for sterility test ..... 1

*Nonsterile:*
- [ ] Medium for sterilization (from Procedure A) .... 550 mL
- [ ] Peristaltic pump (Fig. 10.14) preferably with foot switch ................................................ 1
- [ ] Silicone tubing to fit pump and inlet to filter ...... 50 cm
- [ ] Clamp stand to hold filter .............................. 1

#### Procedure
1. Bring equipment to hood, swab as appropriate, and place in hood.
2. Feed tubing through peristaltic pump.
3. Bring medium to hood and insert upstream end of tubing.
4. Unpack filter and connect to downstream outlet from peristaltic pump.
5. Clamp filter in clamp stand at a suitable height so that bottles for receiving medium can be positioned below the filter with the neck shrouded and removed easily when filled. Use one of the sterile medium bottles to set up the filter, but do not use this bottle, ultimately, for sterile collection.
6. Switch on pump and collect about 20 mL into bottle used for setup.
7. Remove setup bottle, cap it and number it "1."
8. Remove foil from next sterile medium bottle and place under filter bell.
9. Switch on pump.
10. Fill bottle to 100-mL mark.
11. Switch off pump.
12. Remove bottle, cap it, number it, and replace with fresh sterile medium bottle.
13. Repeat steps 8 through 11, filling 5 bottles to 100 mL, and collecting the remainder in the last bottle.
14. Check pH (visually by indicator) and gas the head space in all bottles with 5% CO$_2$ if high, recapping when finished.
15. Replace the aluminum foil over the cap and neck of each bottle to keep it free from dust during storage.
16. Label, date, and initial bottles.
17. Place bottles containing 100 mL medium at 4°C for storage (*see* Section 10.6.4).
18. Place the first and last bottles, sealed, at 37°C and incubate for 1 week (*see* Section 10.6.2).

#### Data
(1) Note pH before and immediately after filtering.
(2) Incubate universal containers or bottles at 37°C for 1 week, and check for contamination.

## Analysis

(1) Explain the difference in pH between vacuum-filtered versus positive-pressure-filtered medium.
(2) Does the pH recover after gassing with $CO_2$?
(3) When would you use one rather than the other?
(4) What filters would you use
  (a) For 5 mL of a crystalline solution?
  (b) For 10 L of medium?
  (c) For 1 L of serum?

## Experimental Variations

There are a number of experimental variables that can be applied to this exercise, for example:

(1) *Stability and storage*: Place one of the 100-mL bottles from positive pressure filtration at $-20°C$, one at $4°C$, one at room temperature ($\sim 20°C$) and one at $37°C$ for two weeks and use in growth curve (Exercise 15) or clonogenic assay (Exercise 20) to check for deterioration.
(2) *Water purity*: Use the three lots of sterile water from Exercise 3 to make up medium, filter sterilize, use later in growth curve or clonogenic assay, and compare.

## 28.3 BASIC CELL CULTURE TECHNIQUES

This section is designed to give the trainee an introduction to cell culture, covering observation, preparation, and sterile manipulation. In most cases an experimental section has been added to make the exercise more interesting and to encourage the trainee to observe and to collect and record data, even from simple manipulations. Standard protocols are cross-referenced for future application of the procedure, while the version in the exercise has additional steps to form the basis of an experiment. More senior staff simply wanting to learn the procedure may choose to go straight to the standard protocol.

## Exercise 7    Observation of Cultured Cells

### Purpose of Procedure
Critical examination of cell cultures.

### Applications
Checking consistency during routine maintenance; evaluation of status of cultures before feeding, subculture, or cryopreservation; assessment of response to new or experimental conditions; detection of overt contamination or misidentification and cross-contamination.

### Training Objectives
Familiarization with appearance of cell cultures of different types and at different densities; awareness of problems of misidentification and cross-contamination; use of microscope and camera; distinction between microbially contaminated and uncontaminated, and between healthy and unhealthy cultures; assessment of growth phase of culture and need for medium replenishment or subculture.

*Supervision:*   Continuous during observation, then intermittent during photography.

*Time:*   30 min.

### Background Information
Morphology, photography (*see* Section 15.5, 15.5.5). **Standard protocols:** Microscopy and photography (*see* Protocols 15.1, 15.6). **Ancillary protocols:** Staining (*see* Protocols 15.2, 15.3); cytocentrifuge (*see* Protocol 15.4); indirect immunofluorescence (*see* Protocol 15.11).

*Demonstration materials or operations:*   Photographic examples of cell morphology, phase contrast of living cultures, as well as fixed and stained preparations; archival photos of cells to be used later in exercises; types of culture vessel suitable for morphological studies such as Petri dishes (*see* Fig. 7.4), chamber slides (*see* Fig. 15.3); cytocentrifuge for suspension cultures (*see* Fig. 15.4).
  Δ *Safety Note.*   No special safety requirements.

### Instructions for Exercise 7

### Outline
Examine and photograph a range of cell lines at different cell densities.

### Equipment and Materials
❑ Range of flask or Petri dish cell cultures at different densities preferably with normal and transformed variants of the same cell (e.g., 3T3 and SV3T3, or BHK21-C13 and BHK21-PyY) at densities including mid-log phase ($\sim$ 50% confluent with evidence of mitoses), confluent (100% of growth area covered and cells packed but not piling up), and postconfluent (cells multilayering and piling up if transformed). If available, include suspension cell cultures at low and high concentrations.
❑ If possible, include examples of contaminated cultures (preferably not Petri dishes to avoid risk of spread) and unhealthy cultures, such as cultures that have gone too long without feeding
❑ Inverted microscope with 4×, 10×, and 20× phase-contrast objectives and condenser
❑ Automatic camera, preferably CCD with monitor but digital SLR with K-mount and photo eyepiece

### Procedure
1. Set up microscope and adjust lighting for Kohler illumination and center phase rings.
2. Bring cultures from incubator. It is best to examine a few flasks at a time, rather than have too many out of the incubator for a prolonged period. Choose a pair, for example, the same cells at low or high density, or a normal and transformed version of the same cell type.
3. Examine each culture by eye, looking for turbidity of the medium, a fall in pH, granularity, vacuolization,

or detached cells. Also try to identify type of cells in monolayer and look for signs of patterning. This can be normal, for example, swirling patterns of fibroblasts at confluence, or transformed, for example, random growth without patterning.

4. Examine at low power (4× objective) by phase-contrast microscopy on inverted microscope, and check cell density and any sign of cell-cell interaction, aggregation, or detachment.

5. Examine at medium (10× objective) and high (20× objective) power and check for the healthy status of the cells, signs of rounding up, contraction of the monolayer, or detachment. Check also for evidence of patterning, indicative of normal fibroblastic growth, or piling up and random overgrowth, evidence of transformation.

6. Check for any sign of microbial contamination.

7. Check homogeneity of culture; evidence of heterogeneity may indicate genetic instability, transformation, or cross-contamination.

8. Look for mitoses and estimate, roughly, their frequency.

9. Photograph each culture, noting the culture details (cell type, date form last passage) and cell density and compare with archival photographs of the same cells to confirm that the cells have not undergone any alteration or cross-contamination.

10. Return cultures to incubator and repeat with a new set.

### Observations and Record

(1) Look for differences in growth pattern, cell density, and morphology in related cultures (Table 28.2). Are there differences from archival photographs that suggest deterioration, alteration of the cells (transformation, heterogeneity, misidentification or cross-contamination)?

(2) Assess health status of cells.

(3) Is there any sign of contamination?

(4) Are cells ready for feeding (*see* Section 12.3.2) or passage (*see* Section 12.4.1)?

(5) Make a numerical estimate of cell density by calculating the area of the 20× objective field and counting the number of cells per field. This will be easiest if a digital camera and monitor are used where the screen can be overlaid with cling film and each cell ticked with a fine felt-tipped marker. Practice will make this "guesstimate" of cell density easier. It becomes critically important if you are unable to carry out a cell count by other means at this stage (*see* Section 12.4.3).

(6) Try to identify and count mitotic cells in these high-power fields.

### Data

*Qualitative:*

(1) Record your observations on morphology, shape, and patterning for all cultures.

(2) Note any contaminations.

(3) Confirm healthy status or otherwise.

*Quantitative:*

(1) Record cell density (cells/cm$^2$) for each culture.

(2) Record mitotic index for each culture.

**TABLE 28.2.** Observation of Culture: Check List

| Status | Criterion | Record | Action indicated (see also Chapter 29) |
|---|---|---|---|
| Appearance | Morphology | | Confirm against archival photographs at same cell density. |
| | Density | | Compare with expected density at this stage of culture. |
| | | | If higher or lower, check and adjust seeding concentration. |
| | Mitoses | | Indicates proliferation. No action required unless absent; then check growth conditions. |
| | Deterioration | | Check for contamination (*see* Section 18.3). |
| | | | Check growth conditions and correct. |
| | | | Discard and replace from stock or from freezer. |
| | Alteration, Transformation | | Discard and replace from frozen stock. |
| | Heterogeneity (evidence of mixed cell types) | | Compare against archival photographs. |
| | | | Check identity and authenticity (*see* Sections 12.1.1, 15.2). |
| | Contamination | | Discard culture and medium and reagents used with it (*see* Section 18.4). |
| | | | Identify and eliminate source if repeated or widespread (*see* Section 18.1, 29.6). |
| Maintenance | Cell density/concentration | | If high, subculture. |
| | pH of medium | | If low, feed or subculture, depending on cell density/concentration. |
| | | | If high, check flasks for leaks, incubator $CO_2$ concentration, evidence of contamination. |
| | | | If all cultures high, check medium preparation. |

*Analysis:*
(1) Account for differences in cell density.
(2) Account for differences in mitotic index.
(3) Compare appearance of cells from normal and transformed cultures and high and low densities and try to explain differences in behavior.

## Exercise 8  Preparing Sterile Medium for Use

### Purpose of Procedure
To prepare complete medium (ready for use) from a working strength (1×) stock and supplements.

### Training Objectives

*Aseptic handling:*  Training in dexterity and sterile manipulation; simple medium preparation.

*Supervision:*  Continuous.

*Time:*  30 min.

*Demonstration of materials and operations:*  Demonstrate how to swab work surface and items brought into hood. Explain the principles of laminar flow and particulate air filtration. Show trainee how to uncap and recap flasks and bottles and how to place the cap on the work surface. Demonstrate holding a pipette, inserting it into a pipette controller, and using it without making nonsterile contact, how to transfer solutions aseptically, sloping bottles and flasks during pipetting (*see* Section 5.1–5.4). Emphasize clearing up and swabbing the hood and checking below the work surface.

### Background Information
Objectives of aseptic technique (*see* Section 5.1); elements of aseptic environment (*see* Section 5.2); sterile handling (*see* Section 5.3); working in laminar flow (*see* Section 5.4); visible microbial contamination (*see* Section 18.3.1). **Standard protocols:** Sterile technique (*see* Protocol 5.1 or Protocol 5.2), preparation of medium from 1× stock (*see* Protocol 10.7)

## Instructions for Exercise 8

### Outline
Check the formulation; if complete, it may be used directly, adding serum if required (*see* Sections 8.6). If the formulation is incomplete (e.g., lacking glutamine), add the appropriate stock concentrate.

**Note.** A supplement (e.g., serum or antibiotics) is a component that is added to the medium and is not in the original formulation. It needs to be indicated in any record or publication. Other additions (e.g., glutamine or $NaHCO_3$) are part of the formulation and are not supplements. They need not be indicated in records or publications unless their concentrations are changed.

### Materials
❑ Medium stock (e.g., Eagle's MEM with 23 mM $NaHCO_3$ without glutamine) .......... 100 mL
❑ Glutamine, 200 mM (will need to be thawed) .... 20 mL
❑ Serum (will need to be thawed), newborn or fetal bovine ................................. 100 mL
❑ Pipettes and other equipment, as listed for aseptic technique (*see* Protocols 5.1, 5.2)
❑ Flask, 25 cm², or Petri dish, 5 or 6 cm.

### Procedure
1. Place glutamine and serum (if frozen) in water bath with clean sterile water.
2. Check formulation of medium, and determine what additions are required (e.g., glutamine).
3. Take medium to hood with any other supplement or addition that is required.
4. Unwrap bottles if they are wrapped in polythene, and swab with 70% alcohol.
5. Uncap bottles.
6. Transfer the appropriate volume of each addition to the stock bottle to make the correct dilution; for 100 mL, use the following ingredients and amounts:
   Glutamine, 200 mM ...............................1 mL
   Serum .................................10 mL (for 10%)
7. Use a different pipette for each addition.
8. Move each new stock to the opposite side of the hood after it has been added, so that you will know that it has been used.
9. Repeat step 6 omitting serum.
10. Remove all additives or supplements from the hood when the medium is complete.
11. Gas the air space with 5% $CO_2$. Do not bubble gas through any medium containing serum, as the medium will froth out through the neck, risking contamination.
12. Recap bottles.
13. Alter labeling to record additions, date and initial.
14. Pipette 5 mL into a 25-cm² flask with a permeable cap or into a 6-cm Petri dish, and incubate overnight in a 5% CO2 incubator, to ensure that the pH equilibrates at the correct value. If it does not, readjust the pH of the medium with sterile 1 M HCl or 1 M NaOH as appropriate and incubate for a further 1 h. Use a 10 µL pipettor to adjust the pH or the flask or Petri dish and multiply up from that to determine the amount need for the main bottle of medium.
15. Pipette a QC sample of 10 mL into a universal container or flask and incubate for 1 week at 37°C.
16. Place the remainder of the medium at 4°C until shown to be sterile (this step may be omitted as experience is gained) and then use for Exercise 9.

### Observation and Analysis
(1) Compare difference in pH between serum-free and serum-containing at step 11 and amount of acid or alkali taken to restore pH to normal (if any).

(2) Check for contamination in QC sample after 24 h, 1 and 2 weeks.

## Exercise 9    Feeding a Monolayer Culture

### Purpose of Procedure
To replace exhausted medium in a monolayer culture with fresh medium.

### Applications
Used to replenish medium between subcultures in rapidly growing cultures, or to change from one type of medium to another.

### Training Objectives
Reinforces aseptic manipulation skills. Introduces one of the basic principles of cell maintenance, that of medium replenishment during propagation cycles. Makes trainee observe culture and become aware of signs of medium exhaustion, such as cell density and/or fall in pH, and also looking for contamination. Awareness of risk of cross-contamination.

*Supervision:* Trainee will require advice in interpreting signs of medium exhaustion and demonstration of medium withdrawal and replenishment.

*Time:*   30 min.

### Background Information
Complete media (*see* Section 16.4); replacement of medium (*see* Section 12.3.2); monitoring for contamination (*see* Section 18.3); cross-contamination (*see* Sections 12.1.1, 15.2, 18.6). *Standard protocol:* Feeding a monolayer (*see* Protocol 12.1). *Ancillary Protocols:* Preparation of complete medium (*see* Protocols 10.7, 10.8, *or* 10.9 *and* Exercise 6, 8); preparation of pH standards (*see* Protocol 8.1); handling dishes or plates (*see* Protocol 5.3).

*Demonstration materials or operations:*   Exercise requires at least three semiconfluent flasks from a continuous cell line such as HeLa or Vero, with details of number of cells seeded and date seeded. Trainee should also be shown how to bring medium from refrigerator, and it should be stressed that a bottle of medium is restricted to one cell line and not shared among other cell lines or operators. Also demonstrate swabbing and laying out hood, use of incubator, retrieving culture from incubator, and observing status of cells by eye and on microscope (*see* Exercise 7, and Protocols 12.1, 15.1). Aspirator with pump for medium withdrawal or discard beaker will be required and the process of medium withdrawal and replacement demonstrated, with gassing with 5% $CO_2$ if necessary.

    Δ *Safety Note.* If human cells are being handled, a Class II biological safety cabinet must be used and waste medium must be discarded into disinfectant (*see* Section 6.8.5; Table 6.7).

## Instructions for Exercise 9

### Outline
Examine the culture by eye and on an inverted microscope. If indicated, such as by a fall in pH, remove the old medium and add fresh medium. Return the culture to the incubator.

### Materials
*Sterile:*
- ❑ Cell cultures: A549 cells (or equivalent rapidly growing monolayer cell line) 4d after seeding at $2 \times 10^4$ cells/mL, 25 cm² flasks............................................. 6
- ❑ Growth media, serum-containing and serum-free, each such as Eagle's 1× MEM with 23 mM $HCO_3$, ±10% FBS, without antibiotics, from Exercise 8..... ~100 mL.
- ❑ Pipettes, graduated, and plugged. If glass, an assortment of sizes, 1 mL, 5 mL, 10 mL, 25 mL, in a square pipette can, or, if plastic, individually wrapped and sorted by size on a rack
- ❑ Unplugged pipettes for aspirating medium if pump or vacuum line is available

*Nonsterile:*
- ❑ Pipette controller or bulb
- ❑ Tubing to receiver connected to vacuum line or to receiver via peristaltic pump
- ❑ Alcohol, 70%, in spray bottle
- ❑ Lint-free swabs or wipes
- ❑ Absorbent paper tissues
- ❑ Pipette cylinder containing water and disinfectant
- ❑ Marker pen with alcohol-insoluble ink
- ❑ Notebook, pen, protocols

### Procedure
1. Prepare the hood by ensuring it is clean and clear and swabbing it with 70% alcohol.
2. Bring the reagents and materials necessary for the procedure, swab bottles with 70% alcohol and place items required immediately in the hood
3. Examine the culture carefully for signs of contamination or deterioration.
4. Check the previously described criteria—pH and cell density or concentration—and, based on your knowledge of the behavior of the culture, decide whether or not the medium needs to be replaced.
5. Take three flasks to the sterile work area and return the other three to the incubator.
6. Uncap the flasks.
7. Take sterile pipette and insert into bulb or pipette controller, or, selecting an unplugged pipette connect to vacuum line or pump.
8. Withdraw the medium from four flasks, and discard into waste beaker. Or, preferably, aspirate medium via a suction line in the hood connected to an external pump.
9. Discard pipette.
10. Uncap the media bottles.

11. Take a fresh pipette and add 5 mL of fresh serum-containing medium to 2 flasks and 5 mL of serum-free medium to 2 flasks. Leave the last 2 flasks without feeding.
12. Discard the pipette.
13. Recap the flasks and the media bottles.
14. Return the cultures to the incubator.
15. Complete record of observations and feeding on record sheet and lab book.
16. Clear away all pipettes and other items, and swab down the work surface.
17. Check cultures after 24, 48, and 72 h. Use for Exercise 11 at 72 h.

### Observations and Analysis
Note cell density and appearance at 72 h, use for cell counting (*see* Exercise 11), and compare flasks for yield.

### Data
Compare appearance (*see* Exercise 7) and cell yield (*see* Exercise 11).

Routine maintenance should be recorded in a record sheet (*see* Table 12.7) and experimental data tabulated in Exercise 11.

## Exercise 10  Preparation of Complete Medium from 10× Stock

### Purpose of Procedure
Addition of concentrated stock and supplements to water to produce a complete medium designed for a specific task.

### Applications
Production of growth medium that will allow the cells to proliferate, maintenance medium that simply maintains cell viability, or differentiation medium that allows cells to differentiate in the presence of the appropriate inducers.

### Training Objectives
Further experience in aseptic handling, increased understanding of the constitution of medium and its supplementation, water quality, and control of pH with sodium bicarbonate.

*Supervision:* A trainee who has responded well to Exercise 8 should need minimum supervision but will require some clarification of the need to add components or supplements before use and the distinction, from using 1× medium, such as in cost and flexibility.

Select Procedures A, B, or C depending on relevance to practices in laboratory.

*Time:* 30 min.

### Background Information
Media (*see* Sections 10.4.3, 10.4.4). **Standard protocol:** Preparation of medium from 10× concentrate (*see* Protocol

10.8). **Ancillary protocols:** Customized medium (*see* Protocol 10.10); preparation of stock medium from powder and sterilization by filtration (*see* Protocol 10.9); preparation of pH standards (*see* Protocol 8.1).

*Demonstration materials or operations:* Set of pH standards (*see* Protocol 8.1). Range of bottles available for medium preparation.

Δ **Safety Note.** No major safety implications unless a toxic (e.g., cholera toxin or cytotoxic drug) or radioactive constituent is being added.

### Instructions for Exercise 10

#### Outline
Add concentrated medium and other constituents, adjust the pH, and use the solution or return it to the refrigerator.

#### Materials
Volumes given are for 500 mL of medium, but stock bottles should contain more than amounts indicated to allow for pipetting.

❑ Pipettes and other items are as listed for aseptic technique (*see* Ex.1), plus sufficient of the following to provide the volumes stipulated:

*A. For sealed culture flask with gas phase of air, low $HCO_3^-$ concentration, atmospheric $CO_2$ concentration, and low buffering capacity.*
*Sterile solutions:*

❑ Premeasured aliquot of UPW .................... 443 mL
❑ Medium, 10× concentrate (e.g., Eagle's MEM with Hanks's salts) ......................... 50 mL
❑ Glutamine, 200 mM ...............................5 mL
❑ NaHCO₃, to give 4 mM final concentration, 7.5% ...............................5 mL
❑ Bovine serum, newborn or fetal ................... 50 mL
❑ NaOH, 1 M ..............................as required

#### Procedure
1. Thaw serum and glutamine and bring to the hood.
2. Swab any bottles that have been in a water bath before placing in the hood.
3. Add constituents as listed in sterile solutions section above. HEPES may be added to increase the buffering capacity, and the flask may be vented to atmosphere for some cell lines at a high cell density if a lot of acid is produced.
4. Add 1 M NaOH to give pH 7.2 at 20°C. When incubated, the medium will rise to pH 7.4 at 37°C, but this figure may need to be checked by a trial titration the first time the recipe is used.

*B. For cultures in open vessels in a $CO_2$ incubator, or under $CO_2$ in sealed flasks, with 5% $CO_2$ and a high bicarbonate concentration.*
*Sterile solutions:*

❑ Premeasured aliquot of sterile water ............. 443 mL

❏ Medium 10× concentrate (e.g., Eagle's
  MEM with Earle's salts) .......................... 50 mL

❏ Glutamine (will need to be thawed),
  to give 2 mM final, 200 mM ......................5 mL

❏ NaHCO₃, to give 26 mM final
  concentration, 7.5% .......................... 14.5 mL

❏ Serum, newborn or fetal bovine.................. 50 mL

❏ NaOH, 1 M ................................. as required

### Procedure

1. Thaw serum and glutamine and bring to the hood.

2. Swab any bottles that have been in a water bath before placing in the hood.

3. Add constituents as listed in sterile solutions section.

4. Add 1 M NaOH to give pH 7.2 at 20°C. When incubated, the medium will rise to pH 7.4 at 37°C, but this figure may need to be checked by a trial titration the first time the recipe is used:

   (a) Dispense 5 mL into each of 5 Petri dishes.

   (b) Add varying amounts of 1N NaOH from 5 to 50 μL to each Petri dish.

   (c) Incubate under 5% CO₂ for a minimum of 2 h and, preferably, overnight.

   (d) Add the proportionate amount of 1 M NaOH to the whole bottle (∼ 100×).

### C. For cultures in open vessels in a CO₂ incubator, or under CO₂ in sealed flasks, with 2% CO₂ and an intermediate bicarbonate concentration.

*Sterile solutions:*

❏ Premeasured aliquot of UPW .................... 443 mL

❏ Medium, 10× concentrate
  (e.g., Ham's F12) ............................... 50 mL

❏ Glutamine (will need to be thawed),
  to give 2 mM final, 200 mM ......................5 mL

❏ NaHCO₃, to give 8 mM final
  concentration, 7.5% .............................9 mL

❏ HEPES, to give 20 mM final
  concentration, 1 M ............................. 10 mL

❏ Serum, newborn or fetal bovine.................. 50 mL

❏ NaOH, 1 N ................................. as required

### Procedure

1. Thaw serum and glutamine and bring to the hood.

2. Swab any bottles that have been in a water bath before placing in the hood.

3. Add constituents as listed in sterile solutions section.

4. Add 1 M NaOH to give pH 7.2 at 20°C. When incubated, the medium will rise to pH 7.4 at 37°C, but this figure may need to be checked by a trial titration the first time the recipe is used.

### Experimental Variations

#### *Water quality (continued from Exercise 3).*

1. Use water from Exercise 3: UPW, deionized water, and tap water.

2. Make up three batches of medium by one of Procedures A, B, or C.

3. Store at 4°C for use in later exercises (Exercise 15 and/or Exercise 20)

#### *Venting flasks.*

1. Prepare medium according to Procedure A.

2. Pipette 5 mL into each of four 25-cm² flasks.

3. Add 10 μL 1-M HEPES to each of two flasks.

4. Seal two flasks, one with HEPES and one without, and slacken the caps (or use gas-permeable caps) on the other two.

5. Incubate at 37°C without CO₂ overnight.

6. Record pH and tabulate against incubation condition.

#### *Analysis of Vented/Unvented Flasks*

(1) Check pH and account for differences.

(2) Which condition is correct for this low-bicarbonate medium?

(3) What effect has HEPES on the stability of pH?

(4) When would venting be appropriate?

### Bicarbonate Concentration

1. Omit the bicarbonate from Procedure B and add varying amounts of sodium bicarbonate as follows:

2. Prepare and label 5 aliquots of 10 mL bicarbonate-free medium in 25-cm² flasks.

3. Add 200 μL, 250 μL, 300 μL, 350 μL, and 400 μL of 7.5% NaHCO₃ to separate flasks.

4. Leave the cap slack (only just engaging on the thread), or use a gas-permeable cap, on the flasks and place at 37°C in a 5% CO₂ incubator.

5. Leave overnight and check pH against pH standards.

6. Record pH and tabulate against volume of NaHCO₃ added.

#### *Analysis*

1. Explain what is happening to change the pH.

2. Calculate the final concentration of bicarbonate in each case and determine the correct amount of NaHCO₃ to use.

3. If none are correct, what would you do to attain the correct pH?

## Exercise 11  Counting Cells by Hemocytometer and Electronic Counter

There are several options in the organization of this exercise. It could be used as an exercise either in the use of the hemocytometer or electronic cell counter, or it could be arranged as a joint exercise utilizing both techniques and comparing the outcomes. One or both these options could be used with Exercise 12 or 13. However, as the initial training in cell counting can make the actual counting process quite slow, it is recommended that cell counting is run as a stand-alone exercise, utilizing cultures set up previously (e.g., from

**TABLE 28.3.** Data Record from Exercise 11, Cell Counting

| Cells per flask at seeding | Hemocytometer or electronic count at harvest | Dilution or sampling fraction[a] | Cell/mL of trypsinate or suspension | Cells harvested per flask | Yield: Cells harvested/ cells seeded |
|---|---|---|---|---|---|
| | | | | | |
| | | | | | |

[a]Electronic counter dilution of 50× (e.g., 0.4 mL cell suspension in 20 mL counting fluid), with counter sample set at 0.5 mL would give a factor of 100. Hemocytometer chamber (Improved Neubauer) counts usually sample 1 mm$^2$ × 0.1 mm deep, or 0.1 mm$^3$. So a factor of 1 × 10$^4$ will give cells/mL (*see* Section 20.1.1).

Exercise 9), and not as a preliminary to another exercise. The combined use of both counting methods will be incorporated in the following description.

### Purpose of Procedure
To quantify the concentration of cells in a suspension.

### Applications
Standardization of cell concentrations at routine subculture; analysis of quantitative growth experiments and cell production via growth curves and cell yields.

### Training Objectives
Quantitative skill. Counting cells and assessment of viability. Evaluation of relative merits of hemocytometer and electronic counting.

*Supervision:* Required during preparation and examination of sample and setting up both counting procedures. Counting samples can proceed unsupervised, although the trainee may require help in analyzing results.

*Time:* 45 min.

### Background Information
Cell counting by hemocytometer (*see* Section 20.1.1); electronic counting (*see* Section 20.1.2); estimation of viability by dye exclusion (*see* Section 21.3.1). **Standard protocols:** Counting by hemocytometer (*see* Protocol 20.1) or electronic counting (*see* Protocol 20.2); viability by dye exclusion (*see* Protocol 21.1).

*Demonstration materials or operations:* Cell cultures used for counting should be provided for the trainee to practice on. The use of the hemocytometer and electronic counter will require demonstration, with appropriate advice on completing calculations at the end. The principles of operation of the electronic counter should also be explained.

Δ **Safety Note.** When human cells are used, handling should be in a Class II microbiological safety cabinet. All plastics and glassware, including the hemocytometer slide and coverslip, should be placed in disinfectant after use, and counting cups and fluid from electronic counting should be disposed of into disinfectant (*see* Section 6.8.5).

### Instructions for Exercise 11

#### Outline
Trypsinize a monolayer culture, or sample a suspension culture, prepare a hemocytometer slide, and add the cells to the counting chamber. Count the cells on a microscope and calculate the cell concentration. Repeat count on electronic cell counter by diluting cells in an electrolyte and placing in sample position.

#### Materials
*Sterile:*
- ❑ Cells for counting, trypsinized monolayers taken from Ex. 9 (previously fed or not) resuspended in 1 mL
- ❑ Growth medium appropriate to cell type ......... 20 mL
- ❑ Trypsin, 0.25% ....................................5 mL
- ❑ D-PBSA ......................................... 10 mL
- ❑ Yellow pipettor tips

*Nonsterile:*
- ❑ Viability stain (e.g., 0.4% Trypan Blue or 1% Naphthalene Black in D-PBSA or HBSS) 1 mL
- ❑ Pipettor, 20 μL or adjustable 100 μL
- ❑ Hemocytometer (Improved Neubauer)
- ❑ Tally counter
- ❑ Microscope (upright or inverted)
- ❑ Counting cups for electronic counter

#### Procedure A. Cell counting by hemocytometer.
1. Sample the cells: (Follow steps 1 to 10 from Ex. 13.)
   (a) Mix the suspension thoroughly to disperse the cells
   (b) Transfer a small sample (e.g., 1 mL) to a vial or universal container. Retain the residue for electronic counting in Procedure B.
   (c) Add an equal volume of viability stain to sample.
2. Prepare the slide:
   (a) Clean the surface of the slide with 70% alcohol, taking care not to scratch the semisilvered surface.
   (b) Clean the coverslip, and, wetting the edges very slightly, press it down over the grooves and semisilvered counting area. The appearance of interference patterns ("Newton's rings"—rainbow

colors between the coverslip and the slide, like the rings formed by oil on water) indicates that the coverslip is properly attached, thereby determining the depth of the counting chamber.

3. Mix the cell sample thoroughly, pipetting vigorously to disperse any clumps, and collect 20 μL into the tip of a pipettor.

4. Transfer the cell suspension immediately to the edge of the hemocytometer chamber, expel the suspension, and let it be drawn under the coverslip by capillarity. Do not overfill or underfill the chamber, or else its dimensions may change, due to alterations in the surface tension; the fluid should run only to the edges of the grooves.

5. Mix the cell suspension, reload the pipettor, and fill the second chamber if there is one.

6. Blot off any surplus fluid (without drawing from under the coverslip), and transfer the slide to the microscope stage.

7. Select a 10× objective, and focus on the grid lines in the chamber. If focusing is difficult because of poor contrast, close down the field iris, or make the lighting slightly oblique by offsetting the condenser.

8. Move the slide so that the field you see is the central area of the grid and is the largest area that you can see bounded by three parallel lines. This area is 1 mm². With a standard 10× objective, this area will almost fill the field, or the corners will be slightly outside the field, depending on the field of view.

9. Count all the cells lying within this 1-mm² area, using the subdivisions (also bounded by three parallel lines) and single grid lines as an aid for counting. Count cells that lie on the top and left-hand lines of each square, but not those on the bottom or right-hand lines, to avoid counting the same cell twice. For routine subculture, attempt to count between 100 and 300 cells per mm²; the more cells that are counted, the more accurate the count becomes. For more precise quantitative experiments, 500 to 1000 cells should be counted.

10. If there are very few cells (< 100/ mm²), count one or more additional squares (each 1 mm²) surrounding the central square.

11. If there are too many cells (> 1000/ mm²), count only five small squares (each bounded by three parallel lines) across the diagonal of the larger (1-mm²) square.

12. Repeat the count, counting only stained cells.

13. If the slide has two chambers, move to the second chamber and do a second set of counts. If not, rinse the slide and repeat the count with a fresh sample.

### Analysis

Subtract the number of stained cells from the total and express the number of unstained cells as a percentage of the total.

Calculate the average of the two counts, and derive the concentration of your sample using the formula

$$c = \frac{n}{v} \times 2$$

where $c$ is the cell concentration (cells/mL), $n$ is the number of cells counted, $v$ is the volume counted (mL), and 2 is the dilution with viability stain. For the improved Neubauer slide, the depth of the chamber is 0.1 mm, and, assuming that only the central 1 mm² is used, $v$ is 0.1 mm³, or $1 \times 10^{-4}$ mL. The formula then becomes

$$c = \frac{n}{10^{-4}} \times 2 \quad \text{or} \quad c = n \times 10^4 \times 2$$

If the cell concentration is high and only the five diagonal squares within the central 1 mm² were counted (i.e., 1/5 of the total), this equation becomes

$$c = n \times 5 \times 10^4 \times 2 \quad \text{or} \quad c = n \times 10^5$$

If the cell concentration is low, count nine 1-mm² squares, five in each chamber of the slide. The expression then becomes

$$c = \frac{n \times 10^4}{9} \quad \text{or} \quad c = n \times 1.1 \times 10^3 \times 2$$

This is the cell concentration in the suspension created after trypsinization (or originally provided). To determine the number of cells per culture flask, multiply by the volume of medium used to resuspend the cells (volume of cell sample provided). To determine the concentration of cells in the culture, divide by the volume of medium in the original flask (5 mL in this case). To determine the density of cells in the culture (for monolayer cells only) divide by the surface area of the flask (25 cm² in this case).

### Procedure B. Electronic cell counting by electrical resistance.

1. Disperse the residual cells from Procedure A by pipetting. The cells must be well mixed and singly suspended.

2. Dilute the sample of cell suspension to 1:50 in 20 mL of counting fluid in a 25-mL beaker or disposable sample cup. An automatic dispenser will speed up this dilution and improve reproducibility.

**Note.** Dispensing counting fluid rapidly can generate air bubbles that will be counted as they pass through the orifice. Consequently the counting fluid should stand for a few moments before counting. If the fluid is dispensed first and the cells added second, this problem is minimized.

3. Mix the suspension well, and place it under the tip of the orifice tube, ensuring that the orifice is covered and that the external electrode lies submerged in the counting fluid in the sample beaker.

4. Check the program settings:

5. Threshold setting(s) (minimum cell size, usually 7.0 μm)
6. Volume to be counted (usually 0.5 mL)
7. Background subtraction (if used)
8. Dilution settings (e.g., 50 if 0.4 mL is counted in 20 mL of D-PBSA).
9. Check the visual analog display:
    (a) To ensure that all cells fall within the threshold setting(s).
    (b) To check for viability or cell debris (indicated by a shoulder on the curve or histogram falling below the normal lower threshold setting).
    (c) To check for aggregation (indicated by particles appearing above the normal size range).
10. Initiate the count sequence.
11. When the count cycle is complete, the size distribution will appear on the analog screen (Fig. 20.4). Switching to the digital screen will give the cell count per mL.

### Experimental Variations

(1) Repeat counts 5 to 10 times with fresh cells with hemocytometer and electronic cell counter, and calculate the mean and standard deviation.
(2) Compare flasks from Exercise 9.

### Data

Calculate cell counts, with viability correction where appropriate, per culture flask, by each counting method. Details of routine maintenance should be recorded in a record sheet (*see* Table 12.8) and experimental data in a separate table (*see* Table 28.3).

### Analysis

(1) Calculate viable cell yield relative to cells seeded into flasks (*see* Ex. 9 for number of cells seeded).
(2) Has refreshing the medium made any difference to cell yield, and if it has, why should it?

## Exercise 12 Subculture of Cells Growing in Suspension

### Purpose of Procedure

Reduction in cell concentration in proportion to growth rate to allow cells to remain in exponential growth.

### Applications

Routine passage of unattached cells such as myeloma, hybridoma, or ascites-derived cultures; expansion of culture for increased cell production and product harvest; setting up replicate cultures for experimental purposes; amplifying stocks for cryopreservation.

### Training Objectives

Familiarization with suspension mode of growth; use of cell counting and viability estimation.

*Supervision:* Initial supervision required to explain principles, but manipulations are simple and, given that the trainee

has already performed at least one method of counting in Exercise 11, should not require continuous supervision, other than intermittent checks on aseptic technique.

*Time:* 30 min.

### Background Information

Propagation in suspension, subculture of suspension culture (*see* Sections 12.4.5, 12.4.6); viability (*see* Section 21.3.1); cell concentration at subculture (*see* Section 12.4.4). **Standard protocol:** Subculture in suspension (*see* Protocol 12.4). **Ancillary protocol:** Scale-up in suspension (*see* Protocol 26.1).

*Demonstration materials or operations:* Trainee will require two suspension cultures, one in late log phase and one in plateau, with details of seeding date and cell concentration, and should be reminded how to add viability determination into hemocytometer counting (*see* Sections 20.1.1, 21.3.1; Protocol 21.1). If stirred culture is to be used rather than static flasks, the preparation of the flasks and the use of the stirrer platform will need to be demonstrated.

Δ **Safety Note.** Where human cells are used, handling should be in a Class II microbiological safety cabinet, and all materials must be disposed of into disinfectant.

### Instructions for Exercise 12

#### Outline

Withdraw a sample of the cell suspension, count the cells, and seed an appropriate volume of the cell suspension into fresh medium in a new flask, restoring the cell concentration to the initial seeding level.

#### Materials

*Sterile*:

❑ Starter culture: HL-60, L1210, or P388, 5 days and 10 days after seeding at $1 \times 10^4$ cells/mL, 25-cm² flasks .. 1
❑ Growth medium, such as Eagle's MEM with Spinner Salts (S-MEM) or RPMI 1640, with 23 mM NaHCO₃, plus 5% calf serum ..................................... 200 mL
❑ Pipettes, graduated, and plugged .......... 1 can of each
   If glass, an assortment of sizes, 1 mL, 5 mL, 10 mL, 25 mL, in a square pipette can, or, if plastic, individually wrapped and sorted by size on a rack or in cans
❑ Unplugged pipettes for aspirating medium if pump or vacuum line available ............................... 1 can
❑ Universal containers or 50-mL centrifuge tubes ........ 4
❑ Stirrer flasks, 500 mL with magnetic pendulum stirrers (Techne, Bellco) ....................................... 2
❑ Culture flasks, 25 cm² ................................. 2

*Nonsterile*:

❑ Pipette controller or bulb
❑ Tubing to receiver connected to vacuum line or to receiver via peristaltic pump
❑ Alcohol, 70%, in spray bottle

❑ Lint-free swabs or wipes
❑ Absorbent paper tissues
❑ Pipette cylinder containing water and disinfectant
❑ Marker pen with alcohol-insoluble ink
❑ Hemocytometer or electronic cell counter
❑ Magnetic stirrer platform
❑ Notebook, pen, protocols

### Procedure

1. Prepare the hood, and bring the reagents and materials to the hood to begin the procedure.
2. Examine the cultures by eye for signs of contamination or deterioration (fall in pH, aggregation, superficial scum, fungal mycelia or spores). This step is more difficult with suspension cultures than with monolayer cells, as the cells are in suspension making the medium cloudy already.
3. Examine on the microscope; cells that are in poor condition are indicated by shrinkage, an irregular outline, and/or granularity. Healthy cells should look clear and hyaline, with the nucleus visible on phase contrast, and are often found in small clumps in static culture.
4. Take the cultures to the sterile work area, remove a sample, and count the cells.
5. Based on the previously described criteria and on your knowledge of the behavior of the culture, decide whether or not to subculture. If subculture is required (Exercise 12 will require subculture), proceed to the next eight steps.
6. Mix the cell suspensions, disperse any clumps by pipetting the cell suspension up and down, and count the cells.
7. Add 50 mL of medium to each of two stirrer flasks.
8. Add 5 mL of medium to each of two 25-cm$^2$ flasks.
9. Add a sufficient number of cells to give a final concentration of $1 \times 10^5$ cells/mL for slow-growing cells (36–48 h doubling time) or $2 \times 10^4$/ mL for rapidly growing cells (12–24 h doubling time). For the cells cited in Materials, seed at a final concentration of $1 \times 10^4$ cells/mL.
10. Gas the air space of the flasks with 5% $CO_2$.
11. Cap the flasks, and take to incubator. Lay the flasks flat, as for monolayer cultures.
12. Cap the stirrer flasks and place on magnetic stirrers set at 60 to 100 rpm, in an incubator or hot room at 37°C. Take care that the stirrer motor does not overheat the culture. Insert a polystyrene foam mat under the bottle if necessary. Induction-driven stirrers generate less heat and have no moving parts.
13. Sample and count the cells in the flasks and stirrers and check viability after 3 days.

### Data

Determine cell concentration and viability in subcultures from log-phase and plateau-phase cells 72 h after subculture. Recording is best done in a table (Table 28.4) and transferred to a spreadsheet.

### Analysis

(1) Calculate the cell yield as described in Table 28.5.
(2) Compare the yield from cells seeded from log and plateau phase.
(3) Compare growth in flasks with that in stirrers.

## Exercise 13   Subculture of Cell Lines Growing in Monolayer

### Purpose of Procedure

Propagating a culture by transferring the cells of a culture to a new culture vessel. This may involve dilution to reseed the same size of culture vessel, or increasing the size of vessel if expansion is required.

### Applications

Routine passage of attached cells such as A549 or HeLa; expansion of culture for increased cell production and product harvest; setting up replicate cultures for experimental purposes; amplifying stocks for cryopreservation.

### Training Objectives

(1) *Assessment of culture*: This exercise requires the trainee to examine and assess the status of a culture. The trainee should note the general appearance, condition, freedom from contamination, pH of the medium, and density of the cells.
(2) *Aseptic handling*: Reinforces skills learned in Exercises 1, 8, and 9.
(3) *Subculture or passage*: This exercise introduces the principle of transferring the culture from one flask to another with dilution appropriate to the expected growth rate. It shows the trainee how to disaggregate cells by the technique of trypsinization, and reinforces how to count cells and assess viability. The trainee is then required to determine the cell concentration and select the correct concentration for reseeding, instilling a concept of quantitation in cell culture and enhancing numeracy skills.
(4) *Growth characteristics*: Familiarization with differences in appearance and cell proliferation between finite and continuous cells lines.

*Supervision:* Provided that the trainee has shown competence in aseptic technique, continuous direct supervision should not be necessary, but the instructor should be on hand for intermittent supervision and to answer questions.

### Background Information

Subculture, criteria for subculture (*see* Section 12.4.1; Figs. 12.2–12.4; Plates 4,5); deterioration and contamination (*see* Figs. 12.1, 18.1); cell counting (*see* Section 20.1); growth cycle and split ratios (*see* Section 12.4.2), cell concentration at subculture (*see* Section 12.4.3; Fig. 12.4); choice of culture

**TABLE 28.4.** Record of Exercise 12: Subculture of Cells Growing in Suspension

| Sample | Volume of cell suspension in culture flask | Cell count from hemocytometer or electronic counter | Dilution or sampling fraction[a] | Cells/mL in flask | Viability (ratio of unstained cells to total) | Dilution factor for viability stain | Viable cells/mL | Cells/flask |
|---|---|---|---|---|---|---|---|---|
| | V | C | D | C × D | R | F | C×D×R×F | C×D×R×F×V |
| Cell counter without viability stain | 20 | 15321 | 100 | 1532100 | 1 | 1 | 1532100 | 30642000 |
| Hemocytometer with viability stain | 20 | 76 | 10000 | 760000 | 0.85 | 2 | 1292000 | 25840000 |
| | | | | | | | | |
| | | | | | | | | |
| | | | | | | | | |
| | | | | | | | | |

[a]For electronic counting, 0.4 mL cell suspension in 20 mL counting fluid is a 50× dilution, and 2× as the counter counts a sample of 0.5 mL of the diluted suspension, giving a factor of 100×. For hemocytometer counting, if the center 1 mm$^2$ is counted, the factor is $1 \times 10^4$; if all 9 fields of 1 mm$^2$ are counted (because the count was low), then the factor is $1 \times 10^4/9$. In practice, it is better to count 5 fields of 1 mm$^2$ on each side of the slide, whereupon the factor becomes $1 \times 10^4/10$, or $1 \times 10^3$.

**TABLE 28.5.** Analysis of Exercise 12

| Sample | Cells/flask at seeding | Cells/flask at next subculture | Yield = cells recovered ÷ cells seeded |
|---|---|---|---|
| | N | C × D × V × R × F | C × D × V × R × F ÷ N |
| Example | 200,000 | 30642000 | 153.21 |
| Log-phase cells | | | |
| Plateau-phase cells | | | |

vessel (see Section 7.3); $CO_2$ and bicarbonate (see Section 8.2.2; Table 8.1); origin of culture cells (see Section 2.7); immortalization (see Section 17.4); and density limitation of cell proliferation (see Section 17.5.2); cross-contamination (see Sections 12.1.1, 15.2, 18.6; Table 12.2; Appendix V). **Standard protocol:** Subculture of monolayer (see Protocol 12.3). **Ancillary protocols:** Using an inverted microscope (see Protocol 15.1); cell counting (see Protocols 20.1, 20.2); preparation of media (see Protocols 10.7, 10.8, 10.9); staining with Giemsa (see Protocol 15.2).

*Demonstration of materials and operations:* The trainee should be shown different types of culture vessel (see Table 7.2; Figs. 7.3–7.8) and photographs of cells, healthy (see Fig. 15.2; Plates 4, 8, 9), unhealthy (see Fig. 12.1), contaminated (see Fig. 18.1a–c), and at different densities (see Fig. 15.2; Plates 4, 8, 9). Instruction should be given in examining cells by phase-contrast microscopy. A demonstration of trypsinization (see Protocol 13.2) will be required.

## Instructions for Exercise 13

### Outline

Remove the medium and rinse the monolayer. Expose the cells briefly to trypsin then remove the trypsin. Incubate the cells then disperse in medium, count, dilute, and reseed the subculture.

### Materials

*Sterile:*

❑ A549 or HeLa continuous cell line culture, 7 days and 14 days after seeding at $2 \times 10^4$ cells/mL, 25-cm$^2$ flasks ... 1

❑ WI-38, MRC-5, or an equivalent normal diploid fibroblast culture, 7 days and 14 days after seeding at $2 \times 10^4$ cells/mL, 25-cm$^2$ flasks ........................ 1

❑ Growth medium, separate media for each cell line, prelabeled with cell line name:
for example, Eagle's 1× MEM with Earle's salts and 23 mM $HCO_3$, without antibiotics ......... 2 × 100 mL

❑ Trypsin, 0.25% in D-PBSA ...................... 10 mL
❑ D-PBSA with 1 mM EDTA ..................... 20 mL
❑ Pipettes, graduated, and plugged. If glass, an assortment of sizes, 1 mL, 5 mL, 10 mL, 25 mL, in a square pipette can, or, if plastic, individually wrapped and sorted by size on a rack
❑ Unplugged pipettes for aspirating medium if pump or vacuum line is available
❑ Universal containers or 50-mL centrifuge tubes ........ 4
❑ Culture flasks, 25 cm$^2$ .................................. 8

*Nonsterile*:
❑ Pipette controller or bulb
❑ Tubing to receiver connected to vacuum line or to receiver via peristaltic pump
❑ Alcohol, 70%, in spray bottle
❑ Lint-free swabs or wipes
❑ Absorbent paper tissues
❑ Pipette cylinder containing water and disinfectant
❑ Hemocytometer or electronic cell counter
❑ Marker pen with alcohol-insoluble ink
❑ Notebook, pen, protocols

**Procedure**

1. Prepare the hood, and bring the reagents and materials to the hood to begin the procedure.
2. Examine the cultures carefully for signs of deterioration or contamination and assess the cell density.
3. Take the fibroblast culture flask to a sterile work area and return the continuous cell line to the incubator. Handle each cell line separately from this point on, repeating this procedure from this step for each cell line.
4. Remove and discard the medium.
5. Add D-PBSA/EDTA prewash (0.2 mL/cm$^2$) to the side of the flasks opposite the cells so as to avoid dislodging cells, rinse the prewash over the cells, and discard.
6. Add trypsin (0.1 mL/cm$^2$) to the side of the flasks opposite the cells. Turn the flasks over and lay them down. Ensure that the monolayer is completely covered. Leave the flasks stationary for 15 to 30 s.
7. Raise the flasks to remove the trypsin from the monolayer and quickly check that the monolayer is not detaching. Using trypsin at 4°C helps prevent premature detachment, if this turns out to be a problem.
8. Withdraw all but a few drops of the trypsin.
9. Incubate, with the flasks lying flat, until the cells round up (check on microscope); when the bottle is tilted, the monolayer should slide down the surface. (This usually occurs after 5–15 min.) Do not leave the flasks longer than necessary, but then again, do not force the cells to detach before they are ready to do so, or else clumping may result.

10. Add medium (0.1–0.2 mL/cm$^2$), and disperse the cells by repeated pipetting over the surface bearing the monolayer.
11. Pipette the suspension up and down sufficiently to disperse the cells into a single cell suspension.
12. Count the cells with a hemocytometer or an electronic particle counter, and record the cell counts.
13. Dilute the cell suspension to $2 \times 10^4$ cells per mL in 25 mL medium.
14. Seed 5 mL of the cell suspension into each of four flasks.
15. Cap the flasks with gas-permeable caps, and return them to the incubator.
16. Repeat this procedure from step 3 for the continuous cell line.
17. Determine cell counts after 7 days in two flasks from each cell line:
    (a) Remove medium and discard.
    (b) Wash cells gently with 2 mL D-PBSA, remove completely, and discard.
    (c) Add 1 mL trypsin to each flask.
    (d) Incubate for 10 min.
    (e) Add 1 mL medium to trypsin and disperse cells by pipetting vigorously to give a single cell suspension.
    (f) Count cells by hemocytometer or electronic cell counter.
18. Calculate number of cells per flask, cells/mL culture medium, and cells/cm$^2$ at time of trypsinization.
19. Fix and stain cells in another flask.

**Data acquisition**

(1) Cell counts at start and in two flasks from each cell line after 1 week (Table 28.6).
(2) Examine and photograph stained flasks in Exercise 14.

**Analysis**

(1) Calculate fold yield:

$$\textit{number of cells recovered} \div \textit{number of cells seeded}$$

(see Table 28.7) and explain the differences between the two cell lines.
(2) Is an intermediate feed required for either of these cells?
(3) Comment on differences in cell morphology between the cell lines.
(4) Why is the fibroblast cell line handled before the continuous cell line and not simultaneously or after?

## Exercise 14 Staining a Monolayer Cell Culture with Giemsa

**Purpose of Procedure**

Staining with a polychromatic stain like Giemsa reveals the morphology characteristic of the fixed cell and can indicate the status and origin of the cells.

**TABLE 28.6.** Record of Exercise 13

| Sample | Volume of trypsinate | Cell count from hemocytometer or electronic counter | Dilution or sampling fraction | Cells/mL in trypsinate | Cells/flask |
|---|---|---|---|---|---|
| | T | C | D | C × D | C × D × T |
| | | | | | |
| | | | | | |
| | | | | | |
| | | | | | |

**TABLE 28.7.** Analysis of Exercise 13

| Sample | Cells per flask at seeding | Cells harvested per flask[a] | Yield = cells seeded ÷ cells harvested |
|---|---|---|---|
| | N | C × D × T | C × D × T ÷ N |
| | | | |
| | | | |
| | | | |
| | | | |

[a]Viability has not been taken into account in this instance as trypsinization, or at least the prewashes before trypsinization, tend to remove most of the nonviable cells when handling a continuous cell line. This is not necessarily the case with an early passage or primary culture, when viability may need to be taken into account (see Recording, Exercise 19).

## Applications

Monitoring cell morphology, usually in conjunction with phase-contrast observations, during routine passage or under experimental conditions. Preparation of permanent record of appearance of the cells for reference purposes. Identification of cell types present in a primary culture. Early indication of cross-contamination.

## Training Objectives

Emphasizes need for observation of cells during and after culture and alerts the user to the significance of the morphology of the cell.

*Supervision:* Minimal.

*Time:* 30 min.

## Background Information

Morphology (*see* Section 15.5); staining (*see* Section 15.5.2).
*Standard protocol:* Giemsa staining (*see* Protocol 15.2).
*Ancillary protocols:* Staining with Crystal Violet (*see* Protocol 15.3); Using an Inverted Microscope (*see* Protocol 15.1); Digital Photography on a Microscope (*see* Protocol 15.6).

*Demonstration materials or operations:* Preparation of stain and staining procedure should be demonstrated and examples provided of previously stained material.

Δ **Safety Note.** Precautions for human cells as in previous exercises until material is fixed. Methanol is both toxic and flammable.

## Instructions for Exercise 14

### Outline

Fix the culture in methanol, stain it directly with undiluted Giemsa, and then dilute the stain 1:5. Wash the culture, and examine it wet.

### Materials

*Nonsterile:*
- ❏ D-PBSA
- ❏ D-PBSA:methanol, 1:1
- ❏ Methanol
- ❏ Undiluted Giemsa stain
- ❏ Deionized water

### Procedure

1. This protocol assumes that a cell monolayer is being used, but fixed cell suspensions can also be used, starting at step 6.
2. Remove and discard the medium.
3. Rinse the monolayer with D-PBSA, and discard the rinse.
4. Add 5 mL of D-PBSA/methanol per 25 cm². Leave it for 2 min and then discard the D-PBSA/methanol.
5. Add 5 mL of fresh methanol, and leave for 10 min.
6. Discard the methanol, and replace it with fresh anhydrous methanol. Rinse the monolayer, and then discard the methanol.
7. At this point, the flask may be dried and stored or stained directly.

8. Add neat Giemsa stain, 2 mL per 25 cm²; rock the flask to make sure that the entire monolayer is covered and remains covered.

9. After 2 min, dilute the stain with 8 mL of water, and agitate it gently for a further 2 min.

10. Displace the stain with water so that the scum that forms is floated off and not left behind to coat the cells. Wash the cells gently in running tap water until any pink cloudy background stain (precipitate) is removed, but stain is not leached out of cells (usually about 10–20 s).

11. Pour off the water, rinse the monolayer in deionized water, and examine the cells on the microscope while the monolayer is still wet. Store the cells dry, and rewet them to re-examine.

### Analysis

(1) Use flasks from Exercise 13 and compare cell morphology of two cell lines.

(2) Photograph before removing medium and fixation (phase-contrast illumination) and after staining (normal bright-field illumination).

## Exercise 15 Construction and Analysis of Growth Curve

### Purpose of Procedure

Familiarization with the pattern of regrowth following subculture; demonstration of the growth cycle in routine subculture and as an analytical tool.

### Applications

Growth curves, cell proliferation assays, cytotoxicity assays, growth stimulation assays, testing media and sera.

### Training Objectives

Setting up experimental replicates; cell counting and viability; plotting and analyzing a growth curve; awareness of differences in doubling times and saturation densities; selecting reseeding concentration.

*Supervision:* Trypsinization and counting should not need supervision, given satisfactory progress in Exercise 11, but some supervision will be necessary while setting up plates.

*Options:* There are two options for this exercise: a simple growth curve of one cell line cultured in flasks as for regular subculture, or use multiwell plates to analyze differences in growth at different densities, between two different cell lines, or under any other selected set of conditions. With flasks, only one cell line at one seeding concentration is feasible without the number of flasks becoming too large for a training exercise. With multiwell plates, harvesting one plate per day, two parameters (e.g., seeding concentration and cell type) can be handled easily, as long as the trainee is made aware of the risks of handling two cell lines simultaneously,

and that no further propagation would be attempted either from the starting flasks or the experimental plates.

*Time:* 1 h on day 0; 30 min each day thereafter up to day 10.

### Background Information

Choice of culture vessel (*see* Section 7.3); replicate sampling (*see* Section 20.8); growth cycle (*see* Section 20.9.2); growth curve, monolayer (*see* Protocol 20.7 for growth curve in flasks to define conditions for routine maintenance *and* Protocol 20.8 in multiwell plates to analyze growth at different seeding densities and/or to compare two different cell lines); microtitration assays (*see* Section 21.3.5). **Standard protocols:** Handling dishes or plates (*see* Protocol 5.3), growth curves in flasks, (*see* Protocol 20.7), multiwell plates, (*see* Protocol 20.8). **Ancillary protocols:** Growth curve in suspension (*see* Protocol 20.9); MTT-based cytotoxicity assay (*see* Protocol 21.4).

*Demonstration materials or operations:* Trainee should be shown the range of multiwell plates available (*see* Table 7.2; Fig. 8.2) and given some indication of their applications. Setting up plates, with the handling precautions to prevent contamination, will need to be demonstrated (*see* Section 5.4.1).

Δ *Safety Note.* Care should be taken when handling human cells (*see* Section 6.8). Particular care is required in handling open plates and dishes because of the increased risk of spillage (*see* Section 5.5.2; Fig. 5.11).

### Instructions for Exercise 15

#### A. Flasks with one cell type at one seeding concentration.

#### Outline

Set up 24 flasks, harvest two flasks at daily intervals for 10 days and count the cells in each. Fix and stain remaining flasks.

#### Materials

Sterile or aseptically prepared:

❑ Monolayer cell culture, A549 or HeLa-S₃, 75-cm² flask, late log phase .............................................. 1
❑ Trypsin, 0.25%, crude, with 10 mM EDTA ..... 10 mL
❑ Growth medium with 4 mM NaHCO₃ ......... 200 mL
❑ D-PBSA (prewash and for cell counting) ........ 500 mL
❑ Flasks, 25 cm² ......................................... 24

#### Procedure

1. Trypsinize the cells as for a regular subculture.
2. Dilute the cell suspension to $2 \times 10^4$ cells/mL in 150 mL of medium.
3. Seed twenty-four 25-cm² flasks.
4. Seal the flasks with regular caps and place in an incubator at 37°C.
5. After 24 h, remove the first two flasks from the incubator, and count the cells.

*Note.* If a hemocytometer is used to count the cells, it may be difficult to use for the lower cell concentrations at the start of the growth curve. Reduce the volume of trypsin to 0.5 mL, disperse the cells carefully, using a pipettor without frothing the trypsin, and transfer the cells to the hemocytometer. Remember to allow for the difference in trypsin volume when calculating the cell number per flask.

6. Repeat sampling at 48 and 72 h.
7. Change the medium at 72 h, or sooner, if indicated by a drop in the pH.
8. Continue sampling daily for rapidly growing cells (i.e., cells with a PDT of 12–24 h), but reduce the frequency of sampling to every 2 days for slowly growing cells (i.e., cells with a PDT> 24 h) until the plateau phase is reached.
9. Keep changing the medium every 3 days, or more frequently as indicated by the fall in pH.
10. Stain the cells in one flask at 2, 5, 7, and 10 days.

***B. Multiwell plates with two cell types at three cell concentrations.***

### Outline
Set up multiwell plates with two cell lines at three different cell concentrations (Fig. 28.1), harvest one plate each day, and count the cells.

### Materials
*Sterile or aseptically prepared*:
❑ Monolayer cell cultures, A549 and MRC-5, 75-cm$^2$ flasks, late log phase ................................... 1 of each
❑ Trypsin, 0.25%, crude, with 10 mM EDTA ...... 10 mL
❑ Growth medium, MEM with Earle's salts and 26 mM NaHCO$_3$ ...................................... 200 mL
❑ D-PBSA (prewash and for cell counting) ....... 1500 mL
❑ Plates, 12 well ...................................... 12

*Nonsterile*:
❑ Plastic box to hold the plates
❑ CO$_2$ incubator or CO$_2$ supply to purge the box with 5% CO$_2$

### Procedure
1. Trypsinize the both flasks of cells separately, as for a regular subculture.
2. Dilute the cell suspension to $1 \times 10^5$ cells/mL, $3 \times 10^4$ cells/mL, and $1 \times 10^4$ cells/mL, in 25 mL of medium for each concentration.
3. Seed twelve 12-well plates with 1 mL per well. Use the left-hand two wells for MRC-5 at each concentration and the right-hand two wells for A549. Seed the $1 \times 10^4$/ mL cell suspensions in the top wells, the $3 \times 10^4$/ mL in the second row, and the $1 \times 10^5$/ mL in the third row (Fig. 28.1; *see also* Fig. 20.7 for other layout options).
4. Place the plates in a humid CO$_2$ incubator or a sealed box gassed with 5% CO$_2$.

5. After 24 h, remove the first plate from the incubator and photograph one well at each concentration for each cell type.
6. Trypsinize and count the cells in the wells:
7. Remove the medium completely from the wells.
8. Add 0.5 mL of trypsin/EDTA to each well, making sure all the cells are covered.
9. Incubate the plate for 15 min.
10. Add 0.5 mL medium with serum, disperse the cells in the trypsin/EDTA/medium, and transfer 0.4 mL of the suspension to 19.6 mL of D-PBSA.
11. Count the cells on an electronic cell counter.

*Note.* A hemocytometer may be used to count the cells, but may be difficult to use for lower cell concentrations. If you use a hemocytometer, reduce the volume of trypsin to 0.1 mL (do not add medium) and disperse the cells carefully, using a pipettor without frothing the trypsin. Transfer the cells to the hemocytometer and count.

12. Repeat sampling at 48 and 72 h, as in steps 5 and 6.
13. Change the medium at 72 h, or sooner, if indicated by a drop in the pH.
14. Continue sampling daily for rapidly growing cells (i.e., cells with a PDT of 12–24 h), but reduce the frequency of sampling to every 2 days for slowly growing cells (i.e., cells with a PDT> 24 h) for 10 days.
15. Keep changing the medium every 1, 2, or 3 days, as indicated by the fall in pH.

*Note.* As two cell lines have been handled together in this exercise, the progeny of these cells (i.e., all cultures derived from them) must be discarded, and on no account should they be propagated for further use.

### Data
Cell counts per well or per flask per day.

### Analysis
(1) Calculate the cells/mL medium in the wells or flasks during culture and cells/cm$^2$ from the cells/well or flask.
(2) Plot each point with the mean for each day on a log scale against days from seeding on a linear scale.
(3) Derive lag time, doubling time, and saturation density.
(4) Which cell concentration would be best for routine subculture?
(5) Account for differences between normal and transformed cell lines.
(6) Why is the cell density higher overall in the plates?

## 28.4  ADVANCED EXERCISES

These exercises are dependent on satisfactory progress in the basic exercises and should not be attempted until that is achieved. Although advanced, they are still of general

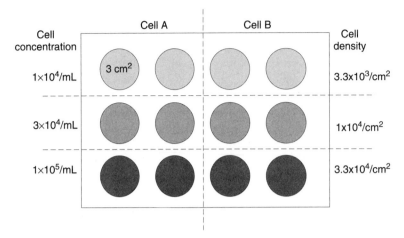

**Fig. 28.1. Layout of 12-Well Plate.** Suggested layout for using three cell concentrations and two different cell types in one plate. Replicate sampling, each day, would require additional plates with the same layout.

application and would be required for anyone claiming general expertise in cell culture. It is, however, possible, if there are time constraints, to defer these exercises if others in the laboratory are already carrying them out and the new trainee or student will not be called upon to perform them. It should be realized, however, if this alternative is adopted, the training cannot be regarded as complete to a reasonable all-round standard until the advanced exercises are performed.

As some basic knowledge is now assumed, variations to the standard protocol will not always be presented in the same detail as for the basic exercises, since it is assumed that a greater degree of experimental planning will be beneficial and a significant part of the training objectives. Whereas the basic exercises are presented in the sequence in which they should be performed, the advanced exercises need not be performed in a specific sequence, and, with a class of students, could be performed in rotation.

### Exercise 16    Cell Line Characterization

Together with mycoplasma detection (*see* Exercise 18), cell line characterization is one of the most important technical requirements in the cell culturist's repertoire. Some form of characterization is essential in order to confirm the identity of cell lines in use, but the techniques selected will be determined by the methodology currently in use in the laboratory. If DNA profiling is available, this single parameter will usually be sufficient to identify individual lines given that comparable data are available for validated stocks of that line [ASN-0002, 2010]; otherwise, more than one technique will be required. Cell lines currently in use will probably have a characteristic already monitored related to the use of the line (e.g., expression of a particular receptor or a specific product, or resistance to a drug), and it may only be necessary to add one other parameter (e.g., chromosomal or isoenzyme analysis). If DNA profiling is not available, it is unlikely that anyone would want to get involved in

setting it up for the sake of a training exercise, and the choice is more likely to be to send the cells to a commercial laboratory for analysis, which has the added advantage that the commercial laboratory will have reference material with which to compare the results. Nevertheless, it is advisable to insert some characterization in the training program to impress upon the student or trainee the importance of cell line authentication, given the widespread use of misidentified cell lines, and the possible consequences (*see* Sections 12.1.1, 15.2, 15.3, 18.6; Table 12.2; Appendix V).

Isoenzyme electrophoresis, which is simply and cheaply performed with a commercially available kit, is suggested as an easily conducted experiment for this exercise.

#### Purpose of Procedure
To confirm the identity of a cell line.

#### Applications
Checking for accidental cross-contamination; quality control of cell lines before freezing and/or initiating a project or program; confirming identity of imported cell lines.

#### Training Objectives
Impress trainee with need for confirmation of cell line identity. Increase awareness of overgrowth, misidentification, and cross-contamination.

*Supervision:*    Preparation of samples and electrophoresis will need supervision, although probably not continuous.

*Time:*    2 h.

#### Background Information
Need for characterization (*see* section 15.1); morphology (*see* Section 15.5); isoenzymes (*see* Section 15.10.1); chromosome content (*see* Section 15.7); DNA profiling (*see* Sections 15.8.2, 15.8.3); antigenic markers (*see* Section 15.11); authentication (*see* Sections 12.1.1, 15.2). **Standard protocol:**

Isoenzyme analysis (*see* Protocol 15.10). *Ancillary Protocols:* Chromosome preparations (*see* Protocol 15.7); DNA profiling (*see* Protocol 15.8); indirect immunofluorescence (*see* Protocol 16.11).

*Demonstration materials or operations:* Use of Authentikit electrophoresis apparatus (*see* Fig. 15.12); examples of DNA profile data (*see* Figs. 15.9, 15.11); examples of karyotypes (*see* Figs. 15.6, 15.7, 15.8).

Δ *Safety Note.* Other than precautions, as before, in the handling of human cell lines, there are no special safety requirements for this exercise.

## Instructions for Exercise 16

### Outline
Cell extracts are prepared, electrophoresed on agarose gels, and developed with chromogenic substrates.

### Procedure
Use standard protocol: Isoenzyme Analysis (*see* Protocol 15.10).

### *Experimental Variations*
Six different cell lines should be examined, chosen from those available in the laboratory, or from the following list: HeLa; KB or Hep-2; Vero; L929, 3T3 or 3T6; BHK21-C13, CHO-K1. Most cell lines from different species can be distinguished by using four isoenzymes: nucleoside phosphorylase, glucose-6-phosphate dehydrogenase, lactate dehydrogenase, and malate dehydrogenase.

### *Data*
Once the gels have been developed with the appropriate chromogenic substrates, they should be photographed or scanned. The gels can be kept.

### *Analysis*
(1) Compare results among different cell lines for each enzyme.
(2) Do you see evidence of cross-contamination or misidentification?
(3) How would you improve resolution among cell lines?

## Exercise 17    Detection of Mycoplasma

### *Purpose of Procedure*
Validation of cell line by proving it to be free of mycoplasmal contamination.

### *Applications*
Routine cell line maintenance; quality control of cell lines before freezing; checking imported cell lines, tissue, and biopsies while in quarantine.

### *Training Objectives*
Awareness of importance of mycoplasma screening. Experience in fluorescence method or PCR for routine screening of cell lines for mycoplasma contamination.

*Supervision:* Setting up cultures and infecting feeder layers should not require supervision, provided that a real mycoplasma contamination is not suspected, in which case the procedure should be carried out in quarantine under strict supervision. Intermittent supervision will be required during mycoplasma staining or DNA extraction and PCR, depending on the experience of the trainee in these areas. Continuous supervision will be required during interpretation of results.

*Time:* 30 min 5 days before start to set up or refeed test culture; 1 h on day 0 to set up indicator cultures; 30 min on day 1 to transfer medium from test culture; 2 to 4 h on day 5 to stain or PCR the cultures and a further 30 min to examine, then or later.

### Background Information
Mycoplasma (*see* Section 18.3.2); validation (*see* Section 6.11). *Standard protocols:* Fluorescence detection of mycoplasma (*see* Protocol 18.2) or detection of mycoplasma contamination by PCR (*see* Protocol 18.3). *Ancillary protocols:* Digital photography (*see* Protocol 15.6).

*Demonstration materials or operations:* Quarantine procedures (*see* Section 18.1.8); use of fluorescence microscope or PCR machine; provision of fixed positive cultures.

Δ *Safety Note.* No special procedures other than standard precautions for human cells (*see* Section 6.8.3). Trainee should be made aware of the severe risks attached to unprotected UV sources and the risk attached to removing working light source from fluorescence microscope.

## Instructions for Exercise 17

### Outline
A test culture is fed with antibiotic-free medium for 5 days, a sample of the medium is transferred to an indicator cell line, known to support mycoplasma growth, and mycoplasma is assayed in the indicator cells by fluorescent DNA staining or PCR.

### Procedure
Use standard protocols: fluorescence detection of mycoplasma (*see* Protocol 18.2) or detection of mycoplasma contamination by PCR (*see* Protocol 18.3).

### *Experimental Variations*
It is difficult to add an experimental element to this exercise except by exploring potential routes of contamination with infected cultures. It is unlikely that any laboratory would wish to undertake this rather hazardous course of action unless special facilities were available.

### *Data*
(1) Results are scored as positive or negative against a fixed positive control (fluorescence) or mycoplasma DNA (PCR).

(2) Records should be kept of all assays and outcomes in a written log or by updating the cell line database.

### Analysis

(1) Mycoplasma-positive specimens will show punctuate or filamentous staining over the cytoplasm (*see* Plate 16*e*, *f*).

(2) Alternatively, electrophoretic migration of PCR product DNA can be compared with incorporated controls (*see* Fig. 18.2).

## Exercise 18  Cryopreservation of Cultured Cells

### Purpose of Procedure

To provide a secure cell stock to protect against contamination, accidental loss though equipment failure or technical error, and genetic and phenotypic instability.

### Applications

Protection of new and existing cell lines; cell banking for archiving and distribution; provision of working cell bank for the lifetime of a project or program; storage of irradiated or mitomycin C-treated feeder cells.

### Training Objectives

Familiarization with cell freezing and thawing procedures and inventory control. Indication of possible variations to improve procedure for difficult cell lines. Comparison or dye exclusion viability with actual cell survival.

*Supervision:* Basic procedures, such as trypsinization, counting, adding preservative, and filling ampoules should not require supervision, but supervision will be required for accessing the liquid nitrogen storage inventory control system, for freezing and transfer of the ampoules to the nitrogen freezer, and for recovery and thawing.

*Time:* 1 h on day 1; 15 min on day 2; 30 min on day 3; 1 h on day 4.

### Background Information

Rationale for freezing (*see* Section 19.1); cooling rate, cryofreezers, and freezer records (*see* Section 19.3); genetic instability (*see* Section 17.3); evolution of cell lines (*see* Section 2.7.2); control of senescence (*see* Section 17.4.1); serial replacement (*see* Section 19.5.2); cell banks (*see* Section 19.6); Cryoprotectants (*see* Section 19.3.3). ***Standard protocols:*** Freezing and thawing cells (*see* Protocol 19.1, 19.2). ***Ancillary protocols:*** Freezing flasks, (*see* Section 19.3.9); vitrification (*see* Protocol 19.3, 19.4); subculture of monolayer (*see* Protocol 12.3); subculture in suspension (*see* Protocol 12.4); cell counting by hemocytometer (*see* Protocol 20.1); electronic cell counting (*see* Protocol 20.2); estimation of viability by dye exclusion (*see* Protocol 21.1).

*Demonstration materials or operations:* Trainee should be shown types of ampoules in regular use, freezing devices (*see* Figs. 19.3–19.5) and types of cryofreezer (*see* Section 19.3.6); should be made aware of the criteria required before freezing (*see* Table 19.1); and should be introduced to the use and upkeep of the freezer inventory control system and record of cell lines (*see* Section 19.5.1) and serial replacement of cultures (*see* Section 19.5.2). The trainee should also be made aware of the importance of monitoring the liquid nitrogen level and temperature of the freezer and the regime for replenishing the liquid nitrogen.

Δ *Safety Note.* Standard precautions if human cell lines are used (*see* Section 6.8.3). In addition, there is a risk of frostbite, asphyxiation, and, where ampoules are stored submerged in liquid nitrogen, explosion (see Section 6.5.6). It is strongly recommended that for the purposes of this exercise and to prevent explosion, ampoules not be submerged in liquid nitrogen but stored in the vapor phase or in a perfused wall freezer.

## Instructions for Exercise 18

### Outline

Cells at a high concentration in medium with preservative (glycerol or DMSO) are cooled slowly (with or without preincubation), frozen slowly, and placed in a liquid nitrogen freezer (Fig. 28.2). They are then thawed rapidly diluted slowly, and reseeded at a high cell density (*see* Fig. 19.9).

### Materials

#### Freezing.

*Sterile or aseptically prepared:*

☐ Culture to be frozen, 75 $cm^2$ flasks, just confluent ..................................................... 2

☐ GM50: growth medium with 50% FBS ..................................... 100 mL

☐ D-PBSA (prewash and count) ..................... 50 mL

☐ Crude trypsin, 0.25% ............................. 10 mL

☐ DMSO freezing medium (DFM): DMSO in GM50, 20% .......................... 10 mL

☐ Glycerol freezing medium (GFM); glycerol in GM50, 20% .......................... 10 mL

☐ Plastic ampoules, 1.2 mL ............................. 12

*Nonsterile:*

☐ Hemocytometer or electronic cell counter .............. 1

☐ Canes for six 1.2-mL plastic ampoules (from nitrogen freezer supplier) ......................................... 2

☐ Tubes of pipe lagging insulation ........................ 2

☐ Cotton or plastic foam plugs for insulation tubes ........ 4

☐ Cardboard tubes to hold canes (from nitrogen freezer supplier) ................................................. 2

☐ Protective gloves, nitrile (for handling DMSO) and cryoprotective (for handling contents of freezer)

☐ Ice bath

#### Thawing.

*Sterile:*

☐ Standard growth medium with 10% FBS ................................. 200 mL

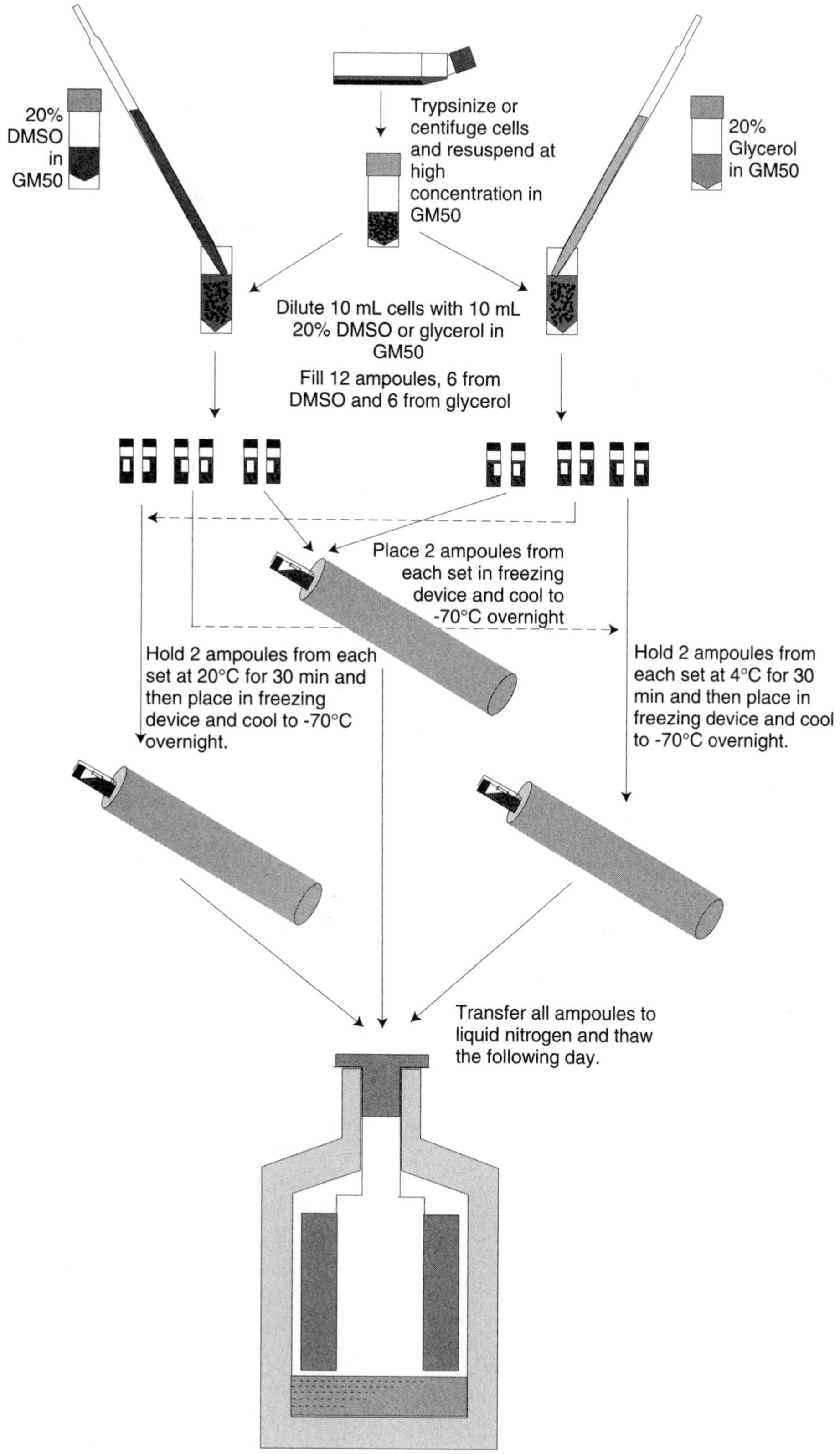

20%
DMSO
in
GM50

Trypsinize or
centifuge cells
and resuspend at
high
concentration in
GM50

20%
Glycerol
in GM50

Dilute 10 mL cells with 10 mL
20% DMSO or glycerol in
GM50

Fill 12 ampoules, 6 from
DMSO and 6 from glycerol

Place 2 ampoules from
each set in freezing
device and cool to
-70°C overnight

Hold 2 ampoules from each
set at 20°C for 30 min and
then place in freezing
device and cool to -70°C
overnight.

Hold 2 ampoules from
each set at 4°C for 30
min and then place in
freezing device and cool
to -70°C overnight.

Transfer all ampoules to
liquid nitrogen and thaw
the following day.

*Fig. 28.2. Options for Freezing Exercise.* This chart suggests experimental variables that can be
added to Exercise 18. It would be feasible to incorporate all of these in one exercise, but it might be
better to choose one variable at a time or per group, if used as a student exercise.

❑ Culture flasks, 25 cm$^2$ ................................12
❑ Pipettes, 1 mL, 10 mL ......................... 20 each
❑ Syringe, 2-mL, and 19-g needle (for glass ampoules)........................................... 1

*Nonsterile:*
❑ Trypan Blue, 0.4% ............................. 10 mL
❑ Alcohol, 70% ................................... 200 mL
❑ Sterilized water at 37°C in a water bath with a rack so that ampoules will not be submerged and the water does not reach the cap of the ampoule (*see* Fig. 19.9, left)

Δ *Safety Note.* It is essential that ampoules not be submerged in liquid nitrogen during storage as they may explode on thawing. Thawing **must** be carried out in a bucket with a lid (*see* Fig. 19.9 right) if there is any chance that the ampoules have been immersed in liquid nitrogen at any time.

❑ Pipettor ............................................. 50 µL
❑ Cryoprotective gloves and face mask
❑ Forceps
❑ Swabs

**Procedure**

*A. Freezing cells.*
1. Make sure the culture satisfies the criteria for freezing and check by eye and on microscope.
2. Trypsinize, resuspend in GM50, and count the cells. If you are using a suspension, count and centrifuge the cells.
3. Dilute cell suspension to $1 \times 10^7$ cells/mL.
4. Divide the cells suspension into two lots of 10 mL each.
5. Dilute each cell suspension 1:1 with DFM or GFM to give $5 \times 10^6$ cells/mL and 10% DMSO or glycerol.
6. Label 6 ampoules for cells in DFM and 6 for cells in GFM, and label 2 of each set A, B, & C.
7. Dispense the cell suspensions into prelabeled ampoules, and cap the ampoules.
8. Place the 4 ampoules labeled 'A' on ice, the 4 labeled B at 37°C, and the 4 labeled C at room temperature (record what this is).
9. After 30 min, clip the 12 ampoules on 2 canes, insert into cardboard tubes, and immediately place in insulation tubes.
10. Place the insulation tubes at −70°C overnight.
11. Check the freezer record before removing the ampoules from the −70°C freezer and identify a suitable location for the ampoules.
12. Remove ampoule canes in cardboard tubes from insulation tubes and transfer the ampoule canes in the cardboard tubes to the gas phase of the liquid N$_2$ freezer, placing the cane and tube into the predetermined canister. This transfer must be done quickly (< 2 min),

as the ampoules will reheat at about 10°C/min, and the cells will deteriorate rapidly if the temperature rises above −50°C.

Δ *Safety Note.* Cryoprotective gloves and a face mask should be used when placing ampoules in a liquid nitrogen freezer. *Do not* submerge the ampoules in liquid nitrogen as they can take up the liquid nitrogen and explode violently on thawing. Submerged ampoules also have a higher risk of contamination.

13. When the ampoules are safely located in the freezer, complete the appropriate entries in the freezer index (*see* Tables 19.2, 19.3).

*B. Thawing frozen cells.*
1. Check the index for the location of the ampoule to be thawed.
2. Collect all necessary materials, prepare the medium, and label the culture flasks.
3. Retrieve the ampoules from the freezer, check the labels and place them in the rack in sterile water at 37°C in the water bath. Avoid getting water up to the cap as this will increase the chance of contamination. A heating block is useful for this, though heat transfer may be slower.

Δ *Safety Note.* A closed lab coat, facemask, and cryoprotective gloves must be worn when removing the ampoule from the liquid nitrogen freezer. Ampoules, including plastic ampoules, must have been stored in the gaseous phase; if stored in the liquid phase, they can inspire the liquid nitrogen and, on thawing, will explode violently. If ampoules have been stored submerged in liquid nitrogen, seek guidance from your supervisor and **use a plastic bucket with a lid for thawing** to contain any explosion (*see* Fig. 19.9).

4. When the ampoule has thawed, double-check the label to confirm the identity of the cells; then swab the ampoules thoroughly with 70% alcohol, and open them one at a time in a laminar-flow hood.
5. Individually transfer the contents of the ampoules to prelabeled culture flasks with a 1-mL pipette.
6. Add medium slowly to the cell suspension: 10 mL over about 2 min added dropwise at the start, and then a little faster, gradually diluting the cells and cryoprotectant.
7. Cap with gas permeable cap and place in CO$_2$ incubator.
8. Stain the dregs in the ampoule with one drop of Trypan Blue and determine cell viability.
9. Trypsinize flasks after 24 h and count cells (this step is for this exercise only and not standard practice in routine freezing).
10. Count the cells in the trypsinate.

### Other Experimental Variations

There are several other experimental variables that can be introduced into this exercise, and the trainee encouraged to devise their own protocol:

(1) Alterations in the freezing rate.

(2) Concentration of cryoprotectant.

(3) Centrifugation after thawing to remove cryoprotectant.

(4) Comparison of vitrification (*see* Protocol 19.3) with slow cooling and freezing.

(5) Removal of preservative by replacing medium the next day or by centrifugation after thawing. If this is selected, then it would be interesting to compare cells from suspension culture (e.g., L5178Y lymphoma, a hybridoma, or HL60) with cells from attached monolayers (e.g., HeLa, A549, Vero, or NRK).

(6) Rapid or slow dilution after thawing.

### Data

(1) Routine records should be completed as for standard freezing (*see* Tables 19.2, 19.3).

(2) Record cell viability from the dregs of each ampoule.

(3) Record the counts from the cultures trypsinized the day after thawing (only for generating survival data; not as a routine).

(4) The results can be tabulated as in Exercises 13 and 14 and the recovery of viable cells calculated on the day of thawing and the recovery of attached cells (for monolayer cultures only) calculated on the day after thawing.

### Analysis

(1) Which preservative is best for your cells?

(2) Does dye exclusion viability agree with recovery after 24 h?

(3) If not, why not?

(4) Is a delay before freezing harmful or beneficial?

(5) Does chilling the cells after adding preservative help?

## Exercise 19   Primary Culture

### Purpose of Procedure

The isolation of cells from living tissue to create a cell culture.

### Applications

Initiation of primary cultures for vaccine production; isolation of specialized cell types for study; chromosomal analysis; development of selective media; provision of short-term cell lines for screening, analytical experiments, or tissue engineering.

### Training Objectives

Awareness of origin and diversity of cultured cells. Appreciation of variations in primary culture methodology.

*Supervision:*   Intermittent.

*Time:*   2 to 4 h day 0; 1 h day 1; 2 h day 3.

### Background

Chick embryos have been selected for this exercise for a variety of reasons. They are readily available with minimal animal care backup and can be dissected without restrictions if less than half-term; full term is 21 days, so 10-day embryos are suitable, if a little small. They are larger, at a given stage, than mouse embryos and give a high yield of cells either from the whole chopped embryo or from isolated organs.

Types of primary culture (*see* Section 11.1); isolation of the tissue (*see* Section 11.2; warm trypsin (*see* Section 11.3.3); cold trypsin (*see* Section 11.3.4). **Standard protocols:** (*see* Protocols 11.1, 11.2, 11.5–11.7). **Ancillary protocols:** mouse embryo (*see* Protocol 11.1), human tissue, (*see* Protocol 11.3), collagenase (*see* Protocol 11.8), mechanical disaggregation (*see* Protocol 11.9), and enrichment of viable cells (*see* Protocol 11.10).

*Demonstration materials or operations:*   Dissection of chick embryos (or alternative tissue source). Photographs of dissection stages (*see* Figs. 11.4, 11.11).

Δ *Safety Note.* Minimal requirements if chick embryo material is used.

## Instructions for Exercise 19

### Outline

An embryo is removed from the egg, dissected, and primary explant and disaggregated cell cultures are set up from the whole chopped embryo and from individual organs using warm and cold trypsinization. This whole exercise can be performed on one embryo, but it is preferable to take fresh embryos for Procedures D and E.

### Materials

*Sterile:*

❑ DBSS: dissection BSS (for dissection; *see* Appendix I) ................................... 200 mL

❑ D-PBSA (dissection medium and trypsin diluent) ......................................... 1000 mL

❑ Growth medium (e.g., 50:50 DMEM:F12 with 20% fetal bovine serum) .......................... 200 mL

❑ Trypsin in D-PBSA (for warm trypsinization), 2.5% ............................ 100 mL

❑ Trypsin in D-PBSA (for cold trypsinization), 0.25% ........................... 100 mL

❑ Small beaker, 20 to 50 mL or egg cup .................. 1

❑ Forceps, straight and curved ............................. 2

❑ Scalpels, #11 blades ..................................... 2

❑ Beaker(for rinsing instruments after alcohol), 100 mL ..................................... 1

❑ Petri dishes (non–tissue-culture grade; for dissection), 9 cm ................................. 12

❑ Culture flasks, 25 cm² .................................... 24

❑ Trypsinization flask: Erlenmeyer flask or stirrer flask, 250 mL .......................................... 1

❑ Magnetic follower, autoclaved in a
  test tube (for trypsinization flask) ........................ 1
❑ Test tubes, preferably glass with screw
  caps (for organ rudiments), 10–15 mL ................. 6
❑ Small Erlenmeyer flask or universal
  container (for cold trypsin), 25–50 mL ................ 1
❑ Pasteur pipettes ......................................... 30
❑ Pipettes, with wide tips, 10 mL ...................... 10

*Nonsterile*:

❑ Embryonated eggs, 10th day of incubation (11th day
  preferable if regulations permit)
❑ Alcohol, 70% (for swabbing egg as well as work surface)
❑ Trypan Blue, 0.4%
❑ Giemsa stain, neat
❑ Swabs
❑ Humid incubator (no additional $CO_2$ above atmospheric
  level)
❑ Pipettor, 100 µL, adjustable
❑ Magnetic stirrer
❑ Hemocytometer or cell counter
❑ Ice bath
❑ Binocular dissecting microscope

### A. Isolation of chick embryos.

### Procedure

*Summary:* The top of the egg is removed and the embryo
carefully lifted out into a Petri dish.

1. Swab the egg with 70% alcohol, and place it with its blunt
   end facing up in a small beaker (Fig. 11.4a).
2. Crack the top of the shell (Fig. 11.4b), and peel the shell
   off to the edge of the air sac with sterile forceps (Fig.
   11.4c).
3. Resterilize the forceps (i.e., dip them in alcohol and rinse
   in sterile D-PBSA), and then use the forceps to peel off
   the white shell membrane to reveal the chorioallantoic
   membrane (CAM) below, with its blood vessels (Fig.
   11.4d, e).
4. Pierce the CAM with sterile curved forceps (Fig. 11.4f),
   and lift out the embryo by grasping it gently under the
   head (Fig. 11.4g, h). Do not close the forceps completely,
   or else the neck will sever; place the middle digit under
   the forceps and use the finger pad to restrict the pressure
   of the forefinger (*see* Fig. 11.4g).
5. Transfer the embryo to a 9-cm Petri dish containing
   20 mL DBSS (Fig. 11.4i).
6. Cut off the head.

### B. Primary explants.

### Procedure

*Summary:* The tissue is chopped finely and rinsed, and the
pieces are seeded onto the surface of a culture flask in a small
volume of medium with a high concentration (i.e., 40–50%)
of serum, such that surface tension holds the pieces in place
until they adhere spontaneously to the surface (Fig. 11.6a).

1. Transfer the head and body to fresh, sterile DBSS.

2. Peel off some skin from the head and chop finely with
   crossed scalpels (*see* Fig. 11.6a, top) into about 1-mm
   pieces.
3. Transfer the pieces with a wide-tipped pipette (wet the
   pipette with BSS first to prevent pieces sticking) to a
   culture flask, about 20 to 30 pieces per 25-cm² flask.
4. Remove most of the fluid, and add 1 mL growth medium
   per 25-cm² growth surface. Tilt the flask gently to spread
   the pieces evenly over the growth surface.
5. Cap the flask with a gas-permeable cap, and place it in a
   $CO_2$ incubator at 37°C for 18–24 h.
6. The following day, check for any signs of outgrowth,
   and add 1 mL medium carefully without dislodging the
   explants.
7. Make up the medium volume gradually to 5 mL per
   25 cm² over the next 3–5 days.
8. Change the medium weekly until a substantial outgrowth
   of cells is observed.

### C. Chick embryo organ rudiments.

### Procedure

*Summary:* Dissect out individual organs or tissues, and place
them, preferably whole, in cold trypsin overnight. Remove
the trypsin, incubate the organs or tissue briefly, and disperse
them in culture medium. Dilute and seed the cultures.

1. Remove an eye and open it carefully, releasing the lens
   and aqueous and vitreous humors (Fig. 11.11c, d).
2. Grasp the retina in two pairs of fine forceps and
   gently peel the pigmented retina off the neural retina
   and connective tissue (Fig. 11.11e). (This step requires
   a dissection microscope for 10-day embryos. A brief
   exposure to 0.25% trypsin in 1 mM EDTA will allow
   the two tissues to separate more easily.) Put the tissue to
   one side of the dish.
3. Pierce the top of the head with curved forceps, and
   scoop out the brain (Fig. 11.11f). Place the brain with
   the retina at the side of the dish.
4. Halve the trunk transversely where the pink color of the
   liver shows through the ventral skin (Fig. 11.11g). If the
   incision is made on the line of the diaphragm, then it
   will pass between the heart and the liver; sometimes the
   liver will go to the anterior instead of the posterior half.
5. Gently probe into the cut surface of the anterior half,
   and draw out the heart and lungs (Fig. 11.11h; tease the
   organs out, and do not cut until you have identified
   them). Separate the heart and lungs and place at the side
   of the dish.
6. Probe the posterior half, and draw out the liver, with
   the folds of the gut enclosed in between the lobes (Fig.
   11.11i). Separate the liver from the gut and place each at
   the side of the dish.
7. Fold back the body wall to expose the inside of the
   dorsal surface of the body cavity in the posterior half.
   The elongated lobulated kidneys should be visible parallel
   to and on either side of the midline.

8. Gently slide the tip of the scalpel under each kidney and tease the kidneys away from the dorsal body wall (Fig. 11.11*j*). (This step requires a dissection microscope for 10-day embryos.) Carefully cut the kidneys free, and place them on one side.

9. Place the tips of the scalpels together on the midline at the posterior end and, advancing the tips forward, one over the other, express the spinal cord as you would express toothpaste from a tube (Fig. 11.11*k*). (This step may be difficult with 10-day embryos.)

10. Turn the posterior trunk of the embryo over, and strip the skin off the back and upper part of the legs (Fig. 11.11*l*). Collect and place this skin on one side.

11. Dissect off muscle from each thigh, and collect this muscle together (Fig. 11.11*m*).

12. Select 6 of these tissues and transfer each to separate test tubes, remove any surplus BSS, add 1 mL of ice-cold 0.25% trypsin, and place these tubes on ice. Make sure that the tissue is immersed in the trypsin.

13. Leave the test tubes for 6 to 18 h at 4°C.

14. Carefully remove the trypsin from the test tubes without disturbing the tissue; tilting and rolling the tube slowly will help.

15. Incubate the tissue in the residual trypsin for 15 to 20 min at 37°C.

16. Add 4 mL of medium to each of two 25-cm² flasks for each tissue to be cultured.

17. Add 2 mL of medium to tubes containing tissues and residual trypsin, and pipette up and down gently to disperse the tissue.

18. Allow any large pieces of undisaggregated tissue to settle.

19. Pipette off the cells in the supernate into the first flask, mix, and transfer 1 mL of diluted suspension to the second flask. This procedure gives two flasks at different cell concentrations and avoids the need to count the cells

20. Change the medium as required (e.g., for brain, it may need to be changed after 24 h, but pigmented retina will probably last 5–7 days), and check for characteristic morphology and function.

21. Fix and stain cultures after 3–7 days as appropriate.

### D. Tissue disaggregation in warm trypsin.

#### Procedure

*Summary:* The headless embryo is chopped and stirred in trypsin for a few hours. The dissociated cells are collected every half hour, centrifuged, and pooled in medium containing serum (Fig. 11.7). If this is done as a class exercise, chopped embryos from several students may be combined in one stirrer flask of trypsin. If done individually, use 6 embryos. Instructions are for one embryo assuming they are combined for trypsinization.

1. Repeat Procedure A with a fresh embryo.

2. Weigh the embryo in a preweighed dry Petri dish with lid.

3. Transfer the body to fresh, sterile DBSS in 9-cm Petri dish.

4. Chop with crossed scalpels (*see* Fig. 11.7) into about 3-mm cubes.

5. Transfer all the pieces with a wide bore pipette to the empty trypsinization flask.

6. Remove most of the residual fluid in the trypsinization flask, and add 180 mL of D-PBSA.

7. Add 20 mL of 2.5% trypsin.

8. Add the magnetic follower to the flask.

9. Cap the flask, and place it on the magnetic stirrer in an incubator or hot room at 37°C.

10. Stir at about 100 rpm for 30 min at 37°C.

11. After 30 min, collect disaggregated cells as follows:
    (a) Allow the pieces to settle.
    (b) Pour off the supernate into a centrifuge tube and place it on ice. Carefully wipe off any medium running down the outside of side arm from which you have poured with a lint-free swab and 70% alcohol.
    (c) Add fresh D-PBSA and trypsin as in step 5 to the pieces remaining in the flask, and continue to stir and incubate for a further 30 min.
    (d) Centrifuge the harvested cells from step 11(b) at approximately 500 *g* for 5 min.
    (e) Resuspend the resulting pellet in 10 mL of medium with serum, and store the suspension on ice.

12. Repeat step 10 four times, when most of the tissue should have disaggregated.

13. Collect and pool the chilled cell suspensions.

14. Remove any large remaining aggregates by filtering through sterile muslin or a proprietary sieve (e.g., *see* Fig. 11.8) or by allowing the aggregates to settle.

15. Count the cells by hemocytometer and note viability.

16. Dilute the cell suspension to $1 \times 10^5$ viable cells/mL in growth medium, and seed three 25 cm² flasks with 5 mL.

17. Check for attached cells the following day.

18. After 3 days, trypsinize and count 2 flasks and fix and stain the third.

### E. Tissue disaggregation in cold trypsin.

#### Procedure

*Summary:* Chop tissue and place in trypsin at 4°C for 6 to 18 h. Incubate after removing the trypsin, and disperse the cells in warm medium (Fig. 11.9).

1. Repeat Procedure A.

2. Transfer the body to fresh, sterile DBSS in a 9-cm Petri dish.

3. Chop with crossed scalpels (*see* Fig. 11.9) into about 3-mm cubes.

4. Transfer the tissue with curved forceps to a 25- or 50-mL flat-bottomed sterile glass vial or Erlenmeyer flask.

5. Allow the pieces to settle.

6. Carefully remove the residual fluid.

7. Add 10 mL 0.25% trypsin.

8. Place the mixture at 4°C for 6 to 18 h.

9. Remove and discard the trypsin carefully, leaving the tissue with only the residual trypsin.

10. Place the vial or flask at 37°C for 20 to 30 min (no stirring is required).

11. Add 10 mL of warm medium and gently pipette up and down until the tissue is completely dispersed.

12. If some tissue does not disperse, then the cell suspension may be filtered through sterile muslin, stainless steel mesh (100–200 μm), or a disposable plastic mesh strainer (Fig. 11.8), or the larger pieces may simply be allowed to settle.

13. Determine the cell concentration in the suspension by hemocytometer and record viability.

14. Dilute the cell suspension to $1 \times 10^5$ cells/mL in growth medium, and seed three 25-cm$^2$ flasks with 5 mL.

15. Check for attached cells the following day.

16. After 3 days, trypsinize and count 2 flasks and fix and stain the third with Giemsa.

### Data

(1) Examine living cultures from organ rudiments after 3 to 5 days and check for morphological differences and signs of contraction in the heart cells.

(2) Cell counts and viability from cells recovered from warm and cold trypsinization.

(3) Trypsinize and count cultures derived from warm and cold trypsin 3 days after seeding.

### Analysis

(1) Try to identify different cell types present in primary cultures from organ rudiments.

(2) How would you propagate these cultures to retain specific cell types?

(3) Calculate and tabulate the recovery of total and viable cells/embryo (Table 28.8).

(4) From the number of cells recovered at the first subculture (after 3 days in this case), calculate the yield of cells per embryo, and as a ratio of the total cells seeded, and viable cells seeded.

(5) Was the dye exclusion staining a good predictor of recovery?

(6) Examine stained cultures for morphological features and compare chopped whole embryo cultures from warm and cold trypsin (D & E).

## Exercise 20    Cloning of Monolayer Cells

### Purpose of Procedure

To dilute cells such that they grow as isolated colonies derived from single cells, demonstrating their proliferative potential and allowing isolation of a specific strain.

### Applications

Isolation of genetic or phenotypic variants; survival assay; growth assay.

### Training Objectives

Introduction to technique of dilution cloning; determination of plating efficiency as a growth or survival parameter; clonal isolation of selected cell types.

*Supervision:* Initial supervision only on day 0, and help later (days 10–14) with identifying clones.

*Time:* 1 h.

### Background Information

Cloning (*see* Section 13.1; Fig. 13.2); plating efficiency (*see* Section 20.10); staining with Crystal Violet (*see* Protocol 15.3; Plate 6a, e). **Standard protocols:** Cloning (*see* Protocol 13.1), plating efficiency (*see* Protocol 20.10). **Ancillary protocols:** Conditioned medium (*see* Protocol 13.2), feeder layers (*see* Protocol 13.3), cloning in suspension (*see* Protocols 13.4, 13.5), isolation of clones (*see* Protocols 13.6–13.8); clonogenic assay (*see* Protocol 21.3).

*Demonstration materials or operations:* Previous cloned cultures stained with Crystal Violet; options for cloning, such as Petri dishes versus microtitration plates; cloning rings for isolation.

Δ **Safety Note.** Minimal risk presented if a nonhuman cell line is used, such as CHO-K1. Use of a growth-arrested feeder layer would, however, require attention to toxicity of mitomycin C or irradiation risk of source, depending on which is used.

### Instructions for Exercise 20

#### Outline

Seed log phase cells at very low density and incubate until colonies form. Fix, stain, and count the colonies to determine plating efficiency.

#### Materials

*Sterile or aseptically prepared:*

❑ CHO cells, 25 cm$^2$ flask, late log phase ................. 3

❑ Medium, 3 batches prepared from UPW, deionized water, and tap water (from Ex. 3): Ham's F12, 5% $CO_2$ equilibrated, 10% FBS .................... 250 mL of each

❑ Trypsin, 0.25%, 1:250, or equivalent.............. 10 mL

❑ Pipettes, 1, 5, 10, and 25 mL .................. 10 of each

❑ Petri dishes, 6 cm, .................................... 30

❑ Universal containers or tubes (for dilution) ............ 20

*Nonsterile:*

❑ D-PBSA .......................................... 500 mL

❑ D-PBSA/MeOH: 50% methanol in D-PBSA ... 500 mL

❑ Methanol ........................................ 500 mL

❑ Crystal Violet .................................... 200 mL

❑ Hemocytometer or electronic cell counter

#### Procedure

1. Trypsinize the first flask of cells to produce a single cell suspension.

**TABLE 28.8.** Record and Analysis of Exercise 19 D, E

| Trypsinization | | Warm | Cold |
|---|---|---|---|
| Age of embryos | | | |
| Number or fraction of embryos used (N) | | | |
| Weight (W) | | | |
| Volume of trypsin | | | |
| Temperature | | | |
| Duration of trypsinization at 4°C | | | |
| Duration of trypsinization at 37°C | | | |
| Volume of medium used to disperse cells after trypsin removal (V) | | | |
| Concentration of cells after dispersal in medium (C) | | | |
| Percentage viability (v) | | | |
| Concentration of viable cells ($C_v = C \times v$) | | | |
| Total number of cells recovered ($n = C \times V$) | | | |
| Number of viable cells recovered ($n_v = C_v \times V$) | | | |
| Total number of cells per embryo ($T = n/N$) | | | |
| Viable cells per embryo ($T_v = n_v/N$) | | | |
| Number of cells per g (n/W) | | | |
| Viable cells per g ($n_v/W$) | | | |
| Number of viable cells seeded per flask (S) | | | |
| Number of cells per flask recovered after 3 days (R) | | | |
| Percentage yield (R/S × 100) | | | |

2. While the cells are trypsinizing:
   (a) Label 12 dishes (on the side of the base), for example, A1, A2, A3, B1, B2, B3, etc
   (b) Label 5 universal containers or tubes for dilution, $10^5, 10^4, 10^3, 10^2$, and 10. Measure out medium for four dilution steps, none in the $10^5$ and 18 mL each of the others, using the UPW medium.
3. When the cells round up and start to detach, disperse the monolayer to a single cell suspension in 5 mL of medium.
4. Count the cells and dilute as follows:
   (a) Dilute trypsinate to $1 \times 10^5/$ mL (approximately 1:10 or 1:20, depending on the number of cells in the flask. The volume is not critical as long as you have a minimum of 16 mL; for example, it might be convenient to dilute, say, $1.8 \times 10^6$ cells/mL in the trypsinate 1 up to 18 mL.
   (b) Dilute 2 mL of the $1 \times 10^5/$ mL suspension to 20 mL (1:10) to give $1 \times 10^4$ cells/mL.
   (c) Dilute 2 mL of the $10^4/$ mL suspension to 20 mL (1:10) to give $1 \times 10^3$ cells/mL.
   (d) Dilute 2 mL of the $10^3/$ mL suspension to 20 mL (1:10) to give $1 \times 10^2$ cells/mL.
   (e) Dilute 2 mL of the $10^2/$ mL suspension to 20 mL (1:10) to give 10 cells/mL.
5. Seed 3 Petri dishes each with 5 mL of medium containing cells from each dilution stage.
6. Place the dishes in a humid 5% $CO_2$ incubator.

7. Repeat the dilution and seeding (steps 4–6) with a second and third set of prelabeled dishes, using the DW or tap water for dilution. If necessary, a fresh flask may be trypsinized to give fresh cells, but, if the first stage has not taken too long, the $1 \times 10^5$ dilution can be used as a starting point.
8. Leave the cultures untouched for 1 week.
9. Collect the culture, discard the lids (hence the reason for labeling the base), wash, fix and stain the clones:
   (a) Remove and discard the medium.
   (b) Rinse the monolayer with D-PBSA, and discard the rinse.
   (c) Add 5 mL of D-PBSA/MeOH per dish for 2 min and then discard the D-PBSA/MeOH.
   (d) Add 5 mL of fresh methanol, and leave for 10 min.
   (e) Discard the methanol, drain the dishes, and allow them to dry.
   (f) Add 5 mL of Crystal Violet and leave for 10 min.
   (g) Rinse the dish in tap water and then in deionized water, and allow the dish to dry.
10. Count the colonies in each dish, excluding those below 50 cells per colony. Magnifying viewers can make counting the colonies easier.

**Data**
Number of colonies per dish.

**Analysis**
(1) Calculate the plating efficiency (PE):

$$PE = colonies\ formed \div cells\ seeded \times 100$$

(2) Compare results with media made from different water qualities.

(3) Why is this a good test for serum?

(4) How would you compare several serum batches?

### *Experimental Variations*

(1) *Quality assurance:* This experiment is designed to demonstrate differences in quality between UPW, DW and tap water, but can be used to test any constituent, such as serum, in which case a range of serum concentrations would be used.

(2) *Seeding concentration:* Once the plating efficiency has been determined, the experiment can be repeated with a narrower range of cell concentration.

(3) *Cytotoxicity:* A simple variable to add into this exercise is the addition of a cytotoxic drug to the cells for 24 h before cloning (*see* Protocol 21.3, Exercise 27). If mitomycin C is used, it becomes a useful preliminary to preparing a feeder layer (*see* Protocol 14.3). An exponential range of concentrations, 0 to 50 μg/mL, would be suitable, such as 0, 0.1, 0.2, 0.5, 1.0, 2.0, 5, 10, 20, 50 μg/mL.

(4) *Feeder layer:* Repeat option 3 with and without feeder layer (*see* Protocol 13.3).

(5) *Isolation of clones:* Isolate and compare morphology of cloned strains.

## 28.5   SPECIALIZED EXERCISES

It is assumed that anyone progressing to the specialized exercises (21–27) will have a specific objective in mind and will select protocols accordingly. The parameters of variability will also be determined by the objectives, so these exercises are not detailed and it is assumed that the student/trainee will, by now, have the skills necessary to design his/her own experiments, following and amending the standard protocols. The training programs included in Table 28.1, however, are thought to have enough general interest to belong to such an extended training program, albeit at a more advanced level.

# CHAPTER 29

# Problem Solving

No matter how well a laboratory is run, problems arise when new personnel, new techniques, or any other new development destabilizes its normal routine. One solution to such problems is to make sure that procedures do not change i.e., to define standard operating procedures (SOPs) and ensure that deviations from these procedures are made only after exhaustive testing of the possible repercussions and with the agreement of all users. However, this is often difficult, particularly in a research environment, where progress demands change and new procedures are introduced continually.

The advice given in the preceding chapters has concentrated on practical, "how to do it" instructions, sometimes with indications of what might go wrong. This chapter attempts to summarize these potential problems under topic headings and adds a few more potential difficulties, queries, and, hopefully, solutions.

*Culture of Animal Cells: A Manual of Basic Technique and Specialized Applications, Sixth Edition,* by R. Ian Freshney
Copyright © 2010 John Wiley & Sons, Inc.

## 29.1  ABNORMAL APPEARANCE OF CELLS

| Nature of problem | Cause | Action | Follow-up | Cross-references |
|---|---|---|---|---|
| **Intracellular granularity or vacuolation** | Cellular deterioration | Check parameters for slow cell growth, above; check growth curve and plating efficiency | Renew stock from freezer or cell bank; check optimization of culture conditions | Section 29.1; Fig. 12.1; Protocols 20.7–20.10 |
| | Phagocytosis | Check on high-power phase microscopy for uptake of neutral red or fluorescent dextran (FITC-dextran, Sigma) | | |
| **Extracellular granularity** | Contamination | Check for uniformity of particle size; plate out medium on nutrient agar if in doubt | Maintain cultures antibiotic free; check by Hoechst 33258 fluorescence; discard if contaminated | Section 18.3 |
| | Precipitate from medium or serum | Check on microscope for variable particle size; plate out on blood agar if in doubt | Check medium for precipitation—may mean a constituent has been lost Check serum for precipitation (not usually harmful) | Section 29.2.2 |
| **Loss of birefringence (edge of cells indistinct)** | Loss of viability, perhaps by drying out | Trypsinize and check viability by dye exclusion and reattachment; discard culture if no cells attach | Check steps of procedure to determine when drying out could occur (e.g., during feeding of a large number of plates) | Section 21.3.1, 21.3.5; Plate 17a |

## 29.2  SLOW CELL GROWTH

### 29.2.1  Problems Restricted to Your Own Stock

| Type of problem | Cause | Action | Follow-up | Cross-references |
|---|---|---|---|---|
| **Contamination** | Overt contamination | Examine visually by naked eye | Discard cells if contaminated | Section 18.3.1, 29.6; Plate 16a–c |
| | | Examine on microscope | Discard cells if contaminated | Section 18.3.1, 29.6; Fig. 19.1a–c |
| | Cryptic contamination | Test for mycoplasma | | Protocol 18.2, 18.3 |
| | | | Discard cells if contaminated | Sections 18.4; 18.5 |
| | | | Check potential routes or causes | Sections 18.1, 28.6; Table 18.1 |

| Type of problem | Cause | Action | Follow-up | Cross-references |
|---|---|---|---|---|
| **Media & reagents** | Test other media and sera | Test different batch | Try different supplier/different product 1×, 10×, powder | Sections 10.6, 29.2 |
| | If medium responsible | | Identify cause/change, supplier/product | Section 29.2 |
| | Trypsin or other dissociation agent | Check batch no. & supplier; changed? | Try new batch or supplier | |
| | | Check exposure: Duration/concentration/temperature | Redefine procedure if necessary | |
| | | Pipetting too vigorous? | Pipette more gently | |
| | | Check toxicity of EDTA | If toxic, remove or change to EGTA | Appendix I: TEGPED |
| | | Check diluent | Wrong diluent/bad batch: replace | |
| **Status of cells** | Test cells: thaw new ampoule & culture separately | Growth curve | Seeding density too low; subculture too frequent; too long in plateau | Sections 12.4.3–12.4.7; Protocols 20.7, 20.8 |

## 29.2.2  Problem More General and Other People Having Difficulty

| Type of problem | Cause | Action | Follow-up | Cross-references |
|---|---|---|---|---|
| **Incubation** | | | | |
| Incubator or warm room temperature control | Faulty thermostat | Check with recording thermometer | Call service engineer | Sections 4.3.1, 4.3.3 |
| | Fan failure | Check with recording thermometer | Call service engineer | Sections 4.3.1, 4.3.3 |
| | Access to frequent | Check with recording thermometer | Change code of practice | Sections 4.3.1, 4.3.3 |
| | Door left open | Check with recording thermometer | Change code of practice | Sections 4.3.1, 4.3.3 |
| Incubator humidity | Water tray empty | Check evaporation rate with preweighed Petri dish of D-PBSA | Check filling routine | Section 4.3.2 |
| | Leakage round doors | Check seals | Call service engineer | Section 4.3.2 |
| | Access to frequent | Check evaporation rate | Control access | Section 4.3.2 |
| $CO_2$ concentration | Access to frequent | Monitor $CO_2$ level | Control access | Section 4.3.2 |
| | $CO_2$ controller | Check pH of standard medium | Call service engineer | |
| | | Monitor $CO_2$ level with $CO_2$ tester (Carborite) | | |
| | | Recalibrate with standard gas mixture | | |
| Media and reagents | Supplier, batch, type, storage | | | Section 29.3 |

| Type of problem | Cause | Action | Follow-up | Cross-references |
|---|---|---|---|---|
| **Recent changes** | | | | |
| Cell culture personnel | New intake, training, or overcrowding | Review changes | Redefine procedures | |
| Preparation | Personnel | Review changes | Redefine procedures | |
| | Procedures or training | Review changes | Redefine procedures | |
| Materials | Chemical contamination | Growth or clonogenic assay | Trace & eliminate contamination | Section 29.5; Protocol 21.3, 21.4 |
| Procedures | Incubation times, speed of operations, sequence of operations, location, or scale | Review procedures | Redefine procedures | |
| Equipment | New equipment | Check performance | Call supplier | |
| | Operational status | Check for faults | Call service engineer | |
| | Location; moved or adjacent item moved | Relocate | | Section 3.2 |
| | Temperature (e.g., incubators; *see above*), water baths, & centrifuges | Check temperature | Institute new practice Call service engineer | |

## 29.3 MEDIUM

## 29.3.1 Formulation, Preparation, and Storage

| Type/scope of problem | Cause | Action | Follow-up | Cross- references |
|---|---|---|---|---|
| **Limited to one individual or general?** | Abnormal procedure | Check among users | Define procedures | Section 6.10 |
| **Medium changed to new type** | Inappropriate formulation | Revert to previous medium | Screen alternative media, sera, if change is required | Section 10.6.3 |
| | | | Match medium to previous type and supplier | |
| | | | Try serum-free formulation | Section 9.5 |
| **Storage** | Duration | Check batch number and delivery date | Revise stock control; define shelf life as 6 months at 4°C | Section 10.6.4 |
| | Glutamine breakdown | Buy medium without glutamine | Define storage conditions | Section 8.4.1 |
| | | Store glutamine frozen and add at use | Test Glutamax | |
| | Temperature | Check temperature continuously on recorder | Maintenance program | |
| | Light | Store in dark | Use only tungsten or low-UV output lamps | Section 10.6.4 |

| Type/scope of problem | Cause | Action | Follow-up | Cross- references |
|---|---|---|---|---|
| **Frequency of changing** | Medium becoming too acid | Check for contamination Feed more often | Revise maintenance schedule or increase buffer in medium | Sections 8.2.3, 12.3.2 |
| **pH fluctuation** | Faulty incubator or $CO_2$ supply | Keep between 7.0 and 7.4 | Call service engineer | Section 29.1.2 |
| pH too high | Incubator $CO_2$ concentration too low | Recalibrate | Call service engineer | |
| | $HCO_3^-$ concentration of medium too low | Prepare fresh medium and equilibrate overnight | Check medium formulation | Protocol 10.8 |
| pH high with DMEM | DMEM requires 10% $CO_2$ | Increase $CO_2$ or reduce $HCO_3^-$ | Purchase or prepare DMEM without $HCO_3^-$ and add for use | Tables 8.1, 8.3 |
| **Osmolality** | Incorrect formulation Incorrect dilution of concentrate Addition of new constituent or test compound | Check osmolality | Check preparation procedure; adjust with sterile NaCl or UPW | Section 8.2.5 |
| **Deficient medium** | Unsuitable formulation | Compare with 1× medium Compare with other media | Consider supplementation with hormones, growth factors, etc. | Section 8.6, 9.4 |
| | Accidental omission | Make fresh batch | Test against previous | |
| | Poor solubility | Ensure all constituents have dissolved before filtration | Revise procedure | Section 10.4.4 |
| | Precipitation on storage | Warm to 37°C; ensure all constituents have dissolved before filtration before use | Discard if precipitate does not dissolve; query supplier | Section 10.6 |
| | Defective component | Replace the components one at a time from an alternative source | Record batch numbers & contact supplier | Section 10.6 |
| | Faulty batch | Compare with previous batch | Try alternative supplier | Section 10.6 |
| **Serum** | Deficient batch | Batch test new serum against old | Select new best batch after overlap period | Section 8.6.1 |
| **Diluent** | If medium is BSS-based, is BSS satisfactory? | Check against fresh 1× medium, bought in complete | Try alternative sources of BSS or a different formulation | |
| | If medium is water based, is ultrapure water satisfactory? | Check against fresh 1× medium, bought in complete | Test UPW | Sections 10.6, 20.10, 21.3.5 |

| Type/scope of problem | Cause | Action | Follow-up | Cross- references |
|---|---|---|---|---|
| **Selecting medium for new culture** | | Screen several media, sera, and serum concentration | | Section 8.6; 10.6.3 |
| | | Batch test serum against old | Select best batch after overlap period | Section 8.6.1 |
| | | Try serum-free | Select by tissue-based criteria (e.g., MCDB 153 for epidermal keratinocytes) | Section 9.5 |
| | | Test hormone or growth factor supplementation | | Table 9.4 |
| | | Try using a feeder layer or conditioned medium | | Sections 8.7.3, 13.2.2, 13.2.3 |

### 29.3.2  Unstable Reagents

| Source of problem | Cause | Action | Follow-up | Cross-references |
|---|---|---|---|---|
| **Glutamine** | Unstable at 37°C; generates ammonia | Store at −20°C | Test and replace with dipeptide (e.g., Glutamax—Invitrogen) | Section 8.4.1 |
| **Serum** | Unstable components (e.g., insulin) | Store at −20°C | Change to serum-free | Sections 9.5, 29.2.1 |
| | Partial thaw | Thaw completely, mix, and refreeze | Aliquot to suit usage | |
| **Other constituents** | Some supplements may be unstable | Aliquot for single use | Use each aliquot once, then discard and do not refreeze | |
| **Trypsin** | Unstable | Aliquot and freeze stock concentrate | Store diluted at 4°C for 2 weeks maximum | |

### 29.3.3  Purity of Medium Constituents

| Source of problem | Cause | Action | Follow-up | Cross-references |
|---|---|---|---|---|
| **Poor water quality** | Unidentified | Collect UPW and test against accredited UPW source and 1× medium | Follow steps below | Protocols 20.7, 20.10; Section 29.2.3 |
| | RO cartridge | Test conductivity and TOC of semipurified water (SPW) | Replace RO cartridge; institute program of replacement | Sections 10.4.1, 10.4.2; Fig. 10.9 |
| | Still | Test the conductivity of SPW | Clean out still; institute cleaning program | Sections 10.4.1, 10.4.2; Fig. 10.9 |
| | Deionizer | Test the conductivity of purified water | Replace deionizer cartridge; institute program of replacement | Sections 10.4.1, 10.4.2; Fig. 10.9 |
| | Charcoal cartridges exhausted | Test TOC (Millipore) | Replace charcoal cartridge; institute program of replacement | Sections 10.4.1, 10.4.2; Fig. 10.9 |
| | Leakage of resin | Replace micropore filter | Institute program of replacement | Sections 10.4.1, 10.4.2; Fig. 10.9 |
| | Contaminated tubing | Check by eye for algae | Clean or replace tubing | Section 10.4.2 |

| Source of problem | Cause | Action | Follow-up | Cross-references |
|---|---|---|---|---|
| **Bicarbonate** | Concentration | Check the conductivity or osmolality of stock against a reference standard solution | Try another batch (make it up or buy it) | |
| | | Check that amount used is correct | Modify medium preparation | Section 8.2.2; Tables 9.1, 9.3, 10.1, 10.2 |
| | | Check $CO_2$ concentration in incubator | Recalibrate or call service engineer | Section 8.2.2 |
| | Precipitate in stock | Check by eye | Replace batch (make it up or buy it) | Appendix I |
| **Antibiotic toxicity** | Concentration too high | Reduce concentration | Omit antibiotic | Section 8.4.7 |
| | Bad batch | Check against fresh batch | Omit antibiotic | Section 8.4.7 |
| | Combinations | Try different combination | Omit antibiotics | Section 8.4.7 |
| | Fungicide (amphotericin B can be toxic) | Change to mycostatin or other antifungal | Use only at primary isolation | Appendix I: DBSS |
| **Antibiotic resistance** | Frequency of use | Change antibiotic | Omit antibiotics | Section 8.4.7 |
| **Serum** | Wrong type | Try new type or serum substitute | Replace with serum-free medium | Section 9.5 |
| | Bad batch | Select new batch & compare with original | Screen batches before reserving | Sections 8.6.1, 8.6.2 |
| | | Check supplier's quality control | Select new supplier | |
| | Concentration too high or too low | Test range of concentrations by growth curve and/or clonal growth | Replace with new batch or serum-free | Protocols 20.7, 20.8, 20.10; Section 9.5 |

## 29.4 SUBSTRATES AND CONTAINERS

| Source of problem | Cause | Action | Follow-up | Cross-references |
|---|---|---|---|---|
| **Plastic cultureware** | Bad batch | Check by plating efficiency relative to control batch | Change batch or supplier | Protocol 20.10 |
| **Glassware** | Inadequate washing, trace contaminants | Check washup procedures; visual check of glassware; test plating efficiency on glass bottles or Petri dishes | Redefine washup procedures; keep chemical glassware separate from tissue culture glassware | Sections 10.3.1, 29.5 |
| | General or restricted to one user | Check differences in procedures and types of glassware used | Redefine washup procedures; change to plastic | Sections 7.1.2, 10.3.1 |
| | General or restricted to one cell type | If only one cell try coating glassware or acid treatment | Change to plastic | Sections 7.1, 7.2, 10.3.1 |
| | Microscope slides | Clean in strong acid | Try coating; use plastic slides (may not be good for fluorescence) | Sections 7.2, 15.5.3 |
| **Caps** | Traces of detergent under liner | Remove liners to wash | Use wadless caps | Section 10.3.3 |

## 29.5  MICROBIAL CONTAMINATION

(See Fig. 29.1; *see also* Table 18.1; Figs. 5.1, 18.1; Plate 16.)

### 29.5.1  Confined to Single User

| Nature of problem | Cause | Action | Follow-up | Cross-references |
|---|---|---|---|---|
| **Sporadic** | | | | |
| Single species | Occasional lapses in aseptic technique | Monitor technique | Re-emphasis training | Sections 5.3–5.5, 28.2, Exercises 1, 28.3, Exercises 8,9 |
| | Personal hygiene | Check the cleanliness of the operator | Hand washing; wear gloves; change laboratory coat in tissue culture; tie back long hair | Sections 5.3–5.5 |
| | | Check if media or reagents unique to user | Test and identify contaminant | Section 18.3 |
| Multispecific | General omissions in sterile technique | Ensure hood is uncluttered; materials have been swabbed before placing in hood; all spillage is mopped up immediately | Is the lab coat changed before commencing culture work? Is the lab coat buttoned? | Sections 5.3–5.5 |
| | | Check with other users of the same hood | Hood may be faulty and need to be serviced | Section 4.2.1 |
| **Repeated** | | | | |
| Single species | Usually a reagent or imported cell line | Check for a unique medium or reagent that no one else uses; Is the cell line unique to that user? | Discard medium, reagent, or cell line; ensure future imported cell lines go through quarantine | Sections 3.2.4, 18.1.7 |
| Multispecific | Chronic failure in aseptic technique | Check for multiple omissions | Review training procedures | Sections 5.3–5.5, 28.2, 28.3 |
| | Adoption of nonstandard procedures | Clarify correct procedure | Review training procedures | Sections 5.3–5.5, 28.2, 28.3 |
| | Overcrowded hood | Demonstrate proper layout | Review training procedures | Section 5.2.1; Fig. 5.4 |
| | Overcrowding in laboratory; too much traffic | Restrict entry | Adopt booking system for use of hoods | |
| | Nonsterile reagents | Exclude nonsterile reagents from hoods | | |
| | Use of laboratory coat from general laboratory or animal house | Check use of laboratory coats | Provide clean coats or gowns for cell culture facility | Section 5.2.4 |

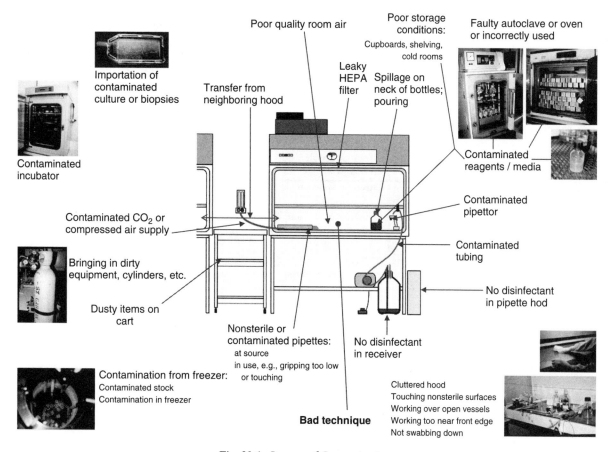

**Fig. 29.1.** Sources of Contamination.

| Nature of problem | Cause | Action | Follow-up | Cross-references |
|---|---|---|---|---|
| **Continuous** | | | | |
| Single species | Chronically contaminated solution | Screen reagents and media in use | Mix medium (antibiotic free) 1:1 with nutrient broth (e.g., L-Broth), & incubate the solution; plate the medium or incubated broth on blood agar, & incubate upside down, with blank controls | Plate 16d |
| | | Check for chronically contaminated cell line | Check the cells by Hoechst staining; mix the cells with broth, and incubate | Protocol 18.2 |
| Multispecific | Total breakdown in aseptic technique | Monitor aseptic technique; check for nonstandard procedures | Rejoin training program | Sections 5.3–5.5, 28.2, 28.3 |
| | Poor location | Reduce crowding, equipment, & traffic | Reorganize layout of aseptic suite | Section 3.2.1 |
| | Crowded hood | Demonstrate layout of hood | | Section 5.2; Fig. 5.4 |
| | Use of nonsterile reagents | Exclude nonsterile reagents from hood | | |

## 29.5.2  Widespread

| Nature of problem | Cause | Action | Follow-up | Cross-references |
|---|---|---|---|---|
| **Sporadic** | | | | |
| Single species | Infrequent use of contaminated reagent or medium; low-level contamination of frequently used medium or reagent | Screen media and reagents | Culture in nutrient broth to amplify and plate out on blood agar | Section 10.6.2; Plate 16*d* |
| | Contaminated incubator | Check incubators; swab & plate out swab if not obvious | Clean out if necessary | Protocol 18.1 |
| Multispecific | High spore count in atmosphere (busy room, building work, seasonal) | Check for spores by exposing bacteriological plates when room is quiet; sample air (*see* Appendix II) with vacuum filter & plate out | Reduce activity generating spores  Check room air filters  Improve atmospheric filtration | Sections 4.2, 29.4.3 |
| | Failure of sterilizing oven | (1) Reduce overcrowding of contents preventing adequate air circulation (2) Check integrity of door seals & any other apertures (3) Check temperature & duration of sterilization cycle: 160°C for 1 h minimum | Check packing of ovens & sterilization cycle; ensure temperature probe or indicator is placed in center of load | Section 10.3, Fig. 10.7 |
| | Failure of steam sterilization | Check autoclaves: (1) For overcrowding of contents preventing adequate steam circulation (2) For temperature & duration of sterilization cycle: 121°C for 15–20 min) | Check packing of autoclave & sterilization cycle; ensure temperature probe & indicator is placed in center of load; ensure that all empty vessels are left open for steam circulation; ensure packaging is steam permeable | Section 10.4; Fig. 4.17; Plate 22*a* |
| | Introduction of new untrained member of staff; not following standard procedures | Check the training of the new staff with supervisor | Update training procedures as necessary | Sections 5.3–5.5, 28.2, 28.3 |
| **Repeated** | | | | |
| Single species | Contaminated reagent or medium | Check the frequency of use of reagents among users to narrow the problem down to common reagents | Test likely candidates by incubating them 1:1 in broth | Section 10.6.2; Plate 16*d* |
| | Contaminated incubator | Check incubators; swab and plate out swab if not obvious | Clean out if necessary | Protocol 18.1 |
| | Faulty or dirty laminar flow hood | | | Section 4.2.1 |

| Nature of problem | Cause | Action | Follow-up | Cross-references |
|---|---|---|---|---|
| Multispecific | Sterilization oven or autoclave failure | (1) Check autoclaves & sterilizing ovens for overcrowding<br>(2) Check for electrical or mechanical failure | (1) Check the printouts & records of the sterilization cycles<br>(2) Check the integrity of the sterilization chamber of ovens | Sections 10.3, 10.4; Fig. 4.17, 10.7; Plate 22*a* |
| | Procedural failure & preparation & sterilization | Check whether new member of staff is following standard procedures | (1) Check the training of new staff<br>(2) Check procedures with the supervisor<br>(3) Check for changes in procedures and/or requirements<br>(4) Redraft SOPs to suit any changed circumstances | Sections 6.3, 28.2 |
| | Contaminated cold store | Check visually & with swab if necessary | (1) Institute cleaning program for cold stores<br>(2) Sore sterile items in sealed packages<br>(3) Review turnover of sterile stocks | Section 10.6.4 |
| | Aseptic technique | Check the aseptic technique of operators; restrict the use of antibiotics | | Sections 5.3–5.5, 8.4.7, 28.2, 28.3; Fig. 12.7 |

## 29.5.3  Air Supply and Laminar-Flow Hoods

| Nature of problem | Cause | Action | Follow-up | Cross-references |
|---|---|---|---|---|
| **Compromised laminar flow hood** | Faulty or dirty laminar-flow hood | (1) Swab work surface and sides of work area with phenolic disinfectant in 70% alcohol<br>(2) Wash below work surface & swab with phenolic disinfectant in 70% alcohol<br>(3) Check other components of ductwork for leaks or contamination<br>(4) Check whether filters are dirty or blocked (e.g., pressure drop across filter)<br>(5) Check integrity of filter with anemometer or nutrient agar plates | Check the cleaning schedules & update as necessary; check the maintenance schedules & update as necessary | Sections 4.2.1, 5.2.1, 5.4 |

| Nature of problem | Cause | Action | Follow-up | Cross-references |
|---|---|---|---|---|
| **Room air quality** | Contaminated room air | Check for spores by exposing bacteriological plates when room is quiet<br><br>Sample air (*see* Appendix II) with vacuum filter and plate out | Reduce activity generating spores; check room air filters; improve atmospheric filtration | Section 3.1.3 |
| | Incoming equipment | Wash & swab down | Check standard procedures | Section 5.5 |
| | Building work or other disturbance nearby | Improve isolation of culture laboratory (e.g., close doors, erect screens, exclude non–tissue-culture staff, clean all equipment and materials entering the room) | | |

## 29.5.4  Specific Contaminants

| Nature of problem | Cause | Action | Follow-up | Cross-references |
|---|---|---|---|---|
| **Bacterial, fungal** | See above | Check for bacterial or fungal contamination of cells or media by using high-power phase contrast | Incubate the cells in broth; plate out the cells on blood agar, and incubate the culture; gram stain the cells, or consult a microbiologist; decontaminate only if vital cell line | Sections 29.4.1–28.4.3 |
| | | | Eradication | Section 18.5.1; Protocol 18.4 |
| **Mycoplasma** | Importing cell lines or biopsies | Quarantine incoming cell lines & primary cultures from biopsies | Confirm screening program operative in house or contracted out (*see* Appendix II: Mycoplasma Testing) | Sections 18.1.7; 18.1.8; Protocols 18.2, 18.3; Plate 16*e, f* |
| | | | Eradication | Section 18.5.2; Protocol 18.5 |
| | Poor aseptic technique | Compare with other users | Re-emphasize precautions | Sections 18.3.2, 29.4.1 |
| | Natural products | Check with other users | Test by fluorescence or PCR | Protocols 18.2, 18.3 |
| | Antibiotic suppression | Culture for test without antibiotics | Test by fluorescence or PCR; restrict the use of antibiotics to primary culture & critical experiments | Protocols 18.2, 18.3 |
| **Viral** | Imported natural products, e.g. sera | Screen with indicator cell lines by TEM or SEM, immunostaining, ELISA, or PCR | Obtain natural products from virus-free sources | |
| | Cell lines | Quarantine; screen by TEM or SEM, immunostaining, or PCR | Obtain cell lines from virus-free sources | Section 18.1.8 |
| | | | Eradication impossible | |

## 29.6  CHEMICAL CONTAMINATION

### 29.6.1  Glassware

| Nature of problem | Cause | Action | Follow-up | Cross-references |
|---|---|---|---|---|
| **Residue after cleaning** | Ineffective washing; resistant soil | Visual examination; add a small volume of BSS with phenol red & look for pH change; clone cells on the glass after sterilization by dry heat; select detergent carefully | Keep tissue culture glassware separate from chemical glassware; ensure that there is no carry-over from the last rinse of a previous chemical wash in a washing machine | Section 10.3.1 |
| **Particulate contamination** | Dust accumulation during storage | Foil cap all open vessels; store in dust-free area or container | Plan storage carefully & rotate stocks | Section 10.6.4 |

### 29.6.2  Pipettes

| Nature of problem | Cause | Action | Follow-up | Cross-references |
|---|---|---|---|---|
| **Residue or blockage after cleaning** | Inadequate flow through during washing; resistant soil (e.g., denatured protein or agar) | Do not allow agar to be used in glass pipettes; collect pipettes into detergent, but rinse in water only; make sure cotton plugs are removed before washing; ensure that pipettes are washed & dried tip uppermost | Check pipettes after washing & before sterilization by visual examination; add a small volume of BSS with phenol red & look for pH change; consider using plastic disposables | Section 10.3.2 |

### 29.6.3  Water Purification

(*See* Sections 10.4.1, 29.2.1.)

### 29.6.4  Cryopreservatives

| Nature of problem | Cause | Action | Follow-up | Cross-references |
|---|---|---|---|---|
| **Contaminated DMSO** | Dissolved plastic or rubber from container | Dispense with glass pipette or polypropylene pipettor tip | Store DMSO in glass or polypropylene with glass or polypropylene cap | Section 19.3.3 |
| **Toxic glycerol** | Deterioration of glycerol on long-term storage | Buy in small amounts that will be used within 3–6 months | Store in dark bottle | Section 19.3.3 |

### 29.6.5  Powders and Aerosols

| Nature of problem | Cause | Action | Follow-up | Cross-references |
|---|---|---|---|---|
| **Dissemination of powders or aerosols** | Handling in draughty area or in laminar flow | Avoid draughts when weighing powders or dispensing liquids; handle toxic chemicals that produce powders or aerosols in a fume hood or chemical safety cabinet | Control traffic of people and equipment into tissue culture laboratory & preparation areas; provide clean laboratory coats for the tissue culture laboratory | Section 6.5.4; Fig. 6.5b |

## 29.7  PRIMARY CULTURE

### 29.7.1  Poor Take in Primary Culture

Primary cultures are usually 50% to 90% viable.

| Nature of problem | Cause | Action | Follow-up | Cross-references |
|---|---|---|---|---|
| **Primary explants do not attach** | Explants too large | Reduce size of explant to 1 mm or less | Practice dissection | Protocols 11.4, 22.9 |
| | Explants low adhesiveness (too fibrous, too differentiated) | Scratch the substrate through the explant; trap explant under a coverslip; embed in a plasma clot | Try coating plastic with collagen laminin, fibronectin, or polylysine; seed onto feeder layer | Section 7.2; Protocols 13.3, 22.9, 22.10 |
| **Incomplete disaggregation** | Inadequate exposure to protease | Incubate the cells in protease for a longer amount of time | Try cold pretreatment before shorter incubation; use an alternative, or additional, protease | Section 11.3.6; Protocol 11.6; Table 12.5 |
| | Cell lysis produces DNA that promotes reaggregation | Add DNAse, 10–20µg/mL, after centrifuging to remove the protease | Improve disaggregation to minimize cell damage (*see above*) | Section 11.3.3; Protocol 12.5, 12.6 |
| **Complete disaggregation but poor attachment** | Cells not very adhesive | If floating cells are viable | Try coating plastic with collagen laminin, fibronectin, or polylysine; seed onto feeder layer | Sections 7.2.1, 7.2.2; Protocols 13.3, 23.4, 24.2 |
| | Cells grow in suspension | Propagate in suspension | Check $Ca^{2+}$ in medium (absent in RPMI 1640) | Section 8.4.3 |
| **Floating cells are mostly nonviable** | Low survival from protease or mechanical digestion | Adjust the concentration to the viable cell count | Remove the nonviable cells | Protocol 11.10 |
| | | Improve disaggregation procedure | Chose cold trypsinization or an alternative protease (e.g., dispase or collagenase) | Sections 11.3.4, 11.3.6, 11.3.7; Protocols 11.6, 11.8 |
| | Tissue necrotic | Dissect off necrotic tissue | Remove the nonviable cells | Protocol 11.10 |
| **Floating cells are nonviable and few cells have attached** | Cell density too low | Increase cell concentration to up to $1 \times 10^6$ cells/mL | Remove the nonviable cells; improve disaggregation procedure (*see above*) | Protocol 11.10 |
| | Cell adhesion is poor | Try coating plastic with matrix, collagen, laminin, fibronectin, or polylysine | Seed onto feeder layer | Sections 7.2.1, 7.2.2; Protocols 13.3, 23.4, 24.2 |
| | Enzymes used are too toxic | Change to a different protease; reduce the exposure time | Try cold pretreatment with protease before incubation | Sections 11.3.4, 11.3.6; Protocol 11.6; Table 12.5 |
| | Medium is very acidic | Check for contamination; reduce cell concentration at seeding | Discard if contaminated | Sections 18.3, 18.4; Plate 16*a–c* |
| | | Add HEPES buffer, & vent the flask | | Section 8.2.3 |
| | Wrong medium used | Try a range of media | Check the literature for media used with your cells | Section 8.6; Tables 8.3, 8.6, 9.1, 9.2, 10.4 |

| Nature of problem | Cause | Action | Follow-up | Cross-references |
|---|---|---|---|---|
| | Supplementation of medium | Use different types or batches of serum | Check the literature | Section 8.6.2 |
| | | Replace the serum with serum-free medium | Check the literature | Section 9.5 |
| | | Use different growth factors or other mitogens | Check the literature | Sections 9.4.4., 9.4.5; Table 9.4 |
| | | Use conditioned medium | Check the literature | Section 8.7.3; Protocol 13.2 |

## 29.7.2  Wrong Cells Selected

| Nature of problem | Cause | Action | Follow-up | Cross-references |
|---|---|---|---|---|
| **Overgrowth by stromal cells** | Stromal cells predominate or favored by medium | Use selective medium | Purify cell population by MACS, FACS, etc. | Sections 9.2.2, 13.6, 14.3, 14.4 |
| | | Use selective substrate | | Sections 7.2.1, 13.8 |
| | | Use a selective feeder layer | | Section 13.8.4; Protocol 24.2 |
| | | | Check for cross-contamination from a feeder layer or a xenograft host | Sections 15.2, 18.6; Protocols 15.7, 15.8, 15.9 |
| **Overgrowth by different, rapidly growing cell type** | Cross-contamination | Check against other cell lines currently in culture | Discard if contaminated | Sections 12.1.1, 15.2, 18.6; Protocol 15.9; Table 12.2; Appendix V |

## 29.7.3  Contamination

| Nature of problem | Cause | Action | Follow-up | Cross-references |
|---|---|---|---|---|
| **Microbial contamination** | Infected tissue | Prewash biopsy with 70% alcohol | Check collection procedure; add antibiotics to collection medium | Section 24.4; Protocol 11.3, step 7; Appendix I: Collection Medium and DBSS |
| | | Eradicate contamination, but only if tissue is irreplaceable | | Section 18.4 |

## 29.8  DIFFERENTIATION

| Nature of problem | Cause | Action | Follow-up | Cross-references |
|---|---|---|---|---|
| **Cells do not differentiate** | Lack of inducers | Apply hormones. Growth factors, etc. | Optimize individual concentrations | Sections 16.7, 16.9 |
| | | Coculture with feeder layer and/or matrix | Screen feeder layers & matrix constituents | Section 16.7.1, 16.7.3 |
| | Wrong cells selected | Use a selective medium in primary culture | Propagate in selective medium | Section 9.2.2; Tables 9.1, 9.2 |
| | | Purify cell population by MACS, FACS, etc. | Propagate in selective medium | Sections 14.3, 14.4, 9.2.2; Tables 9.1, 9.2 |

| Nature of problem | Cause | Action | Follow-up | Cross-references |
|---|---|---|---|---|
| **Loss of product formation** | Lack of inducers | Apply induction conditions as above | | |
| | Genetic instability, wrong clone/subline overgrowing | Thaw fresh stock from freezer or isolate new line | Reclone; assay clones for activity | Sections 13.1, 13.2 |

## 29.9  FEEDING

### 29.9.1  Regular Monolayers

| Nature of problem | Cause | Action | Follow-up | Cross-references |
|---|---|---|---|---|
| **pH falls too quickly** | Contamination | Check for bacterial contamination | Check procedures | Sections 18.3.1, 29.4 |
| | Exhaustion of medium | Feed more frequently | Check suitability of medium | Sections 8.6, 9.5 |
| | Production of lactic acid | Increase buffering capacity with 20 mM HEPES, or by increasing $CO_2/HCO_3^-$ concentration | Use vented flasks | Sections 8.2.2, 8.2.3. |
| **pH rises after feeding** | Flask leaking if incubated in air | Tighten cap | Increase $HCO_3^-$; use $CO_2$ incubator and permeable caps | Section 7.3.3, 8.2.2, 8.2.3 |
| | Film of liquid sealing cap in $CO_2$ incubator | Blot neck with sterile swab or tissue; replace cap with fresh cap | Use gas-permeable caps | |
| | Film of liquid sealing lids of dishes when in $CO_2$ incubator | Remove lid, swab rim with 70% alcohol, replace lid | Check procedures to avoid medium entering capillary space between base & lid | Section 5.5.1; Protocol 5.3 |
| | $HCO_3^-$ concentration too high | Reduce $HCO_3^-$ | Check medium formulation is correct | Section 8.2.2 |
| | $CO_2$ concentration too low | Increase $CO_2$ | Check medium formulation is correct; check $CO_2$ incubator calibration | Sections 4.3.2, 8.2.2 |

### 29.9.2  Cell Cloning

| Nature of problem | Cause | Action | Follow-up | Cross-references |
|---|---|---|---|---|
| **pH too high or too low** | *See above* | | | Sections 28.12.1, 28.5 |
| **Prevention of contamination** | Contamination from humid incubator | Discard lids of dishes or plates, particularly if they have medium on them; swab outside of base with 70% alcohol | Incubate dishes in box and swab box before opening; do not use fan in incubator | Section 5.5.2 |
| | Dirty incubator | Clean out incubator | Revise cleaning program | Protocol 18.1 |

## 29.10   SUBCULTURE

### 29.10.1   Poor Take or Slow Growth

Cell lines are usually 90% to 100% viable cells.

| Nature of problem | Cause | Action | Follow-up | Cross-references |
|---|---|---|---|---|
| **Phase of growth cycle** | Cells subcultured from plateau | Subculture from the exponential or late-exponential phase | Review or repeat growth curve | Protocols 12.3, 20.7–20.9 |
| | Seeding concentration wrong | Adjust seeding concentration to give minimum lag period & late log at time for next subculture | Review or repeat growth curve | Protocols 12.3, 20.7–20.9 |
| | Subculture too frequent | Subculture only in late log phase | Review or repeat growth curve | Protocols 12.3, 20.7–20.9 |
| **Senescence** | Cells have been cultured for too many generations (doublings) | Check generation number | Review length of finite life span of cell line | Sections 2.7.3, 12.4.3, 17.4.1 |
| | | Replace stocks routinely from freezer | Ensure frozen stocks adequate | Section 19.5 |
| | | Immortalize cell line | Determine consequences of immortalization on phenotype & growth properties | Section 17.4; Protocols 17.1, 17.2 |
| **Medium deficiencies** | | | | Section 29.1 |
| **Toxicity** | Chemical contamination | Check individual components of medium by growth curve, microtitration, or plating efficiency | | Section 29.5, Protocols 20.7–20.9, 21.3, 21.4 |

### 29.10.2   Uneven Growth

(*See also* Sections 7.3.5, 29.10.3.)

| Nature of problem | Cause | Action | Follow-up | Cross-references |
|---|---|---|---|---|
| **More cells at edge of dish or well** | Meniscus effect; insufficient medium | Increase volume of medium | Adjust cell concentration so that cell density stays constant | Sections 12.4.4, 20.9.4 |
| **More cells at center of dish** | Swirling of medium when pipetting or to mix cells | Add cells & mix randomly without swirling | | |
| **More cells at center of well** | Scouring of cells from edge of well by washing or feeding | Wash & feed gently | Fixing cells with alcohol before assay helps them to remain attached | Section 21.3.5 |
| **Wells at edge of microtitration late have fewer cells** | "Edge effect": probably evaporation from outer wells | Do not use wells at edge of plate | Use plate sealer instead of lid | Section 21.3.5 |
| **More cells at one side of dish or plate than the other** | Culture vessel not level | Level culture vessels and vent large flasks briefly if sealed & stacked. | Check level of shelves and/or incubator | Fig. 5.10 |

| Nature of problem | Cause | Action | Follow-up | Cross-references |
|---|---|---|---|---|
| **Ribbing or patterning in flasks or concentric circles in dishes** | Vibration from fan motor or door opening and closing | Switch off fan or replace; place a cushion under the flasks, dishes, or plates; restrict access to incubators where cell distribution important (e.g., cloning) | Check incubator stability & maintenance; reallocate incubator space | Sections 7.3.5, 29.10.3; Fig. 7.10 |
| | Uneven heating | Place flasks or dishes on an insulating tile or metal plate | Check the air & temperature distribution in incubator | Sections 4.3.1, 4.3.2; Fig. 7.10 |

## 29.11   CLONING

(*See also* Section 29.1, Slow Cell Growth.)

### 29.11.1   Too Few Colonies per Dish

| Nature of problem | Cause | Action | Follow-up | Cross-references |
|---|---|---|---|---|
| **Poor plating efficiency** | Seeding concentration too low | Increase seeding concentration | Plate at a range of cell concentrations & determine plating efficiency | Section 13.1; Protocol 13.1 |
| | | Enhance plating efficiency | Glucocorticoid, insulin | Section 13.2 |
| | | Use a feeder layer or conditioned medium | Screen different feeder layers or conditioned media for maximum effect | Sections 13.2.2, 13.2.3; Protocols 13.2, 13.3 |
| | Poor handling; taking too long for procedure | Speed up or rationalize procedure | *See above* | |
| | Evaporation in incubator | Replace medium | Check humidity in incubator; check for evaporation by weighing test dishes with medium, & under the same conditions, at start & end of culture period | |
| | Medium inadequate for cloning | Choose a rich medium (e.g., Ham's F12) | Optimize by screening range of rich media | Sections 8.6, 9.5 |
| | $CO_2$ too low | Optimize $CO_2$ level | | Section 8.2.2 |
| | Wrong serum | Use fetal bovine serum, which is usually better than calf or horse serum for cloning; if fetal bovine serum is already being used, increase the concentration | Select serum batch based on the plating efficiency of the cells being used | Section 10.6.3; Protocol 20.10; Plate 6*e* |
| | Plasticware unsuitable | Check source of plasticware and change if necessary | Compare different source of dishes | Section 10.6.3; Protocol 20.10 |
| | Surface charge wrong | Coat the plastic with matrix | Try alternatively charged plastic (e.g., Primaria—BD Biosciences), CellBIND (Corning), FN/V/BSA | Section 7.2.1; Protocols 7.1, 22.9 |

| Nature of problem | Cause | Action | Follow-up | Cross-references |
|---|---|---|---|---|
| | Mycoplasma contamination | Screen for mycoplasma | Eradicate the mycoplasma, but only as a last resort, if the cell line is irreplaceable | Protocols 18.2, 18.3, 18.5; Plate 16e, f |
| **Colonies are too diffuse** | Reduction in cell–cell adhesion; increased cell migration | Use a larger dish to allow space for colonies to spread out | Glucocorticoid at plating (e.g., dexamethasone, $1 - 10\,\mu M$) | Section 13.2.1 |
| | | | Coat dish with matrix | Section 7.2.1; Protocol 22.9 |
| **Poorly attached colonies** | Low adhesiveness of cells or poor substrate | Clone the cells in Methocel in a tissue-culture-grade dish | | Protocol 14.5 |

## 29.11.2   Too Many Colonies per Dish

| Nature of problem | Cause | Action | Follow-up | Cross-references |
|---|---|---|---|---|
| **Colonies overlapping, difficult to count or isolate** | Seeding concentration too high | Reduce the seeding concentration (cells/mL) or seed the same number of cells into a larger dish | Plate at a range of cell concentrations & determine plating efficiency | Section 13.1; Protocol 13.1 |
| | | Grow colonies for shorter time | | |

## 29.11.3   Nonrandom Distribution

(*See also* Section 29.9.2.)

| Nature of problem | Cause | Action | Follow-up | Cross-references |
|---|---|---|---|---|
| **Colonies unevenly spread** | Uneven seeding | Add the cells to medium in a bottle, mix the cell suspension, & then seed the dishes | | |
| | Swirling the dish | Do not swirl the dishes to mix cells or when pipetting | | |
| | Medium not covering all of dish | Make sure that the medium covers all of the bottom of the dish evenly | Ensure that the dishes are level & the incubator is free from vibration | Section 20.9.4 |
| | Incubator vibration | Restrict access of other users to the incubator | Label the box or tray containing the dishes with the phrase, "CLONING, DO NOT MOVE." | |

## 29.12 CROSS-CONTAMINATION AND MISIDENTIFICATION

| Nature of problem | Cause | Action | Follow-up | Cross-references |
|---|---|---|---|---|
| **Cross-contamination** | Accidental mixing, sharing pipettes or reagents, handling more than one cell line at a time, importing non-authenticated stock | Screen all cell lines regularly; follow rules in Section 18.6 | Discard all contaminated lines & replace with authentic stock | Sections 12.1.1, 15.2, 15.8.3, 18.6 |
| | | If detected early may be possible to isolate original line by cloning or physical separation | Confirm identity by DNA profiling or other method | Sections 14.1, 14.3.2, 15.2, 15.8.3 |
| **Misidentification** | Mislabeling of flasks or ampoules, freezer inventory error | Screen all cell lines regularly; check new stocks after thawing; follow rules in Section 18.6 | Confirm identity by DNA profiling or other method; discard all misidentified lines & replace with authentic stock | Sections 12.1.1, 15.2, 15.8.3, 18.6 |

## 29.13 CRYOPRESERVATION

### 29.13.1 Poor Recovery

Thawed cells are usually 50% to 80% viable.

| Nature of problem | Cause | Action | Follow-up | Cross-references |
|---|---|---|---|---|
| **Freezing rate** | Rate at which cells are cooled & frozen not optimal | Change the cooling rate by changing wall thickness of insulated container or using programmable freezer | Optimize cooling curve on programmable freezer | Section 19.3.4 |
| | | | Try vitrification | Section 19.4 |
| **Cell concentration at freezing** | Cells leaky & survive poorly at low concentrations | Increase the cell concentration at freezing (optimum is usually from $1 \times 10^6 - 1 \times 10^7$/cells mL) | | Section 19.3.2 |
| **Cell concentration at thawing** | Cells reseeded at too low a concentration | Increase concentration at freezing (usually to give $> 5\times$ normal seeding concentration on thawing); reduce dilution at thawing (may need to centrifuge to remove preservative) Pool several ampoules & centrifuge to remove preservative | Optimize cell concentration at freezing to give maximum recovery | Section 19.3.8 |
| **Dilution rate after thawing** | Rapid dilution can cause osmotic damage | Add medium slowly to thawed cells | Update procedure | Section 19.3.8 |

| Nature of problem | Cause | Action | Follow-up | Cross-references |
|---|---|---|---|---|
| **Cryoprotectant** | | | | |
| **DMSO** | Contamination from storage container | Check color—should be colorless; do spectroscopic scan and compare with fresh DMSO | Store in glass or polypropylene with glass or polypropylene stopper | |
| | Induction of differentiation | Check literature to confirm cells are not induced to differentiate in DMSO | Centrifuge cells after dilution at thawing | Section 16.7.2 |
| **Glycerol** | Light-induced conversion to acrolein—toxic | Dispense in small volumes that will be used up in a few weeks; store in the dark | | Section 10.6.4 |

## 29.13.2 Changed Appearance after Cryopreservation

(*See also* Section 29.11.)

| Nature of problem | Cause | Action | Follow-up | Cross-references |
|---|---|---|---|---|
| **Mistaken identity** | Poor labeling & inventory control | Check the labeling of the ampoule; if label illegible, discard the ampoule; check the records | Check the authentication | Sections 15.2, 20.3.6, 29.11 |
| **Changed culture conditions since the cells were last grown** | New staff, new procedures, new medium or serum type, batch, or supplier | Revert to previous conditions if possible; screen fresh medium and serum batches | Standardize culture conditions for each cell line | Section 12.4.7 |
| **Contamination** | Leakage of ampoule | Check the seal: cap should be tight, but seal should not be distorted | Confirm correct torsion required to seal cap on ampoule | Section 18.3 |
| | Contamination from water bath on thawing | Do not immerse ampoules; place in a rack only partially submerged; thaw the ampoules in a heating block; swab the ampoules carefully after thawing | Ensure this procedure is only used with ampoules which have not been immersed in liquid nitrogen | Section 19.3.8 |
| | From liquid nitrogen | Do not submerge ampoules in liquid nitrogen | Use vapor-phase storage or a perfused jacket freezer | Section 19.3.6 |

### 29.13.3   Loss of Stock

| Nature of problem | Cause | Action | Follow-up | Cross-references |
|---|---|---|---|---|
| **Explosion** | Liquid nitrogen drawn into ampoule | Freeze in vapor phase or in perfused jacket freezer | Update procedure and safety code | Section 19.3.6 |
| **Last of seed stock removed** | Poor inventory control | Replenish from distribution stock<br>Restock from a reputable cell bank | Check the security of the inventory control; access to seed stock should be restricted to the curator only | Section 19.5.1;<br>Table 19.5; Appendix II: Cell Banks |
| **Depleted user stock** | Normal use | Restock from the distribution stock | | Section 19.5.1 |
| **Depleted distribution stock** | Normal use | Restock from the seed stock | | Section 19.5.1 |

## 29.14   CELL COUNTING

### 29.14.1   Hemocytometer

| Nature of problem | Cause | Action | Follow-up | Cross-references |
|---|---|---|---|---|
| **Variable counts** | Sampling error | Mix the cell suspension thoroughly before sampling; ensure that the cells are singly suspended & not clumped | Revise disaggregation procedure | Protocol 21.1; Table 12.4 |
| | Incorrect use of hemocytometer | Ensure that the coverslip is correctly attached (interference colors should be visible at edges of coverslip) & that the counting chamber is not over- or under filled; a sufficient number of cells should be counted (> 200) | Revise counting procedure & training | Section 20.1.1 |
| **Poor visibility of cells** | Loss of silvering on slide; poor optics | Use phase contrast or a noncentered light path; stain the cells | Renew hemocytometer slide; get phase contrast optics for microscope | |

## 29.14.2 Electronic Counting via Orifice by Resistance

| Nature of problem | Cause | Action | Follow-up | Cross-references |
|---|---|---|---|---|
| **Variable counts** | Sampling error | Mix the cell suspension thoroughly before sampling & ensure the cells are singly suspended & not clumped | Check trypsinization & sampling procedure | Section 20.1.2 |
| **Count cycle slow or fails to start** | Blockage | Run the wash cycle or unblock cycle; rub the orifice of electrical resistance-type counter with the tip of your finger or a fine brush | Soak in detergent for 1–18 h & repeat unblock; repeat wash cycle three times | Section 21.1.2 |
| | Insufficient negative pressure | Check to ensure that the waste reservoir is not full; check the pump (see the manual for a diagnostic test) & connections | If pump failure has occurred; call engineer | |
| **Count is lower than expected, or orifice blocks frequently** | Aggregation of cell suspension | Disperse the cells by pipetting the original sample vigorously, redilute the sample, & proceed; use different disaggregation technique | | Table 12.5 |
| **High background** | Electrode out of beaker or disconnected | Replace the electrode in the beaker; check that the electrode is secure; replace the electrode if the terminal plate is missing | | |
| | Precipitation or contamination in suspension medium | Prepare fresh counting fluid | If trouble persists, filter counting fluid through 0.45-$\mu$m micropore filter | |
| | Interference from other equipment or lights | Fit suppressors; check grounding (earthing) | | |

# CHAPTER 30

# In Conclusion

It has been my intention in the foregoing pages to describe the fundamentals of cell culture in sufficient detail that recourse to the literature is required only to extend your work beyond the basic procedures or to acquire some additional background detail. It is customary, when giving a lecture, to conclude with a summary that highlights the major points raised in the lecture, and that is how I would like to conclude this text.

There are certain requirements that are crucial to successful and reproducible cell culture, and they may be highlighted as follows:

(1) Ensure that your instruction, and the training of anyone who works with you, comes from an experienced source.
(2) Work in a clean, uncrowded, aseptic environment, reserved for culture, and clear up when you have finished.
(3) To avoid the transfer of contamination, including cross-contamination, do not share media, reagents, cultures, or materials with others.
(4) Do not assume that your work is immune to mycoplasma because you have never seen it; test your cells regularly.
(5) Keep adequate records, particularly of changes in procedures, media, or reagents, and keep photographic records of the cell lines that you use.
(6) Work only on cell lines that have been obtained from a properly validated source, such as an international cell bank. Distrust any other cell line, even from the originator, unless you can prove that it is authentic.

(7) Become familiar with the cell lines that you use—their appearance, growth rate, and special characteristics—so that you can respond immediately to any change.
(8) Ensure that your work does not compromise your own safety or that of others working around you.
(9) Preserve cell line stocks in a liquid nitrogen freezer (but not submerged), and replace working stocks regularly.
(10) Protect seed stocks of valuable cell lines, and use other stocks for distribution.
(11) Try to work under conditions that are precisely defined, including minimal use of undefined media supplements, and do not change procedures for trivial reasons.
(12) Do not mix cell culture with other microbiological work.

To cover all of the fascinating aspects of cell and tissue culture would take many volumes and defeat the objective of this book. It has been more my intention to provide sufficient information to set up a laboratory and prepare the necessary materials with which to perform basic tissue culture, and to develop some of the more important techniques required for the characterization and understanding of your cell lines. In collaboration with experts in their fields, I have also included specialized protocols for some of the many applications of cell culture that are possible. This book may not be sufficient on its own, but with help and advice from colleagues and other laboratories, it may make your introduction to tissue culture easier and more satisfying and enjoyable than it otherwise might have been.

# APPENDIX I

# Calculations and Preparation of Reagents

## Calculations

There are a number of simple calculations required to carry out some of the procedures described in this books. Calculations specific to a particular protocol will be found in that protocol, but some general ones are listed below.

### Cell Dilution

To dilute a cell suspension to a lower concentration, the volume of cell suspension (v) required for dilution can be calculated as follows:

$$\frac{\text{Required concentration}}{\text{Starting concentration}} \times \text{Required volume}$$

For example, the volume of cell suspension required to dilute $2.36 \times 10^6$ cells/mL to $5 \times 10^4$ cells/mL in 50 mL would be

$$\frac{5 \times 10^4}{2.36 \times 10^6} \times 50 = 1.06 \text{ mL}$$

that is, dilute 1.06 mL up to 50 mL.

### Cell Population Doublings

The simplest way to calculate cell population doublings in a finite cell line is to dilute at subculture by a power or 2. A split ratio of 2 would be one doubling, a split ratio of 4 would be two doublings, a split ratio of 8 would be three doublings, and so on. The number of population doublings (PD) can be calculated by the formula

$$\text{PD} = \frac{\ln \left( \frac{N_{\text{finish}}}{N_{\text{start}}} \right)}{\ln 2}$$

where ln is the natural logarithm, $N_{\text{start}}$ is the number of cells seeded at the beginning of the growth cycle, and $N_{\text{finish}}$ the number of cells recovered at the end.

For example, for $3.2 \times 10^6$ cells recovered from a seeding of $2 \times 10^5$,

$$\text{PD} = \frac{\ln \left( 3.2 \times \frac{10^6}{2} \times 10^5 \right)}{\ln 2} = \frac{\ln 16}{\ln 2} = \frac{2.7726}{0.6931} = 4$$

## Conversions

### Celsius (centigrade) to Fahrenheit

$$\frac{\text{Temperature in Celsius} \times 9}{5} + 32$$

### Fahrenheit to Celsius

$$\text{Temperature in Celsius (centigrade)}$$
$$= \frac{\text{Temperature in Fahrenheit} - 32}{9} \times 5$$

For example, the temperature in °C at 62°F would be

$$\frac{62 - 32}{9} \times 5 = 16.7$$

### Molarity

Recipes often quote the concentration of constituents in grams, but this will vary depending on the water of crystallization in a salt, or when quoting the weight of free acid or its salt. Quoting concentrations in molarity (M) avoids this problem:

$$M = \frac{\text{Concentration in g/L}}{\text{Molecular mass}}$$

For example, the molarity of 2 g/L of the disodium salt of the dihydrate of EDTA would be

$$\frac{2}{372.2} = 0.005373 \text{ M} \quad or \quad 5.373 \text{ mM} \cong 5 \text{ mM}$$

Conversely, the concentration in g/L is given by multiplying the molarity by the molecular mass:

$$5.373 \times 372.2 = 2.0$$

**Pressure**

Pressure is quoted in different units by different disciplines. Fifteen pounds per square inch (15 psi) = 1 atmosphere (1 atm) = 1 bar = 760 mm. Hg = 30 in. Hg = 100 kiloPascals(100 kPa) = 100 Newtons/meter$^2$ (100 N/m$^2$).

**Radioactivity**

Radioactivity is often expressed in Curies (Ci) but the correct SI unit is the Becquerel (Bq).

$$1 \text{ Ci} = 3.7 \times 10^{10} \text{ Bq} \quad or \quad 37 \text{ GBq}$$

**RPM to *g***

For convenience, most people in a laboratory will quote the speed of the centrifuge when describing a centrifugation step but that will only remain consistent as long as the same centrifuge and rotor are used. Centrifugation is better defined by *g*, the radial acceleration relative to gravity.

$$g = 1.118 \times 10^{-5}$$

$$\times \text{ Radius of the rotor to the center of the tube} \times \text{rpm}^2$$

For example, in a centrifuge rotor rotating at 1500 rpm, with a radius of 18 cm from the center of the spindle to the middle depth of the sample in the tube,

$$g = 1.118 \times 10^{-5} \times 18 \times 1500^2 = 452.79 \cong 450$$

**Preparation of Reagents**

Many of the reagents listed here are available commercially (*see* Appendix II), and often the formulation is provided in the appropriate catalog. Details are provided here for reference or for those who wish to make up the solutions in their own laboratory. The reagents that are listed are those used in several different protocols throughout the book. Specialized reagents used in only one protocol will generally be located within the Materials section of that protocol.

**Note.** Dilutions quoted as, for example, 1:10 or 1:100, are v/v and imply that the final volume is 10 or 100 parts, respectively.

**Acetic/Methanol**

Add 1 part glacial acetic acid to 3 parts methanol. Make up fresh each time used, and keep on ice.

**Alcohol, 70%, v/v**

Ethanol, methanol, or isopropanol have all been used for swabbing surfaces and other topical sterilization. Isopropanol is used clinically and spray cans are commercially available.

| | |
|---|---|
| Alcohol | 700 mL |
| Deionized or RO water | 300 mL |

**Agar 2.5%**

| | |
|---|---|
| Agar | 2.5 g |
| UPW | 100 mL |

(1) Boil to dissolve agar.
(2) Sterilize by autoclaving or boiling for 2 min.
(3) Store at room temperature.

**Amido Black**

*See* Naphthalene Black.

**Amino Acids, Essential**

(*See* Section 8.4.1 and Table 8.3; available as 50× concentrate in 0.1 M HCl from commercial suppliers, e.g., Invitrogen, MP Biomedicals, and Sigma.)

(1) Make up tyrosine and tryptophan together at 50× in 0.1 M HCl and remaining amino acids at 100× in ultrapure water.
(2) Dilute for use (*see* Protocol 10.10).
(3) Sterilize by filtration.
(4) Store in the dark at 4°C.

**Amino Acids, Nonessential**

| Ingredient | g/1 (100×) |
|---|---|
| L-Alanine | 0.89 |
| L-Asparagine H$_2$O | 1.50 |
| L-Aspartic acid | 1.33 |
| Glycine | 0.75 |
| L-Glutamic acid | 1.47 |
| L-Proline | 1.15 |
| L-Serine | 1.05 |
| Water | 1000 mL |

(1) Sterilize by filtration.
(2) Store at 4°C.
(3) Use at a concentration of 1:100.

**Antibiotics**

*See under specific headings* (e.g., penicillin, streptomycin sulfate, kanamycin sulfate, gentamicin, mycostatin).

## Antifoam

(e.g., RD emulsion 9964.40; *see* Appendix II.)

(1) Dispense into aliquots and autoclave to sterilize.
(2) Store at room temperature.
(3) Dilute 0.1 mL/L (i.e., 1:10,000).

## Bactopeptone, 5%

| | |
|---|---|
| Difco Bactopeptone | 5 g |
| Hanks's BSS | 100 mL |

(1) Stir to dissolve.
(2) Dispense in aliquots appropriate to a 1:50 dilution, and autoclave.
(3) Store at room temperature.
(4) Dilute 1:10 for use.

## Balanced Salt Solutions (BSS)

(*See* Table 8.2.)

(1) Dissolve each constituent separately, adding $CaCl_2$ last.
(2) Make up to 1 L.
(3) Adjust to pH 6.5.
(4) Sterilize the solution by autoclaving or filtration. With autoclaving, the pH must be kept below 6.5 to prevent calcium phosphate from precipitating; alternatively, calcium may be omitted and added later. If glucose is included, the solution should be filtered to avoid caramelization of the glucose, or the glucose may be autoclaved separately (*see* Glucose, 20%, in this appendix) at a higher concentration (e.g., 20%) and added later.
(5) With autoclaving, mark the level of the liquid before autoclaving. Store the solution at room temperature, and if evaporation has occurred, make up to mark with sterile ultrapure water before use. If borosilicate glass is used, the bottle may be sealed before autoclaving and no evaporation will occur.

## Broths

(*See* manufacturers' instructions for preparation; *see also* Bactopeptone and tryptose phosphate broth.)

Sterilize by autoclaving.

## Buffered Glycerol Mountant

*See* Mycoplasma.

## Carboxymethylcellulose (CMC)

(1) Weigh out 4 g of CMC and place it in a beaker.
(2) Add 90 mL of Hanks's BSS, and bring the mixture to boil in order to wet the CMC.
(3) Allow the solution to stand overnight at 4°C to clarify.
(4) Make volume up to 100 mL with Hanks's BSS.
(5) Sterilize the solution by autoclaving. The CMC will solidify again, but will redissolve at 4°C.
(6) For use (e.g., to increase the viscosity of the medium in suspension cultures), use 3 mL per 100 mL of growth medium.

## Chick Embryo Extract [Paul, 1975]

(1) Remove embryos from eggs (*see* Protocol 11.2) and place the embryos in 9-cm Petri dishes.
(2) Take out the eyes, using two pairs of sterile forceps.
(3) Transfer the embryos to flat- or round-bottomed 50-mL containers, two embryos to each container.
(4) Add an equal volume of Hanks's BSS to each container.
(5) Using a sterile glass rod that has been previously heated and flattened at one end, mash the embryos in the BSS until they have broken up.
(6) Let the mixture stand for 30 min at room temperature.
(7) Centrifuge the mixture for 15 min at 2000 *g*.
(8) Remove the supernate, and after keeping a sample to check its sterility (*see* Section 10.6.2), dispense the solution into aliquots, freeze and thaw quickly, twice, and store at −20°C.

Extracts of chick and other tissues may also be prepared by homogenization in a Potter homogenizer or Waring blender [Coon and Cahn, 1966].

(1) Homogenize chopped embryos with an equal volume of Hanks's BSS.
(2) Transfer the homogenate to centrifuge tubes, and spin at 1000 *g* for 10 min.
(3) Transfer the supernatant to fresh tubes, and centrifuge for a further 20 min at 10,000 *g*.
(4) Check the sample for sterility (*see* Section 10.6.2).
(5) Dispense into aliquots.
(6) Freeze and thaw quickly, twice, and store at −20°C.

## Citric Acid/Crystal Violet

*See* Crystal Violet.

## CMC

*See* Carboxymethylcellulose.

## Colcemid, 100× Concentrate

| | |
|---|---|
| Colcemid | 100 mg |
| Hanks's BSS | 100 mL |

(1) Stir to dissolve.
(2) Sterilize by filtration.
(3) Dispense into aliquots and store at −20°C

Δ *Safety Note.* Colcemid is toxic; handle it with care by weighing in a fume cupboard and wearing gloves.

## Collagenase

Worthington CLS-grade collagenase or the equivalent (specific activity 1500–2000 U/mg) at 2000 U/mL in Hanks's BSS.

| | |
|---|---|
| Collagenase | 100,000 U |
| Hanks's BSS | 50 mL |

(1) Stir at 37°C for 2 h or at 4°C overnight to dissolve.
(2) Sterilize the solution by filtration, as with serum (*see* Protocol 10.15).

(3) Divide into aliquots, each suitable for 1 to 2 weeks of use.

(4) Store at −20°.

### Collagenase–Trypsin–Chicken Serum (CTC) [Coon and Cahn, 1966]

| Sterile | Volume | Final Concentration |
|---|---|---|
| Ca$^{2+}$- and Mg$^{2+}$-free saline [Moscona, 1952] | 85 mL | |
| Trypsin stock, 2.5%, sterile | 4 mL | 0.1% |
| Collagenase stock, 1%, sterile | 10 mL | 0.1% |
| Chick serum | 1 mL | 1.0% |

Dispense into aliquots and store at −20°C.

### Collection Medium (for Tissue Biopsies)

| | |
|---|---|
| Growth medium | 500 mL |
| Penicillin | 125,000 units |
| Streptomycin | 125 mg |
| Kanamycin | 50 mg |
| or | |
| Gentamycin | 25 mg |
| Amphotericin | 1.25 mg |

Store at 4°C for up to 3 weeks or at −20°C for longer periods.

### Crystal Violet, 0.1%, in 0.1 M Citric Acid

| | |
|---|---|
| Citric acid | 21.0 g |
| Crystal violet | 1.0 g |

(1) Make up to 1000 mL with deionized water.

(2) Stir to dissolve.

(3) To clarify, filter the solution through Whatman No. 1 filter paper.

### Crystal Violet 0.1% in Water

| | |
|---|---|
| Crystal violet | 100 mg |
| Water | 100 mL |

Filter through Whatman No. 1 paper before use.

A 0.1% Crystal Violet solution is available readymade from Merck.

### Dexamethasone (Merck)

1 mg/mL

This reagent comes already sterile in glass vials.

(1) Add 5 mL water by syringe to the vial to dissolve contents.

(2) Remove by syringe when dissolved

(3) Dilute to 1 mg/mL (approximately 2.5 mM).

(4) Divide the solution into aliquots and store at −20°C.

(5) For use, dilute the solution to give 10 to 50 nM (physiological concentration range), 0.1 to 1.0 μM (pharmacological dose range), or 25 to 100 μM (high-dose range).

β-Methasone (GlaxoSmithKline) and methylprednisolone (Sigma) may be prepared in the same way.

### Dissection BSS (DBSS)

(1) To Hanks's BSS without bicarbonate, previously sterilized by autoclaving, add the following (all sterile):

| | |
|---|---|
| Penicillin | 250 U/mL |
| Streptomycin | 250 μg/mL |
| Kanamycin | 100 μg/mL |
| or | |
| Gentamycin | 50 μg/mL |
| Amphotericin B | 2.5 μg/mL |

(2) Store at −20°C.

### D-PBSA

*See* Phosphate-Buffered Saline.

### Dulbecco's PBS without Ca$^{2+}$ and Mg$^{2+}$ (D-PBSA)

(*See* Protocol 10.6.)

### EDTA (Versene)

(1) Prepare as a 10 mM concentrate, 0.374 g/L, in D-PBSA.

(2) Sterilize by autoclaving or filtration.

(3) Dilute 1:10, or 1:5 for use at 1.0 to 2.0 mM, or, exceptionally, 1:2 for use at 5 mM, diluted in D-PBSA or trypsin in D-PBSA.

### EGTA

As for EDTA, but EGTA may be used at higher concentrations because of its lower toxicity.

### Ficoll, 20%

(1) Sprinkle 20 g of Ficoll (GE Heathcare) on the surface of 80 mL UPW

(2) Leave overnight for the Ficoll to settle and dissolve.

(3) Make up to 100 mL in UPW.

(4) Sterilize by autoclaving.

(5) Store at room temperature.

Available commercially combined with metrizoate for lymphocyte preparation (*see* Appendix II: Ficoll-Metrizoate).

### Fixative for Tissue Culture

*See* Acetic/methanol, above.

Alternatively, use pure anhydrous ethanol or methanol (*see* Protocol 15.2), 10% formalin, 1% glutaraldehyde, or 5% paraformaldehyde.

### Gelatin

(From Protocol 22.18; *see also* Subbing Slides in this appendix.)

(1) Prepare a 1% stock solution of gelatin (Sigma) in UHP water

(2) Autoclave at 121°C for 20 min.

(3) Aliquot and store at 4°C.

(4) To prepare the 0.1% working solution, warm 1% gelatin to 37°C until it liquefies and dilute 1:10 in sterile D-PBSA.

## Gentamycin

Stock 50 mg/mL (see Appendix II); dilute to 50 μg/mL for use.

## Gey's Balanced Salt Solution

NaCl . . . . . . . . . . . . . . . . . . . . . . . . . . . . . . . . . . . . 7.00 g
KCl. . . . . . . . . . . . . . . . . . . . . . . . . . . . . . . . . . . . . .0.37 g
$CaCl_2$ . . . . . . . . . . . . . . . . . . . . . . . . . . . . . . . . . . . 0.17 g
$MgCl_2 \cdot 6H_2O$ . . . . . . . . . . . . . . . . . . . . . . . . . . . 0.21 g
$MgSO_4 \cdot 7H_2O$ . . . . . . . . . . . . . . . . . . . . . . . . . .0.07 g
$Na_2HPO_4 \cdot 12H_2O$ . . . . . . . . . . . . . . . . . . . . . 0.30 g
$KH_2PO_4$ . . . . . . . . . . . . . . . . . . . . . . . . . . . . . . . .0.03 g
$NaHCO_3$ . . . . . . . . . . . . . . . . . . . . . . . . . . . . . . . 2.27 g
Glucose . . . . . . . . . . . . . . . . . . . . . . . . . . . . . . . . . 1.00 g
Water, up to . . . . . . . . . . . . . . . . . . . . . . . . . . 1000 mL
$CO_2$ . . . . . . . . . . . . . . . . . . . . . . . . . . . . . . . . . . . . . 5%

## Giemsa Stain

Giemsa stain can be applied undiluted and then diluted with buffer or water (see Protocol 15.2), or diluted in buffer before use. The author has found the first method more successful for cultured cells.

(1) Prepare buffer:
$NaH_2PO_4 \cdot 2H_2O$ . . . . . . . 0.01 M . . . . . . . 1.38 g/L
$Na_2HPO_4 \cdot 7H_2O$ . . . . . . . . . 0.01 M . . . . . . . . 2.68 g/L
Combine in correct proportions to give pH 6.5.
(2) Dilute prepared Giemsa concentrate 1:10 in 100 mL of buffer.
(3) Filter the solution through Whatman No. 1 filter paper to clarify.
(4) Make up a fresh solution each time, because the concentrate precipitates on storage.

## Glucose, 20%

Glucose . . . . . . . . . . . . . . . . . . . . . . . . . . . . . . . . . . . .20 g
Hanks's BSS to . . . . . . . . . . . . . . . . . . . . . . . . . . . 100 mL

(1) Sterilize by autoclaving.
(2) Store at room temperature.

## Glutamine, 200 mM

L-Glutamine . . . . . . . . . . . . . . . . . . . . . . . . . . . . . . 29.2 g
Hanks's BSS . . . . . . . . . . . . . . . . . . . . . . . . . . . . 1000 mL

(1) Dissolve the glutamine in BSS and sterilize by filtration (see Protocols 10.12, 10.13).
(2) Dispense the solution into aliquots and store at −20°C.

## Glutathione

(1) Make 100× stock (i.e., 0.1 M in HBSS or D-PBSA).
(2) Sterilize by filtration.
(3) Dispense into aliquots and store at −20°C.
(4) Dilute to 1 mM for use

## Ham's F12

(See Table 8.3.)

## Hanks's BSS

(See Section 8.3, Table 8.3, and Balanced Salt Solutions, in this appendix.)

## Hanks's BSS without phenol red:

Standard HBSS formulation (see Table 8.2) but omit phenol red.

## HAT Medium

| Drug | Concentration | Dissolve in | Molarity (100 × final) |
|---|---|---|---|
| Hypoxanthine (H) | 136 mg/100 mL | 0.05 N HCl | $1 \times 10^{-2}$ M |
| Aminopterin (A) | 1.76 mg/100 mL | 0.1 N NaOH | $4 \times 10^{-5}$ M |
| Thymidine (T) | 38.7 mg/100 mL | HBSS | $1.6 \times 10^{-3}$ M |

(1) For use in the HAT selective medium (see Protocol 27.6), mix equal volumes of each, sterilize by filtration, and add the mixture to medium at 3% v/v.
(2) Store H and T at 4°C, A at −20°C.

## HB Medium

Add the following to CMRL 1066 medium:

Insulin . . . . . . . . . . . . . . . . . . . . . . . . . . . . . . . . . .5 μg/mL
Hydrocortisone . . . . . . . . . . . . . . . . . . . . . . . . .0.36 μg/mL
β-Retinyl acetate . . . . . . . . . . . . . . . . . . . . . . . . 0.1 μg/mL
Glutamine . . . . . . . . . . . . . . . . . . . . . . . . . . . . . 1.17 mM
Penicillin . . . . . . . . . . . . . . . . . . . . . . . . . . . . . 50 U/mL
Streptomycin . . . . . . . . . . . . . . . . . . . . . . . . . .50 μg/mL
Gentamycin . . . . . . . . . . . . . . . . . . . . . . . . . . . .50 μg/mL
Fungizone . . . . . . . . . . . . . . . . . . . . . . . . . . . . . 1.0 μg/mL
Fetal bovine serum . . . . . . . . . . . . . . . . . . . . . . . . . . 1%

## HBSS

(See Section 8.3, Table 8.2, and Balanced Salt Solutions, in this appendix.)

## Hoechst 33258 [Chen, 1977]

2-[2-(4-Hydroxyphenol)-6-benzimidazoyl]-6-(1-methyl-4-piperazyl)benzimidazole trihydrochloride

(1) Make up 1 mg/mL stock in D-PBSA or HBSS without phenol red.
(2) Store the solution at −20°C. For use, dilute 1:20,000 (1.0 μL in 20 mL) in D-PBSA or HBSS without phenol red at pH 7.0.

Δ **Safety Note.** Because this substance may be carcinogenic, handle it with extreme care. Weigh in a fume cupboard and wear gloves.

## Holtfreter's Buffer

NaCl . . . . . . . . . . . . . . . . . . . . . . . . . . . . . . . . . . . . . . 70 g
KCl . . . . . . . . . . . . . . . . . . . . . . . . . . . . . . . . . . . . . . . .1.0 g
$NaHCO_3$ . . . . . . . . . . . . . . . . . . . . . . . . . . . . . . . . . . .4.0 g

CaCl₂ . . . . . . . . . . . . . . . . . . . . . . . . . . . . . . . . . . . .2.0 g
UPW . . . . . . . . . . . . . . . . . . . . . . . . . . . . . to 1000 mL

(1) Sterilize by filtration.
(2) Store at 4°C.
(3) Dilute 1:20 with sterile UPW for use.

**Kanamycin Sulfate (Kannasyn), 10 mg/mL**

Kanamycin . . . . . . . . . . . . . . . . . . . . . . . . . . . . . .4, 1-g vials
Hanks's BSS . . . . . . . . . . . . . . . . . . . . . . . . . . . . . .400 mL

(1) Add 5 mL of HBSS from a 400-mL bottle of HBSS to each vial.
(2) Leave for a few minutes to dissolve.
(3) Remove the HBSS and kanamycin from the vials, and add them back to the HBSS bottle.
(4) Add another 5 mL of HBSS to each vial to rinse and return to the BSS bottle. Mix well.
(5) Dispense 20-mL aliquots of the solution into sterile containers and store at −20°C.
(6) Test for sterility: Add 2 mL of reagent to 10 mL of sterile medium, free of all other antibiotics, and incubate the solution at 37°C for 72 h.
(7) Use at 100 μg/mL.

**Lactalbumin Hydrolysate 5% (10×)**

Lactalbumin hydrolysate . . . . . . . . . . . . . . . . . . . . . . .5 g
HBSS . . . . . . . . . . . . . . . . . . . . . . . . . . . . . . . . . . .100 mL

(1) Heat to dissolve.
(2) Sterilize by autoclaving.
(3) Use at 0.5%.

**McIlvaines Buffer, pH 5.5**

| | Molarity of stock | Stock conc. | To make 20 mL | To make 100 mL |
|---|---|---|---|---|
| Na₂HPO₄ | 0.2 M | 28.4 g/L | 11.37 mL | 56.85 mL |
| Citric acid | 0.1 M | 21.0 g/L) | 8.63 mL | 43.15 mL |

**Media**
The constituents of some media in common use are listed in Chapters 8 and 9 (see Section 16.4; Tables 8.3, 9.1, 9.2), together with the recommended procedure for their preparation. For those media not described, see Morton [1970], www.ccf.ucsf.edu/Protocols/ccfMediaFBS.asp, or suppliers' catalogs (see Appendix II: Media).

**MEM**
See Media, in this appendix.

**2-Mercaptoethanol**

Stock solution, 5 mM . . . . . . . . . . . . . . . . . . . . . . . .4 μL
HBSS . . . . . . . . . . . . . . . . . . . . . . . . . . . . . . . . . . . .10 mL

(1) Sterilize by filtration in fume cupboard.
(2) Store at −20°C or make up a fresh solution each time.

**Methocel**
See Methylcellulose.

**Methylcellulose (1.8%)**
(1) Weigh out 7.2 g of Methocel, and add it to a 500-mL bottle containing a large magnetic stirrer bar.
(2) Sterilize by autoclaving with the cap loose for penetration of steam.
(3) Add 400 mL of sterile UPW heated to 90°C to wet the Methocel.
(4) Stir at 4°C overnight to dissolve. (The Methocel will gel if the magnet does not keep stirring.)
The resulting solution is now Methocel 2×, and for use, it should be diluted with an equal volume of 2× medium of your choice. It is more accurate to use a syringe (without a needle) than a pipette to dispense Methocel.
(5) For use, add a cell suspension in a small volume of growth medium (see Protocol 13.5)

**Mitomycin C**
Stock solution, 10 μg/mL (50×)
Mitomycin C, 2-mg vial

(1) Measure 20 mL of HBSS into a sterile container.
(2) Remove 2 mL of HBSS by syringe and add it to a vial of mitomycin.
(3) Allow the mixture to dissolve, withdraw the resulting solution, and add it back to the container.
(4) Store for 1 week only at 4°C in the dark. (Cover the container with aluminum foil.)
(5) For longer periods, store at −20°C.
(6) Dilute to 0.25 μg/mL for 18-h exposure or 20 μg/mL for 10-h exposure (see Protocols 13.3, 22.4).

Δ **Safety Note.** Because mitomycin is toxic, reconstitute it in the vial. Work in a fume hood when handling the substance in powder form.

**MTT**
(1) Dissolve 3-(4,5-dimethylthiazol-2-yl)-2,5-diphenyltetrazolium bromide (MTT) 50 mg/mL in D-PBSA.
(2) Sterilize by filtration.
Δ **Safety Note.** MTT is toxic; weigh it in a fume cupboard and wear gloves.

**Mycoplasma Reagents**
Stain (see Hoechst 33258)
Mountant: Glycerol in McIlvaines Buffer pH 5.5

| | | | To make 40 mL |
|---|---|---|---|
| Na₂HPO₄ | 0.4 M | 56.8 g/L | 11.37 mL |
| Citric acid | 0.2 M | 42.0 g/L | 8.63 mL |
| Glycerol | | | 20.00 mL |

(1) Add Vectashield (Vector) to reduce fluorescence fade (*see* manufacturer's instructions)

(2) Check the pH and adjust to 5.5.

### Mycostatin (Nystatin)

Prepare at 2 mg/mL (100×):

Mycostatin ................................... 200 mg
Hanks's BSS .............................. 100 mL

(1) Make up by same method as kanamycin (*see* above, this appendix).

(2) Final concentration, 20 μg/mL.

### N2 Supplement

(*See* Protocol 23.1 *and* Nichols & Ying [2006].)

*For stock solutions*

Insulin (Sigma); dissolve at 25 mg/ml in 0.01 M filter sterilized HCl overnight at 4°C.

Apo-transferrin (Sigma); dissolve at 100 mg/ml in filter sterilized $H_2O$.

Bovine Albumin Fraction V Solution (Gibco, as supplied).

Progesterone (Sigma); dissolve at 0.6 mg/ml in ethanol and filter sterilize.

Putrescine (Sigma); dissolve at 160 mg/ml in $H_2O$ and filter sterilize.

Sodium selenite (Sigma); dissolve at 3 mM in $H_2O$ and filter sterilize.

DMEM:F12 (Invitrogen).

With the exception of DMEM:F12 (stored at 4°C) stock solutions are aliquoted and stored at −20°C.

*For 40 mL N2 supplement add the following stock solutions:*

DMEM:F12 .............................. 27.5 mL
Bovine Albumin Fraction V ..................... 4 mL
Insulin ..................................... 4 mL
(Add 200 μL at a time to prevent precipitation)
Apo-transferrin .............................. 4 mL
Sodium selenite .............................. 40 μL
Putrescine .................................. 400 μL
Progesterone ............................... 132 μL

Aliquot and store at −20°C for up to 3 weeks.

### N2B27 Serum-Free Medium

(*See* Protocol 23.1 *and* Nicholls & Ying [2006].)

DMEM:F12, 1:1 (Invitrogen) with 2 mM L-glutamine, supplemented with Neurobasal Medium (Invitrogen), 2-Mercaptoethanol, 0.1 mM, B27 supplement, 1:100 as supplied (Invitrogen), N2 supplement (1:200, *see* this appendix, above)

(Readymade N2B27 can be purchased from Stem Cell Sciences (www.stemcellsciences.com)

### Naphthalene Black

Prepare at 1% in Hanks's BSS.

Naphthalene Black ........................... 1 g
Hanks's BSS ............................... 100 mL

(1) Dissolve as much as possible of the stain in the HBSS.

(2) Filter the resulting saturated solution through Whatman No.1 filter paper.

### PBS

*See* Phosphate-Buffered Saline.

### PE

*See* Phosphate-Buffered Saline/EDTA.

### Penicillin

Crystapen (benzylpenicillin [sodium]) or equivalent, 1,000,000 units per vial.

Make up as for kanamycin, stock concentration 10,000 U/mL.

Crystapen ........................ 4 vials (4 × $10^6$ U)
HBSS .................................. 400 mL

(1) Store frozen at −20°C in aliquots of 5−10 mL.

(2) Use at 50−100 U/mL.

### Percoll

(1) Readymade and sterile as purchased, Percoll should be diluted with medium or HBSS until the correct density is achieved (*see* Appendix II: Density Meters).

(2) Check the osmolality. Adjusting it to 290 mOsm/Kg will require the diluent to be hypo- or hypertonic, so it is better to dilute a small sample first and check its osmolality, and then scale up.

### Phosphate-Buffered Saline (Dulbecco Solution A; D-PBSA)

(*See* Table 8.2 and Protocol 10.6.)

### Phosphate-Buffered Saline/EDTA, 10 mM (PE)

(1) Make up D-PBSA.

(2) Add EDTA disodium salt, 3.72 g/L, and stir.

(3) Dispense, autoclave, and store at room temperature.

(4) Dilute D-PBSA/EDTA 1:10 to give 1 mM for most applications or 1:2 (5 mM) for high chelating conditions (e.g., trypsinization of CaCo-2 cells).

### Phytohemagglutinin (PHA)

(1) Prepare stock 500 μg/mL (100×) from lyophilized PHA by adding HBSS by syringe to an ampoule.

(2) Dispense into aliquots and store at −20°C.

(3) Dilute 1:100 for use.

### SF12

Ham's F12 (*see* Table 8.3) with additional 2× Eagle's MEM essential amino acids and 1× nonessential amino acids but lacking thymidine, and with 10× folic acid concentration.

### Sodium Bicarbonate

Prepare as 7.5% w/v in UPW. Divide into 50 mL aliquots and sterilize by autoclaving at 121°C for 10 min or filter sterilize. Dilute 3 mL/100 mL medium for use with 5% $CO_2$ and 0.45 mL/100 mL medium for gas phase or air (*see* Section 8.2.2), although the actual amount required will be determined by the pH of the medium at the start and may need to be determined by adding a range of amounts of sodium bicarbonate and checking the pH after incubation at 37°C (*see* Section 10.4.4).

## SSC

(1) Prepare 20x concentrated SSC:

Trisodium citrate (dihydrate), 0.3 M . . . . . . . . . . . 88.3 g
NaCl, 3.0 M . . . . . . . . . . . . . . . . . . . . . . . . . . . . . . 175.3g
Water to . . . . . . . . . . . . . . . . . . . . . . . . . . . . . 1000 mL

(2) Dilute 1x or 2x as required.

## Streptomycin Sulfate

(1) Take 2 mL from a bottle containing 100 mL of sterile HBSS, and add to a 1-g vial of streptomycin.
(2) When streptomycin has dissolved, return the 2 mL to the 98 mL of HBSS.
(3) Dilute 1:200 for use. The final concentration should be 50 µg/mL.

## Subbing Solution for Microscope Slides

Gelatin, 0.1%, chrome alum (chromium potassium sulphate), 0.01%, in deionized or RO water [Kiernan, 1999].

## Trypsin

2.5% w/v in 0.85% (0.14 M) NaCl.
To prepare a 2.5% solution:

Trypsin (e.g., Difco 1:250) . . . . . . . . . . . . . . . . . . . . 25 g
NaCl, 0.85% . . . . . . . . . . . . . . . . . . . . . . . . . . . . . . . . . 1 L

(1) Stir trypsin for 1 h at room temperature or 10 h at 4°C. If the trypsin does not dissolve completely, clarify it by filtration through Whatman No. 1 filter paper.
(2) Sterilize by filtration.
(3) Dispense into 10- to 20-mL aliquots and store at −20°C.
(4) Thaw and dilute 1:10 in D-PBSA or PE for use.
(5) Store diluted trypsin at 4°C for a maximum of 3 weeks.

  **Note.** Trypsin is available as a crude (e.g., Difco 1:250) or purified (e.g., Worthington or Sigma 3× recrystallized) preparation. Crude preparations contain several other proteases that may be important in cell dissociation, but may also be harmful to more sensitive cells. The usual practice is to use crude trypsin, unless the viability of the cells is diminished or reduced growth is observed, in which case purified trypsin may be used. Pure trypsin has a higher specific activity and should therefore be used at a proportionally lower concentration (e.g., 0.01 or 0.05%). Check the specific activity when you purchase trypsin as it varies among suppliers and from batch to batch. The higher the specific activity the lower the concentration that will be required. The traditional specific activity is 1:250 which is 250 USP units/mg. Trypsin at 1:250 is equivalent to 750 BAEE units/mg. Check for mycoplasma when preparing from raw trypsin.

## Trypsin/EDTA

*See* Trypsin, step 4.

## Trypsin, Versene, Phosphate (TVP)

Trypsin (Difco 1:250) . . . . . . . . . . 25 mg (or 1 mL 2.5%)
Phosphate-buffered saline, D-PBSA . . . . . . . . . . . . . 98 mL
Disodium EDTA (2H$_2$O) . . . . . . . . . . . . . . . . . . . . . 37 mg
Chick serum (MP Biomedicals) . . . . . . . . . . . . . . . . 1 mL

(1) Mix D-PBSA and EDTA, autoclave the mixture, and store it at room temperature.
(2) Add chick serum and trypsin before use. If powdered trypsin is used, sterilize it by filtration before adding the serum.
(3) Dispense the solution into aliquots and store at −20°C.

## Tryptose Phosphate Broth

Prepare at 10% in HBSS.

Tryptose phosphate (Difco) . . . . . . . . . . . . . . . . . . . . 100 g
Hanks's BSS . . . . . . . . . . . . . . . . . . . . . . . . . . . . . . . 1000 mL

(1) Stir until dissolved.
(2) Dispense into aliquots of 100 mL and sterilize in the autoclave.
(3) Store at room temperature
(4) Dilute 1:100 (final concentration, 0.1%) for use.

## Tyrode's Solution

NaCl . . . . . . . . . . . . . . . . . . . . . . . . . . . . . . . . . . . . . . 8.00 g
KCl . . . . . . . . . . . . . . . . . . . . . . . . . . . . . . . . . . . . . . . 0.20 g
CaCl$_2$ . . . . . . . . . . . . . . . . . . . . . . . . . . . . . . . . . . . . . 0.20 g
Mg$_2$Cl$_2$ · 6H$_2$O . . . . . . . . . . . . . . . . . . . . . . . . . . . . 0.10 g
NaH$_2$PO$_4$ · H$_2$O . . . . . . . . . . . . . . . . . . . . . . . . . . . 0.05 g
Glucose . . . . . . . . . . . . . . . . . . . . . . . . . . . . . . . . . . . 1.00 g
UPW, make up to . . . . . . . . . . . . . . . . . . . . . . . . . . . . 1 L
Gas phase . . . . . . . . . . . . . . . . . . . . . . . . . . . . . . . . . Air

## Versene

*See* EDTA.

## Viability Stain

*See* Naphthalene Black.
Trypan Blue is available from most tissue culture media suppliers (*see* Appendix II).

## Vitamins

Detailed in media recipes (*see* Tables 8.3, 9.1, 9.2) and available commercially as 100× concentrates.

(1) Make up individually as 1000–10,000× stocks and combine as required to make up a 100× concentrate.
(2) Sterilize by filtration.
(3) Store at −20°C in the dark.

# Sources of Equipment and Materials

The number of suppliers of reagents, equipment and materials used in cell culture is now so extensive that all possible sources are not given, but merely some examples of suppliers for each product. The suppliers' addresses are to be found in Appendix III. Additional suppliers are listed in the BiosupplyNet Source Book, Cold Spring Harbor Laboratory Press, and at www.biosupplynet.com, www.biocompare.com, http://informagen.com/, www.linscottsdirectory.com/, http://www.martex.co.uk/laboratory-supplies/index.htm, http://www.biosciencetechnology.com/, www.cato.com/biotech/bio-prod.html, www.ispex.ca/naccbiologicals.html, www.biospace.com/service_and_supplier.cfm., www.sciquest.com, http://www.coe.montana.edu/che/CompAlph.htm#A.

| Item | Supplier |
|------|----------|
| Accuspin tubes | Sigma Diagnostics |
| Accutase, Accumax | Sigma; TCS; Upstate Biotechnology |
| Acepromazine | Henry Schein |
| Acetonitrile | TAAB |
| Acetylcholine | Sigma |
| Actin, smooth muscle, alpha, antibody | DAKO |
| Activated charcoal | Sigma |
| Adenine (hydrochloride) | Sigma |
| Adipocytes and preadipocytes | BioPredic; Cell Applications; ECACC; Promocell; ReachBio; Zen Biologicals |
| Agar | Invitrogen; BD Biosciences (Difco); Oxoid |
| Agar EM embedding kit | Plano; Polysciences |
| Agarose | Lonza; FMC Bioproducts; Sigma |
| Agarose urea gel | Applied Biosystems |

| Item | Supplier |
|------|----------|
| Agitating water bath | Baker; Grant; *see also* Water baths |
| Air sampling | AES Chemunex; Particle Measuring Systems |
| Air velocity meters | *See* Anemometers |
| Alanyl-L-glutamine dipeptide | *See* Glutamax |
| Albumin (BSA) | Sigma |
| Albumin antibody, rabbit anti-rat | MP Biomedicals |
| Albumin, rat | MP Biomedicals |
| Alcohol disinfectant sprays (IPA) | Oncall Medical supplies |
| Alcohol swabs | BD Biosciences |
| Alcohol-resistant markers | Radleys |
| Alexa Fluor 488 goat anti-mouse IgG | Invitrogen; Molecular Probes |
| Alginate | ISP; NovaMatrix |
| Alkaline phosphatase assay materials | Sigma |
| Alkaline phosphatase reaction immunostaining kit (red) | Vector |
| Alkaline phosphatase staining kit (for stem cells) | Stemgent |
| Alpha medium | *See* Media |
| Alveolar epithelium | SkinEthic |
| Amino acids | JT Baker; Merck; Sigma |

| Item | Supplier |
|---|---|
| Amino acids, essential | *See* MEM essential amino acids |
| Aminoethanol | *See* Ethanolamine |
| Aminopropyltriethoxysilane | Fluka |
| Aminopterin | Sigma |
| Amphotericin B (Fungizone) | Lonza; Invitrogen; MP Biomedicals; PAA; Sigma |
| Ampoules, glass | Wheaton |
| Ampoules, plastic | Alpha Laboratories; CLP; Corning; Fisher Scientific; Greiner; Thermo Fisher (Nunc) |
| Anemometers | Technika; TSI; *see also* Laminar-flow hoods |
| Anesthesia device, veterinary grade | SurgiVet |
| Angiocath (plastic catheter) 20 gauge | BD Biosciences |
| Antibiotics | Invitrogen; Invivogen; Roche; MP Biomedicals; Sigma; *See also* Media |
| Antibodies | GE Healthcare (Amersham); BD Biosciences; Biopool; DAKO; Invitrogen; MP Biomedicals; Peprotech; R&D Systems; Serotec; Roche Diagnostics; Santa Cruz Biotechnology; Sigma; Upstate Biotechnology; Vector |
| Antifoams | Bayer; Dow-Corning; Merck; Sigma |
| Anti-mouse IgG ABC alkaline phosphatase kit | Vector |
| Anti-rabbit IgG ABC peroxidase kit | Vector |
| APAAP complex | DAKO; Vector |
| Apoptosis inhibitor | MP Biomedicals |
| Aprotinin (Trasylol) | Bayer; Serologicals; Sigma |
| Arginine HCl | *See* Chemicals |
| Ascorbic acid | Sigma; Wako |
| Ascorbic-2-phosphate | *See* Chemicals |
| Asparagine | *See* Chemicals |
| Aspartic acid | *See* Chemicals |
| Aspiration pipettes, unplugged | Corning Costar |
| Aspirators (reservoirs) | Bel-Art; Sterilin; Camlab; Corning; Integra; Kimble-Kontes; Techmate |
| Astrocytes and medium | Lonza Biologics; Neuromics |
| Attachment factors | *See* Matrix |
| Authentication of cell lines | ATCC; Bioreliance; Charles River; Cellmark; DSMZ; ECACC; LGC Standards; Orchid Biosciences; *see also* DNA profiling |
| Autoclavable bags and nylon film | Altec; Sterilin; Buck Scientific; Elkay; Jencons; Lab Safety Supply; SPS Medical; Portex; Portland Plastics; Roth |
| Autoclavable media | Sigma; Mediatech |
| Autoclaves | Ace; Astell; Bennet; Global; Integra; LTE; Medorex; Narang; Precision Scientific; SP Industries; Steris; VWR |
| Automatic dispensers | Corning; Genetic Research Instr.; Gilson; Jencons; Matrix Technologies; Michael Smith; MP Biomedicals; Robbins Scientific |
| Automatic pipette | Alpha; BD Biosciences; Roche Diagnostics; Corning; Gilson; MP Biomedicals; Jencons; Labsystems; Rainin |
| Automatic pipette plugger | *See* Pipette plugger |
| Autoradiographic emulsion | *See* Emulsion for autoradiography |
| Autosequencer, ALFred | GE Healthcare (Amersham) |
| Avidin-biotinylated IgG kit | Dako Diagnostics |
| Bacteriocides for water baths | Guest Medical; Henry Schein |
| Bacteriological grade culture dish | BD Biosciences; Fisher |
| Bactopeptone | BD Biosciences (Difco); Invitrogen |
| Balb/c 3T3 A31 cells | ATCC |
| BALB/c nu/nu mice | Nippon CLEA |
| Basic fibroblast growth factor (bFGF) | *See* Growth factors |
| BEGM (bronchial epithelium growth medium) | Lonza |
| Benchcote | Whatman |
| Bench-top autoclaves | Narang; Prestige Medical |
| Betadine | Bruce Medical |
| Bijou containers | Sterilin |
| Biochemicals | Calbiochem; Fluka; Merck; Pierce; Roche Diagnostics; Sigma; U.S. Biochemical |
| Biocoat multiwell plates | BD Biosciences |
| Bioelectrical monitoring | *See* Biosensors |
| Biohazard bags | *See* Autoclavable bags and nylon film |
| Biohazard safety cabinets | *See* Microbiological safety cabinets |

| Item | Supplier |
|---|---|
| Biopsy needles | Stille |
| Bioreactors | Alfa Laval; Bellco; Biotech Instruments; Biovest; Braun; Cellon; Charles River; Corning; Electrolab; Genetic Research Instrumentation; Global Medical; Greiner Bio-one; James Glass; KBI; Medorex; New Brunswick Scientific; PerkinElmer; Sartorius; Sigma; Synthecon; Vivascience; Wave Biotech |
| Bioreactors, disposable | Cellon; Millipore; Sartorius; Xcellerex |
| Bioreactors, mini | Amprotein; Applikon; Hexascreen; Hudson Robotics |
| Biosensors | Analytical Technologies; Applied Biophysics; CellStat; Nova Biomedical; YSI |
| Biotain-MPS serum substitute | Lonza |
| Biotin | Sigma |
| Biotinylated anti-mouse IgG | Vector |
| Biotinylated IgG | Sigma; Vector |
| Blood urea nitrogen reagents | Stanbio Labs |
| BM-cycline | Roche Applied Science |
| BMON software | Ingenieurbüro Jäckel |
| Bone cells | See Osteoblasts |
| Bone marrow | Lonza; ReachBio |
| Bone marrow stromal cells | Cell Applications; ECACC; ReachBio |
| Bone morphogenic proteins (BMPs) | R&D Systems; Wyeth; see also Growth factors |
| Borate buffer | Sigma |
| Bottle brushes | Cole-Parmer |
| Bottle rack | MP Biomedicals |
| Bottles | Applied Scientific; BD Biosciences; Bel-Art; Bellco; Sterilin; Caisson; Camlab; Corning; Fisher; Integra; Kimble Kontes; Thermo Fisher (Nunc); Polytech; Radleys; Schott; Techmate; VWR |
| Bottle-top dispensers | Brand; Jencons; Fisher; Polytech |
| Bouin's fixative | Sigma |
| Bovine hypothalamus | Pel-Freez |
| Bovine pituitaries | Pel-Freez |
| Bovine pituitary extract | Lonza (Clonetics); Cascade; Hammond Cell Technology; Invitrogen; PromoCell |
| Bovine serum (selected batch) | Gemini; Invitrogen; MP Biomedicals; Sigma; SeraLab |
| Bovine serum albumin | GE Healthcare (Amersham); Bayer; BioSource; BD Biosciences; Calbiochem; Lonza (Clonetics); MP Biomedicals; Invitrogen; Intergen; Sigma |
| BrdU DNA cell labeling kit | Invitrogen; Promega; Sigma |
| Bronchial epithelial cells | Cell Applications; ECACC; Lonza; TCS Cellworks; (see also Specialized Cell Cultures) |
| Broths | See Nutrient broths |
| BSA | See Bovine serum albumin |
| BSS, Hanks's, Earle's, etc. | See Media |
| Buffered formalin | Fisher |
| Butanedione monoxime (BDM) | Sigma |
| $CaCl_2$ | See Chemicals |
| Cacodylate buffer | Plano; TAAB |
| Calcium assay kit | Sigma |
| Calcium binding reagent | Sigma |
| Calcium chloride | See Chemicals |
| Calcium/phosphorus combined standard | Sigma |
| Calf serum | See Serum |
| Calf thymus DNA type I | Sigma |
| Cameras | Leica; Olympus; Nikon; Zeiss (see also CCD and Digital Cameras) |
| Cannulae | BD Biosciences; Roboz |
| Carbodiimide, 1-ethyl-3-(dimethylaminopropyl)-, EDC | Pierce |
| Carbon dioxide | Air Products; Cryoservice; Messer; Taylor Wharton |
| Carboxymethyl-cellulose (CMC) | Fisher; Merck |
| Cardiac myocytes/myoblasts | PromoCell; ReachBio |
| Cartilage cells | See Chondrocytes |
| Catheter extension kit, Interlink | Baxter HealthCare |
| Catheters | BD Biosciences |
| CCD cameras | BioWorld; Dage-MTI; Hamamatsu; Leica; Photometrics; Scanalytics; Sony; see also Microscopes |
| CD antibodies | BD Biociences (Pharmingen) |
| cDNA libraries | Clontech |
| Cell banks | ATCC; CellBank Australia; Coriell Cell Repositories (CCR); ECACC; DSMZ; HSRRB; JCRB; RIKEN |

| Item | Supplier |
|---|---|
| Cell counters | Beckman Coulter; Chemometec; Millipore; New Brunswick; Sedna (*see also* Table 20.2) |
| Cell counting solutions | Beckman Coulter; Sedna |
| Cell culture dishes, flasks, and plates | *See* Culture dishes, flasks, and plates |
| Cell culture inserts | *See* Filter well inserts |
| Cell dissociation agents | *See individual enzymes and Trypsin replacements* |
| Cell filter | BD Biosciences; Miltenyi |
| Cell line 5637 | DSMZ |
| Cell line databanks | ATCC; Coriell; DSMZ; ECACC; HSRRB; ICLC |
| Cell lines | *See* Cell banks |
| Cell migration assays | Cell Biolabs |
| Cell proliferation assays | Lumitech (Lonza); Promega; Upstate |
| Cell Quest acquisition software | BD Biosciences |
| Cell scraper | Applied Scientific; Bel-Art; BD Biosciences; Corning; Thermo Fisher (Nunc); Techmate; Techno Plastic |
| Cell separation | Accurate Chemical & Scientific; Applied Immune Sciences; BD Biosciences; BioCarta; Biochrom; Dynal; Kendro; Miltenyi; PerkinElmer; Stem Cell Technologies |
| Cell sizing | Beckman Coulter; Schärfe |
| Cell strainers | BD Biosciences; Dynal; Miltenyi Biotec |
| CellFlo | Spectrum Laboratories |
| Cellgro | *See* Mediatech |
| Centrifugal elutriator | Beckman Coulter |
| Centrifuge tubes | Alpha; Sterilin; BD Biosciences (Falcon); Corning; Du Pont; Eppendorf; Greiner; Omnilab; Thermo Fisher (Nunc); Techno Plastic Products |
| Centrifuges | Beckman; DuPont (Sorval); Fisher; Kendro; Thermo-IEC; Life Sciences International |
| Ceramic rods and scaffolds | Zimmer |
| Ceramics, coral-based | Interpore |
| Chamber slides | Applied Scientific; Bayer; BD Biosciences; Sterilin (Iwaki); Heraeus; Metachem; Thermo Fisher (Nunc); Stem Cell Technologies |
| CHAPS buffer | Calbiochem |
| Chemicals | Calbiochem; Fisher; JT Baker; Merck; Pierce; Research Organics; Roche; Scientific & Chemical Supplies; Sigma; USB; Wako |
| Chick embryo extract | Accurate Chemical & Scientific; Invitrogen |
| Chicken plasma | Invitrogen; MP Biomedicals |
| Chicken serum | Invitrogen; MP Biomedicals; TCS Biologicals |
| CHIRON 99021 | Axon Medchem |
| Chloramphenicol | Sigma |
| Chloroform | *See* Chemicals |
| Chloros | Vernon Morris (*see also* Disinfectants) |
| Cholera toxin | EMD; List Biological; Merck; Sigma |
| Choline chloride | Sigma |
| Chondrocytes | Cell Applications, ECACC, Cell Genix, Promocell, TCS Cellworks |
| Chondroitin sulfate | Sigma |
| Chromosome paints | Applied Imaging; Cambio; Invitrogen; Qbiogene; Vysis |
| Ciprofloxacin | Bayer |
| Clidox disinfectant | Tecniplast (Indulab) |
| Cloning disks, 3 mm | Sigma |
| Cloning rings/cylinders | BelArt; Bellco; Fisher; Scientific Laboratory Supplies |
| Clorox | Polyscience (*see also* Disinfectants) |
| Closed-circuit television (CCTV) | Dage-MTI; Hamamatsu; Leica (*see also* CCD cameras; microscopes) |
| CMC | *See* Carboxymethylcellulose |
| CMF: $Ca^{2+}$, $Mg^{2+}$-free EBSS (Earle's balanced salt solution) | Invitrogen |
| CMF-PBS | *See* Media |
| CMRL-1066 | *See* Media |
| $CO_2$ | Air Products; Cryoservice; Taylor Wharton |
| $CO_2$ automatic changeover unit for cylinders | Air Products and Chemicals Inc.; Gow-Mac; Lab Impex; Nuaire; Thermo-Shandon |
| $CO_2$ controllers | Air Products; Gow-Mac; Lab-Line; Lab Impex; Therma Electron (Forma; Hotpak) |
| $CO_2$ incubators | Barnstead-Thermolyne; Boro Labs; Camlab; Fisher; Heinecke; Kendro; Lab-Impex; Lab-Line; LEEC; Memmert; MP Biomedicals; Napco; New |

| Item | Supplier |
|------|----------|
| | Brunswick Scientific; NuAire; Omnilab; Precision Scientific; Sanyo Gallenkamp; SP Industries; Thermo-Electron; Triple Red; |
| $CO_2$-permeable caps | *See* Culture dishes, flasks, and plates |
| Colcemid | Sigma |
| Collagen | BD Biosciences; Biocolor; Biodesign; Biomedical Technologies; Biosource International; Cambridge Biosciences; Cellon; Chondrex; CR Bard; Davol; Invitrogen; Inamed; Insmed; Lab Vision; MP Biomedicals; Roche; Seikagaku; Sigma; Stratech; Universal Biologicals |
| Collagen, recombinant | FibroGen |
| Collagen sponge, Avitene Ultrafoam | CR Bard; Davol; Inamed; Allergan |
| Collagen type I antibody, rabbit anti-human | Biodesign |
| Collagenase | Chondrex; Invitrogen; Lorne; Roche; Serva; Sigma; Worthington |
| Collagen-coated culture dishes, flasks, plates | BD Biosciences; Corning; Iwaki Glass |
| Colony counters | BioWorld; BioDu Pont (NEN Life Sciences); Cole-Parmer; Don Whitley; Oxford Optronics; Perceptive Instruments; PerkinElmer; Synbiosis; UVP |
| Colony ring marker | Nikon |
| Combi Ring Dish | Renner; Germany |
| Conductivity meters | Corning; Quadrachem; Technika; Thermo Electron |
| Confocal microscope | Bio-Rad; Biotech Instruments; Leica; Nikon; Zeiss |
| Conical centrifuge tube | *See* Centrifuge tubes |
| Continuous roller pump | Cole-Parmer |
| Controlled-atmosphere chamber | Bellco; Vineland; NJ |
| Controlled-rate coolers for liquid $N_2$ freezing | Messer; Thermo Fisher (Nunc); Planer; Thermo Electron; Statebourne; Taylor Wharton |
| Coomassie Blue R | Sigma |
| Copper sulfate | *See* Chemicals |
| Cornwall syringe | BD Biosciences; Cole-Parmer (Chempette); Popper |

| Item | Supplier |
|------|----------|
| Coverslip mountants | *See* Mountants |
| Coverslips, glass | Bayer; Corning; Fisher; Invitrogen; Wheaton |
| Coverslips, plastic | Bayer; Corning; MP Biomedicals; Lux; Invitrogen |
| CPSR serum substitute | Sigma |
| Cryofreezer | *See* Controlled-rate coolers; Freezers, liquid nitrogen |
| Cryogenic freezers | *See* Freezers, Liquid nitrogen |
| Cryogenic vials | *See* Ampoules |
| Cryolabels | Computer Imprintable Label Systems; GA International; Triple Red |
| Cryomarkers | GA International |
| Cryopreservation medium | MP Biomedicals; Invitrogen; JRH Biosciences; Sigma |
| Cryoprotective gloves | Taylor Wharton; Jencons |
| Cryogenic storage | ATCC; ECACC; DSMZ; JCRB; Vindon Scientific; *See also* Cell Banks; Freezers, liquid nitrogen |
| Cryotubes | *See* Ampoules |
| Cryovials | *See* Ampoules |
| Crystal violet | Merck; Fisher |
| Culture bags | Cell Genix; Cellon; Du Pont; MP Biomedicals; PAW BioScience Products (*see* also Media Bags) |
| Culture chambers | Bellco; MP Biomedicals |
| Culture dishes, flasks, and plates | BD Biosciences (Falcon); Sterilin; Corning (Costar); Fisher; Greiner; Invitrogen; Iwaki; MP Biomedicals (Lux); Thermo Fisher (Nunc); Sarstedt; TPP |
| Culture imaging | *See* Culture monitoring systems |
| Culture media, salt solutions, etc. | *See* Media |
| Culture monitoring systems | Chip-man Technologies; Essen Instruments; Pan Biotech |
| Culture slides | *See* Chamber slides |
| Culture tubes | *See* Culture dishes, flasks, and plates |
| Culture vessels | *See* Culture dishes, flasks, and plates |
| Curved forceps | *See* Surgical instruments |
| Custom media | Caisson; *see also* Media |
| Cy5 fluorescent-labeled primers | GE Healthcare (Amersham) |
| Cyclic AMP | Sigma |
| Cyclosporine | Novartis; Sigma |
| Cyprofloxacin | *See* Ciprofloxacin |
| Cysteine | Sigma |
| Cysteine hydrochloride hydrate | *See* Chemicals |
| Cystine | *See* Chemicals |
| Cytobuckets | Thermo Electron Corporation |

| Item | Supplier |
| --- | --- |
| Cytocentrifuge | Bayer; CSP; Electron Microscopy Sciences; Thermo Electron; Sakura Finetek; Wescor |
| Cytofectin GSV | Glen Research |
| Cytogenetic analysis | Cell Culture Characterization Services |
| Cytokeratin antibodies | DAKO; Lab Vision; Santa Cruz Biotechnology; Zymed |
| Cytokines | See Growth factors |
| Cytometer | See Flow cytometers; Scanning cytometer |
| Cytoseal TM 60 fluorescence mountant | Microm; Stephens Scientific; VWR |
| Cytotoxicity assays | ATCC; MatTek; Promega; SkinEthic |
| Cytotoxicity testing | BioReliance; CellStat; see also Cytotoxicity assays |
| Dacron vascular graft | Bard |
| Daigo's T medium | Wako |
| dbcAMP | Sigma |
| DEAE dextran | GE Healthcare (Amersham); Bio-Rad |
| Decontamination (of equipment and facilities) | Anachem; Steris; see also Disinfectants |
| Deionizers | Barnstead Thermolyne; Bellco; Corning; Dow Corning; Elga; High-Q; Millipore; Purite; U.S. Filter; Vivendi Water Systems; VWR |
| Densitometers | Beckman Instruments; Pall Gelman Sciences; Gilford Instruments; Helena; Joyce-Lobel |
| Density marker beads | GE Healthcare (Amersham) |
| Density media | GE Healthcare (Amersham); Genetic Research Instr.; MP Biomedicals; Nycomed; Robbins Scientific; Sigma |
| Density meter | Mettler Toledo; Parr |
| Deoxyribonuclease | Sigma |
| Dermal puncher | Miltey |
| Dermatome | Stortz Instruments; see also Tissue slicers |
| Desmin, mouse, antibody | Sigma |
| Detergents | Alconox; Calbiochem; Decon; MP Biomedicals; Pierce |
| Dexamethasone | Sigma; Merck |
| Dextran | Calbiochem; Fisher; Sigma |
| Dextrose | Fisher |
| Diacetyl fluorescein | Fisher; Sigma |
| Dialysis cassette | Pierce |
| Dialysis tubing | Cole-Parmer; Chemicon; Pierce; Serva; Spectrum |
| Diaminobenzidine | Sigma; Vector |
| Diethylpyrocarbonate (DEPC) | Sigma |
| Digital cameras | Canon; Kodak; Leica; Nikon; Olympus; Polaroid; see also CCD cameras; Microscopes |
| Dimethyl sulfoxide (DMSO) | ATCC; JT Baker; Merck; Sigma |
| Dimethylethylformamide | See Chemicals |
| Dimethylmethylene blue | Polysciences; Sigma |
| Dipeptidyl peptidase IV (CD26) | Biosource; Neomarkers |
| Dishes | See Culture dishes, flasks, and plates |
| Disinfectants | Anachem; BioMedical Products; Day-Impex; Guest Medical; Johnson & Johnson; Lab Impex; Lab Safety Supply; Markson LabSales; MP Biomedicals; Polyscience; Sigma; Steris; Tecniplast; Thomas; Vernon Morris; VWR Scientific |
| Disk filter assembly for sterilization | See Filters |
| Disodium hydrogen orthophosphate | See Chemicals |
| Disodium hydrogen phosphate | See Chemicals |
| Dispase | BD Biosciences; Invitrogen; Roche |
| Dispensers, liquid | Accuramatic; Barnstead; Corning; Gilson; Jencons; Lawson Mardon Wheaton; Matrix Technologies; Mettler Toledo; Michael Smith; MP Biomedicals; Polytech; Popper; Robbins; Thermo Electron; Zinsser |
| Dissecting instruments | Fine Scientific Tools; Fisher; Roboz; Swann-Morton; VWR |
| Dissection microscope | Daigger; Olympus; Leica; Nikon; Zeiss |
| Dithiothreitol | Sigma |
| DMEM with stabilized glutamine | Invitrogen; Biochrom |
| DMEM/F12, 50/50 | Biochrom; Cellgro-Mediatech; Invitrogen; Sigma |
| DNA polymerase premix | Takara Bio |
| DNA preparation kit | Qiagen |
| DNA profiling | Anglia DNA Bioservices; ATCC; Cellmark Diagnostics; Charles River; ECACC; Laboratory Corporation of America; LGC Standards; Orchid Biosciences; see also Authentication |
| DNA sequences | Geron |
| DNA stains | Hoechst; Molecular Probes |
| DNA templates | Ambion |
| DNase | Lorne; Serologicals; Sigma; Worthington |

| Item | Supplier |
|---|---|
| Donkey anti-rabbit IgG | Jackson ImmunoResearch |
| Donor calf serum (DCS) | *See* Serum |
| Donor horse serum | Gemini |
| D-PBSA | *See* Media |
| DPX, Permount | *See* Stains |
| Drug metabolizing kits | BioPredic |
| Dulbecco's modified Eagle's medium (DMEM) | *See* Media |
| Duran glass bottles | Camlab; Fisher; Schott |
| ECGS | *See* Growth factors |
| ECM | *See* Matrix |
| Ecoscint | BS&S; Perkin Elmer |
| EDC, 1-Ethyl-3-(3-Dimethylaminopropyl)-carbodiimide | Pierce |
| EDTA | JT Baker; Merck; Sigma |
| EDTA, sterile | *See* Media |
| Edwards high vacuum grease | Edwards |
| EGF | *See* Growth factors |
| Egg incubators | G.Q.F. Manufacturing |
| EGTA | Sigma |
| Ehrlenmeyer flasks | *See* Glassware |
| EHS matrix (Matrigel, Natrigel) | BD Biosciences |
| Elastase | Roche Diagnostics; Sigma |
| Electrochemical sensors | Finesse Solutions; Sentek; Siemens |
| Electron microscopy | Electron Microscopy Services; TAAB; Structure Probe |
| Electronic cell counter | *See* Cell counters |
| Electronic gas blenders | *See* Gas blenders |
| Electronic thermometers | Comark; Cole-Parmer; Fisher; Harvard; Grant; Labox; Omega; Pierce; Thermo Electron |
| Electrophoresis | GE Healthcare (Amersham); Anachem; Bio-Rad; Invitrogen; Haake; Innovative Chemistry; Life Sciences International; Thermo-Shandon |
| ELISA assays | Assay Designs; R&D Systems |
| ELISA kit for collagen type I | Chondrex |
| ELISA plate readers | *See* Plate readers |
| Embryonic stem cells, mouse & human | Befutur; Millipore; Thermo Fisher |
| Emulsion for autoradiography | GE Healthcare (Amersham) |
| Endofree Maxi Kit | Qiagen |
| Endothelial cells | Amsbio; ATCC; BioPredic; Cell Applications; ECACC; Lonza (Clonetics); PAA; TCS Cellworks; *See also* Specialized cell cultures *and* Table 22.1 |
| Endothelin 1, human, porcine (ET-1) | Sigma |
| Entactin | USB |
| Entellan | Merck |

| Item | Supplier |
|---|---|
| Environmental incubator | *See* Incubator enclosure for microscopes |
| Epidermal growth factor | *See* growth factors |
| Epi-Life serum-free medium | Sigma |
| Epinephrine | Sigma |
| Epithelial cells | ATCC; Cell Applications, ECACC, Invitrogen, Lonza; Millipore, PromoCell; TCS Cellworks; SkinEthic; Zen Biologicals; *See also* Specialized cell cultures *and* Table 22.1 |
| Epithelial membrane antigen (EMA) antibody | Dako |
| Epon | Merck; LR-White; Roth |
| Erlenmeyer flasks, tissue culture grade | *See* Glassware |
| ESGRO stem cell medium | Millipore (Chemicon) |
| Esophageal epithelial cells | TCS Cellworks |
| Estradiol | Sigma |
| Ethanol | *See* Chemicals |
| Ethanolamine | Sigma |
| Ethidium bromide | Sigma |
| Ethylene diamine tetraacetate disodium salt | *See* EDTA |
| Eukitt | Fluka |
| Excell-900 serum substitute | JRH Biosciences |
| Ex-cyte serum substitute | Bayer |
| Extracellular matrix | *See* Matrix |
| F12 | *See* Media |
| F12:DMEM | *See* Media |
| F12H medium | *See* Media |
| FACS | BD Biosciences; Beckman-Coulter |
| FACSCalibur | BD Biosciences |
| Factor VIII antibody | Roche Applied Science |
| Falcon tissue culture flasks | BD Biosciences |
| Fast Green | Sigma |
| Fast red TR salt | SERVA |
| Fast Violet capsule | Sigma |
| Fast-Track software Wavemaker32 ver.6.6 | Instron |
| Fat cells | *See* Adipocytes |
| FBS | *See* Serum |
| FCS | *See* Serum |
| $Fe_2SO_4 \cdot 7H_2O$ | *See* Chemicals |
| Fermentors | *See* Bioreactors |
| Ferrous sulfate | *See* Chemicals |
| Fetal bovine serum (FBS) | *See* Serum |
| FGF, human recombinant | *See* Growth factors |
| Fiber-optic illumination | Fischer Scientific |
| Fibrin sealant, Tisseel VH | Baxter |
| Fibroblast growth factor | *See* Growth factors |
| Fibroblasts | ATCC; BioPredicCell Applications, ECACC; Invitrogen; Lonza; Millipore; PromoCell; TCS Cellworks; Zen |

| Item | Supplier |
|------|----------|
| | Biologicals; *See also* Specialized cell cultures *and* Table 22.1 |
| Fibronectin | Accurate Chemical & Scientific; BD Biosciences; Biomedical Technologies; GE Heathcare (Amersham); Invitrogen (Biosource); Lab Vision; R & D; Sanyo Chemical; Sigma; Stratech; *see also* Matrix |
| Fibronectin, recombinant | Sanyo Chemical Industries |
| Fibronectin/collagen/BSA | Biosource International |
| Ficoll | GE Heathcare (Amersham); MP Biomedicals; Sigma |
| Ficoll–Hypaque | *See* Ficoll-metrizoate |
| Ficoll-metrizoate | GE Heathcare (Amersham); Lonza; MP Biomedicals; Nycomed; Sigma |
| Filter bottom multiwell plates | Thermo Fisher (Nunc); Techmate; Whatman |
| Filter holders | *See* Filters |
| Filter well inserts | BD Biosciences; Sterilin; Corning; Integra; Millipore; Thermo Fisher (Nunc) |
| Filters, mesh or gauze | *See* Mesh filters |
| Filters, sterilization | BD Biosciences; Sterilin; Corning; Millipore; Thermo Fisher (Nunc); Omnilab; Pall Gelman; Sartorius Stedim; Techno Plastic; VWR; Whatman |
| Filtration equipment | *See* Filters, sterilization |
| Fish serum | East Coast Biologics |
| FITC-avidin | Vector Laboratories |
| FITC-conjugated secondary antibody | *See* Antibodies |
| Fixed-bed bioreactors | New Brunswick |
| Flasks | *See* Culture dishes, flasks, and plates |
| FlexiPERM slide™ (8 chambers) | Sartorius |
| Flow cytometers | Accuri; Agilent; BD Biosciences; Beckman Coulter; Genetic Research; Guava Technologies; Miltenyi Biotec |
| Flow meters | *See* Gas-flow meters |
| Fluorescence bleaching inhibitor | *See* Fluorescence fade retardant |
| Fluorescence fade retardant | Calbiochem; Citifluor; Vector |
| Fluorescence mountants | Biomedia; Citifluor; Microm; Stephens Scientific; Vector; VWR |
| Fluorescence-activated cell sorter (FACS) | BD Biosciences |
| FluorSave, fluorescent mounting medium | Calbiochem |
| Folic acid | Sigma |
| Force transducer (muscle culture) | Ingenieurbüro Jäckel |
| Forceps | *See* Dissecting instruments |
| Formaldehyde | *See* Chemicals |
| Formamide | Merck |
| Formic acid | *See* Chemicals |
| Formol saline, buffered | Sigma |
| Four-well plates | *See* Culture dishes, flasks, and plates |
| Fragment Manager software | GE Healthcare (Amersham) |
| Frame grabber card (time-lapse) | Scion |
| Freezer racks and canes | *See* Freezers, liquid nitrogen |
| Freezers, −20°C | Barnstead; Fisher; New Brunswick Scientific; local discount warehouses |
| Freezers, −70°C | Barnstead; Fisher; LabLife Sciences International; New Brunswick; Nuaire; Precision Scientific; Revco |
| Freezers, liquid nitrogen | Aire Liquide; Barnstead-Thermolyne; Boro Labs; Chart Biomed; Cryomed; Genetic Research; Jencons; Messer Cryotherm; Planer; Statebourne; Taylor-Wharton; Thwermo Electron; VWR |
| Freezing medium | *See* Cryopreservation medium |
| FuGENE 6 | Roche |
| Fungicides | BioWhittaker; MP Biomedicals; Invitrogen; Sigma |
| Fungizone | *See* Amphotericin B |
| Gas blenders | Gow-Mac; Signal |
| Gas-flow meters | Muis Controls; Omega Engineering; Titan Enterprises (*see also* Gas blenders) |
| Gas generators | Bioquell; Peak Scientific; Sartec; Texol |
| Gas mixers | *See* Gas blenders |
| Gases | Air Products; British Oxygen; Cryoservice; Matheson Gas; Messer; Taylor-Wharton |
| Gas-permeable caps | *See* Culture dishes, flasks, and plates |
| Gas-permeable tissue culture bags | *See* Media bags and Culture bags |
| Gauze | *See* Mesh filters and gauze |
| Gel/Mount water-based mounting medium | Biomeda |
| Geneticin | Invitrogen |
| Genotoxicity testing | Apredica; Bioreliance; Gentronix |
| Gentamicin | Gemini; Invitrogen; Sigma |
| German glass coverslips | Carolina Biological Supplies |
| Giemsa stain | Fisher; Merck; TCS Biologicals |

| Item | Supplier |
|---|---|
| Gilson Pipetteman | Bellco; Corning; Gilson; Schott; Wheaton |
| Gingival epithelial cells | Millipore |
| Glass coverslips | Chance Propper; Wheaton |
| Glass fiber filters | Millipore; Pall; Whatman |
| Glass universal containers | Camlab |
| Glass-bottomed Petri dishes | Mat Tek |
| Glassware | Bellco; Corning; Kimble-Kontes; Schott; Wheaton |
| Glassware washing machine | Burge; Lancer; Miele; Scientek; Scientific Instrument Centre; Steris Corp. |
| Gloves | Ansell Medical; Applied Scientific; Cryomed; Johnson & Johnson Medical; Lab Safety Supply; Radleys; Renco Corp.; Safeskin; Sentinel Laboratories; Stoelting; Surgicon; VWR Scientific |
| Gloves, nitrile | Ansell; Cole-Parmer; Lab Safety Supplies; Radleys; Sentinel; Stoelting |
| Glucagon | Bedford Laboratories |
| Glucose | *See* Chemicals |
| Glucose, sterile | *See* Media |
| Glutamax | Invitrogen |
| Glutamic acid | *See* Chemicals |
| Glutamine | *See* Chemicals |
| Glutamine, sterile | *See* Media |
| Glutaraldehyde | Fisher Scientific; Plano; Roth; Sigma |
| Glutathione | Sigma |
| Glycerol | Fisher; Mallinckrodt; Merck; Sigma; VWR |
| β-Glycerophosphate | Invitrogen; Sigma |
| Glycine | Sigma |
| Glycine-arginine-glycine-aspartate-serine (GRGDS) peptide | Sigma |
| GM-CSF | *See* Growth factors |
| Goat serum | Sigma |
| GRGDS peptise | Sigma |
| Growth factor reduced Matrigel | BD Biosciences; Fisher Scientific |
| Growth factors and cytokines | Abbiotec; Amgen; Austral Biologicals; BD Biosciences; Biodesign; Biosource; Lonza; Cambridge Bioscience; Collaborative Research (BD Biosciences); GE Healthcare (Amersham); Genzyme; Invitrogen; MP Biomedicals; PeproTech; PerkinElmer; Promega; |
| | R&D Systems; Roche; Serologicals; Sigma; Stratech; UniversalBiologicals; Upstate Biotechnology |
| Guanidine hydrochloride, 8M | Pierce |
| Gyratory shaker | Boeker Scientific; New Brunswick |
| Halothane | Henry Schein |
| Ham's F-10 | *See* Media |
| Ham's F12 | *See* Media |
| Ham's F12:DMEM | *See* Media |
| Hanks's balanced salt solution (HBSS) | *See* Media |
| Harris hematoxylin | Fisher |
| HARV bioreactor | Synthecon |
| HCl | *See* Chemicals |
| Heat-inactivated fetal bovine serum (HIFBS) | *See* Serum |
| Heating blocks | Genetic Research Instr.; Robbins; VP Scientific |
| Hematopoietic cells | Befutur; ReachBio |
| Hematoxylin | Merck; Sigma |
| Hemocytometer | Fisher; Omnilab; Thermo Electron |
| Hen's eggs | Charles River Laboratories |
| Heparan sulfate | Sigma |
| Heparan sulfate proteoglycans (HSPGs) | Sigma |
| HepaRG cells | BioiPredic |
| Heparin | Roche; Sigma |
| Hepatocytes | Admet Technologies; Invitrogen, BioPredic; LGC Standards; Lonza Biologicals, Promocell, ReachBio; TCS Cellworks, Zen Biologicals |
| HEPES | *See* Media |
| HepG2 cells, human hepatoma cell line | *See* Cell banks |
| Hexafluoro-2-propanol (HFIP) | Aldrich |
| HGF/SF, human, recombinant | *See* Growth factors |
| High aspect ratio vessel (HARV) | Synthecon |
| Histidine HCl·H₂O | *See* Chemicals |
| HistoGel™ | Lab Storage Systems, Inc. |
| Histological stains | Merck; MTR Scientific; Sigma |
| Histopaque | Sigma |
| Histostain-SP kit (Streptavidin-peroxidase) | Zymed Laboratories, Inc.; San Francisco; CA; USA |
| HLF cells | Coriell Institute for Medical Research |
| Hoechst 33258 fluorescent mycoplasma stain | MP Biomedicals (complete kit); Polysciences; Sigma |
| Hoechst 33342 DNA stain | Sigma |
| Hollow fiber perfusion culture | Argos; Bellco; Biovest; Cellco; FiberCell; Integra; JM Separations; Spectrum; Unisyn |

| Item | Supplier |
|------|----------|
| Hormones | Sigma |
| Horse serum | *See* Serum |
| HT Tuffryn filters | Pall-Gelman |
| *Htrt* DNA | Geron |
| Human bone marrow stromal cells | Clonetic-Poietics |
| Human epidermal growth factor (hEGF) | *See* Growth factors |
| Human fibronectin | BD Biosciences |
| Human transferrin | Sigma |
| Hyaluronidase | Sigma; Worthington |
| Hydrochloric acid | *See* Chemicals |
| Hydrocortisone and analogs | Biosource; GE Healthcare; Merck; Sigma Upjohn |
| Hydrogels | Nektar |
| Hydrogen peroxide (30% solution) | *See* Chemicals |
| Hydrophobic filters | Millipore; Pall Gelman; Sartorius |
| Hydroxylamine hydrochloride | Pierce |
| Hydroxysuccinimide (NHS) | Pierce |
| Hypaque–Ficoll media | *See* Ficoll-metrizoate |
| Hypochlorite disinfectant | *See* Disinfectants |
| Hypoclearing agent | Kodak |
| Hypoxanthine | Sigma |
| IBMX, 3-isobutyl-1-methylxanthine | Sigma |
| IgG, rat | Sigma |
| Image analysis | Applied Imaging; Bio-Rad; Carl Zeiss; Chip-man Technologies; Essen Instruments; Hamamatsu; Imaging Associates; Imaging Research; Leica; Nikon; Nonlinear Dynamics; Pan Biotech; Perceptive Instruments; PerkinElmer; Scanalytics; Syngene; |
| Immunoanalyzers | Agilent; Guava (*see also* Flow cytometers) |
| Immunoglobulin antibodies, rabbit anti-mouse | Dako Diagnostics |
| Immunoglobulins, mouse anti-rabbit | Dako |
| Incubator enclosure for microscopes | Life Imaging Services |
| Incubators | ATR; Barnstead; Bellco; Binder; Boro Labs; Camlab; Fisher; Genetic Research Instr.; Global Medical; Harvard; Kendro; Infors; Lab-Line; LEEC; LMS; LTE; Memmert; MP Biomedicals; Napco; New Brunswick Scientific; Nuaire; Omnilab; Precision Scientific; Robbins; RS |
| | Biotech; Sanyo Gallenkamp; Scientific Instrument Centre; SciGene; SP Industries; Thermo Electron; Triple Red; VWR |
| Indomethacin | Sigma |
| Injection ports, Interlink system | BD Biosciences |
| Inorganic salts | Merck |
| Insect culture media | BD Biosciences; BioSource; Lonza; Hyclone; Invitrogen; MP Biomedicals; Novagen; Sigma |
| Instruments, dissecting | Fine Scientific Tools; Fisher; Roboz |
| Insulin | Lonza (Clonetics); CP Pharmaceuticals; Eli Lilly; Intergen; Invitrogen; Sigma |
| Insulin/transferrin/selenium (ITS) | BD Biosciences; Fisher Scientific; Invitrogen; Roche Diagnostics; Sigma |
| Interleukins | *See* Growth factors |
| Inverted microscope | Leica; Nikon; Olympus; Zeiss |
| Iodonitrotetrazolium (INT) violet | Sigma |
| Irgacure 2959 photoinitiator for photopatterning | Ciba Specialty Chemicals |
| Iscove's modified Dulbecco's medium (IMDM) | *See* Media |
| Isoenzyme electrophoresis analysis | ATCC; Charles River; DSMZ; ECACC |
| Isoenzyme electrophoresis analysis kit | Innovative Chemistry |
| Isoleucine | *See* Chemicals |
| Isometric force transducer (for muscle culture) | Radnoti |
| Isoproterenol hydrochloride | Sigma |
| Isotonic counting solution | Beckman Coulter; Scharfe; *see also* Media for PBS |
| Isotopes | *See* Radioisotopes |
| ITS Premix serum substitute | MP Biomedicals |
| Kanamycin | Invitrogen; MP Biomedicals; Sigma |
| KCl | *See* Chemicals |
| Keratin 19 (K19) antibody | Dako |
| Keratinocyte medium | Lonza; Cascade; Invitrogen; PromoCell; Sigma |
| Keratinocytes | ATCC, BioPredic; Cell Applications, ECACC, Invitrogen, Lonza, Millipore, Promocell, TCS Cellworks, Zen Biologicals; *see also* Specialized Cell Cultures and Table 22.1 |
| Ketamine | Henry Schein; Sigma |
| Ketanest | Parke-Davis |
| KGF, human recombinant | *See* Growth factors |
| KGM | *See* Keratinocyte medium |

| Item | Supplier |
|---|---|
| KH$_2$PO$_4$ | *See* Chemicals |
| Kodak X-OMAT film | Eastman Kodak |
| Kova slides | Hycor Biomedical |
| Krebs−Henseleit solution | Sigma |
| Labels | Triple Red; *see also* Cryolabels |
| Laboratory glassware washers | *See* Glassware washing machines |
| Laboratory suppliers, general | Camlab; Cole-Parmer; Fisher; Scientific Instrument Centre; Thermo Electron; Triple Red; VWR |
| Lab-Tek 8 chamber tissue culture slides | *See* Chamber slides |
| Lactalbumin hydrolysate | BD Biosciences (Difco); Invitrogen |
| Lactate dehydrogenase (LDH) kit viability assay | Roche Diagnostics; Sigma |
| Laemmli sample buffer | Bio-Rad |
| L-Alanine | *See* Chemicals |
| Laminar-flow hoods/cabinets | AES; Atlas Clean Air; Baker; Bigneat; Bioquell; Biosero; Contamination Control; Envair; Erlab; Germfree Labs; Kendro; Labcaire; LTE; Medical Air Technology (Kendro); MP Biomedicals; Nuaire; Safelab; SP Industries; Thermo Electron |
| Laminin | AMS Biotechnology (recombinant); BD Biosciences; Biomedical Technologies; Invitrogen; Lab Vision; Sigma; Stratech |
| Latex particles (cell sizing standard) | Beckman Coulter |
| Lead II nitrate | *See* Chemicals |
| Leibovitz L-15 medium without glutamine | *See* Media |
| Leighton tubes | Bellco; Corning; Techno Plastic |
| Lentiviral packaging | Addgene |
| Leucine | Sigma-Aldrich |
| LEUCOperm kit | Serotec |
| Levamisol | Serva |
| LHC medium | BioSource |
| Liberase | Roche Applied Science |
| LIF | *See* Growth factors |
| Lifecell tissue culture flask | Nexell Therapeutics |
| Liquid nitrogen freezers | Aire Liquide; Chart Biomed; Cryoservice; Jencons; Messer; Statebourne; Thermo Electron |
| Liquipette | Elkay; Oxford Worldwide |
| Lithium bromide | *See* Chemicals |
| Lithium carbonate | *See* Chemicals |
| Luer stub adaptors | BD Biosciences (Clay Adams) |
| Luer-Lok connectors | Altec; Applied Medical Technol., Harvard Apparatus; Popper |
| Lymphocyte preparation media | GE Healthcare (Amersham); Lonza (BioWhittaker); MP Biomedicals; Nycomed; Robbins Scientific; Sigma |
| Lysine HCl | *See* Chemicals |
| M199 | *See* Media |
| MACS | Dynal; Miltenyi; Stem Cell Technologies |
| Magnesium chloride | *See* Chemicals |
| Magnesium sulfate | *See* Chemicals |
| Magnetic cell sorting: *see* MACS | |
| Magnetic stirrers | Bellco; Camlab; Chemap; Cole-Parmer; Fisher; Hanna; Jencons; Lab-Line; Techne; Thomas; Wheaton |
| Magnifying viewers | Bellco; New Brunswick Scientific |
| Marcaine | NLS Animal Health |
| Marker beads for density centrifugation | GE Healthcare (Amersham) |
| Markers, ampoules | GA International; Radleys |
| Matrigel™ | BD Biosciences; Serva |
| Matrix | Accurate Chemical (assays); BD Biosciences; Biocolor (assays); Biodesign; Biomedical Technologies; BioSource; Cellon; Cook Biotech; Harbor Bio-Products; Imperial; Lab Vision; Matrix Technologies; NovaMatrix; Pierce; Protein Polymer; R&D Systems; Sigma; Stratech; TCS; Tebu-bio |
| May−Grünwald stain solution | Merck |
| McCoy's 5A medium | *See* Media |
| MCDB 105 | *See* Serum-free media |
| MCDB 153 | *See* Serum-free media |
| MCDB-170 and stocks | *See* Serum-free media |
| McIlwain tissue chopper | Campden Instruments |
| Media | ACM Biotech; AES; ATCC; Biochrom; BD Biosciences; Biosource; Caisson; Lonza; Cascade; CellGenix; CellGro-Mediatech; Connaught; Eurobio; Gemini; Genetic Research Instr.; Mediatech; MP Biomedicals; Irvine Scientific; Invitrogen; JRH; PAA; Perbio; PromoCell; Sigma; Sero-Med; StemCell Technologies; Wako |

| Item | Supplier |
| --- | --- |
| Media bags | American Fluoroseal; Cell Genix; DuPont; MP Biomedicals; Sigma |
| Medium 199 | *See* Media |
| Melanocytes | ATCC, BioPredic; Cell Applications; Clonagen; ECACC; Invitrogen; Lonza; Promocell, TCS Cellworks |
| MEM | *See* Media |
| Membrane syringe filters | *See* Filters |
| Mercaptoethanol | *See* Chemicals |
| MES buffered saline | Pierce |
| Mesenchymal stem cells (MSCs) | Befutur; Cell Applications; ECACC; Promocell; TCS Cellworks |
| Mesh filters and gauze | Dynal; Merck; Miltenyi; Sefar; Stanier; Teledyne Tekmar; *see also* Cell strainers |
| Mesothelial cells | Zen Biologicals |
| Metabolic monitoring | *See* Biosensors |
| Methacrylate, poly (2-hydroxyethyl) | Sigma |
| Methacrylate: 3-(trimethoxysilyl) propyl methacrylate | Sigma |
| Methanol | *See* Chemicals |
| Methionine | *See* Chemicals |
| Methocel | Dow Corning |
| Methylcellulose | Dow Corning; Fluka; Fisher; Sigma; Stem Cell Technologies |
| Metrizamide | Nycomed |
| MF-319 photopatterning/photetching developer | Microchem Corp. |
| $Mg^{2+}$ and $Ca^{2+}$-free Hanks's balanced salt solution (HBSS) | *See* Media |
| $MgCl_2 \cdot 6H_2O$ | *See* Chemicals |
| $MgSO_4 \cdot 7H_2O$ | *See* Chemicals |
| Microarray analysis | Affymetrix; Agilent; Bio-Rad; Brinkmann; Chemicon; Imaging Research; Nonlinear Dynamics; Schleicher & Schuell; Stratagene |
| Microbiological safety cabinets | Plas Labs; *see also* Laminar-flow hoods |
| Microcaps, microcapillary tubes | (Drummond) Thermo-Shandon; Fisher |
| Microcarriers | GE Healthcare (Amersham); Sterilin; Bio-Rad; Thermo Fisher (Nunc); Invitrogen; MP Biomedicals; Sigma; SoloHill; TCS |
| Microcentrifuge filter tubes | Corning |
| Microdensitometry | *See* Image analysis; Scanning cytometer |
| Micromanipulators | Brinkmann; Eppendorf; Leica; Nikon; Steris; Stoelting |
| Micrometer slides | Meiji Techno; SPI Supplies |
| Micropipettes | Alpha; Anachem; Applied Scientific; Sterilin; Biohit; Brinkmann; Corning; Elkay; Eppendorf; Genetic Research Instr.; Gilson; Jencons; Lawson Mardon Wheaton; Oxford; Roche; Thermo Electron |
| Microplate reader | *See* Plate readers |
| Micropore filters | *See* Filters, sterilization |
| Microscope incubation chambers | Buck Scientific; Imaging Associates; |
| Microscope slide containers | Corning; Raymond Lamb; Raven; Richardsons |
| Microscope slides | Bellco; Laboratory Sales; Lab-Tek; Microm; Thermo Fisher (Nunc); Propper; Raymond Lamb; Richardsons; VWR |
| Microscopes | Carl Zeiss; Leica; Olympus; Nikon; Prior |
| Microscopes, fluorescence | *See* Microscopes |
| Microtitration equipment | Anachem; Berthold; BioRad; Biospec; BMG Labtech; Brand; Camlab; Corning; Dynex; Elkay; ESA; Genetic Research Instr.; Gilson; Integra; Invitrogen; Molecular Devices; Thermo Fisher (Nunc); Nonlinear Dynamics; PerkinElmer; R & D Systems; Robbins; Techmate; Thermo Electron; VWR; Whatman; Zinsser |
| Microtitration plate centrifugation | Beckman Coulter; Thermo Fisher |
| Microtitration plate homogenizer | *See* MiniBead Beater |
| Microtitration plate readers | *See* Plate readers |
| Microtitration plate reagent dispensers | Thermo Electron Corporation |
| Microtitration plate sealers | *See* Plate sealers |
| Microtitration plates | Applied Biosystems; *see also* Culture dishes, flasks, and plates |
| Microtitration plates with removable wells | Invitrogen |
| Millex-SV syringe-driven filter unit, pore size 5 μm | Millipore |
| Millicell-HA filter holder inserts | Millipore |
| MiniBead Beater microplate homogenizer | Biospec; Daintree |
| Mini-bioreactors | *See* Bioreactors, mini |
| Minimal essential medium | *See* Media |
| Miniperm bioreactor | Greiner Bio-one; Sigma |

| Item | Supplier |
|---|---|
| Mitomycin C | Roche Diagnostics; Sigma |
| Mountants | Biomeda; Citifluor; Merck; Microm; RA Lamb; Statlab; Stephens Scientific; Vector; VWR |
| Mouse early passage cell lines | Lonza |
| Movette pipette | Invitrogen; VWR |
| MRA | MP Biomedicals |
| MSCGM | Lonza (Clonetics, BioWhittaker) |
| MSCs | *See* Mesenchymal stem cells |
| MTT: 3-[4,5-dimethylthiazol-2-yl]-2,5-diphenyl tetrazolium bromide | Sigma |
| MUC-1 (mucin) antibody | Chemicon |
| Multichamber slides | *See* Chamber slides |
| Multipoint pipettors | *See* Micropipettes and pipettors |
| Multiwell filter plates | Whatman |
| Multiwell plates | *See* Culture dishes, flasks, and plates |
| Mycoplasma detection kits | ATCC; Gen-Probe; Irvine Scientific; Lonza, Metachem; MP Biomedicals (*See also* Table 18.2) |
| Mycoplasma elimination | ATCC; DSMZ; ECACC |
| Mycoplasma nutrient agar | BD Biosciences; Sigma |
| Mycoplasma nutrient broth | BD Biosciences |
| Mycoplasma removal agents | Autogen Bioclear; Invivogen; Minerva Biolabs; Lonza; MP Biomedicals; Roche (*See also* Table 18.3) |
| Mycoplasma testing | ATCC; Bionique; BioReliance; Cambio; ECACC; DMSZ; Millipore; Mycoplasma Experience; SGS |
| Mycostatin, Nystatin | BioWhittaker; MP Biomedicals; Invitrogen; Sigma |
| Mylar plastic plate sealers | MP Biomedicals; Invitrogen |
| *myo*-Inositol | Sigma |
| $Na_2HPO_4 \cdot 7H_2O$ | *See* Chemicals |
| NaCl | *See* Chemicals |
| NaCl, 0.9% | Baxter |
| $NaHCO_3$ | *See* Chemicals |
| $NaHCO_3$ (7.5%) sterile | *See* Media |
| NaOH | *See* Chemicals |
| Naphthalene Black | Thermo Fisher; R.A. Lamb; *see also* Stains |
| Naphthol AS-MX | Serva; Sigma |
| NASA bioreactor | Cellon; Synthecon |
| Natrigel | BD Biosciences |
| Neck plug-controlled rate cooler | Taylor Wharton |
| Needles (for syringes) | *See* Syringes |

| Item | Supplier |
|---|---|
| Neural cells | Invitrogen; Lonza; Millipore; Neuromics, TCS Cellworks |
| Neural stem cells | Befutur; Millipore; Neuromics |
| Neutral protease (Dispase) | Roche Diagnostics |
| Neutral Red | TCS Biologicals |
| NGF | *See* Growth factors |
| Niacinamide | Sigma |
| Nigrosin | Sigma; TCS Biologicals |
| Nile Red | Sigma |
| Nitex nylon filter, 100 μm | Teledyne Tekmar (*See also* Mesh filters and gauze) |
| Nitrile gloves | Ansell; Cole-Parmer; Lab Safety Supplies; Radleys; Sentinel; Stoelting |
| Nitrogen freezers | *See* Freezers; liquid nitrogen |
| Nitrophenol standard solution | Sigma |
| Nitrophenyl phosphate | Sigma |
| Nonessential amino acids | Invitrogen; MP Biomedicals; Sigma |
| Normal goat serum | Vector |
| Nuclear Fast Red | Sigma |
| Nucleosides | Sigma |
| Nude mice | Charles River; Nippon CLEA |
| Nutridoma serum substitute | Roche Applied Science |
| Nutrient agar plates | Oxoid |
| Nutrient broths | BD Biosciences (Difco); Invitrogen; MP Biomedicals; TCS |
| Nyaflo membranes | Gelman Sciences |
| Nylon film: *see* Autoclavable bags and nylon film | |
| Nylon mesh filters and gauze | *See* Mesh filters and gauze |
| Nystatin | Sigma; Invitrogen |
| OCG 825-835, 934, photopatterning/photoetching developer | Olin-Ciba-Geigy |
| Oncostatin M | *See* Growth factors |
| Open pulled straws | LEC |
| Opticell culture chamber | Thermo Fisher (BioCrystal) |
| Opti-MEM, reduced serum medium | Invitrogen |
| Optiprep | Sigma |
| Oral epithelium | Millipore; TCS Cellworks |
| Orbital shaker | Bellco |
| Organ baths | Radnoti |
| Organ culture grids and dishes | BD Biosciences |
| Organotypic culture | ACM-Biotech; MatTek; Minucells; SkinEthic |
| ORIGEN™ DMSO freeze medium | IGEN International, Inc. |
| Osmium tetroxide | Plano; Roth; Sigma |
| Osmometer | Advanced Instruments; Analytical Technology; Gonotec; Nova Biomedical; Wescor |

| Item | Supplier |
|---|---|
| Osteoblasts | Cell Applications; ECACC; PromoCell; TCS Cell Works |
| Osteogenic protein 1 (OP-1; BMP-7) | See Growth factors |
| Ovens | Astell; Barnstead; Camlab; Dynalab; Harvard; Kendro; LEEC; LTE; Memmert; Precision Scientific; Scientific Instrument Centre; Thermo Electron |
| Oxygen probe | Microelectrodes |
| p53 antibodies: D0-1, PAb421, PAb240 | Oncogene Science |
| Packaging, sterile | See Sterile packaging |
| Packaging, transportation | Air Packaging Technologies; see also Sample containers for transportation |
| Pancreatic elastase | Sigma |
| Pancreatin | Invitrogen; Sigma |
| Panserin™ 401 serum-free medium | CoaChrom |
| Pantothenic acid | Sigma |
| Papain | Worthington; Sigma |
| Paraformaldehyde | Electron Microscopy Sciences; Merck; Sigma |
| Pastette | Alpha Laboratories; Sterilin; Elkay; Polytech; Richardsons |
| Pasteur pipettes, disposable | Alpha Laboratories; Invitrogen; Sterilin; Elkay; Invitrogen; Oxford Worldwide; Polytech; Richardsons; VWR |
| PBS tablets | Oxoid; see also Media for powder or solution |
| Pen/Strep Fungizone mix | See Media |
| Penicillin | See Media |
| Penicillin/streptamycin | See Media |
| Pentobarbital | Abbott |
| Pepsin | Sigma |
| Peptidase, dipeptidyl, IV (CD26) | Biosource; Neomarkers |
| Peptides: RGD, GRGDS, H-Gly-Arg-Gly-Asp-Ser-OH | Calbiochem |
| Penicillin | See Media |
| Percoll | GE Healthcare (Amersham); Sigma |
| Perfusion cartridges, polycarbonate | Advanced Tissue Sciences |
| Perfusion culture | Amicon; Bioptechs; Cellco; Endotronics; Microgon |
| Peristaltic pumps | See Pumps, peristaltic |
| Permeable caps | See Gas-permeable caps |
| Permount mounting medium | Fisher |
| Peroxidase blocking reagent | Cytomation; Dako |
| Petri dishes | See Culture dishes, flasks, and plates |
| Petriperm dishes | Sartorius |
| PGA scaffolds | Albany International |
| Pharmed tubing | Cole-Parmer |
| Phenol Red | Sigma |
| Phenylalanine | See Chemicals |
| Phenylenediamine (OPD) tablets | Sigma |
| Phenylisothiocyanate | Pierce |
| Phosphocreatine | Sigma |
| Phosphoethanolamine | Sigma |
| Photographic chemicals, films, and papers | Agfa; Eastman Kodak; Fuji; Ilford (and local camera shops) |
| Photoinitiators for photopatterning, Irgacure 2959 | Ciba Specialty Chemicals |
| Photopatterning/photoetching | Microchem Corp; Olin-CIBA-Geigy |
| Photopatterning/photoetching developer, MF-319 | Microchem Corp. |
| Phytohemagglutinin | GE Healthcare (Amersham); Sigma |
| PicoGreen | Molecular Probes |
| Pinning forceps | Fisher Scientific |
| Pipette bulbs | Cole-Parmer; Narang Medical; Sigma |
| Pipette cans | Bellco; Thermo Electron |
| Pipette cylinders or hods | Bel Art; Fisher; Thermo Fisher (Nunc); Radleys |
| Pipette plugger | Bellco; Camlab; Volac |
| Pipette tips, aerosol-resistant | See Pipettors and pipetting aids |
| Pipette washer, drier | Bel Art; Thermo-Shandon; Radleys |
| Pipettes | Alpha Laboratories; BD Biosciences; Bellco; Corning; Fisher; Thermo Fisher (Nunc); Sterilin |
| Pipettes, 100 mL | BD Biosciences |
| Pipettes, wide bore | Bellco |
| Pipettors and pipetting aids | Alpha Laboratories; Anachem; Applied Scientific; Barnstead; Bellco; Sterilin; Brand; Cole-Parmer; Corning; Daiger; Eppendorf; Fisher; Genetic Research Instr.; Gilson; Integra; Jencons; Lawson Mardon Wheaton; Matrix Technologies; Merck; Messer; Mettler-Toledo; MP Biomedicals; Polytech; Radleys; Rainin; Roche; Socorex; Thermo Electron; Whatman |
| Pituitary extract (PE) | Lonza (Clonetics); Cascade; Invitrogen; PromoCell |
| Plasma, equine | MP Biomedicals |
| Plasmids | Addgene |
| Plasmocin | Invivogen |
| Plastic test-tubes | See Centrifuge tubes |
| Plasticware | See Culture dishes, flasks, and plates |

| Item | Supplier |
|---|---|
| Plate readers | Bio-Tek; Berthold; Bio-Rad; BMG; Camlab; Dynex; Fisher; Merck; Messer; Molecular Devices; PerkinElmer; Thermo Electron |
| Plate sealers | Anachem; Brandell; Elkay; Greiner; MP Biomedicals; Porvair |
| Plates | *See* Culture dishes, flasks, and plates |
| Platinum-cured silicone tubing, for air/$CO_2$ exchange | Cole-Parmer |
| Pleated cartridge filter | Pall Gelman |
| Pluronic F-68 | Invitrogen; MP Biomedicals; Serva; Sigma |
| Polaron SC502 sputter coater | Fison |
| Poly HEMA: poly (2-hydroxyethyl) methacrylate | Sigma |
| Poly(ethylene glycol) diacrylate (PEGDA) | Nektar Transforming Therapeutics; Shearwater |
| Polyacrylamide gels | Bio-Rad; GE Healthcare (Amersham) |
| Polybrene | Sigma |
| Polycarbonate tubes & bottles | Thermo Fisher (Nunc) |
| Poly-D-lysine-coated dishes | BD Biosciences (Biocoat) |
| Polylysine, poly-D- and poly-L- | Biomedical Technologies; BD Biosciences; Sigma; Stratech |
| Polymyxin B | Invitrogen |
| Polyolefin heat shrink tubing | Appleton Electronics |
| Polypropylene jar, 30 ml | Thermo Fisher (Nunc) |
| Polystyrene flasks | *See* Culture dishes, flasks, and plates |
| Polyvinyl pyrrolidone | Calbiochem; Sigma; USB |
| Positive photoresists, OCG 825-835 St, Shipley 1813, or Shipley 1818 | Shipley |
| Potassium chloride | *See* Chemicals |
| Potassium dihydrogen orthophosphate | *See* Chemicals |
| Potassium ferricyanide, $K_3Fe(CN)_6$ | *See* Chemicals |
| Potassium ferrocyanide trihydrate, $K_4Fe_3(CN)_6 \cdot 3H_2O$ | *See* Chemicals |
| Potassium phosphate, monobasic | *See* Chemicals |
| Precept tablets | Johnson & Johnson |
| Pressure cooker, bench-top autoclave | Astell Scientific; Harvard Apparatus; LTE; Napco; Valley Forge |
| Pressure monitor | Hewlett-Packard |
| Pressure transducers | Gould; Maxim Medica |
| Pressure vessels | Alpha Laval; Millipore; Pall Gelman; Sartorius Stedim |
| Primaria® flasks and dishes | BD Biosciences |
| Primers | Applied Biosciences; Research Genetics; Riken |
| Progesterone in absolute ethanol | Sigma |
| Programmable freezer | Messer; Planer; Thermo Life Sciences (Cryomed); Statebourne |
| Proline | *See* Chemicals |
| Pronase | Calbiochem; Roche Diagnostics; Sigma |
| Pronectin | Protein Polymer Technologies |
| Propidium iodide | Sigma |
| Prostatic epithelial, fibroblastic, and smooth muscle cells | ATCC; Lonza |
| Protease inhibitors | Intergen; Roche Diagnostics; Sigma |
| Protective clothing | Alexandria Workwear; Lab Safety Supply; Sigma |
| Protein assay | Pierce |
| Proteinase K | Sigma |
| PTFE (Teflon) sheet (for disks) and tube (for cylinders) | Interplast; Plastim |
| Pump, bellows-style | Gorman-Rupp Industries |
| Pumps, peristaltic | Altec; Cole-Parmer; GE Healthcare (Amersham); Gilson; Michael Smith; Verderflex; Watson Marlow |
| Pumps, vacuum | BO; Cole-Parmer; Millipore; Pall Gelman; Varian |
| Putrescine 2HCl | Sigma-Aldrich |
| Pyridoxine HCl | Sigma-Aldrich |
| Qiagen Rneasy Mini Kit | Qiagen |
| QIAshredder | Qiagen |
| QuiAmp DNA mini kit | Qiagen |
| Quinacrine dihydrochloride | Sigma |
| Radioisotopes | GE Healthcare (Amersham); MP Biomedicals; PerkinElmer; Sigma |
| Random hexamers | Invitrogen |
| Rat early passage cell lines | Lonza |
| Rat serum | Pel-Freeze |
| Rat tail collagen | BD Biosciences |
| Rat tails | Biotrol |
| Rats | Charles River; Harlan Sera-Lab |
| RCCS (rotatory cell culture system) | Synthecon; Cellon |
| RD emulsion 9964.40 | Dow Corning |
| RDO decalcifying agent | Apex Engineering |
| Real-time monitoring | *See* Biosensors |
| Recombinant growth factors | Abbiotec; *See also* Growth factors |
| Recombinant matrix proteins: *see* Fibronectin, Laminin | |
| Recording thermometer | Comark; Cole-Parmer; Fisher; Harvard; Grant; Omega; Pierce; Rustrack |
| Refractometers | Beckman; Bellingham & Stanley; Cole-Parmer; Mettler Toledo; Reichert; Thermo Fisher; Technika |
| Repeating pipettor | Brand; Cole-Parmer; Popper; Radleys |

| Item | Supplier |
|---|---|
| Reservoirs | *See* Aspirators |
| Retinoic acid | Sigma |
| Retinol acetate, all *trans* | Sigma |
| Reusable in-line filter assembly | *See* Filters |
| Reverse osmosis | Barnstead; Elga; Millipore; U.S. Filter |
| Reynolds's lead citrate | PI Supplies |
| RGD, GRGDS | Calbiochem |
| Rhodamine B | Sigma |
| Riboflavin | Sigma |
| RNA preparation kit | Qiagen |
| *RNAimage* kits | GenHunter |
| RNAseZap® | Ambion |
| RNeasy mini kit | Qiagen |
| Robotics | Beckman Coulter; Bio-Rad; Cytogration; Gilson; Hudson Robotics; Innovative Cell Technologies; The Autometion Partnership (TAP) |
| Roccal | Henry Schein Rexodent; Pfizer |
| Rocker platform | Bellco |
| Roller bottle rack | Argos; Bellco; Genetic Research Instr.; Integra; Lawson Mardon Wheaton; New Brunswick Scientific; Robbins |
| Roller bottles, glass | Bellco |
| Roller bottles, plastic | Applied Scientific; BD Biosciences; Caisson; Corning; Integra |
| Roller drum | Genetic Research Instr.; New Brunswick |
| Rompun, xylazine hydrochloride | Bayer Vital |
| Rotameters | *See* Gas flow meters |
| Rotary shaker | New Brunswick |
| Rotatory cell culture systems | Cellon; Sigma; Synthecon |
| Round bottom test tubes with cap | Alpha Laboratories; BD Biosciences (Falcon); Bellco; Corning; Kimble-Kontes |
| RPMI 1640 | *See* Media |
| Rubber pipette bulb | Bel Art; Bellco; Sterilin; Cole-Parmer; Fisher; Jencons Scientific; VWR |
| Biological safety cabinets | *See* Laminar-flow hoods |
| Safety products | Air Sea Atlanta; Altec; Bel-Art; Cellutech; Cin-Made; Jencons; Kimberly-Clark; Lab safety Supply; Saf-T-Pak |
| Safranin O | Sigma |
| Sample containers for transportation | Saf-T-Pak; Thermo Fisher (Nunc); Air Sea Atlanta; Cellutech; Cin-Made Corp |

| Item | Supplier |
|---|---|
| | (*see also* www.uos.harvard .edu/ehs/bio_bio_shi.shtml and www.ehs.ucsf.edu/ Safety%20Updates/Bsu/ Bsu5.pdf) |
| SBTI (soybean trypsin inhibitor) | Sigma |
| Scaffolds, tissue engineering | Albany; Cook Biotech; EBI; Minucells; Zimmer |
| Scalpels | *See* Dissecting instruments |
| Scanning cytometer | Beckman Coulter; Compucyte; *see also* Microscopes |
| Scanning electron microscopes (SEM) | ElectroScan; JEOL |
| SCID mice | Charles River |
| Scintillation fluid | BS & S; National Diagnostics; GE Healthcare (Amersham); New England Nuclear (DuPont); PerkinElmer (Packard) |
| Scintillation vials, minivials | PerkinElmer (Packard) |
| Scion-Image software (time-lapse recording) | Scion |
| Scissors | *See* Dissecting instruments |
| Scrynel NYHC nylon gauze | Merck; *see also* Nylon mesh filters |
| Selenious acid | Merck; Sigma |
| SEM (scanning electron microscope) | ElectroScan; JEOL |
| Semipermeable nylon film | *See* Sterile packaging |
| Sequencing gel filter paper | Bio-Rad |
| Sequencing kit | GE Healthcare (Amersham) |
| Serine | *See* Chemicals |
| Serum | ATCC; Biochrom; Biosource; Caisson; Lonza; Invitrogen; Gemini; Globepharm; Harlan Seralab; HyClone; Invitrogen; Irvine; Metachem; MP Biomedicals; PAA; Perbio; PromoCell; Serologicals; Sigma; Sterile Systems; TCS Biologicals |
| Serum replacements | *See* Serum substitutes |
| Serum substitutes | Bayer; Celox; Invitrogen (Biosource); Invitrogen; Irvine; Lonza; Metachem; MP Biomedicals; Protide; Roche Applied sciences; Sigma |
| Serum-free media | AthenaES; Atlanta Biologicals; BD Biosciences; Biosource; Biofluids; Cascade Biologicals; Cell Applications; Cell Culture Services; CellGenix; CellGro; Clonagen; CoaChrom; |

| Item | Supplier |
|---|---|
|  | DuPont (NEN); Hyclone; Hycor; Invitrogen; Irvine; Lonza (Clonetics, Hyclone); Mediatech; Metachem; Millipore; MP Biomedicals; PAA; PeproTech; PromoCell; Roche Applied Science; Sigma (JRH Biosciences); Stem Cell Technologies; Stratech; TCS CellWorks (*see also* Table 9.3); Zen Biologicals |
| Shaking incubator | Camlab; New Brunswick; Radleys |
| Sharps bins | Altec; Cole Parmer; Lab Safety Supply; Stoelting; VWR-Jencons |
| Shipley 354, 1813, 1818; photopatterning developers | Microchem Corp. |
| Sieve with 200-mm mesh | BD Biosciences; Tekmar |
| Sieves | Markson; Retsch; *see also* Mesh filters and Gauze |
| Silica gel | Fisher; Merck |
| Silicone grease | Edwards; British Oxygen; Dow Corning; Girovac |
| Silicone lubricant, inert | Dow Corning |
| Silicone rubber adhesive | Dow Corning; GE Silicones |
| Silicone rubber sheet | Nusil; Silicone Specialty Fabricators |
| Silicone rubber stoppers | Sterilin |
| Silicone tubing | Altec; Sterilin; Dow Corning; Cole-Parmer; Nusil; PAW BioScience Products; Thomas; Watson-Marlow |
| Silicones | Fisher; Merck; Serva; Sigma |
| Siliconizing solution (Sigmacote) | Sigma |
| Silver nitrate | *See* Chemicals |
| SIT serum substitute | Sigma |
| Skeletal muscle growth medium (SKGM) | Lonza (Clonetics) |
| Skeletal myoblasts | Cell Applications; ECACC; Lonza; Promocell; ReachBio; TCS Cellworks; Zen Biologicals |
| Slide boxes, light tight | Bel-Art; Cole-Parmer; Raven; Raymond Lamb |
| Slide boxes, light tight | BD Biosciences; Raven Scientific |
| Slide containers for emulsion (Cyto-Mailer) | Lab-Tek; Fisher; Statlab |
| Slide culture dishes | Vivascience; *see also* Culture dishes, flasks, and plates |
| Slide flasks | Thermo Fisher (Nunc); *see also* Chamber slides |
| Slides, glass | *See* Microscope slides |
| Slow turning lateral vessel (STLV) | Synthecon; Cellon |

| Item | Supplier |
|---|---|
| SLTV (slow turning lateral vessel) | Synthecon; Cellon |
| S-MEM | *See* Media |
| Smooth muscle alpha actin antibody | DAKO |
| Smooth muscle cells | ATCC; Cell Applications; ECACC; Invitrogen; Lonza; Promocell; TCS Cellworks |
| Sodium alginate | ISP Alginate |
| Sodium azide | Sigma |
| Sodium butyrate (NaBt) | Sigma |
| Sodium carbonate | *See* Chemicals |
| Sodium chloride | *See* Chemicals |
| Sodium citrate | *See* Chemicals |
| Sodium deoxycholate | Sigma |
| Sodium formate | Fisher |
| Sodium hydroxide | Sigma |
| Sodium lauryl sulfate (SLS; sodium deoxycholate) | Sigma |
| Sodium orthovanadate, $Na_3VO_4$ | Sigma |
| Sodium phosphate, dibasic heptahydrate | *See* Chemicals |
| Sodium pyruvate | *See* Chemicals |
| Sodium pyruvate, sterile solution | *See* Media |
| Sodium selenate | Sigma |
| Sodium selenite | Sigma |
| Sodium thiosulfate | Sigma |
| SonicSeal slides, 4-well | Thermo Fisher (Nunc) |
| Sorbitol | Sigma |
| Soybean trypsin inhibitor (SBTI) | Sigma |
| Spatulas | Fisher; VWR |
| Specialized cell cultures | ATCC; Amsbio; BD Biosciences; Befutur; Cascade; Cell Applications; CellGenix; Cell Systems; ECACC; Invitrogen; Lonza; Millipore; Neuromics; PromoCell; SkinEthic; TCS Cellworks; SkinEthic; Zen Biologicals (*see also* Table 22.1) |
| Specimen containers | Alpha Laboratories; Corning; Sterilin; Thermo Fisher (Nunc); VWR; *see also* Sample containers for transportation |
| Spectra Max 250 microplate spectrophotometer | Molecular Devices; *see also* Plate readers |
| Spinner flasks | *See* Stirrer flasks |
| Spring scissors | Fine Science Tools Inc.; VWR; *see also* Dissecting instruments |
| SPSS (version 10.0) statistical analysis software | SPSS |
| Sputter coater | Gatan; Fison |
| Stainless steel mesh | BD Biosciences; *see also* Mesh filters and gauze |
| Stains | Fisher; Merck; Molecular Probes; MTR; Sigma; TCS Biologicals |
| Stanzen Petri dishes | Greiner |

| Item | Supplier |
|------|----------|
| Steam-permeable nylon film | *See* Autoclavable nylon film and bags |
| Stem cell factor (SCF) | *See* Growth factors |
| Stem cells | Befutur; Cell Applications; Clonagen; ECACC; Neuromics; Promocell; TCS Cell works; Thermo Fisher; Zen Biologicals |
| Stem cells (isolation, markers, preservation) | Chemicon; Geron; Metachem; Miltenyi; Origen; Roche; Stem Cell Sciences; Universal Biologicals |
| Stereomicroscopes | Leica; Nikon; Olympus; Zeiss |
| Sterile filtration | *See* Filters, sterilization |
| Sterile-indicating tape | *See* Sterility indicators |
| Sterile packaging: cartridge paper, semipermeable nylon film | Applied Scientific; Buck Scientific; KNF Corp; Portex; Roth |
| Sterility indicators | Altec; Alfa Medical; Appleton Woods; Applied Scientific; Bennett; Jencons; Popper; Raven; Roboz; SGM Biotech; SPS Medical; Shamrock; Stoelting; Surgicon |
| Sterilization bags | *See* Sterile packaging |
| Sterilization film | *See* Sterile packaging |
| Sterilizers | *See* Autoclaves; Ovens |
| Sterilizing agents | *See* alcohol, Betadine, disinfectants |
| Sterilizing and drying oven | *See* Ovens |
| Sterilizing filters | *See* Filters, sterilization |
| Sterilizing tape (indicator) | *See* Sterility indicators |
| Stills | Corning; Jencons; Steris |
| Stirrer flasks | Bellco; Corning; Genetic Research Instr.; Integra; Lawson Mardon Wheaton; Techne; *see also* Magnetic stirrers |
| Stirrers | *See* Magnetic stirrers |
| Straws for cryopreservation | LEC |
| Streptavidin-FITC | Dako Diagnostics |
| Streptomycin | *See* Media |
| Sucrose | Fluka; Sigma |
| Sulfo-NHS, *N*-hydroxysulfosuccinimide | Pierce |
| Superscript amplification system | Invitrogen |
| Supplemented keratinocyte defined medium (SKDM) | *See* Keratinocyte growth medium |
| Surgical gauze | Johnson & Johnson; Kendall |
| Surgical Instruments | *See* Dissecting instruments |
| Surgilube | Fougera |
| Suture, Dacron® | Davis and Geck |
| Suture, Dexon® | Davis and Geck |
| Suture, silk | Harvard Apparatus |
| Swiss 3T3 cells | ATCC |
| Sylgard plastic | Dow Corning |

| Item | Supplier |
|------|----------|
| Synoviocytes | Cell Applications; Clonagen; ECACC |
| Syringe filters | *See* Filters |
| Syringe needles, 22 g | *See* Syringes |
| Syringe pump | Harvard Apparatus |
| Syringes | Baxter; BD Biosciences; Gillette; Popper (and general laboratory suppliers) |
| Syringe-tip filters | Microgon; Millipore; Pall-Gelman; Sartorius; *see also* Filters |
| TCA (trichloroacetic acid) | *See* Chemicals |
| TCM, TCH serum substitutes | Celox; MP Biomedicals |
| TEER measurement | Applied Biophysics; BD Biosciences; World Precision Instruments |
| Temperature controllers | *See* Thermostats, proportional controllers |
| Temperature indicator strips | *See* Sterility indicators |
| Temperature recorders | *See* recording thermometers |
| Tension/compression system for muscle culture | Instron |
| Tetramethylbenzidine | Research Diagnostics |
| TGF-β1 | R&D Systems; Research Diagnostics |
| Thermalog sterility indicator | Bennet; Popper |
| Thermanox | Thermo Fisher (Nunc); *see also* Coverslips; plastic |
| Thermometers, recording | Comark; Cole-Parmer; Fisher; Harvard; Grant; Omega; Pierce; Rustrack |
| Thermostats, proportional controllers | Controls & Automation; Fisher; Napco |
| Thiamine-HCl | Sigma |
| Thimerosal | Sigma |
| Thioctic acid | Sigma |
| Thioglycerol | Sigma |
| Three-way stop-cocks | Baxter Healthcare; Sherwood-Davis & Geck |
| Threonine | *See* Chemicals |
| Thymidine | Sigma |
| Time-lapse observation chambers | Bioptechs; Buck Scientific; Carl Zeiss; Intracel; Life Imaging Services; MatTek |
| Time-lapse video | Applied Biosystems; Dage-MTI; Hamamatsu; Imaging Associates; Sutter Instruments; *See also* CCD cameras; video camera; video recorder; Protocol 27.4 |
| Tissue culture flasks | *See* Culture dishes, flasks, and plates |
| Tissue culture inserts | *See* Filter well inserts |
| Tissue culture media | *See* Media |
| Tissue culture plastic flasks and dishes | *See* Culture dishes, flasks, and plates |
| Tissue grinder (pellet pestle mixer) | Kimble-Kontes |
| Tissue sealant, fibrin, Tisseel VH | Baxter |

| Item | Supplier |
| --- | --- |
| Tissue slicers | Alabama Research & Development |
| Tissue Tek-O.T.C embedding compound | Sakura Finetek (Raymond Lamb in UK); Sakura Finetek Europe |
| TOC meter | *See* Total organic carbon meter |
| Tocopherol | Sigma |
| Toluidine Blue | Sigma |
| Total organic carbon (TOC) meter | Millipore; Sartec; Thermo Electron |
| TPA, 12-*O*-tetradecanoyl phorbol-13-acetate | Sigma |
| Transepithelial electrical resistance (TEER) | Applied Biophysics; WPI |
| Transfer pipettes | *See* Pasteur pipettes (disposable); pipettors |
| Transferrin | BD Biosciences; Genway Biotech (recombinant); Sigma |
| Transmission electron microscope | Philips |
| Transplantation chamber (epidermis) | Greiner |
| Transportation containers | Air Packaging Technologies; Air Sea Atlanta; Altec; Bel-Art; Cellutech; Cin-Made; Jencons; Kimberly-Clark; Lab Safety Supply; Thermo Fisher (Nunc); Saf-T-Pak |
| Transwell inserts | Corning; *see also* Filter well inserts |
| Trichlorotrifluoroethane | Sigma |
| Triethylamine | Fisher; Sigma |
| Triiodothyronine | Sigma |
| Tris | *See* Chemicals |
| Tritiated thymidine | NEN Life Sciences (DuPont) |
| Triton X-100 | Sigma; Fisher |
| TRIzol | Invitrogen |
| Trypan blue viability stain | Merck; *see also* Media |
| Trypsin | Sigma; Worthington; *see also* Media |
| Trypsin inhibitors | Biosource; Cascade; Sigma; Serologicals |
| Trypsin replacements/substitutes | Hyclone; Innovative Cell Technologies; Invitrogen; Perbio; Sera-Lab; Sigma |
| Trypsin/ETDA | *See* Media |
| Tryptophan | *See* Chemicals |
| Tryptose phosphate broth | BD Biosciences (Difco); Invitrogen; Oxoid |
| Trypzean | Sigma |
| Tube rotator | New Brunswick Scientific |
| Tuffryn filters | Pall-Gelman |
| Tween 20, 80 | Sigma |
| Tylosin | MP Biomedicals |
| Tyrode's salt solution (TBSS) | *See* Media |
| Tyrosinase (C-19) goat anti-human antibody | Santa Cruz Biotechnology |
| Tyrosine | *See* Chemicals |
| Ultrafoam collagen sponge | Davol |
| Ultra-low attachment plates, 35 mm | Corning (Costar) |
| Ultramicrotome | Leica |
| Ultra-TMB (3, 3′, 5, 5′-tetramethylbenzidine) | Research Diagnostics |
| Ultroser G serum substitute | Invitrogen; *see also* Serum substitutes |
| Universal containers | Sterilin; Camlab; Corning; Elkay; Thermo Fisher (Nunc) |
| Uranyl acetate | Plano; SPI Supplies |
| Urea | *See* Chemicals |
| Urea nitrogen analysis | Stanbio Labs |
| UV light sources | Cole-Parmer; UVP |
| Vacuum pump | *See* Pumps, vacuum |
| Validation of cell lines | ATCC; BioReliance; Cellmark; ECACC; LGC Standards |
| Valine | *See* Chemicals |
| Vectashield | Vector Laboratories |
| Vectashield mounting medium with DAPI | Vector Laboratories |
| Vectastain Elite ABC Kit | Vector Laboratories |
| Vectastain Quick Kit | Vector Laboratories |
| Ventrex serum substitute | JRH Biosciences; *see also* Serum substitutes |
| Versene: *see* EDTA | |
| Viability assays | ATCC; Molecular Probes |
| Viability kit | *See* Media |
| Viability stains | *See* Media |
| Vials for freezing cells | *See* Ampoules |
| Vibratome slicer | Campden Instruments |
| Video camera | *See* CCD cameras; Digital cameras; Microscopes |
| Video systems | *See* Time-lapse video |
| Villin | Novacastra Laboratories; Santa Cruz Biotechnology; Chemicon |
| Vimentin antibody | Sigma |
| Vinblastine | Sigma |
| Vinyl tape | 3M (general laboratory suppliers) |
| Virkon | Day-Impex; Scientific Laboratory Supplies |
| Vitamin B$_{12}$ | Sigma |
| Vitamins | Sigma; *See also* Media |
| Vitamins, sterile solution | *See* media |
| Vitrogen | BD Biosciences; *see also* Collagen |
| Vitronectin | BD Biosciences; Biosource International |
| von Willebrand factor (factor VIII) antibody | Roche Diagnostics |
| Vortex mixer | Gallenkamp; Fisher; general laboratory suppliers |
| Wafer tweezers | Fluoroware |
| Water baths | Camlab; Cole-Parmer; Grant; Polyscience; SciGene; Thermo Electron |

| Item | Supplier |
|---|---|
| Water for tissue irrigation | Baxter |
| Water purification | Applied Biosciences; Barnstead; Elga; Genetic Research Instr.; High-Q; Millipore; Purite; Triple Red; U.S. Filter; Vivendi; Whatman |
| Wave bioreactor | GE Heathcare; Sartorius Stedim; *see also* Culture bags |
| Wavemaker32, software for fatigue testing of ligament | Instron |
| Weise buffer | Merck |
| Wescodyne | Steris |
| Williams medium E | Invitrogen; Sigma |
| Winged infusion set syringes | Merck |
| WST-1 cytotoxicity indicator stain | Serva |
| X-gal | Roche |
| X-ray film | Eastman Kodak; Fuji |
| Xylazine | NLS Animal Health; Sigma |
| Xylene | *See* Chemicals |
| ZnSO$_4$·7H$_2$O | *See* Chemicals |
| Zyderm 2 collagen implant | Insmed; Allergan |

# APPENDIX III

# Suppliers and Other Resources

Addresses, telephone, emails, and fax numbers are not supplied as these vary from one country to another and are available on the websites. Company names tend to change frequently due to mergers and takeovers; if the original URL is still functional (or was at the time of writing), it is still provided as this helps to navigate to the correct site of the new parent company and often makes use of a brand name which has been retained. However, some of these URLs may become redundant in due course so a cross-reference to the parent company is also provided ("*see also* . . ."). Where the URL is already redundant, the new URL is provided against the original name. This list is not intended to be comprehensive but has merely accumulated from companies cited in this and related publications. Many other websites are available that are more comprehensive (*see* introduction to Appendix II).

| Name | URL |
| --- | --- |
| ABB Kent-Taylor | www.tmseurope.co.uk |
| Abbiotec | www.abbiotec.com |
| Abbott Laboratories | www.abbott.com |
| Abcam | www.abcam.com |
| ABD Serotec | www.abdserotec.com |
| Abgene (*see also* Thermo Fisher) | www.abgene.com |
| Accuramatic | www.accuramatic.co.uk |
| Accurate Chemical & Scientific | www.accuratechemical.com |
| Accuri Cytometers | www.accuricytometers.com/welcome/accuri |
| Activox Human Liver Cell Systems | www.activtox.com |
| Adam Equipment | www.adamequipment.com |
| Addgene | www.addgene.org |
| ADMET Technologies | www.admettechnologies.com |
| Advanced Instruments | www.aitests.com |
| AES Chemunex | www.aeslaboratoire.com |
| Affymetrix | www.affymetrix.com |
| Agar Scientific | www.agarscientific.com |
| Agilent Technologies | www.chem.agilent.com |
| Air Liquide | www.dmc.airliquide.com |
| Air Packaging Technologies | www.airbox.com |
| Air Products & Chemicals, | www.airproducts.com |
| Air Sea Atlanta | www.airseaatlanta.com |
| Ajinomoto Company | www.ajiaminoscience.com |
| Alabama Research & Development | www.alspi.com/Slicer.htm |
| Albany International Research | ww3.albint.com |
| Alconox | www.alconox.com |
| Aldrich | www.sigma-aldrich.com |
| Alexis | www.alexis-corp.com |
| Alfa Laval | local.alfalaval.com |
| Alfa Medical | www.sterilizers.com |
| Alpco | www.alpco.com |
| Alpha Laboratories | www.alphalabs.co.uk |

| Name | URL |
| --- | --- |
| Altec Products | www.altecweb.com |
| Ambion | www.ambion.com |
| AME Bioscience | www.amebioscience.com |
| American Association for Cancer Research | www.aacr.org |
| American Fluoroseal | www.toafc.com |
| American Society for Cell Biology | www.ascb.org |
| American Type Culture Collection | www.atcc.org |
| Amersham Biosciences | www4.gelifesciences.com |
| Amgen | www.Amgen.com |
| Amicon: See Millipore | www.millipore.com |
| Amprotein | www.amprotein.com |
| Amsbio | www.amsbio.com |
| Anachem | www.anachem-ltd.com |
| Anachem- Flexus | www.anachem-instruments.co.uk/flexus |
| Analox | www.analox.net |
| Analytical Technology | www.analyticaltechnology.com |
| Anglia DNA Bioservices, | www.angliadna.co.uk |
| Ansell Healthcare | www.ansell.com |
| Anton Paar | www.anton-paar.com |
| Apex Engineering Products | www.rdo-apex.com |
| Appleton Woods | www.appletonwoods.co.uk |
| Applied Biophysics | www.biophysics.com |
| Applied Biosystems | www.appliedbiosystems.com |
| Applied Medical Technology | www.applied-medical.co.uk |
| Applikon | www.applikon.co.uk |
| Apredica | www.apredica.com/Genotox |
| Argos Technologies | www.argos-tech.com |
| Associates of Cape Cod | www.acciusa.com |
| Astell Scientific | www.astell.com |
| Asterand | http://solutions.asterand.com |
| ATCC | www.atcc.org |
| AthenaES | www.athenaes.com |
| Atlanta Biologicals | www.atlantabio.com/ |
| Atlas Clean Air | www.atlascleanair.com |
| ATR: Appropriate Technical Resources | www.atrbiotech.com |
| Austral Biologicals | www.australbiologicals.com |
| Autoclude | www.verderflex.com |
| Autogen Bioclear | www.autogenbioclear.com |
| Axon Medchem | www.axonmedchem.com |
| Axxora LLC | www.axxora.com |
| Baker | www.bakerco.com |
| Barnstead Thermolyne | www.thermoscientific.com |
| Baxter HealthCare | www.baxter.com |
| Bayer | www.bayer.com |
| BD Biosciences | www.bdbiosciences.com |
| Beckman Coulter | www.beckman.com |
| Becton Dickinson | www.bdbiosciences.com |
| Befutur | www.befutur.com |
| Bel-Art Products | www.bel-art.com |
| Bellco Glass | www.bellcoglass.com |
| Bellingham and Stanley | www.bellinghamandstanley.com |
| Bennett Scientific | www.bennett-scientific.com |
| Berthold Technologies | www.bertholdtech.com |
| Bibby Scientific | www.bibby-scientific.com |
| Bigneat | www.bigneat.com |
| Binder GmbH | www.binder-world.com |
| BioCarta | www.biocarta.com |
| Biochrom | www.biochrom.de |
| Biocolor | www.biocolor.co.uk |
| Biocote | www.biocote.com |
| BioCrystal | www.opticell.com |
| Biodesign International | http://meridianlifescience.com |
| Biofluids (see Invitrogen) | www.invitrogen.com |
| Biogene | www.biogene.com |
| Biohit | www.biohit.com |
| BioLife Solutions | www.biolifesolutions.com |
| Biomedical Technologies | www.btiinc.com |
| Biomet | www.biomet.co.uk |
| Bionique Testing Laboratories, | www.bionique.com |
| BioPredic | www.biopredic.com |
| Bioptechs | www.bioptechs.com |
| Bioquell | www.bioquell.com |
| Bio-Rad Laboratories | www.bio-rad.com |
| BioReliance | www.bioreliance.com |
| Biosero | www.bioseroinc.com |
| Biosource International | www.invitrogen.com |
| Biospec | www.biospec.com |
| BioSupplyNet | www.biosupplynet.com |
| Biotech Inst, | www.biotinst.demon.co.uk |
| BioTek Instruments | www.biotek.com |
| Biovest International | www.biovest.com |
| BMG Labtech | www.bmglabtech.com |
| BOC | www.boc-gases.com |
| Boehringer Mannheim | www.roche-applied-science.com |
| Boro Labs (see Labcold) | www.labcold.com |
| Brady Corporation | www.bradyeurope.com |
| Brand | www.brand.de |
| Brandel | www.brandel.com |
| BrandTech | www.brand.de |
| Braun B Biotech (see Sartorius) | www.sartorius-stedim.com |
| Brinkmann Instruments | www.brinkmann.com |

| Name | URL |
| --- | --- |
| British Association for Cancer Research (BACR) | www.bacr.org.uk |
| British Society for Cell Biology | www.bscb.org |
| Bruce Medical Supply | www.brucemedical.com |
| Buck Scientific | www.bucksci.com |
| Burge Scientific | www.arrowmight.co.uk/burge |
| C.R. Bard | www.crbard.com |
| Caisson Laboratories, Inc. | www.caissonlabs.com |
| Calbiochem-Novabiochem | www.merckbiosciences.com |
| Cambio | www.cambio.co.uk |
| Cambrex | www.lonza.com |
| Cambridge BioScience | www.bioscience.co.uk |
| Camlab | www.camlab.co.uk |
| Campden Instruments | www.campden-inst.com |
| Carl Roth | www.carlroth.com |
| Carl Zeiss | www.zeiss.com |
| Cascade Biologics (*see also* Invitrogen) | www.cascadebio.com |
| Cedex (*see* Roche Innovatis) | www.innovatis.com |
| Cell Applications | www.cellapplications.com |
| Cell Biolabs | www.cellbiolabs.com |
| Cell Sciences | www.cellsciences.com |
| CellBank Australia | www.cellbankaustralia.com |
| Cellex | www.biovest.com |
| CellGenix Technologie Transfer | www.cellgenix.com |
| Celliance | www.millipore.com |
| Cellmark | www.cellmark.co.uk |
| Cellon | www.cellon.lu |
| CellStat | www.cellstat.com |
| Cellutech. | www.pollutiononline.com |
| CellzDirect (*see also* Invitrogen) | www.cellzdirect.com |
| Celox | www.celoxmedical.com |
| Celsis In vitro Technologies | www.celsis.com |
| Charles Austen Pumps | www.charlesausten.com |
| Charles River Laboratories | www.criver.com |
| Chart Biomed | www.chartbiomed.com |
| Chemicon (*see also* Millipore) | www.chemicon.com |
| Chemometec | www.chemometec.com |
| Chip-Man Technologies | www.chipmantech.com |
| Chiron | www.chiron.com |
| Chondrex | www.chondrex.com |
| Ciba | www.ciba.com |
| Cin-Made Corp. | www.sonoco.com |
| Clonagen | www.clonagen.com |
| Clonetics | www.lonza.com |
| Coachrom Diagnostica GmbH. | www.coachrom.com |
| Codman | www.codman.com |
| Cole-Parmer Instrument | www.coleparmer.com |
| Comark | www.comarkltd.com |
| CompuCyte | www.compucyte.com |
| Computer Imprintable Label Systems | www.cils-international.com |
| Contamination Control Products | www.ccpcleanroom.com |
| Cook Biotech, | www.cookgroup.com |
| Coriell Cell Repository (CCR) | ccr.coriell.org |
| Corning, Life Sciences | www.corning.com/Lifesciences |
| Costar | www.corning.com/Lifesciences |
| Coulter | www.beckman.com |
| Covance | www.covance.com |
| Cryomed | www.thermoscientific.com |
| Cryoservice | www.cryoservice.co.uk |
| CryoTrack | www.cryotrack.com |
| Cytogration | www.cytogration.com |
| Cytomatrix | www.cytomatrix.com |
| Dage-MTI | www.dagemti.com |
| Daigger | www.daigger.com |
| Dako | www.dako.com |
| Damon IEC | www.thermoscientific.com |
| Davol | www.davol.com |
| Day-Impex | www.day-impex.co.uk |
| Decon Laboratories | www.decon.co.uk |
| Deutsche Sammlung von Mikroorganismen und Zellkulturen (DSMZ) | www.dsmz.de |
| Dianova | www.dianova.de |
| DiaSorin | www.diasorin.com |
| Difco (*see* BD Biosciences) | www.bdbiosciences.com |
| Division of Signal Transduction Therapy, University of Dundee, Scotland | www.lifesci.dundee.ac.uk |
| DNA Bioscience | www.dna-bioscience.co.uk |
| DNA Solutions | www.DNAsolutions.com |
| Domnick Hunter Scientific | www.domnickhunter.com/scientific |
| Don Whitley Scientific | www.dwscientific.co.uk |
| Dow Corning | www.dowcorning.com |
| DSMZ | www.dsmz.com |
| Dynal Biotech | www.invitrogen.com |

| Name | URL |
|---|---|
| Dynalab | www.dynalabcorp.com |
| Dynex Technologies | www.dynextechnologies.com |
| East Coast Biologics | www.eastcoastbio.com |
| Eastman Kodak | www.kodak.com |
| EBI, L.P. | www.biomet.co.uk |
| ECACC | www.hpacultures.org.uk/ collections/ecacc.jsp |
| Econo-med | www.econo-med.com |
| Edwards | www.edwardsvacuum.com |
| Electrolab | www.electrolab.biz |
| Electron Microscopy Services (EMS) | www.emsdiasum.com/ems |
| Elga LabWater | www.elgalabwater.com |
| Elkay Laboratory Products | www.elkay-uk.co.uk |
| EMD Biosciences | www.merckbiosciences.co.uk |
| Envair | www.envair.co.uk |
| Enzo Life Sciences | www.enzo.com |
| Eppendorf | www.eppendorf.com |
| Erlab DFS | www.erlab-dfs.com |
| ESA | www.esainc.com |
| ESACT-UK | www.esactuk.org.uk |
| ESCO | www.escoglobal.com |
| Essen Instruments | www.essen-instruments.com |
| Eurobio | www.eurobio.fr/ |
| European Collection of Cell Cultures (ECACC) | www.hpacultures.org.uk/ collections/ecacc.jsp |
| European Life Scientist Organisation | www.elso.org |
| European Society for Animal Cell Technology (ESACT) | www.esact.org |
| European Society of Toxicology In Vitro (ESTIV) | www.estiv.org |
| European Tissue Culture Society (ETCS) | www.etcs.info |
| European Tissue Culture Society, UK Branch (ETCS-UK) | www.hpacultures.org.uk/ collections/ecacc.jsp |
| Falcon (see BD Biosciences) | www.bdbiosciences.com |
| FiberCell Systems, | www.fibercellsystems.com |
| FibroGen | www.fibrogen.com/ |
| Fine Cut Graphic Imaging | www.finecut.co.uk |
| Fine Science Tools. | www.finescience.com |
| Finesse Solutions | http://finesse.com |
| Fisher Scientific (see also Thermo Fisher) | www.fishersci.com |
| Fison Instruments | www.thermoscientific.com |
| Fitzgerald | www.fitzgerald-fii.com |
| Fluka Chemical | www.sigmaaldrich.com |
| Fluorochem | www.fluorochem.net |
| FMC BioProducts | www.lonzabioscience.com |
| Forma | www.thermoscientific.com |
| Fougera | www.fougera.com |

| | |
|---|---|
| Fresenius | www.fresenius-ag.com |
| Fuji Photo Film | www.fujifilm.com |
| G.Q.F. Manufacturing | www.gqfmfg.com |
| GA International | www.labtag.com |
| Gallenkamp | www.sanyobiomedical.com |
| GE Heathcare | www.gelifesciences.com |
| GE Silicones | www.momentive.com |
| Gemini Bio-Products | www.gembio.com |
| Genetic Research Instrumentation | www.gri.co.uk |
| Genetix | www.genetix.com |
| Gen-Probe | www.gen-probe.com |
| Gentronix | www.gentronix.co.uk |
| Genway Biotech | www.genwaybio.com |
| Germfree Labs | www.germfree.com |
| Geron | www.geron.com |
| Gibco | www.invitrogen.com |
| Gilson | www.gilson.com |
| Glen Research | www.glenres.com |
| Global Medical Instrumentation, (GMI). | www.gmi-inc.com |
| Gonotec | www.gonotec.com |
| Gow-Mac Instrument | www.gow-mac.com |
| Grace Bio-labs | www.gracebio.com |
| Grant Instruments (Cambridge) | www.grant.co.uk |
| Greiner Bio-One GmbH. | www.greinerbioone.com |
| Guava Technologies | www.millipore.com/ flowcytometry |
| Guest Medical | www.guest-medical.co.uk |
| H.V. Skan | www.berlinerglas.com |
| Hamamatsu | www.hamamatsu.com |
| Hamilton | www.hamiltoncompany.com |
| Hammond Cell Technology | https://secure.bluehost.com/ ~hammondc/home/ |
| Hamo: See Steris | www.hamo.com |
| Hanna Instruments | www.hannainst.com |
| Harlan Sera-Lab | www.seralab.co.uk |
| Harvard Apparatus | www.harvardapparatus.com |
| Health Science Research Resources Bank (HSRRB) | www.jhsf.or.jp/English |
| Helena Laboratories | www.helena.com |
| Henry Schein | www.henryschein.com |
| Heraeus | www.thermo.com |
| Hexascreen Culture Technologies | www.hexascreen.com |
| High-Q | www.high-q.com |
| Hirschmann Laborgerate | www.hirschmann-laborgeraete.de |
| Hook & Tucker Zenyx | www.htz.biz |
| Hotpack | www.hotpack.com |
| HSRRB (see Health Science Research Resources Bank) | www.jhsf.or.jp/English |

| Name | URL |
|---|---|
| Hudson Robotics | www.hudsoncontrol.com |
| HyClone (see also Thermo Scientific) | www.hyclone.com |
| Hycor Biomedical | www.hycorbiomedical.com |
| IBM | www.ibm.com |
| ICLC Interlab Cell Line Collection | bioinformatics.istge.it/cldb/descat16.html |
| ICN | www.mpbiomedicals.com |
| IEC | www.thermoscientific.com |
| Ilford | www.ilford.com |
| Image Solutions | www.imsol.co.uk |
| Imaging Associates | www.imas.co.uk |
| Imaging Research Inc. (see GE Healthcare) | www.gehealthcare.com |
| Incucyte | www.essen-instruments.com |
| Infors HT | www.infors-ht.com |
| Innovatis (see also Roche Innovatis) | www.innovatis.com |
| Innovative Cell Technologies | www.innovativecelltech.com |
| Innovative Chemistry | www.innovativechem.com |
| Insight Biotechnology | www.insightbio.com |
| Insmed | www.insmed.com |
| Insmed Aesthetics | www.allergan.com |
| Instron. | www.instron.com |
| Integra Biosciences | www.integra-biosciences.com |
| Interplast | www.interplastinc.com |
| Interpore Cross International | www.interpore.com |
| Intracel | www.intracel.co.uk |
| Invitrogen | www.invitrogen.com |
| Invivogen | www.invivogen.com |
| Irvine Scientific | www.irvinesci.com |
| ISP Alginate | www.ispcorp.com |
| Iwaki (see Sterilin) | www.sterilin.co.uk |
| J.T. Baker (see Mallinckrodt Baker) | www.mallbaker.com |
| Jackson ImmunoResearch | www.jacksonimmuno.com |
| James Glass | www.jamesglass.com |
| Japanese Collection of Research Bioresources (JCRB) | http://cellbank.nibio.go.jp |
| Japanese Tissue Culture Association (JTCA) | http://jtca.umin.jp |
| JCRB | http://cellbank.nibio.go.jp |
| Jencons Scientific | www.jencons.co.uk |
| Jenway | www.jenway.com |
| JEOL | www.jeol.com |
| JM Separations | www.jmseparations.com |
| Johnson & Johnson, Medical | www.jnj.com |
| Jouan | www.thermoscientific.com |
| JRH Biosciences | www.sigmaaldrich.com |
| KBI BioPharma | www.kbibiopharma.com |
| KD Scientific | www.kdscientific.com |
| Keison Products | www.keison.co.uk |
| Kendro | www.thermoscientific.com |
| Kimberley-Clark | www.kc-safety.com |
| Kimble Chase | www.kimble-kontes.com |
| Kinetic Biosystems (KBI Pharma) | www.kbibiopharma.com |
| KNF | www.knf.co.uk |
| Krackeler Scientific | www.krackeler.com |
| Lab Safety Supply | www.labsafety.com |
| Lab Storage Systems | www.labstore.com |
| Lab Vision | www.labvision.com |
| Labcaire Systems | www.labcaire.co.uk |
| Labco Limited | www.labco.co.uk |
| Labcold | www.labcold.com |
| Lab-Line | www.thermoscientific.com |
| LabM | www.labm.com |
| Laboratory Impex Systems | www.lab-impex-systems.co.uk |
| Laboratory Sales (UK) | www.ls-uk.com |
| Labtech International | www.labtech.co.uk |
| Lamb, R.A. | www.ralamb.co.uk |
| Lancer | www.lancer.com |
| Lawson Mardon Wheaton | www.wheatonsci.com |
| LEC Instruments | www.lecinstruments.com |
| LEEC | www.leec.co.uk |
| Leica Microsystems | www.leica-microsystems.com |
| LGC Standards | www.lgcstandards.com |
| Life Imaging Services | www.lis.ch |
| Life Sciences International | www.thermoscientific.com |
| Life Technologies | www.lifetechnologies.com |
| Lifeblood Medical | www.lifebloodmedical.com |
| List Biological Laboratories | www.listlabs.com |
| LMS | www.lms.ltd.uk |
| Lonza Biologics | www.lonza.com |
| Lorne Biochemicals | www.lornelabs.com |
| LTE Scientific | www.lte-scientific.co.uk |
| Ludl Electronic Products | www.ludl.com |
| Lumitech (see Lonza) | www.lonza.com |
| MA Bioservices (see Bioreliance) | www.bioreliance.com |
| Mallinckrodt-Baker. | www.mallbaker.com |
| Marienfeld Laboratory Glassware | www.superior.de |
| Markson LabSales | www.markson.com |
| Marsh Electronics. | www.marshelectronics.com |
| Matrix Technologies (see also Thermo Scientific) | www.matrixtechcorp.com |
| MatTek | www.mattek.com |
| MB Research Laboratories | www.mbresearch.com |

| Name | URL |
|------|-----|
| MBL International | www.mblintl.com |
| MDH | www.bioquell.com |
| Mediatech | www.cellgro.com |
| Medical & Biological Laboratories (MBL) | www.mbl.co.jp |
| Medical Air Technology (MAT) | www.medicalairtechnology.com |
| Medorex | www.medorex.com |
| Meiji Techno | www.meijitechno.com |
| Memmert | www.memmert.com |
| Merck Chemicals | www.merckbio.eu |
| Merck USA | www.vwr.com |
| Cryotherm | www.cryotherm.de |
| Messer | www.messergroup.com |
| Metachem Diagnostics | www.metachem.co.uk |
| Mettler-Toledo | www.mt.com |
| Michael Smith Engineers | www.michael-smith-engineers.co.uk |
| Microflow | www.bioquell.com |
| Microgon | www.spectrapor.com |
| Microm International (see also Thermo Scientific) | www.microm.de |
| Midwest Scientific | www.midsci.com |
| Miele | www.labwashers.com |
| Miles (see Bayer) | www.bayer.com |
| Millipore | www.millipore.com |
| Miltenyi Biotec | www.miltenyibiotec.com |
| Miltex Instruments | www.miltex.com |
| Minerva Biolabs | www.minerva-biolabs.com |
| Minucells and Minutissue Vertriebs | www.minucells.de |
| Molecular Devices | www.moleculardevices.com |
| Molecular Probes (see also Invitrogen) | www.probes.com |
| Momentive | www.momentive.com |
| MP Biomedicals | www.mpbio.com |
| MTR Scientific | www.mtrscientific.com |
| Muis Controls | www.muiscontrols.com |
| MVE | www.chartbiomed.com |
| Mycoplasma Experience | www.mycoplasma-exp.com |
| Nalgene (see also Thermo Fisher) | www.nalgenelabware.com |
| Napco | www.thermoscientific.com |
| Narang Medical | www.narang.com |
| National Cell Culture Center | www.biovest.com |
| National Diagnostics | www.nationaldiagnostics.com |
| Nektar | www.nektar.com |
| NEN Life Sciences | www.perkinelmer.co.uk |
| Neuromics | www.neuromics.com |
| New Brunswick Scientific | www.nbsc.com |
| New England Biolabs | www.neb.com |
| Nexcelom Biosciences | www.nexcelom.com |
| Nikon | www.nikon-instruments.com |
| NLS Animal Health | www.nlsanimalhealth.com |
| Nonlinear Dynamics | www.nonlinear.com |
| Nova Biomedical | www.novabiomedical.com |
| NovaMatrix | www.novamatrix.biz |
| Novartis | www.novartis.com |
| Novo Nordisk | www.novonordisk.com |
| Novocastra (see also Leica) | www.novocastra.co.uk |
| Novozymes Biopharma | www.novozymes.com |
| NuAire | www.nuaire.com |
| Nuclepore | www.whatman.com |
| Nunc (see also Thermo Scientific) | www.nuncbrand.com |
| Nusil Silicone Technology | www.nusil.com |
| Nycomed | www.nycomed.com |
| Olympus | www.olympus-global.com |
| Omega Engineering | www.omega.com |
| Omnilab | www.omnilab.de |
| Oncall Medical Supplies | www.oncallmedicalsupplies.com |
| Oncogene Research Products | www.merckbiosciences.com |
| Oncor (see Qbiogene) | www.qbiogene.com |
| Orchid Cellmark | www.orchidcellmark.com |
| Origen Biomedical | www.origen.com |
| Orion Research | www.thermoscientific.com |
| Oxford Instruments | www.oxinst.com |
| Oxford Worldwide | www.oxfordworldwide.com |
| Oxoid | www.oxoid.com/uk |
| P&T Poultry Supplies and Equipment | www.pandtpoultry.co.uk |
| PAA Laboratories | www.paa.com |
| Pall | www.pall.com/laboratory |
| Pan Biotech | www.pan-biotech.com |
| Partec | www.partec.com |
| Particle Measuring Systems | www.pmeasuring.com |
| PAW BioScience Products | www.pawbioscience.com |
| Peak Scientific Instruments | www.peakscientific.com |
| PelFreeze | www.pelfreez-bio.com |
| PeproTech | www.peprotech.com |
| Peqlab Ltd | www.peqlab.co.uk |
| Perbio Science (see also Thermo Scientific) | www.perbio.com |
| Perceptive Instruments | www.perceptive.co.uk |
| PerkinElmer Life Sciences | www.PerkinElmer.com |
| Pharmacia | www.gelifesciences.com |
| Pharmingen (see also BD Biosciences) | www.pharmingen.com |
| Phoretix International | www.nonlinear.com |

| Name | URL |
| --- | --- |
| Photometrics | www.photomet.com |
| Pierce Chemical (*see also* Thermo Fisher) | www.piercenet.com |
| Planer plc | www.planer.com |
| Plas Labs | www.Plas-Labs.com |
| Plastim | www.plastim.co.uk |
| Polaroid | www.polaroid.com |
| Polyfiltronics | www.whatman.com |
| PolyScience | www.polyscience.com |
| Popper & Sons | www.popperandsons.com |
| Portex Verpakkingen | www.biofolie.com |
| Porvair Sciences | www.porvair-sciences.com |
| Prestige Medical | www.prestigemedical.co.uk |
| Princeton Instruments | www.princetoninstruments.com |
| Prior Scientific | www.prior.com |
| Priorclave | www.priorclave.co.uk |
| Progen Scientific | www.progensci.co.uk |
| Promega | www.promega.com |
| PromoCell | www.promocell.com |
| Protein Polymer Technologies | www.ppti.com |
| Protide Pharmaceuticals | www.protidepharma.com |
| Purite | www.purite.com |
| Qbiogene | www.qbiogene.com |
| Qiagen | www.qiagen.com |
| QMX Laboratories | www.qmxlabs.com |
| Quadrachem Laboratories | www.qclscientific.com |
| R& D Systems | www.rndsystems.com |
| Radleys | www.radleys.co.uk |
| Radnoti Glass Technology | www.radnoti.com |
| Rainin Instrument | www.rainin.com |
| Raven Biological Laboratories | www.ravenlabs.com |
| Raymond A. Lamb (*see also* Thermo Fisher) | www.ralamb.co.uk |
| ReachBio | www.reachbio.com |
| Reichert | www.reichertaI.com |
| Reinnervate | www.reinnervate.com |
| Research Diagnostics | www.fitzgerald-fii.com |
| Research Laboratory Supplies | www.researchsupply.net |
| Research Organics | www.resorg.com |
| Retsch | www.retsch.com |
| Revco Scientific (*see also* Thermo Scientific) | www.revco-sci.com |
| Richardsons of Leicester | www.richardsonsofleicester.co.uk |
| Riken | www.brc.riken.jp |
| Robbins Scientific | www.scigene.com |
| Roboz Surgical Instrument | www.roboz.com |
| Roche Applied Science | www.roche-applied-science.com |
| Roche Innovatis | www.innovatis.com |
| RS Biotech (*see also* New Brunswick Scientific) | www.rsbiotech.com |
| RS Components | www.rs-components.com |
| RTS Life Science | www.rtslifescience.com |
| Rütten Engineering | www.rutten.com |
| Safelab Systems | www.safelab.co.uk |
| Safetec | www.safetec.com |
| Saf-T-Pak | www.saftpak.com |
| Sakura Finetek | www.sakuraus.com |
| Santa Cruz Biotechnology | www.scbt.com |
| Sanyo Chemical Industries | www.sanyo-chemical.co.jp |
| Sanyo Scientific | www.sanyobiomedical.com |
| Sarstedt | www.sarstedt.com |
| Sartec | www.sartec.co.uk |
| Sartorius Stedim Biotech | www.sartorius-stedim.com |
| Scanalytics (*see* BD Biosciences) | www.bdbiosciences.com |
| ScanLaf | www.scanlaf.dk |
| Scharfe Systems: *see also* Innovative Cell Technologies; Roche Innovatis; Sedna Scientific | www.casy-technology.com |
| Schleicher & Schuell | www.schleicher-schuell.com |
| Schott | www.schott.com |
| Scientek. | www.scientek.net |
| Scientific Instrument Centre | www.sic.uk.com |
| Scientific Laboratory Supplies | www.scientific-labs.com |
| SciGene | www.scigene.com |
| Scion | www.scioncorp.com |
| Sedna Scientific | www.sednascientific.com |
| Sefar | www.sefar.com |
| Seikagaku | www.seikagaku.co.jp/english |
| Selznick Scientific Software | www.cclims.com |
| Sentinel Laboratories | www.sentinel-laboratories.com |
| Sera Lab | www.seralab.co.uk |
| Serologicals (*see also* Millipore) | www.serologicals.com |
| Sero-Med | www.seromed.com |
| SERVA Electrophoresis | www.serva.de |
| Severn Biotech | www.severnbiotech.com |
| SGM Biotech | www.sgmbiotech.com |
| SGS SA Life Science | www.sgs.com |
| Shamrock Scientific Speciality Systems | www.shamrocklabels.com |
| Shandon-Scientific | www.thermoscientific.com |
| Siemens Water Technologies | www.water.siemens.com |
| Sigma-Aldrich. | www.sigmaaldrich.com |

| Name | URL |
|---|---|
| Signal Instrument | www.signal-group.com |
| Silicone Specialty Fabricators | www.ssfab.com |
| Silvertree Engineering | www.icespy.com |
| Skan | www.skan.co.uk |
| SkinEthic | www.skinethic.com |
| Smiths Medical International | www.smiths-medical.com |
| Society for In Vitro Biology (SIVB) | www.sivb.org |
| Socorex ISBA SA | www.socorex.ch |
| SoloHill Engineering | www.solohill.com |
| Sonoco | www.sonoco.com |
| Sorvall | www.thermoscientific.com |
| Southern Biotechnology Associates | www.southernbiotech.com |
| Spectrum Laboratories | www.spectrumlabs.com |
| Sper Scientific | www.sperdirect.com |
| SPI Supplies | www.2spi.com |
| SPS Medical | www.spsmedical.com |
| SPSS (see also IBM) | www.spss.com |
| Staniar, J. | www.johnstaniar.co.uk |
| Statebourne Cryogenics | www.statebourne.com |
| Statlab Medical Products | www.statlab.com |
| StemCell Technologies | www.stemcell.com |
| Stemgent | www.stemgent.com |
| Sterile Systems | www.sterilesystems.com |
| Sterilin | www.sterilin.co.uk |
| Steris | www.steris.com |
| Stille | www.stille.se |
| Stoelting | www.stoeltingco.com/physio |
| Stratagene. | www.stratagene.com |
| Stratech Scientific | www.stratech.co.uk |
| Structure Probe | www.2spi.com |
| Stuart Scientific (see Sterilin) | www.sterilin.co.uk |
| Summers Optical | www.emsdiasum.com |
| Surgicon | www.surgicon.com.pk |
| Sutter Instrument | www.sutter.com |
| Swann-Morton | www.swann-morton.com |
| Synbiosis | www.synbiosis.com |
| Syngene | www.syngene.com |
| Synthecon | www.synthecon.com |
| TAAB Laboratories | www.taab.co.uk |
| TAP (see The Automation Partnership) | www.automationpartnership.com |
| Taylor-Wharton Cryogenics | www.taylorwharton.com |
| TCS Biosciences | www.tcsbiosciences.co.uk |
| TCS CellWorks | www.tcscellworkscatalogue.co.uk |
| Tebu-bio APS. | www.tebu-bio.com |
| Tecan Sales Switzerland | www.tecan.com |
| Techmate | www.techmate.co.uk |
| Techne | www.techne.com |
| Technika (see Sper Scientific) | www.sperdirect.com |
| Techno Plastic Products (TPP) | www.tpp.ch |
| Tecniplast UK | www.tecniplast.it |
| Tekmar-Dohrmann | www.teledynetekmar.com |
| Teknomek | www.teknomek.co.uk |
| Teledyne Tekmar | www.teledynetekmar.com |
| Telstar | www.telstar-lifesciences.com |
| Terra Universal | www.terrauniversal.com |
| Texol Products | www.laboratorygas.com |
| The Automation Partnership (TAP) | www.automationpartnership.com/tap/cell_culture |
| The West Group | www.westgroup.co.uk |
| Thermo Electron | www.thermoscientific.com |
| Thermo Fisher Scientific | www.thermofisher.com |
| Thermo Scientific | www.thermoscientific.com |
| Thermolyne | www.thermoscientific.com |
| Thomas Scientific | www.thomassci.com |
| Titan Enterprises | www.flowmeters.co.uk |
| TMS Europe | www.tmseurope.co.uk |
| TPP | www.tpp.ch |
| Triple Red Laboratory Technology | www.triplered.com |
| Tristel Solutions | www.tristel.com |
| TSI | www.tsi.com |
| U.S. Filter | www.water.siemens.com |
| Unibioscreen | www.unibioscreen.com |
| UniEquip | www.uniequip.com |
| Universal Biologicals (Cambridge) | www.universalbiologicals.ltd.uk |
| Upstate Biotechnology | www.millipore.com |
| USB | www.usbweb.com |
| UVP | www.uvp.com |
| V& P Scientific | www.vp-scientific.com |
| Vacuubrand | www.vacuubrand.com |
| Valley Biomedical | www.valleybiomedical.com |
| Varian | www.varianinc.com |
| Vector Laboratories | www.vectorlabs.com |
| Ventria Bioscience | www.ventria.com |
| Verderflex Tube Pumps | www.verderflex.com |
| Verilabs Europe | www.verilabs.nl |
| Vernon Morris | www.vernonmorris.co.uk |
| Vindon Scientific | www.vindon.co.uk |
| Vision BioSystems (see also Leica Microsystems) | www.leica-microsystems.com |
| VisionBiomed | www.visionbiomed.com |
| Vivascience AG | www.sartorius-stedim.de |
| Vivendi Water Systems | www.elgalabwater.com |
| Volac (see also Camlab) | www.sigmaladrich.com |
| VWR International | www.vwr.com |
| Vysis (see also Abbot Molecular) | www.vysis.com |

| Name | URL |
| --- | --- |
| Wako Chemicals | www.wakousa.com |
| Wallac | www.perkinelmer.com |
| Watson-Marlow Bredel Pumps | www.watson-marlow.com |
| Wave Biotech (*see also* Sartorius Stedim) | www.wavebiotech.ch |
| Wescor | www.wescor.com |
| Whatman (*see also* GE Heathcare) | www.whatman.com |
| Wheaton Science Products | www.wheatonsci.com |
| WLD-TEC | www.wld-tec.com |
| World Precision Instruments | www.wpiinc.com |
| Worthington Biochemical | www.worthington-biochem.com |
| WPI (*see* World Precision Instruments) | www.wpiinc.com |
| WTW Measurement Systems | www.wtw.com |
| YSI | www.ysi.com |
| Zeiss (*see* Carl Zeiss) | www.zeiss.com |
| Zen Biologicals | www.zen-bio.com |
| Zimmer | www.zimmer.com |
| Zinsser Analytic | www.zinsser-analytic.com |
| Zymed Laboratories | www.zymed.com |

# Glossary

(Modified after Schaeffer, 1990.)

**Adaptation.** Induction or repression of synthesis of a macromolecule (usually a protein) in response to a stimulus; for example, enzyme adaptation—an alteration in enzyme activity brought about by an inducer or repressor and involving an altered rate of enzyme synthesis or degradation.

**Allograft.** *See* Homograft.

**Amniocentesis.** Prenatal sampling of the amniotic cavity.

**Anchorage dependent.** Requiring attachment to a solid substrate for survival or growth.

**Anemometer.** An instrument for measuring flow rate of air.

**Aneuploid.** Not an exact multiple of the haploid chromosome number (*see* Haploid).

**Apoptosis.** Cell death by a biologically controlled intracellular process involving DNA cleavage and nuclear fragmentation.

**Aseptic.** Free of microbial infection.

**Autocrine.** Receptor-mediated response of a cell to a factor produced by the same cell.

**Autograft.** A graft from one individual transplanted back to the same individual.

**Autoradiography.** Localization of radioisotopes in cells and tissue sections (microautoradiography) and blots from electrophoresis preparations; achieved by exposure of a photographic emulsion placed in close proximity to the specimen.

**Balanced salt solution.** An isotonic solution of inorganic salts present in approximately the correct physiological concentrations; may also contain glucose, but is usually free of other organic nutrients.

**Bioreactor.** Culture vessel for large-scale production of cells, either anchored to a substrate or propagated in suspension. Can also be used for smaller scale three-dimensional culture of constructs for tissue engineering.

**Biostat.** Culture vessel in which physical, physicochemical, and physiological conditions, as well as cell concentration, are kept constant, usually by perfusion, monitoring, and feedback.

**Carcinoma.** A tumor derived from epithelium, usually endodermally or ectodermally derived cells.

**Cell concentration.** Number of cells per mL of medium.

**Cell culture.** Growth of cells dissociated from the parent tissue by spontaneous migration or mechanical or enzymatic dispersal.

**Cell density.** Number of cells per $cm^2$ of substrate.

**Cell fusion.** Formation of a single cell body by the fusion of two other cells, either spontaneously or, more often, by induced fusion with inactivated Sendai virus or polyethylene glycol.

**Cell hybridization.** *See* Hybrid cell.

**Cell line.** A propagated culture after the first subculture.

**Cell strain.** A characterized cell line derived by selection or cloning.

**Centipoise.** Unit of viscosity; 1000 centipoises are equivalent to 1 Pascal-second.

**Centromere.** The point of adhesion between two *chromatids* in a *chromosome*; attaches to spindle during metaphase, telophase, and anaphase of cell division.

**Chemically defined.** Made entirely from pure defined constituents (said of a medium); distinct from "serum free," in which other poorly characterized constituents may be used to replace serum.

**Chromatid.** Paired constituent of a *chromosome* linked by a centromere.

**Chromosome.** A complex of DNA and nucleoproteins forming a defined morphological structure within the nucleus, visible at metaphase during cell division, made up of two morphologically identical *chromatids* joined at the *centromere*, and present in a defined number characteristic for each species.

**Chromosome painting.** Use of specific fluorescent probes to stain defined regions of the chromosome.

**Clone.** A population of cells derived from one cell.

**Commitment.** Irreversible progression from a stem cell to a particular defined lineage endowing the cell with the potential to express a limited repertoire of properties.

**Confluent.** A monolayer of cells in which all cells are in contact with other cells all around their periphery, and no available substrate is left uncovered.

**Constitutive.** Expressed by a cell in the absence of external regulation.

**Contact inhibition.** Inhibition of plasma membrane ruffling and cell motility when cells are in complete contact with other adjacent cells, as in a confluent culture; often precedes, but is not necessarily causally related to, cessation of cell proliferation.

**Continuous cell line or cell strain.** Cell line or strain having the capacity for infinite survival. Previously known as "established" and often referred to as "immortal."

**Construct.** Applied to a genetic recombination of exogenous genetic elements within a plasmid or other vector for transfer of DNA to a recipient cell. Can also be used of recombined cellular elements in an *organotypic* culture for use in tissue engineering.

**Cyclic growth.** Growth from a low cell density to a high cell density with a regular subculture interval; regular repetition of the growth cycle for maintenance purposes.

**Cytokine.** A factor, released by cells, that will induce a receptor-mediated effect on the proliferation, differentiation, or inflammation of other cells; usually a short-range paracrine, rather than systemic, effect.

**Cytostasis.** Cessation of cell proliferation.

**Cytotoxicity.** Cellular damage to one or more metabolic pathways, intracellular processes, or structures resulting in impaired function. Often, but not necessarily, linked to loss of viability.

**Deadaptation.** Reversible loss of a specific property due to the absence of the appropriate inducer (not always defined).

**Dedifferentiation.** Irreversible loss of the specialized properties that a cell would have expressed in vivo. As evidence accumulates that cultures dedifferentiate by a combination of the selection of undifferentiated stromal cells and deadaptation resulting from the absence of the appropriate inducers, the term is going out of favor. It is still correctly applied to mean the progressive loss of differentiated morphology in histological observations of, for example, tumor tissue.

**Density limitation of growth.** Mitotic inhibition correlated with an increase in cell density at confluence.

**Differentiation.** Acquisition of phenotypic properties associated with the fully functional cells in vivo.

**Diploid.** Each chromosome represented as a pair, identical in the autosomes and female sex chromosomes and nonidentical in male sex chromosomes, and corresponding to the chromosome number and morphology of most somatic cells of the species from which the cells are derived.

**Dome.** A hemicystic or blister-like structure in a confluent epithelial monolayer implying ion transport across the monolayer and resulting in the accumulation of water below the monolayer.

**DNA barcoding.** Analysis of polymorphisms in mitochondrial DNA

**DNA fingerprinting.** Binding of multilocus cDNA probes to hypervariable regions of satellite DNA cut by restriction endonucleases and visualized by autoradiography following electrophoresis. Pattern specific to individual from whom, or from which, the DNA was derived.

**DNA profiling.** The generic term for assaying hypervariable regions of satellite DNA, now mainly used to detect the frequency of short tandem repeats (STRs) in microsatellite DNA using PCR of individual loci with defined primers. More sensitive than multilocus *DNA fingerprinting* and readily quantifiable.

**Ectoderm.** The outer germ layer of the embryo, giving rise to the epithelium of the skin.

**Embryonal stem cells.** Totipotent stem cells isolated from the inner cell mass of an early embryo; can be propagated as cell lines with a wide range of differentiation capabilities.

**Embryonic induction.** The interaction (often reciprocal) of cells from two different germ layers, promoting differentiation.

**Endocrine.** Signaling factors, such as hormones, released by one tissue and having an effect on a distant tissue via the systemic vasculature.

**Endoderm.** The innermost germ layer of the embryo, giving rise to the epithelial component of organs such as the gut, liver, and lungs.

**Endothelium.** An epithelium-like cell layer lining spaces within mesodermally derived tissues, such as blood vessels, and derived from the mesoderm of the embryo.

**Enzyme induction.** An increase in synthesis of an enzyme produced by, for example, hormonal stimulation.

**Epithelial.** Describes cells derived from epithelium, but often used more loosely to describe any cells of a polygonal shape with clear, sharp boundaries between them. More correctly, the latter should be referred to as epithelium-like or epithelioid.

**Epithelium.** A covering or lining of cells, as in the surface of the skin or lining of the gut, usually derived from the embryonic endoderm or ectoderm, but sometimes derived from mesoderm, as with kidney tubules and mesothelium lining body cavities.

**ES cells.** *See* Embryonal stem cells.

**Euploid.** Exact multiple of the haploid chromosome set. The correct morphology characteristic of each chromosome pair in the species from which the cells are derived is not implicit in the definition but is usually assumed to be the case; otherwise, we should say "euploid, but with some chromosomal aberrations."

**Explant.** A fragment of tissue transplanted from its original site and maintained in an artificial medium.

**Explantation.** Isolation of tissue for maintenance in vitro, strictly as small fragments with accompanying outgrowth (*see* Primary explant), but often used as a generic term for the isolation of tissue for culture.

**FACS.** *See* Fluorescence-activated cell sorter.

**Fermentor.** Large-scale culture vessel, often used for cells in suspension; derived from same term applied to microbiological culture. *See also* Bioreactor.

**Fibroblast.** A proliferating precursor cell of the mature differentiated fibrocyte.

**Fibroblastic.** Resembling fibroblasts—for example, spindle shaped (bipolar) or stellate (multipolar); usually arranged in parallel arrays at confluence if contact is inhibited. Often the term is used indiscriminately for undifferentiated mesodermal cells, regardless of their relationship to the fibrocyte lineage; implies a migratory type or cell with processes exceeding the nuclear diameter by threefold or more. More correctly, fibroblast-like or fibroblastoid.

**Ficoll–Paque.** Density medium made up of Ficoll combined with a radiopaque iodinated substance, such as sodium metrizoate.

**Finite cell line.** A culture that has been propagated by subculture but is capable of only a limited number of cell generations in vitro before dying out.

**FISH.** *See* Fluorescence in situ hybridization.

**Flow cytometer.** An instrument providing quantitative and qualitative analysis of individual cells in a population by scanning a single cell stream with a laser, or with multiple lasers of different wavelengths, and recording the light that is scattered or the fluorescence that is emitted.

**Fluorescence-activated cell sorter (FACS).** A cell separation device based on electromagnetic sorting of a single cell suspension by means of the scattering of light or the fluorescent properties of individual cells revealed by a laser scanning a single cell stream. *See also* Flow cytometer.

**Fluorescence in situ hybridization (FISH).** Binding of specific fluorescent probes to specific intracellular locations by in situ *hybridization*. *See also* Chromosome painting.

**Generation number.** The number of population doublings (estimated from dilution at subculture) that a culture has undergone since explanation; necessarily contains an approximation of the number of generations in the primary culture.

**Generation time.** The interval from one point in the cell division cycle to the same point in the cycle, one division later; distinct from population-doubling time, which is derived from the total cell count of a population and therefore averages different generation times, including the effect of nongrowing cells.

**Genotype.** The total genetic characteristics of a cell.

**Glycocalyx.** Glycosylated peptides, proteins, and lipids, and glycosaminoglycans attached to the surface of the cell.

**Growth curve.** A semilogarithmic plot of the cell number on a logarithmic scale against time on a linear scale, for a proliferating cell culture; usually divided into the lag phase (the phase before growth is initiated), the log phase (the period of exponential growth), and the plateau (a stable cell count achieved when the culture stops growing at a high cell density).

**Growth cycle.** Growth interval from subculture to the top of the log phase, ready for a further subculture.

**Growth factor.** A factor, released by cells, that induces proliferation in other cells; mostly paracrine in effect, but may be released into the blood by platelets or endothelium.

**Growth medium.** The medium used to propagate a particular cell line; usually a basal medium with additives such as serum or growth factors.

**Haploid.** That chromosome number wherein each chromosome is represented once; in most higher animals, the number present in the gametes and half the number found in most somatic cells.

**Hemicyst.** *See* Dome

**Heterokaryon.** Cell containing two or more genetically different nuclei; usually derived by cell fusion.

**Heteroploid.** A culture in which the cells have chromosome numbers other than diploid and differing from each other.

**Histotypic culture.** A high density culture resembling a tissue-like morphology in vivo. Usually a three-dimensional culture re-created from a dispersed cell culture of one cell type that attempts to regain, by cell proliferation and multilayering or by reaggregation, a tissue-like structure. *See also* Organotypic culture.

**Holding medium.** Medium, usually without serum and growth factors, or with minimal serum, designed to maintain cells in a viable state without proliferation (e.g., for collecting biopsies or maintaining cells at a plateau with no further cell proliferation).

**Homocrine.** Paracrine interaction between cells of the same type.

**Homeothermic.** Able to maintain a constant body temperature despite environmental fluctuations.

**Homograft (Allograft).** A graft derived from a genetically different donor of the same species as the recipient.

**Homokaryon.** Cell containing two or more genetically identical nuclei; usually a product of cell fusion.

**Hybrid cell.** Mononucleate cell that results from the fusion of two different cells, leading to the formation of a synkaryon. *See* Synkaryon.

**Ideogram.** The arrangement of the chromosomes of a cell in order by size and morphology so that the karyotype may be studied and genetically analyzed.

**Immortalization.** The acquisition of an infinite life span. May be induced in finite cell lines by transfection with telomerase, oncogenes, or the large T-region of the SV40 genome, or by infection with SV40 (whole virus) or Epstein−Barr virus (EBV). Immortalization is not necessarily a malignant transformation, although usually a component of malignant transformation.

**Induction.** An increase in effect produced by a given stimulus.

**Infection (other than the commonplace definition).** Transfer of genomic DNA with a retroviral construct containing the DNA sequence under investigation, usually packaged with a promoter sequence and a reporter gene, such as β-galactosidase; the product of an infection may be detected by staining with a chromogenic substrate.

**In situ hybridization.** Binding of specific complementary nucleic acid probes to intracellular locations; cDNA probes for the localization of mRNA sequences, or RNA probes for DNA localization (*see* Chromosome painting). Visualized by radioisotopic labeling of the probe and microautoradiography, or by using a fluorochrome bound to the probe.

**iPS cell.** A pluripotent stem cell induced from genetic manipulation and/or epigenetic regulation of gene expression of adult cells.

**Isograft (syngraft).** A graft derived from a genetically identical or nearly identical donor of the same species as the recipient.

**In ovo.** In the egg—usually the hen's egg.

**In vitro.** Literally "in glass," but used conventionally to mean cultured outside of the host as cell cultures, organ cultures, or short-term organ bath preparations; also used to indicate biochemical and molecular reactions carried out in a test tube, but these reactions are better referred to as *cell free*.

**In vivo.** In the living plant or animal.

**Karyotype.** The distinctive chromosomal complement of a cell.

**Laminar flow.** The flow of a fluid that closely follows the shape of a streamlined surface without turbulence; said of hoods or cabinets characterized by a stable flow of air over the work area so as to minimize turbulence.

**Laminar-flow hood or cabinet.** A workstation with filtered air flowing in a laminar (nonturbulent) manner parallel to or perpendicular to the work surface, to maintain the sterility of the work; the parallel flow is called *horizontal* laminar flow, the perpendicular flow *vertical* laminar flow.

**Leukemia.** Malignant disease of the hematopoietic system, evident as circulating blast cells.

**LIF.** Leukemia inhibitory factor, a cytokine of the interleukin-6 family; used to inhibit differentiation and maintain the stem cell phenotype in ES cells.

**Lipofection.** Transfection of DNA by fusion with lipid-encapsulated DNA.

**Lymphoma.** A solid tumor of lymphoid cells.

**Log phase.** *See* Growth curve.

**Macroautoradiography.** Localization of radioisotopes in whole body sections and blots from electrophoresis preparations, by exposure of a photographic emulsion placed in close proximity to the specimen, usually by placing the film with the blot in a cassette with an intensifier screen.

**MACS (Magnetic-activated cell sorting).** Sorting cells by the magnetic attraction of magnetizable antibody-coated ferritin beads that bind to specific cell surface antigens.

**Malignant.** Invasive or metastatic (i.e., colonizing other tissues); said of a tumor. Usually progressive, leading to the destruction of host cells and, ultimately, death of the host.

**Malignant transformation.** The development of destructive invasion of normal tissue without regulation in space or time; may also lead to metastatic growth (colonization of a distant site with subsequent unregulated invasive growth).

**Manometer.** A U-shaped tube containing liquid, the levels of which in each limb of the U reflect the pressure difference between the ends of the tube.

**Medium.** A mixture of inorganic salts and other nutrients capable of sustaining cell survival in vitro for 24 hours. *Growth medium*: A medium that is used in routine culture such that the cell number increases with time. *Maintenance medium*: A medium that will retain cell survival without cell proliferation (e.g., a low-serum or serum-free medium used with serum-dependent cells). The plural of medium is *media*.

**Mesenchyme.** Loose, often migratory embryonic tissue derived from the mesoderm, giving rise to connective tissue, cartilage, muscle, hemopoietic cells, and so forth, in the adult.

**Mesenchymal stem cells (MSCs).** Stem cells, usually derived from bone marrow, with multipotent differentiation capacity—such as cardiac muscle, neural cells, or hepatocytes—as well as hematopoietic lineages.

**Mesoderm.** A germ layer in the embryo arising between the ectoderm and endoderm and giving rise to mesenchyme, which in turn gives rise to connective tissue, etc. *See* Mesenchyme.

**Microautoradiography.** Localization of radioisotopes in cells and tissue sections, by exposure of a photographic emulsion placed in close proximity to the specimen, usually by dipping it in the melted emulsion. After development, the specimen may be viewed under a microscope.

**Monoclonal.** Derived from a single clone of cells. *Monoclonal antibody*: Antibody produced by a clone of lymphoid cells either in vitro or in vivo. In vitro, the clone is usually derived from a hybrid of a sensitized spleen cell and a continuously growing myeloma cell.

**Morphogenesis.** The development of form and structure of an organism.

**Multipotent stem cell.** A stem cell that can give rise to many different lineages of differentiated cells. Often not distinguished from pluripotent stem cells (*q.v.*) but generally restricted to a stem cell within one tissue type giving rise to multiple lineages within that tissue, such as a colony forming cells within the hematopoietic system.

**Myeloma.** A tumor derived from myeloid cells; used in monoclonal antibody production when the myeloma cell can produce immunoglobulin.

**Neoplastic.** A new, unnecessary proliferation of cells giving rise to a tumor.

**Neoplastic transformation.** The conversion of a nontumorigenic cell into a tumorigenic cell.

**Oncogene.** A gene that, when transfected or infected into normal cells, induces malignant transformation; usually a positively acting gene coding for growth factors, receptors, signal transducers, or nuclear regulators.

**Organ culture.** The maintenance or growth of organ primordia or the whole or parts of an organ in vitro in a way that may allow differentiation and preservation of the architecture or function of the organ.

**Organogenesis.** The development of organs.

**Organotypic culture.** Histotypic culture involving more than one cell type to create a model of the cellular interactions characteristic of an organ in vivo. A reconstruction from dissociated cells or fragments of tissue is implied, as distinct from organ culture, in which the structural integrity of the explanted tissue is retained. *See also* Histotypic culture, *and tissue equivalent.*

**Osmolality.** The concentration of osmotically active particles in an aqueous solution, expressed in osmoles/kg.

**Osmolarity.** The concentration of osmotically active particles in an aqueous solution, expressed in osmoles/L.

**Osmole.** The amount of a substance containing 1 mole of osmotically active particles.

**Paracrine.** An effect of one cell on another adjacent cell mediated by a soluble factor without involvement of the systemic vasculature. *Homocrine*: homotypic paracrine interaction between cells of the same type. *Heterotypic paracrine*: interaction between dissimilar cells.

**Parenchyma.** That part of a tissue carrying out the major function of the tissue, such as the hepatocytes in liver; as distinct from the stroma, such as fibroblastic connective tissue, seen as supporting tissue.

**Pascal.** SI unit of pressure equivalent to 1 newton per square meter.

**Passage.** The transfer or subculture of cells from one culture vessel to another; usually, but not necessarily, involves the subdivision of a proliferating cell population, enabling the propagation of a cell line or cell strain.

**Passage number.** The number of times a culture has been subcultured.

**Pavement-like.** Cells in a regular monolayer or polygonal cells. More correctly, epithelioid or epithelium-like.

**Phenotype.** The aggregate of all the expressed properties of a cell; the product of the interaction of the genotype with the regulatory environment.

**Plateau.** *See* Growth curve.

**Plating efficiency.** The percentage of cells seeded at subculture that gives rise to colonies. If each colony can be said to be derived from one cell, plating efficiency is identical to cloning efficiency. Sometimes the plating efficiency is used loosely to describe the number of cells surviving after subculture, but this is better termed the *seeding efficiency*.

**Ploidy.** Relationship of chromosome number of a given type of cell to that found in normal somatic cells in vivo. *See also* Haploid, Diploid, Euploid, Aneuploid, *and* Heteroploid.

**Pluripotent stem cell.** A stem cell that can give rise to several different unrelated lineages of differentiated cells, following appropriate induction.

**Poikilothermic.** Having a body temperature close to that of the environment and not regulated by metabolism.

**Population density.** The number of monolayer cells per unit area of substrate.

**Population-doubling time.** The interval required for a cell population to double at the middle of the logarithmic phase of growth.

**Precursor cell.** A cell at a stage in a cell differentiation pathway that is assumed to be committed to a particular type of differentiation. Early stages will be proliferative, late stages may not be.

**Primary culture.** A culture started from cells, tissues, or organs taken directly from an organism, and before the first subculture.

**Primary explant.** A fragment of tissue removed from the organism and placed in culture in such as way as to promote its survival and the outgrowth of viable cells.

**Progenitor cells.** Cells that are at an early stage of development, probably proliferating, and not yet expressing differentiated properties. Includes *precursor cells* and *stem cells*.

**Pseudodiploid.** Numerically diploid chromosome number, but with chromosomal aberrations.

**Reagent.** A substance (element, compound, or mixture) that participates in a chemical reaction.

**Quasidiploid.** *See* Pseudodiploid.

**Sarcoma.** A tumor derived from mesodermally derived cells, for example, connective tissue, muscle (*myosarcoma*), or bone (*osteosarcoma*).

**Saturation density.** Maximum number of cells attainable per $cm^2$ (in a monolayer culture) under specified conditions.

**Seeding efficiency.** The percentage of the inoculum that attaches to the substrate within a stated period of time (implying viability, or survival, but not necessarily proliferative capacity).

**Senescence.** Biologically regulated loss of proliferative potential linked to shortening of the *telomeres* of the chromosomes.

**Skin equivalent.** An in vitro reconstruction of skin via organotypic culture, usually incorporating epidermal keratinocytes, collagen, and dermal fibroblasts but usually lacking melanocytes, vascular, neural, and neuroendocrine elements.

**Somatic cell genetics.** The study of cell genetics by the recombination and segregation of genes in somatic cells, usually by fusion.

**Spheroid.** A three-dimensional cluster of cells formed by reaggregation of cells in suspension, usually over a nonadhesive substrate such as agar or agarose.

**Split ratio.** The divisor of the dilution ratio of a cell culture at subculture (e.g., one flask divided into four, or 100 mL up to 400 mL, would be a split ratio of 4).

**Stem cell.** The earliest detectable cell in a lineage with the capacity to generate all the cells in the lineage while maintaining its own population. A *unipotent* stem cell will only give rise to one differentiation pathway, *bipotent* to two, *multipotent* to more than two, *pluripotent* to several, and *totipotent* to all types of differentiated cell. *See also* Embryonal stem cells *and* Mesenchymal stem cells.

**Stroma.** That part of a tissue seen as having a purely supporting role; for example, fibroblastic connective tissue and its vasculature.

**Subconfluent.** Less than confluent; not all of the available substrate is covered.

**Subculture.** *See* Passage.

**Substrate.** The matrix or solid underlay upon which a monolayer culture grows.

**Superconfluent.** Progressing beyond the state in which all the cells are attached to the substrate and multilayering occurs.

**Suppressor gene.** A gene that inhibits the transformed (malignant) phenotype, usually associated with dominant-negative regulation of cell proliferation or cell migration; often suppressor genes are mutated or deleted in transformed cells and cancer.

**Suspension culture.** A culture in which cells will multiply when suspended in growth medium.

**Synkaryon.** A hybrid cell that results from the fusion of the nuclei it carries.

**Telomeres.** Terminal regions of the chromosomes that prevent recombination with other chromosomes and are able to maintain the proliferative capacity of the cell. Progressively shortened during senescence but maintained in stem cells and some tumor cells by telomerase.

**Tetraploid.** Twice the diploid (four times the haploid) number of chromosomes.

**Three-dimensional (3-D) culture.** Culture of cells as aggregates or in a matrix or scaffold such that the cells are in a three-dimensional array, as distinct from the conventional two-dimensional array of standard monolayer culture. *See also* Histotypic culture, Organotypic culture, *and* Spheroids.

**Tissue culture.** Properly, the maintenance of fragments of tissue in vitro, but now commonly applied as a generic term denoting tissue explant culture, organ culture, and dispersed-cell culture, including the culture of propagated cell lines and cell strains.

**Tissue equivalent.** A culture, usually *organotypic* and *three-dimensional*, approximating the histology of the tissue in vivo (though usually lacking vascular, neural, and some other elements). *See also* Skin equivalent.

**Totipotent stem cell.** A stem cell that can give rise to all the tissues of the body.

**Transdifferentiation.** Cells from one lineage acquiring the ability to differentiate into cells of a different lineage.

**Transfection.** The transfer, by artificial means, of genetic material from one cell to another, when less than the whole nucleus of the donor cell is transferred. Transfection is usually achieved by transferring isolated chromosomes, DNA, or cloned genes.

**Transformation.** A permanent alteration of the cell phenotype, presumed to occur via an irreversible genetic change. May be spontaneous, as in the development of rapidly growing continuous cell lines from slow-growing early-passage rodent cell lines, or may be induced by chemical or viral action. Usually produces cell lines that have an increased growth rate, an infinite life span, a lower serum requirement, and a higher plating efficiency and that are often (but not necessarily) tumorigenic.

**Validation (of cell lines).** A process that includes authentication, characterization, and the demonstration of the lack of contamination of a cell line.

**Variant.** A cell line expressing a stable phenotype that is different from the parental culture from which it was derived.

**Viral transformation.** A permanent phenotypic change induced by the genetic and heritable effects of a transforming virus.

**Xenograft.** Transplantation of tissue to a species different from that from which it was derived; often used to describe the implantation of human tumors in athymic (nude), immune-deprived, or immune-suppressed mice.

# APPENDIX V

# Cross-contaminated or Misidentified Cell Lines

Tables V.I and V.II have been generated by collaboration between the author and Amanda Capes-Davis (Founding Manager, CellBank Australia) with feedback from the major cell banks, ATCC, DSMZ, ECACC, JCRB, and RIKEN. It is meant as a preliminary guide to avoiding suspect cell lines. It is also available in updated form on the websites of the cell banks listed above (see Appendix III for URLs), as electronic supplementary material with Capes-Davis et al. [2010], and on www.wiley.com/go/freshney/cellculture.

The "contaminating cell line," in most cases, will have overgrown the claimed original, or will have replaced it by a technical error, and the original cells will no longer exist. Genuine examples of some cell lines, such as RT-4 human bladder carcinoma, may still exist (see Table V.II); their identity would be confirmed by profiling. Attributions in the last two columns indicate where these misidentifications were reported and in no way imply responsibility for the cause by the authors or institutions. Observations made in these lists are based on published reports and details obtained from cell banks, their websites, and Wikipedia. "Reference PubMed ID" refers to the unique ID number assigned by the PubMed database (http://www.ncbi.nlm.nih.gov/pubmed/).

The authors take no credit or responsibility for any of the primary observations and have merely attempted to collate data previously available on other sites. It must be stressed that regardless of the information contained in these tables, all recently acquired cell lines should be tested (e.g., by STR profiling for human cell lines) and compared to reference stock before use.

Cross-contamination has been reported in the cell lines listed in Table V.II but authentic stocks apparently do exist. As with any imported cell line, check identity on receipt or obtain from authenticated stock, such as from a reputable cell bank. Authenticity, or otherwise, can be confirmed following DNA profiling with reference to the standard developed by ATCC SDO workgroup ASN-0002 [2010].

**TABLE V.I.** Cross-contaminated or Misidentified Cell Lines

| Misidentified Cell line | Claimed species | Claimed cell type | Contaminating cell line | Actual species | Actual cell type | Misidentification reported by | Reference PubMed ID |
|---|---|---|---|---|---|---|---|
| 2474/90 | Human | Gastric carcinoma | HT-29 | Human | Colon carcinoma | MacLeod et al., 1999 | 10508494 |
| 2563 (MAC-21) | Human | Lung carcinoma | HeLa | Human | Cervical adenocarcinoma | Nelson-Rees et al., 1981 | 6451928 |
| 2957/90 | Human | Gastric carcinoma | HT-29 | Human | Colon carcinoma | MacLeod et al., 1999 | 10508494 |
| 3051/80 | Human | Gastric carcinoma | HT-29 | Human | Colon carcinoma | MacLeod et al., 1999 | 10508494 |
| 41M | Human | Ovarian carcinoma | OAW 28 | Human | Ovarian carcinoma | Wilson et al., 1996 | 8795574 |
| ACC2 | Human | Salivary gland, adenoid cystic carcinoma | HeLa | Human | Cervical adenocarcinoma | Phuchareon et al., 2009 | 18698025, 19557180 |
| ACC3 | Human | Salivary gland, adenoid cystic carcinoma | HeLa | Human | Cervical adenocarcinoma | Phuchareon et al., 2009 | 18698025, 19557180 |
| ACCM | Human | Salivary gland, adenoid cystic carcinoma | HeLa | Human | Cervical adenocarcinoma | Phuchareon et al., 2009 | 18698025, 19557180 |
| ACCNS | Human | Salivary gland, adenoid cystic carcinoma | Unknown | Mouse | Unknown | Phuchareon et al., 2009 | 19557180 |
| ACCS | Human | Salivary gland, adenoid cystic carcinoma | T-24 | Human | Bladder carcinoma | Phuchareon et al., 2009 | 19557180 |
| ADLC-5M2 | Human | Lung carcinoma | HeLa | Human | Cervical adenocarcinoma | MacLeod et al., 1999 | 10508494 |
| Aedes aegypti, Suitor's clone | Mosquito | Not specified | Unknown | Moth, *Antheraea eucalypti* | Not specified | Nelson-Rees et al., 1981 | 6451928 |
| Aedes vexans culture | Mosquito | Not specified | Unknown | Moth, *Antheraea eucalypti* | Not specified | Nelson-Rees et al., 1981 | 6451928 |
| AG-F | Human | Lymphoma, Hodgkin | CCRF-CEM | Human | Leukemia, acute lymphoblastic, T-cell | Drexler et al., 2003 | 12592342 |
| AKI | Human | Melanoma | HeLa | Human | Cervical adenocarcinoma | Yoshino et al., 2006 | 16643607 |
| ALVA-31 | Human | Prostate carcinoma | PC-3 | Human | Prostate carcinoma | van Bokhoven et al., 2001; Varella-Garcia et al., 2001 | 11304728, 11433521 |
| ALVA-41 | Human | Prostate carcinoma | PC-3 | Human | Prostate carcinoma | van Bokhoven et al., 2001; Pan et al., 2001; Varella-Garcia et al., 2001 | 11304728, 11135436, 11433521 |
| ALVA-55 | Human | Prostate carcinoma | PC-3 | Human | Prostate carcinoma | van Bokhoven et al., 2003 | 14518029 |

| | | | | | | | |
|---|---|---|---|---|---|---|---|
| ALVA-101 | Human | Prostate carcinoma | PC-3 | Human | Prostate carcinoma | van Bokhoven et al., 2003 | 14518029 |
| AMDURII | Human | Skin | LLC-PK1 | Pig | Kidney, normal renal cells | Milanesi et al., 2003 | 14505435 |
| AO | Human | Amnion | HeLa | Human | Cervical adenocarcinoma | Nelson-Rees et al., 1981 | 6451928 |
| ARO81-1 | Human | Thyroid, anaplastic carcinoma | HT-29 | Human | Colon carcinoma | Schweppe et al., 2008 | 18713817 |
| AV3 | Human | Amnion | HeLa | Human | Cervical adenocarcinoma | Nelson-Rees et al., 1981 | 6451928 |
| BCC1/KMC | Human | Basal cell carcinoma | HeLa | Human | Cervical adenocarcinoma | MacLeod et al., 1999 | 10508494 |
| BE-13 | Human | Leukemia, acute lymphoblastic, T-cell | PEER | Human | Leukemia, acute lymphoblastic, T-cell | Drexler et al., 2003 | 12592342 |
| BHP 10-3 | Human | Thyroid, papillary carcinoma | TPC-1 | Human | Thyroid, papillary carcinoma | Schweppe et al., 2008 | 18713817 |
| BHP 14-9 | Human | Thyroid, papillary carcinoma | M14 | Human | Melanoma | Schweppe et al., 2008 | 18713817 |
| BHP 17-10 | Human | Thyroid, papillary carcinoma | M14 | Human | Melanoma | Schweppe et al., 2008 | 18713817 |
| BHP 2-7 | Human | Thyroid, papillary carcinoma | TPC-1 | Human | Thyroid, papillary carcinoma | Schweppe et al., 2008 | 18713817 |
| BHP 5-16 | Human | Thyroid, papillary carcinoma | M14 | Human | Melanoma | Schweppe et al., 2008 | 18713817 |
| BHP 7-13 | Human | Thyroid, papillary carcinoma | TPC-1 | Human | Thyroid, papillary carcinoma | Schweppe et al., 2008 | 18713817 |
| BIC-1 | Human | Esophageal carcinoma | SW-480, SW-620 | Human | Colon Carcinoma | Boonstra et al., 2010 | 20075370 |
| BLIN-1 (also subclone 1E8) | Human | Leukemia, acute lymphoblastic, B-cell precursor | NALM-6 | Human | Leukemia, acute lymphoblastic, B-cell precursor | Drexler et al., 2003 | 12592342 |
| BM-1604 | Human | Prostate carcinoma | DU-145 | Human | Prostate carcinoma | MacLeod et al., 1999 | 10508494 |
| BrCA 5 | Human | Breast carcinoma | HeLa | Human | Cervical adenocarcinoma | Nelson-Rees & Flandermeyer, 1977; Nelson-Rees et al., 1981 | 557237, 6451928 |
| C16 (MRC-5 derivative) | Human | Lung cells, fetal | HeLa | Human | Cervical adenocarcinoma | ECACC website | No PMID |
| CAC2 | Human | Salivary gland, adenoid cystic carcinoma | Unknown | Rat | Unknown | Phuchareon et al., 2009 | 19557180 |
| CaMa (clone 15) | Human | Breast carcinoma | Unknown | Syrian hamster and mouse | Unknown | Nelson-Rees et al., 1981 | 6451928 |

(continued overleaf)

**TABLE V.I.** (*Continued*)

| Misidentified Cell line | Claimed species | Claimed cell type | Contaminating cell line | Actual species | Actual cell type | Misidentification reported by | Reference PubMed ID |
|---|---|---|---|---|---|---|---|
| CaOV | Human | Ovarian carcinoma | HeLa | Human | Cervical adenocarcinoma | Nelson-Rees et al., 1981 | 6451928 |
| CaVe | Human | Gastric carcinoma | HeLa | Human | Cervical adenocarcinoma | Nelson-Rees et al., 1981 | 6451928 |
| Chang liver | Human | Liver, normal hepatic cells | HeLa | Human | Cervical adenocarcinoma | Nelson-Rees et al., 1981 | 6451928 |
| CHB | Human | Astrocytoma | Unknown | Rat | Not specified | Nelson-Rees et al., 1981 | 6451928 |
| CHP-234 | Human | Neuroblastoma | Unknown | Human | Unknown | ATCC website | No PMID |
| Clone 1-5c-4 | Human | Conjunctiva | HeLa | Human | Cervical adenocarcinoma | ECACC website | No PMID |
| Clone-16 | Human | Pancreas, fetal endocrine cells | Unknown | Syrian hamster | Unknown | Matsuba et al., 1988 | 2903855 |
| CMP | Human | Rectal adenocarcinoma | HeLa | Human | Cervical adenocarcinoma | Nelson-Rees et al., 1981 | 6451928 |
| CMPII C2 | Human | Rectal adenocarcinoma | HeLa | Human | Cervical adenocarcinoma | Nelson-Rees et al., 1981 | 6451928 |
| CO (= COLE) | Human | Lymphoma, Hodgkin | CCRF-CEM | Human | Leukemia, acute lymphoblastic, T-cell | Drexler et al., 2003 | 12592342 |
| COLO-677 | Human | Lung carcinoma, small cell | RPMI-8226 | Human | Myeloma | DSMZ website | No PMID |
| COLO-818 | Human | Melanoma | COLO-800 | Human | Melanoma | MacLeod et al., 1999 | 10508494 |
| CoLo-TC | Human | Colon carcinoma | COLO-205 | Human | Colon carcinoma | RIKEN website | No PMID |
| *Culisita inornata* culture | Mosquito | Not specified | Unknown | Moth, *Antheraea eucalypti* | Not specified | Greene et al., 1972; Nelson-Rees et al., 1981 | 4402510, 6451928 |
| D18T | Human | Synovial cell | HeLa | Human | Cervical adenocarcinoma | Nelson-Rees et al., 1981 | 6451928 |
| D98/AH | Human | Not specified | HeLa | Human | Cervical adenocarcinoma | Honma et al., 1992 | 1730567 |
| D98/AH2 Clone B | Human | Not specified | HeLa | Human | Cervical adenocarcinoma | ECACC website | No PMID |
| DAMI | Human | Leukemia, acute myeloid, M7 | HEL | Human | Leukemia, acute myeloid, M6 | MacLeod et al., 1997; Drexler et al., 2003 | 9447816, 12592342 |
| DAPT | Human | Astrocytoma, piloid | HeLa | Human | Cervical adenocarcinoma | Nelson-Rees et al., 1981 | 6451928 |

| | | | | | | | |
|---|---|---|---|---|---|---|---|
| DD | Human | Malignant histiocytosis | K-562 | Human | Leukemia, chronic myeloid, blast crisis | Drexler et al., 2003 | 12592342 |
| Det30A | Human | Breast carcinoma, ascitic fluid | HeLa | Human | Cervical adenocarcinoma | Nelson-Rees et al., 1981 | 6451928 |
| Detroit-6 (Det6) | Human | Sternal marrow cells | HeLa | Human | Cervical adenocarcinoma | Nelson-Rees et al., 1981 | 6451928 |
| Detroit-98 | Human | Sternal marrow cells | HeLa | Human | Cervical adenocarcinoma | Nelson-Rees et al., 1981 | 6451928 |
| Detroit 98/AG | Human | Sternal marrow cells | HeLa | Human | Cervical adenocarcinoma | Nelson-Rees et al., 1981 | 6451928 |
| Detroit 98/AH-2 | Human | Sternal marrow cells | HeLa | Human | Cervical adenocarcinoma | Nelson-Rees et al., 1981 | 6451928 |
| Detroit 98/AH-R | Human | Sternal marrow cells | HeLa | Human | Cervical adenocarcinoma | Nelson-Rees et al., 1981 | 6451928 |
| Detroit 98s | Human | Sternal marrow cells | HeLa | Human | Cervical adenocarcinoma | Nelson-Rees et al., 1981 | 6451928 |
| DJM-1 | Human | Squamous cell carcinoma | BSCC-93 | Human | Skin, squamous cell carcinoma | Yoshino et al., 2006 | 16643607 |
| DRO90-1 | Human | Thyroid, anaplastic carcinoma | A-375 | Human | Melanoma | Schweppe et al., 2008 | 18713817 |
| DuPro-1 | Human | Prostate carcinoma | PC-3 | Human | Prostate carcinoma | van Bokhoven et al., 2003 | 14518029 |
| EB33 | Human | Prostate carcinoma | HeLa | Human | Cervical adenocarcinoma | Nelson-Rees, 1979; Nelson-Rees et al., 1981 | 535908, 6451928 |
| ECTC | Cow | Thyroid, embryonic | Vero | Monkey, African green (Cercopithecus aethiops) | Kidney, normal renal cells | Milanesi et al., 2003 | 14505435 |
| ECV-304 | Human | Endothelium, normal cells | T-24 | Human | Bladder carcinoma | Dirks et al., 1999; MacLeod et al., 1999 | 10614862, 10508494 |
| ED27 | Human | Chorionic villus | HeLa | Human | Cervical adenocarcinoma | Kniss et al., 2002 | 11869090 |
| EEK | Horse | Kidney, embryonic renal cells | NSK | Pig | Kidney, normal renal cells | Milanesi et al., 2003 | 14505435 |
| EH | Human | Leukemia, hairy cell | HK | Human | Leukemia, hairy cell | Drexler et al., 2003 | 12592342 |
| EJ-1 | Human | Bladder carcinoma | T-24 | Human | Bladder carcinoma | Masters et al., 2001 | 11416159 |
| Ej138 | Human | Bladder carcinoma | T-24 | Human | Bladder carcinoma | Azari et al., 2007 | 17254797 |
| EICo | Human | Breast carcinoma | HeLa | Human | Cervical adenocarcinoma | Nelson-Rees et al., 1981 | 6451928 |

*(continued overleaf)*

**TABLE V.I.** (*Continued*)

| Misidentified Cell line | Claimed species | Claimed cell type | Contaminating cell line | Actual species | Actual cell type | Misidentification reported by | Reference PubMed ID |
|---|---|---|---|---|---|---|---|
| EPC | Carp, *Cyprinus carpio* | Epithelial papilloma | Unknown | Fathead minnow, *Pimephales promelas* | Unknown | ATCC website | No PMID |
| EPLC3-2M1 | Human | Lung carcinoma | HeLa | Human | Cervical adenocarcinoma | MacLeod et al., 1999 | 10508494 |
| EPLC-65 | Human | Lung carcinoma | HeLa | Human | Cervical adenocarcinoma | MacLeod et al., 1999 | 10508494 |
| ESP1 | Human | Lymphoma, Burkitt | HeLa | Human | Cervical adenocarcinoma | Nelson-Rees et al., 1981 | 6451928 |
| ETK-1 | Human | Cholangiocarcinoma | SSP-25 | Human | Cholangiocarcinoma | Yoshino et al., 2006 | 16643607 |
| EU-1 | Human | Leukemia, acute lymphoblastic, B-cell precursor | REH | Human | Leukemia, acute lymphoblastic, B-cell precursor | Drexler et al., 2003 | 12592342 |
| EU-7 | Human | Leukemia, acute lymphoblastic, T-cell | CCRF-CEM | Human | Leukemia, acute lymphoblastic, T-cell | Drexler et al., 2003 | 12592342 |
| EUE | Human | Subcutis, fetal | HeLa | Human | Cervical adenocarcinoma | Nelson-Rees et al., 1981 | 6451928 |
| EVLC2 | Human | Umbilical vein endothelium, transfected | Unknown | Human | Non-endothelial? | Unger et al., 2002 | 12453433 |
| F2-4E5 | Human | Thymic epithelium | SK-HEP-1 | Human | Liver carcinoma | MacLeod et al., 1999 | 10508494 |
| F2-5B6 | Human | Thymic epithelium | SK-HEP-1 | Human | Liver carcinoma | MacLeod et al., 1999 | 10508494 |
| F255A4 | Human | Not specified | HeLa | Human | Cervical adenocarcinoma | Nelson-Rees et al., 1981 | 6451928 |
| FB2 | Human | Thyroid, papillary carcinoma | TPC-1 | Human | Thyroid, papillary carcinoma | Ribeiro et al., 2008 | 19087340 |
| FL | Human | Amnion | HeLa | Human | Cervical adenocarcinoma | Nelson-Rees et al., 1981; Ogura et al., 1993 | 6451928, 8397027 |
| Flow 13000 | Human | Lung cells, embryonic fibroblast | MRC-5 | Human | Lung cells, embryonic fibroblast | JCRB website | No PMID |
| Flow 5000 | Human | Lung cells, embryonic fibroblast | Flow 1000 | Human | Lung cells, embryonic fibroblast | JCRB website | No PMID |
| Flow 6000 | Human | Lung cells, embryonic fibroblast | Flow 1000 | Human | Lung cells, embryonic fibroblast | JCRB website | No PMID |

| | | | Flow 3000 | | | JCRB website | No PMID |
|---|---|---|---|---|---|---|---|
| Flow 7000 | Human | Lung cells, embryonic fibroblast | Flow 3000 | Human | Lung cells, embryonic fibroblast | JCRB website | No PMID |
| FQ | Human | Lymphoma, Hodgkin | OMK-210 | Monkey, Owl (*Aotus trivirgatus*) | Kidney, normal renal cells | Nelson-Rees et al., 1981; Drexler et al., 2003 | 6451928, 12592342 |
| FU-RPNT-2 | Human | Kidney, renal primitive neuroectodermal tumor | FU-RPNT-1 | Human | Kidney, renal primitive neuroectodermal tumor | RIKEN website | No PMID |
| G-11 (HBT-3 derivative) | Human | Breast carcinoma | HeLa | Human | Cervical adenocarcinoma | Nelson-Rees et al., 1981 | 6451928 |
| GHE | Human | Astrocytoma | T-24 | Human | Bladder carcinoma | MacLeod et al., 1999 | 10508494 |
| Girardi heart | Human | Heart, normal cells | HeLa | Human | Cervical adenocarcinoma | Nelson-Rees et al., 1981 | 6451928 |
| GM1312 | Human | Myeloma | Correct name, incorrect cell type | Human | EBV+ B-lymphoblastoid cell line | Drexler et al., 2003 | 12592342 |
| GPS-M | Human | Spleen, adult cells | Strain L-M | Mouse | Connective tissue | Nelson-Rees et al., 1981 | 6451928 |
| GPS-PD | Human | Spleen, adult cells | Strain L-M | Mouse | Connective tissue | Nelson-Rees et al., 1981 | 6451928 |
| GREF-X | Human | Liver, hepatic myofibroblast | Unknown | Rat | Unknown | MacLeod et al., 1999 | 10508494 |
| GT3TKB | Human | Gastric carcinoma | RERF-LC-A1 | Human | Lung carcinoma | Yoshino et al., 2006 | 1643607 |
| H249 | Human | Lung carcinoma, small cell | H69 | Human | Lung carcinoma, small cell | Personal communication, M. Liscovitch to R. Nardone | No PMID |
| H-494 | Human | Prostate carcinoma | HeLa | Human | Cervical adenocarcinoma | Williams, 1980 | 6244232 |
| H7D7A and derivatives | Human | Liver, normal cells (SV40-transformed) | HepG2 | Human | Liver, hepatocellular carcinoma | van Pelt et al., 2003 | 12619888 |
| H7D7B and derivatives | Human | Liver, normal cells (SV40-transformed) | HepG2 | Human | Liver, hepatocellular carcinoma | van Pelt et al., 2003 | 12619888 |
| H7D7C and derivatives | Human | Liver, normal cells (SV40-transformed) | HepG2 | Human | Liver, hepatocellular carcinoma | van Pelt et al., 2003 | 12619888 |
| H7D7D and derivatives | Human | Liver, normal cells (SV40-transformed) | HepG2 | Human | Liver, hepatocellular carcinoma | van Pelt et al., 2003 | 12619888 |
| HAG | Human | Thyroid adenoma (goitre) | T-24 | Human | Bladder carcinoma | MacLeod et al., 1999 | 10508494 |

*(continued overleaf)*

**TABLE V.I.** (*Continued*)

| Misidentified Cell line | Claimed species | Claimed cell type | Contaminating cell line | Actual species | Actual cell type | Misidentification reported by | Reference PubMed ID |
|---|---|---|---|---|---|---|---|
| HBC | Human | Breast carcinoma, invasive ductal tumor | Unknown | Rat | Unknown | Nelson-Rees & Flandermeyer, 1977; Nelson-Rees et al., 1981 | 557237, 6451928 |
| HBL-100 | Human | Breast carcinoma | Unknown | Human | Unknown | Yoshino et al., 2006; ATCC website | 16643607 |
| HBT-3 | Human | Breast carcinoma | HeLa | Human | Cervical adenocarcinoma | Nelson-Rees et al., 1981 | 6451928 |
| HBT-39b | Human | Breast carcinoma | HeLa | Human | Cervical adenocarcinoma | Nelson-Rees et al., 1981 | 6451928 |
| HBT-E (HBT-3 clone) | Human | Breast carcinoma | HeLa | Human | Cervical adenocarcinoma | Nelson-Rees et al., 1981 | 6451928 |
| HCE | Human | Cervical carcinoma | HeLa | Human | Cervical adenocarcinoma | Nelson-Rees et al., 1981 | 6451928 |
| HCu-10 | Human | Esophageal squamous cell carcinoma | Hcu-10, Hcu-18, Hcu-22, Hcu-27, Hcu-33, Hcu-37 or Hcu-39 (genetically identical) | Human | Esophageal squamous cell carcinoma | van Helden et al., 1988 | 3167823 |
| HCu-18 | Human | Esophageal squamous cell carcinoma | Hcu-10, Hcu-18, Hcu-22, Hcu-27, Hcu-33, Hcu-37 or Hcu-39 (genetically identical) | Human | Esophageal squamous cell carcinoma | van Helden et al., 1988 | 3167823 |
| HCu-22 | Human | Esophageal squamous cell carcinoma | Hcu-10, Hcu-18, Hcu-22, Hcu-27, Hcu-33, Hcu-37 or Hcu-39 (genetically identical) | Human | Esophageal squamous cell carcinoma | van Helden et al., 1988 | 3167823 |
| HCu-27 | Human | Esophageal squamous cell carcinoma | Hcu-10, Hcu-18, Hcu-22, Hcu-27, Hcu-33, Hcu-37 or Hcu-39 (genetically identical) | Human | Esophageal squamous cell carcinoma | van Helden et al., 1988 | 3167823 |

| Cell line | Species | Origin | Cross-contaminant | Species | Disease | Reference | PMID |
|---|---|---|---|---|---|---|---|
| HCu-33 | Human | Esophageal squamous cell carcinoma | Hcu-10, Hcu-18, Hcu-22, Hcu-27, Hcu-33, Hcu-37 or Hcu-39 (genetically identical) | Human | Esophageal squamous cell carcinoma | van Helden et al., 1988 | 3167823 |
| HCu-37 | Human | Esophageal squamous cell carcinoma | Hcu-10, Hcu-18, Hcu-22, Hcu-27, Hcu-33, Hcu-37 or Hcu-39 (genetically identical) | Human | Esophageal squamous cell carcinoma | van Helden et al., 1988 | 3167823 |
| HCu-39 | Human | Esophageal squamous cell carcinoma | Hcu-10, Hcu-18, Hcu-22, Hcu-27, Hcu-33, Hcu-37 or Hcu-39 (genetically identical) | Human | Esophageal squamous cell carcinoma | van Helden et al., 1988 | 3167823 |
| HCV-29Tmv (HCV-29 derivative) | Human | Bladder, tumorigenic urothelial cells | T-24 | Human | Bladder carcinoma | Christensen et al., 1993 | 8105864 |
| HEC-155 | Human | Uterine adenocarcinoma | HEC-180 | Human | Uterine carcinoma | JCRB website | No PMID |
| HEC-180 | Human | Uterine adenocarcinoma | HEC-155 | Human | Uterine adenocarcinoma | JCRB website | No PMID |
| HEK | Human | Kidney, embryonic renal cells | HeLa | Human | Cervical adenocarcinoma | Nelson-Rees et al., 1981 | 6451928 |
| HEK/HRV (HEK derivative) | Human | Kidney, transformed embryonic renal cells | HeLa | Human | Cervical adenocarcinoma | Nelson-Rees et al., 1981 | 6451928 |
| HEL-R66 | Human | Not specified | Unknown | Monkey, African green (Cercopithecus aethiops) | Unknown | Nelson-Rees et al., 1981 | 6451928, 7236009 |
| HEp-2 (H.Ep.-2) | Human | Laryngeal carcinoma | HeLa | Human | Cervical adenocarcinoma | Nelson-Rees et al., 1981; Chen, 1988 | 6451928, 3180844 |
| Hep-2C | Human | Laryngeal carcinoma | HeLa | Human | Cervical adenocarcinoma | ECACC website | No PMID |
| Hep2 (Clone 2B) | Human | Laryngeal carcinoma | HeLa | Human | Cervical adenocarcinoma | ECACC website | No PMID |

*(continued overleaf)*

**TABLE V.I.** (*Continued*)

| Misidentified Cell line | Claimed species | Claimed cell type | Contaminating cell line | Actual species | Actual cell type | Misidentification reported by | Reference PubMed ID |
|---|---|---|---|---|---|---|---|
| HIMEG-1 | Human | Leukemia, chronic myeloid | HL-60 | Human | Leukemia, acute myeloid, M2 | Drexler et al., 2003 | 12592342 |
| HKB-1 | Human | Lymphoma, Hodgkin | Unknown | Human | Unknown | Drexler et al., 2003 | 12592342 |
| HKMUS | Human | Cervical carcinoma | SKG-II-SF | Human | Cervical carcinoma | RIKEN website | No PMID |
| HKMUS-SF | Human | Cervical carcinoma | SKG-II-SF | Human | Cervical carcinoma | Yoshino et al., 2006 | 16643607 |
| HL111783 | Human | Lung carcinoma | HeLa | Human | Cervical adenocarcinoma | Yoshino et al., 2006 | 16643607 |
| HMV-1 | Human | Melanoma | HeLa | Human | Cervical adenocarcinoma | MacLeod et al., 1999 | 10508494 |
| HPB-MLT | Human | Leukemia, acute lymphoblastic, T-cell | HPB-ALL | Human | Leukemia, acute lymphoblastic, T-cell | Drexler et al., 2003 | 12592342 |
| HPC-36M (HPC-36 derivative) | Human | Prostate carcinoma | HeLa | Human | Cervical adenocarcinoma | Masters et al., 2001 | 11416159 |
| hPTC | Human | Thyroid, papillary cell | Unknown | Pig | Unknown | MacLeod et al., 1999 | 10508494 |
| HSC-41 | Human | Gastric carcinoma | HSC-42 | Human | Gastric carcinoma | JCRB website | No PMID |
| HSG | Human | Salivary gland, submandibular | HeLa | Human | Cervical adenocarcinoma | ECACC website | No PMID |
| HSG-AZA1 | Human | Salivary gland, submandibular | HeLa | Human | Cervical adenocarcinoma | JCRB website | No PMID |
| HSG-AZA3 | Human | Salivary gland, submandibular | HeLa | Human | Cervical adenocarcinoma | JCRB website | No PMID |
| HSGc-C5 | Human | Oral carcinoma | HeLa | Human | Cervical adenocarcinoma | JCRB website | No PMID |
| HS-SULTAN | Human | Myeloma | JIJOYE | Human | Lymphoma, Burkitt | Drexler et al., 2001; Drexler et al., 2003 | 11732505, 12592342 |
| HSY | Human | Salivary gland, parotid | HeLa | Human | Cervical adenocarcinoma | JCRB website | No PMID |
| Hu1734 | Human | Bladder, nonmalignant urothelial cells | HCV-29 | Human | Bladder, nonmalignant urothelial cells | Christensen et al., 1993 | 7905254 |
| Hu456 | Human | Bladder, tumorigenic urothelial cells | T-24 | Human | Bladder carcinoma | Christensen et al., 1993 | 8105864 |
| Hu549 | Human | Bladder, tumorigenic urothelial cells | T-24 | Human | Bladder carcinoma | Christensen et al., 1993 | 8105864 |

| Cell line | Species | Tissue | Cross-contaminant | Species | Tumour type | Reference | PMID |
|---|---|---|---|---|---|---|---|
| Hu609 | Human | Bladder, nonmalignant urothelial cells | J82 | Human | Bladder carcinoma | Christensen et al., 1993; Masters et al., 2001 | 7905254, 11416159 |
| Hu609Tmv (Hu609 derivative) | Human | Bladder, tumorigenic urothelial cells | T-24 | Human | Bladder carcinoma | Christensen et al., 1993 | 8105864 |
| Hu961a, Hu961t (Hu961 derivatives) | Human | Bladder, tumorigenic urothelial cells | T-24 | Human | Bladder carcinoma | Christensen et al., 1993; Masters et al., 2001 | 8105864, 11416159 |
| HuK°39 | Human | Kidney, normal renal cells | HeLa | Human | Cervical adenocarcinoma | Nelson-Rees et al., 1981 | 6451928 |
| HuL-1 | Human | Liver, hepatocellular carcinoma | HeLa | Human | Cervical adenocarcinoma | JCRB website | No PMID |
| Hut | Human | Not specified | HeLa | Human | Cervical adenocarcinoma | Nelson-Rees et al., 1981 | 6451928 |
| IMC-2 | Human | Maxillary carcinoma | HeLa | Human | Cervical adenocarcinoma | MacLeod et al., 1999 | 10508494 |
| IMC-3 | Human | Maxillary carcinoma | HeLa | Human | Cervical adenocarcinoma | Masters et al., 2001 | 11416159 |
| IMC-4 | Human | Maxillary carcinoma | HeLa | Human | Cervical adenocarcinoma | RIKEN website | No PMID |
| Intestine 407 (INT 407; HEI) | Human | Intestinal cells (jejunum/ileum), embryonic | HeLa | Human | Cervical adenocarcinoma | Nelson-Rees et al., 1981 | 6451928 |
| IPDDC-A2 | Human | Astrocytoma | Unknown | Rat | Unknown | ECACC website | No PMID |
| IPTP/98 | Human | Glioblastoma | Unknown | Rat | Unknown | ECACC website | No PMID |
| J-111 | Human | Leukemia, acute myeloid, M5 | HeLa | Human | Cervical adenocarcinoma | Nelson-Rees et al., 1981; Drexler et al., 2003 | 6451928, 12592342 |
| J96 | Human | Leukemia | HeLa | Human | Cervical adenocarcinoma | Nelson-Rees et al., 1981 | 6451928 |
| JCA-1 | Human | Prostate carcinoma | T-24 | Human | Bladder carcinoma | van Bokhoven et al., 2001a, b | 11522622 |
| JHC | Human | Placenta | HeLa | Human | Cervical adenocarcinoma | Nelson-Rees et al., 1981 | 6451928 |
| JHH-1 | Human | Liver, hepatocellular carcinoma | Unknown | Mouse | Unknown | JCRB website | No PMID |
| JHT (JHC derivative) | Human | Placenta | HeLa | Human | Cervical adenocarcinoma | Nelson-Rees et al., 1981 | 6451928 |

*(continued overleaf)*

**TABLE V.I.** (*Continued*)

| Misidentified Cell line | Claimed species | Claimed cell type | Contaminating cell line | Actual species | Actual cell type | Misidentification reported by | Reference PubMed ID |
|---|---|---|---|---|---|---|---|
| JOSK-I | Human | Leukemia, acute myeloid, M4 | U-937 | Human | Lymphoma, histiocytic | MacLeod et al., 1999; Drexler et al., 2003 | 10508494, 12592342 |
| JOSK-K | Human | Leukemia, acute myeloid, M5 | U-937 | Human | Lymphoma, histiocytic | MacLeod et al., 1999; Drexler et al., 2003 | 10508494, 12592342 |
| JOSK-M | Human | Leukemia, chronic myeloid, blast crisis | U-937 | Human | Lymphoma, histiocytic | MacLeod et al., 1999; Drexler et al., 2003 | 10508494, 12592342 |
| JOSK-S | Human | Leukemia, acute myeloid, M5 | U-937 | Human | Lymphoma, histiocytic | MacLeod et al., 1999; Drexler et al., 2003 | 10508494, 12592342 |
| JTC-17 | Human | Skin | HeLa | Human | Cervical adenocarcinoma | Honma et al., 1992; JCRB website | 1730567 |
| JTC-3 | Human | Not specified | HeLa | Human | Cervical adenocarcinoma | Ogura et al., 1997 | 9556756 |
| K051 | Human | Leukemia, acute myeloid, M2 | K-562 | Human | Leukemia, chronic myeloid, blast crisis | Drexler et al., 2003 | 12592342 |
| K1 | Human | Thyroid, papillary carcinoma | GLAG-66 | Human | Thyroid, papillary carcinoma | Ribeiro et al., 2008; Schweppe et al., 2008 | 19087340, 18713817 |
| K2 | Human | Thyroid, papillary carcinoma | GLAG-66 | Human | Thyroid, papillary carcinoma | Ribeiro et al., 2008; Schweppe et al., 2008 | 19087340, 18713817 |
| KAK1 | Human | Thyroid, follicular adenoma | HT-29 | Human | Colon carcinoma | van Staveren et al., 2007; Schweppe et al., 2008 | 17804723, 18713817 |
| KAT10 | Human | Thyroid, papillary carcinoma | HT-29 | Human | Colon carcinoma | van Staveren et al., 2007; Schweppe et al., 2008 | 17804723, 18713817 |
| KAT4 | Human | Thyroid, anaplastic carcinoma | HT-29 | Human | Colon carcinoma | van Staveren et al., 2007; Schweppe et al., 2008 | 17804723, 18713817 |
| KAT5 | Human | Thyroid, papillary carcinoma | HT-29 | Human | Colon carcinoma | Schweppe et al., 2008 | 18713817 |
| KAT50 | Human | Thyroid, differentiated cells | HT-29 | Human | Colon carcinoma | Schweppe et al., 2008 | 18713817 |
| KAT7 | Human | Thyroid, benign follicular hyperplasia | HT-29 | Human | Colon carcinoma | Schweppe et al., 2008 | 18713817 |
| KB | Human | Oral carcinoma | HeLa | Human | Cervical adenocarcinoma | Gartler, 1967; Lavappa et al., 1976; Nelson-Rees et al., 1981 | 4864103, 1250349, 6451928 |
| KCI-MOH1 | Human | Pancreatic carcinoma | HPAC | Human | Pancreatic carcinoma | DSMZ website | No PMID |
| KM-3 | Human | Leukemia, acute lymphoblastic, B-cell precursor | REH | Human | Leukemia, acute lymphoblastic, B-cell precursor | Drexler et al., 2003 | 12592342 |

| | | | | | | | |
|---|---|---|---|---|---|---|---|
| KM3 | Human | Melanoma | Unknown | Rat | Not specified | Moseley et al., 2003 | 12740908 |
| KMS-21-BM | Human | Myeloma | Unknown | Human | Unknown | Drexler et al., 2003 | 12592342 |
| KMT-2 | Human | Umbilical cord blood | KG-1 | Human | Leukemia | Yoshino et al., 2006 | 16643607 |
| KNS-89 | Human | Gliosarcoma | U-251 MG | Human | Glioblastoma | JCRB website | No PMID |
| KOSC-3 | Human | Oral carcinoma | Ca9-22 | Human | Oral carcinoma | JCRB website | No PMID |
| KPB-M15 | Human | Leukemia, chronic myeloid, blast crisis | KYO-1 | Human | Leukemia, chronic myeloid, blast crisis | Drexler et al., 2003 | 12592342 |
| KPL-1 | Human | Breast carcinoma | MCF-7 | Human | Breast carcinoma | DSMZ website | No PMID |
| KP-P1 | Human | Prostate carcinoma | HeLa | Human | Cervical adenocarcinoma | Nelson-Rees et al., 1981 | 6451928 |
| KU-YS | Human | Neuroblastoma | KU-SN | Human | Neuroectodermal tumor | RIKEN website | No PMID |
| L-132 | Human | Lung cells, embryonic | HeLa | Human | Cervical adenocarcinoma | Nelson-Rees et al., 1981 | 6451928 |
| L-41 (J96 derivative) | Human | Leukemia, bone marrow | HeLa | Human | Cervical adenocarcinoma | ECACC website | No PMID |
| LED-Ti | Human | Cervical carcinoma | HeLa | Human | Cervical adenocarcinoma | Nelson-Rees et al., 1981 | 6451928 |
| LLC-15 MB | Human | Breast carcinoma | M14 | Human | Melanoma | Thompson et al., 2004 | 15679051 |
| LR10.6 | Human | Leukemia, acute lymphoblastic, B-cell precursor | NALM-6 | Human | Leukemia, acute lymphoblastic, B-cell precursor | MacLeod et al., 1999; Drexler et al., 2003 | 10508494, 12592342 |
| LT-1 | Frog, grass | Kidney, renal adenocarcinoma | TH and FHM | TH = box turtle; FHM = fathead minnow | TH = heart; FHM = unspecified | Nelson-Rees et al., 1981 | 6451928 |
| LU | Human | Lung cells, fetal | HeLa | Human | Cervical adenocarcinoma | Nelson-Rees et al., 1981 | 6451928 |
| LU 106 | Human | Lung cells, embryonic | HeLa | Human | Cervical adenocarcinoma | Nelson-Rees et al., 1981 | 6451928 |
| Lu-130 | Human | Lung carcinoma | Lu-134A, B | Human | Lung carcinoma | Yoshino et al., 2006 | 16643607 |
| M10T | Human | Synovial cell | HeLa | Human | Cervical adenocarcinoma | Nelson-Rees et al., 1981 | 6451928 |
| MA-104 | Monkey, Rhesus (Macaca mulatta) | Kidney, embryonic renal cells | Vero? | Monkey, African green (Cercopithecus aethiops) | Kidney, normal renal cells | Milanesi et al., 2003 | 14505435 |
| MA-111 | Rabbit | Kidney, newborn | Vero? | Monkey, African green (Cercopithecus aethiops) | Kidney, normal renal cells | Nelson-Rees et al., 1981 | 6451928 |

*(continued overleaf)*

**TABLE V.I.** (*Continued*)

| Misidentified Cell line | Claimed species | Claimed cell type | Contaminating cell line | Actual species | Actual cell type | Misidentification reported by | Reference PubMed ID |
|---|---|---|---|---|---|---|---|
| MA-160 | Human | Prostate adenoma | HeLa | Human | Cervical adenocarcinoma | Nelson-Rees, 1977; Nelson-Rees et al., 1981 | 562836, 6451928 |
| MaTu | Human | Breast carcinoma | HeLa | Human | Cervical adenocarcinoma | MacLeod et al., 1999 | 10508494 |
| MC-4000 | Human | Breast carcinoma | HeLa | Human | Cervical adenocarcinoma | MacLeod et al., 1999 | 10508494 |
| McCoy | Human | Not specified | Strain L | Mouse | Connective tissue | Nelson-Rees et al., 1981 | 6451928 |
| MCF-7/AdrR (NCI/ADR-RES) | Human | Breast carcinoma | OVCAR-8 | Human | Ovarian carcinoma | Liscovitch & Ravid, 2007 | 16504380 |
| MDA-MB-435 | Human | Breast carcinoma | M14 | Human | Melanoma | Ellison et al., 2003; Christgen et al., 2007; Rae et al., 2007 | 12354931, 17786032, 17004106 |
| MDA-N (MDA-MB-435 derivative) | Human | Breast carcinoma, HER2/ERBB2-transfected | M14 | Human | Melanoma | Lorenzi et al., 2009 | 19372543 |
| MDS | Human | Leukemia, chronic myelomonocytic | JURKAT | Human | Leukemia, acute lymphoblastic, T-cell | Drexler et al., 2003 | 12592342 |
| MEL-HO | Human | Melanoma | Unknown | Human | Unknown | MacLeod et al., 1999 | 10508494 |
| ME-WEI | Human | Melanoma | Unknown | Human | Unknown | MacLeod et al., 1999 | 10508494 |
| MGH-U1 (EJ) | Human | Bladder carcinoma | T-24 | Human | Bladder carcinoma | O'Toole et al., 1983; Lin et al., 1985 | 6823318, 4027986 |
| MGH-U2 (HM) | Human | Bladder carcinoma | T-24 | Human | Bladder carcinoma | O'Toole et al., 1983; Lin et al., 1985 | 6823318, 4027986 |
| MHH-225 | Human | Leukemia, acute myeloid, M7 | JURKAT | Human | Leukemia, acute lymphoblastic, T-cell | Drexler et al., 2003 | 12592342 |
| Minnesota EE | Human | Esophageal epithelium | HeLa | Human | Cervical adenocarcinoma | Nelson-Rees et al., 1981 | 6451928 |
| MKB-1 | Human | Leukemia, acute myeloid | CCRF-CEM | Human | Leukemia, acute lymphoblastic, T-cell | MacLeod et al., 1999; Drexler et al., 2003 | 10508494, 12592342 |
| MKN28 | Human | Gastric carcinoma | MKN74 | Human | Gastric carcinoma | JCRB website | No PMID |
| MKN-7 | Human | Gastric carcinoma | Unknown | Human | Lymphoblastoid cells | Suzuki & Sekiguchi, 1999 | No PMID (book chapter) |
| MMAc | Human | Melanoma | Mash-1 | Human | Schwannoma | Yoshino et al., 2006 | 16643607 |
| MOBS-1 | Human | Leukemia, acute myeloid, M5 | U-937 | Human | Lymphoma, histiocytic | Drexler et al., 2003 | 12592342 |

| Cell line | Species | Origin | Cross-contaminant | Tumour type | Reference | PMID |
|---|---|---|---|---|---|---|
| MOLT-15 | Human | Leukemia, acute lymphoblastic, T-cell | CTV-1 | Leukemia, acute myeloid, M5 | MacLeod et al., 1999; Drexler et al., 2003 | 10508494, 12592342 |
| MPanc-96 | Human | Pancreatic carcinoma | AsPC-1 | Pancreatic carcinoma | ATCC website | No PMID |
| MRO87-1 | Human | Thyroid, follicular carcinoma | HT-29 | Colon carcinoma | Schweppe et al., 2008 | 18713817 |
| MS (Monkey Stable) | Monkey | Kidney, normal renal cells | HeLa | Cervical adenocarcinoma | Milanesi et al., 2003 | 14505435 |
| MT-1 | Human | Breast carcinoma | HeLa | Cervical adenocarcinoma | MacLeod et al., 1999 | 10508494 |
| NC-37 | Human | Lymphoblastoid cell line, normal donor | Raji | Lymphoma, Burkitt | JCRB website | No PMID |
| NCC16 | Human | Cervical carcinoma | PHK16-0b | Skin, immortalised keratinocytes | JCRB website | No PMID |
| NCI/ADR-RES (MCF-7/AdrR) | Human | Breast carcinoma | OVCAR-8 | Ovarian carcinoma | Liscovitch & Ravid, 2007 | 16504380 |
| NCI-H1514 | Human | Lung carcinoma | Unknown | Unknown | Durkin et al., 2000; ATCC website | 10949990 |
| NCI-H1622 | Human | Lung carcinoma | Unknown | Unknown | ATCC website | No PMID |
| NCI-H738 | Human | Lung carcinoma | Unknown | Unknown | ATCC website | No PMID |
| NCOL-1 | Human | Intestinal cells, normal colon | LoVo | Colon carcinoma | Melcher et al., 2005 | 15771911 |
| NCTC2544 | Human | Skin epithelium | HeLa | Cervical adenocarcinoma | Nelson-Rees et al., 1981 | 6451928 |
| NCTC3075 | Human | Not specified | HeLa | Cervical adenocarcinoma | Nelson-Rees et al., 1981 | 6451928 |
| ND-1 | Human | Prostate carcinoma | DU 145 | Prostate carcinoma | van Bokhoven et al., 2001 | 11304728 |
| NOI-90 | Human | Lymphoma, non-Hodgkin, natural killer cell | REH | Leukemia, acute lymphoblastic, B-cell precursor | Drexler et al., 2003 | 12592342 |
| NPA87 | Human | Thyroid, papillary carcinoma | M14 | Melanoma | Schweppe et al., 2008 | 18713817 |
| NS-3 | Human | Gastric carcinoma | COLO 201 | Colon carcinoma | JCRB website | No PMID |
| OCUM-6 | Human | Gastric carcinoma | OCUM-11 | Gastric carcinoma | JCRB website | No PMID |
| OE | Human | Endometrium | HeLa | Cervical adenocarcinoma | Nelson-Rees et al., 1981; Ogura et al., 1997 | 6451928, 9556756 |
| OF | Human | Not specified | HeLa | Cervical adenocarcinoma | Ogura et al., 1997 | 9556756 |
| ONCO-DG-1 | Human | Thyroid, papillary carcinoma | OVCAR-3 | Ovarian carcinoma | DSMZ website | No PMID |

*(continued overleaf)*

**TABLE V.I.** (*Continued*)

| Misidentified Cell line | Claimed species | Claimed cell type | Contaminating cell line | Actual species | Actual cell type | Misidentification reported by | Reference PubMed ID |
|---|---|---|---|---|---|---|---|
| OST | Human | Sarcoma (osteosarcoma) | HeLa | Human | Cervical adenocarcinoma | Personal communication, R. Nardone | No PMID |
| OU-AML-1 | Human | Leukemia, acute myeloid, M4 | OCI/AML2 | Human | Leukemia, acute myeloid, M4 | Drexler et al., 2003 | 12592342 |
| OU-AML-2 | Human | Leukemia, acute myeloid, M2 | OCI/AML2 | Human | Leukemia, acute myeloid, M4 | Drexler et al., 2003 | 12592342 |
| OU-AML-3 | Human | Leukemia, acute myeloid, M4 | OCI/AML2 | Human | Leukemia, acute myeloid, M4 | Drexler et al., 2003 | 12592342 |
| OU-AML-4 | Human | Leukemia, acute myeloid, M2 | OCI/AML2 | Human | Leukemia, acute myeloid, M4 | Drexler et al., 2003 | 12592342 |
| OU-AML-5 | Human | Leukemia, acute myeloid, M5 | OCI/AML2 | Human | Leukemia, acute myeloid, M4 | Drexler et al., 2003 | 12592342 |
| OU-AML-6 | Human | Leukemia, acute myeloid, M1 | OCI/AML2 | Human | Leukemia, acute myeloid, M4 | Drexler et al., 2003 | 12592342 |
| OU-AML-7 | Human | Leukemia, acute myeloid, M4 | OCI/AML2 | Human | Leukemia, acute myeloid, M4 | Drexler et al., 2003 | 12592342 |
| OU-AML-8 | Human | Leukemia, acute myeloid, M4 | OCI/AML2 | Human | Leukemia, acute myeloid, M4 | Drexler et al., 2003 | 12592342 |
| OV-1063 | Human | Ovarian carcinoma | Unknown | Human | Unknown | ATCC website | No PMID |
| OVMIU | Human | Ovarian carcinoma | OVSAYO | Human | Ovarian carcinoma | JCRB website | No PMID |
| P1-1A3 | Human | Thymic epithelium | SK-HEP-1 | Human | Liver carcinoma | MacLeod et al., 1999 | 10508494 |
| P1-4D6 | Human | Thymic epithelium | SK-HEP-1 | Human | Liver carcinoma | MacLeod et al., 1999 | 10508494 |
| P39/TSUGANE (P39/TSU) | Human | Leukemia, acute myeloid, M2 | HL-60 | Human | Leukemia, acute myeloid, M2 | Drexler et al., 2003; JCRB website | 12592342 |
| PBEI | Human | Leukemia, acute lymphoblastic, B-cell precursor | NALM-6 | Human | Leukemia, acute lymphoblastic, B-cell precursor | MacLeod et al., 1999; Drexler et al., 2003 | 10508494, 12592342 |
| PC-93 | Human | Prostate carcinoma | HeLa | Human | Cervical adenocarcinoma | van Bokhoven et al., 2003 | 14518029 |
| PEAZ-1 | Human | Prostate carcinoma | HT-1080 | Human | Sarcoma (fibrosarcoma) | | 11433418 |
| PH61-N | Human | Not specified | MIA PaCa-2 | Human | Pancreatic carcinoma | JCRB website | No PMID |
| PLB-985 | Human | Leukemia, acute myeloid, M4 | HL-60 | Human | Leukemia, acute myeloid, M2 | Drexler et al., 2003 | 12592342 |
| PPC-1 | Human | Prostate carcinoma | PC-3 | Human | Prostate carcinoma | Chen, 1993 | 8428522 |
| PSV811 | Human | Skin fibroblast, Werner's syndrome | WI-38 | Human | Lung, normal diploid fibroblasts | JCRB website | No PMID |

| | | | | | | | |
|---|---|---|---|---|---|---|---|
| RAMAK-1 | Human | Muscle synovium | T-24 | Human | Bladder carcinoma | MacLeod et al., 1999 | 10508494 |
| RB | Human | Lymphoma, Hodgkin | OMK-210 | Monkey, Owl (*Aotus trivirgatus*) | Kidney, normal renal cells | Nelson-Rees et al., 1981; Drexler et al., 2003 | 6451928, 12592342 |
| RBHF-1 | Human | Liver, hepatoma | Unknown | Nonhuman | Unknown | MacLeod et al., 1999 | 10508494 |
| RC-2A | Human | Leukemia, acute myeloid, M4 | CCRF-CEM | Human | Leukemia, acute lymphoblastic, T-cell | Drexler et al., 2003 | 12592342 |
| RED-3 | Human | Leukemia, acute myeloid | HL-60 | Human | Leukemia, acute myeloid, M2 | Drexler et al., 2003 | 12592342 |
| REH-6 | Human | Leukemia, acute lymphoblastic, B-cell precursor | Unknown | Mouse | Unknown | Drexler et al., 2003 | 12592342 |
| RERF-LC-OK | Human | Lung carcinoma | Marcus | Human | Astrocytoma | JCRB website | No PMID |
| RM-10 | Human | Leukemia, chronic myeloid, blast crisis | K-562 | Human | Leukemia, chronic myeloid, blast crisis | Drexler et al., 2003 | 12592342 |
| RMG-I | Human | Ovarian carcinoma | RMG-II | Human | Ovarian carcinoma | JCRB website | No PMID |
| RPMI-6666 | Human | Lymphoma, Hodgkin | Correct name, incorrect cell type | Human | EBV+ B-lymphoblastoid cell line | Drexler et al., 2003 | 12592342 |
| RS-1 | Human | Leukemia, acute myeloid, M7 | K-562 | Human | Leukemia, chronic myeloid, blast crisis | Drexler et al., 2003 | 12592342 |
| RY | Human | Lymphoma, Hodgkin | Unknown | Monkey | Unknown | Harris et al., 1981; Drexler et al., 2003 | 7192801, 12592342 |
| SA4 | Human | Sarcoma (liposarcoma) | HeLa | Human | Cervical adenocarcinoma | Nelson-Rees et al., 1981 | 6451928 |
| SAM-1 | Human | Leukemia, chronic myeloid, blast crisis | K-562 | Human | Leukemia, chronic myeloid, blast crisis | Drexler et al., 2003 | 12592342 |
| SBC-2 | Human | Bladder carcinoma | HeLa | Human | Cervical adenocarcinoma | MacLeod et al., 1999 | 10508494 |
| SBC-7 | Human | Bladder carcinoma | HeLa | Human | Cervical adenocarcinoma | MacLeod et al., 1999 | 10508494 |
| SCCTF | Human | Oral carcinoma, squamous cell | SCCKN | Human | Oral carcinoma, squamous cell | Yoshino et al., 2006 | 16643607 |
| SCLC-16H | Human | Lung carcinoma, small cell | SCLC-21/22H | Human | Lung carcinoma, small cell | MacLeod et al., 1999 | 10508494 |
| SCLC-24H | Human | Lung carcinoma, small cell | SCLC-21/22H | Human | Lung carcinoma, small cell | MacLeod et al., 1999 | 10508494 |
| SH-2 | Human | Breast carcinoma | HeLa | Human | Cervical adenocarcinoma | Nelson-Rees et al., 1981 | 6451928 |
| SEG-1 | Human | Esophageal adenocarcinoma | NCI-H460 | Human | Lung carcinoma, large cell | Boonstra et al., 2010 | 20075370 |

*(continued overleaf)*

**TABLE V.I.** (*Continued*)

| Misidentified Cell line | Claimed species | Claimed cell type | Contaminating cell line | Actual species | Actual cell type | Misidentification reported by | Reference PubMed ID |
|---|---|---|---|---|---|---|---|
| SH-3 | Human | Breast carcinoma | HeLa | Human | Cervical adenocarcinoma | Nelson-Rees et al., 1981 | 6451928 |
| SK-MG-1 | Human | Astrocytoma | Marcus | Human | Astrocytoma | JCRB website | No PMID |
| SJPL | Pig | Lung cells, immortalized epithelial | Unknown | Monkey, African green | Unknown | Silversides et al, 2010 | 20200241 |
| SK-GT-5 | Human Esophageal | adenocarcinoma | SK-GT-2 | Human | Gastric carcinoma (gastric fundus) | Boonstra et al, 2010 | 20075370 |
| SK-N-MC | Human | Neuroblastoma | Unknown | Human | Sarcoma (Ewing's) | Staege et al., 2004 | 15548687 |
| SKW-3 | Human | Leukemia, chronic lymphocytic, T-cell | KE-37 | Human | Leukemia, acute lymphoblastic | DSMZ website | No PMID |
| SNB-19 | Human | Glioblastoma | U-251 | Human | Glioblastoma | Azari et al., 2007; ATCC website | 17254797 |
| SPI-801 | Human | Leukemia, acute lymphoblastic, T-cell | K-562 | Human | Leukemia, chronic myeloid, blast crisis | Gignac et al., 1993; Drexler et al., 2003 | 8220135, 12592342 |
| SPI-802 | Human | Leukemia, acute lymphoblastic, T-cell | K-562 | Human | Leukemia, chronic myeloid, blast crisis | Gignac et al., 1993; Drexler et al., 2003 | 8220135, 12592342 |
| SpR | Human | Lymphoma, Hodgkin | OMK-210 | Monkey, owl (*Aotus trivirgatus*) | Kidney, normal renal cells | Nelson-Rees et al., 1981; Drexler et al., 2003 | 6451928, 12592342 |
| SQ-5 | Human | Lung carcinoma | HeLa | Human | Cervical adenocarcinoma | Yoshino et al., 2006 | 16643607 |
| SR-91 | Human | Leukemia, acute lymphoblastic, T cell | AML-193 | Human | Leukemia, acute myeloid, M5 | Drexler et al., 2003 | 12592342 |
| SW-527 | Human | Colon carcinoma | SW-480, SW-620 | Human | Colon carcinoma | Nelson-Rees et al., 1981 | 6451928 |
| SW-598 | Human | Colon carcinoma | SW-480, SW-620 | Human | Colon carcinoma | Nelson-Rees et al., 1981 | 6451928 |
| SW-608 | Human | Colon carcinoma | SW-480, SW-620 | Human | Colon carcinoma | Nelson-Rees et al., 1981 | 6451928 |
| SW-613 | Human | Colon carcinoma | SW-480, SW-620 | Human | Colon carcinoma | Nelson-Rees et al., 1981 | 6451928 |
| SW-732 | Human | Colon carcinoma | SW-480, SW-620 | Human | Colon carcinoma | Nelson-Rees et al., 1981 | 6451928 |
| SW-733 | Human | Colon carcinoma | SW-480, SW-620 | Human | Colon carcinoma | Nelson-Rees et al., 1981 | 6451928 |
| T-1 | Human | Kidney, normal renal cells | HeLa | Human | Cervical adenocarcinoma | Nelson-Rees et al., 1980; Nelson-Rees et al., 1981 | 7394535, 6451928 |
| T-9 (WI-38 derivative) | Human | Lung cells, transformed normal diploid fibroblasts | HeLa | Human | Cervical adenocarcinoma | Nelson-Rees et al., 1981 | 6451928 |

| Cell line | Species | Claimed origin | Cross-contaminant | Species | Actual origin | Reference | PMID |
|---|---|---|---|---|---|---|---|
| T-33 | Human | Leukemia, chronic myeloid, blast crisis | K-562 | Human | Leukemia, chronic myeloid, blast crisis | Drexler et al., 2003 | 12592342 |
| T3M-12 | Human | Lung carcinoma | T3M-1 | Human | Oral carcinoma | RIKEN website | No PMID |
| TCO-1 | Human | Cervical carcinoma | TCS | Human | Cervical carcinoma | Yoshino et al., 2006 | 16643607 |
| TDL-1 | Human | Tonsillar lymphoid cells, non-neoplastic | P3JHR-1 | Human | Lymphoma, Burkitt | Nelson-Rees et al., 1981 | 6451928 |
| TDL-2 | Human | Tonsillar lymphoid cells, non-neoplastic | P3JHR-1 | Human | Lymphoma, Burkitt | Nelson-Rees et al., 1981 | 6451928 |
| TDL-3 | Human | Tonsillar lymphoid cells, non-neoplastic | RPMI 1788 | Human | Lymphoblastoid cell line, normal donor | Nelson-Rees et al., 1981 | 6451928 |
| TDL-4 | Human | Tonsillar lymphoid cells, non-neoplastic | Raji | Human | Lymphoma, Burkitt | Nelson-Rees et al., 1981 | 6451928 |
| TE-12 | Human | Esophageal squamous cell carcinoma | TE-2, TE-3, TE-12 or TE-13 (genetically identical) | Human | Esophageal squamous cell carcinoma | Boonstra et al., 2007 | 17804709 |
| TE-13 | Human | Esophageal squamous cell carcinoma | TE-2, TE-3, TE-12 or TE-13 (genetically identical) | Human | Esophageal squamous cell carcinoma | Boonstra et al., 2007 | 17804709 |
| TE-2 | Human | Esophageal squamous cell carcinoma | TE-2, TE-3, TE-12 or TE-13 (genetically identical) | Human | Esophageal squamous cell carcinoma | Boonstra et al., 2007 | 17804709 |
| TE-3 | Human | Esophageal squamous cell carcinoma | TE-2, TE-3, TE-12 or TE-13 (genetically identical) | Human | Esophageal squamous cell carcinoma | Boonstra et al., 2007 | 17804709 |
| TE671 | Human | Medulloblastoma | RD | Human | Sarcoma (rhabdomyosarcoma) | Stratton et al., 1989; Chen et al., 1989 | 2650908, 2739733 |
| TE671 Subline No.2 | Human | Medulloblastoma | RD | Human | Sarcoma (rhabdomyosarcoma) | ECACC website | No PMID |
| TE-7 | Human | Esophageal adenocarcinoma | TE-2, TE-3, TE-12 or TE-13 (genetically identical) | Human | Esophageal squamous cell carcinoma | Boonstra et al., 2007 | 17804709 |
| TK-1 | Human | Glioblastoma | U-251 MG | Human | Glioblastoma | JCRB website | No PMID |
| TEC61 | Human | Thyroid, endothelium | JEG3 | Human | Choriocarcinoma | Patel et al, 2003 (retraction) | 12388152 |
| TI-1 | Human | Leukemia, acute myeloid, M2 | K-562 | Human | Leukemia, chronic myeloid, blast crisis | Rush et al., 2002; Drexler et al., 2003 | 11871388, 12592342 |

(continued overleaf)

**TABLE V.I.** (*Continued*)

| Misidentified Cell line | Claimed species | Claimed cell type | Contaminating cell line | Actual species | Actual cell type | Misidentification reported by | Reference PubMed ID |
|---|---|---|---|---|---|---|---|
| TMH-1 | Human | Thyroid, benign goitre | IHH-4 | Human | Thyroid, papillary thyroid carcinoma | JCRB website | No PMID |
| TMM | Human | Leukemia, chronic myeloid, blast crisis | Correct name, incorrect cell type | Human | EBV+ B-lymphoblastoid cell line | Drexler et al., 2003 | 12592342 |
| TSU-Pr1 | Human | Prostate carcinoma | T-24 | Human | Bladder carcinoma | van Bokhoven et al., 2001 | 11522622 |
| TuWi | Human | Not specified | HeLa | Human | Cervical adenocarcinoma | Nelson-Rees et al., 1981 | 6451928 |
| U-118 MG | Human | Glioblastoma | U-138 MG | Human | Glioblastoma | ATCC website | No PMID |
| UM-UC-2 | Human | Bladder carcinoma | T-24 | Human | Bladder carcinoma | Chiong et al., 2009 | 19375735 |
| UM-UC-3-GFP (UM-UC-3 derivative) | Human | Bladder carcinoma, GFP-transfected | Unknown, NOT UM-UC-3 | Human | Unknown | Chiong et al., 2009 | 19375735 |
| UTMB-460 | Human | B-cell | CCRF-CEM | Human | Leukemia, acute lymphoblastic, T-cell | Drexler et al., 2003 | 12592342 |
| VM-CUB-III | Human | Bladder carcinoma | VM-CUB-I | Human | Bladder carcinoma | Masters et al., 1986 | 3708594 |
| WiDr | Human | Colon carcinoma | HT-29 | Human | Colon carcinoma | Chen et al., 1987 | 3472642 |
| WISH | Human | Amnion, normal cell | HeLa | Human | Cervical adenocarcinoma | Nelson-Rees et al., 1981 | 6451928 |
| WKD | Human | Conjunctiva | HeLa | Human | Cervical adenocarcinoma | ECACC website | No PMID |
| Wong-Kilbourne | Human | Conjunctiva | HeLa | Human | Cervical adenocarcinoma | Nelson-Rees et al., 1981 | 6451928 |
| WRL 68 | Human | Liver, embryonic cells | HeLa | Human | Cervical adenocarcinoma | ECACC website | No PMID |
| WSU-ALCL | Human | Lymphoma, anaplastic large cell | CCRF-CEM | Human | Leukemia, acute lymphoblastic, T-cell | Drexler et al., 2003 | 12592342 |
| WSU-CLL | Human | Leukemia, chronic lymphocytic | REH | Human | Leukemia, acute lymphoblastic, B-cell precursor | Drexler et al., 2002; Drexler et al., 2003 | 12200708, 12592342 |
| YAA | Human | Monocytes | U-937 | Human | Lymphoma, histiocytic | Drexler et al., 2003 | 12592342 |
| YAP | Human | Monocytes | U-937 | Human | Lymphoma, histiocytic | Drexler et al., 2003 | 12592342 |
| YJ | Human | Leukemia, chronic myelomonocytic | HL-60 | Human | Leukemia, acute myeloid, M2 | Drexler et al., 2003 | 12592342 |

**TABLE V.II.** Cross-contaminated or Misidentified Cell Lines Where Authentic Stock May Exist

| Cell line | Claimed species | Claimed cell type | Contaminating cell line | Actual species | Actual cell type | Misidentification reported by | Reference PubMed ID |
|---|---|---|---|---|---|---|---|
| 207 | Human | Leukemia, acute lymphoblastic, B-cell precursor | REH and SUP-B2 | Human | Leukemia, acute lymphoblastic, B-cell precursor | MacLeod et al., 1999; Drexler et al., 2003 | 10508494, 12592342 |
| BJA-B | Human | Bladder carcinoma | HeLa | Human | Cervical adenocarcinoma | Nelson-Rees & Flandermeyer, 1976 | 1246601 |
| BT-20 | Human | Breast carcinoma | HeLa | Human | Cervical adenocarcinoma | Nelson-Rees et al., 1981 | 6451928 |
| HPB-ALL | Human | Leukemia, acute lymphoblastic, T-cell | JURKAT | Human | Leukemia, acute lymphoblastic, T-cell | Drexler et al., 2003 | 12592342 |
| J-82 | Human | Bladder carcinoma | T-24 | Human | Bladder carcinoma | Masters et al., 1986 | 3708594 |
| KARPAS-45 | Human | Leukemia, acute lymphoblastic, T-cell | Unknown | Human | Unknown | Drexler et al., 1999 | 10516762 |
| KBM-3 | Human | Leukemia, acute myeloid, M4 | HL-60 | Human | Leukemia, acute myeloid, M2 | Drexler et al., 2003 | 12592342 |
| KE-37 | Human | Leukemia, acute lymphoblastic, T-cell | CCRF-CEM | Human | Leukemia, acute lymphoblastic, T-cell | Drexler et al., 1999 | 10516762 |
| L-540 | Human | Lymphoma, Hodgkin | CCRF-CEM | Human | Leukemia, acute lymphoblastic, T-cell | Drexler et al., 2003 | 12592342 |
| MB-02 | Human | Leukemia, acute myeloid, M7 | HU-3 | Human | Leukemia, acute myeloid, M7 | Drexler et al., 1999 | 10516762 |
| RPMI-8402 | Human | Leukemia, acute lymphoblastic, T-cell | Unknown | Human | Unknown | Drexler et al., 1999 | 10516762 |
| RT4 | Human | Bladder carcinoma | HeLa | Human | Cervical adenocarcinoma | Nelson-Rees et al., 1981 | 6451928 |
| U-373 MG | Human | Glioblastoma | U-251 MG | Human | Glioblastoma | Azari et al., 2007; ATCC website | 17254797 |
| U-937 | Human | Lymphoma, histiocytic | Unknown | Human | Unknown | Drexler et al., 1999 | 10516762 |
| UT-7 | Human | Leukemia, acute myeloid, M7 | U-937 | Human | Lymphoma, histiocytic | Drexler et al., 2003 | 12592342 |

# General Textbooks and Relevant Journals

Barlovatz-Meimon, G., & Adolphe, M. (2003). *Culture de cellules animales: Methodologies, applications.* Paris: Editions INSERM. *A collection of technique-oriented chapters on basic and advanced aspects of tissue culture.*

Alberts, B., Johnson, A., Lewis, J., Raff, M., Roberts, K., & Walter, P. (2008). *Molecular biology of the cell.* New York: Garland Science. *Classic textbook on cellular and molecular biology.*

Butler, M., ed. (2004). *Animal Cell Culture and Technology.* London: Bios Scientific Publishers. *A useful introduction to basic biotechnology.*

Cann, A. J., ed. (2004). *Virus culture.* Practical Approach Series, Oxford University Press. *A useful introduction to culture, purification, and assay of viruses.*

Davis, J. M., ed. (2002). *Basic cell culture: A practical approach.* Oxford: IRL Press at Oxford University Press. *Multi-author textbook on basic and apllied aspects, including quality assurance and GLP.*

Doyle, A., Griffiths, J. B., & Newell, D. G., eds. (1998). *Cell and tissue culture: Laboratory procedures.* Chichester, UK: Wiley. *A loose-leaf compendium of general and specialized techniques with regular updates. Very expensive, but a very good source for a wide variety of techniques.*

Freshney, R. I., & Freshney, M. G., eds. (2002). *Culture of epithelial cells,* New York: Wiley-Liss. *Invited chapters on specialized culture of epithelium; technique oriented.*

Freshney, R. I. (1999). *Freshney's culture of animal cells: A multimedia guide.* New York: Wiley-Liss.

Freshney, R. I., & Freshney, M. G. (1996). *Culture of immortalized cells.* New York: Wiley-Liss.

Freshney, R. I., Pragnell, I. B., & Freshney, M. G., eds. (1994). *Culture of hematopoietic cells.* New York: Wiley-Liss. *Second in the series "Culture of Specialized Cells". Invited chapters on specialized techniques.*

Freshney, R. I., Stacey, G. N., & Auerbach, J. M. (2007). *Culture of human stem cells.* Hoboken, NJ: Wiley-Liss. *Protocols for the culture of embryonic, new born and adult stem cells.*

Haynes, L. W., ed. (1999). *The neuron in tissue culture.* Chichester, UK: Wiley.

Helgason, C. D., & Miller, C. L., eds. (2005). *Basic cell culture protocols.* Totowa, NJ: Humana Press. *Wide ranging cover of basic and specialized techniques including mycoplasma detection and authentication.*

Leigh, I. M., Lane, E. B., & Watt, F. M., eds. (1994). *The keratinocyte handbook.* Cambridge: Cambridge University Press.

Leigh, I. M., & Watt, F. M., eds. (1994). *Keratinocyte methods.* Cambridge: Cambridge University Press.

Lodish, H., Berk, A., Kaiser, C. A., Kreiger, M., Scott, M. P., Bretscher, A., & Matsudaira, P. T. (2007). *Molecular cell biology,* New York: Scientific American Books, Freeman.

Masters, J. R. W., ed. (2000). *Animal cell culture, a practical approach,* 3rd ed. Oxford: IRL Press.

Masters, J. R. W., ed. (1991). *Human cancer in primary culture.* London: Kluwer. *Product of a European Tissue Culture Society workshop.*

Masters, J. R. W., & Palsson, B., eds. (1999). *Human cell culture.* Dordrecht: Kluwer.

Mothersill, C., & Austin, B., eds. (2000). *Aquatic invertebrate cell culture.* Chichester, UK: Springer/Praxis. *Review of methods for invertebrate cell culture.*

Papas, D. (2010). *Practical cell analysis.* Chichester, UK, John Wiley & Sons. *Useful chapter on analytical procedures including microscopy, flow cytometry and microfluidics.*

Pfragner, R., & Freshney, R. I., eds. (2004). *Culture of human tumor cells.* Hoboken, NJ: Wiley-Liss. *Invited chapters with culture protocols for several types of human tumors.*

Pollack, R., ed. (1981). *Reading in mammalian cell culture,* Cold Spring Harbor, NY: Cold Spring Harbor Laboratory Press. *A very good compilation of key papers in the early development of the field. Used as a tutorial, for general interest, and for teaching.*

Ravid, K., & Freshney, R. I., eds. (1998). *DNA transfer to cultured cells*. New York: Wiley-Liss. *Invited chapters with practical protocols on DNA transfer technology.*

Shahar, A., de Vellis, J., Vernadakis, A., & Haber, B. (1989). *A dissection and tissue culture manual of the nervous system*. New York: Wiley-Liss. *Useful short protocols; well illustrated.*

Vunjak-Novakovic, G., & Freshney, R. I., eds. (2006). *Culture of cells for tissue engineering*. Hoboken, NJ: Wiley-Liss. *Invited chapters with practical protocols on preparation of cells and matrices for tissue engineering.*

## Useful Journals

### Technique-Oriented Tissue Culture

Biotechnology and Bioengineering
Cell Preservation Technology
Cytotechnology (now incorporating *Methods in Cell Science*)
In Vitro Cell and Development Biology
Tissue Culture Research Communications (Japanese)
Nature Methods

### Cell Biology

Cell
Cell Biology, International Reports
Cellular Biology
Cell Growth & Differentiation
Current Opinion in Cell Biology
European Journal of Cell Biology
Experimental Cell Biology
Experimental Cell Research
Journal of Cell Biology
Journal of Cellular Physiology
Journal of Cell Science
Nature Biotechnology
Nature Cell Biology

### Cancer

British Journal of Cancer
Cancer Research
European Journal of Cancer and Clinical Oncology
International Journal of Cancer
Journal of the National Cancer Institute
Nature Reviews, Cancer

# References

Aaronson, S. A., & Todaro, G. J. (1968). Development of 3T3-like lines from Balb/c mouse embryo cultures: Transformation susceptibility to SV40. *J. Cell Physiol.* **72**:141–148.

Aaronson, S. A., Bottaro, D. P., Miki, T., Ron, D., Finch, P. W., Fleming, T. P., Ahn, J., Taylor, W. G., & Rubin, J. S. (1991). Keratinocyte growth factor: A fibroblast growth factor family member with unusual target cell specificity. *Ann. NY Acad. Sci.* **638**:62–77.

Aaronson, S. A., Todaro, G. J., & Freeman, A. E. (1970). Human sarcoma cells in culture: Identification by colony-forming ability on monolayers of normal cells. *Exp. Cell. Res.* **61**:1–5.

Aasen, T., Raya, A., Maria J Barrero, M. J., Garreta, E., Consiglio, A., Gonzalez, F., Vassena, R., Bilic, J., Pekarik, V., Tiscornia, G., Edel, M.,, Boué, S., & Izpisúa Belmonte, J. C. I. (2008). Efficient and rapid generation of induced pluripotent stem cells from human keratinocytes. *Nat. Biotechnol.* **26**:1276–1284.

Abaza, N. A., Leighton, J., & Schultz, S. G. (1974). Effects of ouabain on the function and structure of a cell line (MDCK) derived from canine kidney; I: Light microscopic observations of monolayer growth. *In Vitro* **10**:172–183.

Abbott, N. J., Hughes, C. C., Revest, P. A., & Greenwood, J. (1992). Development and characterisation of a rat brain capillary endothelial culture: Towards an in vitro blood-brain barrier. *J. Cell Sci.* **103**(Pt 1): 23–37.

Abercrombie, M., & Heaysman, J. E. (1953). Observations on the social behaviour of cells in tissue culture; I: Speed of movement of chick heart fibroblasts in relation to their mutual contacts. *Exp. Cell Res.* **5**:111–131.

Abercrombie, M., & Heaysman, J. E. M. (1954). Observations on the social behaviour of cells in tissue culture; II: "Monolayering" of fibroblasts. *Exp. Cell Res.* **6**:293–306.

Abney, E. R., Williams B. P., & Raff M. C. (1983). Tracing the development of oligodendrocytes from precursor cells using monoclonal antibodies, fluorescence activated cell sorting and cell culture. *Dev. Biol.* **100**:166–171.

ACDP. (2003). Advisory Committee on Dangerous Pathogens. www.advisorybodies.doh.gov.uk/acdp/publications.htm.

ACDP. (2005). Biological agents: Managing the risks in laboratories and healthcare premises. Appendix 3.1, Cell Cultures. www.advisorybodies.doh.gov.uk/acdp/managingtherisks.pdf.

Adams, J. C., & Watt, F. M. (1993). Regulation of development and differentiation by the extracellular matrix. *Development* **117**:1183–1198.

Adams, R. L. P. (1980). Cell culture for biochemists. In Work, T. S., & Burdon, R. H. (eds.), *Laboratory techniques in biochemistry and molecular biology*. Amsterdam: Elsevier/North Holland Biomedical Press.

Adolphe, M. (1984). Multiplication and type II collagen production by rabbit articular chondrocytes cultivated in a defined medium. *Exp. Cell. Res.* **155**:527–536.

Adolphe, M., & Benya, P. D. (1992). Different types of cultured chondrocytes: The in vitro approach to the study of biochemical regulation. In Adolphe, M., ed., *Biological regulation of the chondrocytes*. Boca Raton, FL: CRC Press, pp. 105–139.

Ager, A. (1987). Isolation and culture of high endothelial cells from rat lymph nodes. *J. Cell Sci.* **87**(Pt 1): 133–144.

Agy, P. C., Shipley, G. D., & Ham, R. G. (1981). Protein-free medium for mouse neuroblastoma cells. *In Vitro* **17**: 671–680.

Ahmed, I., Collins, C. A., Lewis, M. P., Olsen, I., & Knowles, J. C. (2004). Processing, characterisation and biocompatibility of iron-phosphate glass fibres for tissue engineering. *Biomaterials* **25**(16): 3223–3232.

Aigner, J., Tegeler, J., Hutzler, P., Campoccia, D., Pavesio, A., Hammer, C., Kastenbauer, E., & Naumann, A. (1998). Cartilage tissue engineering with novel nonwoven structured biomaterial based on hyaluronic acid benzyl ester. *J. Biomed. Mater. Res.* **42**(2): 172–181.

Aizawa, S., Sensui, N., & Yamamuro, Y. (2009). Induction of cholinergic differentiation by 5-azacytidine in NG108-15 neuronal cells. *Neuroreport* **20**:157–160.

Albelda, S. M., Oliver, P. D., Romer, L. H., & Buck, C. A. (1990). EndoCAM: A novel endothelial cell–cell adhesion molecule. *J. Cell. Biol.* **110**:1227–1237.

Alberts, B., Johnson, A., Lewis, J., Raff, M., Roberts, K., & Walter, P. (2008). *Molecular biology of the cell.* New York: Garland Science.

Albrecht, A. M., Biedler, J. L., & Hutchison, D. J. (1972). Two different species of dihydrofolate reductase in mammalian cells differentially resistant to amethopterin and methasquin. *Cancer Res.* **32**:1539–1546.

Alexander, C. L., FitzGerald, U. F., & Barnett, S. C. (2002). Identification of growth factors that promote long-term proliferation of olfactory ensheathing cells and modulate their antigenic phenotype. *Glia* **37**:349–364.

Al-Hajj, M., Becker M. W., Wicha, M., Weissman, I., & Clarke, M. F., (2004). Therapeutic implications of cancer stem cells. *Curr. Opin. Genet. Dev.* **14**:43–47.

Aliabadian M., Kaboli, M., Nijman, V., & Vences, M. (2009). Molecular identification of birds: performance of distance-based DNA barcoding in three genes to delimit parapatric species. *PLoS ONE* **4**(1):e4119.

Alison, M. R., Vig, P., Russo, F., Bigger, B. W., Amofah, E., Themis, M., & Forbes, S. (2004). Hepatic stem cells: From inside and outside the liver? *Cell Prolif.* **37**:1.

Alley, M. C., Scudiero, D. A., Monks, A., Hursey, M. L., Czerwiniski, M. J., Fine, D. L., Abbot, B. J., Mayo, J. G., Shoemaker, R. H., & Boyd, M. R. (1988). Feasibility of drug screening with panels of human tumour cell lines using a microculture tetrazolium assay. *Cancer Res.* **48**:589–601.

Al-Mufti, R., Hambley, H., Farzaneh, F., & Nicolaides, K. H. (2004). Assessment of efficacy of cell separation techniques used in the enrichment of foetal erythroblasts from maternal blood: Triple density gradient vs. single density gradient. *Clin. Lab. Haematol.* **26**(2): 123–128.

Al-Rubeai, M., & Singh, R. P. (1998). Apoptosis in cell culture. *Curr. Opin. Biotechnol.* **9**:152–156.

Al-Rubeai, M., Welzenbach, K., Lloyd, D. R., & Emery, A. N. (1997). A rapid method for evaluation of cell number and viability by flow cytometry. *Cytotechnology* **24**:161–168.

Alvarez-Dolado, M., Pardal, R., Garcia-Verdugo, J. M., Fike, J. R., Lee, H. O., Pfeffer, K., Lois, C., Morrison, S. J., & Alvarez-Buylla, A. (2003). Fusion of bone-marrow-derived cells with Purkinje neurons, cardiomyocytes and hepatocytes. *Nature* **425**:968–973.

Ames, B. N. (1980). Identifying environmental chemicals causing mutations and cancer. *Science* **204**:587–593.

Anazawa T, Balamurugan AN, Bellin M, Zhang HJ, Matsumoto S, Yonekawa Y, Tanaka T, Loganathan G, Papas KK, Beilman GJ, Hering BJ, & Sutherland DE. (2009). Human islet isolation for autologous transplantation: comparison of yield and function using SERVA/Nordmark versus Roche enzymes. *Am. J. Transplant.* **9**:2383–2391.

Anderson, J. E., & Wozniak, A. C. (2004). Satellite cell activation on fibers: Modeling events in vivo. An invited review. *Can. J. Physiol. Pharmacol.* **82**:300–310.

Andersson, L. C., Nilsson, K., & Gahmberg, C. G. (1979a). K562—a human erythroleukemic cell line. *Int. J. Cancer* **23**:143–147.

Andersson, L. C., Jokinen, M., Klein, G., & Nilsson, K. (1979b). Presence of erythrocytic components in the K562 cell line. *Int. J. Cancer.* **24**:5–14.

Andersson, L. C., Jokinen, M., & Gahmberg, C. G. (1979c). Induction of erythroid differentiation in the human leukaemia cell line K562. *Nature* **278**:364–365.

Andreoli, S. P., & McAteer, J. A. (1990). Reactive oxygen molecule-mediated injury in endothelial and renal tubular epithelial cells in vitro. *Kidney Int.* **38**:785–794.

Aninat C, Piton A, Glaise D, Le Charpentier T, Langouët S, Morel F, Guguen-Guillouzo C, Guillouzo A. (2006). Expression of cytochromes P450, conjugating enzymes and nuclear receptors in human hepatoma HepaRG cells. *Drug Metab. Dispos.* **34**:75–83.

Antoniades, H. N., Scher, C. D., & Stiles, C. D. (1979). Purification of human platelet-derived growth factor. *Proc. Natl. Acad. Sci. USA* **76**:1809.

Aoki H, Motohashi T, Yoshimura N, Yamazaki H, Yamane T, Panthier JJ, & Kunisada T. (2005). Cooperative and indispensable roles of endothelin 3 and KIT signalings in melanocyte development. *Dev. Dyn.* **233**:407–417.

Applied Cytometry. (2008). Flow cytometry. www.applied cytometry.com/flow_cytometry.php#cell.

Archer, C. W., Oldfield, S., Redman, S., Haughton, L., Dowthwaite, G., Khan, I., & Ralphs, J. (2007). Isolation, characterization, and culture of soft connective tissue stem/progenitor cells. In Freshney, R. I., Stacey, G. N., & Auerbach, J. M. (eds.), *Culture of human stem cells.* Hoboken: Wiley, pp. 233–248.

Armati, P. J., & Bonner, J. (1990). A technique for promoting Schwann cell growth from fresh and frozen biopsy nerve utilizing D-valine medium. *In Vitro Cell Dev. Biol.* **26**:1116–1118.

Artursson, P., & Magnusson, C. (1990). Epithelial transport of drugs in cell culture II: Effect of extracellular calcium concentration on the paracellular transport of drugs of different lipophilicities across monolayers of intestinal epithelial (Caco-2) cells. *J. Pharmaceut. Sci.* **79**:595–600.

Askanas, V., Bornemann, A., & Engel, W. K. (1990). Immuno-cytochemical localization of desmin at human neuromuscular junctions. *Neurology* **40**:949–953.

Assoian, R. K., & Yung, Y. (2008). A reciprocal relationship between Rb and Skp2: Implications for restriction point control, signal transduction to the cell cycle and cancer. *Cell Cycle* **7**:24–27.

Atala, A., & Lanza, R. P. (2002). *Methods of tissue engineering.* San Diego: Academic Press.

ASN-0002 (2010). American Type Culture Collection Standards Development Organization Workgroup ASN-0002. Cell line misidentification: the beginning of the end. *Nat Rev Cancer* **10**:441–448.

Au, A. M.-J., & Varon, S. (1979). Neural cell sequestration on immunoaffinity columns. *Exp. Cell Res.* **120**:269.

Auerbach, R., & Grobstein, C. (1958). Inductive interaction of embryonic tissues after dissociation and reaggregation. *Exp. Cell Res.* **15**:384–397.

Aunins, J. G., Bader, B., Caola, A., Griffiths, J., Katz, M., Licari, P., Ram, K., Ranucci, C. S., Zhou, W. (2003). Fluid mechanics, cell distribution, and environment in CellCube bioreactors. *Biotechnol. Prog.* **19**:2–8.

Ausubel, F. M., Brent, R., Kingston, R. E., Moore, D. D., Seidman, J. G., Smith, J. A., & Struhl, K. (eds.). (2009). *Current protocols in molecular biology.* Hoboken, NJ: Wiley.

Azari, S., Ahmadi, N., Tehrani, M. J., & Shokri, F. (2007). Profiling and authentication of human cell lines using short tandem repeat (STR) loci: report from the National Cell Bank of Iran. *Biologicals* **35**:195–202.

Babich, H., & Borenfreund, E. (1990). Neutral red uptake. In Doyle, A., Griffiths, J. B., & Newall, D. G. (eds.), *Cell and tissue culture: Laboratory procedures.* Chichester, UK: Wiley, module 4B:7.

Bagley, R. G., Walter-Yohrling, J., Cao, X., Weber, W., Simons, B., Cook, B. P., Chartrand, S. D., Wang, C., Madden, S. L., & Teicher, B. A. (2003). Endothelial precursor cells as a model of tumor endothelium: Characterization and comparison with mature endothelial cells. *Cancer Res.* **63**:5866–5873.

Baker, J. R. J. (2006). Autoradiography: A comprehensive overview. *Royal Microscopical Society microscopy handbooks*, **18**. New York: Oxford Univesrity Press.

Bakolitsa, C., Cohen, D. M., Bankston, L. A., Bobkov, A. A., Cadwell, G. W., Jennings, L., Critchley, D. R., Craig, S. W., & Liddington, R. C. (2004). Structural basis for vinculin activation at sites of cell adhesion. *Nature* **430**:583–586.

Balin, A. K., Goodman, B. P., Rasmussen, H., & Cristofalo, V. J. (1976). The effect of oxygen tension on the growth and metabolism of WI-38 cells. *J. Cell Physiol.* **89**:235–250.

Balkovetz, D. F., & Lipschutz, J. H. (1999). Hepatocyte growth factor and the kidney: It is not just for the liver. *Int. Rev. Cytol.* **186**:225–260.

Ballard, P. L. (1979). Glucocorticoids and differentiation. *Glucocorticoid Horm. Action* **12**:439–517.

Ballard, P. L., & Tomkins, G. M. (1969). Dexamethasone and cell adhesion. *Nature* **244**:344–345.

Balmforth, A. J., Ball, S. G., Freshney, R. I., Graham, D. I., McNamee, B., & Vaughan, P. F. T. (1986). D-1 dopaminergic and beta-adrenergic stimulation of adenylate cyclase in a clone derived from the human astrocytoma cell line G-CCM. *J. Neurochem.* **47**:715–719.

Balls, M., Coecke, S., Bowe, G., Davis, J., Gstraunthaler, G., Hartung, T., Hay, R., Merten, O. W., Price, A., Schechtman, L. M., Stacey, G., & Stokes, W. (2006). The importance of good cell culture practice (GCCP). *ALTEX* **23** suppl: 270–273.

Baltimore, D. (2001). Our genome unveiled. *Nature* **409**:814–816.

Banfalvi, G. (2008). Cell cycle synchronization of animal cells and nuclei by centrifugal elutriation. *Nat. Protoc.* **3**:663–673.

Bansal, R., Stefansson K., & Pfeiffer, S. E. (1992). Proligodendroblast antigen (POA), a developmental antigen expressed by A007/O4-positive oligodendrocyte progenitors prior to the appearance of sulfatide and galactocerebroside. *J. Neurochem.* **58**:2221–2229.

Barber, N., Gez, S., Belov, L., Mulligan, S. P., Woolfson, A., & Christopherson, R. I. (2009). Profiling CD antigens on leukaemias with an antibody microarray. *FEBS Lett.* **583**:1785–1791.

Barbulovic-Nad, I., Au, S. H., & Wheeler, A. R. (2010). A microfluidic platform for complete mammalian cell culture. *Lab Chip.* **10**:1536–1542.

Bard, D. R., Dickens, M. J., Smith, A. U., & Sarck, J. M. (1972). Isolation of living cells from mature mammalian bone. *Nature* **236**:314–315.

Barnes, D., & Sato, G. (1980). Methods for growth of cultured cells in serum-free medium. *Anal. Biochem.* **102**:255–270.

Barnes, W. D., Sirbasku, D. A., & Sato, G. H. (eds.). (1984a). *Cell culture methods for molecular and cell biology; Vol. 1: Methods for preparation of media, supplements, and substrata for serum-free animal cell culture.* New York: Liss.

Barnes, W. D., Sirbasku, D. A., & Sato, G. H. (eds.). (1984b). *Cell culture methods for molecular and cell biology; Vol. 2: Methods for serum-free culture of cells of the endocrine system.* New York: Liss.

Barnes, W. D., Sirbasku, D. A., & Sato, G. H. (eds.). (1984c). *Cell culture methods for molecular and cell biology; Vol. 3: Methods for serum-free culture of epithelial and fibroblastic cells.* New York: Liss.

Barnes, W. D., Sirbasku, D. A., & Sato, G. H. (eds.). (1984d). *Cell culture methods for molecular and cell biology; Vol. 4: Methods for serum-free culture of neuronal and lymphoid cells.* New York: Liss.

Barnett, S. C. (2004). Olfactory ensheathing cells: unique glial cell types? *J. Neurotrauma.* **21**:375–382.

Barnett, S. C., & Chang, L. (2004). Olfactory ensheathing cells: Going solo or in need of a friend? *Trends Neurosci.* **27**(1): 54–60.

Barnett, S. C., & Riddell, J. S. (2004). Olfactory ensheathing cells (OECs) and the treatment of CNS injury: Advantages and possible caveats. *J. Anat.* **24**:57–67.

Barnett, S. C., Hutchins, A.-M., & Noble, M. (1993). Purification of olfactory nerve ensheathing cells of the olfactory bulb. *Dev. Biol.* **155**:337–350.

Barnett, S. C., Riddell, J. S. (2007). Cell transplantation strategies for spinal cord repair: What can be achieved? *Nat. Clin. Pract. Neurol.* **3**:152–161.

Bateman, A. E., Peckham, M. J., & Steel, G. G. (1979). Assays of drug sensitivity for cells from human tumours: in vitro and in vivo tests on a xenografted tumour. *Br. J. Cancer* **40**:81–88.

Batlle-Morera, L., Smith, A., & Nichols, J. (2008). Parameters influencing derivation of embryonic stem cells from murine embryos. *Genesis* **46**:758–767.

Bazill, G. W., Haynes M., Garland J., & Dexter T. M. (1983). Characterisation and partial purification of a haemopoietic cell growth factor in WEHI-3 cell conditioned medium. *Biochem. J.* **210**:747–759.

Beattie GM, Lappi DA, Baird A, Hayek A. (1990). Selective elimination of fibroblasts from pancreatic islet monolayers by basic fibroblast growth factor-saporin mitotoxin. *Diabetes* **39**:1002–1005.

Beddington, R. (1992). Transgenic mutagenesis in the mouse. *Trends Genet.* **8**:10.

Bedrin, M. S., Abolafia, C. M., & Thompson, J. F. (1997). Cytoskeletal association of epidermal growth factor receptor and associated signalling proteins is regulated by cell density in IEC-6 intestinal cells. *J. Cell. Physiol.* **172**:126–136.

Belin, V., & Rousselle, P. (2006). Production of a recombinantly expressed laminin fragment by HEK293-EBNA cells cultured in suspension in a dialysis-based bioreactor. *Protein Expr. Purif.* **48**:43–48.

Bell, E., Sher, S., Hull, B., Merrill, C., Rosen, S., Chamson, A., Asselineau, D., Dubertret, L., Coulomb, B., Lapiere, C., Nusgens, B., & Neveux, Y. (1983). The reconstitution of living skin. *J. Invest. Dermatol.* **81**(1 suppl.):2s–10s.

Benda, P., Lightbody, J., Sato, G., Levine, L., & Sweet, W. (1968). Differentiated rat glial cell strain in tissue culture. *Science* **161**:370.

Benders, A. A. G. M., van Kuppevelt, T. H. M. S. M., Oosterhof, A., & Veerkamp, J. H. (1991). The biochemical and structural maturation of human skeletal muscle cells in culture: The effect of serum substitute, Ultroser. G. *Exp. Cell Res.* **195**:284–294.

Benya, P. D. (1981). Two dimensional CNBr peptide patterns of collagen types I, II and III. *Coll. Relat. Res.* **1**:17–26.

Benya, P. D., Padilla, S. R., & Nimmi, M. E. (1977). The progeny of rabbit articular chondrocytes synthesize collagen type I and III and I trimer, but not type II: Verification by cyanogen bromide peptide analysis. *Biochemistry* **16**:865–872.

Berdichevsky, F., Gilbert, C., Shearer, M., & Taylor-Papadimitriou, J. (1992). Collagen-induced rapid morphogenesis of human mammary epithelial cells: The role of the alpha 2 beta 1 integrin. *J. Cell Sci.* **102**:437–446.

Berky, J. J., & Sherrod, P. C. (eds.). (1977). *Short term in vitro testing for carcinogenesis, mutagenesis and toxicity.* Philadelphia, Franklin Institute Press.

Bernstein, A. (1975). Differentiation of clonal lines of teratocarcinoma cells: Formation of embryoid bodies in vitro. *Proc. Natl. Acad. Sci. USA* **72**:1441–1445.

Bernstine, E. G., Hooper, M. L., Grandchamp, S., & Ephrussi, B (1973). Alkaline phosphatase activity in mouse teratoma. *Proc. Natl. Acad. Sci. USA* **70**:3899–3903.

Berry, M. N., & Friend, D. S. (1969). High yield preparation of isolated rat liver parenchymal cells: A biochemical and fine structural study. *J. Cell Biol.* **43**:506–520.

Bertheussen, K. (1993). Growth of cells in a new defined protein-free medium. *Cytotechnology* **11**:219–231.

Bertoncello, I., Bradley, T. R., & Watt, S. M. (1991). An improved negative immunomagnetic selection strategy for the purification of primitive hemopoietic cells from normal bone marrow. *Exp. Hematol.* **19**:95–100.

Bettger, W. J., Boyce, S. T., Walthall, B. J., & Ham, R. G. (1981). Rapid clonal growth and serial passage of human diploid fibroblasts in a lipid-enriched synthetic medium supplemented with EGF, insulin and dexamethasone. *Proc. Natl. Acad. Sci. USA* **78**:5588–5592.

Bhagavati, S., & Xu, W. (2004). Isolation and enrichment of skeletal muscle progenitor cells from mouse bone marrow. *Biochem Biophys Res Commun.* **318**:119–124.

Bhargava, M., Joseph, A., Knesel, J., Halaban, R., Li, Y., Pang, S., Golberg, I., Setter, E., Donovan, M. A., Zarnegar, R., Faletto, D., & Rosen, E. M. (1992). Scatter factor and hepatocyte growth factor activities, properties, and mechanism. *Cell Growth Differ.* **3**:11–20.

Bickenbach, J. R., & Chism, E. (1998). Selection and extended growth of murine epidermal stem cells in culture. *Exp. Cell Res.* **244**:184–195.

Biedler, J. L. (1976). Chromosome abnormalities in human tumour cells in culture. In Fogh, J. (ed.), *Human tumor cells in vitro.* New York: Academic Press.

Biedler, J. L., & Spengler, B. A. 1976. A novel chromosomal abnormality in human neuroblastoma and anti-folate resistant Chinese hamster cell lines in culture. *J. Natl. Cancer Inst.* **57**:683–695.

Biedler, J. L., Albrecht, A. M., Hutchinson, D. J., & Spengler, B. A. (1972). Drug response, dihydrofolate reductase, and cytogenetics of amethopterin-resistant Chinese hamster cells in vitro. *Cancer Res.* **32**:151–161.

Biernaskie, J. A., McKenzie, I. A., Toma, J. G., & Miller, F. D. (2006). Isolation of skin-derived precursors (SKPs) and differentiation and enrichment of their Schwann cell progeny. *Nat. Protoc.* **1**:2803–2812.

Biggers, J. D., Gwatkin, R. B. C., & Heyner, S. (1961). Growth of embryonic avian and mammalian tibiae on a relatively simple chemically defined medium. *Exp. Cell Res.* **25**:41.

Bignami, A., Dahl, D., & Rueger, D. G. (1980). Glial fibrillary acidic (GFA) protein in normal neural cells and in pathological conditions. In Federoff, S., & Hertz, L. (eds.), *Advances in cellular neurobiology*, Vol. **1**. New York: Academic Press.

Birch, J. R., & Pirt, S. J. (1970). Improvements in a chemically-defined medium for the growth of mouse cells (strain LS) in suspension. *J. Cell Sci.* **7**:661–670.

Birch, J. R., & Pirt, S. J. (1971). The quantitative glucose and mineral nutrient requirements of mouse LS (suspension) cells in chemically-defined medium. *J. Cell Sci.* **8**:693–700.

Birnie, G. D., & Simons, P. J. (1967). The incorporation of $^3$H-thymidine and $^3$H-uridine into chick and mouse embryo cells cultured on stainless steel. *Exp. Cell Res.* **46**:355–366.

Bishop, A. E., Power, R. F., & Polak, J. M. (1988). Markers for neuroendocrine differentiation. *Pathol. Res. Pract.* **183**:119–128.

Bishop, J. M. (1991). Molecular themes in oncogenesis. *Cell* **64**:235–248.

Bissell, D. M., Arenson, D. M., Maher, J. J., & Roll, F. J. (1987). Support of cultured hepatocytes by a laminin-rich gel: Evidence for a functionally significant subendothelial matrix in normal liver. *J. Clin. Invest.* **79**:801–812.

Bjerkvig, R., Laerum, O. D., & Mella, O. (1986a). Glioma cell interactions with fetal rat brain aggregates in vitro, and with brain tissue in vivo. *Cancer Res.* **46**:4071–4079.

Bjerkvig, R., Steinsvag, S. K., & Laerum, O. D. (1986b). Reaggregation of fetal rat brain cells in a stationary culture system; I: Methodology and cell identification. *In Vitro* **22**:180–192.

Blaker, G. J., Birch, J. R., & Pirt, S. J. (1971). The glucose, insulin and glutamine requirements of suspension cultures of HeLa cells in a defined culture medium. *J. Cell Sci.* **9**:529–537.

Blanco, F. J., Geng, Y., & Lotz, M. (1995). Differentiation dependent effects of IL-1 and TGF-β on human articular chondrocyte proliferation are related to nitric oxide synthase expression. *J. Immunol.* **154**:4018–4026.

Blouin, R., Grondin, G., Beaudoin, J., Arita, Y., Daigle, N., Talbot, B. G., Lebel, D., & Morisset, J. (1997). Establishment and immunocharacterization of an immortalized pancreatic cell line derived from the H-2Kb-tsA58 transgenic mouse. *In Vitro Cell Dev. Biol. Anim.* **33**:717–726.

Blow, N. (2009). Cell culture; building a better matrix. *Nat. Methods.* **6**:619–620.

Bobrow, M., Madan, J., & Pearson, P. L. (1972). Staining of some specific regions on human chromosomes, particularly the secondary constriction of no. 9. *Nature* **238**:122–124.

Bochaton-Piallat, M. L., Gabbiani, F., Ropraz, P., & Gabbiani, G. (1992). Cultured aortic smooth muscle cells from newborn and adult rats show distinct cytoskeletal features. *Differentiation* **49**:175–185.

Bockhold, K. J., Rosenblatt, J. D., & Partridge, T. A. (1998). Aging normal and dystrophic mouse muscle: Analysis of myogenicity in cultures of living single fibres. *Muscle Nerve* **21**:173–183.

Bodnar, A. G., Ouellette, M., Frolkis, M., Holt, S. E., Chiu, C.-P., Morin, G. B., Harley, C. B., Shay, J. W., Lichsteiner, S., & Wright, W. E. (1998). Extension of life-span by introduction of telomerase into normal human cells. *Science* **279**:349–352.

Boehnke, K., Mirancea, N., Pavesio, A., Fusenig, N. E., Boukamp, P., & Stark, H.-J. (2007). Effects of fibroblasts and microenvironment on epidermal regeneration and tissue function in long-term skin equivalents. *Eur. J. Cell Biol.* **86**:731–746.

Boehnke, K., et al. (2010). Submitted.

Bögler, O., Wren, D., Barnett, S. C., Land, H., & Noble, M. (1990). Cooperation between two growth factors promotes extended self-renewal and inhibits differentiation of oligodendrocyte-type-2 astrocyte (O-2A) progenitor cells. *Proc. Natl. Acad. Sci. USA* **87**:6368–6372.

Boheler, K. R. (2009). Stem cell pluripotency: A cellular trait that depends on transcription factors, chromatin state and a checkpoint deficient cell cycle. *J. Cell. Physiol.* **221**:10–17.

Bolton, B. J., & Spurr, N. K. (1996). B-lymphocytes. In Freshney, R. I., & Freshney, M. G., (eds.), *Culture of immortalized cells.* New York: Wiley-Liss, pp. 283–298.

Bonaventure, J., Kadhom, N., Cohen-Solal, L., Ng, K. H., Bourguignon, J., Lasselin, C., & Freisinger, P. (1994). Reexpression of cartilage-specific genes by dedifferentiated human articular chondrocytes cultured in alginate beads. *Exp. Cell Res.* **212**:97–104.

Bongso A, Fong CY, Ng SC, Ratnam S. (1994). Isolation and culture of inner cell mass cells from human blastocysts. *Hum. Reprod.* **9**:2110–2117.

Boonstra, J. J., van der Velden, A. W., Beerens, E. C., van Marion, R., Morita-Fujimura, Y., Matsui, Y., Nishihira, T., Tselepis, C., Hainaut, P., Lowe, A. W., Beverloo, B. H., van Dekken, H., Tilanus, H. W., & Dinjens, W. N. (2007). Mistaken identity of widely used esophageal adenocarcinoma cell line TE-7. *Cancer Res.* **67**:7996–8001.

Boonstra, J. J., van Marion, R., Beer, D. G., Lin, L., Chaves, P., Ribeiro, C., Pereira, A. D., Roque, L., Darnton, S. J., Altorki, N. K., Schrump, D. S., Klimstra, D. S., Tang, L. H., Eshleman, J. R., Alvarez, H., Shimada, Y., van Dekken, H., Tilanus, H. W., Dinjens, W. N. (2010). Verification and unmasking of widely used human esophageal adenocarcinoma cell lines. *J Natl Cancer Inst.* **102**:271–274.

Booth, C., O'Shea, J. A., & Potten, C. S. (1999). Maintenance of functional stem cells in isolated and cultured adult intestinal epithelium. *Exp. Cell Res.* **249**:359–366.

Booth, C., & O'Shea, J. A. (2002). Isolation and culture of intestinal epithelial cells. In Freshney, R. I., & Freshney, M. G. (eds.), *Culture of epithelial cells*, 2nd ed. Hoboken, NJ: Wiley-Liss, pp. 303–335.

Booyse, F. M., Sedlak, B. J., & Rafelson, M. E. (1975). Culture of arterial endothelial cells: Characterization and growth of bovine aortic cells. *Thromb. Diathes. Ahemorrh.* **34**:825–839.

Borenfreund, E., Babich, H., & Martin-Alguacil, N. (1990). Rapid chemosensitivity assay with human normal and tumor cells in vitro. *In Vitro Cell Dev. Biol.* **26**:1030–1034.

Bosco, D., Soriano, J. V., Chanson, M., & Meda, P. (1994). Heterogeneity and contact-dependent regulation of amylase release by individual acinar cells. *J. Cell Physiol.* **160**:378–388.

Bossis, I., Voutetakis, A., Matyakhina, L., Pack, S., Abu-Asab, M., Bourdeau, I., Griffin, K. J., Courcoutsakis, N., Stergiopoulos, S., Batista, D., Tsokos, M., & Stratakis, C. A. (2004). A pleiomorphic GH pituitary adenoma from a Carney complex patient displays universal allelic loss at the protein kinase A regulatory subunit 1A (PRKARIA) locus. *J. Med. Genet.* **41**:596–600.

Bossolasco, P., Corti, S., Strazzer, S., Borsotti, C., Del Bo, R., Fortunato, F., Salani, S., Quirici, N., Bertolini, F., Gobbi, A., Deliliers, G. L., Pietro Comi, G., & Soligo, D. (2004). Skeletal muscle differentiation potential of human adult bone marrow cells. *Exp. Cell Res.* **295**:66–78.

Bottenstein, J. E., & Sato, G. (1979). Growth of a rat neuroblastoma cell line in serum free supplemented medium. *Proc. Natl. Acad. Sci. USA* **76**:514–517.

Bottenstein, J. E. (1984). Culture methods for growth of neuronal cells lines in defined media. In Barnes, D. W., Sirbasku, D. A., & Sato, G. H. (eds.), *Methods for serum-free culture of neuronal and lymphoid cells.* New York: Liss, pp. 3–13.

Boukamp, P., Petrusevska, R. T., Breitkreutz, D., Hornung, J., & Markham, A. (1988). Normal keratinisation in a spontaneously immortalised, aneuploid human keratinocyte cell line. *J. Cell Biol.* **106**:761–771.

Bourillot, P. Y., Waltzer, L., Sergeant, A., & Manet, E. (1998). Transcriptional repression by the Epstein-Barr virus EBNA3A protein tethered to DNA does not require RBP-Jkappa. *J. Gen. Virol.* **79**:363–370.

Bouzahzah, B., Nishikawa, Y., Simon, D., & Carr, B. I. (1995). Growth control and gene expression in a new hepatocellular carcinoma cell line, Hep40: Inhibitory actions of vitamin K. *J. Cell Physiol.* **165**:459–467.

Bowman, P. D., Betz, A. L., Ar, D., Wolinsky, J. S., Penney, J. B., Shivers, R. R., & Goldstein, G. (1981). Primary culture of capillary endothelium from rat brain. *In Vitro* **17**:353–362.

Boxman, D. L. A., Quax, P. H. A., Lowick, C. W. G. M., Papapoulos, S. E., Verheijen, J., & Ponec, M. (1995). Differential regulation of plasminogen activation in normal keratinocytes and SCC-4 cells by fibroblasts. *J. Invest. Dermatol.* **104**:374–378.

Boyce, S. T., & Ham, R. G. (1983). Calcium-regulated differentiation of normal human epidermal keratinocytes in chemically defined clonal culture and serum-free serial culture. *J. Invest. Dermatol.* **81**:33s–40s.

Boyd, M. R. (1989). Status of the NCI preclinical antitumor drug discovery screen. *Prin. Prac. Oncol.* **10**:1–12.

Boyd, M., Cunningham, S. H., Brown, M. M., Mairs, R. J., & Wheldon, T. E. (1999). Noradrenaline transporter gene transfer for radiation cell kill by [131I] meta-iodobenzyl-guanidine. *Gene Ther.* **6**:1147–1152.

Boyd, M., Mairs, R. J., Cunningham, S. H., Mairs, S. C., McCluskey, A. G., Livingstone, A., Stevenson, K., Brown, M. M., Wilson, L., Carlin, S., & Wheldon, T. E. (2001). A gene therapy/targeted radiotherapy strategy for radiation cell kill by [131I]MIBG. *J. Gene Med.* **3**:165–172.

Boyd, M., Mairs, R. J., Keith, W. N., Ross, S. C., Welsh, P., Akabani, G., Owens, J., Vaidyanathan, G., Carruthers, R., Dorrens, J., & Zalutsky, M. R. (2004). An efficient targeted radiotherapy/gene therapy strategy utilising human telomerase promoters and radioastatine and harnassing radiation-mediated bystander effects. *J. Gene Med.* **6**:937–947.

Boyd, M., Mairs, S. C., Stevenson, K., Livingstone, A., Clark, A. M., Ross, S. C., & Mairs, R. J. (2002). Transfectant mosaic spheroids: A new model for evaluation of tumour cell killing in targeted radiotherapy and experimental gene therapy. *J. Gene Med.* **4**:1–10.

Boyum, A. (1968a). Isolation of leucocytes from human blood: A two-phase system for removal of red cells with methylcellulose as erythrocyte aggregative agent. *Scand. J. Clin. Lab. Invest.* **21**(suppl 97): 9–29.

Boyum, A. (1968b). Isolation of leucocytes from human blood: Further observations—methylcellulose, dextran and Ficoll as erythrocyte aggregating agents. *Scand. J. Clin. Lab. Invest.* **31**(suppl 97):50.

Braa, S. S., & Triglia, D. (1991). Predicting ocular irritation using three-dimensional human fibroblast cultures. *Cosmetics Toiletries* **106**:55–58.

Braam, S. R., Zeinstra, L., Litjens, S., Ward-van Oostwaard, D., van den Brink, S., van Laake, L., Lebrin, F., Kats, P., Hochstenbach, R., Passier, R., Sonnenberg, A., & Mummery, C. L. (2008). Recombinant vitronectin is a functionally defined substrate that supports human embryonic stem cell self-renewal via alphavbeta5 integrin. *Stem Cells* **26**:2257–2265.

Braaten, J. T., Lee, M. J., Schewk, A., & Mintz, D. H. (1974). Removal of fibroblastoid cells from primary monolayer cultures of rat neonatal endocrine pancreas by sodium ethylmercurithiosalicylate. *Biochem. Biophys. Res. Commun.* **61**:476–482.

Bradford, C. S., Sun, L., & Barnes, D. W. (1994a). Basic FGF stimulates proliferation and suppresses melanogenesis in cell cultures derived from early zebrafish embryos. *Mol. Mar. Biol. Biotech.* **3**:78–86.

Bradford, C. S., Sun, L., Collodi, P., & Barnes, D. W. (1994b). Cell cultures from zebrafish embryos and adult tissues. *J. Tissue Cult. Methods* **16**:99–107.

Bradford, M. (1976). A rapid and sensitive method for the quantitation of microgram quantities of protein utilizing the principle of protein-dye binding. *Anal. Biochem.* **72**:248–254.

Bradley, T. R., & Metcalf, D. (1966). The growth of mouse bone marrow cells in vitro. *Aust. J. Exp. Biol. Med. Sci.* **44**(3): 287–299.

Bradley, A., Evans, M., Kaufman, M. H., & Robertson, E. (1984). Formation of germ-line chimaeras from embryo-derived teratocarcinoma cell lines. *Nature* **309**:255–256.

Bradley, M. O., & Sharkey, N. A. (1977). Mutagenicity and toxicity of visible fluorescent light to cultured mammalian cells. *Nature* **266**:724–726.

Bradley, N. J., Bloom, H. J. G., Davies, A. J. S., & Swift, S. M. (1978a). Growth of human gliomas in immune-deficient mice: A possible model for pre-clinical therapy studies. *Br. J. Cancer* **38**:263.

Bradley, T. R., Hodgson, G. S., & Rosendaal, M. (1978b). The effect of oxygen tension on haemopoietic and fibroblast cell proliferation in vitro. *J. Cell Physiol.* **97**(suppl 1): 517–522.

Bravery, C. A., Batten, P., Yacoub, M. H., & Rose, M. L. (1995). Direct recognition of SLA- and HLA-like class II antigens on porcine endothelium by human T cells results in T cell activation and release of interleukin-2. *Transplantation* **60**:1024–1033.

Breach, M. R. (1968). *Sterilization: Methods and control.* London: Butterworths.

Breen, G. A. M., & De Vellis, J. (1974). Regulation of glycerol phosphate dehydrogenase by hydrocortisone in dissociated rat cerebral cell cultures. *Dev. Biol.* **41**:255–266.

Breitkreutz, D., Stark, H.-J., Mirancea, N., Tomakidi, P., Steinbauer, H., & Fusenig, N. E. (1997). Integrin and basement membrane normalization in mouse grafts of human keratinocytes: Implications for epidermal homeostasis. *Differentiation* **61**:195–209.

Breitman, T. R., Kene, B. R., & Hemmi, H. (1984). Studies of growth and differentiation of human myelomonocytic leukaemia cell lines in serum-free medium. In Barnes, D. W., Sirbasku, D. A., & Sato, G. H. (eds.), *Methods for serum-free culture of neuronal and lymphoid cells.* New York: Liss, pp. 215–236.

Brellier, F., Bergoglio, V., Valin, A., Barnay, S., Chevallier-Lagente, O., Vielh, P., Spatz, A., Gorry, P., Avril, M. F., & Magnaldo, T. (2008). Heterozygous mutations in the tumor suppressor gene PATCHED provoke basal cell carcinoma-like features in human organotypic skin cultures. *Oncogene* **27**:6601–6606.

Bretzel, R. G., Bonath, K., & Federlin, K. (1990). The evaluation of neutral density separation utilizing Ficoll, sodium diatrizoate and Nycodenz and centrifugal elutriation in the purification of bovine and canine islet preparations. *Hormone Metab. Res.* (suppl) **25**:57–63.

Brewer, G. J. (1995). Serum-free B27/Neurobasal medium supports differentiated growth of neurons from the striatum, substantia nigra, septum, cerebral cortex, cerebellum, and dentate gyrus. *J. Neurosci. Res.* **42**:674–683.

Bridger, J. M. (2004). Mammalian artificial chromosomes: modern day feats of engineering—Isambard Kingdom Brunel style. *Cytogenet. Genome Res.* **107**:5–8.

Briggs, R., & King, T. J. (1960). Serial transplantation of embryonic nuclei. *Cold Spring Harb. Symp. Quant. Biol.* **21**:271–290.

Bright, R. K., & Lewis, J. D. (2004). Long-term culture of normal and malignant human prostate epithelial cells. In Pfragner, R., & Freshney, R. I. (eds.), *Culture of human tumor cells*, Hoboken, NJ: Wiley-Liss, pp. 125–144.

Brinch, D. S., & Elvig, S. G. (2001). Evaluation of an in vitro human corneal model as alternative to the in vivo eye irritation testing of enzymes. *Toxicol. Lett.* **123**(suppl.) 1:22.

Brindle, K. M. (1998). Investigating the performance of intensive mammalian cell bioreactor systems using magnetic resonance imaging and spectroscopy. *Biotech. Genet. Eng. Rev.* **15**:499–520.

British Standard BS5726. (2005). Microbiological Safety Cabinets. The Stationery Office, P. O. Box 276, London. www. thenbs.com/PublicationIndex/DocumentSummary.aspx?PubID =76&DocID=274468.

Brito Babapulle, V. (1981). Lateral asymmetry in human chromosomes 1, 3, 4, 15 and 16. *Cytogenet. Cell Genet.* **29**:198–202.

Brockes, J. P., Fields, K. L., & Raff, M. C. (1979). Studies on cultured rat Schwann cells; I: Establishment of purified populations from cultures of peripheral nerve. *Brain Res.* **165**:105–118.

Brook, F. A., & Gardner, R. L. (1997). The origin and efficient derivation of embryonic stem cells in the mouse. *Proc. Natl. Acad. Sci. USA.* **94**:5709–5712.

Brouty-Boyé, D., Kolonias, D., Savaraj, N., & Lampidis, T. J. (1992). Alpha-smooth muscle actin expression in cultured cardiac fibroblasts of newborn rat. *In Vitro Cell Dev. Biol.* **28A**:293–296.

Brower, M., Carney, D. N., Oie, H. K., Gazdar, A. F., & Minna, J. D. (1986). Growth of cell lines and clinical specimens of human nonsmall cell lung cancer in a serum-free defined medium. *Cancer Res.* **46**:798–806.

Brunk, C. F., Jones, K. C., & James, T. W. (1979). Assay for nanogram quantities of DNA in cellular homogenates. *Anal. Biochem.* **92**:497–500.

Brunton, V., Ozanne, B., Paraskeva, C., & Frame, M. (1997). A role for epidermal growth factor receptor, c-Src and focal adhesion kinase in an in vitro model for the progression of colon cancer. *Oncogene* **14**:283–293.

Bruyneel, E. A., Debray, H., de Mets, M., Mareel, M. M., & Montreuil, J. (1990). Altered glycosylation in Madin-Darby canine kidney (MDCK) cells after transformation by murine sarcoma virus. *Clin. Exp. Metastasis* **8**:241–253.

Bryan, D., Sexton, C. J., Williams, D., Leigh, I. M., & McKay, I. (1995). Oral keratinocytes immortalized with the early region of human papillomavirus type 16 show elevated expression of interleukin 6, which acts as an autocrine growth factor for the derived T103C cell line. *Cell Growth Differ.* **6**:1245–1250.

Bryan, T. M., & Reddel, R. R. (1997). Telomere dynamics and telomerase activity in in vitro immortalised human cells. *Eur. J. Cancer* **33**:767–773.

Bryant, P. E., Gray, L. J., & Peresse, N. (2004). Progress towards understanding the nature of chromatid breakage. *Cytogenet Genome Res.* **104**(1–4):65–71.

Brysk, M. M., Santschi, C. H., Bell, T., Wagner, R. F. Jr, Tyring, S. K., & Rajaraman, S. (1992). Culture of basal cell carcinoma. *J. Invest. Dermatol.* **98**:45–49.

Bucana, C. D., Giavazzi, R., Nayar, R., O'Brian, C. A., Seid, C., Earnest, L. E., & Fan, D. (1990). Retention of vital dyes correlates inversely with the multidrug-resistant phenotype of adriamycin-selected murine fibrosarcoma variants. *Exp. Cell Res.* **190**:69.

Buchanan, S. S., Gross, S. A., Acker, J. P., Toner, M., Carpenter, J. F., & Pyatt, D. W. (2004). Cryopreservation of stem cells using trehalose: Evaluation of the method using a human hematopoietic cell line. *Stem Cells Dev.* **13**:295–305.

Buchler, P., Reber, H. A., Buchler, M. W., Friess, H., Lavey, R. S., & Hines, O. J. (2004). Antiangiogenic activity of genistein in pancreatic carcinoma cells is mediated by the inhibition of hypoxia-inducible factor-1 and the down-regulation of VEGF gene expression. *Cancer* **100**:201–210.

Buckingham, M. (1992). Making muscle in mammals. *Trends Genet.* **8**:144–149.

Buehr, M., Meek, S., Blair, K., Yang, J., Ure, J., Silva, J., McLay, R., Hall, J., Ying, Q. L., & Smith, A. (2008). Capture of authentic embryonic stem cells from rat blastocysts. *Cell* **135**:1287–1298.

Buehring, G. C. (1972). Culture of human mammary epithelial cells: Keeping abreast of a new method. *J. Natl. Cancer Inst.* **49**:1433–1434.

Buehring, G. C., Eby, E. A., & Eby, M. J. (2004). Cell line cross-contamination: how aware are mammalian cell culturists of the problem and how to monitor it? *In Vitro Cell. Dev. Biol. Anim.* **40**:211–215.

Buick, R. N., Stanisic, T. H., Fry, S. E., Salmon, S. E., Trent, J. M., & Krosovich, P. (1979). Development of an agar-methyl cellulose clonogenic assay for cells of transitional cell carcinoma of the human bladder. *Cancer Res.* **39**:5051–5056.

Buonassisi, V., Sato, G., & Cohen, A. I. (1962). Hormone-producing cultures of adrenal and pituitary tumor origin. *Proc. Natl. Acad. Sci. USA* **48**:1184–1190.

Burchell, J., & Taylor-Papadimitriou, J. (1989). Antibodies to human milk fat globule molecules. *Cancer Invest.* **17**:53–61.

Burchell, J., Durbin, H., & Taylor-Papadimitriou, J. (1983). Complexity of expression of antigenic determinants recognised by monoclonal antibodies HMFG 1 and HMFG 2 in normal and malignant human mammary epithelial cells. *J. Immunol.* **131**:508–513.

Burchell, J., Gendler, S., Taylor-Papadimitriou, J., Girling, A., Lewis, A., Millis, R., & Lamport, D. (1987). Development and characterisation of breast cancer reactive monoclonal antibodies directed to the core protein of the human milk mucin. *Cancer Res.* **47**:5476–5482.

Burgess, W. H., & Maciag, T. (1989). The heparin-binding fibroblast growth factor family of proteins. *Ann. Rev. Biochem.* **58**:575–606.

Burke, J. F., Price, T. N. C., & Mayne, L. V. (1996). Immortalization of human astrocytes. In Freshney, R. I., & Freshney, M. G. (eds.), *Culture of immortalized cells.* New York: Wiley-Liss, pp. 299–314.

Burrows, M. T. (1912). Rhytmische Kontraktionen der isolierten Herzmuskelzelle Ausserhalb des Organismus. *Münch. Med. Wochenschr.* **59**:1473.

Burwen, S. J., & Pitelka, D. R. (1980). Secretory function of lactating moose mammary epithelial cells cultured on collagen gels. *Exp. Cell Res.* **126**:249–262.

Butler, M. (1991). *Mammalian cell biotechnology.* Oxford: IRL Press at Oxford University Press.

Butler, M. (2004). *Animal cell culture and technology,* 2nd ed. New York: Bios Scientific Publishers.

Butler, M., & Christie, A. (1994). Adaptation of mammalian cells to non-ammoniagenic media. *Cytotechnology* **15**:87–94.

Cable, E. E., & Isom, H. C. (1997). Exposure of primary rat hepatocytes in long-term DMSO culture to selected transition metals induces hepatocyte proliferation and formation of duct-like structures. *Hepatology* **26**:1444–1457.

Calder, C. J., Liversidge, J., & Dick, A. D. (2004). Murine respiratory tract dendritic cells: Isolation, phenotyping and functional studies. *J. Immunol. Methods* **287**:67–77.

Caldon C. E., Daly R. J., Sutherland, R. L., & Musgrove, E. A. (2006). Cell cycle control in breast cancer cells. *J. Cell Biochem.* **97**:261–274.

Campion, D. G. (1984). The muscle satellite cell: A review. *Int. Rev. Cytol.* **87**:225–251.

Cancedda, R., Dozin, B., Giannoni, P. & Quarto, R. (2003). Tissue engineering and cell therapy of cartilage and bone. *Matrix Biol.* **22**:81–91.

Cancela, M. L., Hu, B., & Price, P. A. (1997). Effect of cell density and growth factors on matrix GLA protein expression by normal rat kidney cells. *J. Cell Physiol.* **171**:125–134.

Caniggia, I., Tseu, I., Han, R. N., Smith, B. T., Tanswell, K., & Post, M. (1991). Spatial and temporal differences in fibroblast behavior in fetal rat lung. *Am. J. Physiol. Lung Mol. Cell. Physiol.* **261**:L424–L433.

Canti, R. G. (1928). Cinematograph demonstration of living tissue cells growing in vitro. *Arch. Exp. Zellforsch.* **6**:86–97.

Cantòn, I, D., Cole, D. M., Kemp, E. H., Watson, P. F., Chunthapong, J., Ryan, A. J., MacNeil, S., Haycock, J.W., (2010). Development of a 3D human in vitro skin co-culture model for detecting irritants in real-time. Biotechnol. Bioeng.,. **106**:794–803.

Cao, B., Zheng, B., Jankowski, R. J., Kimura, S., Ikezawa, M., Deasy, B., Cummins, J., Epperly, M., Qu-Petersen, Z., & Huard, J. (2003). Muscle stem cells differentiate into haematopoietic lineages but retain myogenic potential. *Nat. Cell Biol.* **5**:640–646.

Cao, D., Lin, G., Westphale, E. M., Beyer, E. C., & Steinberg, T. H. (1997). Mechanisms for the coordination of intercellular calcium signaling in insulin-secreting cells. *J. Cell Sci.* **110**:497–504.

Capes-Davis A, Theodosopoulos G, Atkin I, Drexler HG, Kohara A, Macleod RA, Masters JR, Nakamura Y, Reid YA, Reddel RR, Freshney RI. Check your cultures! A list of cross-contaminated or misidentified cell lines. *Int. J. Cancer* **127**:1–8.

Caplan, A. I. (1991). Mesenchymal stem cells. *J. Orthop. Res.* **9**:641–650.

Caputo, J. L. (1996). Safety procedures. In Freshney, R. I., & Freshney, M. G. (eds.), *Culture of immortalized cells.* New York: Wiley-Liss, pp. 25–51.

Carlsson, J., & Nederman, T. (1989). Tumour spheroid technology in cancer therapy research. *Eur. J. Cancer Clin. Oncol.* **25**: 1127–1133.

Carmeliet, P. (2005) Angiogenesis in life, disease and medicine. *Nature* **438**:932–936.

Carmichael, J., DeGraff, W. G., Gazdar, A. F., Minna, J. D., & Mitchell, J. B. (1987a). Evaluation of a tetrazolium-based semiautomated colorimetric assay: Assessment of chemosensitivity testing. *Cancer Res.* **47**:936–942.

Carmichael, J., DeGraff, W. G., Gazdar, A. F., Minna, J. D., & Mitchell, J. B. (1987b). Evaluation of a tetrazolium-based semiautomated colorimetric assay: Assessment of radiosensitivity. *Cancer Res.* **47**:943–946.

Carney, D. N., Bunn, P. A., Gazdar, A. F., Pagan, J. A., & Minna, J. D. (1981). Selective growth in serum-free hormone-supplemented medium of tumor cells obtained by biopsy from patients with small cell carcinoma of lung. *Proc. Natl. Acad. Sci. USA* **78**:3185–3189.

Carney, D. N., Gazdar, A. F., Bepler, G., Guccion, J. G., Marangos, P. J., Moodt, T. W., Zweig, M. H., & Minna, J. D. (1985). Establishment and identification of small cell lung cancer cell lines having classic and variant features. *Cancer Res.* **45**:2913–2923.

Carpenter, G., & Cohen, S. (1977). Epidermal growth factor. In Acton, R. T., & Lynn, J. D. (eds.), *Cell culture and its application.* New York: Academic Press, pp. 83–105.

Carr, T., Evans, P., Campbell, S., Bass, P., & Albano, J. (1999). Culture of human renal tubular cells: Positive selection of kallikrein-containing cells. *Immunopharmacology* **44**:161–167.

Carraway, K. L., Fregien, N., Carraway, K. L., III, & Carraway, C. A. (1992). Tumor sialomucin complexes as tumor antigens and modulators of cellular interactions and proliferation. *J. Cell Sci.* **103**:299–307.

Carrel, A. (1912). On the permanent life of tissues outside the organism. *J. Exp. Med.* **15**:516–528.

Carrel, A., & Ebeling, A. H. (1923). Survival and growth of fibroblasts in vitro. *J. Exp. Med.* **38**:487.

Cartwright, T., & Shah, G. P. (1994). Culture media. In Davis, J. M. (ed.), *Basic cell culture: A practical approach.* Oxford: IRL Press at Oxford University Press, pp. 58–91.

Carvalho, K. A., Guarita-Souza, L. C., Rebelatto, C. L., Senegaglia, A. C., Hansen, P., Mendonca, J. G., Cury, C. C., Francisco, J. C., & Brofman, P. R. (2004). Could the coculture of skeletal myoblasts and mesenchymal stem cells be a solution for postinfarction myocardial scar? *Transplant Proc.* **36**:991–992.

Casanova, J. E. (2002). Epithelial cell cytoskeleton and intracellular trafficking. V. Confluence of membrane trafficking and motility in epithelial cell models. *Am. J. Physiol. Gastrointest. Liver. Physiol.* **283**(5):G1015–G1019.

Caspersson, T., Farber, S., Foley, G. E., Kudynowski, J., Modest, E. J., Simonsson, E., Wagh, U., & Zech, L. (1968). Chemical differentiation along metaphase chromosomes. *Exp. Cell Res.* **49**:219–222.

Cavallaro, U., & Christofori, G. (2004). Cell adhesion and signalling by cadherins and Ig-CAMs in cancer. *Nat. Rev. Cancer* **4**:118–132.

Cawkill, D., & Tucker, S. J., (2008). Generating cell Lines for early drug discovery. *BIOForum Europe* **9**:46–47.

CDC/OHS. (1999). *Biosafety in microbiological and biomedical laboratories (BMBL)*, 4th ed. US Department of Health and Human Services, Centers for Disease Control and Prevention, and National Institutes of Health. Washington, DC: US Government Printing Office. www.cdc.gov/OD/ohs/biosfty/bmbl4/bmbl4toc.htm.

CDC-NIOSH. (2009). National Institute for Occupational Safety & Health. www.cdc.gov/NIOSH

Centers for Disease Control. (1988). Update: Universal precautions for prevention of transmission of human immunodeficiency virus, hepatitis B virus, and other blood-borne pathogens in healthcare settings. *MMWR* **37**:377–382, 387, 388.

Ceriani, R. L., Taylor-Papadimitriou, J., Peterson, J. A., & Brown, P. (1979). Characterization of cells cultured from early lactation milks. *In Vitro* **15**:356–362.

Chambard, M., Mauchamp J., & Chabaud, O. (1987). Synthesis and apical and basolateral secretion of thyroglobulin by thyroid cell monolayers on permeable substrate: Modulation by thyrotropin. *J. Cell Physiol.* **133**:37–45.

Chambard, M., Vemer, B., Gabrion, J., Mauchamp, J., Bugeia, J. C., Pelassy, C., & Mercier, B. (1983). Polarization of thyroid cells in culture: Evidence for the basolateral localization of the iodide "pump" and of the thyroid-stimulating hormone receptor–adenyl cyclase complex. *J. Cell Biol.* **96**:1172–1177.

Chang, H., & Baserga, R. (1977). Time of replication of genes responsible for a temperature sensitive function in a cell cycle specific to mutant from a hamster cell line. *J. Cell Physiol.* **92**:333–343.

Chang, S. E., Keen, J., Lane, E. B., & Taylor-Papadimitriou, J. (1982). Establishment and characterisation of SV40-transformed human breast epithelial cell lines. *Cancer Res.* **42**:2040–2053.

Chaproniere, D. M., & McKeehan, W. L. (1986). Serial culture of adult human prostatic epithelial cells in serum-free medium containing low calcium and a new growth factor from bovine brain. *Cancer Res.* **46**:819–824.

Chen C.-S., Toda, K.-I., Maruguchi, Y., Matsuyoshi, N., Horriguchi, Y., & Imamura, S. (1997). Establishment and characterization of a novel in vitro angiogenesis model using a microvascular endothelial cell line, F-2C, cultured in chemically defined medium. *In Vitro Cell Dev. Biol. Anim.* **33**:796–802.

Chen, T. R., Dorotinsky, C., Macy, M., & Hay, R. (1989). Cell identity resolved. *Nature* **340**:106.

Chen, T. R., Drabkowski, D., Hay, R. J., Macy, M., & Peterson, W., Jr. (1987). WiDr is a derivative of another colon adenocarcinoma cell line. HT-29. *Cancer Genet. Cytogenet.* **27**:125–134.

Chen, T. R. (1988). Re-evaluation of HeLa, HeLa S3, and Hep-2 karyotypes. *Cytogenet. Cell Genet.* **48**:19–22.

Chen, G., & Palmer, A. F. (2009). Hemoglobin-based oxygen carrier and convection enhanced oxygen transport in a hollow fiber bioreactor. *Biotechnol. Bioeng.* **102**:1603–1612.

Chen, T. C., Curthoys, N. P., Lagenaur, C. F., & Puschett, J. B. (1989). Characterization of primary cell cultures derived from rat renal proximal tubules. *In Vitro Cell Dev. Biol.* **25**:714–722.

Chen, T. R. (1977). In situ detection of mycoplasm contamination in cell cultures by fluorescent Hoechst 33258 stain. *Exp. Cell Res.* **104**:255.

Chen, Y. C., Lee, D. C., Hsiao, C. Y., Chung, Y. F., Chen, H. C., & Thomas, J. P. (2009). The effect of ultra-nanocrystalline diamond films on the proliferation and differentiation of neural stem cells. *Biomater.* **30**:3428–3435.

Cherry, R. S., & Papoutsakis, E. T. (1990). Understanding and controlling fluid-mechanical injury of animal cells in bioreactors. In Spier, R. E., & Griffiths, J. B., *Animal cell biotechnology*, Vol. 4. London: Academic Press, pp. 71–121.

Cho, J.-K., & Bikle, D. D. (1997). Decrease of Ca-ATPase activity in human keratinocytes during calcium-induced differentiation. *J. Cell Physiol.* **172**:146–154.

Chiong, E., Dadbin, A., Harris, L. D., Sabichi, A. L., & Grossman, H. B. (2009). The use of short tandem repeat profiling to characterize human bladder cancer cell lines. *J. Urol.* **181**:2737–2748.

Chopin, D., Barei-Moniri, R., Maille, P., Le Frere-Belda, M. A., Muscatelli-Groux, B., Merendino, N., Lecerf, L., Stoppacciaro, A., & Velotti, F. (2003). Human urinary bladder transitional cell carcinomas acquire the functional Fas ligand during tumor progression. *Am. J. Pathol.* **162**:1139–1149.

Chopra, D. P., & Xue-Hu, I. C. (1993). Secretion of alpha-amylase in human parotid gland epithelial cell culture. *J. Cell Physiol.* **155**:223–233.

Chopra, D. P., Grignon, D. J., Joiakim, A., Mathieu, P. A., Mohamed, A., Sakr, W. A., Powell, I. J., & Sarkar, F. H. (1996). Differential growth factor responses of epithelial cell cultures derived from normal human prostate, benign prostatic hyperplasia and primary prostate carcinoma. *J. Cell Physiol.* **169**:269–280.

Chou, T. C. (2006). Theoretical basis, experimental design, and computerized simulation of synergism and antagonism in drug combination studies. *Pharmacol. Rev.* **58**:621–681.

Christensen, R., Kolvraa, S., Blaese, R. M., Jensen, T. G. (2000). Development of a skin-based metabolic sink for phenylalanine by overexpression of phenylalanine hydroxylase and GTP cyclohydrolase in primary human keratinocytes. *Gene Ther.* **7**:1971–1978.

Christensen, B., Hansen, C., Kieler, J., & Schmidt, J. (1993). Identity of non-malignant human urothelial cell lines classified as transformation grade I (TGrI) and II (TGrII). *Anticancer Res.* **13**:2187–2191.

Christgen, M., & Lehmann, U. (2007). MDA-MB-435: The questionable use of a melanoma cell line as a model for human breast cancer is ongoing. *Cancer Biol. Ther.* **6**:1355–1357.

Chung, S., Shin, B. S., Hwang, M., Lardaro, T., Kang, U. J., Isacson, O., & Kim, K. S. (2006). Neural precursors derived from embryonic stem cells, but not those from fetal ventral mesencephalon, maintain the potential to differentiate into dopaminergic neurons after expansion in vitro. *Stem Cells.* **24**:1583–1593.

Cioni, C., Filoni, S., Aquila, C., Bernardini, S., & Bosco, L. (1986). Transdifferentiation of eye tissue in anuran amphibians: Analysis of the transdifferentiation capacity of the iris of *Xenopus laevis* larvae. *Differentiation* **32**:215–220.

Clarke, G. D., & Ryan, P. J. (1980). Tranquilizers can block mitogenesis in 3T3 cells and induce differentiation in Friend cells. *Nature* **287**:160–161.

Clayton, R. M., Bower, D. J., Clayton, P. R., Patek, C. E., Randall, F. E., Sime, C., Wainwright, N. R., & Zehir, A. (1980). Cell culture in the investigation of normal and abnormal differentiation of eye tissues. In Richards, R. J., & Rajan, K. T. (eds.), *Tissue culture in medical research (II)*. Oxford: Pergamon Press, pp. 185–194.

Cobbold, P. H., & Rink, T. J. (1987). Fluorescence and bioluminescence measurement of cytoplasmic free $Ca^{2+}$. *Biochem. J.* **248**:313.

Cohen, C., Roguet, R., Cottin, M., Olive, C., Leclaire, J., & Rougier, A. (1997). The use of Episkin, a reconstructed epidermis, in the evaluation of the protective effect of sunscreens against chemical lipoperoxidation induced by UVA. *J. Invest. Dermatol.* **108**:77.

Cohen, S. (1962). Isolation of a mouse submaxillary gland protein accelerating incisor eruption and eyelid opening in the newborn animal. *J. Biol. Chem.* **237**:1555–1562.

Cole, J., Fox, M., Garner, R. C., McGregor, D. B., & Thacker, J. (1990). Gene mutation assays in cultured mammalian cells. In Kirkland, D. J. (ed.), *Basic mutagenicity tests: UKEMS recommended procedures, Part 1 (revised)*. Cambridge: Cambridge University Press, pp. 87–114.

Cole, R. J., & Paul, J. (1966). The effects of erythropoietin on haem synthesis in mouse yolk sac and cultured foetal liver cells. *J. Embryol. Exp. Morphol.* **15**:245–260.

Cole, S. P. C. (1986). Rapid chemosensitivity testing of human lung tumour cells using the MTT assay. *Cancer Chemother. Pharmacol.* **17**:259–263.

Coleman, R. M., Case, N. D., & Guldberg, R. E. (2007). Hydrogel effects on bone marrow stromal cell response to chondrogenic growth factors. *Biomaterials* **28**:2077–2086.

Collen, D., Stassen, J. M., Marafino, B. J., Jr., Builder, S., De Cock, F., Ogez, J., Tajiri, D., Pennica, D., Bennett, W. F., Salwa, J., et al. (1984). Biological properties of human tissue-type plasminogen activator obtained by expression of recombinant DNA in mammalian cells. *J. Pharmacol. Exp. Ther.* **231**(1): 146–152.

Collins, C. H., & Kennedy, D. A. (1999). Evolution of microbiological safety cabinets. *Br. J. Biomed. Sci.* **56**:161–169.

Collins, S. J., Gallo, R. C., & Gallagher, R. E. (1977). Continuous growth and differentiation of human myeloid leukaemic cells in suspension culture. *Nature* **270**:347–349.

Collodi, P. (1998). DNA transfer to blastula-derived cultures from zebrafish. In Ravid, K., & Freshney, R. I. (eds.), *DNA transfer to cultured cells*. New York: Wiley-Liss, pp. 69–92.

Collodi, P., & Barnes, D. W. (1990). Mitogenic activity from trout embryos. *Proc. Natl. Acad. Sci. USA* **87**:3498–3502.

Collodi, P., Kamei, Y., Ernst, T., Miranda, C., Buhler, D. R., & Barnes, D. W. (1992). Culture of cells from zebrafish (*Brachydanio rerio*) embryo and adult tissues. *Cell Biol. Toxicol.* **8**:43–61.

Compton, C. C., Warland, G., Nakagawa, H., Opitz, O. G., & Rustgi, A. K. (1998). Cellular characterization and successful transfection of serially subcultured normal human esophageal keratinocytes. *J. Cell Physiol.* **177**:274–281.

Conkie, D., Affara, N., Harrison, P. R., Paul, J., & Jones, K. (1974). In situ localization of globin messenger RNA formation. II. After treatment of Friend virus-transformed mouse cells with dimethyl sulfoxide. *J Cell Biol.* **63**:414–419.

Conrad, S., Renninger, M., Hennenlotter J., Wiesner, T., Just, L., Bonin, M., Aicher, W., Bühring, H-J., Mattheus, U., Mack, A., Wagner, H.-J., Minger, S., Matzkies, M., Reppel, M., Hescheler, J., Sievert, K.-D., Stenzl, A., & Skutella, T. (2008). Generation of pluripotent stem cells from adult human testis. *Nature* **456**:344–349.

Cooke, J. A., & Minger, S. L. (2007). Human embryonal stem cell lines: Derivation and culture. In Freshney, R. I., Stacey, G. N., & Auerbach, J. M. (eds.), *Culture of human stem cells*. Hoboken, NJ: Wiley, pp. 24–59.

Coolen, N. A., Verkerk, M., Reijnen, L., Vlig, M., van den Bogaerdt, A. J., Breetveld, M., Gibbs, S., Middelkoop, E., Ulrich, M. M. (2007). Culture of keratinocytes for transplantation without the need of feeder layer cells. *Cell Transplant.* **16**:649–661.

Coon, H. D. (1968). Clonal cultures of differentiated rat liver cells. *J. Cell Biol.* **39**:29a.

Coon, H. G., & Cahn, R. D. (1966). Differentiation in vitro: Effects of Sephadex fractions of chick embryo extract. *Science* **153**:1116–1119.

Cooper, G. W. (1965). Induction of somite chondrogenesis by cartilage and notochord: A correlation between inductive activity and specific stages of cytodifferentiation. *Dev. Biol.* **12**:185–212.

Cooper, J. K., Syke, G., King, S., Cottrill, K., Ivanova, N. V., Hanne, R., & Ikonomi, P. (2007). Species identification in cell culture: A two-pronged molecular approach. *In Vitro Cell. Dev. Biol. Anim.* **43**:344–351.

Cooper, P. D., Burt, A. M., & Wilson, J. N. (1958). Critical effect of oxygen tension on rate of growth of animal cells in continuous suspended culture. *Nature* **182**:1508–1509.

Cooper, S., & Broxmeyer, H. (1994). Purification of murine granulocyte-macrophage progenitor cells (CFC-GM) using counterflow centrifugal elutriation. In Freshney, R. I., Pragnell, I. B., & Freshney, M. G. (eds.), *Culture of haemopoietic cells*. New York: Wiley-Liss, pp. 223–234.

Coriell, L. L., Tall, M. G., & Gaskill, H. (1958). Common antigens in tissue culture cell lines. *Science* **128**:198.

Corlu, A., Ilyin, G., Cariou, S., Lamy, I., Loyer, P., & Guguen-Guillouzo, C. (1997). The coculture: A system for studying the regulation of liver differentiation/proliferation activity and its control. *Cell Biol. Toxicol.* **13**:235–242.

Corlu, A., Kneip, B., Lhadi, C., Leray, G., Glaise, D., Baffet, G., Bourel, D., & Guguen-Guillouzo, C. (1991). A plasma membrane protein is involved in cell contact-mediated regulation of tissue-specific genes in adult hepatocytes. *J. Cell Biol.* **115**:505–515.

Cou, J. Y. (1978). Establishment of clonal human placental cells synthesizing human choriogonadotropin. *Proc. Natl. Acad. Sci. USA* **75**:1854–1858.

Coulombel, L., Eaves, A. C., & Eaves, C. J. (1983). Enzymatic treatment of long-term human marrow cultures reveals the preferential location of primitive hemopoietic progenitors in the adherent layer. *Blood* **62**:291–297.

Courtenay, V. D., Selby, P. I., Smith, I. E., Mills, J., & Peckham, M. J. (1978). Growth of human tumor cell colonies from biopsies using two soft-agar techniques. *Br. J. Cancer* **38**:77–81.

Coutinho, L. H., Gilleece, M. H., de Wynter, E. A., Will, A., & Testa, N. G. (1992). Clonal and long-term cultures using human bone marrow. In Testa, N. G., & Molineaux, G. (eds.), *Haemopoiesis: A practical approach*. Oxford: IRL Press at Oxford University Press, pp. 75–106.

Cox, P., & Chrisochoidis, M. (2003). Directive 2003/15/EC of the Eurtopean Parliament and of the Council. http://eur-lex.europa.eu/LexUriServ/site/en/oj/2003/l_066/l_066200303 11en00260035.pdf.

Crabb, I. W., Armes, L. G., Johnson, C. M., & McKeehan, W. L. (1986). Characterization of multiple forms of prostatropin (prostate epithelial growth factor) from bovine brain. *Biochem. Biophys. Res. Commun.* **136**:1155–1161.

Crabtree, B., & Subramanian, V. (2007). Behavior of endothelial cells on Matrigel and development of a method for a rapid and reproducible in vitro angiogenesis assay. *In Vitro Cell. Dev. Biol. Anim.* **43**:87–94.

Creasey, A. A., Smith, H. S., Hackett, A. I., Fukuyama, K., Epstein, W. L., & Madin, S. H. (1979). Biological properties of human melanoma cells in culture. *In Vitro* **15**:342.

Croce, C. M. (1991). Genetic approaches to the study of the molecular basis of human cancer. *Cancer Res.* **51** (suppl):5015s–5018s.

Cronauer, M. V., Eder, I. E., Hittmair, A., Sierek, G., Hobisch, A., Culig, Z., Thurnhur, M., Bartsch, G., & Klocker, H. (1997). A reliable system for the culture of human prostatic cells. *In Vitro Cell Dev. Biol. Anim.* **33**:742–744.

Crouch, E. C., Stone, K. R., Bloch, M., & McDivitt, R. W. (1987). Heterogeneity in the production of collagens and fibronectin by morphologically distinct clones of a human tumor cell line: Evidence for intratumoral diversity in matrix protein biosynthesis. *Cancer Res.* **47**(22): 6086–6092.

Cruikshank, C. N. D., & Lowbury, E. J. L. (1952). Effect of antibiotics on tissue cultures of human skin. *Br. Med. J.* **2**:1070.

Csoka, K., Nygren, P., Graf, W., PAAhlman, L., Glimelius, B., & Larsson, R. (1995). Selective sensitivity of solid tumors to suramin in primary cultures of tumor cells from patients. *Int. J. Cancer* **63**:356–360.

Cuttitta, F., Carney, D. N., Mulshine, J., Moody, T. W., Fedorko, J., Fischler, A., & Minna, J. D. (1985). Bombesin-like peptides can function as autocrine growth factors in human small-cell lung cancer. *Nature* **316**:823.

Dairkee, S. H., Deng, G., Stampfer, M. R., Waldman, F. M., & Smith, H. S. (1995). Selective cell culture of primary breast carcinoma. *Cancer Res.* **55**:2516–2519.

Dairkee, S. H., Paulo, S. C., Traquina, P., Moore, D. H., Ljung, B. M., & Smith, H. S. (1997). Partial enzymatic degradation of stroma allows enrichment and expansion of primary breast tumor cells. *Cancer Res.* **57**:1590–1596.

Dame, M. K., & Varani, J. (2008). Recombinant collagen for animal product-free dextran microcarriers. *In Vitro Cell. Dev. Biol. Anim.* **44**:407–414.

Dame, M. K., Spahlinger, D. M., DaSilva, M., Perone, P., Dunstan, R., & Varani, J., (2008). Establishment and characteristics of Gottingen minipig skin in organ culture and monolayer cell culture: Relevance to drug safety testing. *In Vitro Cell. Dev. Biol. Anim.* **44**:245–252.

Dame, M. K., Bhagavathula, N., Mankey, C., Dasilva, M., Paruchuri, T., Aslam, M. N.,& Varani, J., (2010). Human colon tissue in organ culture: preservation of normal and neoplastic characteristics. *In Vitro Cell Dev Biol Anim.* **46**:114–122.

Damon, D. H., Lobb, R. R., D'Amore, P. A., & Wagner, J. A. (1989). Heparin potentiates the action of acidic fibroblast growth factor by prolonging its biological half-life. *J. Cell Physiol.* **138**:221–226.

Danen, E. H., & Yamada, K. M. (2001). Fibronectin, integrins, and growth control. *J. Cell. Physiol.* **189**:1–13.

Danes, B. S., & Paul, J. (1961). Environmental factors influencing respiration of strain L cells. *Exp. Cell Res.* **24**:344–355.

Dangles, V., Femenia, F., Laine, V., Berthelmy, M., LeRhun, D., Poupon, M. F., Levy, D., & Schwartz-Cornil, I. (1997). Two and three-dimensional cell structures govern epidermal growth factor survival function in human bladder carcinoma cell lines. *Cancer Res.* **57**:3360–3364.

Danielson, K. G., McEldrew, D., Alston, J. T., Roling, D. B., Damjanov, A., Damjanov, I., Daska, I., & Spinner, N. (1992). Human colon carcinoma cell lines from the primary tumor and a lymph node metastasis. *In Vitro Cell Dev. Biol.* **A28**:7–10.

Darcy, K. M., Shoemaker, S. F., Lee, P.-P. H., Vaughan, M. M., Black, J. D., & Ip, M. M. (1995). Prolactin and epidermal growth factor regulation of the proliferation, morphogenesis, and functional differentiation of normal rat mammary epithelial cells in three dimensional primary culture. *J. Cell Physiol.* **163**:346–364.

Darling, J. L. (2004). In Vitro culture of malignant brain tumors. In Pfragner, R., & Freshney, R. I. (eds.), *Culture of human tumor cells.* Hoboken, NJ: Wiley-Liss, pp. 349–372.

Darnell, J. E. (1982). Variety in the level of gene control in eukaryotic cells. *Nature* **297**:365–371.

Dart, J. (2003). Corneal toxicity: The epithelium and stroma in iatrogenic and factitious disease. *Eye* **17**:886–892.

Darzynkiewicz, Z., Li, X., & Gong, J. (1994). Assays of cell viability: Discrimination of cells dying by apoptosis. *Methods Cell Biol.* **41**:15–38.

Davies, B., Brown, P. D., East, N., Crimmin, M. J., & Balkwill, F. R. (1993). A synthetic matrix metalloproteinase inhibitor decreases tumour burden and prolongs survival of mice bearing ovarian carcinoma xenografts. *Cancer Res.* **53**:2087–2091.

Davis, J. B., & Stroobant, P. (1990). Platelet-derived growth factors and fibroblast growth factors are mitogens for rat Schwann cells. *J. Cell. Biol.* **110**:1353–1360.

Davison, P. M., Bensch, R., & Karasek, M. A. (1983). Isolation and long-term serial cultivation of endothelial cells from microvessels of the adult human dermis. *In Vitro* **19**:937–945.

Dawe, C. J., & Potter, M. (1957). Morphologic and biologic progression of a lymphoid neoplasm of the mouse in vivo and in vitro. *Am. J. Pathol.* **33**:603.

De Larco, J. E., & Todaro, G. J. (1978). Epithelioid and fibroblastoid rat kidney cell clones: Epidermal growth factor receptors and the effect of mouse sarcoma virus transformation. *J. Cell Physiol.* **94**:335–342.

De Leij, L., Poppema, S., Nulend, I. K., Haar, A. T., Schwander, E., Ebbens, F., Postmus, P. E., & The, T. H. (1985). Neuroendocrine differentiation antigen on human lung carcinoma and Kulchitski cells. *Cancer Res.* **45**:2192–2200.

De Ridder, L., & Mareel, M. (1978). Morphology and $^{125}$I-concentration of embryonic chick thyroids cultured in an atmosphere of oxygen. *Cell Biol. Int. Rep.* **2**:189–194.

De Ridder, L. (1997). Autologous confrontation of brain tumor derived spheroids with human dermal spheroids. *Anticancer Res.* **17**:4119–4120.

De Ridder, L., & Calliauw, L. (1990). Invasion of human brain tumors in vitro: Relationship to clinical evolution. *J. Neurosurg.* **72**:589–593.

De Ridder, L., & Mareel, M. (1978). Morphology and $^{125}$I-concentration of embryonic chick thyroids cultured in an atmosphere of oxygen. *Cell Biol. Int. Rep.* **2**:189–194.

De Silva, R., Moy, E. L., & Reddel, R. R. (1996). Immortalization of human bronchial epithelial cells. In Freshney, R. I., & Freshney, M. G. (eds.), *Culture of immortalized cells.* New York: Wiley-Liss, pp. 121–144.

De Silva, R., Whitaker, N. J., Rogan, E. M., & Reddel, R. R. (1994). HPV-16 E6 and E7 genes, like SV40 early region genes, are insufficient for immortalization of human mesothelial and bronchial epithelial cells. *Exp. Cell Res.* **213**:418–427.

De Vitry, F., Camier, M., Czernichow, P., Benda, P., Cohen, P., & Tixier-Vidal, A. (1974). Establishment of a clone of mouse hypothalamic neurosecretory cells synthesizing neurophysin and vasopressin. *Proc. Natl. Acad. Sci. USA* **71**:3575–3579.

De Vonne, T. L., & Mouray, H. (1978). Human $\alpha_2$-macroglobulin and its antitrypsin and antithrombin activities in serum and plasma. *Clin. Chim. Acta* **90**:83–85.

De Vos, A., & Van Steirteghem, A. (2001). Aspects of biopsy procedures prior to preimplantation genetic diagnosis. *Prenat Diagn.* **21**:767–780.

De Vries, J. E., Dinjens, W. N. M., De Bruyne, G. K., Verspaget, H. W., van der Linden, E. P. M., de Bruine, A. P., Mareel, M. M., Bosman, F. T., & ten Kate, J. (1995). in vivo and in vitro invasion in relation to phenotypic characteristics of human colorectal carcinoma cells. *Br. J. Cancer* **71**:271–277.

De Wever, O., Westbroek, W., Verloes, A., Bloemen, N., Bracke, M., Gespach, C., Bruyneel, E., & Mareel, M. (2004). Critical role of N-cadherin in myofibroblast invasion and migration in vitro stimulated by colon-cancer-cell-derived TGF-β or wounding. *J. Cell Sci.* **117**:4691–4703.

de Wynter, E., Allen, T., Coutinho, L., Flavell, D., Flavell, S. U., & Dexter, T. J. (1993). Localisation of granulocyte macrophage colony-stimulating factor in human long-term bone marrow cultures. *J. Cell Sci.* **106**:761–769.

Degos, L. (1997). Differentiation in acute promyelocytic leukemia: European experience. *J. Cell Physiol.* **173**:285–287.

Del Vecchio, P., & Smith, J. R. (1981). Expression of angiotensin converting enzyme activity in cultured pulmonary artery endothelial cells. *J. Cell Physiol.* **108**:337–345.

Del Vecchio, S., Stoppelli, M. P., Carriero, M. V., Fonti, R., Massa, O., Li, P. Y., Botti, G., Cerra, M., d'Aiuto, G., Espositi, G., & Salvatore, M. (1993). Human urokinase receptor concentration in malignant and benign breast tumours by in vitro quantitative autoradiography: Comparison with urokinase levels. *Cancer Res.* **53**:3198–3206.

DeMars, R. (1957). The inhibition of glutamine of glutamyl transferase formation in cultures of human cells. *Biochim. Biophys. Acta* **27**:435–436.

Demers, G. W., Halbert, C. L., & Galloway, D. A. (1994). Elevated wild-type p53 protein levels in human epithelial cell lines immortalized by the human papillomavirus type 16 E7 gene. *Virology* **198**:169–174.

DeMichele, M. A., Davis, A. L., Hunt, J. D., Landreneau, R. J., & Siegfried, J. M. (1994). Expression of mRNA for three bombesin receptor subtypes in human bronchial epithelial cells. *Am. J. Resp. Cell Mol. Biol.* **11**:66–74.

Deng, J., Steindler, D. A., Laywell, E. D., & Petersen, B. E. (2003). Neural trans-differentiation potential of hepatic oval cells in the neonatal mouse brain. *Exp. Neurol.* **182**:373–382.

Dennis, C. (2003). Developmental biology: Synthetic sex cells. *Nature* **424**:364–366.

Dennis, C., Gallagher, R., & Campbell, P. (2001). The Humane Genome. *Nature* **209**:813–858.

Desai, K. V., Kavanaugh, C. J., Calvo, A., & Green, J. E. (2002). Chipping away at breast cancer: Insights from microarray studies of human and mouse mammary cancer. *Endocr. Relat. Cancer.* **9**:207–220.

Deshpande, A. K., Baig, M. A., Carleton, S., Wadgoankar, R., & Siddiqui, M. A. Q. (1993). Growth factors activate cardiogenic differentiation in avian mesodermal cells. *Mol. Cell Differ.* **1**:269–284.

Detrisac, C. J., Sens, M. A., Garvin, A. J., Spicer, S. S., & Sens, D. A. (1984). Tissue culture of human kidney epithelial cells of proximal tubule origin. *Kidney Int.* **25**:383–390.

Dexter, T. J., Spooncer, E., Simmons, P., & Allen, T. D. (1984). Long-term marrow culture: An overview of technique and experience. In Wright, D. G., & Greenberger, J. S. (eds.), *Long-term bone marrow culture*. New York: Liss, Kroc Foundation Series 18, pp. 57–96.

Dexter, T. M., Allen, T. D., Scott, D., & Teich, N. M. (1979). Isolation and characterisation of a bipotential haematopoietic cell line. *Nature* **277**:417–474.

Dexter, T. M., Testa, N. G., Allen, T. D., Rutherford, S., & Scolnick, E. (1981). Molecular and cell biological aspects of erythropoiesis in long-term bone marrow cultures. *Blood* **58**:699–707.

Dick, J. E. (2009). Looking ahead in cancer stem cell research. *Nat. Biotechnol.* **27**:44–46.

Dickson, J. D., Flanigan, T. P., & Kemshead, J. T. (1983). Monoclonal antibodies reacting specifically with the cell surface of human astrocytes in culture. *Biochem. Soc. Trans.* **11**:208.

Dickson, M. A., Hahn, W. C., Ino, Y., Ronfard, V., Wu, J. Y., Weinberg, R. A., Louis, D. N., Li, F. P., & Rheinwald, J. G. (2000). Human keratinocytes that express hTERT and also bypass a p16(INK4a)-enforced mechanism that limits life span become immortal yet retain normal growth and differentiation characteristics. *Mol. Cell Biol.* **20**:1436–1447.

Dimitroulakos, J., Squire, J., Pawlin, G., & Yeger, H. (1994). NUB-7: A stable-1-type human neuroblastoma cell line inducible along N- and S-type cell lineages. *Cell Growth Differ.* **5**:373–384.

Dinnen, R., & Ebisuzaki, K. (1990). Mitosis may be an obligatory route to terminal differentiation in the Friend erythroleukemia cell. *Exp. Cell Res.* **191**:149–152.

Dirks, W. G., MacLeod, R. A. F., & Drexler, H. G. (1999). ECV3O4 (endothelial) is really T24 (bladder carcinoma): cell line cross-contamination at source. In Vitro Cell. Dev. *Biol. Anim.* **35**:558–559.

Ditadi, A., de Coppi, P., Picone, O., Gautreau, L., Smati, R., Six, E., Bonhomme, D., Ezine, S., Frydman, R., Cavazzana-Calvo, M., & André-Schmutz, I. (2009). Human and murine amniotic fluid c-Kit+Lin- cells display hematopoietic activity. *Blood* **113**:3953–3960.

Dobbie, I., & Zicha, D. (2002). Microscopy of living cells. In Davis, J. M., (ed.), *Basic cell culture*. New York: Oxford University Press, pp. 107–133.

Dobbs, L. G., & Gonzalez, R. F. (2002). Isolation and culture of pulmonary alveolar epithelial type II cells. In Freshney, R. I., & Freshney, M. G. (eds.), *Culture of epithelial cells*, 2nd ed. Hoboken, NJ: Wiley-Liss, pp. 278–301.

Dobbs, L. G., Pian, M., Dumars, S., Maglio, M., & Allen, L. (1997). Maintenance of the differentiated type II cell phenotype by culture with an apical air surface. *Am. J. Physiol.* **273**:L347–L354.

Dodson, M. V., Mathison, B. A., & Mathison, B. D. (1990). Effects of medium and substratum on ovine satellite cell attachment, proliferation and differentiation. *In Vitro Cell Differ. Dev.* **29**(1): 59–66.

Doering, C. B., Healey, J. F., Parker, E. T., Barrow, R. T., & Lollar, P. (2002). High level expression of recombinant porcine coagulation factor VIII. *J. Biol. Chem.* **277**:38345–38349.

DoH. (2006). Health technical memorandum 07-01: Safe management of healthcare waste. www.dh.gov.uk/en/Publications andStatistics/Publications/PublicationsPolicyAndGuidance/DH _063274.

Doherty, P., Ashton, S. V., Moore, S. E., & Walsh, F. S. (1991). Morphoregulatory activities of NCAM and N-cadherin can be accounted for by G protein-dependent activation of L- and N-type neuronal $Ca^{2+}$ channels. *Cell* **67**:21–33.

Dolbeare, F., & Selden, J. R. (1994). Immunochemical quantitation of\ bromodeoxyuridine: Application to cell-cycle kinetics. *Method Cell Biol.* **41**:297–316.

Domogatskaya, A., Rodin, S., Boutaud, A., & Tryggvason, K. (2008). Laminin-511 but not -332, -111, or -411 enables mouse embryonic stem cell self-renewal in vitro. *Stem Cells* **26**:2800–2809.

Dorrell, C., Abraham, S. L., Lanxon-Cookson, K. M., Canaday, P. S., Streeter, P. R., & Grompe, M. (2008). Isolation of major pancreatic cell types and long-term culture-initiating cells using novel human surface markers. *Stem Cell Res.* **1**:183–194.

Dotto, G. P., Parada, L. F., & Weinberg, R. A. (1985). Specific growth response of *ras*-transformed embryo fibroblasts to tumour promoters. *Nature* **318**:472–475.

Doucette, J. R. (1984). The glial cells in the nerve fibre layer of the rat olfactory bulb. *Anat. Rec.* **210**:285–391.

Doucette, J. R. (1990). Glial influences on axonal growth in the primary olfactory system. *Glia* **3**:433–449.

Douglas, G. C., Vandevoort, C., Kumar, P., Chang, T. C., and Golos, T. G. (2009). Trophoblast Stem cells: Models for investigating trophectoderm differentiation and placental development. *Endocr Rev.* **30**:228−240.

Douglas, J. L., & Quinlan, M. P. (1994). Efficient nuclear localization of the Ad5 E1A 12S protein is necessary for immortalization but not cotransformation of primary epithelial cells. *Cell Growth Differ.* **5**:475−483.

Douglas, W. H. J., McAteer, J. A., Dell'Orco, R. T., & Phelps, D. (1980). Visualization of cellular aggregates cultured on a three-dimensional collagen sponge matrix. *In Vitro* **16**:306−312.

Dowling, P., Walsh, N., & Clynes, M. (2008). Membrane and membrane-associated proteins involved in the aggressive phenotype displayed by highly invasive cancer cells. *Proteomics* **8**:4054−4065.

Doyle, A., Griffiths, J. B., & Newall, D. G. (eds.). (1990−1999). *Cell and tissue culture: Laboratory procedures.* Chichester, UK: Wiley.

Doyle, A., Morris, C., & Mowles, J. M. (1990). Quality control. In Doyle, A., Hay, R., & Kirsop, B. E. (eds.), *Animal cells, living resources for biotechnology.* Cambridge: Cambridge University Press, pp. 81−100.

Dracopoli, N. C., Haines, J. L., Korf, B. R., Morton, C. C., Seidman, C. E., Rosenzweig, A., Seidman, J. G., & Smith, D. R. (eds.). (2004). *Short protocols in human genetics.* Hoboken, NJ: Wiley.

Draper, J. S., Pigott, C., Thomson, J. A., & Andrews, P. W. (2002). Surface antigens of human embryonic stem cells: Changes upon differentiation in culture. *J Anat.* **200**:249−258.

Drejer, J., Larsson, O. M., & Schousboe, A. (1983). Characterization of uptake and release processes for D- and L-aspartate in primary cultures of astrocytes and cerebellar granule cells. *Neurochem. Ref.* **8**:231−243.

Drexler, H. G. (2004). Establishment and culture of human leukemia-lymphoma cell lines. In Pfragner, R., & Freshney. R. I. (eds.), *Culture of human tumor cells.* Hoboken, NJ: Wiley-Liss, pp. 319−348.

Drexler, H. G., & Minowada, J. (2000). Human leukemia-lymphoma cell lines: historical perspective, state of the art and future prospects. In Masters, J. R. W., & Palsson, B. O., (eds.), *Human cell culture, Vol. III Cancer cell lines, Part 3: Leukemia and lymphomas.* Dordrecht: Kluwer, pp. 1−88.

Drexler, H. G., Dirks, W. G., Matsuo, Y., & MacLeod, R. A. (2003). False leukemia-lymphoma cell lines: An update on over 500 cell lines. *Leukemia* **17**:416−426.

Drexler, H. G., Gignac, S. M., Hu, Z.-B., Hopert, A., Fleckenstein, E., Viges, M., & Uphoff, C. C. (1994). Treatment of mycoplasma contamination in a large panel of cell cultures. *In Vitro Cell Dev. Biol.* **30A**:344−347.

Driever, W., & Rangini, Z. (1993). Characterization of a cell line derived from zebra-fish (*Brachydanio rerio*). *In Vitro Cell Dev. Biol. Anim.* **29A**:749−754.

Driever, W., Stemple, D., Schier, A., & Solnica-Krezel, L. (1994). Zebrafish: Genetic tools for studying vertebrate development. *Trends Genet.* **10**:152−159.

Du, Y., & Funderburgh, J. L. (2007). Culture of human corneal stem cells. In Freshney, R. I., Stacey, G. N., & Auerbach, J. M. (eds), *Culture of human stem cells.* Hoboken, NJ: Wiley, pp. 251−280.

Duffy, M. J., Reilly, D., O'Sullivan, C., O'Higgins, N., Fennelly, J. J., & Andreasen, P. (1990). Urokinase-plasminogen activator, a new and independent prognostic market in breast cancer. *Cancer Res.* **50**:6827−6829.

Duksin, D., Maoz, A., & Fuchs, S. (1975). Differential cytotoxic activity of anticollagen serum on rat osteoblasts and fibroblasts in tissue culture. *Cell* **5**:83−86.

Dulbecco, R. (1952). Production of plaques in monolayers of tissue cultures by single particles of an animal virus. *Proc. Natl. Acad. Sci. USA* **38**:747.

Dulbecco, R., & Elkington, J. (1973). Conditions limiting multiplication of fibroblastic and epithelial cells in dense cultures. *Nature* **246**:197−199.

Dulbecco, R., & Freeman, G. (1959). Plaque formation by the polyoma virus. *Virology* **8**:396−397.

Dulbecco, R., & Vogt, M. (1954). Plaque formation and isolation of pure cell lines with poliomyelitis viruses. *J. Exp. Med.* **199**:167−182.

Duncan, E. L., De Silva, R., & Reddel, R. R. (1996). Immortalization of human mesothelial cells. In Freshney, R. I., & Freshney, M. G. (eds.), *Culture of immortalized cells.* New York: Wiley-Liss, pp. 239−258.

Dunham, L. J., & Stewart, H. L. (1953). A survey of transplantable and transmissible animal tumors. *J. Natl. Cancer Inst.* **13**:1299−1377.

Dunn, J. C., Yarmush, M. L., Koebe, H. G., & Tompkins, R. G. (1989). Hepatocyte function and extracellular matrix geometry: long-term culture in a sandwich configuration. *FASEB J.* **3**:174−177.

Dunn, G. A., & Zicha, D. (1993). Phase-shifting interference microscopy applied to the analysis of cell behaviour. *Symp. Soc. Exp. Biol.* **47**:91−106.

Dupin, E., Glavieux, C., Vaigot, P., & le Douarin, N. M. (2000). Endothelin 3 induces the reversion of melanocytes to glia through a neural crest-derived glial-melanocytic progenitor. *Proc. Natl. Acad. Sci. USA* **97**:7882−7887.

Dupin, E., Real, C., Glavieux-Pardanaud, C., Vaigot, P., & le Douarin, N. M. (2003). Reversal of developmental restrictions in neural crest lineages, transition from Schwann cells to glial-melanocytic precursors in vitro. *Proc. Natl. Acad. Sci. USA* **100**:5229−5233.

Dutt, K., Harris-Hooker, S., Ellerson, D., Layne, D., Kumar, R., & Hunt, R. (2003). Generation of 3D retina-like structures from a human retinal cell line in a NASA bioreactor. *Cell Transplant.* **12**:717−731.

Eagle, H. (1955). The specific amino acid requirements of mammalian cells (stain L) in tissue culture. *J. Biol. Chem.* **214**:839.

Eagle, H. (1959). Amino acid metabolism in mammalian cell cultures. *Science* **130**:432.

Eagle, H. (1973). The effect of environmental pH on the growth of normal and malignant cells. *J. Cell Physiol.* **82**:1−8.

Eagle, H., Foley, G. E., Koprowski, H., Lazarus, H., Levine, E. M., & Adams, R. A. (1970). Growth characteristics of virus-transformed cells. *J. Exp. Med.* **131**:863−879.

Earle, W. R., Schilling, E. L., Stark, T. H., Straus, N. P., Brown, M. F., & Shelton, E. (1943). Production of malignancy in vitro; IV: The mouse fibroblast cultures and changes seen in the living cells. *J. Natl. Cancer Inst.* **4**:165−212.

Easty, D. M., & Easty, G. C. (1974). Measurement of the ability of cells to infiltrate normal tissues in vitro. *Br. J. Cancer* **29**:36–49.

Ebendal, T., (1976). The relative roles of contact inhibition and contact guidance in orientation of axons extending on aligned collagen fibrils in vitro. *Exp. Cell Res.* **98**:159–169.

Ebert, A. D., Yu, J., Rose, F. F. Jr, Mattis, V. B., Lorson, C. L., Thomson, J. A., & Svendsen, C. N. (2009). Induced pluripotent stem cells from a spinal muscular atrophy patient. *Nature* **457**:277–280.

EBRA. (2009). European Biomedical Research Association: http://www.ebra.org.

Echarti, C., & Maurer, H. R. (1989). Defined, serum-free culture conditions for the GM-microclonogenic assay using agar-capillaries. *Blut* **59**:171–176.

Echarti, C., & Maurer, H. R. (1991). Lymphokine-activated killer cells: Determination of their tumor cytolytic capacity by a clonogenic microassay using agar capillaries. *J. Immunol. Methods* **143**:41–47.

ECVAM. (2008). Scientifically validated methods. http://ecvam.jrc.it/.

Edelman, G. M. (1973). Nonenzymatic dissociations; B: Specific cell fractionation of chemically derivatized surfaces. In Kruse, P. F. Jr., & Patterson, M. K. Jr. (eds.), *Tissue culture methods and applications*. New York: Academic Press, pp. 29–36.

Edelson, J. D., Shannon, J. H., & Mason, R. J. (1988). Alkaline phosphatase: A marker of alveolar type II cell differentiation. *Am. Rev. Respir. Dis.* **138**:1268–1275.

Edington, K. G., Berry, I. J., O'Prey, M., Burns, J. E., Clark, L. J., Mitchell, R., Robertson, G., Soutar, D., Coggins, L. W., & Parkinson, E. K. (2004). Multistage head and neck squamous cell carcinoma. In Pfragner, R., & Freshney. R. I. (eds.), *Culture of human tumor cells*. Hoboken, NJ: Wiley-Liss, pp. 261–288.

Edvardsson, L., & Olofsson, T. (2009). Real-time PCR analysis for blood cell lineage specific markers. *Methods Mol Biol.* **496**:313–322.

Edwards, A. M., Silva, E., Jofre, B., Becker, M. I., & De Ioannes, A. E. (1994). Visible light effects on tumoral cells in a culture medium enriched with tryptophan and riboflavin. *J. Photochem. Photobiol. B* **24**:179–186.

Edwards, P. A. W., Easty, D. M., & Foster, C. S. (1980). Selective culture of epithelioid cells from a human squamous carcinoma using a monoclonal antibody to kill fibroblasts. *Cell Biol. Int. Rep.* **4**:917–922.

Edwards, R. G. (1996). The history of assisted human conception with especial reference to endocrinology. *Exp. Clin. Endocrinol. Diabetes* **104**(3): 183–204.

Eisinger, M., Lee, J. S., Hefton, J. M., Darzykiewicz, A., Chiao, J. W., & Deharven, E. (1979). Human epidermal cell cultures: Growth and differentiation in the absence of dermal components or medium supplements. *Proc. Natl. Acad. Sci. USA* **76**:5340.

Elliget, K. A., & Lechner, J. F. (1992). Normal human bronchial epithelial cell cultures. In Freshney, R. I. (ed.), *Culture of epithelial cells*. New York: Wiley-Liss, pp. 181–196.

Ellison, G., Klinowska, T., Westwood, R. F. R., Docter, E., French, T., & Fox, J. C. (2003). Further evidence to support the melanocytic origin of MDA-MB-435. *J. Clin. Pathol.* **55**:294–299.

Elvin, P., Wong, V., & Evans, C. W. (1985). A study of the adhesive, locomotory and invasive behaviour of Walker 256 carcinosarcoma cells. *Exp. Cell Biol.* **53**:9–18.

EMAP. (2003). 3D embryo atlas project: http://genex.hgu.mrc.ac.uk/

Emura, M., Katyal, S. L., Ochiai, A., Hirohashi, S., & Singh, G. (1997). In Vitro reconstitution of human respiratory epithelium. *In Vitro Cell Dev. Biol. Anim.* **33**:602–605.

Enders, J. F., Weller, T. J., & Robbins, F. C. (1949). Cultivation of Lansing strain of poliomyelitis in cultures of various human embryonic tissues. *Science* **109**:85.

Eng, L. F., & Bigbee, J. W. (1979). Immunochemistry of nervous-system specific antigens. In Aprison, M. H. (ed.), *Advances in neurochemistry*. New York: Plenum Press, pp. 43–98.

Engel, L. W., Young, N. A., Tralka, T. S., Lippman, M. E., O'Brien, S. J., & Joyce, M. J. (1978). Establishment and characterization of three new continuous cell lines derived from breast carcinomas. *Cancer Res.* **38**:3352–3364.

Entwistle, A. (1998). A comparison between the use of a high-resolution CCD camera and 35mm film for obtaining coloured micrographs. *J. Microsc.* **192**:81–89.

Eppig, J. J., & O'Brien, M. J. (1996), Development in vitro of mouse oocytes from primordial follicles. *Biol. Reprod.* **54**:197–207.

Epstein, M. A., & Barr, Y. M. (1964). Cultivation in vitro of human lymphoblasts from Burkitt's malignant lymphoma. *Lancet* **1**:252.

Ermis, A., Muller, B., Hopf, T., Hopf, C., Remberger, K., Justen, H. P., Welter, C., & Hanselmann, R. (1998). Invasion of human cartilage by cultured multicellular spheroids of rheumatoid synovial cells: A novel in vitro model system for rheumatoid arthritis. *J. Rheumatol.* **25**:208–213.

Ernsberger, U. (2008). The role of GDNF family ligand signalling in the differentiation of sympathetic and dorsal root ganglion neurons. *Cell Tissue Res.* **333**:353–371.

Eroglu, A., Russo, M. J., Bieganski, R., Fowler, A., Cheley, S., Bayley, H., & Toner, M. (2000). Intracellular trehalose improves the survival of cryopreserved mammalian cells. *Nat. Biotechnol.* **18**:163–167.

Eschenhagen, T., & Zimmerman, W. H. (2006). Engineered heart tissue. In Vunjak-Novakovic, G. & Freshney, R. I. *Culture of Cells for Tissue Engineering*. Hoboken: John Wiley & Sons, 260–291.

Espiñeira, M., González-Lavín, N., Vieites, J. M., & Santaclara, F. J. (2009). Development of a method for the genetic identification of commercial bivalve species based on mitochondrial 18S rRNA sequences. *J Agric Food Chem.* **57**:495–502.

Ethier, S. P., Mahacek, M. L., Gullick, W. J., Frank, T. S., & Weber, B. L. (1993). Differential isolation of normal luminal mammary epithelial cells and breast cancer cells from primary and metastatic sites using selective media. *Cancer Res.* **53**:627–635.

EuroStemCell. (2009). www.eurostemcell.org

Evans, M. J., & Kaufman, M. H. (1981). Establishment in culture of pluripotential cells from mouse embryos. *Nature* **292**:154–156.

Evans, T., Rosenthal, E. T., Youngblom, J., Distel, D., & Hunt, T. (1983). Cyclin: a protein specified by maternal mRNA in sea urchin eggs that is destroyed at each cleavage division. *Cell* **33**:389–396.

Evans, V. J., & Bryant, J. C. (1965). Advances in tissue culture at the National Cancer Institute in the United States of America. In Ramakrishnan, C. V. (ed.), *Tissue culture*. The Hague: Junk, pp. 145–167.

Evans, V. J., Bryant, J. C., Fioramonti, M. C., McQuilkin, W. T., Sanford, K. K., & Earle, W. R. (1956). Studies of nutrient media for tissue C cells in vitro; I: A protein-free chemically defined medium for cultivation of strain L cells. *Cancer Res.* **16**:77.

Evans, V. J., & Sanford, K. K. (1978). Development of defined media for studies on malignant transformation in culture. In Katsuta, H. (ed.), *Nutritional requirements of cultured cells*. Baltimore: University Park Press, pp 149–192.

Evans-Molina, C., Vestermark, G. L., & Mirmira, R. G. (2009). Development of insulin-producing cells from primitive biologic precursors. *Curr. Opin. Organ Transplant.* **14**:56–63.

Fael Al-Mayhani, T. M., Ball, S. L., Zhao, J. W., Fawcett, J., Ichimura, K., Collins, P. V., & Watts, C. (2009). An efficient method for derivation and propagation of glioblastoma cell lines that conserves the molecular profile of their original tumours. *J. Neurosci. Methods*, **176**:192–199.

Fantini, J., Galons, J. P., Abadie, B., Canioni, P., Cozzone, P. J., Marvali, J., & Tirard, A. (1987). Growth in serum-free medium of human colonic adenocarcinoma cell lines on microcarriers: A two-step method allowing optimal cell spreading and growth. *In Vitro* **23**:641–646.

Fata, J. E., Werb, Z., & Bissell, M. J. (2004). Regulation of mammary gland branching morphogenesis by the extracellular matrix and its remodeling enzymes. *Breast Cancer Res.* **6**:1–11.

FDA. (2003). Good laboratory practice for nonclinical laboratory studies. CFR Title 21, Vol 1. Part 58. www.access.gpo.gov/nara/cfr/waisidx_03/21cfr58_03.html.

Federoff, S. (1975). In Evans, V. J., Perry, V. P., & Vincent, M. M. (eds.), *Manual of the Tissue Culture Association* **1**: 53–57.

Federoff, S., & Richardson, A. (2001). *Protocols for neural cell culture*. Clifton, NJ: Humana Press.

Fell, H. B., & Robison, R. (1929). The growth, development and phosphatase activity of embryonic avian femora and limb buds cultivated in vitro. *Biochem. J.* **23**:767–784.

Fenton, J. I., & Hord, N. G. (2004). Flavonoids promote cell migration in nontumorigenic colon epithelial cells differing in Apc genotype: Implications of matrix metalloproteinase activity. *Nutr. Cancer.* **48**:182–188.

Fergusson, R. J., Carmichael, J., & Smyth, J. F. (1980). Human tumour xenografts growing in immunodeficient mice: A useful model for assessing chemotherapeutic agents in bronchial carcinoma. *Thorax* **41**:376–380.

Fernandes, T. G., Diogo, M. M., Clark, D. S., Dordick, J. S., & Cabral, J. M. (2009). High-throughput cellular microarray platforms: applications in drug discovery, toxicology and stem cell research. *Trends Biotechnol.* **27**:342–349.

Fernig, D. G., & Gallagher, J. T. (1994). Fibroblast growth factors and their receptors: An information network controlling tissue growth, morphogenesis and repair. *Prog. Growth Factor Res.* **5**:353–377.

Finbow, M. E., & Pitts, J. D. (1981). Permeability of junctions between animal cells. *Exp. Cell Res.* **131**:1–13.

Fiorino, A. S., Diehl, A. M., Lin, H. Z., Lemischka, I. R., & Reid, L. M. (1998). Maturation-dependent gene expression in a conditionally transformed liver progenitor cell line. *In Vitro Cell Dev Biol Anim.* **34**:247–258.

Fischer, A. (1925). *Tissue culture. Studies in experimental morphology and general physiology of tissue cells in vitro*. London: Heinemann.

Fischer, G. A., & Sartorelli, A. C. (1964). Development, maintenance and assay of drug resistance. *Methods Med. Res.* **10**:247.

Fisher, H. W., Puck, T. T., & Sato, G. (1958). Molecular growth requirements of single mammalian cells: The action of fetuin in promoting cell attachment of glass. *Proc. Natl. Acad. Sci. USA* **44**:4–10.

Fitzsimmons, S. A., Ireland, H., Barr, N. I., Cuthbert, A. P., Going, J. J., Newbold, R. F., & Parkinson, E. K. (2003). Human squamous cell carcinomas lose a mortality gene from chromosome 6q14.3 to q15. *Oncogene* **22**:1737–1746.

Flier, J. S. (1995). The adipocyte: Storage depot or node on the energy information superhighway? *Cell* **80**:15–18.

Flynn, D., Yang, J., & Nandi, S. (1982). Growth and differentiation of primary cultures of mouse mammary epithelium embedded in collagen gel. *Differentiation* **22**:191–194.

Focus on Alternatives (2009). Serum-free media for cell culture: www.focus-alternative.org.

Fogh, J. (1977). Absence of HeLa cell contamination in 169 cell lines derived from human tumors. *J. Natl. Cancer Inst.* **58**:209–214.

Fogh, J., & Trempe, G. (1975). In Fogh, J. (ed.), *Human tumor cells in vitro*. New York: Academic Press, pp. 115–159.

Folkman, I., & Moscona, A. (1978). Role of cell shape in growth control. *Nature* **273**:345–349.

Folkman, I., Haudenschild, C. C., & Zetter, B. R. (1979). Long-term culture of capillary endothelial cells. *Proc. Natl. Acad. Sci. USA* **76**:5217–5221.

Folkman, J. (1992). The role of angiogenesis in tumor growth. *Sem. Cancer Biol.* **3**:65–71.

Folkman, J., & D'Amore, P. A. (1996). Blood vessel formation: what is its molecular basis? Minireview. *Cell* **87**:1153–1155.

Folkman, J., & Haudenschild, C. (1980). Angiogenesis in vitro. *Nature* **288**:551–556.

Fong, C. Y., Peh, G. S., Gauthaman, K., & Bongso, A. (2009). Separation of SSEA-4 and TRA-1-60 labelled undifferentiated human embryonic stem cells from a heterogeneous cell population using magnetic-activated cell sorting (MACS) and fluorescence-activated cell sorting (FACS). *Stem Cell Rev Rep.* **5**:72–80.

Fontana, A., Hengarner, H., de Tribolet, N., & Weber, E. (1984). Glioblastoma cells release interleukin 1 and factors inhibiting interleukin 2-mediated effects. *J. Immunol.* **132**(4): 1837–1844.

Food and Drug Administration (2007). GMP: www.fda.gov/CDRH/DEVADVICE/32.html.

Foreman, J., & Pegg, D. E. (1979). Cell preservation in a programmed cooling machine: The effect of variations in supercooling. *Cryobiology* **16**:315–321.

Forsten-Williams, K., Chu, C. L., Fannon, M., Buczek-Thomas, J. A., & Nugent, M. A. (2008). Control of growth factor networks by heparan sulfate proteoglycans. *Ann. Biomed. Eng.* **36**:2134–2148.

Frame, M., & Norman, J. (2008). A tal(in) of cell spreading. *Nat. Cell Biol.* **10**:1017–1019.

Frame, M., Freshney, R. I., Shaw, R., & Graham, D. I. (1980). Markers of differentiation in glial cells. *Cell Biol. Int. Rep.* **4**:732.

Frame, M. C., Freshney, R. I., Vaughan, P. E. T., Graham, D. I., & Shaw, R. (1984). Interrelationship between differentiation and malignancy-associated properties in glioma. *Br. J. Cancer* **49**:269–280.

Franceschini, I. A., & Barnett, S. C. (1996). Low-affinity NGF-receptor and E-N-CAM expression define two types of olfactory nerve ensheathing cells that share a common lineage. *Dev. Biol.* **173**:327–343.

Franklin, J. M., Gilson, J. M., Franceschini, I. A., & Barnett, S. C. (1996). Schwann cell-like myelination following transplantation of an olfactory-nerve-ensheathing-cell line into areas of demyelination in the adult CNS. *Glia* **17**:217–224.

Franklin, R. J. M, & Barnett, S. C. (1997). Do olfactory glia have advantages over Schwann cells for CNS repair? *J. Neurosci. Res.* **50**:1–8.

Franklin, R. J. M., & Barnett, S. C. (2000). Olfactory ensheathing cells and CNS regeneration: The sweet smell of success? *Neuron* **28**:1–4.

Franssen, E. H., De Bree, F. M., Essing, A. H., Ramon-Cueto, A., & Verhaagen. J. (2008). Comparative gene expression profiling of olfactory ensheathing glia and Schwann cells indicates distinct tissue repair characteristics of olfactory ensheathing glia. *Glia* **56**:1285–1298.

Fredin, B. L., Seiffert, S. C., & Gelehrter, T. D. (1979). Dexamethasone-induced adhesion in hepatoma cells: The role of plasminogen activator. *Nature* **277**:312–313.

Freed, L. E., & Vunjak-Novakovic, G. (2002). Spaceflight bioreactor studies of cells and tissues. *Adv. Space Biol. Med.* **8**:177–95.

Freedman, V. H., & Shin, S. (1974). Cellular tumorigenicity in nude mice: Correlation with cell growth in semi-solid medium. *Cell* **3**:355–359.

Freshney, R. I., & Morgan, D. (1978). Radioisotopic quantitation in microtitration plates by an autofluorographic method. *Cell Biol. Int. Rep.* **2**:275–380.

Freshney, R. I. (1972). Tumour cells disaggregated in collagenase. *Lancet* **2**:488–489.

Freshney, R. I. (1976). Separation of cultured cells by isopycnic centrifugation in metrizamide gradients. In Rickwood, D. (ed.), *Biological separations*. Washington, DC: Information Retrieval, pp. 123–130.

Freshney, R. I. (1978). Use of tissue culture in predictive testing of drug sensitivity. *Cancer Top.* **1**:5–7.

Freshney, R. I. (1980). Culture of glioma of the brain. In Thomas, D. G. T., & Graham, D. I. (eds.), *Brain tumours: Scientific basic, clinical investigation and current therapy*. London: Butterworths, pp. 21–50.

Freshney, R. I., & Freshney, M. G. (2002). In Freshney, R. I. (ed.), *Culture of epithelial cells*, 2nd ed. Hoboken, NJ: Wiley-Liss, chs. 6, 10, 11, 12, 13.

Freshney, R. I., & Freshney, M. G. (eds.). (1996). *Culture of immortalized cells*. New York: Wiley-Liss.

Freshney, R. I., & Hart, E. (1982). Clonogenicity of human glia in suspension. *Br. J. Cancer* **46**:463.

Freshney, R. I., Celik, F., & Morgan, D. (1982a). Analysis of cytotoxic and cytostatic effects. In Davis, W., Malvoni, C., & Tanneberger, St. (eds.), *The control of tumor growth and its biological base*, Fortschritte in der Onkologie, Band 10. Berlin: Akademie, pp. 349–358.

Freshney, R. I., Hart, E., & Russell, J. M. (1982b). Isolation and purification of cell cultures from human tumours. In Reid, E., Cook, G. M. W., & Moore, D. J. (eds.), *Cancer cell organelles; methodological surveys (B): Biochemistry*, Vol. **2**. Chicester, UK: Horwood, pp. 97–110.

Freshney, R. I., Morgan, D., Hassanzadah, M., Shaw, R., & Frame, M. (1980a). Glucocorticoids, proliferation and the cell surface. In Richards, R. J., & Rajan, K. T. (eds.), *Tissue culture in medical research (II)*. Oxford: Pergamon Press, pp. 125–132.

Freshney, R. I., Paul, J., & Kane, I. M. (1975). Assay of anticancer drugs in tissue culture: Conditions affecting their ability to incorporate $^3$H-leucine after drug treatment. *Br. J. Cancer* **31**:89–99.

Freshney, R. I., Pragnell, I. B., & Freshney, M. G., (eds.). (1994). *Culture of haemopoietic cells*. New York: Wiley-Liss.

Freshney, R. I., Sherry, A., Hassanzadah, M., Freshney, M., Crilly, P., & Morgan, D. (1980b). Control of cell proliferation in human glioma by glucocorticoids. *Br. J. Cancer* **41**:857–866.

Freshney, R. I. (1985). Induction of differentiation in neoplastic cells. *Anticancer Res.* **5**:111–130.

Freyer, J. P., & Sutherland, R. M. (1980). Selective dissociation and characterization of cells from different regions of multicell tumour spheroids. *Cancer Res.* **40**:3956–3965.

Friend, C., Patuleia, M. C., & Nelson, J. B. (1966). Antibiotic effect of tylosine on a mycoplasma contaminant in a tissue culture leukemia cell line. *Proc. Soc. Exp. Biol. Med.* **121**:1009.

Friend, C., Scher, W., Holland, J. G., & Sato, T. (1971). Hemoglobin synthesis in murine virus-induced leukemic cells in vitro; 2: Stimulation of erythroid differentiation by dimethyl sulfoxide. *Proc. Natl. Acad. Sci. USA* **68**:378–382.

Friend, K. K., Dorman, B. P., Kucherlapati, R. S., & Ruddle, F. H. (1976). Detection of interspecific translocations in mouse–human hybrids by alkaline Giemsa staining. *Exp. Cell Res.* **99**:31–36.

Froud, S. J. (1999). The development, benefits and disadvantages of serum-free media. *Dev. Biol. Stand.* **99**:157–166.

Fry, J., & Bridges, J. W. (1979). The effect of phenobarbitone on adult rat liver cells and primary cell lines. *Toxicol. Lett.* **4**:295–301.

Fu, V. X., Schwarze, S. R., Reznikoff, C. A., & Jarrard, D. F. (2004). The establishment and characterization of bladder cancer cultures in vitro. In Pfragner, R., & Freshney. R. I. (eds.), *Culture of human tumor cells*, Hoboken, NJ: Wiley-Liss, pp. 97–123.

Fujita, H., Asahina, A., Mitsui, H., & Tamaki, K. (2004). Langerhans cells exhibit low responsiveness to double-stranded RNA. *Biochem. Biophys. Res. Commun.* **319**:832–839.

Fuller, B. B., & Meyskens, F. L. (1981). Endocrine responsiveness in human melanocytes and melanoma cells in culture. *J. Natl. Cancer Inst.* **66**:799–802.

Fullerton, N. E., Boyd, M., Mairs, R. J., Keith, W. N., Alderwish, O., Brown, M. M., Livingstone, A., & Kirk, D. (2004). Combining a targeted radiotherapy and gene therapy approach for adenocarcinoma of prostate. *Prostate Cancer Prostatic Dis.* **7**:355–363.

Furue, M., & Saito, S. (1997). Synergistic effect of hepatocyte growth factor and fibroblast growth factor-1 on the branching morphogenesis of rat submandibular gland epithelial cells. *Tissue Cult. Res. Commun.* **16**:189–194.

Fusenig, N. E., & Boukamp, P. (1998). Multiple stages and genetic alterations in immortalization, malignant transformation, and tumor progression of human skin keratinocytes. *Mol. Carcinog.* **23**:144–158.

Fusenig, N. E. (1994a). Epithelial–mesenchymal interactions regulate keratinocyte growth and differentiation in vitro. In Leigh, I. M., Lane, E. B., & Watt, F. M. (eds.), *Keratinocyte handbook*, Vol. **1**. Cambridge: Cambridge University Press, pp. 71–94.

Fusenig, N. E. (1994b). Cell culture models: Reliable tools in pharmacotoxicology? In Fusenig, N. E., & Graf, H. (eds.), *Cell culture in pharmaceutical research*. Berlin: Springer, pp. 1–7.

Fusenig, N. E., Limat, A., Stark, H.-J., & Breitkreutz, D. (1994). Modulation of the differentiated phenotype of keratinocytes of the hair follicle and from epidermis. *J. Dermatol. Sci.* **7**:142–151.

Galluzzi, L., Aaronson, S. A., Abrams, J., Alnemri, E. S., Andrews, D. W. , Baehrecke, E. H., et al. (2009). Guidelines for the use and interpretation of assays for monitoring cell death in higher eukaryotes. *Cell Death Differ.* **16**(8): 1093–1107.

Ganmore, I., Smooha, G., & Izraeli, S. (2009). Constitutional aneuploidy and cancer predisposition. *Hum. Mol. Genet.* **18**(R1):R84–93.

Gao, R., Ustinov, J., Pulkkinen, M. A., Lundin, K., Korsgren, O., & Otonkoski, T. (2003). Characterization of endocrine progenitor cells and critical factors for their differentiation in human adult pancreatic cell culture. *Diabetes* **52**:2007–2015.

Gard, A. L., & Pfeiffer, S. E. (1990). Two proliferative stages of the oligodendrocyte lineage (A2B5+O4−and O4+GalC−) under different mitogenic control. *Neuron* **5**:615–625.

Gardner, D. K., & Lane, M. (2003). Towards a single embryo transfer. *Reprod. Biomed. Online* **6**(4): 470–481.

Garnier, D., Loyer, P., Ribault, C., Guguen-Guillouzo, C., & Corlu, A. (2009). Cyclin-dependent kinase 1 plays a critical role in DNA replication control during rat liver regeneration. . *Hepatology* **50**:1946–1956.

Garrido, T., Riese, H. H., Aracil, M., & Perez-Aranda, A. (1995). Endothelial cell differentiation into capillary-like structures in response to tumour cell conditioned medium: A modified chemotaxis chamber assay. *Br. J. Cancer* **71**:770–775.

Gartler, S. M. (1967). Genetic markers as tracers in cell culture. 2nd Decennial Review Conference on Cell, Tissue and Organ Culture. *NCI Monograph* 26, pp. 167–195.

Gartner, S., & Kaplan, H. S. (1980). Long-term culture of human bone marrow cells. *Proc. Natl. Acad. Sci. USA* **77**:4756–4759.

Gaudernack, T., Leivestad, T., Ugelstad, J., & Thorsby, E. (1986). Isolation of pure functionally active CD8+ T cells. Positive selection with monoclonal antibodies directly conjugated to monosized magnetic microspheres. *J. Immunol. Methods* **90**:179–187.

Gaush, C. R., Hard, W. L., & Smith, T. F. (1966). Characterization of an established line of canine kidney cells (MDCK). *Proc. Soc. Exp. Biol. Med.* **122**:931–933.

Gazdar, A. F., Carney, D. N., & Minna, J. D. (1983). The biology of nonsmall cell lung cancer. *Semin. Oncol.* **10**:3–19.

Gazdar, A. F., Carney, D. N., Russell, E. K., Sims, H. L., Baylin, S. B., Bunn, P. A., Guccion, J. G., & Minna, J. D. (1980). Establishment of continuous, clonable cultures of small cell carcinoma of the lung which have amine precursor uptake and decarboxylation properties. *Cancer Res.* **40**:3502–3507.

Gazdar, A. F., Zweig, M. H., Carney, D. N., Van Steirteghen, A. C., Baylin, S. B., & Minna, J. D. (1981). Levels of creatine kinase and its BB isoenzyme in lung cancer specimens and cultures. *Cancer Res.* **41**:2773–2777.

Gazdar, A. F. Oie, H. K., Shackleton, C. H., Chen, T. R., Triche, T. J., Myers, C. E., Chrousos, G. P., Brennan, M. F., Stein, C. A., & La Rocca, R. V. (1990). Establishment and characterization of a human adrenocortical carcinoma cell line that expresses multiple pathways of steroid biosynthesis. *Cancer Res.* **50**:5488–5496.

Geijsen, N., Horoschak, M., Kim, K., Gribnau, J., Eggan, K., & Daley, G. Q. (2004). Derivation of embryonic germ cells and male gametes from embryonic stem cells. *Nature* **427**:148–154.

Germain, L., Rouabhia, M., Guignard, R., Carrier, L., Bouvard, V., & Auger, F. A. (1993). Improvement of human keratinocyte isolation and culture using thermolysin. *Burns* **19**:99–104.

Gerper, P., Whang-Peng, J., & Monroe, J. H. (1969). Transformation and chromosome changes induced by Epstein-Barr virus in normal human leukocyte cultures. *Proc. Natl. Acad. Sci. USA* **63**(3): 740–747.

Gey, G. O., Coffman, W. D., & Kubicek, M. T. (1952). Tissue culture studies of the proliferative capacity of cervical carcinoma and normal epithelium. *Cancer Res.* **12**:364–365.

Ghigo, D., Priotto, C., Migliorino, D., Geromin, D., Franchino, C., Todde, R., Costamagna, C., Pescarmona, G., & Bosia, A. (1998). Retinoic acid-induced differentiation in a human neuroblastoma cell line is associated with an increase in nitric oxide synthesis. *J Cell Physiol.* **174**:99–106.

Ghosh, C., & Collodi, P. (1994). Culture of cells from zebrafish (*Brachydanio rerio*) blastula-stage embryos. *Cytotechnology* **14**:21–26.

Ghosh, D., Danielson, K. C., Alston, J. T., & Heyner, S. (1991). Functional differential of mouse uterine epithelial cells grown on collagen gels or reconstituted basement membranes. *In Vitro Cell Dev. Biol.* **27A**:713–719.

Giard, D. J., Aaronson, S. A., Todaro, G. J., Arnstein, P., Kersey, J. H., Dosik, K., & Parks, W. P. (1972). In vitro cultivation of human tumors: Establishment of cell lines derived from a series of solid tumors. *J. Natl. Cancer Inst.* **51**:1417.

Gignac, S. M., Steube, K., Schleithoff, L., Janssen, J. W., MacLeod, R. A., Quentmeier, H., & Drexler, H. G. (1993). Multiparameter approach in the identification of cross-contaminated leukemia cell lines. *Leukemia Lymphoma* **10**(4–5): 359–368.

Gilbert, S. F., & Migeon, B. R. (1975). D-Valine as a selective agent for normal human and rodent epithelial cells in culture. *Cell* **5**:11–17.

Gilbert, S. F., & Migeon, B. R. (1977). Renal enzymes in kidney cells selected by D-valine medium. *J. Cell Physiol.* **92**:161–168.

Gilchrest, B. A., Albert, L. S., Karassik, R. L., & Yaar, M. (1985). Substrate influences human epidermal melanocyte attachment and spreading in vitro. In Vitro Cell Dev. Biol., **21**:114.

Gillette, R. W. (1987). Alternatives to pristane priming for ascitic fluid and monoclonal antibody production. *J. Immunol. Methods* **99**:21–23.

Gillis, S., & Watson, J. (1981). Interleukin-2 dependent culture of cytolytic T cell lines. *Immunol. Rev.* **54**:81–109.

Gilot, D., Loyer, P., Corlu, A., Glaise, D., Lagadic-Gossmann, D., Atfi, A., Morel, F., Ichijo, H., & Guguen-Guillouzo, C. (2002). Liver protection from apoptosis requires both blockage of initiator caspase activities and inhibition of ASK1/JNK pathway via glutathione S-transferase regulation. *J. Biol. Chem.* **277**:49220–49229.

Giovanella, B. C., Stehlin, J. S., & Williams, L. J. (1974). Heterotransplantation of human malignant tumors in "nude" mice; II: Malignant tumors induced by injection of cell cultures derived from human solid tumors. *J. Natl. Cancer Inst.* **52**:921.

Giron, J. A., Lange, M., & Baseman, J. B. (1996). Adherence, fibronectin-binding and induction of cytoskeleton reorganization in cultured human-cells by *Mycoplasma penetrans*. *Infect. Immun.* **64**:197–208.

Gjerset, R., Yu, A., & Haas, M. (1990). Establishment of continuous cultures of T-cell acute lymphoblastic leukemia cells at diagnosis. *Cancer Res.* **50**:10–14.

Glaise, D., Ilyin, G. P., Loyer, P., Cariou, S., Bilodeau, M., Lucas, J., Puisieux, A., Ozturk, M., & Guguen-Guillouzo, C. (1998). Cell cycle gene regulation in reversibly differentiated new human hepatoma cell lines. *Cell Growth Diff*. **9**:165–176.

Glavin, G. B., Szabo, S., Johnson, B. R., Xing, P. L., Morales, R. E., Plebani, M., & Nagy, L. (1996). Isolated rat gastric mucosal cells: Optimal conditions for cell harvesting, measures of viability and direct cytoprotection. *J. Pharmacol. Exp. Therapeut.* **276**:1174–1179.

Gluzman, Y. (1981). SV40-transformed simian cells support the replication of early SV40 mutants. *Cell* **23**:175–182.

Gobet, R., Raghunath, M., Altermatt, S., Meuli-Simmen, C., Benathan, M., Dietl, A., & Meuli, M. (1997). Efficacy of cultured epithelial autografts in pediatric burns and reconstructive surgery. *Surgery* **121**:546–661.

Goding, C. R., & Fisher, D. E. (1997). Meeting review: Regulation of melanocyte differentiation and growth. *Cell Growth Differ.* **8**:935–940.

Goh, S., Ong, P. F., Song, K. P., Riley, T. V., and Chang, B. J. (2007). The complete genome sequence of Clostridium difficile phage phiC2 and comparisons to phiCD119 and inducible prophages of CD630. *Microbiology* **53**:676–685.

Goldberg, B. (1977). Collagen synthesis as a marker for cell type in mouse 3T3 lines. *Cell* **11**:169–172.

Goldman, B. I., & Wurzel, J. (1992). Effects of subcultivation and culture medium on differentiation of human fetal cardiac myocytes. *In Vitro Cell Dev. Biol.* **28A**:109–119.

Goldstein, S., Murano, S., Benes, H., Moerman, E. J., Jones, R. A., & Thweatt, R. (1989). Studies on the molecular genetic basis of replicative senescence in Werner syndrome and normal fibroblasts. *Exp. Gerontol.* **24**(5–6): 461–468.

Goldwasser, E. (1975). Erythropoietin and the differentiation of red blood cells. *Fed. Proc.* **34**:2285–2292.

Gomez-Lechon, M. J., Jover, R., Donato, T., Ponsoda, X., Rodriguez, C., Stenzel, K. G., Klocke, R., Paul, D., Guillen, I., Bort, R., & Castell, J. V. (1998). Long-term expression of differentiated functions in hepatocytes cultured in three-dimensional collagen matrix. *J. Cell. Physiol.* **177**:553–562.

Gomm, J. J., Coope, R. C., Browne, P. J., & Coombes, R. C. (1997). Separated human breast epithelial and myoepithelial cells have different growth factor requirements in vitro but can reconstitute normal breast lobuloalveolar structure. *J. Cell Physiol.* **171**:11–19.

Good Laboratory Practice Regulations (1999). Statutory Instrument 1999 No. 3106. www.opsi.gov.uk/si/si1999/19993106.htm.

Good, N. E., Winget, G. D., Winter, W., Connolly, T. N., Izawa, S., & Singh, R. M. M. (1966). Hydrogen ion buffers and biological research. *Biochemistry* **5**:467–477.

Goodwin, G., Shaper, J. H., Abezoff, M. D., Mendelsohn, G., & Baylin, S. B. (1983). Analysis of cell surface proteins delineates a differentiation pathway linking endocrine and nonendocrine human lung cancers. *Proc. Natl. Acad. Sci. USA* **80**:3807–3811.

Gordon, K. E., Ireland, H., Roberts, M., Steeghs, K., McCaul, J. A., MacDonald, D. G., & Parkinson, E. K. (2003). High levels of telomere dysfunction bestow a selective disadvantage during the progression of human oral squamous cell carcinoma. *Cancer Res.* **63**:458–467.

Gospodarowicz, D. (1974). Localization of fibroblast growth factor and its effect alone and with hydrocortisone on 3T3 cell growth. *Nature* **249**:123–127.

Gospodarowicz, D., & Moran, J. (1974). Growth factors in mammalian cell cultures. *Annu. Rev. Biochem.* **45**:531–558.

Gospodarowicz, D., Delgado, D., & Vlodavsky, I. (1980). Permissive effect of the extracellular matrix on cell proliferation in vitro. *Proc. Natl. Acad. Sci. USA* **77**:4094–4098.

Gospodarowicz, D., Greenburg, G., & Birdwell, C. R. (1978b). Determination of cell shape by the extracellular matrix and its correlation with the control of cellular growth. *Cancer Res.* **38**:4155–4171.

Gospodarowicz, D., Greenburg, G., Bialecki, H., & Zetter, B. R. (1978a). Factors involved in the modulation of cell proliferation in vivo and in vitro: The role of fibroblast and epidermal growth factors in the proliferative response of mammalian cells. *In Vitro* **14**:85–118.

Goto, S., Miyazaki, K., Funabiki, T., & Yasumitsu, H. (1999). Serum-free culture conditions for analysis of secretory proteinases during myogenic differentiation of mouse C2C12 myoblasts. *Anal Biochem.* **272**:135–142.

Gottwald, E., Lahni, B., Thiele, D., Giselbrecht, S., Welle, A., & Weibezahn, K. F. (2008). Chip-based three-dimensional cell culture in perfused micro-bioreactors. *J. Vis. Exp.* **15**:564.

Gougos, A., & Letarte, M. (1990). Primary structure of endoglin, an RGD-containing glycoprotein of human endothelial cells. *J Biol Chem* **265**:8361–8364.

Gower, W. R., Risch, R. M., Godellas, C. V., & Fabri, P. J. (1994). HPAC, a new human glucocorticoid sensitive pancreatic ductal adenocarcinoma cell line. *In Vitro Cell Dev. Biol.* **30A**:151–161.

Goy, A., Ramdas, L., Remache, Y. K., Gu, J., Fayad, L., Hayes, K. J., Coombes, K. R., Barkoh, B. A., Katz, R., Ford, R., Cabanillas, F., & Gilles, F. (2003). Establishment and characterization by gene expression profiling of a new diffuse large B-cell lymphoma cell line, EJ-1, carrying t(14;18) and t(8;14) translocations. *Lab Invest.* **83**:913–916.

Grafström, R. C. (1990a). In vitro studies of aldehyde effects related to human respiratory carcinogenesis. *Mutation Res.* **238**:175–184.

Grafström, R. C. (1990b). Carcinogenesis studies in human epithelial tissues and cells in vitro: Emphasis on serum-free culture conditions and transformation studies. *Acta Physiol. Scand.* **140**:93–133.

Grafström, R. C. (2002). Human oral epithelium. In Freshney, R. I. & Freshney, M. G. (eds.), *Culture of epithelial cells*, 2nd ed. Hoboken, NJ: Wiley-Liss, pp. 195–255.

Graham, F. L., & Van der Eb, A. J. (1973). A new technique for the assay of infectivity of human adenovirus 5 DNA. *Virology* **52**:456–461.

Graham, F. L., Smiley, J., Russell, W. C., & Nairn, R. (1977). Characteristics of a human cell line transformed by DNA from human adenovirus type. *J. Gen. Virol.* **36**:59–72.

Graham, G. J. Freshney, M. G., Donaldson, D., & Pragnell, I. B. (1992). Purification and biochemical characterisation of human and murine stem cell inhibitors. *Growth Factors* **7**:151–160.

Grampp, G. E., Sambanis, A., & Stephanopoulos, G. N. (1992). Use of regulated secretion in protein production from animal cells: An overview. *Adv. Biochem. Eng. Biotechnol.* **46**:35–62.

Grandolfo, M., d'Andrea, P., Paoletti, S., Martina, M., Silvestrini, G., Bonucci, E., & Vittur, F. (1993). Culture and differentiation of chondrocytes entrapped in alginate beads. *Calcif. Tissue Int.* **52**:131–138.

Granner, D. K., Hayashi, S., Thompson, E. B., & Tomkins, G. M. (1968). Stimulation of tyrosine aminotransferase synthesis by dexamethasone phosphate in cell culture. *J. Mol. Biol.* **35**:291–301.

Graziadei, P. P. C., & Monti Graziadei, G. A. (1979). Neurogenesis and neuron regeneration in the olfactory system of mammals; I: Morphological aspects of differentiation and structural organisation of the olfactory sensory neurons. *J. Neurocytol.* **8**:1–18.

Graziadei, P. P. C., & Monti Graziadei, G. A. (1980). Neurogenesis and neuron regeneration in the olfactory system of mammals; III: Deafferentation and reinnervation of the olfactory bulb following section of the *fila olfactoria* in rat. *J. Neurocytol.* **9**:145–162.

Graziano, A., d'Aquino, R., Tirino, V., Desiderio, V., Rossi, A., & Pirozzi, G. (2008). The stem cell hypothesis in head and neck cancer. *J. Cell. Biochem.* **103**:408–412.

Greco, B., & Recht, L. (2003). Somatic plasticity of neural stem cells: Fact or fancy? *J. Cell. Biochem.* **88**:51–56.

Green, A. E., Athreya, B., Lehr, H. B., & Coriell, L. L. (1967). Viability of cell cultures following extended preservation in liquid nitrogen. *Proc. Soc. Exp. Biol. Med.* **124**:1302–1307.

Green, D. F., Hwang, K. H., Ryan, U. S., & Bourgoignie, J. J. (1992). Culture of endothelial cells from baboon and human glomeruli. *Kidney Int* **41**:1506–1516.

Green, H. (1977). Terminal differentiation of cultured human epidermal cells. *Cell* **11**:405–416.

Green, H., & Kehinde, O. (1974). Sublines of mouse 3T3 cells that accumulate lipid. *Cell* **1**:113–116.

Green, H., & Thomas, J. (1978). Pattern formation by cultured human epidermal cells: Development of curved ridges resembling dermatoglyphs. *Science* **200**:1385–1388.

Green, H., Kehinde, O., & Thomas, J. (1979). Growth of cultured human epidermal cells into multiple epithelia suitable for grafting. *Proc. Natl. Acad. Sci. USA* **76**:5665–5668.

Greenberger, J. S. (1980). Self-renewal of factor-dependent haemopoietic progenitor cell lines derived from long-term bone marrow cultures demonstrate significant mouse strain genotypic variation. *J. Supramol. Struct.* **13**:501–511.

Greene, L. A., & Tischler, A. S. (1976). Establishment of a noradrenergic clonal line of rat adrenal pheochromocytoma cells which respond to nerve growth factor. *Proc Natl Acad Sci USA,* **73**:2424–2428.

Gregory, P. A., Bracken, C. P., Bert, A. G., & Goodall, G. J. (2008). MicroRNAs as regulators of epithelial-mesenchymal transition. *Cell Cycle* **7**:3112–3118.

Gregory, C. A., & Prockop, D. J. (2007). Fundamentals of culture and characterization of mesenchymal stem/progenitor cells (MSCs) from bone marrow stroma. In Freshney, R. I., Stacye, G. N., & Auerbach, J. M. (eds.), *Culture of Human Stem Cells.* Hoboken, NJ: Wiley, pp. 208–232.

Greider, C. W., & Blackburn, E. H. (1996). Telomeres, telomerase, and cancer. *Sci. Am.* **274**:92–97.

Griffiths, B. (2000). Scaling up of animal cell cultures. In Masters, J. R. W. (ed.), *Animal cell culture: A practical approach.* Oxford: Oxford University Press, pp. 19–67.

Griffiths, B. (2001). Scale-up of suspension and anchorage-dependent animal cells. *Mol. Biotechnol.* **17**:225–238.

Griffiths, J. B., & Pirt, G. J. (1967). The uptake of amino acids by mouse cells (Strain LS) during growth in batch culture and chemostat culture: The influence of cell growth rate. *Proc. R. Soc. Biol.* **168**:421–438.

Gripon, P., Rumin, S., Urban, S., Le Seyec, J., Glaise, D., Cannie, I., Guyomard, C., Lucas, J., Trepo, C., & Guguen-Guillouzo, C. (2002). Infection of a human hepatoma cell line by hepatitis B virus. *Proc. Natl. Acad. Sci. USA* **99**(24): 15655–15660.

Grizzle, W. E., & Polt, S. H. (1988). Guidelines to avoid personal contamination by infective agents in research laboratories that use human tissue. *J. Tissue Cult. Methods* **11**:191–200.

Gross, S. K., Lyerla, T. A., Williams, M. A., & McCluer, R. H. (1992). The accumulation and metabolism of glycosphingolipids in primary kidney cell cultures from beige mice. *Mol. Cell Biochem.* **118**:61–66.

Groves, A., Barnett, S. C., Franklin, R. J. M., Crang, A. J., Mayer, M., Blakemore, W. & Nobel, M. (1993). Repair of demyelinated lesions by transplants of purified 0-2A progenitor cells. *Nature* **362**:453–455.

Gryseels, T. (2008). Considering cell culture automation in upstream bioprocess development. *Bioprocess Intl.* **6**:12–16.

Gugel, E. A., & Sanders, M. E. (1986). Needle-stick transmission of human colonic adenocarcinoma. *New Engl. J. Med.* **315**:1487.

Guguen-Guillouzo, C. (2002). Isolation and culture of animal and human hepatocytes. In Freshney, R. I., and Freshney, M. G. (eds.), *Culture of epithelial cells,* 2nd ed. Hoboken, NJ: Wiley-Liss, pp. 337–379.

Guguen-Guillouzo, C., & Guillouzo, A. (1986). Methods for preparation of adult and fetal hepatocytes. In Guguen-Guillouzo, C., & Guillouzo, A. (eds.), *Isolated and cultured hepatocytes.* Paris: Les Éditions INSERM, John Libbey Eurotext, pp. 1–12.

Guguen-Guillouzo, C., Campion, J. P., Brissot, P., Glaise, D., Launois, B., Bourel, M., & Guillouzo A. (1982). High yield preparation of isolated human adult hepatocytes by enzymatic profusion of the liver. *Cell Biol. Int. Rep.* **6**:625–628.

Guguen-Guillouzo, C., Clement, B., Baffet, G., Beaumont, C., Morel-Chany, E., Glaise, D., & Guillouzo, A. (1983). Maintenance and reversibility of active albumin secretion by adult rat hepatocytes co-cultured with another liver epithelial cell type. *Exp. Cell Res.* **143**:47–54.

Guilbert, L. I., & Iscove, N. N. (1976). Partial replacement of serum by selenite, transferrin, albumin and lecithin in haemopoietic cell cultures. *Nature* **263**:594–595.

Guillouzo, A., & Guguen-Guillouzo, C. (2008). Evolving concepts in liver tissue modeling and implications for in vitro toxicology. *Expert Opin Drug Metab Toxicol.* **4**:1279–1294.

Guiraud, J. M., Beuron, F., Sion, B., Brassier, G., Faivre, J., Thieulant, M. L., & Duval, J. (1991). Human prolactin producing pituitary adenomas in three dimensional culture. *In Vitro Cell Dev. Biol.* **27a**:188–190.

Gullino, P. M., & Knazek, R. A. (1979). Tissue culture on artificial capillaries. In Jakoby, W. B., & Pastan, I. (eds.), *Methods in Enzymology, Vol. 58: Cell Culture.* New York, Academic Press, pp. 178–184.

Gumbiner, B. M., & Yamada, K. M. (1995). Cell-to-cell contact and extracellular matrix. *Curr. Opin. Cell Biol.* **7**:615–618.

Gumbiner, B. (1992). Epithelial morphogenesis. *Cell* **69**:385–387.

Gumbiner, B., & Simons, K. (1986). A functional assay for proteins involved in establishing an epithelial occluding barrier: Identification of a uvomorulin-like polypeptide. *J. Cell Biol.* **102**:457–468.

Guner, M., Freshney, R. I., Morgan, D., Freshney, M. G., Thomas, D. G. T., & Graham, D. I. (1977). Effects of dexamethasone and betamethasone on in vitro cultures from human astrocytoma. *Br. J. Cancer* **35**:439–447.

Gupta, K., Ramakrishnan, S., Browne, P. V., Solovey, A., & Hebbel, R. P. (1997). A novel technique for culture of human dermal microvascular endothelial cells under either serum-free or serum-supplemented conditions: Isolation by panning and stimulation with vascular endothelial growth factor. *Exp. Cell Res.* **230**:244–251.

Gurdon, J. B., & Melton, D. A. (2008). Nuclear reprogramming in cells. *Science* **322**:1811–1815.

Gurdon, J. B. (1960). Factors responsible for the abnormal development of embryos obtained by nuclear transplantation in *Xenopus laevis. J. Embryol. Exp. Morphol.* **8**:327–340.

Gustafson, C. J., Eldh, J., & Kratz, G. (1998). Culture of human urothelial cells on a cell-free dermis for autotransplantation. *Eur. Urol.* **33**:503–506.

Hacker, D. L., De Jesus, M., & Wurm, F. M. (2009). 25 years of recombinant proteins from reactor-grown cells: Where do we go from here? *Biotechnol. Adv.* **27**:1023–1027.

Hafny, B. E. L., Bourre, J.-M., & Roux, F. (1996). Synergistic stimulation of gamma-glutamyl transpeptidase and alkaline phosphatase activities by retinoic acid and astroglial factors in immortalised rat brain microvessel endothelial cells. *J. Cell. Physiol.* **167**:451–460.

Halaban, R. (2004). Culture of melanocytes from normal, benign, and malignant lesions. In Pfragner, R., & Freshney, R. I. (eds.), *Culture of human tumor cells.* Hoboken, NJ: Wiley-Liss, pp. 289–318.

Halaban, R., & Alfano, F. D. (1984). Selective elimination of fibroblasts from cultures of normal human melanocytes. *In Vitro* **20**:447–450.

Hallermeyer, K., & Hamprecht, B. (1984). Cellular heterogeneity in primary cultures of brain cells revealed by immunocytochemical localisation of glutamyl synthetase. *Brain Res.* **295**:1–11.

Halleux, C., & Schneider, Y.-J. (1994). Iron absorption by Caco-2 cells cultivated in serum-free medium as in vitro model of the human intestinal epithelial barrier. *J. Cell. Physiol.* **158**:17–28.

Ham, R. G. (1963). An improved nutrient solution for diploid Chinese hamster and human cell lines. *Exp. Cell Res.* **29**:515.

Ham, R. G. (1965). Clonal growth of mammalian cells in a chemically defined synthetic medium. *Proc. Natl. Acad. Sci. USA* **53**:288.

Ham, R. G. (1984). Growth of human fibroblasts in serum-free media. In Barnes, D. W., Sirbasku, D. A., & Sato, G. H. (eds.), *Cell culture methods for molecular and cell biology*, Vol. **3**. New York: Liss, pp. 249–264.

Ham, R. G., & McKeehan, W. L. (1978). Development of improved media and culture conditions for clonal growth of normal diploid cells. *In Vitro* **14**:11–22.

Ham, R. G., & McKeehan, W. L. (1979). Media and growth requirements. In Jakoby, W. B., & Pastan, I. H. (eds.), *Methods in enzymology; vol. 58: Cell culture.* New York: Academic Press, pp. 44–93.

Hamburger, A. W., & Salmon, S. E. (1977). Primary bioassay of human tumor stem cells. *Science* **197**:461–463.

Hamburger, A. W., Salmon, S. E., Kim, M. B., Trent, J. M., Soehnlen, B., Alberts, D. S., & Schmidt, H. J. (1978). Direct cloning of human ovarian carcinoma cells in agar. *Cancer Res.* **38**:3438–3444.

Hamilton, T. C., Young, R. C., McKoy, W. M., Grotzinger, K. R., Green, J. A., Chu, E. W., Whang-Peng, J., Rogan, A. M., Green, W. R., & Ozols, R. F. (1983). Characterization of a human ovarian carcinoma cell line (NIH:OVCAR-3) with androgen and estrogen receptors. *Cancer Res.* **43**:5379–5389.

Hamilton, W. G., & Ham, R. G. (1977). Clonal growth of Chinese hamster cell lines in protein-free media. *In Vitro* **13**:537–547.

Hammond, S. L., Ham, R. G., & Stampfer, M. R. (1984). Serum free growth of human mammary epithelial cells: Rapid clonal growth in defined medium and extended serial passage with pituitary extract. *Proc. Natl. Acad. Sci. USA* **81**:5435–5439.

Häner, K., & Henzi, T. Pfefferli, M. Künzli, E. Salicio, V. Schwaller, B. (2010). A bipartite butyrate-responsive element in the human calretinin (CALB2) promoter acts as a repressor in colon carcinoma cells but not in mesothelioma cells. *J. Cell. Biochem.* **109**:519–531.

Hanks, J. H., & Wallace, R. E. (1949). Relation of oxygen and temperature in the preservation of tissues by refrigeration. *Proc. Exp. Biol. Med.* **71**:196.

Hansson, B., Rönnbäck, L., Persson, L. I., Lowenthal, A., Noppe, M., Alling, C., & Karlsson, B. (1984). Cellular composition of primary cultures from cerebral cortex, striatum, hippocampus, brain-stem and cerebellum. *Brain Res.* **300**:9–18.

Hansson, M. G. (2009). Ethics and biobanks. *Br. J. Cancer* **100**:8–12.

Hardman, P., Klement, B. J., & Spooner, B. S. (1990). Growth and morphogenesis of embryonic mouse organs on Biopore membrane. *In Vitro Cell Dev. Biol.* **26**:119–120.

Harrington, W. N., & Godman, G. C. (1980). A selective inhibitor of cell proliferation from normal serum. *Proc. Natl. Acad. Sci. USA Biol. Sci.* **77**:423–427.

Harris, H., & Watkins, J. F. (1965). Hybrid cells from mouse and man: Artificial heterokaryons of mammalian cells from different species. *Nature* **205**:640–646.

Harris, L. W., & Griffiths, J. B. (1977). Relative effects of cooling and warming rates on mammalian cells during the freeze–thaw cycle. *Cryobiology* **14**:662–669.

Harris, M. (1959). Essential growth factor in serum dialysate for chick skeletal muscle fibroblasts. *Proc. Soc. Exp. Biol. Med.* **102**:468.

Harrison, R. G. (1907). Observations on the living developing nerve fiber. *Proc. Soc. Exp. Biol. Med.* **4**:140–143.

Hart, I. R., & Fidler, I. J. (1978). An in vitro quantitative assay for tumor cell invasion. *Cancer Res.* **38**:3218–3224.

Hartley, R. S., & Yablonka-Reuveni, Z. (1990). Long-term maintenance of primary myogenic cultures on a reconstituted basement membrane. *In Vitro Cell Dev. Biol.* **26**:955–961.

Hartung, T., & Daston, G. (2009). Are in vitro tests suitable for regulatory use? Toxicol Sci., **111**:233-7.

Hashimoto, N. (2004). Stem cell systems in skeletal muscle. *Tanpakushitsu Kakusan Koso* **49**:741–748.

Hass, R., Gunji, H., Hirano, M., Weichselbaum, R., & Kufe, D. (1993). Phorbol ester-induced monocytic differentiation is associated with G2 delay and down regulation of cdc25 expression. *Cell Growth Differ.* **4**:159–166.

Hassell, J. R., Robey, P. G., Barrach, H. J., Wilczek, J., Rennard, S. I., & Martin, G. R. (1980). Isolation of a heparan sulfate-containing proteoglycan from basement membrane. *Proc. Natl. Acad. Sci. USA* **77**:4494–4498.

Hassell, T., Gleave, S., & Butler, M. (1991). Growth inhibition in cell culture. *Appl. Biochem. Biotechnol.* **30**:30–41.

Haudenschild, C. C., Zahniser, D., Folkman, J., & Klagsbrun, M. (1976). Human vascular endothelial cells in culture. *Exp. Cell Res.* **98**:175–183.

Hauschka, S. D., & Konigsberg, I. R. (1966). The influence of collagen on the development of muscle clones. *Proc. Natl. Acad. Sci. USA* **55**:119–126.

Häuselmann, H. J., Fernandes, R. J., Mok, S. S., Schmid, T. M., Block, J. A., Aydelotte, M. B., Kuettner, K. E., & Thonar, E. J. M. A. (1994). Phenotypic stability of bovine articular chondrocytes after long-term culture in alginate beads. *J. Cell Sci.* **107**:17–27.

Häuselmann, H. J., Masuda, K., Hunziker, E. B., Neidhart, M., Mok, S. S., Michel, B. A., & Thonar, E. J.-M. A. (1996). Adult human chondrocytes cultured in alginate form a matrix similar to native human cartilage. *Am. J. Physiol. Cell Physiol.* **40**:C742–C752.

Hay, R. J. (1979). Identification, separation and culture of mammalian tissue cells. In Reid, E. (ed.), *Methodological surveys in biochemistry; Vol. 8, Cell Populations.* London: Ellis Horwood, pp. 143–160.

Hay, R. J. (2000). Cell line preservation and characterization. In Masters, J. R. W. (ed.). *Animal cell culture, a practical approach.* Oxford: IRL Press at Oxford University Press, pp. 95–148.

Hay, R. J., & Cour, I. (1997). Testing for microbial contamination: Bacteria and fungi. In Doyle, A., Griffiths, J. B., & Newall, D. G. (eds.), *Cell and tissue culture: Laboratory procedures.* Chichester, UK: Wiley, module 7A:2.

Hay, R. J., & Strehler, B. L. (1967). The limited growth span of cell strains isolated from the chick embryo. *Exp. Gerontol.* **2**:123.

Hay, R. J., Miranda-Cleland, M., Durkin, S., & Reid, Y. A. (2000). Cell line preservation and authentication. In Masters, J. R. W. (ed.), *Animal cell culture.* Oxford: Oxford University Press, pp. 69–103.

Hayashi, I., & Sato, G. H. (1976). Replacement of serum by hormones permits growth of cells in a defined medium. *Nature* **259**:132–134.

Hayflick, L., & Moorhead, P. S. (1961). The serial cultivation of human diploid cell strains. *Exp. Cell Res.* **25**:585–621.

Haynes, L. W. (1999). *The neuron in tissue culture.* Chichester, UK: Wiley.

Heald, K. A., Hail, C. A., & Downing, R. (1991). Isolation of islets of Langerhans from the weanling pig. *Diabetes Res.* **17**:7–12.

Heffelfinger, S. C., Hawkins, H. H., Barrish, J., Taylor, L., & Darlington, G. (1992). SK HEP-1: A human cell line of endothelial origin. *In Vitro Cell Dev. Biol.* **28A**:136–142.

Heinzman, J. M., Brower, S. L., & Bush, J. E. (2008). Comparison of angiogenesis-related factor expression in primary tumor cultures under normal and hypoxic growth conditions. *Cancer Cell Int.* **10**:11.

Heldin, C. H., Westermark, B., & Wasteson, A. (1979). Platelet-derived growth factor: Purification and partial characterization. *Proc. Natl. Acad. Sci. USA* **76**:3722–3726.

Hemmati-Brivaniou, A., Kelly, O. G., & Melton, D. A. (1994). Follistatin, an antagonist of activin, is expressed in the Spemann organizer and displays direct neuralizing activity. *Cell* **77**:283–295.

Henkens, T., Papeleu, P., Elaut, G., Vinken, M., Rogiers, V., & Vanhaecke, T. (2007). Trichostatin A, a critical factor in maintaining the functional differentiation of primary cultured rat hepatocytes. *Toxicol. Appl. Pharmacol.* **218**:64–71.

Heyderman, E., Steele, K., & Ormerod, M. G. (1979). A new antigen on the epithelial membrane: Its immunoperoxidase localisation in normal and neoplastic tissue. *J. Clin. Pathol.* **32**:35–39.

Heyworth, C. M., & Spooncer, E. (1992). In vitro clonal assays for murine multipotential and lineage restricted myeloid progenitor cells. In Testa, N. G., & Molineux, G. (eds.), *Haemopoiesis: A practical approach.* Oxford: IRL Press at Oxford University Press, pp. 37–54.

HFEA. (2009). Human Fertilisation and Embryology Authority: www.hfea.gov.uk.

Higuchi, K. (1977). Cultivation of mammalian cell lines in serum-free chemically defined medium. *Methods Cell. Biol.* **14**:131.

Hill, B. T. (1983). An overview of clonogenic assays for human tumour biopsies. In Dendy, P. P., & Hill, B. T. (eds.), *Human tumour drug sensitivity testing in vitro.* New York: Academic Press, pp. 91–102.

Hilwig, I., & Gropp, A. (1972). Staining of constitutive heterochromatin in mammalian chromosomes with a new fluorochrome. *Exp. Cell. Res.* **75**:122–126.

Hince, T. A., & Roscoe, J. P. (1980). Differences in pattern and level of plasminogen activator production between a cloned cell line from an ethylnitrosourea-induced glioma and one from normal adult rat brain. *J. Cell Physiol.* **104**:199–207.

Hino, H., Tateno, C., Sato, H., Yamasaki, C., Katayama, S., Kohashi, T., Aratani, A., Asahara, T., Dohi, K., & Yoshizato, K. (1999). A long-term culture of human hepatocytes which show a high growth potential and express their differentiated phenotypes. *Biochem. Biophys. Res. Commun.* **256**:184–189.

Hirai, Y., Takebe, K., Takashina, M., Kobayashi, S., & Takeichi, M. (1992). Epimorphin: A mesenchymal protein essential for epithelial morphogenesis. *Cell* **69**:471–481.

Hlubinova, K., Feldsamova, A., & Prachar, J. (1994). Evaluation of two methods for elimination of mycoplasma. *In Vitro Cell Dev. Biol.* **30A**:21–22.

Ho, L., Greene, C. L., Schmidt, A. W., & Huang, L. H. (2004). Cultivation of HEK 293 cell line and production of a member of the superfamily of G-protein coupled receptors for drug discovery applications using a highly efficient novel bioreactor. *Cytotechnology* **45**(3): 117–123.

Hoffman, R. A., & Houck, D. W. (1997). Cell separation using flow cytometric cell sorting. In Recktenwald, D., & Radbruch, A. (eds.), *Cell separation methods and applications.* New York: Dekker, pp. 237–270.

Hofmann, S., Kaplan, D., Vunjak-Novakovic, G., & Meinel, L. (2006). Tisue engineering of bone. In Vunjak-Novokovic, G., & Freshney, R. I. (eds.), *Culture of cells for tissue engineering.* Hoboken, NJ: Wiley, pp. 325–373.

Hogenboom, D. O., Dimarzio, C. A., Gaudette, T. J., Devaney, A. J. & Lindberg, S. C. (1998). Three-dimensional images generated by quadrature interferometry. *Opt. Lett.* **23**:783–785.

Hogge, D. E., Lansdorp, P. M., Reid, D., Gerhard, B., & Eaves, C. J. (1996). Enhanced detection, maintenance, and differentiation of primitive human hematopoietic cells in cultures containing murine fibroblasts engineered to produce human steel factor, interleukin-3, and granulocyte colony-stimulating factor. *Blood* **88**:3765–3773.

Hoheisel, D., Nitz, T., Franke, H., Wegener, J., Hakvoort, A., Tilling, T., & Galla, H. J. (1998). Hydrocortisone reinforces the blood-brain properties in a serum free cell culture system. *Biochem. Biophys. Res. Commun.* **247**:312–315.

Hohla, F., Schally, A. V., Kanashiro, C. A., Buchholz, S., Baker, B., Kannadka, C., Moder, A., Aigner, E., Datz, C., & Halmos, G. (2007). Growth inhibition of non-small-cell lung carcinoma by BN/GRP antagonist is linked with suppression of K-Ras, COX-2, and pAkt. *Proc. Natl. Acad. Sci. USA* **104**:18671–18676.

Holbrook, K. A., & Hennings, H. (1983). Phenotypic expression of epidermal cells in vitro: A review. *J. Invest. Dermatol.* **81**:11s–24s.

Hollenberg, M. D., & Cuatrecasas, P. (1973). Epidermal growth factor: Receptors in human fibroblasts and modulation of action by cholera toxin. *Proc. Natl. Acad. Sci. USA* **70**:2964–2968.

Holley, R. W., Armour, R., & Baldwin, J. H. (1978). Density-dependent regulation of growth of BSC-1 cells in cell culture: Growth inhibitors formed by the cells. *Proc. Natl. Acad. Sci. USA* **75**:1864–1866.

Holt, S. E., Shay, J. W., & Wright, W. E. (1996). Refining the telomere–telomerase hypothesis of ageing and cancer. *Nat. Biotechnol.* **14**:836–839.

Home Office, UK. (2005). Code of practice for the humane killing of animals under Schedule 1. http://scienceandresearch. homeoffice.gov.uk/animal-research/publications-and-reference /publications/code-of-practice/code-of-practice-housing-care/ humane_killing.pdf?view = Binary.

Home Office, UK. (2009). Research and testing using animals. www.homeoffice.gov.uk/science-research/animal-testing.

Hoober, J. K., & Cohen, S. (1967). Epidermal growth factor. I. The stimulation of protein and ribonucleic acid synthesis in chick embryo epidermis. *Biochim. Biophys. Acta* **138**:347–356.

Honma, M., Kataoka, E., Ohnishi, K., Ohno, T., Takeuchi, M., Nomura, N., & Mizusawa, H. (1992). A new DNA profiling system for cell line identification for use in cell banks in Japan. In Vitro Cell Dev Biol. *Animal* **27A**(1):24–28.

Hopps, H., Bernheim, B. C., Nisalak, A., Tjio, J. H., & Smadel, J. E. (1963). Biologic characteristics of a continuous cell line derived from the African green monkey. *J. Immunol.* **91**:416–424.

Horibata, K., & Harris, A. W. (1970). Mouse myelomas and lymphomas in culture. *Exp. Cell Res.* **60**:61–77.

Horita, A., & Weber, L. J. (1964). Skin penetrating property of drugs dissolved in dimethylsulfoxide (DMSO) and other vehicles. *Life Sci.* **3**:1389–1395.

Hornsby, P. J., Yang, L., Lala, D. S., Cheng, C. Y. & Salmons, B. (1992). A modified procedure for replica plating of mammalian cells allowing selection of clones based on gene expression. *Biotechniques* **12**:244–251.

Horster, M. (1979). Primary culture of mammalian nephron epithelia: Requirements for cell outgrowth and proliferation from defined explanted nephron segments. *Pflugers Arch.* **382**:209–215.

Hotamisligil, G. S., Arner, P., Caro, J. F., Atkinson, R. L., & Speigelman, B. M. (1995). Increased adipose tissue expression of tumor necrosis factor-α in human obesity and insulin resistance. *J. Clin. Invest.* **95**:2409–2415.

Howard, M., Kessler, S., Chused, T., & Paul, W. E. (1981). Long term culture of normal mouse B lymphocytes. *Proc. Natl. Acad. Sci. USA* **78**:5788–5792.

Howe, M., Zhao, J., Bodenburg, Y., McGuckin, C. P., Forraz, N., Tilton, R. G., Urban, R. J., & Denner, L. (2009). Oct-4A isoform is expressed in human cord blood-derived CD133 stem cells and differentiated progeny. *Cell Prolif.* **42**:265–275.

Hoyer, L. W., de los Santos, R. P., & Hoyer, J. R. (1973). Antihemophilic factor antigen: Localization in endothelial cells by immunofluorescence microscopy. *J. Clin. Invest.* **52**:2737–2744.

Hrzenjak, A., Moinfar, F., Kremser, M. L., Strohmeier, B., Staber, P. B., Zatloukal, K., & Denk, H. (2006). Valproate inhibition of histone deacetylase 2 affects differentiation and decreases proliferation of endometrial stromal sarcoma cells. *Mol. Cancer Ther.* **5**:2203–2210.

HSE. (1998). Simple guide to the provision and use of work equipment regulations 1998. http://www.hse.gov.uk/pubns/ indg291.pdf.

HSE. (2008a). Infections at work and genetically modified organisms (GMOs)—Information on microbiological containment and biosafety. www.hse.gov.uk/biosafety/

HSE. (2008b). Radiation protection publications. http://www.hse. gov.uk/radiation/ionising/publications.htm.

HSE. (2009). Infection at work—Information sources. www.hse. gov.uk/biosafety/information.htm.

HSE-COSHH. Control of substances hazardous to health (COSHH). www.hse.gov.uk/coshh/index.htm.

Hsieh-Bonassera, N. D., Wu, I., Lin, J. K., Schumacher, B. L., Chen, A. C., Masuda, K., Bugbee, W. D., & Sah, R. L. (2009). Expansion and Redifferentiation of Chondrocytes from Osteoarthritic Cartilage: Cells for Human Cartilage Tissue Engineering. *Tissue Eng.* Part A. doi: 10.1089.

Hu, G.-F., Riordan, J. F., & Vallee, B. L. (1997). A putative angiogenin receptor in angiogenin-responsive human endothelial cells. *Proc. Natl. Acad. Sci. USA* **94**:204–209.

Huangfu, D., Osafune, K., Maehr, R., Guo, W., Eijkelenboom, A., Chen, S.-B., Muhlestein, W., & Melton, D. A. (2008). Induction of pluripotent stem cells from primary human fibroblasts with only Oct4 and Sox2. *Nat. Biotechnol.* **26**:1269–1275.

Hughes, P., Marshall, D., Reid, Y., Parkes, H., & Gelber, C. (2007). The costs of using unauthenticated, over-passaged cell lines: How much more data do we need? *Biotechniques* **43**:575–586.

Huleihel, M., Abuelhija, M., & Lunenfeld, E. (2007). In vitro culture of testicular germ cells: Regulatory factors and limitations. *Growth Factors* **25**:236–252.

Hull, R. N., Cherry, W. R., & Weaver, G. W. (1976). The origin and characteristics of a pig kidney cell strain, LLC-PKI. *In Vitro* **12**:670–677.

Hume, D. A., & Weidemann, M. J. (1980). *Mitogenic lymphocyte transformation*. Amsterdam: Elsevier/North Holland Biomedical Press.

Hume. R., Bell, J., Cossar, D., Giles, M., Hallas, A., Kelly, R. (1996). Differential release of prostaglandins by organ cultures of human fetal trachea and lung. *In Vitro Cell Dev. Biol. Anim.* **32**:24–29.

Hurley, C. K., & Johnson, A. H. (2001). HLA type of EBV-transformed human B cell lines. *Curr. Protoc. Immunol.* App. 1G.

Hurt, E. M., Kawasaki, B. T., Klarmann, G. J., Thomas, S. B., & Farrar, W. L. (2008). CD44+CD24- prostate cells are early cancer progenitor/stem cells that provide a model for patients with poor prognosis. *Br. J. Cancer* **98**:756–765.

Huschtscha, L. I., & Holliday, R. (1983). Limited and unlimited growth of SV40-transformed cells from human diploid MRC-5 fibroblasts. *J. Cell Sci.* **63**:77–99.

Hwang, R. F., Moore, T., Arumugam, T., Ramachandran, V., Amos, K. D., Rivera, A., Ji, B., Evans, D. B., & Logsdon, C. D. (2008). Cancer-associated stromal fibroblasts promote pancreatic tumor progression. *Cancer Res.* **68**:918–926.

Hynes, R. O. (1973). Alteration of cell-surface proteins by viral transformation and by proteolysis. *Proc. Natl. Acad. Sci. USA* **70**:3170–3174.

Hynes, R. O. (1992). Integrins: Versatility, modulation, and signaling in cell adhesion. *Cell* **69**:11–25.

Hyvonen, T., Alakuijala, L., Andersson, L., Khomutov, A. R., Khomutov, R. M. & Eloranta, T. O. (1988). 1-Amino-oxy-3-aminopropane reversibly prevents the proliferation of cultured baby hamster kidney cells by interfering with polyamine synthesis. *J. Biol. Chem.* **263**:1138–1144.

ICH. (2005). The International Conference on Harmonisation of Technical Requirements for Registration of Pharmaceuticals for Human Use, Q5A, Q5D. www.ich.org/cache/compo/363-272-1.html.

Ichikawa, Y., Pluznik, D. H., & Sachs, L. (1966). In vitro control of the development of macrophage and granulocyte colonies. *Proc. Natl. Acad. Sci. USA* **56**:488–495.

Ieda, M., Tsuchihashi, T., Ivey, K. N., Ross, R. S., Hong, T. T., Shaw, R. M., & Srivastava, D. (2009). Cardiac fibroblasts regulate myocardial proliferation through beta1 integrin signaling. *Dev. Cell* **6**:233–244.

Iguchi, H., Mizumoto, K., Shono, M., Kono, A., & Takiguchi, S. (2004). Pancreatic cancer-derived cultured cells: genetic alterations and application to an experimental model of pancreatic cancer metastasis. In Pfragner, R., & Freshney. R. I. (eds.), *Culture of human tumor cells.* Hoboken, NJ: Wiley-Liss, pp. 81–96.

Ikonomou, L., Schneider, Y. J., & Agathos, S. N. (2003). Insect cell culture for industrial production of recombinant proteins. *Appl. Microbiol. Biotechnol.* **62**:1–20.

Illmensee, K., & Mintz, B. (1976). Totipotency and normal differentiation of single teratocarcinoma cells cloned by injection into blastocysts. *Proc. Natl. Acad. Sci. USA* **73**:549–553.

Ilsie, A. W., & Puck, T. T. (1971). Morphological transformation of Chinese hamster cells by dibutyryl adenosine cycline 3′:5′-monophosphate and testosterone. *Proc. Natl. Acad. Sci. USA* **2**:358–361.

Imaizumi, T., Lankford, K. L., Waxman, S. G., Greer, C. A., & Kocsis, J. D. (1998). Transplanted olfactory ensheathing cells remyelinate and enhance axonal conduction in the demyelinated dorsal columns of the rat spinal cord. *J. Neurosci.* **18**:6176–6185.

Inamatsu, M., Matsuzaki, T., Iwanari, H., & Yoshizato, K. (1998). Establishment of rat dermal papilla cell lines that sustain the potency to induce hair follicles from afollicular skin. *J. Invest. Dermatol.* **111**:767–775.

Inokuchi, S., Handa, H., Imai, T., Makuuchi, H., Kidokoro, M., Tohya, H., Aizawa, S., Shimamura, K., Ueyama, Y., Mitomi, T., & Sawada, Y. (1995). Immortalisation of human oesophageal epithelial cells by a recombinant SV40 adenovirus vector. *Br. J. Cancer* **71**:819–825.

Inoué, S., & Spring, K. R. (1997). *Video microscopy*, 2nd ed. New York: Plenum Press.

International Human Genome Sequencing Consortium. (2001). Initial sequencing and analysis of the human genome. *Nature* **409**:860–921.

Ireland, G. W., Dopping-Hepenstal, P. J., Jordan, P. W., & O'Neill, C. H. (1989). Limitation of substratum size alters cytoskeletal organization and behaviour of Swiss 3T3 fibroblasts. *Cell Biol. Int. Rep.* **13**:781–790.

Iscove, N. N., Guilbert, L. W., & Weyman, C. (1980). Complete replacement of serum in primary cultures of erythropoietin-dependent red cell precursors (CFU-E) by albumin, transferrin, iron, unsaturated fatty acid, lecithin and cholesterol. *Exp. Cell Res.* **126**:121–126.

Iscove, N., & Melchers, F. (1978). Complete replacement of serum by albumin, transferrin and soybean lipid in cultures of lipopolysaccharide-reactive B lymphocytes. *J. Exp. Med.* **147**:923–933.

Isom, H. C., Secott, T., Georgoff, I., Woodworth, C., & Mummaw, J. (1985). Maintenance of differentiated rat hepatocytes in primary culture. *Proc. Natl. Acad. Sci. USA.*, **82**:3252-6.

Itagaki, A., & Kimura, G. (1974). TES and HEPES buffers in mammalian cell cultures and viral studies: Problems of carbon dioxide requirements. *Exp. Cell Res.* **83**:351–360.

Iyer, V. R., Eisen, M. B., Ross, D. T., Schuler, G., Moore, T., Lee, J. C. F., Trent, J. M., Staudt, L. M., Hudson, J. Jr., Boguski, M. S., Lashkari, D., Shalon, D., Botstein, D., & Brown, P. O. (1999). The transcriptional program in the response of human fibroblasts to serum. *Science* **283**:83–87.

Izutsu, K. T., Fatherazi, S., Belton, C. M., Oda, D., Cartwright, F. D., & Kenny, G. E. (1996). Mycoplasma orale infection affects K and Cl currents in the HSG salivary gland cell line. *In Vitro Cell Dev. Biol. Anim.* **32**:361–365.

Jackman, J., & O'Connor, P. M. (2001). Methods for synchronizing cells at specific stages of the cell cycle. *Curr. Protoc. Cell. Biol.* Chapter 8:Unit 8.3.

Jackson, J. P., Tonge, P. D., & Andrews, P. W. (2007). Techniques for neural differentiation of human EC and ES cells. In Freshney, R. I., Stacey, G. N., & Auerbach, J. M. (eds.), *Culture of human stem cells*. Hoboken, NJ: Wiley, pp. 61–91.

Jacobs, J. P. (1970). Characteristics of a human diploid cell designated MRC-5. *Nature* **227**:168–170.

Jacobs, J. P., Garrett, A. J., & Meron, R. (1979). Characteristics of a serially propagated human diploid cell designated MRC-9. *J. Biol. Stand.* **7**:113–122.

Jaffe, E. A., Nachman, R. L., Becker, G. C., & Ninick, C. R. (1973). Culture of human endothelial cells derived from umbilical veins. *J. Clin. Invest.* **52**:2745–2756.

Jain, R. K. Schlenger, K., Hockel, M., & Yuan, F. (1997). Quantitative angiogenesis assays: Progress and problems. *Nat. Med.* **3**:1203–1208.

Janssen, M. E., Kim, E., Liu, H., Fujimoto, L. M., Bobkov, A., Volkmann, N., & Hanein, D. (2006). Three-dimensional structure of vinculin bound to actin filaments. *Mol. Cell.* **21**:271–281.

Janssen, M. I., van Leeuwen, M. B., Scholtmeijer, K., van Kooten, T. G., Dijkhuizen, L., & Wosten, H. A. (2003). Coating with genetic engineered hydrophobin promotes growth of fibroblasts on a hydrophobic solid. *Biomaterials* **23**:4847–4854.

Jat, P. S., Noble, M. D., Ataliotis, P., Tanaka, Y., Yannoutsos, N., Larsen, L., & Kioussis, D. (1991). Direct derivation of conditionally immortal cell lines from an H-2Kb-tsA58 transgenic mouse. *Proc. Natl. Acad. Sci. USA* **88**:5096–5100.

Jeffreys, A. J., Wilson, V., & Thein, S. L. (1985). Individual specific "fingerprints" of human DNA. *Nature* **316**:76–79.

Jeng, Y.-J., Watson, C. S., & Thomas, M. L. (1994). Identification of vitamin D-stimulated phosphatase in IEC-6 cells, a rat small intestine crypt cell line. *Exp. Cell Res.* **212**:338–343.

Jenkins, N. (ed.). (1992). *Growth factors, a practical approach.* Oxford: IRL Press at Oxford University Press.

Jessell, T. M., & Melton, D. A. (1992). Diffusible factors in vertebrate embryonic induction. *Cell* **68**:257–270.

Jin, D. I., Lee, S. H., Choi, J. H., Lee, J. S., Lee, J. E., Park, K. W., & Seo, J. S. (2003). Targeting efficiency of a-1,3-galactosyl transferase gene in pig fetal fibroblast cells. *Exp. Mol. Med.* **35**:572–577.

Jinard, F., Sergent-Engelen, T., Trouet, A., Remacle, C., & Schneider, Y.-J. (1997). Compartment coculture of porcine arterial endothelial and smooth muscle cells on a microporous membrane. *In Vitro Cell Dev. Biol. Anim.* **33**:92–103.

Jones, R. J., Matsui, W. H., & Smith, B. D. (2004). Cancer stem cells: Are we missing the target? *J. Natl. Cancer Inst.* **96**:583–585.

Joukov, V., Kaipainen, A., Jeltsch, M., Pajusola, K., Olofsson B., Kumar, V., Eriksson, U., & Alitalo, K. (1997). Vascular endothelial growth factors VEGF-B and VEGF-C. *J. Cell Physiol.* **173**:211–215.

Jozan, S., Roche, H., Cheutin, F., Carton, M., & Salles, B. (1992). New human ovarian cell line OVCCR1/sf in serum-free medium. *In Vitro Cell Dev. Biol.* **28A**:687–689.

Judson, R., & Blelloch, R. (2009). Embryonic stem cell-specific microRNAs promote induced pluripotency. *Nat. Biotechnol.* **27**:459–461.

Kaartinen, L., Nettesheim, P., Adler, K. B., & Randell, S. H. (1993). Rat tracheal epithelial cell differentiation in vitro. *In Vitro Cell Dev. Biol.* **29A**:481–492.

Kahn, P., & Shin, S.-L. (1979). Cellular tumorigenicity in nude mice: Test of association among loss of cell-surface fibronectin, anchorage independence, and tumor-forming ability. *J. Cell Biol.* **82**:1.

Kaji, K., Norrby, K., Paca, A., Mileikovsky, M., Mohseni, P., & Woltjen, K. (2009). Virus-free induction of pluripotency and subsequent excision of reprogramming factors. *Nature* **458**:715–716.

Kaltenbach, J. P., Kaltenbach, M. H., & Lyons, W. B. (1958). Nigrosin as a dye for differentiating live and dead ascites cells. *Exp. Cell Res.* **15**:112–117.

Kamata, Y., Shiraga, H., Tai, A., Kawamoto, Y., & Gohda, E. (2007). Induction of neurite outgrowth in PC12 cells by the medium-chain fatty acid octanoic acid. *Neuroscience* **146**:1073–1081.

Kaminska, B., Kaczmarek, L., & Grzelakowska-Sztabert, B. (1990). The regulation of $G_0-S$ transition in mouse T lymphocytes by polyamines. *Exp. Cell Res.* **191**:239–245.

Kanczkowski, W., Zacharowski, K., Wirth, M. P., Ehrhart-Bornstein, M., & Bornstein, S. R. (2009). Differential expression and action of Toll-like receptors in human adrenocortical cells. *Mol. Cell Endocrinol.* **300**:57–65.

Kane, D. A., Warga, R. M., & Kimmel, C. B. (1992). Mitotic domains in the early embryo of the zebrafish. *Nature* **360**:735–737.

Kao, F. T., & Puck, T. T. (1968). Genetics of somatic mammalian cells; VII: Induction and isolation of nutritional mutants in Chinese hamster cells. *Proc. Natl. Acad. Sci. USA* **60**:1275–1281.

Kao, W.-Y., & Prockop, D. I. (1977). Proline analogue removes fibroblasts from cultured mixed cell populations. *Nature* **266**:63–64.

Karenberg, J. R., & Freelander, E. F. (1974). Giemsa technique for the detection of sister chromatid exchanges. *Chromosoma* **48**:355–360.

Kasper, S. (2008). Stem cells: The root of prostate cancer? *J. Cell. Physiol.* **216**:332–336.

Katdare, M., Osborne, M., & Telang, N. T. (2004). Soy isoflavone genistein modulates cell cycle progression and induces apoptosis in HER-2/neu oncogene expressing human breast epithelial cells. *Int. J. Oncol.* **21**:809–815.

Kato, R., Maeda, T., Akaike, T., & Tamai, I. (2009). Characterization of nucleobase transport by mouse Sertoli cell line TM4. Biol Pharm Bull., **32**:450–455.

Kawa, S., Kimura, S., Hakomori, S., & Igarashi, Y. (1997). Inhibition of chemotactic motility and trans-endothelial migration of human neutrophils by sphingosine 1-phosphate. *FEBS Lett.* **420**:196–200.

Kawada, H., Fujita, J., Kinjo, K., Matsuzaki, Y., Tsuma, M., Miyatake, H., Muguruma, Y., Tsuboi, K., Itabashi, Y., Ikeda, Y., Ogawa, S., Okano, H., Hotta, T., Ando, K., & Fukuda, K. (2004). Nonhematopoietic mesenchymal stem cells can be mobilized and differentiate into cardiomyocytes after myocardial infarction. *Blood* **104**:3581–3587.

Kawasaki, E. S. (2004). Microarrays and the gene expression profile of a single cell. *Ann. NY Acad. Sci.* **1020**:92–100.

Kédinger, M., Simon-Assmann, P., Alexandre, E., & Haffen, K. (1987). Importance of a fibroblastic support for in vitro differentiation of intestinal endodermal cells and for their response to glucocorticoids. *Cell Differ.* **20**:171–182.

Keen, M. J., & Rapson, N. T. (1995). Development of a serum-free culture medium for the large scale production of recombinant protein from a Chinese hamster ovary cell line. *Cytotechnology* **17**:153–163.

Keenen, B., de la Serna, I.L. (2008). Chromatin remodeling in Embryonic stem cells: Regulating the balance between pluripotency and differentiation. *J. Cell. Physiol.* **219**:1–7.

Keilova, H. (1948). The effect of streptomycin on tissue cultures. *Experientia* **4**:483.

Keith, W. N. (2003). In situ analysis of telomerase RNA gene expression as a marker for tumor progression. *Methods Mol. Med.* **75**:163–176.

Keles, G. E., Berger, M. S., Lim, R., & Zaheer, A. (1992). Expression of glial fibrillary acidic protein in human medulloblastoma cells treated with recombinant glia maturation factor-beta. *Oncol. Res.* **4**:431–437.

Kelley, D. S., Becker, J. E., & Potter, V. R. (1978). Effect of insulin, dexamethasone, and glucagon on the amino acid transport ability of four rat hepatoma cell lines and rat hepatocytes in culture. *Cancer Res.* **38**:4591–4601.

Kelly, J. H., & Sussman, N. L. (2000). A fluorescent cell-based assay for cytochrome P-450 isozyme 1A2 induction and inhibition. *J. Biomol. Screen.* **5**:249–254.

Kempson, S. A., McAteer, J. A., Al-Mahrouq, H. A., Dousa, T. P., Dougherty, G. S., & Evan, A. P. (1989). Proximal tubule characteristics of cultured human renal cortex epithelium. *J. Lab. Clin. Med.* **113**:285–296.

Kendig, D. M., & Tarloff, J. B. (2006). Inactivation of lactate dehydrogenase by several chemicals: Implications for in vitro toxicology studies. *Toxicol In Vitro* **21**(1): 125–132.

Kenworthy, P., Dowrick, P., Baillie-Johnson, H., McCann, B., Tsubouchi, H., Arakaki, N., Daikuhara, Y., & Warn, R. M. (1992). The presence of scatter factor in patients with metastatic spread to the pleura. *Br. J. Cancer* **66**:243–247.

Kern, P. A., Knedler, A., & Eckel, R. H. (1983). Isolation and culture of microvascular endothelium from human adipose tissue. *J. Clin. Invest.* **71**:1822–1829.

Kerr, D. A., Llado, J., Shamblott, M. J., Maragakis, N. J., Irani, D. N., Crawford, T. O., Krishnan, C., Dike, S., Gearhart, J. D., & Rothstein, J. D. (2003). Human embryonic germ cell derivatives facilitate motor recovery of rats with diffuse motor neuron injury. *J. Neurosci.* **23**:5131–5140.

Kew, O., Sutter, R., de Gourville, E., Dowdle, W., & Pallansch, M. (2005). Vaccine-derived polioviruses and the endgame strategy for global polio eradication. *Ann. Rev. Microbiol.* **59**:587–635.

Khan, M. Z., Spandidos, D. A., Kerr, D. J., McNicol, A. M., Lang, J. C., de Ridder, L., & Freshney, R. I. (1991). Oncogene transfection of mink lung cells: effect on growth characteristics In vitro and in vivo. *Anticancer Res.* **11**:1343–1348.

Khetani, S. R., & Bhatia, S. N. (2008). Microscale culture of human liver cells for drug development. *Nat. Biotechnol.* **26**:120–126.

Kibbey, M. C., Royce, L. S., Dym, M., Baum, B. J., & Kleinman, H. K. (1992). Glandular-like morphogenesis of the human submandibular tumour cell line A253 on basement membrane components. *Exp. Cell Res.* **198**:343–351.

Kiefer, J., Alexander, A., & Farach-Carson, M. C. (2004). Type I collagen-mediated changes in gene expression and function of prostate cancer cells. *Cancer Treat Res.* **118**:101–124.

Kiernan, J. A. (1999). Strategies for preventing detachment of sections from glass slides. *Microscopy Today* **6**:22–24.

Kikuchi, K., Sumaru, K., Edahiro, J., Ooshima, Y., Sugiura, S., Takagi, T., & Kanamori, T. (2009). Stepwise assembly of micropatterned co-cultures using photoresponsive culture surfaces and its application to hepatic tissue arrays. *Biotechnol. Bioeng.* **103**:552–561.

Kimhi, Y. H., Palfrey, C., & Spector, I. (1976). Maturation of neuroblastoma cells in the presence of dimethyl sulphoxide. *Proc. Natl. Acad. Sci. USA* **73**:462–466.

Kingsbury, A., Gallo, V., Woodhams, P. L., & Balazs, R. (1985). Survival, morphology and adhesion properties of cerebellar interneurons cultured in chemically defined and serum-supplemented medium. *Dev. Brain Res.* **17**:17–25.

Kinoshita, T., & Miyajima, A. (2002). Cytokine regulation of liver development. *Biochim. Biophys. Acta.* **1592**:303–312.

Kinsella, J. L., Grant, D. S., Weeks, B. S., & Kleinman, H. K. (1992). Protein kinase C regulates endothelial cell tube formation on basement membrane matrix, Matrigel. *Exp. Cell Res.* **199**:56–62.

Kirkland, S. C., & Bailey, I. G. (1986). Establishment and characterisation of six human colorectal adenocarcinoma cell lines. *Br. J. Cancer* **53**:779–785.

Klagsbrun, M., & Baird, A. (1991). A dual receptor system is required for basic fibroblast growth factor activity. *Cell* **67**:229–231.

Klein, B., Pastink, A., Odijk, H., Westerveld, A., & van der Eb, A. J. (1990). Transformation and immortalization of diploid xeroderma pigmentosum fibroblasts. *Exp. Cell Res.* **191**:256–262.

Kleinman, H. K., Philp, D., & Hoffman, M. P. (2003). Role of the extracellular matrix in morphogenesis. *Curr Opin Biotechnol.* **14**:526–532.

Kleinman, H. K., McGarvey, M. L., Hassell, J. R., Star, V. L., Cannon, F. B., Laurie, G. W., & Martin, G. R. (1986) Basement membrane complexes with biological activity. *Biochemistry* **25**:312–318.

Kleinsmith, L. J., & Pierce, G. B. (1964). *Cancer Res.* **24**:1544–1551.

Klevjer-Anderson, P., & Buehring, G. C. (1980). Effect of hormones on growth rates of malignant and nonmalignant human mammary epithelia in cell culture. *In Vitro* **16**:491–501.

Klingel, S., Rothe, G., Kellermann, W., & Valet, G. (1994). Flow cytometric determination of cysteine and serine proteinase activities in living cells with rhodamine 110 substrates. *Methods Cell Biol.* **41**:449–459.

Klinger, R. Y., Niklason, L. E. (2006). Tissue-engineered blood vessels. In Vunjak-Novakovic, G., & Freshney, R. I. (eds.), *Culture of cells for tissue engineering*. Hoboken, NJ: Wiley, pp. 294–322.

Klug, C. A., & Jordan, C. T. (2002). *Hematopoietic stem cell protocols*. Clifton, NJ: Humana Press.

Knazek, R. A., Gullino, P., Kohler, P. O., & Dedrick, R. (1972). Cell culture on artificial capillaries: An approach to tissue growth in vitro. *Science* **178**:65–67.

Knazek, R. A., Kohler, P. O., & Gullino, P. M. (1974). Hormone production by cells grown in vitro on artificial capillaries. *Exp. Cell Res.* **84**:251.

Knedler, A., & Ham, R. G. (1987). Optimized medium for clonal growth of human microvascular endothelial cells with minimal serum. *In Vitro* **23**(7): 481–491.

Kneuchel, R., & Masters, J. R. W. (1999). Bladder cancer. In Masters, J. R. W., & Palsson, B. (eds.), *Human cell culture*, Vol. **1**. Dordrecht: Kluwer, pp. 213–230.

Knight, D. J., & Breheny, D. (2002). Alternatives to animal testing in the safety evaluation of products. *Altern. Lab. Anim.* **30**:7–22.

Kniss, D. A., Xie, Y., Li, Y., Kumar, S., Linton, E. A., Cohen, P., Fan-Havard, P., Redman, C. W., & Sargent, I. L. (2002). ED(27) trophoblast-like cells isolated from first-trimester chorionic villi are genetically identical to HeLa cells yet exhibit a distinct phenotype. *Placenta* **23**:32–43.

Knott, C. L., Kuus-Reichel, K., Liu, R., & Wolfert, R. L. (1997). Development of antibodies for diagnostic assays. In Price, C., & Newman, D. (eds.), *Principles and practice of immunoassay*, 2nd ed. New York: Stockton Press, pp. 36–64.

Knowles, B. B., Howe, C. C., & Aden, D. P. (1980). Human hepatocellular carcinoma cell lines secrete the major plasma proteins and hepatitis B surface antigen. *Science* **209**:497–499.

Kohler, G., & Milstein, C. (1975). Continuous cultures of fused cells secreting antibody of predefined specificity. *Nature* **256**:495–497.

Kohlhepp, E. A., Condon, M. E., & Hamburger, A. W. (1987). Recombinant human interferon-α enhancement of retinoic acid induced differentiation of HL-60 cells. *Exp. Hematol.* **15**:414–418.

Kokkinos, M. I., Wafai, R., Wong, M. K., Newgreen, D. F., Thompson, E. W., & Waltham, M. (2007). Vimentin and epithelial-mesenchymal transition in human breast cancer–observations in vitro and in vivo. *Cells Tissues Organs* **185**:191–203.

Kondo, T., & Raff, M. (2000). Oligodendrocyte precursor cells reprogrammed to become multipotential CNS stem cells. *Science* **289**:1754–1757.

Kondo, T., & Raff, M. (2004). Chromatin remodeling and histone modification in the conversion of oligodendrocyte precursors to neural stem cells. *Genes Dev.* **18**:2963–1272.

Kondo, S., Kooshesh, F., & Sauder, D. N. (1997). Penetration of keratinocyte-derived cytokines into basement membrane. *J. Cell Physiol.* **171**:190–195.

Koopman, L. A., Szuhai, K., van Eendenburg, J. D., Bezrookove, V., Kenter, G. G., Schuuring, E., Tanke, H., & Fleuren, G. J. (1999). Recurrent integration of human papillomaviruses 16, 45, and 67 near translocation breakpoints in new cervical cancer cell lines. *Cancer Res.* **59**:5615–5624.

Kopper, L., & Hajdu, M. (2004). Tumor stem cells. *Pathol. Oncol. Res.* **10**:69–73.

Koren, H. S., Handwerger, B. S., & Wunderlich, J. R. (1975). Identification of macrophage-like characteristics in a murine tumor cell line. *J. Immunol.* **114**:894–897.

Korenberg, J. R., Chen, X. N., Adams, M. D., & Venter, J. C. (1995). Toward a cDNA map of the human genome. *Genomics* **29**:364–370.

Kortesmaa, J., Yurchenco, P., & Tryggvason, K. (2000). Recombinant Laminin-8 (alpha4 beta1 gamma1); production, purification, and interactions with integrins. *J. Biol. Chem.* **275**:14853–14859.

Koschier, F. J., Roth, R. N., Wallace, K. A., Curren, R. D., & Harbell, J. W. (1997). A comparison of three-dimensional human skin models to evaluate the dermal irritation of selected petroleum products. *In Vitro Toxicol.* **10**:391–405.

Kreisberg, J. L., Sachs, G., Pretlow, T. G. E., & McGuire, R. A. (1977). Separation of proximal tubule cells from suspensions of rat kidney cell by free-flow electrophoresis. *J. Cell. Physiol.* **93**:169–172.

Kroemer, G., Galluzzi, L., Vandenabeele, P., Abrams, J., Alnemri, E. S., Baehrecke, E. H., Blagosklonny, M. V., El-Deiry, W. S., Golstein, P., Green, D. R., Hengartner, M., Knight, R. A., Kumar, S., Lipton, S. A., Malorni, W., Nuñez, G., Peter, M. E., Tschopp, J., Yuan, J., Piacentini, M., Zhivotovsky, B., & Melino, G. (2009). Classification of cell death: Recommendations of the Nomenclature Committee on Cell Death 2009. *Cell Death Differ.* **16**:3–11.

Krunic, D., Moshir, S., Greulich-Bode, K. M., Figueroa, R., Cerezo, A., Stammer, H., Stark, H. J., Gray, S. G., Nielsen, K. V., Hartschuh, W., & Boukamp, P. (2009). Tissue context-activated telomerase in human epidermis correlates with little age-dependent telomere loss. *Biochim. Biophys. Acta.* **1792**:297–308

Kruse, P. F. Jr., Keen, L. N., & Whittle, W. L. (1970). Some distinctive characteristics of high density perfusion cultures of diverse cell types. *In Vitro* **6**:75–78.

Kruse, R. H., Puckett, W. H., & Richardson, J. H. (1991). Biological safety cabinetry. *Clin. Microbiol. Rev.* **4**:207–241.

Kujoth, G. C., & Fahl, W. E. (1997). c-sis/platelet-derived growth factor-B promoter requirements for induction during the 12-O-tetradecanoylphorbol-13-acetate-mediated megakaryoblastic differentiation of K562 human erythroleukemia cells. *Cell Growth Differ.* **8**:963–977.

Kuriharcuch, W., & Green, H. (1978). Adipose conversion of 3T3 cells depends on a serum factor. *Proc. Natl. Acad. Sci. USA* **75**:6107–6110.

Kurtz, J. W., & Wells, W. W. (1979). Automated fluorometric analysis of DNA, protein, and enzyme activities: Application of methods in cell culture. *Anal. Biochem.* **94**:166.

Kuslak, S. L., & Marker, P. C. (2007). Fibroblast growth factor receptor signaling through MEK-ERK is required for prostate bud induction. *Differentiation* **75**:638–651.

Labarca, C., & Paigen, K. (1980). A simple, rapid, and sensitive DNA assay procedure. *Anal. Biochem.* **102**:344–352.

Labarge, M. A., Petersen, O. W., & Bissell, M. J. (2007). Culturing mammary stem cells. In Freshney, R. I., Stacey, G. N., & Auerbach, J. M. (eds.), *Culture of human stem cells*. Hoboken, NJ: Wiley, pp. 282–302.

Lacroix, M. (2008). Persistent use of "false" cell lines. *Int. J. Cancer* **122**:1–4.

Laferte, S., & Loh, L. C. (1992). Characterization of a family of structurally related glycoproteins expressing beta 1-6-branched asparagine-linked oligosaccharides in human colon carcinoma cells. *Biochem. J.* **283**:192–201.

Lag, M., Becher, R., Samuelsen, J. T., Wiger, R., Refsnes, M., Huitfeldt, H. S., & Schwarze, P. E. (1996). Expression of CYP2B1 in freshly isolated and proliferating cultures of epithelial rat lung cells. *Exp. Lung Res.* **22**:627–649.

Lamb, R. F., Hennigan, R. F., Katsanakis, K. D., Turnbull, K., MacKenzie, E. D., Birnie, G. D., & Ozanne, B. W. (1997). AP-1-mediated invasion requires increased expression of the hyaluronan receptor, CD44. *Mol. Cell Biol.* **17**:963–976.

Lan, S., Smith, H. S., & Stampfer, M. R. (1981). Clonal growth of normal and malignant human breast epithelia. *J. Surg. Oncol.* **18**:317–322.

Lane, E. B. (1982). Monoclonal antibodies provide specific intramolecular markers for the study of tonofilament organisation. *J. Cell Biol.* **92**:665–673.

Langdon, S. P. (2004). Characterization and authentication of cancer cell lines: An overview. *Methods Mol. Med.* **88**:33–42.

Lange, W., Brugger, W., Rosenthal, F. M., Kanz, L., & Lindemann, A. (1991). The role of cytokines in oncology. *Int. J. Cell Clon.* **9**:252–273.

Lao, L., Tan, H., Wang, Y., & Gao, C. (2008). Chitosan modified poly(L-lactide) microspheres as cell microcarriers for cartilage tissue engineering. *Colloids Surf. B Biointerfaces* **66**:218–225.

Larsen, M. C., Brake, P. B., Pollenz, R. S., & Jefcoate, C. R. (2004). Linked expression of Ah receptor, ARNT, CYP1A1, and CYP1B1 in rat mammary epithelia, in vitro, is each substantially elevated by specific extracellular matrix interactions that precede branching morphogenesis. *Toxicol. Sci.* **82**:46–61.

Lasfargues, E. Y. (1973) Human mammary tumors. In Kruse, P., & Patterson, M. K. (eds.), *Tissue culture methods and applications.* New York: Academic Press, pp. 45–50.

Laslett, A. L., Filipczyk, A. A., & Pera, M. F. (2003). Characterization and culture of human embryonic stem cells. *Trends Cardiovasc. Med.* **13**:295–301.

Lasnitzki, I. (1992). Organ culture. In Freshney, R. I. (ed.), *Animal cell culture, a practical approach.* Oxford: IRL Press at Oxford University Press, pp. 213–261.

Latt, S. A. (1973). Microfluorometric detection of DNA replication in human metaphase chromosomes. *Proc. Natl. Acad. Sci. USA* **70**:3395–3399.

Latt, S. A. (1981). Sister chromatid exchange formation. *Annu. Rev. Genet.* **15**:11–55.

Laug, W. E., Tokes, Z. A., Benedict, W. F., & Sorgente, N. (1980). Anchorage independent growth and plasminogen activator production by bovine endothelial cells. *J. Cell Biol.* **84**:281–293.

Laurent, L., Wong, E., Li, G., Huynh, T., Tsirigos, A., Ong, C. T., Low, H. M., Kin Sung, K. W., Rigoutsos, I., Loring, J., & Wei, C. L., (2010). Dynamic changes in the human methylome during differentiation. *Genome Res.* **20**:320–331.

Lavappa, K. S., Macy, M. L., & Shannon, J. E. (1976). Examination of ATCC stocks for HeLa marker chromosomes in human cell lines. *Nature* **259**:211–213.

Lavappa, K. S. (1978). Survey of ATCC stocks of human cell lines for HeLa contamination. *In Vitro* **14**(5): 469–475.

Lavoie, J. F., Biernaskie, J. A., Chen, Y., Bagli, D., Alman, B., Kaplan, D. R., & Miller, F. D. (2009). Skin-derived precursors differentiate into skeletogenic cell types and contribute to bone repair. *Stem Cells Dev.* **8**:893–906.

Law, L. W., Dunn, T. B., Boyle, P. J., & Miller, J. H. (1949). Observations on the effect of a folic acid antagonist on transplantable lymphoid leukemia in mice. *J. Natl. Cancer Inst.* **10**:179–192.

Le Douarin, N. M., Creuzet, S., Couly, G., & Dupin, E. (2004). Neural crest cell plasticity and its limits. *Development* **131**:4637–4650.

Le Page, C., Ouellet, V., Madore, J., Ren, F., Hudson, T. J., Tonin, P. N., Provencher, D. M., & Mes-Masson, A. M. (2006). Gene expression profiling of primary cultures of ovarian epithelial cells identifies novel molecular classifiers of ovarian cancer. *Br. J. Cancer* **94**:436–445.

Le Poole, I. C., van den Berg, F. M., van den Wijngaard, R. M., Galloway, D. A., van Amstel, P. J., Buffing, A. A., Smiths, H. L., Westerhof, W., & Das, P. K. (1997). Generation of a human melanocyte cell line by introduction of HPV16 E6 and E7 genes. *In Vitro Cell Dev. Biol. Anim.* **33**:42–49.

Le Roith, D., & Raizada, M. K. (eds.). (1989). *Molecular and cellular biology of insulin-like growth factors and their receptors.* New York: Plenum.

Leake, R. E., Freshney, R. I., & Munir, I. (1987). Steroid responses in vivo and in vitro. In Green, B., & Leake, R. E., (eds.), *Steroid hormones: A practical approach.* Oxford: IRL Press at Oxford University Press, pp. 205–218.

Lebeau, M. M., & Rowley, J. D. (1984). Heritable fragile sites in cancer. *Nature* **308**:607–608.

Lechardeur, D., Schwartz, B., Paulin, D., & Scherman, D. (1995). Induction of blood–brain barrier differentiation in a rat brain-derived endothelial cell line. *Exp. Cell Res.* **220**:161–170.

Lechner, J. F., & LaVeck, M. A. (1985). A serum free method for culturing normal human bronchial epithelial cells at clonal density. *J. Tissue Cult. Methods* **9**:43–48.

Lechner, J. F., Haugen, A., Autrup, H., McClendon, I. A., Trump, B. F., & Harris, C. C. (1981). Clonal growth of epithelial cells from normal adult human bronchus. *Cancer Res.* **41**:2294–2304.

Leder, A., & Leder, P. (1975). Butyric acid, a potent inducer of erythroid differentiation in cultured erythroleukemic cells. *Cell* **5**:319–322.

Lee, M. Y., Chou, C. Y., Tang, M. J., & Shen, M. R. (2008). Epithelial-mesenchymal transition in cervical cancer: correlation with tumor progression, epidermal growth factor receptor overexpression, and snail up-regulation. *Clin. Cancer Res.* **14**:4743–4750.

Lee, D. R., Kaproth, M. T. & Parks, J. E. (2001). In vitro production of haploid germ cells from fresh or frozen-thawed testicular cells of neonatal bulls. *Biol. Reprod.* **65**:873–878.

Lee, L. E. J., Dayeh, V. R., Schirmer, K., & Bols, N. C. (2009). Applications and potential uses of fish gill cell lines: examples with RTgill-W1. *In Vitro Cell Dev. Biol. Anim.* **45**:127–134.

Lee, T. H., Baik, M. G., Im, W. B., Lee, C. S., Han, Y. M., Kim, S. J., Lee, K. K., & Choi, Y. J. (1996). Effects of EHS matrix on expression of transgenes in HC11 cells. *In Vitro Cell Dev. Biol. Anim.* **32**:454–456.

Lefort, N., Feyeux, M., Bas, C., Féraud, O., Bennaceur-Griscelli, A., Tachdjian, G., Peschanski, M., & Perrier, A. L. (2008). Human embryonic stem cells reveal recurrent genomic instability at 20q11.21. *Nat. Biotechnol.* **26**:1364–1366.

Lehle, K., Buttstaedt, J., & Birnbaum, D. E. (2003). Expression of adhesion molecules and cytokines in vitro by endothelial cells seeded on various polymer surfaces coated with titaniumcarboxonitride. *J. Biomed. Mater. Res.* **65A**:393–401.

Leibo SP, Mazur P. (1971). The role of cooling rates in low-temperature preservation. *Cryobiology* **8**:447–452.

Leibovitz, A. (1963). The growth and maintenance of tissue cell cultures in free gas exchange with the atmosphere. *Am. J. Hyg.* **78**:173–183.

Leigh, I. M., & Watt, F. M. (1994). *Keratinocyte methods.* Cambridge: Cambridge University Press.

Leighton, J. (1991). Radial histophysiologic gradient culture chamber rationale and preparation. *In Vitro Cell Dev. Biol.* **27A**:786–790.

Leighton, J., Mark, R., & Rush, G. (1968). Patterns of three-dimensional growth in collagen coated cellulose sponge: Carcinomas and embryonic tissues. *Cancer Res.* **28**:286–296.

Lemare, F., Steimberg, N., Le Griel, C., Demignot, S., & Adolphe, M. (1998). Dedifferentiated chondrocytes cultured in alginate beads: Restoration of the differentiated phenotype and of the metabolic response to interleukin-1β. *J. Cell Physiol.* **176**:303–313.

Lennon, D., & Caplan, A. (2006). Mesenchymal stem cells. In Vunjak-Novakovic, G., & Freshney. R. I. (eds.), *Culture of cells for tissue engineering.* Hoboken,NJ- Wiley-Liss, pp. 23–60.

Léobon, B., Garcin, I., Menasche, P., Vilquin, J. T., Audinat, E., & Charpak, S. (2003). Myoblasts transplanted into rat infarcted myocardium are functionally isolated from their host. *Proc. Natl. Acad. Sci. USA* **100**:7808–7811.

LePage, K. T., Dickey, R. W., Gerwick, W. H., Jester, E. L., & Murray, T. F. (2005). On the use of neuro-2a neuroblastoma cells versus intact neurons in primary culture for neurotoxicity studies. *Crit. Rev. Neurobiol.* **17**:27–50.

Lesuffleur, T., Barbat, A., Dussaulx, E., & Zweibaum, A. (1990). Growth adaptation to methotrexate of HT-29 human colon carcinoma cells is associated with their ability to differentiate into columnar absorptive and mucus-secreting cells. *Cancer Res.* **50**:6334–6343.

Lever, J. (1986). Expression of differentiated functions in kidney epithelial cell lines. *Min. Elec. Metab.* **12**:14–19.

Levi-Montalcini, R. (1966). The nerve growth factor: its mode of action on sensory and sympathetic nerve cells. *Harvey Lect.* **60**:217–259.

Levi-Montalcini, R. C. P. (1979). The nerve-growth factor. *Sci. Am.* **240**:68.

Levin, D. B., Wilson, K., Valadares de Amorim, G., Webber, J., Kenny, P., & Kusser, W. (1995). Detection of p53 mutations in benign and dysplastic nevi. *Cancer Res.* **55**:4278–4282.

Levine, E. M., & Becker, B. G. (1977). Biochemical methods for detecting mycoplasma contamination. In McGarrity, G. T., Murphy, D. G., & Nichols, W. W. (eds.), *Mycoplasma infection of cell cultures.* New York: Plenum Press, pp. 87–104.

Lewitzky, M., & Yamanaka, S. (2007). Reprogramming somatic cells towards pluripotency by defined factors. *Curr. Opin. Biotechnol.* **18**:467–473.

Ley, K. D., & Tobey, R. A. (1970). Regulation of initiation of DNA synthesis in Chinese hamster cells; II: Induction of DNA synthesis and cell division by isoleucine and glutamine in $G_1$-arrested cells in suspension culture. *J. Cell Biol.* **47**:453–459.

Li, P., Tong, C., Mehrian-Shai, R., Jia, L., Wu, N., Yan, Y., Maxson, R. E., Schulze, E. N., Song, H., Hsieh, C. L., Pera, M. F., & Ying, Q. L. (2008). Germline competent embryonic stem cells derived from rat blastocysts. *Cell* **135**:1299–1310.

Li, A. P., Roque, M. M., Beck, D. J., & Kaminski, D. L. (1992). Isolation and culturing of hepatocytes from human livers. *J. Tissue Cult. Methods* **14**:139–146.

Li, N. F., Broad, Lu, S., Y. J., Yang, J. S., Watson, R., Hagemann, T., Wilbanks, G., Jacobs, I., Balkwill, F., Dafou, D., & Gayther, S. A. (2007). Human ovarian surface epithelial cells immortalized with hTERT maintain functional pRb and p53 expression. *Cell Prolif.* **40**:780–794.

Li, Y. Sauvé, Y. Li, D. Lund, R. D., & Raisman, G. (2003b) Transplanted olfactory ensheathing cells promote regeneration of cut adult rat optic nerve axons. *J. Neurosci.* **23**:7783–7788.

Li, Y. M., Schacher, D. H., Liu, Q., Arkins, S., Rebeiz, N., McCusker, R. H. Jr., Dantzer, R., & Kelley, K. W. (1997). Regulation of myeloid growth and differentiation by the insulin-like growth factor I receptor. *Endocrinology.* **138**:362–368.

Li, Y., Decherchi, P., & Raisman, G. (2003a) Transplantation of olfactory ensheathing cells into spinal cord lesions restores breathing and climbing. *J. Neurosci.* **23**:727–731.

Li, Y., Field, P. M., & Raisman, G. (1997). Repair of adult rat corticospinal tract by transplants of olfactory ensheathing cells. *Science* **277**:2000–2002.

Li, Y., Foster, W., Deasy, B. M., Chan, Y., Prisk, V., Tang, Y., Cummins, J., & Huard, J. (2004). Transforming growth factor-beta1 induces the differentiation of myogenic cells into fibrotic cells in injured skeletal muscle: A key event in muscle fibrogenesis. *Am. J. Pathol.* **164**:1007–1019.

Lidington, E. A., Rao, R. M., Marelli-Berg, F. M., Jat, P. S., Haskard, D. O., & Mason, J. C. (2002). Conditional immortalization of growth factor-responsive cardiac endothelial cells from H-2K(b)-tsA58 mice. *Am. J. Physiol. Cell. Physiol.* **282**:C67–C74.

Liebermann, D., & Sachs, L. (1978). Nuclear control of neurite induction in neuroblastoma cells. *Exp. Cell Res.* **113**:383–390.

Lillie, I. H., MacCallum, D. K., & Jepsen, A. (1980). Fine structure of subcultivated stratified squamous epithelium grown on collagen rafts. *Exp. Cell. Res.* **125**:153–165.

Lim, G., Karaskova, J., Vukovic, B., Bayani, J., Beheshti, B., Bernardini, M., Squire, J. A., & Zielenska, M. (2004). Combined spectral karyotyping, multicolor banding, and microarray comparative genomic hybridization analysis provides a detailed characterization of complex structural chromosomal rearrangements associated with gene amplification in the osteosarcoma cell line MG-63. *Cancer Genet. Cytogenet.* **153**:158–164.

Limat, A., Breitkreutz, D., Thiekötter, G., Klein, E. C., Braathen, L. R., Hunziker, T., & Fusenig, N. E. (1995). Formation of a regular neo-epidermis by cultured human outer root sheath cells grafted on nude mice. *Transplantation* **59**:1032–1038.

Limat, A., Hunziker, T., Boillat, C., Bayreuther, K., & Noser, F. (1989). Post-mitotic human dermal fibroblasts efficiently support the growth of human follicular keratinocytes. *J. Invest. Dermatol.* **92**:758–762.

Limat, A., Mauri, D., & Hunziker, T. (1996). Successful treatment of chronic leg ulcers with epidermal equivalents generated from cultured autologous outer root sheath cells. *J. Invest. Dermatol.* **107**:128–135.

Lin, C. W., Lin, J. C., & Prout, G. R. Jr. (1985). Establishment and characterization of four human bladder tumor cell lines and sublines with different degrees of malignancy. *Cancer Res.* **45**:5070–5079.

Lin, T., Ambasudhan, R., Yuan, X., Li, W., Hilcove, S., Abujarour, R., Lin, X., Hahm, H. S., Hao, E., Hayek, A., & Ding, S. (2009). A chemical platform for improved induction of human iPSCs. *Nature Methods.* **6**:805–808.

Linser, P., & Moscona, A. A. (1980). Induction of glutamine synthetase in embryonic neural retina-localization in Muller fibers and dependence on cell interaction. *Proc. Natl. Acad. Sci. USA* **76**:6476–6481.

Liotta, L. (1987). The role of cellular proteases and their inhibitors in invasion and metastasis: Introductionary overview. *Cancer Metastasis Rev.* **9**:285–287.

Lippincott-Schwartz, J., & Manley, S. (2009). Putting super-resolution fluorescence microscopy to work. *Nat. Methods* **6**:21–23.

Lippincott-Schwartz, J., Glickman, J., Donaldson, J. G., Robbins, J., Kreis, T. E., Seamon, K. B., Sheetz, M. P., & Klausner, R. D. (1991). Forskolin inhibits and reverses the effects of Brefeldin A on Golgi morphology by a cAMP-independent mechanism. *J. Cell Biol.* **112**:567.

Lipscomb, E. A., & Mercurio, A M. (2005). Mobilization and activation of a signaling competent alpha6beta4integrin underlies its contribution to carcinoma progression. *Cancer Metastasis Rev.* **24**:413–423.

Liscovitch, M., & Ravid, D. (2007). A case study in misidentification of cancer cell lines: MCF-7/AdrR cells (re-designated NCI/ADR-RES) are derived from OVCAR-8 human ovarian carcinoma cells. *Cancer Lett.* **245**:350–352.

Littlefield, J. W. (1964a). Selection of hybrids from matings of fibroblasts in vitro and their presumed recombinants. *Science* **145**:709–710.

Littlefield, J. W. (1964b). Three degrees of guanylic acid pyrophosphorylase deficiency in mouse fibroblasts. *Nature* **203**:1142–1144.

Litwin, J. (1973). Titanium disks. In Kruse, P. F., & Patterson, M. K. (eds.), *Tissue culture methods and applications.* New York: Academic Press, pp. 383–387.

Liu, J. Y., Hafner, J., Dragieva, G., & Burg, G. (2004). Bioreactor microcarrier cell culture system (Bio-MCCS) for large-scale production of autologous melanocytes. *Cell Transplant.* **13**:809–816.

Liu, L., Delbe, J., Blat, C., Zapf, J., & Harel, L. (1992). Insulin like growth factor binding protein (IGFBP-3), an inhibitor of serum growth factors other than IGF-I and -II. *J. Cell Physiol.* **153**:15–21.

Liu, M., Xu, J., Souza, P., Tanswell, B., Tanswell, A. K., & Post, M. (1995). The effect of mechanical strain on fetal rat lung cell proliferation: Comparison of two- and three-dimensional culture systems. *In Vitro Cell Dev. Biol. Anim.* **31**:858–866.

Liu, T. F., Cohen, K. A., Willingham, M. C., Tatter, S. B., Puri, R. K., & Frankel, A. E. (2003). Combination fusion protein therapy of refractory brain tumors: demonstration of efficacy in cell culture. *J. Neurooncol.* **65**:77–85.

Lohmann, V., Körner, F., Koch, J., Herian, U., Theilmann, L., & Bartenschlager, R. (1999). Replication of subgenomic hepatitis C virus RNAs in a hepatoma cell line. *Science* **285**:110–113.

Lokesh Battula, V., Bareiss, P. M., Treml, S., Conrad, S., Albert, I., Hojak, S., Abele, H., Schewe, B., Just, L., Skutella, T., & Bühring, H.-J. (2007). Human placenta and bone marrow derived MSC cultured in serum-free, b-FGF-containing medium express cell surface frizzled-9 and SSEA-4 and give rise to multilineage differentiation. *Differentiation* **75**:279–291.

Lopez-Casillas, F., Wrana, J. L., & Massague, J. (1993). Betaglycan presents ligand to the TGF-β signalling receptor. *Cell* **73**:1435–1444.

Lorenzi, P. L., Reinhold, W. C., Varma, S., Hutchinson, A. A., Pommier, Y., Chanock, S. J., & Weinstein, J. N. (2009). DNA fingerprinting of the NCI-60 cell line panel. *Mol. Cancer Ther.* **8**:713–724.

Lotan, R., & Lotan, D. (1980). Stimulation of melanogenesis in a human melanoma cell line by retinoids. *Cancer Res.* **40**:33–45.

Lou, Z., Wang, Y., Nawoschik, S., Riddell, D., Martone, R., & Dunlop, J. (2008). Implementing CompacT SelecT automated cell culture in support of Alzheimer's disease-targeted drug discovery. *JALA* **13**:3.

Lounis, H., Provencher, D., Godbout, C., Fink, D., Milot, M.-J., & Mes-Masson, A.-M. (1994). Primary cultures of normal and tumoral ovarian epithelium: A powerful tool for basic molecular studies. *Exp. Cell Res.* **215**:303–309.

Lovelock, J. E., & Bishop, M. W. H. (1959). Prevention of freezing damage to living cells by dimethyl sulphoxide. *Nature* **183**:1394–1395.

Lowe, K. C., Davey, M. R., & Power, J. B. (1998). Perfluorochemicals: Their applications and benefits to cell culture. *Trends Biotechnol.* **16**:272–277.

Loyer, P., Cariou, S., Glaise, D., Bilodeau, M., Baffet, G., & Guguen-Guillouzo, C. (1996). Growth factor dependence of progression through G1 and S phases of adult rat hepatocytes in vitro: Evidence of a mitogen restriction point in mid-late G1. *J. Biol. Chem.* **271**(19): 11484–11492.

Luikart, S. D., Maniglia, C. A., Furcht, L. T., McCarthy, J. B., & Oegama, T. R. (1990). A heparan sulphate-containing fraction of bone marrow stroma induces maturation of HL-60 cell in vitro. *Cancer Res.* **50**:3781–3785.

Lundqvist, M., Mark, J., Funa, K., Heldin, N. E., Morstyn, G., Weddell, B., Layton, J., & Oberg, K. (1991). Characterisation of a cell line (LCC-18) from a cultured human neuroendocrine-differentiated colonic carcinoma. *Eur. J. Cancer* **12**:1662–1668.

Lutolf, M. P., & Hubbel, J. A. (2005). Synthetic biomaterials as instructive extracellular microenvironments for morphogenesis in tissue engineering. *Nat. Biotechnol.* **23**:47–55.

Lutz, M. P., Gaedicke, G., & Hartmann, W. (1992). Large-scale cell separation by centrifugal elutriation. *Anal. Biochem.* **200**:376–380.

Ma, T., Grayson, W. L., Fröhlich, M., & Vunjak-Novakovic, G. (2009). Hypoxia and stem cell-based engineering of mesenchymal tissues. *Biotechnol. Prog.* **25**:32–42.

Maas-Szabowski, N., & Fusenig, N. E. (1996). Interleukin-1-induced growth factor expression in postmitotic and resting fibroblasts. *J. Invest. Dermatol.* **107**:849–855.

Maas-Szabowski, N., Fusenig, N. E., & Shimotoyodome, A. (1999). Keratinocyte growth regulation in fibroblast co-cultures via a double paracrine mechanism. *J. Cell Sci.* **112**:1843–1853.

Maas-Szabowski, N., Stark, H. J., & Fusenig, N. E. (2000). Keratinocyte growth regulation in defined organotypic cultures through IL-1-induced KGF expression in resting fibroblasts. *J. Invest. Dermatol.* **114**:1075–1084.

Maas-Szabowski, N., Stark, H. J., & Fusenig, N. E. (2002). Cell interaction and epithelial differentiation. In Freshney, R. I. & Freshney, M. G. (eds.), *Culture of epithelial cells,* 2nd ed. Hoboken, NJ: Wiley-Liss, pp. 31–63.

MacDonald, C. M., Freshney, R. I., Hart, E., & Graham, D. I. (1985). Selective control of human glioma cell proliferation by specific cell interaction. *Exp. Cell Biol.* **53**:130–137.

Macé, K., Gonzalez, F. J., McConnell, I. R., Garner, R. C., Avanti, O., Harris, C. C., & Pfeifer, A. M. A. (1994). Activation of promutagens in a human bronchial epithelial cell line stably expressing human cytochrome P450 1A2. *Mol. Carcinogen.* **11**:65–73.

Maciag, T., Cerondolo, J., Ilsley, S., Kelley, P. R., & Forand, R. (1979). Endothelial cell growth factor from bovine hypothalamus: Identification and partial characterization. *Proc. Natl. Acad. Sci. USA* **76**:5674–5678.

Macieira-Coelho, A. (1973). Cell cycle analysis; A: Mammalian cells. In Kruse, P. F., & Patterson, M. K. (eds.), *Tissue culture methods and applications*. New York: Academic Press, pp. 412–422.

Macklis, J. D., Sidman, R. L., & Shine, H. D. (1985). Cross-linked collagen surface for cell culture that is stable, uniform, and optically superior to conventional surfaces. *In Vitro* **21**:189–194.

MacLeod, R. A., Dirks, W. G., Matsuo, Y., Kaufmann, M., Milch, H., & Drexler, H. G. (1999). Widespread intraspecies cross-contamination of human tumor cell lines arising at source. *Int. J. Cancer* **83**:555–563.

Macpherson, I. (1973). Soft agar technique. In Kruse, P. F., & Patterson, M. K. (eds.), *Tissue culture methods and applications*. New York: Academic Press, pp. 276–280.

Macpherson, I., & Bryden, A. (1971). Mitomycin C treated cells as feeders. *Exp. Cell Res.* **69**:240–241.

Macpherson, I., & Montagnier, L. (1964). Agar suspension culture for the selective assay of cells transformed by polyoma virus. *Virology* **23**:291–294.

Macpherson, I., & Stoker, M. (1962). Polyoma transformation of hamster cell clones: An investigation of genetic factors affecting cell competence. *Virology* **16**:147.

Macy, M. (1978). Identification of cell line species by isoenzyme analysis. *Manual Am. Tissue Cult. Assoc.* **4**:833–836.

Maddugoda, M. P., Crampton, M. S., Shewan, A. M., & Yap, A. S. (2007). Myosin VI and vinculin cooperate during the morphogenesis of cadherin cell contacts in mammalian epithelial cells. *J. Cell Biol.* **178**:529–540.

Magee, J. C., Stone, A. E., Oldham, K. T., & Guice, K. S. (1994). Isolation, culture, and characterization of rat lung microvascular endothelial cells. *Am. J. Physiol. Lung. Mol. Cell. Physiol.* **267**:L433–L441.

Mahdavi, V., & Hynes, R. O. (1979). Proteolytic enzymes in normal and transformed cells. *Biochim. Biophys. Acta* **583**:167–178.

Maherali, N., Ahfeldt, T., Rigamonto, A., Utikal, J., Cowan, C., & Hochedlinger, K. (2008). A high-efficiency system for the generation of human induced pluripotent stem cells. *Cell Stem Cell* **3**:340–345.

Mairs, R. J., & Wheldon, T. E. (1996). Experimental tumour therapy with targeted radionuclides in multicellular tumour spheroids. In Hagen, U., Jung, H., & Streffer, C. (eds.), *Radiation research, 1895–1995* (Proc. 10th Intl. Congr. Radiation Research, Wurzburg, Germany, Vol. 2. Universitatsdruckerei, H. Sturtz A. G., Wurtzburg. pp. 819–826. ISBN: 300000842X).

Majore, I., Moretti, P., Hass, R., & Kasper, C. (2009). Identification of subpopulations in mesenchymal stem cell-like cultures from human umbilical cord. *Cell Commun. Signal.* **7**:6.

Mal, A., & Harter, M. L. (2003). MyoD is functionally linked to the silencing of a muscle-specific regulatory gene prior to skeletal myogenesis. *Proc. Natl. Acad. Sci. USA* **100**:1735–1739.

Maltese, W. A., & Volpe, I. J. (1979). Induction of an oligodendroglial enzyme in C-6 glioma cells maintained at high density or in serum-free medium. *J. Cell Physiol.* **101**:459–470.

Malvestiti, F., Colombo, D., Perego, D., Rodeschini, O., Finelli, P., Larizza, L., & Giardino, D. (2009). Fluorescence in situ hybridization dissection of a chronic myeloid leukemia case

bearing the apparently balanced translocations (9;22)(q34;q11.2) and (11;11)(p15;q13). *Cancer Genet. Cytogenet.* **188**:42–47.

Management of Health and Safety at Work Regulations. (1999). SI. 1999 No. 3242. www.opsi.gov.uk/si/si1999/19993242.htm.

Mangi, A. A., Noiseux, N., Kong, D., He, H., Rezvani, M., Ingwall, J. S., & Dzau, V. J. (2003). Mesenchymal stem cells modified with Akt prevent remodeling and restore performance of infarcted hearts. *Nat. Med.* **9**:1195–1201.

Maniatis, T., Hardison, R. C., Lacy, E., Lauer, J., O'Connell, C., Quon, D., Sim, G. K., & Efstradiadis, A. (1978). The isolation of structural genes from libraries of eukaryotic DNA. *Cell* **15**:687–701.

Maramorosch, K. (1976). *Invertebrate tissue culture.* New York: Academic Press.

Marchionni, M. A., Goodearl, A. D., Chen, M. S., Bermingham-McDonogh, O., Kirk, C., Hendricks, M., Danehy, F., Misumi, D., Sudhalter, J., Kobayashi, K., Wroblewski, D., Lynch, C., Baldassare, M., Hiles, I., Davis, J. B., Hsuan, J. J., Totty, N. F., Otsu, M., McBurney, R. N., Waterfield, M. D., Stroobant, P., & Gwynne, D. (1993). Glial growth factors are alternatively spliced erbB2 ligands expressed in the nervous system. *Nature* **362**:312–318.

Marcus, M., Lavi, U., Nattenberg, A., Ruttem, S., & Markowitz, O. (1980). Selective killing of mycoplasmas from contaminated cells in cell cultures. *Nature* **285**:659–660.

Mardh, P. H. (1975). Elimination of mycoplasmas from cell cultures with sodium polyanethol sulphonate. *Nature* **254**:515–516.

Mareel, M. M., Bruynell, E., & Storme, G. (1980). Attachment of mouse fibrosarcoma cells to precultured fragments of embryonic chick heart. *Virchows Arch. B Cell Pathol.* **34**:85–97.

Mareel, M., Kint, J., & Meyvisch, C. (1979). Methods of study of the invasion of malignant C3H-mouse fibroblasts into embryonic chick heart in vitro. *Virchows Arch. B Cell Pathol.* **30**:95–111.

Marelli-Berg, F. M., Peek, E., Lidington, E. A., Stauss, H. J., & Lechler, R. I. (2000). Isolation of endothelial cells from murine tissue. *J. Immunol. Methods* **244**:205–215.

Marh, J., Tres, L. L., Yamazaki, Y., Yanagimachi, R. & Kierszenbaum, A. L. (2003). Mouse round spermatids developed in vitro from preexisting spermatocytes can produce normal offspring by nuclear injection into in vivo-developed mature oocytes. *Biol. Reprod.* **69**:169–176.

Marijnissen, W. J., van Osch, G. J., Aigner, J., van der Veen, S. W., Hollander, A. P., Verwoerd-Verhoef, H. L. & Verhaar, J. A. (2002). Alginate as a chondrocyte-delivery substance in combination with a non-woven scaffold for cartilage tissue engineering. *Biomaterials* **23**:1511–1517.

Mariotti, E., Mirabelli, P., Di Noto, R., Fortunato, G., & Salvatore, F. (2008). Rapid detection of mycoplasma in continuous cell lines using a selective biochemical test. *Leuk Res.* **32**:323–326.

Mark, J. (1971). Chromosomal characteristics of neurogenic tumours in adults. *Hereditas* **68**:61–100.

Markovic, O., & Markovic, N. (1998a). Cell cross-contamination in cell cultures: The silent and neglected danger. *In Vitro Cell Dev. Biol. Anim.* **34**:1–8.

Markovitz, D., Goff, S., & Bank, A (1988a). A safe packaging line for gene transfer: separating viral genes on two different plasmids. *J. Virol.* **62**:1120–1124.

Markowitz, D., Goff, S., & Bank, A. (1988b). Construction and use of a safe and efficient amphotropic packaging cell line. *Virology* **167**:400–406.

Marks, P. A., Richon, V. M., Kiyokawa, H., & Rifkind, R. A. (1994). Inducing differentiation of transformed cells with hybrid polar compounds: A cell cycle-dependent process. *Proc. Natl. Acad. Sci. USA* **91**:10251–10254.

Markus, G., Takita, H., Camiolo, S. M., Corsanti, J., Evers, J. L., & Hobika, J. H. (1980). Content and characterization of plasminogen activators in human lung tumours and normal lung tissue. *Cancer Res.* **40**:841–848.

Marques da Silva, C., Miranda Rodrigues, L., Passos da Silva Gomes, A., Mantuano Barradas, M., Sarmento Vieira, F., Persechini, P. M., & Coutinho-Silva, R. (2008). Modulation of P2X7 receptor expression in macrophages from mineral oil-injected mice. **Immunobiology 213**:481–492.

Marsh, J. W., Donovan, M., Burholt, D. R., George, L. D., & Kornblith, P. L. (2004). Metastatic lung disease to the central nervous system: In vitro response to chemotherapeutic agents. *J. Neurooncol.* **66**:81–90.

Marshall, C. J. (1991). Tumor suppressor genes. *Cell* **64**:313–326.

Marte, B. M., Meyer, T., Stabel, S., Standke, G. J. R., Jaken, S., Fabbro, D., & Hynes, N. E. (1994). Protein kinase C and mammary cell differentiation: Involvement of protein kinase C alpha in the induction of beta-casein expression. *Cell Growth Differ.* **5**:239–247.

Martin, S. M., Schwartz, J. L., Giachelli, C. M., & Ratner, B. D. (2004). Enhancing the biological activity of immobilized osteopontin using a type-1 collagen affinity coating. *J. Biomed. Mater. Res.* **70A**:10–19.

Martin, G. R. (1975). Teratocarcinomas as a model system for the study of embryogenesis and neoplasia. *Cell* **5**:229–243.

Martin, G. R. (1978). Advantages and limitations of teratocarcinoma stem cells as models of development. In Johnson, M. H. (ed.), *Development in mammals*, Vol. **3**. Amsterdam: North Holland, p. 225.

Martin, G. R. (1981). Isolation of a pluripotent cell line from early mouse embryos cultured in medium conditioned by teratocarcinoma stem cells. *Proc. Natl. Acad. Sci. USA* **78**:7634–7638.

Martin, G. R., & Evans, M. J. (1974). The morphology and growth of a pluripotent teratocarcinoma cell line and its derivatives in tissue culture. *Cell* **2**:163–172.

Martin, G. R., & Evans, M. J. (1975). Differentiation of clonal lines of teratocarcinoma cells: formation of embryoid bodies in vitro. *Proc. Natl. Acad. Sci. USA* **72**:1441–1445.

Maskell, R., & Green, M. (1995). Applications of the comet assay technique. *Int. Micr. Lab.* **6**:2–5.

Massague, J., Cheifetz, S., Laiho, M., Ralph, D. A., Weis, F. M., & Zentella, A. (1992). Transforming growth factor-beta. *Cancer Surveys* **12**:81–103.

Masson, E. A., Atkin, S. L., & White, M. C. (1993). D-valine selective medium does not inhibit human fibroblast growth in vitro. *In Vitro Cell Dev. Biol. Anim.* **29A**:912–913.

Masters JR, Hepburn PJ, Walker L, Highman WJ, Trejdosiewicz LK, Povey S, Parkar M, Hill BT, Riddle PR, Franks LM. (1986). Tissue culture model of transitional cell carcinoma: characterization of twenty-two human urothelial cell lines. *Cancer Res.* **46**:3630–3636.

Masters, J. R. W., & Palsson, B. (eds.) (1999–2000). *Human cell culture*, Vols. 1 & 2, Dordrecht: Kluwer.

Masters, J. R. W., Thomson, J. A., Daly-Burns, B., Reid, Y. A., Dirks, W. G., Packer, P., Toji, L. H., Ohno, T., Tanabe, H., Arlett, C. F., Kelland, L. R., Harrison, M., Virmani, A., Ward, T. H., Ayres, K. L., & Debenham, P. G. (2001). STR profiling provides an international reference standard for human cell lines. *Proc. Natl. Acad. Sci. USA* **98**:8012–8017.

Masters, J. R., Bedford, P., Kearney, A., Povey, S., & Franks, L. M. (1988). Bladder cancer cell line cross-contamination: Identification using a locus-specific minisatellite probe. *Br. J. Cancer* **57**(3): 284–286.

Masuda, K., & Sah, R. (2006). Tissue engineering of articular cartilage. In Vunjak-Novakovic, G., Freshney. R. I. (eds.), *Culture of cells for tissue engineering*. Hoboken, NJ: Wiley-Liss, pp. 157–189.

Masui, T., Lechner, J. F., Yoakum, G. H., Willey, J. C., & Harris, C. C. (1986b). Growth and differentiation of normal and transformed human bronchial epithelial cells. *J. Cell Physiol.* **4** (suppl): 73–81.

Masui, T., Wakefield, L. M., Lechner, J. F., LaVeck, M. A., Sporn, M. B., & Harris, C. C. (1986a). Type beta transforming growth factor is the primary differentiation-inducing serum factor for normal human bronchial epithelial cells. *Proc. Natl. Acad. Sci. USA* **83**:2438–2442.

Mather, J. (1979). Testicular cells in defined medium. In Jakoby, W. B., & Pastan, I. H. (eds.), *Methods in enzymology; Vol. 57: Cell culture*. New York: Academic Press, p. 103.

Mather, J. P. (1998). Making informed choices: Medium, serum, and serum-free medium; How to choose the appropriate medium and culture system for the model you wish to create. *Methods Cell Biol.* **57**:19–30.

Mather, J. P., & Sato, G. H. (1979a). The growth of mouse melanoma cells in hormone supplemented, serum-free medium. *Exp. Cell Res.* **120**:191.

Mather, J. P., & Sato, G. H. (1979b). The use of hormone supplemented serum-free media in primary cultures. *Exp. Cell Res.* **124**:215.

Matsui, A., Zsebo, K., & Hogan, B. L. M. (1992). Derivation of pluripotential embryonic stem cells from murine primordial germ cells in culture. *Cell* **70**:841–847.

Matsumoto, S., Arakawa, Y., Ohishi, M., Yanaihara, H., Iwanaga, T., & Kurokawa, N. (2008). Suppressive action of pituitary adenylate cyclase activating polypeptide (PACAP) on proliferation of immature mouse Leydig cell line TM3 cells. *Biomed Res.* **29**:321–330.

Mayne, L. V., Price, T. N. C., Moorwood, K., & Burke, J. F. (1996). Development of immortal human fibroblast cell lines. In Freshney, R. I., & Freshney, M. G. (eds.), *Culture of immortalized cells*. New York: Wiley-Liss, pp. 77–93.

Mayne, L. V., Priestly, A., James, M. R., & Burke, J. F. (1986). Efficient immortalisation and morphological transformation of human fibroblasts with SV40 DNA linked to a dominant marker. *Exp. Cell Res.* **162**:530–538.

McAteer, J. A., Kempson, S. A., & Evan, A. P. (1991). Culture of human renal cortex epithelial cells. *J. Tissue Cult. Methods* **13**:143–148.

McCall, E., Povey, J., & Dumonde, D. C. (1981). The culture of vascular endothelial cells on microporous membranes. *Thromb. Res.* **24**:417–431.

McCarthy, K. D., & de Vellis, J. (1980). Preparation of separate astroglial and oligodendroglial cell cultures from rat cerebral tissue. *J. Cell Biol.* **85**:890–902.

McCormack, S. A., Viar, M. J., Tague, L., & Johnston, L. R. (1996). Altered distribution of the nuclear receptor rar (β) accompanies proliferation and differentiation changes caused by retinoic acid in Caco-2 cells. *In Vitro Cell Dev. Biol. Anim.* **32**:53–61.

McCormick C., & Freshney, R. I. (2000). Activity of growth factors in the IL-6 group in the differentiation of human lung adenocarcinoma. *Br. J. Cancer* **82**:881–890.

McCormick, C., Freshney, R. I., & Speirs, V. (1995). Activity of interferon alpha, interleukin 6 and insulin in the regulation of differentiation in A549 alveolar carcinoma cells. *Br. J. Cancer* **71**:232–239.

McCoy, T. A., Maxwell, M., & Kruse, P. F. Jr. (1959). Amino acid requirements of the Novikoff hepatoma in vitro. *Proc. Soc. Exp. Biol. Med.* **100**:115–118.

McDevitt, M. A., Condon, M., Stamberg, J., Karp, J. E., & McDiarmid, M. (2007). Fluorescent in situ hybridization (FISH) in bone marrow and peripheral blood of leukemia patients: Implications for occupational surveillance. *Mutat Res.* **629**:24–31.

McDonald, E. R. 3rd, & El-Deiry, W. S. (2000). Ontrol as a basis for cancer drug development. *Int. J. Oncol.* **16**:871–886.

McDouall, R. M., Yacoub, M., and Rose, M. L. (1996). Isolation, culture, and characterisation of MHC class II-positive microvascular endothelial cells from the human heart. *Microvasc Res* **51**:137–152.

McFarland, D. C., Liu, X., Velleman, S. G., Zeng, C., Coy, C. S., & Pesall, J. E. (2003). Variation in fibroblast growth factor response and heparan sulfate proteoglycan production in satellite cell populations. *Comp. Biochem. Physiol. C Toxicol. Pharmacol.* **134**:341–351.

McGarrity, G. J. (1976). Spread and control of mycoplasmal infection of cell cultures. *In Vitro* **12**:643–648.

McGarrity, G. J. (1982). Detection of mycoplasmic infection of cell cultures. In Maramorosch, K. (ed.), *Advances in cell culture*, Vol. **2**. New York: Academic Press, pp. 99–131.

McGowan, J. A. (1986). Hepatocyte proliferation in culture. In Guillouzo, A., & Guguen-Guillouzo, C. (eds.), *Isolated and cultured hepatocytes*. Paris: Editions Inserm, John Libbey Eurotext, pp. 13–38.

McGregor, D. B., Edwards, I, Riach, C. J., Cattenach, P., Martin, R., Mitchell, A. & Caspary, W. J. (1988). Studies of an S9 based metabolic activation system used in the mouse lymphoma L51768Y cell mutation assay. *Mutagenesis* **3**:485–490.

McGuckin, C. P., Forraz, N., Allouard, Q., & Pettengell, R. (2004). Umbilical cord blood stem cells can expand hematopoietic and neuroglial progenitors in vitro. *Exp. Cell Res.* **295**:350–359.

McGuire, P. G., & Orkin, R. W. (1987) Isolation of rat aortic endothelial cells by primary explant techniques and their phenotypic modulation by defined substrata. *Lab. Invest.* **57**:94–105.

McKeehan, W. L. (1977). The effect of temperature during trypsin treatment on viability and multiplication potential of single normal human and chicken fibroblasts. *Cell Biol. Int. Rep.* **1**:335–343.

McKeehan, W. L., & Ham, R. G. (1976a). Stimulation of clonal growth of normal fibroblasts with substrata coated with basic polymers. *J. Cell Biol.* **71**:727–734.

McKeehan, W. L., & Ham, R. G. (1976b). Methods for reducing the serum requirement of growth in vitro of non-transformed diploid fibroblasts. *Dev. Biol. Standard.* **37**:97–98.

McKeehan, W. L., & McKeehan, K. A. (1979). Oxocarboxylic acids, pyridine nucleotide-linked oxidoreductases and serum factors in regulation of cell proliferation. *J. Cell Physiol.* **101**:9–16.

McKeehan, W. L., Adams, P. S., & Rosser, M. P. (1982). Modified nutrient medium MCDB 151 (WJAC401), defined growth factors, cholera toxin, pituitary factors, and horse serum support epithelial cell and suppress fibroblast proliferation in primary cultures of rat ventral prostate cells. *In Vitro* **18**:87–91.

McKeehan, W. L., Adams, P. S., & Rosser, M. P. (1984). Direct mitogenic effects of insulin, epidermal growth factor, cholera toxin, unknown pituitary factors and possibly prolactin, but not androgen, on normal rat prostate epithelial cells in serum-free primary cell culture. *Cancer Res.* **44**:1998–2010.

McKeehan, W. L., Hamilton, W. G., & Ham, R. G. (1976). Selenium is an essential trace nutrient for growth of WI-38 diploid human fibroblasts. *Proc. Natl. Acad. Sci. USA* **73**:2023–2027.

McKeehan, W. L., McKeehan, K. A., Hammond, S. L., & Ham, R. G. (1977). Improved medium for clonal growth of human diploid cells at low concentrations of serum protein. *In Vitro* **13**:399–416.

McLean, J. S., Frame, M. C., Freshney, R. I., Vaughan, P. F. T., & Mackie, A. E. (1986). Phenotypic modification of human glioma and non-small cell lung carcinoma by glucocorticoids and other agents. *Anticancer Res.* **6**:1101–1106.

Meeker, A. K., & De Marzo, A. M. (2004). Recent advances in telomere biology: Implications for human cancer. *Curr. Opin. Oncol.* **16**:32–38.

Mège, R. M., Gavard, J., & Lambert, M. (2006). Regulation of cell–cell junctions by the cytoskeleton. *Curr. Opin. Cell Biol.* **18**:541–548.

Mege, R. M., Matsuzaki, F., Gallin, W. J., Goldberg, J. I., Cummingham, B. A., & Edelman, G. M. (1989). Construction of epithelioid sheets by transfection of mouse sarcoma cells with cDNAs for chicken cell adhesion molecules. *Proc. Natl. Acad. Sci. USA* **85**:7274–7278.

Melera, P. W., Wolgemuth, D., Biedler, J. L., & Hession, C. (1980). Antifolate-resistant Chinese hamster cells: Evidence from independently derived sublines for the overproduction of two dihydrofolate reductases encoded by different mRNAs. *J. Biol. Chem.* **255**:319–322.

Menasche, P. (2004). Skeletal myoblast transplantation for cardiac repair. *Expert Rev. Cardiovasc. Ther.* **2**:21–28.

Menasche, P., Hagege, A. A., Scorsin, M., Pouzet, B., Desnos, M., Duboc, D., Schwartz, K., Vilquin, J. T., & Marolleau, J. P. (2001). Myoblast transplantation for heart failure. *Lancet*, **357**:279–280.

Ment, L. R., Stewart, W. B., Scaramuzzino, D., & Madri, J. A. (1997). An in vitro three-dimensional coculture model of cerebral microvascular angiogenesis and differentiation. *In Vitro Cell Dev. Biol. Anim.* **33**:684–691.

Merten, O. W. (1999). Safety issues of animal products used in serum-free media. *Dev. Biol. Stand.* **99**:167–180.

Messing, E. M., Fahey, I. L., deKernion, I. B., Bhuta, S. M., & Bubbers, I. E. (1982). Serum-free medium for the in vitro growth of normal and malignant urinary bladder epithelial cells. *Cancer Res.* **42**:2392–2397.

Metcalf, D. (1969). Studies on colony formation in vitro by mouse bone marrow cells. I. Continuous cluster formation and relation of clusters to colonies. *J. Cell Physiol.* **74**:323–332.

Metcalf, D. (1990). The colony stimulating factors. *Cancer* **65**:2185–2195.

Meyskens, F. L., & Fuller, B. B. (1980). Characterization of the effects of different retinoids on the growth and differentiation of a human melanoma cell line and selected subclones. *Cancer Res.* **40**:2194–2196.

Michalopoulos, G., & Pitot, H. C. (1975). Primary culture of parenchymal liver cells on collagen membranes: Morphological and biochemical observations. *Exp. Cell Res.* **94**:70–78.

Michler-Stuke, A., & Bottenstein, J. (1982). Proliferation of glial-derived cells in defined media. *J. Neurosci. Res.* **7**:215–228.

Midgley, C. A., Craig, A. L., Hite, J. P., & Hupp, T. R. (1998). Baculovirus expression and the study of the regulation of the tumor suppressor protein p53. In Ravid, K., & Freshney, R. I. (eds.), *DNA transfer to cultured cells.* New York: Wiley-Liss, pp. 27–54.

Miklic, S., Juric, D. M., & Caman-Krzan, M. (2004). Differences in the regulation of BDNF and NGF synthesis in cultured neonatal rat astrocytes. *Int. J. Dev. Neurosci.* **22**:119–130.

Mikulits, W., Dolznig, H., Edelmann, H., Sauer, T., Deiner, E. M., Ballou, L., Beug, H., & Mullner, E. W. (1997). Dynamics of cell cycle regulators: Artefact-free analysis by recultivation of cells synchronized by centrifugal elutriation. *DNA Cell Biol.* **16**:849–859.

Milanesi, E., Ajmone-Marsan, P., Bignotti, E., Losio, M. N., Bernardi, J., Chegdani, F., Soncini, M., & Ferrari, M. (2003). Molecular detection of cell line cross-contaminations using amplified fragment length polymorphism DNA fingerprinting technology. *In Vitro Cell. Dev. Biol. Anim.* **39**:124–130.

Miller, A. D., Garcia, J. V., von Suhr, N., Lynch, C. M., Wilson, C., & Eiden, M. V. (1991). Construction and properties of retrovirus packaging cells based on gibbon ape leukemia virus. *J. Virol.* **65**:2220–2224.

Miller, A. D., Garcia, J. V., von Suhr, N., Lynch, C. M., Wilson, C., & Eiden, M. V. (1991). Construction and properties of retrovirus packaging cells based on gibbon abe leukemia virus. *J. Virol.* **65**:2220–2224.

Miller, G. G., Walker, G. W. R., & Giblack, R. E. (1972). A rapid method to determine the mammalian cell cycle. *Exp. Cell Res.* **72**:533–538.

Miller, G., Lisco H., Kohn, H. I., Stitt, D., & Enders, J. F.. (1971). Establishment of cell lines from normal adult human blood leukocytes by exposure to Epstein–Barr virus and neutralization by human sera with Epstein–Barr virus antibody. *Proc. Soc. Exp. Biol. Med.* **137**:1459–1465.

Miller, R. G., & Phillips, R. A. (1969). Separation of cells by velocity sedimentation. *J. Cell Physiol.* **73**:191–201.

Mills, J. B., Rose, K. A., Sadagopan, N., Sahi, J., & de Morais, S. M. (2004). Induction of drug metabolism enzymes and MDR1 using a novel human hepatocyte cell line. *J. Pharmacol. Exp. Ther.* **309**:303–309.

Mills, K. J., Volberg, T. M., Nervi, C., Grippo, J. F., Dawson, M. I., & Jetten, A. M. (1996). Regulation of retinoid-induced differentiation in embryonal carcinoma PCC4, aza1R cells: Effects of retinoid-receptor selective ligands. *Cell Growth Differ.* **7**:327–337.

Milo, G. E., Ackerman, G. A., & Noyes, I. (1980). Growth and ultrastructural characterization of proliferating human keratinocytes in vitro without added extrinsic factors. *In Vitro* **16**:20–30.

Mirskey, R., Dubois, C., Morgan, L., & Jessen, K. R. (1990). O4 and A007 sufatide antibodies bind to embryonic Schwann cells prior to the appearance of galactocerebroside: Regulation by axon-Schwann cell signals and cyclic AMP. *Development* **109**:105–116.

Mitaka, T., Norioka, K.-I., & Mochizuki, Y. (1993). Redifferentiation of proliferated rat hepatocytes cultures in L15 medium supplemented with EGF and DMSO. *In Vitro Cell Dev. Biol.* **29A**:714–722.

Mitaka, T., Sattler, C. A., Sattler, G. L., Sargent, L. M., & Pitot, H. C. (1991). Multiple cell cycles occur in rat hepatocytes cultured in the presence of nicotinamide and epidermal growth factor. *Hepatology* **13**:21–30.

Mitalipov, S., & Wolf, D. (2009). Totipotency, pluripotency and nuclear reprogramming. *Adv. Biochem. Eng. Biotechnol.* **114**:185–199.

Mitelman, F., Johansson, B., & Mertens, F. (eds.). (2009). Mitelman database of chromosome aberrations in cancer. http://cgap.nci.nih.gov/Chromosomes/Mitelman.

Mitelman, F., (ed.) (1995). *ISCN. An international system for human cytogenetic nomenclature.* Basel: S. Karger.

Miura, M., Gronthos, S., Zhao, M., Lu, B., Fisher, L. W., Robey, P. G., & Shi, S. (2003). SHED: Stem cells from human exfoliated deciduous teeth. *Proc. Natl. Acad. Sci. USA* **100**:5807–5812.

Mohammad, R. M., Li, Y., Mohamed, A. N., Pettit, G. R., Adsay, V., Vaitkevicius, V. K, Al-Katib, A. M., & Sarkar, F. H. (1999). Clonal preservation of human pancreatic cell line derived from primary pancreatic adenocarcinoma. *Pancreas* **19**:353–361.

Moll, R., Franke, W. W., & Schiller, D. L. (1982). The catalog of human cytokeratins: Patterns of expression in normal epithelia, tumours and cultured cells. *Cell* **31**:11–24.

Monaghan, S. R., Kent, M. L., Watral, V. G., Kaufman, R. J., Lee, L. E., & Bols, N. C. (2009). Animal cell cultures in microsporidial research: their general roles and their specific use for fish microsporidia. *In Vitro Cell Dev. Biol. Anim.* **45**:135–147.

Monroy, B., Honiger, J., Darquy, S., & Reach, G. (1997). Use of polyethyleneglycol for porcine islet cryopreservation. *Cell Transplant.* **6**:613–621.

Monstein, H. J., Ohlsson, B., & Axelson, J. (2001). Differential expression of gastrin, cholecystokinin-A and cholecystokinin-B receptor mRNA in human pancreatic cancer cell lines. *Scand. J. Gastroenterol.* **36**:738–743.

Montagnier, L. (1968). Correlation entre la transformation des cellule BHK21 et leur resistance aux polysaccharides acides en milieu gélifié. *CR Acad. Sci.* **D267**:921–924.

Montesano, R., Matsumoto, K., Nakamura, T., & Orci, L. (1991). Identification of a fibroblast-derived epithelial morphogen as hepatocyte growth factor. *Cell* **67**:901–908.

Montesano, R., Soriano, J. V., Pepper, M. S., & Orci, L. (1997). Induction of epithelial branching tubulogenesis in vitro. *J. Cell Physiol.* **173**:152–161.

Moore, G. E., Gerner, R. E., & Franklin, H. A. (1967). Culture of normal human leukocytes. *J. Am. Med. Assoc.* **199**:519–524.

Morgan, D., Freshney, R. I., Darling, J. L., Thomas, D. G. T., & Celik, F. (1983). Assay of anticancer drugs in tissue culture: Cell cultures of biopsies from human astrocytoma. *Br. J. Cancer* **47**:205–214.

Morgan, J. E., Beauchamp, J. R., Pagel, C. N., Peckham, M., Ataliotis, P., Jat, P. S., Noble, M. D., Farmer, K., & Partridge, T A.. (1994). Myogenic cell lines derived from transgenic mice carrying a thermolabile T antigen: A model system for the derivation of tissue-specific and mutation-specific cell lines. *Dev. Biol.* **162**:486–498.

Morgan, J. E., Moore, S. E., Walsh, F. S., & Partridge, T. A. (1992). Formation of skeletal muscle in vivo from the mouse C2 cell line. *J. Cell Sci.* **102**:779–787.

Morgan, J. G., Morton, H. J., & Parker, R. C. (1950). Nutrition of animal cells in tissue culture; I: Initial studies on a synthetic medium. *Proc. Soc. Exp. Biol. Med.* **73**:1.

Morton, H. J. (1970). A survey of commercially available tissue culture media. *In Vitro* **6**:89–108.

Moscona, A. A. (1952). Cell suspension from organ rudiments of chick embryos. *Exp. Cell Res.* **3**:535.

Moscona, A. A., & Piddington, R. (1966). Stimulation by hydrocortisone of premature changes in the developmental pattern of glutamine synthetase in embryonic retina. *Biochim. Biophys. Acta* **121**:409–411.

Moseley, G. W., Elliott, J., Wright, M. D., Partridge, L. J., & Monk, P. N. (2003). Interspecies contamination of the KM3 cell line: implications for CD63 function in melanoma metastasis. *Int. J. Cancer* **105**:613–616.

Mosmann, T. (1983). Rapid colorimetric assay for cellular growth and survival: Application to proliferation and cytotoxicity assays. *J. Immunol. Methods* **65**:55–63.

Mothersill, C., & Austin, B. (2000). *Aquatic invertebrate cell culture.* Chichester, UK: Springer/Praxis.

Moulton, D. G. (1974). Dynamics of cell populations in the olfactory epithelium. *Ann. NY Acad. Sci.* **237**:52–61.

Mowles, J. (1988). The use of ciprofloxacin for the elimination of mycoplasma from naturally infected cell lines. *Cytotechnology* **1**:355–358.

MRC. (2008). Responsibility in the use of animals in bioscience research: Expectations of the major research council and charitable funding bodies. www.mrc.ac.uk/Utilities/Documentrecord/index.htm?d=MRC001897.

MRC. (2009). Ethics and research governance. www.mrc.ac.uk/Newspublications/Publications/Ethicsandguidance/MRC003291.

Muffler, S., Stark, H.-J., Amoros, M., Falkowska-Hansen, B., Boehnke, K., Bühring, H. J., Marmé, A., Bickenbach, J. R., & Boukamp P. (2008). A stable niche supports long-term maintenance of human epidermal stem cells in organotypic cultures. *Stem Cells* **10**:2506–25015.

Muirhead, E. E., Rightsel, W. A., Pitcock, J. A., & Inagami, T. (1990). Isolation and culture of juxtaglomerular and renomedullary interstitial cells. *Methods Enzymol.* **191**:152–167.

Mullender, M., El Haj, A. J., Yang, Y., van Duin, M. A., Burger, E. H., & Klein-Nulend, J. (2004). Mechanotransduction of bone cells in vitro: Mechanobiology of bone tissue. *Med. Biol. Eng. Comput.* **42**:14–21.

Muller, U., Wang, D., Denda, S., Meneses, J. J., Pedersen, R. A., & Reichardt, L. F. (1997). Integrin alpha 8 beta 1 is critically important for epithelial–mesenchymal interactions during kidney morphogenesis. *Cell* **88**:603–613.

Mullin, J. M., Marano, C. W., Laughlin, K. V., Nuciglio, M., Stevenson, B. R., & Soler, A. P. (1997). Different size limitations for increased transepithelial paracellular solute flux across phorbol ester and tumour necrosis factor treated epithelial cell sheets. *J. Cell Physiol.* **171**:226–233.

Munthe-Kaas, A. C., & Seglen, P. O. (1974). The use of metrizamide as a gradient medium for isopycnic separation of rat liver cells. *FEBS Lett.* **43**:252–256.

Murakami, H. (1984). Serum-free cultivation of plasmacytomas and hybridomas. In Barnes, D. W., Sirbasku, D. A., & Sato, G. H. (eds.), *Methods for serum-free culture of neuronal and lymphoid cells.* New York: Liss, pp. 197–206.

Murakami, H., & Masui, H. (1980). Hormonal control of human colon carcinoma cell growth in serum-free medium. *Proc. Natl. Acad. Sci. USA* **77**:3464–3468.

Murao, S., Gemmell, M. A., & Callaghan, M. F. (1983). Control of macrophage cell differentiation in human promyelocytic HL-60 leukemia cells by 1,25-dihydroxyvitamin $D_3$ and phorbol-12-myristate-13-acetate. *Cancer Res.* **43**:4989–4996.

Murnane, J. P. (2006). Telomeres and chromosome instability. *DNA Repair (Amst.)* **5**:1082–1092.

Murphy, C. L., & Polak, J. M. (2004). Control of human articular chondrocyte differentiation by reduced oxygen tension. *J. Cell. Physiol.* **199**:451–459

Murphy, S. J., Watt, D. J., & Jones, G. E. (1992). An evaluation of cell separation techniques in a model mixed cell population. *J. Cell Sci.* **102**:789–798.

Naeyaert, J. M., Eller, M., Gordon, P. R., Park, H.-Y., & Gilchrest, B. A. (1991). Pigment content of cultured human melanocytes does not correlate with tyrosinase message level. *Br. J. Dermatol.* **125**:297–303.

Nakagawa, M., Koyanagi, M., Tanabe, K., Takahashi, K., Ichisaka, T., Aoi, T., Okita, K., Mochiduki, Y., Takizawa, N., & Yamanaka, S. (2007). Generation of induced pluripotent stem cells without Myc from mouse and human fibroblasts. *Nat. Biotechnol.* **26**:101–106.

Nardone R. M. (2008). Curbing rampant cross-contamination and misidentification of cell lines. *Biotechniques* **45**:221–227.

Nardone, R. M., Todd, G., Gonzalez, P., & Gaffney, E. V. (1965). Nucleoside incorporation into strain L cells: Inhibition by pleuropneumonia-like organisms. *Science* **149**:1100–1101.

NASA. (1997). Bioreactors workshop on regulation of cell and tissue differentiation: Proceedings. *In Vitro Cell. Dev. Biol. Anim.* **33**:325–405.

Nehls, V., & Drenckhahn, D. (1995). A novel, microcarrier-based in vitro assay for rapid and reliable quantification of three-dimensional cell migration and angiogenesis. *Microvasc Res.* **50**:311–322.

Nehring, D., Gonzalez, R., Pörtner, R., & Czermak, P. (2006). Experimental and modelling study of different process modes for retroviral production in a fixed bed reactor. *J. Biotechnol.* **122**:239–253.

Nelson, J. B., ed (1960). Biology of the pleuropneumonia-like organisms. *Ann. NY Acad. Sci.* **79**:305.

Nelson-Rees, W. A. (1979). Prostatic acid phosphatase in cell lines MA 160, EB33 and HeLa cells. *In Vitro* **15**:935.

Nelson-Rees, W. A., Daniels, D., & Flandermeyer, R. R. (1981). Cross-contamination of cells in culture. *Science* **212**:446–452.

Nelson-Rees, W., & Flandermeyer, R. R. (1977). Inter- and intraspecies contamination of human breast tumor cell lines HBC and BrCa5 and other cell cultures. *Science* **195**:1343–1344.

Neugut, A. I., & Weinstein, I. B. (1979). Use of agarose in the determination of anchorage-independent growth. *In Vitro* **15**:351.

Neumann. T., Hauschka, S. D., & Sanders, J. E. (2003). Tissue engineering of skeletal muscle using polymer fiber arrays. *Tissue Eng.* **9**:995–1003.

Neves, A. A., Medcalf, N., & Brindle, K. (2003). Functional assessment of tissue-engineered meniscal cartilage by magnetic resonance imaging and spectroscopy. *Tissue Eng.* **9**:51–62.

Newman, M. B., Davis, C. D., Kuzmin-Nichols, N., & Sanberg, P. R. (2003). Human umbilical cord blood (HUCB) cells for central nervous system repair. *Neurotox Res.* **5**:355–368.

Neyfakh, A. A. (1987). Use of fluorescent dyes as molecular probes for the study of multidrug resistance. *Exp. Cell Res.* **174**:168.

Nichols, J., & Ying, Q. L. (2006). Derivation and propagation of embryonic stem cells in serum- and feeder-free culture. *Methods Mol. Biol.* **329**:91–98.

Nichols, W. W., Murphy, D. G., & Christofalo, V. J. (1977). Characterization of a new diploid human cell strain, IMR-90. *Science* **196**:60–63.

Nicola, N. A. (1987). Hemopoietic growth factors and their interactions with specific receptors. *J. Cell Physiol. (Suppl.)* **5**:9–14.

Nicolson, G. L. (1976). Trans-membrane control of the receptors on normal and tumor cells; II: Surface changes associated with transformation and malignancy. *Biochim. Biophys. Acta* **458**:1–72.

Nicosia, R. F., & Ottinetti, A. (1990). Modulation of microvascular growth and morphogenesis by reconstituted basement membrane gel in three-dimensional cultures of rat aorta: A comparative study of angiogenesis in Matrigel, collagen, fibrin, and plasma clot. *In Vitro Cell Dev. Biol.* **26**:119–128.

Nicosia, R. F., Tchao, R., & Leighton, J. (1983). Angiogenesis-dependent tumor spread in reinforced fibrin clot culture. *Cancer Res.* **43**:2159–2166.

Nielsen, V. (1989). Vibration patterns in tissue culture vessels. *Nunc Bulletin 2* (May 1986; rev. March 1989). Roskilde, Denmark: A/S Nunc.

Nienow, A. W. (2006). Reactor engineering in large scale animal cell culture. *Cytotechnology* **50**:9–33.

NIH. (2007). Bioethics resources on the web. http://bioethics.od.nih.gov/IRB.html.

NIH OBA. (2009). Animal experiments covered under the NIH Guidelines for Research Involving Recombinant DNA molecules. http://oba.od.nih.gov/oba/ibc/FAQs/Animal%20Experiments%20Covered%20under%20the%20NIH%20Guidelines.pdf.

Niklason, L. E., Abbott, W., Gao, J., Klagges, B., Hirschi, K. K., Ulubayram, K., Conroy, N., Jones, R., Vasanawala, A., Sanzgiri, S., & Langer, R. (2001). Morphological and mechanical characteristics of engineered bovine arteries. *J. Vasc. Surg.* **33**(3): 628–638.

Nilos, R. M., & Makarski, J. S. (1978). Control of melanogenesis in mouse melanoma cells of varying metastatic potential. *J. Natl. Cancer Inst.* **61**:523–526.

Nilsson, J.-A., Zheng, X., Sundqvist, K., Liu, Y., Atzori, L., Elfwing, Å, Arvidson, K., & Grafström, R. C. (1998). Toxicity of formaldehyde to human oral fibroblasts and epithelial cells: Influences of culture conditions and role of thiol status. *J. Dental Res.* **77**:1896–1903.

Nims, R. W., Shoemaker, A. P., Bauternschub, M. A., Rec, L. J., & Harbell, J. W. (1998). Sensitivity of isoenzyme analysis for the detection of interspecies cell line cross-contamination. *In Vitro Cell Dev. Biol. Anim.* **34**:35–39.

Nishikawa, S., Goldstein, R. A., & Nierras, C. R. (2008). The promise of human induced pluripotent stem cells for research and therapy. *Nat. Rev. Mol. Cell. Biol.* **9**:725–729.

Nishizuka, S., Winokur, S. T., Simon, M., Martin, J., Tsujimoto, H., & Stanbridge, E. J. (2001). Oligonucleotide microarray expression analysis of genes whose expression is correlated with tumorigenic and non-tumorigenic phenotype of HeLa × human fibroblast hybrid cells. *Cancer Lett.* **165**:201–209.

Noble, M., & Barnett, S. C. (1996). Production and growth of conditionally immortal primary glial cell cultures and cell lines. In Freshney, R. I., & Freshney, M. G. (eds.), *Culture of immortalized cells*. New York: Wiley-Liss, pp. 331–366.

Noble, M., & Murrey, K. (1984). Purified astrocytes promote the in vitro division of a bipotential glial progenitor cell. *EMBO J.* **3**:2243–2247.

Norwood, T. H., Zeigler, C. J., & Martin, G. M. (1976). Dimethyl sulphoxide enhances polyethylene glycol-mediated somatic cell fusion. *Somatic Cell Genet.* **2**:263–270.

Nowak, J. A., & Fuchs, E. (2009). Isolation and culture of epithelial stem cells. *Methods Mol. Biol.* **482**:215–232.

NRC. (2009). *Biosafety in the laboratory: Prudent practices for the handling and disposal of infectious materials*. Washington, DC: National Academy Press. www.nap.edu/catalog.php?record_id=1197.

NSF. (1993). NSF/ANSI Standard 49. www.cdc.gov/OD/OHS/biosfty/bmbl/table3.htm.

Nuffield Council on Bioethics. (2009). Human tissue: Ethical and legal issues in. www.nuffieldbioethics.org.

Nurse, P. (1990). Universal control mechanism regulating onset of M-phase. *Nature* **344**:503–508.

Nuwaysir, E. F., Bittner, M., Trent, J., Barrett, J. C., & Afshari, C. A. (1999). Microarrays and toxicology: The advent of toxicogenomics. *Mol. Carcinog.* **24**:153–159.

O'Brien, S. J., Shannon, J. E., & Gail, M. H. (1980). Molecular approach to the identification and individualization of\ human and animal cells in culture: Isozyme and allozyme genetic signatures. *In Vitro* **16**:119–135.

O'Hare, M. J., Ellison, M. L., & Neville, A. M. (1978). Tissue culture in endocrine research: Perspectives, pitfalls, and potentials. *Curr. Top. Exp. Endocrinol.* **3**:1–56.

Obata, Y., Kono, T. & Hatada, I. (2002). Gene silencing: maturation of mouse fetal germ cells in vitro. *Nature* **418**: 497.

O'Brien, S. J., Menninger, J. C., & Nash, W. G. (2006). Atlas of mammalian chromosomes. Hoboken, NJ: Wiley.

OECD. (2009). Good laboratory practice. www.oecd.org/department/0,2688,en_2649_34381_1_1_1_1_1,00.html.

Oft, M., Akhurst, R. J., & Balmain, A. (2002). Metastasis is driven by sequential elevation of H-ras and Smad2 levels. *Nat. Cell Biol.* **4**:487–494.

Ogura, H., Fujii, R., Hamano, M., Kuzuya, M., Nakajima, H., Ohata, R., & Mori, T. (1997). Detection of HeLa cell contamination: Presence of human papillomavirus 18 DNA as HeLa marker in JTC-3, OG and OE cell lines. *Jpn. J. Med. Sci. Biol.* **50**:161–167.

Ogura, H., Yoshinouchi, M., Kudo, T., Imura, M., Fujiwara, T., & Yabe, Y. (1993). Human papillomavirus type-18 DNA in so-called HEP-2-cells. KB-cells and F1-cells: Further evidence that these cells are HeLa-cell derivatives. *Cell Mol. Biol.* **39**:463–467.

Oh, C. K., Kwon, Y. W., Kim, Y. S., Jang, H. S., & Kwon, K. S. (2003). Expression of basic fibroblast growth factor, vascular endothelial growth factor, and thrombospondin-1 related to microvessel density in nonaggressive and aggressive basal cell carcinomas. *J. Dermatol.* **30**:306–313.

Oh, Y. S., Kim, E. J., Schaffer, B. S., Kang, Y. H., Binderup, L., MacDonald, R. G., & Park, J. H. (2001). Synthetic low-calcaemic vitamin $D_3$ analogues inhibit secretion of insulin-like growth factor II and stimulate production of insulin-like growth factor-binding protein-6 in conjunction with growth suppression of HT-29 colon cancer cells. *Mol. Cell Endocrinol.* **183**:141–149.

Ohmichi, H., Koshimizu, U., Matsumoto, K., & Nakamura, T. (1998). Hepatocyte growth factor (HGF) acts as a mesenchyme-derived morphogenic factor during fetal lung development. *Development* **125**:1315–1324.

Ohno, T., Saijo-Kurita, K., Miyamoto-Eimori, N., Kurose, T., Aoki, Y., & Yosimura, S. (1991). A simple method for in situ freezing of anchorage-dependent cells including rat liver parenchymal cells. *Cytotechnology* **5**:273–277.

OHRP. (2002). Guidance for investigators and institutional review boards regarding research involving human embryonic stem cells, germ cells, and cell-derived test articles. Office for Human Research Protections, Department of Health and Human Services. www.hhs.gov/ohrp/humansubjects/guidance/stemcell .pdf.

OHRP. (2010). Common Rule. Federal Policy for the Protection of Human Subjects http://www.dhhs.gov/ohrp/policy/common.html.

Oie, H. K., Russell, E. K., Carney, D. N., & Gazdar, A. F. (1996). Cell culture methods for the establishment of the NCI series of lung cancer cell lines. *J. Cell Biochem.* **24** (suppl): 24–31.

OLAW. (2002). Institutional animal care and use committee guidebook. Office of Laboratory Animal Welfare and Applied Research Ethics National Association http://grants.nih.gov/grants/olaw/GuideBook.pdf;

Olie, R. A., Looijenga, L. H. J., Dekker, M. C., de Jong, F. H., van Dissel-Emiliani, F. M. F., de Rooij, D. G., van der Holt, B., & Oosterhuis, J. W. (1995). Heterogeneity in the in vitro survival and proliferation of human seminoma cells. *Br. J. Cancer* **71**:13–17.

Olsson, I., & Ologsson, T. (1981). Induction of differentiation in a human promyelocytic leukemic cell line (HL-60). *Exp. Cell Res.* **131**:225–230.

Ootani, A., Li, X., Sangiorgi, E., Ho, Q. T., Ueno, H., Toda, S., Sugihara, H., Fujimoto, K., Weissman, I. L., Capecchi, M. R., & Kuo, C. J. (2009). Sustained in vitro intestinal epithelial culture within a Wnt-dependent stem cell niche. *Nat. Med.* **15**:701–706.

Orellana, S. A., Neff, C. D., Sweeney, W. E. & Avner, E. D. (1996). Novel Madin Darby canine kidney cell clones exhibit unique phenotypes in response to morphogens. *In Vitro Cell Dev. Biol. Anim.* **32**:329–339.

Orly, J., Sato, G., & Erickson, G. F. (1980). Serum suppresses the expression of hormonally induced function in cultured granulosa cells. *Cell* **20**:817–827.

Osborne, C. K., Hamilton, B., Tisus, G., & Livingston, R. B. (1980). Epidermal growth factor stimulation of human breast cancer cells in culture. *Cancer Res.* **40**:2361–2366.

Osborne, H. B., Bakke, A. C., & Yu, J. (1982). Effect of dexamethasone on HMBA-induced Friend cell erythrodifferentiation. *Cancer Res.* **42**:513–518.

Osborne, R., Durkin, T., Shannon, H., Dornan, E., & Hughes, C. (1999). Performance of open-fronted microbiological safety cabinets: The value of operator protection tests during routine servicing. *J. Appl. Microbiol.* **86**:962–970.

OSHA. (2009). *The essential health & safety manual.* www.osha-occupational-health-and-safety.com.

Ostertag, W., & Pragnell, I. B. (1978). Changes in genome composition of the Friend virus complex in erythroleukaemia cells during the course of differentiation induced by DMSO. *Proc. Natl. Acad. Sci. USA* **75**:3278–3282.

Ostertag, W., & Pragnell, I. B. (1981). Differentiation and viral involvement in differentiation of transformed mouse and rat erythroid cells. *Curr. Topics Microbiol. Immunol.* **94/95**:143–208.

O'Toole, C. M., Povey, S., Hepburn, P., & Franks, L. M. (1983). Identity of some human bladder cancer cell lines. *Nature* **301**:429–430.

Owens, R. B., Smith, H. S., & Hackett, A. J. (1974). Epithelial cell culture from normal glandular tissue of mice: Mouse epithelial cultures enriched by selective trypsinisation. *J. Natl. Cancer Inst.* **53**:261–269.

Paddenberg, R., Wulf, S., Weber, A., Heimann, P., Beck, L., & Mannherz, H. G. (1996). Internucleosomal DNA fragmentation in cultured cells under conditions reported to induce apoptosis may be caused by mycoplasma endonucleases. *Eur. J. Cell Biol.* **71**:105–119.

Pantel, K., Dickmanns, A., Zippelius, A., Klein, C., Shi, J., Hoechtlen-Vollmar, W., Schlimok, G., Weckermann, D., Oberneder, R., Fanning, E., & Rietmüller, G. (1995). Establishment of micrometastatic cell lines: A novel source of tumor cell vaccines. *J. Natl. Cancer Inst.* **87**:1162–1168.

Paquet-Durand, F., Tan, S, & Bicker, G. (2004). Turning teratocarcinoma cells into neurons: rapid differentiation of NT-2 cells in floating spheres. *Brain Res. Dev. Brain Res.* **142**:161–167.

Paquette, J. S., Tremblay, P., Bernier, P. V., Auger, F. A., Laviolette, M., Germain, L., Boutet, M., Boulet, L. P., & Goulet, F. (2003). Production of tissue-engineered three-dimensional human bronchial models. *In Vitro Cell. Dev. Biol. Anim.* **39**:213–220.

Paraskeva, C., & Williams, A. C. (1992). The colon. In Freshney, R. I. (ed.), *Culture of epithelial cells.* New York: Wiley-Liss, pp. 82–105.

Paraskeva, C., Buckle, B. G., & Thorpe, P. E. (1985). Selective killing of contaminating human fibroblasts in epithelial cultures derived from colorectal tumors using an anti-Thy-1 antibody-ricin conjugate. *Br. J. Cancer* **51**:131–134.

Paraskeva, C., Buckle, B. G., Sheer, D., & Wigley, C. B. (1984). The isolation and characterisation of colorectal epithelial cell lines at different stages in malignant transformation from familial polyposis coli patients. *Int. J. Cancer* **34**:49–56.

Pardee, A. B., Cherington, P. V., & Medrano, E. E. (1984). On deciding which factors regulate cell growth. In Barnes, D. W., Sirbasku, D. A., & Sato, G. H. (eds.), *Methods for serum-free culture of epithelial and fibroblastic cells*. New York: Liss, pp. 157–166.

Park, H. Y., Perez, J. M., Laursen, R., Hara, M., & Gilchrest, B. A. (1999). Protein kinase C-beta activates tyrosinase by phosphorylating serine residues in its cytoplasmic domain. *J. Biol. Chem.* **274**:16470–16478.

Park, H. Y., Russakovsky, V., Ohno, S., & Gilchrest, B. A. (1993). The β isoform of protein kinase C stimulates melanogenesis by activating tyrosinase in pigment cells. *J. Biol. Chem.* **268**:11742–11749.

Park, J. G., & Gazdar, A. F. (1996). Biology of colorectal and gastric cancer cell lines. *J. Cell Biochem.* **24** (suppl.): 131–141.

Park, J.-G., Ku, J.-L., Kim, H.-S., Park, S.-Y., & Rutten, M. J (2004). Culture of normal and malignant gastric epithelium. In Pfragner, R., & Freshney. R. I. (eds.), *Culture of human tumor cells*. Hoboken, NJ: Wiley-Liss, pp. 23–66.

Parker, R. C. (1938). *Methods of tissue culture*. London: Hamish Hamilton Medical Books.

Parker, R. C. (1961). *Methods of tissue culture*, 3rd ed. London: Pitman Medical.

Parker, R. C., Castor, L. N., & McCulloch, E. A. (1957). Altered cell strains in continuous culture. Special publications. *NY Acad. Sci.* **5**:303–313.

Parker, R. C., Healy, G. M., & Fisher, D. C. (1954). Nutrition of animal cells in tissue culture. VII. Use of replicate cell cultures in the evaluation of synthetic media. *Canad. J. Biochem. Physiol.* **32**:306.

Parkinson, E. K., & Yeudall, W. A. (2002). The epidermis. In Freshney, R. I., & Freshney, M. G. (eds.), *Culture of epithelial cells*, 2nd ed. New York: Wiley-Liss, pp. 65–94.

Parums, D. V., Cordell, J. L., Micklem, K., Heryet, A. R., Gatter, K. C., and Mason, D. Y. (1990). JC70: A new monoclonal antibody that detects vascular endothelium associated antigen on routinely processed tissue sections. *J. Clin. Pathol.* **43**:752–757

Pasch, J., Schiefer, A., Heschel, I., Dimoudis, N., & Rau, G. (2000). Variation of the HES concentration for the cryopreservation of keratinocytes in suspensions and in monolayers. *Cryobiology* **41**:89–96.

Patel, K., Moore, S. E., Dickinson, G., Rossell, R. J., Beverley, P. C., Kemshead, J. T. & Walsh, F. S. (1989). Neural cell adhesion molecule (NCAM) is the antigen recognised by monoclonal antibodies of similar specificity in small-cell lung carcinoma and neuroblastoma. *Int. J. Cancer* **44**:573–578.

Patel, V. A., Logan, A., Watkinson, J. C., Uz-Zaman, S., Sheppard, M. C., Ramsden, J. D., & Eggo, M. C. (2009). Retraction of Patel Logan, A., Watkinson, J. C., Uz-Zaman, S., Sheppard, M. C., Ramsden, J. D., & Eggo, M. C. Isolation and characterization of human thyroid endothelial cells. *Am J Physiol Endocrinol Metab*. **284**:E168–E176. *Am J Physiol Endocrinol Metab.*, **297**:E562.

Paul, J. (1975). *Cell and tissue culture*. Edinburgh: Churchill Livingstone.

Paul, J., Conkie, D., & Freshney, R. I. (1969). Erythropoietic cell population changes during the hepatic phase of erythropoiesis in the foetal mouse. *Cell Tissue Kinet.* **2**:283–294.

Paul, J., & Pearson, E. S. (1960). The action of insulin on the metabolism of cell cultures. *J. Endocrin.* **21**:287–294.

Pavelic, K., Antonic, M., Pavelic, L., Pavelic, J., Pavelic, Z., & Spaventi, S. (1992). Human lung cancers growing on extracellular matrix: Expression of oncogenes and growth factors. *Anticancer Res.* **12**:2191–2196.

Peat, N., Gendler, S. J., Lalani, N., Duhig, T., & Taylor-Papadimitriou, J. (1992). Tissue-specific expression of a human polymorphic epithelial mucin (MUC1) in transgenic mice. *Cancer Res.* **52**:1954–1960.

Pedersen, N., Mortensen, S., Sorensen, S. B., Pedersen, M. W., Rieneck, K., Bovin, L. F., & Poulsen, H. S. (2003). Transcriptional gene expression profiling of small cell lung cancer cells. *Cancer Res.* **63**(8): 1943–1953.

Peehl, D. M. (2002). Human prostatic epithelial cells. In Freshney, R. I. & Freshney, M. G., (eds.), *Culture of epithelial cells*, 2nd ed., Hoboken, NJ: Wiley-Liss, pp. 171–194.

Peehl, D. M., & Ham, R. G. (1980). Clonal growth of human keratinocytes with small amounts of dialysed serum. *In Vitro* **16**:526–540.

Peehl, D. M., & Stanbridge, E. J. (1981). Anchorage-independent growth of normal human fibroblasts. *Proc. Natl. Acad. Sci. USA* **78**:3053–3057.

Pegolo, G., Askanas, V., & Engel, W. K. (1990). Expression of muscle-specific isozymes of phosphorylase and creatine kinase in human muscle fibers cultured aneurally in serum-free, hormonally/chemically enriched medium. *Int. J. Dev. Neurosci.* **8**:299–308.

Peppelenbosch, M. P., Tertoolen, L. G. J., DeLaat, S. W., & Zivkovic, D. (1995). Ionic responses to epidermal growth factor in zebrafish cells. *Exp. Cell Res.* **218**:183–188.

Pereira, M. E. A., & Kabat, E. A. (1979). A versatile immunoadsorbent capable of binding lectins of various specificities and its use for the separation of cell populations. *J. Cell Biol.* **82**:185–194.

Pereira-Smith, O., & Smith, J. (1988). Genetic analysis of indefinite division in human cells: Identification of four complementation groups. *Proc. Natl. Acad. Sci. USA* **85**:6042–6046.

Perin, L., Giuliani, S., Jin, D., Sedrakyan, S., Carraro, G., Habibian, R., Warburton, D., Atala, A., & De Filippo, R. E. (2007). Renal differentiation of amniotic fluid stem cells. *Cell Prolif*. **40**:936–948.

Perl, A.-K., Wilgenbus, P., Dahl, U., Semb, H., & Christofori, G. (1998). A causal role for E-cadherin in the transition from adenoma to carcinoma. *Nature* **392**:190–193.

Perry, P., & Wolf, S. (1974). New Giemsa method for the differential staining of sister chromatids. *Nature* **251**:156–158.

Persad, S., & Dedhar, S. (2003). The role of integrin-linked kinase (ILK) in cancer progression. *Cancer Metastasis Rev.* **22**:375–384.

Pertoft, H., & Laurent, T. C. (1982). Sedimentation of cells in colloidal silica (Percoll). In Pretlow, T. G., & Pretlow, T. P. (eds.), *Cell separation, methods and selected applications*, Vol. **1**. New York: Academic Press, pp. 115–152.

Peters, D. M., Dowd, N., Brandt, C., & Compton, T. (1996). Human papilloma virus E6/E7 genes can expand the lifespan of human corneal fibroblasts. *In Vitro Cell Dev. Biol. Anim.* **32**:279–284.

Petersen, O. W., van Deurs, B., Nielsen, K. V., Madsen, M. W., Laursen, I., Balslev, I., & Briand, P. (1990). Differential tumorigenicity of two autologous human breast carcinoma cell lines, HMT-3909S1 and HMT-3909S8, established in serum-free medium. *Cancer Res.* **50**:1257–1270.

Petersen, O. W., van Deurs, B., Nielsen, K. V., Madsen, M. W., Laursen, I., Balslev, I., & Briand, P. (1990). Differential tumorigenicity of two autologous human breast carcinoma cell lines, HMT-3909S1 and HMT-3909S8, established in serum-free medium. *Cancer Res.* **50**:1257–1270.

Petersen, D. F., Anderson, E. C., & Tobey, R. A. (1968). Mitotic cells as a source of synchronized cultures. In Prescott, D. M. (ed.), *Methods in cell physiology*. New York: Academic Press, pp. 347–370.

Petersen, O. W., Gudjonsson, T., Villadsen, R., Bissell, M. J., & Ronnov-Jessen, L. (2003). Epithelial progenitor cell lines as models of normal breast morphogenesis and neoplasia. *Cell Prolif.* **36** (suppl.)1: 33–44.

Petit, B., Masuda, K., d'Souza, A., Otten, L., Pietryla, D., Hartmann, D. J., Morris, N. P., Uebelhart, D., Schmid, T. M., & Thonar, E. J.-M. A. (1996). Characterization of crosslinked collagens synthesized by mature articular chondrocytes cultures in alginate beads: Comparison of two distinct matrix compartments. *Exp. Cell Res.* **225**:151–161.

Pevny, L., & Rao, M. S. (2003). The stem-cell menagerie. *Trends Neurosci.* **26**:351–359.

Pfeffer, L. M., & Eisenkraft, B. L. (1991). The antiproliferative and antitumour effects of human alpha interferon on cultured renal carcinomas correlate with the expression of a kidney-associated differentiation antigen. *Interferons Cytokines* **17**:30–31.

Pfragner, R., & Freshney, R. I. (eds.) (2004). *Culture of human tumor cells*. Hoboken, NJ: Wiley-Liss.

Phillips, P. D., & Cristofalo, V. J. (1988). Classification system based on the functional equivalency of mitogens that regulate WI-38 cell proliferation. *Exp. Cell Res.* **175**:396–403.

Phucharoen, J., Ohta, Y., Woo, J. M., Eisele, D. W., & Tetsu, O. (2009). Genetic profiling reveals cross-contamination and misidentification of 6 adenoid cystic carcinoma cell lines: ACC2, ACC3, ACCM, ACCNS, ACCS and CAC2. *PLoS One* **4**:e6040.

Pickering, S. J., Braude, P. R., Patel, M., Burns, C. J., Trussler, J., Bolton, V., & Minger, S. (2003). Preimplantation genetic diagnosis as a novel source of embryos for stem cell research. *Reprod Biomed Online* **7**:353–364.

Pignata, S., Maggini, L., Zarrilli, R., Rea, A., & Acquaviva, A. M. (1994). The enterocyte-like differentiation of the Caco-2 tumour cell line strongly correlates with responsiveness to cAMP and activation of kinase A pathway. *Cell Growth Differ.* **5**:967–973.

Pignatelli, M., & Bodmer, W. F. (1988). Genetics and biochemistry of collagen binding-triggered glandular differentiation in a human colon carcinoma cell line. *Proc. Natl. Acad. Sci. USA* **85**:5561–5565.

Pinnel, S. R., Murad, S., & Darr, D. (1987). Induction of collagen synthesis by ascorbic acid: A possible mechanism. *Arch Dermatol.* **123**:1684–1686.

Pitot, H., Periano, C., Morse, P., & Potter, V. R. (1964). Hepatomas in tissue culture compared with adapting liver in vitro. *Natl. Cancer Inst. Monogr.* **13**:229–245.

Piyathilake, C. J., Frost, A. R, Manne, U., Weiss, H., Bell, W. C., Heimburger, D. C., & Grizzle, W. E. (2002). Differential expression of growth factors in squamous cell carcinoma and precancerous lesions of the lung. *Clin. Cancer Res.* **8**:734–744.

Pizzonia, J. H., Gesek, F. A., Kennedy, S. M., Coutermarach, B. A., Bacskal, B. J., & Friedman, P. A. (1991). Immunomagnetic separation, primary culture, and characterisation of cortical thick ascending limb plus distal convoluted tubule cells from mouse kidney. *In Vitro Cell Dev. Biol.* **27A**:409–416.

Planas-Silva, M. D., & Weinberg, R. A. (1997). The restriction point and control of cell proliferation. *Curr. Opin. Cell Biol.* **9**:768–772.

Planz, B., Wang, Q., Kirley, S. D., Lin, C. W., & McDougal, W. S. (1998). Androgen responsiveness of stromal cells of the human prostate: Regulation of cell proliferation and keratinocyte growth factor by androgen. *J. Urol.* **160**:1850–1855.

Platsoucas, C. D., Good, R. A., & Gupta, S. (1979). Separation of human lymphocyte-T subpopulations (T-mu, T-gamma) by density gradient electrophoresis. *Proc. Natl. Acad. Sci. USA* **76**:1972.

Ploss, A., Khetani, S. R., Jones, C. T., Syder, A. J., Trehan, K., Gaysinskaya, V. A., Mu, K., Ritola, K., Rice, C. M., & Bhatia, S. N. (2010). Persistent hepatitis C virus infection in microscale primary human hepatocyte cultures. *Proc Natl Acad Sci U S A.* **107**:3141–3145.

Plumb, J. A., Milroy, R., & Kaye, S. B. (1989). Effects of the pH dependence of 3-(4,5-dimethylthiazol-2-yl)-2,5-diphenyltetrazolium bromide-formazan absorption on chemosensitivity determined by a novel tetrazolium-based assay. *Cancer Res.* **49**:4435–4440.

Polak, J. M., & Van Noorden, S. (2003). *Immunocytochemistry*. Oxford: Bios Scientific Publishers.

Pollack, M. S., Heagney, S. D., Livingston, P. O., & Fogh, J. (1981). HLA-A, B, C & DR alloantigen expression on forty-six cultured human tumor cell lines. *J. Natl. Cancer Inst.* **66**:1003–1012.

Pollack, R. (1981). *Readings in mammalian cell culture*, 2nd ed. Cold Spring Harbor, NY: Cold Spring Harbor Laboratory Press.

Pollard, J. W., & Walker, J. M. (eds.). (1990). *Animal cell culture: Methods in molecular biology*, Vol. **5**. Clifton, NJ: Humana Press.

Pollard, T. D., & Borisy, G. G. (2003). Cellular motility driven by assembly and disassembly of actin filaments. *Cell* **112**:453–465.

Pollock, G. S., Franceschini, I. A., Graham, G., & Barnett, S. C. (1999). Neuregulin is a mitogen and survival factor for olfactory bulb ensheathing cells and a related isoform is produced by astrocytes. *Eur. J. Neurosci.* **11**:769–780.

Pollock, M. F., & Kenny, G. E. (1963). Mammalian cell cultures contaminated with pleuro-pneumonia-like organisms; III: Elimination of pleuro-pneumonia-like organisms with specific antiserum. *Proc. Soc. Exp. Biol. Med.* **112**:176–181.

Polnaszek, N., Kwabi-Addo, B., Wang, J., & Ittmann, M. (2004). FGF17 is an autocrine prostatic epithelial growth factor and is upregulated in benign prostatic hyperplasia. *Prostate* **60**:18–24.

Pontecorvo, G. (1975). Production of mammalian somatic cell hybrids by means of polyethylene glycol treatment. *Somat. Cell Genet.* **1**:397–400.

Pontén, J., & Macintyre, E. H. (1968). Long term culture of normal and neoplastic human glia. *Acta Pathol. Microbiol. Scand.* **74**(4): 465–486.

Pontén, J. (1975). Neoplastic human glia cells in culture. In Fogh, J. (ed.), *Human tumor cells in vitro*. New York: Plenum Press, pp. 175–185.

Pontén, J., & Westermark, B. (1980). Cell generation and aging of nontransformed glial cells from adult humans. In Fedorof, S., & Hertz, L. (eds.), *Advances in cellular neurobiology*, Vol. **1**. New York: Academic Press, pp. 209–227.

Pontier, S. M., & Muller, W. J. (2009). Integrins in mammary-stem-cell biology and breast-cancer progression: A role in cancer stem cells? *J. Cell Sci.* **122**:207–214.

Poot, M., Hoehn, H., Kubbies, M., Grossmann, A., Chen, Y., & Rabinovitch, P. S. (1994). Cell-cycle analysis using continuous bromodeoxyuridine labeling and Hoechst 33358-ethidium bromide bivariate flow cytometry. *Methods Cell Biol.* **41**:327–340.

Pörtner, R., & Platas Barradas, O. B. J. (2007). Cultivation of mammalian cells in fixed-bed reactors. *Anim. Cell Biotechnol. Methods Protocols* **24**:353–369.

Portnoy, J., Curran-Everett, D., & Mason, R. J. (2004). Keratinocyte growth factor stimulates alveolar type II cell proliferation through the extracellular signal-regulated kinase and phosphatidylinositol 3-OH kinase pathways. *Am. J. Respir. Cell Mol. Biol.* **30**:901–907.

Post, M., Floros, J., & Smith, B. T. (1984). Inhibition of lung maturation by monoclonal antibodies against fibroblast-pneumocyte factor. *Nature* **308**:284–286.

Potten, C. S., Booth, C., Tudor, G. L., Booth, D., Brady, G., Hurley, P., Ashton, G., Clarke, R., Sakakibara, S., & Okano, H. (2003). Identification of a putative intestinal stem cell and early lineage marker; musashi-1. *Differentiation* **71**:28–41.

Powell, C. A., Smiley, B. L., Mills, J. & Vandenburgh, H. H., (2002). Mechanical stimulation improves tissue-engineered human skeletal muscle. *Am. J. Physiol. Cell Physiol* **283**:C1557–C1565.

Powers, D. A. (1989). Fish as model systems. *Science* **246**:352–357.

Pragnell, I. B., Wright, E. G., Lorimore, S. A., Adam, J., Rosendaal, M., de Lamarter, J. F., Freshney, M., Eckmann, L., Sproul, A., & Wilkie, N. (1988). The effect of stem cell proliferation regulators demonstrated with an in vitro assay. *Blood* **72**:196–201.

Prasad, K. N., Edwards-Prasad, J., Ramanujam, S., & Sakamoto, A. (1980). Vitamin E increases the growth inhibitory and differentiating effects of tumour therapeutic agents on neuroblastoma and glioma cells in culture. *Proc. Soc. Exp. Biol. Med.* **164**:158–163.

Preissmann, A., Wiesmann, R., Buchholz, R., Werner, R. G., & Noe, W. (1997). Investigations on oxygen limitations of adherent cells growing on macroporous microcarriers. *Cytotechnology* **24**:121–134.

Pretlow, T. G., & Pretlow, T. P. (1989). Cell separation by gradient centrifugation methods. *Methods Enzymol.* **171**:462–482.

Pretlow, T. G., Delmoro, C. M., Dilley, G. G., Spadafora, C. G., & Pretlow, T. P. (1991). Transplantation of human prostatic carcinoma into nude mice in Matrigel. *Cancer Res.* **51**:3814–3817.

Prince, G. A., Jenson, A. B., Billups, L. C., & Notkins, A. L. (1978). Infection of human pancreatic beta cell cultures with mumps virus. *Nature* **271**:158–161.

Przyborski, S. (2007). Derivation and culture of human embryonal carcinoma cell lines. In Freshney, R. I., Stacey, G. N., & Auerbach, J. M. (eds.), *Culture of human stem cells*. Hoboken, NJ: Wiley, pp. 134–158.

Puck, T. T., & Marcus, P. I. (1955). A rapid method for viable cell titration and clone production with HeLa cells in tissue culture: The use of X-irradiated cells to supply conditioning factors. *Proc. Natl. Acad. Sci. USA* **41**:432–437.

Puck, T. T., Cieciura, S. J., & Robinson, A. (1958). Genetics of somatic mammalian cells; III: Long term cultivation of euploid cells from human and animal subjects. *J. Exp. Med.* **108**:945–956.

Pun KM, Albrecht C, Castella V, Fumagalli L. (2009). Species identification in mammals from mixed biological samples based on mitochondrial DNA control region length polymorphism. *Electrophoresis* **30**:1008–1014.

Punchard, N., Watson, D., Thomson, R., & Shaw, M. (1996). Production of immortal human umbilical vein endothelial cells. In Freshney, R. I., & Freshney, M. G. (eds.), *Culture of immortalized cells*. New York: Wiley-Liss, pp. 203–238.

Quarles, J. M., Morris, N. G., & Leibovitz, A. (1980). Carcinoembryonic antigen production by human colorectal adenocarcinoma cells in matrix-perfusion culture. *In Vitro* **16**:113–118.

Quax, P. H., Frisdal, E., Pedersen, N., Bonavaud, S., Thibert, P., Martelly, I., Verheijen, J. H., Blasi, F., & Barlovatz-Meimon, G. (1992). Modulation of activities and RNA level of the components of the plasminogen activation system during fusion of human myogenic satellite cells in vitro. *Dev. Biol.* **151**:166–175.

Quintanilla, M., Brown, K., Ramsden, M., & Balmain, A. (1986). Carcinogen specific mutation and amplification of Ha-ras during mouse skin carcinogenesis. *Nature* **322**:78–79.

Quon, M. J. (1998). Transfection of rat adipose cells by electroporation. In Ravid, K., & Freshney, R. I. (eds.), *DNA transfer to cultured cells*. New York: Wiley-Liss, pp. 93–110.

Rabito, C. A., Tchao, R., Valentich, J., & Leighton, J. (1980). Effect of cell substratum interaction of hemicyst formation by MDCK cells. *In Vitro* **16**:461–468.

Radisic, M., Park, H., Chen, F., Salazar-Lazzaro, J. E., Wang, Y., Dennis, R., Langer, R., Freed, L. E., & Vunjak-Novakovic, G. (2006). Biomimetic approach to cardiac tissue engineering: Oxygen carriers and channeled scaffolds. *Tissue Eng.* **12**:2077–2091.

Radisky, D. C., Hirai, Y., Bissell, M. J. (2003). Delivering the message: Epimorphin and mammary epithelial morphogenesis. *Trends Cell Biol.* **13**:426–434.

Rae, J. M., Creighton, C. J., Meck, J. M., Haddad, B. R., Johnson, M. D. (2007). MDA-MB-435 cells are derived from M14 melanoma cells: A loss for breast cancer, but a boon for melanoma research. *Breast Cancer Res. Treat.* **104**:13–19.

Raff, M. C. (1990). Glial cell diversification in the rat optic nerve. *Science* **243**:1450–1455.

Raff, M. C., Abney, E., Brockes, J. P., & Hornby-Smith, A. (1978). Schwann cell growth factors. *Cell* **15**:813–822.

Raff, M. C., Fields, K. L., Hakomori, S. L., Minsky, R., Pruss, R. M., & Winter, J. (1979). Cell-type-specific markers for distinguishing and studying neurons and the major classes of glial cells in culture. *Brain Res.* **174**:283–309.

Rainey, W. E., Saner, K., & Schimmer, B. P. (2004). Adrenocortical cell lines. *Mol. Cell Endocrinol.* **228**:23–38.

Rainov, N. G., & Ren, H. (2003). Clinical trials with retrovirus mediated gene therapy: What have we learned? *J. Neurooncol.* **65**:227–36.

Raiser, D. M., & Kim, C. F. (2009). Sca-1 and cells of the lung: A matter of different sorts. *Stem Cells* **27**(3): 606–611.

Raisman, G. (1985). Specialized neuroglial arrangement may explain the capacity of vomeronasal axons to reinnerveate central neurons. *Neuroscience* **14**:237–254.

Raisman, G. (2001) Olfactory ensheathing cells: Another miracle cure for spinal cord injury? *Nat. Rev. Neurosci.* **5**:369–375.

Rak, J., Mitsuhashi, Y., Erdos, V., Huang, S.-N., Filmus, J., & Kerbel, R. S. (1995). Massive programmed cell death in intestinal epithelial cells induced by three-dimensional growth conditions: Suppression by mutant c-H-ras oncogene expression. *J. Cell Biol.* **131**:1587–1598.

Ramaekers, F. C. S., Puts, J. J. G., Kant, A., Moesker, O., Jap, P. H. K., & Vooijs, G. P. (1982). Use of antibodies to intermediate filaments in the characterization of human tumors. *Cold Spring Harbor Symp. Quant. Biol.* **46**:331–339.

Ramon-Cueto, A., Plant, G. W., Avila, J., & Bunge, M. B. (1998). Long-distance axonal regeneration in the transected adult rat spinal cord is promoted by olfactory ensheathing glia transplants. *J. Neurosci.* **18**:3803–3815.

Ranscht, B., Clapshaw, P. A., Price, J., Noble, M., & Seifert, W. (1982). Development of oligodendrocytes and Schwann cells studied with monoclonal antibody against galactocerebroside. *Proc. Natl. Acad. Sci. USA* **79**:2709–2713.

Rattner, A., Sabido, O., Massoubre, C., Rascle, F., & Frey, J. (1997). Characterization of human osteoblastic cells: Influence of the culture conditions. *In Vitro Cell Dev. Biol. Anim.* **33**:757–762.

Ravid, K., & Freshney, R. I. (eds.). (1998). *DNA transfer to cultured cells*. New York: Wiley-Liss.

Ray, M. E., Yang, Z. Q., Albertson, D., Kleer, C. G., Washburn, J. G., Macoska, J. A., & Ethier, S. P. (2004). Genomic and expression analysis of the 8p11–12 amplicon in human breast cancer cell lines. *Cancer Res.* **64**:40–47.

Raz, A. (1982). B16 melanoma cell variants: Irreversible inhibition of growth and induction of morphologic differentiation by anthracycline antibiotics. *J. Natl. Cancer Inst.* **68**:629–638.

Reaven, G. M. (1995). The fourth musketeer: From Alexandre Dumas to Claude Bernard. *Diabetologia* **38**:3–13.

Recktenwald, D. (ed.) (1997). *Cell separation methods and applications*. New York: Decker.

Reddel, R., Ke, Y., Gerwin, B. I., McMenamin, M. G., Lechner, J. F., Su, R. T., Brash, D. E., Park, J. B., Rhim, J. S., & Harris, C. C. (1988). Transformation of human bronchial epithelial cells by infection with SV40 or adenovirus-12 SV40 hybrid virus or transfection via strontium phosphate coprecipitation with a plasmid containing SV40 early region genes. *Cancer Res.* **48**:1904–1909.

Reed, S. I. (2003). Ratchets and clocks: The cell cycle, ubiquitylation and protein turnover. *Nat. Rev. Mol. Cell Biol.* **4**(11): 855–864.

Reel, J. R., & Kenney, F. T. (1968). "Superinduction" of tyrosine transaminase in hepatoma cell cultures: Differential inhibition of synthesis and turnover by actinomycin D. *Proc. Natl. Acad. Sci. USA* **61**:200–206.

Reeves, M. E. (1992). A metastatic tumour cell line has greatly reduced levels of a specific homotypic cell adhesion molecule activity. *Cancer Res.* **52**:1546–1552.

Reid, L. M., & Rojkind, M. (1979). New techniques for culturing differentiated cells. Reconstituted basement membrane rafts. In Jakoby, W. B., & Pastan, I. H. (eds.) *Methods in enzymology; Vol 57: cell culture*. New York: Academic Press, pp. 263–278.

Reitzer, L. J., Wice, B. M., & Kennel, D. (1979). Evidence that glutamine, not sugar, is the major energy source for cultured HeLa cells. *J. Biol. Chem.* **254**:2669–2677.

Relou, I. A., Damen, C. A., van der Schaft, D. W., Groenewegen, G., & Griffioen, A. W. (1998). Effect of culture conditions on endothelial cell growth and responsiveness. *Tissue Cell* **30**:525–530.

Repesh, L. A. (1989). A new in vitro assay for quantitating tumor cell invasion. *Invas. Metast.* **9**:192–208.

Rheinwald, J. G., & Beckett, M. A. (1981). Tumorigenic keratinocyte lines requiring anchorage and fibroblast support cultured from human squamous cell carcinomas. *Cancer Res.* **41**:1657–1663.

Rheinwald, J. G., & Green, H. (1975). Serial cultivation of strains of human epidermal keratinocytes: The formation of keratinizing colonies from single cells. *Cell* **6**:331–344.

Rheinwald, J. G., & Green, H. (1977). Epidermal growth factor and the multiplication of cultured human keratinocytes. *Nature* **265**:421–424.

Richler, C., & Yaffe, D. (1970). The in vitro cultivation and differentiation capacities of myogenic cell lines. *Dev. Biol.* **23**:1–22.

Richmond, A., Lawson, D. H., Nixon, D. W. & Chawla, R. K. (1985). Characterization of autostimulatory and transforming growth factors from human melanoma cells. *Cancer Res.* **45**:6390–6394.

Rickwood, D., & Birnie, G. D. (1975). Metrizamide, a new density gradient medium. *FEBS Lett.* **50**:102–110.

Ribeiro. F. R., Meireles, A. M., Rocha, A. S., & Teixeira, M. R. (2008). Conventional and molecular cytogenetics of human non-medullary thyroid carcinoma: characterization of eight cell line models and review of the literature on clinical samples. *BMC Cancer* **16**(8):371.

Rieder, C. L, & Cole, R. W. (2002). Cold-shock and the mammalian cell cycle. *Cell Cycle.* **1**:169–175.

Rieske, P., Augelli, B. J., Stawski, R., Gaughan, J., Azizi, S. A., & Krynska, B. (2009). A population of human brain cells expressing phenotypic markers of more than one lineage can be induced in vitro to differentiate into mesenchymal cells. *Exp. Cell Res.* **315**:462–473.

Rindler, M. J., Chuman, L. M., Shaffer, L., & Saier, M. H. Jr. (1979). Retention of differentiated properties in an established dog kidney epithelial cell line (MDCK). *J. Cell Biol.* **81**:635–648.

Rippon, H. J., & Bishop, A. E. (2004). Embryonic stem cells. *Cell Prolif.* **37**:23–34.

Rizzino, A. (2002). Embryonic stem cells provide a powerful and versatile model system. *Vitam Horm.* **64**:1–42.

Robertson, E., Bradley, A., Kuehn, M., & Evans, M. (1986). Germline transmission of genes introduced into cultured pluripotential cells by retroviral vector. *Nature.* **323**:445–448.

Robertson, K. M., & Robertson, C. N. (1995). Isolation and growth of human primary prostate epithelial cultures. *Methods Cell Sci.* **17**:177–185.

Robey, R. W., To, K. K., Polgar, O., Dohse, M., Fetsch, P., Dean, M., & Bates, S. E. (2009). ABCG2: A perspective. *Drug Deliv. Rev.* **61**:3–13.

Rodbell, M. (1964). Metabolism of isolated fat cells: Effects of hormones on glucose metabolism and lipolysis. *J. Biol. Chem.* **239**:375–380.

Rodriguez, A. M., Elabd, C., Delteil, F., Astier, J., Vernochet, C., Saint-Marc, P., Guesnet, J., Guezennec, A., Amri, E. Z., Dani, C., & Ailhaud, G. (2004). Adipocyte differentiation of multipotent cells established from human adipose tissue. *Biochem. Biophys. Res. Commun.* **315**:255–263.

Rofstad, E. K. (1994). Orthotopic human melanoma xenograft model systems for studies of tumor angiogenesis, pathophysiology, treatment sensitivity and metastatic pattern. *Br. J. Cancer* **70**:804–812.

Rogers, A. W. (1979). *Techniques of autoradiography*, 3d ed. Amsterdam: Elsevier/North Holland Biomedical Press.

Roguet, R., Cohen, C., & Rougier, A. (1994). A reconstituted human epidermis to assess cutaneous irritation, photoirritation and photoprotection in vitro. *In Vitro Skin Toxicol.* **10**:141–149.

Rojkind, M., Gatmaitan, Z., Mackensen, S., Giambrone, M. A., Ponce, P., & Reid, L. M. (1980). Connective tissue biomatrix: Its isolation and utilization for long term cultures of normal rat hepatocytes. *J. Cell Biol.* **87**:255–263.

Romano, P., Manniello, A., Aresu, O., Armento, M., Cesaro, M., & Parodi, B. (2009). Cell line data base: Structure and recent improvements towards molecular authentication of human cell lines. *Nucl. Acids Res.* **37**:D925–D932.

Rooney, S. A., Young, S. L., & Mendelson, C. R. (1995). Molecular and cellular processing of lung surfactant. *FASEB J.* **8**:957–967.

Roper, S., & Hemberger, M. (2009). Defining pathways that enforce cell lineage specification in early development and stem cells. *Cell Cycle* **8**:1515–1525.

Rosenblatt, J. D., Lunt, A. I., Parry, D. J., & Partridge, T. A. (1995). Culturing satellite cells from living single muscle fiber explants. *In Vitro Cell. Dev. Biol.* **31**:773–779.

Rosenblatt, J. D., Parry, D. J., & Partridge, T. A. (1996). Phenotype of adult mouse muscle myoblasts reflects their fiber type of origin. *Differentiation* **60**:39–45.

Rosenfeld, M. A., Yoshimura, K., Trapnell, B. C., Yoneyama, K., Rosenthal, E. R., Dalemenas, W., Fukayama, M., Bargon, J., Stier, L. E., Stratford-Perricaudet, L., Perricaudet, M., Guggino, W. B., Pavirani, A., Lecocq, J.-P., & Crystal, R. G. (1992). In vivo transfer of the human cystic fibrosis transmembrane conductance regulator gene to the airway epithelium. *Cell* **68**:143–155.

Rosenman, S. J., & Gallatin, W. M. (1991). Cell surface glycoconjugates in intercellular and cell-substratum interactions. *Semin. Cancer Biol.* **2**:357–366.

Rossi, G. B., & Friend, C. (1967). Erythrocytic maturation of (Friend) virus-induced leukemic cells in spleen clones. *Proc. Natl. Acad. Sci. USA* **58**:1373–1380.

Rothblat, G. H., & Morton, H. E. (1959). Detection and possible source of contaminating pleuropneumonia-like organisms (PPLO) in culture of tissue cells. *Proc. Soc. Exp. Biol. Med.* **100**:87.

Rothfels, K. H., & Siminovitch, L. (1958). An air drying technique for flattening chromosomes in mammalian cells growth in vitro. *Stain Technol.* **33**:73–77.

Rotman, B., & Papermaster, B. W. (1966). Membrane properties of living mammalian cells as studied by enzymatic hydrolysis of fluorogenic esters. *Proc. Natl. Acad. Sci. USA* **55**:134–141.

Rous, P., & Jones, F. A. A. (1916). A method of obtaining suspensions of living cells from the fixed tissues, and for the plating out of individual cells. *J. Exp. Med.* **23**:555.

Rowley S. D., & Byrne, D. V. (1992). Low-temperature storage of bone marrow in nitrogen vapor-phase refrigerators: Decreased temperature gradients with an aluminum racking system. *Transfusion* **32**:750–754.

Rudland, P. S. (1992). Use of peanut lectin and rat mammary stem cell lines to identify a cellular differentiation pathway for the alveolar cell in the rat mammary gland. *J. Cell Physiol.* **153**:157–168.

Ruff, M. R., & Pert, C. B. (1984). Small cell carcinoma of the lung: Macrophage-specific antigens suggest hematopoietic stem cell origin. *Science* **225**:1034–1036.

Rules and Guidance for Pharmaceutical Manufacturers and Distributors, (2007). www.mhra.gov.uk/Publications/Regulatoryguidance/Medicines/index.htm.

Rundlett, S. E., Gordon, D. A., & Miesfeld, R. L. (1992). Characterisation of a panel of rat ventral prostate epithelial cell lines immortalized in the presence or absence of androgens. *Exp. Cell Res.* **203**:214–221.

Ruoff, N. M., & Hay, R. J. (1979). Metabolic and temporal studies on pancreatic exocrine cells in culture. *Cell Tissue Res.* **204**:243–252.

Rush, L. J., Heinonen, K., Mrozek, K., Wolf, B. J., Abdel-Rahman, M., Szymanska, J., Peltomaki, P., Kapadia, F., Bloomfield, C. D., Caligiuri, M. A., & Plass, C. (2002). Comprehensive cytogenetic and molecular genetic characterization of the TI-1 acute myeloid leukemia cell line reveals cross-contamination with K-562 cell line. *Blood* **99**:1874–1876.

Russo, J., Calaf, G., & Russo, I. H. (1993). A critical approach to the malignant transformation of human breast epithelial cells with chemical carcinogens. *Crit. Rev. Oncog.* **4**(**4**): 403–417.

Rutzky, L. P., Tomita, J. T., Calenoff, M. A., & Kahan, B. D. (1979). Human colon adenocarcinoma cells; III: In vitro organoid expression and carcino-embryonic antigen kinetics in hollow fiber culture. *J. Natl. Cancer Inst.* **63**:893–902.

Rygaard, K., Moller, C., Bock, E., & Spang-Thomsen, M. (1992). Expression of cadherin and NCAM in human small cell lung cancer cell lines and xenografts. *Br. J. Cancer* **65**:573–577.

Sá, R., Neves, R., Fernandes, S., Alves, C., Carvalho, F., Silva, J., Cremades, N., Malheiro, I., Barros, A., & Sousa, M. (2008). Cytological and expression studies and quantitative analysis of the temporal and stage-specific effects of follicle-stimulating hormone and testosterone during cocultures of the normal human seminiferous epithelium. *Biol. Reprod.* **79**:962–975.

Saalbach, A., Aust, G., Haustein, U. F., Herrmann, K., & Anderegg, U. (1997). The fibroblast-specific MAb AS02: A novel tool for detection and elimination of human fibroblasts. *Cell Tissue Res.* **290**:593–599.

Sachs, L. (1978). Control of normal cell differentiation and the phenotypic reversion of malignancy in myeloid leukaemia. *Nature* **274**:535–539.

Safford, K. M., & Rice, H. E. (2007). Tissue culture of adipose-derived stem cells. In Freshney, R. I, Stacey, G. N., & Auerbach, J. M. (eds.), *Culture of human stem cells*. Hoboken, NJ: Wiley, pp. 303–315.

Saier, M. H. (1984). Hormonally defined, serum free medium for a proximal tubular kidney epithelial cell line, LLC-PKI. In Barnes, W. D. (ed.), *Methods for serum free culture of epithelial and fibroblastic cells*. New York: Liss, pp. 25–31.

Sakaguchi, Y., Sekiya, I., Yagishita, K., Ichinose, S., Shinomiya, K., & Muneta, T. (2004). Suspended cells from trabecular bone by collagenase digestion become virtually identical to mesenchymal stem cells obtained from marrow aspirates. *Blood* **104**:2728–2735.

Sakaguchi, Y., Sekiya, I., Yagishita, K., & Muneta, T. (2005) Comparison of human stem cells derived from various mesenchymal tissues: Superiority of synovium as a cell source. *Arthritis Rheum.* **52**:2521–2529.

Sambrook, J. F., & Russell, D. (2006). *Condensed protocols from Molecular Cloning: A Laboratory Manual*. Cold Spring Harbor Press.

Sanchez-Ramos, J. R. (2002). Neural cells derived from adult bone marrow and umbilical cord blood. *J. Neurosci. Res.* **69**:880–893.

Sanford, K. K., Earle, W. R., & Likely, G. D. (1948). The growth in vitro of single isolated tissue cells. *J. Natl. Cancer Inst.* **9**:229.

Sanford, K. K., Earle, W. R., Evans, V. J., Waltz, H. K., & Shannon, I. E. (1951). The measurement of proliferation in tissue cultures by enumeration of cell nuclei. *J. Natl. Cancer Inst.* **11**:773.

Sanford, K. K., Westfall, B. B., Fioramonti, M. C., McQuilkin, W. T., Bryant, J. C., Peppers, E. V., & Earle, W. R. (1955). The effect of serum fractions on the proliferation of strain L mouse cells in vitro. *J. Natl. Cancer. Inst.* **16**:789.

Santamaria. D., & Ortega, S., (2006). Cyclins and CDKS in development and cancer: Lessons from genetically modified mice. *Front. Biosci.* **11**:1164–1188.

Santiago-Walker, A., Li, L., Haass, N. K., & Herlyn, M. (2009). Melanocytes: From morphology to application. *Skin Pharmacol Physiol.* **22**:114–121.

Sarang, Z., Haig, Y., Hansson, A., Vondracek, M., WärngAArd, L., & Grafström, R. C. (2003). Microarray assessment of fibronectin, collagen and integrin expression and the role of fibronectin-collagen coating in the growth of normal, SV40T-antigen-immortalized and malignant human oral keratinocytes. *ATLA* **31**:575–585.

Sasaki, M., Honda, T., Yamada, H., Wake, N., & Barrett, J. C. (1996). Evidence for multiple pathways to cellular senescence. *Cancer Res.* **54**:6090–6093.

Sato, G. H., & Yasumura, Y. (1966). Retention of differentiated function in dispersed cell culture. *Trans. NY Acad. Sci.* **28**:1063–1079.

Sattler, G. A., Michalopoulos, G., Sattler, G. L., & Pitot, H. C. (1978). Ultrastructure of adult rat hepatocytes cultured on floating collagen membranes. *Cancer Res.* **38**:1539–1549.

Saunders, N. A., Bernacki, S. H., Vollberg, T. M., & Jetten, A. M. (1993). Regulation of transglutaminase type-I expression in squamous differentiating rabbit tracheal epithelial cells and human epidermal keratinocytes: Effects of retinoic acid and phorbol esters. *Mol. Endocrinol.* **7**:387–398.

Scanlon, E. F., Hawkins, R. A., Fox, W. W., & Smith, W. S. (1965). Fatal homotransplanted melanoma (a case report). *Cancer* **18**:782–789.

Schaeffer, W. I. (1990). Terminology associated with cell, tissue and organ culture, molecular biology and molecular genetics. *In Vitro Cell Dev. Biol.* **26**:97–101.

Schäfer-Korting M, Bock, U., Diembeck, W., Düsing, H. J., Gamer, A., Haltner-Ukomadu, E., Hoffmann, C., Kaca, M., Kamp, H., Kersen, S., Kietzmann, M., Korting, H. C., Krächter, H. U., Lehr, C. M., Liebsch, M., Mehling, A., Müller-Goymann, C., Netzlaff, F., Niedorf, F., Rübbelke, M. K., Scähfer, U., Schmidt, E., Schreiber, S., Spielmann, H., Vuia, A., & Weimer, M. (2008). The use of reconstructed human epidermis for skin absorption testing: Results of the validation study. *Altern Lab Anim.* **36**:161–187.

Schaffer, K., Herrmuth, H., Mueller, J., Coy, D. H., Wong, H. C., Walsh, J. H., Classen, M., Schusdziarra, V., & Schepp, W. (1997). Bombesin-like peptides stimulate somatostatin release from rat fundic D cells in primary culture. *Am. J. Physiol. Gastrointest. Liver Physiol.* **273**:G686–G695.

Schamblott, M. J., Axelman, J., Sterneckert, J., Christoforou, N., Patterson, E. S., Siddiqi, M. A., Kahler, H., Ifeanyi, L. A., & Gearhart, J. D. (2002). Stem cell culture: Pluripotent stem cells. In. Atala, A., & Lanza, R. P. (eds.), *Methods of tissue engineering*. San Diego: Academic Press, pp. 411–420.

Schamblott, M. J., Axelman, J., Wang, S., Bugg, E. M., Littlefield, J. W., Donovan, P. J., Blumenthal, P. D., Huggins, G. R., & Gearhart, J. D (1998). Derivation of pluripotent stem cells from cultured human primordial germ cells. *Proc. Natl. Acad. Sci. USA* **95**:13726–13731.

Scher, W., Holland, J. G., & Friend, C. (1971). Hemoglobin synthesis in murine virus-induced leukemic cells in vitro; I: Partial purification and identification of hemoglobins. *Blood* **37**:428–437.

Schimmelpfeng, L., Langenberg, U., & Peters, I. M. (1968). Macrophages overcome mycoplasma infections of cells in vitro. *Nature* **285**:661.

Schlechte, W., Brattain, M., & Boyd, D. (1990). Invasion of extracellular matrix by cultured colon cancer cells: Dependence on urokinase receptor display. *Cancer Commun.* **2**:173–179.

Schlessinger, J., Lax, I., & Lemmon, M. (1995). Regulation of growth factor activation by proteoglycans: What is the role of the low affinity receptors? *Cell* **83**:357–360.

Schmidt, R., Reichert, U., Michel, S., Shrott, B., & Boullier, M. (1985). Plasma membrane transglutaminase and cornified envelope competence in cultured human keratinocytes. *FEBS Lett.* **186**:204.

Schneider, E. L., & Stanbridge, E. I. (1975). A simple biochemical technique for the detection of mycoplasma contamination of cultured cells. *Methods Cell Biol.* **10**:278–290.

Schoenlein, P. V., Shen, D.-W., Barrett, J. T., Pastan, I. T., & Gottesman, M. M. (1992). Double minute chromosomes carrying the human multidrug resistance 1 and 2 genes are generated from the dimerization of submicroscopic circular DNAs in colchicine-selected KB carcinoma cells. *Mol. Biol. Cell* **3**:507–520.

Schor, S. L. (1994). Cytokine control of cell motility modulation and mediation by the extracellular matrix. *Prog. Growth Factor Res.* **5**:223–248.

Schousboe, A., Thorbek, P., Hertz, L., & Krogsgaard-Larsen, P. (1979). Effects of GABA analogues of restricted conformation on GABA transport in astrocytes and brain cortex slices and on GABA receptor binding. *J. Neurochem.* **33**:181–189.

Schulman, H. M. (1968). The fractionation of rabbit reticulocytes in dextran density gradients. *Biochim. Biophys. Acta* **148**:251–255.

Schwartz Albiez, R., Heidtmann, H.-H., Wolf, D., Schirrmacher, V., & Moldenhauer, G. (1991). Three types of human lung tumour cell lines can be distinguished according to surface expression of endogenous urokinase and their capacity to bind exogenous urokinase. *Br. J. Cancer* **65**:51–57.

Schwartz, D., & Rotter, V. (1998). P53-dependent cell cycle control: response to genotoxic stress. *Semin. Cancer Biol.* **8**(5): 325–336.

Schweppe, R. E., Klopper, J. P., Korch, C., Pugazhenthi, U., Benezra, M., Knauf, J. A., Fagin, J. A., Marlow, L. A., Copland, J. A., Smallridge, R. C., Haugen, B. R. (2008). Deoxyribonucleic acid profiling analysis of 40 human thyroid cancer cell lines reveals cross-contamination resulting in cell line redundancy and misidentification. *J. Clin. Endocrinol. Metab.* **3**:4331–4341.

Schwob, J. (2002) Neuronal regeneration and the peripheral olfactory system. *Anat. Rec.* **269**:33–49.

Scott, W. N, McCool, K., & Nelson, J. (2000) Improved method for the production of gold colloid monolayers for use in the phagokinetic track assay for cell motility. *Anal. Biochem.* **287**:343–344.

Scotto, K. W., Biedler, I. L., & Melera, P. W. (1986). Amplification and expression of genes associated with multidrug resistance in mammalian cells. *Science* **232**:751–755.

Seeds, N. W. (1971). Biochemical differentiation in reaggregating brain cell culture. *Proc. Natl. Acad. Sci. USA* **68**:1858–1861.

SEFREC. (2005). Interactive database about serum free medias and cell lines. www.sefrec.com.

Seglen, P. O. (1975). Preparation of isolated rat liver cells. *Methods Cell Biol.* **13**:29–83.

Seidel, J. O., Pei, M., Gray, M. L., Langer, R., Freed, L. E., & Vunjak-Novakovic, G. (2004). Long-term culture of tissue engineered cartilage in a perfused chamber with mechanical stimulation. *Biorheology* **41**:445–458.

Seifert, H. H., Meyer, A., Cronauer, M. V., Hatina, J., Müller, M., Rieder, H., Hoffmann, M. J., Ackermann, R., Schulz, W. A. (2007). A new and reliable culture system for superficial low-grade urothelial carcinoma of the bladder. *World J. Urol.* **25**:297–302.

Seifert, W., & Müller, H. W. (1984). Neuron-glia interaction in mammalian brain: Preparation and quantitative bioassay of a neurotropic factor (NTF) from primary astrocytes. In Barnes, D. W., Sirbasku, D. A., & Sato, G. H. (eds.), *Methods for serum-free culture of neuronal and lymphoid cells.* New York: Liss, pp. 67–78.

Seigel, G. A. (1996). Establishment of an E1a-immortalised retinal cell culture. *In Vitro Cell Dev. Biol. Anim.* **32**:66–68.

Selby, P. J., Thomas, M. J., Monaghan, P., Sloane, J., & Peckham, M. J. (1980). Human tumour xenografts established and serially transplanted in mice immunologically deprived by thymectomy, cytosine arabinoside and whole-body irradiation. *Br. J. Cancer* **41**:52.

Selmi, T. A., Verdonk, P., Chambat, P., Dubrana, F., Potel, J. F., Barnouin, L., & Neyret, P. (2008). Autologous chondrocyte implantation in a novel alginate-agarose hydrogel: Outcome at two years. *J. Bone Joint Surg. Br.* **90**:597–604.

Serakinci, N., Christensen, R., Graakjaer, J., Cairney, C. J., Keith, W. N., Alsner, J., Saretzki, G., & Kolvraa, S. (2007). Ectopically hTERT expressing adult human mesenchymal stem cells are less radiosensitive than their telomerase negative counterpart. *Exp. Cell Res.* **313**:1056–1067.

Sérandour, A. L., Loyer, P., Garnier, D., Courselaud, B., Théret, N., Glaise, D., Guguen-Guillouzo, C., & Corlu, A. (2005). TNFalpha-mediated extracellular matrix remodeling is required for multiple division cycles in rat hepatocytes. *Hepatology* **41**:478–486.

Seruya, M., Shah, A., Pedrotty, D., du Laney, T., Melgiri, R., McKee, J. A., Young, H. E., & Niklason, L. E. (2004). Clonal population of adult stem cells: Life span and differentiation potential. *Cell Transplant.* **13**:93–101.

Shah, G. (1999). Why do we still use serum in the production of biopharmaceuticals? *Dev. Biol. Stand.* **99**:17–22.

Shall, S. (1973). Sedimentation in sucrose and Ficoll gradients of cells grown in suspension culture. In Kruse, P. F., & Patterson, M. K. (eds.), *Tissue culture methods and applications.* New York: Academic Press, pp. 198–204.

Shall, S., & McClelland, A. J. (1971). Synchronization of mouse fibroblast LS cells grown in suspension culture. *Nat. New Biol.* **229**:59–61.

Shaner, N. C., Steinbach, P. A. & Tsien, R. Y. (2005). A guide to choosing fluorescent proteins. *Nat. Methods* **2**:905–909.

Shansky, J., Chromiak, J., Del Tatto, M., & Vandenburgh, H. (1997). A simplified method for tissue engineering skeletal muscle organoids in vitro. *In Vitro Cell Dev. Biol. Anim.* **33**:659–661.

Shansky, J., Ferland, P., McGuire, S., Powell, C., DelTatto, M., Nackman, M., Hennessey, J., & Vandenburgh, H. H. (2006). Tissue engineering human skeletal muscle for clinical applications. In Vunjak-Novakovic, G., & Freshney, R. I. (eds.), *Culture of cells for tissue engineering.* Hoboken, NJ: Wiley-Liss, pp. 239–258.

Shapiro, H. (2003). *Practical flow cytometry.* Hoboken, NJ: Wiley.

Shaw, R. G., Johnson, A. R., Schulz, W. W., Zahlten, R. N., & Combes, B. (1984). Sinusoidal endothelial cells from normal guinea pig liver: Isolation, culture and characterization. *Hepatology* **4**:591–602.

Shay, J. W., & Wright, W. E. (1989). Quantitation of the frequency of immortalization of normal human diploid fibroblasts by SV40 large T antigen. *Exp. Cell Res.* **184**:109–118.

Shay, J. W., Wright, W. E., Brasiskyte, D., & der Haegen, A. (1993). E6 of human papillomavirus type 16 can overcome the M1 stage of immortalization in human mammary epithelial cells but not in human fibroblasts. *Oncogene* **8**:1407–1413.

Shea, C. M., Edgar, C. M., Teinhorn, T. A., Louis, C. & Gerstenfeld, L. C. (2003). BMP treatment of C3H10T1/2 mesenchymal stem cells induces both chondrogenesis and osteogenesis. *J. Cell. Biochem.* **90**:1112–1127.

Shiga, M., Kapila, Y. L., Zhang, Q., Hayami, T., & Kapila, S. (2003). Ascorbic acid induces collagenase-1 in human periodontal ligament cells but not in MC3T3-E1 osteoblast-like cells: potential association between collagenase expression and changes in alkaline phosphatase phenotype. *J. Bone Miner. Res.* **18**:67–77.

Shih, C., Shilo, B. Z., Goldfarb, M. P., Dannenberg, A., & Weinberg, R. A. (1979). Passage of phenotypes of chemically transformed cells via transfection of DNA and chromatin. *Proc. Natl. Acad. Sci. USA* **76**:5714–5718.

Shih, S. J., Dall'Era, M. A., Westphal, J. R., Yang, J., Sweep, C. G., Gandour-Edwards, R., & Evans, C. P. (2003). Elements regulating angiogenesis and correlative microvessel density in benign hyperplastic and malignant prostate tissue. *Prostate Cancer Prostatic Dis.* **6**:131–137.

Shipley, G. D., & Ham, R. G. (1983). Multiplication of Swiss 3T3 cells in a serum-free medium. *Exp. Cell Res.* **146**:249–260.

Shipman, C. (1973). Control of culture pH with synthetic buffers. In Kruse, P., & Patterson, M. K. (eds.), *Tissue culture methods and applications*. New York: Academic Press, p. 710.

Simara, P., Koutna, I., Stejskal, S., Krontorad, P., Rucka, Z., Peterkova, M., & Kozubek, M. (2009). Combination of mRNA and protein microarray analysis in complex cell profiling. *Neoplasma* **56**:141–149.

Simon-Assmann, P., Turck, N., Sidhoum-Jenny, M., Gradwohl, G., & Kedinger, M. (2007). In vitro models of intestinal epithelial cell differentiation. *Cell Biol Toxicol.* **23**:241–256.

Simon-Assmann, P., Kédinger, M., & Haffen, K. (1986). Immunocytochemical localization of extracellular matrix proteins in relation to rat intestinal morphogenesis. *Differentiation* **32**:59–66.

Simonsen, J. L., Rosada, C., Serakinci, N., Justesen, J., Stenderup, K., Rattan, S. I., Jensen, T. G., & Kassem, M. (2002). Telomerase expression extends the proliferative life-span and maintains the osteogenic potential of human bone marrow stromal cells. *Nat. Biotechnol.* **20**:592–596.

Simonsen, J. L., Rosada, C., Serakinci, N., Justesen, J., Stenderup, K., Rattan, S. I., Jensen, T. G., & Kassem, M. (2002). Telomerase expression extends the proliferative life-span and maintains the osteogenic potential of human bone marrow stromal cells. *Nat. Biotechnol.* **20**:560–561.

Sinha, M. K., Buchanan, C., Raineri-Maldonado, C., Khazanie, P., Atkinson, S., DiMarchi, R., & Caro, J. F. (1990). IGF-II receptors and IGF-II-stimulated glucose transport in human fat cells. *Am. J. Physiol.* **258**:E534–E542.

Skehan, P., Storeng, R., Scudiero, D., Monks, A., McMahon, J., Vistica, D., Warren, J. T., Bokesch, H., Kenney, S., & Boyd, M. R. (1990). New colorimetric cytotoxicity assay for anticancer-drug screening. *J. Natl. Cancer Inst.* **82**:1107–1112.

Skobe, M., & Fusenig, N. E. (1998). Tumorigenic conversion of immortal human keratinocytes through stromal cell activation. *Proc. Natl. Acad. Sci. USA* **95**:1–6.

Skloot, R. (2010). The Immortal Life of Henrietta Lacks. New York, Crown Publishers.

Smagghe, G., & Goodman, C. L. (2009). Applications and future directions in invertebrate and fish cell culture. *In Vitro Cell Dev. Biol. Anim.* **45**:91–92.

Smagghe, G., & Goodman, C. L. (2009). Applications and future directions in Invertebrate and fish cell culture. *In Vitro Cell. Dev. Biol. Anim.* **45**:91–92.

Smagghe, G., Goodman, C. L., & Stanley, D. (2009). Insect cell culture and applications to research and pest management. *In Vitro Cell. Dev. Biol. Anim.* **45**:93–105.

Smith, A. (2006). A glossary for stem-cell biology. *Nature* **441**:1060.

Smith, A. D., Datta, S. P., Smith, G. H., Campbell, P. N., Bentley, R., & McKenzie, H. A. (eds.). (1997). *Oxford dictionary of biochemistry and molecular biology*. Oxford: Oxford University Press.

Smith, A. G. (2001). Embryo-derived stem cells: of mice and men. *Ann. Rev. Cell Dev. Biol.* **17**:435–462.

Smith, A. G., Heath, J. K., Donaldson, D. D., Wong, G. G., Moreau, J., Stahl, M., & Rodgers, D. (1988). Inhibition of pluripotential stem cell differentiation by purified polypeptides. *Nature* **336**:688–690.

Smith, H. S., Lan, S., Ceriani, R., Hackett, A. J., & Stampfer, M. R. (1981). Clonal proliferation of cultured non-malignant and malignant human breast epithelia. *Cancer Res.* **41**:4637–4643.

Smith, H. S., Owens, R. B., Hiller, A. J., Nelson-Rees, W. A., & Johnston, J. O. (1976). The biology of human cells in tissue culture; I: Characterization of cells derived from osteogenic sarcomas. *Int. J. Cancer* **17**:219–234.

Smith, M. D., Summers, M. D., & Frazer, M. J. (1983). Production of human beta interferon in insect cells infected with a baculovirus expression vector. *Mol. Cell Biol.* **3**:2156–2165.

Smith, P. K., Krohn, R. I., Hermanson, G. T., Mallia, A. K., Gartner, F. H., Provenzano, M. D., Fujimoto, E. K., Goeke, N. M., Olson, B. J., & Klenk, D. C. (1985). Measurement of protein using bicinchoninic acid. *Anal. Biochem.* **150**:235–239.

Smith, S., & de Lange, T. (1997). TRFI, a mammalian telomeric protein. *Trends Genet.* **113**:21–26.

Smith, W. L., & Garcia-Perez, A. (1985). Immunodissection: Use of monoclonal antibodies to isolate specific types of renal cells. *Am. J. Physiol.* **248**:F1–F7.

Smola, H., Stark, H.-J., Thiekötter, G., Mirancea, N., Krieg, T., & Fusenig, N. E. (1998). Dynamics of basement membrane formation by keratinocyte–fibroblast interactions in organotypic skin culture. *Exp. Cell Res.* **239**:399–410.

Smola, H., Thiekötter, G., & Fusenig, N. E. (1993). Mutual induction of growth factor gene expression by epidermal–dermal cell interaction. *J. Cell Biol.* **122**:417–429.

Smyth, M. J., Rodney L Sparks, R. L., & Wharton, W. (1993). Proadipocyte cell lines: Models of cellular proliferation and differentiation. *J. Cell Sci.* **106**:1–9.

Snyder, E. Y., Deitcher, D. L., Walsh, C., Arnold Aldea, S., Hartwieg, E. A., & Cepko, C. L. (1992). Multipotent neural cell lines can engraft and participate in development of mouse cerebellum. *Cell* **68**:33–51.

Sokol, F., Kuwert, E., Wiktor, T. J., Hummeler, K., & Koprowski, H. (1968). Purification of rabies virus grown in tissue culture. *J. Virol.* **2**:836–849.

Solassol, J., Crozet, C., & Lehmann, S. (2003). Prion propagation in cultured cells. *Br. Med. Bull.* **66**:87–97.

Solter, D., Skreb, N., & Damjanov, I. (1970). Extrauterine growth of mouse egg-cylinders results in malignant teratoma. *Nature* **227**:503–504.

Sommer, C. A., Sommer, A. G., Longmire, T. A., Christodoulou, C., Thomas, D. D., Gostissa, M., Alt, F. W., Murphy, G. J., Kotton, D. N., & Mostoslavsky, G. (2010). Excision of reprogramming transgenes improves the differentiation potential of iPS cells generated with a single excisable vector. *Stem Cells* **28**:64–74.

Sommer, I., & Schachner, M. (1981). Cells that are O4-antigen-positive and O1-antigen-negative differentiate into O1 antigen-positive oligodendrocytes. *Neurosci. Lett.* **29**:183–188.

Sonoyama, W., Yamaza, T., Gronthos, S., & Shi, S. (2007). Multipotent stem cells in dental pulp. In Freshney, R. I., Stacey, G. N., & Aurebach, J. M. (eds.), *Culture of human stem cells*. Hoboken, NJ: Wiley, pp. 187–206.

Sordillo, L. M., Oliver, S. P., & Akers, R. M. (1988). Culture of bovine mammary epithelial cells in D-valine modified medium: Selective removal of contaminating fibroblasts. *Cell Biol. Int. Rep.* **12**:355–364.

Soriano, V., Pepper, M. S., Nakamura, T., Orci, L., & Montesano, R. (1995). Hepatocyte growth factor stimulates extensive development of branching duct-like structures by cloned mammary gland epithelial cells. *J. Cell Sci.* **108**:413–430.

Sorieul, S., & Ephrussi, B. (1961). Karyological demonstration of hybridization of mammalian cells in vitro. *Nature* **190**:653–654.

Sorour, O., Raafat, M., El-Bolkainy, N., & Mohamad, R. (1975). Infiltrative potentiality of brain tumors in organ culture. *J. Neurosurg.* **43**:742–749.

Soule, H. D., Maloney, T. M., Wolman, S. R., Teterson, W. D., Brenz, R., McGrath, C. M., Russo, J., Pauley, R. J., Jones, R. F., & Brooks, S. C. (1990). Isolation and characterization of a spontaneously immortalized human breast epithelial cell line, MCF-10. *Cancer Res.* **50**:6075–6086.

Soule, H. D., Vasquez, J., Long, A., Albert, S., & Brennan, M. (1973). A human cell line from a pleural effusion derived from a breast carcinoma. *J. Natl. Cancer Inst.* **51**:1409–1416.

Southam, C. M. (1958). Homotransplantation of human cell lines. *Bull. NY Acad. Med.* **34**:416–423.

Southern, P. J., & Berg, P. (1982). Transformation of mammalian cells to antibiotic resistance with a bacterial gene under control of the SV40 early region promoter. *J. Mol. App. Genet.* **1**:327–341.

Southgate, J., Masters, J. W. R., & Trejdosiewicz, L. K. (2002). In Freshney, R. I., & Freshney, M. G., (eds.), *Culture of epithelial cells*, 2nd ed. New York: Wiley-Liss, pp. 383–399.

Speirs, V. (2004). Primary culture of human mammary tumor cells. In Pfragner, R., & Freshney. R. I. (eds.), *Culture of human tumor cells*. Hoboken, NJ: Wiley-Liss, pp. 205–219.

Speirs, V., Green, A. R., & White, M. C. (1996). Collagenase III: A superior enzyme for complete disaggregation and improved viability of normal and malignant human breast tissue. *In Vitro Cell Dev. Biol. Anim.* **32**:72–74.

Speirs, V., Ray, K. P., & Freshney, R. I. (1991). Paracrine control of differentiation in the alveolar carcinoma, A549, by human foetal lung fibroblasts. *Br. J. Cancer* **64**:693–699.

Spinelli, W., Sonnenfeld, K. H., & Ishii, N. (1982). Effects of phorbol ester tumor promoters and nerve growth factor on neurite outgrowth in cultured human neuroblastoma cells. *Cancer Res.* **42**:5067–5073.

Splinter, T. A. W., Beudeker, M., & Beek, A. V. (1978). Changes in cell density induced by isopaque. *Exp. Cell Res.* **111**:245–251.

Spooncer, E., Eliason, J., & Dexter, T. M. (1992). Long-term mouse bone marrow cultures. In Testa, N. G., & Molineux, G. (eds.), *Haemopoiesis: A practical approach.* Oxford: IRL Press at Oxford University Press, pp. 55–74.

Spremulli, E. N., & Dexter, D. L. (1984). Polar solvents: A novel class of antineoplastic agents. *J. Clin. Oncol.* **2**:227–241.

Sredni, B., Sieckmann, D. G., Kumagai, S. H., Green, I., & Paul, W. E. (1981). Long term culture and cloning of non-transformed human B-lymphocytes. *J. Exp. Med.* **154**:1500–1516.

Staab, C. A., Vondracek, M., Custodio, H., Johansson, K., Nilsson, J. A., Morgan, P., Hoog, J. O., Cotgreave, I., & Grafstrom, R. C. (2004). Modelling of normal and premalignant oral tissue by using the immortalised cell line, SVpgC2a: A review of the value of the model. *Altern. Lab. Anim.* **32**:401–405.

Stacey, G. N., Bolton, B. J., Morgan, D., Clark, S. A., & Doyle, A. (1992). Multilocus DNA fingerprint analysis of cell-banks: Stability studies and culture identification in human B-lymphoblastoid and mammalian cell lines. *Cytotechnology* **8**:13–20.

Stacey, G. N., Masters, J. R. M., Hay, R. J., Drexler, H. G., MacLeod, R. A. F., & Freshney, R. I. (2000). Cell contamination leads to inaccurate data: We must take action now. *Nature* **403**:356.

Stadtfeld, M., Nagaya, M., Utikal, J., Weir, G., & Hochedlinger, K. (2008). Induced pluripotent stem cells generated without viral integration. *Science* **322**:945–949.

Staege, M. S., Hutter, C., Neumann, I., Foja, S., Hattenhorst, U. E., Hansen, G., Afar, D., & Burdach, S. E. (2004). DNA microarrays reveal relationship of Ewing family tumors to both endothelial and fetal neural crest-derived cells and define novel targets. *Cancer Res.* **64**:8213–8221.

Stampfer, M. R., Yaswen, P., & Taylor-Papadimitriou, J. (2002). Culture of human mammary epithelial cells. In Freshney, R. I. & Freshney, M. G. (eds.), *Culture of epithelial cells*, 2nd ed. Hoboken, NJ: Wiley-Liss, pp. 95–135.

Stampfer, M., Halcones, R. G., & Hackett, A. J. (1980). Growth of normal human mammary cells in culture. *In Vitro* **16**:415–425.

Stanbridge, E. J., & Doersen, C.-J. (1978). Some effects that mycoplasmas have upon their injected host. In McGarrity, G. J., Murphy, D. G., & Nichols, W. W. (eds.), *Mycoplasma infection of cell cultures.* New York: Plenum Press, pp. 119–134.

Stanley, M. A. (2002). Culture of human cervical epithelial cells. In Freshney, R. I. & Freshney M. G. (eds.), *Culture of epithelial cells*, 2nd ed. Hoboken, NJ: Wiley-Liss, pp. 138–169.

Stanley, M. A., & Parkinson, E. (1979). Growth requirements of human cervical epithelial cells in culture. *Int. J. Cancer* **24**:407–414.

Stanners, C. P., Eliceri, G. L., & Green, H. (1971). Two types of ribosome in mouse–hamster hybrid cells. *Nat. New Biol.* **230**:52–54.

Stanton, B. A., Biemesderfer, D., Wade, J. B., & Giebisch, G. (1981). Structural and functional study of the rat distal nephron: Effects of potassium adaptation and depletion. *Kidney Int.* **19**:36–48.

Stark, H.-J., Boehnke, K., Mirancea, N., Willhauck, M., Pavesio, A., Fusenig, N. E., & Boukamp, P. (2006). Epidermal homeostasis in long-term scaffold-enforced skin equivalents. *J. Invest. Dermatol. Sympos. Proc.* **11**:93–105.

Stark, H.-J., Willhauck, M. J., Mirancea, N., Boehnke, K., Nord, I., Breitkreutz, D., Pavesio, A., Boukamp, P., & Fusenig, N. E. (2004). Authentic fibroblast matrix in dermal equivalents normalizes epidermal histogenesis and dermo-epidermal junction in organotypic co-culture. *Eur. J. Cell Biol.* **83**:631–645.

Stark, H. J., Baur, M., Breitkreutz, D., Mirancea, N., & Fusenig, N. E. (1999). Organotypic keratinocyte cocultures in defined medium with regular epidermal morphogenesis and differentiation. *J. Invest. Dermatol.* **112**:681–691.

States, B., Foreman, J., Lee, J., & Segal, S. (1986). Characteristics of cultured human renal cortical epithelia. *Biochem. Med. Metab. Biol.* **36**:151–161.

Steele, M. P., Levine, R. A., Joyce-Brady, M., & Brody, J. S. (1992). A rat alveolar type II cell line developed by adenovirus 12 SE1A gene transfer. *Am. J. Resp. Cell Mol. Biol.* **6**:50–56.

Steinberg, M. L. (1996). Immortalization of human epidermal keratinocytes by SV40. In Freshney, R. I., & Freshney, M. G. (eds.), *Culture of immortalized cells.* New York: Wiley-Liss, pp. 95–120.

Stern, P., West, C., & Burt, D. (2004). Culture of cervical carcinoma cell lines. In Pfragner, R., & Freshney, R. I. (eds.), *Culture of human tumor cells*. Hoboken, NJ: Wiley-Liss, pp. 179–204.

Stevens, L. C. (1970). Experimental production of testicular teratomas in mice of strains 129, A/He, and their F1 hybrids. *J. Natl. Cancer Inst.* **44**, 923–929.

Stevens, L. C. (1968). The development of teratomas from intratesticular grafts of tubal mouse eggs. *J. Embryol. Exp. Morphol.* **20**:329–341.

Stewart, C. E. H., James, P. L., Fant, M. E., & Rotwein, P. (1996). Overexpression of insulin-like growth factor-II induces accelerated myoblast differentiation. *J. Cell. Physiol.* **169**:23–32.

Stoker, M. G. P. (1973). Role of diffusion boundary layer in contact inhibition of growth. *Nature* **246**:200–203.

Stoker, M. G. P., & Rubin, H. (1967). Density dependent inhibition of cell growth in culture. *Nature* **215**:171–172.

Stoker, M., O'Neill, C., Berryman, S., & Waxman, B. (1968). Anchorage and growth regulation in normal and virus transformed cells. *Int. J. Cancer* **3**:683–693.

Strange, R., Li, F., Fris, R. R., Reichmann, E., Haenni, B., & Burri, P. H. (1991). Mammary epithelial differentiation in vitro: Minimum requirements for a functional response to hormonal stimulation. *Cell Growth Differ.* **2**:549–559.

Strangeways, T. S. P. & Fell, H. B. (1925). Experimental studies on the differentiation of embryonic tissues growing in vivo and in vitro. I. The development of the undifferentiated limb-bud (a) when subcutaneously grafted into the post-embryonic chick and (b) when cultivated in vitro. *Proc. Roy. Soc, London* **B99**:340–366.

Strangeways, T. S. P. & Fell, H. B. (1926). Experimental studies on the differentiation of embryonic tissues growing in vivo and in vitro; II: The development of the isolated early embryonic eye of the fowl when cultivated in vitro. *Proc. Roy. Soc, London* **B100**:273–283.

Stratton, M. R., Darling, J., Pilkington, G. J., Lantos, P. L., Reeves, B. R., & Cooper, C. S. (1989). Characterization of the human cell line TE671. *Carcinogenesis* **10**:899–905.

Strickland, S., & Beers, W. H. (1976). Studies on the role of plasminogen activator in ovulation: In vitro response of granulosa cells to gonadotropins, cyclic nucleotides, and prostaglandins. *J. Biol. Chem.* **251**:5694–5702.

Stryer, L. (1995). *Biochemistry*, 4th ed. New York: Freeman.

Styles, J. A. (1977). A method for detecting carcinogenic organic chemicals using mammalian cells in culture. *Br. J. Cancer* **36**:558.

Su, X., Sorenson, C. M., & Sheibani, N. (2003). Isolation and characterization of murine retinal endothelial cells. *Mol. Vis.* **1**:171–178.

Subramanian, M., Madden, J. A., & Harder, D. R. (1991). A method for the isolation of cells from arteries of various sizes. *J. Tissue Cult. Methods* **13**:13–20.

Suemori, H., Yasuchika, K., Hasegawa, K., Fujioka, T., Tsuneyoshi, N., & Nakatsuji, N. (2006). Efficient establishment of human embryonic stem cell lines and long-term maintenance with stable karyotype by enzymatic bulk passage. *Bio. Biophys. Res. Commun.* **345**:926–932.

Sugimoto, M., Tahara, H., Ide, T., & Furuichi, Y. (2004). Steps involved in immortalization and tumorigenesis in human B-lymphoblastoid cell lines transformed by Epstein–Barr virus. *Cancer Res.* **64**:3361–3364.

Suh, H., Song, M. J., & Park, Y. N. (2003). Behavior of isolated rat oval cells in porous collagen scaffold. *Tissue Eng.* **9**:411–20.

Sumer, H., Jones, K. L., Liu, J., Rollo, B. N., van Boxtel, A. L., Pralong, D., & Verma, P. J. (2009). Transcriptional changes in somatic cells recovered from ES-somatic heterokaryons. *Stem Cells Dev.* **18**:1361–1368.

Sun, L., Bradford, C. S., & Barnes, D. W. (1995a). Feeder cell cultures for zebrafish embryonal cells in vitro. *Mol. Mar. Biol. Biotech.* **4**:43–50.

Sun, L., Bradford, C. S., Ghosh, C., Collodi, P., & Barnes, D. W. (1995b). ES-like cell cultures derived from early zebrafish embryos. *Mol. Mar. Biol. Biotech.* **4**:193–199.

Sundqvist, K., Liu, Y., Arvidson, K., Ormstad, K., Nilsson, L., ToftgAArd, R., & Grafström, R. C. (1991). Growth regulation of serum-free cultures of epithelial cells from normal human buccal mucosa. *In Vitro Cell Dev. Biol.* **A27**:562–568.

Sutherland, H. J., Eaves, C. J., Lansdorp, P. M., Thacker, J. D., & Hogge, D. E. (1991). Differential regulation of primitive human hematopoietic cells in long-term cultures maintained on genetically engineered murine stromal cells. *Blood* **78**:666–672.

Sutherland, R. M. (1988). Cell and micro environment interactions in tumour microregions: The multicell spheroid model. *Science* **240**:117–184.

Suva, D., Garavaglia, G., Menetrey, J., Chapuis, B, Hoffmeyer, P, Bernheim, L., & Kindler, V. (2004). Non-hematopoietic human bone marrow contains long-lasting, pluripotential mesenchymal stem cells. *J. Cell. Physiol.* **198**:110–118.

Suzuki, T., Saha, S., Sumantri, C., Takagi, M., & Boediono, A. (1995). The influence of polyvinylpyrrolidone on freezing of bovine IVF blastocysts following biopsy. *Cryobiology* **32**:505–510.

Suzuki, T., & Sekiguchi, M. (1999). Gastric cancer. In Masters J. R. W., & Palsson, B. (eds.), *Human cell culture*, Vol. **2**., Dordrecht: Kluwer Academic Publishers, pp. 257–291.

Sykes, J. A., Whitescarver, J., Briggs, L., & Anson, J. H. (1970). Separation of tumor cells from fibroblasts with use of discontinuous density gradients. *J. Natl. Cancer Inst.* **44**:855–864.

Szabowski, A., Maas-Szabowski, N., Andrecht, S., Kolbus, A., Schorpp-Kistner, M., Fusenig, N. E., & Angel, P. (2000). c-Jun and JunB antagonistically control cytokine-regulated mesenchymal-epidermal interaction in skin. *Cell* **103**:745–755.

Szpirer, C., Szpirer, J., Klinga-Levan, K., Stahl, F., & Levan, G. (1996). The rat: An experimental animal in search of a genetic map. *Folia Biol. (Prague)* **42**:175–226. http://ratmap.gen.gu.se/Idiogram.html.

Takahashi, K., Tanabe, M., Ohnuki, M., Narita, T., Ichisaka, K., Tomoda, S., & Yamanaka, S. (2007). Induction of pluripotent stem cells from adult human fibroblasts by defined factors. *Cell* **131**:861–872.

Takahashi, M., Galligan, C., Tessarollo, L., & Yoshimura, T. (2009). Monocyte chemoattractant protein-1 (MCP-1), not MCP-3, is the primary chemokine required for monocyte recruitment in mouse peritonitis induced with thioglycollate or zymosan A. *J. Immunol.* **183**:3463–3471.

Takahashi, H., Yanagi, Y., Tamaki, Y., Muranaka, K., Usui, T., & Sata, M. (2004). Contribution of bone-marrow-derived cells to choroidal neovascularization. *Biochem. Biophys. Res. Commun.* **320**:372–375.

Takahashi, K., & Okada, T. S. (1970). Analysis of the effect of "conditioned medium" upon the cell culture at low density. *Dev. Growth Differ.* **12**:65–77.

Takahashi, K., Suzuki, K., Kawahara, S., & Ono, T. (1991). Effects of lactogenic hormones on morphological development and growth of human breast epithelial cells cultivated in collagen gels. *Jpn. J. Cancer Res.* **82**:553.

Takeda, K., Minowada, J., & Bloch, A. (1982). Kinetics of appearance of differentiation-associated characteristics in ML-1, a line of human myeloblastic leukaemia cells, after treatment with TPA, DMSO, or Ara-C. *Cancer Res.* **42**:5152–5158.

Tarella, C., Ferrero, D., Gallo, E., Luyca Pagliardi, G., & Ruscetti, F. W. (1982). Induction of differentiation of HL-60 cells by dimethylsulphoxide: Evidence for a stochastic model not linked to the cell division cycle. *Cancer Res.* **42**:445–449.

Tashjian, A. H., Jr. (1979). Clonal strains of hormone-producing pituitary cells. In Jakoby, W. B., & Pastan, I. H. (eds.), *Methods in enzymology; Vol. 57: Cell culture.* New York: Academic Press, pp. 527–535.

Tashjian, A. H., Yasamura, Y., Levine, L., Sato, G. H., & Parker, M. (1968). Establishment of clonal strains of rat pituitary tumor cells that secrete growth hormone. *Endocrinology* **82**:342–352.

Taub, M. L., Yang, S. I., & Wang, Y. (1989). Primary rabbit proximal tubule cell cultures maintain differentiated functions when cultured in a hormonally defined serum-free medium. *In Vitro Cell Dev. Biol.* **25**:770–775.

Taub, M.. (1984). Growth of primary and established kidney cell cultures in serum-free media. In Barnes, D. W., Sirbasku, D. A., & Sato, G. H. (eds.), *Methods for serum-free culture of epithelial and fibroblastic cells.* New York: Liss, pp. 3–24.

Taylor, D. A. (2001). Cellular cardiomyoplasty with autologous skeletal myoblasts for ischemic heart disease and heart failure. *Curr. Control Trials Cardiovasc. Med.* **2**:208–210.

Taylor, J. H. (1958). Sister chromatid exchanges in tritium labeled chromosomes. *Genetics* **43**:515–529.

Taylor-Papadimitriou, J., Purkiss, P., & Fentiman, I. S. (1980). Choleratoxin and analogues of cyclic AMP stimulate the growth of cultured human epithelial cells. *J. Cell Physiol.* **102**:317–322.

Taylor-Papadimitriou, J., Stampfer, M., Bartek, J., Lewis, A., Boshell, M., Lane, E. B., & Leigh, I. M. (1989). Keratin expression in human mammary epithelial cells cultured from normal and malignant tissue: Relation to in vivo phenotypes and influence of medium. *J. Cell Sci.* **94**:403–413.

Tedder, R. S., Zuckerman, M. A., Goldstone, A. H., Hawkins, A. E., Fielding, A., Briggs, E. M., Irwin, D., Blair, S., Gorman, A. M., Patterson, K. G., Linch, D. C., Heptonstall, J., & Brink, N. S. (1995). Hepatitis B transmission from contaminated cryopreservation tank. *Lancet* **346**:137–140.

Telling, R. C., & Ellsworth, R. (1965). Submerged culture of hamster kidney cells in a stainless steel vessel. *Biotechnol. Bioeng.* **7**:4-1–434.

Temin, H. M. (1966). Studies on carcinogenesis by avian sarcoma viruses; III: The differential effect of serum and polyanions on multiplication of uninfected and converted cells. *J. Natl. Cancer Inst.* **37**:167–175.

Terasaki, T., Kameya, T., Nakajima, T., Tsumuraya, M., Shimosato, Y., Kato, K., Ichinose, H., Nagatsu, T., & Hasegawa, T. (1984). Interconversion of biological characteristics of small cell lung cancer cells depending on the culture conditions. *Gann* **75**:1689–1699.

Tesei, A., Zoli, W., Arienti, C., Storci, G., Granato, A. M., Pasquinelli, G., Valente, S., Orrico, C., Rosetti, M., Vannini, I., Dubini, A., Dell'Amore, D., Amadori, D., & Bonafè, M. (2009). Isolation of stem/progenitor cells from normal lung tissue of adult humans. *Cell Prolif.* **42**:298–308.

Testa, N. G., & Molineux, G. (eds.). (1993). *Haemopoiesis.* Oxford: Oxford University Press.

Thelwall, P. E., Neves, A. A., & Brindle, K. M. (2001). Measurement of bioreactor perfusion using dynamic contrast agent-enhanced magnetic resonance imaging. *Biotechnol. Bioeng.* **75**:682–690.

Thomas, D. G. T., Darling, J. L., Paul, E. A., Mott, T. C., Godlee, J. N., Tobias, J. S., Capra, L. G., Collins, C. D., Mooney, C., Bozek, T., Finn, G. P., Arigbabu, S. O., Bullard, D. E., Shannon, N., & Freshney, R. I. (1985). Assay of anti-cancer drugs in tissue culture: Relationship of relapse free interval (FRI) and in vitro chemosensitivity in patients with malignant cerebral glioma. *Br. J. Cancer* **51**:525–532.

Thomas, R. J., Chandra, A., Liu, Y., Hourd, P. C., Conway, P. P., & Williams, D. J. (2007). Manufacture of a human mesenchymal stem cell population using an automated cell culture platform. *Cytotechnology* **55**:31–39.

Thomas, S., Gray, E., & Robinson, C. J. (1997). Response of HUVEC and EAhy926 and fibroblast growth factors. *In Vitro Cell Dev. Biol. Anim.* **33**:492–494.

Thompson, E. B., Tomkins, J. M., & Curran, J. F. (1966). Induction of tyrosine α-ketoglutarate transaminase by steroid hormones in a newly established tissue culture cell line. *Proc. Natl. Acad. Sci. USA* **56**:296–303.

Thompson, L. H., & Baker, R. M. (1973). Isolation of mutants of cultured mammalian cells. In Prescott, D. (ed.), *Methods in cell biology*, Vol. **6**. New York: Academic Press, pp. 209–281.

Thomson, A. A., Foster, B. A., & Cunha, G. R. (1997). Analysis of growth factor and receptor mRNA levels during development of the rat seminal vesicle and prostate. *Development* **124**:2431–2439.

Thomson, A. W. (ed.). (1991). *Cytokine handbook.* London: Academic Press.

Thompson, E. W., Waltham, M., Ramus, S. J., Hutchins, A. M., Armes, J. E., Campbell, I. G., Williams, E. D., Thompson, P. R., Rae, J. M., Johnson, M. D., & Clarke, R. (2004). LCC15-MB cells are MDA-MB-435: A review of misidentified breast and prostate cell lines. *Clin. Exp. Metastasis.* **21**:535–541.

Thomson, J. A., Itskovitz-Eldor, J., Shapiro, S. S., Waknitz, M. A., & Swiergiel, J. J. (1998). Embryonic stem cell lines derived from human blastocysts. *Science* **282**:1145–1147.

Till, J. E., & McCulloch, E. A. (1961). A direct measurement of the radiation sensitivity of normal mouse bone marrow cells. *Radiat. Res.* **14**:213–222.

Tobey, R. A., Anderson, E. C., & Peterson, D. F. (1967). Effect of thymidine on duration of G1 in chinese hamster cells. *J. Cell Biol.* **35**:53–67.

Todaro, G. J., & DeLarco, I. E. (1978). Growth factors produced by sarcoma virus-transformed cells. *Cancer Res.* **38**:4147–4154.

Todaro, G. J., & Green, H. (1963). Quantitative studies of the growth of mouse embryo cells in culture and their development into established lines. *J. Cell Biol.* **17**:299–313.

Toft, A., Scott, D. Barnett, S. C., & Riddell, J. S. (2007). Electrophysiological evidence that olfactory cell transplants improve function after spinal cord injury. *Brain* **130**:970–984.

Tomakidi, P., Fusenig, N. E., Kohl, A., & Komposch, G. (1997). Histomorphological and biochemical differentiation capacity in organotypic co-cultures of primary gingival cells. *J. Periodont. Res.* **32**:388–400.

Topley, P., Jenkins, D. C., Jessup, E. A., & Stables, J. N. (1993). Effect of reconstituted basement membrane components on the growth of a panel of human tumour cell lines in nude mice. *Br. J. Cancer* **67**:953–958.

Torday, J. S., & Kourembanas, S. (1990). Fetal rat lung fibroblasts produce a TGF-β homologue that blocks type II cell maturation. *Dev. Biol.* **13**:35–41.

Totoiu, M. O., Nistor, G. I., Lane, T. E., & Keirstead, H. S. (2004). Remyelination, axonal sparing, and locomotor recovery following transplantation of glial-committed progenitor cells into the MHV model of multiple sclerosis. *Exp. Neurol.* **187**:254–265.

Tozer, B. T., & Pirt, S. J. (1964). Suspension culture of mammalian cells and macromolecular growth promoting fractions of calf serum. *Nature* **201**:375–378.

Trapp, B. D., Honegger, P., Richelson, E., & Webster, H. de F. (1981). Morphological differentiation of mechanically dissociated fetal rat brain in aggregating cell cultures. *Brain Res.* **160**:235–252.

Triglia, D., Braa, S. S., Yonan, C., & Naughton, G. K. (1991). Cytotoxicity testing using neutral red and MTT assays on a three-dimensional human skin substrate. *Toxic. In Vitro* **5**:573–578.

Trotter, J., & Schachner, M. (1988). Cells positive for the O4 surface antigen isolated by cell sorting are able to differentiate into oligodendrocytes and type-2 astrocytes. *Dev. Brain Res.* **46**:115–122.

Trowell, O. A. (1959). The culture of mature organs in a synthetic medium. *Exp. Cell Res.* **16**:118–147.

Troyer, D. A., & Kreisberg, J. I. (1990). Isolation and study of glomerular cells. *Methods Enzymol.* **191**:141–152.

Tsao, M. C., Walthall, B. I., & Ham, R. G. (1982). Clonal growth of normal human epidermal keratinocytes in a defined medium. *J. Cell Physiol.* **110**:219–229.

Tsao, S.-W., Mok, S. C., Fey, E. G., Fletcher, J. A., Wan, T. S. K., Chew, E.-C., Muto, M. G., Knapp, R. C., & Berkowitz, R. S. (1995). Characterisation of human ovarian surface epithelial cells immortalized by human papilloma viral oncogenes (HPV-E6E7 ORFs). *Exp. Cell Res.* **218**:499–507.

Tsuruo, T., Hamilton, T. C., Louis, K. G., Behrens, B. C., Young, R. C., & Ozols, R. F. (1986). Collateral susceptibility of adriamycin-, melphalan-, and cisplatin-resistant human ovarian tumor cells to bleomycin. *Jpn. J. Cancer Res.* **77**:941–945.

Tumilowicz, J. J., Nichols, W. W., Cholon, J. J., & Greene, A. E. (1970). Definition of a continuous human cell line derived from neuroblastoma. *Cancer Res.* **30**:2110–2118.

Turner, R. W. A., Siminovitch, L., McCulloch, E. A., & Till, J. E. (1967). Density gradient centrifugation of hemopoietic colony-forming cells. *J. Cell Physiol.* **69**:73–81.

Turnpenny, L., & Hanley, N. (2007). Culture of the human germ cell lineage. In Freshney, R. I. Stacey, G. N. & Auerbach, J. M. (eds.), *Culture of human stem cells*. Hoboken, NJ: Wiley, pp. 107–132.

Tuszynski, M. H., Roberts, J., Senut, M. C., U, H. S., & Gage, F. H. (1996). Gene therapy in the adult primate brain: Intraparenchymal grafts of cells genetically modified to produce nerve growth factor prevent cholinergic neuronal degeneration. *Gene Therapy* **3**:305–314.

Tveit, K. M., & Pihl, A. (1981). Do cells lines in vitro reflect the properties of the tumours of origin? A study of lines derived from human melanoma xenografts. *Br. J. Cancer* **44**:775–786.

Tweats, D. J., & Scott, A. D., Westmoreland, C., & Carmichael, P. L. (2007). Determination of genetic toxicity and potential carcinogenicity in vitro: Challenges post the Seventh Amendment to the European Cosmetics Directive. *Mutagenesis* **22**:5–13.

Twentyman, P. R. (1980). Response to chemotherapy of EMT6 spheroids as measured by growth delay and cell survival. *Eur. J. Cancer* **42**:297–304.

Unkless, I., Dano, K., Kellerman, G., & Reich, E. (1974). Fibrinolysis associated with oncogenic transformation: Partial purification and characterization of cell factor, a plasminogen activator. *J. Biol. Chem.* **249**:4295–4305.

Uphoff, C. C., & Drexler, H. G. (1999). Detection of mycoplasma contamination in cell cultures by PCR analysis. *Hum. Cell* **12**:229–236.

Uphoff, C. C., & Drexler, H. G. (2002). Comparative antibiotic eradication of mycoplasma infections from continuous cell lines. *In Vitro Cell Dev. Biol. Anim.* **38**:86–89.

Uphoff, C. C, & Drexler, H. G. (2002). Comparative PCR analysis for detection of mycoplasma infections in continuous cell lines. *In Vitro Cell. Dev. Biol. Anim.* **38**:79–85.

Uphoff, C. C., & Drexler, H. G. (2004). Elimination of mycoplasma from infected cell lines using antibiotics. *Methods Mol. Med.* **88**:327–334.

Uphoff, C. C., Gignac, S. M., & Drexler, H. G. (1992). Mycoplasma contamination in human leukemia cell lines; I: Comparison of various detection methods. *J. Immunol. Methods* **149**:43–53.

Ure, J. M., Fiering, S., & Smith, A. G. (1992). A rapid and efficient method for freezing and recovering clones of embryonic stem cells. *Trends Genet.* **8**:6.

US EPA. (2009). Good laboratory practices standards. www. epa.gov/oecaerth/monitoring/programs/fifra/glp.html.

US/NRC. (2008). Medical, industrial, and academic uses of nuclear materials. www.nrc.gov/materials/medical.html.

UWMC Cytogenetics Laboratory. (2003). www.pathology. washington.edu/galleries/Cytogallery/

Uzgare, A. R., Xu, Y., & Isaacs, J. T. (2004). In vitro culturing and characteristics of transit amplifying epithelial cells from human prostate tissue. *J. Cell. Biochem.* **91**:196–205.

Uzgare, A. R., Xu, Y., & Isaacs, J. T. (2004). In vitro culturing and characteristics of transit amplifying epithelial cells from human prostate tissue. *J. Cell. Biochem.* **91**:196–205.

Vachon, P. H., Perreault, N., Magny, P., & Beallieu, J.-F. (1996). Uncoordinated transient mosaic patterns of intestinal hydrolase expression in differentiating human enterocytes. *J. Cell Physiol.* **166**:198–207.

Vago, C. (ed.). (1971). *Invertebrate tissue culture*, Vol. **1**. New York: Academic Press.

Vago, C. (ed.). (1972). *Invertebrate tissue culture*, Vol. **2**. New York: Academic Press.

Vaheri, A., Ruoslahti, E., Westermark, B., & Pontén, J. (1976). A common cell-type specific surface antigen in cultured human glial cells and fibroblasts: Loss in malignant cells. *J. Exp. Med.* **143**:64–72.

van Bokhoven, A., Varella-Garcia, M., Korch, C., Johannes, W. U., Smith, E. E., Miller, H. L., Nordeen, S. K., Miller, G. J., & Lucia, M. S. (2003). Molecular characterization of human prostate carcinoma cell lines. *Prostate*, **57**:205–225.

van Bokhoven, A., Varella-Garcia, M., Korch, C., & Miller, G. J. (2001a). TSU-Pr1 and JCA-1 cells are derivatives of T24 bladder carcinoma cells and are not of prostatic origin. *Cancer Res.* **61**:6340–6344.

van Bokhoven, A., Varella-Garcia, M., Korch, C., Hessels, D., & Miller, G. J. (2001b). Widely used prostate carcinoma cell lines share common origins. *Prostate* **47**:36–51.

Van Diggelen, O., Shin, S., & Phillips, D. (1977). Reduction in cellular tumorigenicity after mycoplasma infection and elimination of mycoplasma from infected cultures by passage in nude mice. *Cancer Res.* **37**:2680–2687.

Van Helden, P. D., Wiid, I. J., Albrecht, C. F., Theron, E., Thornley, A. L., & Hoal-van Helden, E. G. (1988). Cross-contamination of human esophageal squamous carcinoma cell lines detected by DNA fingerprint analysis. *Cancer Res.* **48**:5660–5662.

van Pelt, J. F., Decorte, R., Yap, P. S., & Fevery, J. (2003). Identification of HepG2 variant cell lines by short tandem repeat (STR) analysis. *Mol. Cell Biochem.* **243**:49–54.

Vanhoecke, B., Derycke, L., Van Marck, V., Depypere, H., De Keukeleire, D., & Bracke, M. (2005). Antiinvasive effect of xanthohumol, a prenylated chalcone present in hops (*Humulus lupulus* L.) and beer. *Int. J. Cancer* **117**:889–895.

Van Roozendahl, C. E. P., van Ooijen, B., Klijn, J. G. M., Claasen, C., Eggermont, A. M. M., Henzen-Logmans, S. C., & Foekens, J. A. (1992). Stromal influences on breast cancer cell growth. *Br. J. Cancer* **65**:77–81.

Vanparys, P. (2002). ECVAM and pharmaceuticals. *Altern. Lab. Anim.* **30**(suppl. 2): 221–223.

van Staveren, W. C., Solís, D. W., Delys, L., Duprez, L., Andry, G., Franc, B., Thomas, G., Libert, F., Dumont, J. E., Detours, V., & Maenhaut, C. (2007). Human thyroid tumor cell lines derived from different tumor types present a common dedifferentiated phenotype. *Cancer Res.* **67**:8113–120.

Varella-Garcia, M., Boomer, T., & Miller, G. J. (2001). Karyotypic similarity identified by multiplex-FISH relates four prostate adenocarcinoma cell lines: PC-3, PPC-1, ALVA-31, and ALVA-41. *Genes Chromosomes Cancer* **31**:303–315.

Varga, K., Goldstein, R. F., Jurkuvenaite, A., Chen, L., Matalon, S., Sorscher, E. J., Bebok, Z., & Collawn, J. F. (2008). Enhanced cell-surface stability of rescued DeltaF508 cystic fibrosis transmembrane conductance regulator (CFTR) by pharmacological chaperones. *Biochem J.* **410**:555–564.

Varga Weisz, P. D., & Barnes, D. W. (1993). Characterization of human plasma growth inhibitory activity on serum-free mouse embryo cells. *In Vitro Cell Dev. Biol.* **A29**:512–516.

Varon, S., & Manthorpe, M. (1980). Separation of neurons and glial cells by affinity methods. In Fedoroff, S., & Hertz, L. (eds.), *Advances in cellular neurobiology*, Vol. **1**. New York: Academic Press, pp. 405–442.

Vaziri, H., & Benchimol, S. (1998). Reconstitution of telomerase activity in normal human cells leads to elongation of telomeres and extended replicative life-span. *Curr. Biol.* **8**:279–282.

Velcich, A., Palumbo, L., Jarry, A., Laboisse, C., Racevskis, J., & Augenlicht, L. (1995). Patterns of expression of lineage-specific markers during the in vitro-induced differentiation of HT29 colon carcinoma cells. *Cell Growth Differ.* **6**:749–757.

Venitt, S. (1984). *Mutagenicity testing: A practical approach.* Oxford: IRL Press.

Venkatasubramanian, J., Sahi, J., & Rao, M. C. (2000). Ion transport during growth and differentiation. *Ann. N Y Acad. Sci.* **915**:357–372.

Venteicher, A. S., Abreu, E. B., Meng, Z., McCann, K. E., Terns, R. M., Veenstra, T. D., Terns, M. P., & Artandi, S. E. (2009). A human telomerase holoenzyme protein required for Cajal body localization and telomere synthesis. *Science* **323**:644–648.

Verbruggen, G., Veys, E. M., Wieme, N., Malfait, A. M., Gijselbrecht, L., Nimmegeers, J., Almquist, K. F., & Broddelez, C. (1990). The synthesis and immobilization of cartilage-specific proteoglycan by human chondrocytes in different concentrations of agarose. *Clin. Exp. Rheumatol.* **8**:371–378.

Vermeulen, L., Sprick, M. R., Kemper, K., Stassi, G., & Medema, J. P. (2008). Cancer stem cells: Old concepts, new insights. *Cell Death Differ.* **15**:947–958.

Verschraegen, C. F, Hu, W., Du, Y., Mendoza, J., Early, J., Deavers, M., Freedman, R. S., Bast, R. C. Jr., Kudelka, A. P., Kavanagh, J. J., & Giovanella, B. C. (2003). Establishment and characterization of cancer cell cultures and xenografts derived from primary or metastatic Mullerian cancers. *Clin. Cancer Res.* **9**:845–852.

Vescovi, A., Gritti, A., Cossu, G., & Galli, R. (2002). Neural stem cells: Plasticity and their transdifferentiation potential. *Cells Tissues Organs* **171**:64–76.

Vescovi, A. L., Galli, R., & Reynolds, B. A. (2006). Brain tumour stem cells. *Nat. Rev. Cancer* **6**:425–436.

Vierbuchen, T., Ostermeier, A., Pang, Z. P., Kokubu, Y., Südhof, T. C., & Wernig, M. (2010). Direct conversion of fibroblasts to functional neurons by defined factors. *Nature.* **463**:1031–1032.

Vierick, J. L., McNamara, P., & Dodson, M. V. (1996). Proliferation and differentiation of progeny of ovine unilocular fat cells (adipofibroblasts). *In Vitro Cell Dev. Biol. Anim.* **32**:564–572.

Vilamitjana-Amedee, J., Bareile, R., Rouais, F., Caplan, A. I., & Harmand, M. F. (1993). Human bone marrow stromal cells express an osteoblastic phenotype in culture. *In Vitro Cell Dev. Biol.* **A29**:699–707.

Vincan, E., Whitehead, R. H., & Faux, M. C. (2008). Analysis of Wnt/FZD-mediated signalling in a cell line model of colorectal cancer morphogenesis. *Methods Mol. Biol.* **468**:263–273.

Virant-Klun, I., Zech, N., Rožman, P., Vogler, A., Cvjetičanin, B., Klemenc, P., Maličev, E., & Meden-Vrtovec, H. (2008). Putative stem cells with an embryonic character isolated from the ovarian surface epithelium of women with no naturally present follicles and oocytes. *Differentiation* **76**:843–856.

Vireque, A. A., Camargo, L. S., Serapião, R. V., Rosa, E., Silva, A. A., Watanabe, Y. F., Ferreira, E. M., Navarro, P. A., Martins, W. P., & Ferriani, R. A. (2009). Preimplantation development and expression of Hsp-70 and Bax genes in bovine blastocysts derived from oocytes matured in alpha-MEM supplemented with growth factors and synthetic macromolecules. *Theriogenology* **71**(4): 620–627.

Visco, V., Bava, F. A., d'Alessandro, F., Cavallini, M., Ziparo, V., & Torrisi, M. R. (2009). Human colon fibroblasts induce differentiation and proliferation of intestinal epithelial cells through the direct paracrine action of keratinocyte growth factor. *J. Cell Physiol.* **220**:204–213.

Vistica, D. T., Skehan, P., Scudiero, D., Monks, A., Pittman, A., & Boyd, M. R. (1991). Tetrazolium-based assays for cellular viability: A critical examination of selected parameters affecting formazan production. *Cancer Res.* **51**:2515–2520.

Vlodavsky, I., Lui, G. M., & Gospodarowicz, D. (1980). Morphological appearance, growth behavior and migratory activity of human tumor cells maintained on extracellular matrix versus plastic. *Cell* **19**:607–617.

Vogel, F. R., & Powell, M. F. (1995). A compendium of vaccine adjuvants and excipients. In Powell, M. F., & Newman, M. (eds.), *Vaccine design: The subunit and adjuvant approach.* New York: Plenum Press, pp. 141–228.

von Briesen, H., Andreesen, R., Esser, R., Brugger, W., Meichsner, C., Becker, K, & Rubsamen-Waigmann, H. (1990). Infection of monocytes/macrophages by HIV in vitro. *Res Virol.* **141**:225–231.

Von der Mark, K. (1986). Differentiation, modulation and dedifferentiation of chondrocytes. *Rheumatology* **10**:272–315.

Von Hoff, D. D., Clark, G. M., Weis, G. R., Marshall, M. H., Buchok, J. B., Knight, W. A., & Lemaistre, C. F. (1986). Use of in vitro dose response effects to select antineoplastics for high dose or regional administration regimens. *J. Clin. Oncol.* **4**:18–27.

Vondracek, M., Weaver, D., Sarang, Z., Hedberg, J. J., Willey, J., WärngAArd, L. and Grafström, R. C. (2002). Transcript profiling of enzymes involved in detoxification of xenobiotics and reactive oxygen in human normal and Simian virus 40 T antigen-immortalized oral keratinocytes. *Int. J. Cancer* **99**:776–782.

Vonen, B., Bertheussen, K., Giaever, A. K., Florholmen, J., & Burhol, P. G. (1992). Effect of a new synthetic serum replacement on insulin and somatostatin secretion from isolated rat pancreatic islets in long term culture. *J. Tissue Cult. Methods* **14**:45–50.

Vouret-Craviari, V., Boulter, E., Grall, D., Matthews, C., & Van Obberghen-Schilling, E. (2004). ILK is required for the assembly of matrix-forming adhesions and capillary morphogenesis in endothelial cells. *J Cell Sci.* **117**:4559–4569.

Voyta, J. C., Via, D. P., Butterfield, C. W., & Zetter, B. R. (1984). Identification and isolation of endothelial cells based on their increased uptake of acetylated–low density lipoprotein. *J. Cell Biol.* **99**:2034–2040.

Vries, J. E., Benthem, M., & Rumke, P. (1973). Separation of viable from nonviable tumor cells by flotation on a Ficoll-triosil mixture. *Transplantation* **5**:409–410.

Vunjak-Novakovic, G. (2006). Tissue engineering: Basic considerations. In Vunjak-Novakovic, G., & Freshney, R. I. (eds.), *Culture of cells for tissue engineering.* Hoboken, NJ: Wiley-Liss, pp. 132–155.

Vunjak-Novakovic, & G., Freshney, R. I. (eds.). (2006). *Culture of cells for tissue engineering.* Hoboken, NJ: Wiley-Liss.

Wada, T., Dacy, K. M., Guan, X.-P., & Ip, M. M. (1994). Phorbol 12-myristate 13 acetate stimulates proliferation and ductal morphogenesis and inhibits functional differentiation of normal rat mammary epithelial cells in primary culture. *J. Cell Physiol.* **158**:97–109.

Waleh, N. S., Brody, M. D., Knapp, M. A., Mendonca, H. L., Lord, E. M., Koch, C. J., Laderoute, K. R., & Sutherland, R. M. (1995). Mapping of the vascular endothelial growth factor-producing hypoxic cells in multicellular tumour spheroids using a hypoxia-specific marker. *Cancer Res.* **55**:6222–6226.

Walston, J., Silver, K., Bogardus, C., Knowler, W. C., Celi, F. S., Austin, S., Manning, B., Strosberg, A. D., Stern, M. P., Raben, N., Sorkin J. D., Roth, J., & Shuldiner, A. R. (1995). Time of onset of non–insulin-dependent diabetes mellitus and genetic variation in the beta 3-adrenergic-receptor gene. *N. Engl. J. Med.* **333**:343–347.

Walter, H. (1975). Partition of cells in two-polymer aqueous phases: A method for separating cells and for obtaining information on their surface properties. In Prescott, D. M. (ed.), *Methods in cell biology.* New York: Academic Press, pp. 25–50.

Walter, H. (1977). Partition of cells in two-polymer aqueous phases: A surface affinity method for cell separation. In Catsimpoolas, N. (ed.), *Methods of cell separation.* New York: Plenum Press, pp. 307–354.

Wang, H. C., & Fedoroff, S. (1972). Banding in human chromosomes treated with trypsin. *Nat. New Biol.* **235**:52–53.

Wang, J., Torbenson, M., Wang, Q., Ro, J. Y., & Becich, M. (2003). Expression of inducible nitric oxide synthase in paired neoplastic and non-neoplastic primary prostate cell cultures and prostatectomy specimen. *Urol. Oncol.* **21**:117–122.

Wang, R. J, & Nixon, B. R. (1978). Identification of hydrogen peroxide as a photoproduct toxic to human cells in tissue-culture medium irradiated with "daylight" fluorescent light. *In Vitro.* **14**:715–22.

Ward, J. P., & King, J. R. (1997). Mathematical modelling of avascular tumour growth. *IMA J. Math. App. Med. Biol.* **14**:39–69.

Watanabe, T., Kondo, K., & Oishi, M. (1991). Induction of in vitro differentiation of mouse erythroleukemia cells by genistein, an inhibitor of tyrosine kinases. *Cancer Res.* **51**:764–768.

Watt, F. M. (2001). Stem cell fate and patterning in mammalian epidermis. *Curr. Opin. Genet. Dev.* **11**:410–417.

Watt, F. M. (2002). The stem cell compartment in human interfollicular epidermis. *J Dermatol Sci.* **28**:173–180.

Waymouth, C. (1959). Rapid proliferation of sublines of NCTC clone 929 (Strain L) mouse cells in a simple chemically defined medium (MB752/1). *J. Natl. Cancer Inst.* **22**:1003.

Waymouth, C. (1970). Osmolality of mammalian blood and of media for culture of mammalian cells. *In Vitro* **6**:109–127.

Waymouth, C. (1974). To disaggregate or not to disaggregate: Injury and cell disaggregation, transient or permanent? *In Vitro* **10**:97–111.

Waymouth, C. (1979). Autoclavable medium AM 77B. *J. Cell Physiol.* **100**:548–550.

Waymouth, C. (1984). Preparation and use of serum-free culture media. In Barnes, W. D., Sirbasku, D. A., & Sato, G. H. (eds.), *Cell culture methods for molecular and cell biology; Vol. 1: Methods for preparation of media, supplements, and substrata for serum-free animal cell culture.* New York: Liss, pp. 23–68.

Webber, D. J., & Minger, S. L. (2004). Therapeutic potential of stem cells in central nervous system regeneration. *Curr. Opin. Investig. Drugs* **5**:714–719.

Wehder, L., Arndt, S., Murzik, U., Bosserhoff, A. K., Kob, R., von Eggeling, F., & Melle, C. (2009). Annexin A5 is involved in migration and invasion of oral carcinoma. *Cell Cycle* **8**:1552–1558.

Weibel, E. R., & Palade, G. E. (1964). New cytoplasmic components in arterial endothelia. *J. Cell Biol.* **23**:101–102.

Weichselbaum, R., Epstein, I., & Little, J. B. (1976). A technique for developing established cell lines from human osteosarcomas. *In Vitro* **12**:833–836.

Weinberg, R. A. (ed.). (1989). *Oncogenes and the molecular origins of cancer.* Cold Spring Harbor, NY: Cold Spring Harbor Laboratory Press.

Weiss, M. C., & Green, H. (1967). Human–mouse hybrid cell lines containing partial complements of human chromosomes and functioning human genes. *Proc. Natl. Acad. Sci. USA* **58**:1104–1111.

Wells, D. L., Lipper, S. L., Hilliard, J. K., Stewart, J. A., Holmes, G. P., Herrmann, K. L., Kiley, M. P., & Schonberger, L. B. (1989). *Herpesvirus simiae* contamination of primary rhesus monkey kidney cell cultures: CDC recommendations to minimize risks to laboratory personnel. *Diagn. Microbiol. Infect. Dis.* **12**:333–335.

Welm, B., Behbod, F., Goodell, M. A., & Rosen, J. M. (2003). Isolation and characterization of functional mammary gland stem cells. *Cell Prolif.* **36**(s1): 17–32.

Wessells, N. K. (1977). *Tissue interactions and development.* Menlo Park, CA: Benjamin.

Wessels, D., Titus, M., & Soll, D. R. (1996). A *Dictyostelium* myosin I plays a crucial role in regulating the frequency of pseudopods formed on the substratum. *Cell Motil. Cytoskeleton* **33**:64–79.

Westerfield, M. (1993). *The zebrafish book: A guide for the laboratory use of zebrafish (Brachydanio rerio).* Eugene: University of Oregon Press.

Westermark, B., & Wasteson, A. (1975). The response of cultured human normal glial cells to growth factors. *Adv. Metab. Disord.* **8**:85–100.

Westermark, B. (1974). The deficient density-dependent growth control of human malignant glioma cells and virus-transformed glialike cells in culture. *Int. J. Cancer* **12**:438–451.

Westermark, B. (1978). Growth control in miniclones of human glial cells. *Exp. Cell. Res.* **111**:295–299.

Westermark, B., Pontén, J., & Hugosson, R. (1973). Determinants for the establishment of permanent tissue culture lines from human gliomas. *Acta Pathol. Microbiol. Scand.* **A81**:791–805.

Wewetzer, K., Verdu, E., Angelov, D. N., & Navarro, X. (2002) Olfactory ensheathing glia and Schwann cells: two of a kind? *Cell Tissue Res.* **309**:337–345.

Whitehead, R. H., & Robinson, P. S. (2009). Establishment of conditionally immortalized epithelial cell lines from the intestinal tissue of adult normal and transgenic mice. *Am. J. Physiol. Gastrointest. Liver Physiol.* **296**:G455–G460.

Whitehead, R. H. (2004). Establishment of cell lines from colon carcinoma. In Pfragner, R., & Freshney. R. I. (eds.), *Culture of human tumor cells.* Hoboken, NJ: Wiley-Liss, pp. 67–80.

Whitlock, C. A., Robertson, D., & Witte, O. N. (1984). Murine B cell lymphopoiesis in long term culture. *J. Immunol. Methods* **67**:353–369.

WHO. (1989). Guidelines on the sterilization and disinfection methods effective against human immunodeficiency virus (HIV). WHO AIDS Series 2, 2nd ed. Geneva: WHO. http://whqlibdoc.who.int/aids/WHO_AIDS_2.pdf.

Whur, P., Magudia, M., Boston, I., Lockwood, J., & Williams, D. C. (1980). Plasminogen activator in cultured Lewis lung carcinoma cells measured by chromogenic substrate assay. *Br. J. Cancer* **42**:305–312.

Wienberg, J., & Stanyon, R. (1997). Comparative painting of mammalian chromosomes. *Curr. Opin. Genet. Dev.* **7**:784–791.

Wiktor, T. J., Fernandes, M. V., & Koprowski, H. (1964). Cultivation of rabies virus in human diploid cell strain WI-38. *J. Immunol.* **93**:353–366.

Wilkenheiser, K. A., Vorbroker, D. K., Rice, W. R., Clark, J. C., Bachurski, C. J., Oie, H. K., & Whitsett, J. E. (1991). Production of immortalized distal respiratory epithelial cell lines from surfactant protein C/simian virus 40 large tumor antigen transgenic mice. *Proc. Natl. Acad. Sci. USA* **90**:11029–11033.

Wilkins, L. Gilchrest, B. A., Szabo, G., Weinstein, R., & Maciag, T. (1985). The stimulation of normal human melanocyte proliferation in vitro by melanocyte growth factor from bovine brain. *J. Cell Physiol.* **122**:350.

Willey, J. C., Moser, C. E. Jr., Lechner, J. F., & Harris, C. C. (1984). Differential effects of 12-O-tetradecanoylphorbol-13-acetate on cultured normal and neoplastic human bronchial epithelial cells. *Cancer Res.* **44**:5124–5126.

Williams, B. P., Abney, E. R., & Raff, M. C. (1985). Macroglial cell development in embryonic rat brain: Studies using monoclonal antibodies, fluorescence-activated cell sorting and cell culture. *Dev. Biol.* **112**:126–134.

Williams, G. M., & Gunn, J. M. (1974). Long-term cell culture of adult rat liver epithelial cells. *Exp. Cell Res.* **89**:139–142.

Williams, M. J., & Clark, P. (2003). Microscopic analysis of the cellular events during scatter factor/hepatocyte growth factor-induced epithelial tubulogenesis. *J. Anat.* **203**:483–503.

Williams, R. D. (1980). Human urologic cancer cell lines. *Invest. Urol.* **17**:359–363.

Williams, R. L., Hilton, D. J, Pease, S., Willson, T. A., Stewart, C. L., Gearing, D. P., Wagner, E. F., Metcalf, D., Nicola, N. A., & Gough, N. M. (1988). Myeloid leukaemia inhibitory factor maintains the developmental potential of embryonic stem cells. *Nature* **336**:684–687.

Willmarth, N. E., Albertson, D. G., & Ethier, S. P. (2004). Chromosomal instability and lack of cyclin E regulation in hCdc4 mutant human breast cancer cells. *Breast Cancer Res.* **6**:R531–R539.

Wilson, A. P., Dent, M., Pejovic, T., Hubbold, L., & Radford, H. (1996). Characterisation of seven human ovarian tumour cell lines. *Br. J. Cancer* **74**:722–727.

Wilson, A. P. (2004). The development of human ovarian epithelial tumor cell lines from solid tumors and ascites. In Pfragner, R., & Freshney, R. I. (eds.), *Culture of human tumor cells.* Hoboken, NJ: Wiley-Liss, pp. 145–178.

Wilson, A. P., Dent, M., Pejovic, T., Hubbold, L., & Rodford, H. (1996). Characterisation of seven human ovarian tumour cell lines. *Br. J. Cancer* **74**:722–727.

Wilson, C., Scullin P., Worthington, J., Seaton, A., Maxwell, P., O'Rourke, D., Johnston, P. G., McKeown, S. R., Wilson, R. H., O'Sullivan, J. M., & Waugh, D. J. J. (2008). Dexamethasone potentiates the antiangiogenic activity of docetaxel in castration-resistant prostate cancer. *Br. J. Cancer* **99**:2054–2064.

Wilson, P. D., Dillingham, M. A., Breckon, R., & Anderson, R. J. (1985). Defined human renal tubular epithelia in culture: Growth, characterization, and hormonal response. *Am. J. Physiol.* **248**:F436–F443.

Wilson, P. D., Schrier, R. W., Breckon, R. D., & Gabow, P. A. (1986). A new method for studying human polycystic kidney disease epithelia in culture. *Kidney Int.* **30**:371–378.

Wise, C. (2002). *Epithelial cell culture protocols.* Clifton, NJ: Humana Press.

Wistuba, I. I., Bryant, D., Behrens, C., Milchgrub, S., Virmani, A. K., Ashfaq, R, Minna, J. D., & Gazdar, A. F. (1999). Comparison of features of human lung cancer cell lines and their corresponding tumors. *Clin. Cancer Res.* **5**:991–1000.

Witkowski, J. A. (1990). The inherited character of cancer: An historical survey. *Cancer Cells* **2**:229–257.

Wolf, D. P., Meng, L., Ely, J. J., & Stouffer, R. L. (1998). Recent progress in mammalian cloning. *J. Assist. Reprod. Genet.* **15**:235–239.

WMA. (2008). Declaration of Helsinki-Ethical Principles for Medical Research Involving Human Subjects: http://www.wma.net/en/30publications/10policies/b3/index.html.

Wolff, E. T., & Haffen, K. (1952). Sur une méthode de culture d'organes embryonnaires in vitro. *Tex. Rep. Biol. Med.* **10**:463–472.

Wolswijk, G., & Noble, M. (1989). Identification of an adult-specific glial progenitor cell. *Development* **105**:387–400.

Woltjen, K., Michael, I. P., Mohseni, P., Desai, R., Mileikovsky, M., Hämäläinen, R., Cowling, R., Wang, W., Liu, P., Gertsenstein, M., Kaj, K., Sung, H.-K., & Nagy, A. (2009). piggyBac transposition reprograms fibroblasts to induced pluripotent stem cells. *Nature* **458**:766–770.

Wootton, M., Steeghs, K., Watt, D., Munro, J., Gordon, K., Ireland, H., Morrison, V., Behan, W., & Parkinson, E. K. (2003). Telomerase alone extends the replicative lifespan of human skeletal muscle cells without compromising genomic stability. *Hum. Gene Ther.* **14**:1473–1487.

Wright, K. A., Nadire, K. B., Busto, P., Tubo, R., McPherson, J. M., & Wentworth, B. M. (1998). Alternative delivery of keratinocytes using a polyurethane membrane and the implications for its use in the treatment of full-thickness burn injury. *Burns* **24**:7–17.

Wu, H., Friedman, W. J., & Dreyfus, C. F. (2004). Differential regulation of neurotrophin expression in basal forebrain astrocytes by neuronal signals. *J. Neurosci. Res.* **76**:76–85.

Wu, R. (2004). Growth of human lung tumor cells in culture. In Pfragner, R., & Freshney. R. I. (eds.), *Culture of human tumor cells.* Hoboken, NJ: Wiley-Liss, pp. 1–21.

Wu, Y. J., Parker, L. M., Binder, N. E., Beckett, M. A., Sinard, J. H., Griffiths, C. T., & Rheinwald, J. G. (1982). The mesothelial keratins: A new family of cytoskeletal proteins identified in cultured mesothelial cells and nonkeratinizing epithelia. *Cell* **31**:693–703.

Wuarin, L., Verity, M. A., & Sidell, N. (1991). Effects of interferon-gamma and its interaction with retinoic acid on human neuroblastoma differentiation. *Int. J. Cancer* **48**:136–141.

Wurmser, A. E., Nakashima, K., Summers, R. G., Toni, N., D'amour, K. A., Lie, C. C., & Gage, F. H. (2004). Cell fusion-independent differentiation of neural stem cells to the endothelial lineage. *Nature* **430**:350–356.

Wurster-Hill, D., Cannizzaro, L. A., Pettengill, O. S., Sorenson, G. D., Cate, C. C., & Maurer, L. H. (1984). Cytogenetics of small cell carcinoma of the lung. *Cancer Genet. Cytogenet.* **13**:303–330.

Wyllie, F. S., Bond, J. A., Dawson, T., White, D., Davies, R., & Wynford-Thomas, D. (1992). A phenotypically and karyotypically stable human thyroid epithelial line conditionally immortalized by SV40 large T antigen. *Cancer Res.* **52**:2938–2945.

Wysocki, L. J., & Sata, V. L. (1978). "Panning" for lymphocytes: A method for cell selection. *Proc. Natl. Acad. Sci. USA* **75**:2844–2848.

Xing, J. G., El-Sweisi, W., Lee, L. E. J., Cullodi, P., Seymour, C., Mothersill, C., & Bols, N. C. (2009). Development of a zebrafsih spleen cell line, ZSSJ, and its growth arrest by gamma radiation and capacity to act as feeder cells. *In Vitro Cell. Dev. Biol. Anim.* **45**:163–174.

Yaffe, D. (1968). Retention of differentiation potentialities during prolonged cultivation of myogenic cells. *Proc. Nat. Acad. Sci. USA* **61**:477–483.

Yamada, K. M., & Geiger, B. (1997). Molecular interactions in cell adhesion complexes. *Curr. Opin. Cell Biol.* **9**:76–85.

Yamada, T., Placzek, M., Tanaka, H., Dodd, J., & Jessell, T. M. (1991). Control of cell pattern in the developing nervous system: Polarizing activity of the floor plate and notochord. *Cell* **64**:635–647.

Yamaguchi, N., Yamamura, Y., Koyama, K., Ohtsuji, E., Imanishi, J., & Ashihara, T. (1990). Characterization of new pancreatic cancer cell lines which propagate in a protein-free chemically defined medium. *Cancer Res.* **50**:7008–7014.

Yan, G., Fukabori, Y., Nikolaropoulost, S., Wang, F., & McKeehan, W. L. (1992). Heparin binding keratinocyte growth factor is a candidate stromal to epithelial cell andromedin. *Mol. Endocrinol.* **6**:2123–2128.

Yanai, N., Suzuki, M., & Obinata, M. (1991). Hepatocyte cell lines established from transgenic mice harboring temperature-sensitive simian virus 40 large T-antigen gene. *Exp. Cell Res.* **197**:50–56.

Yang, J., Mani, S. A., Donaher, J. L., Ramaswamy, S., Itzykson, R. A., Come, C., Savagner, P., Gitelman, I., Richardson, A., & Weinberg, R. A. (2004). Twist, a master regulator of morphogenesis, plays an essential role in tumor metastasis. *Cell* **117**:927–39.

Yasamura, Y., Tashijian, A. H., & Sato, G. (1966). Establishment of four functional clonal strains of animal cells in culture. *Science* **154**:1186–1189.

Yeager. T. R., DeVries, S., Farrard, D. F., Kao, C., Nakada, S. Y., Monn, T. D., Bruskewitz, R., Stadler, W. M., Meisner, L. F., Gilchrest, K. W., Newyton, M. A., Waldman, F. M., & Reznikoff, C. A. (1998). Overcoming cellular senescence in human cancer pathogenesis. *Genes Dev.* **12**:163–174.

Yen, A., Coles, M., & Varvayanis, S. (1993). 1,25-dihydroxy vitamin $D_3$ and 12-O-tetradecanoyl phorbol-13-acetate synergistically induce monocytic cell differentiation: FOS and RB expression. *J. Cell Physiol.* **156**:198–203.

Yeoh, G. C. T., Hilliard, C., Fletcher, S., & Douglas, A. (1990). Gene expression in clonally derived cell lines produced by in vitro transformation of rat fetal hepatocytes: Isolation of cell lines which retain liver-specific markers. *Cancer Res.* **50**:75–93.

Yerganian, G., & Leonard, M. J. (1961). Maintenance of normal in situ chromosomal features in long-term tissue cultures. *Science* **133**:1600–1601.

Yeung, J., Wai, E., & So, A. C. (2009). Identification and characterization of hematopoietic stem and progenitor cell populations in mouse bone marrow by flow cytometry. *Methods Mol. Biol.* **538**:1–15.

Yevdokimova, N., & Freshney, R. I. (1997). Activation of paracrine growth factors by heparan sulphate induced by glucocorticoid in A549 lung carcinoma cells. *Br. J. Cancer* **76**:261–289.

Ying, Q. L., Nichols, J., Chambers, I., & Smith, A. (2003). BMP induction of Id proteins suppresses differentiation and sustains embryonic stem cell self-renewal in collaboration with STAT3. *Cell* **115**:281–292.

Ying, Q. L., Wray, J., Nichols, J., Batlle-Morera, L., Doble, B., Woodgett, J., Cohen, P., & Smith, A. (2008). The ground state of embryonic stem cell self-renewal. *Nature* **453**:519–523.

Yoshida, M., & Beppu, T. (1990). In Doyle, A., Griffiths, J. B., & Newell, D. G. (eds.), *Cell and tissue culture: Laboratory procedures.* Chichester, UK: J Wiley, module 4E2.

Yoshida, S. (2009). Casting back to stem cells. *Nat. Cell Biol.* **11**:118–120.

Yoshino, K., Iimura, E., Saijo, K., Iwase, S., Fukami, K., Ohno, T., Obata, Y., & Nakamura Y. (2006). Essential role for gene profiling analysis in the authentication of human cell lines. *Hum Cell* **19**:43–48.

Young, H. E., Duplaa, C., Romero-Ramos, M., Chesselet, M. F., Vourc'h, P., Yost, M. J., Ericson, K., Terracio, L., Asahara, T., Masuda, H., Tamura-Ninomiya, S., Detmer, K., Bray, R. A., Steele, T. A., Hixson, D., el-Kalay, M., Tobin, B. W., Russ, R. D., Horst, M. N., Floyd, J. A., Henson, N. L., Hawkins, K. C., Groom, J., Parikh, A., Blake, L., Bland, L. J., Thompson, A. J., Kirincich, A., Moreau, C., Hudson, J., Bowyer, F. P. 3rd, Lin, T. J., & Black, A. C. Jr. (2004). Adult reserve stem cells and their potential for tissue engineering. *Cell. Biochem. Biophys.* **40**:1–80.

Yu, J., Vodyanik, M. A., Kim Smuga-Otto, K., Antosiewicz-Bourget, J., Frane, J. L., Shulan Tian, S., Nie, J., Jonsdottir, G. A., Ruotti, V., Stewart, R., Slukvin, I. I., James, A., & Thomson, J. A. (2007). Induced pluripotent stem cell lines derived from human somatic cells. *Science* **318**:1917–1920.

Yuhas, J. M., Li, A. P., Martinez, A. O., & Ladman, A. J. (1977). A simplified method for production and growth of multicellular tumour spheroids (MTS). *Cancer Res.* **37**:3639–3643.

Yuspa, S. H., Koehler, B., Kulesz-Martin, M., & Hennings, H. (1981). Clonal growth of mouse epidermal cells in medium with reduced calcium concentration. *J. Invest. Dermatol.* **76**:144–146.

Yusufi, A. N. K., Szczepanska-Konkel, M., Kempson, S. A., McAteer, J. A., & Dousa, T. P. (1986). Inhibition of human renal epithelial $Na^+/Pi$ cotransport by phosphonoformic acid. *Biochem. Biophys. Res. Commun.* **139**:679–686.

Zaroff, L., Sato, G. H., & Mills, S. E. (1961). Single-cell platings from freshly isolated mammalian tissue. *Exp. Cell Res.* **23**:565–575.

Zborowski, M., & Chalmers, J. J. (2007). *Magnetic cell separation: Laboratory techniques in biochemistry and molecular biology 32.* Amsterdam: Elsevier.

Zeltinger, J., & Holbrook, K. A. (1997). A model system for long-term serum-free suspension organ culture of human fetal tissues: Experiments on digits and skin from multiple body regions. *Cell & Tissue Res.* **290**:51–60.

Zeng, C., Pesall, J. E., Gilkerson, K. K., & McFarland, D. C. (2002). The effect of hepatocyte growth factor on turkey satellite cell proliferation and differentiation. *Poult Sci.* **81**:1191–1198.

Zhang, X., Jiang, G., Cai, Y., Monkley, S. J., Critchley, D. R., & Sheetz, M. P. (2008). Talin depletion reveals independence of initial cell spreading from integrin activation and traction. *Nat. Cell Biol.* **10**:1062–1068.

Zhang, Y., Proenca, R., Maffei, M., Barone, M., Leopold, L., & Friedman, J. M. (1994). Positional cloning of the mouse obese gene and its human homologue. *Nature* **372**:425–432.

Zhao, H., Dreses-Werringloer, U., Davies, P., & Marambaud, P. (2008). Amyloid-beta peptide degradation in cell cultures by mycoplasma contaminants. *BMC Res Notes.* **1**:38.

Zhao, R., & Daley, G. Q. (2008). From fibroblasts to iPS cells: Induced pluripotency by defined factors. *J. Cell. Biochem.* **105**:949–955.

Zhou, S., Schuetz, J. D., Bunting, K. D., Colapietro, A.-M., Sampath, J., Morris, J. J., Lagutina, I., Grosveld, G. C., Osawa, M., Nakauchi, H., & Sorrentino, B. P. (2001). The ABC transporter Bcrp1/ABCG2 is expressed in a wide variety of stem cells and is a molecular determinant of the side-population phenotype. *Nat. Med.* **7**:1028–1034.

Zhu, C., & Joyce, N. C. (2004). Proliferative response of corneal endothelial cells from young and older donors. *Invest. Ophthalmol. Vis. Sci.* **45**:1743–1751.

Zhu, S. Y., Cunningham, M. L., Gray, T. E., & Nettesheim, P. (1991). Cytotoxicity, genotoxicity and transforming activity of 4-(methylnitrosamino)-1-(3-pyridyl)-1-butanone (NNK) in rat trachea epithelial cells. *Mutation Res.* **261**(4): 249–259.

Zicha, D., & Dunn, G. A. (1995). An image processing system for cell behaviour studies in subconfluent cultures. *J. Microsc.* **179**:11–21.

Zimmermann, H., Hillgartner, M., Manz, B., Feilen, P., Brunnenmeier, F., Leinfelder, U., Weber, M., Cramer, H., Schneider, S., Hendrich, C., Volke, F., & Zimmermann, U. (2003). Fabrication of homogeneously cross-linked, functional alginate microcapsules validated by NMR-, CLSM- and AFM-imaging. *Biomaterials* **24**:2083–2096.

Zoubiane, G. S., Valentijn, A., Lowe, E. T., Akhtar, N., Bagley, S., Gilmore, A. P., & Streuli, C. H. (2003). A role for the cytoskeleton in prolactin-dependent mammary epithelial cell differentiation. *J. Cell Sci.*: **117**:271–280.

Zuk, P. A., Zhu, M., Ashjian, P., De Ugarte, D. A., Huang, J. I., Mizuno, H., Alfonso, Z. C., Fraser, J. K., Benhaim, P., & Hedrick, M. H. (2002). Human adipose tissue is a source of multipotent stem cells. *Mol. Biol. Cell.* **13**:4279–4295.

zur Nieden, N. I., Kempka, G., & Ahr, H. J. (2003). In vitro differentiation of embryonic stem cells into mineralized osteoblasts. *Differentiation* **71**:18–27.

Zwain, I. H., Morris, P. L., & Cheng, C. Y. (1991). Identification of an inhibitory factor from a Sertoli clonal cell line (TM4) that modulates adult rat Leydig cell steroidogenesis. *Mol. Cell Endocrinol.* **80**:115–126.

# INDEX

*Culture of Animal Cells: A Manual of Basic Technique and Specialized Applications, Sixth Edition*, by R. Ian Freshney
Copyright © 2010 John Wiley & Sons, Inc.